Introduction to Plant Biochemistry

by

T. W. GOODWIN

Johnston Professor of Biochemistry,
University of Liverpool

and

E. I. MERCER

Senior Lecturer in The Department of Biochemistry and Agricultural Biochemistry,
University College of Wales, Aberystwyth

Second Edition

PERGAMON PRESS

Member of Maxwell Macmillan Pergamon Publishing Corporation

OXFORD · NEW YORK · BEIJING · FRANKFURT
SÃO PAULO · SYDNEY · TOKYO · TORONTO

U.K.	Pergamon Press plc, Headington Hill Hall, Oxford OX3 0BW, England
U.S.A.	Pergamon Press Inc., Maxwell House, Fairview Park, Elmsford, NY 10523, U.S.A.
PEOPLE'S REPUBLIC OF CHINA	Pergamon Press, Room 4037, Qianmen Hotel, Beijing, People's Republic of China
FEDERAL REPUBLIC OF GERMANY	Pergamon Press GmbH, Hammerweg 6, D-6242 Kronberg, Federal Republic of Germany
BRAZIL	Pergamon Editora Ltda, Rua Eça de Queiros, 346, CEP 04011, Paraiso, São Paulo, Brazil
AUSTRALIA	Pergamon Press Australia Pty Ltd, P.O. Box 544, Potts Point, N.S.W. 2011, Australia
JAPAN	Pergamon Press, 5th Floor, Matsuoka Central Building, 1-7-1 Nishishinjuku, Shinjuku-ku, Tokyo 160, Japan
CANADA	Pergamon Press Canada Ltd., Suite No. 271, 253 College Street, Toronto, Ontario, Canada M5T 1R5

First edition 1972
Reprinted 1975
Second edition 1983
Reprinted 1985, 1986, 1988 (with corrections), 1990 (with corrections)

Library of Congress Cataloging in Publication Data

Goodwin, T. W. (Trevor Walworth)
Introduction to plant biochemistry.
(Pergamon international library of science, technology, engineering, and social studies)
Includes index.
1. Botanical chemistry. I. Mercer, E. I. (Eric Ian)
II. Title. III. Series.
QK861.G66 1982 581.19′2 81–15903

British Library Cataloguing in Publication Data

Goodwin, T. W.
Introduction to plant biochemistry.—2nd ed.
—(Pergamon international library)
1. Botanical chemistry
I. Title II. Mercer, E. I.
581.19′2 QK861

ISBN 0-08-024922-1 (Hardcover)
ISBN 0-08-024921-3 (Flexicover)

Typeset by Cotswold Typesetting Ltd., Gloucester
Printed in Great Britain by BPCC Wheatons Ltd, Exeter

**THE PERGAMON TEXTBOOK
INSPECTION COPY SERVICE**

An inspection copy of any Pergamon Textbook will gladly be
sent to academic staff without obligation for their consideration for course adoption or recommendation.
Copies may be retained for a period of 60 days from receipt
and returned if not suitable. When a particular title is adopted
or recommended for adoption for class use and the recommendation results in a sale of 12 or more copies the inspection
copy may be retained with our compliments.
The Publishers will be pleased to receive suggestions for
revised editions and new titles.

Other Pergamon publications of related interest

Books

COOMBS *et al.*   Techniques in Bioproductivity and Photosynthesis, 2nd Edition

FAHN   Plant Anatomy. 4th Edition

HADER & TEVINI   General Photobiology

MAYER & POLJAKOFF-MAYBER   The Germination of Seeds, 4th Edition

WAREING & PHILLIPS   Growth and Differentiation in Plants, 3rd Edition

Journals (*free specimen copies gladly sent on request*)

BIOCHEMICAL SYSTEMATICS AND ECOLOGY

CURRENT ADVANCES IN PLANT SCIENCE

ENVIRONMENTAL AND EXPERIMENTAL BOTANY

JOURNAL OF THERMAL BIOLOGY

PHOTOCHEMISTRY AND PHOTOBIOLOGY

PHYTOCHEMISTRY

SOIL BIOLOGY AND BIOCHEMISTRY

Preface to the Second Edition

THE RAPIDITY with which advances have been made in biochemistry during the last decade are particularly impressive in the field of plant biochemistry. This is due to no small degree to the developments in many biochemical techniques which now allow intransigent plant material to be examined with the same degree of sophistication as *E. coli* or liver. Both the expansion in factual information and the development of biochemical concepts in plants has inevitably meant that a second edition of *Introduction to Plant Biochemistry* has had to be a new and much longer book. Our original aim however has remained the same: to provide for students who have already studied general biochemistry for one year, a thorough and rigorous voyage through the major fields of plant biochemistry.

The only chapter which stands largely unchanged is the first, in which the philosophy of our approach is outlined. Chapter 2 is now confined to bioenergetics; the section on enzyme kinetics which was present in this chapter in the first edition has been omitted. We regret this but the enormous expansion of the topic did not allow us to treat it adequately in the space available; however, several excellent specialist texts are now available.

Two characteristics of the text should be emphasized; firstly where appropriate, if sufficient information is available, the metabolic activities of plants have been compared with those of animals or bacteria. Secondly the number of summarizing diagrams has been increased so as to provide a birds eye view of the details being considered in the text.

Two appendices are included: one explaining the intricacies of the Cahn–Ingold–Prelog nomenclature which is now widely used to specify absolute configuration at chiral centres; the second gives a comprehensive list of units and dimensions based on the SI unit system.

We thank many friends and colleagues for the help they have given us in many ways during the preparation of the second edition; in particular Dr. D. S. Jones for the time he spent over the drafts of Chapter 10, Dr. R. N. Jones for his help with part of Chapter 3, Prof. P. F. Wareing, Prof. M. A. Hall and Dr. R. Horgan for their advice during the writing of Chapter 15, Miss B. T. Foulkes and Mrs. K. E. van Mechelen for typing the manuscript, Mrs. C. M. A. Mercer, Miss R. I. Baloch and Mr. I. A. Khalil for their invaluable help with the preparation of the indexes and Mr. M. J. Richardson of Pergamon Press for his understanding and encouragement.

We would welcome any comments which would help to improve the acceptability of future editions of the book.

Liverpool
Aberystwyth

T. W. G.
E. I. M.

Preface to the First Edition

IN WRITING this book we have assumed that the student who reads it will already have had at least a one-year course of general biochemistry such as that covered by E. E. Conn and P. K. Stumpf's deservedly popular *Outlines of Biochemistry*. This assumption has allowed us to omit much introductory material whilst emphasizing certain aspects which from our experience give students trouble, and to concentrate on the aim of presenting a precise view of plant biochemistry without making the book oversized. This also fits in with one aspect of our general philosophy of the teaching of biochemistry which is that all biologists should very early on in their university studies be exposed to an introductory course in biochemistry taught in a Biochemistry Department; this provides a solid foundation on which experts in associated subjects can later on discuss the specialized aspects of biochemistry with which they are particularly concerned.

The book is therefore intended for use in the second or third year of a general honours course in biochemistry, for second- or third-year honours students in other disciplines who wish to take a 'unit' of plant biochemistry as an additional subject, or for honours botany students whose physiology courses are biochemically orientated. It should also be useful to natural product chemists and students in the more applied aspects of biochemistry, such as Food Science and Agricultural Biochemistry.

We are grateful to many colleagues for helpful discussion and particularly to Dr. L. J. Rogers, who read and criticized the text in its early form; many improvements are due to his comments. Dr. Mary Hallaway and Dr. F. W. Hemming also kindly read the manuscript in its later stages, and made many valuable suggestions. In addition we are greatly indebted to Miss Joan Peel for typing the manuscript.

We hope that users of the book will let us have their comments, so that any future editions can be improved.

Liverpool
Aberystwyth

T. W. G.
E. I. M.

Contents

CHAPTER 1

Introductory Chapter

LIVING organisms, whether they are plants, animals or microbes, are made up basically of the same chemical components. Biochemistry is the study of the way in which these components are synthesized and utilized by the organisms in the life processes. The unity of life, the common ancestry of living things, and the similarity in chemical make-up, suggest that very similar chemical processes are carried out in all organisms. Comparative biochemistry has shown that while this is true in essence it is untrue in detail; there is a basic pattern of chemical processes common to all organisms but major or minor modifications to it have been evolved by different organisms to suit their particular requirements.

In considering the biochemistry of plants, therefore, it is necessary to deal not only with processes which are unique to plants but also with processes which are carried out, with only minor differences, by all living organisms. Thus many of the topics discussed in this book are to a greater or lesser degree based on the comparative approach. Occasionally the use of this approach makes a virtue of necessity because of lack of information in some fields of plant biochemistry compared with that available in equivalent areas of animal or microbial biochemistry. Frequently biochemical pathways are first elucidated in detail in animal or microbial tissues and then just one or two critical experiments on plant tissues are carried out to see whether the same pathway operates in the plant kingdom. There are several possible reasons for this approach, some of which are mentioned during the course of the book.

Members of the plant kingdom may be divided into two groups according to the nutrients they require; they are either autotrophic or heterotrophic. Autotrophic organisms require only simple inorganic nutrients; from these they can synthesize all the complex organic molecules needed for their growth and reproduction. Heterotrophic organisms are unable to live solely on simple inorganic nutrients, although small quantities of these are essential; their major nutrients consist of complex organic materials.

Apart from the short period during germination when they are growing heterotrophically on their seed stores, all green (higher) plants are autotrophic. Their nutrients are CO_2, H_2O, NO_3^-, SO_4^{2-} and PO_4^{3-}, plus trace amounts of other elements. These simple inorganic materials supply the six elements, carbon, hydrogen, oxygen, nitrogen, sulphur and phosphorus, which make up the bulk of the tissue components, the proteins (C, H, O, N, S, P), carbohydrates (C, H, O), lipids (C, H, O, N, S, P), nucleic acids (C, H, O, N, P), etc. Carbon dioxide supplies both carbon and oxygen, water supplies only hydrogen, nitrate supplies only nitrogen, sulphate supplies both sulphur and oxygen and phosphate supplies both phosphorus and oxygen. However, the oxygen of sulphate and phosphate remains bound to the sulphur or phosphorus atoms throughout its metabolism.

These facts give a clue to the major biochemical difference between plants and animals: it is in their biosynthetic ability. Not only can plants make all the universally required molecules such as DNA, RNA and proteins from inorganic nutrients, a property shared with some photosynthetic bacteria, but they show enormous versatility and prodigality in synthesizing complex materials which have no immediately obvious functions; these include many bizarre amino acids (Chapter 9), the terpenoids (Chapter 11), alkaloids (Chapter 13) and the flavonoids (Chapter 14), all of which have been widely studied from the biosynthetic point of view.

Plants are far more versatile in this aspect of their metabolism than animals although like animals their catabolic processes are fairly circumscribed. Microorganisms, of course, excel in their versatility in degrading organic molecules, particularly those from plants which find their way to the soil. Higher plants appear to have failed to bring about one important synthesis, that of vitamin B_{12}; but they have devised pathways which bring about the same results as certain vitamin B_{12}-requiring pathways in microorganisms, without the necessity for invoking this vitamin as an enzyme co-factor. So we see that from such simple materials as CO_2 and inorganic N, P and S (plus the trace elements), green plants can synthesize all the complex organic molecules which make up their tissues. However, these synthetic processes require energy. This energy comes from the sun. Green plants have the ability to trap light energy from the sun and utilize it to build up carbohydrates from CO_2 and H_2O; this process is called photosynthesis (see Chapter 5). The light energy has been used to form chemical bonds between carbon atoms, hydrogen atoms and oxygen atoms derived from molecules of CO_2 and H_2O. In effect the light energy has been trapped in the newly formed carbohydrate molecules as chemical bond energy. This chemical bond energy cannot be released or made available to do 'work' until the chemical bond is broken. The carbohydrates then (and all other organic molecules) can be regarded as repositories of 'trapped' energy. The green plant utilizes a considerable proportion of the carbohydrate for structural purposes; the cellulose of plant cell walls is the most widespread and abundant carbohydrate on this planet. However, the green plant also derives energy from a portion of the carbohydrate by breaking it down oxidatively in a controlled and gradual way to CO_2 and water in the process called respiration (Chapter 6). A large part of the energy which is released as chemical bonds are broken in this process is conserved in a chemically available form as adenosine 5′-triphosphate (ATP); this is the available form utilized throughout the living world, indeed ATP has been called the 'universal biological energy currency'. The efficiency of this trapping process is 40–50% and the energy which is not trapped is lost as heat. The trapped energy is built into the ATP molecule and ATP is, therefore, said to be a 'high energy compound'. The

energy of this ATP is then used in a number of different ways, such as in the synthesis of the other tissue components, proteins, lipids, nucleic acids, etc., and for the transport of substances across membranes (e.g. the uptake of NO_3^-, SO_4^{2-}, PO_4^{3-} and other ions from the soil into the root hairs). Thus it can be seen that the energy which enables green plants to live on simple inorganic nutrients is entirely derived from the sun.

Many of the lower forms of plant life, e.g. fungi, are heterotrophic. They are not green and cannot trap the sun's energy to enable them to convert simple inorganic nutrients into the complex organic tissue components. They must live on preformed complex organic molecules; preformed, that is, by autotrophic organisms of which green plants are quantitatively the most important. These organic molecules provide the two essentials for the life of heterotrophic organisms, the C, H, O, N, S, and P necessary for the manufacture of their own tissue components and the energy, conserved as chemical bond energy, with which their synthetic and other life processes are driven. The heterotrophic forms of plant life, therefore, degrade complex organic molecules into simpler molecules which can be introduced into the oxidative respiratory pathway to produce utilizable energy in the form of ATP, and a considerable number of different, small molecules which are intermediates in the respiratory pathway and which can be utilized as starting materials for the plant's own synthetic processes.

The difference between the biochemical capabilities of autotrophic green plants and the heterotrophic lower forms of plant life stems from the immediate origin of the energy which drives the synthetic processes essential to life. The green plant possesses the unique chemical machinery for converting light energy into chemical bond energy whilst the heterotrophic plant does not; however, both autotrophe and heterotrophe have the ability to obtain energy by oxidation of organic molecules and to utilize it in the same way in the form of ATP.

Animals are heterotrophic organisms; they derive energy and C, H, O, N, S and P from the preformed organic molecules which constitute their food. Their biochemistry, therefore, is, in considerable measure, basically comparable with that of heterotrophic plants, but they have also evolved unique biochemical activities necessary for their continued existence.

Bacteria, on the other hand, vary, some being autotrophic and some heterotrophic. Heterotrophic bacteria which grow aerobically (that is in the presence of oxygen) derive their energy by oxidation of exogenous preformed organic substances, and thus their basic biochemistry is somewhat similar to that of animals and heterotrophic plants; anaerobic bacteria have devised other, generally less efficient, means of deriving available energy, but it is still in the form of ATP. Autotrophic bacteria may be divided into two classes, the phototrophic bacteria which utilize light energy, and the chemoautotrophic bacteria, which derive their energy in a unique way by the oxidation of exogenous inorganic substances such as H_2S. The phototrophic bacteria are subdivided into photolithotrophic bacteria (e.g. green and purple sulphur bacteria) whose growth is dependent on exogenous inorganic hydrogen donors and photo-organotrophic bacteria (e.g. purple, non-sulphur bacteria) whose growth is dependent on exogenous organic hydrogen donors. The biochemistry of photolithotrophic bacteria is related to that of green plants.

A recurrent theme in this introduction has been the origin of the energy which motivates the life process. The central role of energy in biochemistry cannot be over-emphasized. The student should be thoroughly acquainted with the fundamental principles of energetics. Without such an understanding we feel that the biochemistry of plants or indeed any other organism becomes insecurely based. For this reason we have introduced biochemical energetics very early in this book (Chapter 2) and have, where knowledge permits, emphasized the energy relations of reactions discussed in later chapters. In Chapter 3 we have attempted to correlate structure within the plant cell with function; this should be valuable in introducing plant cell morphology to the non-botanist and in pin-pointing biological function to the non-biochemist. Subsequent chapters deal with the metabolism of carbohydrates, lipids, proteins and nucleic acids; since these are the main chemical components of all living tissues we have not confined ourselves too strictly to the plant but have introduced relevant comparisons with animals and bacteria. In spite of gaps in our knowledge it is in this type of study that the 'unity and diversity' of biochemistry is most clearly demonstrated. Plant cells contain nuclei in which the general pattern of biochemical genetics is the same as in any other cell; they contain mitochondria which respire in the same way as mitochondria from animals and they contain ribosomes which synthesize protein in very much the same way as do bacterial and animal ribosomes; however, their unique activity, photosynthesis, is carried out in a unique organelle, the chloroplast. The existence of a vacuole, not present in animals or bacteria, poses problems of transport and storage in plant cells which are unique and biochemically virtually unexplored; finally, the need for the plant to stand 'on its own feet' has resulted in the formation of a unique cell wall.

Later chapters are confined to topics which are more or less exclusive to the plant. We have thought it beneficial to the appreciation of the mechanism of the light phase of photosynthesis to emphasize the basic physics and physical chemistry involved. Some topics, because of our own research interests, may have been over-emphasized (although we have tried hard to strike a balance) whilst others, for example that on plant-growth substances, because we have attempted to concentrate on the essential biochemical aspects rather than deal with the complexities unearthed by the plant physiologists, may be under-emphasized. Nevertheless our aim throughout has been to give an integrated introductory picture of the biochemical processes which go on in plant tissues.

The generally accepted trivial names of enzymes are used in the text. However, the student should become familiar with the systematic names and numbers of enzymes recommended by the *Commission on Enzymes of the International Union of Biochemistry*. Accordingly the first time an enzyme is mentioned in a chapter it is given a superscript number which corresponds to the number of that enzyme in the list of systematic names at the end of the chapter.

CHAPTER 2

Bioenergetics

A. BIOENERGETICS

1. Free energy

An external source of energy is required to join atoms together to form stable molecules. The most fundamental example of this in biochemistry is the utilization of the radiant energy of the sun by photosynthetic tissues of plants to 'fix' atmospheric CO_2 and to convert it, through the formation of covalent linkages, into the complex carbon compounds which make up a living plant. Such reactions, which require an external source of energy in order to occur, are termed *endergonic*. Conversely, when specific bonds are broken there is a net release of energy (i.e. the energy liberated by cleavage of the bonds is greater than that needed to be introduced into the system to effect this cleavage). This is essentially what happens in respiration. Such reactions are said to be *exergonic*. In order to understand the energy changes involved in biochemical reactions, these concepts must be put into more precise terms.

The energy liberated or consumed in a reaction which can be used in or supplied to another system is termed the *free energy change* (ΔG) of the reaction.

ΔG can be defined as that portion of the total energy change which is available to do work as the reaction proceeds towards equilibrium at constant temperature, pressure and volume. In an exergonic reaction [eq. (2.1)] ΔG is negative because there is an overall loss of free energy in the conversion of a molecules of A and b molecules of B into c molecules of C and d molecules of D.

$$aA + bB \rightarrow cC + dD + \text{'Free energy'} \quad (2.1)$$

The reverse reaction, the reaction of c molecules of C and d molecules of D to produce a molecules of A and b molecules of B [eq. (2.2)] is endergonic because there is an overall gain in free energy (i.e. energy has to be taken up) and ΔG is positive.

$$\text{'Free energy'} + cC + dD \rightarrow aA + bB. \quad (2.2)$$

Clearly there must be a situation where $\Delta G = 0$; this occurs when reactions (2.1) and (2.2) are in equilibrium.

In the reversible reaction [eq. (2.3)] which is the sum of eqs. (2.1) and (2.2), it can be shown that the

$$aA + bB \rightleftharpoons cC + dD \quad (2.3)$$

apparent equilibrium constant,† K'_{eq} [eq. (2.4), where [A], [B], [C] and [D] are the concentrations

$$K'_{eq} = \frac{[C]^c[D]^d}{[A]^a[B]^b} \quad (2.4)$$

of A, B, C and D at equilibrium and where the concentration of water, if it is a participant in the reaction, is taken as 1], is related to ΔG by eq. (2.5), in which R is the gas constant ($8.314 \text{ J K}^{-1} \text{ mol}^{-1}$),

---

† K'_{eq} is used rather than the thermodynamic equilibrium constant, K_{eq}, because concentrations and not activities of the reactions have been used; K_{eq} and K'_{eq} are only equal at infinite dilution.

T is the absolute temperature in degrees Kelvin ($K = 273.15 + °C$), a, b, c and d are the number of molecules of each chemical species participating in the balanced reaction and [A], [B], [C] and [D] are their concentrations.

$$\Delta G = -RT \log_e K'_{eq} + RT \log_e \frac{[C]^c[D]^d}{[A]^a[B]^b} \quad (2.5)$$

$$\underbrace{\phantom{-RT \log_e K'_{eq}}}_{(1)} \quad \underbrace{\phantom{RT \log_e \frac{[C]^c[D]^d}{[A]^a[B]^b}}}_{(2)}$$

From eq. (2.5) it is apparent that ΔG is equal to the difference between terms (2) and (1). Term (1) involves K'_{eq} which is the ratio of products/reactants (i.e. $[C]^c[D]^d/[A]^a[B]^b$) in the reaction [eq. (2.3)] when it is in equilibrium. Term (2) defines the actual products/reactants ratio that is maintained throughout the reaction; it is, therefore, a measure of the extent to which the reaction is displaced from equilibrium.

Now in both chemistry and biochemistry it is useful to be able to compare the ΔG value of different reactions. To do this the ΔG value of each reaction must be determined under identical conditions; these are referred to as standard conditions. The standard conditions agreed upon by chemists are that (i) all participants of the reaction be maintained at a concentration of 1 M; this also applies to water if it happens to be a participant in the reaction (e.g. as in a hydrolysis), (ii) if one of the participants is a gas it be maintained at a partial pressure of 1 atmosphere and (iii) the temperature be maintained at 25°C (298.15 K).

When the standard values are inserted into eq. (2.5), term (2) becomes $RT \log_e 1$ which equals 0; thus term (2) is eliminated and ΔG equals $-RT \log_e K'_{eq}$. ΔG measured under these conditions is called the *standard free energy change* and is designated $\Delta G°$ [eq. (2.6)].

$$\Delta G° = -RT \log_e K'_{eq}. \quad (2.6)$$

$\Delta G°$ may be defined in two ways. Firstly, it can be defined as the difference between the sum of the free energies of the products and the sum of the free energies of the reactants when each product and reactant is present in its standard state; this is given by eq. (2.7) for reaction (2.3) where $G°_A$, $G°_B$, etc.,

$$\Delta G° = [cG°_C + dG°_D] - [aG°_A + bG°_B] \quad (2.7)$$

are the intrinsic free energies of the participants in their standard state. Secondly, $\Delta G°$ can be defined as the amount of free energy absorbed or lost per mole when A and B are changed into C and D at 25°C and 1 atmosphere pressure, with concentrations of A, B, C and D being maintained throughout at a concentration of 1 M.

$\Delta G°$ values, however, have a major disadvantage so far as the biochemist is concerned. This follows from the fact that hydrogen ions (H^+) participate in many of the enzyme-catalysed reactions occurring in living tissues. We have already seen that $\Delta G°$ describes the free energy change in a reaction where all the participants have a concentration of 1 M. However, a concentration of 1 M H^+ gives a pH of 0, a pH at which almost all enzymes are denatured. There would, therefore, be no reaction to study under these conditions. To get around this difficulty biochemists have replaced $\Delta G°$ with $\Delta G°'$ which is defined as the free energy change per mole of reactant at 25°C and 1 atmosphere pressure when the concentrations of all reactants and products are maintained at 1 M *with the exception of H^+ which is maintained at 10^{-7} M* (which is the H^+ concentration of pH 7.0). $\Delta G°'$ is thus the standard free energy change at pH 7.0 [eq. (2.8)].

$$\left.\begin{array}{l} \Delta G°' = -RT \log_e K'_{eq} \\ \text{or} \\ \Delta G°' = -RT \, 2.303 \log_{10} K'_{eq} \end{array}\right\} \text{at pH 7.0} \quad (2.8)$$

Also embodied in this definition of $\Delta G°'$ is the fact that if a participant in the reaction exists, at pH 7.0, as a mixture of its unionized and ionized forms or as a mixture of two differently ionized forms, then it is the sum of these different forms that is maintained at 1 M.

An important point to grasp about $\Delta G°'$ values is that they are pH-dependent; that is to say they are only correct at the pH at which they were determined, namely pH 7.0, and that their value changes with changing pH. The magnitude of this change in reaction (2.9) can be calculated from eq. (2.10) provided that no reactants or products ionize within

$$aA + bB \rightleftharpoons cC + dD + nH^+ \quad (2.9)$$

the pH range concerned or, if they do, they balance each other out by being on opposite sides of eq. (2.9) and have identical pK_a values. In eq. (2.10) ΔpH is equal to $pH_{(+7)}$ minus $pH_{(=7)}$ and n has the value of

n in eq. (2.9). Note that when H^+ is on the left-hand side of eq. (2.9) the sign of the last term in eq. (2.10) becomes positive and that when, in eq. (2.10), ΔpH is equal to 7, the term $\Delta G^{\circ\prime}_{(pH \neq 7)}$ becomes ΔG°.

$$\Delta G^{\circ\prime}_{(pH \neq 7)} = \Delta G^{\circ\prime}_{(pH = 7)} - RT\, 2.303 \cdot n \cdot \Delta pH \quad (2.10)$$

If the reactants and/or products of reaction (2.9) do ionize within the pH range of interest and their pK_a values are different from one another, equation (2.10) is inappropriate for the determination of the change in the $\Delta G^{\circ\prime}$ value that accompanies a change in pH. An alternative equation can be derived to cope with this situation. This equation includes the acid dissociation constants (K_a values) of all the ionizing species participating in the reaction. There are occasions, however, when one or more of these K_a values are not known, thus making it impossible to use the equation.

The $\Delta G^{\circ\prime}$ values of hydrolysis of a number of biologically important compounds are given in Table 2.1.

The standard free energy change at pH 7.0 ($\Delta G^{\circ\prime}$) is related to the actual free energy change (ΔG) of reaction (2.3) at that pH by eqs. (2.11). When using these equations at pH 7.0 it is not necessary to include $[H^+]$ or $[H_2O]$ even though they may be participants in the reaction (i.e. they may be A, B, C or D); moreover, it is understood that the concentration of each participant (e.g. [A]) includes all its unionized or ionized forms.

$$\left. \begin{array}{l} \Delta G = \Delta G^{\circ\prime} + RT \log_e \dfrac{[C]^c[D]^d}{[A]^a[B]^b} \\ \text{or} \\ \Delta G = \Delta G^{\circ\prime} + RT\, 2.303 \log_{10} \dfrac{[C]^c[D]^d}{[A]^a[B]^b} \end{array} \right\} \quad (2.11)$$

It is the ΔG value that indicates the direction in which and the extent to which a reversible reaction [eq. (2.3)] will proceed. If the ΔG value is negative the reaction will proceed from left to right as written; if it is positive the reaction will proceed from right to left. Thus the sign (+ or −) of the ΔG value indicates the directionality of the reaction. Whether the reaction is proceeding from left to right or vice versa, it is proceeding towards a state of minimum energy which, of course, occurs when a state of equilibrium exists and ΔG equals zero. The extent to which the reaction proceeds is indicated by the magnitude of the ΔG value, because it is this that indicates how far the system is from equilibrium. It should be noted that a reaction whose $\Delta G^{\circ\prime}$ (or ΔG°) is positive can still proceed from left to right as written provided that the ratio of the concentrations of reactants [i.e. [A] and [B] in eq. (2.3)] to the concentrations of products [i.e. [C] and [D] in eq. (2.3)] is sufficiently great to give a negative ΔG when eq. (2.11) is applied.

The ΔG value gives no information whatsoever about the rate of the reaction (i.e. the rate at which it

Table 2.1. *Standard free energy* ($\Delta G^{\circ\prime}$) *of hydrolysis of some biologically important compounds*

Compound	Type of bond hydrolysed	$\Delta G^{\circ\prime}$ kJ mol^{-1}
Phosphoenolpyruvate	enolic phosphate	−61.9
1,3-diphosphoglycerate	acyl phosphate	−49.3
ATP ($\rightarrow$ADP + P$_i$)†	pyrophosphate	−30.5
ATP ($\rightarrow$AMP + PP$_i$)	pyrophosphate	−31.8
ADP ($\rightarrow$AMP + P$_i$)	pyrophosphate	−27.2
Pyrophosphate ($\rightarrow$2P$_i$)	pyrophosphate	−33.5
UDP-glucose ($\rightarrow$UDP + G)	pyrophosphate ester	−33.5
Acetyl-CoA	thioester	−32.2
Glucose 1-phosphate	phosphate ester	−20.9
Glucose 6-phosphate	phosphate ester	−13.8
Fructose 6-phosphate	phosphate ester	−15.9
sn-Glycerol 3-phosphate	phosphate ester	− 9.2

† Measured in the presence of 20 mM Mg^{2+}.

approaches equilibrium); indeed it does not even indicate whether the reaction will proceed at all under the prevailing conditions. There are many reactions with very large negative ΔG values which do not proceed at a detectable rate within the temperature range of living organisms without the aid of catalysts *in vitro* or enzymes *in vivo*. A very good example of this is D-glucose which can sit in a bottle on a laboratory shelf for years in the presence of oxygen without being oxidized to CO_2 and H_2O even though the $\Delta G^{\circ\prime}$ of this reaction [eq. (2.12)] is -2870 kJ mol^{-1}. This reaction proceeds extremely

$$C_6H_{12}O_6 + 6O_2 \rightarrow 6CO_2 + 6H_2O \quad (2.12)$$
$$(\Delta G^{\circ\prime} = -2870 \text{ kJ mol}^{-1})$$

well at high temperatures; the glucose burns. This indicates that despite the very negative $\Delta G^{\circ\prime}$, the reaction will proceed, in the absence of catalysts or enzymes, only if there is an input of energy in the form of heat. The energy which is required to get the reaction going is called the activation energy (see Section A.3). The oxidation of D-glucose to CO_2 and H_2O, however, proceeds at a considerable rate in the process known as respiration (see Chapter 6) in most living organisms without the input of this 'extra' energy. The explanation of this is that the enzymes catalysing the multistep process markedly lower the activation energy of the individual reactions and thereby the overall activation energy.

The ΔG, ΔG° and $\Delta G^{\circ\prime}$ values of reactions are additive in any sequence of consecutive reactions. For instance, in Fig. 2.1 the reaction sequence $A \rightarrow E$ via B, C and D consists of a series of consecutive reactions ($A \rightarrow B$, $B \rightarrow C$, $C \rightarrow D$ and $D \rightarrow E$) which are linked or coupled together by common intermediates (B, C and D). The ΔG, ΔG° and $\Delta G^{\circ\prime}$ of the overall reaction sequence is the sum of the ΔG, ΔG° and $\Delta G^{\circ\prime}$ values respectively of the individual reactions, each being given its proper sign. These relationships are described in general terms in eq. (2.13) where the subscripts 1, 2, 3, etc., refer to the individual reactions.

$$\left.\begin{array}{l} \Delta G_{\text{overall}} = \Delta G_1 + \Delta G_2 + \Delta G_3, \ldots, \text{etc.} \\[4pt] \Delta G^{\circ}_{\text{overall}} = \Delta G^{\circ}_1 + \Delta G^{\circ}_2 + \Delta G^{\circ}_3, \ldots, \text{etc.} \\[4pt] \Delta G^{\circ\prime}_{\text{overall}} = \Delta G^{\circ\prime}_1 + \Delta G^{\circ\prime}_2 + \Delta G^{\circ\prime}_3, \ldots, \text{etc.} \end{array}\right\} \quad (2.13)$$

Since the overall ΔG, ΔG° or $\Delta G^{\circ\prime}$ values of the

reaction sequence $A \rightarrow E$ are solely measures of the difference between the free energy contents of A and E under different sets of circumstances, it follows that this difference must be the same regardless of the route taken or the number of individual reactions involved as is shown in Fig. 2.1.

It is apparent from Fig. 2.1 that individual reactions in a reaction sequence can have a positive $\Delta G^{\circ\prime}$ (e.g. $C \rightarrow D$, $\Delta G^{\circ\prime} = +10$ kJ mol^{-1}) provided that the overall $\Delta G^{\circ\prime}$ has a negative value. Similarly individual reactions in a reaction sequence may have a positive ΔG° value provided that the overall ΔG° is negative. This is, however, not true of ΔG values; all the reactions in a reaction sequence must have a negative ΔG value because, by definition, only if they have a negative ΔG do they proceed in a left to right direction, the direction of the reaction sequence itself. This apparent paradox becomes clear when one remembers that $\Delta G^{\circ\prime}$ and ΔG° values give the free energy change under standard conditions whereas the ΔG value is free energy change under conditions which may be far removed from the standard. Thus in the case of reaction $C \rightarrow D$ (Fig. 2.1), if the [C]/[D] ratio obtaining during the conversion of $A \rightarrow E$ is 125/1, the ΔG, calculated by means of eq. (2.11), is -1.97 kJ mol^{-1}. Such a ratio is not unlikely because C is being continuously formed from B and D is being continually removed by conversion to E. There are innumerable examples amongst the metabolic reaction sequences of living organisms of reactions with positive $\Delta G^{\circ\prime}$ values being pulled or pushed in the left to right direction by being coupled to reactions with large negative $\Delta G^{\circ\prime}$ values; in fact the concept of coupled reactions operating in the manner described is very important in biochemistry. An oft-quoted example occurs in the TCA cycle (see Chapter 6) where the conversion of malate into oxaloacetate ($\Delta G^{\circ\prime} = +27.9$ kJ mol^{-1}) is pulled in that direction by the conversion of oxaloacetate and acetyl-CoA into citrate ($\Delta G^{\circ\prime} = -31.4$ kJ mol^{-1}), thus giving an overall $\Delta G^{\circ\prime}$ of -3.5 kJ mol^{-1}. Another example, in which the consecutive nature of the process is less clear, is that of the reaction catalysed by hexokinase[1] [eq. (2.14)]. This reaction can be considered to be the sum of two other reactions [eqs. (2.15) and

D-glucose + ATP $\rightleftharpoons$

D-glucose 6-phosphate + ADP $\quad (2.14)$

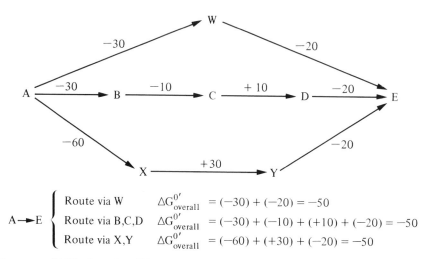

$$
A \rightarrow E \begin{cases} \text{Route via W} & \Delta G^{0'}_{\text{overall}} = (-30) + (-20) = -50 \\ \text{Route via B,C,D} & \Delta G^{0'}_{\text{overall}} = (-30) + (-10) + (+10) + (-20) = -50 \\ \text{Route via X,Y} & \Delta G^{0'}_{\text{overall}} = (-60) + (+30) + (-20) = -50 \end{cases}
$$

Fig. 2.1. The additive nature of $\Delta G^{\circ'}$ values. (A and E represent the compounds at the start and finish respectively of reaction sequences which proceed via compounds W or B, C, D or X, Y. The numbers are the $\Delta G^{\circ'}$ values of the individual reactions in kJ mol^{-1}.)

(2.16)] which are coupled together by having orthophosphate (P_i) as their common intermediate. The $\Delta G^{\circ'}$ values of the left to right reactions of eqs. (2.15) and (2.16) are -30.5 and $+13.8$ kJ mol^{-1}

$$\text{ATP} + H_2O \rightleftharpoons \text{ADP} + P_i \tag{2.15}$$

$$P_i + \text{D-glucose} \rightleftharpoons \text{D-glucose 6-phosphate} + H_2O \tag{2.16}$$

respectively. The overall $\Delta G^{\circ'}$ for the two reactions operating in sequence is therefore -16.7 kJ mol^{-1} which therefore must also be the $\Delta G^{\circ'}$ of the left to right reaction of eq. (2.14). Thus we may say that the mildly endergonic phosphorylation of D-glucose has been driven by the considerably exergonic dephosphorylation of ATP. However, when making that statement it is important to bear in mind that the splitting of reaction (2.14) into two reactions [eqs. (2.15) and (2.16)] operating in sequence does not imply that this is the mechanism of reaction (2.14). In fact it most certainly is not the mechanism; the likely mechanism is shown in Fig. 2.2 and does not involve the hydrolysis of ATP or the release of free P_i. Nevertheless it is perfectly permissible from a thermodynamic viewpoint to regard it as such, reactions (2.15) and (2.16) being one theoretically possible route by which orthophosphate may be transferred from ATP to D-glucose. Since the overall $\Delta G^{\circ'}$ is the same regardless of the route taken (see Fig. 2.1) this is a perfectly correct way of arriving at the $\Delta G^{\circ'}$ of eq. (2.14).

What can be effectively regarded as the $\Delta G^{\circ'}$ of hydrolysis of ATP ($\rightarrow$ADP + P_i or $\rightarrow$AMP + PP_i) is frequently the driving force which gets metabolic processes going and keeps them going. This is accomplished by reactions that are coupled to ATP hydrolysis in the same sense that reaction (2.14) is. For instance for fatty acids to enter the β-oxidation process [see Chapter 6, Section E.2(ii)(a)] they must be activated by conversion to their acyl-CoA derivatives. Since the formation of fatty acyl-CoA from free fatty acid and coenzyme A [eq. (2.17)] is endergonic ($\Delta G^{\circ'} = +32.2$ kJ mol^{-1}) it has to be driven by coupling it to the hydrolysis of ATP [$\rightarrow$AMP + PP_i; eq. 2.18)] which is exergonic ($\Delta G^{\circ'} = -31.8$ kJ mol^{-1}) but not decisively so. The reaction is pulled to completion by the exergonic hydrolysis of the pyrophosphate ($\Delta G^{\circ'} = -33.5$ kJ mol^{-1}) produced in reaction (2.18) [eq. (2.19)], thus giving an overall $\Delta G^{\circ'}$ of -33.1 kJ mol^{-1}.

$$R \cdot COOH + HS \cdot CoA \rightleftharpoons$$
$$R \cdot CO \cdot S \cdot CoA + H_2O \tag{2.17}$$

$$\text{ATP} + H_2O \rightleftharpoons \text{AMP} + PP_i \tag{2.18}$$

$$PP_i + H_2O \rightleftharpoons P_i + P_i \tag{2.19}$$

The mechanism of coupling of reactions (2.17) and (2.18) is shown in Fig. 2.3, where it is apparent that the cleavage of pyrophosphate from ATP yields a fatty acyl-adenylate (i.e. fatty acyl-AMP) which

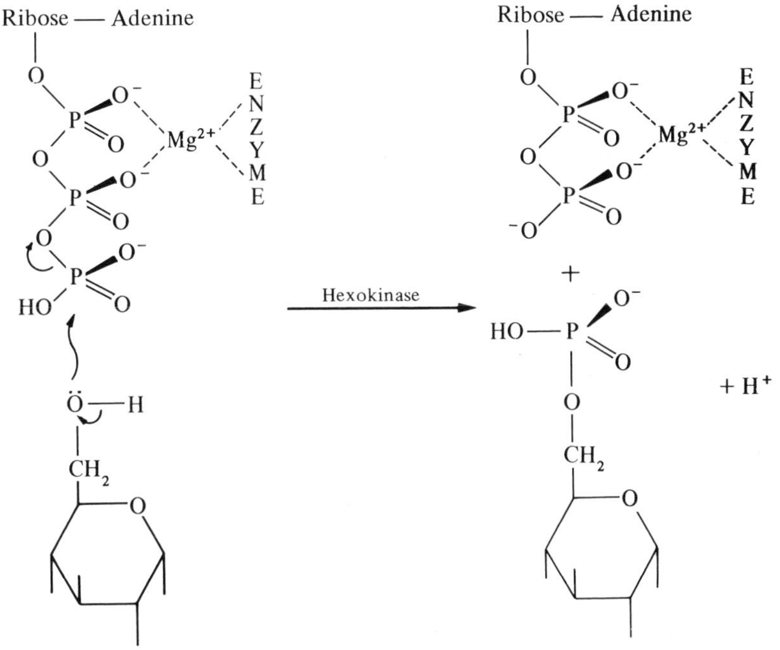

Fig. 2.2. Likely mechanism of the hexokinase-catalysed phosphorylation of D-glucose.

then reacts with coenzyme A to give the fatty acyl-CoA and AMP. Thus there are two sequential reactions linked by fatty acyl-adenylate as their common intermediate.

The biosynthesis of proteins from amino acids (see Chapter 10) is begun by a reaction which is both mechanistically and energetically similar to the formation of fatty acyl-CoA as is clear from Fig. 2.3. The amino acid is converted into its amino acyl-adenylate which then reacts with the 3'-hydroxyl group of the terminal nucleotide in a transfer RNA molecule to form an amino acyl-tRNA and AMP.

ATP plays a central role in metabolism by being used in coupled reactions of the kind described above to drive endergonic reactions in the required direction. In these processes it is cleaved to ADP or AMP. Since the amount of ATP in living tissues is finite it would soon be totally converted into ADP and AMP unless there were processes designed to regenerate it. The most important of these processes in plants is photophosphorylation, a process which is an integral part of photosynthesis (see Chapter 5). ATP is also regenerated in respiration (see Chapter 6) by processes known as substrate-level phosphorylation and oxidative phosphorylation. The

energy ($\Delta G^{\circ\prime} = +30.5 \text{ kJ mol}^{-1}$) required to drive eq. (2.20) from left to right as written is derived, in

$$\text{ADP} + \text{P}_i \rightleftharpoons \text{ATP} + \text{H}_2\text{O} \qquad (2.20)$$

the case of photophosphorylation and oxidative phosphorylation, indirectly from the flow of electrons from a negative electrode potential (see Section 5) to a positive electrode potential via a chain of oxidation-reduction systems. In the case of substrate-level phosphorylation, eq. (2.20) is coupled to very exergonic reactions such as the cleavage of phosphate from phosphoenol pyruvate ($\Delta G^{\circ\prime}$ of hydrolysis = $-61.9 \text{ kJ mol}^{-1}$) or 1,3-diphosphoglyceric acid ($\Delta G^{\circ\prime}$ of hydrolysis = $-49.3 \text{ kJ mol}^{-1}$).

Because of the importance of ATP in metabolism some comment should be made about (i) its $\Delta G^{\circ\prime}$ and its intracellular ΔG value and (ii) the structural basis for its relatively high $\Delta G^{\circ\prime}$ of hydrolysis. In this book the $\Delta G^{\circ\prime}$ of hydrolysis of ATP ($\rightarrow$ADP) is given as $-30.5 \text{ kJ mol}^{-1}$. However, the published values vary somewhat. This variation is due to (i) the impracticability of measuring the equilibrium constant of reaction (2.15) because the equilibrium is so far to the right that analytical methods are not

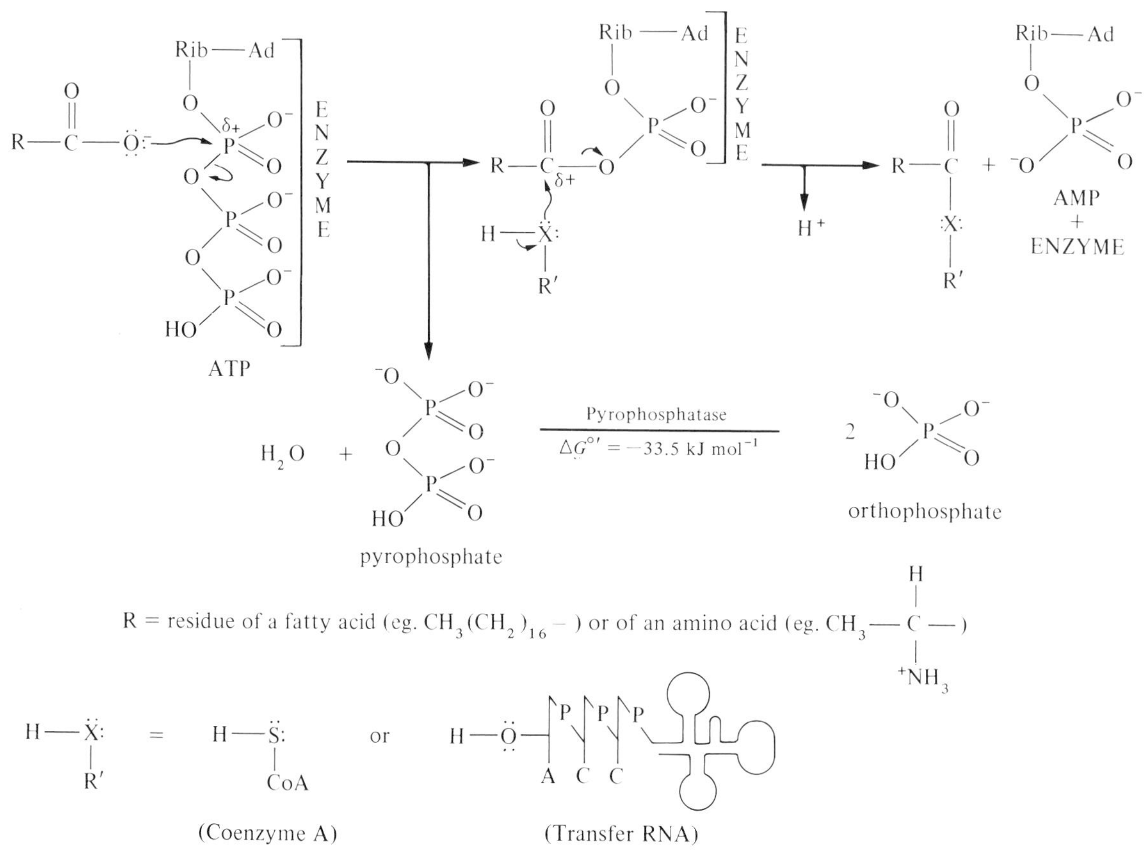

Fig. 2.3. The coupling of ATP cleavage to the generation of fatty acyl-CoA and amino acyl-tRNA.

sufficiently accurate to measure the exact equilibrium concentrations of ATP, ADP and P_i (remember that the $\Delta G^{\circ\prime}$ can be calculated with eq. (2.8) if the equilibrium constant at pH 7 is known), (ii) analytical difficulties in determining accurate equilibrium constants of coupled reactions (e.g. reaction (2.14) [proceeding L→R] and reaction (2.16) [proceeding R→L]) from which the $\Delta G^{\circ\prime}$ of hydrolysis of ATP can be determined indirectly and (iii) the fact that measurements by different researchers have been made under slightly different conditions.

The actual free energy change of hydrolysis, ΔG, of ATP *in vivo* depends upon several factors, the temperature, the pH, the concentrations of ATP, ADP and P_i and the concentration of metal ions, particularly Mg^{2+}, which are known to complex reversibly with the various ionic species of ATP, ADP and P_i. The variation of the ΔG with pH is due to the change in the relative proportions of the various ionic species of ATP (i.e. ATP^{3-}, ATP^{4-}), ADP (i.e. ADP^{2-}, ADP^{3-}) and P_i (i.e. $H_2PO_4^-$, HPO_4^{2-}) as the pH changes. Since the affinity of Mg^{2+} ions is different for each of these ionic species and since this affinity changes with pH, it is apparent that the interrelationship of pH, Mg^{2+} concentration and the ΔG of hydrolysis of ATP is complex. It is also clear that *in vivo* the ΔG of hydrolysis of ATP is unlikely to be constant for there are likely to be slight differences in pH, Mg^{2+} (or other metal ion) concentration and in the concentrations of ATP, ADP and P_i (sum of all ionic forms of each) from tissue to tissue, cell type to cell type and intracellular compartment to intracellular compartment. A value of about -52.3 kJ mol^{-1} has been estimated for the ΔG of hydrolysis of ATP in the cytosol of certain animal cells; this figure takes account of the cytosol pH, Mg^{2+} concentration and ATP, ADP and P_i concentrations.

The structural basis for the relatively high free

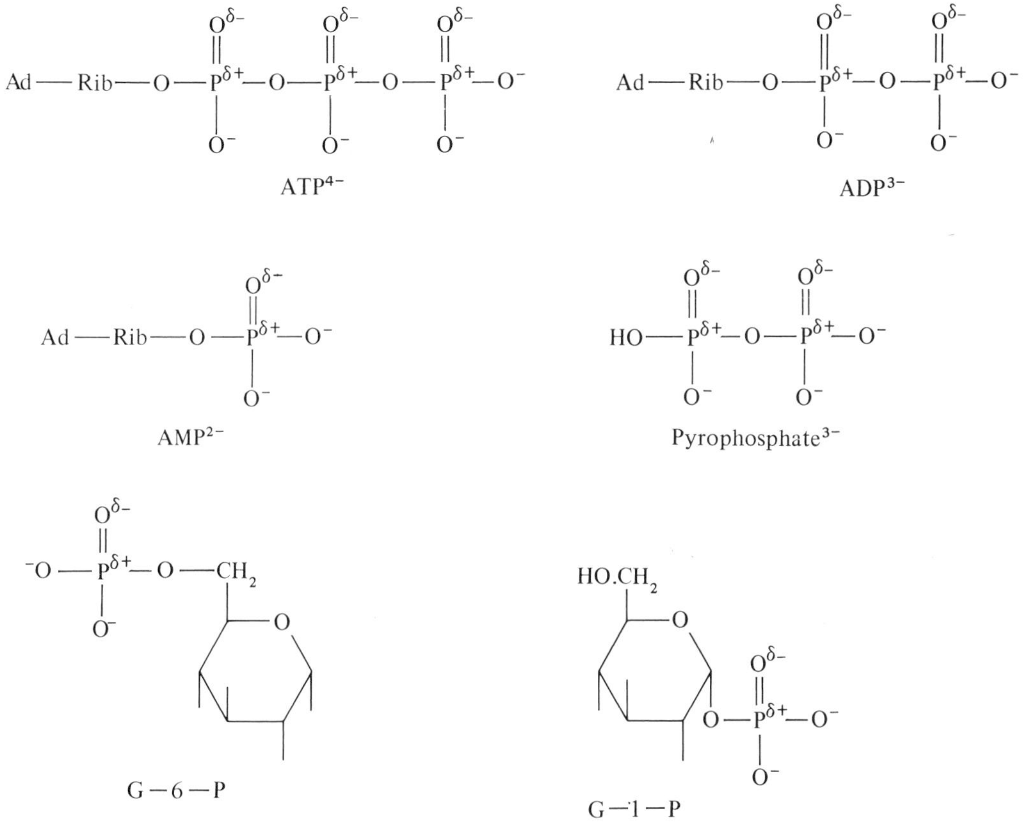

Fig. 2.4. Electrostatic repulsion between like charges in pyrophosphate esters but not in orthophosphate esters.

energy change of hydrolysis of ATP is the instability due to charge repulsion amongst the three inter-linked phosphate residues. This is at its greatest in the ionized forms, ATP^{3-} and ATP^{4-}, which are present in roughly equal amounts at pH 7. In these ions each phosphorus atom carries an oxygen atom with a full negative charge owing to ionization and a doubly-bonded oxygen atom. The latter both carry a fractional negative charge because of the tendency of the electrons in the P=O bond to be drawn closer to the more electronegative oxygen atom; for the same reason each phosphorus atom carries a fractional positive charge ($\delta+$) (see Fig. 2.4). The presence of these charges means that the ATP molecule must possess sufficient internal energy to overcome the electrostatic repulsion between like charges. However, when the terminal phosphate or pyrophosphate is cleaved by hydro-lysis to yield ADP or AMP respectively this energy is released and contributes to the negative ΔG of the

reaction. The same reasoning accounts for the relatively high $\Delta G^{\circ\prime}$ of hydrolysis ($-27.2 \text{ kJ mol}^{-1}$) of ADP ($\rightarrow$AMP) and of pyrophosphate ($-33.5 \text{ kJ mol}^{-1}$). The lack of electrostatic repul-sion between like charges in AMP and phosphate monoesters like D-glucose 1-phosphate and D-glucose 6-phosphate (see Fig. 2.4) explains their low $\Delta G^{\circ\prime}$ of hydrolysis (see Table 2.1).

2. Enthalpy and entropy

The free energy change, ΔG, of reaction (2.1) is related to two other thermodynamic properties of the reactants, A and B, and the products, C and D, by eq. (2.21), in which ΔH is the change in enthalpy, T is the absolute temperature (degrees Kelvin) and

$$\Delta G = \Delta H - T \Delta S \qquad (2.21)$$

ΔS is the change in entropy. Enthalpy is frequently called heat content because the change in enthalpy (ΔH) of a system, like reaction (2.1), is manifested as the heat which is exchanged between the isothermal system and its surroundings at constant pressure. The ΔH of a reaction can be measured by carrying out the reaction in a bomb calorimeter; it is obtained from eq. (2.22) where ΔU is the observed heat change, R is the gas constant, T is the absolute temperature and n is the increase in moles of gaseous products over reactants.

$$\Delta H = \Delta U + nRT \qquad (2.22)$$

Entropy is a measure of the randomness or disorderliness of the system. The more random or disordered the system, the higher is the entropy and conversely the less random or more ordered the system, the lower is the entropy. Thus if a reaction results in a change from a relatively ordered state to a less ordered state there is an increase in entropy When a reaction is in equilibrium $\Delta G = 0$ and $\Delta H = T \Delta S$. If ΔH has a large negative value it is apparent from eq. (2.21) that ΔG is nearly always negative (i.e. the reaction is exergonic). Under these circumstances ΔS can be negative (i.e. have a low value), thus allowing for the synthesis of ordered structures (e.g. proteins, polysaccharides) from less ordered structures (e.g. amino acids, mono-saccharides).

3. Activation energy

The law of mass action states that the rate of a chemical reaction is proportional to the concentration of the reactants, so that in reaction (2.23) the velocity, v, can be expressed as $v = k_r[A][B]$ where k_r is a constant called the rate constant.

$$A + B \rightarrow C + D \qquad (2.23)$$

In 1889 the Swedish chemist, Arrhenius, examined the available data on the effect of temperature on the rates of chemical reactions and concluded that the rate constant k_r of a reaction and the absolute temperature, T (degrees Kelvin), were related in a manner defined by eq. (2.24), which has become known as the Arrhenius equation.

$$\frac{d \log_e k_r}{dT} = \frac{E}{RT^2} \qquad (2.24)$$

In this equation R is the gas constant ($8.314 \text{ J K}^{-1} \text{ mol}^{-1}$) and E is a constant known as the activation energy of the reaction and expressed in J mol^{-1}.

The integrated form of the Arrhenius equation [eq. (2.25)] is more useful since it is an equation of a straight line. It follows therefore that if $\log_{10} k_r$ is

or
$$\left. \begin{array}{c} \log_e k_r = \dfrac{-E}{RT} + A \\[2mm] \log_{10} k_r = \dfrac{-E}{2.303\,RT} + A' \end{array} \right\} \qquad (2.25)$$

(A and A' are constants of integration.)

plotted as ordinate (y-axis) against $1/T$ as abscissa (x-axis) a straight line with a negative slope, whose value is equal to $-E/2.303R$, will result. Thus from the slope the value of the activation energy, E, can be calculated.

The significance of E is that it is a measure of the energy that molecules [A and B in eq. (2.23)] must acquire before they can undergo a reaction [i.e. be converted into C and D in eq. (2.23)] under a given set of conditions. This was originally interpreted in terms of the Collision Theory developed by Arrhenius and van't Hoff. However, this theory is strictly only applicable to reactions between gases and has been superseded by the Transition State Theory developed by Eyring. In simple terms this theory states that reaction (2.23) will only take place when molecules A and B acquire sufficient energy to form a transition state complex AB^+ whose potential energy is greater than that of A plus B. The transition state complex AB^+ can then decompose to yield the products of the reaction, C and D, whose potential energy is, of course, less than that of A and B since reaction (2.23) must be exergonic. This is depicted graphically in Fig. 2.5. The rate of the reaction is proportional to the concentration of AB^+. The activation energy is the amount of energy required to bring all the molecules in 1 mole of A and B at a given temperature to the top of the energy barrier separating A and B from C and D.

There are two common ways of increasing the rate of a chemical reaction. The first is by increasing

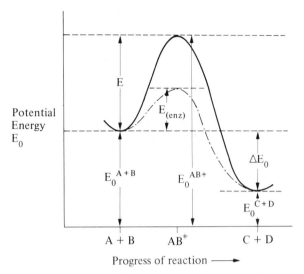

Fig. 2.5. Graph showing energy of activation of an uncatalysed reaction (E) and of an enzyme-catalysed reaction (E_{enz}).

the temperature; this increases the thermal motion of the reactants [A and B in eq. (2.23)] and thus raises the average potential energy of the reactant molecules bringing a greater proportion of them to the top of the energy barrier. The second way is by adding a catalyst. This combines transiently with the reactants to form a transition state with a lower activation energy; a catalyst therefore functions by lowering the energy barrier. Since enzymes are catalysts, they, too, increase the rate of chemical reactions by lowering the activation energy. With this lower activation energy a greater proportion of the reactant molecules [A and B in eq. (2.23)] have sufficient energy to reach the top of the energy barrier even though their average potential energy has not changed. The effect of an inorganic catalyst and an enzyme on activation energy is exemplified by the decomposition of hydrogen peroxide to water and oxygen. The activation energy of the uncatalysed reaction is about 75 kJ mol^{-1}; the activation energy of the reaction when catalysed by colloidal platinum is about 54 kJ mol^{-1} and the activation energy of the reaction when catalysed by the enzyme catalase[2] is about 29 kJ mol^{-1}. Enzymes are the most efficient catalysts known. This is fortunate because they represent the only way that the activation energy barrier can be surmounted in living organisms which can only live within a fairly narrow temperature range.

4. Redox potentials

Respiration is basically the oxidation of a substrate with the liberation of free energy which can be used for various endergonic cellular processes; the fundamental process is thus the transfer of electrons from a substrate to oxygen. The free energy changes associated with such processes can be calculated from the redox potentials of the two components (substrate and oxygen).

The redox potential (or reduction-oxidation potential or reduction potential) of a compound is a quantitative measure of its ability to gain or lose electrons (i.e. its electron affinity). If a solution of a ferric salt is placed in contact with a chemically inert platinum electrode, the ferric (Fe^{3+}) ions tend to gain electrons and become reduced to ferrous (Fe^{2+}) ions; this leaves the platinum positively charged. The reverse situation will occur when a platinum electrode is placed in a solution of a ferrous salt. In a mixture of the two salts an equilibrium will be set up [eq. (2.26)]. The potential

$$Fe^{3+} + e^- \rightleftharpoons Fe^{2+} \qquad (2.26)$$

of the platinum electrode at equilibrium is designated E and is related to the concentration of the oxidant (Fe^{3+}) and the reductant (Fe^{2+}) and to the prevailing temperature by eq. (2.27), where n is the

$$E = E_0 + \frac{RT}{nF} \log_e \frac{[\text{oxidant}]}{[\text{reductant}]} \qquad (2.27)$$

number of moles of electrons per mole of reactant transferred in the redox reaction, R is the gas constant (8.314 J K^{-1} mol^{-1}), T is the absolute temperature and F is the Faraday (96,487 coulombs mol^{-1}). It is apparent from eq. (2.27) that when the concentrations of oxidant and reductant are equal, the second term on the right-hand side of the equation becomes zero and E becomes equal to E_0; the significance of the term E_0 will be made apparent later.

It is impossible, in practice, to measure the *absolute* potential of a reduction-oxidation system (redox system) or half-cell such as that described in eq. (2.26). It is, however, possible to measure the potential relative to another half-cell, that is to determine by how much the potential of one half-cell is different from that of another. This is accomplished by connecting the two half-cells together by means of an agar-KCl bridge,† so forming an electrical cell. If the electrodes of the two half-cells are now connected by a metal wire electrons will flow from one half-cell to the other. The direction of electron flow depends upon the relative affinities (i.e. the redox potentials) of the two half-cells for electrons; the electrons will flow from the half-cell with the lesser affinity for electrons to the half-cell with the greater affinity for electrons. The half-cell with the lesser affinity for electrons will therefore assume the reducing role in the reduction-oxidation reaction resulting from the interaction of the two half-cells, whilst the half-cell with the greater affinity for electrons will assume the oxidizing role. The 'intensity' of the electron flow from the reducing half-cell to the oxidizing half-cell is termed the electromotive force (e.m.f.) of the electrical cell. The electromotive force is, in effect, an expression of the difference in the electron affinities of the two half-cells which constitute the electrical cell. Since the redox potential of a half-cell is a measure of its electron affinity, the e.m.f. of the electrical cell is the arithmetic difference between the redox potentials of the two half-cells, that is the *potential difference* between the two half-cells. The e.m.f. of an electrical cell can be measured directly and is expressed in volts.‡

It has been found convenient to relate the redox potential of all half-cells to that of a common standard half-cell. The half-cell chosen as the standard is the hydrogen half-cell which is usually called the standard hydrogen electrode. It consists of a platinum strip coated with platinum black immersed in a solution of hydrogen ions at unit activity (i.e. a concentration of 1 M, pH = 0) saturated with hydrogen gas at a pressure of 1 atmosphere (i.e. all the components of the electrode are in their standard states). It is necessary to define the

concentration of H_2 (pressure) and H^+ (pH) because both affect the reaction [eq. (2.28)] which

$$2H^+ + 2e^- \rightleftharpoons H_2 \qquad (2.28)$$

gives rise to the redox potential. The redox potential of the standard hydrogen electrode is arbitrarily given the value of 0.000 volt at all temperatures.

In order to determine the redox potential of a given half-cell it must be suitably joined to the standard hydrogen electrode and the e.m.f. of the resulting electrical cell measured. Since the redox potential of the standard hydrogen electrode is zero and the e.m.f. is the potential difference between the two half-cells concerned, the redox potential of the half-cell in question is equal to the e.m.f. of the cell.

By measuring the redox potentials of half-cells relative to that of the standard hydrogen electrode a scale of electrode potentials is set up. This is called the hydrogen scale and is measured in volts. On this scale a half-cell which is more strongly reducing (i.e. has a lesser affinity for electrons) than the standard hydrogen electrode has a negative redox potential (E is $-$ve); if such a half-cell were joined to the standard hydrogen electrode electrons would flow from that half-cell to the standard hydrogen electrode. Conversely a half-cell which is more strongly oxidizing (i.e. has a greater affinity for electrons) than the standard hydrogen electrode has a positive redox potential (E is $+$ve); if such a half-cell were joined to the standard hydrogen electrode electrons would flow from the standard hydrogen electrode to that half-cell.

From eq. (2.27) it was seen that the redox potential (E) of a half-cell depends upon (a) the concentrations of oxidant and reductant, and (b) the temperature. Therefore in order to facilitate the comparison of one half-cell with another on the hydrogen scale it is necessary to define certain standard conditions under which these redox potentials shall be measured. These conditions are (i) the concentrations of oxidant and reductant shall be present at unit activity (which is for most practical purposes a concentration of 1 M), (ii) the temperature shall be constant (usually 25° unless stated otherwise), (iii) the pressure shall be 1 atmosphere. When a half-cell which fulfils these requirements is coupled with the standard hydrogen electrode the e.m.f. measured gives the *standard redox potential* of that half-cell. The standard redox potential is

---

† The agar-KCl bridge enables electrical neutrality (i.e. equality of positive and negative charges) to be maintained in both half-cells which constitute the electrical cell by allowing positively or negatively charged ions to migrate from one half-cell to the other as required.

‡ A volt is that unit of potential difference (e.m.f.) which produces a current of one ampere (1 amp ≡ standard flow of electrons) through a conductor which has a resistance (to the passage of electrons) of one ohm.

designated E_0. It is usually the E_0 values of half-cells which are tabulated in the literature for reference purposes. Knowing the E_0 and the concentrations of oxidant and reductant (or even the ratio of their concentrations) one can calculate, using eq. (2.27), the actual redox potential E (often designated E_h) of a given half-cell.

The term midpoint potential, designated E_m, is frequently to be found in the literature. Its value is numerically equal to that of E_0, yet E_0 and E_m are not synonyms because they are determined under different sets of conditions. As stated above the term E_0 implies that the concentrations of oxidant and reductant in the half-cell in question are each 1 M during the measurement. However, it is not practicable to obtain 1 M concentrations of oxidant and reductant in the case of many biochemically important redox systems (e.g. membrane-bound cytochromes); the best that can be done is to measure the redox potential when the concentrations of oxidant and reductant are equal but very much lower than 1 M. The resulting redox potential which is called the midpoint potential, E_m, is nevertheless equal to the E_0 value because it is clear from eq. (2.27) that whenever the oxidant and reductant of a redox system are present in equal concentrations, regardless of whether they are each 1 M, the term $\log_e$ [oxidant]/[reductant] equals 0 and $E = E_0$.

(i) EFFECT OF pH

When the E_0 value for the Fe^{3+}/Fe^{2+} half-cell is measured at different pH values (pH 0–5) it is found that there is little change. This lack of change of E_0 with pH is not, however, found with half-cells in which hydrogen ions participate in the chemical reaction. Therefore in the case of these cells the hydrogen ion concentration has to be taken into consideration when the redox potential is determined; eq. (2.27) is modified to accommodate the hydrogen ion concentration, giving eq. (2.29).

$$E = E_0 + \frac{RT}{nF} \log_e \frac{[\text{oxidant}]}{[\text{reductant}]}$$

$$+ \frac{RT}{F} \log_e [H^+] \qquad (2.29)$$

Equation (2.29) is usually simplified by replacing E_0 and the term involving H^+ by E_0' whose value is given by eq. (2.30).

$$E_0' = E_0 + \frac{RT}{F} \log_e [H^+] \qquad (2.30)$$

This gives eq. (2.31).

$$E = E_0' + \frac{RT}{nF} \log_e \frac{[\text{oxidant}]}{[\text{reductant}]} \qquad (2.31)$$

When the concentration of the oxidant is equal to that of the reductant in eq. (2.31) E is equal to E_0'. The E_0' value of a half-cell is therefore obtained by arranging for the concentrations of oxidant and reductant to be equal at a known pH, coupling it to the standard hydrogen electrode and measuring the e.m.f. The E_0' value of a half-cell is thus defined as the redox potential at a stated pH and temperature when the concentrations of oxidant and reductant are equal; it is the *standard redox potential or midpoint potential at a stated pH other than 0* (if the pH is not stated it is assumed to be pH 7; if the temperature is not stated it is assumed to be 25°C). E_0' values, like E_0 values, are tabulated for reference purposes for they provide a yardstick by which the redox potentials of different half-cells may be compared on the hydrogen scale. Knowing the E_0' and the concentrations of oxidant and reductant (or even the ratio of their concentrations) one can calculate, using eq. (2.31), the actual redox potential (E or E_h) of a given half-cell. The E_0' values of some biologically important half-cells are given in Table 2.2.

Further examination of eq. (2.29) reveals that when the concentration of the oxidant is equal to that of the reductant the second term is eliminated giving eq. (2.32).

$$E = E_0 + \frac{RT}{F} \log_e [H^+]$$

$$= E_0 + 2.303 \frac{RT}{F} \log_{10} [H^+] \qquad (2.32)$$

Since $pH = -\log_{10} [H^+]$, eq. (2.32) can be modified to become eq. (2.33) which, after substituting the appropriate constants, simplifies to eq. (2.34).

$$E = E_0 + \left[\left(2.303 \frac{RT}{F} \right) \times (-pH) \right] \qquad (2.33)$$

$$E = E_0 - 0.059 \text{ pH (at 25°C)} \qquad (2.34)$$

From eq. (2.34) it can be seen that the redox potential (E) of a half-cell at pH 0 is equal to E_0 but

Table 2.2. E_0' values of some biologically important redox systems

Redox system (half-cell)	E_0' (V) at pH 7
Acetate $+ CO_2 + 2H^+ + 2e$/pyruvate	-0.699
Succinate $+ CO_2 + 2H^+ + 2e$/α-oxyglutarate $+ H_2O$	-0.673
Acetate $+ 2H^+ + 2e$/acetaldehyde	-0.600
3-Phosphoglycerate $+ 2H^+ + 2e$/3-phosphoglyceraldehyde $+ H_2O$	-0.55
Gluconate $+ 2H^+ + 2e$/glucose $+ H_2O$	-0.45
Ferredoxin $-Fe^{3+} + e$/ferredoxin $-Fe^{2+}$	-0.43
$CO_2 + 2H^+ + 2e$/formate	-0.42
$H^+ + e$/$\frac{1}{2}H_2$	-0.413
Acetyl-CoA $+ 2H^+ + 2e$/acetaldehyde $+ CoA$	-0.41
Pyruvate $+ CO_2 + 2H^+ + 2e$/malate	-0.33
$NAD^+ + 2H^+ + 2e$/NADH $+ H^+$	-0.32
$NADP^+ + 2H^+ + 2e$/NADPH $+ H^+$	-0.32
Lipoate $(-S-S-) + 2H^+ + 2e$/lipoate $(-SH\ HS-)$	-0.29
1,3-Diphosphoglyceric acid $+ 2H^+ + 2e$/3-phosphoglyceraldehyde $+ P_i$	-0.29
$FMN + 2H^+ + 2e$/FMNH$_2$	-0.22
$FAD + 2H^+ + 2e$/FADH$_2$	-0.18
Cytochrome b_6 ($\equiv b_{563}$) $-Fe^{3+} + e$/cytochrome $b_6 - Fe^{2+}$	-0.18
Oxaloacetate $+ 2H^+ + 2e$/malate	-0.175
Acetaldehyde $+ 2H^+ + 2e$/ethanol	-0.163
Fumarate $+ 2H^+ + 2e$/succinate	$+0.030$
Cytochrome b_{559} LP$-Fe^{3+} + e$/cytochrome b_{559} LP$-Fe^{2+}$	$\sim +0.06$
Ubiquinone $+ 2H^+ + 2e$/ubiquinol	$+0.10$
Cytochrome $c-Fe^{3+} + e$/cytochrome $c-Fe^{2+}$	$+0.25$
Cytochrome $a-Fe^{3+} + e$/cytochrome $a-Fe^{2+}$	$+0.29$
Cytochrome $f-Fe^{3+} + e$/cytochrome $f-Fe^{2+}$	$+0.365$
Cytochrome b_{559} HP$-Fe^{3+} + e$/cytochrome b_{559} HP$-Fe^{2+}$	$\sim +0.37$
Plastocyanin$_{(ox)} + e$/plastocyanin$_{(red)}$	$+0.37$
P700$^{+\cdot} + e$/P700	$+0.43$
Cytochrome $a_3-Fe^{3+} + e$/cytochrome a_3-Fe^{2+}	$+0.55$
$\frac{1}{2}O_2 + 2H^+ + 2e$/H$_2$O (or $\frac{1}{2}O_2 + H_2O + 2e$/2OH$^-$)	$+0.817$

at the physiological pH of 7 it differs from E_0 by -0.413 volt (i.e. -0.059×7). This means that the redox potential of the standard hydrogen electrode, which is arbitrarily taken as 0.000 volt at pH 0, becomes -0.413 volt at pH 7 and 25°C; or, in other words, the E_0' value of the hydrogen electrode (at pH 7 and 25°C) is -0.413 volt.

(ii) THE OXYGEN ELECTRODE

As respiration is concerned with the transfer of electrons to oxygen it is important to know the standard redox potential of the oxygen electrode. The reaction taking place at the oxygen electrode can be described in a number of ways but from the biochemist's point of view that given in eq. (2.35) is the most appropriate.

$$\tfrac{1}{2}O_2 + H_2O + 2e^- \rightleftharpoons 2OH^- \qquad (2.35)$$

The standard oxygen electrode, in which oxygen

and hydroxyl ions are present in their standard state (O_2 at 1 atmosphere; OH$^-$ at a concentration of 1 M = pH 14), has a redox potential of $+0.401$ volt on the hydrogen scale (i.e. if the standard hydrogen electrode were joined to the standard oxygen electrode the e.m.f. would be 0.401 volt). The redox potential of the standard oxygen electrode is thus $+0.401$ volt at pH 14. The redox potential at pH 0 (which is the standard redox potential, E_0) can be calculated from eq. (2.34), $E_0 = [0.401 + (0.059 \times 14)] = +1.23$ volts. Similarly the redox potential at pH 7 (which is the E_0' value) can also be calculated from eq. (2.34), $E_0' = [1.23 - (0.059 \times 7)] = +0.817$ volt.

5. Free energy changes in a redox reaction

The standard free energy change at defined pH

$(\Delta G^{\circ\prime})$ in an isothermal reduction-oxidation reaction can be calculated from eq. (2.36) where n is the

$$\Delta G^{\circ\prime} = -nF\,\Delta E_0' \qquad (2.36)$$

number of moles of electrons[†] transferred per mole of reactant through a potential difference of $\Delta E_0'$ ($\Delta E_0' = E_0'$ of the oxidizing redox couple or half-cell minus the E_0' of the reducing redox couple or half-cell) and F is the Faraday (96,487 coulombs mol^{-1}). This gives $\Delta G^{\circ\prime}$ in terms of volt coulombs per mole or $J\,mol^{-1}$ since 1 volt coulomb equals 1 joule.

Therefore from eq. (2.36) the $\Delta G^{\circ\prime}$ value of a redox reaction can be calculated provided that the E_0' values of the two redox couples (half-cells) are known. To take an example: at pH 7 and 25°C the E_0' of the $NAD^+/NADH$ redox couple is -0.320 volt and the E_0' of the $\frac{1}{2}O_2/OH^-$ redox couple is $+0.817$ volt; this gives a $\Delta E_0'$ of $+0.817 - (-0.320) = 1.137$ volts which, substituted in eq. (2.36) gives a $\Delta G^{\circ\prime}$ for the reaction indicated [eq. (2.37)] of $-2 \times 96,487 \times 1.137 = 219,411\ J\ mol^{-1}$. In this calculation $n = 2$ because 2 moles of electrons are transferred per mole of reactant.

$$NADH + H_2O + \tfrac{1}{2}O_2 \rightleftharpoons NAD^+ + H^+ + 2OH^-$$
$$(2.37)$$

In this example, since E_0' values have been used, the concentration of the oxidant equals that of the reductant in both redox couples (see definition of E_0'). However, in the living cell this situation is unlikely to occur, and the true ΔG may be considerably different from the $\Delta G^{\circ\prime}$ calculated above.

The true ΔG of the reaction can be calculated provided that the E_0' values and the steady-state ratios of oxidant and reductant of both redox couples are known by making use of eq. (2.31) and eq. (2.38), which is a more general expression of eq. (2.36).

$$\Delta G = -nF\,\Delta E \quad \text{or} \quad -nF\,\Delta E_h \qquad (2.38)$$

[†] Since 1 mole of any compound contains 6.023×10^{23} molecules (Avogadro's number), 1 mole of electrons contains 6.023×10^{23} electrons.

For example, if the ratio of $NAD^+/NADH$ were 100:1 and that of $\frac{1}{2}O_2/OH^-$ were 2:1, the true redox potentials (E or E_h) of the two redox couples can be calculated from eq. (2.31) as follows:

$$E = E_0' + 2.303\,\frac{RT}{nF}\log_{10}\frac{[NAD^+]}{[NADH]}$$

$$= -0.320 + \left(\frac{2.303 \times 8.314 \times 298.15}{2 \times 96,487}\right)^{[‡]} \times \log_{10}\frac{100}{1}$$

$$= -0.320 + (0.0296) \times 2$$

$$= -0.261 \text{ volt (at 25°C)}$$

and similarly

$$E = E' + 2.303\,\frac{RT}{nF}\log_{10}\frac{[\frac{1}{2}O_2]}{[OH^-]}$$

$$= +0.817 + \left(\frac{2.303 \times 8.314 \times 298.15}{2 \times 96,487}\right) \times \log_{10}\frac{2}{1}$$

$$= +0.817 + (0.0296) \times 0.301$$

$$= +0.826 \text{ volt (at 25°C)}$$

The ΔE for the reduction-oxidation reaction [eq. (2.37)] is therefore $+0.826 - (-0.261) = +1.087$ volts. This value, substituted in eq. (2.38), gives a true ΔG (*not* $\Delta G^{\circ\prime}$) of $-209,763\ J\ mol^{-1}$.

SUGGESTIONS FOR FURTHER READING

Morris, J. G. (1973) *A Biologist's Physical Chemistry*, 2nd edition. Edward Arnold Ltd., London.

Segel, I. H. (1976) *Biochemical Calculations*, 2nd edition. Wiley, New York, London.

ENZYMES

1. ATP: D-hexose 6-phosphotransferase, EC 2.7.1.1.
2. Pyrophosphate phosphohydrolase, EC 3.6.1.1.

[‡] The term $2.303RT/nF$ is approximately equal to $0.059/n$ at 25°C; at 30°C it equals $0.06/n$.

CHAPTER 3

Structure and Function of the Plant Cell

CONTENTS

A. GENERAL MORPHOLOGY OF A TYPICAL PLANT CELL

Although this section purports to describe the structure of a typical plant cell it should be emphasized that there is no such thing as a typical plant cell. Plants are made up of many different types of cells which have different structures and functions. These different cell types can be divided into two broad groups; the first is responsible for performing all the metabolic activities of the plant, while the second group is metabolically inactive and functions either as a mechanical support or for conduction of fluids through the plant. Since the cells of the second group developed from meta-bolically active cells they cannot be regarded as typical plant cells. Thus for the purposes of this section one must select a metabolically active cell. Perhaps the most convenient choice is a mature parenchymatous cell since such cells not only contain all the biochemically important cell or-ganelles but also make up about 80% of the cell complement of a higher plant.

A parenchymatous cell can be regarded as being made up of two primary components, the cell wall and the protoplast (Fig. 3.1). The cell wall com-pletely surrounds the protoplast and provides mechanical support for it. It is composed of two layers; the outer layer is called the primary wall whilst the inner layer, nearest the protoplast and usually composed of several sublayers, is known as the secondary wall. Sandwiched between the primary walls of adjacent cells is an intercellular layer known as the middle lamella.

The protoplast is the metabolically active compo-nent of the cell and consists of two parts, the cytoplasm and the vacuole. The cytoplasm is a fluid

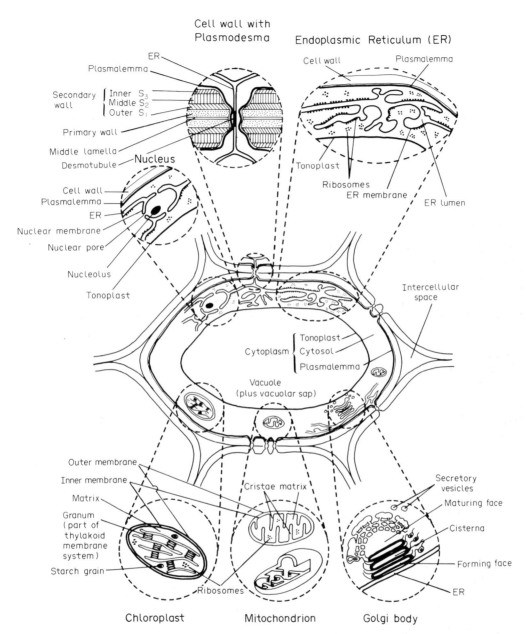

Fig. 3.1. Structure of the plant cell.

material of varying viscosity which is usually present as a thin film immediately within the cell wall and completely enclosing the vacuole. The cytoplasm may, however, traverse the vacuole as strands, or as sheets which subdivide the vacuole into several distinct compartments; the cell may thus have one large vacuole or two or more smaller ones. Vacuoles normally contain an aqueous fluid which is called cell sap or vacuolar sap. Each vacuole is bounded by a membrane known as the tonoplast which separates the vacuolar sap from the cytoplasm. The outer edge of the cytoplasm is bounded by a membrane known as the plasmalemma or plasma membrane.

Embedded in the ground phase of the cytoplasm, the cytosol, are numerous particulate and membranous inclusions. The most extensive of these is the endoplasmic reticulum which consists of a ramified network of interconnected, membrane-bound tubules and vesicles. The endoplasmic reticular membrane is continuous with the nuclear membrane but not with the plasmalemma or tonoplast.

The largest of the particulate inclusions is the nucleus which is frequently spherical with a diameter typically in the range 6–8 nm. It is surrounded by a double membrane which is perforated by nuclear pores and is continuous with the membranes of the endoplasmic reticulum. Within the nuclear membrane is the nucleoplasm, embedded in which are the chromosomes which carry the genetic information of the cell. The most obvious feature in the nucleoplasm is the nucleolus, a region where ribosomes are formed.

The next largest of the particulate inclusions are the chloroplasts which are one type of the general group of intracellular organelles known as plastids; they contain the green chlorophyll pigments and the yellow carotenoid pigments and are specialized for photosynthesis. They are typically lens-shaped with a diameter of up to 10 μm and are surrounded by a double membrane. They are present to the extent of 200–400 per cell.

More numerous but somewhat smaller are the mitochondria which are centres of intracellular oxidation. They can change shape with ease but most commonly take the shape of a short rod with hemispherical ends having a length rarely exceeding 6 μm.

The Golgi apparatus has the property of being both particulate and membranous. It consists of a number, often quite large, of Golgi bodies or dictyosomes. Each dictyosome is composed of a stack of flattened vesicles known as cisternae which has a diameter of 1–2 μm. The dictyosomes are usually closely associated with endoplasmic reticular membranes or the nuclear membrane and are surrounded by tiny spherical vesicles which appear to have budded off the cisternae.

The most numerous of all the particulate inclusions are the ribosomes, which occur in the cytoplasm, either free or attached to endoplasmic reticular or nuclear membranes, in the chloroplasts and the mitochondria. They are roughly spherical organelles; those in the cytoplasm are slightly larger (largest diam. 26 nm) than those in the chloroplasts and mitochondria (largest diam. 22 nm). They are composed of RNA and protein and are concerned in protein synthesis.

Microbodies constitute another group of particulate inclusions. They are roughly spherical with a diameter of 0.2–1.5 μm and are surrounded by a single membrane. They include peroxisomes and glyoxysomes. The former occur in photosynthetic cells and are frequently closely associated with chloroplasts and mitochondria. They play a role in photorespiration. The latter are much more restricted in the type of cell in which they are to be found; typically they are present in the cells of the endosperm or cotyledons of fat-storing seeds.

Spherosomes are also particulate inclusions restricted to the cells of lipid-storing tissues. They are roughly spherical with a diameter of 0.4–3 μm and are surrounded by a single membrane whose structure is still a matter of debate. It is their function to synthesize, store and ultimately mobilize lipid which usually takes the form of triglyceride.

Although the protoplasts of neighbouring cells are separated from one another by their cell walls and the middle lamella, they are connected by thin strands of cytoplasm known as plasmodesmata. In fact it is probable that all living cells in higher plants are interconnected by plasmodesmata; they also occur in pteridophytes, bryophytes and many algae and have also been reported as occurring in fungi and some blue-green algae (cyanobacteria), although this is less well documented. In higher plant cells plasmodesmata occur with a frequency of 1–15 per μm^2 of wall surface. A plasmodesma typically consists of a cylindrical hole, about 16–20 nm in diameter, through the cell wall–middle lamella–cell wall sandwich, which may be 0.1–1.0 μm thick, separating two protoplasts. The hole is lined by a cylinder of plasmalemma which is continuous with the plasmalemma of both protoplasts. The lumen of the plasmalemmal cylinder, which is 7–10 nm in diameter, is filled with cytoplasm and often contains a minute tubule, known as a desmotubule, running axially through it. The desmotubule is frequently seen to be connected at either end to elements of the endoplasmic reticulum of the two interconnected protoplasts; it is thus possible that the endoplasmic reticular lumen of the two protoplasts are continuous with one another via the desmotubule. If this is so then substances may pass from one protoplast to the other through the cytoplasmic ground substance and through the desmotubule. The dimensions of these two passageways prevent the transference of organelles or large macromolecules but clearly water and small dissolved solutes may pass through.

The importance of the intercellular spaces must not be overlooked. These spaces arise in the developmental phase of cell growth. In the meristematic regions, a cell is in contact with its neighbours over the whole of its surface but during differentiation and development the primary walls separate at the corners so forming intercellular spaces. These spaces around each cell become interlinked so forming, in mature tissue, an intercellular space system which ends at the stomatal openings. Air diffuses through the stomatal openings, so permeating the whole system and ensuring that each cell in the tissue receives an adequate supply of oxygen and/or carbon dioxide for its metabolic activities.

The existence of plasmodesmata and intercellular spaces subdivides the plant into two major compartments, known as the symplast and the apoplast. The symplast is the living part of the plant made up of the interconnected protoplasts bounded by what is in effect the one continuous plasmalemma of the plant. The apoplast is the non-living part of the plant external to the plasmalemma and composed of the cell walls, the intercellular spaces and the lumen of dead structures such as xylem vessels. Both compartments are utilized for the transport of materials through the plant.

B. STRUCTURE AND FUNCTION OF THE CELLULAR COMPONENTS

1. The cell wall

The structure and function of the cell wall are considered in detail in Chapter 4. However, it is appropriate to point out here that it is the possession of a cell wall which is one of the key ways in which a plant cell differs from an animal cell. This difference frequently makes the study of the metabolic activities of plant cells more difficult than those of animal cells.

A technique much favoured by biochemists is to burst open cells, liberate their contents and then isolate the various membranous and particulate components in order to study separately their metabolic activities. Ideally this involves subjecting the tissue to shearing forces which are sufficiently powerful to break open the outer covering of the cells and yet insufficiently powerful to disrupt the intracellular organelles. The outer covering of the animal cell is the plasma membrane. This is readily broken by low shearing forces which cause little disruption of the intracellular organelles. However, the outer covering of the plant cell is the cell wall which is a tough structure requiring much greater shearing forces to break it open. The use of these greater shearing forces causes much greater disruption of intracellular organelles; the extent of this disruption frequently adversely affects their metabolic function. This is just one of the reasons why biochemical investigations in plants have often lagged behind those on animals.

2. Membranes

(i) GENERAL STRUCTURE AND PROPERTIES

The most satisfactory model of membrane structure at present is the fluid mosaic model put forward by Singer and Nicholson in 1972 (see Fig. 3.2). This postulates that the membrane consists of a bilayer of amphipathic lipids (e.g. phospholipids, glycolipids, sterols) with globular proteins embedded in it.

The lipid molecules of the bilayer, being amphipathic, have within their structure hydrophilic

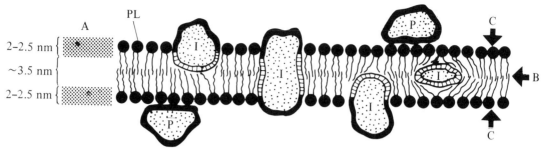

Fig. 3.2. Cross-section through the fluid mosaic model of a cell membrane (after Singer and Nicholson, 1972).
A = Probable relationship of the dark–light–dark-layered effect, produced when membranes are stained with OsO_4 and
    $KMnO_4$, with the model.
B = Line of fracture produced by the technique of freeze fracture.
C = Surfaces seen when the technique of deep etching is applied after freeze fracture.
 I = Integral protein.
P = Peripheral protein.
PL = Phospholipid.
▬▬▬ = Polar (hydrophilic) protein surface.
☐☐☐ = Non-polar (hydrophobic) protein surface.

(or polar) and hydrophobic (or non-polar) moieties. This is exemplified by phosphatidylcholine, a phospholipid common to most membranes (see Fig. 3.3). Phosphatidylcholine is composed of a relatively small, compact 'head' which is hydrophilic because of the presence of a positively charged nitrogen atom and a negatively charged oxygen atom, and two relatively long 'tails' which are hydrocarbon in character and therefore hydrophobic.

In the bilayer the amphipathic molecules are arranged such that their hydrophobic 'tails' point towards each other and perhaps intermingle at the tips to some extent. Both surfaces of the bilayer are in consequence composed of the hydrophilic 'heads'. The bilayer is therefore stable in an aqueous environment because its surfaces readily associate with water.

The proteins which are embedded in the lipid bilayer are globular. They may be divided into two types, peripheral and integral. The peripheral proteins can be readily removed from the membrane by solutions of high ionic strength (e.g. 1 M NaCl); it is therefore assumed that they are bound only or mainly to the polar surface of the lipid bilayer by electrostatic or hydrogen bonds. Cytochrome c, occurring on the outer face of the inner mitochondrial membrane (see Section B.5), is a typical peripheral membrane protein. The integral proteins, however, are deeply embedded in the lipid bilayer. They may be divided into three types: (i) those

which penetrate the bilayer from one side or the other but do not pass right through it, (ii) those which pass right through the bilayer and thus have part of their structure exposed to the aqueous environment on either side of the bilayer and (iii) those which are totally embedded in the hydrophobic core of the bilayer. Those parts of the surface of the integral proteins which are in contact with the hydrophobic core of the lipid bilayer are hydrophobic whilst those parts which are exposed to the aqueous environment are hydrophilic. Hydrophobic surfaces are the result of a predominance of amino acids with non-polar side chains (e.g. leucine, valine) in that particular region whilst hydrophilic surfaces occur when amino acids with polar side chains (e.g. glutamic acid, lysine) predominate.

The membrane must not be thought of as a static structure. Both the lipid and protein molecules are free to move and indeed many are in constant motion. However, they may only move readily in the plane of the membrane, a process known as lateral diffusion. The movement from one side of the membrane to the other, termed transverse diffusion or 'flip-flop', is a very slow process. A phospholipid molecule, for instance, may diffuse laterally (i.e. within its own layer of the bilayer) at a rate of about $2~\mu m~sec^{-1}$ which indicates that the viscosity of the membrane is about 100 times that of water. On the other hand, that same phospholipid may flip-flop

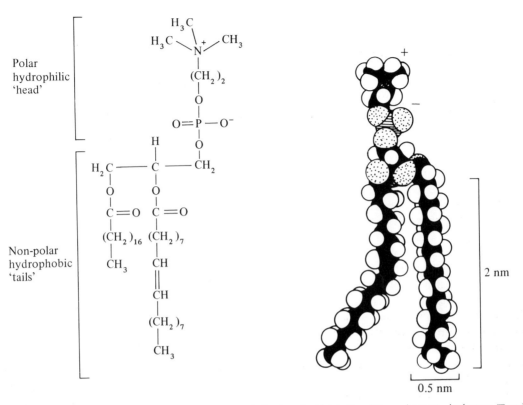

Fig. 3.3. Chemical structure and a space-filling molecular model of a phosphatidylcholine. (■ = carbon; □ = hydrogen; ▨ = nitrogen; ▨ = oxygen; ▤ = phosphorus.)

only once in several hours; the reason for the slow rate of flip-flop is that it involves the passage of the polar head through the hydrophobic interior of the bilayer. Proteins diffuse laterally much more slowly than phospholipids because they are much larger; experiments with animal membranes have indicated a rate of several $\mu m\ min^{-1}$.

It is evident from the fluid mosaic model that membranes may differ from one another whilst still having the same basic structure. They may differ in the type and relative proportions of the lipids in the bilayer, in the type of proteins present and in the ratio of total lipid to total protein. Such differences have an important bearing on the function of the membrane. For instance, the plasma membrane of the Schwann cell which winds tightly around the axons of neurones in animals to form myelin has the highest lipid:protein ratio (4:1, w/w) of any known membrane; this is consistent with the role of myelin

as an electrical insulator. At the other extreme the plasma membrane in many bacteria has a lipid:protein ratio of 1:3 (w/w) due to the presence of many enzyme systems. Most membranes, however, have a lipid:protein ratio of about 1:1 (w/w). A further example stems from the fact that the nature of the fatty-acid residues in the lipids of the bilayers has an important bearing on the temperature range in which the membrane can function. The lipid bilayer undergoes a physical change from a relatively fluid, flexible liquid crystalline state which is compatible with normal membrane function to a solid gel-like structure which is not, when the temperature drops below a fairly narrow range known as the phase transition range. Lipids with unsaturated fatty-acid residues have a lower phase transition range than those with saturated fatty acids. This thermal phase transition of membranes has an important consequence for organisms, like plants, which cannot control their internal temperature; clearly they must choose the type of fatty-acid residues of their membrane lipids to suit the temperature range of the environment in which they exist.

Natural membranes are known to have two other properties which are consistent with the fluid mosaic model. Firstly they are asymmetric, their outer surface has different properties from their inner surface. This can obviously be best accounted for by the presence of different proteins on or in the two surfaces. This membrane asymmetry is very obvious in the plasma membrane of animal cells, the outer surface of which possesses glycoproteins [see Chapter 7, Section D.2(vi)] which constitute recognition sites for a range of different substances. The presence of such recognition sites on the outer surface of the plasmalemma of the plant cell is not yet established, indeed it has been argued that the need for such external recognition sites is made unnecessary in vegetative cells by the presence of the cell wall. This mention of membrane recognition sites leads us to the second property of natural membranes which is that they are able to recognize each other. It has been observed that certain membranes within the cell are able to fuse with some internal membranes but not with others. For instance, two mitochondria can fuse with each other as can two vacuoles and a Golgi vesicle with the plasmalemma; however, a mitochondrion cannot fuse with the nucleus nor can a vacuole with the plasmalemma. The most reasonable explanation for such specificity is that the different species of cellular membrane have recognition sites for each other. It is presumed that these recognition sites are protein in character.

One of the key functions of a membrane is to control the passage of substances across it. Natural membranes allow some substances through but not others; they are said to be selectively permeable. Moreover, the different membranes of the cell have different selective permeabilities; in this way the different cellular compartments are able to have different complements of chemical components and different functions. This selective permeability is based upon 'carrier' systems, probably protein in character, which are capable of recognizing a particular compound and transporting it across the membrane. These carrier systems are able to operate against a concentration gradient; to do this they require energy which is frequently derived from the hydrolysis of ATP. Some substances are, however, able to pass across a membrane by simple diffusion, water being the prime example; many lipids are also able to do this by virtue of their solubility in the lipid bilayer. Substances crossing a membrane by simple diffusion do so down a concentration or electrochemical gradient.

Solid material not in solution may pass across membranes by endo- or exocytosis. In these processes the material is engulfed by the membrane forming a small vesicle which is then detached on the other side; the vesicle then opens to release its contents. The term endocytosis is usually confined to the passage of solid material across the plasma membrane into the cytoplasm. The reverse process in which substances are ejected from the cell is called exocytosis.

When most biological membranes are examined by electron microscopy after staining with osmium tetroxide or potassium permanganate they exhibit a characteristic three-layered structure consisting of two darkly stained lines, 2–2.5 nm wide, on either side of a clear unstained central region about 3.5 nm wide (see Fig. 3.2). It is assumed that the dark lines correspond to the hydrophilic heads of the lipid bilayer and the clear central region to the two sets of hydrophobic tails.

The internal and external surfaces of membranes can be seen in the electron microscope when the techniques of freeze-fracture and deep-etching are used. Cells or fragments of membrane are rapidly frozen at the temperature of liquid nitrogen ($-195.9°C$). The specimen is then fractured by a sharp blow from a microtome knife. The plane of fracture in the specimen follows the plane of greatest weakness which is the plane in the middle of the lipid bilayer because it contains least ice. This follows from two facts: (i) the hydrophobicity of the lipid tails excludes virtually all water from the central region of the bilayer and (ii) the strength of the frozen specimen is largely due to the ice it contains. Thus electron microscopic examination of the surface produced by freeze-fracture gives a picture of the interior of the membrane. If, after freeze-fracture, the ice is sublimed away, a process called deep-etching, the exterior surface can also be seen in the electron microscope.

(ii) THE PLASMALEMMA

The plasmalemma bounds the outer surface of the protoplast. It differs from the endoplasmic reticular membranes by appearing slightly thicker and in having no adhering ribosomes. The plasmalemma of one cell is continuous with that of adjacent cells via the plasmodesmata. Its outer surface is covered with partly embedded, square-packed spherical granules about 15 nm in diameter which are believed to be the enzyme complexes responsible for the synthesis of the cellulose microfibrils of the cell wall [see Chapter 4, Section D.1(i)]. Small vesicles derived from the Golgi cisternae which are believed to contain polysaccharides destined for the cell wall matrix appear to be directed to the plasmalemma by microtubules (see Section B.9). The vesicles fuse with the plasmalemma and then deposit their contents into the cell wall by exocytosis. During this process the membrane surrounding the vesicles is incorporated into and becomes indistinguishable from the plasmalemma.

(iii) THE ENDOPLASMIC RETICULUM

The endoplasmic reticulum is composed of an extensive network of interconnected, membrane-bound tubules and vesicles which extend throughout the cytoplasm. The endoplasmic reticular membrane is continuous with the nuclear membrane but not with the plasmalemma or the tonoplast. The endoplasmic reticular membrane of adjacent cells is connected through the plasmodesmata.

Some parts of the endoplasmic reticular membrane have ribosomes (see Section B.3) adhering to their cytoplasmic surfaces whilst other parts do not; the former are referred to as 'rough endoplasmic reticulum', the latter 'smooth endoplasmic reticulum'. The ribosomes of the rough endoplasmic reticulum are in the process of translating an *m*RNA molecule into a protein which, when complete, ends up in the lumen of the endoplasmic reticulum. This happens because the protein in question has at its N-terminal end (the end synthesized first) a sequence of twenty or so hydrophobic amino acids which, because of this characteristic, is able to pass readily through the endoplasmic reticular membrane dragging the rest of the molecule behind it. Thus the protein, during the period of its synthesis, serves to anchor the ribosome forming it to the endoplasmic reticular membrane.

The endoplasmic reticular lumen is a distinct compartment within the cell which is connected to that of adjacent cells via plasmodesmata.

The endoplasmic reticulum is a dynamic structure. It exhibits both growth and turnover. At cell division it undergoes partial fragmentation. It has a close relationship with the Golgi apparatus [see Section B.2(v)].

A broad spectrum of enzyme systems is built into the endoplasmic membrane which consequently plays an important part in the metabolism of the cell.

When a cell is disrupted by homogenization the endoplasmic reticular membrane, the plasmalemma and the tonoplast are broken into small fragments which close up to form small sacs 50–150 nm in diameter. When the homogenate is subjected to differential centrifugation (see Section C) these membranous sacs are sedimented between 40,000g and 105,000g and constitute what biochemists call the microsomal fraction of the cell. Since the endoplasmic reticular membrane has a much greater area than that of the plasmalemma and tonoplast, the majority of the membranous sacs constituting the microsomal fraction are derived from it. Consequently biochemists regard the microsomal fraction as essentially a preparation of endoplasmic reticular membrane. Some of the microsomal sacs have ribosomes adhering to their outer surface and have clearly been derived from regions of rough endoplasmic reticulum. Partial separation of the microsomal fraction into its various membranous components has been achieved by density gradient centrifugation (see Fig. 3.19).

(iv) THE TONOPLAST AND VACUOLES

A vacuole is a compartment within the protoplast containing an aqueous solution and separated from the cytoplasm by a membrane known as the tonoplast. The number of vacuoles in a cell and their size vary enormously. In a mature parenchymatous cell about 90% of the volume of the protoplast is taken up by a few, sometimes only one, large vacuoles. A meristematic cell, on the other hand, has a very large number of tiny vacuoles.

The vacuolar contents are very varied and differ from plant to plant and from cell type to cell type.

There is only one compound common to all, namely water. Most of the other vacuolar contents are dissolved in the water, often in high concentration; solute concentrations of 0.4–0.6 M have been measured. These solutes include inorganic ions such as Na^+, K^+, Ca^{2+}, Mg^{2+}, Cl^-, SO_4^{2-}, PO_4^{3-}, carbohydrates such as sucrose (e.g. vacuoles in the stem cells of sugar cane), other sugars and mucilages, organic acids, such as malic acid in CAM plants (see Chapter 5, Section D.4), and other TCA cycle acids, phenolic compounds such as flavonoids (e.g. in many flower petals) and tannins (see Chapter 14), and nitrogenous compounds such as amino acids, amides, peptides, proteins (some of which are enzymes), betalains (e.g. betanin in beetroot) and alkaloids. The vacuoles of some cells also contain substances which are not dissolved; these may occur as suspensions, amorphous deposits or crystals and be inorganic (e.g. calcium oxalate) or organic (e.g. protein) in character.

The tonoplast which surrounds each vacuole is the same thickness as the plasmalemma and looks very like it in many ways. However, it is different as is shown by the fact that, unlike the plasmalemma, it does not stain with phosphotungstic acid. It is able to accumulate substances within the vacuole by active transport processes mediated by specific permeases and by fusion with membrane-bound vesicles, containing the substances in question, frequently derived from the Golgi apparatus.

Vacuoles appear to have three main functions in the cell. Firstly they maintain the cell in a turgid condition; this results from the fact that the high concentration of osmotically active substances actively accumulated in the vacuole draws water into the vacuole which therefore increases in volume and pushes the protoplast tightly against the surrounding cell wall. This increases the rigidity of the cell. Thus the skeletal system of soft plant tissues is provided by the co-operation of cell walls and vacuole.

Secondly vacuoles have a storage function. Some of the substances stored appear to be waste products which have been excreted from the cytoplasm into the vacuole. Others, however, have a distinct function. Some are food reserves; for instance the vacuoles of the cells of developing seeds store large quantities of protein and are often called protein bodies or aleurone grains. Other stored substances serve to attract members of the animal kingdom (e.g. the flavonoids in flower petals which attract insects to assist in pollination) or repel them (e.g. unpalatable or poisonous compounds like alkaloids which deter animals from eating the plant). Still other stored substances are enzymes which confer on vacuoles their third function, that of acting as lysosomes.

Animal lysosomes contain a battery of enzymes with an acid pH (~ 5) optimum which are capable of catalysing the hydrolysis of all the main cellular components, nucleic acids, proteins, carbohydrates and lipids. Their function is to digest organic matter brought into the cell by endocytosis and degenerating intracellular components. The products of digestion are released into the cytoplasm and re-used. Many of the hydrolases characteristic of animal lysosomes have been found in the vacuoles of cells from a variety of members of the plant kingdom ranging from algae and fungi to higher plants. Moreover, these enzymes have been shown to digest cellular material entering the vacuoles by the vesicular route. It therefore appears clear that vacuoles are able to function as lysosomes.

(v) THE GOLGI APPARATUS

The Golgi apparatus of the cell consists of a number of distinct units called Golgi bodies or dictyosomes. In some cells there is only one dictyosome which therefore constitutes the whole Golgi apparatus.

Each dictyosome is built up of a stack of flattened membrane-bound sacs, about 1–2 μm in diameter, called cisternae (see Fig. 3.4). The number of cisternae in a dictyosome is variable but is usually in the range 4–8. The cisternae may have a simple disc-like shape with a smooth, though dilated, perimeter or may have a more complex structure in which the peripheral region of the disc is extensively perforated to form a network of branched tubules whose ends are frequently dilated. The cisternae are surrounded by a variable number of small vesicles which appear to have been nipped off the dilated peripheral region.

Within the dictyosome the membranes of successive cisternae are close but do not actually touch owing to the presence of a film about 10 nm thick of intervening material of unknown composition.

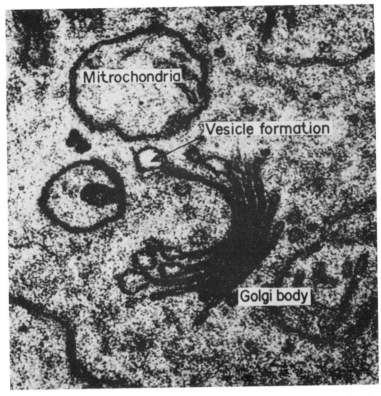

Fig. 3.4. An electron micrograph of a Golgi dictyosome from a bean root tip cell after permanganate fixation (× 50,000). (Original kindly supplied by Dr. R. Barton, UCW, Aberystwyth.)

Since complete dictyosomes can be isolated by the homogenization and differential centrifugation techniques of the biochemist, it is evident that this material cements the cisternae together.

Dictyosomes usually lie with the bottom-most cisterna of the stack close to and parallel with a section of endoplasmic reticular membrane. A close examination of dictyosomes positioned in this way shows that the further a cisterna is from the endoplasmic reticular membrane the larger and more numerous are the peripherally attached or associated vesicles and the more conspicuous are the cisternal contents. Between the endoplasmic reticular membrane and the nearest cisterna are a number of small vesicles which are known as transitional vesicles. It is clear that these vesicles are derived from the adjacent endoplasmic reticular membrane and that they ultimately coalesce to form a Golgi cisterna. At the other end of the dictyosome the top-most cisterna gradually breaks up into small vesicles which move through the cytoplasm to fuse with the plasmalemma or tonoplast.

The dictyosome is therefore a dynamic structure; new cisternae are continually being produced from the endoplasmic reticulum at one end of the cisternal stack which is known as the forming face whilst the oldest cisterna is gradually being lost in the form of vesicles at the other end which is known as the maturing face. Thus a given cisterna gradually moves its way up the stack until it reaches the maturing face and breaks up into vesicles. During this process it undergoes differentiation and becomes metabolically active. The formation of a new cisterna must be accompanied by the production of the layer of intercisternal material which cements it to the cisterna above. At the other end of the dictyosome the loss of the mature cisterna as vesicles must be accompanied by the break down of the cementing material.

It is believed that a considerable range of compounds are synthesized in the Golgi cisternae. The enzymes catalysing these syntheses are produced by the ribosomes of the rough endoplasmic reticulum [see Section B.2(iii)]. These enzymes, after formation, evidently pass through the lumen of the endoplasmic reticulum until they reach the dictyosome formation site and are incorporated into the transitional vesicles and hence into a cisterna.

The compounds produced in the cisternae leave the dictyosome packed in the vesicles produced at the maturing face and from the periphery of the cisternae nearest to it. The vesicles then pass through the cytoplasm, possibly under the direction of microtubules (see Section B.9), and fuse with either the plasmalemma or the tonoplast, dumping their contents in the process outside the protoplast (as, for example, in the case of the polysaccharides of the cell wall matrix) or into the vacuole respectively.

(vi) The endomembrane concept

This concept was developed by Morré and Mollenhauer in the early 1970s. It proposes that certain membranes of the cell are integrated into a developmental and functional continuum known as the endomembrane system. The components of this system are membranes of the endoplasmic reticulum, the nucleus, the Golgi apparatus and the Golgi vesicles along with the plasmalemma and the tonoplast. Also included are the outer membranes of the mitochondrion and the chloroplast. Not included are the inner membranes of the mitochondrion and the chloroplast.

The concept suggests that all new membrane produced within the endomembrane system is formed by the endoplasmic reticulum or by the nuclear membrane which is, of course, continuous with the endoplasmic reticulum. Membrane material is then transferred from the endoplasmic reticulum to the Golgi apparatus by means of the transitional vesicles [see Section B.2(v)]. From the Golgi apparatus membrane material is transferred to the plasmalemma and tonoplast in the form of Golgi vesicles. There is thus an outward flow of membrane from the endoplasmic reticulum to the plasmalemma and tonoplast via the Golgi apparatus. The concept also suggests that there is an inward flow of membrane material from the plasmalemma to the tonoplast via vesicles formed during endocytosis. During these flows it is postulated that membrane differentiation occurs.

The evidence for the outward membrane flow comes from two sources: (i) electron micrographs showing all the postulated membrane continuities and (ii) pulse-chase experiments using radioactively labelled precursors of membrane components. In the latter the radioactively labelled precursor is supplied to the tissue for a short time and is then replaced by unlabelled precursor. The intracellular location of radioactivity is then determined by examining aliquots of the tissue immediately after the pulse of labelled precursor and after various intervals of time during the chase with unlabelled precursor. It has been found that the radioactivity resides at first solely in the endoplasmic reticulum and then successively in the dictyosomes, Golgi vesicles close to the dictyosomes, Golgi vesicles close to the plasmalemma and tonoplast and finally in the plasmalemma and tonoplast.

The outer membranes of the mitochondrion and chloroplast are included in the endomembrane system because both organelles are thought to have originated in the evolutionary sense from prokaryotic cells which invaded the ancestral plant cell by endocytosis and began a symbiotic relationship with it [see Sections B.5 and B.6(ii)]. It follows from this that the outer membrane of the mitochondrion and chloroplast corresponds to the membrane of the endocytotic vacuole which originated from the plasmalemma of the host cell. It would therefore seem reasonable to postulate that the outer membrane of these organelles is formed by the same mechanism as the host plasmalemma. The inner membrane of the mitochondrion and chloroplast according to this hypothesis originates from the plasma membrane of the endosymbiont, a bacterium and a blue-green alga respectively, and would therefore not be expected to be formed by the same mechanism as the host plasmalemma.

3. Ribosomes

Ribosomes are small organelles composed of RNA and protein that play a vital role in protein synthesis (see Chapter 10). Plant cells contain three distinct kinds of ribosomes which occur in different intracellular compartments, namely the cytoplasm (cytosol), the mitochondrion and the chloroplast. The mitochondrial and chloroplastic ribosomes are smaller (70S) than the cytoplasmic ribosomes (80S); they are in fact the same size as the ribosomes of prokaryotic cells.

Plant cytoplasmic ribosomes are oblate or prolate spheroids with long and short axes of 26 and 19 nm respectively. They are composed of two subunits, a large one with a sedimentation constant of 60S and a small one of 40S, which readily dissociate at low Mg^{2+} concentrations. The small subunit fits over the large subunit leaving a tunnel between the two which can accommodate the mRNA, aminoacyl-tRNA and other factors during protein synthesis. The large subunit contains three rRNA molecules, one of 25S (MW 1,300,000), one of 5S (MW 40,000) and one of 5.8S (MW 50,000) whilst the small subunit contains one rRNA of 18S (MW 700,000). Both subunits also contain proteins; 45–50 proteins are present in the 60S subunit and about 30 in the 40S subunit.

The structure of mitochondrial and chloroplastidic ribosomes are discussed in Sections B.5 and B.6 respectively.

The rRNA molecules of the cytoplasmic ribosomes are formed by transcription of nuclear genes in the nucleolus (see Section B.4) whilst rRNA molecules of mitochondrial and chloroplastidic ribosomes are formed by transcription of mitochondrial and chloroplastidic genes respectively. The protein components of cytoplasmic ribosomes are coded by nuclear genes and synthesized in the cytoplasm. Present evidence indicates that the majority of the protein components of mitochondrial and chloroplastidic ribosomes are formed in the cytoplasm by translation of mRNA transcribed from nuclear genes in some cases and organellar genes in others.

The cytoplasmic ribosomes may be found in the cytosol or attached to endoplasmic reticular membranes so producing the so-called 'rough' endoplasmic reticulum. The ribosomes are anchored to the endoplasmic reticulum by the proteins they are in the process of synthesizing (see Chapter 4, Section D.3).

4. The nucleus

The nucleus is the largest of the cytoplasmic inclusions, its diameter ranging from about 2 μm in yeast to about 8 μm in higher plant cells. It is surrounded by a double membrane, consisting of two unit membranes, each 8 nm thick, separated by a space, about 15 nm wide, known as the perinuclear space. It is perforated by pores which have a diameter in the range 30–100 nm and tend to be octagonal in shape. The inner and outer membranes are fused together at the annulus of each pore. About 8% of the surface area of the nuclear membrane is occupied by pores. There are, therefore, about 200 pores in a yeast nucleus and about 3000 in a higher plant nucleus. The pores allow the transport of substances between the cytosol and the nucleus. Nuclear imports include the nucleotide precursors of DNA and RNA, histones and ribosomal proteins, whilst nuclear exports include mRNA, tRNA and ribosomal subunits.

The nuclear membranes are continuous with those of the endoplasmic reticulum. Thus the perinuclear space is continuous with the lumen of the endoplasmic reticular tubules and vesicles.

In fungi and many other lower plants the nuclear membrane remains intact during cell division but does divide and become partitioned between the daughter nuclei. In higher plants the nuclear membrane, like the endoplasmic reticulum, undergoes fragmentation at cell division and is reformed towards the end of the process around the daughter nuclei.

Within the nuclear membrane there is the nucleoplasm, which by virtue of the nuclear pores is continuous with the cytosol. Embedded in the nucleoplasm are the chromosomes and the nucleolus. The chromosomes cannot be distinguished in the non-dividing cell but can be readily seen by light microscopy during mitotic or meiotic divisions. The nucleolus, on the other hand, is usually the only feature of the nucleus of a non-dividing cell that can be distinguished by light microscopy. However, it disappears during the first stage of cell division and reappears towards the end of the process in each of the daughter nuclei.

The chromosomes contain most of the information for the regulation of the activities of the cell (some information is contained in the mitochondrial and chloroplastidic DNA—see Sections B.5 and B.6). The number of chromosomes in the somatic cells of a given plant species is constant but this number varies enormously from species to species, ranging from four to several hundred. In each somatic cell there are normally two chromosomes of each type, so that each nucleus has two identical sets of chromosomes and is said to be diploid. However, the somatic cells of some higher plants have more than two sets of chromosomes and are said to be polyploid. Polyploidy results from endomitotic divisions. The germ cells have only one set of chromosomes and are said to be haploid.

Chromosomes are composed of DNA and histone proteins in roughly equal quantities. Histones are relatively small proteins which carry a net positive charge at pH 7 due to the presence of unusually large numbers of lysine and arginine residues. Five different histones have been found in chromosomes, designated I (MW $\sim$21,000, lys $\geqslant$ arg), IIb1 (MW $\sim$15,000, lys $\eqsim$ arg), IIb2 (MW $\sim$14,000, lys $\eqsim$ arg), III (MW $\sim$15,500, lys $<$ arg) and IV (MW $\sim$11,000, lys $<$ arg). The structure of histone I is almost species-specific whilst that of the other four shows very little change from species to species. These positively charged histones are held by electrostatic linkages to the negatively charged internucleotide phosphate residues of the DNA to form the complex nucleoprotein that is the chromosome. Each chromosome contains one extremely long DNA molecule and several million histone molecules.

The molecular architecture of the chromosome is different at different stages of the cell cycle. The mitotic cycle of meristematic cells is usually divided into four phases: M, G_1, S and G_2. The length of the cycle depends upon the quantity of DNA (which is related to the number and size of the chromosomes) in the nuclei of the plant species or variety; the more DNA the longer is the cycle. Figure 3.5 shows the lengths of the different phases of the mitotic cycle in the onion species Allium cepa. Here the M or mitotic phase, when the chromosomes are visible and actively dividing, takes 2 hr and corresponds to prophase, metaphase, anaphase and telophase of mitosis. Collectively the G_1, S and G_2 phases take 15 hr and correspond to interphase of mitosis. During the S or synthetic phase chromosome replication occurs; there is thus a doubling of DNA content of the nucleus. During the two periods between the M and S phases there is no active synthesis of DNA. These periods therefore correspond to gaps (hence G_1 and G_2) in the mitotic cycle when the chromosomes are, in terms of replicatory activity, quiescent. They are, however, fully active in their cell regulatory capacity. During

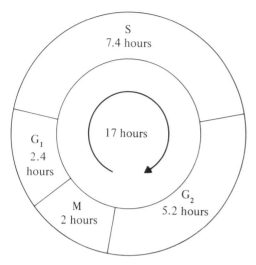

Fig. 3.5. Diagrammatic representation of the mitotic cycle and its component phases. The times of the different phases are those determined for the onion species *Allium cepa*. The timing of the cycle and its component parts is achieved by labelling the DNA in a batch of interphase root meristem nuclei with [^{3}H]thymidine and then following its progress at intervals of time through the division cycle by radioautography. Interphase = $G_1 + S + G_2$; M = prophase, metaphase, anaphase and telophase of mitosis. (Courtesy of Dr. R. N. Jones, UCW, Aberystwyth.)

the G_1 phase each chromosome consists of a single chromatid containing a single DNA molecule whereas, during the G_2 phase, it consists of two identical chromatids, each with an identical DNA molecule.

During the M phase the chromosomes are very much shorter and fatter than they are during interphase ($G_1 + S + G_2$). The DNA must therefore be much more tightly coiled or folded within the M phase chromosome than in the interphase chromosome. Any attempt to explain the molecular architecture of chromosomes must take account of the gross structural changes seen in the mitotic cycle and in the process known as meiosis which is responsible for the production of germ cells. It must also take account of the role that DNA plays in providing information for cell regulation.

Although the molecular architecture of the more complex M phase chromosomes is largely unknown, that of the interphase chromosomes, which are usually referred to as chromatin, is gradually becoming clearer. The key clue came in 1973 when electron micrographs were obtained showing chromatin fibres in a more 'stretched-out' form than

usual and looking rather like beads on a string. The beads, now called nucleosomes, have been shown to consist of a disc-like core composed of two each of histones IIb1, IIb2, III and IV around which are coiled 1.75 turns of the DNA double helix, corresponding to 145 consecutive nucleotide pairs (about 80 nucleotide pairs per turn). The nucleosomes are linked together by the DNA double helix. The gap between adjacent nucleosomes appears in electron micrographs to be 2–14 nm. One molecule of the species-specific histone I is bound to each inter-nucleosome DNA segment (see Fig. 3.6).

Nucleosomes, however, only represent first-order DNA packaging. The natural chromatin fibre is 2–3 times the diameter of the nucleosome. Clearly, therefore, the nucleosome 'string of beads' must undergo some form of further coiling. The precise nature of this is uncertain. One suggestion is that it is coiled into a cylinder with about six nucleosomes per turn. Since chromatin is itself a very 'stretched-out' chromosome structure, it is clear that it must undergo several more orders of packaging to attain the much shorter and fatter chromosome forms seen during the M phase.

The division phase of the mitotic cell cycle (see Fig. 3.7) begins with prophase when the already-replicated chromosomes shorten by some form of coiling and become readily distinguishable by light microscopy. It is apparent that each chromosome is a double structure, being composed of two chromatids lying close together throughout their length and being joined at, or near to, a relatively uncoiled region called the centromere.

The centromere may be at any position within the chromosome, but for individual chromosomes its location is fixed and unvarying. The centromere is of considerable importance since it is responsible for the mobility of the chromosomes. Throughout prophase the chromosomes continue to undergo further orders of coiling which cause them to shorten considerably. As this shortening reaches its maximum, the nucleolus detaches itself, disorganizes and finally disappears. At the same time the nuclear membrane breaks down.

Early in metaphase a structure known as the nuclear spindle is formed. This has the shape of a rugby football and is composed of protein microtubules (see Section B.9) some of which run from pole to pole whilst others, termed half-spindle fibres,

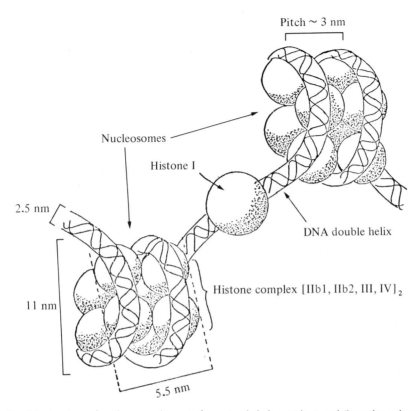

Fig. 3.6. Possible structure of nucleosomes in part of an extended chromatin strand (interphase chromosome).

extend from one pole to the equator of the spindle. In those fungi and lower plants which do not disorganize their nuclear membranes during cell division, the spindle forms within the nuclear membrane.

During metaphase the chromosomes move towards the equator of the spindle and attach themselves by their centromeres to the spindle fibres. It appears that the half-centromeres corresponding to the two chromatids of a given chromosome become attached to half-spindle fibres originating from opposite poles. At anaphase the centromere of each chromosome completes its division into daughter centromeres (formerly half-centromeres) which are then pulled to opposite poles by contraction of the half-spindle fibres to which they are attached. This separates the chromatids and pulls them to opposite ends of the spindle. There are now two identical sets of daughter chromosomes at opposite ends of the spindle.

During telophase a new nuclear membrane is formed around both sets of daughter chromosomes, which simultaneously reverse their coiling behaviour and return to their interphase structure (chromatin). The formation of the daughter nuclei is closely followed by division of the cytoplasm. This commences with the conversion of the spindle fibres in the equatorial region into phragmoplast fibres and is followed by the formation of the cell plate which initially occurs within the confines of the spindle but soon extends to the outer wall. The cell plate becomes the middle lamella of the newly forming cell wall which will separate the daughter cells. Completion of the wall is accomplished by the laying down of a primary wall on the cell plate by both daughter cells; the cell plate is thus sandwiched between the two primary walls and becomes the middle lamella (see Chapter 4, Sections C.1 and E).

Mitosis is concerned with the cycle of chromosome duplication and division, and is synchronous with the cell cycle itself which duplicates and divides the other cellular components. Daughter cells thus formed are genetically identical with one another and with the parent cell from which they arose, in terms of number and type of chromosomes. Indeed the mechanism of heredity depends upon the fidelity of DNA replication and chromosome duplication at mitosis.

Plant cells normally cease their cell and chromosome cycles of division when growth and development are complete and most are arrested in the G_1-phase following the final mitosis. It is not uncommon, however, in differentiated tissues, to find that the cells and chromosomes fail to cease division synchronously and for the chromosomes to pass through one or more additional phases of DNA duplication without passing through the M-phase at all. These cycles of DNA duplication without division are called endomitotic cycles.

There are two different types of endomitotic cycles, one leading to endopolyploidy, the other to polyteny. Their mechanistic relationship to conventional mitosis is seen in Fig. 3.8.

Endopolyploidy occurs when the chromatids of chromosomes, formed during the S-phase, separate from one another without the breakdown of the nuclear membrane and the M-phase. This results in a tetraploid nucleus. The chromatids may then undergo further cycles of duplication and separation without ensuing M-phases; this results in highly polyploid nuclei.

Polyteny occurs when the chromatids, formed during the S-phase, replicate once or several times without separating from each other and without breakdown of the nuclear membrane and the M-phase. This results in multistranded or polytene

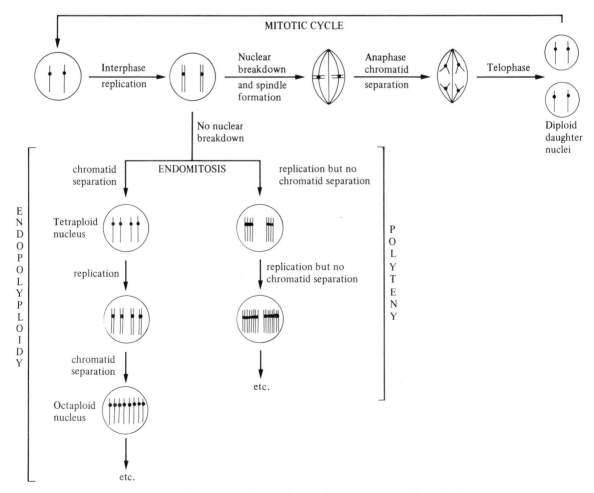

Fig. 3.8. Relationship between mitosis and endomitosis (endopolyploidy and polyteny).

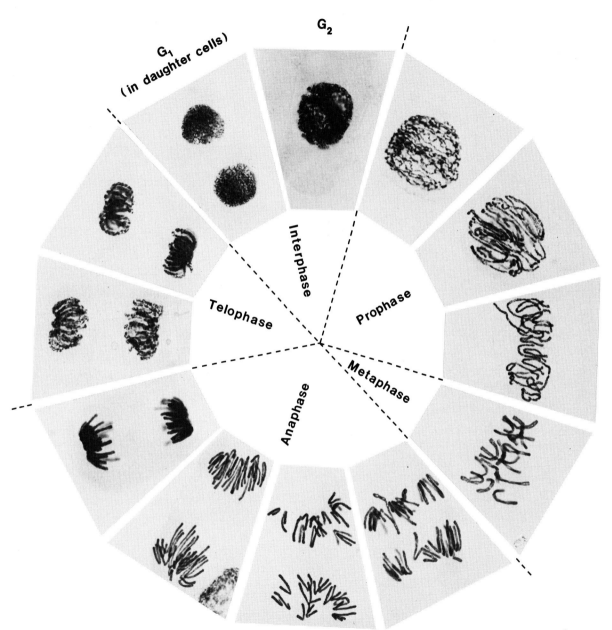

Fig. 3.7. The events of mitosis in the onion species *Allium cepa*, which is diploid with 16 chromosomes, seen in the light microscope after staining with acetocarmine. (Photographs kindly provided by Dr. R. N. Jones, UCW, Aberystwyth.)

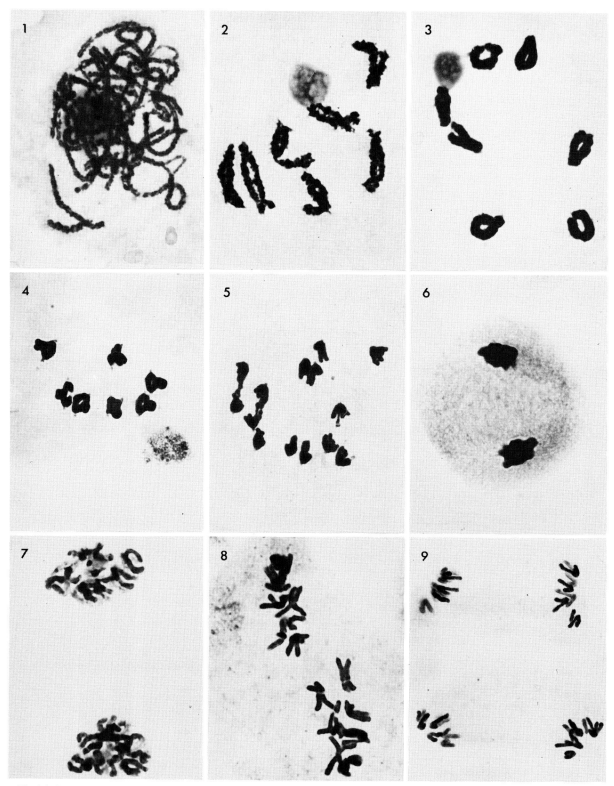

Fig. 3.9. Some of the stages of meiosis in Rye, *Secale cereale*, which is diploid with 14 chromosomes, seen in the light microscope after staining with acetocarmine. 1 = pachytene (Prophase I), 2 = Diplotene (Prophase I), 3 = Diakinesis, 4 = Metaphase I, 5 = Anaphase I, 6 = Telophase I, 7 = Prophase II, 8 = Metaphase II, 9 = Anaphase II. (Photographs kindly provided by Dr. R. N. Jones, UCW, Aberystwyth.)

chromosomes, each strand being a chromatid. Polyteny is less common than endopolyploidy but is found in some of the nuclei associated with the developing embryos of flowering plants.

The physiological significance of endomitotic reduplication of chromosomes is not well understood. It is usually regarded as a means of gene amplification within the cell. This may be particularly important in those somatic cells which are required to produce large quantities of certain metabolites in short periods of time. This is exemplified by the endopolyploid cells of the tapetum which nourish the developing pollen mother cells. The relative functional advantages of endopolyploidy and polyteny are not clear.

Cells with endopolyploid or polytene nuclei are normally excluded from the germ line and therefore pose no problems to heredity and the formation of haploid gametes. However, some plants are naturally polyploid throughout their somatic tissue and their germ-line cells. This form of numerical chromosome variation is known in over 35% of flowering plant species and in some groups, such as grasses, up to 70% of species are either polyploid or have polyploid varieties. There are no difficulties at mitosis in having more than a diploid complement of chromosomes because each member of the set behaves quite independently of, but in synchrony with, the others. However, this is not so at meiosis as will be seen later.

Meiosis (Gk. *meioum*—to reduce) is a developmental modification of mitosis and is complementary to fertilization in the cycle of sexual reproduction. It takes place in the pollen mother cells of the anthers and the megaspore mother cell in the embryo sac to produce haploid cells that initiate the gametophyte phase of the life-cycle. Meiosis differs from mitosis in that the parent cell divides twice but the chromosomes divide only once, thus producing four daughter cells each with a reduced number of chromosomes (half the number in the case of a diploid parent cell).

Both of the meiotic cell divisions (see Fig. 3.9) can be conveniently divided into four phases: pro-, meta-, ana- and telophase. Prophase of the first cell division (prophase I) is subdivided into the leptotene, zygotene, pachytene and diplotene stages and terminates with diakinesis.

During the interphase that precedes the first meiotic division the nuclear DNA is replicated; the S-phase is longer than in mitosis and the G_2-phase is absent.

In diploid meiosis the chromosomes first become visible in the light microscope, by coiling and shortening, as single-stranded structures (not double as in prophase of mitosis) during the leptotene stage. However, the duplicated nature of the chromosomes has been revealed by electron microscopy. During the zygotene stage homologous chromosomes come together to form pairs which are known as bivalents. The mechanism of pairing is not fully understood but it is thought to be initiated by attachment of homologous chromosome ends at closely adjacent sites on the inside of the nuclear membrane.

During the pachytene stage the bivalents undergo further shortening. Electron microscopic examination of the pachytene bivalent has revealed the presence of a tripartite structure known as the synaptonemal complex which runs between and parallel to the paired chromosomes. It is about $0.2\ \mu m$ wide and is composed of protein molecules in the form of a central element and two lateral elements. The synaptonemal complex is thought to be implicated in the pairing and crossing-over events that take place between homologous chromosomes, but the mechanisms by which these events are accomplished are by no means clear.

During the diplotene stage the synaptonemal complex disintegrates and the pairing attraction between the homologous chromosomes constituting the bivalent lapses and they begin to separate. The separating chromosomes can at this stage be seen by light microscopy to be composed of two chromatids. The chromosomes ultimately become separated throughout their length except at one or more points where 'crossing over' has taken place. A 'cross-over' point (chiasma) occurs when a chromatid from each chromosome of the bivalent breaks at a corresponding point and then the broken ends from non-sister chromatids reunite.

The precise mechanism of crossing over is not known at the molecular level but a number of models have been suggested which have in common a scheme of chain separation following the breakage of DNA half-helices and then the rejoining of the unwound DNA chains to give a region of hybrid DNA consisting of a part of the polynucleotide chain from one chromatid and a part from one of the homologous, non-sister chromatids. The models also include a DNA repair system which recognizes and corrects any illegitimate base pairings which may have arisen when the hybrid DNA spans a site of heterozygosity within a gene. During this phase of meiosis a very small amount of DNA synthesis has been detected (representing about 0.3% of the total meiotic DNA synthesis) which may well be due to this DNA repair process.

The process of crossing over is the basis of the genetic exchange in heterozygous genotypes which results in recombination of alleles at different loci on homologous chromosomes.

Prophase I terminates with diakinesis in which the bivalents reach their maximum degree of contraction and adjacent chromatid segments between the chiasmata rotate so that they come to lie at right angles to each other, instead of in the same plane as formerly.

Early in metaphase I the nuclear membrane breaks down and is replaced by a spindle. The bivalents arrange themselves on the spindle in such a way that homologous centromeres lie on opposite sides of, and equidistant from, the equator.

At anaphase I the attraction between sister chromatids lapses allowing the chromosomes of each bivalent to move to opposite ends of the spindle. Each anaphase I chromosome thus consists of two chromatids which are new combinations of parts from the chromatids of the homologous chromosome pair (bivalent) from which it was derived. The two chromatids are joined only at their centromere.

Telophase I then follows. The chromosomes uncoil and become a diffuse mass of chromatin again and in some species (e.g. *Tradescantia*, maize) there may be an interphase stage with a nuclear membrane forming round the telophase chromosomes. In this interphase, occurring between the two divisions of meiosis, there is no DNA synthesis. In other plants there is no telophase I or interphase and the anaphase I chromosomes pass directly to prophase of the second division (prophase II). Whichever of these procedures occurs the second division follows rapidly upon the first.

In prophase II the chromosomes coil up once more and, as in mitosis, it is apparent that they are each composed of two chromatids. However, in contrast to mitosis, the chromatids of each chromosome are not genetically identical (because of crossing over) and are only joined together at the centromere.

In metaphase II spindles are formed in each of the two daughter cells or around the two sets of chromosomes. The chromosomes move towards the equator of the spindle and attach themselves by their centromere to the spindle fibres.

In anaphase II the centromeres divide and the chromatids are separated to opposite poles. Following telophase II nuclear membranes are assembled around the four sets of chromosomes and cytoplasmic division takes place giving the four haploid products of meiosis. These form the microspores and megaspores which develop into the male and female gametophytes in the pollen grains of the anther and in the embryo sac respectively. In the limited number of cell divisions which take place in these gametophytes, to give the sperm nuclei of the male gametes and the egg cell in the female, the nuclei possess only a haploid set of chromosomes and in this respect they differ in their structure and organization from the nuclei of the cells of the rest of the plant which corresponds to the sporophyte phase of the life cycle.

In the description of mitosis given earlier reference was made to the frequent occurrence of polyploidy among higher plant species and it was pointed out that the additional sets of chromosomes posed no difficulties during the mitosis of somatic cell division. This is not true of meiosis; in polyploid individuals the mechanics of meiosis are seriously affected. Chromosomes pair up as multivalents of three, four or more depending upon the level of ploidy, although only two of the homologues actually come together over any one chromosome region. Crossing over takes place and the multivalents arrange themselves on the equator of the spindle at metaphase I. The segregation of chromosomes to the poles at anaphase is then highly irregular; the level of irregularity depends upon the degree of ploidy and also upon the number and position of the chiasmata within the multivalent and upon the configuration assumed by the multivalent on the equator. Equal separation of anaphase I chromosomes can be more efficient in even-numbered rather than odd-numbered polyploids but in general terms autopolyploids (which have multiple sets of *homologous* chromosomes) have a high degree of infertility. For this reason many have evolved vegetative means of reproduction by bulbs, corms or even vivipary where the place of the seed is taken by a bulbil or some equivalent vegetative structure. Allopolyploids (which arise by species hybridization and subsequent chromosome doubling and which have *homoeologous* chromosomes) are, however, able to effect proper bivalent formation and a more efficient meiosis.

The DNA of the chromosomes contains information for the regulation of cellular activity. The basic units of information are genes, of which there are a very large number in a chromosome. Each gene consists of a linear sequence of nucleotide residues which constitutes no more than a tiny fraction of the DNA molecule of the chromosome. Within the chromosomes there is only one copy of the majority of the genes. However, there are multiple copies of some genes. The multiplicity, in some cases, runs into thousands. The genes for cytoplasmic ribosomal RNA (*r*RNA) and transfer RNA (*t*RNA) are examples of reiterated genes.

The genes are transcribed within the nucleus by DNA-directed RNA polymerases[1] to form RNA molecules which, after enzymic modification, become messenger RNA (*m*RNA), *r*RNA and *t*RNA molecules.

The *m*RNA and *t*RNA molecules pass from the nucleus into the cytosol where they play key roles in protein synthesis (see Chapter 10). The nucleotide sequences of the *m*RNAs are translated into the amino acid sequences of specific proteins by the ribosomes (see Section B.3). The *t*RNAs assist in this process by transferring amino acids to the appropriate *m*RNA codons.

Three of the highly reiterated cytoplasmic *r*RNA genes form a sequential cluster in equivalent locations on a homologous pair of chromosomes and are known as the nucleolar organizer. Transcription of these genes produces a 45S-precursor *r*RNA molecule which undergoes enzymic cleavage and modification to yield three of the four *r*RNA molecules (18S, 25S and 5.8S) found in ribosomes; it seems likely that the gene for the 5S *r*RNA is on a different chromosome. Ribosomal proteins, synthesized in the cytosol and translocated into the nucleus via the nuclear pores, bind specifically and in a defined sequence to the *r*RNA molecules to form the 40S ribosomal subunit, containing the 18S *r*RNA, and the 60S subunit which contains the 25S, 5.8S and 5S *r*RNAs. The subunits then pass into the cytosol via the nuclear pores and aggregate to form functional 80S ribosomes.

The region within the nucleus where ribosomal subunit formation takes place is known as the nucleolus. The nucleolus is usually the only feature of the nucleus of a non-dividing cell that can be seen by light microscopy. It is frequently spherical and consists of a dense irregular central region surrounded by a more translucent region. Electron microscopy shows that the latter consists of loosely packed particles which, in size and density, resemble ribosomal subunits. It also shows that the denser central region is composed of closely packed fibrillar material which appears to be chromosomal in character and may be the nucleolar organizer.

The nucleolus cannot be seen during cell division. It begins to disorganize at the onset of prophase and has completely disappeared by the end of prophase. Two nucleoli begin to be formed in the second half of telophase, one in each of the reforming nuclei of the two daughter cells. They are much smaller than the predivision structure and appear to consist only of the densely packed fibrillar material. They increase in size and have completed their reorganization by the end of G_1. They remain fully organized during the S- and G_2-phases of the mitotic cycle.

5. Mitochondria

Mitochondria are cytoplasmic organelles which are specifically stained by the dye Janus green B. They are centres of intracellular oxidation and contain the enzymes of the tricarboxylic acid cycle, the respiratory electron transport chain and of oxidative phosphorylation as well as many others.

Examination of living higher plant cells by phase-contrast light microscopy has revealed that mitochondria undergo considerable changes of shape as they are moved around by cytoplasmic streaming; they are sometimes globular, cylindrical or branched and sometimes they split into portions or fuse with one another. When they are globular their diameter is 0.5–1.5 μm and when they are cylindrical their length rarely exceeds 6 μm. In higher plant cells, however, the most common shape taken up by mitochondria is that of a short rod with roughly hemispherical ends.

The number of mitochondria per higher plant cell is in the hundreds or even thousands, depending upon the size and type of cell, for instance, the central cells of the root cap of maize have about 200 when young and about 2500 when mature. For a given cell type the number per unit volume of cytoplasm is roughly constant; in the case of metabolically active cells, such as phloem companion cells, about a fifth of the cytoplasmic volume is taken up by mitochondria.

Some algae (e.g. *Chlorella*) and the yeast, *Saccharomyces*, have only one large mitochondrion per cell; this is a much-branched structure with many lobes which extends throughout the cytoplasm and is usually referred to as a mitochondrial reticulum. Again the ratio between mitochondrial and cytoplasmic volume is constant; during growth the *Chlorella* cell undergoes a 6-fold increase in volume whilst throughout this period the mitochondrion occupies about 2.5% of the cytoplasmic volume. *Chlamydomonas*, a unicellular alga, has 10–15 multi-lobed mitochondria.

Mitochondria are surrounded by a double-membraned envelope. The inner membrane is thrown into a series of folds or invaginations known as cristae. The cristae of many, but not all, algae and fungi are tubular rather than plate-like. Tubular cristae are seen in higher plant mitochondria, particularly those of phloem companion cells, but generally they are in a minority. Although quite a range of different shapes have been seen, the cristae of higher plant mitochondria generally open out from a narrow neck to a globular or plate-like structure.

Except where it invaginates to form cristae, the inner membrane follows the contours of the outer membrane, leaving an intermembrane space of fairly constant width which is continuous with the intracristal space. Although both the intermembrane and intracristal spaces are apparent in electron micrographs of sections fixed with osmium tetroxide and potassium permanganate, they are not apparent in freeze-etched electron micrographs. Consequently there may be much less space between the mitochondrial membranes *in vivo* than is usually supposed.

The mitochondrial envelope encloses a finely granular highly proteinaceous matrix within which there are two types of electron-dense particles, namely ribosomes and calcium-containing granules, and fibrillar regions known as nucleoids.

The mitochondrial ribosomes are smaller than those in the cytoplasm (cytosol) of the cell. They typically have a sedimentation constant of 70S and are composed of a large subunit (50S) and a small subunit (30S). The large subunit contains at least two rRNA molecules, one of 23S and one of 5S, whilst the small subunit contains one rRNA molecule of 16S; all these rRNA molecules are coded by the mitochondrial DNA. Both subunits also contain a considerable number of different species of proteins, the majority of which are coded by nuclear DNA.

The calcium-containing granules contain mainly calcium phosphate. The calcium is taken into the mitochondrion as Ca^{2+} in a process energetically coupled to electron transport. This accumulation of Ca^{2+} is accompanied by the uptake of an equivalent amount of phosphate.

The fibrillar material in the nucleoids contains DNA. Mitochondria contain several copies of a circular, histone-free DNA molecule (chromosome). Mitochondrial DNA molecules from various plant sources have 15,000–75,000 base pairs and so could code for 16–80 proteins each having 300 amino acids (MW 40,000). Since it is known that some of this DNA codes for ribosomal RNA, it is clear that the mitochondrial DNA does not have the capacity to code for the 100 or so different proteins estimated to be present in mitochondria. Thus most mitochondrial proteins must be coded by nuclear DNA.

The mitochondrial matrix contains all the enzymes of the tricarboxylic acid (TCA) cycle [see Chapter 6, Section E.1(ii)] with the exception of the flavoprotein, succinate dehydrogenase[2], which is built into the inner surface of the inner membrane probably where it is folded into cristae. There are grounds for thinking that the TCA cycle enzymes are structurally organized, possibly as a loose multi-enzyme complex, in the matrix adjacent to the membrane-bound succinate dehydrogenase.

The inner membrane, particularly the cristae, is characterized by the presence of stalked particles on the surface facing the matrix. The particles have a spherical head about 9 nm in diameter linked to the membrane by a short stalk. They are part of a complex structural unit which contains an ATP-ase[3]. ATP-ases are capable of catalysing the hydrolysis of ATP to ADP and orthophosphate and the formation of ATP from ADP and orthophosphate. The ATP-ase of the inner membrane has the latter function; it couples the free energy decline brought about by electron flow down the respiratory electron transport chain to the phosphorylation of ADP by a mechanism which is imperfectly understood but which is probably best reflected by Mitchell's chemiosmotic hypothesis [see Chapter 6, Section B.3(ii)]. The ATP-ase complex is composed not only of the spherical head and stalk but also of a unit built into, and quite possibly extending right through the thickness of, the inner membrane to which the stalk is attached. The headpiece is known as the F_1 component and consists of at least five different proteins which collectively have an MW of 360,000; the ATP-ase activity appears to be located in the two heaviest of these proteins. The intramembrane part of the complex is known as the F_0 component and consists of at least four different proteins with a collective MW of 77,000. A further protein is required to bind F_1 to F_0 and is presumably located in the stalk. The F_1 proteins are coded by nuclear DNA whilst the F_0 proteins are coded by mitochondrial DNA.

The components of the respiratory electron transport chain are built into the inner membrane as structural units. It has been estimated that about 10% of the area of the inner membrane is taken up by these units. The individual components of the chain occupy different positions within the thickness of the membrane. There are two NADH dehydrogenases, both of which contain flavoproteins. One is located on the inner side of the membrane and catalyses the reoxidation of NADH produced in the matrix of the mitochondrion during the operation of the TCA cycle, for example, whilst the other, peculiar to plants, is located on the outer face of the membrane and catalyses the reoxidation of NADH produced outside the mitochondrion. The TCA cycle enzyme succinate dehydrogenase, another flavoprotein, is located on the inner face of the inner membrane. These flavoproteins pass electrons via iron–sulphur proteins to ubiquinone which is probably located in the lipid core of the membrane. From ubiquinone the electrons are passed via centrally located cytochromes b to cytochromes c which are located on the outer face of the inner membrane. The components of cytochrome oxidase[4], cytochromes a and a_3, are located across the inner membrane with the a_3 component nearest the matrix.

The outer membrane has a higher lipid content than the inner and has a totally different complement of enzymes some of which are reminiscent of those of the endoplasmic reticulum.

The two mitochondrial membranes also have different permeability properties. The outer membrane is freely permeable to low molecular weight compounds and to a number of proteins whereas the inner membrane presents a barrier to many ions, low molecular weight compounds and proteins and possesses specific trans-membrane transport systems.

New mitochondria arise by division of existing mitochondria. In most cells mitochondrial division follows their growth and differentiation. It is possible that division is initiated when a critical DNA content is reached. Division is accomplished by the invagination of the inner membrane. This invagination, instead of forming a crista, extends and eventually divides the matrix compartment into two distinct compartments both of which are enclosed within the same outer membrane. A constriction then forms between the compartments and separation is ultimately achieved. The mechanism by which both daughter mitochondria receive nucleoid regions is not known.

The ultimate origin of mitochondria is an interesting question. Increases in our knowledge of mitochondrial structure and genetics have shown striking similarities with bacteria and it has been suggested that they evolved from symbiotic prokaryotic organisms.

6. Plastids

Plastids constitute a family of organelles which are peculiar to plant cells. One or other of the various types of plastid are present in all plant cells except those of the fungi. The family is composed of proplastids, chloroplasts, chromoplasts, amyloplasts and etioplasts. Proplastids are the tiny precursors of all the other members of the family and are particularly abundant in meristematic cells. Chloroplasts house the photosynthetic apparatus and are usually green. Chromoplasts are pigmented yellow to red by carotenoids but do not photosynthesize. Amyloplasts are not pigmented and are specialized for the synthesis and storage of starch grains. Etioplasts are normally a transitory stage; they are formed when the development of proplastids into chloroplasts is interrupted by lack of light. The term leucoplast occurs in the literature; it refers not to a distinct type of plastid but to all the non-pigmented plastids.

All plastids are surrounded by a double membrane, have a system of internal membranes which in some types of plastid are extensive but which in others are vestigial, chromosomes, composed of histone-free DNA, located in regions known as nucleoids, and ribosomes. Plastoglobuli (osmiophilic globules) are also widespread.

(i) PROPLASTIDS

Proplastids are the small, colourless or pale green, undifferentiated plastids occurring in the meristematic cells of the shoot and the root. They are roughly ellipsoid or spherical structures with a diameter of 1–1.5 μm. They have little internal structure. The inner membrane of the bounding envelope occasionally extends into the homogenous matrix in the form of invaginations. There may also be one or two isolated vesicles and thylakoids [see Section 6(ii)]. Small starch grains may be present in the proplastids of the root meristem.

Proplastids are present to the extent of 7–20 per cell in the shoot meristem and 20–40 per cell in the root meristem. They are not confined to the cells of vascular plants; they are present in the apical cells of multicellular algae, e.g. *Lomentaria baileyana*.

(ii) CHLOROPLASTS

In higher plants chloroplasts are found mainly in the palisade and spongy mesophyll cells of the leaf. They are also present in the guard cells of the leaf epidermis. The other epidermal cells, in many species, have only a few, small chloroplasts, the exception being shade plants like *Helxine soleroli* whose epidermal cells are packed with large chloroplasts. Chloroplasts are also found in the cells of all other green tissues. They are absent from the meristematic cells of the shoot and from the cells of the root.

The number of chloroplasts varies with cell type and plant species but generally increases with cell size. In the palisade and mesophyll cells of a spinach leaf there are typically 300–400 and 200–300 chloroplasts per cell respectively. They are usually found around the periphery of the cell close to the plasmalemma.

All chloroplasts, by definition, contain the pigment chlorophyll. They are, however, not always green; in brown and red algae their green colour is masked by other pigments.

The chloroplasts of vascular plants are shaped like a bi-convex, plano-convex or concavo-convex lens with a circular or ellipsoid outline whose long diameter is within the range 3–10 μm. Those bryophyte species which have been examined, e.g. the liverwort *Anthoceros*, have one large chloroplast (long diam. ∼40 μm) per mature cell in the gametophyte; their internal structure is similar to that of vascular plant chloroplasts. The chloroplasts of eukaryotic algae exhibit a wide variation in size, shape and number per cell. The chloroplasts of *Euglena* spp. are lens-shaped, those of *Spirogyra* spp. are helical, those of *Zygnema* spp. are star-shaped whilst those of *Oedogonium* spp. are in the form of an irregular network. The chloroplasts of algae of the Chrysophyta, Cryptophyta and Phaeophyta are enclosed within a double-membraned sac of endoplasmic reticulum known as the 'chloroplast endoplasmic reticulum'. Blue-green algae (or Cyanobacteria) are prokaryotic and do not have chloroplasts; their photosynthetic apparatus is present in thylakoids (see later) which lie free in the cytosol.

The internal structure of all chloroplasts (see Fig. 3.10) is characterized by a system of membranes sometimes referred to as the 'chloroplast lamellae', embedded in a hydrophilic, proteinaceous matrix or stroma. The basic subunit of this membranous structure is a sac or vesicle surrounded by a single membrane to which the name thylakoid (from the Greek, 'sack-like') was given in 1960 by Menke. The vesicle is completely flattened so that in cross-section it appears as a pair of parallel membranes, closely applied to each other and joined at the ends. The arrangement of thylakoids is different in the

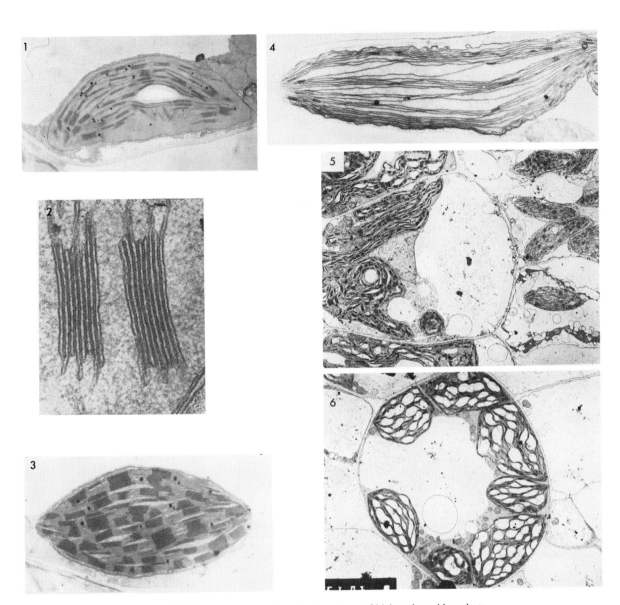

Fig. 3.10. Electron micrographs of various types of higher plant chloroplasts.

1. Cross-section of a typical chloroplast from a C_3 plant, *Spinacia oleraceae* ($\times 18,000$); the clear zone in the centre is a starch grain.
2. Cross-section of two grana from a chloroplast of *Spinacia oleraceae* ($\times 170,000$).
3. Cross-section of a chloroplast from a mesophyll cell of a C_4 plant, *Zea mays* ($\times 20,000$); note the presence of grana. The chloroplasts of mesophyll cells from all three types of C_4 plants (see Chapter 5, Section D.3) have similar structures. They differ from those of C_3 plants in having a more pronounced system of vesicles, the peripheral reticulum (Pr), located immediately within the envelope; this could be associated with the rapid transport of substances into and out of the chloroplast.
4. Cross-section of a chloroplast from a bundle sheath cell of the NADP-malic enzyme type of C_4 plant, *Zea mays*, which is characterized by the paucity of grana ($\times 15,000$).
5. Cross-section of a bundle sheath cell from an NAD-malic enzyme type of C_4 plant, *Sporobolus giganteus*, in which the chloroplasts are characteristically arranged in a centripetal manner (i.e. they congregate at the side of the cell nearest the vascular bundle). Note that grana are present in the chloroplasts; the clear areas are starch grains ($\times 15,000$).
6. Cross-section of a bundle sheath cell from a PEP carboxykinase type of C_4 plant, *Sporobolus aeroides*, in which the chloroplasts are characteristically arranged in a centrifugal manner (i.e. they congregate at the side of the cell remote from the vascular bundle). Note that grana are present in the chloroplasts; the numerous clear areas are starch grains ($\times 5000$).

(The original of 2 was kindly supplied by Prof. K. Mühlethaler, Zurich; the remaining originals were kindly supplied by Dr. J. Coombs, Tate & Lyle Ltd., Reading, and Mr. A. D. Greenwood, Imperial College, London, and reproduced with the permission of Elsevier/North Holland Biomedical Press.)

chloroplasts of different members of the plant kingdom. The simplest arrangement occurs in the red algae (Rhodophyta) where the thylakoids lie separately and roughly parallel to each other. Attached to their outer surface is a regular array of particles, about 30–40 nm in diameter, which contain phycobiliprotein molecules and are called phycobilisomes [see Chapter 5, Section C.2(iii)]. The next simplest arrangement occurs in the Cryptomonad algae where the thylakoids are associated loosely in stacks of two. In all other algae, with the exception of some of the Chlorophyta, the thylakoids occur in stacks of three. In many species of the Chlorophyta the thylakoid arrangement is like that of higher plant chloroplasts.

An early concept of the internal structural organization of the higher plant chloroplast suggested that there were two sizes of thylakoid, the large or 'stroma' thylakoid which approached the diameter of the chloroplast itself and the small or 'grana' thylakoid, which was 30–60 nm in diameter. These were arranged as shown in Fig. 3.11, A. According to this concept the tiny green grains seen in chloroplasts by the light microscopists and termed grana consisted of stacks of small thylakoids, often alternating with large thylakoids. There may be 40–60 grana per chloroplast and usually 5–20 thylakoids per granum. This concept suggests that the lumen of each thylakoid is not connected to that of any other thylakoid; this is now known to be incorrect.

It was then found that the membranes between the grana were perforated; they could not therefore be simply the component membranes of large thylakoids. The size of the perforations varied considerably; frequently they were so large that the membranes interconnecting the grana were no more than narrow tubes which became known as frets (see Fig. 3.11, B). It was then realized that the frets branch and by so doing interconnect the lumen of the granal thylakoids (see Fig. 3.11, C); this meant that the thylakoid system constitutes a single, complex cavity separated from the stroma by the thylakoid membrane system.

Further electron microscopic examination showed structures which suggested that a fret may intersect a granum at an angle to the plane of the granal thylakoids and be connected to them in sequence (see Fig. 3.11, D). These fret connections always seem to describe a right-handed helix ascending the granal stack. Even closer examination revealed that several frets, rather than just one, may ascend the granal stack and interconnect with the component thylakoids (see Fig. 3.11, E).

The way in which this structure is formed depends upon whether the chloroplast develops directly from a proplastid or an etioplast. In Nature direct chloroplast development from a proplastid occurs when the tissue is exposed to normal daylight conditions whilst indirect development via an etioplast occurs in tissues which undergo their early development in the dark.

In the former process the inner membrane of the proplastid invaginates into the stroma at several points over a period of time. These invaginations are eventually nipped off and become the flattened vesicles we know as thylakoids. As differentiation proceeds stacking of the thylakoids occurs. The mechanism by which stacks of thylakoids arise has to take account of the fact that the intra-thylakoid spaces of the different thylakoids of a granum are in communication with each other. Two different suggestions by which this may occur are shown in Fig. 3.12.

In the latter case the membranous tubules of the prolamellar body [see Section B.6(iii)] are reorganized into the thylakoid system in the light.

Built into the thylakoid membranes are the pigment systems and electron carriers which carry out the light phase of photosynthesis. However, the arrangement of these components within the membranes is still uncertain (see Chapter 5, Section C.4).

Dispersed in the stroma of the chloroplast are other structures. Plastoglobuli are droplets, usually small (diam. 50–220 nm), rich in lipid particularly plastoquinone, without a bounding membrane, which occur close to the thylakoid membranes. They are particularly numerous when thylakoid membranes are breaking down such as in a senescing chloroplast and are least numerous during the formation of thylakoids. This, taken with the fact that plastoquinone is a thylakoid component, suggests that they are formed from some of the lipids liberated when thylakoids break down.

Starch grains are usually present in the stroma lying close to the thylakoid membranes. They disappear if the plant is kept in the dark for about 12–24 hr but reappear after a few hours in the light.

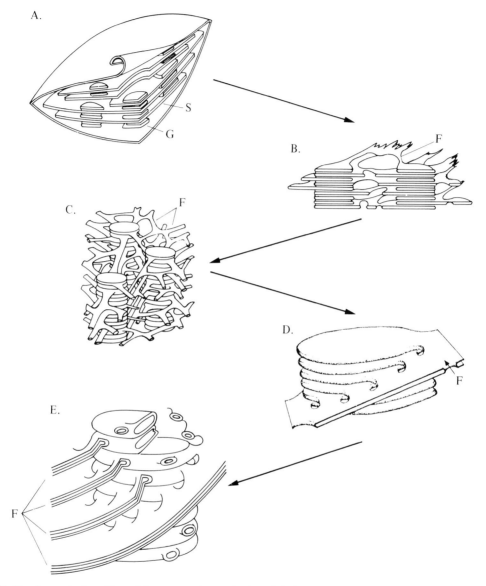

Fig. 3.11. Development of our ideas of the structural organization of the internal membranes of the higher plant chloroplast.
A. An early representation of the internal structure of the chloroplast, cut open to show the small, granal thylakoids (G) and the large, stromal, thylakoids (S) (redrawn from Eriksson, Walles and von Wettstein, *Ber. deut. bot. Ges.* **74**, 221, 1961).
B. A later representation showing that the intergranal membranes are perforated, producing narrow or broad membrane-bound channels, known as frets (F), interconnecting the grana (redrawn, with permission, from J. Heslop-Harrison, *Planta* **58**, 273, 1962).
C. A representation showing much-branched frets which allow for intra and intergranal connections (reproduced by permission of Prof. T. E. Weier).
D. A single fret ascends a stack of granal discs at an angle, describing a right-handed helix, and connects to each disc in turn (redrawn, with permission, from J. Heslop-Harrison, *Planta* **60**, 243, 1963).
E. A single granum showing that several frets can ascend the stack, as in D; thus each granal disc has multiple fret connections (adapted, with permission, from *Ultrastructure and Biology of Plant Cells* by B. E. S. Gunning and M. W. Steer, Publ. Edward Arnold Ltd., 1975).

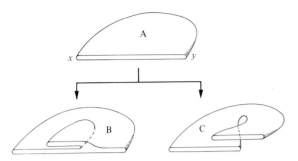

Fig. 3.12. Methods by which thylakoid stacking could arise. A = single thylakoid cross-sectioned between x and y; B = conjunctive stacking; C = disjunctive stacking. (Redrawn after Wehrmeyer and Röbbelen.)

They are usually ovoid and up to 1.5 μm in length. Mesophyll cells contain 1–5 starch grains per chloroplast but much larger numbers are often present in the bundle sheath chloroplasts of C_4 plants.

Ribosomes occur free in the stroma and bound to the outer surface of the thylakoid membranes. They are smaller than the ribosomes in the cytosol of the cell having sedimentation constants of about 70S. The lengths of their long and short axes are 22 and 17 nm respectively. At Mg^{2+} concentrations less than 10 mM they dissociate into large (50S) and small (30S) subunits. Chloroplast ribosomes contain four RNA molecules, one of 23S (MW 1,100,000), one of 5S (MW 40,000) and one of 4.5S which are present in the large subunit and one of 16S (MW 560,000) which is located in the small subunit. Protein constitutes about 46% of the chloroplast ribosome; this is made up of about seventy-five different protein species.

Higher plant and algal chloroplasts contain 20–50 identical chromosomes which are located within specific regions of the stroma known as nucleoids, the number and size of which increase with chloroplast size. Each chromosome contains a histone-free, circular DNA molecule with a circumference of 40–46 μm (MW 83–95 × 10^6) in higher plants and 36–62 μm (MW 75–128 × 10^6) in algae. This length of DNA, allowing for known repeating segments, has been estimated to be capable of coding for about 126 proteins of MW 40,000. However, relatively few chloroplastidic proteins have so far been shown to be coded by chloroplastidic DNA. The large subunit of ribulose 1,5-bisphosphate carboxylase-oxygenase[5] [see Chapter 5, Section D.2(i)] is known to be coded by a chloroplastidic gene whilst the small subunit is coded by a nuclear gene. The higher plant chloroplast chromosome contains two genes each for the four types of chloroplast ribosomal RNA; physical mapping shows that the genes occur in the sequence 16S, 23S, 4.5S and 5S with a spacer of about 2200 base pairs between the first two. It also contains genes, probably one of each, for the tRNA of each amino acid. To date the genetic function of only about 12% of the chloroplast DNA sequence has been accounted for.

Chloroplasts contain a DNA polymerase[6] and the plastidic DNA is replicated in a semiconservative manner within the chloroplast itself. They possess all the machinery for protein synthesis; however, many chloroplastidic proteins are coded by nuclear genes, synthesized by cytoplasmic ribosomes and then imported. The formation of chlorophyll from δ-aminolaevulinic acid occurs in the chloroplast as does the formation of the haem of chloroplastidic cytochromes, the phycobilins (in certain algae) and the carotenoid biosynthetic pathway certainly from MVA and possibly from CO_2. Fatty acids such as palmitic, oleic and linoleic acids are formed from acetyl-CoA in the chloroplast, as are galactolipids (see Chapter 8).

(iii) Etioplasts

The prefix 'etio' in etioplast indicates a plastid which is characteristic of an etiolated leaf (i.e. a leaf of a plant grown in the dark). Etioplasts are formed naturally in the primary leaves or cotyledons of germinating seedlings before they emerge from the soil into the light and in the differentiating meristematic cells at the base of leaves of members of the Gramineae where the coleoptile or older leaves cut out most of the light.

Etioplasts can be regarded as stages in the development of chloroplasts. They contain many but not all of the chloroplastidic proteins. They contain only tiny amounts of carotenoids, protochlorophyll and plastid quinones but the composition of the colourless polar lipid fraction is very similar to that of chloroplasts. They have ribosomes and DNA.

They are irregularly ellipsoidal with a long diameter of about 3 μm. The double-membraned envelope contains a uniform proteinaceous stroma in which are embedded 1–4 prolamellar bodies (see Fig. 3.13). These are large quasicrystalline structures composed of interconnected membranous tubules in a regular array. They are called prolamellar bodies to mark the fact they are the precursors of the lamellae of the chloroplast; when etiolated leaves are exposed to the light the etioplasts rapidly develop into chloroplasts and the prolamellar bodies are transformed into the thylakoid membrane system.

The prolamellar bodies are composed of basic repeating units which most commonly consist of four short lengths of tubule joined together at one point in a tetrahedral fashion (cf. the four bonds of a carbon atom). Each of the four 'legs' of a given tetrahedron may then link to a leg of another tetrahedron. In this manner a three-dimensional lattice is built up.

(iv) CHROMOPLASTS

Chromoplasts are the carotenoid-containing plastids that give the yellow, orange and red colours to many fruits (e.g. tomato) and flower petals and to some roots (e.g. carrot).

They usually develop from chloroplasts and so have roughly the same size and shape as chloroplasts. They have most of the structural features of the chloroplast; the photosynthetic membrane system has, however, been replaced by structures rich in carotenoids.

Five different types of chromoplast have been recognized on the basis of their internal structure. The simplest type is globular and possesses only homogenous plastoglobuli (diam. 150 nm) which are thought to contain the carotenoids; it is typical of most flower petals coloured by carotenoids. The membranous type has up to twenty-five sets of concentric membranes which contain the carotenoids; it is found in daffodil petals. The tubulous type, found in the fruit of *Capsicum annuum*, possesses many fibres (diam. 15–80 nm) which are thought to contain carotenoprotein complexes. The reticulo-tubular type, found in the spadix appendix of *Typhonium divaricatum*, contains a ramifying network of branched, non-parallel tubules. The crystalline type contains carotenoid in the form of crystals; this type of chromoplast is usually formed when only hydrocarbon carotenoids (e.g. β-carotene, lycopene) are present. They occur, for instance, in tomato fruit where the carotene (mainly lycopene) occurs in crystalline tubules (15–48 μm long, 1–2 μm diam.) built up of concentric lamellae.

No physiological function is known for chromoplasts. It has been suggested that their bright colour in flower petals and fruit may attract insects and birds for pollination and seed dispersal respectively.

(v) AMYLOPLASTS

Amyloplasts are mature plastids which are almost totally filled with starch in the form of grains. They are found in storage tissues such as the cotyledons or endosperm of seeds and tubers. They are also present as the differentiated cells of the root, particularly the root cap.

Amyloplasts are surrounded by the usual plastid double membrane. The starch grains constitute the main internal feature. There may be just one large (100 μm diam.) grain as in the potato tuber amyloplasts or up to about eight as in root cap amyloplasts. The grains are embedded in the stroma which also contains DNA in nucleoid areas and a few ribosomes. The structure and composition of the grains is discussed in Chapter 7, Section D.1(i).

The function of amyloplasts in storage tissues is to synthesize starch from imported sucrose produced in the photosynthetic tissues (see Chapter 5, Section D.6), to store it and then to mobilize it [see Chapter 7, Section D.1(i)] when the plant needs carbohydrate as, for instance, in germination. The amyloplasts in the root cap, however, have a totally different function; they play an essential role in gravity perception.

(vi) INTERCONVERSION OF PLASTIDS

Chloroplasts, etioplasts, chromoplasts and amyloplasts can all develop directly from proplastids. The reverse process, that is the simplification of plastid structure back from a specialized form to the proplastid form, is conditional upon cell division and is seen, for instance, when a meristem is formed in a differentiated tissue.

Etioplasts develop into chloroplasts in the light. Amyloplasts are commonly formed as intermediate stages in the development of etioplasts and chloro-

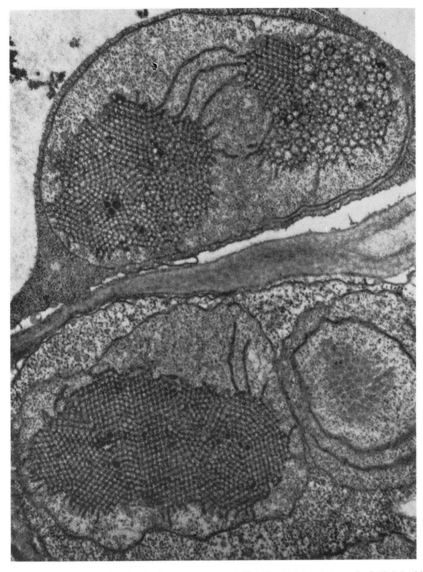

Fig. 3.13. Electron micrograph of prolamellar bodies in mature mesophyll cells of etiolated *Avena* leaf. (Original kindly supplied by Professor B. E. S. Gunning.)

plasts. Even the apparently permanent amyloplasts of potato tuber will develop a green thylakoid system if illuminated. Chromoplasts can be formed from chloroplasts; the reverse process has been observed in carrots and oranges.

(vii) DIVISION OF PLASTIDS

In meristems the division of proplastids matches cell division but in other tissues this relationship is not so close. For instance, plastid division continues in young leaves well after cell division has ceased. However, each cell type within a given plant species appears to have a characteristic final number of plastids; many factors, both internal (e.g. DNA content of the cell nucleus—polyploidy increases plastid numbers) and external (e.g. intensity and wavelength of light), influence this number.

The division of proplastids and chloroplasts is well known but little is known about the division of other plastid types. Two methods of higher plant chloroplast division have been described. One involves the development of a constriction in the mid-region which narrows until the chloroplast separates into two daughters of equal size. The other involves the invagination of the inner membrane in the mid-region of the chloroplast ultimately dividing it into two (cf. mitochondrial division). The mechanism which ensures an equal distribution of plastid DNA into the daughter chloroplasts is unknown.

(viii) ORIGIN OF PLASTIDS

The population of plastids growing and dividing within the cells of a plant is similar to a population of symbiotic, unicellular microorganisms multiplying in the cells of a host. In 1905 Mereschkowski suggested that chloroplasts in algae and higher plants arose in evolution from symbiotic blue-green algae (cyanobacteria); Ris and Plaut revived the suggestion in 1962.

It is a plausible and attractive hypothesis for there are many similarities between blue-green algae and chloroplasts, particularly those of the red algae. The thylakoids of blue-green algae lie free in the cytoplasm of the cell as they do in the stroma of the chloroplast; moreover they are single and are covered with phycobilisomes as are those of the red algae. Blue-green algae carry out an oxygen-

evolving photosynthesis using essentially the same electron transport and CO_2-fixation pathways as do chloroplasts. The ribosomes of blue-green algae are of the same size (70S) as those of chloroplasts. However, the most compelling evidence in favour of the theory lies in the homologies between the primary sequences of certain proteins and nucleic acids in chloroplasts and the corresponding molecules in blue-green algae.

7. Microbodies

Microbodies are distinctive intracellular organelles which, during the past 15 years, have been shown by electron microscopy to be present in most eukaryotic members of the plant kingdom. They are typically spherical or oblate in shape with a diameter within the range 0.2–1.5 μm and are bounded by a single membrane. The matrix is usually finely granular but occasionally crystalline or fibrillar inclusions are present. They characteristically contain the enzyme catalase and therefore react positively in the histochemical test with 3,3'-diaminobenzidine. Hydrogen peroxide-generating oxidases are also present.

The two most-studied types of microbody are peroxisomes and glyoxysomes. Peroxisomes are present in the photosynthetic cells of higher plant leaves. Glyoxysomes are present in the cells of the endosperm or cotyledons of fat-storing seeds and in the fat-rich cells of the aleurone layer of barley and wheat seeds and the scutellum of maize seeds. Microbodies are also present in the non-photosynthetic tissues of higher plants, in gymnosperms, pteridophytes, bryophytes, fungi and in algae.

Glyoxysomes were first isolated by Beevers in 1967 from the endosperm of germinating castor bean seeds. Peroxisomes were first isolated from spinach leaves a year later by Tolbert. The original isolation procedures involved the homogenization of the plant tissue in 0.5 M sucrose solution buffered at pH 7.5, followed by centrifugation at low speed ($\sim$250g for 15 min) to remove cell debris and then at 10,000g for 30 min. The resulting sedimented pellet, which contained mitochondria, proplastids and microbodies, was then resuspended in 1 M sucrose solution buffered at pH 7.5 and subjected to centrifugation at $65,000g$ for 3–5 hr on a linear gradient ranging from 30% (w/w) to 60% (w/w) sucrose in 1 mM EDTA at pH 7.5. This produced two major bands in the centrifuge tube composed of mitochondria (density = 1.19 g ml^{-1}) and microbodies (density = 1.25 g ml^{-1}) with a minor band of proplastids (density = 1.23 g ml^{-1}) in between. However, microbodies are fragile and it has since been found that gentler homogenization and fewer manipulations increase the yield. Consequently it is now usual to subject the homogenate directly to centrifugation on the sucrose gradient.

Peroxisomes in the leaf photosynthetic cells of C_3 plants (see Chapter 5, Section D.2) are about one-third and one-half as numerous as chloroplasts and mitochondria respectively and are frequently seen to be closely associated with these organelles. This is consistent with their functional association with chloroplasts and mitochondria in photorespiration (see Chapter 6, Section F.1). Photorespiration, in fact, is the only process that peroxisomes are known to participate in at the present time. Of the enzymes shown to be present in peroxisomes, namely glycollate oxidase[7], glutamate:glyoxylate aminotransferase, serine:glyoxylate aminotransferase[8], glutamate:oxaloacetate aminotransferase[9], hydroxypyruvate reductase[10], catalase[11] and urate oxidase[12], only the last is not involved in photorespiration. The function of urate oxidase in the peroxisome is unknown.

Although photorespiration is much less evident in C_4 plants (see Chapter 5, Section D.3), peroxisomes are, nevertheless, found in their leaf photosynthetic cells. However, they are more abundant in the bundle sheath cells than in the mesophyll cells. This would be consistent with a photorespiratory role because glycollate, the substrate of photorespiration, is produced only in the bundle sheath chloroplasts. It is possible that the low rate of photorespiration in these plants is due to the re-fixation of the CO_2 produced from glycollate into oxaloacetate in the mesophyll chloroplasts.

Peroxisomes have also been shown to be present in the leaves of several CAM plants (see Chapter 5, Section D.4).

Glyoxysomes develop in the cells of the endosperm or cotyledons of fat-storing seeds such as castor bean or sunflower during germination and are present only while the growth of the seedling is being supported by the stored fat. They develop in a similar manner in the fat-rich cells of the aleurone layer of seeds such as barley and wheat which store starch in their endosperm and in the endosperm cells of jojoba seeds which store wax esters [see Chapter 8, Section C.1(iii)].

They are frequently seen to be closely associated with the fat-storing organelles, the spherosomes (see Section 8). This is consistent with their function which is to assist in the conversion of the stored fat into carbohydrate. The glyoxysomes allow the acetyl-CoA produced by β-oxidation of the fatty acyl residues of the stored fat to escape oxidation to CO_2 in the TCA cycle and to be used for carbohydrate synthesis. They operate a series of reactions known as the glyoxylate cycle which plays a key role in enabling them to do this. The enzymes catalysing the reactions of the glyoxylate cycle are citrate synthase[13], aconitase[14], isocitrate lyase[15], malate synthase[16] and malate dehydrogenase[17]. Other glyoxysome enzymes also participating in the conversion of fat into carbohydrate are a monoglyceride-specific lipase, acyl-CoA synthetase[18], those of the β-oxidation spiral (see Chapter 6, Section G.3) and catalase.

Glyoxysomes also possess a number of enzymes which have little obvious connection with the conversion of fat into carbohydrate. They are glycollate oxidase, hydroxypyruvate reductase, urate oxidase, allantoinase[19], serine:glyoxylate aminotransferase, glutamate:glyoxylate aminotransferase, glutamate:oxaloacetate aminotransferase, L-histidine:ammonia lyase[20], L-phenylalanine:ammonia lyase[21], L-tyrosine:ammonia lyase, D-amino acid oxidase[22] and L-amino acid oxidase[23]. The role of these enzymes in glyoxysome function is at present unknown.

The microbodies in tissues other than leaves and fat-storing seeds are less well understood. Indeed those of non-photosynthetic higher plant tissues have been referred to as non-specialized microbodies. The microbodies of autotrophically grown green algae might be expected to resemble functionally leaf peroxisomes since several species produce glycollate. However, it is not yet clear whether the enzymes of the glycollate pathway are present in the algal microbodies. Similarly the microbodies of algae, which are capable of heterotrophic growth on

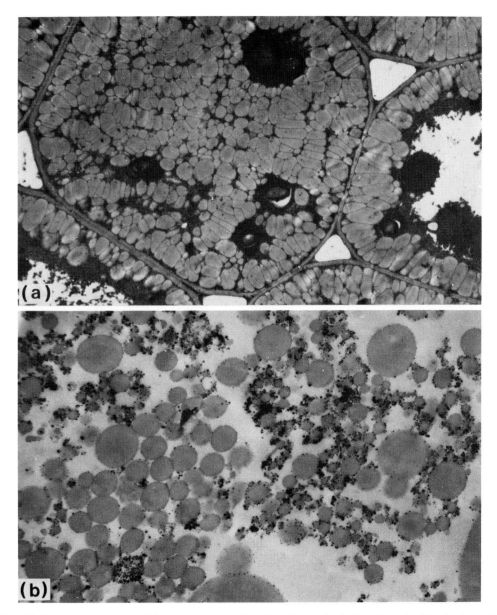

Fig. 3.14. Electron micrographs of the spherosomes in the cells of the seeds of abyssinian kale (*Crambe abyssinica*).
(a) Spherosomes in mature seeds packed so tightly that their normal spherical shape has been distorted ($\times 7500$).
(b) Spherosomes in an isolated fat fraction ($\times 15,000$).
(Original photos kindly provided by Dr. B. W. Nichols and Mr. C. G. Smith of Unilever Research and reproduced with permission from *European J. Biochem.* **43**, 281–290, 1974.)

acetate as the sole carbon source, might be expected to resemble glyoxysomes. However, in several algae the enzymes of the glyoxylate cycle, which are induced under these conditions, are soluble and therefore not present in microbodies. The glyoxylate cycle enzymes of *Euglena*, on the other hand, do appear to be located in glyoxysomes.

The origin and development of microbodies within the cell is of considerable interest. Since they contain no genetic or biosynthetic machinery, which would give them a degree of autonomy, they must be formed and regulated by other components of the cell. Most biochemical and ultrastructural evidence suggests that they are formed from the endoplasmic reticulum. It has been suggested that the blind end of an endoplasmic reticular vesicle is expanded, differentiated and eventually nipped off as a microbody. The synthesis of the membrane of the microbody is therefore brought about by endoplasmic reticular enzymes. The enzymes of the microbody are synthesized by ribosomes attached to the adjacent endoplasmic reticulum. Their synthesis and transfer to the microbody matrix is presumably analogous to that of export proteins in animal cells (see Chapter 4, Section D.3).

In many fat-storing seeds (e.g. sunflower) the fat is stored in cotyledons which emerge from the ground during germination and become functional leaves for a short time. While fat is being consumed the cotyledons contain glyoxysomes. When the cotyledons have emerged from the ground and fat consumption is complete no glyoxysomes are present but typical leaf peroxisomes are. This transition from glyoxysomes to peroxisomes normally occurs when the cotyledons are exposed to the light as they emerge from the ground. The mechanism of the transition is not yet clear. Two models have been suggested. One, proposed by Newcomb and known as the one-population hypothesis, suggests that the glyoxysomes change into peroxisomes. The other proposed by Beevers and known as the two-population hypothesis suggests that glyoxysomes disappear during the transition and are replaced by newly generated peroxisomes.

8. Lipid-rich organelles

Some plant tissues store lipid. The best-known examples are the endosperm of castor bean seeds, the cotyledons of sunflower, rape and peanut seeds, the aleurone layer of barley and wheat seeds and the mesocarp of the avocado pear. In each of these examples the stored lipid is triglyceride but this is not always the case; the cotyledons of jojoba seeds, for instance, store wax esters. However, in all these tissues the stored lipid occurs in special organelles which are abundantly distributed throughout the cytoplasm of the cells. In the triglyceride-storing tissues these organelles have been variously called spherosomes, oleosomes or oil bodies whilst those of the jojoba seed have been called wax bodies. Throughout this book they will be called spherosomes.

Spherosomes are generally spherical or oblate and have diameters in the range 0.4–3 μm (see Fig. 3.14). They are surrounded by a membrane, the structure of which is still a matter of controversy. There are those who believe it to be a single unit membrane as described in Section B.2(i) and there are others who believe that it is a half-unit membrane composed of one layer of the bimolecular leaflet of lipid plus the proteins normally embedded in it.

The two postulated membrane structures are directly related to the two current theories of spherosome ontogeny, both of which agree on the endoplasmic reticular origin of the spherosome. Frey-Wyssling in 1963 suggested that they are formed by nipping off the blind end of an endoplasmic reticular vesicle. This vesicle will have been differentiated into a triglyceride-synthesizing structure by receipt of the required enzymes. The synthesis and transfer of these enzymes to the vesicle is presumably analogous to that of export proteins in animal cells (see Chapter 4, Section D.3). The enzymes then catalyse the synthesis of triglycerides within the vesicle from appropriate precursors imported from the cytosol; it is well established that spherosomes can form triglycerides from acetyl-CoA, malonyl-CoA and *sn*-glycerol-3-phosphate [see Chapter 8, Sections C.1(i) and C.2(i)]. The triglycerides so formed accumulate in and eventually fill the spherosome. According to this theory the spherosome membrane would be structurally like that of the endoplasmic reticulum, namely a unit membrane [see Fig. 3.15(a)].

An alternative ontogenic mechanism was postu-

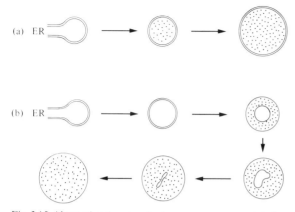

Fig. 3.15. Alternative theories of spherosome ontogeny; (a) after Frey-Wyssling, 1964; (b) after Schwarzenbach, 1971. (ER = endoplasmic reticulum; a double line represents a unit membrane whilst a single line represents half a unit membrane; dotted areas represent triglyceride.)

lated in 1971 by Schwarzenbach [see Fig. 3.15(b)]. This again involves the nipping off of an endoplasmic reticular vesicle presumably after differentiation and receipt of the triglyceride synthesizing enzymes. These enzymes are built into the vesicle membrane. The triglycerides formed as a result of their catalytic activity accumulate in between the two layers of the bimolecular leaflet of lipid of the vesicle membrane. This causes the two halves of the membrane to be forced apart. Eventually the inner half disappears leaving the outer half, presumably containing the triglyceride-synthesizing enzymes, surrounding the accumulated triglyceride. Support for half-unit membrane has come from a study of peanut cotyledon spherosomes which, in the electron microscope, have a single-line membrane, 2–3.5 nm in width (cf. the double-line unit membrane, ~8.5 nm in width).

The chemical composition of spherosomes isolated from different plant species shows small variation. Generally triglyceride constitutes more than 80% of the weight of the spherosome and in some cases as much as 98.5%. Protein and phospholipid are the other two components; presumably these represent the triglyceride-synthesizing enzymes and the outer membrane respectively.

Spherosomes are formed during seed development; the lipid synthesized and stored in them is intended to support the growth of the plant in the period between the germination of the seed and the

development of adequate photosynthetic capability. Accordingly during germination the triglyceride or wax ester is hydrolysed into its component fatty acids and alcoholic component and these then pass to other parts of the cell for further utilization [see Section B.7; Chapter 7, Section B.2 and Chapter 8, Section C.1(iii)]. Triglyceride-storing spherosomes contain a lipase with an acid pH optimum which is very active during germination. The spherosomes of jojoba seeds presumably contain a wax ester hydrolase.

9. Microtubules

Although microtubules were discovered in animal and algal cells in the late 1950s, they were not seen in higher plant cells until 1962 when glutaraldehyde fixation was introduced. They consist of straight cylinders of undetermined length (frequently at least several microns) with an external diameter of 24–25 nm. The wall of the cylinder has a thickness of 5–6 nm and the diameter of the apparently hollow lumen is about 12 nm. Microtubules are sometimes surrounded by a 'clear halo' 5–10 nm wide, the structure and significance of which is not known.

Cross-sections of microtubules show that the wall is composed of thirteen circular subunits each with a diameter of about 5 nm. Longitudinal views reveal a series of filaments made up of two types of subunit which occur alternately along their length. The subunits are globular with a diameter of about 5 nm; since they overlap one another slightly and are of two types, the filament has an 8-nm periodicity. The thirteen filaments which constitute the microtubule wall exhibit a small longitudinal displacement relative to each other which gives the helical structure seen in Fig. 3.16.

Microtubules are composed of molecules of tubulin, a globular protein (MW 120,000) containing two subunits of equal size but different structure (α and β). Tubulin molecules are arranged in a helical fashion to form the microtubule wall. The α and β subunits are the two different structural components of the filaments. The filaments are of little structural significance for the microtubule is constructed by the successive addition of tubulin molecules to a developing helix rather than by the

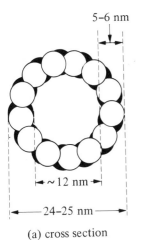

(a) cross section

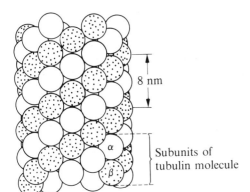

Subunits of
tubulin molecule

(b) side view

Fig. 3.16. Model of a microtubule (after J. Bryan, *Fed. Proc.* **33**, 152, 1974).

aggregation of thirteen pre-formed filaments. Each tubulin molecule binds one molecule of GTP tightly and a second one loosely.

Three broad categories of microtubules occur in plant cells, the microtubules which constitute the nuclear spindle in mitosis (see Section B.4), the microtubules which are found in the cytoplasm and the microtubules that are structural components of flagellae and cilia (which are confined to motile cells). The latter are much more stable structures than the first two categories which break down and re-form with ease apparently in tune with specific requirements of the cell.

It is believed that the labile types of microtubules are in a dynamic equilibrium with a pool of free tubulin molecules [eq. (3.1)].

$$\text{Tubulin (associated as microtubules)} \rightleftharpoons \text{tubulin (free molecules)} \quad (3.1)$$

Thus the tubules can be formed at one site in the cell by shifting the equilibrium of eq. (3.1) to the left and then when their task is complete be disassembled by shifting the equilibrium to the right. This would explain the fact that microtubules do not move as intact structures within the cell but 'disappear' from one site and 'reappear' at another. Clearly there must be some signal to bring about these shifts in the equilibrium of eq. (3.1) and this has led to the concept of microtubule 'initiation sites' in the cell. The physical and chemical nature of these signals is as yet unknown. During the assembly of microtubules from free tubulin molecules, one GTP molecule is hydrolysed to GDP and orthophosphate per tubulin molecule.

The dynamic equilibrium depicted in eq. (3.1) is severely displaced by the alkaloids colchicine, vinblastine and vincristine (Fig. 3.17). Colchicine binds to one of the subunits of free tubulin molecules with displacement of GTP. This pulls the equilibrium to the right by diminishing the concentration of tubulin molecules available for microtubule formation. Thus colchicine prevents the formation of microtubules and promotes the disassembly of those already present. One of the most striking effects of colchicine is the inhibition of mitosis by disruption of the mitotic spindle; dividing cells treated with colchicine appear to be blocked at metaphase and polyploid progeny result. Vinblastine and vincristine also bind to free tubulin molecules but at a different site from colchicine; in this case, however, tubulin is precipitated in the cytoplasm in crystalline form. Because these alkaloids interfere with the formation of mitotic spindles and thus block cell division, they have been used to treat rapidly growing cancers.

In addition to their function in the mitotic spindle and in the flagellae and cilia of motile cells, microtubules are thought to be concerned with the determination of the shape of non-spherical, naked cells, e.g. algae such as *Ochromonas* and the male gametes of algae, liverworts, mosses, ferns and cycads, and with cell-wall formation in non-naked cells. When the microtubules of the naked cells are disrupted by colchicine a spherical shape results which persists until the colchicine is removed and

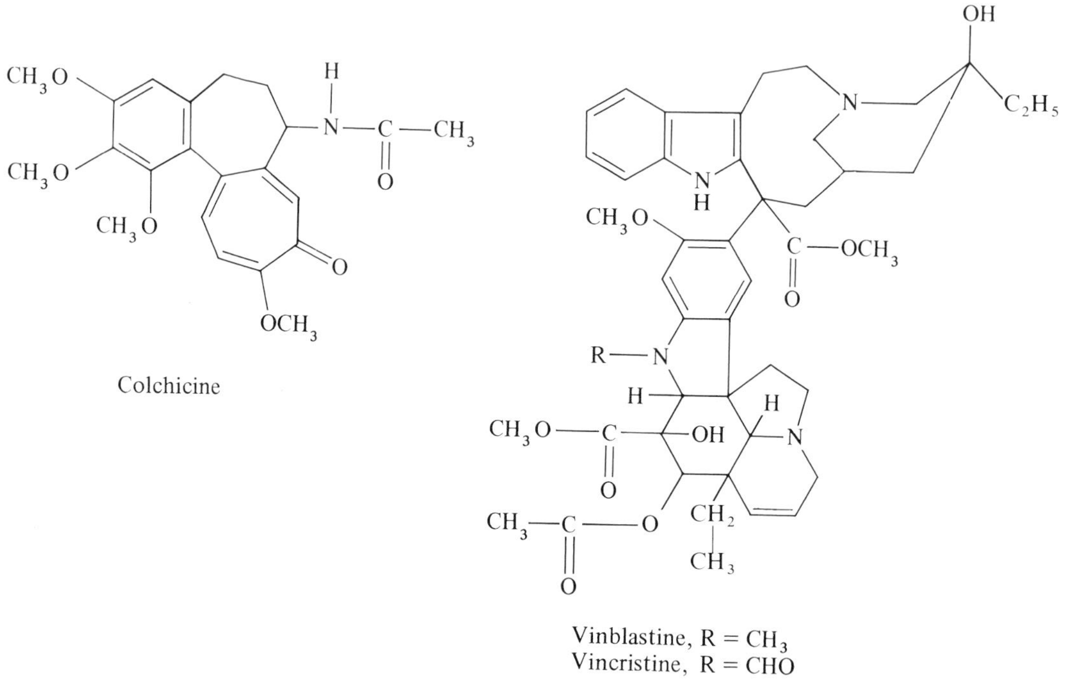

Fig. 3.17. Alkaloid inhibitors of microtubule formation.

Colchicine

Vinblastine, R = CH₃
Vincristine, R = CHO

the microtubules re-form. In non-naked cells the orientation of cellulose microfibrils as they are initially laid down in the cell wall (see Chapter 4, Section C) matches the orientation of microtubules in the cytoplasm; moreover, disruption of the microtubules by colchicine leads to the random orientation of newly laid down microfibrils and often to an abnormal wall structure. This has led to the idea that microtubules determine the orientation of newly formed microfibrils. If this idea is correct it requires that the orienting information be transmitted across the plasmalemma because the microtubules are in the cytoplasm whilst the microfibril-synthesizing enzyme complex [see Chapter 4, Section D.1(i)] is built into the outer surface of the plasmalemma. It has also been suggested that microtubules direct the Golgi vesicles carrying cell-wall matrix polysaccharide through the cytoplasm to those parts of the plasmalemma immediately adjacent to active cell wall synthesis.

C. ISOLATION OF CELLULAR COMPONENTS

Much of our knowledge of the functions and enzymic properties of the subcellular components just described has been derived from a study of pure preparations of these components. These preparations are made by disrupting the cell wall and the cytoplasmic membranes (plasmalemma, endoplasmic reticulum and tonoplast) and then separating the various subcellular components by differential centrifugation, a technique which exploits differences in their size and density. This procedure results in rather crude subcellular fractions which are each enriched with one particular component and are by no means pure. These fractions are then purified by density gradient centrifugation. The identity of the type of organelle isolated in a particular fraction is checked by microscopic examination and by looking for the presence of a compound or enzyme characteristic of it; the latter are known as markers. Moreover, the purity of a fraction is often confirmed by checking for the absence of markers characteristic of likely contaminants.

1. Cell-disruption techniques

Several methods have been used for the disruption of plant cells. All of these are mechanical because the presence of the tough cell wall renders inadequate such techniques as ultrasonic vibration and cycles of freezing and thawing which are successful with many animal tissues.

The most common method employs a blender whose high-speed blades exert large shearing forces on the cells. The tissue is immersed in an equal weight of homogenizing medium (see Section C.2) and blended for 0.5–2 min at full speed; provided the tissue is not too fibrous this results in the disruption of most of the cells. Unfortunately this procedure is too severe for some of the more delicate cellular components; for instance, it destroys a high proportion of nuclei and microbodies and fragments many chloroplasts.

A gentler method employs a Potter–Elvehjem homogenizer fitted with a loose-fitting glass pestle which is rotated at 200–400 revolutions per minute. The tissue is chopped into small pieces and then added to the homogenizer tube with an equal volume of homogenization medium. The rotating pestle is then worked slowly up and down the tube about 20 times during the course of 1–2 min. This forces the tissue between the pestle and the static wall of the tube several times and breaks the cells.

Probably the gentlest procedure of all is hand grinding with a mortar and pestle. The tissue is chopped into small pieces, mixed with an equal volume of homogenization medium and sometimes a little washed, silver sand to act as an abrasive and then ground for 2–3 min.

All these cell disruption procedures are carried out at 2–4°C to minimize autolytic changes. These changes result from the bringing together of enzymes and substrates normally compartmentalized from one another.

2. Homogenization media

The ideal homogenization medium should be capable of maintaining the morphological and functional integrity of those cellular organelles not disrupted by the grinding procedures (e.g. mitochondria) and of preventing deleterious changes which would otherwise result from the mixing of the contents of normally separate cellular compartments (e.g. the vacuolar contents with the cytosol).

In an effort to fulfil these requirements homogenization media usually contain, (i) 0.25 M sucrose (or mannitol) so as to be isotonic with the cytosol, (ii) a pH 7–8 buffer (often Tris) at a concentration of 50 mM to neutralize the acidic vacuolar contents, (iii) a sulphydryl compound such as dithiothreitol or mercaptoethanol at a concentration of 10 mM to minimize the inactivation by oxidation of those enzymes which have a cysteine residue at their active site, (iv) magnesium ions at a concentration of 10 mM to prevent the dissociation of cytoplasmic and organellar ribosomes, (v) calcium ions at a concentration of 1 mM to prevent the clumping of nuclei and (vi) polyvinylpyrrolidone to precipitate out the tannins released from the vacuole which would otherwise bind to enzymes and inactivate them.

3. Centrifugation

The initial procedure, termed differential centrifugation (see Fig. 3.18), is used to separate the components of the homogenate into a number of fractions (usually five), namely nuclei, chloroplasts, mitochondria, microsomes [small membranous vesicles derived artefactually mainly from the endoplasmic reticulum, see Section B.2(iii)] and supernatant (a mixture of the soluble phase of the cytoplasm mixed with the contents of the vacuole and any other ruptured organelle).

These fractions are, however, impure. The fraction containing nuclei also contains cell debris, chloroplasts and chloroplast fragments as major contaminants. The chloroplast fraction is contaminated by nuclei and nuclear fragments. The mitochondrial fraction is contaminated by proplastids and, provided the homogenization has been

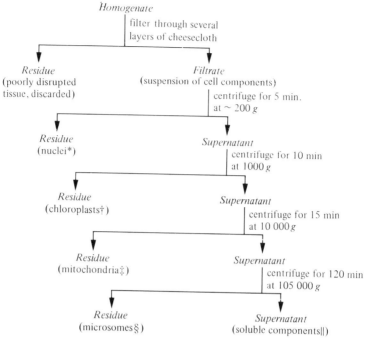

Fig. 3.18. A generalized procedure for separating a tissue homogenate into crude fractions enriched in a particular cell component.
* Contaminated with cell debris, chloroplasts and chloroplast fragments.
† Contaminated with nuclei and nuclear fragments.
‡ Contaminated with proplastids and microbodies.
§ Consists of membranous vesicles derived artefactually from endoplasmic reticulum (mainly), plasmalemma and tonoplast; ribosomes adhere to vesicles derived from rough endoplasmic reticulum.
‖ Consists of the soluble phase of the cytoplasm (cytosol) mixed with the soluble contents of the vacuole and any other ruptured organelle.

gentle, microbodies (peroxisomes or glyoxysomes depending upon the tissue). The microsomal fraction is often contaminated with mitochondria.

All the particulate fractions can be further purified by gently resuspending in the homogenization medium and centrifuging for a second time at the g force at which they were originally sedimented. However, it is more usual to purify them by density gradient centrifugation. In this technique the cell fraction is resuspended in homogenization medium and then carefully layered on to the top of a sucrose gradient in a centrifuge tube. The gradient is made by successively adding layers of sucrose solution of decreasing concentration (and therefore density) to the centrifuge tube so gently that they do not mix to any great extent. The tube may then be allowed to stand at 2–4°C for about an hour to allow diffusion to smooth the steps in the gradient. After the resuspended cell fraction has been added, the tube is centrifuged in a swinging-bucket rotor. The cell components are pulled through the sucrose gradient by the centrifugal force and separate into distinct zones according to their relative densities. Depending upon the nature of the gradient and the length of the centrifugal period a given zone may or may not reach that part of the gradient which has a density equal to its own or may reach the bottom of the tube. The various zones separated in this way may be removed by careful pipetting. This technique exploits the difference in density of the various cell components. These differences are shown in Fig. 3.19 along with a curve relating the concentration of a sucrose solution to its density.

Nuclei can be isolated from the crude nuclear fraction by centrifugation at $24,000g$ for 60 min on a single layer of 60% (w/w) sucrose solution which has a density of 1.296 g ml^{-1}. Nuclei, which have a density of 1.32 g ml^{-1}, can move much further through this solution than chloroplasts or chloroplast fragments (density <1.24 g ml^{-1}).

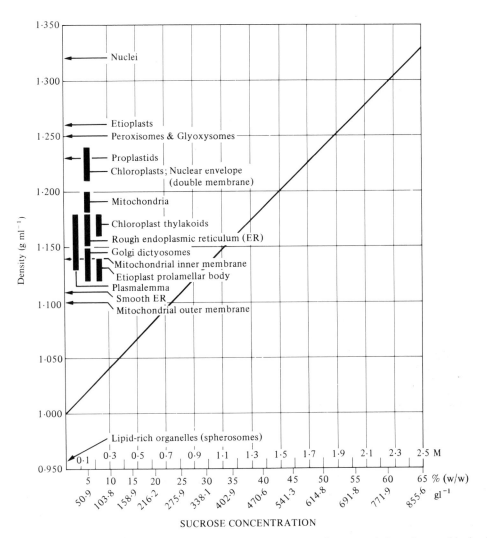

Fig. 3.19. Density of cellular components and a graph relating the concentration of sucrose solution, often used in density gradient centrifugation, to its density at 0°C.

Chloroplasts can be freed from contamination by nuclei by centrifugation at 150,000 g for 90 min on a sucrose gradient ranging from 10% (w/w); density 1.041 g ml^{-1}) to 55% (w/w; density 1.266 g ml^{-1}). Nuclei reach the bottom of the tube whilst the chloroplasts move no further than the region of the gradient having a density of 1.24 g ml^{-1}.

The separation of the crude mitochondrial fraction into mitochondria, proplastids and microbodies by density gradient centrifugation is described in Section B.7.

4. Checking the identity of isolated cell components

The identity of cell components isolated by differential and density gradient centrifugation should be checked before they are used. This is accomplished in two ways, firstly by microscopic examination often after treatment with a specific stain and secondly by assaying for a characteristic compound or enzyme. A list of the latter is given in Table 3.1.

Table 3.1. *Commonly used biochemical markers of plant cell components*

Organelle	Suborganelle	Marker
1. Nucleus	Intact	DNA
	Envelope	Attached DNA
	Nucleolus	RNA polymerase I
2. Chloroplast	Intact	Chlorophyll, ribulose bisphosphate carboxylase (C_3 plants), pyruvate:orthophosphate dikinase[25] (C_4 plants), NADP:triose phosphate dehydrogenase
	Thylakoid	Chlorophyll
	Matrix	Ribulose bisphosphate carboxylase (C_3 plants), pyruvate:orthophosphate dikinase (C_4 plants), NADP:triose phosphate dehydrogenase[26]
3. Mitochondrion	Intact	Cytochrome oxidase, fumarase[24], succinate dehydrogenase, succinate:cytochrome c reductase
	Outer membrane	Antimycin A-insensitive NADH:cytochrome c reductase
	Inner membrane	Cytochrome oxidase, succinate dehydrogenase, succinate:cytochrome c reductase
	Matrix	Fumarase
4. Proplastid	Intact	Ribulose bisphosphate carboxylase, fatty acid synthase, nitrite reductase[27]
5. Peroxisome	Intact	Catalase, hydroxypyruvate reductase
6. Glyoxysome	Intact	Catalase, isocitrate lyase, malate synthase
7. Microsomes	—	Phospholipid-synthesizing enzymes, Antimycin A-insensitive NAD(P)H:cytochrome c reductase
8. Vacuole	Intact	Ribonuclease, phosphodiesterase, pigments characteristic of plant species

5. Isolation of particularly fragile cell components

Golgi dictyosomes have been successfully isolated by homogenizing the tissue very gently and then treating the homogenate with glutaraldehyde which stabilizes the protein-rich membranes and enables them to withstand the ensuing centrifugation.

Vacuoles have been isolated from a few tissues. This can usually only be accomplished when they are small and numerous. Yeast vacuoles are prepared by first removing the cell wall by incubation with snail gut enzyme and then lysing the resulting protoplast with 0.7 M mannitol. The vacuoles survive this osmotic shock and can be isolated by density gradient centrifugation.

6. Isolation of chloroplasts by a non-aqueous method

The tissue is freeze-dried and then disrupted in a hand-operated glass tissue grinder in the presence of an ice-cold mixture of cyclohexane and carbon tetrachloride with a density of 1.32 g ml^{-1}. The resulting homogenate is then filtered through several layers of cheese cloth to remove large particles and then subjected to density gradient centrifugation. The gradient is composed of a bottom layer of cyclohexane:carbon tetrachloride mixture of density 1.36 g ml^{-1}, followed by successive layers of density 1.32, 1.30 and 1.29 g ml^{-1}. After layering the filtered homogenate on to the top of the gradient, the tube is centrifuged at 12,000g for 15 min in a swinging bucket rotor. The chloroplasts migrate to the surface of the gradient leaving other materials lower down the tube.

Chloroplasts prepared in this way have the advantage that their enzyme complement is complete because there is none of the leaching out of stromal enzymes which accompanies aqueous isolation procedures. However, their outer membrane is frequently missing and a proportion of their lipids are removed; sometimes small amounts of other organelles (e.g. mitochondria) may be adsorbed on to them.

SUGGESTIONS FOR FURTHER READING

General texts

Robards, A. W. (ed.) (1974) *Dynamic Aspects of Plant Ultrastructure.* McGraw-Hill Book Co. (UK) Ltd.
Gunning, B. E. S. and Steer, M. W. (1975) *Ultrastructure and the Biology of Plant Cells.* Edward Arnold Ltd.
Reid, E. (ed.) (1979) *Plant Organelles.* John Wiley & Sons, Chichester, New York, Sydney, Toronto.

Specialist texts and reviews

Tolbert, N. E. (1971) 'Microbodies-peroxisomes and glyoxysomes', *Ann. Rev. Plant Physiol.* **22**, 45–74.
Jordan, E. G. (1971) *The Nucleolus.* Oxford Univ. Press.
Cocking, E. C. (1972) 'Plant cell protoplasts-isolation and development', *Ann. Rev. Plant. Physiol.* **23**, 29–50.
John, B. and Lewis, K. R. (1972) *Somatic Cell Division.* Oxford Univ. Press.
John, B. and Lewis, K. R. (1973) *The Meiotic Mechanism.* Oxford Univ. Press.
Hepler, P. K. and Palevitz, B. A. (1974) 'Microtubules and microfilaments', *Ann. Rev. Plant Physiol.* **25**, 309–362.
Robards, A. W. (1975) 'Plasmodesmata', *Ann. Rev. Plant Physiol.* **26**, 13–29.
Morré, D. J. (1975) 'Membrane biogenesis', *Ann. Rev. Plant Physiol.* **26**, 441–481.
Beevers, H. (1975) 'Organelles from castor bean seedlings: biochemical roles in gluconeogenesis and phospholipid biosynthesis', in *Recent Advances in the Chemistry and Biochemistry of Plant Lipids* (Galliard, T. and Mercer, E. I., eds.). Academic Press.
Barber, J. (ed.) (1976) *The Intact Chloroplast (Topics in Photosynthesis*, Vol. I). Elsevier/North-Holland Biomedical Press.
Nagl, W. (1976) 'Nuclear organization', *Ann. Rev. Plant Physiol.* **27**, 39–69.
Yeoman, M. M. (ed.) (1976) *Cell Division in Higher Plants.* Academic Press.
Matile, P. (1978) 'Biochemistry and function of vacuoles', *Ann. Rev. Plant Physiol.* **29**, 193–213.
Kirk, J. T. O. and Tilney-Bassett, R. A. E. (1978) *The Plastids; their chemistry, structure, growth and inheritance*, 2nd edition. Elsevier/North-Holland Biomedical Press.
Beevers, H. (1979) 'Microbodies in higher plants', *Ann. Rev. Plant Physiol.* **30**, 159–193.
Quail, P. H. (1979) 'Plant cell fractionation', *Ann. Rev. Plant Physiol.* **30**, 425–484.
Wool, I. G. (1979) 'The structure and function of eukaryotic ribosomes', *Ann. Rev. Biochem.* **48**, 719–754.
Roberts, K. and Hyams, J. S. (eds.) (1979) *Microtubules.* Academic Press.

ENZYMES

1. Nucleosidetriphosphate: RNA nucleotidyltransferase, EC 2.7.7.6.
2. Succinate: (acceptor) oxidoreductase, EC 1.3.99.1.
3. ATP phosphohydrolase, EC 3.6.1.3.

4. Ferrocytochrome c: oxygen oxidoreductase, EC 1.9.3.1.
5. 3-Phospho-D-glycerate carboxy-lyase (dimerizing), EC 4.1.1.39.
6. Deoxynucleosidetriphosphate: DNA deoxynucleotidyl-transferase, EC 2.7.7.7.
7. Glycollate: oxygen oxidoreductase, EC 1.1.3.1.
8. L-Serine: glyoxylate aminotransferase, EC 2.6.1.45.
9. L-Aspartate: 2-oxoglutarate aminotransferase, EC 2.6.1.1.
10. D-Glycerate: NAD$^+$ oxidoreductase, EC 1.1.1.81.
11. Hydrogen peroxide: hydrogen peroxide oxidoreductase, EC 1.11.1.6.
12. Urate: oxygen oxidoreductase, EC 1.7.3.3.
13. Citrate oxaloacetate-lyase (pro-3S-CH$_2$COO$^-$ →acetyl CoA), EC 4.1.3.7.
14. Citrate (isocitrate) hydro-lyase, EC 4.2.1.3.
15. $threo$-D$_s$-Isocitrate glyoxylate-lyase, EC 4.1.3.1.

16. L-Malate glyoxylate-lyase (CoA-acetylating), EC 4.1.3.2.
17. L-Malate: NAD$^+$ oxidoreductase, EC 1.1.1.37.
18. Acid: CoA ligase (AMP forming), EC 6.2.1.3.
19. Allantoin amidohydrolase, EC 3.5.2.5.
20. L-Histidine ammonia-lyase, EC 4.3.1.3.
21. L-Phenylalanine ammonia-lyase, EC 4.3.1.5.
22. D-Amino acid: oxygen oxidoreductase (deaminating), EC 1.4.3.3.
23. L-Amino acid: oxygen oxidoreductase (deaminating), EC 1.4.3.2.
24. L-Malate hydro-lyase, EC 4.2.1.2.
25. ATP: pyruvate, orthophosphate phosphotransferase, EC 2.7.9.1.
26. D-Glyceraldehyde 3-phosphate: NADP$^+$ oxidoreductase, EC 1.2.1.13.
27. NAD(P)H: nitrite oxidoreductase, EC 1.6.6.4.

CHAPTER 4

The Plant Cell Wall

CONTENTS

A. GENERAL DESCRIPTION

Outside the plasmalemma of the majority of plant cells there exists a tough coat known as the cell wall. The cell wall is important biologically for two reasons. Firstly, it determines, to a great extent, the morphology, and, to some extent, the function of the cell. Secondly, since it forms the limiting envelope of the cell, it may be directly involved in regulating cell expansion.

When first formed the cell wall is very thin but becomes progressively thicker during growth of the cell owing to the deposition of new layers of cell wall material. These new layers are formed immediately adjacent to the plasmalemma; indeed, as will be apparent later, the plasmalemma itself and enzymes partially embedded in it are intimately concerned in the formation of the cell wall. This means that as new layers are formed the older ones are pushed further and further away from the protoplast. In many cell walls the thickening is not uniform; the classic example of this are the pits seen in the walls of the tracheids of both angiosperms and gymnosperms.

The chemical and physical structure of the wall varies widely from plant group to plant group and from cell type to cell type. There are, however, two general characteristics which are common to all plant cell walls; firstly, they are not chemically homogeneous but are composed of several different materials; and secondly, they are not physically homogeneous but are built up of distinct layers.

The most important chemical components of all plant cell walls are polysaccharides. These can be subdivided into two broad categories, those which exist within the wall in crystalline form and those which do not. The polysaccharides of the former category are very long, unbranched molecules which, within the cell wall, are aggregated together in bundles called microfibrils. Within most types of microfibrils the long axes of the polysaccharides are parallel to each other [the exceptions are those of the xylan seaweeds, see Section B.1(ii)(c)]; moreover, the majority of the molecules, constituting the core of the structure, are so regularly arranged that they form a crystal lattice. The microfibrils are embedded in a matrix composed of the cell wall polysaccharides which are not crystalline. The matrix polysaccharides are more often than not multi-branched molecules containing several different species of monosaccharide residues. The matrix is often described as amorphous but this cannot always be the case, for it is clear that some relatively linear matrix polysaccharides lie parallel to the microfibrils. The polysaccharide structure of the cell wall has been likened to that of reinforced concrete with the microfibrils being the equivalent of the steel rod reinforcing structure and the matrix polysaccharides the equivalent of the concrete surrounding it.

The next most abundant organic component of the cell walls of higher plants is lignin which is a complex, highly ramified polymer of phenylpropane (C_6–C_3) residues. It is a major constituent of the cell walls of the supporting tissues of higher plants (e.g. xylem) but also occurs to a lesser extent in the cell walls of other tissues. It does not appear to be present in algal cell walls. When present, lignin forms a network which permeates the matrix of the cell wall and is an important structural component of the wall.

It is now well established that proteins are present in the cell wall. Some are enzymes whilst others are thought to have a structural role.

Certain specialized cells have additional chemical components. Those cell walls of higher plants which are exposed to the atmosphere (e.g. epidermal cells) have a cuticle, composed of layers of cutin and wax, superimposed on the normal wall. Inorganic incrustations occur in the cell walls of some land plants and many algae.

Water is an extremely important constituent of all plant cell walls, though this is often forgotten in many descriptions of cell-wall chemistry. Water is able to permeate the matrix of the wall but not the crystalline regions of the microfibrils. The amount of water present varies from wall to wall and depends upon the nature of the matrix polysaccharides, with which it forms close intermolecular associations and gel-like structures, and the degree of lignification (the more lignin, the less water is present).

All plant cell walls are made up of distinct layers. All the layers have the same basic structure, being composed of microfibrils embedded in a polysaccharide matrix. However, they may differ from one another in several respects, the most important of which are (i) their thickness, (ii) the ratio of the microfibrillar component to the matrix component, (iii) the orientation of the microfibrils within the matrix relative to the long axis of the cell, (iv) the nature of the matrix polysaccharides, (v) the degree of lignification and (vi) the water content.

In higher plants the cell wall is typically composed of three layers which are called the middle lamella, the primary wall and the secondary wall. The middle lamella forms an amorphous intercellular layer between the primary walls of adjacent cells; it is the first layer to be formed when a cell divides and is, therefore, the initial partition between the newly formed daughter cells. A primary wall is then formed upon the middle lamellae from each side by the daughter cells, so sandwiching it between them. The primary walls continue to grow and to increase in area and thickness while the daughter cells grow in size; the primary walls, therefore, expand to accommodate the increasing size of the daughter cells. Towards the end of their growth, deposition of the secondary wall on top of the primary wall begins and continues after growth has stopped. The extent to which the secondary wall thickens depends upon the type of cell into which the daughter cells have differentiated. In many parenchyma cells the secondary wall remains thin though their protoplasts remain alive for long periods. In other cells, the secondary wall thickens so much that the cell lumen is almost filled, after which the protoplast, which is responsible for its synthesis and deposition, dies and is autolysed; this is exemplified by some phloem fibres which appear as rods, each with a narrow, central, empty, threadlike lumen running down its length. In most other elongated higher plant cells, e.g. tracheids, xylem vessels, collenchyma cells, many phloem fibres, the secondary wall becomes appreciably thick but nevertheless leaves a fairly large lumen when the protoplast eventually degenerates. When the secondary wall becomes very thick it is normally made up of several different layers which differ from one another mainly in the relative orientation of their microfibrils.

In algae cell walls are composed of many layers or lamellae. It is not usual, however, to use the terms primary wall and secondary wall in relation to algal cell walls. Indeed in many algae, e.g. the Cladophorales, there is no obvious difference between the wall in that part of a cell which is growing and that part which is not growing; hence there is no distinction between a primary wall layer and a secondary wall layer and the terms, therefore, do not apply.

The importance of cell walls, in both their usefulness to man and in the carbon economy of the biosphere, should not be overlooked. It has been estimated that about 10^{12} tons of cell wall material are synthesized and destroyed annually; of this cellulose constitutes 10^{11} tons which makes it the most abundant organic compound in the biosphere. The energy equivalent of this cell-wall material is about 1.8×10^{19} kJ. Wood, which is composed entirely of cell walls, is used by man in huge amounts in the form of timber for construction purposes and to produce paper. To this must be added the cotton and other vegetable fibre industries and the food manufacturers who use cell-wall constituents such as alginic acid and pectins as food additives.

B. CHEMICAL COMPOSITION

1. Polysaccharides

(i) EXTRACTION AND FRACTIONATION OF CELL-WALL POLYSACCHARIDES

Over the years many methods have been used to extract and fractionate cell-wall polysaccharides. However, the method developed by Jermyn and Isherwood (1959) has now become the standard procedure. The cell-wall material, e.g. wood, algal cell walls, is dried either in an oven at 105°C or with acetone and then extracted for 12 hr with boiling water. The extract, which, when higher plant tissue is used, contains a group of matrix polysaccharides called pectins along with some soluble protein, is centrifuged off. The residue is then subjected to mild chlorination; this removes any lignin and tends to break any linkages between microfibrillar polysaccharides and the remaining matrix polysaccharides which are collectively called hemicelluloses. This is then followed by extraction with 4 M KOH at room temperature for several hours. The extract, which is then centrifuged off, contains the alkali-soluble hemicelluloses. The residue is the microfibrillar component.

This procedure separates the cell-wall polysaccharides into three solubility classes: the pectins, the hemicelluloses and the microfibrillar polysaccharides. Within the first two of these classes are a number of different polysaccharide species which vary from plant type to plant type. There are, therefore, no standard methods whereby these classes may be separated into their component polysaccharide subclasses; each biological material requires handling in its own way. Various fractionation procedures are used for this, such as fractional precipitation by electrolytes or organic solvents, column chromatography and treatment with specific enzymes.

An alternative approach to this problem is that of Albersheim who has used as his biological material cell walls isolated from suspension-cultured cells, in particular those of sycamore. This approach has two advantages. Firstly, the cultured cells are homogeneous and possess only primary walls; thus the cell-wall preparation obtained is of pure primary walls from a dicotyledonous plant. This contrasts with most higher plant cell-wall preparations which are often from a mixture of cell types (as in wood) and usually include both primary and secondary walls. The second advantage is that these cells secrete into the culture medium polysaccharides which are structurally related to the wall matrix polysaccharides. These secreted polysaccharides constitute an easily available source of material for structural studies and the development of analytical methods. In some ways this approach is rather limited since it can only examine primary walls but this is more than compensated for by the penetrating observations that can be, and indeed have been, made.

(ii) MICROFIBRILLAR POLYSACCHARIDES

The most common microfibrillar polysaccharide found in the plant kingdom is cellulose. However, three other microfibrillar polysaccharides do occur. These are chitin, which is present in the majority of fungi (the remaining fungi have cellulose), the β-1,4-mannans which occur in the green algal families, Codiaceae, Dasycladaceae and Derbesiaceae, and the β-1,3-xylans which occur in the green algal families, Bryopsidaceae, Caulerpaceae, Udotaceae and Dichotomosiphonaceae. The algae, *Porphyra* and *Bangia*, which belong to the primitive order Bangiales of the red algae possess both β-1,4-mannans and β-1,3-xylans. However, only the β-1,3-xylans occur in the cell wall as microfibrils; the β-1,4-mannan is amorphous and is present in the cuticle.

(a) *Cellulose*

Cellulose molecules are unbranched chains of D-glucopyranose residues linked by $\beta1\rightarrow4$ glycosidic bonds. They vary considerably in length but the average number of glucose residues is about 8000 per molecule. The hydroxyl groups at carbon atoms 2, 3 and 6 are unsubstituted. The repeating unit in the cellulose molecule is the cellobiose residue; its conformation is shown in Fig. 4.1.

Cellulose molecules within plant cell walls are organized into biological units of structure known as microfibrils. A microfibril consists of a bundle of cellulose molecules each arranged with its long axis parallel to that of the others. The cross-section of the bundle is usually oval. The cellulose molecules constituting the core of the microfibril are arranged in perfect three-dimensional array and thus constitute a crystal lattice. Surrounding this crystalline core is a region of cellulose molecules which run parallel to those in the core but which do not have perfect three-dimensional order; this is called the paracrystalline phase. Amongst the cellulose molecules of the paracrystalline cortex of the microfibril are some matrix polysaccharides, usually hemicelluloses; these are least abundant at the surface of the crystalline core but become more and more numerous as the outermost region of the cortex is reached. It has been suggested that it is this mixture of cellulose and hemicellulose chains that prevents the cellulose molecules of the cortex from forming a crystal lattice. Water molecules are able to penetrate the paracrystalline cortex but not the crystalline core of the microfibril.

The microfibrils within a given cell wall have a range of lengths as do the cellulose molecules which are their main component. They are frequently many microns long. Since an average cellulose molecule, composed of 8000 glucose residues, has a length, within the crystal lattice of the microfibril core, of 4.12 μm, it is evident that relatively few cellulose molecules will run the whole length of the micro-

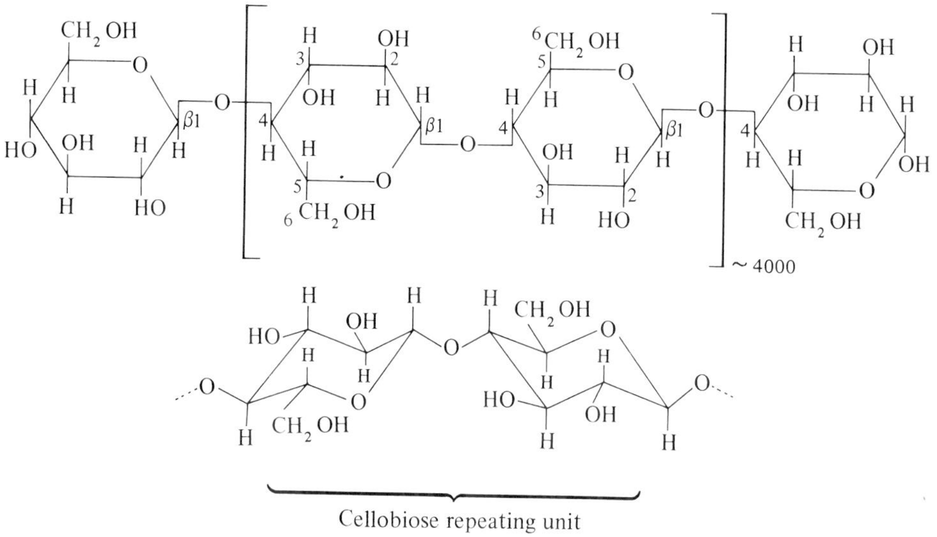

Fig. 4.1. Structure and conformation of the cellulose molecule.

fibril. Most will not. There will therefore be regions along the length of the microfibril where some cellulose molecules end, others begin and the majority pass straight through. In these regions the crystal lattice is distorted to some extent.

Within the plant kingdom there appear to be two types of cellulose microfibril which differ in their cross-sectional area. The cross-section of the micro-fibrils of all higher plants and most lower plants have external long and short diameters of 8.5 and 4.5 nm respectively. The long and short diameters of the crystalline core of these microfibrils are 5 and 3 nm respectively. The cross-section of the crystal-line core is estimated to cut across about fifty cellulose molecules. The cross-sectional area of the cortical region of the microfibril is estimated to cut across a further 100 polysaccharide molecules (cellulose plus hemicelluloses). The microfibrils of certain green algae, in particular *Valonia* and members of the Cladophorales, have microfibrils with a much larger cross-sectional area. The exter-nal long and short diameters of these microfibrils are 18.5 nm and 11.5 nm respectively with crystal-line core long and short diameters of 17 and 11 nm respectively. The cross-section of the core is esti-mated to cut across about 500 cellulose molecules whilst that of the cortex is estimated to cut across about 160 molecules. The latter are all poly-saccharides composed of glucose residues, pre-sumably cellulose, since microfibrils from these algae yield only glucose on hydrolysis.

It should be noted that older concepts of micro-fibrillar structure involving micelles and elementary fibrils (or protofibrils) are no longer tenable.

The structure of the cellulose crystal lattice, which constitutes the core of the microfibril, has been studied extensively by X-ray diffraction analysis. Each cellulose chain within the crystal lattice has a zig-zag conformation (see Fig. 4.2). Successive glucose residues are rotated through 180° with respect to each other, thus permitting a hydrogen bond to form between the hydroxyl group at C–3 of one glucose residue and the pyranose ring oxygen atom of the next glucose residue. This impedes the rotation of adjacent glucose residues around the glycosidic bond which links them together and produces a stiff, band-like molecule with the puck-ered (chair conformation) faces of the pyranose rings all in the same plane.

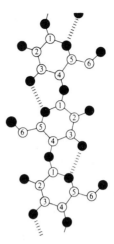

Fig. 4.2. Part of a cellulose chain showing the 'bent' conformation proposed by Hermans (1949). Open circles represent carbon atoms, filled circles represent oxygen atoms and barred lines represent hydrogen bonds.

The unit cell of the crystal lattice (see Fig. 4.3) is composed of cellobiose residues from five cellulose molecules. The length of the longitudinal b-axis of the unit cell is 1.03 nm, equivalent to the length of the cellobiose residue; the lengths of the a- and c-axes are 0.835 nm and 0.79 nm respectively. The angle between the a- and c-axes is 84°. Four out of the five cellobiose residues are arranged at the vertical corners of the cell, whilst the fifth occupies

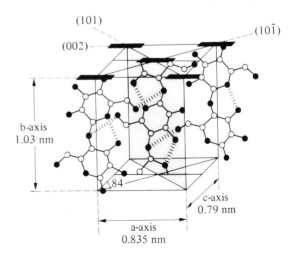

Fig. 4.3. Unit cell of the crystalline core of a higher plant cellulose microfibril (according to Meyer and Misch, 1936). (Only three of the five cellobiose residues are shown. The centrally placed residue is shown as being upside down with respect to the others; whether this is so is now being seriously questioned. The shaded (101) plane lies parallel to the surface of the plasmalemma.)

the centre; the puckered faces of the pyranose rings are 'flat' in the plane bounded by the a- and b-axes (the 002 plane). The central cellobiose residue of the unit cell is the odd one out. To account for the presence of certain X-ray reflections and the absence of others it must in some way be different from the four corner residues. In the model of the unit cell proposed by Meyer and Misch in 1936 the central cellobiose residue was different in two ways. Firstly, it was out of phase with respect to the other four by a quarter of the fibre period, namely $1.03/4 = 0.257$ nm, and secondly it was upside down, thus giving an antiparallel arrangement of cellulose chains in the microfibril. Preston in 1974 maintained that the crystallographic evidence upon which the second difference was based was weak and that the general argument that led Meyer to this conclusion was not strong either. It is Preston's opinion that the displacement of the central cellobiose residue by a quarter of the fibre period is alone sufficient to account for the X-ray diffraction pattern. Many attempts have been made using a variety of techniques to check whether or not the central cellobiose residue is upside down but at the present time there is insufficient information to come to an unequivocal conclusion. Until recently, however, there has been general acceptance of the antiparallel arrangement largely because so many other polymers are known to occur in this form. The doubts that have been expressed against it have

usually come from biochemists who find it difficult to correlate the antiparallel arrangement with what is now known about the biosynthesis of microfibrils [see Section D.1(i)]. Within the unit cell adjacent cellobiose residues are linked by hydrogen bonding, which occurs principally between the C-5 hydroxyl group of one residue and the glycosidic bond oxygen atom of the adjacent residue. Within the cell wall the crystal lattice of the microfibril is oriented such that the 101 plane (shaded in Fig. 4.3) is parallel to the surface of the plasmalemma.

(b) β-1,4-Mannans

The cell walls of the so-called 'mannan weeds', which include the green algal genera *Acetabularia*, *Batophora*, *Codium*, *Cymopolia*, *Dasycladus*, *Derbesia*, *Halicoryne* and *Neomeris*, are remarkably simple and uniform. After removal of adhering cytoplasm and mineral incrustation only the boiling-water treatment of the standard fractionation [Section B.1(i)] extracts a significant quantity of polysaccharide from the cell wall: on hydrolysis this yields mainly D-galactose and D-glucose. The insoluble residue which results from the extraction procedure constitutes about 90% of the wall polysaccharide and has proved to be an unbranched polysaccharide composed of D-mannopyranose residues linked together by β1→4 glycosidic linkages (see Fig. 4.4). The β-1,4-mannan molecules

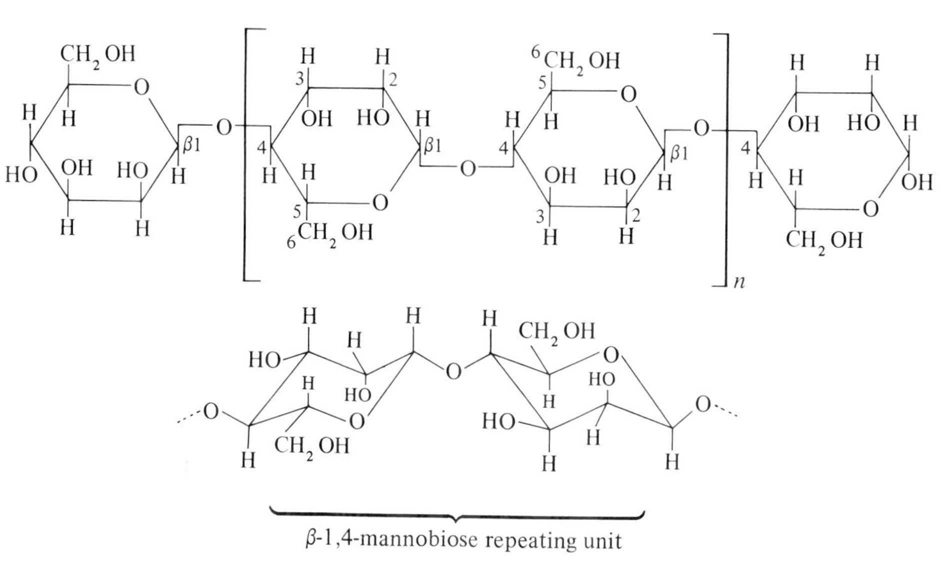

β-1,4-mannobiose repeating unit

Fig. 4.4. Structure and conformation of the β-1,4-mannan molecule.

are aggregated together within the cell walls of these algae as microfibrils which are structurally analogous to cellulose microfibrils.

(c) β-1,3-Xylans

The cell walls of the so-called 'xylan weeds' are composed of microfibrils embedded in a matrix that is rich in glucans. The microfibrils are made up of β-1,4-xylan molecules, which are unbranched chains of D-xylopyranose residues linked together by β1→3 glycosidic linkages (see Fig. 4.5). The ratio of xylan to glucan in this group of algae is about 4:1 except for the genus *Bryopsis* in which the ratio is about 1:1. The glucan of these walls is readily removed by the boiling-water treatment of the standard fractionation procedure. However, the xylan (of the microfibrils) which remains after extraction with boiling water is soluble in fairly dilute alkali; this is quite unlike cellulose or the β-1,4-mannans and more characteristic of the hemicelluloses of higher plant cell walls which often include polysaccharides rich in D-xylopyranose.

The internal structure of the β-1,3-xylan microfibrils is quite different from that of cellulose or β-1,4-mannan microfibrils. The structure is still crystalline but the crystal lattice is quite different. This arises because the β1→3 glycosidic linkages of the xylan cause its chain to curve. X-ray diffraction studies indicate that within the microfibril each

xylan molecule is coiled to turn a right-handed helix which has six xylopyranose residues per turn in a pitch of 1.836 nm and whose axis is parallel to that of the microfibril; the axial rise of each xylopyranose unit is therefore 0.306 nm. Three such xylan helices are intertwined and held together by hydrogen bonds formed between the C–2 hydroxyl groups of the individual chains. These project inwards towards the axis of the triple helix and form a triad of hydrogen bonds (see Fig. 4.6) which is repeated every 0.306 nm. The complete xylan microfibril is composed of a considerable number of these triple helices lying parallel to each other.

(d) Chitin

Chitin constitutes the microfibrillar component of the cell walls of most fungi; it is also the principal component of the hard exoskeleton of many invertebrates. Chitin molecules are long, unbranched chains of N-acetyl-D-glucosamine residues linked by β1→4 glycosidic linkages (see Fig. 4.7). They are arranged in the microfibril in a manner analogous to that of the cellulose microfibril. X-ray diffraction studies indicate that the microfibril is highly crystalline and that the repeating unit along the fibre axis of the unit cell is the chitobiose residue which is held in the zig-zag conformation characteristic of cellulose (see Fig. 4.2) by hydrogen bonding between

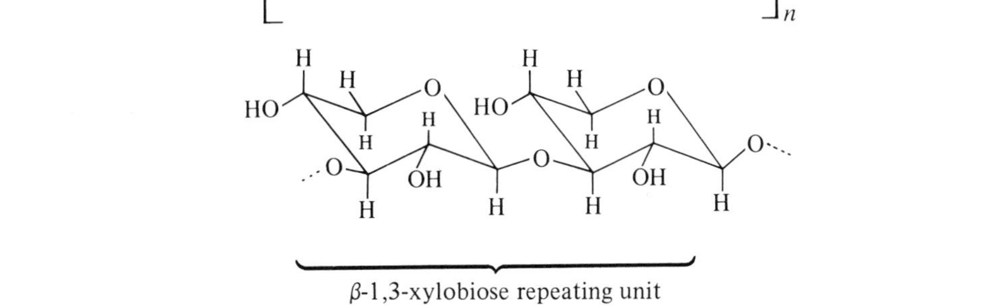

β-1,3-xylobiose repeating unit

Fig. 4.5. Structure and conformation of the β-1,3-xylan molecule.

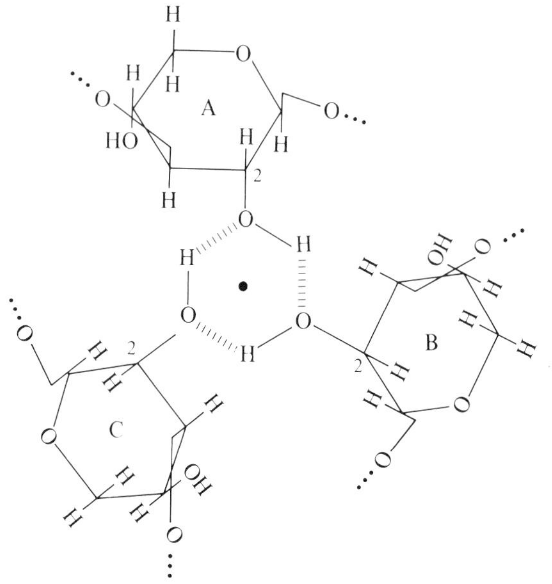

Fig. 4.6. The triad of hydrogen bonds holding the xylan triple helix together, viewed along its axis. (A, B and C are D-xylopyranose residues occurring at the same level in the triple helix; each is in a different xylan chain: ııı = hydrogen bonds; ● = axis of triple helix.)

the C–3 hydroxyl group of one N-acetyl-D-glucosamine residue and the glycosidic oxygen of the next. Within the unit cell adjacent cellobiose residues run in opposite directions. This indicates that within the microfibril itself the chitin molecules are antiparallel. The X-ray diffraction data make this conclusion inescapable in contrast to the data for cellulose microfibrils [see Section B.1(ii)(a)].

(iii) MATRIX POLYSACCHARIDES

The polysaccharides of the plant cell wall are generally subdivided into two classes, the hemicelluloses and the pectins. This subdivision is based upon solubility differences. The pectins are a group of polysaccharides which are extracted from the cell wall by prolonged treatment with boiling water, whilst the hemicelluloses are a group of poly-saccharides which are extracted by 4 M KOH solution at room temperature. Within both groups are a range of polysaccharides which are almost always made up of two or more different mono-saccharide species and frequently highly branched. Moreover, the range of polysaccharides within the two groups is different for different plant groups and

even for different wall layers. There is increasing evidence to suggest that a number of the poly-saccharides which are presently regarded as distinct polymeric species are, in fact, covalently linked to other polysaccharide and non-polysaccharide com-ponents of the cell wall. Figure 4.8 gives the structures of the monosaccharides which appear in matrix polysaccharides.

(a) Hemicelluloses

The name given to this group of polysaccharides is not a good one. It was proposed in 1891 by Schulze to describe a group of polysaccharides which were relatively easily extracted from various plant tissues and which he thought were precursors of cellulose, i.e. half-built cellulose molecules, hence the name hemi (Gk., half) cellulose. It is now, of course, very clear that they are not cellulose precursors and we are stuck with a name that implies a relationship that does not exist.

The hemicelluloses may be subdivided into three subgroups: the xylans, the mannans and the galac-tans. Each subgroup contains polysaccharides with considerable structural diversity and is named after the predominant monosaccharide.

Xylans. It is convenient to include within this subclass all those hemicellulose polysaccharides which are rich in D-xylopyranose (I) residues whether these occur in the main chain or only as frequent branches attached to it. The structure of the most abundant xylan of hardwoods (pre-dominantly the thick secondary walls of dicoty-ledonous angiosperm trees) is shown in Fig. 4.9(a). It is a linear chain of D-xylopyranose residues linked together by β1→4 glycosidic linkages; seven out of every ten xylose residues are acetylated, mostly at C–3 but occasionally at C–2. 4-O-Methyl-α-D-glucuronic acid (II) is linked to some of the xylose residues by an α1→2 glycosidic linkage. This type of xylan also occurs in coniferous wood (predominantly the thick secondary walls of gymnosperm trees) although it is not the main hemicellulose; moreover, there is a slight structural difference in that an α-L-arabofuranose (III) may be linked to some of the xylose residues by α1→3 glycosidic linkages. Ferns also contain this type of xylan.

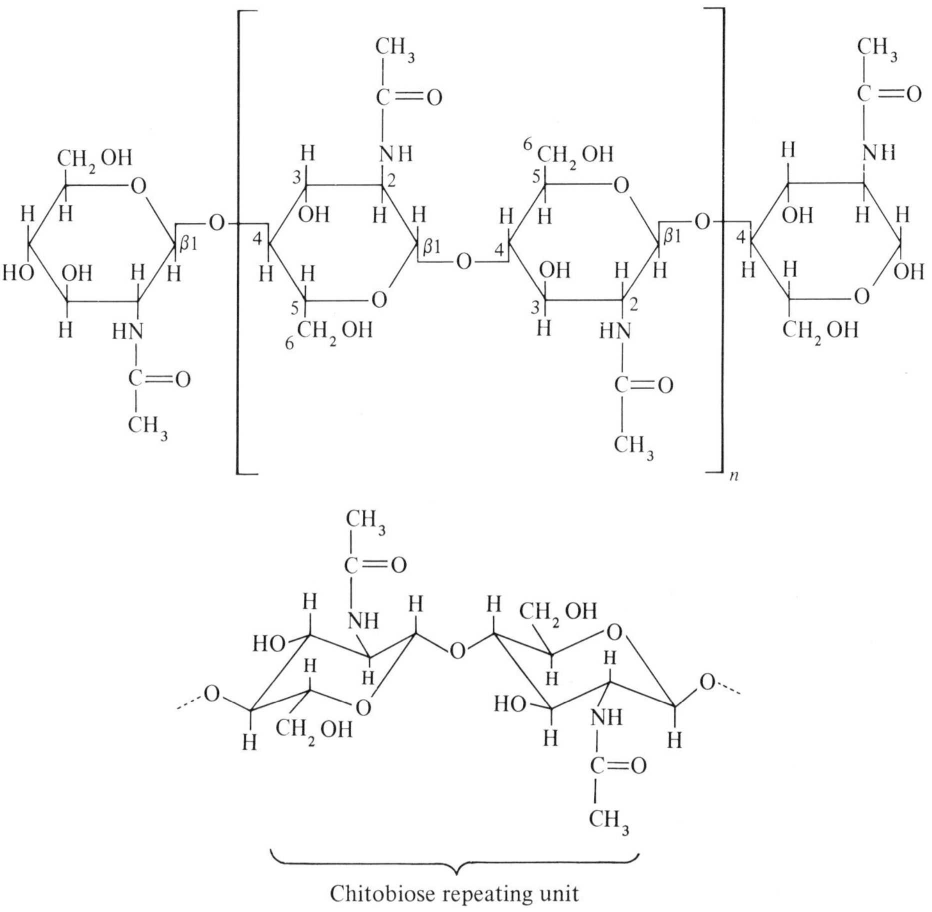

Fig. 4.7. Structure and conformation of the chitin molecule.

A totally different type of xylan has been found by Albersheim in the primary cell wall of suspension-cultured sycamore cells. It is the only hemicellulose component of these cell walls and it is thought likely that this is true of the primary walls of all dicotyledonous plants. In this xylan D-xylopyranose residues are present in branches rather than in the main polysaccharide chain which is composed of D-glucopyranose residues (IV) linked together by $\beta1 \rightarrow 4$ glycosidic linkages. The majority of the branches attached to the glucan chain are single D-xylopyranose residues linked to it by an $\alpha1 \rightarrow 6$ glycosidic linkage. Some branches are, however, trisaccharides composed of α-L-fuco-pyranose (V), α-D-galactopyranose (VI) and α-D-xylopyranose; again it is the xylose which is attached to the glucan chain, by an $\alpha1 \rightarrow 6$ linkage.

Some glucose residues carry two branches, namely the trisaccharide just mentioned and an L-arabo-pyranose linked to it by a $1 \rightarrow 2$ glycosidic linkage. The structure of this xylan is shown in Fig. 4.9(b). There is good evidence to indicate that in the dicotyledonous primary wall it is glycosidically linked, via the glucose residue at the reducing end of the glucan chain, to C–2 of a D-galactopyranose residue in an arabinan-galactan, a neutral pectin.

Mannans. Mannans constitute the main hemi-cellulose of coniferous woods (soft woods). Hydrolysis of this type of mannan yields D-mannose, D-glucose and D-galactose in the ratio 3:1:1. It consists of a main chain composed of β-D-mannopyranose (VII) and β-D-glucopyranose re-sidues in the ratio 3:1 linked together by $\beta1 \rightarrow 4$

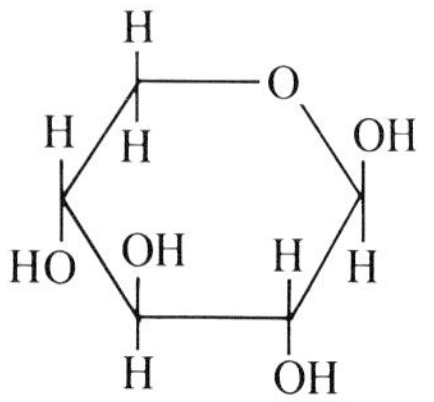

β-D-xylopyranose (I)

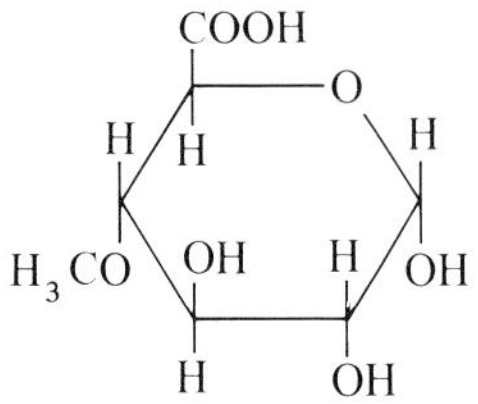

4-*O*-methyl-α-D-
glucuronic acid (II)

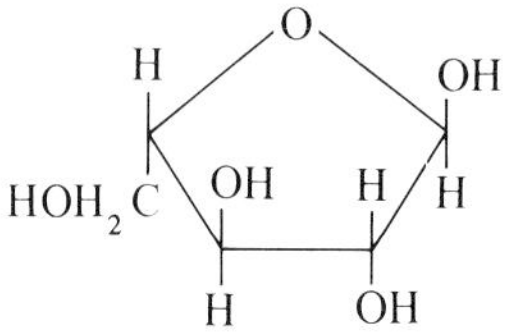

α-L-arabofuranose (III)

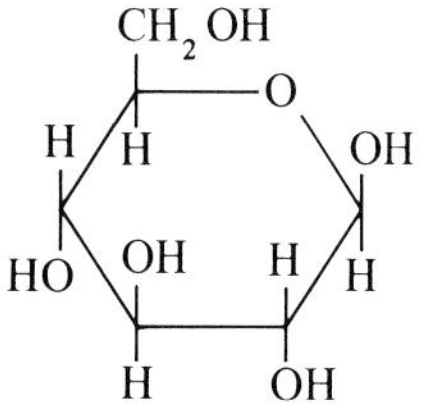

β-D-glucopyranose (IV)

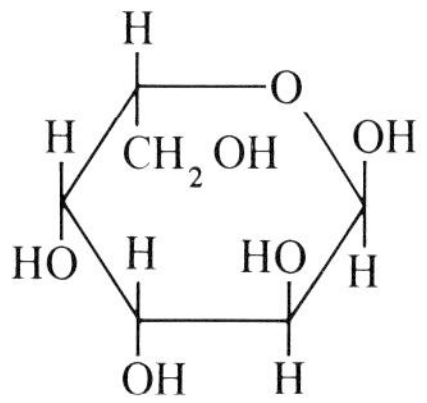

α-L-fucopyranose (V)

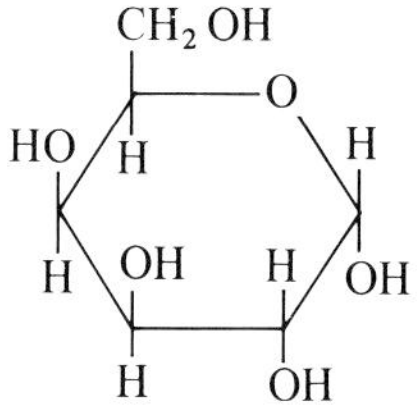

α-D-galactopyranose (VI)

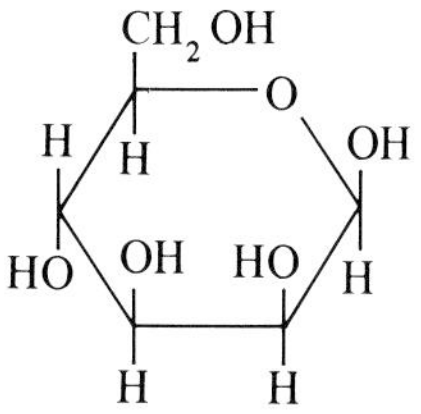

β-D-mannopyranose (VII)

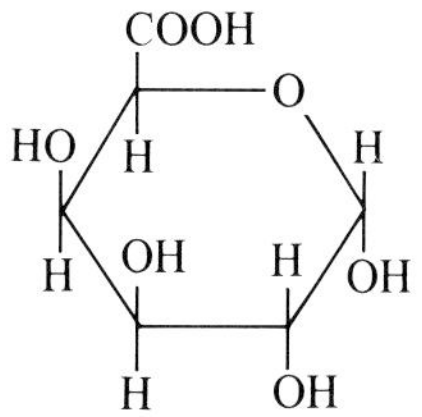

α-D-galacturonic acid (VIII)

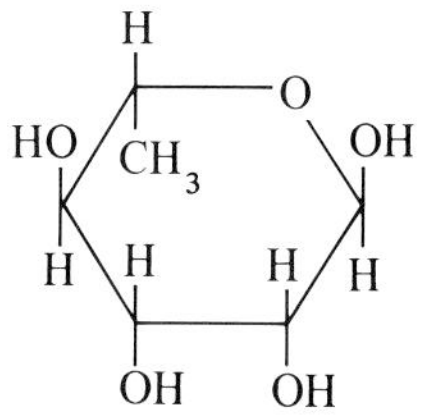

α-L-rhamnopyranose (IX)

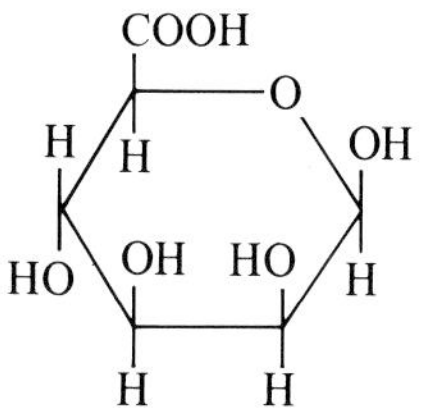

β-D-mannuronic acid (X)

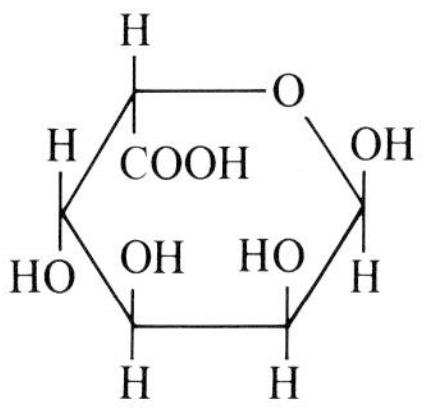

α-L-guluronic acid (XI)

Fig. 4.8. Structure of the monosaccharide building units of the cell-wall matrix polysaccharides.

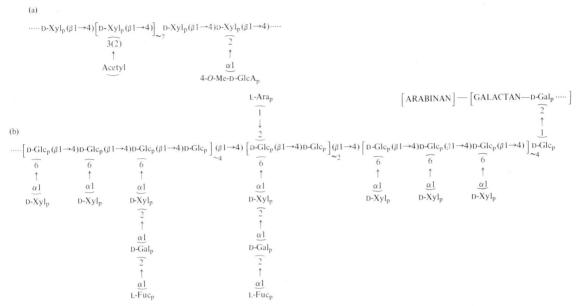

Fig. 4.9. Partial structures of xylan components of the class of hemicelluloses. (D-Xyl$_p$ = D-xylopyranose; 4-O-Me-D-GlcA$_p$ = 4-O-methyl-D-glucuronic acid; D-Glc$_p$ = D-glucopyranose; L-Ara$_p$ = L-arabopyranose; D-Gal$_p$ = D-galactopyranose; L-Fuc$_p$ = L-fucopyranose.)

glycosidic linkages. Single β-D-galactopyranose residues are linked by $\beta1\rightarrow6$ linkages to some of the mannose residues of the main chain. The hydroxyl groups at C–2 (mainly) and C–3 of some of the mannose residues are acetylated (see Fig. 4.10). Glucomannans are the main hemicelluloses in ferns and in *Psilotum* and *Equisetum*. Mannans, containing about 5% glucose, occur as hemicellulose in the mannan weeds [see Section B.1(ii)(b)].

Galactans. The galactan subgroup of the hemicelluloses is mainly composed of arabinogalactans. They are particularly abundant in some larches but also occur in small amounts in other soft- and hardwoods. The arabinogalactan molecule consists of a main chain of β-galactopyranose units linked to each other by $\beta1\rightarrow3$ glycosidic linkages, attached to

which are branches composed of D-galactopyranose- or L-arabofuranose-containing disaccharides (see Fig. 4.11). Arabinogalactans with a similar structure but with some hydroxyl groups sulphated occur in some algae such as *Cladophora*, *Caulerpa* and *Codium*.

(b) *Pectins*

Pectins are not only important cell-wall matrix polysaccharides, but also occur in some plant juices. The most abundant component of the pectin polysaccharides are polyuronic acids. In the case of higher plants these are polymers built up mainly of α-D-galacturonic acid residues. However, in the marine brown algae the polyuronic acid is structurally different and is usually called alginic acid. In higher

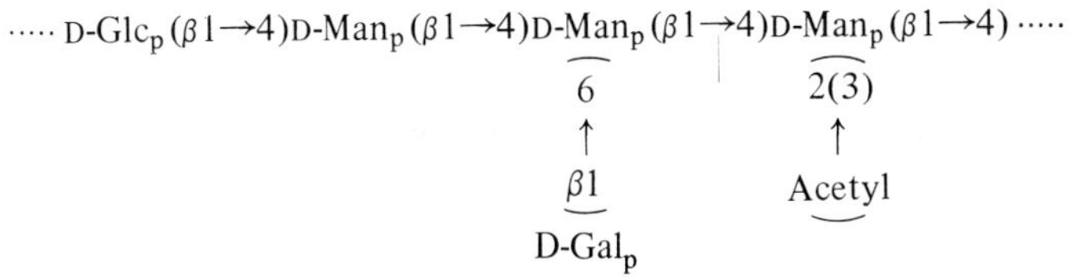

$$\cdots\cdots \text{D-Glc}_p(\beta1\rightarrow4)\text{D-Man}_p(\beta1\rightarrow4)\text{D-Man}_p(\beta1\rightarrow4)\text{D-Man}_p(\beta1\rightarrow4)\cdots\cdots$$

$$\begin{array}{cc} \underline{6} & \underline{2(3)} \\ \uparrow & \uparrow \\ \beta1 & \text{Acetyl} \\ \text{D-Gal}_p & \end{array}$$

Fig. 4.10. Partial structure of the mannan component of the class of hemicelluloses. (D-Glc$_p$ = D-glucopyranose; D-Man$_p$ = D-mannopyranose; D-Gal$_p$ = D-galactopyranose.)

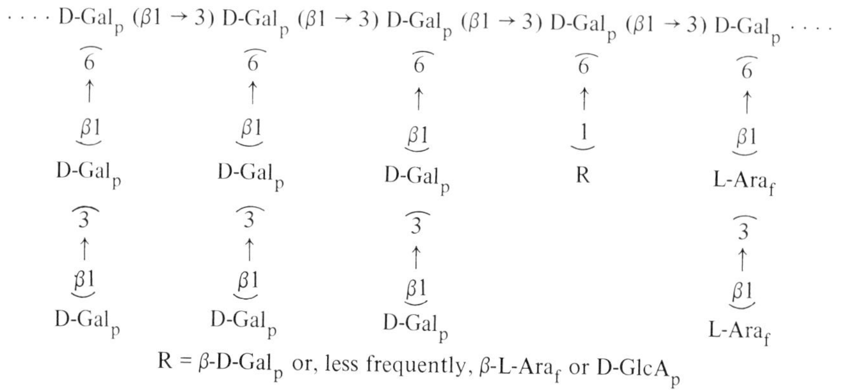

$\cdots\cdots$ D-Gal$_p$ ($\beta1 \to 3$) D-Gal$_p$ ($\beta1 \to 3$) D-Gal$_p$ ($\beta1 \to 3$) D-Gal$_p$ ($\beta1 \to 3$) D-Gal$_p$ $\cdots\cdots$

```
        6            6            6           6            6
        ↑            ↑            ↑           ↑            ↑
       β1           β1           β1          1           β1
     D-Gal_p      D-Gal_p      D-Gal_p       R         L-Ara_f

        3            3            3                        3
        ↑            ↑            ↑                        ↑
       β1           β1           β1                       β1
     D-Gal_p      D-Gal_p      D-Gal_p                  L-Ara_f
```

R = β-D-Gal$_p$ or, less frequently, β-L-Ara$_f$ or D-GlcA$_p$

Fig. 4.11. Partial structure of the arabinogalactan component of the class of hemicelluloses. (D-Gal$_p$ = D-galactopyranose; L-Ara$_f$ = L-arabofuranose; D-GlcA$_p$ = D-glucuronic acid.)

plants the galacturonic acid polymers are accompanied by the less abundant and neutral arabinans and galactans.

Polyuronic acids. The polyuronic acids of higher plant cell walls are mainly composed of unbranched chains in which the most abundant residue is D-galacturonic acid (VIII). Albersheim has shown that there are three different polyuronic acids in the primary wall of suspension cultured sycamore cells, two of which conform to this pattern and one which does not. The two which do conform are a homogalacturonan composed of at least twenty-five D-galacturonic acid residues linked by $\alpha1\to4$ glycosidic linkages and a high molecular weight rhamnogalacturonan whose structure is shown in Fig. 4.12. The rhamnogalacturonan consists of segments composed of about eight α-1,4-linked D-galacturonic residues separated from one another by a trisaccharide made up of L-rhamnopyranose $(1\to4)$ D-galacturonopyranose $(\alpha1\to2)$ L-rhamnopyranose. The terminal D-galacturonic residue of

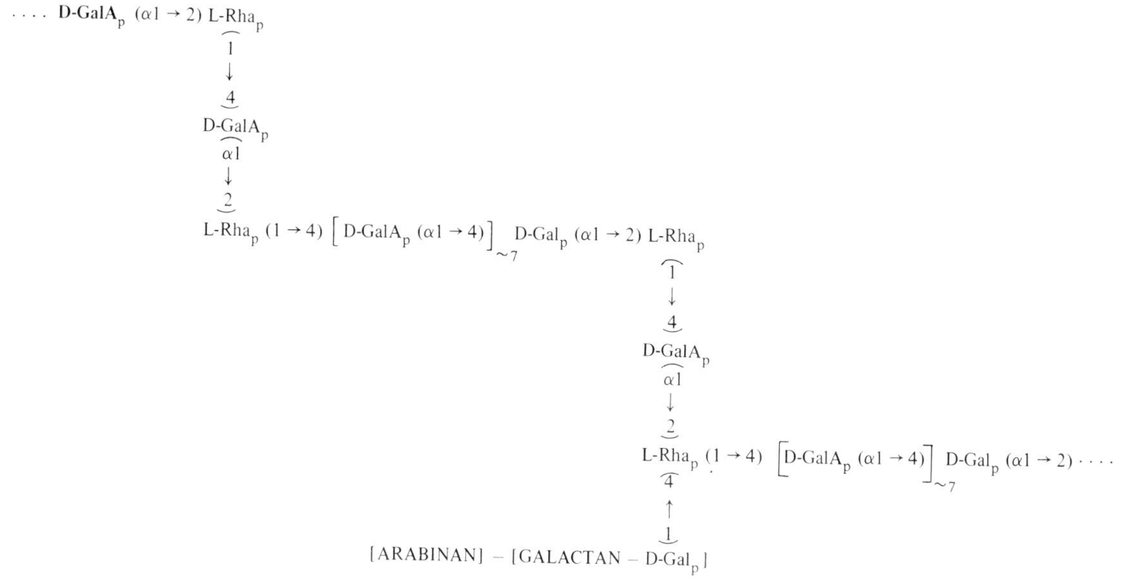

```
···· D-GalA_p (α1 → 2) L-Rha_p
                          1
                          ↓
                          4
                       D-GalA_p
                          α1
                          ↓
                          2
    L-Rha_p (1 → 4) [ D-GalA_p (α1 → 4) ]_{~7} D-Gal_p (α1 → 2) L-Rha_p
                                                                    1
                                                                    ↓
                                                                    4
                                                                 D-GalA_p
                                                                    α1
                                                                    ↓
                                                                    2
                                               L-Rha_p (1 → 4) [D-GalA_p (α1 → 4)]_{~7} D-Gal_p (α1 → 2) ····
                                                  4
                                                  ↑
                                                  1
                            [ARABINAN] – [GALACTAN – D-Gal_p]
```

Fig. 4.12. Partial structure of the rhamnogalacturonan of the primary wall of sycamore showing the way it links to the neutral pectins. (D-GalA$_p$ = D-galacturonic acid; L-Rha$_p$ = L-rhamnopyranose; D-Gal$_p$ = D-galactopyranose.)

the pure galacturonan segments is linked to the L-rhamnose (IX)-containing trisaccharide by an $\alpha1\rightarrow2$ glycosidic link. The presence in the rhamnogalacturonan of the L-rhamnose residues and their attendant $\alpha1\rightarrow2$ linkages causes the chain to zig-zag. The polyuronic acid which does not conform to the usual pattern is also classed as a rhamnogalacturonan but appears to be a much-branched structure containing, in addition to D-galacturonic acid and L-rhamnose, a variety of other monosaccharides, some of which are quite rare. It is not yet possible to draw even a partial structure for it.

It is clear that in the primary wall of the sycamore cells the rhamnogalacturonan is glycosidically linked to one of the pectin polysaccharides, namely the galactan (see Fig. 4.12). The linkage in question is $\alpha1\rightarrow4$ and involves the D-galactopyranose residue at the reducing end of the galactan chain and an L-rhamnose residue in the rhamnogalacturonan.

The carboxyl group in each D-galacturonic acid residue may exist within the wall in any one of three states: it may be esterified with methanol to give a carboxymethyl group, it may form salts with certain metal ions, particularly calcium, or it may remain unmodified.

Since the D-galacturonic acid-containing polysaccharides are the major components of the pectins, their properties are in effect the properties of the whole group of pectin polysaccharides. Indeed the galacturonans are often referred to as pectin and their calcium salts as calcium pectate. Because of this it may offend some cell-wall experts to see the alginic acids of brown algae classified as pectins. Our justification for so classifying them is that they are also polyuronic acids and that they appear to fulfil the same role in the cell wall of brown algae as do the galacturonans in the higher plant cell wall.

Alginic acids are built up from two types of uronic acid, namely β-D-mannuronic acid (X) and α-L-guluronic acid (XI), which occur in ratios ranging from 2:1 to 1:2. Within the alginic acid molecule there appear to be stretches composed solely of $\beta1\rightarrow4$-linked D-mannuronic acid residues, stretches composed solely of $\alpha1\rightarrow4$-linked L-guluronic acid residues and stretches where the two uronic acids occur in almost alternating sequence. The carboxyl groups of the residues frequently form salts with Na^+, Ca^{2+} and Mg^{2+} ions.

Alginic acid is of considerable commercial importance being used as a thickener and stabilizer and, when suitably derivatized, as an emulsifier. The galacturonan of higher plants is also of commercial value. It is used in the food industry, under the name pectin, as a setting agent. An aqueous solution of 'pectin', when heated with sugar under acid conditions (pH 2–3.5), solidifies to a clear solid jelly on cooling. This is called a pectin–sugar–acid jelly. The formation of this type of jelly is the basis of jam-making. The fruit, containing pectic substances in the parenchymatous cell walls and cell sap, is boiled with a small amount of water and then sugar is added. The boiling serves to concentrate the pectic substances. The sugar, along with the lowered pH due to the natural fruit acids, causes gel formation (the setting of the jam) on cooling. When pectins are dispersed in water they exist as negatively charged, hydrophilic, colloidal particles stabilized by water layers held by the negative electrostatic field. The stability of the colloid is decreased by the addition of sugar whose dehydrating action disturbs the pectin–water balance, and by the addition of hydrogen ions which further upset the balance by reducing the negative charge on the pectin. This results in gel formation owing to an aggregation of clumps of pectin molecules.

Arabinans and galactans. These polysaccharides are usually much less abundant pectin components than the polyuronic acids. Moreover, they are neutral in contrast to the acidic polyuronic acids. Arabinans are branched polymers of L-arabofuranose residues. They are composed of a linear chain of $\alpha1\rightarrow5$-linked L-arabofuranose residues, alternate residues of which have an $\alpha1\rightarrow3$-linked L-arabofuranose, single-unit side chain (see Fig. 4.13). Galactans are largely unbranched chains of $\beta1\rightarrow4$-linked D-galactopyranose residues. In the primary cell wall of sycamore cells the arabinan and galactan are so closely associated that they were initially described as a single polysaccharide called an arabinogalactan. It is thought that at least some of the arabinan chains are linked to galactan chains by an $\alpha1\rightarrow4$ glycosidic linkage as shown in Fig. 4.13. The galactan or the galactan moiety of the arabinogalactan is also linked to the rhamnogalacturonan and the xyloglucan in the sycamore primary wall as shown in Fig. 4.13.

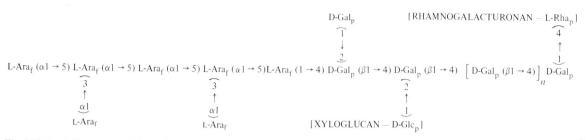

Fig. 4.13. Partial structures of the arabinan and galactan of the primary wall of sycamore cells, showing the way they link together and the way they link to xyloglucan and to rhamnogalacturonan. (L-Ara$_f$ = L-arabofuranose; D-Gal$_p$ = D-galactopyranose; L-Rha$_p$ = L-rhamnopyranose; D-Glc$_p$ = D-glucopyranose.)

2. Lignin

In general the occurrence of lignin in the plant kingdom is associated with the presence of supporting and conducting tissues (e.g. xylem); the lignin is an important constituent of the cell walls of these tissues. Thus lignin is found in vascular plants such as lycopods, ferns, gymnosperms and angiosperms but is absent from non-vascular plants such as fungi and algae. The mosses are anomalous in that they do not have cells characteristic of xylem and yet contain lignin-like compounds; however, there are some doubts as to whether these are true lignins.

Lignin is not confined to the cell walls of conducting tissues but also occurs in the cell walls of pith, roots, fruits, buds, bark and cork. Lignification of the cell wall occurs after the laying down of the polysaccharide components of the wall and towards the end of the growing period of the cell. The distribution of lignin within the wall is not uniform; in general those layers that were laid down first become the most heavily lignified. Thus the middle lamella and the primary wall undergo the greatest lignification and the secondary wall the least.

The purpose of lignification appears to be two-fold. Firstly, it strengthens the cell wall by forming a ramified network throughout the matrix, thus anchoring the cellulose microfibrils more firmly. This strengthening of the cell wall has been considered in the evolutionary context by Neish who felt that the development of the ability to lignify cell walls made possible the formation of supporting and conducting tissues which in turn enabled plants to grow into large, upright structures. Secondly, lignin protects the microfibrils of the wall from chemical, physical and biological attack.

(i) EXTRACTION PROCEDURES

Lignin is extremely difficult to extract from the cell wall. Until relatively recently all extraction procedures involved the use of very aggressive chemical reagents, e.g. 40% HCl, a 3:1 mixture of 36% HCl and 80% H_3PO_4, 2% H_2SO_4 followed by ammoniacal CuO, at elevated temperatures. These procedures caused considerable modification of the extracted lignin; moreover, the nature of the modification depended upon the extracting reagent. An additional complication was the self-condensation of lignin molecules which occurred at the elevated temperatures used. These modified lignins were usually given names which indicated their methods of extraction, e.g. HCl lignin, Cuproxam lignin.

These difficulties were eased in 1939 when Brauns showed that about 5% of the total lignin of wood could be extracted if freshly prepared sawdust was extracted with 95% aqueous ethanol at room temperature for several days. This combination of a neutral, relatively inert solvent and low temperature apparently caused no structural changes and enabled an unmodified lignin to be extracted. Lignin extracted in this way is called 'Brauns lignin' or 'native lignin'. It is thought to be identical to 'protolignin', which is the name usually given to lignin as it exists in the cell wall. The yield of native lignin can be improved if wood which has been decayed by a brown rot fungus (e.g. *Poria vaillantii*) is extracted by the Brauns method. Brown rot fungi secrete a cocktail of enzymes which degrade the cellulose and other polysaccharide components of the wall but do not attack the lignin which thereby becomes more accessible to the extracting solvent. The brown rot fungi are so named because the colour of the rotted area of wood is brown. This

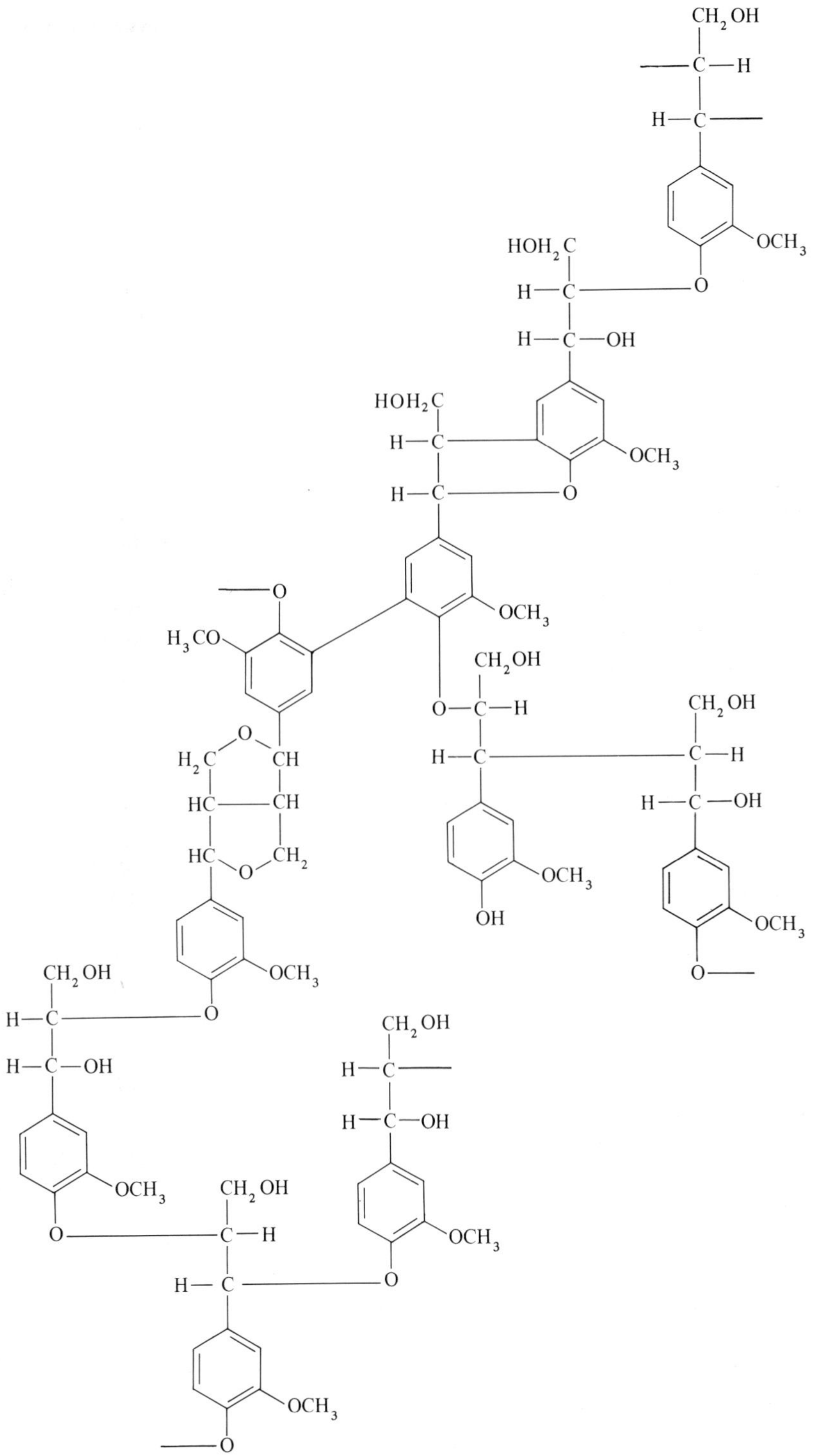

Fig. 4.14. Hypothetical partial structure of gymnosperm lignin.

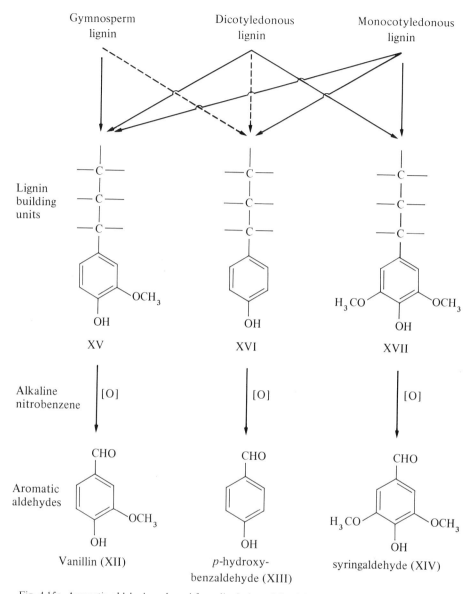

Fig. 4.15a. Aromatic aldehydes released from lignin by mild oxidation with alkaline nitrobenzene.

distinguishes them from the white rot fungi which degrade all the components of wood, including lignin, and leave a white rotted area.

(ii) STRUCTURE

The term lignin covers a group of closely related, high-molecular-weight polymers whose main, if not only, building unit is a phenylpropane (C_6–C_3) residue. The structural variations of the phenyl-propane residue are few but there are numerous ways in which they can be linked together. The order in which the different phenylpropane units and their linkages occur in the polymer is random. The polymerization process (see Section D.2) occurs by a random condensation of free radicals and involves no template (cf. the mRNA template in protein synthesis, see Chapter 10, Section E.4). Because of this the structure of each lignin molecule may well be unique; it is therefore not possible to write down the structure of lignin. One can, however, draw a hypothetical, partial structure which depicts the different types of phenylpropane building units and the different ways in which they are linked together, as has been done in Fig. 4.14 for gymnosperm lignin.

The lignin of monocotyledonous angiosperms, dicotyledonous angiosperms and gymnosperms are structurally different from one another. This difference stems from the structure of the phenylpropane building units. This is apparent from the different aromatic aldehydes produced when native lignin from these sources is subjected to mild oxidation with nitrobenzene in alkaline medium. Gymnosperm lignin yields mostly vanillin (XII) but also a little p-hydroxybenzaldehyde (XIII), dicotyledonous lignin yields mostly vanillin and syringaldehyde (XIV) but also a little p-hydroxybenzaldehyde whilst monocotyledonous lignin yields all three aldehydes. Nitrobenzene oxidation yields aromatic aldehydes from benzenoid nuclei with a hydroxyl group *para* to an alphatic side chain when the α-carbon atom of the side chain bears a hydroxyl or sulphonic acid group, or is involved in the formation of a carbonyl group or a double bond. This indicates that vanillin, p-hydroxybenzaldehyde and syringaldehyde arise from phenylpropane building units XV, XVI and XVII respectively (see Fig. 4.15a). Gymnosperm lignin therefore appears to arise by the polymerization of many phenylpropane units of type XV with a few of type XVI; similarly dicotyledonous lignin is formed by polymerization of types XV and XVII with a few of type XVI and monocotyledonous lignin by polymerization of all three types.

The C_3 side chain of the phenylpropane units usually has a *trans* double bond between C–1 and C–2 and a hydroxyl group on C–3. The C–3 primary alcohol group is, however, often replaced by an esterified carboxyl group in the lignin of grasses.

There is increasing evidence that lignin is covalently linked to polysaccharides of the cell-wall matrix; however, the nature of this linkage is unknown.

3. Proteins

Protein was first shown to be associated with cell walls (in this case the primary wall of meristematic cells) in 1924. However, this and later findings were challenged as being due to cytoplasmic contamination. More recently improved methods of cell-wall preparation have shown beyond doubt that cell walls, particularly those of growing cells, contain proteins which represent 5–10% of the cell wall. These cell-wall proteins may be divided into two classes, enzymes and structural proteins.

(i) ENZYMES

A considerable range of hydrolases have been found in cell walls, including invertase[1], various glucanases, pectin methylesterase[2], ATP-ase, DNA-ase, RNA-ase and various phosphatases. Several oxidases are also present, including ascorbic acid oxidase and the laccase involved in lignin formation.

None of these enzymes are concerned with *de novo* synthesis of cell-wall polysaccharides for it is now clear that such synthesis takes place in the Golgi bodies or at the outer surface of the plasmalemma (see Section D.1). Some of them may be concerned with the making and breaking of the glycosidic bonds which hold the various species of cell-wall polysaccharides together, perhaps facilitating growth and extension. Pectin methylesterase catalyses the hydrolysis of the ester bond in the —$COOCH_3$ group of the galacturonic residues of the polyuronic acid component of the pectins, thereby producing carboxyl groups which may then form salt linkages with Ca^{2+} ions.

One of the wall oxidases is almost certainly concerned with the production of the free radicals of the lignin building units which then randomly polymerize within the substance of the wall, so forming the lignin network (see Section D.2).

(ii) STRUCTURAL PROTEINS

A glycoprotein, exceptionally rich in the amino acid 4-*trans*-hydroxy-L-proline (see Fig. 4.15b), relatively rich in the amino acids L-serine, L-threonine, L-alanine and L-aspartic acid, and rich in the monosaccharides L-arabofuranose and D-galactopyranose, has been found in cell walls, particularly primary cell walls. A partial structure of this glycoprotein has been elucidated (see Fig. 4.15a). There is no evidence at present that the glycoprotein is covalently linked to cell-wall polysaccharides; however, this does not rule out the

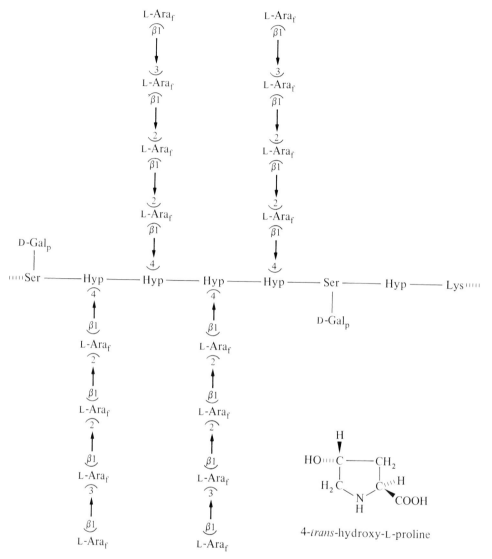

Fig. 4.15b. A partial structure of the cell wall structural glycoprotein. (D-Gal$_p$ = D-galactopyranose; L-Ara$_f$ = L-arabofuranose; Ser = L-serine; Hyp = 4-*trans*-hydroxy-L-proline; Lys = L-lysine.)

possibility that it is strongly bound to the poly-saccharides by non-covalent bonds. The glyco-protein is thought to be concerned with cell-wall extension and accordingly has been called exten-sin. The evidence for this is tenuous.

4. Water

Water is an important structural component of

the cell wall. It forms part of the gel structure of the pectins. Because of this changes in the water content of the wall cause reversible changes in the texture of the cell-wall matrix as the pectin gel changes to a viscous solution. It also reduces the hydrogen bonding between the cellulose microfibrils and the hemicellulose chains that are considered [see Section B.1(ii)(a)] to coat them; thus changes in the water content of the wall cause changes in the degree of adherence between the microfibrils and the matrix.

Water also has a solvent and chemical role to play within the wall. As a solvent it has obvious effects on the permeability of the wall to other molecules and ions; the more water in the wall the more permeable it is. As a chemical, it is a reactant species in the hydrolysis of interchain glycosidic bonds which is thought to occur during cell-wall growth.

As growth of the cell ceases the space occupied by water in the wall becomes progressively filled with lignin, thus making the matrix, and therefore the whole wall, much more rigid.

5. Incrusting substances

(i) CUTICULAR SUBSTANCES (cutin and suberin)

The outermost surface (i.e. that exposed to the atmosphere) of the cell walls of epidermal cells are covered with a hydrophobic cuticle. The principal function of the cuticle is to reduce the excessive losses and gains of water by the underlying tissue which would take place in its absence. It probably also protects the tissue from chemical, physical and biological attack to some degree.

The cuticle is generally composed of three layers (see Fig. 4.16). The outermost is a surface layer of wax, often called the epicuticular wax. The wax of this layer is often crystalline and gives rise to characteristic patterns. Beneath the epicuticular wax is a layer of cutin embedded in wax. The innermost layer is a mixture of cutin, wax, cell-wall polysaccharides and possibly traces of protein. This layer adheres to the middle lamella of the epidermal cells.

Cutin is the principal polymer of the cuticle. It consists of a complex mixture of hydroxy fatty acids which are linked together by ester bonds to give a three-dimensional network. The majority of the fatty acids that are the cutin building units contain sixteen or eighteen carbon atoms and are of three main types: (i) saturated or monounsaturated fatty acids, (ii) ω-hydroxy saturated or monounsaturated fatty acids and (iii) di- or trihydroxy saturated fatty acids which have one hydroxyl group on the ω-carbon atom and the other(s) usually on carbon atoms 9 and 10. The ester bonds which link the fatty acids together to form cutin occur between the carboxyl group of one fatty acid and the ω- or 'in chain' hydroxyl group of another fatty acid (see Fig. 4.17).

The epicuticular wax is a complex mixture containing mixtures of esters of fatty acids and long-chain alcohols, free long-chain fatty acids and hydroxy fatty acids, free long-chain alcohols (primary, secondary and α,ω-diols), long-chain alkanes and long-chain ketones.

Suberin is the polymeric substance which replaces cutin in the material which covers the epidermal cell layer of the underground parts of plants (e.g. roots, tubers). It is built on the same principal as cutin but differs in respect of its building units which are mainly (i) α,ω-saturated or monounsaturated dicarboxylic acids, ranging from C_{16} to C_{22} or more with octadec-9-en-1,18-dioic acid predominating, (ii) ω-hydroxy saturated or unsaturated fatty acids and (iii) α,ω-long-chain diols.

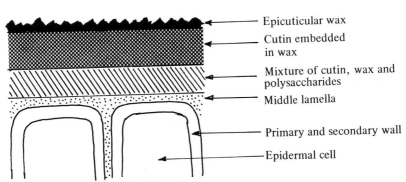

Fig. 4.16. Schematic representation of a cross-section through the cuticle of a leaf.

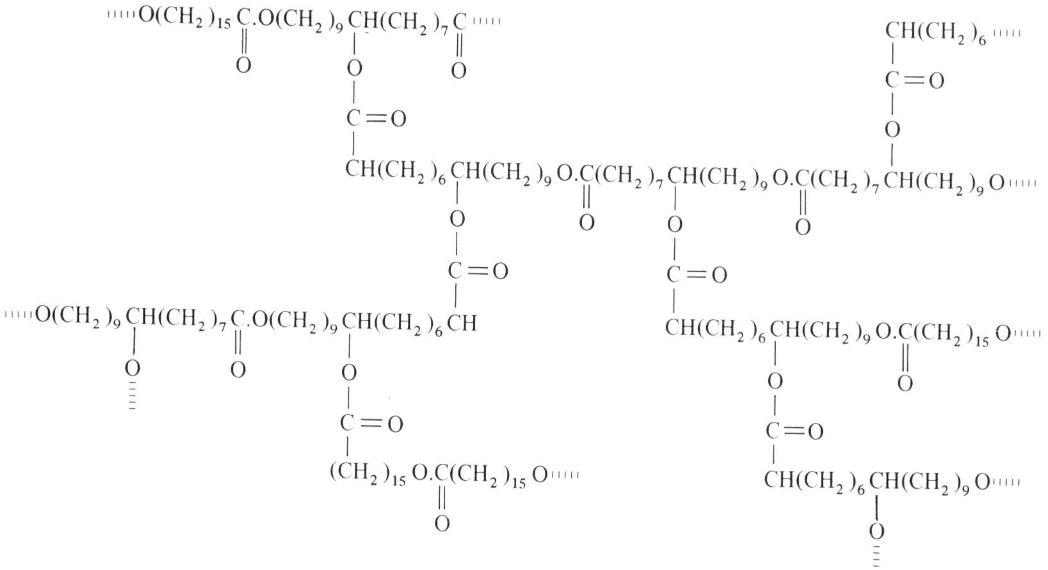

$$\text{''''O(CH}_2\text{)}_{15}\overset{\text{O}}{\underset{\|}{\text{C}}}.\text{O(CH}_2\text{)}_9\underset{\underset{\|}{\text{O}}}{\text{CH(CH}_2\text{)}_7}\overset{\text{O}}{\underset{\|}{\text{C}}}\text{''''}$$

$$\text{C}=\text{O}$$

$$\underset{\underset{\|}{\text{O}}}{\text{CH(CH}_2\text{)}_6}\underset{\underset{\|}{\text{O}}}{\text{CH(CH}_2\text{)}_9}\text{O.}\underset{\text{O}}{\overset{\|}{\text{C}}}\text{(CH}_2\text{)}_7\underset{\underset{\|}{\text{O}}}{\text{CH(CH}_2\text{)}_9}\text{O.}\underset{\text{O}}{\overset{\|}{\text{C}}}\text{(CH}_2\text{)}_7\underset{\underset{\|}{\text{O}}}{\text{CH(CH}_2\text{)}_9}\text{O ''''}$$

Fig. 4.17. Partial structure of a cutin molecule.

(ii) INORGANIC COMPOUNDS

Incrustations of minerals, mainly calcium carbonate and calcium silicate, are found in the cell walls of some plants. These substances may also appear in the cytoplasm or vacuole of the cell in the form of crystals. Heavy mineral deposits are found in the walls of the hair cells of borages, crucifers and cucurbits as well as in the cell walls of many green algae. Silicious cell walls occur commonly amongst the Equisetaceae, Cyperaceae and Gramineae (Poaceae).

C. PHYSICAL COMPOSITION

1. Higher plant cell walls

As indicated in the introduction, the higher plant cell wall is typically composed of three layers, called the middle lamella, the primary wall and the secondary wall.

The middle lamella is an amorphous layer which occurs between the primary walls of adjacent cells. It is the first layer to be formed when a cell divides and is, therefore, the initial partition between the newly formed daughter cells. Each daughter cell subsequently deposits a primary wall on to the middle lamella which is therefore sandwiched between the primary walls of the two daughter cells. Thus it is impossible to say to which of the daughter cells the middle lamella 'belongs'. It is a product of the parent cell from which they arose and was instrumental in their creation. It is the only true intercellular layer in plants and as such probably should not be regarded as part of the cell wall. It is a region which is free of cellulose microfibrils and is thought to be composed mainly of pectins. In the softer tissues of the plant (e.g. pith, cambium) the middle lamella remains unlignified, whilst in woody tissues it becomes heavily permeated with lignin.

The primary wall is a flexible structure which is capable of growth and extension. It can thus accommodate the increasing size of the protoplast. During the growth period of the cell the primary wall is composed of cellulose microfibrils embedded in matrix polysaccharides. Towards the end of the growth period a secondary wall may or may not be deposited upon it, depending on the cell type. If a secondary wall is formed, the primary wall becomes heavily lignified; if it does not form, the primary wall may or may not remain unlignified. The principal matrix polysaccharides of the primary wall of dicotyledonous plants are pectins; this does not, however, appear to be the case with monocotyledonous plants, where the rather sparse information that is available indicates that pectins are only a minor component of the primary wall.

Albersheim in 1978 put forward a tentative and preliminary model of the primary wall of dicotyledonous plants based on his extensive studies with suspension-cultured sycamore cells (see Fig. 4.18). The model is designed not to be spatially or quantitatively accurate but simply to give an idea how the microfibrils and the various matrix polysaccharides appear to coexist in the primary wall. Hemicellulosic xyloglucans are believed to completely coat the cellulose microfibrils (although only a few of them are shown in Fig. 4.18) and are held to

them by hydrogen bonds. The reducing ends of some of the xyloglucans are believed to be linked, glycosidically, by their reducing ends to an arabinan–galactan molecule (labelled arabinogalactan in Fig. 4.18) which is in turn glycosidically linked to a rhamnogalacturonan molecule. Each of the latter can link to several arabinan–galactan molecules.

During the growth period of the cell, the cellulose microfibrils have the ability to change their orientation with respect to each other within the matrix of the primary wall. If the microfibrils were unable to move over each other in this way as the primary wall grows it would be necessary to postulate the insertion of new microfibrils into the wall in order to maintain an intact layer. The fluidity of the matrix which allows this microfibrillar reorientation is lost when lignin replaces much of the water in the matrix at the end of the growth period; this does not matter because the need for microfibrillar reorientation

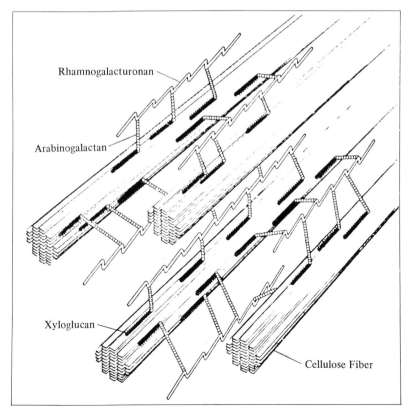

Rhamnogalacturonan

Arabinogalactan

Xyloglucan

Cellulose Fiber

Fig. 4.18. A tentative and preliminary model of the primary cell wall of dicotyledonous plants. (The model is not intended to be spatially or quantitatively accurate. Reproduced, with permission, from P. Albersheim (1978) *Biochemistry of Carbohydrates, II, International Review of Biochemistry*, Vol. 16, ed. D. J. Manners, Publ. University Park Press, Baltimore.)

disappears when the cell wall no longer needs to increase in area.

In the primary cell wall of meristematic cells the microfibrils form a random network within the matrix. As the cell elongates the microfibrils in the walls parallel to the axis of elongation begin to exhibit a more organized pattern; those in the end walls of the cell remain as a random network. On the inner surface of the primary wall the microfibrils are transversely oriented (i.e. they are at right angles to the axis of elongation), whilst at the outer surface they are more or less longitudinally oriented (i.e. they are parallel to the long axis of the cell). Moreover, there is a progressively increasing tendency to change from a transverse to a longitudinal orientation through the thickness of the wall from the inner to the outer surfaces. This arrangement, proposed by Roelofsen and Houwink in 1953, is called the *simple multinet* structure. The way in which it arises depends upon the fluidity of the wall matrix which allows microfibrillar reorientation during wall extension; this is explained in Section E.

In addition to the simple multinet structure the primary walls of many cells develop axial ribs of longitudinally oriented microfibrils between regions of simple multinet. There are two types of rib, *integral ribs*, which extend through the thickness of the wall, and *outer ribs*, which are superimposed on the outer surface of the wall.

The primary wall, therefore, has two distinct microfibrillar components, the simple multinet and the longitudinal ribs. The primary walls of different types of cells reflect variations in the relative contributions of these two components; thus, in hair cells the longitudinal ribs are absent and the wall consists entirely of multinet, parenchyma cells have both multinet and longitudinal ribs, whilst in collenchyma primary cell walls the longitudinal ribs are the predominant feature.

The primary wall is punctured in places, particularly in the cross wall which results from cell division, by plasmodesmata. Plasmodesmata in primary walls are usually associated with an area of reduced deposition of cell-wall material, known as a primary pit field.

Whilst some plant cells have no further cell-wall layer(s) deposited upon the primary wall, many others do. These extra layers constitute the secondary wall. Their basic composition is the same as that of the primary wall, in that cellulose microfibrils are embedded in a polysaccharide matrix. However, there are usually three differences between the primary and secondary walls, firstly in the orientation of the microfibrils, secondly in the composition of the matrix and thirdly in the number of lamellae (i.e. sublayers) that are present. In general the simple multinet of the primary wall is replaced by a helical arrangement of microfibrils in the secondary wall. In the matrix of the secondary wall it is usually hemicelluloses rather than pectins which predominate. Whilst the primary wall is a single, discrete layer, the secondary wall is composed of up to three lamellae (sublayers). The outermost of these (i.e. the lamella next to the primary wall) is called the S_1 lamella, the middle, the S_2 lamella and the innermost, the S_3 lamella. The S_2 lamella is frequently the thickest layer of the whole cell wall, so that its composition and physical properties largely determine the overall characteristics of the wall as a whole. The S_2 lamella is so thick because it is made up of sublamellae (subsublamellae).

The S_1, S_2 and S_3 lamellae are distinguished from each other by differences in the pitch of the helically arranged microfibrils. In the S_1 lamella there are two helices of microfibrils, one right-handed and one left-handed; both make an angle with respect to the long axis of the cell of about 50°. In each of the sublamellae of the S_2 lamella there is a single helix of microfibrils which makes an angle of about 10° with respect to the long axis of the cell. In the S_3 lamella there is a single helix of microfibrils which is in the opposite sense to those of the S_2 sublamellae and which makes an angle of about 80° with respect to the long axis of the cell.

Although the S_1, S_2 and S_3 layers are usually present in the secondary wall (see Fig. 4.19) there are many exceptions. In some cells the secondary wall is composed of the S_1 layer only, in others it is composed of the S_1 and S_2 layers only, whilst in rare cases there is a quite different arrangement. For instance in bamboo fibres the secondary wall is composed of S_1 and S_2 lamellae alternating three or four times.

Lignification of the secondary wall does occur but never so extensively as that of the primary wall and middle lamella; the S_3 lamella is frequently unlignified.

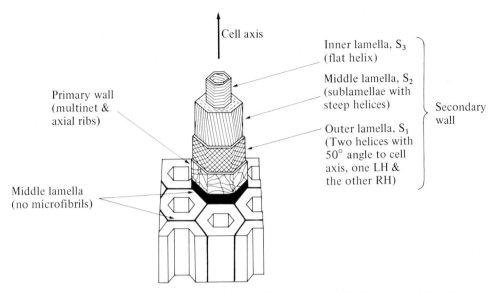

Cell axis

Inner lamella, S₃
(flat helix)

Middle lamella, S₂
(sublamellae with
steep helices)

Secondary
wall

Outer lamella, S₁
(Two helices with
50° angle to cell
axis, one LH &
the other RH)

Primary wall
(multinet &
axial ribs)

Middle lamella
(no microfibrils)

Fig. 4.19. Structure of cell wall of wood fibre (old xylem vessel) (after Wardrop and Bland).

2. Algal cell walls

Algal cell walls are built on the same pattern as those of higher plants in that they consist of microfibrils embedded in a polysaccharide matrix. Most algae have cellulosic microfibrils but others have microfibrils composed of β-1,4-mannans or β-1,3-xylans [see Sections B.1(ii)(b) and (c)]. Algal cell walls are never lignified. They are built up of many layers but these cannot be subdivided into a primary and a secondary wall.

A particularly beautiful example of an algal cell wall is that of *Valonia*, which has been studied in detail and shown to be representative of many algae with cellulose walls. *Valonia* is a tropical marine alga. The young organism consists of a macroscopic, multinucleate vesicle, anchored basally by rhizoids. The wall of the mature vesicle is about 0.04 mm thick and consists of about 1600 lamellae each of which is about 20 nm thick. The lamella thickness corresponds to the thickness of one or two *Valonia* microfibrils [which have a greater cross-sectional area than those of higher plants—see Section B.1(ii)(a)]. What makes the lamellae distinct from one another is the orientation of the microfibrils within them. Three types of orientation are present, all of them helical. The helices all run

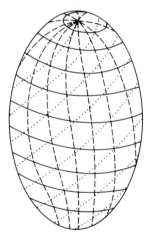

Fig. 4.20. The three different helical arrangements of cellulose microfibrils in the lamellae of the cell wall of *Valonia*. (Redrawn, with permission, from *The Physical Biology of Plant Cell Walls* by R. D. Preston, Publ. Chapman & Hall, 1974.)

from one pole of the vesicle to the other (see Fig. 4.20). In one set of lamellae the microfibrils describe a left-handed helix which makes an angle of about 80° to the long axis of the vesicle. In another set the microfibrils describe a left-handed helix which makes an angle of about 10° to the long axis of the vesicle. In the third set the microfibrils describe a right-handed helix which bisects the angle between the other two. The lamellae with the left-handed helices are the most abundant in the wall, each constituting about 600 out of the total of 1600; the remaining 400 lamellae have the right-handed helix. Each lamella of the algal wall is adjacent to a lamella with a different helical arrangement of microfibrils. The wall has, however, one lamella that does not have one of the three helical arrangements of microfibrils. This is the first lamella to be laid down around a previous naked aplanospore; in it the relatively small number of microfibrils are randomly arranged. Since all the other lamellae are laid down successively within this layer, it becomes the outermost lamella of the wall.

D. BIOSYNTHESIS OF CELL-WALL COMPONENTS

1. Polysaccharides

The biosynthesis of the individual polysaccharide components of the cell wall are dealt with in some detail in Chapter 7, Section D.2. The ensuing sections [D.1(i) and (ii)] of this chapter will not reiterate what has been written in Chapter 7 but will concentrate largely upon the site of synthesis of the various cell-wall polysaccharides and the means by which they are incorporated into the wall.

(i) MICROFIBRILLAR POLYSACCHARIDES
Cellulosic microfibrils are almost certainly formed at the outer surface of the plasmalemma. The main evidence for this is that electron microscopy has shown that a considerable proportion of the outer surface of the plasmalemma of a range of higher plant cell types (e.g. conifer cambial cells, parenchyma cells of broad bean (*Vicia faba*) stem internodes) and algae (e.g. *Valonia, Chaetomorpha*) is covered with partly embedded, square-packed spherical granules. From some of these granules fibres with the known dimensions of microfibrils emanate. The diameter of the granules is always somewhat larger than the bigger diameter of the oval cross-section of the microfibril known to be produced; thus the diameter of the granules in the cellulosic green algae is about 30 nm [microfibril large diameter equals 18.5 nm, see Section B.1(ii)

(a)] whilst that of higher plant granules is about 15 nm [microfibril large diameter equals 8.5 nm, see Section B.1(ii)(a)].

It is generally thought that these granules are aggregates of the enzyme cellulose synthase which catalyses eq. (7.44) (see Chapter 7). The primer in eq. (7.44) is the non-reducing end of a cellulose molecule which is already a member of a microfibril, thus the end of a microfibril provides upwards of fifty non-reducing ends in the case of higher plants and upwards of 500 in the case of the cellulosic algae [provided that all the cellulose molecules are running in the same direction within the microfibril, see Section B.1(ii)(a)]. The cellulose synthase molecules present in each granule, which must be about as numerous as the cellulose molecules within the microfibril, each catalyse the transfer of a glucose residue from GDP-D-glucose to a non-reducing end that is sufficiently close. Thus the whole microfibril is elongated. It is thought that such an elongation takes place each time a microfibrillar end comes sufficiently close to a granule. Since the granules are likely to change position within the membrane [lateral movement of membrane proteins is known to occur, see Chapter 3, Section B.2(i)] and since the microfibrils themselves change position in the wall (at least, they do during the growth of the cell), it is clear that several different granules are likely to have been involved in the lengthening of a given microfibril.

The square packing of the granules is considered to impose restrictions upon the number of possible directions that microfibrils can 'grow'. Whatever orientation a microfibril has within the wall when its 'end' comes into contact with a granule, its subsequent direction of elongation will be along one or other of the sides of the square created by the granular packing or, less probably, along one of the diagonals (see Fig. 4.21). This prediction of microfibrillar directionality within the wall is often seen to be borne out in Nature. Thus the plasmalemmal granules appear to have two roles; firstly, to catalyse the elongation of microfibrils, and secondly, to impose an initial directionality upon them. This directionality is modified in the growing cell by the stresses imposed by wall extension.

The substrate required for microfibril extension, GDP-D-glucose, is synthesized within the protoplast by the reactions described in Chapter 7,

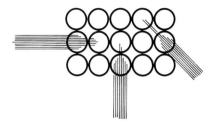

Fig. 4.21. Diagrammatic representation of square-packed array of granules partly embedded in the outer surface of the plasmalemma with cellulose microfibrils emanating from some of them.

Section B.3(i)(b). It must then be exported through the plasmalemma to reach the cellulose synthase-containing granules. How it does this is not yet clear; it may be accomplished by a specific active transport process or by exocytosis.

It is apparent that the microfibril elongation mechanism just described is very satisfactory provided that all the cellulose chains within the microfibril run in the same direction, so that they all have their non-reducing ends at the same end of the microfibril. It is much more difficult to see how it could operate if the cellulose molecules have the antiparallel arrangement that has still to be disproved [see Section B.1(ii)(a)].

Little is known of the formation of the β-1,4-mannan and β-1,3-xylan microfibrils of some algae or the chitin microfibrils of most fungi.

(ii) MATRIX POLYSACCHARIDES

The matrix polysaccharides (pectins and hemicelluloses) are synthesized in the cisternae of the Golgi bodies. Pure preparations of Golgi cisternae have been shown to contain the synthase enzymes which catalyse the reactions that are characteristic of pectin and hemicellulose formation [see Chapter 7, Sections D.2(ii) and (iii)].

This relationship between the Golgi bodies and the biosynthesis of the matrix polysaccharides has been elegantly demonstrated by the autoradiographic-electron microscopic studies of Northcote which were carried out on sections of the root cap of wheat after the administration of D-[U-³H]glucose. After 5 min exposure of the root to the D-[U-³H]glucose, radioactivity was located almost entirely within the Golgi bodies. Analysis of the radioactive material showed it to be a poly-

saccharide or mixture of polysaccharides composed of D-galacturonic acid, D-galactose and L-arabinose residues; it therefore closely resembled pectin. When roots were incubated in a solution of D-[U-³H]glucose for 15 min and then in a solution of unlabelled D-glucose for different lengths of time (10, 20 and 60 min) the radioactive polysaccharide was found firstly in the Golgi cisternae, then in the Golgi vesicles immediately surrounding the cisternae, then in Golgi vesicles close to the plasmalemma and fused with it, and finally in the cell wall itself. This time sequence of labelling clearly indicates that matrix polysaccharides (predominantly pectins in this case) are synthesized in the Golgi cisternae and are then transported to the plasmalemma in vesicles nipped off the Golgi cisternae. The vesicles then fuse with the plasmalemma and empty their contents into the cell wall by exocytosis.

This fits in beautifully with what is known about the formation of a new cell wall after a cell division (see Section E). The cell plate, which is the initial partition between the newly formed daughter cells, is formed by the coalescing of Golgi vesicles which contain pectins. The cell plate becomes the middle lamella which is sandwiched between the primary walls of the daughter cells and is known to be a region rich in pectin.

Since there is a change in nature of the matrix polysaccharides from predominantly pectin to predominantly hemicellulose in most higher plant cell walls as active growth comes to an end it must be assumed that there is some mechanism for controlling the relative rates of production of pectins and hemicelluloses within the Golgi cisternae. It is not yet clear what this is.

2. Lignin

The process of lignification involves a series of enzyme-catalysed reactions leading from CO_2 via carbohydrates, phenylpropanoid amino acids (i.e. phenylalanine, tyrosine) and cinnamic acid derivatives to cinnamyl alcohol derivatives. A proportion of these are then converted, by an enzymic dehydrogenation, into free radicals which initiate a random, non-enzymic polymerization process giving lignin.

Much of what we know of lignin biosynthesis has

come by administering radioactively labelled putative precursors to plants at a time when active lignification is occurring. Sometimes the labelling of the precursor was general (e.g. [U-^{14}C]- or [G-^{14}C]glucose)† but frequently it was located solely in one or two specific atoms (e.g. [carboxy-^{14}C]cinnamic acid). The purpose of this line of experimentation was (i) to find out which compounds could be incorporated into lignin; generally labelled compounds could be used for this and (ii) to find out whether a specific atom of a known precursor ended up at a specific location or became randomized within the lignin molecule; label at a specific location would indicate the direct incorporation of the precursor into lignin whilst random labelling would indicate that the precursor had been incorporated indirectly, probably via a small degradation product, and could not therefore be regarded as a true lignin precursor. The experimental procedure, after the labelled, putative precursor had been administered to the plant for a sufficient time, was to extract the lignin, radioassay it and then degrade it in such a way that specific atoms or groups of atoms could be radioassayed. Most of these biosynthetic studies have been carried out on grasses and other herbaceous plants since they are simpler to handle experimentally than trees which synthesize most of the lignin in the biosphere.

Carbon dioxide is readily incorporated into lignin as would be expected. However, labelling experiments with $^{14}CO_2$ have also shown that there is no significant metabolic turnover of lignin; once lignin is formed it remains within the cell wall until the plant decays.

Glucose, which is an early product of photosynthesis, is almost as good a general carbon source within the plant as CO_2. It is hardly surprising, therefore, to find that ^{14}C from labelled D-glucose is readily incorporated into lignin. The distribution of labelled atoms in vanillin derived from lignin biosynthesized in the presence of [1-^{14}C]- and [6-^{14}C]glucose suggested that the aromatic rings of lignin were formed via shikimic acid.

Shikimic acid has been found in the leaves and

stems of many species of plant. Moreover, it has been shown by administering [2,6-^{14}C]shikimic acid to young sugar-cane plants that the carbon skeleton of this compound is incorporated unchanged into the aromatic rings of lignin. The biosynthesis of shikimic acid from D-glucose, which has been elucidated mainly by work on certain bacteria and fungi, is shown in Fig. 4.22; present evidence indicates that the same pathway operates in plants. The key starting materials are phosphoenolpyruvate (PEP) and D-erythrose 4-phosphate (E-4-P) which are derived from D-glucose via glycolysis and the oxidative pentose phosphate pathway (see Chapter 6, Sections E.1 and F.3) respectively. Both of these compounds are lignin precursors.

The reaction between PEP and E-4-P is closely analogous to an aldol condensation. The mechanism shown in Fig. 4.22 takes account of the experimental fact that when PEP labelled with ^{18}O in the C–O–P bridge is used all the ^{18}O is lost in the eliminated orthophosphate and none remains in the product, 7-phospho-2-keto-3-deoxy-D-arabinoheptonic acid. The cyclization of the latter to give 5-dehydroquinic acid probably involves the enolic intermediate, shown in parenthesis in Fig. 4.22, formed by the elimination of orthophosphate; the enzyme catalysing this reaction requires NAD^+ which is thought to be concerned with the transient oxidation of the C–5 hydroxyl group to a carbonyl group which would facilitate the elimination of the orthophosphate. The dehydration of 5-dehydroquinic acid yields 5-dehydroshikimic acid which is then reduced by NADPH to shikimic acid.

Shikimic acid is the known precursor of the aromatic amino acids, L-phenylalanine and L-tyrosine. These compounds are phenylpropane (C_6–C_3) derivatives as are the building units of lignin and, not surprisingly, they were soon shown to be lignin precursors. Phenylalanine is an excellent precursor in all plants but tyrosine is only really effective in the grasses.

The first step in the conversion of shikimic acid into phenylalanine and tyrosine is its phosphorylation (see Fig. 4.22). The resulting 5-phosphoshikimic acid then condenses with a molecule of PEP to form 3-enoylpyruvyl-shikimic acid 5-phosphate. The mechanism of this reaction, shown in Fig. 4.23, takes account of the experimental fact that

---

† The symbol 'U' in [U-^{14}C] is the abbreviation of the word 'uniformly' and indicates that the ^{14}C is uniformly distributed amongst all six carbon atoms; the symbol 'G' in [G-^{14}C]glucose is the abbreviation of the word 'generally' and indicates that the ^{14}C is distributed amongst all six carbon atoms but not necessarily uniformly.

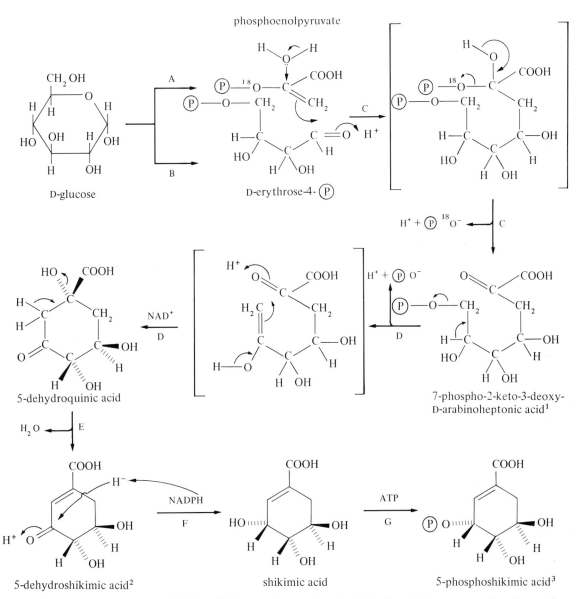

Fig. 4.22. Formation of 5-phosphoshikimic acid from D-glucose. (A = glycolysis; B = oxidative pentose phosphate pathway; C = phospho-2-keto-3-deoxyheptonate aldolase, EC 4.1.2.15; D = 5-dehydroquinate synthase; E = 5-dehydroquinate dehydratase, EC 4.2.1.10; F = shikimate dehydrogenase, EC 1.1.1.25; G = shikimate kinase, EC 2.7.1.71; 1 = 3-deoxy-D-arabinoheptulosonate-7-phosphate; 2 = 3-dehydroshikimic acid; 3 = 3-phosphoshikimic acid; 1, 2 and 3 = alternative names and numbering.)

when the reaction takes place in ^3H_2O tritium is incorporated into the methylene group. This strongly suggests the addition–elimination mechanism shown.

The subsequent reactions in the pathway are designed to (i) insert two more double bonds in the ring, so aromatizing it and (ii) reposition the C_3 side chain. The second double bond is inserted into the

ring by the elimination of orthophosphate (reaction B, Fig. 4.23), so generating chorismic acid. The repositioning of the C_3 side chain occurs next (reaction C, Fig. 4.23) with the formation of prephenic acid. Prephenic acid may then undergo decarboxylation and dehydration under the catalytic influence of prephenate dehydratase[3] (reaction D, Fig. 4.23), so forming phenylpyruvic acid.

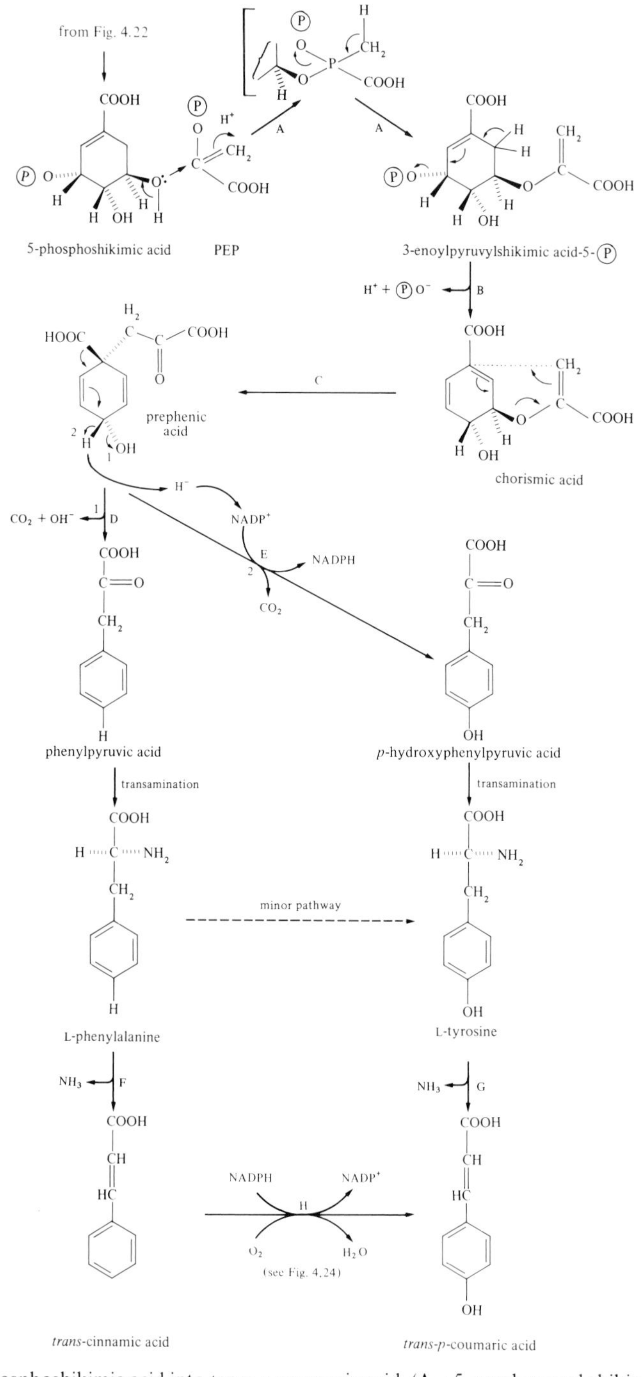

Fig. 4.23. Conversion of 5-phosphoshikimic acid into *trans-p*-coumaric acid. (A = 5-enoylpyruvyl shikimic acid 5-phosphate synthase; B = chorismate synthase, EC 4.6.1.4; C = chorismate mutase, EC 5.4.99.5; D = prephenate dehydratase, EC 4.2.1.51; E = prephenate dehydrogenase, EC 1.3.1.13; F = L-phenylalanine ammonia lyase, EC 4.3.1.5; G = L-tyrosine ammonia lyase (grasses only); H = cinnamate 4-hydroxylase, EC 1.14.13.11.)

Alternatively it may undergo decarboxylation and dehydrogenation under the catalytic influence of the $NADP^+$-requiring prephenate dehydrogenase[4] (reaction E, Fig. 4.23) so forming p-hydroxyphenylpyruvic acid. The decarboxylations in these two reactions introduce the third double bond into the ring so aromatizing it. Both the products of reactions D and E (Fig. 4.23) can undergo transamination from L-glutamic acid, for example, to yield L-phenylalanine and L-tyrosine respectively. There is also evidence to indicate that prephenic acid can be transaminated directly to form pretyrosine which then undergoes aromatization to give L-tyrosine (see Chapter 9, Section E.5) in bluegreen algae.

It is doubtful whether the hydroxylation of L-phenylalanine to yield L-tyrosine occurs to any great extent in plants. There have been reports that some plant enzymes are capable of catalysing this reaction but it seems likely that L-phenylalanine is not their *in vivo* substrate.

Both L-phenylalanine and L-tyrosine can be deaminated to form *trans*-cinnamic acid and *trans*-p-coumaric acid respectively (see Fig. 4.23). The deamination of L-phenylalanine occurs in all plants and is catalysed by the enzyme phenylalanine ammonia lyase[5]; in this reaction the α-amino group and the β *pro*-S hydrogen are removed leaving a *trans* $\alpha\beta$-double bond. The deamination of L-tyrosine appears to occur only in grasses and is catalysed by L-tyrosine ammonia lyase. Both *trans*-cinnamic acid and *trans*-p-coumaric acid are excellent precursors of lignin in all plants, their carbon skeletons being incorporated intact. However, it is *trans*-p-coumaric acid, with its 4-hydroxyl group, which is the more immediate precursor. In grasses *trans*-p-coumaric acid may arise in two ways: (i) from prephenic acid via p-hydroxyphenylpyruvic acid and L-tyrosine or (ii) from prephenic acid via phenylpyruvic acid, L-phenylalanine and *trans*-cinnamic acid. In all other plants only the second of these routes occurs. The second route involves the hydroxylation of *trans*-cinnamic acid at C–4. This reaction is catalysed by a cytochrome P450-mixed function oxygenase, cinnamate 4-hydroxylase, which, in common with all such enzymes, requires O_2 and NADPH. It has been studied in detail by Zenk who has shown that the introduction of the hydroxyl group causes the hydrogen originally present at C–4 to migrate to the immediately adjacent carbon atom. This hydroxylation-induced migration of hydrogen is called the NIH shift because it was originally discovered by researchers at a laboratory of the National Institutes of Health in the U.S.A. who used substrates labelled at C–4 with ^2H or ^3H. The likely mechanism of this reaction is shown in Fig. 4.24. The enzyme system is a microsomal one and has been studied in young *Catalpa* leaf tissue, etiolated pea and sorghum shoots.

The biosynthetic route from *trans*-p-coumaric acid to the lignin building units p-coumaryl alcohol, coniferyl alcohol and sinapyl alcohol is shown in Fig. 4.25. This route involves up to three steps: (i) the introduction of a hydroxyl group at C–3 and sometimes also at C–5 of the aromatic ring, (ii) the methylation of the one or two extra hydroxyl groups (but not that at C–4) and (iii) the reduction of the carboxyl group of the C_3-side chain to a primary alcohol group. These three steps occur in that order. The methyl donor for the methylation of the 3- and 5-hydroxyl groups is S-adenosylmethionine and the enzyme catalysing the reaction is a methyltransferase which appears to be specific to the intermediates of the lignin biosynthetic pathway. The reduction of the carboxyl group is a three-step process. The first step is the formation of the high-energy coenzyme A thioester derivatives of p-coumaric, ferulic and sinapic acids; the enzyme, 4-coumarate:CoA ligase[7], catalyses the activation of all three acids. The second step is the reduction of the coenzyme A thioesters to their corresponding aldehydes; this is catalysed in each case by cinnamyl-CoA:NADPH oxidoreductase. The third step is the reduction of the aldehydes to the corresponding primary alcohols, p-coumaryl, coniferyl and sinapyl alcohols; this reaction is catalysed in each case by cinnamyl alcohol:$NADP^+$ oxidoreductase. All the named compounds in Fig. 4.25 have been shown to be readily incorporated into lignin with their carbon skeletons intact.

The lignin building units p-coumaryl alcohol, coniferyl alcohol and sinapyl alcohol are often stored in the cambial cells of plants as their β-glucosides (see Fig. 4.26). When required they are liberated by a β-glucosidase which is also located in the cambial cells.

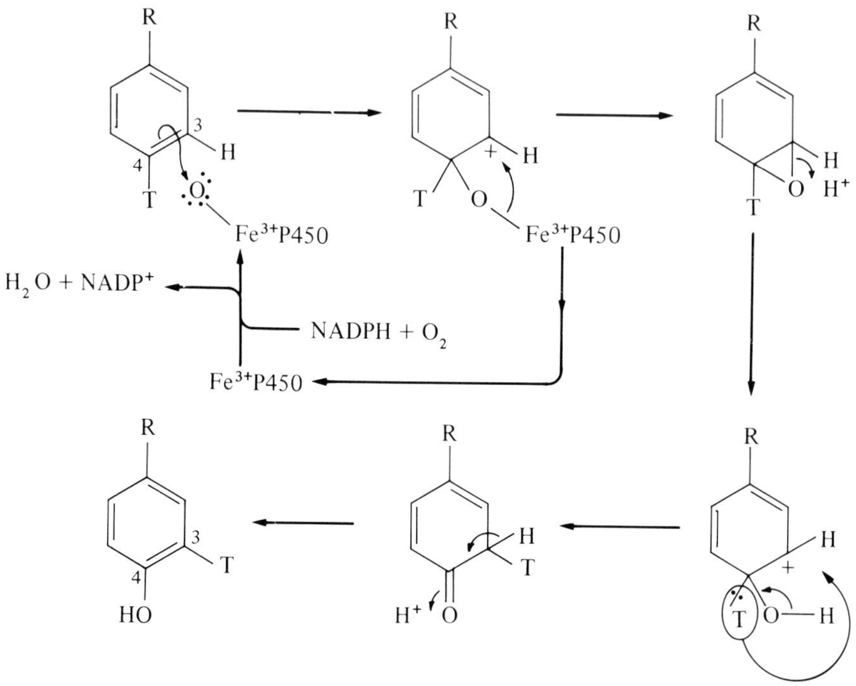

Fig. 4.24. Hydroxylation-induced migration of hydrogen; the N.I.H. shift.

It was Erdtman in 1939 who first suggested that lignin was probably formed by the polymerization of such compounds as coniferyl alcohol. This was confirmed in the 1950s by Freudenberg who found that a lignin-like polymer was produced *in vitro* by incubating coniferyl alcohol, under aerobic conditions, with a phenol oxidase obtained from mushrooms. This enzyme was later shown to be a laccase (so named because the first such enzyme was obtained from the Japanese lac tree, *Rhus vernicifera*), which is a copper-containing monophenol monooxygenase and catalyses the general reaction (4.1).

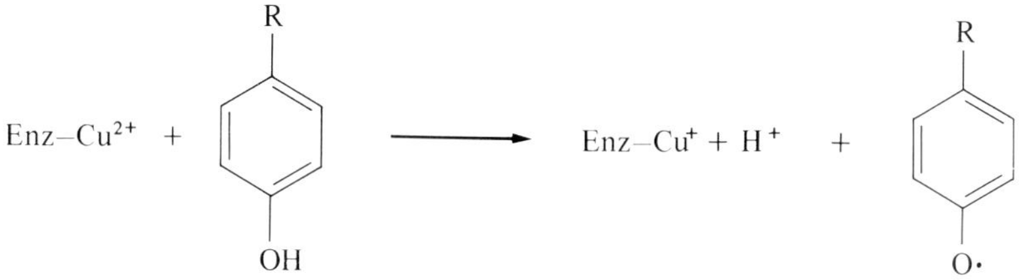

$$2 \ Enz\text{-}Cu^+ + 2H^+ + \tfrac{1}{2}O_2 \rightarrow 2 \ Enz\text{-}Cu^{2+} + H_2O$$
$$(4.2)$$

A similar enzyme, also capable of causing the formation of a lignin-like polymer from coniferyl alcohol, has been found in cambial cells.

The reactions bringing about the lignin-like polymer *in vitro* were investigated by stopping the incubation after decreasing periods of time when dimeric and trimeric derivatives of coniferyl alcohol accumulated. These were isolated and their structures determined. The structures of three of the dimers are shown in Fig. 4.27. They indicate three

The Cu$^+$ of the enzyme prosthetic group is then regenerated by the action of molecular oxygen as shown in eq. (4.2).

ways in which two coniferyl alcohol residues may link together. A comparison of their structures with the partial structure of lignin shown in Fig. 4.14

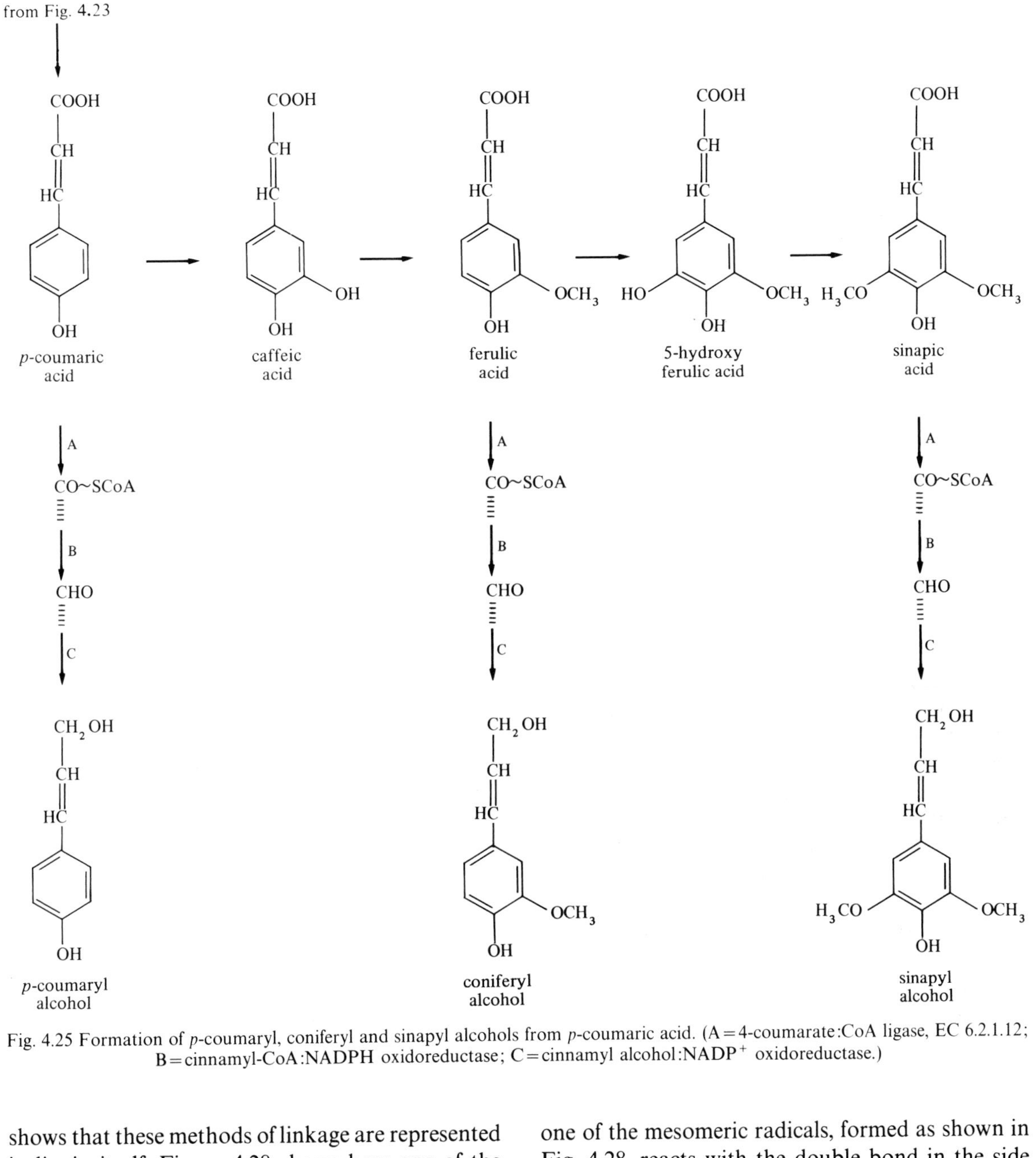

from Fig. 4.23

Fig. 4.25 Formation of p-coumaryl, coniferyl and sinapyl alcohols from p-coumaric acid. (A = 4-coumarate:CoA ligase, EC 6.2.1.12; B = cinnamyl-CoA:NADPH oxidoreductase; C = cinnamyl alcohol:NADP$^+$ oxidoreductase.)

shows that these methods of linkage are represented in lignin itself. Figure 4.28 shows how one of the dimers, guaiacyl glycerol-β-coniferyl ether, is formed by the coupling of two of the three mesomeric free radicals produced by the action of the laccase [eq. (4.1)] and subsequent spontaneous hydration of the resulting quinone methide.

Some of the likely reactions involved in the formation of lignin itself are shown in Fig. 4.29. Here

one of the mesomeric radicals, formed as shown in Fig. 4.28, reacts with the double bond in the side chain of a coniferyl alcohol molecule to produce a dimeric free radical. Being still a radical, this can react with a double bond of another coniferyl alcohol molecule forming a trimeric free radical or, as is shown in Fig. 4.29, form a stable trimer by reacting with a monomeric free radical. The stable trimer may then react with another monomeric free *p*-coumaryl-β-D-glucoside — Coniferin — Syringin

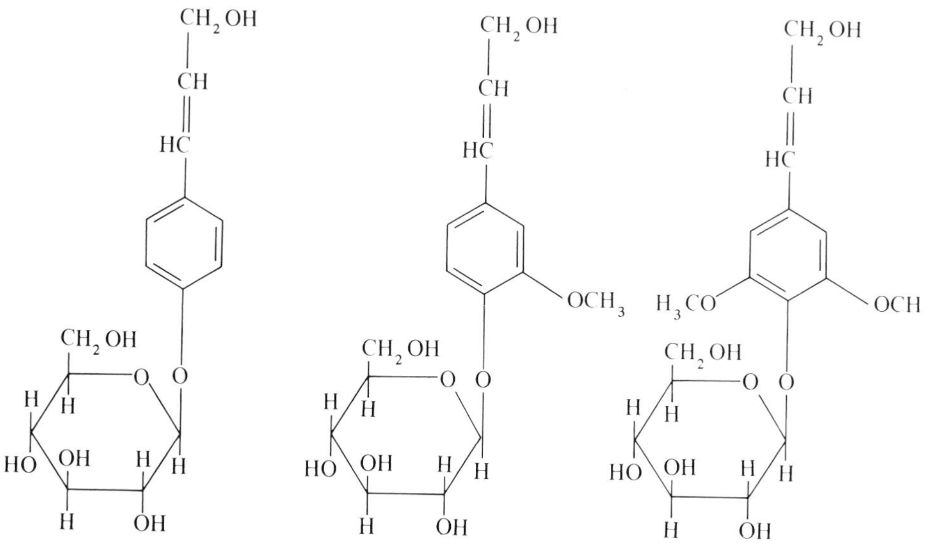

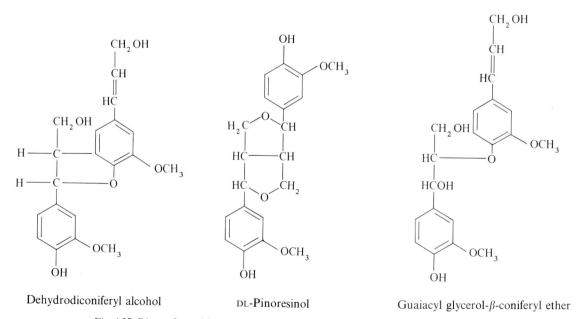

Fig. 4.26 Glucosides of *p*-coumaryl, coniferyl and sinapyl alcohols.

Dehydrodiconiferyl alcohol — DL-Pinoresinol — Guaiacyl glycerol-β-coniferyl ether

Fig. 4.27. Dimers formed by *in vitro* enzyme-catalysed dehydrogenation of coniferyl alcohol.

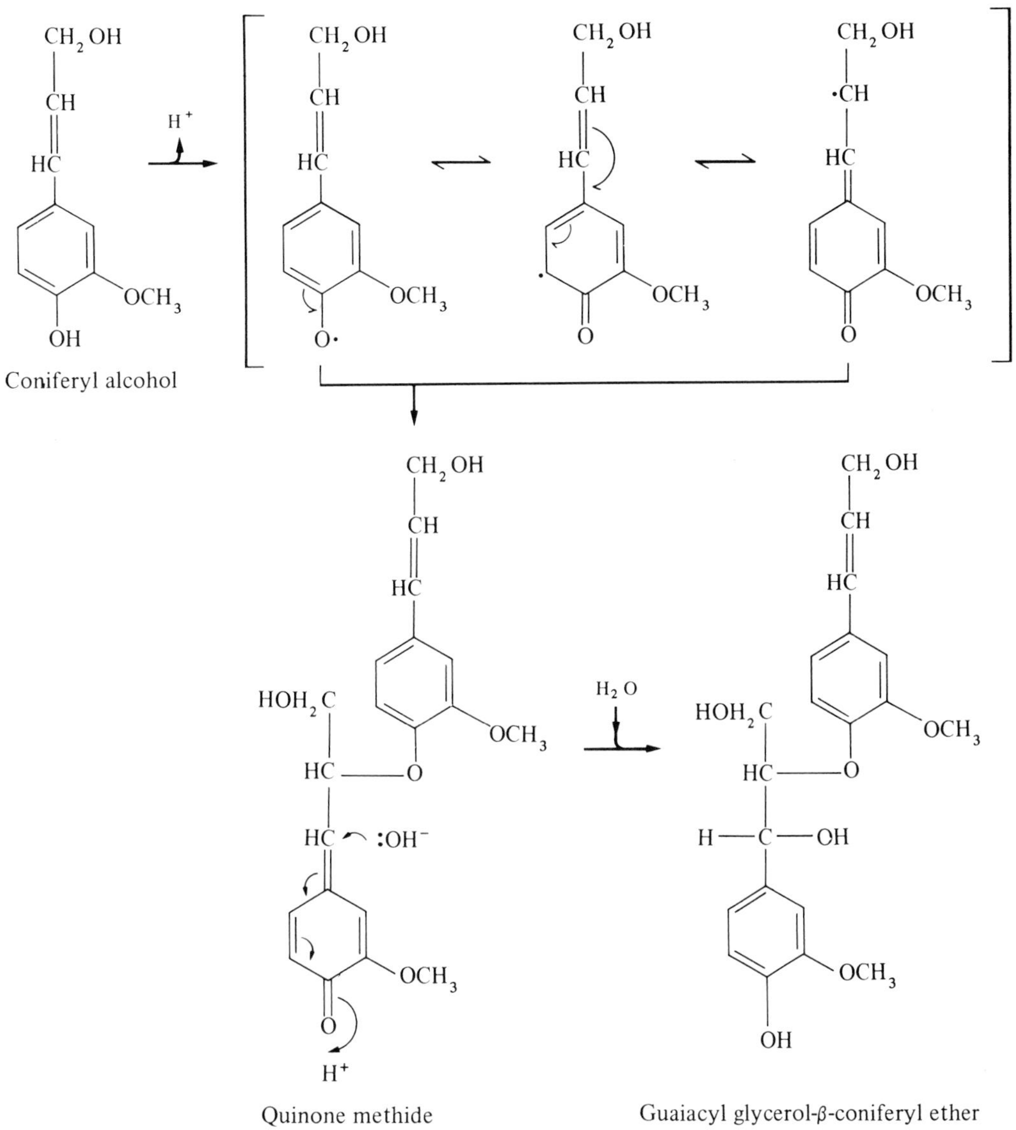

Fig. 4.28. Formation of mesomeric free radicals by dehydrogenation of coniferyl alcohol and one example of how they may couple together.

radical to give a tetrameric free radical and so the process can go on indefinitely. Furthermore, hydration such as is shown in Fig. 4.28 may take place at any time. The polymerization process therefore involves the random, non-enzymic reactions of free radicals and stable molecules and produces a huge net-like molecule which permeates the matrix of the cell wall.

The structure of lignin from various classes of plants [see Section B.2 (ii)] indicates that gymno-sperm lignin is largely formed by the random reactions of coniferyl alcohol with the mesomeric free radicals formed from it by the laccase. Dicotyledonous angiosperm lignin is formed by the random reactions of coniferyl and sinapyl alcohols with the mesomeric free radicals formed from them by the laccase whilst monocotyledonous angiosperm lignin is even more complex, being formed in a similar manner from coniferyl, sinapyl and p-coumaryl alcohols.

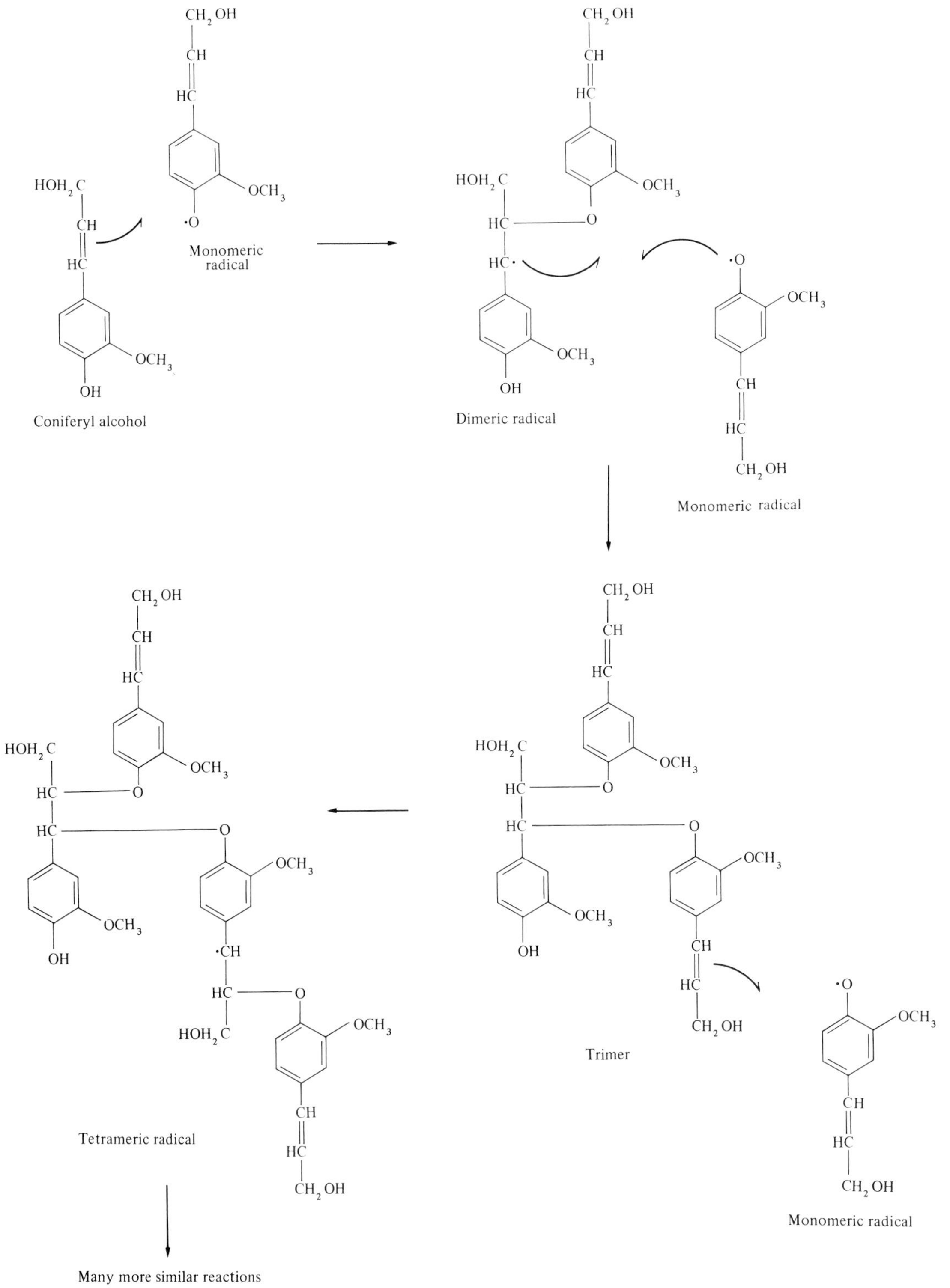

Fig. 4.29. An illustration of how a gymnosperm lignin polymer may be formed by the interaction of coniferyl alcohol and the free radicals formed from it.

The biosynthesis of coniferyl, sinapyl and *p*-coumaryl alcohols takes place in the protoplast of the cell but their laccase-catalysed dehydrogenation to yield monomeric free radicals and the subsequent polymerization process occur within the cell wall. The cell wall is known to contain a number of oxidases [see Section B.3 (i)], one of which is almost certainly the laccase concerned in lignin formation. Clearly the three alcohols must diffuse from the cytoplasm through the plasmalemma and deep into the substance of the wall where they meet the laccase. Since lignification does not occur throughout the growth and development of the cell wall but only begins towards the end of the active growth of the cell, there must be some signal which starts it off. Although it is not known what this signal is, there are indications that it may exert its effect by controlling the synthesis (and possibly the activity) of certain key enzymes, such as 4-coumarate:CoA ligase, operating at the level of the C_6–C_3 intermediates.

3. Proteins

Little is known of the biosynthesis of the specific proteins, structural or enzymic, of the cell wall. However, the little direct experimental knowledge that we have, taken with what is generally known about the biosynthesis of proteins that are exported from the cell, enable a fairly clear, but nevertheless tentative, picture to be painted.

In animal cells proteins which are destined for secretion are synthesized on rough endoplasmic reticulum. This is a consequence of the fact that such proteins are initially synthesized with a sequence of very hydrophobic amino acids at the N-terminal end. This sequence is not part of the secreted protein and is therefore removed before secretion takes place. Since the biosynthesis of a protein begins at its N-terminal end, this hydrophobic segment is formed first as the ribosome proceeds down the *m*RNA molecule from 5′ end to 3′ end in the translational phase (see Chapter 10, Section E). The hydrophobic nature of the N-terminal segment enables it to dissolve in and pass through the endoplasmic reticular membrane. This, of course, anchors the ribosome and the *m*RNA to the membrane and is

the explanation of how rough endoplasmic reticulum arises. The hydrophobic segment, having passed through the membrane, causes the rest of the protein molecule, as it is synthesized, to follow. Therefore the protein molecule, when complete, ends up in the lumen of an endoplasmic reticular tubule along which it moves, presumably passively, until it reaches an endoplasmic reticular vesicle. Now it is from the membranes of such vesicles that the Golgi cisternae are formed. Thus ultimately the protein finds itself in a Golgi cisterna. Within the lumen of endoplasmic reticular tubules and vesicles and the Golgi cisternae are other proteins which got there by the same process. One of these is an enzyme which catalyses the hydrolytic removal of the hydrophobic segment from the protein. The protein is now in the form in which it is secreted. Secretion takes place in a manner analogous to that in which cell-wall matrix polysaccharides are deposited into the wall. The protein is packed into Golgi vesicles which travel to the plasma membrane, fuse with it and empty their contents to the exterior by exocytosis.

Assuming that this process also occurs in plant cells, and there is no reason to suppose that it does not, then this is clearly the means by which the enzymes of the cell wall get there. It may also explain how the synthases catalysing the formation of the matrix polysaccharides get into the Golgi cisternae; however, there would have to be some means of keeping them there, since they are not found, in active form at least, in the cell wall itself.

There is evidence to indicate that the Golgi bodies are concerned in the glycosylation of the hydroxyproline-rich glycoprotein extensin [see Section B.3 (ii)] and its transport to the wall. It may be assumed that the protein moiety of extensin is synthesized and translocated to the Golgi vesicles in the manner previously described. There it is glycosylated by polysaccharide synthases. The resulting glycoprotein is then packed into Golgi vesicles and deposited into the cell wall along with the matrix polysaccharides.

4. Incrusting substances

Although considerable progress has been made in

elucidating the mechanism of biosynthesis of the components of wax and the monomers of cutin [see Chapter 8, Section C.2(ii)], little is known of the mechanism of cutin formation or its whereabouts in the cell. As would be expected, however, all the synthetic processes involved in wax and cutin formation take place in the epidermal layer of cells only.

Little is known of the mechanism of mineral deposition in cell walls. Mineralization of the wall is, however, peculiar to a minority of plants. Since it increases with age, it has been suggested that it is the way in which the cell rids itself of excess minerals.

E. FORMATION AND GROWTH OF THE CELL WALL

The primary wall begins to develop during anaphase of mitotic nuclear division (see Chapter 3, Section B.4) with the formation of the cell plate. As the chromosomes move to opposite poles of the nuclear spindle, pieces of endoplasmic reticulum and ribosomes accumulate in the equatorial region. During telophase the spindle breaks down and the two groups of chromosomes begin to be enveloped in nuclear membranes. This is coincident with a ramification of the endoplasmic reticular elements in the equatorial region which, along with the spindle fibres, form a structure called the phragmoplast. The phragmoplast, formed originally within what were the confines of the spindle, grows outwards towards the exterior walls of the cell. During this process a large number of small vesicles, about 20–50 nm in diameter, appear within the phragmoplast. It now seems clear that these vesicles are derived from Golgi bodies and carry material for formation of the cell wall. The vesicles appear to be directed to this region of the cell by microtubules which were present in the mitotic spindle and which appear in great concentration at, and perpendicular to, the edges of the developing wall. The vesicles coalesce to form a structure known as the cell plate which eventually reaches the side walls of the cell and divides it into two daughter cells. The cell plate constitutes the middle lamella of the newly forming cell wall. It is pierced at several points by strands of membrane which form the plasmodesmata. The cell plate is composed mainly of pectic polysaccharides. However, it shows positive birefringence† as soon as the vesicles have coalesced indicating that cellulose is being deposited on it in the form of microfibrils. This occurs on both sides indicating that each of the newly formed daughter cells is beginning to lay down a primary wall on the cell plate which then becomes the structure known as the middle lamella.

Growth of the primary wall is continuous, due to the laying down of more cellulose microfibrils and non-cellulosic polysaccharide matrix material. The microfibrils are laid down on the inner (cytoplasmic) surface of the primary wall. In the case of primary walls which are parallel to the direction of growth the microfibrils are laid down in a transverse, parallel array. As the cell elongates the transversely oriented microfibrils are pulled into a more longitudinal orientation rather like the individual coils of an extending spring, so explaining the mode of formation of the simple multinet (see Section C.1). In the case of primary walls which are transverse to the direction of growth (cross walls) the distribution of microfibrils is random. Concomitant with the formation of microfibrils is the deposition of non-cellulosic polysaccharide matrix material. During the growth phase there is a gradual change in the nature of the non-cellulosic polysaccharides deposited in the wall. This appears to be mainly due to an increase in the proportion of hemicellulosic polysaccharides synthesized but may also involve transglycosylation reactions within the wall thereby altering the nature of the various interpolysaccharide connections.

Secondary thickening of the wall involves further deposition of microfibrils and non-cellulosic polysaccharides, almost entirely hemicelluloses, in a distinct fashion and often in distinct regions of the wall. There is evidence that the matrix polysaccharides are synthesized in the Golgi bodies and transported to the cell wall in the Golgi vesicles. Microtubules (about 25 nm in diameter) appear to direct the vesicles to the sites of matrix deposition; it is significant that during the even, all-round growth of the primary wall the microtubules are distributed

---

† When polarized light is passed through a system containing asymmetrical molecules arranged in an orderly pattern it is diffracted; the system is then said to exhibit birefringence. Birefringence is, therefore, an indication of order or disorder in aggregates of asymmetric molecules.

round the inner surfaces of the plasmalemma whilst they are grouped at the sites of thickening during the growth of the secondary wall.

Lignin begins to be deposited in the various layers of the wall, including the middle lamella, towards the end of the growth of the cell and frequently continues throughout the lifetime of the protoplast.

There is a continuous development of the cell wall starting during the final stages of cell division, continuing through the phases of differentiation and growth and ending with the cessation of the cell's metabolic activity.

SUGGESTIONS FOR FURTHER READING

Freudenberg, K. and Neish, A. C. (1968) *Constitution and Biosynthesis of Lignin*. Springer-Verlag, Berlin, Heidelberg, New York.

Martin, J. T. and Juniper, B. E. (1970) *The Cuticles of Plants*. St. Martins Press, New York.

Sarkanen, K. V. and Ludwig, C. H., eds. (1971) *Lignins: occurrence, formation, structure and reactions*. Wiley-Interscience, New York.

Higuchi, T. (1971) 'Formation and biological degradation of lignins', *Adv. Enzymol.* **34**, 207–283.

Shafizadeh, F. and McGinnis, G. D. (1971) 'Morphology and biogenesis of cellulose and plant cell walls', *Adv. Carbohydrate Chem. Biochem.* **26**, 297–349.

Cleland, R. (1971) 'Cell wall extension', *Ann. Rev. Plant Physiol.* **22**, 197–222.

Northcote, D. H. (1972) 'Chemistry of the plant cell wall', *Ann. Rev. Plant. Physiol.* **23**, 113–132.

Towle, G. A. and Whistler, R. L. (1973) 'Hemicelluloses and gums', in *Phytochemistry*, Vol. I, *The Process and Products of Photosynthesis* (Miller, L. P., ed.). Van Nostrand Reinhold Co., New York.

Teng, J. and Whistler, R. L. (1973) 'Cellulose and chitin', *ibid*.

Doesburg, J. J. (1973) 'The pectic substances', *ibid*.

Schubert, W. J. (1973) 'Lignin' in *Photochemistry*, Vol. III, *Inorganic Elements and Special Groups of Chemicals* (Miller, L. P., ed.). Van Nostrand Reinhold Co., New York.

Martin, J. T. (1973) 'Cutins and suberin', *ibid*.

Preston, R. D. (1974) *The Physical Biology of Plant Cell Walls*. Chapman & Hall, London.

Albersheim, P. (1974) 'Primary cell wall and control of elongation growth', in *Plant Carbohydrate Biochemistry*, (Pridham, J. B., ed.). *Ann. Proc. Phytochem. Soc.* No. 10, pp. 145–164.

Northcote, D. H. (1974) 'Sites of synthesis of the polysaccharides of the cell wall', *ibid*., pp. 165–181.

Stafford, H. A. (1974) 'The metabolism of aromatic compounds', *Ann. Rev. Plant Physiol.* **25**, 459–486.

Grisebach, H. and Hahlbrock, K. (1974) 'Enzymology and regulation of flavonoid and lignin biosynthesis in plants and plant cell suspension cultures', *Recent Adv. Phytochem.* **8**, 21–52.

Grisebach, H. (1977) 'The structure, biosynthesis and degradation of wood', *Recent Adv. Phytochem.* **11**, 141–184.

Grisebach, H. (1977) 'Biochemistry of lignification', *Naturwissenschaften* **64**, 619–625.

Albersheim, P. (1978) 'Concerning the structure and biosynthesis of the primary cell walls of plants', in *Biochemistry of Carbohydrates*, Vol. II (Manners, D. J., ed.), *International Review of Biochemistry*, Vol. 16 (Kornberg, H. L. and Phillips, D. C., series eds.). pp. 127–150. University Park Press, Baltimore.

Turvey, J. R. (1978) 'Biochemistry of algal polysaccharides', *ibid*., pp. 151–177.

Hahlbrock, K. and Grisebach, H. (1979) 'Enzymic controls in the biosynthesis of lignin and flavonoids', *Ann. Rev. Plant Physiol.* **30**, 105–130.

Darvill, A., McNeil, M. and Albersheim, P. (1979) 'The primary cell walls of flowering plants', in *The Plant Cell* (Tolbert, N. E. ed.). Academic Press, New York.

ENZYMES

1. β-D-Fructofuranoside fructohydrolase, EC 3.2.1.26.
2. Pectin pectyl-hydrolase, EC 3.1.1.11.
3. Prephenate hydro-lyase (decarboxylating), EC 4.2.1.51.
4. Prephenate:$NADP^+$ oxidoreductase (decarboxylating), EC 1.3.1.13.
5. L-Phenylalanine:ammonia lyase, EC 4.3.1.5.
6. Cinnamate, NADPH:oxygen oxidoreductase (4-hydroxylating), EC 1.14.13.11.
7. 4-Coumarate:CoA ligase, EC 6.2.1.12.

CHAPTER 5

Photosynthesis

A. INTRODUCTION

Photosynthesis is a process in which carbon dioxide is converted into carbohydrates [reaction (5.1)]; it is carried out in the light by all organisms which possess chlorophyll (see Section B).

$$CO_2 \rightarrow (CH_2O)_n \qquad (5.1)$$

Although photosynthesis is a multistep process whose complexities have yet to be fully elucidated, the outlines of its mechanism can be arrived at by making use of a number of chemical and biochemical principles and a few intelligent guesses.

Firstly, it is at once apparent that CO_2 is at a higher level of oxidation (i.e. has less H) than carbohydrate. Therefore to convert CO_2 into $(CH_2O)_n$ hydrogen must be added; in other words, the process must involve the reduction of CO_2 [reaction (5.2) where [H] is reducing power].

$$CO_2 + [H] \rightarrow (CH_2O)_n \qquad (5.2)$$

It is possible to go even further than this because it is known that such a reduction is effected by the addition of electrons [reaction (5.3)].

$$CO_2 + e^- (+H^+) \rightarrow (CH_2O)_n \qquad (5.3)$$

There must, therefore, be a biological oxidation-reduction system (redox system) to provide the electrons for this reduction. It would appear likely that the E_0' of this system is more negative than -0.4 V which is the approximate E_0' of the $CO_2/(CH_2O)_n$ system; this would allow the electrons to flow exergonically down an electrochemical gradient. The photosynthetic process can, at this stage of the argument, be described by reaction (5.4) in which X^- is the reduced form of a redox system X/X^- which has an E_0' more negative than -0.4 V.

$$CO_2 \xrightarrow[X^- \quad X]{\quad H^+ \quad} (CH_2O)_n \qquad (5.4)$$

Secondly, it can be predicted that the conversion of CO_2 into carbohydrate will require the input of a large amount of energy. This follows from the fact that respiration, the process in which carbohydrate is broken down into CO_2 (i.e. the reverse of photosynthesis), releases a large amount of free energy [reaction (5.5)]. Thus an equivalent amount of free energy must be available to convert CO_2 into carbohydrate.

$$(CH_2O)_n + O_2 \xrightarrow[\text{respiration}]{\text{aerobic}} CO_2 + H_2O + \text{ENERGY}$$

$$\Delta G^{\circ\prime} = -2872 \text{ kJ mol}^{-1} \text{ glucose} \qquad (5.5)$$

Thirdly, the ways in which this input of energy is used in the process can be guessed at with a fair degree of accuracy. To keep reaction (5.4) going the oxidized form, X, of the redox system has to be reduced back to X^-. The electron carrying out this reduction must ultimately come from the reduced form of another redox system which can be called the 'electron donor'. Ideally the electron donor should have two properties: (i) it should have an E_0' more negative than that of the X/X^- system so that the electron carrying out the reduction of X to X^- can flow exergonically down an electrochemical gradient and (ii) it should be abundant and ubiquitous in the biosphere. Unfortunately no naturally occurring redox system has both of these properties. There are many which are locally abundant and there is one (oxygen/water) which is both abundant and almost ubiquitous, but all have an E_0' more positive than the X/X^- system (e.g. E_0'

of $O_2/H_2O = +0.82$ V). There are some redox systems which have a sufficiently negative E_0' but none of them are present in the biosphere in the required quantity. The choice is a clear one; the over-riding property must be that of abundance. Thus the electron donor will have the disadvantage of having an E_0' more positive than the X/X^- system. This means that a redox system with a greater electron-donating power is to receive an electron from a system with a lesser electron-donating power or, to put it another way, an electron is to be forced up an electrochemical gradient. This is analogous to pushing water back up a waterfall; it can be done but only with the input of energy. Therefore one can guess that one of the uses to which the predicted influx of energy is put is to force an electron up an electrochemical gradient from the reduced form of the electron donor to X. The second likely use of the predicted energy input is in the generation of ATP from ADP and ortho-phosphate. This follows from the fact that phosphorylated intermediates are known to participate in carbohydrate metabolism (see Chapter 7) including that associated with photosynthesis (see Section D). The production of phosphorylated intermediates immediately implicates ATP as the phosphorylating agent. The ADP produced in the phosphorylation reactions must be converted back into ATP so as to keep the process going; this is an energy-requiring process ($\Delta G^{\circ\prime} = +30.5$ kJ mol^{-1} ADP). Thus the photosynthetic process at this stage is depicted in Fig. 5.1.

Fourthly, it is easy to guess where the energy comes from because photosynthesis will only proceed in the light. The energy which drives photosynthesis is, therefore, light energy. It follows from

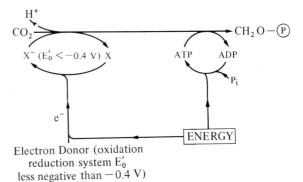

Fig. 5.1. Simple basic concept of photosynthesis.

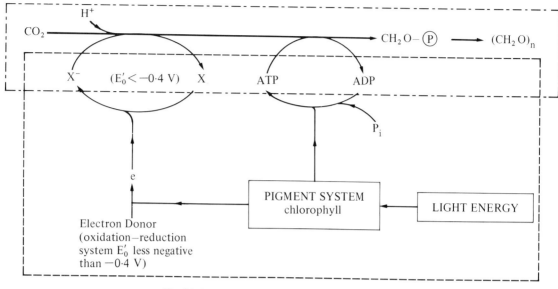

Fig. 5.2. More sophisticated concept of photosynthesis.
–––––––– = Light phase
––– –– –– = Dark phase

this that the photosynthetic organism must have a mechanism for trapping light energy and handing it on to other participants in the photosynthetic process. This light-trapping agent is a pigment system whose main component is chlorophyll. Thus the photosynthetic process can now be simply outlined as shown in Fig. 5.2. From this scheme (Fig. 5.2) several requirements of photosynthesis become apparent:

1. the conversion of CO_2 into carbohydrate requires (a) reducing power and (b) ATP;
2. light is required *only* for the production of the reducing power and ATP;
3. light is *not* required for the conversion of CO_2 into carbohydrate *provided that reducing power and ATP are available.*†

These requirements cover the observations made in 1905 by Blackman that photosynthesis involved both light-dependent reactions and light-independent (dark) reactions. Blackman came to this conclusion largely because temperature affected the rate of photosynthesis in the same manner as it did chemical reactions (i.e. within the limits of enzyme stability an increase in temperature increased the photosynthetic rate) although photochemical reactions, such as must take place in

†This is not strictly true, see Section D.2 (iv).

photosynthesis, are not affected by temperature. From Fig. 5.2 it can be seen that the light phase involves the production of reducing power and ATP whilst the dark phase involves the use of the reducing power and ATP in the conversion of CO_2 into carbohydrate.

B. PHOTOSYNTHETIC ORGANISMS

Photosynthesis is carried out by all organisms which possess any of the several species of chlorophyll. These include all those members of the plant kingdom, ranging from algae to higher plants, that are green, for their colour is due to the presence of chlorophyll. They also include many species which contain chlorophyll but, in spite of this, are not green because the chlorophyll is masked by other pigments (e.g. brown algae, red algae, copper beech trees). In addition some bacteria are also photosynthetic. These comprise the order Rhodospirillales which is divided into the suborders, Rhodospirillineae and Chlorobineae. The Rhodospirillineae are made up of two families, the Chromataceae (formerly known as the Thiorhodaceae or purple sulphur bacteria) and the

Rhodospirillaceae (formerly known as the Athio-rhodaceae or purple non-sulphur bacteria). The Chlorobineae contain only one family, the Chloro-biaceae (formerly known as the Chlorobacteriaceae or green sulphur bacteria).

Since photosynthetic bacteria do not fall within the scope of this book, bacterial photosynthesis will not be dealt with, save to say that, although it is described by the scheme shown in Fig. 5.2, it differs from that carried out by members of the plant kingdom in a number of ways. It utilizes, for instance, only one pigment system instead of two. The chlorophylls in that pigment system are differ-ent; they are species of bacteriochlorophyll and/or chlorobium chlorophyll. The nature of the light phase is also different. In some species the reductant (X^- in Fig. 5.2) is generated by using some of the ATP, formed in the light phase, to reverse electron flow in the respiratory electron transport chain (or an electron transport chain using some of the components of the respiratory chain). In other species X^- is generated in a plant-like manner but using electron donors other than water as the ultimate electron source. Moreover, photosynthetic bacteria do not produce oxygen as an end product of photosynthesis.

C. THE LIGHT PHASE OF PHOTOSYNTHESIS

1. Introduction

As can be seen from Fig. 5.2 the light phase of photosynthesis involves the use of light energy to generate ATP and reducing power. The reducing power in photosynthetic organisms of the plant kingdom is NADPH which is the reduced com-ponent of the redox system $NADP^+/NADPH$ ($E_0' = -0.32$ V). The electrons required to convert $NADP^+$ into NADPH in the light phase are derived from water. Water is the reduced component of the redox system $\frac{1}{2}O_2/H_2O$ ($E_0' = +0.82$ V); water is thus the electron donor of Fig. 5.2. When electrons are taken from water to reduce $NADP^+$ oxygen, the oxidized component of the $\frac{1}{2}O_2/H_2O$ system, is generated. Oxygen is therefore produced as a by-product of photosynthesis in the plant kingdom.

In order to transfer electrons from water to $NADP^+$ energy is required. The magnitude of this energy requirement can be calculated from eq. (5.6) (see also Chapter 2).

$$\Delta G^{\circ\prime} = -nF\Delta E_0' \qquad (5.6)$$

where n = the number of moles of electrons transfer-red per mole of reactant, which in this case equals 2 because $NADP^+/NADPH$ is a two-electron redox system,

F = the Faraday = 96,487 coulombs mol^{-1},

$\Delta E_0' = E_0'$ of the oxidizing redox system minus the E_0' of the reducing redox system which in this case equals $-0.32 - (+0.82) = -1.14$ V,

$$\therefore \Delta G^{\circ\prime} = -2 \times 96,487 \times (-1.14)$$
$$= +219,990 \text{ J mol}^{-1} \text{ or } +220 \text{ kJ mol}^{-1}$$
$$\text{(approx.)}$$

Thus about 220 kJ of energy are required to produce 1 mole of NADPH utilizing electrons derived from water. The energy that is used for this purpose takes the form of photons (or quanta) of red light (~ 700 nm). One mole of 700 nm photons (i.e. an Einstein) has energy equal to about 171 kJ (see Table 5.2). Clearly at least 2 moles of photons of 700 nm light are required to produce 1 mole of NADPH; in fact as will be seen later (Section C.6) 4 moles are required, giving a stoichiometry of 4 photons per NADPH generated (and per H_2O used up and $\frac{1}{2}O_2$ evolved). One reason for the apparently excessive photon requirement is that ATP has to be generated as well as NADPH. To generate 1 mole of ATP by phosphorylating ADP about 30.5 kJ of energy are required.

The flow of electrons from water to $NADP^+$ is not direct but passes through a number of redox systems which are interposed between them. Two pigment systems known as Pigment System I (PS I) and Pigment System II (PS II) are inserted into this electron transport chain at two different points and serve to introduce the energy which is required to force the electrons up the electrochemical gradient.

The two pigment systems are the engines which drive the light phase of photosynthesis. Both are composed of an array of pigment molecules, usually called the antenna pigments, which harvest incom-ing photons of light and then pass the resulting energy, in the form of molecular excitons, on to a 'reaction centre' where a special form of chlorophyll is raised to an excited state (Chl*). In this state it is extremely reactive and is a sufficiently powerful reducing agent to pass an electron to (i.e. to reduce) the oxidized form of redox system (A/A^-). The A^- which results then passes the electron on, via other components of the electron transport chain, ultimately to $NADP^+$. Having donated an electron, the special form of chlorophyll is now a cationic free radical (Chl^+). It is reduced back to ground-state chlorophyll by receipt of an electron from the reduced form of another redox system (D/D^-); the resulting D is reduced back to D^- by an electron ultimately derived from water (see Fig. 5.3).

2. Photosynthetic pigments

The pigments involved in the light phase of photosynthesis in members of the plant kingdom fall into three well-defined classes: the chlorophylls, the carotenoids and the phycobilins. All three classes occur in the photosynthetic pigment system as chromoproteins, that is to say, as pigment–protein complexes. In the case of chlorophylls and carotenoids the pigment–protein linkage is relatively weak, being composed of non-covalent bonds. These bonds are easily broken; hence the chlorophylls and carotenoids can be extracted simply by macerating plant tissue with an organic solvent such as acetone or alcohol. However, the linkage between phycobilins and protein is covalent; thus phycobilins occur in the photosynthetic tissue as distinct molecular species called phycobiliproteins. The phycobiliproteins are water soluble and are readily extracted with water or dilute salt solutions from macerated tissue. The cleavage of the phycobilin–protein linkage requires quite vigorous hydrolysing conditions.

(i) CHLOROPHYLLS

Four different chlorophylls have been extracted from plants; their structures and distribution are shown in Fig. 5.4 and Table 5.1 respectively.

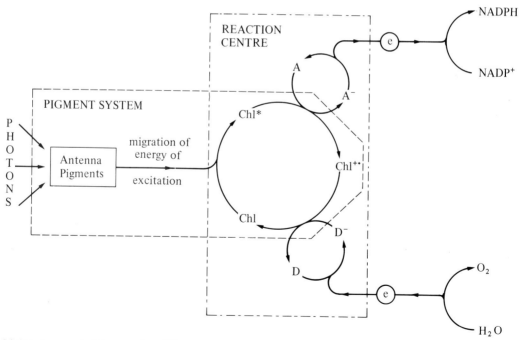

Fig. 5.3. Simple concept of the operation of Pigment Systems I and II. (PS I and PS II are inserted at two different points in the chain of redox systems carrying electrons from H_2O to $NADP^+$. They each have a different acceptor (A/A^-) and donor (D/D^-) redox system. The figure makes no attempt to describe the stoichiometry or the mechanism of the redox reactions involved; in fact, two electrons are required to reduce each molecule of $NADP^+$ to NADPH and these are transferred in the form of a hydride ion (see Chapter 6, Section B.2[ii](a)) from the redox system that is the immediate reductant of $NADP^+$.)

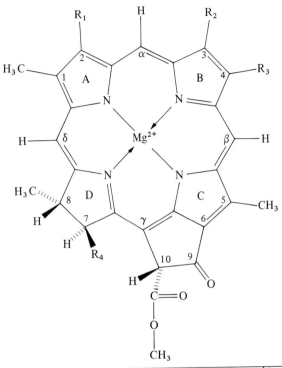

Textual No.	Chlorophyll Species	R_1	R_2	R_3	R_4	Loss of C-7 & C-8 hydrogens giving a 7,8 double bond
I	a	$-CH = CH_2$	$-CH_3$	$-CH_2 CH_3$	X (see below)	No
II	b	$-CH = CH_2$	$-CHO$	$-CH_2 CH_3$	X	No
III	c_1	$-CH = CH_2$	$-CH_3$	$-CH_2 CH_3$	$-CH = CH.COOH$	Yes
IV	c_2	$-CH = CH_2$	$-CH_3$	$-CH = CH_2$	$-CH = CH.COOH$	Yes
V	d	$-CHO$	$-CH_3$	$-CH_2 CH_3$	X	No

$$X = -CH_2 CH_2 \underset{\underset{O}{\|}}{C} - O - CH_2 \quad \text{---}$$

$$\underbrace{\hspace{3cm}}_{\text{propionic acid}} \quad \underbrace{\hspace{5cm}}_{\text{phytyl}}$$

Fig. 5.4. Structure of plant chlorophylls. (The arrangement of double bonds in the macrocyclic ring of chlorophyll is not fixed. Resonance of the conjugated double-bond system makes it possible to write down a number of mesomeric structures, one of which is shown above. The positions of the two co-ordinate bonds ($\rightarrow$) and the two covalent bonds between the Mg^{2+} and pyrrole nitrogens, whilst being appropriate to the mesomer shown, are also not fixed for the same reason. The two positive charges on the Mg^{2+} are balanced by two negative charges which are shared amongst the four pyrrole nitrogens; thus chlorophyll is electrically neutral.)

Chlorophyll a (I) is found in all photosynthetic members of the plant kingdom. Moreover, within each species it is the most abundant chlorophyll. Chlorophyll b (II) is almost as widespread, but is absent from all the algae except the Chlorophyceae and the Euglenophyceae. Two types of chlorophyll c, namely c_1 (III) and c_2 (IV), are known. These occur together in the Phaeophyceae, Chrysophyceae and Bacillariophyceae but only chlorophyll c_2 is present in the Cryptophyceae. Chlorophyll

Table 5.1. Distribution of the major photosynthetic pigments in the plant kingdom

Organism	Chlorophylls					Phycobiliproteins		Carotenoids	
	a	b	c_1	c_2	d	Phycoerythrin	Phycocyanin	Carotenes	Xanthophylls
Higher plants, pteridophytes and bryophytes	+	+	−	−	−	−	−	β-carotene α-carotene	lutein violaxanthin neoxanthin
Chlorophyceae (green)	+	+	−	−	−	−	−	β-carotene	lutein violaxanthin neoxanthin
Euglenophyceae	+	+	−	−	−	−	−	β-carotene	neoxanthin diadinoxanthin
Phaeophyceae (brown)	+	−	+	+	−	−	−	β-carotene	fucoxanthin violaxanthin
Chrysophyceae (yellow-brown)	+	−	+	+	−	−	−	β-catotene	fucoxanthin
Xanthophyceae (yellow-green)	+	−	−	−	−	−	−	β-carotene	neoxanthin diadinoxanthin
Bacillariophyceae (diatoms)	+	−	+	+	−	−	−	β-carotene	neoxanthin diadinoxanthin fucoxanthin
Cryptophyceae	+	−	−	+	−	+	+	α-carotene β-carotene	alloxanthin
Rhodophyceae (red)	+	−	−	−	+	+ + +	+[c]	α-carotene β-carotene	lutein zeaxanthin
Cyanophyceae[a] (blue-green)	+	−	−	−	−	+	+ + +[c]	β-carotene	echinenone myxoxanthophyll zeaxanthin
Prochlorophyta[b] (genus: Prochloron)	+	+	−	−	−	−	−	β-carotene	zeaxanthin

(The bracket spanning the algal rows from Chlorophyceae to Prochlorophyta is labelled **Algae**.)

+ and − = presence and absence respectively.

[a] The taxonomic position of the blue-green algae is at present in a state of flux: some authors believe that they should be classed as bacteria because they are prokaryotic and therefore refer to them as cyanobacteria; other authors believe that they should remain within the algal classification because their mechanism of photosynthesis is more plant-like than bacteria-like (they produce oxygen and have two photosystems). Being prokaryotes they do not have chloroplasts.

[b] Prochloron constitutes the only known genus within the new algal division Prochlorophyta. It is symbiotically associated with colonies of didemnid ascidians and lives in warm marine habitats. It is prokaryotic but its photosynthetic mechanism is plant-like in that it evolves oxygen and has two photosystems which are built into the membranes of paired thylakoids.

[c] Allophycocyanin is also present in small quantities.

d (V) has been extracted from a number of species of Rhodophyceae but whether it exists in these organisms *in vivo* is still uncertain.

(ii) CAROTENOIDS

Many plant carotenoids are known. Although they are important photosynthetic pigments they do have other unrelated functions. Thus carotenoids are to be found (i) in non-photosynthetic plant species, e.g. fungi, (ii) in non-photosynthetic tissues or organelles in photosynthetic species, e.g. the petals, anthers and pollen of some flowers, the eyespot of *Euglena* spp.

The structures and distribution of the main carotenoids concerned in photosynthesis are shown in Fig. 5.5 and Table 5.1 respectively. Higher plants have the carotenes, α- and β-carotene (VII and VI) and the xanthophylls, lutein (IX), violaxanthin (X) and neoxanthin (XII). The limited information available suggests that pteridophytes and bryophytes have essentially the same carotenoids as

Textual
No.

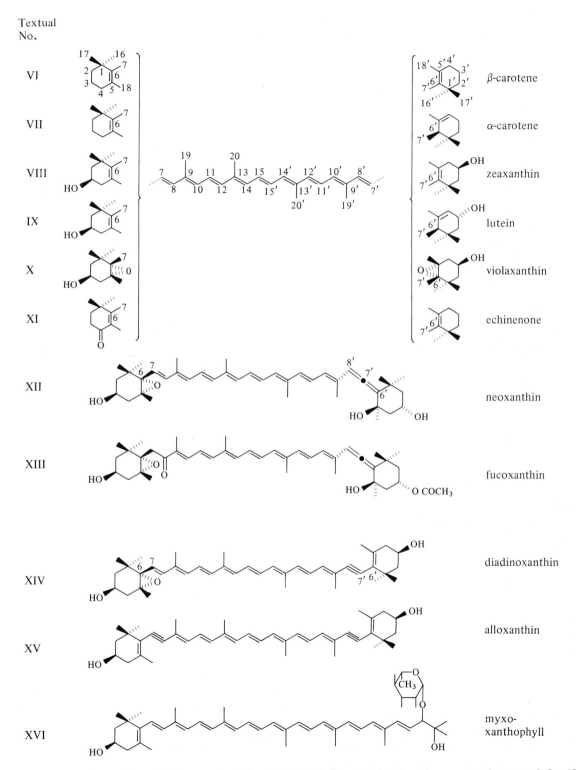

Fig. 5.5. Structures of some of the main plant carotenoids (see Table 5.1). Carotenoids VI–XI have the same central segment; Cs 7 and 7′ are bonded to Cs 6 and 6′ of the left-hand and right-hand rings respectively, at the same level in the figure.

higher plants. The Chlorophyceae have the typical higher plant carotenoids. The rest of the algal families have quite a variety of carotenoids. Notable amongst these are (i) fucoxanthin (XIII), which is present in such quantity in Phaeophyceae that it gives them their characteristic brown coloration, and (ii) the carotenoid glycosides which are characteristic of the blue-green algae, e.g. myxoxanthophyll (XVI) the 2'-O-rhamnoside of the aglycone, myxol.

(iii) PHYCOBILINS

The phycobilins, present *in vivo* as phycobiliproteins, occur in three algal families, the Rhodophyceae, Cyanophyceae and Cryptophyceae. The phycobiliproteins are divided into two main groups on the basis of their colour, the red phycoerythrins and the blue phycocyanins. The distribution of these compounds is shown in Table 5.1. The structures of the corresponding bilins, phycoerythrobilin (XVII) and phycocyanobilin (XVIII) and the phycoerythrobilin–protein linkage (XIX) are shown in Fig. 5.6. It should be noted that although the phycobilins are called 'open tetrapyrroles' and are usually depicted as linear structures to contrast them with the 'closed tetrapyrrolic' or porphyrin structure of the chlorophylls, the true shape of the molecule is more accurately represented by the 'open circle' shown in structure XIX.

Phycoerythin and phycocyanin are composed of two different protein subunits, designated α (MW $\sim 19,000$) and β (MW $\sim 21,000$), which occur in a 1:1 stoichiometric ratio. Each subunit carries covalently bound phycobilin. Depending upon its source a subunit may carry 1–4 molecules of phycobilin. These are usually of one type only (i.e. phycoerythrobilin or phycocyanobilin) giving a red or a blue subunit. However, some subunits have both types of phycobilin, usually with one type predominating. *In vivo* the subunits occur as aggregates of $(\alpha,\beta)_n$ where n is commonly 3 or 6 giving MWs of $\sim 134,000$ and $\sim 268,000$ respectively.

Allophycocyanin, so named because it was originally thought to be a special form of phycocyanin, is present in small amounts in the Rhodophyceae and Cyanophyceae. Unlike the other phycobiliproteins it has only one type of protein subunit (MW $\sim 15,500$); each subunit carries one molecule of phycocyanin. Six subunits are aggregated together *in vivo* giving particles of MW $\sim 100,000$.

In the algal cells the phycobiliproteins are aggregated together in special granules called phycobilisomes which occur in a regular array on the surface of the thylakoid membranes.

Chlorophyll a is the most important photosynthetic pigment for it is present in all photosynthetic members of the plant kingdom. The other forms of chlorophyll (i.e. b, c_1, c_2 and d), the carotenoids and the phycobilins are frequently called accessory pigments.

3. Absorption spectra of the photosynthetic pigments

The three classes of photosynthetic pigment absorb light of different wavelengths; the chlorophylls absorb in the blue (~ 450 nm) and red (650–700 nm) regions of the visible spectrum whilst the carotenoids and phycobilins absorb in the ranges 400–500 nm and 500–650 nm respectively. The spectra shown in Fig. 5.7 are of representative pigments taken after extraction, purification and solution in an appropriate solvent.

It is found, however, that the absorption maxima of many of these pigments are not the same *in vivo* as they are after extraction from the tissue. This is particularly true of the chlorophylls; chlorophyll a, for example, has a single sharp absorption peak in the red region of the spectrum after extraction and solution in an appropriate solvent (662 nm in diethyl ether but the value depends upon the solvent) whilst *in vivo* there is a much more diffuse red peak which has been resolved by curve analysis into several components (see Fig. 5.8). Each of these components represents the absorption spectrum of a different form of chlorophyll a. There now appear to be at least six different species of chlorophyll a *in vivo*, namely chlorophylls $a_{700}, a_{692}, a_{684}, a_{677}, a_{670}$ and a_{662}, present in markedly different amounts. Curve analysis (Fig. 5.8) also shows that there may be two different forms of chlorophyll b *in vivo*; the most abundant form absorbs in the red region at 650 nm whilst the minor form absorbs at 640 nm. Chlorophyll c_2 has a single red absorption peak *in vivo* at 633 nm. Carotenoids also have different

Textual No

XVII Phycoerythrobilin

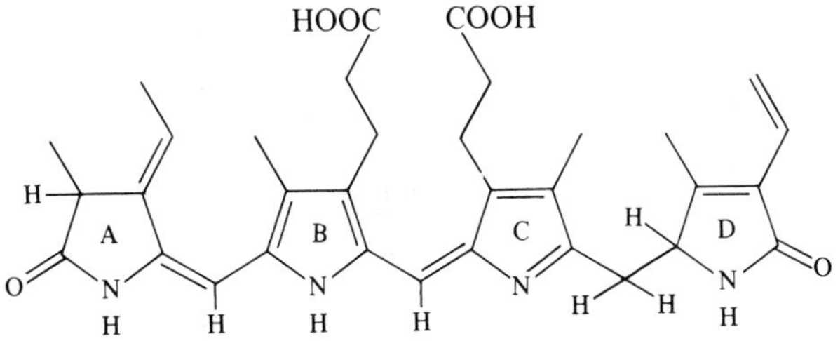

XVIII Phycocyanobilin

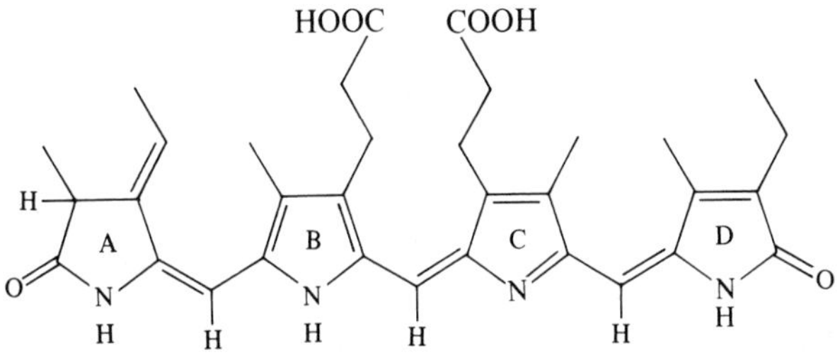

XIX

cysteine residue Phycoerythrobilin-protein complex (a phycoerythrin)

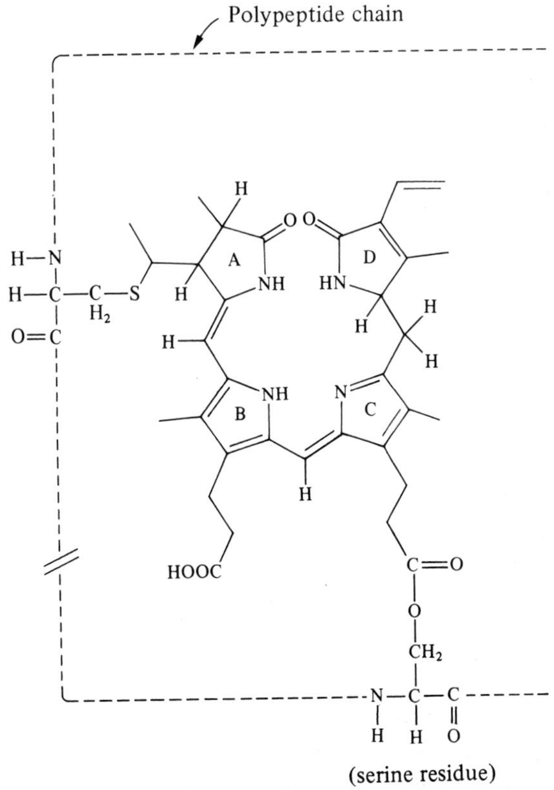

(serine residue)

Fig. 5.6. Structures of phycobilins.

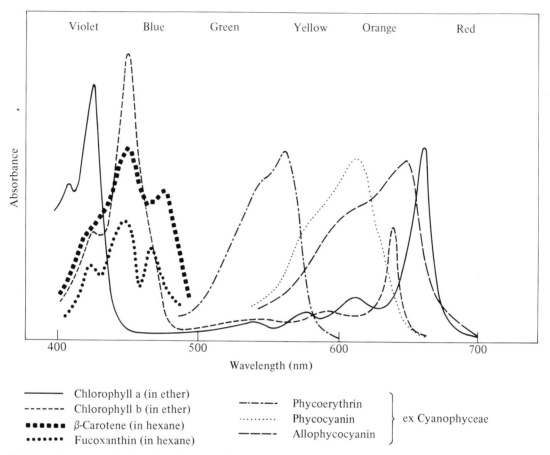

Fig. 5.7. Absorption spectra of selected members of the three groups of photosynthetic pigments, the chlorophylls, the carotenoids and the phycobilins.

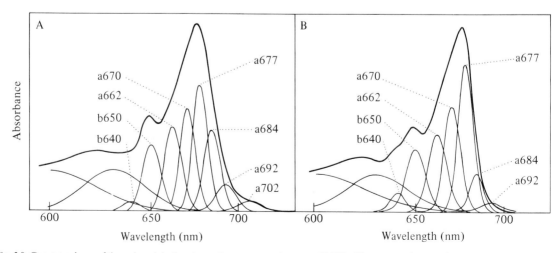

Fig. 5.8. Curve analyses of the red peak in the absorption spectra, taken at $-196°C$, of fragments of stroma lamellae (spectrum A) which constitute a chloroplast fraction enriched in PS I and of fragments of grana lamellae (spectrum B) which constitute a chloroplast fraction enriched in PS II (adapted from French, Brown and Lawrence, 1972).

absorption maxima. Fucoxanthin, for example, which has maxima at 427, 450 and 476 nm in hexane, has an *in vivo* absorption in the range 500–590 nm.

4. Pigment systems I and II

There is now a great deal of evidence to indicate that the photosynthetic pigments are organized into two pigment systems, PS I and PS II, and that it is as part of these systems that they participate in the light phase of photosynthesis. According to the classical view PS I and PS II are both functionally and physically distinct. However, there are some indications that this separation may not be absolute; there is, for example, evidence that some of the light harvested by PS II may 'spill' over to PS I (but not *vice versa*).

A partial separation of PS I and PS II has been achieved by disrupting chloroplast preparations either mechanically (e.g. by forcing them through a needle valve) or with detergents (e.g. dilute solutions of digitonin or Triton X-100). Centrifugation of the resulting homogenate at 3000g yields a clear, green supernatant which can be separated by centrifugation at 60,000g on a 10–50% (w/v) sucrose density gradient into an upper, light fraction and a lower, heavy fraction. The former has been shown to be composed mainly of fragments of stroma thylakoid membranes which constitute a fraction enriched in PS I. The latter fraction is composed of fragments of the granal stack membranes which, although they contain both PS I and PS II, nevertheless constitute a fraction enriched in PS II. Pure PS II fractions have not yet been obtained.

Analysis of the light (PS I-containing) fraction shows that it contains about 200 antenna (light harvesting) chlorophylls, about fifty carotenoids (mainly carotenes), one cytochrome f, one plastocyanin, two cytochrome b_{563} (= cytochrome b_6) and one or two membrane-bound ferredoxin molecules per molecule of chlorophyll a_{700} (= P700). The non-pigment components of this fraction are members of the chain of redox systems which carry electrons in the light-phase process; moreover, they are the systems which are closest to PS I on both the electron donor and recipient side. PS I itself appears

to be composed of units each with one molecule of P700, constituting the reaction centre (see Section C.1 and Fig. 5.3), and about 250 antenna pigments. Curve analysis of the red peak of the absorption spectrum of this fraction taken at −196°C (Fig. 5.8, A) indicates that the antenna pigments include chlorophylls a_{692}, a_{684}, a_{677}, a_{670}, a_{662}, b_{650} and b_{640}. Other work, however, suggests that there is virtually no chlorophyll b in PS I. For instance, gel electrophoresis of sodium dodecyl-sulphate (SDS) extracts of chloroplasts separates a chlorophyll–protein complex (known as CPI) which contains P700 and chlorophyll a but no chlorophyll b and appears to be the heart of PS I. This complex has a MW of 110,000 and has 14 chlorophyll molecules built into it. It is clearly essential for the operation of the light phase because it is lacking in *Scenedesmus* mutant 8, a mutant which cannot live photosynthetically. Since PS I absorbs light maximally in the red at wavelengths longer than 680 nm, it is likely that chlorophylls a_{677}, a_{684} and a_{692} along with the carotenes constitute the bulk of the antenna pigments in this system, handing the energy of excitation on to P700 at the reaction centre.

Analysis of the heavy (PS II-enriched) fraction shows that it contains about 200 antenna chlorophylls, about fifty carotenoids (mainly xanthophylls), an unidentified primary electron donor (Z), an unidentified primary electron acceptor (Q), about four plastoquinones, two cytochrome b_{559} molecules and six atoms of manganese per molecule of chlorophyll a_{680} (= P680) (or chlorophyll a_{690} = P690 in some species). Again the non-pigment components are members of the photosynthetic electron transport chain. PS II itself appears to consist of units each with one molecule of P680 (or P690), constituting the reaction centre, and about 250 antenna molecules. Curve analysis of the red peak of the absorption spectrum of this fraction taken at −196°C (Fig. 5.8, B) is not particularly informative about the identity of the antenna chlorophylls of PS II because the fraction also contains PS I; however, the virtual absence of a 700-nm absorption peak clearly shows that P700 is only to be found in PS I.

Other work suggests that virtually all of the chlorophyll b and xanthophylls of higher plants and green algae are not present in PS II but occur in another pigment–protein complex known as the 'light harvesting chlorophyll protein' (LHCP). The LHCP, which also contains some chlorophyll *a*, appears to be capable of transferring the energy of the light photons it absorbs to both PS I and PS II and would therefore be responsible for the light 'spill over' mentioned earlier. PS II thus consists of chlorophylls a_{662}, a_{670} and a_{677} acting as antenna pigments handing energy on to chlorophyll a_{680} at the reaction centre. A chlorophyll–protein complex consistent with this composition and quite distinct from CPI and LHCP has been found by gel electrophoresis of SDS-extracts of chloroplasts. The chloroplasts of plants grown at relatively low light intensities (> 3 klx) are particularly rich in LHCP whilst those of plants grown at much higher light intensities (~ 20 klx) have much less. This suggests that at low light intensities the antenna pigments of PS I and PS II are incapable of absorbing enough light photons to maintain the photosynthetic rate and have to be assisted by LHCP which is therefore present in quantity: on the other hand, at reasonably high light intensities they require little assistance and consequently have much less LHCP.

A protein component of LHCP has been shown to be coded by a nuclear gene and synthesized by cytoplasmic ribosomes. This contrasts with the CPI protein which is coded by a chloroplastidic gene and synthesized by chloroplastidic ribosomes. If the protein of the PS II chlorophyll–protein complex also proves to be coded and synthesized within the chloroplast, the endosymbiont hypothesis for the evolutionary origin of the chloroplast would suggest that PS I and PS II originated from the blue-green algal 'invader' and the LHCP was developed by the invaded host cell sometime later to boost light trapping efficiency; this suggestion also fits in with the fact that blue-green algae have chlorophyll *a*, as do PS I and PS II of higher plants, but no chlorophyll *b* which appears to be confined to LHCP in higher plants.

In Phaeophyceae, Chrysophyceae, Bacillariophyceae and Cryptophyceae chlorophyll c_1 and/or c_2 apparently take over the role of chlorophyll *b*.

In Rhodophyceae and Cyanophyceae the phycobilins present in the phycobilisomes have a light-harvesting role supplementing that of the antenna pigments.

The location of PS I and PS II in the thylakoid membranes of the chloroplast is not yet clear. The stroma thylakoid membranes are particularly rich in PS I. The thylakoid membranes of the granal stack contain both PS I and PS II. The arrangement of PS I and PS II within the membranes is uncertain. For some time it was thought that PS I and PS II were asymmetrically distributed across the membrane with PS I located in the outer half and PS II in the inner half. Support for this comes from a combination of the results of freeze-fracture electron micrographs of chloroplast membranes and studies of the photochemical activity and ultrastructure of particles obtained by mechanical or detergent disruption of chloroplasts. The former reveal 11 nm diam. and 17 nm diam. particles located in the hydrophobic faces of the outer and inner halves of the membrane respectively whilst the latter strongly suggests that PS I is associated with the smaller particles and PS II with the larger. However, how the freeze-fracture electron micrographs relate to the structure of the thylakoid membrane is controversial; it has been suggested, for instance, that the 11 nm and 17 nm diam. particles might arise from two different sets of membrane proteins, both of which span the membrane, that are sheared during the freeze-fracture process. Later work has indicated that PS II (but not PS I) spans the membrane. Moreover, other studies emphasize the fact that the membrane has a fluid matrix which can allow lateral movements of macromolecular complexes, presumably including those seen by electron microscopy.

The nature of P700 and P680 (or P690) which lie at the heart of the PS I and PS II reaction centres respectively are under active investigation. Each reaction centre is presumed to be a complex of the pigment plus the redox system which is the immediate donor of electrons to it and the redox system which is the immediate recipient of electrons from it. To date such reaction centres have not been isolated. However, results from electron paramagnetic resonance studies indicate that the unpaired spin [i.e. the unbalanced spin due to the unpaired electron in oxidized P700 (i.e. P700$^+$)] is shared by two special chlorophyll *a* molecules. This indicates that P700 is composed of two molecules of chlorophyll *a* linked together in some way; these have been called the 'chlorophyll special pair' (Chl_{sp}). The precise nature of Chl_{sp} is not known. However, several structural models have been proposed. The most satisfactory of these, that of Shipman and his colleagues, suggests that two chlorophyll *a* molecules are held together by two molecules of a bifunctional ligand which must have (i) a lone pair of electrons with which to form a co-ordinate bond with Mg^{2+} of chlorophyll *a* and (ii) the ability to form hydrogen bonds. Ligands of the general type R—X—H, where R=H or an alkyl group and X=O, NH or S, are suitable. Thus water, H—O—H, could link the two chlorophyll *a* mole-cules together, with O providing the lone pair for co-ordination with Mg^{2+} of one of the chlorophyll *a* molecules and H bonding to the carbonyl oxygen of the other. This arrangement is shown in Fig. 5.9. It sets the porphyrin macrocyles at a π–π stacking distance of 0.36 nm. This not only brings the π molecular orbitals of the conjugated double bond systems of the two chlorophyll molecules into contact but also allows them to overlap optimally. It is the extent of this π overlap that allows the unpaired spin to be delocalized over (i.e. shared between) the two chlorophyll molecules that con-stitute the special pair cationic free radical (i.e. P700$^{+\cdot}$). An interesting feature of this model is that it does not rule out the possibility that the two crosslinking bifunctional ligands could be located in

a protein. Ligands of the general class R—X—H occur in amino acid side chains; thus it is possible that the OH groups of serine or threonine, the NH_2 group of lysine or the SH group of cysteine could be instrumental in forming the Chl_{sp} *in vivo*.

5. Mechanism of pigment system function

Having considered the make-up of PS I and PS II we are now in a position to consider the question of how they trap light energy and pass it on to other molecules. To begin with it will be useful to review some relevant physicochemical aspects of light and light absorption.

Light has both wave and particle attributes; it is transmitted in waves and is absorbed and emitted in packets called photons or quanta. The energy of a photon is equal to the product of its frequency (v) and Planck's constant (h) (6.6262×10^{-34} J s) [eq. (5.7)]:

$$E_{photon} = hv \quad \text{where}$$

$$v = \frac{c \text{ (velocity of light} = 2.9979 \times 10^8 \text{ m s}^{-1})}{\lambda \text{ (wavelength in m)}}$$

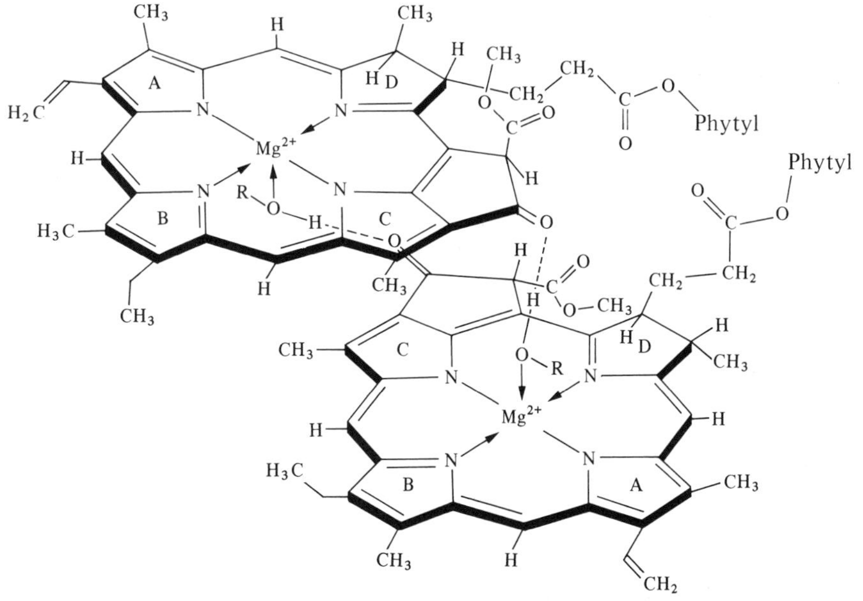

Fig. 5.9. Proposed structure of the chlorophyll *a* (special pair) which appears to be the P700 of PS I (*in vivo* R may be the residue of an amino acid which is built into a protein component of PS I). (Redrawn from Shipman, L. L., Cotton, T. M., Norris, J. R. and Katz, J. J., 1976.)

$$E_{photon} = \frac{hc}{\lambda}$$

$$= \frac{6.6262 \times 10^{-34} \times 2.9979 \times 10^8}{\lambda \text{ (in nm)} \times 10^{-9}}$$

$$= \frac{1.9865 \times 10^{-16}}{\lambda \text{ (in nm)}} \text{ J} \qquad (5.7)$$

The energy of a photon is often also expressed in electron volts (eV). Since 1 eV has an energy of 1.6022×10^{-19} J it follows that the energy of a photon expressed in eV is given by eq. (5.8):

$$E_{photon} = \frac{1.9865 \times 10^{-16}}{\lambda \text{ (in nm)}} \times \frac{1}{1.6022 \times 10^{-19}}$$

$$= \frac{1.2399 \times 10^3}{\lambda \text{ (in nm)}} \text{ eV} \qquad (5.8)$$

According to Einstein's law of photochemical equivalence a molecule can react, in a photochemical reaction, only after absorbing one photon. It follows, therefore, that one mole can react only after absorbing 6.0222×10^{23} photons (6.0222×10^{23} is Avogadro's number (N) which is the number of molecules in a mole of any substance). Now the energy of 6.0222×10^{23} photons is equal to $6.0222 \times 10^{23} \, hv$ or $6.0222 \times 10^{23} \, hc/\lambda$; this quantity of energy is called an Einstein. By definition, therefore, an Einstein is the amount of radiant energy of the appropriate wavelength that must be absorbed by 1 mole of a substance before it can react in a photochemical reaction. The energy of an Einstein is given by eq. (5.9).

$$E_{Einstein} = \frac{Nhc}{\lambda}$$

$$= \frac{6.0222 \times 10^{23} \times 6.6262 \times 10^{-34} \times 2.9979 \times 10^8}{\lambda \text{ (in nm)} \times 10^{-9}}$$

$$= \frac{1.1963 \times 10^8}{\lambda \text{ (in nm)}} \text{ J} \qquad (5.9)$$

The relationships between wavelength, frequency, etc., and energy of light involved in photosynthesis are given in Table 5.2. It will be noted that the energy of a photon (or an Einstein) varies directly with the frequency (v) and inversely with the wavelength (λ); thus the greater the frequency and the shorter the wavelength, the larger the energy.

Since it is the function of the pigment systems to absorb light energy, we must now consider what happens when a pigment molecule absorbs a photon.

The individual atoms constituting a pigment molecule are held together by covalent bonds (usually single or double) which are formed by the overlapping of atomic orbitals of adjacent atoms to form molecular orbitals containing two electrons which are identical in every way save that they have opposite spins. In the case of a carbon–carbon single bond the overlap is between two sp$_3$ hybrid atomic orbitals each of which originally contained one electron. The resulting molecular orbital is symmetrical about the C–C axis, and contains two electrons of opposite spins; it is called a sigma (σ) bond [see Fig. 5.10(a)]. In the case of a carbon–carbon double bond the overlap is between two sets of atomic orbitals. One set are sp$_2$ hybrids which overlap to give a σ-bond whilst the other set are unhybridized p-orbitals which overlap to give a π-bond. Like the σ-bond, the π-bond contains two electrons of opposite spins. Thus the C=C bond is composed of a σ-bond and a π-bond [see Fig. 5.10(b)]; this is also true of other types of double bond.

According to molecular orbital theory the combination of x atomic orbitals gives rise to x molecular orbitals. Thus the linear combination of two sp$_3$ or of two sp$_2$ hybrid atomic orbitals gives rise to two molecular orbitals which are called the bonding (σ) and antibonding (σ^*) molecular orbitals [see Fig. 5.10(c)]. Similarly the linear combination of two p atomic orbitals gives rise to a bonding (π) and an antibonding (π^*) molecular orbital [see Fig. 5.10(d)]. The antibonding molecular orbitals (σ^* and π^*) have higher electronic energy levels than their bonding counterparts.

Atomic orbitals which are not involved in bond formation (i.e. combination to form molecular orbitals) are described as non-bonding (n) orbitals; these are exemplified by the lone electron pairs of oxygen and nitrogen. The electrons occupying n-orbitals generally have energies between that of the highest bonding and lowest antibonding molecular orbital. The relative electronic energies of these orbitals are shown in Fig. 5.11.

In almost every stable organic molecule there is an even number of electrons. When the molecule is

Table 5.2. Relationship between spectroscopic units and energy

Wavelength (nm)[a]	Wavenumber[b] (cm^{-1})	Frequency (THz)[c]	Energy of photon		Einstein[f] (kJ)[g]
			(aJ)[d]	(eV)[e]	
700	14,286	428.3	0.2838	1.7712	170.9
692	14,451	433.2	0.2871	1.7917	172.9
690	14,493	434.5	0.2879	1.7969	173.4
684	14,620	438.3	0.2904	1.8126	174.9
680	14,706	440.9	0.2921	1.8233	175.9
677	14,771	442.8	0.2934	1.8314	176.7
670	14,925	447.5	0.2965	1.8505	178.6
662	15,106	452.9	0.3001	1.8729	180.7
650	15,385	461.2	0.3056	1.9074	184.0
640	15,625	468.4	0.3104	1.9372	186.9
633	15,798	473.6	0.3138	1.9587	188.9
615	16,260	487.5	0.3230	2.0160	194.5
580	17,241	516.9	0.3425	2.1377	206.3
560	17,857	535.3	0.3547	2.2140	213.6
500	20,000	599.6	0.3973	2.4797	239.3
475	21,053	631.1	0.4182	2.6102	251.8
450	22,222	666.2	0.4414	2.7552	265.8
400	25,000	749.5	0.4966	3.0996	299.1

[a] 1 nm = 10^{-9} m.
[b] Wavenumber = 1/wavelength (cm) = 1/wavelength (nm) $\times 10^{-7}$.
[c] 1 THz = 10^{12} Hz or 10^{12} cycles sec^{-1} (v = velocity of light (m s^{-1})/wavelength (nm) $\times 10^{-9}$ = 2.9979×10^8/wavelength (nm) $\times 10^{-9}$ = 2.9979×10^{17}/wavelength (nm)).
[d] 1 aJ = 10^{-18} J.
[e] An electron volt (eV) is the energy acquired by an elementary charge (= 1.6022×10^{-19} C) when it moves through a potential of 1 volt; since C × V = J, an eV has an energy of 1.6022×10^{-19} J.
[f] An Einstein is a unit of energy equal to one mole of photons, i.e. 6.0222×10^{23} photons = 6.0222×10^{23} × energy of one photon of a given wavelength; it is usually given in J.
[g] 1 kJ = 10^3 J.

in its lowest energy state, the ground state, the electrons are present, as pairs with opposite spins, in the bonding (σ, π) and non-bonding (n) molecular orbitals. When a photon of appropriate energy is absorbed an electron is promoted from a bonding (σ, π) or non-bonding (n) molecular orbital to an antibonding (σ^*, π^*) molecular orbital; the molecule is then said to be in an excited state. Figure 5.11 shows that four such promotions are possible; $\sigma \to \sigma^*$, $\pi \to \pi^*$, $n \to \sigma^*$ and $n \to \pi^*$. The energy of the photon bringing about any one of these transitions must be at least equal to the energy difference between the two orbitals involved, the general order of which is $\sigma \to \sigma^* > n \to \sigma^* > \pi \to \pi^* > n \to \pi^*$ (see Fig. 5.11). Since the energy of a photon is inversely proportional to its wavelength, the wavelength of a photon bringing about a $\sigma \to \sigma^*$ transition is shorter than that bringing about a $n \to \pi^*$ transition (see Table 5.3). The transitions shown in Table 5.3

require photons that have a wavelength shorter than that of the visible light (range 400–740 nm) which is absorbed by the photosynthetic pigments. It would appear, at first sight, that none of these transitions can explain the light-absorption characteristics of the photosynthetic pigments. However, it is found that the wavelength of photons bringing about the $\pi \to \pi^*$ transition becomes longer as the number of double bonds *in conjugation* in the molecule becomes greater (see Table 5.4). This results from the increase in the number of delocalized π-electrons which accompanies an increase in conjugation. The more double bonds there are in conjugation, the more π-orbitals there are in the system and the greater is the energy range from the lowest to the highest energy π-orbital; this is equally true of the π^*-orbitals. Since the average energy of the π-orbitals and of the π^*-orbitals does not change appreciably as conjugation increases it

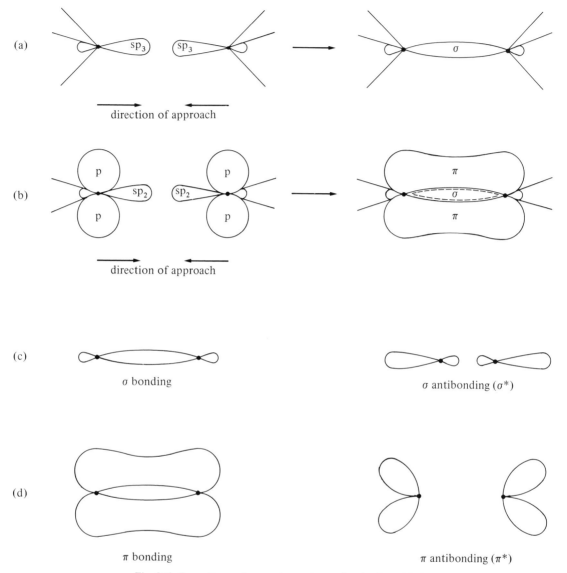

Fig. 5.10. Formation and approximate shape of molecular orbitals.

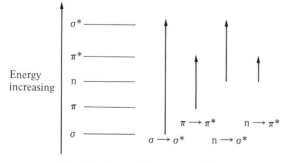

Fig. 5.11. Types of electron transitions.

follows that the energy of the highest energy π-orbital increases and that of the lowest energy π^*-orbital decreases with increasing conjugation (see Fig. 5.12). Thus the promotion of an electron from the highest energy π-orbital to the lowest energy π^*-orbital requires less energy as the number of conjugated double bonds increases. Consequently the photons bringing about these promotions have increasingly long wavelengths. By the time there are seven or more double bonds in conjugation the photons required for the $\pi \rightarrow \pi^*$ transitions are in the

Table 5.3. Energies of photons bringing about different types of electronic transitions

Compound	Bond or electron	Transition	Photon		
			Wavelength (nm)	Energy (J)	Energy (eV)
Ethane	C—C (σ bond)	$\sigma \rightarrow \sigma^*$	135	1.47×10^{-18}	9.18
Formaldehyde	$\bar{e}$ of a lone pair on the oxygen atom	$n \rightarrow \sigma^*$	160	1.24×10^{-18}	7.75
Ethylene	C=C (π bond)	$\pi \rightarrow \pi^*$	175	1.14×10^{-18}	7.09
Formaldehyde	$\bar{e}$ of a lone pair on the oxygen atom	$n \rightarrow \pi^*$	285	6.97×10^{-19}	4.35

N.B. The wavelengths of the photons bringing about the transitions shown above apply specifically to the compounds named and do not apply to all compounds.

Table 5.4. Relationship between absorption maxima and the number of conjugated double bonds

Carotenoid	Number of conjugated double bonds	Wavelength (nm) of max. absorption[a]
Butadiene[b]	2	217
Phytoene	3	286
Phytofluene	5	347
ζ-Carotene	7	400
Neurosporene	9	439
Lycopene	11	472
3,4-Didehydrolycopene	13	492
3,4,3',4'-tetrahydro-lycopene	15	510

[a]Carotenoid spectra taken in light petroleum.
[b]Butadiene is not a carotenoid.
Data taken from Davies, B. H. (1976) in *Chemistry and Biochemistry of Plant Pigments* (Goodwin, T. W., ed.), Vol. 2, p. 125, Academic Press, London and New York.

visible region of the spectrum. Since all the photosynthetic pigments have seven or more conjugated double bonds it is clear that $\pi \rightarrow \pi^*$ transitions account in great measure for their ability to absorb visible light.

The picture of the excited-state molecule that has been painted so far is now complicated by the fact that every electronic energy level, whether bonding, antibonding or non-bonding, is subdivided into vibrational energy levels. These are also quantized, requiring a quantum of energy to raise an electron from one vibrational energy level to the next

highest. However, the energy difference between adjacent vibrational energy levels is very much smaller than that between electronic energy levels, being 0.075–0.5 eV (equivalent to photons of wavelength 16,500–2500 nm which occur in the infrared region of the electromagnetic spectrum). The vibrational energy levels are caused by the atoms within a molecule vibrating back and forth with respect to each other. This vibration affects the distribution of the electrons about the atomic nuclei and consequently influences the electronic energy of the molecular orbital in question.

We saw earlier that when a photon of visible light is absorbed by a photosynthetic pigment an electron in the conjugated double bond system undergoes a $\pi \rightarrow \pi^*$ transition and the molecule is said to have been raised from its ground state to an excited state. The excited state is more energetic than the ground state by the amount of energy introduced by the absorbed photon. In the ground state molecule the electron which is to be excited in the $\pi \rightarrow \pi^*$ transition is one of a pair (with opposite spins) residing in the lowest vibrational energy level of the electronic energy level of the π molecular orbital in question. When the photon is absorbed the electron is promoted to one of the higher vibrational energy levels of the electronic energy level of the π^* molecular orbital. This promotion does not alter the direction of its spin. The time taken for this process is 10^{-15} sec. The promoted electron then drops down from the higher vibrational level to the lowest vibrational level of the π^* electronic energy level. This occurs in a stepwise manner from vibrational level

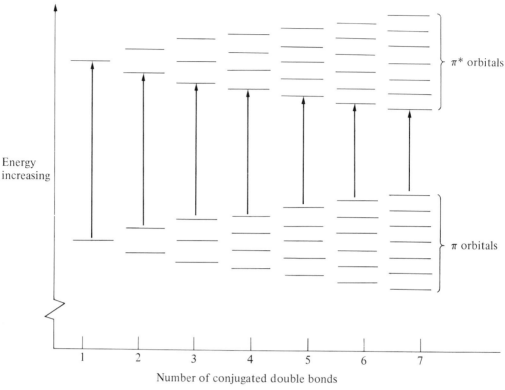

Energy increasing

Number of conjugated double bonds

Fig. 5.12. Effect of conjugation on the π and π^* energy levels. (The vertical arrows represent $\pi \rightarrow \pi^*$ transitions which decrease in energy as conjugation increases; thus the wavelengths of the photons effecting the transitions increase as conjugation increases.)

to vibrational level until the lowest is reached; it is called the vibrational cascade. A quantum of energy equal in size to the energy difference between adjacent vibrational levels is lost at each downward step of the cascade. These quanta are of infrared (or heat) energy. They are not lost by emission but are absorbed by other molecules which collide with the excited state molecule and are then distributed amongst the remaining molecules in the system by further collisions. The vibrational cascade is therefore a non-radiative process. It takes 10^{-9}–10^{-12} sec. The pigment molecule is now in the excited state. Since the promoted electron has not changed the direction of its spin during this process, its spin is opposite to that of the electron which has remained behind in the π molecular orbital. The spins of these two electrons therefore still cancel each other out in the excited state molecule just as they did in the ground-state molecule; the spins are said to be 'paired'. Now a molecule in which all the electron spins are paired has no magnetic moment

and is called a singlet. Thus in the $\pi \rightarrow \pi^*$ transition just described a ground-state singlet (S_0) has been converted into an excited state singlet.

It is possible for a molecule to absorb photons of different wavelengths (and therefore energies). Chlorophyll is a prime example, absorbing in the red and the blue regions of the visible spectrum. It can thus absorb a 'red photon' ($\lambda \sim 662$ nm; $E = 3.00 \times 10^{-19}$ J or 1.87 eV) or a 'blue photon' ($\lambda \sim 430$ nm; $E = 4.62 \times 10^{-19}$ J or 2.88 eV) and be converted into two different excited-state singlets. These are distinguished by referring to them as the first- and second-excited-state singlets, S_1 and S_2, of which S_1 is the least energetic. The processes showing the formation of S_1 and S_2 are shown in Fig. 5.13.

The lifetime of the excited-state singlet is very short. In the context of the excited state the term 'lifetime' is defined as the time required for the number of molecules in that state to decrease by $1/e$ (where $e = 2.71828$) which roughly equals 0.37 or

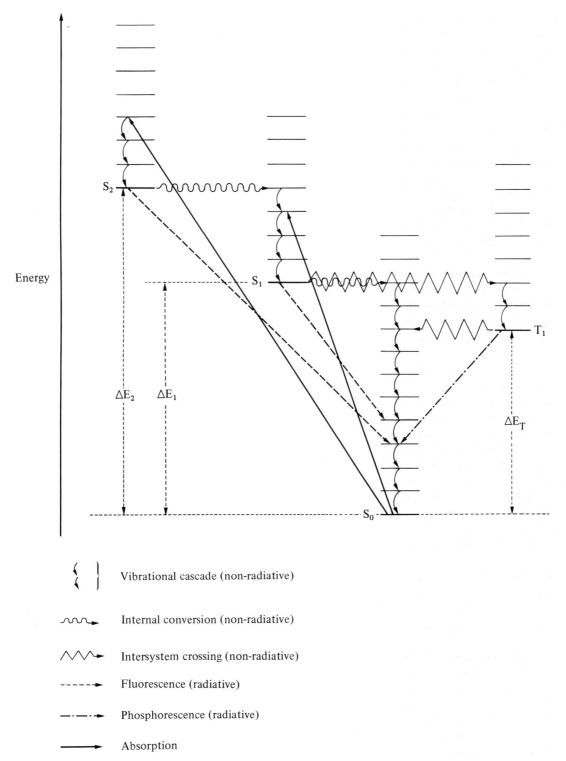

Vibrational cascade (non-radiative)

Internal conversion (non-radiative)

Intersystem crossing (non-radiative)

Fluorescence (radiative)

Phosphorescence (radiative)

Absorption

Fig. 5.13. Modified Jablonski diagram.

37%. This means that the electron in the lowest vibrational level of the π^* electronic energy level drops back to the lowest vibrational level of the π electronic energy level in a very short time, about 10^{-10}–10^{-6} sec. When this happens energy equal to ΔE_1 (or ΔE_2) in Fig. 5.13 is dissipated. There are several ways in which this can happen:

1. *By a combination of internal conversion and the vibrational cascade* (see Fig. 5.13): here the promoted electron in S_2 is transferred to an equivalent energy level in the set of vibrational levels associated with S_1. This is a radiationless process called internal conversion and occurs in 10^{-12}–10^{-10} sec. The electron then drops down the vibrational cascade to yield S_1; this is also a radiationless process. This is then followed by a second internal conversion from S_1 to an equivalent energy level in the S_0 set of vibrational levels and then a second vibrational cascade to yield S_0. In the overall process the dissipated energy (ΔE_2) is redistributed amongst the surrounding molecules.

2. *By fluorescence* (see Fig. 5.13): here the promoted electron in S_2 or S_1 drops back to one of the energy levels in the set of vibrational levels associated with S_0, emitting a photon as it does so; this process is called fluorescence. The electron then drops down the vibrational cascade to S_0. The energy of the photon emitted in fluorescence is less than that which formed the excited state by an amount equal to that lost in the excited state and ground state vibrational cascades. It therefore has a longer wavelength than the absorbed photon. Fluorescence occurs in 10^{-9}–10^{-5} sec.

3. *By phosphorescence* (see Fig. 5.13): here the promoted electron undergoes intersystem crossing. In this process the spin of the electron is reversed. This is a very unlikely event and occurs relatively rarely; it may be caused by the collision of the excited-state molecule with another molecule. The promoted electron and its former partner now have identical spins which give the molecule a magnetic moment which can be oriented in three different ways relative to arbitrarily chosen axes. The molecule thus has three states which are, however, only distinguishable in a strong magnetic field. These three states are, therefore, normally treated as

one which is termed the triplet state. Intersystem crossing normally involves some loss of energy due to vibrational cascade as is shown in the conversion of S_1 into T_1 in Fig. 5.13. The electron in T_1 cannot now return to S_0 until its spin is reversed once more. Since this occurs infrequently, the lifetime of the excited-state triplet is much longer than that of the excited-state singlet; it is usually in the range 10^{-5}–10^{-3} sec but in some cases may be as long as 10 sec. When eventually the electron in T_1 has its spin reversed it falls back to one of the energy levels in the set of vibrational levels associated with S_0, emitting a photon as it does so. This process is called phosphorescence. The electron then drops down the vibrational cascade to S_0. The energy of the photon emitted in phosphorescence is less than that of the photon which formed the equivalent excited singlet and is usually less than that of the photon emitted from that singlet by fluorescence because of the losses incurred in the various vibrational cascades. Hence the wavelength of the phosphorescent photon is longer than the absorbed photon and the fluorescent photon.

4. *By energy transfer*: molecules in the excited state may revert to the ground state by transferring their energy of excitation (i.e. ΔE_1, ΔE_2 or ΔE_T, Fig. 5.13) to a molecule of the same or another species which is in the ground state. The acceptor molecule then attains the excited state. In any such energy transfer the overall spin angular momentum must not change. Thus for an excited-state donor molecule (D) and a ground-state acceptor molecule (A), three modes of electronic energy transfer are possible and are shown in eqs. (5.10), (5.11) and (5.12); the arrows in parentheses indicate the electron spins.

$$D_{S_1}(\uparrow\downarrow) + A_{S_0}(\uparrow\downarrow) \rightarrow D_{S_0}(\uparrow\downarrow) + A_{S_1}(\uparrow\downarrow) \quad (5.10)$$

$$D_{T_1}(\uparrow\uparrow) + A_{S_0}(\uparrow\downarrow) \rightarrow D_{S_0}(\uparrow\downarrow) + A_{T_1}(\uparrow\uparrow) \quad (5.11)$$

$$D_{T_1}(\uparrow\uparrow) + A_{T_0}(\downarrow\downarrow) \rightarrow D_{S_0}(\uparrow\downarrow) + A_{S_1}(\uparrow\downarrow) \quad (5.12)$$

Since excited triplets have been shown not to participate in normal photosynthetic pigment system function, only energy transfer by the first of the above modes [eq. (5.10)] need concern us.

The most likely mechanism for this energy transfer is by molecular excitons, although this has not been unequivocally demonstrated. A

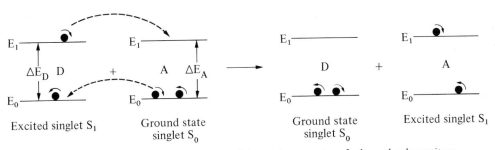

Fig. 5.14. Simplistic idea of the mechanism of electronic energy transfer by molecular excitons.

simplistic idea of the mechanism of this process is given in Fig. 5.14, where the electron in the antibonding molecular orbital at energy level E_1 in the donor molecule D (an excited singlet) is transferred to the equivalent orbital in the acceptor molecule A (a ground-state singlet). This transfer is balanced by the concomitant transfer of an E_0 electron (with the correct spin) from A to the equivalent orbital of D. Molecule D is now a ground-state singlet and molecule A an excited singlet. Put another way the energy of excitation (equal to the energy difference between E_1 and E_0) has been transferred from D to A. Energy transfer by this means is only possible if (i) molecules D and A are very close to each other (as they would be in PS I and PS II) and (ii) ΔE_D is equal to or greater than ΔE_A (Fig. 5.14). If ΔE_D is greater than ΔE_A the electron transferred from energy level E_1 of D would enter a vibrational level of equivalent energy in the comparable electronic energy level of A; the electron would then drop down the vibrational cascade, losing some energy, to yield the S_1 of A. This energy loss prevents A transferring its energy of excitation back to D, a process which is of course possible if, as is shown in Fig. 5.14, ΔE_D is equal to ΔE_A.

5. *By photochemistry:* here the excited singlet uses its energy of excitation to do chemical work. In the case of chlorophyll in PS I and PS II this chemical work is to donate an electron to the oxidized form of a redox system so reducing it (see Fig. 5.3). It is able to do this because in its excited form it is a much more powerful reducing agent than it was in its ground state.

We have now seen five different ways in which a pigment molecule in an excited singlet state may lose its energy of excitation. Which of these processes occur in the operation of PS I and PS II? Clearly

energy transfer occurs. This is the mechanism by which the energy of excitation is taken from the antenna pigments to P700 and P680 at the reaction centres. This process will involve many transfers of the type outlined in Fig. 5.14. During this sequence internal conversions and vibrational cascades will occur but presumably few excited molecules will subside beyond the S_1 level in these processes. Once the energy of excitation has reached the reaction centres, photochemistry occurs. Fluorescence and phosphorescence are wasteful processes and do not occur to any great extent in PS I and PS II *in vivo*. However, chlorophyll fluoresces strongly *in vitro*. Its fluorescence is always in the red region of the spectrum regardless of whether a red or a blue photon has been absorbed. This indicates that the chlorophyll S_2 singlet, formed by the absorption of a blue photon, subsides to the S_1 singlet by internal conversion and the vibrational cascade (see Fig. 5.13) before fluorescence occurs. From this it is concluded that chlorophyll S_2 singlets formed in PS I and PS II *in vivo* drop to S_1 singlets before energy transfer takes place. The small amount of fluorescence that does occur during the operation of the pigment systems is solely from chlorophyll a S_1 singlets regardless of the wavelength of the photon absorbed by the systems. This is very strong evidence of energy transfer from antenna molecules to chlorophyll a.

Energy transfer in PS I and PS II is a 'downhill' process in that the size of the energy quantum progressively decreases as it migrates from the antenna pigments to the P700 and P680 at the reaction centres. This follows from the fact that ΔE_D (Fig. 5.14) is equal to or greater than ΔE_A. Thus the antenna pigments absorb more energetic (shorter wavelength) photons than the pigment at the reaction centre. Consequently once the energy of excitation has reached P700 in PS I and P680 in PS II there can be no 'back' energy transfer and only photochemistry can productively follow. The energy-transfer sequences in PS I and PS II which are consistent with this concept are shown in Fig. 5.15. The process of light absorption and energy transfer in PS I is outlined in Fig. 5.16; a similar process probably operates in PS II.

The path taken by the migrating energy of excitation in PS I and PS II is thought to resemble a 'random walk' that continues until the energy quantum is transferred to P700 or P680 at the reaction centres. The reaction centres may be dispersed randomly amongst all the antenna molecules of the pigment system such that any quantum of excitation energy can wander anywhere within the system and reach any of the reaction centres. Alternatively each reaction centre might be confined to and served by a fixed number of antenna molecules such that a quantum of excitation has

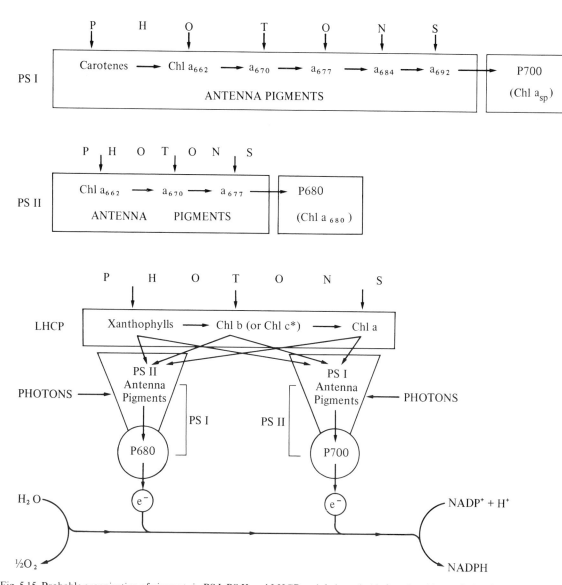

Fig. 5.15. Probable organization of pigments in PS I, PS II and LHCP and their probable functional interrelationship. (PS I = Pigment system I; PS II = Pigment system II; LHCP = Light harvesting chlorophyll protein; *chlorophyll c replaces chlorophyll b in Phaeophyceae, Chrysophyceae, Bacillariophyceae and Cryptophyceae.)

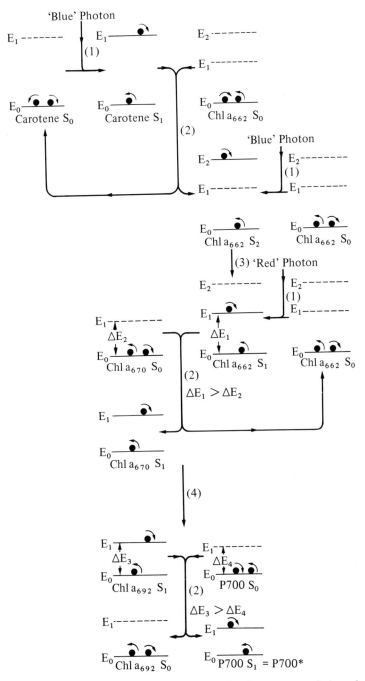

Fig. 5.16. Photon absorption and energy transfer in PS I (1 = photon absorption; 2 = energy transfer by molecular excitons; 3 = internal conversion and vibrational cascade; 4 = energy transfer through the sequence Chl $a_{670} \rightarrow$ Chl $a_{677} \rightarrow$ Chl $a_{684} \rightarrow$ Chl a_{692}).

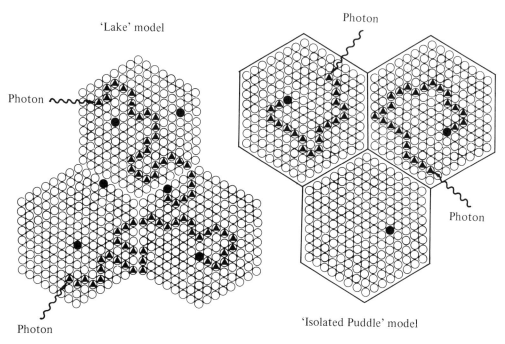

Fig. 5.17a. 'Lake' and 'Isolated Puddle' models of excitation energy flow in photosynthetic pigment systems.
○ = antenna pigment.
● = P700 or P680 at the reaction centre.
▲ = antenna pigment molecule in an excited state having received a quantum of energy by absorption of a photon or by energy transfer.

access to one reaction centre within the pigment system. These two alternatives are called the 'lake model' and the 'isolated puddle model' (see Fig. 5.17a); it is not yet clear which model is operative.

The photochemical reaction at the reaction centres must occur very rapidly indeed, otherwise the P700 and P680 S_1 singlets (i.e. P700* and P680*) would lose their energy of excitation by fluorescence. It has been shown that the photochemical reaction in the bacterial pigment system occurs within about 6×10^{-12} sec which is several orders of magnitude faster than fluorescence. It is clear that for the photochemical reaction to occur as rapidly as this the P700 and P680 must each be located in the chloroplast thylakoid membrane very near to the redox system that they donate an electron to; it is equally likely that the redox system that reduces the resulting P700$^{+\cdot}$ and P680$^{+\cdot}$ is also physically close to P700 and P680.

With the reduction of the redox systems (X and Q, see Section C.6) which receive electrons from P700*

and P680* in the photochemical reactions, the part played by light in the process of photosynthesis is over. All the other reactions that take place in the generation of NADPH and ATP and their subsequent use in converting CO_2 into carbohydrate are not dependent upon light [but see Section D.2(iv)]. It could be said with a great degree of truth that the process of photosynthesis is over once X^- and Q^- have been generated, a process which takes less than a millionth of a second.

In the foregoing account of pigment system function carotenoids have been shown to play a light-harvesting and energy-transferring role. However, they also have another, even more important, function in that they prevent photodynamic damage.

Photodynamic damage is caused by oxygen molecules in their first excited singlet state, Ox_{S_1} in Fig. 5.17b, often referred to as singlet oxygen. Singlet oxygen is very reactive and is capable of oxidizing a whole range of organic compounds, thereby rendering them unfit for their normal physiological function (i.e. damaging them). Singlet oxygen is reactive because, (i) it is in an excited state and is therefore more energetic than the ground state and (ii) it is a singlet (i.e. the promoted electron and its former partner have opposite spins) which permits its reaction with other singlet molecules. In these two respects singlet oxygen differs markedly from the oxygen of the atmosphere which is (i) in the ground state and (ii) a triplet (i.e. it has two electrons with identical spins). The fact that oxygen is a triplet means that its reaction with most organic compounds (which are singlets) to give singlet products is a spin-forbidden process; this explains why materials composed of organic compounds (and this includes living organisms) can exist at the climatic temperatures of this planet in an atmosphere of 20.95% (by vol.) oxygen without being oxidized.

Singlet oxygen is formed from atmospheric oxygen (i.e. the ground-state triplet, Ox_{T_0}) when energy is supplied. This energy is usually supplied in the form of light in the presence of a photosensitizer. Such a photosensitizer is chlorophyll. We have seen that when a chlorophyll molecule absorbs a photon of light of the appropriate wavelength it is transformed into its first or second excited singlet state (Chl_{S_1} or Chl_{S_2}); the latter then subsides to the former by internal conversion and the vibrational cascade (see Fig. 5.13). The vast majority of the Chl_{S_1} molecules are used in the light phase of photosynthesis but some are transformed by intersystem crossing into excited state triplets (Chl_{T_1}). The reaction between atmospheric oxygen and Chl_{T_1} is a spin-permitted process because both are triplets [see eq. (5.12)]. Such a reaction results in the formation of two singlets, Chl_{S_0} and singlet oxygen. The latter then causes photodynamic damage by oxidizing susceptible substrates in the vicinity (see Fig. 5.17b).

Carotenoids are able to prevent photodynamic damage occurring within the photosynthetic apparatus in three different ways (mechanisms 1, 2 and 3 in Fig. 5.17b). In the first of these, carotenoids quench (by about 90%) the first excited triplet state of the chlorophyll photosensitizer. The carotenoid ground-state singlet (Car_{S_0}) reacts with Chl_{T_1} in a reaction of the type given in eq. (5.11) to yield a chlorophyll ground-state singlet (Chl_{S_0}) and a carotenoid first excited-state triplet (Car_{T_1}). The latter then subsides to Car_{S_0} by intersystem crossing and the vibrational cascade, the energy of excitation being lost as heat. In this quenching mechanism the carotenoid competes with oxygen (Ox_{T_0}) for Chl_{T_1}.

In the second mechanism carotenoids quench (by 99.999%) singlet oxygen directly. The Car_{S_0} reacts with the singlet oxygen in a reaction which is the reverse of the type in eq. (5.12) to give Ox_{T_0} and Car_{T_1}. The latter then subsides to Car_{S_0} as before.

In the third mechanism carotenoids act as a preferred substrate for oxidation by any singlet oxygen that may have escaped mechanisms 1 and 2. This method is probably only of minor importance because it has been shown that only one in every thousand carotenoid molecules involved in quenching is oxidized.

Quantitatively mechanism 2 is probably the most important means of protection against photodynamic damage. Only carotenoids with nine or more conjugated double bonds are capable of quenching in this way because only in such molecules is the triplet energy level near or below that of singlet oxygen.

The importance of carotenoids in protecting plants against photodynamic damage is clearly shown by the fact that any mutation markedly lowering the level of carotenoids or their number of conjugated double bonds is lethal. Such mutants can only be grown heterotrophically in the dark; as soon as they are exposed to the light in the presence of oxygen they die. At first sight it might be thought that they could survive in the presence of light in an atmosphere devoid of oxygen but this is not so because higher plant and algal photosynthesis produces oxygen from water as a by-product.

6. Generation of NADPH and ATP by non-cyclic electron flow

The route taken by electrons derived from water in getting to NADPH in the light phase of photosynthesis is shown in Fig. 5.18. This process is called *non-cyclic photophosphorylation*, *non-cyclic* because the electrons follow a non-cyclic track, *photo-* because it is light energy that drives them along this track and *phosphorylation* because as they

Photodynamic damage

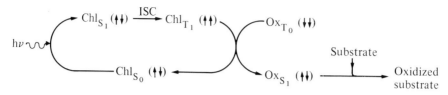

Action of carotenoids in preventing photodynamic damage

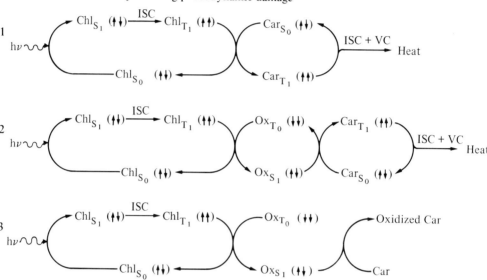

Fig. 5.17b. Mechanism by which photodynamic damage occurs and the ways in which carotenoids prevent it. [Chl = a chlorophyll species; Ox = oxygen; Car = a carotenoid species with at least 9 conjugated double bonds; $h\nu$ = a photon of light of the appropriate wavelength; ISC = intersystem crossing; VC = vibrational cascade; S_0 = ground-state singlet; S_1 = first excited-state singlet; T_0 = ground-state triplet; T_1 = first excited-state triplet; the arrows in parentheses indicate electron spins—see eqs. (5.10), (5.11) and (5.12).]

are driven along the track ADP is phosphorylated to yield ATP.

The route can be conveniently divided into five sections, the two pigment systems and three electron transport chains. The latter are chains of redox systems; one receives electrons from P700* and passes them to $NADP^+$, a second, referred to subsequently as the intermediate electron transport chain, receives electrons from P680* and passes them to $P700^{+\cdot}$ and a third, referred to subsequently as the oxygen-evolution system, takes electrons from water and passes them to $P680^{+\cdot}$.

(i) THE P700*–$NADP^+$ ELECTRON TRANSPORT CHAIN

The identity of the first redox system of this chain is unknown. It is usually called X. It has been

detected by a characteristic electron-spin resonance signal which accompanies photooxidation of P700 at low temperatures. An indirect estimate of its E'_0 has suggested a value as negative as -0.73 V. Thus to reduce X to X^- P700* must be a very powerful reducing agent with an E'_0 even more negative than -0.73 V.

X^- then reduces ferredoxin. Chloroplasts appear to contain three different ferredoxins, a soluble one which has an E'_0 of -0.43 V and a 2Fe–2S centre (see Fig. 5.19) and two membrane-bound species called Fd_{590} and Fd_{540} because their E'_0 values are 0.59 V and 0.54 V respectively. Fd_{590} and Fd_{540} have 4Fe–4S centres. All the ferredoxins are one-electron redox systems and do not carry protons along with the electron. There is evidence to suggest that X^- reduces one of the membrane-bound

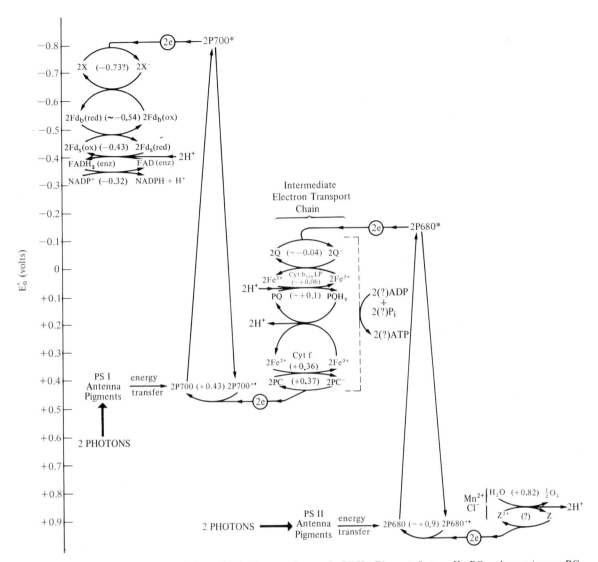

Fig. 5.18. Non-cyclic photophosphorylation. (PS I—Pigment System I; PS II—Pigment System II; PQ—plastoquinone; PC—plastocyanin; Cyt b_{559} LP—low potential cytochrome b_{559}; Cyt f—cytochrome f; Fd_b—membrane-bound ferredoxin; Fd_s—soluble ferredoxin; FAD(enz)—ferredoxin: NADP oxidoreductase: Q,X and Z—unidentified redox systems; NB.Z^{4+} must be formed before O_2 can be evolved (see Fig. 5.23).]

ferredoxins, possibly Fd_{540} as indicated in Fig. 5.18, in non-cyclic photophosphorylation and that this then reduces the soluble ferredoxin. Fd_{590} is thought to function in cyclic photophosphorylation (see Section C.7).

Soluble ferredoxin forms a 1:1 complex in the chloroplast with the next redox system, the enzyme ferredoxin:NADP oxidoreductase[1], an FAD-containing flavoprotein. It is thought that this complex is loosely bound to the outer surface of the

thylakoid membrane (see Fig. 5.20). For soluble ferredoxin to reduce the FAD of ferredoxin:NADP oxidoreductase two electrons and two H^+ ions are required; the latter are taken from the aqueous environment (balancing the two produced from water) whilst the former are derived from two molecules of reduced soluble ferredoxin. The reduced flavoprotein then reduces $NADP^+$ producing NADPH and H^+. Because the NADP is a two-electron redox system, two electrons are required for

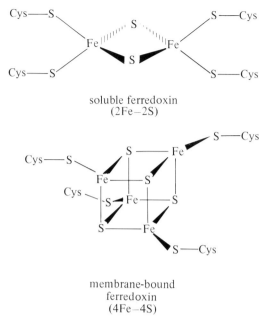

soluble ferredoxin
(2Fe–2S)

membrane-bound
ferredoxin
(4Fe–4S)

Fig. 5.19. Iron-sulphur centres of chloroplast ferredoxins.

the production of one molecule of NADPH. Thus a pair of electrons must flow down this electron transport chain, being derived from two P700* entities which have been excited as a result of the absorption, and subsequent energy transfer, of two photons by two PS I antenna pigment molecules.

(ii) THE INTERMEDIATE ELECTRON TRANSPORT CHAIN

When the two P700* entities have each donated an electron to X they become P700$^{+\cdot}$ cationic free radicals. In order to keep non-cyclic photophosphorylation going they must each receive an electron to reduce them to ground state P700. Now P700 (ground state)/P700$^{+\cdot}$ is a one-electron redox system with an E_0' of $+0.43$ V. Thus P700$^{+\cdot}$ can be reduced by the more electronegative plastocyanin, a one-electron redox system with an E_0' of $+0.37$ V. Plastocyanin is a protein of MW 21,000 containing two atoms of copper per molecule; it is blue in the oxidized form. It is present in the chloroplast at a concentration equal to 0.2% of the total chlorophyll.

Oxidized plastocyanin is then reduced by cytochrome f (Latin *frons*, leaf), a typical c-type cytochrome with an E_0' of $+0.36$ V. Like all cytochromes it is a one-electron redox system, the

porphyrin iron changing from Fe^{2+} in the reduced form to Fe^{3+} in the oxidized form. Cytochrome f is present in the chloroplast at a concentration equal to 0.25% of the total chlorophyll.

Oxidized cytochrome f is in turn reduced by plastoquinol, the reduced form of the two-electron plastoquinone/plastoquinol redox system (see Fig. 5.21) which has an E_0' of about $+0.1$ V. As well as the two electrons, the PQ/PQH$_2$ system also carries two H$^+$; these are of great significance in the generation of ATP. When PQ is reduced by two molecules of ferrocytochrome b_{559} (low potential), the redox system which precedes it in the electron transport chain, two H$^+$ ions are taken from the aqueous matrix of the chloroplast. When the resulting PQH$_2$ reduces two molecules of ferricytochrome f, two H$^+$ ions are released into the lumen of the thylakoid. Thus as two electrons flow down the intermediate electron transport chain through the PQ/PQH$_2$ system two H$^+$ ions are pumped across the thylakoid membrane. The resulting H$^+$ gradient (i.e. high H$^+$, low OH$^-$ inside the thylakoid; low H$^+$, high OH$^-$ outside) is thought to be the driving force behind the conversion of ADP into ATP, a process catalysed by an anisotropic ATP-ase located in the thylakoid membrane (see Fig. 5.20). This chemiosmotic mechanism for linking electron flow with the phosphorylation of ADP is, in essence, the same as that of oxidative phosphorylation in mitochondria [see Chapter 6, Section B.3 (ii)]. As can be seen from Fig. 5.20 it would be expected that two molecules of ATP would be generated for each pair of electrons passing through the PQ/PQH$_2$ system; this is consistent with experimental values of the ATP/2e ratio (i.e. the number of ATPs formed per NADPH generated) which are now consistently greater than 1 and less than 2.

Plastoquinone is present in the chloroplast in much greater quantity than any of the other redox systems, amounting to 5–10% of the total chlorophyll. It is the only component of the intermediate electron transport chain which can account for the large pool of electron acceptors kinetically observed to occur between PS I and PS II.

Plastoquinone is reduced by a b-type cytochrome, known as cytochrome b_{559}. This appears to exist in two interconvertible forms which have different E_0' values. The low potential form, which

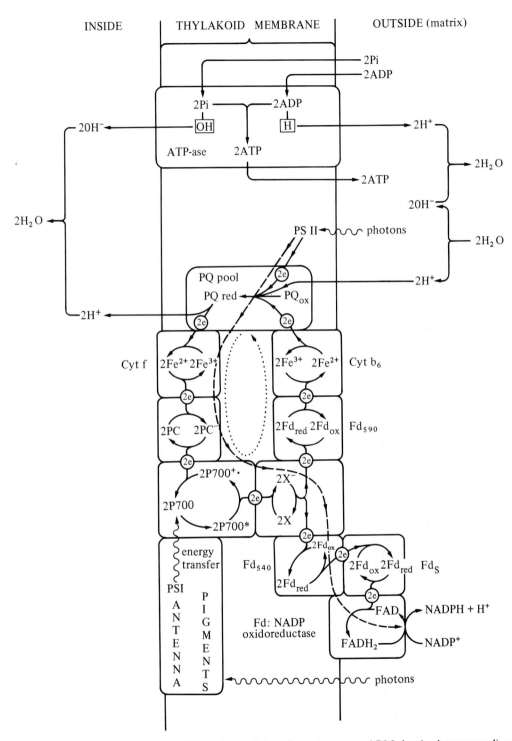

Fig. 5.20. A possible arrangement in the thylakoid membrane of the redox systems around PS I showing how non-cyclic and cyclic electron flow may occur and how ATP may be coincidentally generated. (— — — = non-cyclic flow; · · · · · · = cyclic I flow.)

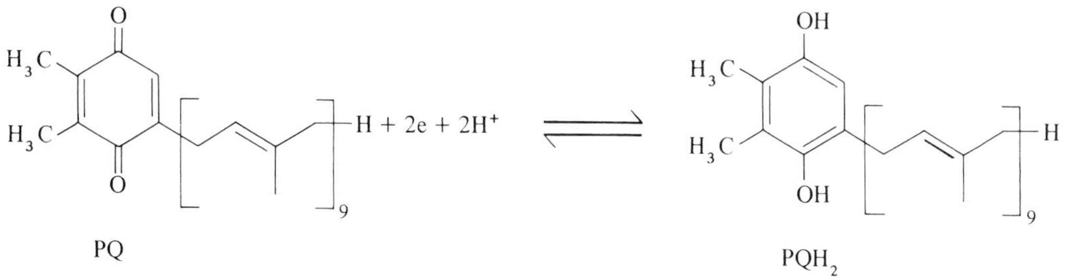

Fig. 5.21. Plastoquinone, a two-electron redox system.

reduces plastoquinone, has an E_0' value of about $+0.06$ V. The high potential form, which under special conditions may reduce P680$^{+\cdot}$ (see Fig. 5.24), has an E_0' of $+0.37$ V and does not participate in non-cyclic photophosphorylation.

Ferricytochrome b_{559} (LP) is, in turn, reduced by Q^-, the reduced form of an unidentified redox system (Q/Q^-) with an E_0' of about -0.04 V which is the immediate acceptor of electrons from P680*.

In the intermediate electron transport chain an electron pair drops from an E_0' of about -0.04 V to one of $+0.37$ V yielding about 79 kJ mol^{-1} [see eq. (5.6)] which is apparently conserved as a H$^+$ ion gradient and used to generate 2ATP from 2ADP and 2P$_i$, a process that requires about 61 kJ mol^{-1}. The purpose of the intermediate electron transport chain is thus to generate ATP by a process akin to that in the mitochondrion. From this it follows that it is the function of PS II to boost electrons from the potential of water ($E_0' = +0.82$ V) to a potential sufficiently negative to allow them to enter the top of the intermediate electron transport chain. When the electrons have dropped down the potential gradient of the intermediate electron transport chain and generated their ATP, they must then be boosted to a sufficiently negative potential to reduce NADP$^+$ ($E_0' = -0.32$ V); it is the function of PS I to do this.

(iii) THE OXYGEN-EVOLUTION SYSTEM

When the two P680* entities have each donated an electron to Q they become P680$^{+\cdot}$ cationic free radicals and in order to keep the process going they must each receive an electron to reduce them to ground state P680. The two electrons which do this are ultimately derived from water (the reduced form of the redox system H$_2$O/$\frac{1}{2}$O$_2$ whose E_0' is $+0.82$ V) and oxygen is produced as a by-product. This is

possible because P680$^{+\cdot}$ is the oxidized form of the redox system P680 (ground state)/P680$^{+\cdot}$ which has a relatively less electro-negative E_0' of about $+0.9$ V; electrons can therefore drop down a potential gradient from H$_2$O at $+0.82$ V to P680$^{+\cdot}$ at $+0.9$ V.

The components which participate in the transfer of electrons from H$_2$O to P680$^{+\cdot}$ have not been identified. Mn^{2+} and Cl$^-$ are known to be required but how they function is not clear. The reason for lack of progress in this area is that the oxygen-evolution system is very difficult to handle experimentally. It is very fragile and is dependent upon the structural integrity of the thylakoid membranes; it is therefore inactivated by the detergents which are used in the laboratory to fractionate the components of the membrane-bound electron transport chains.

It has been suggested that there is a redox system, usually called Z, operating between H$_2$O and P680$^{+\cdot}$. The identity of Z is unknown but there is evidence that it must accumulate four positive charges before a molecule of oxygen can be evolved. This comes from the results of the experiment shown in Fig. 5.22 in which spinach chloroplasts, after 40 min in the dark, were subjected to intense flashes of light, each of 10^{-5} sec duration, every second. The amount of oxygen evolved at each flash was then determined polarographically. Figure 5.22 shows that the oxygen yield in the sequence of flashes oscillates with a period of four; thus the oxygen yield of flashes 1 and 2 are very low, that of flash 3 is very high and that of flash 4 is of intermediate size. The oscillations are damped, their amplitude progressively decreasing as the number of flashes increases. Nevertheless the periodicity of four is very clear with maximum oxygen yield occurring on the 3rd, 7th, 11th, etc., flashes. This strongly suggests a

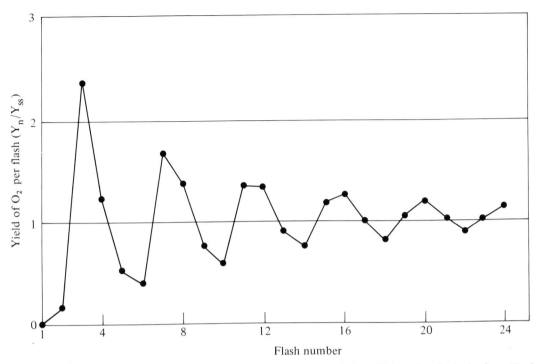

Fig. 5.22. Yields of oxygen produced by isolated spinach chloroplasts exposed to flashes of light at 1-sec intervals after a 40-min dark period. The light flashes were of 10^{-5} sec duration and of equal intensity; they were strong enough to excite all the light-trapping centres in the preparation simultaneously. Individual flash yield values (Y_n) have been normalized to a steady-state yield (Y_{ss}) (adapted from Fowler and Kok, 1974).

four-step process. It has been suggested that each step indicates the absorption of one photon which results in the accumulation of one positive charge on Z. Therefore the first photon would produce Z^+, the second Z^{2+}, the third Z^{3+} and the fourth Z^{4+} after which a molecule of O_2 is evolved. This explanation, however, requires a slight modification because as formulated it would give oxygen yield maxima on the 4th, 8th, 12th, etc., flashes rather than the 3rd, 7th, 11th as is observed. The modification assumes that in the dark there are two stable species, namely Z and Z^+, present in the ratio 1:3. Thus the first flash will convert Z^+ to Z^{2+}, the second Z^{2+} to Z^{3+} and the third Z^{3+} to Z^{4+}, after which the periodicity of four is established. The operation of such a mechanism is shown in Fig. 5.23 where $Z.P680.Q$ represents the three components of the PS II reaction centre in close physical contact. The absorption of four photons by PS II leads, in a stepwise manner, to the formation of four molecules of reduced cytochrome b_{559} (LP) and $Z^{4+}.P680.Q$. The latter then reacts with two molecules of water, abstracting four electrons to nullify the four positive charges so regenerating $Z.P680.Q$ and coincidentally releasing one molecule of oxygen and four H^+ ions.

The stoichiometry of non-cyclic photophosphorylation is given by eqs. (5.13) and (5.14) which define the number of photons which must be absorbed to generate one molecule of NADPH and to evolve one molecule of O_2 respectively. The only slight doubt in these equations is the number of molecules of ATP generated; however, there is now good reason to believe that 2ATPs are generated per electron pair rather than one as was once thought.

$$4 \text{ photons} + H_2O \rightarrow NADPH + 2ATP + \tfrac{1}{2}O_2$$
$$(5.13)$$

$$8 \text{ photons} + 2H_2O \rightarrow 2NADPH + 4ATP + O_2$$
$$(5.14)$$

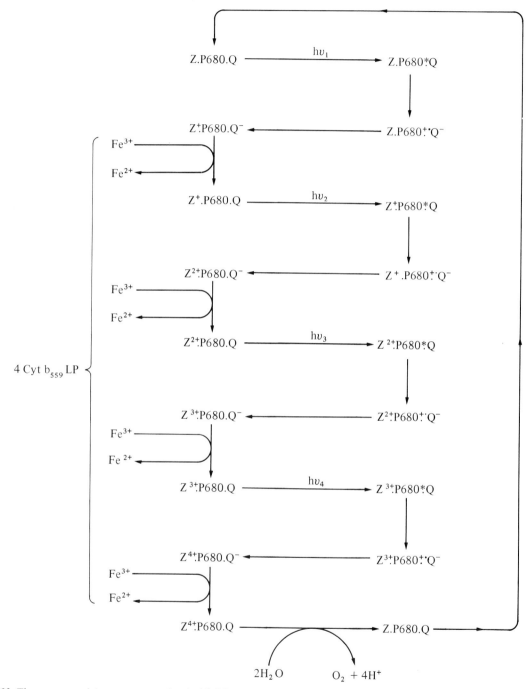

Fig. 5.23. The oxygen-evolving system associated with PS II. (Four positive charges must accumulate on Z, the electron donor to $P680^+$, before a molecule of oxygen can be evolved.)

7. Cyclic electron flow

It has been known for some time that a cyclic electron flow around PS I can be produced artificially with the aid of added redox systems. These systems are reduced when light is absorbed by PS I, possibly by picking up electrons from X^-. They then in turn promote the reduction of $P700^+$ thereby reforming P700 (ground state) and completing the cycle. The redox systems which induce this artificial cyclic flow may be divided into three classes on the basis of how they reduce $P700^+$. Some, with E_0' values more negative than 0.0 V, reduce plastoquinone; the electrons then flow from PQH_2 down the rest of the intermediate electron transport chain to $P700^+$. Others, with E_0' values more positive than 0.0 V, by-pass PQ and reduce either plastocyanin and then $P700^+$ or $P700^+$ directly. When plastoquinone is on the cyclic route a hydrogen ion gradient is generated across the thylakoid membrane and this in turn drives the phosphorylation of ADP. The process is then called cyclic photophosphorylation. The first demonstration of this artificially induced cyclic photophosphorylation came in 1955 when Arnon used flavin mononucleotide (FMN/FMNH$_2$; $E_0' = -0.22$ V) as the added redox system.

Since then cyclic photophosphorylation has been shown to occur naturally in chloroplasts. The precise route taken by the electrons in this process is not yet clear. It has been suggested that the flow is from X^- to Fd_{590} (one of the two membrane-bound ferredoxins mentioned in Section C.6) and then to plastoquinone via cytochrome b_6 (also known as cytochrome b_{563}) ($E_0' = -0.18$ V). From the resulting PQH_2 the route to $P700^+$ is via cytochrome f and plastocyanin. Since this electron transport chain includes plastoquinone, electron flow is accompanied by the generation of ATP; presumably in the ratio 2ATP/2e. The outline of this process is shown in Fig. 5.24. Its possible structural organization within the thylakoid membrane in relation to that of the noncyclic route around PS I is shown in Fig. 5.20. This process has been called Cyclic I by Govindjee (1975) to distinguish it from the cyclic flow of electrons which occurs under special conditions around PS II and which is called Cyclic II.

Cyclic II is thought to involve the transfer of electrons from P680* back to $P680^+$ via Q and the low- and high-potential forms of cytochrome b_{559} (see Fig. 5.24). There is no evidence to suggest that electron transport via the Cyclic II route is accompanied by the generation of ATP; it is not therefore a cyclic phosphorylation process and its physiological significance is not clear.

What is the function of cyclic photophosphorylation (i.e. Cyclic I)? In answering this question one must bear in mind the fact that cyclic photophosphorylation generates only ATP. Since it does not generate NADPH it cannot support the dark phase photosynthesis by itself. It must therefore play a supportive role to non-cyclic photophosphorylation. Since it is likely that non-cyclic photophosphorylation produces ATP and NADPH in the ratio 2:1, rather than 1:1 as was once thought, the ATP and NADPH requirements of the dark phase of photosynthesis can probably be met by the operation of this process unaided. However, non-cyclic photophosphorylation will not be able to supply the additional ATP required for the other synthetic processes which go on in the chloroplast, e.g. starch, lipid, pigment and nucleic acid synthesis. Nor will it be able to supply ATP for export from the chloroplast to support extrachloroplastidic syntheses (note that ATP cannot cross the chloroplast membranes but can be exported indirectly by means of the 3-phosphoglyceric acid–dihydroxyacetone-phosphate shuttle, see Section D.7). It is thought that it is the function of cyclic photophosphorylation to produce the extra ATP required for these processes.

8. Reagents which affect photosynthetic electron flow

Such reagents can be divided into three classes, those that can draw off electrons from the photosynthetic electron flow (electron acceptors), those that can introduce electrons into the flow (electron donors) and those that can interrupt flow (inhibitors). Some of these reagents have proved useful as herbicides.

(i) ELECTRON ACCEPTORS

Methyl viologen, benzyl viologen, anthraquinone-2-sulphonate and diaminodurene receive

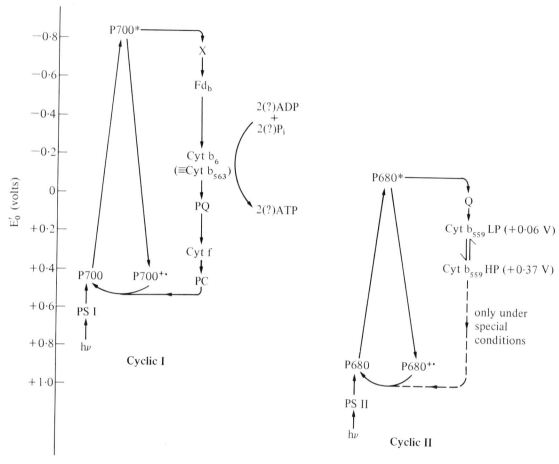

Fig. 5.24. Pathways of cyclic electron flow (only Cyclic I, otherwise known as cyclic photophosphorylation, is known to generate ATP).

electrons from PS I at a position preceding ferredoxin; this means that they are probably reduced by X^-.

Ferricyanide and dichlorophenolindophenol accept electrons from a component of the intermediate electron transport chain, probably in the region of cytochrome f and plastocyanin.

(ii) ELECTRON DONORS

Hydroquinone, hydroxylamine, phenylenediamine, diphenylcarbazide and manganous ions donate electrons into the electron flow on the H_2O side of cytochrome b_{559} LP. The reduced forms of dichlorophenolindophenol (DCPIP), N,N,N',N'-tetramethyl-p-phenylenediamine (TMPD) and 2,3,5,6-tetramethyl-p-phenylenediamine (or diaminodurene, DAD) donate electrons to a compo-

nent of the intermediate electron transport chain, although not necessarily the same one.

(iii) INHIBITORS OF ELECTRON FLOW

2'-Phosphoadenosine diphosphate ribose (NADP minus the nicotinamide moiety) inhibits ferredoxin:NADP oxidoreductase[1] and therefore stops the final step in noncyclic electron flow.

Disalicylidenepropanediamine (DSPD) inhibits electron flow at the ferredoxin level.

2,5-Dibromo-3-methyl-6-isopropyl-p-benzo-quinone (DBMID) acts as a plastoquinone competitor and blocks electron flow between cytochrome b_{559}(LP) and cytochrome f.

Substituted ureas such as 3-(p-chlorophenyl)-1,1-dimethylurea (CMU) and 3-(3,4-dichlorophenyl)-1,1-dimethylurea (DCMU), triazines such as 2-

chloro-4-ethylamino-6-isopropylamino-*s*-triazine) (atrazine) and 2-chloro-4,6-bis(chloro-ethylamino)-*s*-triazine (simazine) and 2*n*-heptyl-4-hydroxy-quinoline-*N*-oxide (HOQNO) block electron flow between Q^- and cytochrome b_{559} LP.

(iv) Herbicides (see Fig. 5.25)

Triazines such as atrazine (XX) and simazine (XXI) were introduced as herbicides in 1952. These compounds can be absorbed by leaves and by roots. However, absorption by leaves is not followed by translocation within the plant whereas root absorption is, via the apoplastic system. The compounds are reversibly bound to clay minerals in the soil which therefore limits the quantity available for root absorption; this disadvantage is compensated for by absorption by the leaves following topical application. These compounds inhibit plant growth but their main herbicidal action is in blocking non-cyclic photophosphorylation by stopping electron flow between Q^- and cytochrome b_{559} LP.

The substituted ureas, CMU (Monuron) (XXII) and DCMU (Diuron) (XXIII), are herbicides which are generally applied to the soil, from which they are absorbed passively by plant roots and translocated via the apoplastic system. Some, however, also show activity when applied to leaves. They also kill plants by blocking non-cyclic photophosphorylation at the same point as the triazines.

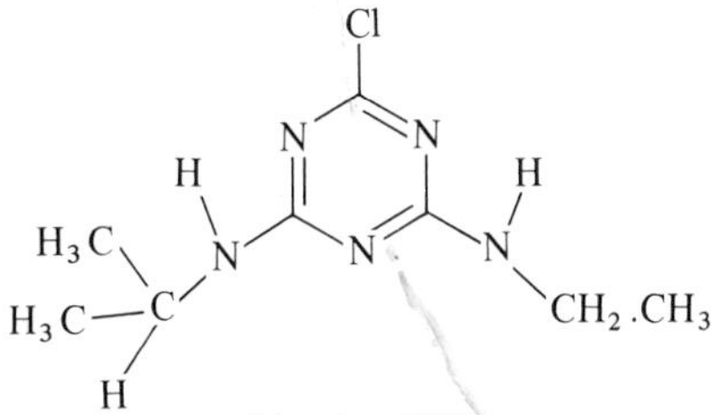

Atrazine (XX)

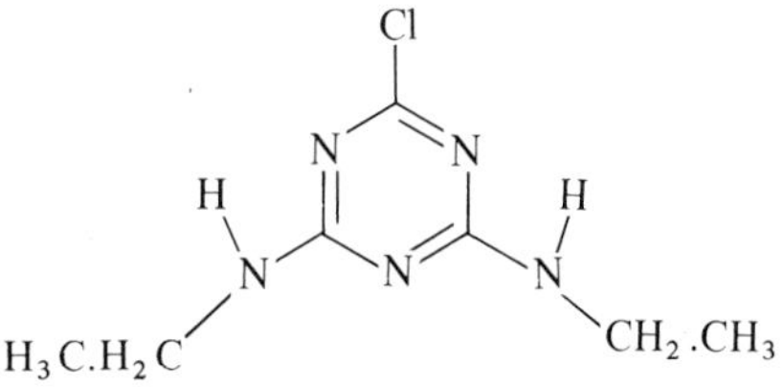

Simazine (XXI)

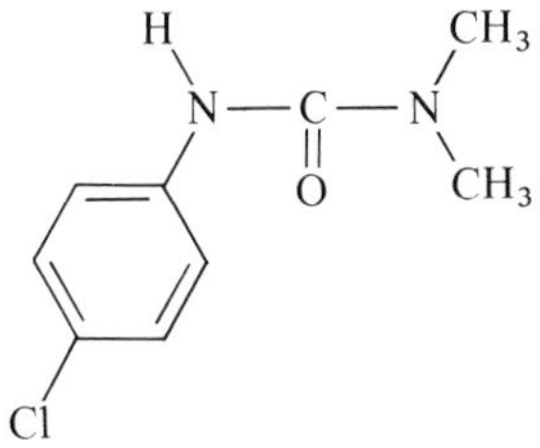

Monuron (CMU; XXII)

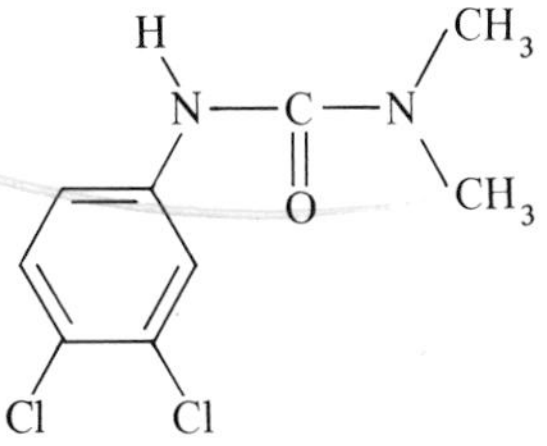

Diuron (DCMU; XXIII)

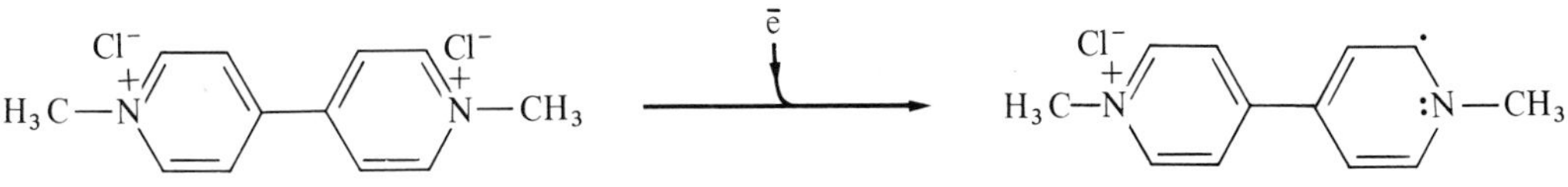

Paraquat (XXIV)

Reduced Paraquat (XXV) (planar free radical in which the unpaired electron is delocalized over all twelve nuclear atoms)

Fig. 5.25. Structures of some herbicides which interrupt photosynthetic electron flow.

The bipyridylium herbicides such as paraquat (1,1'-dimethyl 4,4'-bipyridylium dichloride, XXIV) were introduced in the 1950s. Paraquat is, in effect, methyl viologen, a redox system with an E'_0 of -0.44 V, which accepts electrons from PS I at a position preceding the ferredoxins. It therefore taps electrons from the P700*–NADP$^+$ transport chain and markedly reduces the rate of NADPH production. However, its phytotoxic effect is too rapid to be fully explained by the inhibition of NADPH generation. It appears that the planar free radical (XXV) which is produced when paraquat accepts an electron from X^- [eq. (5.15)] in turn, reduces oxygen to the superoxide anionic free radical ($O_2^{-\cdot}$) [eq. (5.16)]. This is then converted into hydrogen peroxide and oxygen by the enzyme superoxide dismutase[2] in the chloroplast [eq. (5.17)]. The hydrogen peroxide then reacts with the unsaturated fatty acyl moieties of the acyl lipids of the chloroplast membranes in a destructive manner. This results in a disruption of the chloroplast membranes and thus the integrity of the whole photosynthetic apparatus.

$$X^- + \text{Para } Q_{(ox)} \rightarrow X + \text{Para } Q_{(red)} \qquad (5.15)$$

$$\text{Para } Q_{(red)} + O_2 \rightarrow \text{Para } Q_{(ox)} + O_2^{-\cdot} \qquad (5.16)$$

$$2O_2^{-\cdot} + 2H^+ \rightarrow H_2O_2 + O_2 \qquad (5.17)$$

Paraquat can be taken up by the roots and translocated in the apoplastic system although its mobility is more limited than the substituted ureas and triazines mentioned above. It is very tightly bound to clay minerals in soil because of its positive charges. It is very toxic to animals. There have been several human fatalities caused by its accidental ingestion. Death results within 2–3 weeks from severe pulmonary fibrosis and attendant liver and kidney damage.

D. THE DARK PHASE OF PHOTOSYNTHESIS

1. Variations in the mechanism of the dark phase of photosynthesis

In the dark phase of photosynthesis the reduced pyridine nucleotide and ATP produced in the light phase are used to convert CO_2 into carbohydrate. In higher plants three different mechanisms of

carrying out this process are now recognized. It would, however, be more accurate to describe these as one basic mechanism with two variants.

The basic mechanism is the photosynthetic carbon reduction cycle or Calvin cycle (named after its discoverer, Melvin Calvin); more recently it has been called the C_3 pathway or the C_3 mode of photosynthesis and those plants which use it exclusively have been called C_3 plants. Such plants generally grow in the temperate regions of the world, the optimum day temperature for CO_2 assimilation in them being 15–25°C.

The two variants are the C_4 pathway (or C_4 mode of photosynthesis) and the process known as crassulacean acid metabolism (CAM). Plants which use the C_4 pathway generally grow in tropical environments and are called C_4 plants. Those which use the CAM pathway frequently grow in arid desert regions and are called CAM plants.

C_3 plants convert CO_2 into carbohydrate by the Calvin cycle only. C_4 plants and CAM plants also use the Calvin cycle in the conversion of CO_2 into carbohydrate but the process also involves additional enzyme-catalysed steps. The nature of these additional steps, the time during the day when they occur and their cellular location distinguish C_4 and CAM pathways from each other.

The morphology of the leaves of these three plant types is directly related to the CO_2 assimilatory mechanism that they use. C_3 and C_4 plants have a wide and overlapping range of stomatal concentration (5000–30,000 stomata per cm^2 depending upon the species). However, the stomata of C_4 plants have a higher resistance to gaseous diffusion than those of C_3 plants. The stomata of CAM plants are much less numerous (100–800 stomata per cm^2 depending upon the species) and also have a high resistance to gaseous diffusion. The leaves of C_4 plants are characterized by an extensive network of air spaces which allows air to bathe a high proportion of the photosynthetic cells; few photosynthetic cells are more than 2 or 3 cell layers from the air. C_4 leaves are also typified by the concentric layers of the two different types of photosynthetic cells present, the mesophyll and bundle sheath cells which surround the vascular bundles [see Section D.3(i)]. Although cross-sections of the leaves of C_3 plants indicate a much less open structure than is present in C_4 leaves, sections taken parallel to the vascular bundles show that inter-cellular spaces are nevertheless plentiful. The internal structure of the leaves of CAM plants has few air spaces and photosynthetic cells are frequently 7–9 cell layers away from the air. There is thus a high resistance to gaseous diffusion into, out of and within CAM leaves; the purpose of this is clearly to minimize water loss but inevitably results in poorer access of CO_2 to the photosynthetic cells.

2. The Calvin cycle (C_3 pathway)

(i) MECHANISM OF THE CALVIN CYCLE

The key reaction in the Calvin cycle is that in which CO_2 is introduced [Fig. 5.26; reaction 1 and eq. (5.18)]; CO_2 reacts with D-ribulose 1,5-diphos-phate (RuDP) to form two molecules of D-3-phosphoglyceric acid (3-PGA) under the catalytic influence of the enzyme ribulose 1,5-bisphosphate carboxylase-oxygenase (RuDP-carboxylase)[3]. This reaction is complex in that CO_2 adds on to a 5C sugar to form a 6C sugar intermediate which then splits into two 3C sugars; the suggested mechanism is given in eq. (5.19). This indicates that RuDP is firstly tautomerized to the corresponding enol which is then carboxylated at C–2 to form a β-keto acid intermediate. For this carboxylation to take place, C–2 must be made nucleophilic. This is probably accomplished by the removal of a proton from the 3-OH group so giving an enolate anion which can undergo β-carboxylation [eq. (5.19A)]. That 2-carboxy-3-ketoarabinitol 1,5-diphosphate is an intermediate is indicated by (i) the fact that RuDP carboxylase can catalyse its hydrolytic cleavage to

$$(5.18)$$

| D-ribulose 1,5-diphosphate (RuDP) | enediol of RuDP | 2-carboxy-3-keto-D-pentitol 1,5-diphosphate | D-3-phospho-glyceric acid (2 molecules) |

$$(5.19)$$

| enolate anion of RuDP | 2-carboxy-3-keto D-pentitol 1,5-diphosphate | D-3-phospho-glyceric acid (2 molecules) |

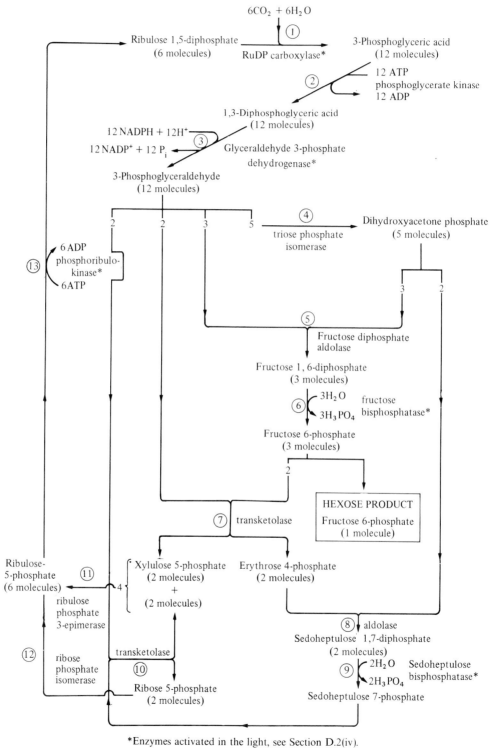

*Enzymes activated in the light, see Section D.2(iv).

Fig. 5.26. Photosynthetic carbon reduction cycle or Calvin cycle.

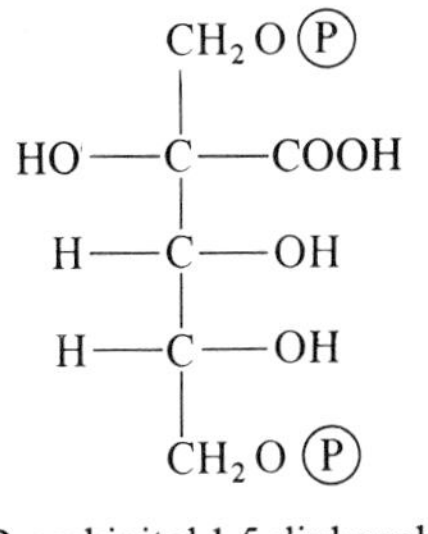

$$
\begin{array}{c}
CH_2O\,\text{\textcircled{P}} \\
| \\
HO-C-COOH \\
| \\
H-C-OH \\
| \\
H-C-OH \\
| \\
CH_2O\,\text{\textcircled{P}}
\end{array}
$$

Fig. 5.27. 2-Carboxy-D-arabinitol 1,5-diphosphate, an inhibitor of RuDP carboxylase.

two molecules of 3-PGA and (ii) the fact that 2-carboxy-D-arabinitol 1,5-diphosphate (Fig. 5.27) inhibits RuDP carboxylase almost irreversibly. The 2-carboxy-3-ketoarabinitol 1,5-diphosphate is then hydrolytically cleaved between C–2 and C–3 [eq. (5.19B)].

CO_2, rather than HCO_3^-, is the true carboxylation substrate of RuDP carboxylase. Label from $H^{14}CO_3^-$ can nevertheless be incorporated into 3-PGA by RuDP carboxylase by virtue of the equilibrium shown in eq. (5.20).

$$CO_2 + H_2O \rightleftharpoons H_2CO_3 \rightleftharpoons H^+ + HCO_3^- \quad (5.20)$$

The question of whether CO_2 or HCO_3^- is the true substrate of a carboxylation enzyme can be decided by use of $HC^{18}O_3^-$. If CO_2 is the true substrate the product of the reaction will contain only two ^{18}O atoms whereas it will contain three if

HCO_3^- is the true substrate. It should be noted that the equilibrium between dissolved CO_2 and HCO_3^- is not attained instantaneously in pure water. The enzyme carbonic anhydrase[4] catalyses the attainment of this equilibrium. This enzyme is widely distributed in plants and is known to occur in the chloroplast.

RuDP carboxylase requires Mg^{2+} in order to catalyse eq. (5.18). There is evidence that the initial step in the reaction mechanism is the production of an active enzyme–CO_2–Mg^{2+} complex [eq. (5.21)] which then reacts with RuDP. Mg^{2+} has been implicated in the formation and stabilization of the β-keto acid intermediate, 2-carboxy-3-keto-D-arabinitol 1,5-diphosphate.

It was once thought that RuDP carboxylase also required Cu^{2+} for its catalytic activity; it is now known that this is not the case.

RuDP carboxylase has another catalytic activity; it can also behave as RuDP oxygenase (hence its full name, D-ribulose 1,5-diphosphate carboxylase-oxygenase). Thus it can catalyse the cleavage of RuDP in the presence of O_2 into 3-PGA and phosphoglycollic acid [eq. (5.22)].

It is possible that the mechanism of this reaction involves an attack by O_2 on the enolate anion [see eq. (5.19A)] producing a peroxide which is cleaved hydrolytically [eq. (5.23)]. It is not clear how RuDP carboxylase activates oxygen [see Chapter 8, Section C.1(ii)(d)]. The presence of the transition metal ion, Cu^{2+}, in the enzyme has been reported.

$$
\text{Enz} + CO_2 \underset{\text{slow}}{\rightleftharpoons} \text{Enz}-CO_2 \overset{Mg^{2+}}{\underset{\text{fast}}{\rightleftharpoons}} \text{Enz}-CO_2-Mg^{2+} \quad (5.21)
$$
$$
\text{(inactive)} \qquad\qquad \text{(inactive)} \qquad\qquad \text{(active)}
$$

$$
\begin{array}{c}
CH_2O\,\text{\textcircled{P}} \\
| \\
C=O \\
| \\
H-C-OH \\
| \\
H-C-OH \\
| \\
CH_2O\,\text{\textcircled{P}}
\end{array}
+ O_2 \xrightarrow{Mg^{2+}}
\begin{array}{c}
CH_2O\,\text{\textcircled{P}} \\
| \\
COOH
\end{array}
+
\begin{array}{c}
COOH \\
| \\
H-C-OH \\
| \\
CH_2O\,\text{\textcircled{P}}
\end{array}
\quad (5.22)
$$

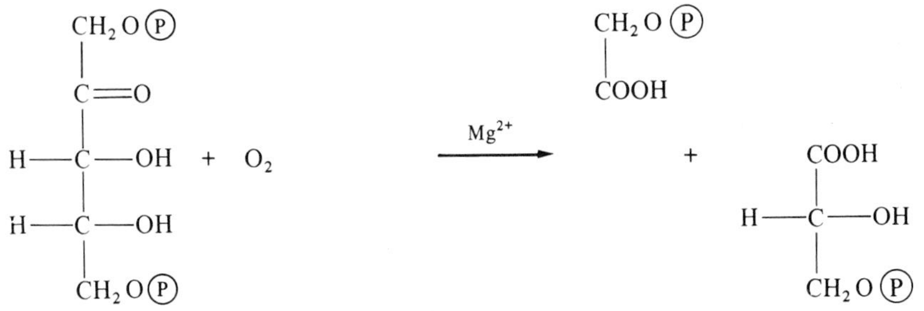

D-ribulose 1,5-diphosphate

phospho-glycollic acid

3-phospho-glyceric acid

By analogy with many oxygenases it might be thought that the Cu^{2+} would enable the oxygen to participate in the reaction by forming a complex with it. However, recent reports indicate that Cu^{2+} is not necessary for the oxygenase activity of RuDP carboxylase.

analytical ultracentrifugation and was called Fraction 1 protein. It has a MW of about 5.5×10^5 and consists of eight large subunits (MW 51,000–58,000) and eight small subunits (MW 12,000–18,000) which are thought to be arranged in the manner shown in Fig. 5.28. The large subunits

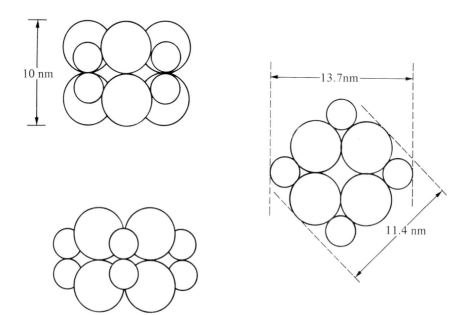

$$\text{(5.23)}$$

enolate anion
of RuDP

peroxide

The phosphoglycollic acid produced in eq. (5.22) is metabolized in photorespiration (see Chapter 6, Section F.1).

RuDP carboxylase constitutes about 50% of the soluble proteins in leaves and as such must be the most abundant protein in nature. It was originally distinguished from the other soluble leaf proteins by

are coded for by chloroplastidic DNA with nuclear DNA codes for the small subunits. The formation of the large and small subunits is discussed in Chapter 10, Section E.9(viii) and represented diagrammatically in Fig. 10.25. The large subunits are catalytically active in the absence of the small subunits; present evidence indicates that the

Fig. 5.28. Possible structure of ribulose 1,5-bisphosphate carboxylase-oxygenase (the positions of the small subunits are uncertain). [Adapted from Baker, T. S., Eisenberg, D. and Eiserling, F. A. (1977).]

same active site is involved in carboxylase and oxygenase functions. It would appear that the small subunits have a regulatory function.

It used to be believed that the catalytic activity of RuDP carboxylase was too low to account for observed photosynthetic CO_2 fixation rates. This belief was based upon the incompatibility of measured $K_{m(CO_2)}$ values of 200–500 μM with the fact that the maximum concentration of CO_2 in the chloroplast is about 10 μM (the conc. of CO_2 in water which is in equilibrium with air, containing 0.03% CO_2 by vol, at 1 atmosphere pressure and 25°C is 10 μM). However, it has now been shown that RuDP carboxylase is very unstable after its release from chloroplasts. If it is assayed within 30–60 sec of its release the $K_{m(CO_2)}$ is found to be 11–18 μM but 3 min later the $K_{m(CO_2)}$ is $\sim 500 \mu M$. Clearly the early estimates of the $K_{m(CO_2)}$ of RuDP carboxylase did not take account of its instability and the new values are more representative of the *in vivo* $K_{m(CO_2)}$. Since the $K_{m(CO_2)}$ values are close to the likely chloroplastidic CO_2 concentrations, RuDP carboxylase can account for photosynthetic CO_2 fixation rates.

The $K_{m(O_2)}$ for the oxygenase activity of RuDP carboxylase is 200–400 μM. This overlaps the maximum concentration, 260 μM, of oxygen in the chloroplast (the conc. of O_2 in water which is in equilibrium with air, containing 20.95% O_2 by vol, at 1 atmosphere pressure and 25°C is 260 μM).

The carboxylase activity of RuDP carboxylase is competitively inhibited by O_2; the $K_{i(O_2)}$ is 200–400 μM. Similarly the oxygenase activity is competitively inhibited by CO_2; the $K_{i(CO_2)}$ is 20–40 μM. Thus the relative rates of the two reactions *in vivo* will be dependent upon the concentration of CO_2 and O_2 in the chloroplast stroma.

3-PGA, resulting from eq. (5.18), is then phosphorylated to give 1,3-diphosphoglyceric acid (1,3-diPGA); ATP generated in the light phase of photosynthesis is the phosphorylating agent [eq. (5.24); reaction 2, Fig. 5.26].

Phosphoglycerate kinase[5] catalyses the reaction which, unlike most kinase-catalysed reactions, is freely reversible. However, during daylight the continuous regeneration of ATP by photophosphorylation drives the reaction in the direction of 1,3-diPGA. This reaction is also seen in glycolysis [see Chapter 6, Section E.1(i)] but there it operates in the reverse direction with the purpose of producing ATP rather than 1,3-diPGA.

The 1,3-diPGA is now reduced to 3-phosphoglyceraldehyde (3-PGAld) by the NADPH produced in the light phase of photosynthesis [eq. (5.25); reaction 3, Fig. 5.26].

The reaction is catalysed by glyceraldehyde-3-phosphate dehydrogenase[6]. Again this reaction participates in glycolysis in the reverse direction; however, there it is catalysed by the cytoplasmic

$$
\begin{array}{ccc}
\text{COOH} & & \text{C} \overset{O}{\underset{O\,\circled{P}}{\diagup}} \\
| & & | \\
\text{H—C—OH} \quad + \quad \text{ATP} & \rightleftharpoons & \text{H—C—OH} \quad + \quad \text{ADP} \qquad (5.24) \\
| & & | \\
\text{CH}_2\text{O}\,\circled{P} & & \text{CH}_2\text{O}\,\circled{P}
\end{array}
$$

3-phospho-
glyceric acid

1,3-diphospho-
glyceric acid

$$
\begin{array}{ccc}
\text{C} \overset{O}{\underset{O\,\circled{P}}{\diagup}} & & \text{CHO} \\
| & & | \\
\text{H—C—OH} \quad + \quad \text{NADPH} + \text{H}^+ & \rightleftharpoons & \text{H—C—OH} \quad + \quad \text{NADP}^+ + \text{H}_3\text{PO}_4 \\
| & & | \\
\text{CH}_2\text{O}\,\circled{P} & & \text{CH}_2\text{O}\,\circled{P} \qquad (5.25)
\end{array}
$$

1,3-diphospho-
glyceric acid

3-phospho-
glyceraldehyde

NAD-linked glyceraldehyde-3-phosphate dehydrogenase[7] [see Chapter 6, Section E.1(i)].

The 3-PGAld can be utilized in the Calvin cycle in four different ways. The first is shown in eq. (5.26) (reaction 4, Fig. 5.26) in which it is isomerized to dihydroxyacetone phosphate (DiHOAcP) under the catalytic influence of triosephosphate isomerase[8]. This reaction is also seen in glycolysis, where it is catalysed by a cytoplasmic triosephosphate isomerase.

$$
\begin{array}{ccc}
\text{CHO} & & \text{CH}_2\text{OH} \\
| & & | \\
\text{H—C—OH} & \rightleftharpoons & \text{C=O} \\
| & & | \\
\text{CH}_2\text{O}\,\textcircled{P} & & \text{CH}_2\text{O}\,\textcircled{P}
\end{array}
\qquad (5.26)
$$

3-phospho-   dihydroxy-
glyceraldehyde   acetone phosphate

The second way in which 3-PGAld is used is shown in eq. (5.27) (reaction 5, Fig. 5.26). Here it condenses with DiHOAcP to form D-fructose 1,6-diphosphate (F-1,6-diP). This reaction is catalysed by fructose diphosphate aldolase[9]. It also operates in glycolysis where it is catalysed by a cytoplasmic FDP aldolase.

The F-1,6-diP is then dephosphorylated by fructose bisphosphatase[10] to yield D-fructose 6-phosphate [eq. (5.28); reaction 6, Fig. 5.26]. It is allosterically stimulated by F-1,6-diP and requires Mg^{2+} for activity.

$$
\begin{array}{ccc}
\text{CH}_2\text{O}\,\textcircled{P} & & \\
| & & \text{CH}_2\text{O}\,\textcircled{P} \\
\text{C=O} & & | \\
| & & \text{C=O} \\
\text{CH}_2\text{OH} & & | \\
 & & \text{HO—C—H} \\
+ & \rightleftharpoons & | \\
 & & \text{H—C—OH} \\
\text{CHO} & & | \\
| & & \text{H—C—OH} \\
\text{H—C—OH} & & | \\
| & & \text{CH}_2\text{O}\,\textcircled{P} \\
\text{CH}_2\text{O}\,\textcircled{P} & &
\end{array}
\qquad (5.27)
$$

D-fructose
1,6-diphosphate

The third way in which 3-PGAld is utilized is shown in eq. (5.29) (reaction 7, Fig. 5.26). Here it reacts with an equivalent amount of the F-6-P formed in eq. (5.28) to produce equal amounts of D-xylulose 5-phosphate (Xu-5-P) and D-erythrose 4-phosphate (E-4-P). This reaction is catalysed by transketolase[11] which requires thiamine pyrophosphate (TPP) and Mg^{2+} for activity. Transketolase also participates in the oxidative pentose phosphate pathway (see Chapter 6, Section F.3) and catalyses the general reaction in which a ketol moiety ($CH_2OH \cdot CO \cdot$) is transferred from a D-ketose sugar with a $3S,4R (\equiv threo)$ configuration to a D-aldose sugar. The TPP acts as a go-between in the reaction, receiving the ketol group from the ketose donor and handing it on to the recipient aldose; the ketol group is transiently covalently bound via its carbonyl carbon atom to C–2 of the thiazole ring of TPP during the reaction.

$$
\begin{array}{ccccc}
\text{CH}_2\text{O}\,\textcircled{P} & & & \text{CH}_2\text{OH} & \\
| & & & | & \\
\text{C=O} & & & \text{C=O} & \\
| & & & | & \\
\text{HO—C—H} & & & \text{HO—C—H} & \\
| & + \text{H}_2\text{O} & \rightleftharpoons & | & + \text{H}_3\text{PO}_4 \quad (5.28) \\
\text{H—C—OH} & & & \text{H—C—OH} & \\
| & & & | & \\
\text{H—C—OH} & & & \text{H—C—OH} & \\
| & & & | & \\
\text{CH}_2\text{O}\,\textcircled{P} & & & \text{CH}_2\text{O}\,\textcircled{P} &
\end{array}
$$

D-fructose   D-fructose
1,6-diphosphate   6-phosphate

$$
\begin{array}{c}
CH_2OH \\
| \\
C=O \\
| \\
HO-C-H \\
| \\
H-C-OH \\
| \\
H-C-OH \\
| \\
CH_2O\,\textcircled{P}
\end{array}
\;+\;
\begin{array}{c}
CHO \\
| \\
H-C-OH \\
| \\
CH_2O\,\textcircled{P}
\end{array}
\;\underset{}{\overset{TPP,\;Mg^{2+}}{\rightleftharpoons}}\;
\begin{array}{c}
CHO \\
| \\
H-C-OH \\
| \\
H-C-OH \\
| \\
CH_2O\,\textcircled{P}
\end{array}
\;+\;
\begin{array}{c}
CH_2OH \\
| \\
C=O \\
| \\
HO-C-H \\
| \\
H-C-OH \\
| \\
CH_2O\,\textcircled{P}
\end{array}
\qquad (5.29)
$$

D-fructose 6-phosphate   3-phospho-glyceraldehyde   D-erythrose 4-phosphate   D-xylulose 5-phosphate

The E-4-P formed in eq. (5.29) then reacts with an equal amount of DiHOAcP formed in eq. (5.26) to yield D-sedoheptulose 1,7-diphosphate (Su-1,7-diP) [eq. (5.30); reaction 8, Fig. 5.26]. The reaction is probably catalysed by fructose diphosphate aldolase[9], which has a fairly broad specificity requiring only that the D-ketose diphosphate produced (or cleaved, in the reverse reaction) has a 3S, 4R configuration.

The Su-1,7-diP is then dephosphorylated to the 7-monophosphate (Su-7-P) by sedoheptulose bis-phosphatase[34] (reaction 9, Fig. 5.26). Su-7-P then reacts with an equal amount of 3-PGAld to form equimolar amounts of Xu-5-P and D-ribose 5-phosphate (R-5-P) [eq. (5.31); reaction 10, Fig. 5.26]. This is the fourth way in which 3-PGAld is used in the Calvin cycle. The enzyme catalysing the reaction is again transketolase[11]. The reaction also occurs in the oxidative pentose phosphate pathway.

$$
\begin{array}{c}
CH_2O\,\textcircled{P} \\
| \\
C=O \\
| \\
CH_2OH \\
+ \\
CHO \\
| \\
H-C-OH \\
| \\
H-C-OH \\
| \\
CH_2O\,\textcircled{P}
\end{array}
\;\rightleftharpoons\;
\begin{array}{c}
CH_2O\,\textcircled{P} \\
| \\
C=O \\
| \\
HO-C-H \\
| \\
H-C-OH \\
| \\
H-C-OH \\
| \\
H-C-OH \\
| \\
CH_2O\,\textcircled{P}
\end{array}
\qquad (5.30)
$$

D-sedoheptulose 1,7-diphosphate

The Xu-5-P formed in reactions 7 and 10, Fig. 5.26 [eqs. (5.29) and (5.31)], then undergoes epimerization at C–3 under the catalytic influence of

$$
\begin{array}{c}
CH_2OH \\
| \\
C=O \\
| \\
HO-C-H \\
| \\
H-C-OH \\
| \\
H-C-OH \\
| \\
H-C-OH \\
| \\
CH_2O\,\textcircled{P}
\end{array}
\;+\;
\begin{array}{c}
CHO \\
| \\
H-C-OH \\
| \\
CH_2O\,\textcircled{P}
\end{array}
\;\overset{TPP,\;Mg^{2+}}{\rightleftharpoons}\;
\begin{array}{c}
CHO \\
| \\
H-C-OH \\
| \\
H-C-OH \\
| \\
H-C-OH \\
| \\
CH_2O\,\textcircled{P}
\end{array}
\;+\;
\begin{array}{c}
CH_2OH \\
| \\
C=O \\
| \\
HO-C-H \\
| \\
H-C-OH \\
| \\
CH_2O\,\textcircled{P}
\end{array}
\qquad (5.31)
$$

D-sedoheptulose 7-phosphate   3-phospho-glyceraldehyde   D-ribose 5-phosphate   D-xylulose 5-phosphate ribulose phosphate 3-epimerase[12] to form D-ribulose 5-phosphate (Ru-5-P), again a reaction which occurs in the oxidative pentose phosphate pathway [eq. (5.32); reaction 11, Fig. 5.26].

$$
\begin{array}{ccc}
\mathrm{CH_2OH} & & \mathrm{CH_2OH} \\
| & & | \\
\mathrm{C}\!=\!\mathrm{O} & & \mathrm{C}\!=\!\mathrm{O} \\
| & & | \\
\mathrm{HO}\!-\!\mathrm{C}\!-\!\mathrm{H} & \rightleftharpoons & \mathrm{H}\!-\!\mathrm{C}\!-\!\mathrm{OH} \\
| & & | \\
\mathrm{H}\!-\!\mathrm{C}\!-\!\mathrm{OH} & & \mathrm{H}\!-\!\mathrm{C}\!-\!\mathrm{OH} \\
| & & | \\
\mathrm{CH_2O}\,\circ\!\mathrm{P} & & \mathrm{CH_2O}\,\circ\!\mathrm{P}
\end{array}
\qquad (5.32)
$$

D-xylulose 5-phosphate   D-ribulose 5-phosphate

The R-5-P formed in reaction 10, Fig. 5.26, is also isomerized to Ru-5-P [eq. (5.33), reaction 12, Fig. 5.26]. The enzyme concerned is ribose phosphate isomerase[13]. This reaction also occurs in the oxidative pentose phosphate pathway.

$$
\begin{array}{ccc}
\mathrm{CHO} & & \mathrm{CH_2OH} \\
| & & | \\
\mathrm{H}\!-\!\mathrm{C}\!-\!\mathrm{OH} & & \mathrm{C}\!=\!\mathrm{O} \\
| & & | \\
\mathrm{H}\!-\!\mathrm{C}\!-\!\mathrm{OH} & \rightleftharpoons & \mathrm{H}\!-\!\mathrm{C}\!-\!\mathrm{OH} \\
| & & | \\
\mathrm{H}\!-\!\mathrm{C}\!-\!\mathrm{OH} & & \mathrm{H}\!-\!\mathrm{C}\!-\!\mathrm{OH} \\
| & & | \\
\mathrm{CH_2O}\,\circ\!\mathrm{P} & & \mathrm{CH_2O}\,\circ\!\mathrm{P}
\end{array}
\qquad (5.33)
$$

D-ribose 5-phosphate   D-ribulose 5-phosphate

Finally the Ru-5-P formed in reactions 11 and 12, Fig. 5.26, is phosphorylated at the expense of ATP generated by photophosphorylation, producing Ru-1,5-diP [eq. (5.34); reaction 13, Fig. 5.26]. The enzyme catalysing the reaction is phosphoribulokinase[14].

With the production of Ru-1,5-diP in reaction 13, Fig. 5.26, the starting material of the Calvin cycle has been regenerated. In effect Ru-1,5-diP has been carboxylated, taken through a series of reactions and then regenerated. It is difficult at first sight to see how any sugar has been made in this cycle, until its stoichiometry is considered.

(ii) STOICHIOMETRY OF THE CALVIN CYCLE

It is obvious that to produce one molecule of hexose (6C sugar) 6 molecules of CO_2 are required. Let us, therefore, introduce into the cycle 6 molecules of CO_2 and see how a molecule of hexose can be produced. The introduction of 6 molecules of CO_2 requires 6 molecules of Ru-1,5-diP to act as acceptors and the product will be 12 molecules of 3-PGA which are then phosphorylated at the expense of 12 ATPs to form 12 molecules of 1,3-diPGA. The 12 molecules of 1,3-diPGA are then reduced to 12 molecules of 3-PGAld, using 12 molecules of NADPH in the process.

Now 3-PGAld can be utilized in the cycle in four different ways (reactions 4, 5, 7 and 10, Fig. 5.26). Five of the 12 molecules of 3-PGAld are converted into 5 molecules of DiHOAcP; 3 are condensed with 3 of the 5 molecules of DiHOAcP to produce 3 molecules of F-1,6-diP which are then dephosphorylated to 3 molecules of F-6-P. There now remain 4 molecules of 3-PGAld and 2 molecules of DiHOAcP.

If one of the 3 molecules of F-6-P is removed, it represents the product of the condensation of 6 molecules of CO_2. This leaves 2 molecules of F-6-P.

So from 2 molecules of F-6-P, 4 molecules of 3-PGAld and 2 molecules of DiHOAcP, 6 molecules

$$
\begin{array}{ccccc}
\mathrm{CH_2OH} & & & \mathrm{CH_2O}\,\circ\!\mathrm{P} & \\
| & & & | & \\
\mathrm{C}\!=\!\mathrm{O} & & & \mathrm{C}\!=\!\mathrm{O} & \\
| & & & | & \\
\mathrm{H}\!-\!\mathrm{C}\!-\!\mathrm{OH} & +\ \mathrm{ATP} \longrightarrow & \mathrm{H}\!-\!\mathrm{C}\!-\!\mathrm{OH} & +\ \mathrm{ADP} & \qquad(5.34) \\
| & & | & \\
\mathrm{H}\!-\!\mathrm{C}\!-\!\mathrm{OH} & & \mathrm{H}\!-\!\mathrm{C}\!-\!\mathrm{OH} & \\
| & & | & \\
\mathrm{CH_2O}\,\circ\!\mathrm{P} & & \mathrm{CH_2O}\,\circ\!\mathrm{P} &
\end{array}
$$

D-ribulose 5-phosphate   D-ribulose 1,5-diphosphate

of Ru-1,5-diP have to be regenerated so that the cycle can continue. This is possible because we have the 30 fixed carbon atoms needed for 6 5C molecules. The rest of the cycle is, therefore, devoted to the rearrangement of these carbon atoms to produce the 6 pentose molecules.

The 2 molecules of F-6-P react with 2 of the 4 remaining molecules of 3-PGAld, yielding 2 molecules of Xu-5-P and 2 molecules of E-4-P. The 2 molecules of E-4-P then condense with the 2 remaining molecules of DiHOAcP producing 2 molecules of Su-1,7-diP which are then dephosphorylated to Su-7-P. Apart from the latter there now remain 2 molecules of 3-PGAld and 2 molecules of Xu-5-P.

The 2 molecules of Su-7-P condense with the 2 molecules of 3-PGAld to produce 2 molecules of Xu-5-P (making a total of 4) and 2 molecules of Ru-5-P.

All 30 carbon atoms are now in the form of pentose (4 molecules of Xu-5-P and 2 molecules of R-5-P) and are now isomerized to 6 molecules of Ru-1,5-diP.

Finally, 6 molecules of ATP are used to phosphorylate the 6 molecules of Ru-5-P so producing the 6 molecules of Ru-1,5-diP required to begin the cycle again.

Thus in this cyclic process 6 molecules of CO_2 are in effect converted into 1 molecule of F-6-P at the expense of 12 molecules of NADPH and 18 molecules of ATP. The overall reaction can be given as:

$$6CO_2 + 12NADPH + 12H^+ + 18ATP + 11H_2O$$
$$\downarrow$$
$$F\text{-}6\text{-}P + 12NADP^+ + 18ADP + 17P_i$$

It may be said that to produce the equivalent of 1 molecule of hexose sugar, (a) 6 molecules of CO_2 must be introduced and the cycle turned once, or (b) the cycle must be turned 6 times.

It will be noted from the overall reaction that for every molecule of CO_2 utilized, 2 molecules of NADPH and 3 molecules of ATP are required. However, this is the stoichiometry of the Calvin cycle itself and does not take account of the effect of photorespiration (see Chapter 6, Section F.1) which causes the loss of photosynthetically fixed carbon. In C_3 plants, growing in their normal environments, this loss is 20–40%. Thus the net numbers of

NADPH and ATP molecules required to fix 1 molecule of CO_2 in the photosynthetic cell of a C_3 plant (i.e. the resultant of photosynthesis minus photorespiration) are very much higher than 2 and 3 respectively.

(iii) Evidence for the Calvin cycle

(a) *Evidence from studies of the variation of labelling of intermediates with time when* $^{14}CO_2$ *is administered to a photosynthesizing organism.* The rationale of the approach is as follows. In a plant carrying out photosynthesis under steady-state conditions CO_2 is being converted into hexose through a series of intermediates, i.e.

$$CO_2 \rightarrow A \rightarrow B \rightarrow C \rightarrow D \rightarrow hexose$$

If $^{14}CO_2$ is introduced into this system it will take a finite time for some of the label to be incorporated into hexose; during this period all the intermediates will also be labelled. If the time period over which the $^{14}CO_2$ is administered is gradually reduced over a series of experiments it will eventually be found that the label has not had sufficient time to get to hexose but has reached, say, intermediate C; thus C and all intermediates before it will be labelled. By reducing the period of $^{14}CO_2$ administration still further a situation will eventually be reached where the label has only got as far as A, i.e. only one compound will be labelled. This compound must be the first compound in the series, the carboxylation product. Such experiments will, therefore, identify the carboxylation product and also give an indication of the sequence of intermediates in the formation of hexose from the rate at which they become labelled.

Calvin carried out a series of these experiments using the green algae, *Chlorella* and *Scenedesmus*. The algal cells were allowed to grow under constant conditions of light and CO_2. Then $^{14}CO_2$ was introduced for a given period and the cells were killed by dropping them into boiling alcohol which also served to extract the labelled compounds. The compounds in the alcoholic extract were separated from each other and identified by paper chromatography. The radioactivity in these compounds was then determined. When the algae were exposed to $^{14}CO_2$ for 30 sec, ^{14}C was found in 3-PGA, triose phosphates and hexose phosphates. When the period of exposure was shortened to 5 sec, however, ^{14}C was found only in 3-PGA. This suggested that the carboxylation product was 3-PGA and that triose phosphates and hexose phosphates were then formed.

(b) *Evidence from the labelling pattern of the carbon chain in the intermediates.* After a short exposure to $^{14}CO_2$, the labelled 3-PGA, triose phosphates and hexose phosphates were separated from one another and isolated. Then the molecules were degraded in such a way that each carbon could be obtained separately and assayed for radioactivity.

It was found that 3-PGA was labelled in the carboxyl group. Since this was the first compound to be labelled, this finding suggested that CO_2 condensed with a 2C compound. However, a long search for such a 2C compound failed and led to a re-examination of the early products of photosynthetic CO_2 fixation. Amongst these were found the pentose phosphates and Su-7-P and Su-1,7-diP. This in turn led to speculation, later proved, that CO_2 condensed with a pentose phosphate (Ru-1,5-diP) to produce a 6C compound which was then cleaved into two molecules of 3-PGA.

It was found that C-3 and C-4 of F-6-P were labelled heavily compared with the other carbon atoms indicating that hexoses were formed by condensation of two trioses. The labelling pattern of other intermediates could also be explained by the operation of the carbon reduction cycle.

(c) *Evidence from changes in the steady-state levels of intermediates.* The carbon reduction cycle can be abbreviated to the scheme shown below. From this it can be seen that to regenerate Ru-1,5-diP from 3-PGA, NADPH and ATP, produced in the light phase of photosynthesis, are required. Thus in the absence of light the supply of NADPH and ATP is

cut off and the level of 3-PGA will rise and the level of Ru-1,5-diP will fall. If light is now supplied, NADPH and ATP can be produced again and the accumulated 3-PGA can be converted into triose phosphates and on to Ru-1,5-diP; there will thus be a drop in the high level of 3-PGA and a rise in the low level of Ru-1,5-diP.

The following experiment was carried out by Calvin using *Chlorella*. Photosynthesis in the presence of $^{14}CO_2$ was allowed to proceed until steady-state conditions obtained, when any change in the concentration of intermediates could be measured by determining their radioactivity. The results of turning off the light and then turning it on again upon the levels of 3-PGA and Ru-1,5-diP are shown in Fig. 5.29. Turning off the light brings about an immediate rise in the level of 3-PGA and a fall in the level of Ru-1,5-diP. Ru-1,5-diP remains at a low level throughout the dark period but the 3-PGA level falls steadily from its high value probably owing to its conversion into other metabolites. When the light is turned on again there is a rapid drop in the level of 3-PGA and a rapid rise in the level of Ru-1,5-diP until the steady-state levels are again reached. The response of these compounds to the presence or absence of light indicates that they are related in a cyclic manner.

(d) *Evidence from the isolation of enzymes.* All the enzymes of the Calvin cycle have been demonstrated in chloroplasts. Those enzymes which are peculiar to the cycle are found only in the chloroplasts.

However, the quantity of some of these enzymes in isolated chloroplasts is not high enough to account for the known capacity of intact leaves for photosynthetic carbon fixation. This should not be regarded as negative evidence but rather a consequence of the difficulty in isolating undamaged chloroplasts. The outer of the two chloroplast membranes

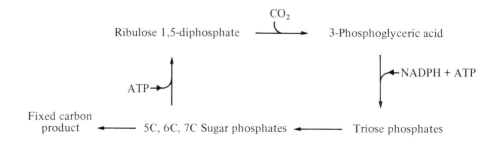

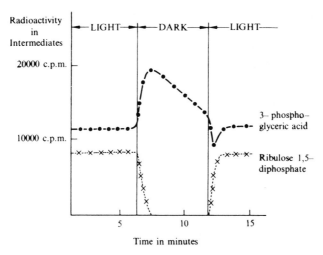

Fig. 5.29a. Effect of periods of light and dark on the levels of 3-phosphoglyceric acid and ribulose 1,5-diphosphate.

is often lost during isolation procedures and this allows the loss of soluble enzymes from the stroma.

(iv) ROLE OF LIGHT IN THE ACTIVATION OF 'DARK PHASE' ENZYMES

Up to this point in the chapter the assimilation of CO_2 into carbohydrate has been described as the 'dark phase' of photosynthesis, the implication being that light plays no part in the Calvin cycle other than to generate the ATP and NADPH required to drive it. Recent work has, however, shown that this is not the case, for it has become clear that light activates at least five of the Calvin cycle enzymes, namely RuDP carboxylase, glyceraldehyde 3-phosphate dehydrogenase, fructose bisphosphatase, sedoheptulose bisphosphatase and phosphoribulokinase. Thus it is probably now incorrect to speak of the 'dark phase' photosynthesis; the term has, however, been retained in this book so that continuity with older textbooks may be maintained and possible confusion avoided.

The presence or absence of light promotes a number of changes within the chloroplast stroma (where most of the Calvin cycle enzymes reside) which in turn promote activation or deactivation of the five enzymes mentioned previously (see Fig. 5.29b). For instance, light causes H^+ ions to be

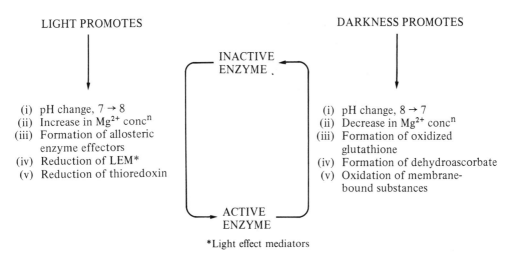

Fig. 5.29b. Light- and dark-promoted changes which in turn promote the activation or deactivation of some chloroplastidic enzymes.

pumped from the stroma of the chloroplast into the thylakoid lumen (see Fig. 5.20); this causes the pH of the stroma to rise from about 7 to 8. In the absence of light H^+ ions leak back into the stroma and its pH falls from 8 back to 7. Light also causes Mg^{2+} ions to be pumped from the thylakoid lumen into the stroma. These ionic changes in the stroma help to activate RuDP carboxylase, fructose bisphosphatase, sedoheptulose bisphosphatase and phosphoribulokinase because the new H^+ and Mg^{2+} concentrations are optimal for these enzymes.

Light also promotes the formation of compounds which are positive allosteric effectors of some of these enzymes. ATP and NADPH, formed by light-driven non-cyclic photophosphorylation, are positive allosteric effectors of RuDP carboxylase and glyceraldehyde 3-phosphate dehydrogenase. Moreover, several of the intermediates of the Calvin cycle occurring between F-1,6-diP and Ru-5-P are allosteric effectors of RuDP carboxylase; the stromal concentration of these compounds will rise as light generates ATP and NADPH and initiates the fixation of CO_2.

Light also promotes the reduction of compounds which have been termed 'light effect mediators' (LEMs) which in turn activate some of these enzymes. The identity of all of these LEMs is not known. However, it is envisaged that they are proteins containing cyst(e)ine residues that occur in the oxidized, disulphide (cystine or —S—S—) state in the dark and the reduced, sulphydryl (cysteine or —SH HS—) state in the light. Moreover, it is believed that the light-induced transition from oxidized to reduced state is brought about by electrons tapped-off from non-cyclic photophosphorylation shortly after the operation of P700. It is also believed that the reduced LEM interacts directly with the inactive form of the enzyme, reducing some key grouping (possibly a —S—S— moiety) in it and thereby producing its active form.

The LEM about which most is known is thioredoxin, a low molecular weight, hydrogen-carrying, cyst(e)ine protein. It is reduced in the light by the reduced form of soluble ferredoxin (generated by non-cyclic photophosphorylation) under the catalytic influence of the enzyme ferredoxin-thioredoxin reductase, a chromophore-less protein of MW 38,000 (see Fig. 5.29c). Reduced thioredoxin activates glyceraldehyde 3-phosphate dehydrogenase, fructose bisphosphatase, sedoheptulose bisphosphatase and phosphoribulokinase. Chloroplast thioredoxin exists in two forms designated f and m; thioredoxin f is the form which activates the enzymes just mentioned. Thioredoxin m is particularly effective in activating the NADP-specific malate dehydrogenase which is a key enzyme in the mesophyll chloroplasts of those plants that carry out the 'NADP-malic enzyme-dependent' C_4 mode of photosynthesis [see Section D.3(i) and Fig. 5.30].

Another light-activated enzyme, probably activated by a LEM, is pyruvate, orthophosphate dikinase[20] which is peculiar to photosynthetic CO_2 fixation in C_4 and CAM plants [see Sections D.3(i) and D.4(i) respectively].

The enzyme ADP-glucose pyrophosphorylase, a key enzyme in the chloroplastidic synthesis of starch, is light modulated via metabolic effectors (see Fig. 5.36). It is allosterically activated by 3-PGA and allosterically inhibited by orthophosphate. Since light increases the intrachloroplastidic content of 3-PGA and decreases that of orthophosphate the enzyme is activated in the light and therefore able to play its part in the conversion of hexose phosphate into starch (see Section D.6).

3. The C_4 mode of photosynthesis (C_4 pathway)

(i) MECHANISM OF THE C_4 PATHWAY

Although the C_4 mode of photosynthesis was discovered in two tropical grasses (sugar cane and maize) it occurs in many other species of Gramineae, some species of Cyperaceae and in some species of the dicotyledonous families, Aizoaceae, Amaranthaceae, Chenopodaceae, Compositae, Euphorbiaceae, Nyctaginaceae, Portulacaceae and Zygophyllaceae. These C_4 plants grow optimally at high light intensities and at day temperatures of 30–35°C. They are characterized by having (i) high photosynthetic rates (40–80 mg CO_2 fixed per dm^2 of leaf surface per hour, cf. 15–40 for C_3 plants), (ii) high growth rates (4–5 g of dry wt produced per dm^2 of leaf surface per day, cf. 0.5–2 for C_3 plants), (iii) low photorespiration rates, (iv) low rates of water loss in relation to quantity of dry matter produced (250–350 g H_2O per g dry wt, cf. 450–950 for C_3 plants) and (v) an unusual leaf structure. The latter

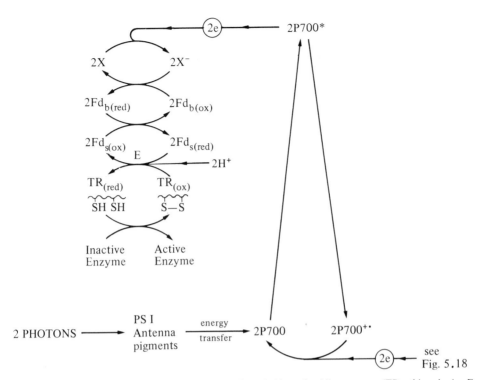

Fig. 5.29c. The role of thioredoxin in the light-dependent activation of chloroplastic enzymes. (TR = thioredoxin; E = ferredoxin-thioredoxin reductase; all other abbreviations have the same meaning as those in Fig. 5.18.)

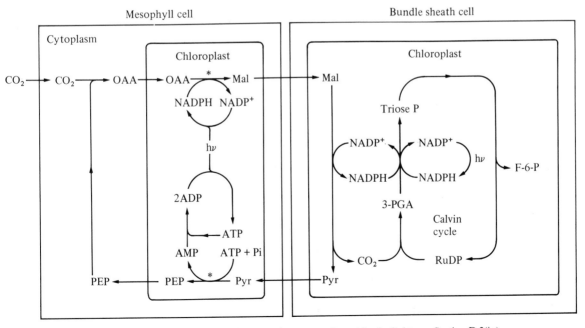

*The enzymes catalysing these reactions are activated in the light, see Section D.2(iv)

Fig. 5.30. C_4 photosynthesis in those plant species that utilize the NADP-malic enzyme to catalyse the decarboxylation reaction.

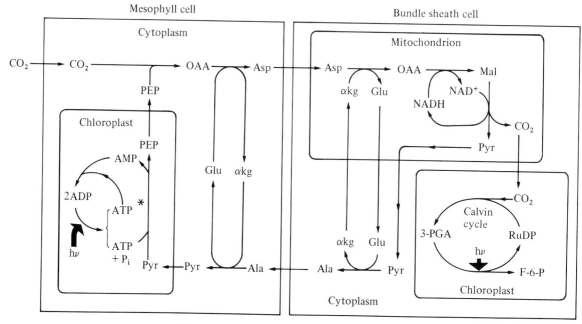

*Enzyme activated in the light, see Section D.2(iv)

Fig. 5.31. C_4 photosynthesis in those plant species that utilize the NAD-malic enzyme to catalyse the decarboxylation reaction.

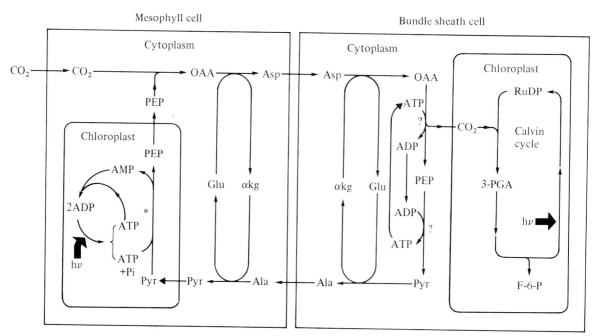

*Enzyme activated in the light, see Section D.2(iv)

Fig. 5.32. C_4 photosynthesis in those plant species that utilize PEP carboxykinase to catalyse the decarboxylation reaction.

is typified by (i) open air spaces extending from the atmosphere to a very high proportion of photosynthetic cells, thereby allowing efficient CO_2 uptake, (ii) a double spiral of bundle-sheath cells tightly packed around the vascular bundles, (iii) mesophyll cells loosely arranged in spirals around the layer of bundle-sheath cells and (iv) the presence of many plasmodesmata between the bundle-sheath cells and the mesophyll cells. Chloroplasts are present in the bundle-sheath cells and the mesophyll cells but are quite different in appearance. This difference is most marked in the advanced C_4 plants, the Panicoid grasses (see Fig. 3.10) where the bundle-sheath chloroplasts have no grana or grana that are much reduced in size. The first proof that the mechanism of photosynthetic carbon fixation in such plants was different from that of C_3 plants came in 1965 when Kortschak and his coworkers showed that the C_4 dicarboxylic acids, L-malic acid and L-aspartic acid, were labelled first when sugarcane leaves were allowed to photosynthesize in $^{14}CO_2$ and that there was a distinct lag before ^{14}C appeared in 3-PGA and the remaining Calvin cycle intermediates. This work was taken up and greatly extended by Hatch and Slack during the next few years. They initially examined the labelling pattern of the carbon chain of the compounds labelled by the photosynthetic fixation of $^{14}CO_2$ and found (i) that the ^{14}C present in oxaloacetic acid (OAA), malic acid and aspartic acid was almost entirely located in C-4, (ii) that 3-PGA was labelled only in C-1 and (iii) that the labelling pattern of the hexose monophosphates was consistent with their having been formed from 3-PGA. These findings suggested that a C_3 compound was initially carboxylated to form a C_4 dicarboxylic acid which then transferred its C-4 by a 'transcarboxylation reaction' to an acceptor molecule with the formation of 3-PGA; the acceptor molecule would thus provide C-2 and C-3 of 3-PGA with C-1 coming from C-4 of the dicarboxylic acid. The transcarboxylation reaction, however, proved elusive and it was eventually shown that C-4 of the C_4 dicarboxylic acid is removed as CO_2 in a decarboxylation reaction. The CO_2 was then incorporated into C-1 of 3-PGA by RuDP carboxylase, the Calvin cycle carboxylation enzyme. Thus the C_4 pathway was shown to involve two carboxylation reactions. It was then demonstrated that the two carboxylation reactions are

physically separated from one another. The initial carboxylation, that of the C_3 compound which is now known to be phosphoenolpyruvate (PEP), takes place in the mesophyll cells whilst the decarboxylation of the C_4 dicarboxylic acid and the second carboxylation, catalysed by RuDP carboxylase, take place in the bundle sheath cells.

The broad outline of the mechanism of C_4 photosynthesis is now clear; only the detail remains to be filled in. It is apparent that there are minor variations in this mechanism in different C_4 plants. Three such variations have been discovered to date and all differ principally in the way the C_4 dicarboxylic acid is decarboxylated in the bundle-sheath cells. Some species (e.g. *Zea mays*, *Saccharum officinarum*, *Sorghum sudanense*) utilize the NADP-malic enzyme[15] for this reaction. A second group (e.g. *Atriplex spongiosa*, *Portulaca oleracea*, *Amaranthus edulis*) utilize the NAD-malic enzyme[16] whilst a third group (e.g. *Panicum maximum*, *Chloris gayana*, *Sporobolus fimbriatus*) use PEP carboxykinase[17]. The three different pathways are shown in Figs. 5.30, 5.31 and 5.32.

Air passes through the open stomata of the leaf into the extensive intercellular spaces and bathes the mesophyll cells. CO_2 then passes into the cytoplasm of these cells where it dissolves and ionizes probably under the catalytic influence of carbonic anhydrase [eq. (5.35)].

The resulting HCO_3^- is then used by PEP carboxylase[18] to carboxylate PEP with the formation of OAA [eq. (5.36)]; HCO_3^-, rather than CO_2, in contrast to RuDP carboxylase[3], is the substrate of PEP carboxylase[18].

Interconversion of OAA, malic acid and aspartic acid then rapidly occurs in all C_4 species. Thus all three C_4 dicarboxylic acids are rapidly labelled in the presence of $^{14}CO_2$. However, net utilization of C_4 dicarboxylic acids occurs via malic acid in NADP-malic enzyme species (Fig. 5.30) and via aspartic acid in NAD-malic enzyme (Fig. 5.31) and PEP carboxykinase species (Fig. 5.32).

In NADP-malic enzyme species the OAA passes into the mesophyll chloroplasts where NADP-malate dehydrogenase[19] catalyses its reduction to malic acid using NADPH generated by the light phase of photosynthesis [eq. (5.37)]. NADP-malate dehydrogenase is activated in the presence of light. The malic acid then passes from the mesophyll

$$^{14}CO_2 \; + \; H_2O \; \rightleftharpoons \; H_2^{\,14}CO_3 \; \rightleftharpoons \; H^+ \; + \; H^{14}CO_3^- \qquad (5.35)$$

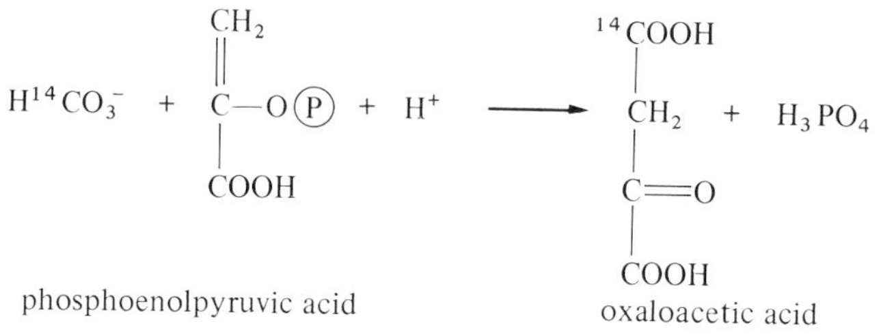

$$
H^{14}CO_3^- \; + \;
\begin{array}{c}
CH_2 \\
\parallel \\
C\!-\!O\,\circled{P} \\
\vert \\
COOH
\end{array}
\; + \; H^+ \; \longrightarrow \;
\begin{array}{c}
^{14}COOH \\
\vert \\
CH_2 \\
\vert \\
C\!=\!O \\
\vert \\
COOH
\end{array}
\; + \; H_3PO_4 \qquad (5.36)
$$

phosphoenolpyruvic acid   oxaloacetic acid

$$
\begin{array}{c}
^{14}COOH \\
\vert \\
CH_2 \\
\vert \\
C\!=\!O \\
\vert \\
COOH
\end{array}
\; + \; NADPH \; + \; H^+ \; \rightleftharpoons \;
\begin{array}{c}
^{14}COOH \\
\vert \\
CH_2 \\
\vert \\
H\!-\!C\!-\!OH \\
\vert \\
COOH
\end{array}
\; + \; NADP^+ \qquad (5.37)
$$

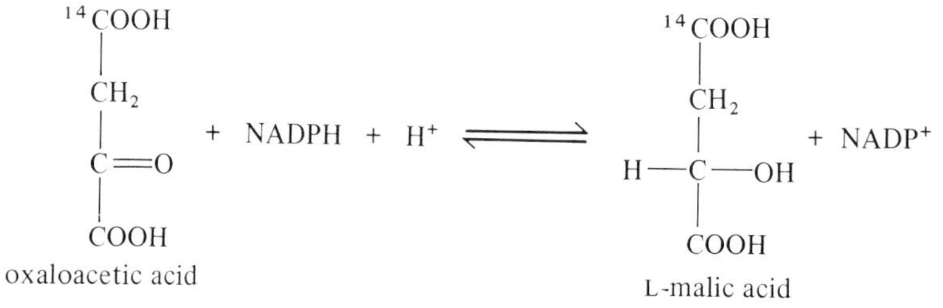

oxaloacetic acid   L-malic acid

chloroplasts into the chloroplasts of the bundle-sheath cells probably by way of the cytoplasm of the plasmodesmata which connect the two types of cell. It is then decarboxylated by the NADP-specific malic enzyme[15] to yield CO_2 and pyruvic acid [eq. (5.38)].

The CO_2 becomes the substrate for RuDP carboxylase[3] and is incorporated into 3-PGA [eq. (5.18)] which is then converted into F-6-P by the operation of the Calvin cycle (see Fig. 5.26). The NADPH produced in eq. (5.38) is probably reoxi-

dized by using it as the reductant in the conversion of 1,3-diPGA into 3-PGAld [eq. (5.25)] in the Calvin cycle thereby supplementing the NADPH produced for that purpose in the light phase of photosynthesis. The pyruvic acid produced in eq. (5.38) passes back from the bundle-sheath chloroplasts into the mesophyll chloroplasts again probably by way of the cytoplasm of the plasmodesmata. It is then converted into PEP by the enzyme pyruvate, orthophosphate dikinase[20] which catalyses eq. (5.39); this enzyme is activated in the presence of light.

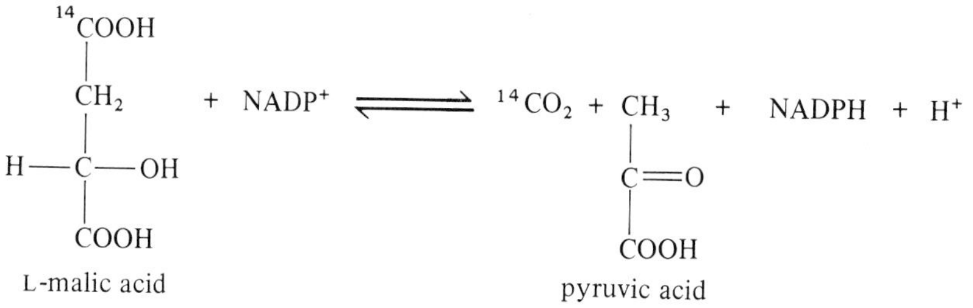

$$
\begin{array}{c}
^{14}COOH \\
\vert \\
CH_2 \\
\vert \\
H\!-\!C\!-\!OH \\
\vert \\
COOH
\end{array}
\; + \; NADP^+ \; \rightleftharpoons \; ^{14}CO_2 \; + \;
\begin{array}{c}
CH_3 \\
\vert \\
C\!=\!O \\
\vert \\
COOH
\end{array}
\; + \; NADPH \; + \; H^+ \qquad (5.38)
$$

L-malic acid   pyruvic acid

$$
\begin{array}{c}
CH_3 \\
\vert \\
C\!=\!O \\
\vert \\
COOH
\end{array}
\; + \; ATP \; + \; P_i \; \rightleftharpoons \;
\begin{array}{c}
CH_2 \\
\parallel \\
C\!-\!O\,\circled{P} \\
\vert \\
COOH
\end{array}
\; + \; AMP \; + \; PP_i \qquad (5.39)
$$

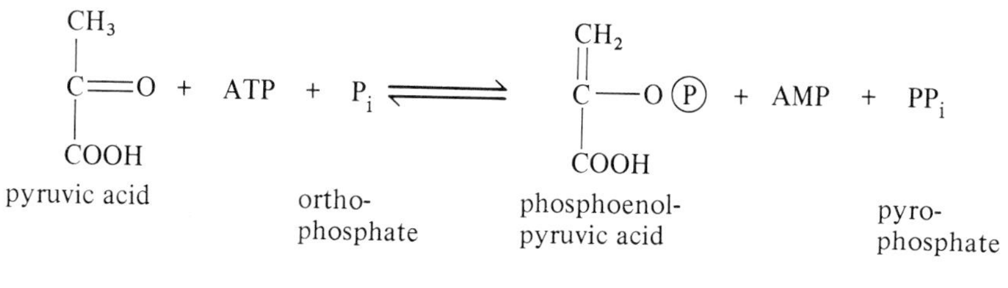

pyruvic acid   ortho-   phosphoenol-   pyro-
                           phosphate   pyruvic acid   phosphate

$$\text{Enzyme} + \text{ATP} + \text{P}_i \rightleftharpoons \text{enzyme-P} + \text{AMP} + \text{PP}_i \tag{5.40}$$

$$\text{Enzyme-P} + \text{Pyruvic acid} \rightleftharpoons \text{Enzyme} + \text{PEP} \tag{5.41}$$

$$\begin{array}{c} \text{O} \quad\quad \text{O} \\ \| \quad\quad \| \\ {}^-\text{O}-\text{P}-\text{O}-\text{P}-\text{O}^- \\ | \quad\quad | \\ \text{O}^- \quad\quad \text{O}^- \end{array} + \text{H}_2\text{O} \rightleftharpoons 2\ \begin{array}{c} \text{O} \\ \| \\ \text{HO}-\text{P}-\text{O}^- \\ | \\ \text{O} \end{array} \tag{5.42}$$

pyrophosphate   orthophosphate

Present evidence indicates that the mechanism of the reaction described by eq. (5.39) involves two steps [eqs. (5.40) and (5.41)].

The pyrophosphate produced in eq. (5.39) is hydrolysed to orthophosphate by pyrophosphatase[21] [eq. (5.42)].

The AMP produced in eq. (5.39) is phosphorylated by ATP under the catalytic influence of adenylate kinase[22] [eq. (5.43)].

$$\text{AMP} + \text{ATP} \rightleftharpoons 2\text{ADP} \tag{5.43}$$

The resulting ADP is then converted back into ATP in the light phase of photosynthesis. The mesophyll chloroplasts are particularly rich in pyrophosphatase and adenylate kinase. The overall process involved in the conversion of pyruvic acid into PEP is shown in Fig. 5.33 and from this it is apparent that two ATPs are required to drive it. The PEP produced in eq. (5.39) diffuses from the chloroplast into the cytoplasm of the mesophyll cell and becomes the substrate for PEP carboxylase[18] [eq. (5.36)]; the cycle of C_4 photosynthesis is thus complete.

In those C_4 plant species which utilize the NAD-

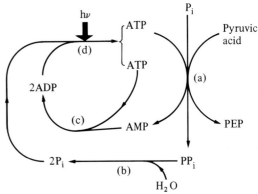

(a) = pyruvate, orthophosphate dikinase (activated by light)
(b) = pyrophosphatase
(c) = adenylate kinase
(d) = light phase of photosynthesis

Fig. 5.33. Mechanism of conversion of pyruvic acid into PEP in mesophyll chloroplasts.

malic enzyme[16] for the decarboxylation reaction (Fig. 5.31) the OAA produced in eq. (5.36) is transaminated by the cytoplasmic L-aspartate aminotransferase[23] utilizing L-glutamic acid as the amino-donor [eq. (5.44)].

$$\begin{array}{c} {}^{14}\text{COOH} \\ | \\ \text{CH}_2 \\ | \\ \text{C}=\text{O} \\ | \\ \text{COOH} \end{array} + \begin{array}{c} \text{COOH} \\ | \\ \text{CH}_2 \\ | \\ \text{CH}_2 \\ | \\ \text{H}-\text{C}-\text{NH}_2 \\ | \\ \text{COOH} \end{array} \rightleftharpoons \begin{array}{c} {}^{14}\text{COOH} \\ | \\ \text{CH}_2 \\ | \\ \text{H}-\text{C}-\text{NH}_2 \\ | \\ \text{COOH} \end{array} + \begin{array}{c} \text{COOH} \\ | \\ \text{CH}_2 \\ | \\ \text{CH}_2 \\ | \\ \text{C}=\text{O} \\ | \\ \text{COOH} \end{array} \tag{5.44}$$

oxaloacetic acid   L-glutamic acid   L-aspartic acid   α-ketoglutaric acid

The resulting aspartic acid then passes from the cytoplasm of the mesophyll cell to the mitochondria of the bundle-sheath cells probably via the plasmodesmata. There OAA is re-formed by reversal of the transamination reaction [eq. (5.44)]. The OAA is then reduced by the mitochondrial NAD-specific malate dehydrogenase[24] to L-malic acid [eq. (5.45)].

The malic acid is decarboxylated by the mitochondrial NAD-malic enzyme[16] to yield pyruvic acid and CO_2; the requirement of this reaction [eq. (5.46)] for NAD^+ balances the NAD^+ production of eq. (5.45).

The CO_2 diffuses from the mitochondria to the chloroplasts where it is incorporated into carbohydrate by the Calvin cycle. The pyruvic acid diffuses into the cytoplasm where it is transaminated by L-alanine aminotransferase[25] utilizing L-glutamic acid as the amino donor [eq. (5.47)].

Provided glutamic acid can pass from the mitochondria into the cytoplasm and α-ketoglutaric acid can move in the opposite direction, the two transamination reactions would balance one another. The alanine produced in eq. (5.47) passes from the cytoplasm of the bundle sheath cell to that of the mesophyll cell probably via the plasmodesmata. It is then converted into pyruvic acid by L-alanine aminotransferase utilizing the α-ketoglutaric acid produced in eq. (5.44) as the amino-acceptor; this reaction is the reverse of that

$$
\begin{array}{c}
^{14}COOH \\
| \\
CH_2 \\
| \\
C=O \\
| \\
COOH
\end{array}
+ NADH + H^+ \rightleftharpoons
\begin{array}{c}
^{14}COOH \\
| \\
CH_2 \\
| \\
H-C-OH \\
| \\
COOH
\end{array}
+ NAD^+ \qquad (5.45)
$$

oxaloacetic acid   L-malic acid

$$
\begin{array}{c}
^{14}COOH \\
| \\
CH_2 \\
| \\
H-C-OH \\
| \\
COOH
\end{array}
+ NAD^+ \rightleftharpoons {}^{14}CO_2 +
\begin{array}{c}
CH_3 \\
| \\
C=O \\
| \\
COOH
\end{array}
+ NADH + H^+ \qquad (5.46)
$$

L-malic acid   pyruvic acid

$$
\begin{array}{c}
CH_3 \\
| \\
C=O \\
| \\
COOH
\end{array}
+
\begin{array}{c}
COOH \\
| \\
CH_2 \\
| \\
CH_2 \\
| \\
H-C-NH_2 \\
| \\
COOH
\end{array}
\rightleftharpoons
\begin{array}{c}
CH_3 \\
| \\
H-C-NH_2 \\
| \\
COOH
\end{array}
+
\begin{array}{c}
COOH \\
| \\
CH_2 \\
| \\
CH_2 \\
| \\
C=O \\
| \\
COOH
\end{array}
\qquad (5.47)
$$

pyruvic acid   L-glutamic acid   L-alanine α-ketoglutaric acid

$$\begin{array}{c} {}^{14}\text{COOH} \\ | \\ \text{CH}_2 \\ | \\ \text{C}=\text{O} \\ | \\ \text{COOH} \end{array} \quad + \quad \text{ATP} \rightleftharpoons {}^{14}\text{CO}_2 \; + \quad \begin{array}{c} \text{CH}_2 \\ || \\ \text{C}-\text{O}\,\textcircled{P} \\ | \\ \text{COOH} \end{array} \quad + \quad \text{ADP} \qquad (5.48)$$

oxaloacetic
acid

phosphoenol
pyruvic acid

shown in eq. (5.47). The pyruvic acid then passes into the mesophyll chloroplasts for conversion into PEP by the mechanism shown in Fig. 5.33, thus completing the cycle of C_4 photosynthesis.

In those C_4 plant species which utilize PEP carboxykinase[17] for the carboxylation reaction (Fig. 5.32) the reaction sequence is very similar to that of the species utilizing the NAD-malic enzyme. It differs principally in that the OAA produced by transamination of the L-aspartic acid imported from the mesophyll cells is decarboxylated by the enzyme PEP carboxykinase to yield CO_2 and PEP [eq. (5.48)].

The intracellular locations of PEP carboxykinase and that of L-aspartate aminotransferase, which catalyses the preceding reaction, are at present not known. Nor is the fate of PEP, although it is reasonable to assume that it is converted into pyruvate which is then transaminated to yield L-alanine and translocated to the mesophyll cells. These uncertainties are indicated in Fig. 5.32 by question marks.

The CO_2 produced in eq. (5.48) is utilized as the substrate for the Calvin cycle in the chloroplasts of the bundle-sheath cells. Alanine from the bundle-sheath cells is converted into PEP in the mesophyll cells by the same sequence of reactions as are used by the NAD-malic enzyme-utilizing species.

It will be seen from Figs. 5.30, 5.31 and 5.32 that there is a marked difference between the functions of the mesophyll chloroplasts and those of the bundle-sheath cells. The mesophyll chloroplasts do not have a functional Calvin cycle because RuDP carboxylase and several other of the necessary enzymes are not present. They do, however, have the enzymes required for the conversion of 3-PGA into triose phosphates. In spite of their inability to convert CO_2 into carbohydrate by the Calvin cycle,

the mesophyll chloroplasts do operate the light phase of photosynthesis. The ATP so generated is used to convert pyruvate into PEP. The NADPH is used in the NADP-malic enzyme species to reduce OAA to malic acid. In the NAD-malic enzyme species (Fig. 5.31) and PEP carboxykinase species (Fig. 5.32) NADP-specific malate dehydrogenase activity in mesophyll chloroplasts is very low and absent respectively; thus the photogenerated NADPH is unlikely to be used to reduce OAA to malic acid (which is still generated in these species despite its not being on the C_4 pathway). It is quite possible that the photogenerated NADPH is used in the conversion of 3-PGA into triose phosphates. The 3-PGA in question is formed in the bundle-sheath chloroplasts and is translocated to the mesophyll chloroplasts for reduction. The resulting triose phosphates are either used for starch synthesis (see Section F.6) or returned to the bundle-sheath chloroplasts for conversion into F-6-P by the Calvin cycle. Such fluxes of 3-PGA and triose phosphates between the mesophyll and bundle-sheath chloroplasts are known to occur.

The bundle-sheath chloroplasts have a fully operative Calvin cycle and light phase of photosynthesis to produce the ATP and NADPH necessary to drive it.

(ii) STOICHIOMETRY OF THE C_4 MODE OF PHOTOSYNTHESIS

Figure 5.34 summarizes the path of carbon in the C_4 mode of photosynthesis. The operation of the Calvin cycle in the bundle-sheath cell is identical with that in the chloroplasts of C_3 plants. Thus the stoichiometry is the same and 3ATP and 2NADPH are required for the fixation of each CO_2. In the C_4 mode an additional 2ATP is required for the conversion of pyruvic acid into PEP in the meso-

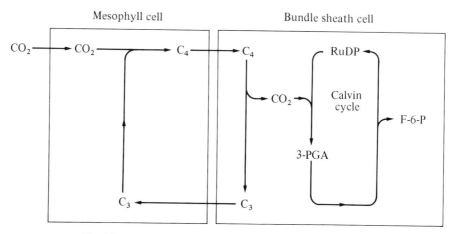

Fig. 5.34. Summary of the path of carbon in C_4 photosynthesis.

phyll cells (see Fig. 5.33). In the PEP-carboxykinase species it is assumed that the ATP required by the decarboxylation reaction in the bundle-sheath cells is regenerated when PEP is converted into pyruvate (cf. glycolysis). There is, however, no extra requirement for NADPH. That this is so needs a little explanation in the case of the NADP-malic enzyme species (Fig. 5.30). Here it is apparent that one molecule of photogenerated NADPH is utilized per CO_2 fixed during the reduction of OAA to malic acid in the mesophyll chloroplast. However, during the decarboxylation of that same malic acid in the bundle-sheath chloroplast one molecule of NADPH is generated per CO_2 produced. This NADPH is then utilized along with one photogenerated NADPH, to provide the 2NADPH required to fix one CO_2 in the Calvin cycle.

Therefore the C_4 mode of photosynthesis requires 5ATP and 2NADPH for the fixation of each CO_2 [cf. 3ATP and 2NADPH per CO_2 in C_3 plants— Section D.2(ii)]. Thus the C_4 mode of photosynthesis is less efficient than the C_3 mode. This conclusion, though true, does not accurately reflect the overall situation in the two types of plant because no account has been taken of the effect of photorespiration (see Chapter 6, Section F.1). The photorespiration rate in C_4 plants is very much less than in C_3 plants. Thus the net numbers of molecules of ATP and NADPH required to fix one molecule of CO_2 (i.e. the resultant of photosynthesis minus photorespiration) are considerably lower in

C_4 plants than in C_3 plants. In the overall sense, therefore, C_4 plants are more efficient assimilatory organisms than C_3 plants.

(iii) PURPOSE OF THE C_4 MODE OF PHOTOSYNTHESIS

At first sight the reactions peculiar to C_4 photosynthesis, namely the synthesis of a C_4 compound by carboxylation of PEP, its translocation from mesophyll to bundle-sheath cell and its subsequent decarboxylation to regenerate CO_2, seem pointless and wasteful of 2ATP per CO_2 fixed [see Section D.3(ii)]. However, they must have a purpose and the elucidation of this purpose has exercised the minds of plant biochemists for some time.

The general consensus is that the reactions peculiar to C_4 photosynthesis serve to concentrate CO_2 in the bundle-sheath chloroplasts. Such a concentration has been demonstrated experimentally. This then raises the question as to why it is necessary for the C_4 plant to do this.

The initial answer to this question, now known to be incorrect, was that it enabled the low affinity of RuDP carboxylase to be circumvented. However, this explanation was based upon measurements of 200–500 μM for the $K_{m(CO_2)}$ of RuDP carboxylase, such values being inconsistent with the high rate of CO_2 fixation observed in C_4 plants. More recent measurements, confirmed in several laboratories, show that the $K_{m(CO_2)}$ of RuDP carboxylase, freshly released from broken chloroplasts, is 11–18 μM which is consistent with the high CO_2 fixation rate.

It is now thought that CO_2 concentration is required because C_4 plants have a higher stomatal resistance to gaseous diffusion than C_3 plants. This causes a much greater CO_2 concentration gradient, during steady-state photosynthesis, between the atmosphere and the mesophyll cell surface in C_4 plants than that in C_3 plants. The CO_2 concentration in the mesophyll cell is high enough in C_3 plants to allow RuDP carboxylase to operate satisfactorily but would be far too low to allow this in C_4 plants. However, the CO_2 concentration in C_4 plants, low though it is, is high enough for the satisfactory operation of PEP carboxylase. Hence PEP carboxylase is used to raise the low concentration of CO_2 in the mesophyll cell to a concentration in the bundle-sheath cell high enough to allow RuDP carboxylase to function satisfactorily.

The higher stomatal resistance to gaseous diffusion seen in C_4 plants, just used to explain the need in these plants to concentrate CO_2, in turn poses the question as to what advantage this gives them over C_3 plants. It is this. A higher stomatal resistance reduces the loss of water from the plant through transpiration. This saving in water allows the C_4 plant to grow in hotter, drier environments than C_3 plants, but its penalty is that CO_2 uptake is more difficult. This difficulty in CO_2 uptake, in turn, leads to the requirement for a CO_2-concentrating mechanism. From this line of reasoning C_4 photosynthesis is an evolutionary adaptation to hotter, drier conditions.

Another reason which has been put forward in explanation of the advantage to the C_4 plant of concentrating CO_2 in the bundle-sheath cells is that it depresses photorespiration, thereby reducing the considerable loss (20–40%) of photosynthetically fixed carbon seen in C_3 plants. Photorespiration depends upon the fact that RuDP carboxylase also has RuDP oxygenase activity; the latter catalyses the oxygen-dependent cleavage of RuDP into 3-PGA and phosphoglycollic acid [see Section D.2(i)]. Since CO_2 and O_2 are competitive substrates for RuDP carboxylase, a higher $CO_2:O_2$ ratio resulting from CO_2 concentration within the bundle-sheath chloroplast will favour the carboxylase activity and depress the oxygenase activity, thereby decreasing photorespiration.

4. Crassulacean acid metabolism (CAM)

(i) MECHANISM OF CAM

Plants which exhibit crassulacean acid metabolism are succulents. Many belong to the dicotyledonous family Crassulaceae, e.g. *Kalanchoe* spp., *Sedum* spp., but others belong to the monocotyledonous families, Agavaceae, Bromeliaceae, Liliaceae and Orchidaceae, to the dicotyledonous families, Aizoaceae, Asclepiadaceae, Bataceae, Cactaceae, Caryophyllaceae, Chenopodiaceae, Compositae, Convolvulaceae, Euphorbiaceae, Plantaginaceae, Portulacaceae and Vitaceae and even to the Pteridophyte family, Polypodiaceae. The sparse occurrence of CAM amongst present-day primitive plants suggests that it developed late in evolution; this is supported by the fact that $^{13}C/^{12}C$-isotope discrimination ratios (see Section D.5) indicate that fossil plant carbon was exclusively assimilated by C_3 photosynthesis. CAM plants exhibit the following characteristics: (i) normally their stomata are open during the night (i.e. in the dark) and closed during the day (these stomatal movements are the reverse of those of non-CAM plants), (ii) CO_2 is fixed during the hours of darkness in chloroplast-containing cells of photosynthetic leaf or stem tissue and considerable quantities of free L-malic acid are synthesized, (iii) this malic acid is stored in the large vacuoles which are characteristic of the cells of CAM plants, (iv) during the ensuing hours of daylight the malic acid is decarboxylated and the resulting CO_2 converted into sucrose and storage glucan (e.g. starch) by light-driven, C_3 photosynthesis, (v) during the next period of darkness some of the storage glucan present in the cell is catabolized to provide an acceptor molecule for the dark CO_2-fixation reaction. Thus, resulting from (ii), (iv) and (v), CAM tissues show a diurnal cycle in which the level of malate rises and the level of storage glucan falls during the night and the converse occurs during the day.

The biochemistry underlying these CAM characteristics is shown in Fig. 5.35. During the night CO_2 enters the tissue through the open stomata. Once within the cells of the tissue the CO_2 reacts with phosphoenolpyruvate (PEP) under the catalytic influence of PEP carboxylase[18] ($K_{m(CO_2)} = 0.02$ mM;

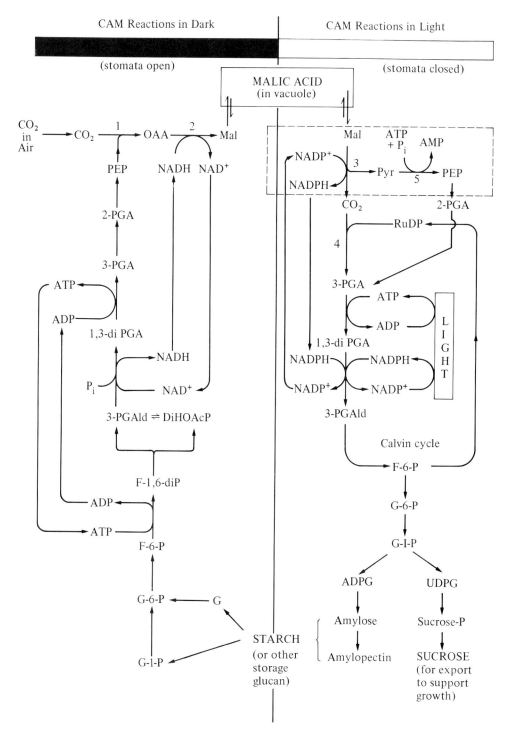

Fig. 5.35. Pathways of crassulacean acid metabolism during darkness and daylight. (1 = PEP carboxylase; 2 = malate dehydrogenase; 3 = NADP malic enzyme; 4 = RuDP carboxylase; 5 = pyruvate, orthophosphate dikinase, activated by light—see Section D.2(iv). Reactions within dashed rectangle occur as such in CAM plants of Crassulaceae, Cactaceae and Agavaceae but not in other families, see text.)

$K_{m(PEP)} = 0.19$ mM) to form oxaloacetic acid (OAA) [eq. (5.36)]. The intracellular location of PEP carboxylase is not known with certainty; in Fig. 5.35 it is shown as being cytoplasmic but some studies have shown that it may be chloroplastic. The enzyme is allosterically inhibited by malate ($K_i = 1$ mM), a property which is thought to play an important role in the regulation of CAM as described later. Malate dehydrogenase[24] then catalyses the reduction of OAA by NADH and the resulting malic acid is actively transported across the tonoplast membrane into the vacuole, thus keeping the cytoplasmic malic acid concentration below the $K_{i(malate)}$ of PEP carboxylase. The production of malic acid at the expense of CO_2 and PEP proceeds throughout the night but its rate slackens off as dawn approaches. The source of PEP for the carboxylation reaction is the storage glucan produced during the previous light period. This is catabolized to glucose 1-phosphate which is then converted via the glycolysis pathway to PEP. The $NAD^+/NADH$ balance of the cell during the night is maintained by the reciprocal requirements of malate dehydrogenase and glyceraldehyde 3-phosphate dehydrogenase[7].

During the following day malic acid passes from the vacuole back into the cytoplasm and is decarboxylated to provide CO_2 for light-driven C_3 photosynthesis. Two different decarboxylation mechanisms are used in CAM although they apparently never occur together in the same plant species. Members of the Crassulaceae, Cactaceae and Agavaceae decarboxylate malic acid directly utilizing the NADP-malic enzyme[15] which produces CO_2 and pyruvate. Members of the Liliaceae, Bromeliaceae and Asclepiadaceae do not possess the NADP-malic enzyme but instead have a very active PEP carboxykinase[17]. In these families malic acid is firstly oxidized to OAA by a malate dehydrogenase. It is not yet clear whether this is a cytoplasmic NAD-malate dehydrogenase[24] or the light-activated, chloroplastic NADP-malic dehydrogenase[19]. The OAA is then converted into CO_2 and PEP with the utilization of ATP by PEP carboxykinase[17] whose intracellular location is as yet uncertain. In the family Euphorbiaceae some species use the PEP carboxykinase[17] decarboxylating mechanism whilst others use the NADP-malic enzyme[15]. The CO_2 produced by malic acid

decarboxylation is then used for conversion into carbohydrate by the Calvin cycle enzymes located in the chloroplast. It is fixed into 3-phosphoglycerate (3-PGA) by RuDP carboxylase[3]. The properties of the RuDP carboxylase of the CAM plant, *Kalanchoe daigremontiana* ($K_{m(CO_2)} = 18$ μM; $K_{m(O_2)} = 490$ μM), are very similar to those of plants utilizing C_3 and C_4 photosynthesis. Fructose 6-phosphate is tapped off the Calvin cycle and used for the synthesis of storage glucan, e.g. starch. The starch is stored in the chloroplast during the day and then utilized as a source of PEP for the dark carboxylation reaction during the following night. DiHOAcP is also tapped off the Calvin cycle. It passes from the chloroplast to the cytoplasm where it is converted, via F-6-P, into sucrose which is then exported from the cell and utilized for growth of the plant.

The pyruvate produced by the NADP-malic enzyme-catalysed decarboxylation of malic acid and the PEP produced by the PEP carboxykinase-malic acid decarboxylation mechanism are probably utilized for carbohydrate synthesis during the day. The pyruvate is firstly converted into PEP by pyruvate, orthophosphate dikinase[20]. This enzyme is, however, not found in those CAM plants which utilize PEP carboxykinase in malic acid decarboxylation since PEP is produced directly. PEP in both types of CAM plant is then converted via reversed glycolysis into 3-PGA which is then utilized in the Calvin cycle.

(ii) REGULATION OF CAM

The way in which CAM is controlled is of great interest and is the subject of much research. There are apparently two types of CAM regulation which may be termed 'short-term regulation' and 'long-term regulation'. The former refers to the regulation of CAM during the daily cycle whilst the latter refers to regulation over periods longer than a day (e.g. a week, several weeks or a season).

(a) *Short-term regulation of CAM*

Several hypothetical models have been suggested over the years to explain how CAM is regulated during the daily cycle. All have been based on experimental observations but all have necessarily included a degree of speculation which, in many cases, has been disproved by subsequent experimentation. Probably the most attractive hypothesis at the moment is that put forward by Kluge (1975) which can be summarized as follows. Towards the end of the day PEP carboxylase begins to generate malic acid by the dark fixation of CO_2. At this time the vacuolar concentration of malic acid is low, having been depleted during the light period. Thus malic acid can be readily actively transported into the vacuole. This has two effects: firstly, it keeps the concentration of malic acid in the cytoplasm low thus preventing the allosteric inhibition of PEP carboxylase[18] and secondly, it increases the concentration of malic acid in the vacuole. The latter causes the osmotically driven uptake of water into the vacuole. This in turn increases the turgor pressure. When a critical turgor pressure is reached the net influx of malate into the vacuole which has obtained thus far is switched to net efflux. The malic acid so released floods into the cytoplasm and inhibits PEP carboxylase[18], so switching off dark fixation of CO_2 into malic acid. The malic acid now in the cytoplasm is decarboxylated and the resulting CO_2 (and 3C compound) utilized in C_3 photosynthesis. This process will eventually use up the malic acid produced by dark CO_2 fixation and so lower the cytoplasmic malic acid concentration to below the $K_{i(malate)}$ of PEP carboxylase[18]. This allows PEP carboxylase[18] to function once again and the cycle to repeat itself.

The most speculative aspect of this hypothesis is the sudden pressure-dependent switch in the properties of the tonoplast. However, one such pressure-dependent switch has been demonstrated; in the alga *Valonia* the net influx of K^+ changes to net efflux as the turgor pressure increases over a narrow range.

A feature of this hypothetical regulatory mechanism is that it will operate regardless of whether there is or is not a rhythmical change of light and dark periods or a diurnal change of temperature. Thus the long, cool night/short, hot day regime, regarded as most favourable for CAM, would not be essential if this hypothesis were correct. The availability of light and water are the critical factors. Thus a considerable boost to the veracity of this hypothesis would be the demonstration that CAM could proceed under constant light and constant temperature. Such a demonstration has been made.

(b) *Long-term regulation of CAM*

There is little doubt that the most important factor in the long-term regulation of CAM is the availability of water. When the supply of water is limited or non-existent other factors such as day and night temperatures, thermoperiod or photoperiod appear to have no more than a minor influence on CAM. When, however, the supply of water is adequate the latter factors can assume a greater regulatory influence as is seen in those CAM plants growing in tropical or semitropical regions.

The former situation is most clearly demonstrated in the cacti which can survive and grow in some of the most arid desert regions in the world. During the daytime when it is extremely hot the cacti close their stomata. This, taken with their low surface area to volume ratio and highly impervious, heavily waxed cuticle, virtually prevents any gaseous exchange between the plant and its environment. Thus evaporative water loss is almost zero. So, of course, is CO_2 influx but this does not matter because CO_2 is being provided for C_3 photosynthesis by malic acid decarboxylation. During the night, when evaporative demand is much lower, the stomata may or may not open. Whether they open depends upon the availability of water. If there has been recent rainfall the stomata open and CO_2 can be taken in and malic acid formed; during the ensuing period of daylight this malic acid is used for C_3 photosynthesis, there is a net production of carbohydrate and growth of the cactus can take place. If, however, there has been no rainfall for some time, as is usual in such regions, the stomata remain tightly shut since even the lower evaporative water loss of the night could not be made up by root absorption. Thus there can be no CO_2 uptake during the night and therefore no net carbohydrate production during the following day. Under these conditions there is no growth; the best that the cactus can achieve is to survive. This it can do for periods of months and years. During these periods it cannot utilize in the normal way the light energy that it cannot avoid absorbing, i.e. by fixing CO_2 by photosynthesis and utilizing the resulting carbohydrate to support growth. It has to dissipate the light energy in a non-productive cycle of carboxylation and decarboxylation which utilizes CAM, photorespiration and normal dark respiration. During the day malic acid is decarboxylated to provide CO_2 for C_3 photosynthesis until it is depleted. When this occurs photorespiration provides the CO_2 for C_3 photosynthesis. During the night the starch formed during the preceding day is converted into CO_2 and PEP by the normal respiratory processes of glycolysis and operation of the TCA cycle. The CO_2 and PEP are then converted back into malate. The diurnal fluctuation of malic acid (7.5–15 μmol per g fresh wt) under these conditions is very much lower than that under conditions of water availability (50–100 μmol per g fresh wt). This recycling of CO_2 through the CAM and the other pathways is frequently referred to as 'idling' and persists throughout the period of drought. With the coming of the first rainfall the cactus takes up water, its tissues rehydrate and the stomata recommence their night-time opening thus allowing the net intake of CO_2 and the transition from idling to productive CAM.

Some CAM plants grow in regions where water availability is less of a problem, for instance those in more tropical regions. These plants appear to shift from CAM to C_3 photosynthesis during periods of water abundance and then back again to CAM as the water supply diminishes. This shift from CAM to C_3 photosynthesis is evidenced by a change in the $^{13}C/^{12}C$-isotope discrimination ratio (see Section D.5) and is characterized by the reversal of the stomatal opening and shutting pattern, i.e. from night opening and day shutting which is typical of CAM to night shutting and day opening. Whilst the key factor in shifting a CAM plant which is operating in the C_3 mode back to CAM is the onset of drought, it is clear from experiments carried out in growth chambers that other factors such as an increased diurnal temperature fluctuation, short photoperiods or increased salinity can also induce this transition.

It thus appears that CAM is an adaptation which allows plants to survive and grow in extremely arid environments. Moreover, evidence is accumulating which strongly suggests that this adaptation is inducible in some CAM plants. There would therefore appear to be two types of CAM plants,

obligate or constitute CAM plants (e.g. the cacti) and facultative or inducible CAM plants (e.g. *Mesembryanthemum crystallinium*).

5. Variation in the $^{13}C/^{12}C$ isotope ratio in the carbon fixed during the C_3, C_4 and CAM modes of photosynthesis

The natural abundances of the two stable isotopes of carbon are $^{13}C = 1.108\%$ and $^{12}C = 98.892\%$. Thus 1.108% of the CO_2 molecules of the atmosphere will each have a ^{13}C atom whilst the rest will have a ^{12}C atom. It has been experimentally demonstrated that RuDP carboxylase discriminates against CO_2 molecules containing ^{13}C to a considerable degree. This means that $^{13}CO_2$ has a smaller chance of being converted into 3-PGA by RuDP carboxylase[3] than has $^{12}CO_2$. Thus 3-PGA, and the substances produced from it, has a lower $^{13}C/^{12}C$ ratio than has the CO_2 from which it was formed. It has also been shown that PEP carboxylase[18] does not discriminate against $^{13}CO_2$ to anything like the same extent as RuDP carboxylase[3]. Since the C_3 mode of photosynthesis utilizes RuDP carboxylase[3] only, whilst the C_4 mode and CAM utilize PEP carboxylase[18] and RuDP carboxylase[3], it can be seen that differences in the $^{13}C/^{12}C$ ratio in the carbon fixed by these processes are likely to be seen.

The change in $^{13}C/^{12}C$ ratio, from that in CO_2 to that in the fixed carbon in the plant, is commonly expressed as the $\delta^{13}C\%_0$ which is given by the following expression:

$$\delta^{13}C(\%_0) = \left[\frac{^{13}C/^{12}C \text{ in the fixed C of the plant}}{^{13}C/^{12}C \text{ in a standard}^*} - 1 \right] \times 10^3$$

From this expression it can be seen that if there has been no change in the $^{13}C/^{12}C$ ratio during photosynthetic carbon fixation $\delta^{13}C(\%_0)$ will equal zero. If, however, there has been discrimination against $^{13}CO_2$ the value of $\delta^{13}C(\%_0)$ will be a negative number; moreover, the greater the discrimination the more negative will be the number.

*Commonly limestone ($CaCO_3$).

A δ^{13}C value of $-3\%_0$ has been measured for PEP carboxylase[18] at 25°C and pH 8.5. In contrast to this the δ^{13}C value for RuDP carboxylase at 24°C and pH 8.2 is $-33.7\%_0$, clearly demonstrating the greater degree of discrimination against $^{13}CO_2$ exerted by RuDP carboxylase[3]. Significantly the δ^{13}C value for RuDP carboxylase[3] becomes less negative as the temperature rises (i.e. δ^{13}C at 37°C and pH 8.2 = $-18.3\%_0$).

Values of δ^{13}C in the range -23 to $-34\%_0$ have been found for plants operating the C_3 mode of photosynthesis whilst values in the range -10 to $-18\%_0$ are typical of C_4 plants. These δ^{13}C ranges tie in nicely with (i) the identity of the primary carboxylase (i.e. RuDP carboxylase[3] in C_3 plants; PEP carboxylase[18] in C_4 plants) and (ii) the fact that C_3 plants have a low temperature optimum for photosynthesis whilst C_4 plants have a high temperature optimum. The higher temperature at which the generally tropical C_4 plants photosynthesize clearly reduces the discrimination against $^{13}CO_2$ by RuDP carboxylase[3], which in these plants is the secondary carboxylase. The δ^{13}C values obtained for CAM plants fall into the range -14 to $-33\%_0$. The lower value is clearly within the δ^{13}C range of C_4 plants whilst the higher value falls into that of C_3 plants. This has been interpreted as indicating that many CAM plants are able to change from the C_3 mode of photosynthesis to CAM and *vice versa* in response to changing environmental conditions. It has, for instance, been shown that CAM plants living in arid regions have less negative δ^{13}C($\%_0$) values than those in regions with adequate water. The same effect is also seen within the same CAM species; when *Aloe vera* plants, grown under conditions where moisture was not a limiting factor, were transferred from an environment designed to induce CAM (i.e. a regime of 9 hr light at 30°C/15 hr dark at 15°C) to one which was not (i.e. continuous light at 30°C) the δ^{13}C changed from $-21\%_0$ to $-29.1\%_0$.

The δ^{13}C($\%_0$) values obtained from fossil plant material representative of seven geological periods from the Pliocene to the Permian are consistently in the range which is typical of C_3 plants today. This, along with the sparse occurrence of CAM amongst present-day primitive plants, has been taken to indicate that C_3 photosynthesis has been operative as a major pathway for the fixation of atmospheric

CO$_2$ since plants came on to the land. However, these results do not positively exclude the possibility that C_4 photosynthesis or CAM was in existence during the same geological period. Nevertheless it is tempting to take them as supporting evidence for the thesis that C_4 photosynthesis and CAM evolved later than and as an adaptation of C_3 photosynthesis.

6. Starch and sucrose metabolism in photosynthetic cells

Photosynthetically fixed carbon is ultimately converted into two main carbohydrate products, sucrose and starch, in photosynthetic cells. Sucrose is the form in which most of the photosynthetically fixed carbon is exported from the cell. It is translocated to those parts of the plant which have need of it, such as the regions of active growth and developing seeds or tubers. Starch is formed as a temporary storage form of fixed carbon during very active periods of photosynthesis; it accumulates within the chloroplasts as starch grains. During the subsequent dark period it is converted into sucrose and exported from the cell.

The way in which sucrose and starch are formed in the photosynthetic cells of C_3 plants and the way in which starch is converted into sucrose are shown in Figs. 5.36 and 5.37 respectively. Our knowledge of these pathways is based upon (i) the observed formation of starch grains within chloroplasts, (ii) the known distribution of the relevant enzymes in the chloroplast and the cytoplasm and (iii) the movements of photosynthetic intermediates between the chloroplast and the rest of the cell (see Fig. 5.38).

For starch formation F-6-P is tapped off the Calvin cycle and converted into ADP-glucose via G-6-P and G-1-P. The enzymes glucose phosphate isomerase[26], phosphoglucomutase[27] and G-1-P adenylyltransferase[28] catalyse the reactions involved. The latter enzyme is allosterically activated by F-6-P and 3-PGA and inhibited by orthophosphate. The ADP-glucose is then converted into amylose; the primer required by this reaction is maltose or a short-chain dextrin produced during the preceding dark period by the action of amylases

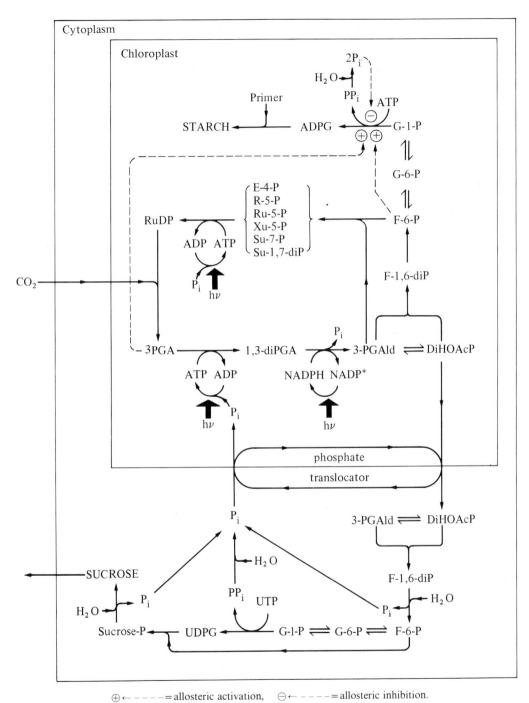

Fig. 5.36. Synthesis of sucrose and starch in the photosynthetic cell of a C_3 plant.

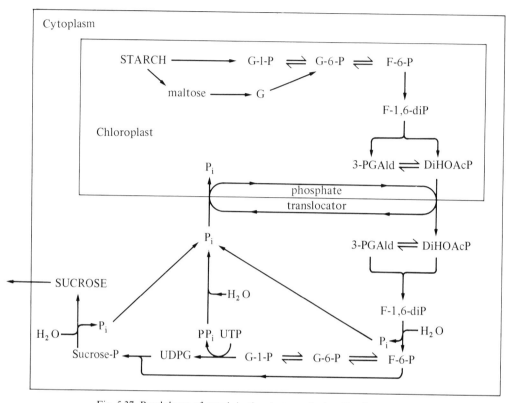

Fig. 5.37. Breakdown of starch in the photosynthetic cells of C_3 plants.

on the starch grains formed during the light period before that. A considerable proportion of the amylose is then converted into amylopectin.

Sucrose is synthesized in the cytoplasm. The principal fixed carbon export from the chloroplast is DiHOAcP and it is from this that sucrose is synthesized. Although 3-PGA is also known to pass from the chloroplast into the cytoplasm it is doubtful whether it contributes greatly to sucrose synthesis because this would require a massive export of ATP and reducing equivalents from the chloroplasts to support its initial conversion to triose phosphate. The outflow of DiHOAcP from the chloroplast results in a considerable loss of phosphate; if this loss were to go on uncorrected there would eventually be insufficient orthophosphate in the chloroplast to allow the photo-generation of ATP. To get over this difficulty there is a 'phosphate translocator' built into the inner membrane of the chloroplast; this ensures that for every phosphate lost to the chloroplast as DiHOAcP, one orthophosphate ion is transferred

back from the cytoplasm. The DiHOAcP is then converted into G-1-P by the cytoplasmic glycolysis enzymes (see Chapter 6, Section E.1). The G-1-P is then converted into UDP-glucose by G-1-P uridylyltransferase[29]. Sucrose is then formed by the successive actions of sucrose phosphate synthase[30] and sucrose phosphatase[31] [see Chapter 7, Section C.1(i)].

It is perhaps appropriate to note that DiHOAcP is also the starting-point for the synthesis of lipids and amino acids.

Breakdown of the starch in the chloroplastidic starch grains occurs when photosynthesis slows down or stops. Attack by starch phosphorylase[32] produces G-1-P whilst attack by the amylases, α-1,6-glucosidic bond-cleaving enzymes and α-glucosidase[33] produces glucose. Some of the maltose and short-chain dextrins, produced in the latter process, probably act as primers for starch synthesis during the succeeding light period. The glucose and G-1-P are then converted by chloroplastidic enzymes into DiHOAcP which is then exported to

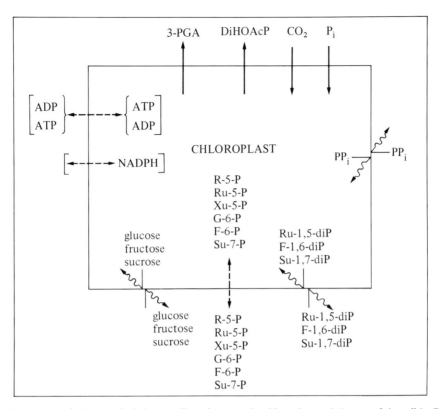

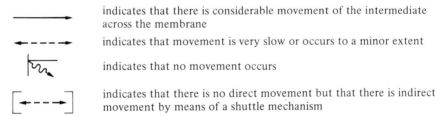

Fig. 5.38. Movements of photosynthetic intermediates between the chloroplast and the rest of the cell in C_3 plants.

⟶ indicates that there is considerable movement of the intermediate across the membrane

⟵ – – –➤ indicates that movement is very slow or occurs to a minor extent

indicates that no movement occurs

[⟵ – – –➤] indicates that there is no direct movement but that there is indirect movement by means of a shuttle mechanism

the cytoplasm for sucrose synthesis, and subsequent export from the cell, or for lipid or amino acid formation within the cell.

The metabolism of starch and sucrose in the photosynthetic cells of C_4 plants has many similarities to and some differences from that in C_3 plants. Starch grains are known to form in both mesophyll and bundle-sheath chloroplasts in spite of the fact that only in the bundle-sheath chloroplasts is there CO_2 fixation into carbohydrate. It is thought that the origin of the fixed carbon used for starch synthesis in the mesophyll chloroplasts is 3-PGA produced in the bundle-sheath chloroplasts and then translocated to the mesophyll chloroplasts.

The 3-PGA is converted into triose phosphates as described in Section D.2(i) and from there to starch in the manner shown in Fig. 5.36. Starch and sucrose synthesis occur in the bundle-sheath cells as shown in Fig. 5.36; sucrose is exported to other parts of the plant via the immediately adjacent vascular tissue. Starch breakdown to triose phosphates presumably occurs in both mesophyll and bundle-sheath chloroplasts as is shown in Fig. 5.37.

Less is known of the intracellular location of the enzymes concerned in the metabolism of starch and sucrose in CAM plants. It is likely that starch is formed in the chloroplast and sucrose in the cytoplasm during the light period as is shown in

Fig. 5.36. The breakdown of starch during the dark period to provide PEP for the dark carboxylation reaction is known to be initiated by starch phosphorylase[32] and/or amylases and glucokinase. The glycolysis pathway then takes the resulting glucose phosphate to PEP. At present it is not clear whether part or the whole of the glycolysis sequence takes place in the chloroplast or the cytoplasm.

7. Transport of light-generated ATP into the cytoplasm—the 3-phosphoglyceric acid/dihydroxyacetone phosphate shuttle

ATP, generated in the light phase of photosynthesis (see Sections C.6 and C.7), is needed to drive endergonic reactions occurring in the cytoplasm of the photosynthetic cell as well as those occurring within the chloroplast. Thus ATP is required to pass from the chloroplast to the cytoplasm. This presents a problem because the membranes of the chloroplast envelope are not freely permeable to ATP or ADP.

How is this problem solved? The answer does not appear to lie with the adenine nucleotide translocator (or antiport) for which there is evidence in the envelope of spinach chloroplasts. The ATP/ADP exchange rates that have been measured for this translocator are far too low to account for the observed light-dependent transport of ATP from the chloroplast to the cytoplasm. It is, however, currently believed that the problem is solved in an indirect manner by a shuttle mechanism known as the 3-phosphoglyceric acid/dihydroxyacetone phosphate shuttle (3-PGA/DiHOAcP shuttle) which is outlined in Fig. 5.39.

The shuttle commences with the 3-PGA→DiHOAcP section of the photosynthetic carbon reduction cycle [see Section D.2(i), eqs. (5.24), (5.25) and (5.26)] in which ATP and NADPH, generated in the light phase of photosynthesis, are used to convert 3PGA into 1,3-diPGA and 1,3-diPGA into 3-PGAld respectively. Some of the DiHOAcP resulting from the isomerization of 3-

PGAld is exported from the chloroplast to the cytoplasm instead of proceeding round the rest of the photosynthetic carbon reduction cycle. The passage of DiHOAcP across the membranes of the chloroplast envelope is mediated by the phosphate translocator (also known as the DiHOAcP or 3-PGA/P_i antiport: an antiport is a specific membrane transport protein that mediates the concerted movement of two different molecular or ionic species in opposite directions) which is believed to reside in the inner membrane. As the phosphate translocator transports one molecule of DiHOAcP out of the chloroplast it simultaneously transports one orthophosphate ion into the chloroplast.

In the cytoplasm the exported DiHOAcP is converted back into 3-PGA by glycolysis enzymes (see Fig. 6.35). DiHOAcP is isomerized by triose phosphate isomerase to 3-PGAld which is then oxidatively phosphorylated by glyceraldehyde phosphate dehydrogenase to the high-energy acyl phosphate compound, 1,3-diPGA. The latter is then used by phosphoglycerate kinase to phosphorylate ADP and concomitantly produce 3-PGA. The ATP produced in this reaction not only replaces the ATP used up in the 3-PGA→1,3-diPGA reaction of the photosynthetic carbon reduction cycle but, more importantly, replaces it in the cytoplasm; thus ATP has, in effect, been transferred from the chloroplast to the cytoplasm. The shuttle is completed by the return of the 3-PGA to the chloroplast, and thereby to the photosynthetic carbon-reduction cycle, via the phosphate translocator. Thus in this cyclical process light-generated ATP is transferred to the cytoplasm in the form of triose phosphate; however, there is no net loss of 3-PGA from the chloroplast.

The NADH formed in the cytoplasm during the operation of the shuttle is oxidized back to NAD$^+$ by using it to reduce oxaloacetic acid (OAA) to L-malic acid under the catalytic influence of the NAD-dependent cytoplasmic malate dehydrogenase. The resulting L-malic acid is then converted back into OAA by means of another shuttle, the malate/OAA shuttle. The L-malic acid is transported into the chloroplast via the dicarboxylate translocator (or antiport) in exchange for OAA. Once in the chloroplast the L-malic acid is oxidized to OAA by the chloroplastidic NAD-dependent- or NADP-dependent malate dehydrogenase. The NADPH that results from the operation of the latter enzyme

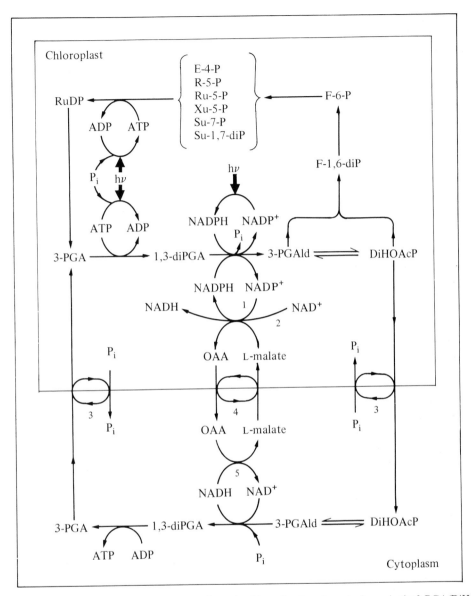

Fig. 5.39. Mechanism of transport of light-generated ATP from the chloroplast into the cytoplasm via the 3-PGA/DiHOAcP shuttle. [1 = Chloroplastidic NADP-dependent malate dehydrogenase, EC 1.1.1.82; 2 = chloroplastidic NAD-dependent malate dehydrogenase, EC 1.1.1.37; 3 = phosphate translocator (antiport); 4 = dicarboxylate translocator (antiport); 5 = cytoplasmic NAD-dependent malate dehydrogenase, EC 1.1.1.37; hν = photons driving the light phase of photosynthesis.]

in this reaction may be used directly as the reductant in the 1,3-PGA→3-PGAld reaction of the photo-synthetic reduction cycle.

In the dark the two shuttle systems described above can reverse direction. It is believed that they then supply the chloroplasts with ATP in the manner shown in Fig. 5.40. This is consistent with the experimental observation that in the absence of light chloroplasts *in vivo* have a much higher ATP content than isolated chloroplasts.

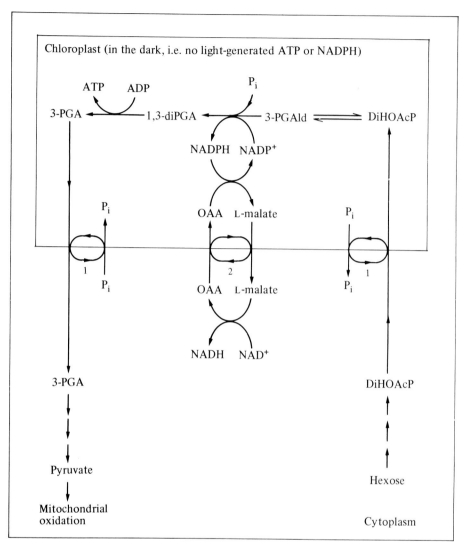

Fig. 5.40. Mechanism of generation of ATP in the chloroplast in the dark from hexose-derived dihydroxyacetone phosphate imported from the cytoplasm. (1 = phosphate translocator; 2 = dicarboxylate translocator.)

SUGGESTIONS FOR FURTHER READING

The light phase of photosynthesis

Trebst, A. (1974) 'Energy conservation in photosynthetic electron transport of chloroplasts', *Ann. Rev. Plant Physiol.* **25**,
423–458.

Radmer, R. and Kok, B. (1975) 'Energy capture in photosynthesis: photosystem II', *Ann. Rev. Biochem.* **44**, 409–433.

Bolton, J. R. and Warden, J. T. (1976) 'Paramagnetic intermediates in photosynthesis', *Ann. Rev. Plant Physiol.* **27**,
375–383.

Several chapters in *Topics in Photosynthesis*, Vol. 1 (*The Intact
Chloroplast*) and Vol. 2 (*Primary Processes in Photo-
synthesis*, 1976 and 1977) (Barber, J., ed.). Elsevier,
Amsterdam, New York, Oxford.

Cramer, W. A. and Whitmarsh, J. (1977) 'Photosynthetic
cytochromes', *Ann. Rev. Plant Physiol.* **38**, 133–172.

Junge, W. (1977) 'Membrane potentials in photosynthesis', *Ann.
Rev. Plant Physiol.* **28**, 503–536.

Avron, M. (1977) 'Energy transduction in chloroplasts', *Ann. Rev.
Biochem.* **46**, 143–155.

Blankenship, R. E. and Parson, W. W. (1978) 'The photochemical
electron transfer reactions of photosynthetic bacteria and
plants', *Ann. Rev. Biochem.* **47**, 635–653.

Katz, J. J., Shipman, L. L. and Norris, J. R. (1979) 'Structure and function of photoreaction-centre chlorophyll', in *Chlorophyll Organisation and Energy Transfer in Photosynthesis*, Ciba Foundation Symposium 61 (new series) (Wolstenholme, G. and Fitzsimons, D. W., eds.), pp. 1–34. Excerpta Medica, Amsterdam, Oxford, New York.

Sauer, K. (1979) 'Photosynthesis—the light reaction', *Ann. Rev. Physical Chem.* **30**, 155–178.

The dark phase of photosynthesis

Black, C. C., Jr. (1973) 'Photosynthetic carbon fixation in relation to net CO_2 uptake', *Ann. Rev. Plant Physiol.* **24**, 253–286.

Kelly, G. J., Latzko, E. and Gibbs, M. (1976) 'Regulatory aspects of photosynthetic carbon metabolism', *Ann. Rev. Plant Physiol.* **27**, 181–205.

Jensen, R. G. and Bahr, J. T. (1977) 'Ribulose 1,5-bisphosphate carboxylase-oxygenase', *Ann. Rev. Plant Physiol.* **28**, 379–400.

Laetsch, W. M. (1974) 'The C_4 syndrome: a structural analysis', *Ann. Rev. Plant Physiol.* **25**, 27–52.

Hatch, M. D. (1976) 'The C_4 pathway of photosynthesis: mechanism and function', in *CO_2 Metabolism and Plant Productivity* (Burris, R. H. and Black, C. C., eds.), pp. 59–81. University Park Press, Baltimore.

Edwards, G. E., Huber, S. C., Ku, S. B., Rathnam, C. K. M., Gutierrez, M. and Mayne, B. C. (1976) 'Variation in photochemical activities of C_4 plants in relation to CO_2 fixation', *ibid.*, pp. 83–112.

Black, C. C., Goldstein, L. D., Ray, T. B., Kestler, D. P. and Mayne, B. C. (1976) 'The relationship of plant metabolism to internal leaf and cell morphology and to the efficiency of CO_2 assimilation', *ibid.*, pp. 113–139.

Kluge, M. (1976) 'Models of CAM regulation', *ibid.*, pp. 205–216.

Osmond, C. B. (1976) 'CO_2 assimilation and dissimilation in the light and dark in CAM plants', *ibid.*, pp. 217–233.

Crews, C. E., Williams, S. L., Vines, H. M. and Black, C. C. (1976) 'Changes in the metabolism and physiology of CAM plants grown in controlled environments', *ibid.*, pp. 235–250.

Ting, I. P. (1976) 'CAM in natural ecosystems in relation to annual CO_2 uptake patterns and water utilization', *ibid.*, pp. 251–268.

Osmond, C. B. (1978) 'Crassulacean acid metabolism: a curiosity in context', *Ann. Rev. Plant Physiol.* **29**, 379–414.

Troughton, J. H. (1971) 'Aspects of the evolution of the photosynthetic carboxylation reaction in plants', in *Photosynthesis and Photorespiration* (Hatch, M. D., Osmond, C. B. and Slatyer, R. O., eds.), pp. 124–129. Wiley-Interscience.

Heber, U. (1974) 'Metabolite exchange between chloroplasts and cytoplasm', *Ann. Rev. Plant Physiol.* **25**, 393–421.

Buchanan, B. B. (1980) 'Role of light in the regulation of chloroplast enzymes', *Ann. Rev. Plant Physiol.* **31**, 341–374.

Lorimer, G. H. (1981) 'The carboxylation and oxygenation of ribulose 1,5-bisphosphate', *Ann. Rev. Plant Physiol.* **32**, 349–383.

ENZYMES

1. NADPH: ferredoxin oxidoreductase, EC 1.6.7.1.
2. Superoxide: superoxide oxidoreductase, EC 1.15.1.1.
3. 3-Phospho-D-glycerate carboxy-lyase (dimerizing), EC 4.1.1.39.
4. Carbonate hydro-lyase, EC 4.2.1.1.
5. ATP: 3-phospho-D-glycerate 1-phosphotransferase, EC 2.7.2.3.
6. D-Glyceraldehyde 3-phosphate: $NADP^+$ oxidoreductase (phosphorylating), EC 1.2.1.13.
7. D-Glyceraldehyde 3-phosphate: NAD^+ oxidoreductase (phosphorylating), EC 1.2.1.12.
8. D-Glyceraldehyde 3-phosphate ketol-isomerase, EC 5.3.1.1.
9. D-Fructose 1,6-diphosphate D-glyceraldehyde 3-phosphate-lyase, EC 4.1.2.13.
10. D-Fructose 1,6-diphosphate 1-phosphohydrolase, EC 3.1.3.11.
11. Sedoheptulose 7-phosphate: D-glyceraldehyde 3-phosphate glycolaldehydetransferase, EC 2.2.1.1.
12. D-Ribulose 5-phosphate 3-epimerase, EC 5.1.3.1.
13. D-Ribose 5-phosphate ketol-isomerase, EC 5.3.1.6.
14. ATP: D-ribulose 5-phosphate 1-phosphotransferase, EC 2.7.1.19.
15. L-Malate: $NADP^+$ oxidoreductase (oxaloacetate decarboxylating), EC 1.1.1.40.
16. L-Malate: NAD^+ oxidoreductase (decarboxylating), EC 1.1.1.39.
17. ATP: oxaloacetate carboxy-lyase (transphosphorylating), EC 4.1.1.49.
18. Orthophosphate: oxaloacetate carboxy-lyase (phosphorylating), EC 4.1.1.31.
19. L-malate: $NADP^+$ oxidoreductase, EC 1.1.1.82.
20. ATP: pyruvate, orthophosphate phosphotransferase, EC 2.7.9.1.
21. Pyrophosphate phosphohydrolase, EC 3.6.1.1.
22. ATP: AMP phosphotransferase, EC 2.7.4.3.
23. L-Aspartate: 2-oxoglutarate aminotransferase, EC 2.6.1.1.
24. L-Malate: NAD^+ oxidoreductase, EC 1.1.1.37.
25. L-Alanine: 2-oxoglutarate aminotransferase, EC 2.6.1.2.
26. D-Glucose 6-phosphate ketol-isomerase, EC 5.3.1.9.
27. α-D-Glucose 1,6-bisphosphate:α-D-glucose 1-phosphate phosphotransferase, EC 2.7.5.1.
28. ATP: α-D-glucose 1-phosphate adenylyltransferase, EC 2.7.7.27.
29. UTP: α-D-glucose 1-phosphate uridylyltransferase, EC 2.7.7.9.
30. UDP glucose: D-fructose 6-phosphate 2-α-glucosyltransferase, EC 2.4.1.14.
31. Sucrose 6^F-phosphate phosphohydrolase, EC 3.1.3.24.
32. 1,4-α-D-Glucan: orthophosphate α-glucosyltransferase, EC 2.4.1.1.
33. α-D-glucoside glucohydrolase, EC 3.2.1.20.
34. Sedoheptulose 1,7-bisphosphate 1-phosphohydrolase, EC 3.1.3.37.

CHAPTER 6

Respiration

CONTENTS

A. INTRODUCTION

Respiration is an oxidative process in which complex molecules such as carbohydrates and fats are broken down into carbon dioxide and a high proportion of the liberated free energy (see Chapter 2, Section A.1) is conserved by using it to generate ATP from ADP and orthophosphate.

The overall process is complex and can be subdivided into a number of component reaction sequences. The function of most of these reaction sequences (e.g. glycolysis, the tricarboxylic acid cycle, the fatty acid β-oxidation spiral) is to generate high-energy-reduced compounds (e.g. NADH, succinate) at the expense of the substrates they oxidize. These high-energy reductants are then reoxidized indirectly by molecular oxygen by means of a sequence of redox systems (see Chapter 2, Section A.4) known as the *terminal electron transport chain* which is located in the inner membrane of the mitochondrion (see Chapter 3, Section B.5).

The terminal electron transport chain removes a pair of electrons from the high-energy reductant, which has a relatively negative E_0' value (e.g. -0.32 V for NADH), and passes them along the sequence of redox systems whose E_0' values become progressively more positive, until they eventually reach oxygen and reduce it to water at an E_0' value of $+0.817$ V. The movement of the electron pair down a potential gradient (i.e. from a negative E_0' value to a positive E_0' value) is, like the flow of water down a waterfall, an exergonic process in which free energy is released.

The quantity of free energy released is dependent upon (i) the magnitude of the potential difference, $\Delta E_0'$ (analogous to the height of the waterfall), and (ii) the number of electron pairs passing down the potential gradient (analogous to the quantity of water flowing down the waterfall). This quantity has been calculated, in Chapter 2, Section A.5 using eq. (2.36), to be 219,411 J when one mole of electron pairs passes from NADH to oxygen.

This liberated free energy is conserved with about 40% efficiency by using it to drive the endergonic phosphorylation of ADP [eq. (6.1)]. The flow of electrons down the terminal electron transport chain is, therefore, coupled to phosphorylation and the

$$ADP + H_3PO_4 \rightarrow ATP + H_2O \qquad (6.1)$$
$$\Delta G^{\circ\prime} = +30.5 \text{ kJ mol}^{-1}$$

resulting process is known as *oxidative phosphorylation*.

Oxidative phosphorylation completes the oxidative breakdown of complex molecules and is, therefore, the final reaction sequence of respiration.

The purpose of respiration is the generation of ATP (see Fig. 6.1). ATP is often said to be the energy currency of the cell. The free energy made available when it is hydrolysed to ADP and orthophosphate [the reverse of eq. (6.1)] is used to drive the endergonic reactions of the pathways responsible for the synthesis of all the chemical components of the plant cell.

It is important to realize, however, that respiration is not the only way that ATP can be generated in most plants. As has been described in Chapter 5, ATP is generated during the light phase of photosynthesis. The difference between photosynthetic and respiratory ATP generation lies in the source of free energy to drive eq. (6.1). In photosynthesis it is derived from the sun whilst in respiration it comes from the oxidation of complex molecules. Since photosynthesis can only take place in the light and in those cells which possess the photosynthetic apparatus, it follows that the ATP generated in the plant during the hours of darkness and in non-photosynthetic tissues arises from respiration.

Since oxidative phosphorylation is the final stage which is common to the respiratory catabolism of all complex molecules, a major part of this chapter (Sections B, C and D) will be devoted to it. Section E

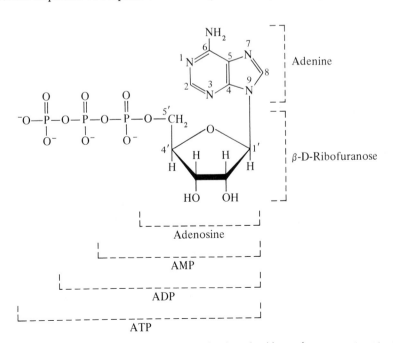

Fig. 6.1. Structure of adenosine 5′-triphosphate (ATP) and its related nucleotides and component parts. (ADP = adenosine 5′-diphosphate; AMP = adenosine 5′-monophosphate.)

will deal with the processes generating the high-energy reductants which are its substrates. Section F will deal with processes which have some similarity to aspects of respiration but which, unlike true respiration, are not linked to ATP generation.

B. OXIDATIVE PHOSPHORYLATION

1. Isolation of 'coupled' mitochondria

Oxidative phosphorylation takes place in the inner membrane of the mitochondrion (see Chapter 3, Section B.5). Thus meaningful investigations into the mechanism of this process depend upon the isolation of intact mitochondria uncontaminated by other cellular components. Moreover, the isolated mitochrondria should be capable of oxidative phosphorylation; such mitochondria are said to be 'coupled'. The term 'coupled' indicates that the processes of electron transport and ADP phosphorylation, which together constitute oxidative phosphorylation, are joined together such that one cannot occur without the other. This is the state of affairs which obtains *in vivo*; electron transport is obligatorily coupled to phosphorylation. However, the coupling of the two processes is fragile and easily damaged during the isolation of mitochondria. Consequently, unless great care is taken, uncoupled mitochondria are isolated; these mitochondria will oxidize NADH and succinate at the expense of oxygen via the electron transport chain but will not simultaneously phosphorylate ADP.

It is particularly difficult to isolate intact, uncontaminated, coupled mitochondria from plants for a number of reasons, including: (i) the problem of disrupting tough cell walls of plant tissues without also disrupting the mitochondria, (ii) control of pH is often difficult because the acidic contents of the vacuole are released during homogenization, (iii) many inhibitory substances such as polyphenols are released from the vacuoles during homogenization and (iv) several enzymes such as polyphenol oxidase[1] and the non-specific acylhydrolase [see Chapter 8, Section C.3(iii)(a)] are liberated during homogenization, which catalyse the formation of inhibitory substances.

In spite of these difficulties good mitochondria can be produced by judicious choice of plant tissue and homogenization medium and by careful use of mild cell-disruption procedures followed by differential and density gradient centrifugation (see Chapter 3, Sections B.7 and C).

A simple and effective test of the 'intactness' of mitochondria is to measure their ability to reduce *added* cytochrome c [see Section B.2(ii)(e)] in the presence of succinate. Cytochrome c cannot penetrate the inner mitochondrial membrane and thus cannot reach the enzyme succinate:cytochrome c reductase. Therefore any recorded reduction of cytochrome c is the result of the activity of succinate:cytochrome c reductase liberated from fragmented mitochondria.

2. Nature of the terminal electron transport chain

(i) GENERAL COMPOSITION AND OPERATION OF THE TERMINAL ELECTRON TRANSPORT CHAIN

The terminal electron transport chain is composed of a number of substances, known as redox compounds, which can shuttle readily back and forth between their oxidized and reduced forms. The ability of each of these redox compounds to gain electrons (i.e. for the oxidized form to be converted into the reduced form) is quantitatively expressed by its standard redox potential, E_0' (see Chapter 2, Section A.4). Since electron pairs flow in the terminal electron transport chain from a reductant with an E_0' value of -0.32 V to the oxidant oxygen (i.e. the redox system $\frac{1}{2}O_2/H_2O$) with an E_0' value of $+0.817$ V, it follows that the redox compounds must be arranged in order of decreasing negative or increasingly positive E_0' (see Figs. 6.2 and 6.3).

With the exception of ubiquinone (UQ) all the components of the chain are proteins with characteristic prosthetic groups (from the Greek, *prosthesis*, and now having the meaning 'additional part'). They are of three kinds: the flavoproteins which have prosthetic groups of flavin mononucleotide (FMN) or flavin adenine dinucleotide (FAD), the cytochromes which have haem prosthetic groups and the iron–sulphur proteins in which the prosthetic group consists of non-haem iron ligated with inorganic sulphur or cysteine sulphur. Ubiquinone, on the other hand, is a lipid consisting of a benzoquinonoid nucleus bound to a long, hydrophobic isoprenoid side chain.

The components of the terminal electron transport chain can also be classified in two other ways: (i) into 'electron carriers' and 'hydrogen carriers' and (ii) into 'one-electron' and 'two-electron redox compounds'.

Whether a component is classified as an electron carrier or a hydrogen carrier has an important bearing on its function in oxidative phosphorylation [see Section B.3(ii)]. What is the difference between these two types of carrier? Since the purpose of the terminal electron transport chain is to transfer (or assist in the *carrying* of) electrons from the high-energy reductant to oxygen, all its components must be electron carriers but some of them carry protons (i.e. H^+) in addition to electrons; these components are called hydrogen carriers to distinguish them from the electron carriers which carry electrons only. The flavoproteins and ubiquinone are hydrogen carriers; the cytochromes and iron–sulphur proteins are electron carriers.

Whether a component is classified as a one-electron or a two-electron redox compound has an important bearing on the stoichiometry of the redox reaction it participates in during the electron transport process. One-electron redox compounds have a one-electron difference between their oxidized and reduced forms; it therefore takes the addition of only one electron to convert the oxidized form into the reduced form and the removal of one electron to carry out the reverse process. Two-electron redox compounds have a two-electron difference between the oxidized and reduced states. The flavoproteins and ubiquinone are two-electron redox compounds; the cytochromes are one-electron compounds. The iron–sulphur proteins are more difficult to classify but are perhaps best regarded as one-electron redox compounds.

How does this classification as one-electron or two-electron redox compounds affect their ability to function in the terminal electron transport chain? Reference has already been made in this chapter to the passage of an *electron pair* down the electron transport chain to oxygen. A pair of electrons are involved because both of the high-energy reductants being oxidized as a result of the operation of the chain, NADH and succinate, are two-electron redox systems. Thus when NADH is oxidized to NAD^+

and succinate is oxidized to fumarate two electrons enter the electron transport chain. These two electrons are then handed from component to component, dropping down the electrical potential gradient in a stepwise manner, until they reach the last one, cytochrome oxidase, which uses them, along with two protons, to reduce one atom of oxygen to one molecule of water. Thus each component of the electron transport chain is required to handle two electrons at a time. This is no problem for the two-electron redox compounds. The one-electron redox compounds, however, cannot cope and consequently have to operate in pairs; thus two molecules of each of the cytochromes are required to transport the electron pair received from one molecule of ubiquinone to one atom of oxygen.

The majority of the components of the terminal electron transport chain are built into the inner membrane of the mitochondrion. The remainder are loosely bound to the outer surface of the inner membrane; these are cytochrome c in animal cells and cytochrome c_{547}, its functional equivalent, in higher plant cells. These two cytochromes are therefore 'peripheral proteins' (see Chapter 3, Section B.2(i)).

The sequence in which the components of the terminal electron transport chains of animal and plant mitochondria occur is given in Figs. 6.2 and 6.3 respectively and their probable arrangement in the respective inner membranes is shown in Figs. 6.4 and 6.5.

The terminal electron transport chain of animal mitochondria has been studied more intensively than that of plant mitochondria and is consequently better understood. Because of this it will, in this chapter, be taken as the model against which the plant system will be compared.

The animal terminal electron transport chain receives electron pairs at a number of flavoprotein-containing enzyme complexes. One of them, FP_1 in Figs. 6.2 and 6.4, receives electron pairs solely from NADH; it, along with several iron–sulphur protein molecules, constitutes an enzyme complex known as NADH dehydrogenase[2] which catalyses the transfer of an electron pair (and H^+) from a molecule of NADH to a molecule of ubiquinone, being itself first reduced and then oxidized in the process. The NADH oxidized by NADH dehydrogenase is generated in the mitochondrial matrix only, as a

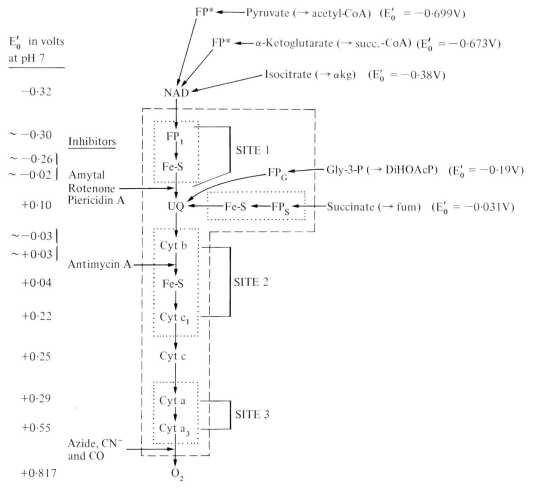

Fig. 6.2. Terminal electron transport chain and sites of energy conservation in animal mitochondria. (FP* = flavoprotein component of the pyruvate- and α-ketoglutarate dehydrogenase complexes; FP_1 = flavoprotein component of NADH dehydrogenase; FP_S = flavoprotein component of succinate dehydrogenase; FP_G = FAD-linked glycerol-3-phosphate dehydrogenase; Fe–S = iron–sulphur protein; Cyt a, $-a_3$, $-b$, $-c$ and $-c_1$ = a, b and c-type cytochromes; SITE 1, -2 and -3 = sites of energy conservation; succ-CoA = succinyl-CoA; αkg = α-ketoglutarate; Gly-3-P = sn-glycerol 3-phosphate; DiHOAcP = dihydroxyacetone phosphate; all components within the dashed line are built into or bound to the inner mitochondrial membrane; all components within dotted lines constitute enzyme complexes.)

result of the catalytic activities of such enzymes as pyruvate dehydrogenase[3] and isocitrate dehydrogenase[4], α-ketoglutarate dehydrogenase[5], and malate dehydrogenase[6] of the tricarboxylic acid cycle. Other flavoproteins receive electron pairs directly from reduced substrates. For instance, FP_S in Figs. 6.2 and 6.4, along with several iron–sulphur protein molecules, constitutes an enzyme complex known as succinate dehydrogenase[7] which catalyses the transfer of an electron pair (plus $2H^+$) from a molecule of succinate in the mitochondrial matrix

to a molecule of ubiquinone; the succinate is thus oxidized to fumarate and the ubiquinone reduced to ubiquinol. Succinate dehydrogenase is a tricarboxylic acid cycle enzyme [see Section E.1(ii)]; it is the only enzyme of this cycle to be built into the inner mitochondrial membrane. Another flavoprotein, FP_G in Figs. 6.2 and 6.4, is located in the outer surface of the inner mitochondrial membrane and catalyses the transfer of an electron pair (plus $2H^+$) from a molecule of sn-glycerol 3-phosphate in the intermembrane space to a molecule of ubiquinone;

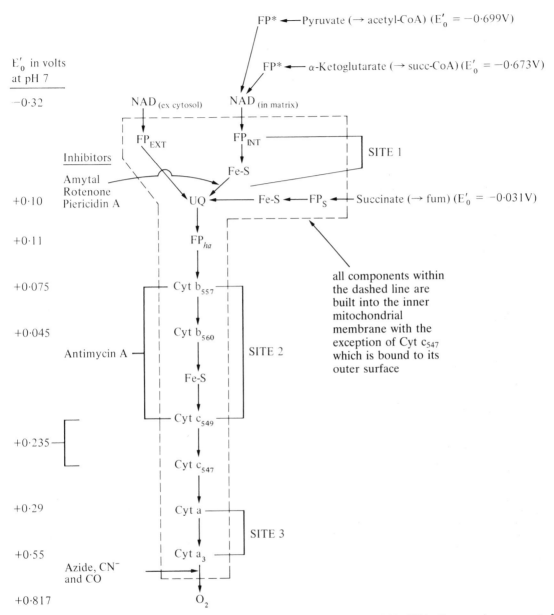

Fig. 6.3. Terminal electron transport chain and sites of energy conservation in plant mitochondria. (FP* = flavoprotein component of the pyruvate and α-ketoglutarate dehydrogenase complexes; FP_{EXT} and FP_{INT} = flavoprotein NADH dehydrogenases located in the external and internal surfaces respectively of the inner mitochondrial membrane; FP_S = succinate dehydrogenase; FP_{ha} = flavoprotein with a high (i.e. +ve) E'_0 and a large absorbance change on redox change; Fe–S = iron–sulphur protein; Cyt a, -b and -c = cytochromes a, -b and -c, 3-figure subscript numbers give the wavelength in nm of the absorption peak of the α-band of the related cytochrome at room temp.; SITE 1, -2 and -3 = sites of free energy conservation.)

the sn-glycerol 3-phosphate is oxidized to dihydroxy-acetone phosphate and the ubiquinone reduced to ubiquinol. This flavoprotein is the enzyme glycerol-3-phosphate dehydrogenase[8]; it is part of a shuttle mechanism designed to get electron pairs from

NADH generated in the cytosol into the terminal electron transport chain. This is required because neither NADH nor NAD^+ can diffuse across the inner mitochondrial membrane.

The reduced ubiquinone (ubiquinol) produced by

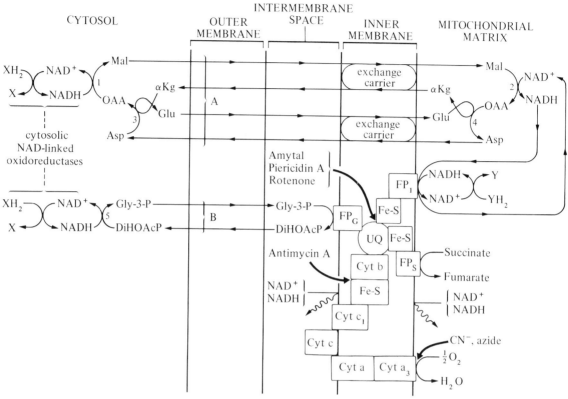

Fig. 6.4. Probable structural organization of the components of the terminal electron transport chain in the inner membrane of the animal mitochondrion. (Abbreviations as in Fig. 6.2 except for: 1 and 2 = cytosolic and mitochondrial malate dehydrogenases respectively; 3 and 4 = cytosolic and mitochondrial aspartate aminotransferases respectively; 5 = cytosolic NAD-linked glycerol-3-phosphate dehydrogenase; Mal = L-malate; OAA = oxaloacetate; Asp = L-aspartic acid; Glu = L-glutamic acid; A = malate-oxaloacetate-aspartate shuttle; B = glycerol 3-phosphate-dihydroxyacetone phosphate shuttle.)

the activities of these flavoprotein enzyme complexes is then reoxidized by an enzyme complex known as ubiquinol-cytochrome c reductase[9] which consists of cytochrome b, cytochrome c_1 and iron–sulphur protein. This enzyme system catalyses the transfer of two electrons (but not two H^+) from one molecule of ubiquinol to two molecules of cytochrome c, being first reduced and then oxidized in the process; the two H^+ ions pass into the aqueous milieu of the cell probably on the cytosolic side of the inner mitochondrial membrane (see Section B.3).

The two molecules of reduced cytochrome c (ferrocytochrome c) are then reoxidized by oxygen (formally one atom of oxygen or $\frac{1}{2}O_2$) under the catalytic influence of an enzyme complex known as cytochrome oxidase[10] or cytochrome aa_3. Cytochrome oxidase is also referred to as the 'terminal oxidase' because its activities terminate the electron transport chain by passing the electron pairs flowing down it, to oxygen. The two electrons passed to $\frac{1}{2}O_2$ by cytochrome are joined by two H^+ ions, taken from the mitochondrial matrix, to produce one molecule of water. Cytochrome oxidase is composed of two a-type cytochromes known as cytochrome a and cytochrome a_3; the former spans the outer half of the inner mitochondrial membrane whilst the latter spans the inner half (see Fig. 6.4). Thus it is cytochrome a which oxidizes the two molecules of ferrocytochrome c; it then hands the electrons on to cytochrome a_3 which uses them to reduce the $\frac{1}{2}O_2$ to H_2O. Both components of cytochrome oxidase, therefore, undergo reduction and oxidation in passing the electron pair across the width of the inner mitochondrial membrane from ferrocytochrome c to oxygen.

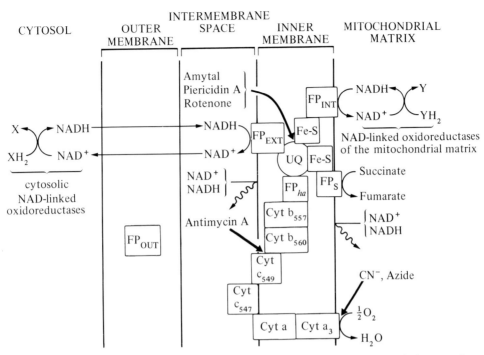

Fig. 6.5. Probable structural organization of the components of the terminal electron transport chain in the inner membrane of the plant mitochondrion. (Abbreviations are as in Fig. 6.3; ∿— = membrane is impermeable to....; FP_{OUT} = a flavoprotein NADH dehydrogenase in the outer membrane which does not normally pass electrons on to the terminal electron transport chain.)

The flow of electron pairs down the animal terminal electron transport chain can be interrupted at various points by inhibitors as is shown in Figs. 6.2 and 6.4; the structures of some of these inhibitors is shown in Fig. 6.6. The activity of NADH dehydrogenase is blocked by amytal, rotenone and piericidin A. Amytal is a barbiturate and like all barbiturates blocks NADH dehydrogenase but at relatively high concentrations; the sedative action of barbiturates appears to be brought about by other actions on neural membranes although a lowering of respiration may increase this action. Rotenone is extracted from the roots of certain tropical plants (*Lonchoncarpus nicou*, *Derris elliptica*). It forms a tight complex with NADH dehydrogenase and appears to block the transfer of electrons from the iron–sulphur protein components to ubiquinone; it is inhibitory (65%) at a concentration of 33 nmol per g of mitochondrial protein. Its poisonous effects have long been used by South American Indians to help them catch fish. It is also now employed as an insecticide. It is relatively non-toxic to mammals because it is only poorly absorbed; however, it

readily passes through the gills of fish and the breathing tubes of insects. Piericidin A is an antibiotic produced by a *Streptomyces* sp. and, being a structural analogue of ubiquinone, competes with it for electron pairs passing down the electron transport chain. It is inhibitory (50%) at a concentration of 20 nmol per g of mitochondrial protein. The antimycins A are a group of antibiotics, also produced by a *Streptomyces* sp., which inhibit the ubiquinol–cytochrome c reductase by blocking the passage of electrons from cytochrome b to cytochrome c_1, probably between cytochrome b and the iron–sulphur protein component. They are inhibitory (50%) with beef heart submitochondrial particles at a concentration of 50 nM. Cytochrome oxidase is inhibited by cyanide, azide and carbon monoxide; in all three cases it is the cytochrome a_3 component that reacts with the inhibitor. Cyanide and azide form co-ordination complexes with the Fe^{3+} of the ferricytochrome a_3 whilst carbon monoxide forms a similar complex with the Fe^{2+} of the ferrocytochrome a_3. Cyanide, azide and carbon monoxide are inhibitory (50%) with crude heart

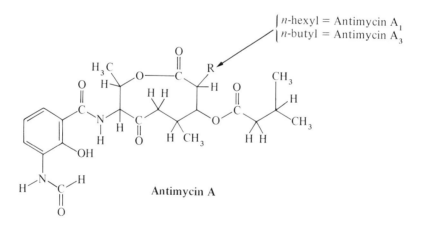

Fig. 6.6. Some inhibitors of the terminal electron transport chain.

mitochondrial preparations at concentrations of 0.5 μM, 0.7 μM and 40 μM respectively.

The terminal electron transport chain of higher plants, like that of animals, receives electron pairs at a number of flavoprotein enzyme complexes. One of them FP_{INT} (Figs. 6.3 and 6.5) appears to fulfil the same role as the FP_1 of the animal chain in that it receives electron pairs solely from NADH generated in the mitochondrial matrix. It is the flavoprotein component of an enzyme system that can be called the 'internal NADH dehydrogenase' because it is built into the inner surface of the inner mitochondrial membrane; the other components of the enzyme system are iron–sulphur proteins. Plant mitochondria differ from those of animals in having a second NADH dehydrogenase; this is built into the outer surface of the inner mitochondrial membrane and can be called the 'external NADH dehydrogenase'. It appears to be composed only of a flavoprotein, FP_{EXT} (see Figs. 6.3 and 6.5), and catalyses the transfer of an electron pair (and H^+) from a molecule of NADH located in the intermembrane space to a molecule of ubiquinone. It is therefore responsible for the reoxidation of NADH generated in the cytosol; although NADH and NAD^+ cannot diffuse across the inner mitochondrial membrane, they can readily do so across the outer mitochondrial membrane which is readily permeable to molecules with a molecular weight less than about 10^4. The possession of the 'external NADH dehydrogenase' means that plant cells have no need to resort to shuttle mechanisms to get electron pairs from NADH generated outside the mitochondrion into the terminal electron transport chain (see Section C). Plant mitochondria have a succinate dehydrogenase built into the inner surface of the inner mitochondrial membrane; like that of animal mitochondria, it consists of a flavoprotein (FP_S) and iron–sulphur proteins. It catalyses the transfer of two electrons (plus $2H^+$) from a molecule of succinate, formed during the operation of the tricarboxylic acid cycle, to ubiquinone.

The ubiquinol produced by the activities of these flavoprotein enzyme complexes is then reoxidized by an enzyme complex which is the functional equivalent of the animal ubiquinol–cytochrome c reductase. It appears to be composed of a flavoprotein (FP_{ha}), two b-type cytochromes, b_{557} and b_{560}, iron–sulphur protein and a c-type cytochrome, c_{549}. It catalyses the transfer of two electrons (but not two H^+) from a molecule of ubiquinol to two molecules of cytochrome c_{547}, its component members being successively reduced and oxidized in the process; the two H^+ ions presumably pass into the aqueous milieu of the intermembrane space as in animal mitochondria. Cytochrome c_{547} appears to be the plant equivalent of the animal cytochrome c since it is also readily extracted from plant mitochondria with salt solutions; by analogy with cytochrome c it is therefore assumed to be a peripheral protein bound to the outer surface of the inner mitochondrial membrane.

The two molecules of ferrocytochrome c_{547} [the prefix 'ferro' indicates that the cytochrome is in the reduced state; see Section B.2(ii)(e)] are then reoxidized by $\frac{1}{2}O_2$ under the catalytic influence of cytochrome oxidase. The plant cytochrome oxidase shows subtle differences from that of animals but in most respects it is very similar in composition, membrane location, function and sensitivity to inhibitors.

The flow of electron pairs down the plant electron transport chain is interrupted by the same inhibitors as inhibit the animal system (see Figs. 6.3 and 6.5). However, only the 'internal NADH dehydrogenase' is inhibited by amytal, rotenone and piericidin A; the 'external NADH dehydrogenase' is unaffected. This ties in with the fact that the internal enzyme contains a number of iron–sulphur proteins whose activities are blocked by these inhibitors whilst the external enzyme does not. Antimycin A inhibits the plant ubiquinol–cytochrome c_{547} reductase, probably at or soon after the b-type cytochromes. Cyanide, azide and carbon monoxide inhibit the cytochrome a_3 component of the plant cytochrome oxidase in the same manner as that of animals.

(ii) COMPONENTS OF THE TERMINAL ELECTRON TRANSPORT CHAIN

(a) *The pyridine nucleotide-linked sector*

Two structurally similar pyridine nucleotides, nicotinamide adenine dinucleotide (NAD) and nicotinamide adenine dinucleotide phosphate (NADP), occur in Nature; their structures are shown in Fig. 6.7. They function as coenzymes to many oxidoreductase enzymes, accepting or donating the equivalent of two electrons and one proton per molecule from or to the substrate of the particular enzyme.

Neither NAD nor NADP is bound as a prosthetic group to the majority of the oxidoreductases that use them as coenzymes; there are, however, a few oxidoreductases, notably glyceraldehyde-3-phosphate dehydrogenase[11], operating in glycolysis [see Section E.1(i)(a)] which do have a tightly bound pyridine nucleotide. Most of the NAD and NADP in the cell is therefore in an unbound state dissolved in the aqueous milieu. There is, however, compartmentalization of the NAD and NADP

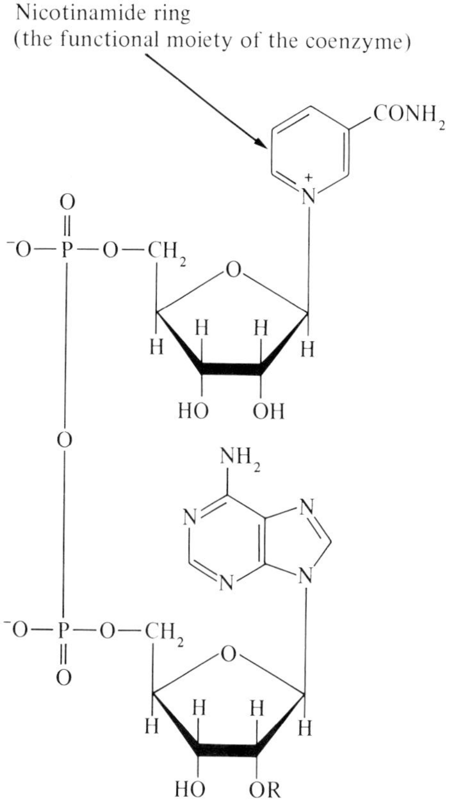

Nicotinamide ring
(the functional moiety of the coenzyme)

Fig. 6.7. Structure of the pyridine nucleotides; NAD^+, $R = H$; $NADP^+$, $R = PO_3^{2-}$.

that utilizes the oxidized form of NAD, namely NAD^+, as the recipient of electrons from the substrate being oxidized. The resulting NADH, the reduced form of NAD, then has to be converted back into NAD^+ to keep the catabolic sequence going. This is due to the fact that the NAD (i.e. NAD^+ plus NADH) concentration within the various intracellular compartments is very low. Thus if there were no reoxidation of NADH, all the NAD would very soon be in the reduced form and there would be no NAD^+ to serve the oxidoreductases of the catabolic process. Under aerobic conditions the terminal electron transport chain, either directly or indirectly, is responsible for the reoxidation of the NADH.

NADP, on the other hand, generally acts as a reducing agent (or reductant) in anabolic (or biosynthetic) processes, which are reductive in an overall sense. Thus one or more reactions in a catabolic sequence may be catalysed by an oxidoreductase that utilizes the reduced form of NADP, namely NADPH, as the donor of electrons to the substrate being reduced. This simultaneously produces $NADP^+$ which has to be reduced back to NADPH to keep the anabolic sequence going, for precisely the same reason that NADH had to be reoxidized. The means by which the $NADP^+$ is reduced depends upon the type of cell in which it is present. In non-photosynthetic cells, occurring in plants or animals, the $NADP^+$ is reduced by two reactions of the non-reversible phase of the oxidative pentose phosphate pathway (see Section F.3). In the chloroplasts of photosynthetic cells, in the light, it is reduced in the light phase of photosynthesis (see Chapter 5, Section C.6).

The way in which the pyridine nucleotides act as oxidants or reductants in reactions catalysed by oxidoreductases is shown in Fig. 6.9. The functional part of these compounds is the nicotinamide ring. The oxidized forms (NAD^+ and $NADP^+$) carry a positive charge which is usually shown as being located on the nitrogen atom of the nicotinamide ring. However, resonance within the ring system allows three different mesomeric structures to be written down, each with the positive charge located at a different place. One of these, shown in Fig. 6.9, has the positive charge located at C–4 and it is the form of $NAD(P)^+$ which most conveniently shows how its reduction to NAD(P)H takes place. The

within the cell. For example, neither the oxidized nor the reduced form of either pyridine nucleotide is able to pass into or out of the mitochondrion. This is due to impermeability of the inner mitochondrial membrane to these compounds; this membrane is noted for its selective permeability and further examples will become apparent later in the chapter. The outer mitochondrial membrane is relatively permeable as has been mentioned earlier and will allow the oxidized and reduced forms of the pyridine nucleotides to diffuse across it. The consequences of this is that the NAD of the mitochondrial matrix, for instance, cannot mix with that of the cytosol or *vice versa*.

NAD and NADP have different functions in the cell (see Fig. 6.8). In general terms NAD acts as an oxidizing agent (or oxidant) in catabolic processes which are, of course, oxidative in an overall sense. Thus one or more reactions in a catabolic sequence of reactions may be catalysed by an oxidoreductase

Role of NAD

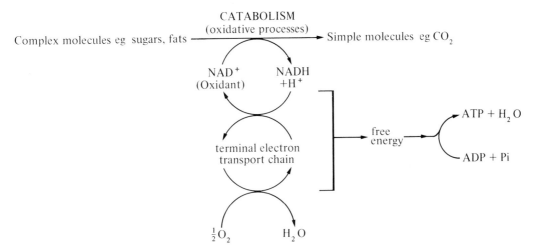

Role of NADP

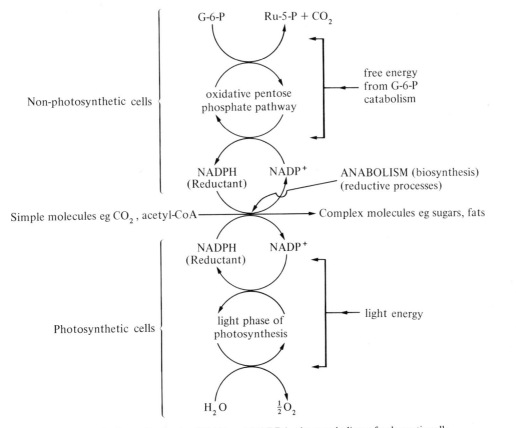

Fig. 6.8. Generalized role of NAD and NADP in the metabolism of eukaryotic cells.

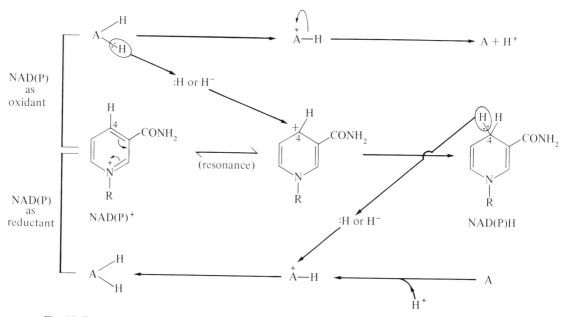

Fig. 6.9. Participation of the pyridine nucleotides, NAD and NADP, as the coenzymes of oxidoreductases.

oxidoreductase catalyses the transfer of a hydrogen atom with both its bonding electrons, as a hydride ion, H^-, from the substrate molecule (AH_2 in Fig. 6.9) to the positively charged C–4 of $NAD(P)^+$ so forming NAD(P)H; the electron pair of H^- become the bonding pair of the newly formed C–4→H covalent bond. The substrate, having lost H^- from one of its atoms, becomes a positively charged ion, AH^+, which stabilizes itself by losing H^+ to the surrounding aqueous medium. In the reverse reaction the oxidoreductase catalyses the transfer of H^- from C–4 of NAD(P)H to the protonated substrate. Thus the difference between $NAD(P)^+$ and $NAD(P)H$ is a hydride ion which is the equivalent of two electrons and one proton H^+. Hence it can be seen why NAD and NADP are classed as 'hydrogen carriers' and 'two-electron redox compounds' [see Section B.2(i)].

Oxidoreductases utilizing NAD or NADP as their coenzyme are stereospecific with respect to the hydrogen at C–4 of the nicotinamide ring of these compounds. The reduced form carries two hydrogen atoms at C–4 which project away from each other and from the plane of the ring itself. The nicotinamide ring can be thought of as having two surfaces or faces; the surface on which the conventional numbering of the component atoms of the

ring appears to the observer to be proceeding from 1 to 6 in an anticlockwise direction has been arbitrarily designated as the A-face whilst that on which the 1→6 numbering sequence is clockwise has been designated as the B-face. The hydrogen on C–4 that projects upwards from the A-face of the ring is usually referred to as H_A; it is the 4 pro-R hydrogen according to the Cahn–Ingold–Prelog convention for specifying absolute configuration. Similarly the hydrogen that projects upwards from the B-face is referred to as H_B and is the 4 pro-S hydrogen. Oxidoreductases catalysing the oxidation of NADH and NADPH can distinguish between H_A and H_B and remove one or other of them with absolute specificity. The oxidoreductases that remove H_A are said to be A-specific whilst those that remove H_B are said to be B-specific. This stereospecificity also extends to the oxidation of NAD^+ and $NADP^+$. A-specific oxidoreductases catalyse the addition of H^- to C–4 on the A-face of the ring such that the newly inserted hydrogen becomes H_A and the single hydrogen already present on C–4 becomes H_B. Similarly B-specific oxidoreductases catalyse the insertion of H_B. Figure 6.10 lists a number of A- and B-specific oxidoreductases and shows this stereospecificity with respect to the pyridine nucleotides in relation to the insertion

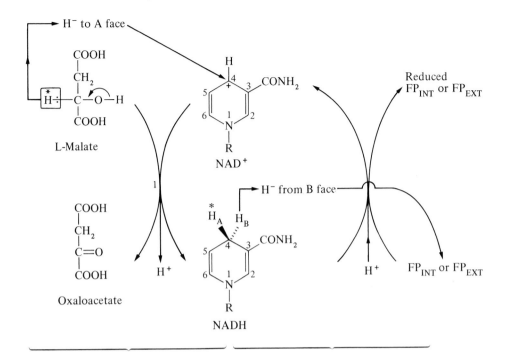

'A' specific oxidoreductases

1. Malate dehydrogenase (see above; EC 1.1.1.37)

2. Isocitrate dehydrogenase (EC 1.1.1.41)

3. FP_{OUT} (NADH dehydrogenase from outer mitochondrial membrane; see Fig. 6.5)

4. Alcohol dehydrogenase (EC 1.1.1.1)

5. Glyoxylate reductase (EC 1.1.1.26)

6. NADP 'photoreductase' (ex spinach chloroplasts)

'B' specific oxidoreductases

1. FP_{INT} and FP_{EXT} (NADH dehydrogenases from inner mitochondrial membrane, see above & Fig. 6.5)

2. Glycerol-3-phosphate dehydrogenase (EC 1.1.1.8)

3. Glyceraldehyde-3-phosphate dehydrogenase (EC 1.2.1.12)

4. Glucose-6-phosphate dehydrogenase (EC 1.1.1.49)

5. Glutamate dehydrogenase (EC 1.4.1.3)

6. Farnesyltransferase (polymeric form (EC 2.5.1.21)

Fig. 6.10. The stereospecificity of NAD(P)-linked oxidoreductase enzymes with respect to the 'A'- and 'B'-faces of the nicotinamide ring of NAD(P). (H_A = the H atom on C–4 which projects upwards from the A-face of the nicotinamide ring $\equiv$ 4 pro-R; H_B = the H atom that projects upwards from the B-face $\equiv$ 4 pro-S; the A-face of the nicotinamide ring is defined as the face on which the numbering appears anticlockwise; on the B-face the numbering is clockwise.)

of an electron pair from malate into the higher plant electron transport chain. Malate, in the mitochondrial matrix, is converted into oxaloacetate under the catalytic influence of malate dehydrogenase; NAD^+, also in the mitochondrial matrix, is the oxidant and is consequently reduced to NADH. Malate dehydrogenase is an A-specific oxidoreductase and so transfers a H^- ion to the A-face of

NAD^+. The resulting NADH is then reoxidized by the flavoprotein component of the internal NADH dehydrogenase that is built into the inner surface of the inner mitochondrial membrane. The internal NADH dehydrogenase, like the external NADH dehydrogenase (see Fig. 6.10), is a B-specific oxidoreductase and so removes a H^- ion from the B-face of NADH. Thus, in this instance, the electron pair taken from malate in the form of H^- is not the same as that which is inserted into the terminal electron transport chain.

The absorption spectra of the oxidized and reduced forms of the pyridine nucleotides are different. The oxidized forms, NAD^+ and $NADP^+$, have a single peak with a maximum at 260 nm. In the reduced forms, NADH and NADPH, the 260-nm peak is slightly less persistent† and is joined by a smaller, broader peak with a maximum at 340 nm (see Fig. 6.11). This difference in spectra

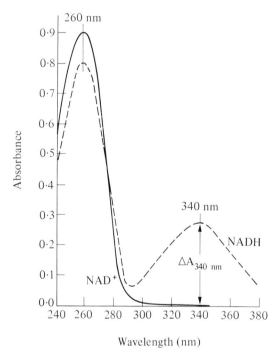

Fig. 6.11. Absorption spectra of NAD^+ and NADH.

means that the formation or disappearance of NAD(P)H in a reaction can be followed spectrophotometrically by monitoring at 340 nm. Similarly the rate and extent of the oxidation or reduction of the pyridine nucleotides can be monitored by recording the change in absorbance at

†The term 'persistence' is the difference, measured in absorbance units, in intensity of an absorption peak from that of the nearest minimum on the curve.

340 nm (ΔA_{340}) of the assay mixture over a period of time.

(b) Flavoproteins

Flavoproteins participate in the terminal electron transport chains of animals and plants as components of NADH dehydrogenases (e.g. FP_1 in animals, FP_{INT} and FP_{EXT} in plants), of succinate dehydrogenase (FP_S) and of oxidoreductases utilizing other substrates (e.g. FP_G in animals, FP_{ha} in plants). In these enzymes they are often associated with iron–sulphur proteins [see Section B.2(ii)(c)].

The prosthetic group of these flavoproteins is either flavin mononucleotide (FMN) or flavin adenine dinucleotide (FAD), whose structures are shown in Fig. 6.12. Both contain a riboflavin residue which consists of a 7,8-dimethylisoalloxazine ring linked from position 10 to the 1' carbon of the alcohol, D-ribitol; because this linkage is not a glycosidic linkage it is technically incorrect to call the prosthetic groups a mononucleotide (FMN) or a dinucleotide (FAD) but, nevertheless, the names are firmly established in the biochemical literature.

Most flavoproteins utilize FAD as their coenzyme; however, the NADH dehydrogenases utilize FMN. In most flavoproteins the prosthetic group is not covalently bound to the apoprotein and can be removed by dialysis. However, in some the prosthetic group is covalently bound; FP_S of succinate dehydrogenase is one of these and linkage between the FAD and the apoprotein is shown in Fig. 6.13.

The functional part of FMN and FAD consists of the two nitrogen-containing rings of the 7,8-dimethylisoalloxazine residue; nitrogen atoms 1 and 5 each receive the equivalent of one electron and one proton when reduction occurs. Thus both flavin prosthetic groups are classed as 'hydrogen carriers' and 'two-electron redox systems'. The mechanism by which reduction and oxidation of these systems occur is not yet clear. Figure 6.12(a) shows a formal mechanism in which a hydride ion (from NADH for instance) is transferred to N–5 and, following electronic rearrangement, a proton is accepted from the aqueous environment by N–1. However, experimental evidence appears to be against such a mechanism. It may well be that a semiquinone is involved in the process [see Fig. 6.12(b)]; this free

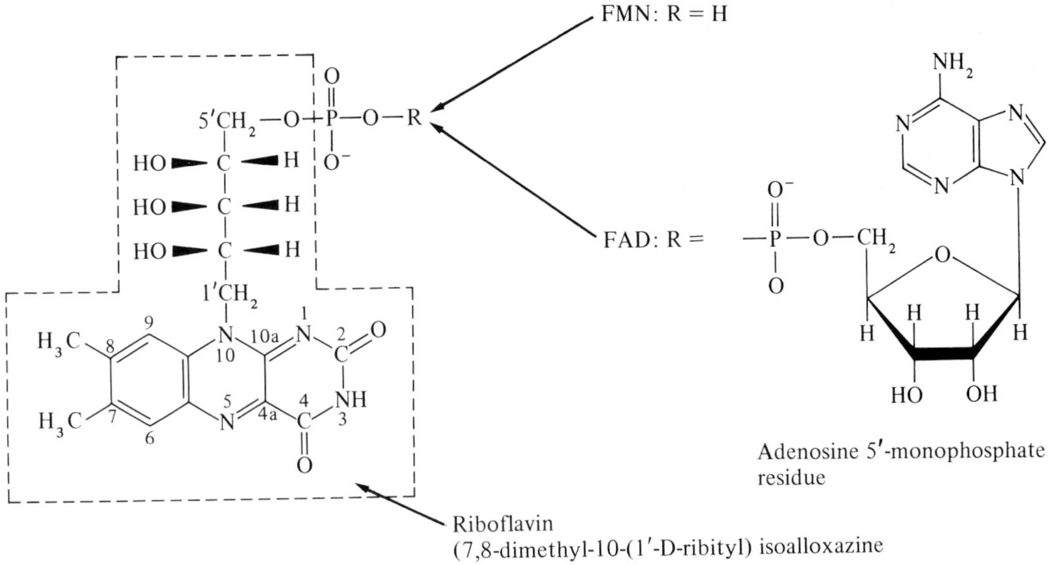

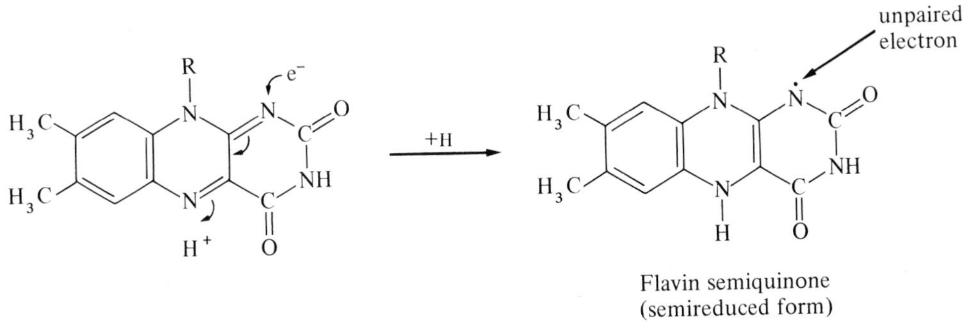

Fig. 6.12. Structure of the flavin coenzymes, flavin mononucleotide (FMN) and flavin adenine dinucleotide (FAD), and the possible mechanisms of their reduction.

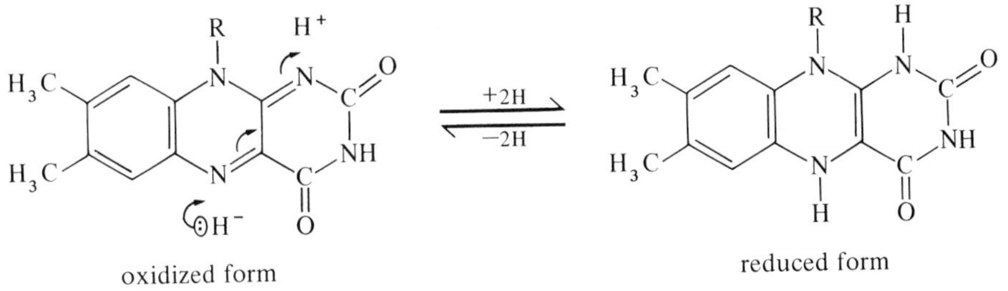

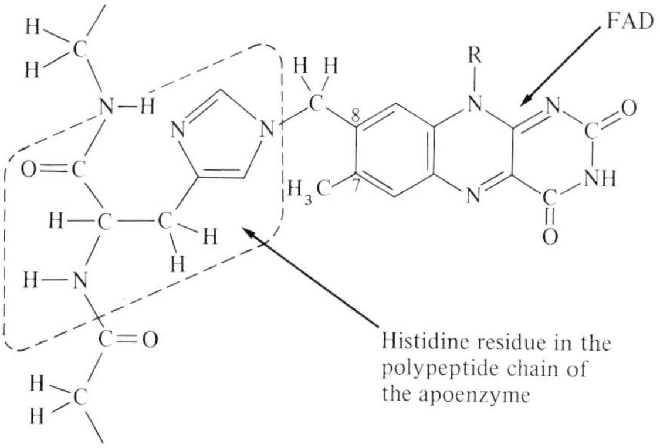

Fig. 6.13. Covalent binding of FAD to the apoenzyme in succinate dehydrogenase (EC 1.3.99.1).

radical has been detected by electron spin resonance spectroscopy.†

The flavoproteins of the NADH dehydrogenases are stereospecific with respect to the hydrogen at C–4 of NADH. Both FP_{INT} and FP_{EXT} of the internal and external NADH dehydrogenases of higher plants (see Fig. 6.5) are B-specific oxidoreductases. The outer mitochondrial membrane of higher plants also has a flavoprotein-containing NADH dehydrogenase, designated FP_{OUT} in Fig. 6.5; this flavoprotein is an A-specific oxidoreductase. This NADH dehydrogenase of the outer mitochondrial membrane does not transfer electron pairs to the terminal electron transport chain *in vivo* but can be induced to do so in the laboratory if cytochrome c is added and the mitochondria are suitably damaged.

---

†Electron spin resonance (ESR) or electron paramagnetic resonance (EPR) spectroscopy detects molecules or atoms which have unpaired electrons and are thus paramagnetic. Since most molecules have all their electrons paired and are diamagnetic, EPR spectroscopy is an effective way of detecting radicals, such as the flavin semiquinone, which may be transiently formed in a reaction. It is also an effective way of studying iron–sulphur proteins [see Section B.2(ii)(c)] which, in the reduced form, are paramagnetic and give a marked and characteristic signal. The EPR spectrometer measures the absorption of microwaves by a sample placed in a magnetic field which is altered at a constant microwave frequency (often 9.5 GHz). The spectrum is characterized by g values which are a function of microwave frequency and magnetic field. A g value is a measure of the ratio of an electron's magnetic moment to its spin angular momentum. EPR spectra are recorded as the first derivative of microwave absorption (dA/dH) against the strength of the magnetic field (H) (see Fig. 6.15). EPR signals are observed only at low temperatures, those of liquid nitrogen or liquid helium.

Flavoproteins as a group span a wide range of redox potentials, from E_0' values of -0.49 V to $+0.19$ V. Why is this so? In free FMN and FAD the E_0' value depends upon the relative amounts of resonance stabilization of the oxidized and reduced forms; an examination of the structures of these forms in Fig. 6.12 (a) shows that the oxidized form has the longer system of conjugated double bonds and so will have more resonance stabilization than the reduced form. Anything that alters the relative amounts of resonance stabilization in the oxidized and reduced forms will also alter the E_0' value. An important factor that can and does alter the resonance stabilization of the two forms is the strength of their binding to the apoprotein of the flavoprotein; this is, of course, a function of the structure of the apoprotein. If the oxidized form is bound weakly to the apoprotein whilst the reduced form is bound strongly the apoprotein-bound flavin has a greater tendency to remain in the reduced form than the unbound flavin. It, therefore, has a more positive E_0' value than the unbound flavin. Conversely, if the oxidized form is bound more tightly to the apoprotein than the reduced form, the apoprotein-bound flavin has a greater tendency to remain in the oxidized form and the E_0' value is more negative than that of the unbound flavin. Thus, Nature has, by varying the structure of the apoproteins, been able to modify the E_0' value of the $FMN/FMNH_2$ and $FAD/FADH_2$ redox systems in both electropositive and electronegative directions and so produce a range of flavoproteins with E_0' values tailored to function in a broad E_0' spectrum of metabolic niches.

As with the pyridine nucleotides the absorption spectra of the oxidized and reduced forms of FMN and FAD are different (see Fig. 6.14). This difference is most marked in the 300–500-nm section of the visible spectrum where the oxidized forms of the flavins have two absorption peaks (at 375 nm and 450 nm for FMN) whilst the reduced forms do not. Thus the rate and extent of the reduction or oxidation of flavoproteins can be measured by recording the change in absorbance at either 375 nm or 450 nm (or their equivalents, since the position of these maxima is slightly altered by the apoprotein).

The study of flavoproteins is technically difficult and as a result there is still confusion as to which of

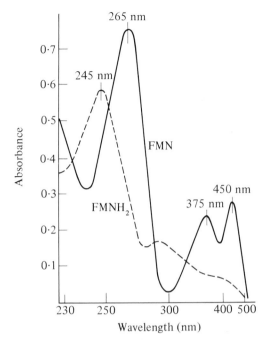

Fig. 6.14. Absorption spectra of FMN and $FMNH_2$.

the flavoproteins that have been identified in higher plant mitochondria are involved in the terminal electron transport chain or in what position in the electron flow sequence they function. Flavoproteins frequently show changes in fluorescence and absorption when they undergo redox changes. Moreover, the extent of these changes varies for different flavoproteins. These differences, taken with differences in their E_0' values (actually measured as E_m values—see Chapter 2, Section A.4), allow the various flavoproteins to be distinguished from one another. Using these criteria at least four flavoproteins have been identified in plant mitochondria. They are designated FP_{lf}, FP_{la}, FP_{ma} and FP_{ha} where f and a denote high fluorescence and absorption respectively and l, m and h denote low (i.e. $-ve$), mid and high (i.e. $+ve$) E_0' values. The E_m values of these flavoproteins, measured at pH 7.2, are -0.155, -0.070, $+0.020$ and $+0.110$ V respectively. The use of electron flow inhibitors has shown that FP_{ha} is a component of the main terminal electron transport chain operating between ubiquinone and the cytochromes b (see Figs. 6.3 and 6.5) whilst FP_{ma} is a component of the cyanide-resistant electron transport chain operating between ubiquinone and the cyanide-insensitive ter-

minal oxidase (see Section D and Fig. 6.34). The cyanide-resistant electron transport chain branches from the main electron transport chain at the ubiquinone stage. At the present time it is not clear what the functions of FP_{lf} and FP_{la} are. Since the internal and external NADH dehydrogenases and succinate dehydrogenase are known to have flavoprotein components (designated FP_{INT}, FP_{EXT} and FP_S in this book) it is clear that more research is required in this area. The situation is further complicated by evidence suggesting that NADH generated by the malic enzyme[12] of the mitochondrial matrix, catalysing eq. (6.2), is reoxidized by a different flavoprotein-containing NADH dehydrogenase from that used by NADH resulting from the activity of malate dehydrogenase.

$$\text{L-Malate} + NAD^+ \rightleftharpoons \text{pyruvate} + CO_2$$
$$+ NADH + H^+ \qquad (6.2)$$

It should be noted that flavoproteins are also components of enzymes other than those of the mitochondrial electron transport chain. For instance, a flavoprotein called NADH:dihydrolipoamide oxidoreductase[13] is one of the three different enzymic components of both the pyruvate dehydrogenase complex and the α-ketoglutarate dehydrogenase complex which catalyse key reactions in the mitochondrial matrix [see Figs. 6.2, 6.3 and Section E.1(ii)]. And a flavoprotein catalyses a reductive step in the multistep process brought about by yeast fatty acid synthase [see Chapter 8, Section C.1(i)(b)].

(c) *Iron–sulphur proteins*

Proteins containing iron may be divided into three classes: (i) haemoproteins [which include the cytochromes, see Section B.2(ii)(e)], (ii) iron–sulphur proteins (the subject of this section) and (iii) other iron-containing proteins (which include the oxygenases).

Iron–sulphur proteins (Fe–S proteins) are proteins which contain within their structure a lattice composed of interlinked iron and sulphur atoms which is usually referred to as an iron–sulphur cluster. The iron atoms involved in the cluster are not associated with haem, whilst the sulphur atoms are of two types, inorganic sulphur and cysteine sulphur (i.e. the sulphur of L-cysteine residues that are part of the protein). The inorganic sulphur atoms are relatively easily removed from the Fe–S cluster particularly under acid conditions and are called the labile-sulphur atoms. Iron–sulphur clusters are classified according to the number of iron and labile-sulphur atoms in the cluster; hence there are two types of ferredoxin in higher plant chloroplasts [see Chapter 5, Section C.6(i) and Fig. 5.19] which are designated as [2Fe–2S]- and [4Fe–4S]-ferredoxins. Some iron–sulphur proteins have more than one Fe–S cluster.

A single Fe–S cluster can only accept or donate one electron, one of the iron atoms changing back and forth between Fe^{3+} and Fe^{2+}. Iron–sulphur proteins are therefore classed as 'one-electron redox systems'. Since they cannot accept or donate hydrogen they are also classified as 'electron carriers'.

The molecular weights of iron–sulphur proteins cover a wide range, starting with bacterial ferredoxins at 6000, rising to 12,000 for the chloroplastidic ferredoxins and extending to much higher values for the more complex iron–sulphur proteins. The wide range of molecular weights indicates that the structure of the protein moiety also varies considerably from one type of Fe–S protein to another. Since the nature of the protein modifies the affinity of the Fe–S cluster(s) for electrons, it is hardly surprising that Fe–S proteins, as a class, cover a wide range of E_0' values (-0.42 V to $+0.35$ V).

When Fe–S proteins are reduced at very low temperatures the change in the valency state of the iron can be detected by EPR spectroscopy (see footnote on p. 178); under these conditions the Fe^{2+} form exhibits a very strong EPR signal, the g-value of which is characteristic of the Fe–S protein. The EPR spectrum of a [2Fe–2S]-ferredoxin is shown in Fig. 6.15 with a g-value of 1.94.

Little is known of the structures of the Fe–S proteins of the animal or plant terminal electron transport chains, in fact they are usually referred to as Fe–S centres. The NADH dehydrogenase of animal mitochondria has at least four Fe–S centres designated N–1, N–2, N–3 and N–4 with g-values of 1.94, 1.92, 1.89 and 1.86 respectively; N–1 and N–2 have E_0' values of ~ -0.30 and -0.02 V respectively and the N–3 + N–4 complex has an E_0' value of ~ -0.25 V. An NADH dehydrogenase, probably

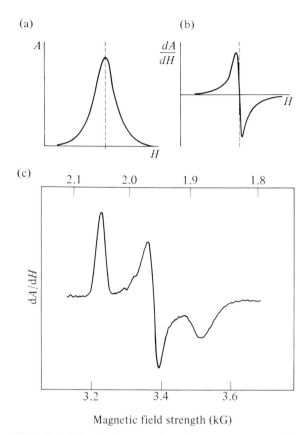

Fig. 6.15. (a) Plot of microwave absorption (A) against magnetic field strength (H). (b) Plot of the first derivative of microwave absorption (dA/dH) against magnetic field strength. (c) A typical EPR spectrum (i.e. a first derivative spectrum) of a reduced iron–sulphur protein (ferredoxin from *Porphyra umbilicalis*) at 22 K (kindly supplied by Dr. L. J. Rogers, U.C.W., Aberystwyth).

the internal enzyme, from Jerusalem artichoke (*Helianthus tuberosus*) and *Arum maculatum* appendix mitochondria have at least three Fe–S centres, two of which (g-values of 1.93 and 1.92) appear to be the equivalent of the animal N–1 and N–2 respectively whilst the third (g-value of 1.87) appears to be the equivalent of the animal N–3 + N–4 complex. Two other Fe–S centres have been detected in *Arum maculatum* appendix mitochondria which appear to correspond to the S–1 and S–2 Fe–S centres of animal mitochondrial succinate dehydrogenase.

(d) *Ubiquinone*

The term ubiquinone is the designation of a

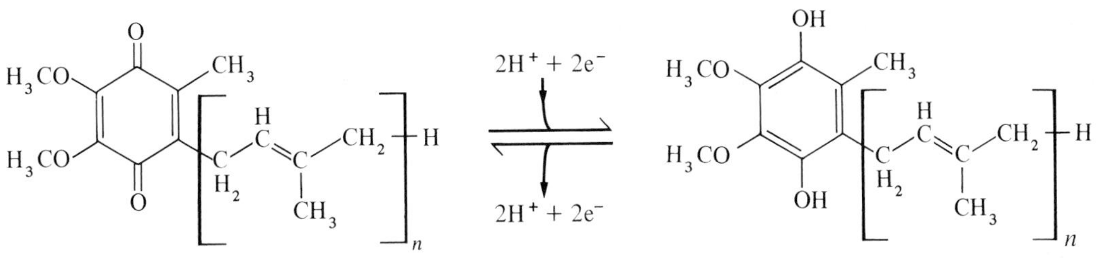

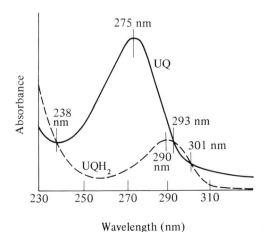

Fig. 6.16. Structures of the oxidized and reduced forms of ubiquinone ($n=9$ or 10 in higher plants).

family of compounds all having the same quinonoid nucleus but differing from one another in the number of isoprene units in the lipophilic side chain; in animal mitochondria the number is 10, in higher plant mitochondria it is 9 or 10 and in bacteria it is 6.

Ubiquinone is reduced by the equivalent of two electrons and two protons to ubiquinol (see Fig. 6.16); it is thus a 'two-electron' redox compound and a 'hydrogen carrier'. The absorption spectra of the oxidized and reduced forms are different from one another (see Fig. 6.17) but because the absorption maxima occur in the ultraviolet region of the spectrum they are largely masked in mitochondrial membrane preparations by the ultraviolet absorption of other components. This makes it difficult to study *in situ*.

Ubiquinone is a small molecule compared with the protein components of the terminal electron transport chain; moreover, being a lipid, it is freely

soluble and therefore mobile in the lipid core of the inner mitochondrial membrane. It is present in the inner membrane in relatively large amounts; the molecular ratio of ubiquinone to any of the various cytochrome species is about 5:1. These two properties probably explain its role at a 'junction point' in the electron transport chain, receiving electron pairs from several different donors and handing them on to the cytochrome sequence.

(e) Cytochromes

Cytochromes fall into the class of iron-containing proteins known as haemoproteins. The term 'haem' (an alternative spelling is 'heme') is defined by IUB Enzyme Commission as 'any tetrapyrrolic chelate of iron'; the terms 'ferrohaem' and 'ferrihaem' refer to Fe^{2+} and Fe^{3+} oxidation states in haem. The term haemoprotein refers to a protein containing haem as a prosthetic group. Cytochromes are defined as haemoproteins whose characteristic mode of action involves the transfer of reducing equivalents associated with a reversible change in oxidation state of the prosthetic group; formally, this redox change involves a single-electron reversible equilibrium between the Fe^{2+} and Fe^{3+} states of the haem iron. Thus cytochromes are classed as 'one-electron redox compounds'; since they cannot accept or donate hydrogen, they are also classed as 'electron carriers'.

Cytochromes were discovered in animal cells by MacMunn in 1886; he called them histo- or myohaematins because he thought they were related to haem and haematin. He assigned them a respiratory function. His results were so severely criticized by Hoppe–Seyler, who was probably the

Fig. 6.17. Absorption spectra of ubiquinone (UQ) and ubiquinol (UQH$_2$) in ethanol.

most influential physiological chemist of the day, that they were forgotten until 1925 when Keilin rediscovered the cytochromes. Keilin was studying the respiratory activity of insect muscles and showed, with a hand spectroscope, that they contained substances with absorption bands like those of reduced haem compounds (see Fig. 6.18). These bands disappeared when the tissue was made aerobic and reappeared as it became anaerobic.

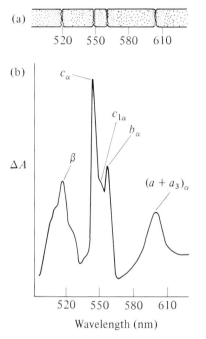

Fig. 6.18. (a) Absorption bands of reduced cytochromes of animal mitochondria seen in a hand spectroscope. (b) The difference spectrum between the cytochromes of anaerobic and aerobic cells. The α-bands of cytochromes c, c_1, b and $a + a_3$ can be seen as well as the composite β-band of all the cytochromes.

Keilin showed that these compounds were also present in many other tissues of animal, plant and bacterial origin. He gave them the name 'cytochromes' and suggested that they functioned by carrying electrons from fuel molecules to oxygen. He divided them into three classes, a, b and c, according to the position of their absorption bands when in the reduced state.

The reduced form of cytochromes characteristically have three absorption bands which appear as peaks in the recorder trace of a spectrophotometer; these are referred to as the α, β and γ (or Soret) bands. The α and β bands largely disappear

when the cytochrome is in the oxidized form whilst the γ band is shifted about 10 nm to shorter wavelengths and loses in persistence. The α peaks of reduced cytochromes taken as a group mostly lie in the range 545–600 nm whilst the β and γ peaks mostly lie in the ranges 520–535 nm and 415–445 nm respectively. The position of the α peak within the range 545–600 nm has, until relatively recently, been used to classify cytochromes as a-, b- or c-type; in a-type cytochromes the α peak is usually in the range 590–600 nm whilst for b- and c-type cytochromes it is usually in the ranges 555–565 nm and 545–555 nm respectively.

A particularly good method of finding the positions of the α peaks and thereby identifying the cytochromes in a tissue preparation is to record the difference spectrum between the reduced and oxidized forms. This is accomplished by placing the reduced (anaerobic) tissue preparation in the test cell of a suitable spectrophotometer and the oxidized (aerobic) tissue preparation in the reference cell and then taking the spectrum. Such a difference spectrum, taken of the cytochromes of the animal terminal electron transport chain, is given in Fig. 6.18 (b). It clearly shows the α peaks of cytochromes b, c_1, c and cytochrome oxidase $(a + a_3)$ along with a peak arising from the overlapping β peaks of all these cytochromes.

We now recognize that there are four classes of cytochrome rather than Keilin's three; these are termed cytochromes a, $-b$, $-c$ and $-d$. This classification is based upon the nature of the haem moiety and whether or not it is covalently bound to its apoprotein. However, the position of the α peak in the absorption spectrum of the reduced cytochrome is still of importance. Moreover, Keilin's a-, b- and c-type cytochromes fit into the new a, b and c classes.

The structures of the haem prosthetic groups of cytochromes a, b, c and d are shown in Fig. 6.19. All have a porphyrin ring structure composed of four pyrrole rings, designated A, B, C and D, that are linked by methine (—CH=) bridges, designated α, β, γ and δ. The nitrogen atoms of the pyrrole rings are bonded to a centrally placed iron atom which can undergo a reversible redox change, $Fe^{2+} \rightleftharpoons Fe^{3+}$. The two carbon atoms of each pyrrole ring that are not involved in methine bridges carry substituents whose nature is different for different cytochromes. The prosthetic group may or may not

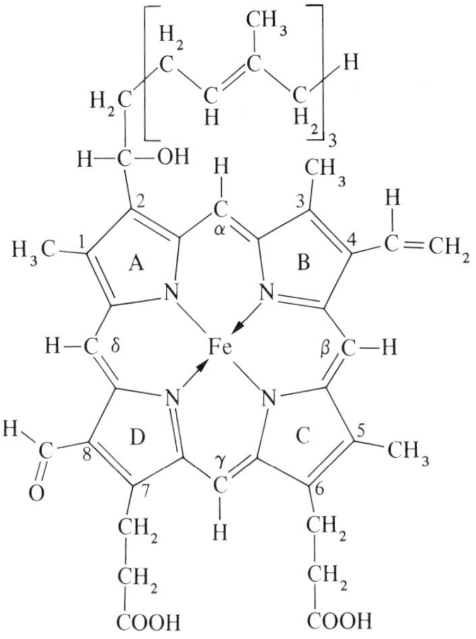

Haem *a* (non-covalently bound prosthetic group of cytochromes *a*)

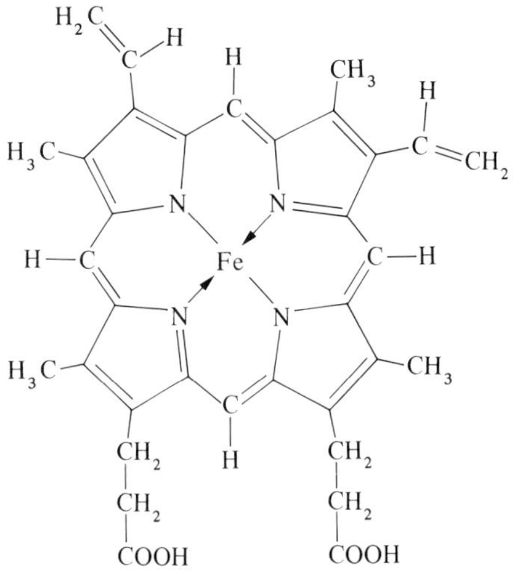

Protohaem (non-covalently bound prosthetic group of cytochromes *b*)

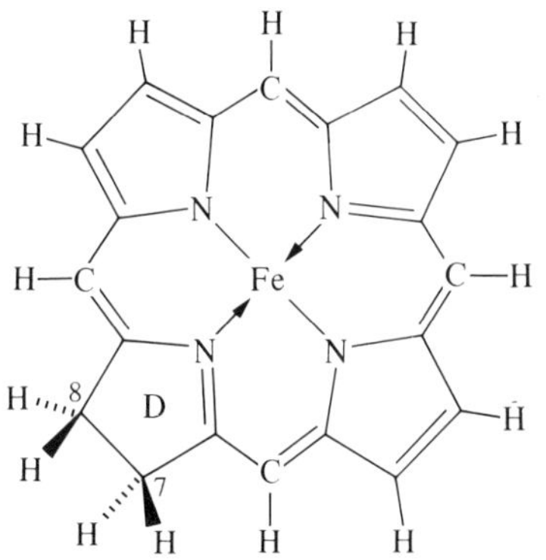

Chlorin (cytochromes *d* have non-covalently bound prosthetic groups that are derivatives of chlorin in which the H atoms at positions 1–8 are replaced by other groupings. Note that chlorophylls are even more complex chlorin derivatives)

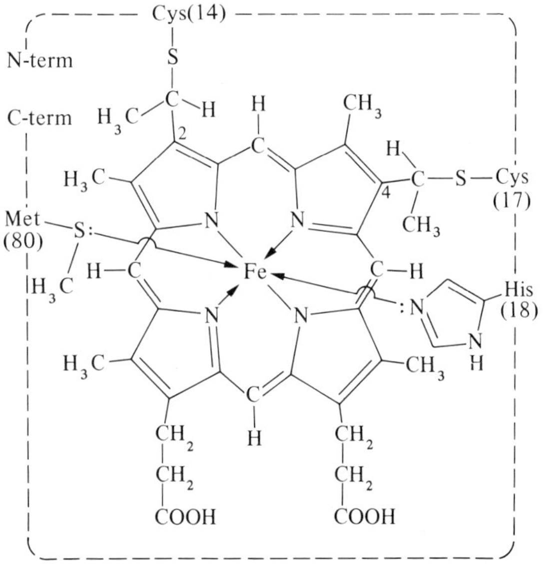

Protohaem covalently bound to the apoenzyme by thioether linkages at positions 2 & 4 as in cytochrome *c*. Cytochromes *c* are characterized by covalent linkages between haem side chains and the apoenzyme.

Fig. 6.19. Structures of the haem prosthetic groups of the four major groups of cytochromes, *a*, *b*, *c* and *d*. (Note that the term 'haem' refers to 'any tetrapyrrolic chelate of iron' and that the haem iron exists as Fe^{3+} in oxidized cytochromes and as Fe^{2+} in reduced cytochromes. Note also that the arrangement of double bonds in the porphyrin macrocycle is not fixed. Resonance of the conjugated double-bond system makes it possible to write down a number of mesomeric structures, one of which is shown above. The position of the two co-ordinate bonds ($\rightarrow$) and the two covalent bonds between the Fe and the pyrrole nitrogens, whilst being appropriate for the mesomer shown, are also not fixed for the same reason.)

be covalently bound to the apoprotein depending upon the cytochrome type.

Cytochromes a are defined as having a non-covalently bound haem a prosthetic group; haem a is characterized by the C_{15} isoprenoid residue linked to the usual C_2 substituent at C–2 (ring A) and a formyl group at C–8 (ring D). Cytochromes b are defined as having a non-covalently bound protohaem (or related haem without a formyl group) prosthetic group. Cytochromes c include all cytochromes in which the haem prosthetic group is covalently linked to the apoprotein via any of its side chains on positions 1–8. Cytochromes d are defined as having a non-covalently bound prosthetic group in which the degree of conjugation of double bonds is less than in porphyrin; a typical example is the chlorin type which has two substituents on both C–7 and C–8 in contrast to the porphyrin type which has only one.

Originally as new cytochromes were discovered they were given a subscript number (e.g. c_1, a_3) to distinguish them from others previously known. Whilst this nomenclature is retained for well-established cytochromes, all recently discovered cytochromes have been given a name based upon the wavelength (nm) of the α-band determined at room temperature (e.g. cytochrome c_{557}).

Cytochromes as a group cover a wide range of E_0' values, from about 0.0 V to about +0.6 V. The E_0' value of a cytochrome is a function of (i) the structure of its prosthetic group, (ii) the structure of its apoprotein and (iii) the method of binding between the prosthetic group and the apoprotein.

The only cytochrome about which there is a good deal of structural information is animal mitochondrial cytochrome c; this is due to the fact that it is easily extracted from mitochondria by salt solutions and is thus readily purified. Its apoprotein has a molecular weight of about 13,000 with 104 amino acids. The protohaem prosthetic group is linked by thioether bonds from positions 2 and 4 to cysteine residues 14 and 17 in the apoprotein and by coordinate bonds from the sulphur of methionine 80 and an imidazole nitrogen of histidine 18 to the central iron atom. X-ray crystallography has shown that the protohaem sits in a deep cleft in the apoprotein and is thus almost surrounded by the polypeptide chain; it is thought that this is also true of other cytochromes. This structure poses the problem of how an incoming electron reaches the haem Fe and how an outgoing electron leaves it. Various explanations have been put forward but none has yet been fully accepted.

Two c-type cytochromes, c_{547} and c_{549}, have been detected in higher plant mitochondria. Both have an E_0' value of about +0.235 V. Most of the experimental evidence indicates that c_{547} is more closely associated than c_{549} with cytochrome oxidase in the operation of the terminal electron transport chain. This ties in with the fact that c_{547} is readily extracted from plant mitochondria by salt solutions whilst c_{549} remains membrane-bound; c_{547} thus appears to be the plant equivalent of animal cytochrome c which is known to hand electrons directly to cytochrome oxidase.

Our knowledge of b-type cytochromes in both plants and animals is somewhat confused at present. It was believed until 1974 that plant mitochondria contained three b-type cytochromes, b_{557} ($E_0' = +0.075$ V), b_{560} ($E_0' = +0.045$ V) and b_{566} ($E_0' = -0.070$ V). Since then evidence has been presented to suggest the presence of two further cytochromes b with α absorption peaks close to 558 nm. The precise functions of all these b-type cytochromes are far from clear because experimental findings have been confusing and contradictory. Present opinion places b_{557} and b_{560} in the terminal electron transport chain after FP_{ha} and before cytochrome c_{549} (see Figs. 6.3 and 6.5).

The cytochrome oxidase of higher plant mitochondria is considered to be very similar to that of animal mitochondria. Subtle differences are, however, indicated by the fact that at low temperatures the α and γ (Soret) absorption bands of plant cytochrome oxidase each split into doublets (α at 589 and 597 nm; γ at 438 and 445 nm) whereas those of animal cytochrome oxidase do not. A great deal of work has been carried out on yeast and beef heart cytochrome oxidases in the past decade and has increased our knowledge of its structure, biosynthesis and mechanism of function considerably.

Yeast cytochrome oxidase is composed of two haems a, two copper atoms and seven different proteins designated I–VII with molecular weights ranging from 40,000 to 4500. The three largest proteins (I–III) are extremely hydrophobic and are coded by mitochondrial genes and synthesized on mitochondrial ribosomes. The four smallest proteins (IV–VII) are relatively hydrophilic and are coded by nuclear genes and synthesized by cytosolic ribosomes.

Beef heart cytochrome oxidase is composed of two haems *a*, two copper atoms and six different proteins, designated I (MW 40,000), II (MW 22,500), III (MW 15,000), IV (MW 11,200), V (MW 9800) and VI (MW 7300). The whole complex is cylindrical, has a length of 8.3 nm and lies perpendicular to the plane of the inner membrane. Since the width of the membrane lipid bilayer is about 4 nm cytochrome oxidase extends considerably on either side of it. The arrangement of the different proteins within the cylinder has recently been elucidated; III is located at the mitochondrial matrix end, II, V and VI at the cytosolic end and I and IV are in the middle with IV adjacent to III. It is presumed that one of the haems *a* is located in III and the other in the II, V and VI complex with the two atoms of copper placed centrally. The proteins at the cytosolic end plus one haem *a* must correspond to the cytochrome *a* component of cytochrome oxidase whilst those at the matrix end plus one haem *a* must correspond to the cytochrome a_3 component. The atoms of copper, which are known to undergo redox changes ($Cu^+ \rightleftharpoons Cu^{2+}$) during the operation of cytochrome oxidase, may serve to transfer electrons between *a* and a_3.

(f) *Lipids*

The electron and hydrogen carriers of the terminal electron transport chain are organized into functional units within the inner mitochondrial membrane and are consequently in intimate contact with the components of the lipid bilayer. These lipids, whilst not participating directly in electron flow, do appear to be essential to the functioning of the electron transport chain, probably by providing the ideal environment for its activity. This is borne out by the observation that removal of the phospholipids from mitochondria destroys the activity of the electron transport chain particularly at the cytochrome oxidase level but that this can be restored by adding them back. Moreover, the lipid composition of the inner mitochondrial membrane is significantly different from that of the outer membrane;

diphosphatidylglycerol, for instance, is localized in the inner membrane [see Chapter 8, Section B.3(i) (a)].

(iii) EVIDENCE FOR THE SEQUENCE OF COMPONENTS IN THE TERMINAL ELECTRON TRANSPORT CHAIN

The probable sequence of electron and hydrogen carriers in the electron transport chain in animals is outlined in Fig. 6.2 and that in higher plants in Fig. 6.3. These sequences, particularly that of higher plants, cannot be taken as the final picture but are likely to be a reasonably close approximation to the truth. Variations in detail can be expected as researchers probe even more deeply into the problem.

The sequences indicated in Figs. 6.2 and 6.3 are the result of a number of different lines of investigation, of which the following are the most significant.

(a) Sophisticated spectrophotometric techniques, pioneered by Britton Chance, have revealed the time sequence of reduction of the components of the electron transport chain. This depends upon (i) the fact that the oxidized and reduced forms of most of the components of the chain are different from each other and from those of the other components (see Figs. 6.11, 6.14, 6.17 and 6.18), (ii) the fact that the passage of an electron pair down the chain will take a finite length of time and (iii) the fact that as an electron pair passes down the chain each carrier will be reduced in turn.

(b) The same spectrophotometric instrumentation used in (a) has been used to determine 'cross-over' points in the presence of electron transport inhibitors such as those indicated in Figs. 6.2 and 6.3. If such an inhibitor is added to a mitochondrial preparation in a steady state of respiration then electrons will accumulate in the components of the electron transport chain on the input side of the block; these components will therefore become fully reduced. Simultaneously electrons will continue to flow to oxygen from those components on the output side of the block; these components will therefore become fully oxidized. There is thus a sharp transition from reduced to oxidized components at the position of the block; this transition is called a cross-over point (see Fig. 6.20). The position of the cross-over point of a particular inhibitor is

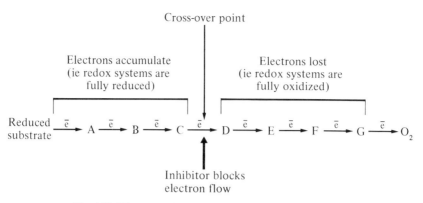

Fig. 6.20. Diagrammatic representation of a cross-over point.

determined by taking the difference spectrum between normal and inhibited mitochondria; a positive change in absorbance (ΔA) at the critical wavelength of a chain component indicates that it has become more reduced whereas a negative ΔA indicates that it has become more oxidized. The determination of the cross-over points for a range of inhibitors tells a good deal about the sequence in which the components of the chain operate.

(c) The appearance or disappearance of ESR signals under appropriate conditions can indicate the position of a particular iron–sulphur protein in the chain.

(d) The redox potential (E) of the components of the chain must be consistent with a flow of electrons from a negative (low) potential to a positive (high) potential. It is important to realize that the redox potential of each component that obtains *in vivo* takes account of (i) the steady-state ratio of the oxidized and reduced forms and (ii) the prevailing pH [see Chapter 2, eq. (2.29)].

(e) Important evidence has come from experiments involving the fragmentation and reconstitution of the electron transport chain, pioneered by D. E. Green using animal mitochondria. A considerable proportion of the area of the inner mitochondrial membrane is taken up by the electron transport complexes. When the inner membrane is disrupted these complexes themselves are frequently disrupted and the resulting fragments can be isolated. Since fragmentation occurs preferentially at the weakest points which are, of course, common to all the complexes, relatively few fragments of the electron transport chain are produced. This greatly

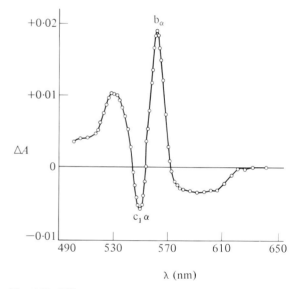

Fig. 6.21. Difference spectrum between animal mitochondria respiring normally and in the presence of the inhibitor, antimycin A. The positive ΔA indicates that cytochrome b has become more reduced and the negative ΔA indicates that cytochrome c_1 has become more oxidized.

facilitates their isolation. Fragmentation is usually brought about by treatment with detergents such as sodium cholate, digitonin or Triton X-100, or by ultrasonication. Both methods cleave the cytochrome sequence by removing the cytochrome c; it has been mentioned previously that cytochrome c is not built into the inner membrane but is loosely bound to its outer surface and can be washed off by the aqueous salt solutions that are used in both these methods. Detergents act by solubilizing the inner membrane lipid bilayer; they will also solubilize ubiquinone and thus cleave the electron transport chain at the junction point of the flavoprotein complexes and the cytochrome sequence. Thus the main fragments to be formed when animal inner mitochondrial membranes are treated by these methods are those which are shown in Fig. 6.2 as residing in the dotted rectangles. These fragments have been analysed in two ways: firstly, their chemical composition has been determined and secondly, their catalytic behaviour has been examined. In this way it has been shown, for instance, that one of the fragments is composed of cytochromes b and c_1 and iron–sulphur protein and is only capable of catalysing the oxidation of ubiquinol by cytochrome c. Complete terminal electron transport chain activity has been obtained by mixing together equimolar amounts of the four fragments shown in Fig. 6.2 with rather greater amounts of ubiquinone and cytochrome c.

3. Coupling of phosphorylation to electron transport

(i) SITES OF PHOSPHORYLATION

As an electron pair passes down the terminal electron transport chain the free energy released is used to phosphorylate ADP and thus form ATP. This raises the question of how many ADP molecules are phosphorylated per electron pair passing down the chain. This question was first answered in 1941 when Ochoa measured the P/O ratio in animal mitochondria. The P/O ratio is the number of moles of orthophosphate used up [in phosphorylating ADP, eq. (6.1)] per g atom (i.e. half mole) of oxygen used up (by being reduced to H_2O). Since the number of moles of orthophosphate used up equals the number of moles of ATP formed and the reduction of one mole of oxygen atoms ($= 1$ g atom $O_2 = 0.5$ mole O_2) requires one mole of electron pairs ($= 2$ moles of electrons), the P/O ratio is a measure of the ATP/$2e^-$ ratio.

Since 1941 the P/O ratio has been determined many times. Its magnitude depends upon the origin of the electron pairs. If they come from NADH the P/O ratio approximates to 3 indicating that three molecules of ADP are phosphorylated as an electron pair travels down the electron transport chain from NADH to $\frac{1}{2}O_2$. If the electron pair originates

from succinate or from sn-glycerol 3-phosphate the P/O ratio approximates to 2 indicating that two molecules of ADP are phosphorylated as an electron pair travels from succinate or sn-glycerol 3-phosphate to $\frac{1}{2}O_2$. These P/O ratios of 3 or 2 are dependent upon the use of fully coupled mitochondria (see Section B.1) in their determination; failure to use coupled mitochondria results in much lower P/O ratios.

The determination of the P/O ratios led to the concept of phosphorylation sites. Since three molecules of ATP were generated for every electron pair passing from NADH to $\frac{1}{2}O_2$ it was inferred that there were three sites in the electron transport chain at which a molecule of ADP was phosphorylated. Moreover, the P/O ratio of 2 for succinate oxidation suggested that one of the phosphorylation sites (site 1, Fig. 6.2) was located between NADH and UQ and preceded the input of electron pairs from succinate into the electron transport chain. This has been confirmed experimentally by use of antimycin A to block electron flow between cytochromes b and c_1 (see Fig. 6.2) and ferricyanide (ferricyanide/ferrocyanide constitute a redox system with an E_0' of $+0.36$ V) as an artificial recipient of electrons from UQH_2; reoxidation of NADH by this modified electron transport chain gave a P/O ratio of 1. A similar experiment with plant mitochondria has shown the same location for phosphorylation site 1 (see Fig. 6.3).

The location of phosphorylation site 3 between ferrocytochrome c and oxygen was first determined in animals by artificially introducing electrons into the electron transport chain at the level of cytochrome c and finding that the P/O ratio was 1. The electron donor in this experiment was L-ascorbic acid which reduced ferricytochrome c non-enzymatically in the presence of tetramethyl-p-phenylenediamine; the sequence of electron flow is shown in Fig. 6.22. Subsequently it became possible to use ferrocytochrome c directly as an electron donor to the electron transport chain; again the P/O ratio was 1. Similar experiments with plant mitochondria show the same location for phosphorylation site 3 (see Fig. 6.3).

Confirmation of the location of phosphorylation sites 1 and 3 and the elucidation of the location of site 2 have come from experiments using the spectrophotometric technique of 'cross-over' points

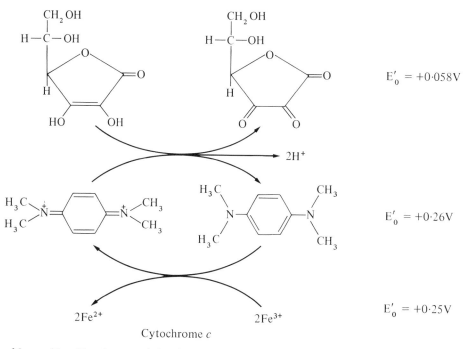

$E'_0 = +0{\cdot}058V$

$E'_0 = +0{\cdot}26V$

$E'_0 = +0{\cdot}25V$

Cytochrome *c*

Fig. 6.22. Use of L-ascorbic acid and tetramethyl-*p*-phenylendiamine to introduce electrons artificially into the terminal electron transport chain at the cytochrome *c* level.

[see Section B.2(iii)]. The results of such an experiment which demonstrates the presence of a phosphorylation site in mung bean mitochondria between cytochromes *b* and *c* are shown in Fig. 6.23. The upper and lower traces in Fig. 6.23 show the degree of reduction of cytochromes *b* and *c* respectively in the presence and absence of ADP when the mitochondria are oxidizing L-malate via endogenous NAD^+. Cytochrome *b* is more oxidized in the presence of ADP than in its absence. Conversely cytochrome *c* is more reduced in the presence of ADP than in its absence. This indicates that when ATP formation is blocked (as it is in the absence of ADP) electron flow from cytochrome *b* to cytochrome *c* is blocked. Thus a phosphorylation site (site 2) exists between these two cytochromes. With the demonstration of the multiplicity of cytochromes *b* and *c* in plant mitochondria further work is needed to pinpoint it exactly. However, present opinion places it in the position shown in Fig. 6.3.

The three phosphorylation sites in both animal and plant mitochondria are therefore located in analogous positions. Note that in neither is there a phosphorylation site in succinate dehydrogenase

which catalyses the transfer of an electron pair from succinate to UQ; this explains the P/O ratio of 2 for succinate oxidation. There is, however, one important difference between the plant and animal terminal electron transport chains; plants have an external NADH dehydrogenase whilst animals do not. When electrons flow to oxygen via the external NADH dehydrogenase the P/O ratio has been found to be 2 (not 3 as with the internal NADH dehydrogenase); this indicates that there is no phosphorylation site associated with the external NADH dehydrogenase.

Since the $\Delta G^{\circ\prime}$ of ADP phosphorylation is $+30.5$ kJ mol^{-1} the output of free energy at each of the phosphorylation sites must be significantly greater than 30.5 kJ mol^{-1}. Moreover, since the $\Delta G^{\circ\prime}$ of a redox reaction is a function of the potential difference ($\Delta E'_0$) between the reactants [see Chapter 2, eq. (2.36)], the potential through which the electron pair must drop at each phosphorylation site must be greater than 0.158 V. It can be seen from Figs. 6.2 and 6.3 that the three phosphorylation sites not only correspond with drops in potential greater than 0.158 V but also are the only parts of the

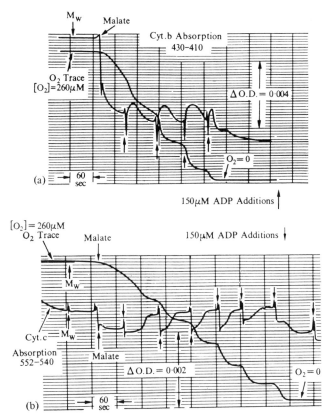

Fig. 6.23. Cross-over method of determining the location of a phosphorylation site. The two sets of traces demonstrate the presence of such a site between cytochromes *b* and *c* in mung bean mitochondria. The spectrophotometric trace in (a) shows that cytochrome *b* is more oxidized in the presence of ADP than in its absence whilst that in (b) shows that cytochrome *c* is more reduced in the presence of ADP than in its absence. (Data of W. D. Bonner: reproduced from *Plant Biochemistry*, 2nd edition, eds. J. Bonner and J. E. Varner, with permission of the authors and the Academic Press.)

electron transport chain to have potential drops of this magnitude.

Current opinion is that these potential drops associated with the phosphorylation sites occur within the enzyme complexes of the electron transport chain. Thus site 1 in animal mitochondria is associated with NADH dehydrogenase which is composed of FP_1 and several Fe–S centres collectively covering an E_0' range of ~ -0.30 V to -0.02 V. The potential drop within this NADH dehydrogenase is therefore about 0.28 V which corresponds to a $\Delta G^{\circ\prime}$ of -54.03 kJ per mole of electron pairs. Note that the E_0' values of the most electronegative and most electropositive components of the NADH dehydrogenase are very close to those of NADH (the electron donor) and UQ (the electron acceptor) respectively. This is thought to

facilitate the efficient transfer of electrons from NADH to the NADH dehydrogenase and from the latter to UQ. This matching of the E_0' values of the components of the enzyme complex to its electron donor and its electron acceptor is not peculiar to NADH dehydrogenase; as can be seen from Fig. 6.24 it also applies to the ubiquinol–cytochrome *c* reductase and cytochrome oxidase enzyme complexes. Figure 6.24 also gives the potential drops occurring within these complexes and the resulting output of free energy.

The efficiency of free energy conservation of the three phosphorylation sites ranges from about 60% to 72% (see Fig. 6.24) whilst the overall efficiency of oxidative phosphorylation is 41.7%. These figures must be regarded as approximate since they are calculated from standard free energy changes ($\Delta G^{\circ\prime}$)

Substrate	Enzyme Complex	E_0' (V)		$\Delta E_0'$ (V)	$\Delta G^{0\prime}$ (kJmol^{-1})[*]	% Efficiency[†]
		Substrate	Enzyme Component			
NAD		-0.32				
$2e^-$	FP$_1$		~ -0.30			
	Fe-S			~ 0.28	-54.03	56.45
	Fe-S		-0.02			
UQ		$+0.10$				
$2e^-$	Cyt b		~ 0.00			
	Fe-S			~ 0.22	-42.45	71.85
	Cyt c_1		$+0.22$			
2 Cyt.c		$+0.25$				
$2e^-$	Cyt a		$+0.29$			
				0.26	-50.17	60.79
	Cyt a_3		$+0.55$			
$\frac{1}{2}O_2$		$+0.817$				

*Calculated using the equation $\Delta G^{\circ\prime} = n F \Delta E_0'$ (see Chapter 2, section A5)
†%Efficiency of free energy conservation as ATP calculated from $\Delta G^{\circ\prime}$ values

Fig. 6.24. Standard potential differences, free energy changes and percentage efficiencies of free energy conservation associated with the phosphorylation sites in animal mitochondria.

rather than the free energy changes (ΔG) that actually obtain in the mitochondrion; however, they are a reasonable approximation.

(ii) MECHANISM OF PHOSPHORYLATION

Over the past three decades a great deal of effort has been put into the elucidation of the molecular mechanism by which the free energy released by electron transport is used to phosphorylate ADP. This effort has led to a much better understanding of the problem but has not yet provided the complete solution.

During this period three hypothetical mechan-isms have come to the fore, the chemical-coupling, the conformational-coupling and the chemiosmotic hypotheses.

The chemical-coupling hypothesis was proposed first. It postulates that each of the free energy-yielding redox reactions of the electron transport chain is coupled to the free energy-requiring ADP phosphorylation reaction by a common high-energy intermediate compound. The high-energy intermediate is formed by the redox reaction and is then used in the ADP phosphorylation reaction. Thus it is supposed that much of the free-energy output of the redox reaction is conserved in the high-energy intermediate and is then used to drive

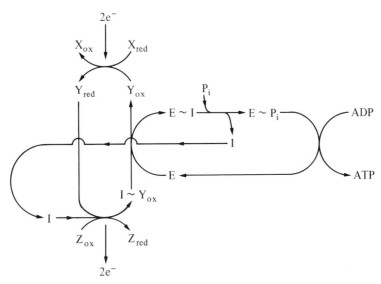

Fig. 6.25. A simple conception of the chemical-coupling hypothesis. (X, Y and Z = redox compounds in the terminal electron transport chain; I = a coupling factor, possibly a protein; E = the enzyme ATPase; ~ indicates a high-energy compound.)

the endergonic phosphorylation of ADP. In this way much of the free energy of the redox reaction is conserved in ATP. One version of this hypothesis is shown diagrammatically in Fig. 6.25.

The chemical-coupling hypothesis is very attractive since it is simple to understand and is in principle no different from substrate-level phosphorylations (see Section E.1) which also depend upon the participation of a common high-energy intermediate. However, it has lost favour in recent years because (i) none of the postulated coupling factors has yet been found and (ii) it does not explain the experimental observation that the inner mitochondrial membrane must be intact and form a closed vesicle for oxidative phosphorylation to occur.

The conformational-coupling hypothesis postulates that the free energy produced by the redox reactions is conserved as a change in the conformation of a protein component of the inner mitochondrial membrane. This change would be from a low-energy conformation to a high-energy conformation and would involve the breaking of a number of weak bonds, such as hydrogen bonds or hydrophobic bonds, in the molecule and the making of others. It is known that such energy-dependent conformational changes do occur in proteins; in fact cytochrome c undergoes a small change when it is oxidized and reduced. It is thought that changes

such as these in some of the components of the enzyme complexes of the electron transport chain might be transmitted to one of the proteins of the F_1 component of the ATP-ase complex of the inner membrane (see Chapter 3, Section B.5). The energy of this now energized ATP-ase is then thought to drive the phosphorylation reaction. As this occurs the ATP-ase reverts to its low-energy conformation. Several interesting mechanisms have been put forward to explain how conformational changes in the ATP-ase could bring about the phosphorylation of ADP but none has much experimental backing.

The conformational-coupling hypothesis is also very attractive, although it is, in effect, only a variant of the chemical-coupling hypothesis. It postulates the generation of a high-energy state within a given molecule brought about by the formation of several weak bonds whilst the chemical-coupling hypothesis postulates the generation of a high-energy molecule brought about by the joining together of two components by a single covalent bond. It may be thought of as being the mechanistic opposite of muscle contraction; in the latter the free energy released by the hydrolysis of ATP causes a dramatic change in the conformation of muscle proteins whereas in the conformational-coupling hypothesis it is the change in protein conformation that provides the free energy for ATP formation. An indication that conformational changes in membrane proteins are associated with oxidative phosphorylation is indicated by the dramatic change in the appearance of mitochondria when they change from the resting state to the active state (see Fig. 6.26). In the resting state, known as state 4 according to the definition of Chance and Williams, the mitochondria are severely depleted in ADP and so cannot produce ATP by oxidative phosphorylation. Under these circumstances the mitochondria have an orthodox appearance; the inner membrane is closely applied to the outer membrane except where it forms cristae [see Fig. 6.26(a)]. In the active state, known as state 3 according to Chance and Williams, the mitochondria have a more than adequate supply of ADP and can form ATP rapidly by oxidative phosphorylation. Under these circumstances the mitochondria are said to exhibit the condensed conformation [see Fig. 6.26(b)] in which the inner membrane has pulled away from the outer membrane and has become more tightly folded. This causes the matrix to be reduced in volume by about 50% by expulsion of water. As the level of ADP falls, by conversion into ATP, the rate of oxidative phosphorylation drops; the mitochondria gradually revert to state 4 and the orthodox conformation. It should be noted that the orthodox and condensed conformations are extreme states produced in the laboratory; the orthodox conformation corresponds to the state in which oxidative phosphorylation is at a minimum whilst the condensed conformation corresponds to a state in which it is at a maximum. In vivo mitochondria exist in a state somewhere between these extremes; this allows the rate of oxidative phosphorylation to be controlled, being increased or decreased according to the cell's immediate energy needs.

The chemiosmotic hypothesis differs from the two other hypotheses in postulating that the free energy produced by the redox reactions of the electron transport chain is used to generate a high-energy state rather than a high-energy compound or a high-energy conformation within a molecule. This high-energy state takes the form of an electrochemical gradient of H^+ ions across the inner mitochondrial membrane.

According to the chemiosmotic hypothesis, as electron pairs pass down the electron transport chain, H^+ ions are pumped across the inner mitochondrial membrane from the matrix to the intermembrane space. This requires that those redox reactions of the electron transport chain that involve H^+ ions as well as electrons are so arranged within the membrane that when H^+ ions are taken up they are abstracted only from the matrix and when they are given out they are discharged only into the intermembrane space. These redox reactions are those that involve the class of electron transport chain components known as hydrogen carriers (e.g. flavoproteins, UQ) [see Section B.2(i)]. The hydrogen-carrying capability of some of the components of the electron transport chain is of fundamental importance to the chemiosmotic hypothesis; this is not so with the other two hypotheses.

It is further postulated that the electrochemical gradient of H^+ ions across the inner membrane constitutes the driving force for the ATP-ase catalysed phosphorylation of ADP (see Fig. 6.27). This driving force has been termed 'proton motive force' and is composed of two elements. These are (i) a pH difference or gradient across the inner membrane of about 1.5 pH units which is due to the difference in H^+ ion concentration in the aqueous solution present in the intermembrane space and the matrix and (ii) an electrical potential difference or gradient of about 0.15 V due to the accumulation of H^+ ions on the outer surface of the inner membrane and of anions (e.g. OH^-) on the inner surface. This is expressed mathematically in eq. (6.3).

Proton motive force $(V) =$

$$\Delta\psi + \left(\frac{-2.303RT}{F} \cdot \Delta pH \right) \quad (6.3)$$

where $\Delta\psi$ = electrical potential difference
$\quad\quad (= \sim 0.15 \text{ V})$,
$\quad\Delta pH = pH_{(outside)} - pH_{(matrix)} (= \sim -1.5)$,
$\frac{-2.303RT}{F} = -0.0592$ at 25°C.

How does the electrochemical gradient of H^+ ions drive the endergonic phosphorylation of ADP? As can be seen from eq. (6.1) the formation of ATP from ADP and orthophosphate involves the removal of the elements of water between the two reactants; in fact it has been demonstrated that ADP loses H^+ and orthophosphate loses OH^-. This reaction is catalysed by the ATP-ase built into

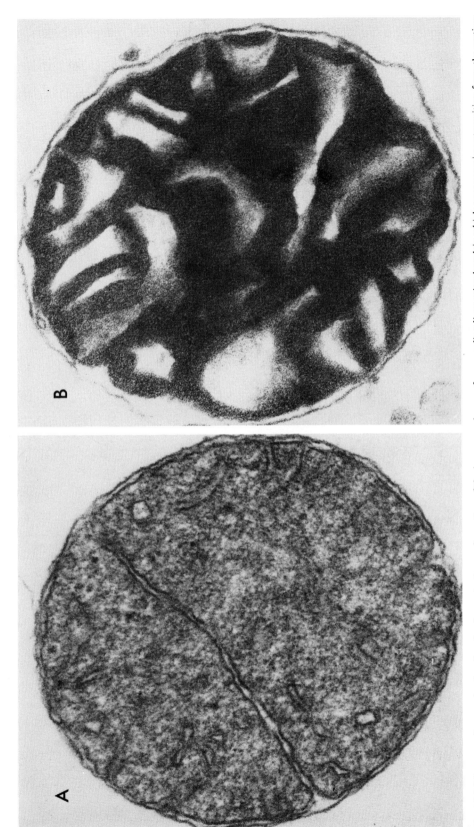

Fig. 6.26. Electron micrographs showing the change in the conformation of the inner membrane of mammalian liver mitochondria which accompanies a transition from the resting respiratory state (state 4, ADP very low) (A) to the active respiratory state (state 3, ADP high) (B). Kindly supplied by Dr. E. A. Munn, Institute of Animal Pysiology, Babraham, Cambridge.)

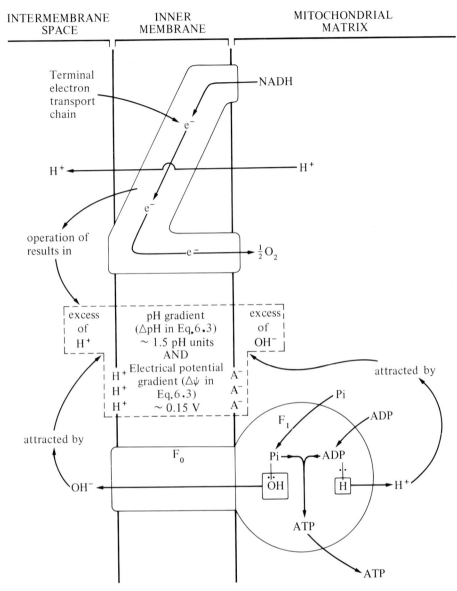

Fig. 6.27. Possible mechanism by which the pH gradient (ΔpH) and electrical potential gradient ($\Delta\psi$) are utilized by the ATP-ase of the stalked particles of the inner mitochondrial membrane to generate ATP from ADP and orthophosphate. (A$^-$ = negatively charged ions, e.g. OH$^-$, accumulating on the matrix side of the inner membrane.)

the inner mitochondrial membrane. The ATP-ase is a multi-protein complex consisting of a spherical headpiece known as the F_1 component, a short stalk and an intramembrane part known as the F_0 component. The F_1 headpiece projects into the mitochondrial matrix and the ATP-ase activity appears to be located in it (see Chapter 3, Section B.5). The chemiosmotic hypothesis postulates that

when ATP and orthophosphate, derived from the matrix, bind to the ATP-ase active site the H$^+$ and OH$^-$, removed in the formation of ATP, are pulled in opposite directions by the electrochemical gradient. The H$^+$ ions move into the matrix, attracted by the excess of OH$^-$ and the relative electronegativity, whilst the OH$^-$ ions move into the intermembrane space, attracted by the excess H$^+$ and the relative electropositivity; in both cases H_2O is formed. It is suggested that the opposite directionality of ion movement is facilitated by the nature of the ATP-ase active site which is supposed to have vectoral character, releasing H^+ in the direction of the matrix and OH^- in the direction of the intermembrane space. The ATP formed in this process is released into the matrix (see Fig. 6.27).

There is a good deal of experimental evidence in favour of the chemiosmotic hypothesis. It is a prerequisite of the hypothesis that the inner mito-chondrial membrane be intact and form a closed vesicle otherwise the H^+ ions pumped across would leak back; for the same reason the inner membrane should not allow the passive diffusion of H^+ ions. It has been demonstrated that the inner membrane is impermeable to H^+ ions and that it has to be intact for oxidative phosphorylation to occur. Furthermore, it has been demonstrated *in vitro* that as electrons flow down the electron transport chain, H^+ ions are transported across the inner membrane from matrix to the surrounding medium (they can readily pass through the outer mitochondrial mem-brane) which becomes more acidic than the matrix by 1.5 pH units. Indirect evidence has indicated that electron flow down the chain is capable of generat-ing an electrical potential of 0.15 V across the inner membrane with the outer surface positive with respect to the inner. Finally the connection between the electrochemical gradient of H^+ ions and the phosphorylation of ADP has been made by de-monstrating that when the appropriate H^+ ion gradient is artificially established across the inner membrane, in the absence of electron flow, ADP and orthophosphate are converted into ATP.

Despite this rather compelling evidence in its favour, there are, however, difficulties with the chemiosmotic hypothesis. The first of these stems from nature of the phosphorylation sites [see Section B.3(i)] which are located in the positions shown in Figs. 6.2 and 6.3. It has been ex-perimentally shown that as an electron pair passes down the electron transport chain from NADH to $\frac{1}{2}O_2$ six H^+ ions are pumped across the inner membrane. Since there are three phosphorylation sites between NADH and oxygen this means that for one molecule of ADP to be phosphorylated, two H^+ must be pumped across the inner membrane, i.e. the H^+/P ratio is 2. It follows from this that at each

phosphorylation site two H^+ ions must be pumped across the inner membrane as an electron pair passes through it. As has been mentioned earlier, the only redox systems that can participate in the pumping of H^+ are the hydrogen carriers, the flavoproteins and UQ. Thus the position of the hydrogen carriers in the electron transport chain should, according to the chemiosmotic hypothesis, dictate where the phosphorylation sites are. As will be seen from Fig. 6.28, which shows the animal electron transport chain, only the first phosphoryl-ation site (site 1 in Fig. 6.2) coincides with a hydrogen carrier and the pumping of $2H^+$ across the inner membrane. The second H^+ pumping site in Fig. 6.28 does not coincide with site 2 in Fig. 6.2. Moreover, the third phosphorylation site (site 3 in Fig. 6.2) is not associated with any hydrogen carrier and therefore cannot pump H^+ ions across the inner membrane. In fact the animal electron transport chain, as envisaged in Fig. 6.2, which is known to have three phosphorylation sites between NADH and oxygen, has only two H^+ pumping sites. In order to get over this discrepancy a third H^+ pumping site has been postulated in the ubiquinol–cytochrome c reductase complex; this is shown as the X/XH_2 system in Fig. 6.28. It is thought that this system might be UQ operating at a second site in the electron transport chain. If this were the case then this site would be coincident with phosphorylation site 2.

A second difficulty with the chemiosmotic hy-pothesis lies in the fact that it has not yet been established whether the electrochemical gradient of H^+ ions is an obligatory step in oxidative phos-phorylation, as is required by the hypothesis, or is simply the result of a side reaction caused by the reversible vectoral hydrolysis of a high-energy intermediate compound generated by the operation of the electron transport chain. The latter would therefore be the obligatory intermediate in oxidative phosphorylation, as suggested in the chemical coupling hypothesis, and would be in equilibrium with the electrochemical gradient of H^+ ions. These alternatives are presented diagrammatically in Fig. 6.29.

In spite of these difficulties the majority biochem-ical opinion at the present time is that the chemios-motic hypothesis most nearly approximates to the actual mechanism of oxidative phosphorylation. As

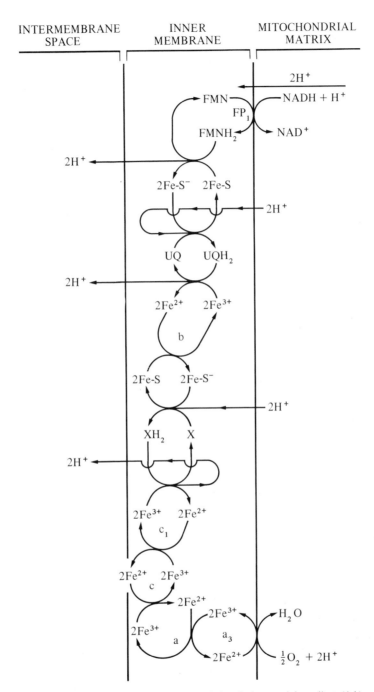

Fig. 6.28. Possible mechanism by which the pH gradient (ΔpH) and electrical potential gradient ($\Delta\psi$) are set up across the inner mitochondrial membrane by the operation of the terminal electron transport chain in animal cells. (Note: it has been necessary to postulate that an extra hydrogen carrier, X/XH_2, operates between cytochromes b and c_1 in order to account for the fact that three pairs of H^+ ions are pumped across the membrane per oxygen atom reduced to water; it has been suggested that UQ is this carrier and thus operates in two positions in the chain.)

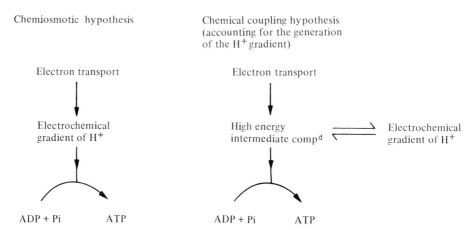

Fig. 6.29. Diagrammatic representation of the question 'Is the electrochemical gradient of H$^+$ ions an obligatory intermediate in oxidative phosphorylation?'.

a mechanism it is particularly attractive because it can also account for photophosphorylation in chloroplasts [see Chapter 5, Section C.6(ii)] and photosynthetic bacteria.

(iii) Uncoupling of oxidative phosphorylation

The phosphorylation of ADP is obligatorily coupled to electron transport. Thus *in vivo* there cannot be electron flow down the terminal electron transport chain without the conversion of ADP into ATP or *vice versa*.

Certain compounds are capable of disengaging this link between electron transport and ADP phosphorylation. They are called uncoupling agents. The first of these to be described was 2,4-dinitrophenol in 1948. Since then others have been discovered. They seem to fall into two distinct classes. The first class are typically lipid-soluble compounds with an acidic moiety and at least one aromatic ring; it therefore includes 2,4-dinitrophenol. The structures of representatives of this class are shown in Fig. 6.30. The second class are called ionophores and only uncouple in the presence of specific cations, usually K$^+$ or Na$^+$. The structures of a representative sample of this class are shown in Fig. 6.31.

Uncoupling agents allow substrates (e.g. NADH, succinate) to be oxidized via the terminal electron transport chain with the consumption of oxygen but without the concomitant phosphorylation of ADP. Both classes of uncoupler bring this about by

discharging the high-energy intermediate or state generated by electron transport. The way they do this fits in best with the chemiosmotic hypothesis.

Uncoupling agents of the 2,4-dinitrophenol class allow H$^+$ ions to flow through the inner mitochondrial membrane into the matrix so collapsing the electrochemical gradient of H$^+$ generated by electron transport. They are able to do this because they can dissolve in the lipid core of the membrane and therefore diffuse across the membrane; since they have an acidic group which can ionize to yield H$^+$ ions they can equilibrate the H$^+$ ion concentration on either side of the inner membrane.

Ionophores collapse the electrochemical gradient of H$^+$ ions by allowing cations to flow through the inner mitochondrial membrane into the matrix. Valinomycin and nonactin are antibiotics produced by several *Streptomyces* spp. They both have a macrocyclic ring structure which allows them to co-ordinately bind a cation in the centre of the molecule. They both bind K$^+$ much more strongly than Na$^+$. In valinomycin the K$^+$ is co-ordinately bound to six carbonyl oxygen atoms giving the octahedral arrangement shown in Fig. 6.31. In nonactin the K$^+$ is co-ordinately bound to eight oxygen atoms, four from carboxyl groups and four from the heterocyclic rings. When the K$^+$ is bound in these molecules it is tucked away in the centre of an essentially hydrophobic molecule; it has been rendered lipid soluble. As such it can diffuse across the lipid core of the inner mitochondrial membrane. Since the K$^+$–antibiotic complex is positively

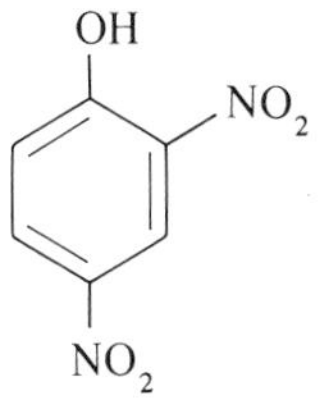

2,4-Dinitrophenol

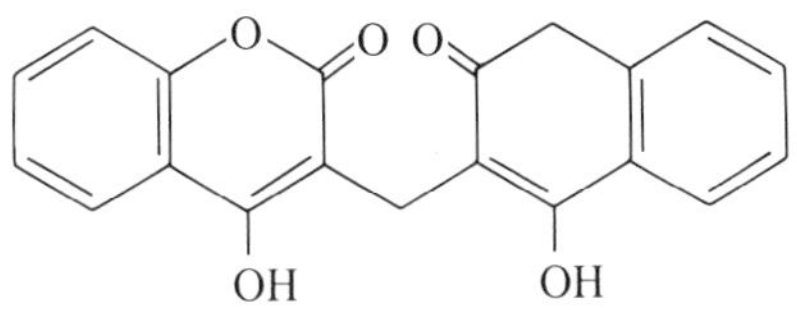

Dicoumarol

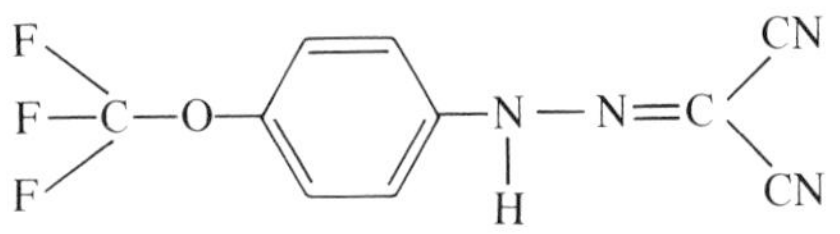

Carbonylcyanide *p*-trifluoromethoxyphenylhydrazone

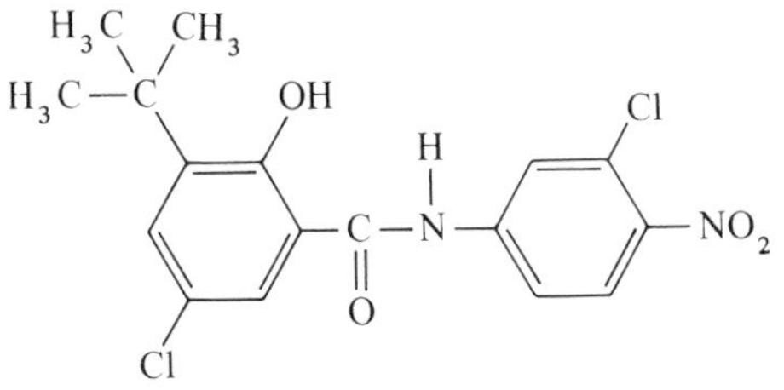

5-Chloro-3-*tert*-butyl-2'-chloro-4'-nitrosalicylanilide

Fig. 6.30. Lipid-soluble uncouplers of oxidative phosphorylation.

charged it is pulled into the mitochondrial matrix by the electrochemical gradient of H^+ ions, simultaneously collapsing it. Valinomycin and nonactin are said to be cation-carrying ionophores since they actually carry the cation across the inner membrane. This contrasts with the ionophore, gramicidin A, which allows cations to cross the inner membrane by forming a channel through which

they pass. Gramicidin A is the main component of gramicidin D, an antibiotic produced by *Bacillus brevis*. K^+ and Na^+ ions are able to pass with equal facility through the channel it makes in the inner membrane, dragged in the direction of the matrix by the electrochemical gradient of H^+ ions, simultaneously collapsing it.

(iv) Inhibition of oxidative phosphorylation

The structures of three important inhibitors of oxidative phosphorylation are shown in Fig. 6.32. Oligomycins are an antibiotic mixture produced by an actinomycete similar to *Streptomyces diastatochromogenes*. They inhibit oxidative phosphorylation by blocking the activity of the ATP-ase in the inner mitochondrial membrane; they appear to do this by binding to one of the proteins of the F_0 component of the ATP-ase.

Atractylic acid is a poisonous glucoside present in the rhizomes of the plant *Atractylis gummifera*. Bongkrekic acid is one of two toxic antibiotics produced by *Pseudomonas cocovenenans* when grown on partly defatted coconut (the other is toxoflavin). It is named from 'bongkrek', a moulded coconut product from Indonesia which becomes highly poisonous after *P. cocovenenans* outgrows the mould. Both these compounds inhibit oxidative phosphorylation by blocking the ADP–ATP antiport [see Section B.3(v)] in the inner mitochondrial membrane. Thus ADP cannot pass from the cytosol to the mitochondrial matrix for phosphorylation and ATP cannot pass in the opposite direction. Since ADP phosphorylation is obligatorily linked to electron transport, once all the ADP in the matrix has been converted into ATP, oxidative phosphorylation comes to a stop.

(v) Other uses for the high-energy state generated by electron transport

It is likely that *in vivo* mitochondria use a significant fraction of the high-energy state generated by electron transport for purposes other than the generation of ATP. One of these is the active transport of certain ions and metabolites across the inner mitochondrial membrane.

Although the outer mitochondrial membrane is freely permeable to compounds with a molecular

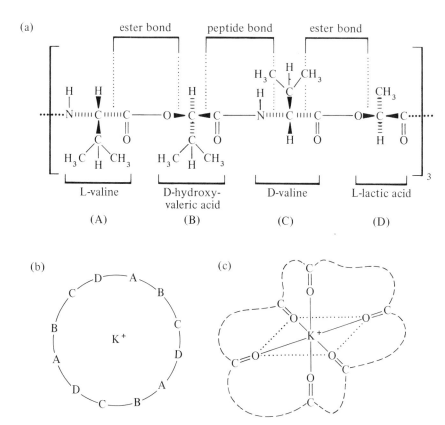

Valinomycin (a) the valinomycin repeating unit, (b) the cyclical arrangement of the repeating unit in valinomycin, (c) the octahedral binding of K^+ in the centre of the valinomycin molecule by six carbonyl oxygen atoms.

Nonactin

Gramicidin A

Fig. 6.31. Ionophoric uncouplers of oxidative phosphorylation.

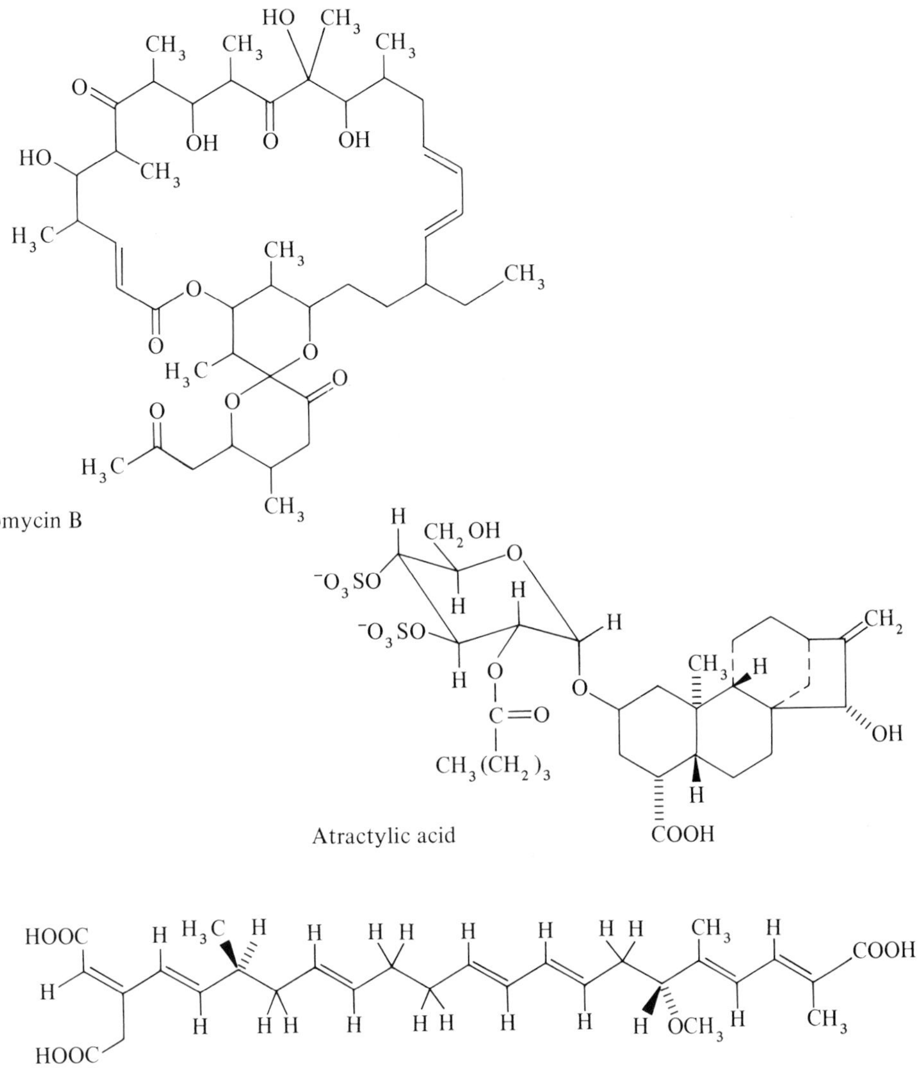

Oligomycin B

Atractylic acid

Bongkrekic acid

Fig. 6.32. Some inhibitors of oxidative phosphorylation.

weight less than about 10,000, the inner membrane is selectively permeable. For example, it is impermeable to K^+, Na^+, Cl^-, Br^-, I^-, NAD^+, NADH, $NADP^+$, NADPH, AMP, CoASH and acyl-CoAs. However, it is permeable to OH^-, $H_2PO_4^-$, ADP, ATP, pyruvate, dicarboxylic acids such as succinate, L-malate, fumarate and α-ketoglutarate, tricarboxylic acids such as citrate and amino acids such as L-aspartate and L-glutamate.

The passage of these ions and metabolites across the inner membrane is mediated by specific transport systems which are variously called exchange carriers, porters or translocases. They are probably composed of specialized groups of proteins. Each system is specific to certain ions or metabolites. Most promote a 1:1 exchange across the membrane and are frequently called antiport systems; the most studied is the ADP–ATP antiport which simultaneously transports one molecule of ADP and one of ATP in opposite directions.

These antiport systems are passive, the transported ions or metabolites moving down concentration gradients until equilibrium is reached. However, they can also transport actively (i.e. against concentration gradients) when they are coupled to the high-energy state generated by electron transport. It has been shown that the electrochemical gradient of H^+ ions generated by electron transport can be used to drive the ions or metabolites through their specific antiport systems against concentration gradients. The way this operates for two very important antiports, the ADP^{3-}–ATP^{4-} and the $H_2PO_4^-$–OH^- antiports, is shown in Fig. 6.33. The ADP^{3-}–ATP^{4-} exchange is

said to be driven by the difference in electrical charge on the two compounds and the difference in potential [$\Delta\psi$ of eq. (6.3)] across the inner membrane. The $H_2PO_4^-$–OH^- exchange is said to be driven by the pH gradient [ΔpH of eq. (6.3)] across the inner membrane. The $H_2PO_4^-$–OH^- antiport is blocked by HS-group inhibitors such as N-ethylmaleimide whilst the ADP^{3-}–ATP^{4-} antiport is blocked by atractylic and bongkrekic acids [see Section B.3(iv)]. The $H_2PO_4^-$–OH^- antiport is the means by which the orthophosphate formed by ATP hydrolysis in the cytosol is returned to the mitochondrial matrix to phosphorylate ADP. The ADP^{3-}–ATP^{4-} antiport is the means by which ATP, generated in the mitochondrial matrix by oxidative phosphorylation, is transported to the

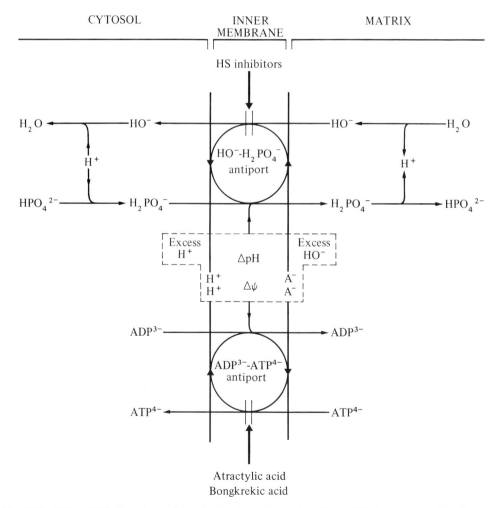

Fig. 6.33. The ADP–ATP and HO–P_i antiports driven by the electrochemical gradient of H^+ ions generated by electron transport.

cytosol for use, and ADP, formed in the cytosol by ATP hydrolysis, is returned to the mitochondrial matrix for phosphorylation.

Other antiports in the inner mitochondrial membrane exchange dicarboxylic acids and tricarboxylic acids.

C. OXIDATION OF EXOGENOUS NADH

The inner membrane of plant mitochondria, unlike its animal counterpart, is equipped with an NADH dehydrogenase built into its external surface. This enzyme is a flavoprotein (FP_{EXT} in Fig. 6.5). It oxidizes NADH present in the intermembrane space and reduces the UQ of the terminal electron transport chain. Its function is to allow NADH generated by oxidative reactions in the cytosol to be reoxidized by oxygen via the terminal electron transport chain. For this to happen NADH must diffuse through the outer mitochondrial membrane; it can readily do this because the outer membrane is readily permeable to compounds with a molecular weight below about 10,000. The importance of the external NADH dehydrogenase is that because of the impermeability of the inner mitochondrial membrane to either NADH or NAD^+ [see Section B.3(v)], the NADH cannot get to the NADH dehydrogenase built into the inner surface of this membrane.

The internal NADH dehydrogenase is composed of a flavoprotein and Fe–S proteins [see Section B.2 (ii)] and is common to both plant and animal mitochondria. Its function in plants is to allow NADH generated by oxidative reactions in the mitochondrial matrix to be reoxidized by oxygen via the terminal electron transport chain. In animals, however, this enzyme has to reoxidize NADH generated in the cytosol as well as in the mitochondrial matrix because there is no external NADH dehydrogenase. This presents the problem of how the cytosol-produced NADH gets to the internal NADH dehydrogenase. The solution is shown in Fig. 6.4; although the NADH itself cannot pass across the inner membrane its 'reducing equivalent' can, by means of shuttle mechanisms.

The first of these shuttle mechanisms (A in Fig. 6.4) makes use of two pairs of isoenzymes, the cytosolic and mitochondrial malate dehydrogenases and the cytosolic and mitochondrial aspartate aminotransferases[14] along with two exchange carriers [see Section B.3(v)], the malate-α-ketoglutarate antiport and the aspartate-glutamate antiport. The cytosolic NADH is reoxidized by the cytosolic malate dehydrogenase and the resulting L-malate passes across the inner mitochondrial membrane, in exchange for α-ketoglutarate, via the malate-α-ketoglutarate antiport. The L-malate is oxidized to oxaloacetate by the mitochondrial malate dehydrogenase using NAD^+ in the matrix as the oxidizing agent. This produces NADH which is then reoxidized by the internal NADH dehydrogenase. Thus the reducing equivalent of NADH passes across the inner membrane in the form of L-malate. The shuttle is completed by conversion of the oxaloacetate formed in the mitochondrial matrix to aspartate which passes out across the inner membrane in exchange for glutamate via the aspartate-glutamate antiport. In the cytosol the aspartate is converted back into oxaloacetate. Although this shuttle is shown in Fig. 6.4 as taking reducing equivalents into the mitochondrion, it is reversible and can, under certain circumstances, take reducing equivalents out of the mitochondrion.

The second shuttle mechanism (B in Fig. 6.4) is simpler and unidirectional. It involves the NAD-specific glycerol-3-phosphate dehydrogenase[15] present in the cytosol and the flavoprotein glycerol-3-phosphate dehydrogenase (FP_G in Fig. 6.4) built into the outer surface of the animal inner mitochondrial membrane. The former enzyme oxidizes cytosolic NADH at the expense of dihydroxyacetone phosphate. The resulting sn-glycerol 3-phosphate diffuses into the mitochondrial intermembrane space where it is converted back into dihydroxyacetone phosphate by FP_G. The dihydroxyacetone phosphate then diffuses back into the cytosol to complete the cycle. Thus the reducing equivalent of NADH reaches the UQ component of the terminal electron transport chain via sn-glycerol 3-phosphate and FP_G without ever crossing the inner mitochondrial membrane. Since the reducing equivalent by-passes the first phosphorylation site (site 1 in Fig. 6.2) only two ATPs are generated per NADH utilizing the shuttle.

Since plant mitochondria have an external NADH dehydrogenase they have no need of such shuttle mechanisms to handle exogenous NADH (i.e. NADH produced outside the mitochondrion). The external NADH dehydrogenase by-passes the first phosphorylation site (site 1 in Fig. 6.3) which is associated with the internal NADH dehydrogenase. Thus the P/O ratio for the mitochondrial oxidation of exogenous NADH is 2. The oxidation of exogenous NADH is not inhibited by amytal, piericidin A or rotenone in contrast to that of endogenous NADH; however, the oxidation of both exogenous and endogenous NADH is inhibited by antimycin A. This shows that although different NADH dehydrogenases are involved in handling exogenous and endogenous NADH the route from UQ to oxygen is identical.

D. CYANIDE-RESISTANT RESPIRATION

Plant mitochondria differ from animal mitochondria in that they can possess two different routes by which electrons can be transported from substrates like NADH and succinate to oxygen. The first route is that discussed at length in Section B.2 and depicted in Figs. 6.3 and 6.5; this route is blocked by cyanide which inhibits cytochrome oxidase. The second route is not blocked by cyanide and is thus called the cyanide-resistant electron transport chain; similarly the catabolic process in which it forms the terminal electron transport chain is called cyanide-resistant respiration.

The structure of the cyanide-resistant electron transport chain and its relationship with the 'conventional' cyanide-sensitive one is shown in Fig. 6.34. It is apparent from Fig. 6.34 that the two electron transport chains are not totally distinct in that electron pairs enter both of them via the internal and external NADH dehydrogenases and succinate dehydrogenase and then pass to UQ. It is after UQ that the two chains separate. In the 'conventional' chain the electrons pass in sequence via FP_{ha}, the b- and c-type cytochromes and cytochrome oxidase to oxygen. In the cyanide-resistant chain the electron pairs pass from UQ to flavoprotein FP_{ma} [$\Delta E_0' = +0.02$ V; see Section B.2 (ii)(b)] and from there to a cyanide-resistant terminal oxidase which hands them on to oxygen. The nature of the terminal oxidase (X in Fig. 6.34) is

not yet clear. However, it is generally believed to be an iron-containing protein which is neither a haemoprotein nor an iron–sulphur protein on the following evidence: (i) it is inhibited by hydroxamates (e.g. mCLAM whose structure is given in Fig. 6.34) which are known to chelate transition metals strongly, (ii) this inhibition is counteracted by Fe^{2+}, (iii) hydroxamates do not chelate the iron of iron–sulphur proteins. There is some evidence to suggest that the cyanide-resistant terminal oxidase catalyses the reduction of oxygen so as to produce H_2O_2 rather than H_2O; the H_2O_2 so formed in the mitochondria would be rapidly converted into H_2O and O_2 by catalase[16].

The evidence that the cyanide-resistant route branches off the conventional electron transport chain immediately after UQ may be summarized as, (i) antimycin A, which blocks between cytochromes b_{557} and c_{549}, does not block cyanide-resistant respiration, (ii) amytal, piercidin A and rotenone, which block between Fe–S of the internal NADH dehydrogenase and UQ, also block cyanide-resistant respiration, (iii) oxidation of endogenous NADH in the presence of antimycin A or cyanide gives a P/O ratio of 1 indicating that phosphorylation site 1 is coupled to this particular electron flow, (iv) oxidation of exogenous NADH or succinate in the presence of antimycin A or cyanide gives a P/O ratio of 0 proving that there is no phosphorylation site between UQ and the cyanide-resistant terminal oxidase.

The way the plant controls the flow of electrons through the two electron transport chains is not clear but there is evidence to indicate that the intensity of the electron flux through the cytochrome system of the conventional chain is a regulating factor.

The physiological significance of cyanide-resistant respiration is not clear. It is doubtful that its function is to allow plants to survive in the presence of cyanide even though many plants release HCN from cyanogenic glucosides [see Chapter 7, Section B.3(i)(c)] when they are injured. This follows from the fact that those plant tissues which have the most active cyanide-resistant respiration have no cyanogenic glucosides.

Cyanide-resistant respiration is thought to be responsible for the climacteric in fruits (i.e. the marked increase in respiration during and just prior

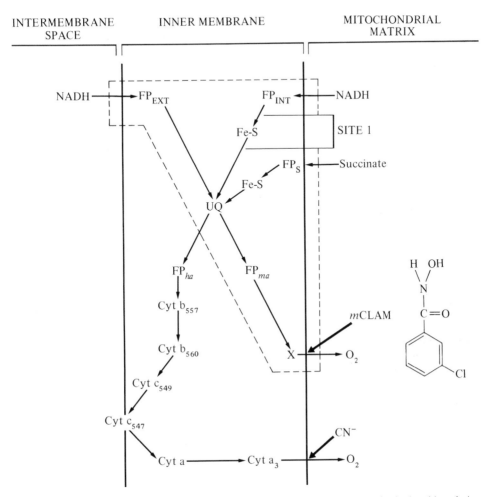

Fig. 6.34. Probable structure of the electron transport chain operating in 'cyanide-resistant respiration' and its relationship with the normal cyanide-sensitive terminal electron transport chain. (FP_{ma} = flavoprotein with a mid-range E'_0 and a large absorbance change on redox change; X = CN^--insensitive terminal oxidase with a non-haem iron centre; m-CLAM = m-chlorobenzhydroxamic acid; all other abbreviations are as in Fig. 6.3.)

to ripening). The climacteric is induced by ethylene and it is thought that ethylene brings this about by stimulating cyanide-resistant respiration.

Certain germinating seeds exhibit cyanide-resistant respiration during the early stages of water imbibition.

The best understood role of cyanide-resistant respiration is, however, in generating heat in thermogenic tissues. Thermogenicity is seen in the flowers or inflorescences of certain plants such as the water lily, *Victoria*, and the arum lilies, *Arum maculatum*, *Sauromatum guttatum*, *Symplocarpus foetidus* (skunk cabbage) and *Philodendron selloum*.

In skunk cabbage the inflorescence is a spadix covered with tiny hermaphrodite flowers which produce heat and a foul smell. In *Arum* and *Sauromatum* the inflorescence has differentiated into staminate and pistillate flowers plus a special, sterile finger-like organ known as the appendix or osmophore which produces heat and smell. The heat is produced in these inflorescences to volatilize the odiferous compounds formed in them. The latter are frequently amines or indoles and serve to attract pollinating insects. In species with an appendix the temperature within this organ reaches a maximum several hours before the shedding of pollen; this is consistent with the fact that in these species obligatory cross pollination is combined with proterogyny (i.e. the maturation of the pistil before the stamens). The production of heat is due to cyanide-resistant respiration in the mitochondria of the thermogenic tissues of the inflorescences. These tissues require ATP for their normal metabolic processes. This ATP is produced by cyanide-resistant respiration rather than conventional respiration. Since the P/O ratio of cyanide-resistant respiration is 1 in contrast to 3 for conventional respiration, the cells have to oxidize three times as much fuel to get the required amount of ATP. Assuming that oxidation is via NADH, this will produce about 4.5 times as much waste heat (i.e. energy that is not conserved as ATP) than normal. It is this extra heat produced as a result of having to burn more fuel to get the same amount of ATP that constitutes thermogenicity.

The thermogenicity of these inflorescences is analogous to that of brown adipose tissue in animals where the heat production is thought to be due to the uncoupling of phosphorylation from the electron transport chain.

The amount of heat produced by plant inflorescences is quite considerable. The temperature of the appendix of *Arum italicum* can be as high as 51°C with an air temperature of 15°C whilst that of the skunk cabbage spadix can reach 30°C when the air temperature is as low as 5°C.

E. PROCESSES GENERATING SUBSTRATES FOR OXIDATIVE PHOSPHORYLATION

1. Oxidation of carbohydrates

(i) GLYCOLYSIS

(a) *Mechanism of glycolysis*

Glycolysis is the main pathway of carbohydrate catabolism. It is a process in which monosaccharides are broken down to pyruvic acid, two molecules of which are formed per monosaccharide residue. In plants D-glucose and D-fructose are the main monosaccharides catabolized by glycolysis although others capable of conversion into these sugars or their phosphate esters (see Chapter 7, Section B.3) can also be handled. Glycolysis is an anaerobic process but not obligatorily so; it can proceed in the absence or in the presence of oxygen. Since anaerobiosis rarely obtains in higher plants, glycolysis normally takes place in the presence of oxygen in higher plant cells.

Glycolysis is a key metabolic component of the respiratory process which generates energy in the form of ATP in cells where photosynthesis is not taking place. Particularly important amongst these are the cells of germinating seedlings and the non-photosynthetic cells of mature plants.

The D-glucose entering the glycolysis sequence is derived from starch or sucrose whilst the D-fructose is derived from sucrose. The starch in question is either stored in seeds or stored temporarily in the chloroplasts of mature plants; its enzyme-catalysed breakdown to glucose or glucose 1-phosphate is discussed in Chapter 7, Section D.1(i). The sucrose is imported via the phloem from photosynthetic or storage tissues and hydrolysed by invertase[17].

Glycolysis occurs in the cytosol of the cells. The possibility of the presence of a complete glycolysis sequence in the chloroplasts of photosynthetic cells in addition to the cytosolic sequence has been put forward. Although many glycolytic enzymes, including phosphofructokinase[18], phosphoglycerate phosphomutase[23], enolase[24] and pyruvate kinase[28], have been shown to be present in the chloroplast stroma, present opinion is still against the operation of a complete glycolysis sequence in this organelle, though less strongly than a few years ago.

The glycolysis pathway is shown in Fig. 6.35. Although the purpose of glycolysis, as a component of respiration, is to generate ATP, the initial stages involve the expenditure of ATP. Both D-glucose and D-fructose require to be converted to their 6-phosphates; this is accomplished by hexokinase[19] using ATP as the phosphorylating agent. This step is avoided in the case of D-glucose 1-phosphate (G-1-P) derived from starch by the action of phosphorylase[20]. ATP is also expended in converting D-fructose 6-phosphate (F-6-P) into D-fructose 1,6-diphosphate (F-1,6-diP). This ATP expenditure is recouped in the later stages of glycolysis.

F-1,6-diP is cleaved into two triose phosphates, dihydroxyacetone phosphate (DiHOAcP) and gly-

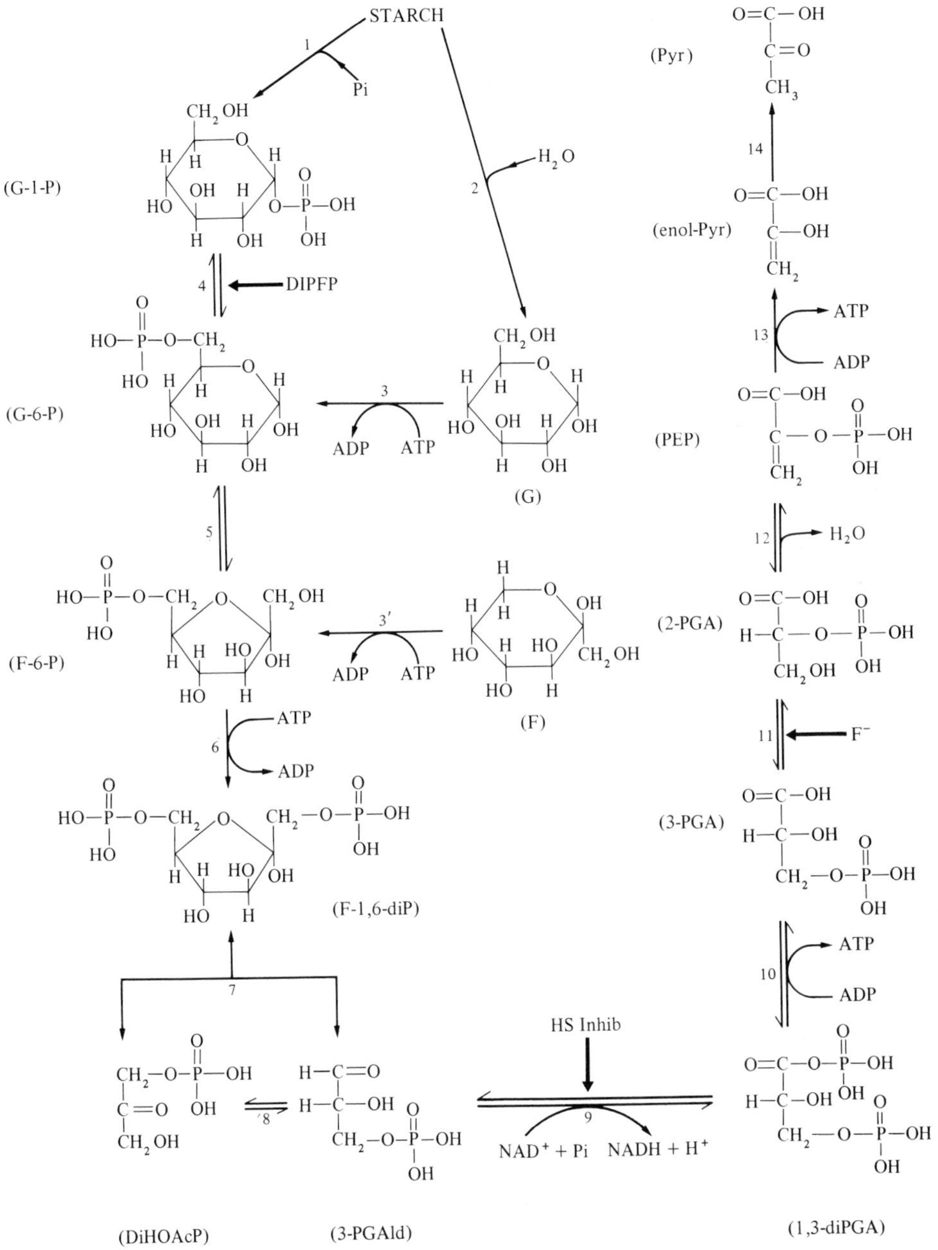

STARCH

Pi

(G-1-P)

CH$_2$OH
H H O H
HO OH H O
O—P—OH
H OH OH

4 ◄— DIPFP

(G-6-P)

O
HO—P—O—CH$_2$
HO H H O H
HO OH H OH
H OH

ADP ATP

—H$_2$O

2

(G)

CH$_2$OH
H H O H
HO OH H OH
H OH

3

(Pyr)

O=C—OH
C=O
CH$_3$

14

(enol-Pyr)

O=C—OH
C—OH
CH$_2$

13

ATP

ADP

(PEP)

O=C—OH
C—O—P—OH
CH$_2$ OH

(F-6-P)

O
HO—P—O—CH$_2$ O CH$_2$OH
HO
H H HO OH
HO H

5

3'

ADP ATP

(F)

H O
H H OH
HO H HO CH$_2$OH
HO H

6

ATP

ADP

(2-PGA)

O=C—OH O
H—C—O—P—OH
CH$_2$OH OH

12

H$_2$O

(3-PGA)

O=C—OH
H—C—OH O
CH$_2$—O—P—OH
OH

11

F⁻

(F-1,6-diP)

O
HO—P—O—CH$_2$ O CH$_2$—O—P—OH
HO OH
H H HO OH
HO H

7

10

ATP

ADP

(1,3-diPGA)

O
O=C—O—P—OH
H—C—OH OH O
CH$_2$—O—P—OH
OH

HS Inhib

9

NAD⁺ + Pi NADH + H⁺

(DiHOAcP)

CH$_2$—O—P—OH
C=O OH
CH$_2$OH

8

(3-PGAld)

H—C=O
H—C—OH O
CH$_2$—O—P—OH
OH

Fig. 6.35. The glycolysis pathway. (G = D-glucose; F = D-fructose; G-1-P = D-glucose 1-phosphate; G-6-P = D-glucose 6-phosphate; F-6-P = D-fructose 6-phosphate; F-1,6-diP = D-fructose 1,6-diphosphate; DiHOAcP = dihydroxyacetone phosphate; 3-PGAld = glyceraldehyde 3-phosphate; 1,3-PGA = 1,3-diphosphoglyceric acid; 3-PGA = 3-phosphoglyceric acid; 2-PGA = 2-phosphoglyceric acid; PEP = phosphoenol pyruvic acid; enol-Pyr = enolpyruvic acid; Pyr = pyruvic acid; reactions 1–13 are catalysed by the following enzymes: 1 = phosphorylase, EC 2.4.1.1; 2 = α- and β-amylases, EC 3.2.1.1 and EC 3.2.1.2, oligo-1,6-glucosidase, EC 3.2.1.10 and α-glucosidase, EC 3.2.1.20; 3 and 3' = hexokinase, EC 2.7.1.1; 4 = phosphoglucomutase, EC 2.7.5.1; 5 = glucosephosphate isomerase, EC 5.3.1.9; 6 = phosphofructokinase, EC 2.7.1.11; 7 = fructose bisphosphate aldolase, EC 4.1.2.13; 8 = triosephosphate isomerase, EC 5.3.1.1; 9 = glyceraldehyde-phosphate dehydrogenase, EC 1.2.1.12; 10 = phosphoglycerate kinase, EC 2.7.2.3; 11 = phosphoglycerate phosphomutase, EC 5.4.2.1; 12 = enolase, EC 4.2.1.11, 13 = pyruvate kinase, EC 2.7.1.40; 14 = spontaneous reaction; ◄— = inhibition by; DIPFP = diisopropylfluorophosphate; HS Inhib = inhibitors of HS-enzymes, e.g. iodoacetamide; → = physiologically irreversible reaction; ⇌ or ↔ = physiologically reversible reaction; N.B. since two molecules of 3-PGAld are formed from one molecule of glucose the number of reactant molecules in all reactions from 9 onwards is doubled.)

ceraldehyde 3-phosphate (3-PGAld) by aldolase[21]; DiHOAcP is derived from carbon atoms 1-3 of F-1,6-diP and 3-PGAld from carbon atoms 4-6. The DiHOAcP is then converted into 3-PGAld by triosephosphate isomerase[22]; thus two molecules of 3-PGAld are produced from each hexose residue entering the glycolysis pathway.

The 3-PGAld is then converted into 1,3-diphosphoglyceric acid (1,3-diPGA) by glyceraldehyde-phosphate dehydrogenase[11]. In this reaction a low-energy phosphate ester is converted into a very high-energy acyl phosphate; the $\Delta G^{\circ\prime}$ of hydrolysis of the acyl phosphate moiety of 1,3-diPGA is -49.3 kJ mol^{-1}. Glyceraldehyde-phosphate dehydrogenase has been crystallized from yeast; it is tetramer of four identical subunits. Each subunit has a molecular weight of 35,000 and has one NAD^{+} tightly bound to it; thus this enzyme is an exception to the general rule that NAD-linked oxidoreductases do not have bound NAD [see Section B.2(ii)(a)]. The enzyme also possesses —SH groups which are essential to its catalytic activity; alkylating agents such as iodoacetamide irreversibly inhibit it by reacting with the —SH groups as shown in eq. (6.4).

$$\text{ENZ—SH} + \text{ICH}_2\text{CONH}_2 \rightarrow$$
$$\text{ENZ—S—CH}_2\text{CONH}_2 + \text{HI} \quad (6.4)$$

The mechanism by which glyceraldehyde-phosphate dehydrogenase is thought to catalyse the conversion of 3-PGAld into 1,3-diPGA is shown in Fig. 6.36.

The 1,3-diPGA is used in the next reaction of glycolysis to generate ATP; the phosphate of the high-energy acyl phosphate moiety is transferred to ADP so generating ATP and 3-phosphoglyceric acid (3-PGA). The $\Delta G^{\circ\prime}$ of this reaction is $-49.3 + (+30.5) = -18.8$ kJ mol^{-1}.

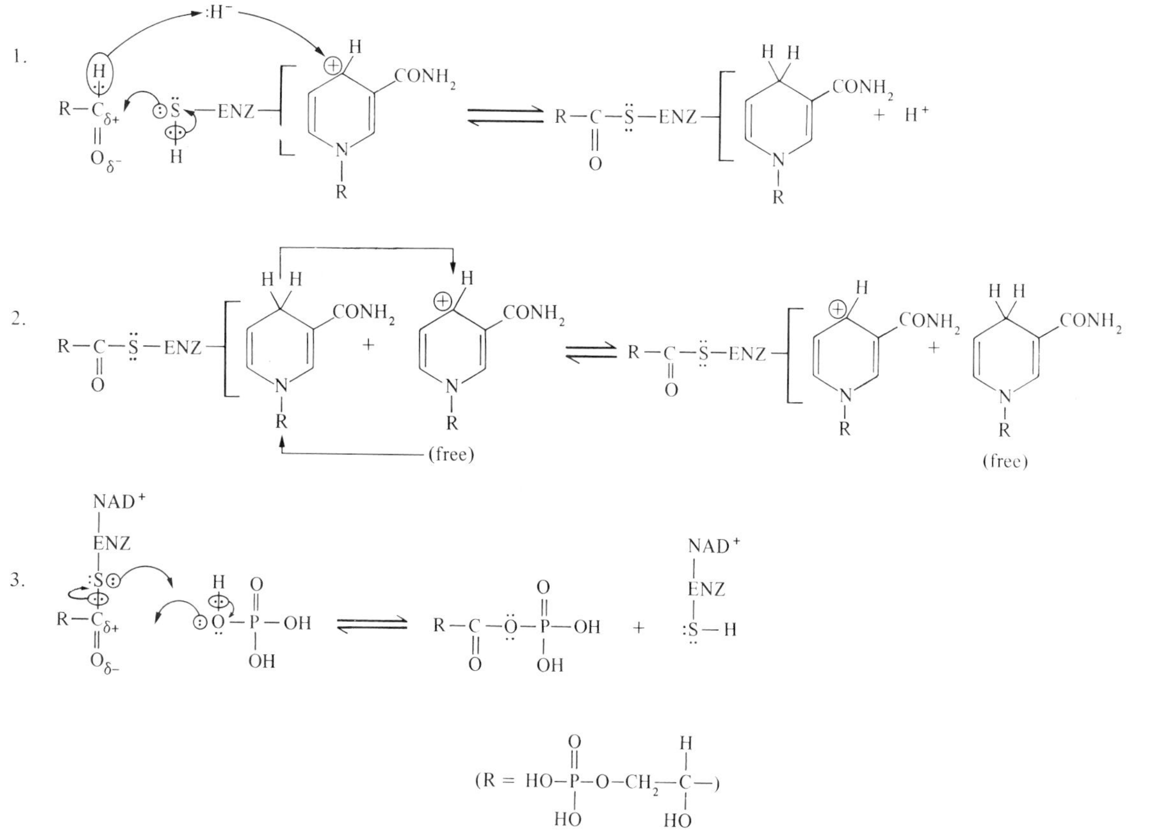

Fig. 6.36. Conversion of glyceraldehyde 3-phosphate into 1,3-diphosphoglyceric acid by glyceraldehyde-phosphate dehydrogenase. (The enzyme has an SH group at its active site and tightly bound NAD^{+}; the NADH formed in reaction 1 is much less tightly bound and exchanges with NAD^{+} dissolved in the aqueous milieu in reaction 2.)

The 3-PGA is then isomerized to 2-phosphoglyceric acid (2-PGA) by phosphoglycerate phosphomutase[23], an enzyme which requires Mg^{2+} for activity. It is inhibited by fluoride ions in the presence of phosphate because the Mg^{2+} is removed as largely unionized magnesium fluorophosphate. The transfer of the phosphate residue from the C–3 to the C–2 hydroxyl group enables 2-PGA to be dehydrated by enolase[24] with the formation of phosphoenolpyruvate (PEP), a high-energy enolic phosphate ($\Delta G^{\circ\prime}$ of hydrolysis = -61.9 kJ mol^{-1}).

The PEP is used in the final reaction of glycolysis to generate ATP; the phosphate of the high-energy enolic phosphate is transferred to ADP so generating ATP and enolpyruvic acid which spontaneously tautomerizes to pyruvic acid. The $\Delta G^{\circ\prime}$ of this reaction is $-61.9 + (+30.5) = -31.4$ kJ mol^{-1}; this high negative value makes the reaction virtually impossible to reverse under physiological circumstances. For different reasons the hexokinase and phosphofructokinase-catalysed reactions are also physiologically irreversible. All the other reactions of glycolysis are reversible. In the process known as gluconeogenesis (see Chapter 7, Section B.2) a modified glycolysis pathway is made to run backwards; the modifications are the by-passing of the physiologically irreversible reactions by use of different enzymes.

The glycolysis pathway, therefore, produces two molecules of pyruvic acid from every hexose residue entering it along with two molecules of NADH and four molecules of ATP. The latter are produced by substrate-level phosphorylation; this contrasts with oxidative phosphorylation in which ATP is generated at the level of the terminal electron transport chain. Under aerobic conditions the two molecules of NADH are reoxidized by the external NADH dehydrogenase of the inner mitochondrial membrane (see Sections B.2 and B.3, Figs. 6.3 and 6.5) with the production of two ATP molecules per NADH. Thus under the aerobic conditions that normally prevail in higher plant cells the gross production of ATP is eight molecules per hexose residue. However, the net production of ATP is lower than this because of the utilization of ATP in the kinase-catalysed reactions (3, 3′ and 6 in Fig. 6.35). When D-glucose or D-fructose are the starting compounds the net production is $8 - 2 = 6$ ATP molecules per hexose; when G-1-P is the starting

material it is $8 - 1 = 7$ ATP molecules per glucose residue.

Under anaerobic conditions the NADH generated in reaction 9 (Fig. 6.35) cannot be reoxidized by the terminal electron transport chain. However, to keep glycolysis going it must be reoxidized. The way this is accomplished depends upon the organism. When yeast cells are made to grow anaerobically the NADH is reoxidized in a process leading to the formation of ethanol (see Fig. 6.37). The pyruvate formed by glycolysis is decarboxylated to acetaldehyde by pyruvate decarboxylase[25], an enzyme using thiamine pyrophosphate (TPP) (see Fig. 6.38) as its coenzyme. The acetaldehyde is then reduced to ethanol under the catalytic influence of alcohol dehydrogenase[26] using NADH as the reductant. Thus the NADH is converted back into NAD^+ and ethanol accumulates as a by-product. The overall process, glucose to ethanol, is called alcoholic fermentation and is the basis of the beer and wine-making industries.

The reoxidation of NADH produced by anaerobic glycolysis in animal tissues such as overworked muscle is accomplished by using it to reduce pyruvic acid into L-lactic acid (see Fig. 6.37) under the catalytic influence of lactate dehydrogenase[27]. Thus L-lactate is the end product of glycolysis under these circumstances. Since anaerobic glycolysis occurs only rarely in plants the problem of NADH reoxidation arises equally rarely; however, when it does arise, as in the case of potato tubers maintained under anaerobic conditions, the L-lactate-forming reoxidation procedure is adopted. This is also true of some green algae (e.g. a few strains of *Chlorella*, *Scenedesmus* D_3 and *Ankistrodesmus braunii*) when grown anaerobically.

(b) *Regulation of glycolysis*

Two enzymes regulate the rate of glycolysis, phosphofructokinase and pyruvate kinase[28], with the former probably playing the major role.

Phosphofructokinase is ideally suited to regulate glycolysis. It catalyses a non-equilibrium reaction, it acts on the first committed step of the reaction sequence and it is allosterically controlled by a number of important metabolites. The latter include (i) phosphorylated intermediates of glycolysis such as 3-PGA, 2-PGA and PEP, (ii) adenine nucleotides

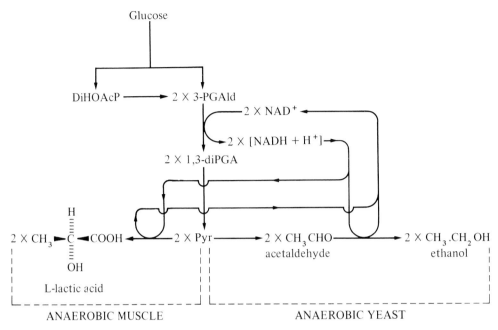

Fig. 6.37. Reoxidation of the NADH produced by glycolysis under anaerobic conditions in muscle and yeast.

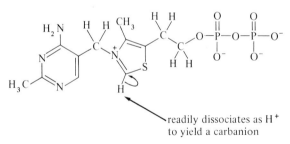

readily dissociates as H^+
to yield a carbanion

Fig. 6.38. Thiamine pyrophosphate, the bound coenzyme of α-keto acid decarboxylases (e.g. pyruvate decarboxylase, EC 4.1.1.1), α-keto acid dehydrogenase complexes (e.g. pyruvate dehydrogenase, EC 1.2.4.1 and α-ketoglutarate dehydrogenase, EC 1.2.4.2) and transketolase, EC 2.2.1.1.

and orthophosphate, (iii) the initial metabolites of the oxidative pentose phosphate pathway, e.g. 6-phosphogluconic acid, and (iv) the initial metabolite of the tricarboxylic acid cycle, citric acid. Although there are some differences in the regulatory control of plant phosphofructokinase from that of animals, notably in the effect of adenine nucleotides, the same general picture emerges. The enzyme is allosterically inhibited by PEP, ATP, 6-phosphogluconic acid and citric acid; this is entirely sensible since it means that when the cell is

rich in energy, as will be the case when the concentration of these compounds is high, the oxidation of carbohydrate via glycolysis is switched off. The enzyme is allosterically stimulated by orthophosphate which will have a higher concentration when the cell is poor in energy.

Pyruvate kinase also catalyses a non-equilibrium reaction. However, it has not so far been possible to demonstrate allosteric kinetics for the plant enzyme. Nevertheless it is inhibited by ATP and citrate and stimulated by its substrate, ADP. These phenomena are likely to be of regulatory significance and operate synergistically with control of glycolysis via phosphofructokinase.

(ii) THE TRICARBOXYLIC ACID CYCLE (TCA CYCLE)

(a) *Mechanism of the TCA cycle*

Under the aerobic conditions which normally obtain in plant tissues the pyruvate formed by glycolysis diffuses into the mitochondrion and is oxidized completely to CO_2 with the production of more ATP. The oxidation of pyruvate involves two stages: (i) the oxidative decarboxylation of pyruvate

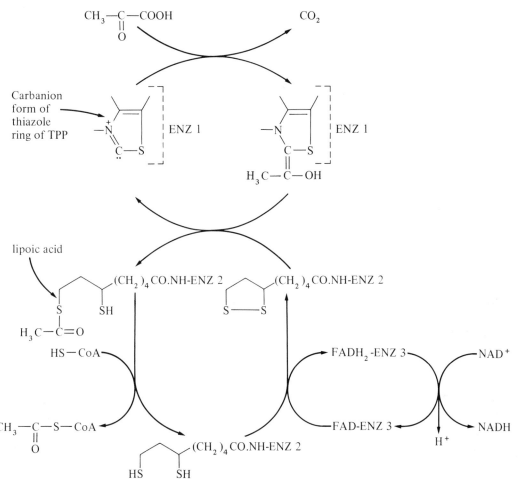

Fig. 6.39. Stages in the conversion of pyruvic acid into acetyl-CoA by the multienzyme pyruvate dehydrogenase complex. (ENZ 1 = pyruvate:lipoamide oxidoreductase (decarboxylating and acceptor-acetylating), EC 1.2.4.1; ENZ 2 = acetyl-CoA: dihydro-lipoamide S-acetyltransferase, EC 2.3.1.12; ENZ 3 = NADH:lipoamide oxidoreductase, EC 1.6.4.3.)

to acetyl-CoA and (ii) the oxidation of the acetyl moiety of acetyl-CoA in the TCA cycle.

The oxidative decarboxylation of pyruvate is catalysed by the pyruvate dehydrogenase complex. This is a rather complex multistep process (see Fig. 6.39) whose overall stoichiometry is given in eq. (6.5). The pyruvate dehydrogenase complex is a structural unit composed of multiple copies of

$$CH_3.C.COOH + CoA.SH + NAD^+ \rightleftharpoons$$
$$\underset{O}{\overset{\|}{}}$$

$$CH_3.C.S.CoA + CO_2 + NADH + H^+$$
$$\underset{O}{\overset{\|}{}}$$

$$\Delta G^{\circ\prime} = -9.4 \text{ kJ mol}^{-1} \quad (6.5)$$

three distinct enzymes each of which catalyses a different step in the multistep process. The first enzyme to operate is pyruvate:lipoamide oxidored-uctase[3] (ENZ 1, Fig. 6.39) which uses TPP as its coenzyme; the TPP is dissociable from the apo-enzyme. This enzyme catalyses the removal of the carboxyl group of pyruvate as CO_2 followed by the transfer of the resulting 2C unit to the terminal sulphur atom of the oxidized form of the lipoic acid coenzyme covalently bound to the second enzyme of the complex. During this reaction the 2C unit is temporarily bound to the thiazole ring of TPP as an acetol residue. The initial decarboxylation is physiologically irreversible and thereby confers irreversibility on the overall reaction [eq. (6.5)].

The second enzyme, acetyl-CoA:dihydrolipo-amide S-acetyltransferase[29] (ENZ 2, Fig. 6.39), then transfers the acetyl residue from its lipoic acid coenzyme to the sulphydryl sulphur of coenzyme A (see Fig. 8.14), so forming acetyl-CoA. This, however, leaves the lipoic acid coenzyme in its reduced form. It is reoxidized by the third enzyme, the FAD-specific NADH:lipoamide oxidoreductase[13] (ENZ 3, Fig. 6.39) using NAD^+ as the oxidant. The resulting NADH is reoxidized by the internal NADH dehydrogenase of the terminal electron transport chain with the generation of three ATPs.

Arsenite (AsO_3^-) inhibits the pyruvate dehydrogenase complex irreversibly because it forms a cyclic complex with the dihydrolipoic acid of ENZ 2 [eq. (6.6)].

$$(6.6)$$

The TCA cycle (see Fig. 6.40) commences with the condensation of the acetyl residue of acetyl-CoA with oxaloacetic acid (OAA) to form the tricarboxylic acid, citric acid; the reaction is catalysed by citrate synthase[30]. This enzyme is stereospecific in its action; it dictates that the acetyl residue attacks the si-face of oxaloacetate and takes up the 3-pro-S position in citric acid.

Citric acid is then converted into threo-D_S-isocitric acid ($2R, 3S$-isocitric acid; one of the four optical isomers this acid) by the enzyme aconitase[31]. This involves the trans abstraction of the elements of water to yield cis-aconitic acid followed by the trans addition of the elements of water to yield threo-D_S-isocitric acid. This is another example of stereospecific enzyme action. Aconitase treats the prochiral centre of citrate (C–3) in a chiral manner such that the removal of H^+ and the addition of OH^- solely involves the pro-R CH_2COOH group (i.e. the CH_2COOH group arising from OAA and not that from acetyl-CoA). Aconitase is inhibited by fluorocitrate. This is formed when fluoroacetyl-CoA, formed from fluoroacetate ($FCH_2 \cdot COO^-$), reacts with OAA under the catalytic influence of citrate synthase. Fluoroacetate is present in several plants

in Australia, South Africa and South America. These plants are intensely poisonous to cattle. The fluoroacetate in the plants is evidently compartmentalized from the TCA cycle otherwise they also would succumb.

The threo-D_S-isocitric acid is then oxidatively decarboxylated by isocitrate dehydrogenase[4] to α-ketoglutaric acid; oxalosuccinic acid is thought to be an enzyme-bound intermediate in the reaction. Plants contain two isocitrate dehydrogenases, one is specific for NAD^+, the other for $NADP^+$. The former is found only in the mitochondria and is the TCA cycle enzyme, the latter is found in the cytosol and in the mitochondria but is not concerned with the catabolic operation of the TCA cycle. In the conversion of threo-D_S-isocitric acid into α-ketoglutarate the C-3 carboxyl group, derived from OAA, is lost as CO_2 and NADH is generated.

The α-ketoglutaric acid is then oxidatively decarboxylated to succinyl-CoA. This reaction is catalysed by the α-ketoglutarate dehydrogenase complex[5] which is very like the pyruvate dehydrogenase complex. It is composed of multiple copies of three enzymes which are the analogues of those of pyruvate dehydrogenase; in fact the NADH:lipoamide oxidoreductases of the two complexes appear to be identical. Thus the steps of the oxidative decarboxylation of α-ketoglutaric acid are identical to those of pyruvate (see Fig. 6.39) save that a succinol residue is bound to the TPP of ENZ 1, a succinyl residue is bound to the lipoic acid of ENZ 2 and succinyl-CoA is the product. In this reaction the carboxyl group next to the carbonyl moiety is lost as CO_2; this carboxyl group is also derived from OAA. NADH is generated. The reaction is physiologically irreversible and inhibited by arsenite for the same reasons as is the oxidative decarboxylation of pyruvic acid. It is, in fact, the only physiologically irreversible reaction of the TCA cycle; this is of considerable significance because it confers directionality on the cycle as a whole.

The succinyl-CoA is then converted into succinic acid by succinate thiokinase[32]. The free energy which results from the cleavage of the high-energy thioester bond is conserved in this reaction by the coupled phosphorylation of ADP. The plant succinate thiokinase generates ATP rather than GTP as is the case with the animal enzyme. This is the

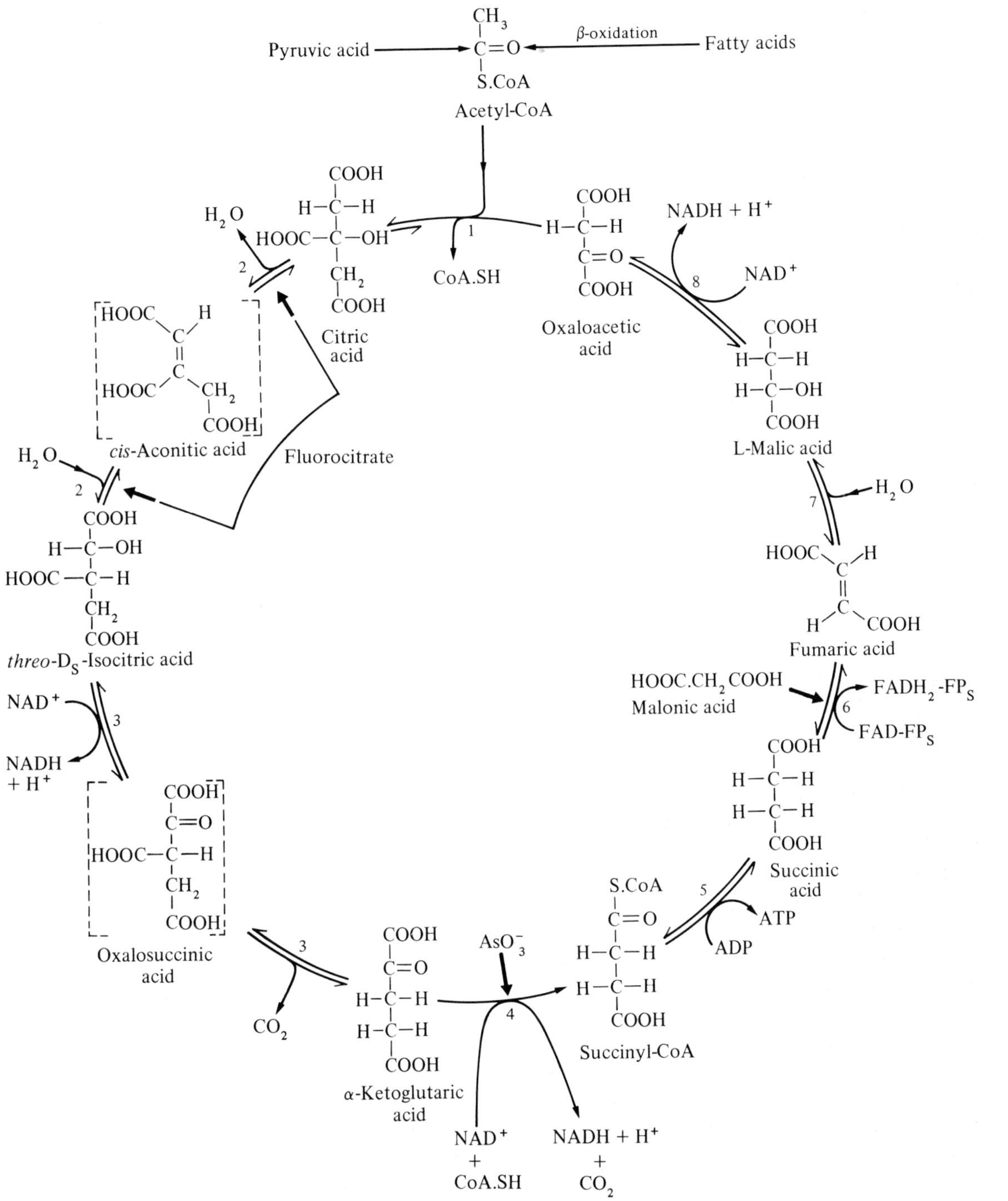

Fig. 6.40. The tricarboxylic acid cycle (TCA cycle). (Reactions 1–8 are catalysed by the following enzymes: 1 = citrate synthase, EC 4.1.3.7; 2 = aconitase, EC 4.2.1.3; 3 = isocitrate dehydrogenase, EC 1.1.1.41; 4 = α-ketoglutarate dehydrogenase complex, EC 1.2.4.2; 5 = succinate thiokinase, EC 6.2.1.5; 6 = succinate dehydrogenase, EC 1.3.99.1; 7 = fumarate hydratase, EC 4.2.1.2; 8 = malate dehydrogenase, EC 1.1.1.37; ◄ = inhibition by; → = physiologically irreversible reaction.)

only substrate-level phosphorylation associated with the TCA cycle.

Of the four carbon atoms of succinate, two, forming a CH_2COOH unit, are derived from the acetyl unit whilst the other two, also forming a CH_2COOH unit, are derived from the two central carbons of OAA. At this point in the TCA cycle the 2C unit from acetyl-CoA loses its identity because the next enzyme, succinate dehydrogenase, cannot distinguish between the two CH_2COOH units of succinate.

Succinate dehydrogenase catalyses the *trans* elimination of two hydrogen atoms from succinic acid with the formation of fumaric acid. It is an FAD-containing flavoprotein with Fe–S centres (see Section B.2) and is built into the inner mitochondrial membrane. It is the only TCA enzyme not to be located in the mitochondrial matrix. During the course of the reaction the FAD is reduced to $FADH_2$; the latter is reoxidized at the expense of oxygen via the terminal electron transport chain with the concomitant generation of 2ATPs. In fact succinate dehydrogenase can be regarded as a component enzyme of terminal electron transport as well as of the TCA cycle.

Fumaric acid is then hydrated by fumarate hydratase[33] with the production of L-malic acid. This is a stereospecific reaction, H^+ being added to the *re* face of fumaric acid and OH^- to the *si* face.

The TCA cycle is completed by the conversion of L-malic acid into OAA by malate dehydrogenase which uses NAD^+ as the oxidant. The $\Delta G^{\circ\prime}$ of this reaction is $+27.9$ kJ mol^{-1} which corresponds to an equilibrium constant of 1.29×10^{-5}. This unfavourable equilibrium is overcome *in vivo* by coupling the reaction to that catalysed by citrate synthase (see Chapter 2, Section A.1).

The TCA cycle is therefore a process in which acetyl residues of acetyl-CoA are oxidized to CO_2. It should be noted that the carbon atoms emerging as CO_2 are not derived from the acetyl unit just introduced into the cycle; they arise from the carboxyl groups of the molecule of OAA that condensed with that acetyl unit. Thus the newly introduced acetyl unit becomes a C—COOH residue in the OAA regenerated by the cycle. Utilization of the OAA in subsequent turns of the cycle will, however, cause the carbon atoms of the acetyl unit to be lost as CO_2.

During one turn of the TCA cycle three molecules of NADH are generated. These are reoxidized by the internal NADH dehydrogenase of the terminal electron transport chain; the 3 electron pairs then pass down the chain to oxygen and 3 ATPs are generated per pair. Thus 9 ATPs are formed. To these are added the two formed as a result of the succinate dehydrogenase reaction making a total of 11 ATPs formed by oxidative phosphorylation. When the one ATP formed by substrate level phosphorylation is added, the total number of ATPs generated per turn of the cycle (or per acetyl unit) is 12.

We are now in a position to calculate how many ATPs are formed when one hexose unit is oxidized to CO_2 via glycolysis, the pyruvate dehydrogenase complex and the TCA cycle. One hexose unit yields two molecules of pyruvate which in turn yield two acetyl residues. The latter yield 24 ATPs when oxidized in the TCA cycle. Their formation from pyruvate yields a further 6 ATPs. The aerobic glycolysis of one molecule of free hexose yields 6 ATPs whilst that of G-1-P yields 7 ATPs [see Section E.1(i)(a)]. Thus the complete oxidation of a molecule of free hexose yields 36 ATPs whilst that of G-1-P yields 37 ATPs. The efficiency of free energy conservation for the complete oxidation of D-glucose is therefore $36 \times 30.5 \times 100/2870 = 38.26\%$.

(b) *Regulation of the TCA cycle*

Regulation of the TCA cycle *in vivo* is difficult to establish in both plants and animals. The likely control point in the TCA cycle proper is the citrate synthase-catalysed step. However, control may also be exerted at the pyruvate dehydrogenase-catalysed step since this generates the substrate for the cycle.

The citrate synthase of mitochondria is inhibited by ATP. This is a logical form of control since at high levels of ATP the ATP-generating-TCA cycle will be shut down but will speed up again as the ATP levels fall. It is interesting to note that the citrate synthase of the glyoxysome is not inhibited by ATP; this fits in with the fact that the glyoxylate cycle [see Chapter 8, Section C.1(iii)] is not concerned with ATP generation.

The pyruvate dehydrogenase complex of potato tubers has been shown to be strongly inhibited by

NADH. The inhibition is competitive (with NAD^+) suggesting that the $NAD^+/NADH$ ratio in the mitochondrion may exert fine control over the TCA cycle.

2. Oxidation of fats (triglycerides)

In many animals fat is the major fuel to be stored. Most have a special tissue, adipose tissue, devoted to fat storage and metabolism; in a 70-kg man about 15 kg of fat is present in this tissue, sufficient to enable him to survive about 40 days of starvation without dramatically depleting his body protein.

In general plants do not use fat as a fuel and therefore do not store it; the level of triglycerides in the vegetative tissues is very low. Therefore fat cannot be said to be a major source of substrates (e.g. NADH) for oxidative phosphorylation in plants.

There is, however, one important exception to this generalization. Fat is stored in considerable amounts in the seeds of certain plants [e.g. castor bean; see Chapter 8, Section C.1(iii)]. When these seeds germinate the fat is used to support growth until photosynthesis is fully established. During this time most of the fat is converted into carbohydrates (see Chapter 7, Section B.2), proteins and other lipids. These conversion processes require ATP to drive them. This ATP is generated by oxidative phosphorylation, the substrates (e.g. NADH) for which come mainly from the first part of the conversion of fat into carbohydrate, involving the glyoxylate cycle and part of the TCA cycle (see Chapter 7, Section B.2). It is also likely that some fatty acids derived from the fat are oxidized directly to CO_2, via β-oxidation [see Section E.2(ii)(a)] and the TCA cycle, in the mitochondrion in order to generate substrates for oxidative phosphorylation. Thus, during the germination of fat-storing seeds, fats are used for ATP production.

The catabolism of triglycerides involves three stages: (i) the hydrolytic cleavage of the three ester bonds to yield glycerol and fatty acids, (ii) the catabolism of glycerol and (iii) the catabolism of fatty acids. Stage (i) is catalysed by lipases and is described in Chapter 8, Section C.2(i)(b). Stages (ii) and (iii) are described below; however, certain aspects of stage (iii) are also touched on in Chapter 8, Section C.1(ii).

(i) OXIDATION OF GLYCEROL

Glycerol is firstly phosphorylated by ATP under the catalytic influence of glycerol kinase[34]. The resulting sn-glycerol 3-phosphate is then oxidized to DiHOAcP by glycerol-3-phosphate dehydrogenase which uses NAD^+ as the oxidant. The DiHOAcP can then be catabolized to pyruvic acid via the glycolysis pathway (or utilized for glucose synthesis by gluconeogenesis: see Chapter 7, Section B.2). The conversion of glycerol into pyruvic acid yields 2 ATPs by substrate level phosphorylation and 2 NADHs which when reoxidized by the terminal electron transport chain via the external NADH dehydrogenase yield a further 4 ATPs. The pyruvic acid can diffuse into the mitochondrion and be oxidized to $3CO_2$ [see Section E.1(ii)(a)] with the production of a further 15 ATPs by oxidative phosphorylation. Thus the gross yield of ATP per molecule of glycerol completely oxidized is 21; this is reduced to a net yield of 20 ATPs per glycerol by the glycerol kinase-catalysed reaction.

(ii) OXIDATION OF FATTY ACIDS

Quantitatively the most important catabolic route for fatty acids is known as the β-oxidation pathway or β-oxidation spiral. This process is linked to ATP production in that it produces (i) substrates (e.g. NADH) that are reoxidized via the terminal electron transport chain with the generation of ATP, and (ii) acetyl-CoA which, provided it is formed in the mitochondrion rather than the glyoxysome, may be catabolized via the TCA cycle with the generation of twelve ATPs per acetyl residue [see Section E.1(ii)(a)].

Fatty acids may also be catabolized by the α-oxidation pathway; however, the general consensus of opinion is that this pathway is not linked to ATP production. It will, nevertheless, be described in this section, for the sake of completeness.

(a) β-Oxidation of fatty acids

The enzymes of the β-oxidation spiral occur in mitochondria and in glyoxysomes.

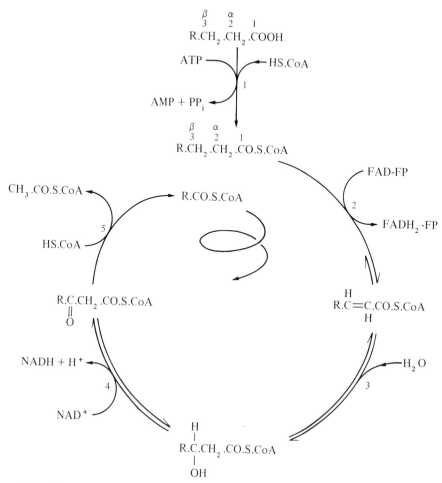

Fig. 6.41. The fatty acid β-oxidation spiral. (1 = acyl-CoA synthase, EC 6.2.1.3; 2 = acyl-CoA dehydrogenase, EC 1.3.99.3; 3 = enoyl-CoA hydratase, EC 4.2.1.17; 4 = 3-hydroxyacyl-CoA dehydrogenase, EC 1.1.1.35; 5 = acetyl-CoA acyltransferase, EC 2.3.1.16.)

The β-oxidation spiral capable of handling saturated fatty acids is shown in Fig. 6.41. Before a fatty acid can be catabolized in the spiral it must be activated by conversion into its coenzyme A derivative. This is accomplished by acyl-CoA synthetase[35] which catalyses the reaction shown in eq. (6.7); the mechanism of this reaction is described in Fig. 2.3. The AMP resulting from this reaction reacts

$$R—COOH + HS.CoA + ATP \rightleftharpoons$$

$$\underset{\substack{\| \\ O}}{R—C—S.CoA} + AMP + PP_i$$

$$(6.7)$$

with ATP under the catalytic influence of adenylate

kinase[36] to form two ADPs (eq. 6.8) which are converted back into ATP by substrate level or

$$AMP + ATP \rightleftharpoons ADP + ADP \qquad (6.8)$$

oxidative phosphorylation. Thus the activation of a fatty acid molecule effectively utilizes two molecules of ATP.

The first reaction of the β-oxidation spiral is catalysed by acyl-CoA dehydrogenase[37]. In it the fatty acyl-CoA is oxidized to a *trans*-α,β-dehydrofatty acyl-CoA (*trans*-Δ^2-dehydrofatty acyl-CoA). Acyl-CoA dehydrogenase is an FAD-containing flavoprotein. During its catalytic activity the FAD is reduced to $FADH_2$. The way in which the $FADH_2$ is reoxidized depends upon whether β-oxidation is taking place in the mitochondrion or the glyoxysome. In the former it is reoxidized by oxygen via the terminal electron transport chain with the generation of two ATPs. In the latter it is probably reoxidized directly by oxygen with the production of H_2O_2 [see Chapter 8, Section C.1 (iii)].

The *trans-α,β*-dehydrofatty acyl-CoA is then stereospecifically hydrated by enoyl-CoA hydratase[38] with the production of L(+)-β-hydroxyfatty acyl-CoA (L(+)-3-hydroxyfatty acyl-CoA). This enzyme can also catalyse the hydration of *cis-α,β*-dehydrofatty acyl-CoA but produces the D(−)-β-hydroxyfatty acyl-CoA. The latter cannot be handled by the enzyme catalysing the next step of the β-oxidation spiral. The significance of this will become apparent when the β-oxidation of unsaturated fatty acids is described.

The L(+)-β-hydroxyfatty acyl-CoA is then oxidized to β-ketofatty acyl-CoA by β-hydroxyacyl-CoA dehydrogenase[39] which uses NAD^+ as the oxidant. The way the resulting NADH is reoxidized depends upon the location of the β-oxidation spiral. If it is operating in the mitochondrion the NADH is reoxidized by oxygen via the terminal electron transport chain with the generation of three ATPs. If it is operating in the glyoxysome the NADH is reoxidized via the external NADH dehydrogenase

(see Section C). This results in the generation of two ATPs via the terminal electron transport chain. The participation of shuttle systems cannot, however, be ruled out.

The final step of the β-oxidation spiral involves the thiolytic cleavage of the α,β-bond of the β-ketofatty acyl-CoA with the production of acetyl-CoA and a fatty acyl-CoA that is two carbon atoms shorter than the one that entered the spiral. The enzyme catalysing this reaction is acetyl-CoA acyltransferase[40].

The shortened fatty acyl-CoA then re-enters the spiral at the acyl-CoA dehydrogenase-catalysed step (reaction 2, Fig. 6.41) and proceeds through it, cleaving off another 2C unit from its carboxyl end as acetyl-CoA. The fatty acyl-CoA continues round and round the spiral until it has been completely degraded to acetyl-CoA. This will, however, only occur if the fatty acyl-CoA has an even number of carbon atoms. If the fatty acyl-CoA has an odd number of carbon atoms the final turn of the spiral produces a molecule of propionyl-CoA (from the ω, ω-1 and ω-2 carbon atoms) as well as a molecule of acetyl-CoA. The propionyl-CoA cannot be catabolized by the β-oxidation spiral; the way in which it is handled is described later (see Fig. 6.42).

The fate of the acetyl-CoA produced at each turn

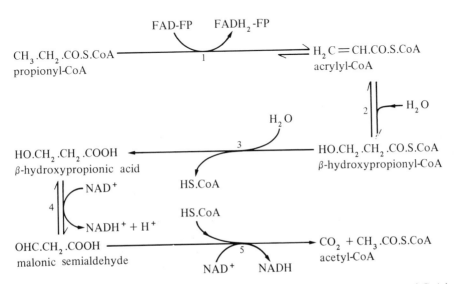

Fig. 6.42. Catabolism of propionyl-CoA in higher plants. (1 = acyl-CoA dehydrogenase, EC 1.3.99.3; 2 = enoyl-CoA hydrogenase, EC 4.2.1.17; 3 = β-hydroxypropionyl-CoA hydrolase activity; 4 = 3-hydroxypropionate dehydrogenase, EC 1.1.1.59; 5 = malonyl semialdehyde dehydrogenase (acetylating), EC 1.2.1.18.)

of the spiral depends upon whether β-oxidation is taking place in the mitochondrion or the glyoxysome. In the latter it enters the glyoxylate cycle in which two acetyl residues are effectively used to generate a molecule of succinate [see Chapter 8, Section C.1(iii)]. The succinate then passes from the glyoxysome to the mitochondrion where it is converted into OAA by reactions (6–8) of the TCA cycle (see Fig. 6.40). The OAA is converted into PEP which is used for carbohydrate formation (see Fig. 7.1). The NADH generated in the L-malate→OAA steps of the glyoxylate and TCA cycles can be used for ATP generation by oxidative phosphorylation.

In the mitochondrion the acetyl-CoA can be oxidized by the TCA cycle with the production of twelve ATPs per acetyl residue [see Section E.1(ii) (a)].

Therefore the mitochondrial oxidation of fatty acids to CO_2 via the β-oxidation spiral and TCA cycle leads to the generation of a great deal of ATP. For instance, the complete oxidation of one molecule of palmitic acid (16C) yields 129 ATPs. This figure is arrived at as follows. Seven turns of the β-oxidation spiral degrade palmitic acid to eight molecules of acetyl-CoA. Since each turn yields five ATPs, this gives a total of thirty-five ATPs from β-oxidation. The oxidation of eight acetyl-CoAs via the TCA cycle yields $8 \times 12 = 96$ ATPs. Thus the gross ATP production is 131 ATPs per palmitic acid molecule. However, this is reduced to a net figure of 129 ATPs per molecule because two ATPs are used up in the activation of the palmitic acid. Thus, given that the $\Delta G^{\circ\prime}$ for the complete oxidation of palmitic acid to CO_2 and H_2O is -9782 kJ mol^{-1} and that the $\Delta G^{\circ\prime}$ for the phosphorylation of ADP is $+30.5$ kJ mol^{-1} the efficiency of energy conservation is $129 \times 30.5 \times 100/9782 = 40.2\%$.

The $\Delta G^{\circ\prime}$ values of the four steps of the β-oxidation spiral, taken in sequence, are -20, -3.1, $+15.7$ and -27.8 kJ mol^{-1}. Thus the shortening of an acyl-CoA by one 2C unit is thermodynamically favourable, the overall $\Delta G^{\circ\prime}$ being -35.2 kJ mol^{-1} (equivalent to a K_{eq} of 1.48×10^6).

The propionyl-CoA produced during the β-oxidation of fatty acids with an odd number of carbon atoms is further catabolized by the pathway shown in Fig. 6.42. The first two steps are catalysed by the β-oxidation enzymes, acyl-CoA dehydrogenase and enolase. The thioester bond in the resulting β-hydroxypropionyl-CoA is then hydrolysed. The β-hydroxypropionic acid is reduced to malonic semialdehyde by the NAD-specific 3-hydroxypropionate dehydrogenase[41]. Finally the malonic semialdehyde is oxidatively decarboxylated to yield acetyl-CoA.

This pathway differs from that in animals, where propionyl-CoA is carboxylated to S-methylmalonyl-CoA which is then isomerized by a specific racemase[42] to R-methylmalonyl-CoA. The latter undergoes an intramolecular rearrangement, catalysed by the cobalamine-enzyme, methylmalonyl-CoA mutase[43], to form succinyl-CoA which then enters the TCA cycle. Since plants do not synthesize cobalamin (Vitamin B_{12}) they cannot produce methylmalonyl-CoA mutase and therefore this method of catabolizing propionyl-CoA is not open to them.

The most common unsaturated fatty acids in plants have a single cis-Δ^9 double bond or a methylene-interrupted system of cis double bonds (see Chapter 8, Section B.1). Fatty acids of this type can be catabolized by the β-oxidation spiral provided that it is supplemented by two extra enzymes. These are a $\Delta^{3\text{-}cis} \to \Delta^{2\text{-}trans}$-enoyl-CoA isomerase[44] and a D($-$)-3-hydroxyacyl-CoA epimerase. The operation of these enzymes in the β-oxidation of linoleic acid is shown in Fig. 6.43. The former enzyme catalyses the switching of the cis-Δ^3 double bond of the $\Delta^{3(cis),6(cis)}$-dodecadienoyl-CoA, produced after three turns of the β-oxidation spiral, to a trans-Δ^2 double bond so that a substrate for enoyl-CoA hydratase is formed; this double bond was the cis-Δ^9 double bond of linoleic acid. The latter enzyme converts the D($-$)-β-hydroxyoctanoyl-CoA, produced when enoyl-CoA hydratase hydrates the cis-Δ^2 double bond of cis-Δ^2-octenoyl-CoA (formerly the cis-Δ^{12} double bond of linoleic acid), into L($+$)-β-hydroxyoctanoyl-CoA so that β-hydroxyacyl-CoA dehydrogenase can function.

It should be noted that for each double bond present in an unsaturated fatty acid, two fewer ATPs are formed when it is completely oxidized than would be formed from the corresponding saturated fatty acid. This follows from the fact that when either of the additional enzymes is required to operate in a given turn of the β-oxidation spiral the $FADH_2$-producing step (reaction 2, Fig. 6.41) is bypassed. Thus the two ATPs that would have resulted

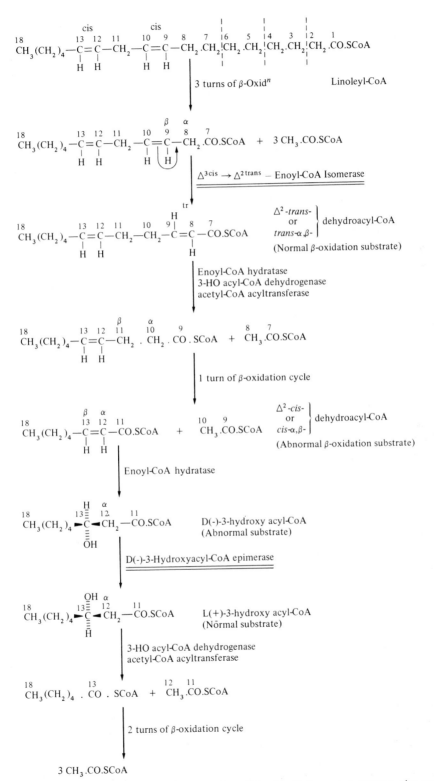

Fig. 6.43. Catabolism of linoleic acid (this involves the operation of the β-oxidation spiral enzymes plus two extra enzymes, $\Delta^{3\text{-}cis} \rightarrow \Delta^{2\text{-}trans}$-enoyl-CoA isomerase and D(−)-3-hydroxyacyl-CoA epimerase).

from the reoxidation of $FADH_2$ via the terminal electron transport chain are not produced.

(b) α-Oxidation of fatty acids

As the name implies, α-oxidation involves the oxidation of the α-carbon atom (i.e. C–2) of a fatty acid. The process brings about the removal of the carboxyl group as CO_2, with the oxidized α-carbon atom replacing it as the carboxyl group of a fatty acid one carbon atom shorter than the original.

Two α-oxidation systems in plants have been described, one occurring in the cotyledons of germinating peanuts, the other in the leaves of pea and castor bean plants. The former appeared to utilize a peroxidase and required a H_2O_2-generating system whilst the latter required molecular O_2 rather than H_2O_2. However, more recent work in Stumpf's laboratory has shown that a common α-oxidation system occurs in both seeds and leaves (see Fig. 6.44).

The first step in this process is catalysed by an FAD-flavoprotein enzyme. The initial attacking species is probably the enzymic FAD in its free radical semiquinone form ($XH\cdot$ in Fig. 6.44; see also Fig. 6.12). This stereospecifically removes H_D from the α-carbon of a free fatty acid. The resulting reduced flavoprotein (XH_2 in Fig. 6.44) then reacts with O_2 to form the adduct XH_2O_2. The latter reacts with the fatty acid free radical resulting from

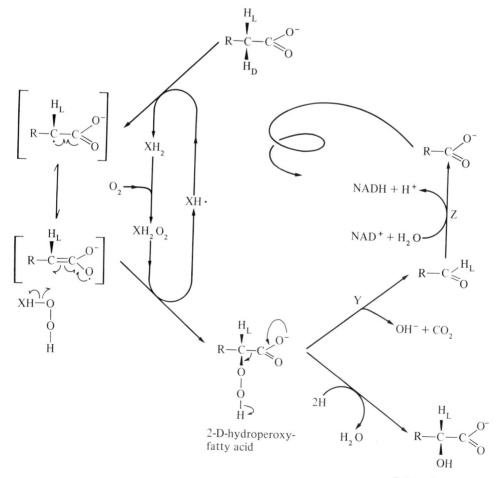

Fig. 6.44. Proposed mechanism of α-oxidation in plants (adapted from Shine and Stumpf, 1974). (X appears to be an FAD-flavoprotein enzyme; Y appears to be an anhydrodecarboxylase; Z is an NAD-specific aldehyde dehydrogenase.)

the initial attack and generates a 2-D-hydroperoxyfatty acid. This may then undergo anhydrodecarboxylation to yield a fatty aldehyde, as a continuation of α-oxidation, or it may be reduced to form a 2-D-hydroxyfatty acid as an end product.

The fatty aldehyde, which is one carbon shorter than the original fatty acid, is then oxidized by an NAD-specific aldehyde dehydrogenase[45] to its corresponding fatty acid. This completes one turn of the α-oxidation spiral. Further turns successively reduce it by one carbon atom at a time. Although it has not yet been demonstrated, the reoxidation of the NADH produced in α-oxidation might be expected to take place via the terminal electron transport chain with the generation of ATP. However, as was stated earlier, the present consensus is that α-oxidation is not linked to ATP production.

The physiological significance of α-oxidation is discussed in Chapter 8, Section C.1(ii)(b).

3. Oxidation of amino acids

Oxidation is an essential and important part of amino acid metabolism in animals because of the protein intake in food. In plants, on the other hand, the synthesis of amino acids for growth is by far the most important aspect of their metabolism. Little, if anything, is known of their oxidation in plants and if it does occur it can only play a very minor part in providing energy for the plant.

F. 'RESPIRATION-LIKE' PROCESSES NOT LINKED TO ATP GENERATION

1. Photorespiration

Photorespiration is a catabolic process, occurring only in the light, in which O_2 is consumed and CO_2 is evolved. Its substrate is glycollate formed from phosphoglycollate in the chloroplast by a specific phosphatase[46] (see Fig. 6.45). The phosphoglycollate is in turn derived from D-ribulose 1,5-diphosphate (RuDP) by the enzyme RuDP carboxylase-oxygenase[47] operating in its oxygenase mode [see

Chapter 5, eqs. (5.22) and (5.23)]. Glycollate is thus derived from Cs 1 and 2 of RuDP. If it is formed in the presence of $^{18}O_2$ one of its carboxyl oxygen atoms is labelled with ^{18}O. If it is formed in the presence of $^{14}CO_2$ its two carbon atoms are equally labelled with ^{14}C; this is what would be expected from its RuDP origin since the operation of the Calvin cycle in the presence of $^{14}CO_2$ produces RuDP equally labelled in Cs 1 and 2.

The further metabolism of the glycollic acid involves two other intracellular organelles, the peroxisome and the mitochondrion (see Chapter 3, Sections B.7 and B.5). It passes from the chloroplast into the peroxisome where glycollate oxidase[48] catalyses its oxidation to glyoxylate. Glycollate oxidase is an FMN-containing flavoprotein which uses one molecule of O_2 and produces one molecule of H_2O_2 for each molecule of glycollate oxidized. The H_2O_2 is converted into H_2O and O_2 by catalase in the peroxisome. The glyoxylate is then transaminated to yield glycine; this reaction may be catalysed by glutamate:glyoxylate aminotransferase or serine:glyoxylate aminotransferase, both of which are present in the peroxisome.

The glycine then passes from the peroxisome to the mitochondrion. Here two molecules of glycine are converted into one molecule each of L-serine, CO_2 and NH_3 in a two-step process. In the first step one molecule of glycine is cleaved by the enzyme glycine synthase[49] into three parts; the carboxyl- and amino groups are eliminated as CO_2 and NH_3 respectively whilst the methylene carbon becomes bound to tetrahydrofolate (THF) as $N^{5,10}$-methylene THF. This reaction requires NAD^+ as an oxidant; the resulting NADH is re-oxidized by the mitochondrial electron transport chain with the attendant generation of ATP. In the second step serine hydroxymethyltransferase[50], a pyridoxal phosphate-requiring enzyme, catalyses the formation of L-serine from the methylene carbon of $N^{5,10}$-methylene THF and a second molecule of glycine.

The ammonia produced in the glycine synthase-catalysed step is probably conserved within the cell by conversion to L-glutamate. The mitochondrion contains glutamate dehydrogenase[51] which could catalyse this conversion. However, its $K_{m(NH_4^+)}$ is so high (5–40 mM) that before any significant conversion could take place the NH_4^+ concentration within

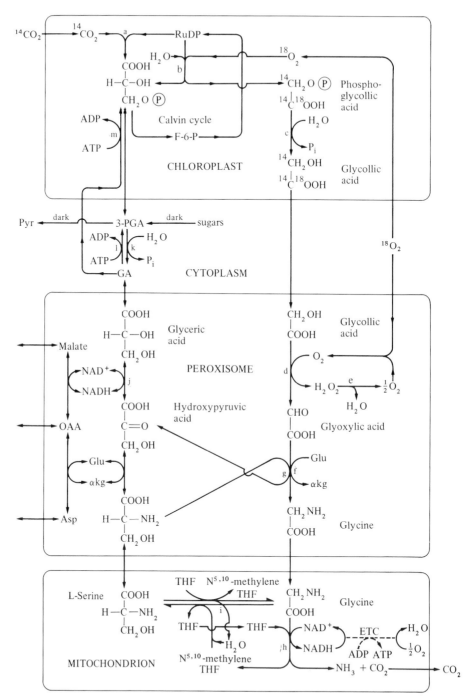

Fig. 6.45. Proposed pathways of photorespiration. (RuDP = D-ribulose 1,5-diphosphate; 3-PGA = 3-phosphoglyceric acid; GA = glyceric acid; Glu = L-glutamic acid; αkg = α-ketoglutaric acid; OAA = oxaloacetic acid; Pyr = pyruvic acid; THF = tetrahydrofolic acid; ETC = terminal electron transport chain; a = carboxylase activity of RuDP carboxylase-oxygenase, EC 4.1.1.39; b = oxygenase activity of RuDP carboxylase-oxygenase, EC 4.1.1.39; c = phosphoglycollate phosphatase, EC 3.1.3.18; d = glycollate oxidase, EC 1.1.3.1; e = catalase, EC 1.11.1.6; f = glutamate:glyoxylate aminotransferase; g = serine:glyoxylate aminotransferase; h = glycine synthase, EC 2.1.2.10; i = serine hydroxymethyltransferase, EC 2.1.2.1; j = hydroxypyruvate reductase, EC 1.1.1.81; k = phosphoglycerate phosphatase, EC 3.1.3.38; l = glycerate kinase, EC 2.7.1.31 (cytoplasmic); m = glycerate kinase, EC 2.7.1.31 (chloroplastidic).)

the mitochondrion would have to rise to a toxic level. It is therefore much more likely that NH_4^+ passes from the mitochondrion to the chloroplast where the coupling together of the light-driven reactions catalysed by glutamine synthetase[52] ($K_{m(NH_4^+)} < 0.5$ mM) and glutamate synthase[53] (see Chapter 9, Section B.4) produces the same result.

The L-serine passes back to the peroxisome where it is transaminated to hydroxypyruvate. The latter is then reduced by the NAD-requiring hydroxypyruvate reductase[54] to glyceric acid which may then pass into the chloroplast where a specific glycerate kinase[55] catalyses its phosphorylation to 3-PGA thus completing a cycle of reactions whose stoichiometry is given in eq. (6.9).

$$2 \text{ RuDP} + 3O_2 + 3H_2O \rightarrow 3 \text{ 3-PGA} + CO_2 + P_i$$

$\underbrace{\qquad\qquad}$

(10C, 4P)   (9C, 3P)   (1C) (1P)

                                              (6.9)

Two of the three molecules of 3-PGA in eq. (6.9) result directly from reaction b (Fig. 6.45) whilst the third arises from the catabolism of the two molecules of phosphoglycollate produced in the same reaction.

The cycle is made up of two distinct parts. That from RuDP to glycine is physiologically irreversible whilst that from glycine to 3-PGA is reversible. The cycle is probably driven in the direction indicated in eq. (6.9) by the directionality of the irreversible reactions. However, when photorespiration is blocked, as occurs, for example, when the oxygenase activity of RuDP carboxylase-oxygenase is inhibited at high $CO_2 : O_2$ ratios, serine and glycine can be formed from 3-PGA by the reactions of the reversible part of the cycle.

The magnitude of photorespiration is difficult to assess accurately. It is usually done by measuring the CO_2 compensation point which is defined as the constant CO_2 concentration which is ultimately reached when a plant is kept in a system which is closed to the atmosphere but which is under constant illumination. At the CO_2 compensation point CO_2 uptake (due to photosynthesis) equals CO_2 evolution (due to photorespiration). The lower the value of the CO_2 compensation point is, the less photorespiration is taking place. It is found that C_3 plants (see Chapter 5, Section D.2) have CO_2 compensation points in the range 40–60 ppm (cf.

CO_2 concentration in air $= 0.03\%$ by vol. or 300 ppm) whilst C_4 plants (see Chapter 5, Section D.3) often have values which are less than 10 ppm. This indicates that photorespiration is very active in C_3 plants but is relatively insignificant in C_4 plants. It has been estimated that as a result of photorespiration C_3 plants lose 20–40% of the carbon fixed by photosynthesis. Such losses in C_4 plants are very small. This would indicate that C_4 plants grow more efficiently than C_3 plants. This is particularly unfortunate in view of the fact that all the major agricultural crop species with the exception of maize, millet, sorghum and sugar cane are C_3 plants.

What the function of photorespiration is remains obscure. It appears to be a totally wasteful process and yet it may be that future research will show that it has a useful role. Various suggestions as to what this role may be have been made. One of the most intriguing of these is that it constitutes a complex terminal oxidase system for consuming excess ATP and NADPH which may have been unavoidably generated in the light phase of photosynthesis under high light intensities. The fact that plants which are totally free of photorespiration have not so far been found certainly suggests that it is a useful process. However, in spite of this and because of the promise of more efficient agricultural production in its absence, methods of diminishing the extent of photorespiration in C_3 plants are under active investigation. Attempts are being made to alter, mutagenically and chemically, the enzyme RuDP carboxylase-oxygenase in such a way that its oxygenase activity is diminished and/or its carboxylase activity is increased. This may prove difficult because the same active site is involved in the carboxylase and the oxygenase functions [see Chapter 5, Section D.2(i)]. Inhibitors of glycollate oxidase, e.g. α-hydroxy-2-pyridinemethanesulphonate, 2-hydroxy-3-butynoic acid, are also being investigated as means of reducing photorespiration.

2. Microsomal electron transport

The endoplasmic reticular membranes [see Chapter 3, Section B.2(iii)] have a number of electron transport chains built into them which utilize oxygen but are not linked to the phosphorylation of ADP.

A particularly important one is composed of a flavoprotein, cytochrome P450 and possibly Fe–S protein. The flavoprotein, termed NADPH-cytochrome P450 reductase, catalyses the transfer of a pair of electrons from NADPH to cytochrome P450 which uses them, along with molecular oxygen, to catalyse hydroxylation reactions. There appears to be a range of cytochromes P450, each with a different substrate specificity.

Cytochromes P450 are enzymes with a non-covalently bound protohaem prosthetic group (see Fig. 6.19); during their catalytic activity the haem Fe undergoes redox change and coordinately binds a molecule of oxygen. They are given the name P450 because the carbon monoxide coordination complex of their reduced form absorbs at 450 nm.

Cytochromes P450 fall into the category of monooxygenases (also known as mixed function oxygenases and hydroxylases) which catalyse reactions of the type given in eq. (6.10). Thus one atom of

$$\text{Substrate} + O_2 + \text{co-substrate-H}_2 \rightarrow$$
$$\text{(reduced form)}$$
$$\text{substrate-OH} + H_2O + \text{co-substrate} \quad (6.10)$$
$$\text{(oxidized form)}$$

the oxygen molecule is incorporated into the substrate hydroxyl group whilst the other is incorporated into water. The reduced co-substrate (usually NADPH) provides the pair of electrons required for the reaction and is therefore oxidized in the process.

The details of the mechanism of the reaction catalysed by cytochromes P450 are not yet clear but present evidence indicates a reaction sequence like that shown in Fig. 6.46. The function that a cytochrome P450 is considered to perform in reaction (6.10) is to facilitate the cleavage of the bond between the two oxygen atoms in the O_2 molecule and thereby to produce a reactive oxygen species known as an oxene. The cleavage of the oxygen bond is difficult; the oxygen molecule has a high dissociation enthalpy ($\Delta H = +493.7 \text{ kJ mol}^{-1}$). However, when O_2 is coordinately bound to the haem Fe, the receipt of two electrons (from NADPH) by the P450 (steps 2 and 5, Fig. 6.46) enables the O—O bond to become strongly polarized. Whilst in this state reaction H^+ from the

aqueous milieu allows H_2O to be eliminated (step 8, Fig. 6.46) with the formation of the oxene. The oxene is electrophilic and reacts with the bound substrate (steps 9 and 10, Fig. 6.46). The hydroxylated substrate is then released from the hydrophobic binding site. This regenerates P450–Fe^{3+} which can then bind another substrate molecule (step 1, Fig. 6.46) and repeat the process.

Cytochrome P450-containing electron transport chains are responsible for a range of hydroxylations in plants. A good example is the first step of the sterol demethylation process in which the 14α-methyl group is converted into a 14α-CH_2OH group.

Another example of a microsomal electron transport chain involving a monooxygenase is that which carries out the desaturation of fatty acids [see Chapter 8, Section C.1(i)(d)].

3. The oxidative pentose phosphate pathway

Although glycolysis is the major catabolic route for hexoses, some plant cells have, in addition to glycolysis, an alternative route which is called the oxidative pentose phosphate pathway or the hexose monophosphate shunt. The latter name is less frequently used nowadays but is particularly appropriate because the word shunt indicates that it is an offshoot or by-pass of another, more direct pathway; in fact the latter pathway is the glycolysis sequence itself. As can be seen from Fig. 6.47 the shunt begins at G-6-P in the glycolysis sequence and ultimately rejoins it at F-6-P and 3-PGAld. The enzymes catalysing the reactions of the shunt are present in the cytosol of plant cells, as are those of glycolysis.

The oxidative pentose phosphate pathway can be divided into two distinct phases. Phase 1, from G-6-P to D-ribulose 5-phosphate (Ru-5-P), is physiologically irreversible whilst phase 2, from Ru-5-P to F-6-P and 3-PGAld, is reversible.

Phase 1 commences with the oxidation of G-6-P to D-glucono-δ-lactone 6-phosphate; this reaction is catalysed by glucose-6-phosphate dehydrogenase[56] which uses $NADP^+$ as the oxidant. The δ-lactone is then hydrolysed by 6-phosphogluconolactonase[57] to 6-phosphogluconic acid which is then

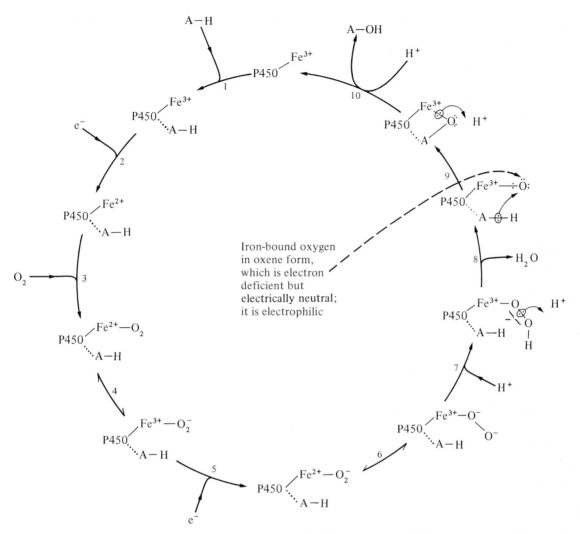

Fig. 6.46. Postulated mechanism for the hydroxylation of a substrate (A–H) by a cytochrome P450. (.... = non-covalent binding of the substrate to a hydrophobic site on the P450 apoprotein; ↔ = resonance.)

oxidatively decarboxylated to Ru-5-P by 6-phos-phogluconate dehydrogenase[58] using NADP$^+$ as the oxidant. Thus in phase 1 two molecules of NADPH are generated per molecule of G-6-P.

In phase 2 Ru-5-P is isomerized to D-ribose 5-phosphate (R-5-P) and D-xylulose 5-phosphate (Xu-5-P) by ribosephosphate isomerase[59] and ribu-losephosphate 3-epimerase[60] respectively. R-5-P and Xu-5-P then react under the catalytic influence of transketolase[61] to form 3-PGAld and D-sedoheptulose 7-phosphate (Su-7-P). These com-pounds in turn react together in the presence of

transaldolase[62] to form F-6-P and D-erythrose 4-phosphate (E-4-P). E-4-P and Xu-5-P may then react together in the presence of transketolase to form 3-PGAld and F-6-P which can then be further metabolized via the glycolysis pathway.

The oxidative pentose phosphate pathway has three main functions in plants. Firstly, it generates NADPH which is used as a reductant in biosynthe-tic processes in circumstances when NADPH is not being generated by photosynthesis; it is therefore particularly important in non-photosynthetic tissues, e.g. in differentiating tissues, germinating

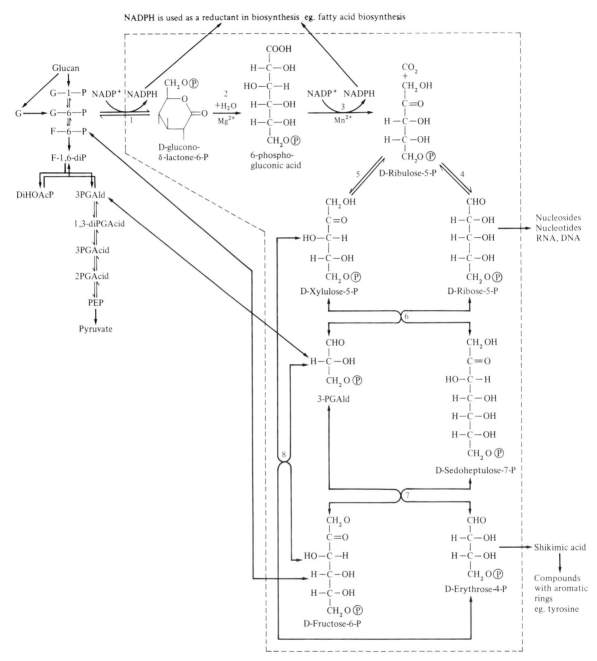

Fig. 6.47. The oxidative pentose phosphate pathway. (1 = glucose-6-phosphate dehydrogenase, EC 1.1.1.49; 2 = 6-phosphogluconolac-tonase, EC 3.1.1.31; 3 = 6-phosphogluconate dehydrogenase, EC 1.1.1.44; 4 = ribosephosphate isomerase, EC 5.3.1.6; 5 = ribulosephos-phate 3-epimerase, EC 5.1.3.1; 6 and 8 = transketolase, EC 2.2.1.1; 7 = transaldolase, EC 2.2.1.2; the reactions within the dashed line constitute the oxidative pentose phosphate pathway.)

seeds, and during the hours of darkness. Secondly, it generates R-5-P which is required for the biosynthesis of nucleotides and nucleic acids. Thirdly, it generates E-4-P which is required for the synthesis of shikimic acid, the precursor of aromatic rings (see Chapter 4, Section D.2).

The reversibility of phase 2 of the oxidative pentose phosphate pathway is particularly useful since it (i) enables any excess pentose phosphate, produced during NADPH generation, to be returned to the glycolysis sequence and (ii) allows R-5-P and E-4-P to be formed from 3-PGAld and F-6-P without generating NADPH should it be necessary.

The oxidative pentose phosphate pathway, by forming reduced pyridine nucleotide (NADPH), would at first sight appear to be generating substrate for oxidative phosphorylation. However, the NADPH is not used for this purpose; it is used as a reductant in biosynthetic reactions [see Section B.2 (ii)(a)].

SUGGESTIONS FOR FURTHER READING

Electron transport and oxidative phosphorylation

Baltscheffsky, H. and Baltscheffsky, M. (1974) 'Electron transport phosphorylation', *Ann. Rev. Biochem.* **43**, 871–897.
Meeuse, B. J. D. (1975) 'Thermogenic respiration in aroids', *Ann. Rev. Plant Physiol.* **26**, 117–126.
Palmer, J. M. (1976) 'The organization and regulation of electron transport in plant mitochondria', *Ann. Rev. Plant Physiol.* **27**, 133–157.
Boyer, P. D., Chance, B., Ernster, L., Mitchell, P., Racker, E. and Slater, E. (1977) 'Oxidative phosphorylation and photophosphorylation', *Ann. Rev. Biochem.* **46**, 955–1026.
Solomos, T. (1977) 'Cyanide-resistant respiration in higher plants', *Ann. Rev. Plant Physiol.* **28**, 279–297.
Wiskich, J. T. (1977) 'Mitochondrial metabolite transport', *Ann. Rev. Plant Physiol.* **28**, 45–69.

Carbohydrate oxidation

Turner, J. F. and Turner, D. H. (1975) 'The regulation of carbohydrate metabolism', *Ann. Rev. Plant Physiol.* **26**, 159–186.

Photorespiration

Tolbert, N. E. (1971) 'Microbodies—peroxisomes and glyoxysomes', *Ann. Rev. Plant Physiol.* **22**, 45–74.

Kelly, G. J., Latzko, E. and Gibbs, M. (1976) 'Regulatory aspects of photosynthetic carbon metabolism', *Ann. Rev. Plant Physiol.* **27**, 181–205.
Tolbert, N. E. and Ryan, F. J. (1976) 'Glycollate biosynthesis and metabolism during photorespiration', in *CO₂ Metabolism and Plant Productivity* (Burris, R. H. and Black, C. C., eds.), pp. 141–159. University Park Press, Baltimore.
Chollet, R. (1977) 'The biochemistry of photorespiration', *TIBS* **2**, 155–159.

ENZYMES

1. Benzenediol: oxygen oxidoreductase, EC 1.10.3.2.
2. NADH: (acceptor) oxidoreductase, EC 1.6.99.3.
3. Pyruvate: lipoamide oxidoreductase (decarboxylating and acceptor-acetylating), EC 1.2.4.1.
4. *threo*-D$_S$-Isocitrate: NAD$^+$ oxidoreductase (decarboxylating), EC 1.1.1.41.
5. 2-Oxoglutarate: lipoamide oxidoreductase (decarboxylating and acceptor succinylating), EC 1.2.4.2.
6. L-Malate: NAD$^+$ oxidoreductase, EC 1.1.1.37.
7. Succinate: (acceptor) oxidoreductase, EC 1.3.99.1.
8. *sn*-Glycerol-3-phosphate: (acceptor) oxidoreductase, EC 1.1.99.5.
9. Ubiquinol: ferricytochrome *c* oxidoreductase, EC 1.10.2.2.
10. Ferrocytochrome *c*: oxygen oxidoreductase, EC 1.9.3.1.
11. D-Glyceraldehyde-3-phosphate: NAD$^+$ oxidoreductase (phosphorylating), EC 1.2.1.12.
12. L-Malate: NAD$^+$ oxidoreductase (decarboxylating), EC 1.1.1.38.
13. NADH: lipoamide oxidoreductase, EC 1.6.4.3.
14. L-Aspartate: 2-oxoglutarate aminotransferase, EC 2.6.1.1.
15. *sn*-Glycerol-3-phosphate: NAD$^+$ 2-oxidoreductase, EC 1.1.1.8.
16. Hydrogen peroxide: hydrogen peroxide oxidoreductase, EC 1.11.1.6.
17. β-D-Fructofuranoside fructohydrolase, EC 3.2.1.26.
18. ATP: D-Fructose 6-phosphate 1-phosphotransferase, EC 2.7.1.11.
19. ATP: D-hexose 6-phosphotransferase, EC 2.7.1.1.
20. 1,4-α-D-Glucan: orthophosphate α-D-glucosyltransferase, EC 2.4.1.1.
21. D-Fructose-1,6-bisphosphate D-glyceraldehyde-3-phosphate lyase, EC 4.1.2.13.
22. D-Glyceraldehyde-3-phosphate ketol-isomerase, EC 5.3.1.1.
23. D-Phosphoglycerate 2,3-phosphomutase, EC 5.4.2.1.
24. 2-Phospho-D-glycerate hydro-lyase, EC 4.2.1.11.
25. 2-Oxo-acid carboxy-lyase, EC 4.1.1.1.
26. Alcohol: NAD$^+$ oxidoreductase, EC 1.1.1.1.
27. L-Lactate: NAD$^+$ oxidoreductase, EC 1.1.1.27.
28. ATP: pyruvate 2-*O*-phosphotransferase, EC 2.7.1.40.
29. Acetyl-CoA: dihydrolipoamide *S*-acetyltransferase, EC 2.3.1.12.
30. Citrate: oxaloacetate-lyase (pro-3*S*-CH₂COO$^-$ →acetyl-CoA), EC 4.1.3.7.
31. Citrate (isocitrate) hydro-lyase, EC 4.2.1.3.
32. Succinate: CoA ligase (ADP-forming), EC 6.2.1.5.
33. L-Malate hydro-lyase, EC 4.2.1.2.
34. ATP: glycerol 3-phosphotransferase, EC 2.7.1.30.
35. Acid: CoA ligase (AMP-forming), EC 6.2.1.3.
36. ATP: AMP phosphotransferase, EC 2.7.4.3.
37. Acyl-CoA: (acceptor) oxidoreductase, EC 1.3.99.3.

38. L-3-Hydroxyacyl-CoA hydro-lyase, EC 4.2.1.17.
39. L-3-Hydroxyacyl-CoA: NAD$^+$ oxidoreductase, EC 1.1.1.36.
40. Acyl-CoA: acetyl-CoA C-acyltransferase, EC 2.3.1.16.
41. 3-Hydroxypropionate: NAD$^+$ oxidoreductase, EC 1.1.1.59.
42. Methylmalonyl-CoA racemase, EC 5.1.99.1.
43. Methylmalonyl-CoA CoA-carbonylmutase, EC 5.4.99.2.
44. $\Delta^{3\text{-}cis} \rightarrow \Delta^{2\text{-}trans}$-Enoyl-CoA isomerase, EC 5.3.3.7.
45. Formate: NAD$^+$ oxidoreductase, EC 1.2.1.3.
46. 2-Phosphoglycollate phosphohydrolase, EC 3.1.3.18.
47. 3-Phospho-D-glycerate carboxy-lyase (dimerizing), EC 4.1.1.39.
48. Glycollate: oxygen oxidoreductase, EC 1.1.3.1.
49. 5,10-Methylene-tetrahydrofolate: ammonia hydroxymethyltransferase (carboxylating, reducing), EC 2.1.2.10.
50. 5,10-Methylene-tetrahydrofolate: glycine hydroxymethyltransferase, EC 2.1.2.1.
51. L-Glutamate: NAD$^+$ oxidoreductase (deaminating), EC 1.4.1.2.
52. L-Glutamate: ammonia ligase (ADP-forming), EC 6.3.1.2.
53. L-Glutamate: NAD$^+$ oxidoreductase (transaminating), EC 1.4.1.14.
54. D-Glycerate: NAD(P)$^+$ 2-oxidoreductase, EC 1.1.1.81.
55. ATP: D-glycerate 3-phosphotransferase, EC 2.7.1.31.
56. D-Glucose-6-phosphate: NADP$^+$ oxidoreductase, EC 1.1.1.49.
57. 6-Phospho-D-gluconate-δ-lactone lactonohydrolase, EC 3.1.1.31.
58. 6-Phospho-D-gluconate: NADP$^+$ 2-oxidoreductase (decarboxylating), EC 1.1.1.44.
59. D-Ribose-5-phosphate ketol-isomerase, EC 5.3.1.6.
60. D-Ribulose-5-phosphate 3-epimerase, EC 5.1.3.1.
61. Sedoheptulose-7-phosphate: D-glyceraldehyde-3-phosphate glycolaldehydetransferase, EC 2.2.1.1.
62. Sedoheptulose-7-phosphate: D-glyceraldehyde-3-phosphate dihydroxyacetonetransferase, EC 2.2.1.2.

CHAPTER 7

Carbohydrate Biosynthesis

CONTENTS

A. INTRODUCTION

In the mature photosynthesizing plant atmospheric carbon dioxide is converted, in the process of photosynthesis, into D-fructose 6-phosphate (F-6-P) (see Chapter 5, Section D) from which a whole range of monosaccharides and monosaccharide derivatives can be formed (see Section B.3). Some of these monosaccharide derivatives are the precursors from which oligo- and polysaccharides are formed. Probably the most important of the oligosaccharides formed is sucrose which is the means by which fixed carbon and energy are transported around the plant. Two classes of polysaccharides are synthesized, the structural polysaccharides (e.g. cellulose, pectins) and the storage polysaccharides (e.g. starch, fructosans). Structural polysaccharides are required for the formation of cell walls; thus the synthesis of structural polysaccharides is to be found in all those regions of the plant where tissue growth is occurring. Storage polysaccharides are synthesized under circumstances where they form temporary or permanent stores of fixed carbon and energy. An example of a temporary store is the starch laid down as starch grains in the chloroplast during a period of active photosynthesis (see Chapter 5, Section D.6) and which is mobilized and exported in the form of sucrose during the next dark period. More permanent stores of polysaccharides are built up during the growth of the plant in the seeds which develop after the fertilization of the flowers and in the various vegetative propagatory structures (e.g. stem tubers, corms, bulbs) that may also be produced. The function of the stored polysaccharide in these structures is to supply the immature, initially non-photosynthetic plant which emerges upon germination, with the fixed carbon and energy it requires to allow it to grow and develop until it is fully photosynthetic.

Carbohydrate biosynthesis in the recently germinated plant is similar but not identical to that in the mature, photosynthesizing plant. Sucrose is synthesized and transported to the meristematic regions where it is utilized to synthesize the structural polysaccharides which are required for the formation of cell walls around the rapidly dividing and growing cells. However, it is not utilized for the synthesis of storage polysaccharides since the emphasis is upon growth and development rather than the laying down of reserves for the next generation. The source of the sucrose in the recently germinated plant is the reserve material stored in the seed or vegetative propagatory structure. This is broken down and then utilized to form the monosaccharide derivatives required for sucrose synthesis. In the majority of cases the reserve material is a polysaccharide; this is usually starch but fructosans are stored in the vegetative, propagatory structures of members of the Compositae, Campanulaceae and Gramineae. Here enzymes catalyse the breakdown of the polysaccharide to the component monosaccharide from which the derivatives of D-glucose and D-fructose, required for sucrose synthesis, are formed. Some seeds, however, store fats (triglycerides or triacylglycerols) (see Chapter 8) instead of a polysaccharide. During germination these fats are broken down and converted into phosphate esters of D-glucose in a process called *gluconeogenesis* which involves the participation of many different enzymes operating in a closely integrated manner. The D-glucose phosphates are then converted into the monosaccharide derivatives required for sucrose synthesis.

The integration of what detailed knowledge we have of the individual reactions involved in the biosynthesis of carbohydrates into an overall picture which can be related to the life cycle of the plant is the result of experiments designed to determine the fate of photosynthetically fixed CO_2. Such experiments have involved exposing a single at-

tached leaf to $^{14}CO_2$ and following the distribution of the resulting fixed ^{14}C throughout the whole plant. When a young actively growing tobacco leaf is exposed to $^{14}CO_2$, almost all the fixed ^{14}C is retained within that same leaf and used for polysaccharide and protein synthesis (i.e. used for its own growth); only when the leaf is half grown does exportation of carbon as sucrose become significant. However, when the leaf is fully expanded, about half the photosynthetically fixed carbon is immediately exported as sucrose, mainly to the stem and roots where it is predominantly converted into polysaccharides; the remaining half of the assimilated carbon remains in the leaf mainly as starch and sugar. Only about 2% of the carbon exported from the fully expanded leaf is transported to other leaves, which are always above and on the same side of the stem as the $^{14}CO_2$-treated leaf, and only about 0.1% of the exported carbon reaches the growing tip. These findings indicate that the carbon necessary for the growth of a leaf during the early stages of its development is supplied by the more mature and actively photosynthesizing leaves below, each of which contributes a small amount. The apex also appears to receive its carbon in the form of small contributions from all actively photosynthesizing leaves. Similar experiments have been carried out with monocotyledonous plants, in particular the cereals, rice, rye and barley. The pattern of ^{14}C distribution was essentially the same as that observed in the tobacco plant, the main difference being that after ear-emergence those lower leaves which are still capable of photosynthesis stopped contributing carbon to the apex. The carbon laid down as starch in the developing grains of the ear was assimilated by the ear itself and the topmost leaves and sheaths.

The importance of sucrose as the main transport material of higher plants is well established. Analysis of the sieve tube sap of plants exposed to $^{14}CO_2$ has revealed that sucrose is the major ^{14}C-labelled component. The storage carbohydrate of seeds, tubers and rhizomes is converted into sucrose during germination or sprouting; moreover, added sucrose stimulates the growth of rye seedlings. These findings indicate that during germination and sprouting, polysaccharide food stores are converted into sucrose which is then translocated to the rapidly developing shoots and roots.

B. BIOSYNTHESIS OF CARBOHYDRATES

1. Synthesis from carbon dioxide

The ultimate source of all the carbon of carbohydrates in plants is the CO_2 of the atmosphere. All plants which are capable of photosynthesis operate the Calvin cycle in which CO_2 is converted into carbohydrate; this cycle is driven by ATP and NADPH formed as a result of the absorption of light energy of the sun. The key reaction of the Calvin cycle is the carboxylation of D-ribulose 1,5-diphosphate (RuDP) with the production of two molecules of 3-phosphoglyceric acid (3-PGA) [eq. (7.1)] which is catalysed by RuDP carboxylase-oxygenase[1]; the mechanism of this reaction is discussed in Chapter 5, Section D.2(i).

Although all photosynthetically capable plants operate the Calvin cycle, only the C_3 plants (see Chapter 5, Section D.2) fix atmospheric CO_2 directly into carbohydrate by this means. C_4 plants and CAM plants (see Chapter 5, Sections D.1, D.3, D.4) initially carboxylate phosphoenolpyruvate (PEP) to form oxaloacetic acid [eq. (7.2)], the reaction is catalysed by PEP carboxylase[2] and bicarbonate (HCO_3^-), derived from CO_2, is the carboxylating agent. Oxaloacetic acid, or malic acid derived from it, is subsequently decarboxylated to yield the CO_2 which then becomes the substrate for reaction (7.1). The advantage that this extra carboxylation step confers on C_4 and CAM plants is discussed in Chapter 5, Sections D.3(iii) and D.4.

The 3-PGA formed in reaction (7.1) is converted into the triose phosphates, 3-phosphoglyceraldehyde (3-PGAld) and dihydroxyacetone phosphate (DiHOAcP), and a number of tetrose, pentose, hexose and heptose phosphates during the operation of the Calvin cycle. Some of these intermediates are tapped off the cycle and may undergo a considerable number of interconversions (see Section B.3) before being converted into glycosides or oligo- or polysaccharides (see Sections C and D).

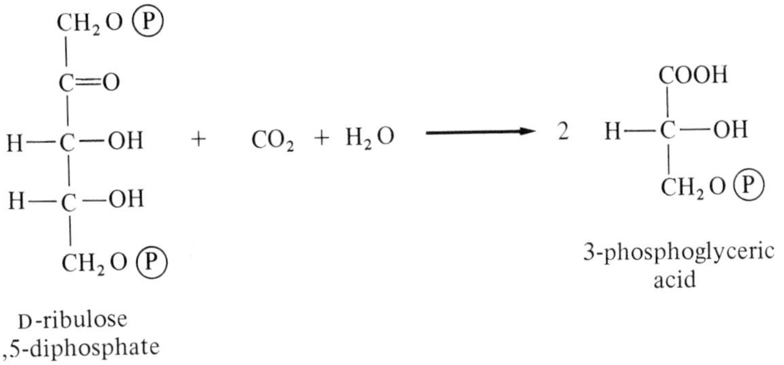

$$(7.1)$$

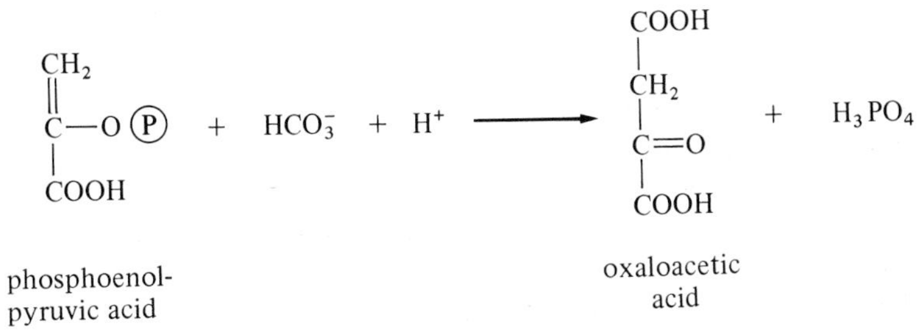

D-ribulose
1,5-diphosphate

3-phosphoglyceric
acid

$$(7.2)$$

phosphoenol-
pyruvic acid

oxaloacetic
acid

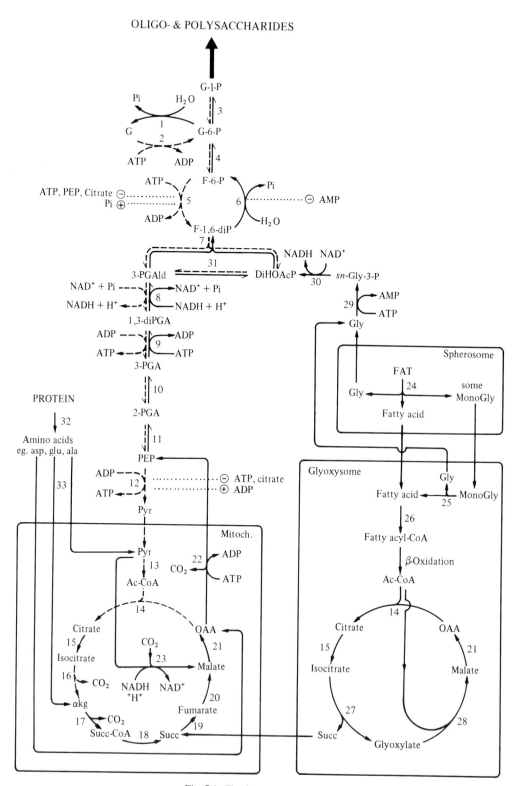

Fig. 7.1. (For legend see p. 231.)

2. Synthesis by gluconeogenesis

The term gluconeogenesis literally means the formation of glucose from new; in this context the 'new sources' of carbon for glucose synthesis do not arise directly from carbohydrate or CO_2. The most-investigated 'new source' in plants is the fat which is stored in the endosperm of some seeds (e.g. castor bean seed, *Ricinus communis*). The endosperm of the castor bean is not just a collection of dead parenchymatous cells packed with reserve material that

is broken down by enzymes secreted into it from the surrounding layer of aleurone cells, as it is in germinating cereal seeds. The cells of the castor bean endosperm, although they do not divide, are very active metabolically. They contain, in addition to the usual complement of eukaryotic intracellular organelles, spherosomes which contain the stored fat. As soon as germination of the seed begins glyoxysomes (see Chapters 3 and 8) develop and gluconeogenesis from fat commences (see Fig. 7.1). The spherosomes contain a lipase with an acid pH

Fig. 7.1 (*opposite*). Synthesis of carbohydrates by gluconeogenesis.

──────> steps in gluconeogenesis
─ ─ ─ ─> steps in carbohydrate breakdown via glycolysis and the TCA cycle
·······⊖ inhibition of enzyme
·······⊕ activation of enzyme

Abbreviations: G = D-glucose; G-1-P = D-glucose 1-phosphate; G-6-P = D-glucose 6-phosphate; F-6-P = D-fructose 6-phosphate; F-1,6-diP = D-fructose 1,6-diphosphate; 3-PGAld = 3-phosphoglyceraldehyde; DiHOAcP = dihydroxyacetone phosphate; 1,3-diPGA = 1,3-diphosphoglyceric acid; 3-PGA = 3-phosphoglyceric acid; 2-PGA = 2-phosphoglyceric acid; PEP = phosphoenol-pyruvate; Pyr = pyruvate; Ac-CoA = acetyl-coenzyme A; αkg = α-oxoglutarate; Succ-CoA = succinyl-coenzyme A; Succ = succinate; OAA = oxaloacetate; asp = asparate; glu = glutamate; ala = alanine; Gly = glycerol; *sn*-Gly-3-p = *sn*-glycerol 3-phosphate; MonoGly = monoglyceride; Mitoch = mitochondrion.

Enzymes:
1 = Glucose 6-phosphatase (EC 3.1.3.9)
2 = Hexokinase (EC 2.7.1.1)
3 = Phosphoglucomutase (EC 2.7.5.1)
4 = Phosphohexose isomerase (EC 5.3.1.9)
5 = Phosphofructokinase (EC 2.7.1.11)
6 = Fructose 1,6-diphosphatase (EC 3.1.3.11)
7 = Aldolase (EC 4.1.2.13)
8 = Glyceraldehyde 3-phosphate dehydrogenase (EC 1.2.1.12)
9 = Phosphoglycerate kinase (EC 2.7.2.3)
10 = Phosphoglyceromutase (EC 2.7.5.3)
11 = Enolase (EC 4.2.1.11)
12 = Pyruvate kinase (EC 2.7.1.40)
13 = Pyruvate dehydrogenase complex
14 = Citrate synthase (EC 4.1.3.7)
15 = Aconitate hydratase (EC 4.2.1.3)
16 = Isocitrate dehydrogenase (EC 1.1.1.41)
17 = αOxoglutarate dehydrogenase complex
18 = Succinyl-CoA hydrolase (EC 3.2.1.3)
19 = Succinate dehydrogenase (EC 1.3.99.1)
20 = Fumarate hydratase (EC 4.2.1.2)
21 = Malate dehydrogenase (EC 1.1.1.37)
22 = Phosphoenolpyruvate carboxykinase (ATP) (EC 4.1.1.49)
23 = Malic enzyme (EC 1.1.1.39)
24 = Lipase (acid pH optimum) (EC 3.1.1.3)
25 = Lipase (alkaline pH optimum) (EC 3.1.1.3)
26 = Acyl-CoA synthetases
27 = Isocitrate lyase (EC 4.1.3.1)
28 = Malate synthase (EC 4.1.3.2)
29 = Glycerol kinase (EC 2.7.1.30)
30 = Glycerol 3-phosphate dehydrogenase (EC 1.1.1.8)
31 = Triose phosphate isomerase (EC 5.3.1.1)
32 = Proteolytic enzymes
33 = Transaminases or other catabolic processes

optimum bound to the inner surface of their single outer membrane. This catalyses the hydrolysis of the fat stored within the spherosome producing free fatty acids, glycerol and some monoglycerides (monoacylglycerols). The free fatty acids and monoglycerides pass into the glyoxysome where a membrane-bound, monoglyceride-specific lipase with an alkaline pH optimum catalyses the hydrolysis of the monoglycerides as they enter the organelle. The glycerol which results from the action of these lipases passes into the cytosol where it is phosphorylated to *sn*-glycerol 3-phosphate by glycerol kinase[3]. This is then oxidized by *sn*-glycerol 3-phosphate dehydrogenase[4] to dihydroxyacetone phosphate (DiHOAcP) which can be utilized for glucose formation as will be apparent later. The free fatty acids are converted into their coenzyme A thioester derivatives by acyl-CoA synthetases within the glyoxysome. The fatty acyl-CoAs are then degraded by the β-oxidation enzymes and the resulting acetyl-CoA is converted into succinate by the operation of the glyoxylate cycle [see Chapter 8, Section C.1(iii)].

The succinate cannot be further metabolized in the glyoxysome because the relevant enzymes are not present. It therefore leaves the glyoxysome and enters the mitochondrion where it is converted into OAA via fumarate and L-malate by the appropriate TCA cycle enzymes. The OAA is then converted into phosphoenolpyruvate (PEP) in a reaction catalysed by PEP carboxykinase[5] [eq. (7.3)], whose intracellular location is uncertain. Since the affinity of PEP carboxykinase for OAA is high whereas that for CO_2 is low, the formation of PEP in reaction (7.3) is greatly favoured. PEP is then converted into D-glucose 1-phosphate (G-1-P) by a slightly modified glycolysis sequence [see Chapter 6, Section E.1 (i)] operating in reverse. In the presence of high intracellular ATP/ADP and NADH/NAD$^+$ ratios all the reactions of the glycolysis sequence from

G-1-P to PEP, save that catalysed by phosphofructokinase[6], can be driven backwards. During the conversion of fat into carbohydrate in the endosperm of the germinating castor bean the ATP/ADP and NADH/NAD$^+$ ratios are high owing to the operation of β-oxidation in the glyoxysome and the enzymes succinate dehydrogenase[7] and malate dehydrogenase[8] (reactions 19 and 21, Fig. 7.1) in the mitochondrion, all of which generate NADH or $FADH_2$. Some of the NADH and all the $FADH_2$ can be used to generate ATP by oxidative phosphorylation (see Chapter 6, Sections B and C) in the mitochondrion. In fact these processes provide more than sufficient reductant and ATP to drive not just the reaction sequence PEP→G-1-P but also the sequence OAA→sucrose in the castor bean endosperm. The phosphofructokinase-catalysed reaction of glycolysis is not reversed in gluconeogenesis since a low ATP/ADP ratio, which does not exist under these circumstances, would be required. The reaction is by-passed by the straightforward hydrolysis catalysed by fructose 1,6-diphosphatase[9] (reaction 6, Fig. 7.1).

It is apparent from Fig. 7.1 that apart from the replacement of phosphofructokinase with fructose 1,6-diphosphatase the enzymes involved in the gluconeogenesis sequence PEP to G-1-P are those of glycolysis, a process in which G-1-P is catabolized to PEP and then to pyruvate. Because of this there must be some means within the cell of controlling these two contrary processes. There are two obvious ways in which this could be accomplished. The first would involve two distinct sets of enzymes separated from one another by being in different intracellular compartments. In one compartment would be a cocktail of enzymes dedicated to glycolysis, namely all those catalysing reactions 3–12 (Fig. 7.1) with the exception of fructose 1,6-diphosphatase. In the other compartment would be essentially the same cocktail of enzymes

$$
\begin{array}{c}
\text{COOH} \\
| \\
\text{CH}_2 \\
| \\
\text{C}=\text{O} \\
| \\
\text{COOH}
\end{array}
+ \text{ATP} \rightleftharpoons \text{CO}_2 +
\begin{array}{c}
\text{CH}_2 \\
\| \\
\text{C}-\text{O}\,\textcircled{P} \\
| \\
\text{COOH}
\end{array}
+ \text{ADP} \qquad (7.3)
$$

oxaloacetic acid

phosphoenol-pyruvic acid

save that pyruvate kinase would not be present and phosphofructokinase would have been replaced by fructose 1,6-diphosphatase; this cocktail would be dedicated to gluconeogenesis. Then depending upon the requirements of the cell one or other of these two sets of enzymes would operate.

The second way would have all the enzymes mentioned above in the same compartment and control over whether glycolysis or gluconeogenesis occurred would be effected by altering the catalytic activities of the three key enzymes phosphofructokinase, pyruvate kinase[10] and fructose 1,6-diphosphatase. This could be accomplished if phosphofructokinase and pyruvate kinase were allosterically inhibited by high levels of ATP whilst fructose 1,6-diphosphatase was allosterically stimulated by them. The control could be accentuated if phosphofructokinase and pyruvate kinase were allosterically stimulated by high levels of orthophosphate, ADP or AMP whilst fructose 1,6-diphosphatase was allosterically inhibited by them.

Whichever of these methods of control was operative the result would be the same. When the ATP/ADP ratio was low glycolysis would be switched on and gluconeogenesis would be switched off. This would be logical since ATP needs to be generated when the ATP/ADP ratio is low, and glycolysis, followed by the TCA cycle, will generate the ATP whereas gluconeogenesis would utilize it. On the other hand, when the ATP/ADP ratio was high glycolysis would be switched off and gluconeogenesis would be switched on. This again is logical since there is no need to generate ATP when the ATP/ADP ratio is high.

It is not absolutely clear which of these two control mechanisms operates in the endosperm of the germinating castor bean. It has been suggested that the glycolysis enzymes are present in the cytosol, as is usual, whilst the enzymes of gluconeogenesis are located in the plastids; against this, there is some evidence of allosteric control of the key enzymes. Such allosteric control has been demonstrated in a number of other plant tissues known to carry out gluconeogenesis. In general phosphofructokinase is inhibited by high levels of ATP, PEP and citrate and is stimulated (or the inhibition removed) by high levels of orthophosphate and sometimes D-fructose 6-phosphate (F-6-P); pyruvate kinase is inhibited by high levels of ATP and citrate and stimulated by ADP or orthophosphate and fructose 1,6-diphosphatase is inhibited by AMP.

Returning to the castor bean endosperm, it is clear that the G-1-P generated by gluconeogenesis is converted to sucrose within the cells of the endosperm by the process described in Section C.1(i). The sucrose passes to the developing castor bean seedling where it supports growth and development until photosynthesis can take over.

As can be seen from Fig. 7.1 gluconeogenesis from protein can also occur. Many of the amino acids produced when proteins are enzymically hydrolysed can be catabolized to pyruvate or compounds which can enter the TCA cycle. Obvious examples are alanine, aspartic acid and glutamic acid. Pyruvate, derived from alanine, is converted into L-malic acid by the NAD-malic enzyme[11] within the mitochondrion [eq. (7.4)]. The malic acid is then converted into OAA by the operation of the TCA cycle. Glutamic acid enters the TCA cycle after conversion into α-oxoglutarate and then follows the cycle round to yield OAA. Aspartic acid is converted into OAA directly. The OAA derived from the amino acids is then converted into G-1-P by the route described previously.

Although gluconeogenesis from protein amino acids undoubtedly does occur in plant tissues it probably has little quantitative significance.

$$
\begin{array}{c}
CH_3 \\
| \\
C{=}O \\
| \\
COOH
\end{array}
\;+\; CO_2 \;+\; NADH \;+\; H^+ \;\rightleftharpoons\;
\begin{array}{c}
COOH \\
| \\
CH_2 \\
| \\
H{-}C{-}OH \\
| \\
COOH
\end{array}
\;+\; NAD^+ \qquad (7.4)
$$

pyruvic acid   L-malic acid

3. Monosaccharide derivatives and interconversions

The monosaccharides, formed by photosynthesis (see Section B.1 and Chapter 5) or by gluconeogenesis (Section B.2), can be converted into other monosaccharides; indeed monosaccharides are able to undergo a considerable number of interconversions. These interconversions are important in sugar breakdown processes because all monosaccharides have to be converted into glucose, fructose or one of their phosphate esters before they can enter the glycolysis pathway. They are also important in forming the monosaccharide precursors of the oligo- and polysaccharides. The reactants in the monosaccharide interconversions are usually monosaccharide derivatives although some interconversions of free monosaccharides do occur; the latter are, however, more common in the lower forms of plant life. Two types of derivatives are particularly important in these interconversions, phosphate esters and nucleoside diphosphate (NDP) esters.

(i) FORMATION OF MONOSACCHARIDE DERIVATIVES

(a) *Monosaccharide phosphates*

Monosaccharide phosphates inevitably arise during photosynthesis and gluconeogenesis. However, free monosaccharides such as those produced during starch degradation in germinating seeds can be converted into their phosphate esters by a reaction [e.g. eq. (7.5)], catalysed by kinases, in which the phosphate group is almost always introduced into a primary alcohol group. Exceptions are L-arabinose (X), D-glucuronic acid (V) and D-galacturonic acid (VI), which are converted into their 1-phosphates.

Although ketopentoses (e.g. D-xylulose [XIII]) and ketohexoses (e.g. D-fructose [VIII]) have two primary alcohol groups, only that on the highest numbered carbon (i.e. C–5 or C–6) is phosphorylated by plant kinases.

The most widespread kinase in plants is hexokinase[12] which has a fairly low specificity towards its monosaccharide substrate, being able to catalyse the phosphorylation of D-glucose (I) (eq. 7.5), D-fructose (VII), D-mannose (III) and D-glucosamine (IV); it is, however, absolutely specific towards the phosphorylating agent, ATP, as are all kinases. Many kinases are specific for particular monosaccharides. Examples which catalyse the initial phosphorylation of monosaccharides in plants are D-glucuronokinase[13], D-galactokinase, D-galacturonokinase[14], D-xylulokinase[15], L-arabokinase[16] and D-fructokinase[17].'

Two examples of kinases catalysing the further phosphorylation of already-phosphorylated monosaccharides are phosphofructokinase[6] and phosphoribulokinase[18].

All kinases are activated by Mg^{2+} ions.

(b) *Nucleoside diphosphate monosaccharides (NDP-sugars)*

A nucleoside diphosphate monosaccharide is a compound in which a sugar is attached by a glycosidic linkage to the terminal phosphate of a nucleoside diphosphate, e.g. uridine diphosphate glucose (UDP-G or UDP-D-glucose) [eq. (7.6)]. Many such compounds are known in which a considerable range of monosaccharides or their uronic acid derivatives are most commonly linked to uridine-, cytosine-, adenosine- and guanosine diphosphates (UDP, CDP, ADP or GDP).

α-D-glucose

$(\Delta G^{\circ\prime} = -16.7 \text{ kJ mol}^{-1})$

α-D-glucose
6-phosphate

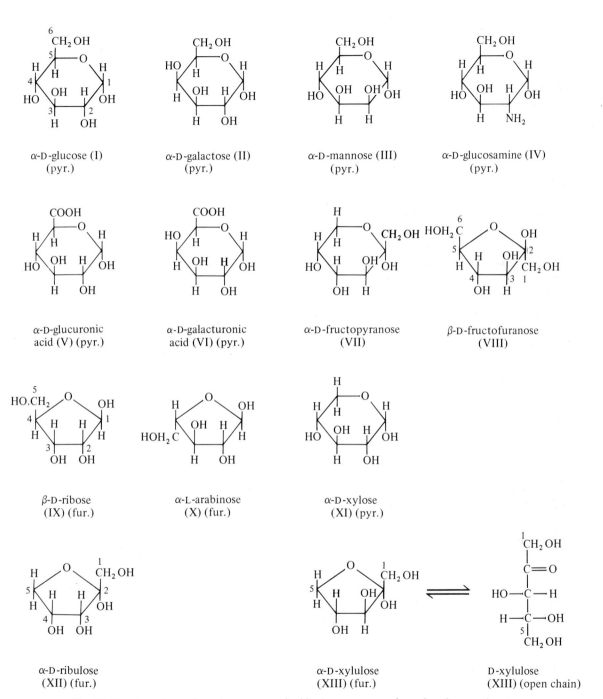

Fig. 7.2. Structure of some important monosaccharides (pyr. = pyranose form; fur = furanose form).

Uridine diphosphate glucose
(UDPG or UDP-D-glucose)

$$(7.6)$$

The NDP-sugars are involved not only in monosaccharide interconversions but also in oligo- and polysaccharide biosynthesis (see Sections C and D). They are formed, not from free monosaccharides, but from their 1-phosphates according to the general eq. (7.7) in which the eliminated pyrophosphate is derived from the terminal two phosphate moieties of the nucleoside triphosphate (NTP).

$$\text{Glycose-1-P} + \overset{*\ *}{\text{P-P}}\text{-P-R-N} \rightleftharpoons$$
$$\text{NTP}$$

$$\text{Glycose-P-P-R-N} + \overset{*\ *}{\text{P-P}} \quad (7.7)$$
$$\text{NDP-sugar} \qquad \text{pyro-}$$
$$\text{phosphate}$$

The generalized name given to the enzymes catalysing reaction (7.7) is NTP:glycose 1-phosphate nucleotidyltransferase; if, for example, NTP were UTP and glycose 1-phosphate were D-glucose 1-phosphate the systematic name of the enzyme would be UTP:D-glucose 1-phosphate uridylyltransferase (EC 2.7.7.9). These enzymes are, however, frequently given the trivial name 'pyrophosphorylase', e.g. UDP-glucose pyrophosphorylase.

The equilibrium constant of reaction (7.7) is about unity indicating that the reaction is freely reversible. *In vivo* the reaction is pulled in the direction of UDP-G synthesis by the continuous removal of the pyrophosphate under the influence of pyrophosphatase[19] which catalyses its exergonic ($\Delta G^{\circ\prime} = -33.5\ \text{kJ mol}^{-1}$) hydrolysis to orthophosphate. The $\Delta G^{\circ\prime}$ of hydrolysis of UDP-G ($\rightarrow$UDP+G) is $-33.5\ \text{kJ mol}^{-1}$ in contrast to $-20.9\ \text{kJ mol}^{-1}$ for that of the G-1-P from which it was formed; UDP-G is thus more suited than G-1-P (or G-6-P, $\Delta G^{\circ\prime}$ of hydrolysis = $-13.8\ \text{kJ mol}^{-1}$) to be a glycosyl donor in oligo- and polysaccharide

biosynthesis. The same argument applies to all NDP-sugars.

It will be noted that the substrate for the pyrophosphorylases is the glycose 1-phosphate rather than the glycose 6-phosphate which is the usual product of the kinase-catalysed reactions [see Section B.3(i)(a)]. The necessary internal repositioning of the phosphate group is brought about by the catalytic activity of a group of enzymes known as phosphoglycomutases [see Section B.3(ii)(a)]. Thus the formation of UDP-G, for example, from D-glucose is a four-step process [eqs. (7.8), (7.9), (7.10) and (7.11)].

$$\text{D-glucose} + \text{ATP} \rightleftharpoons \text{D-glucose 6-phosphate} + \text{ADP}$$
$$\text{(hexokinase)} \qquad\qquad\qquad\qquad (7.8)$$

$$\text{D-glucose 6-phosphate} \rightleftharpoons \text{D-glucose 1-phosphate}$$
$$\text{(phosphoglucomutase)} \qquad\qquad\qquad (7.9)$$

$$\text{D-glucose 1-phosphate} + \text{UTP} \rightleftharpoons$$
$$\text{UDP-D-glucose} + \text{PP}_i \quad (7.10)$$
$$\text{(UTP: D-glucose 1-phosphate uridylyltransferase)}$$

$$\text{Pyrophosphate} + \text{H}_2\text{O} \rightarrow 2 \text{ orthophosphate} \quad (7.11)$$
$$\text{(pyrophosphatase)}$$

Although monosaccharide 1-phosphates are by far the most common substrates of the pyrophosphorylase enzymes, a monosaccharide uronic acid 1-phosphate, D-glucuronic acid 1-phosphate, is the substrate of UTP:D-glucuronic acid uridylyltransferase which catalyses the formation of UDP-D-glucuronate in mung bean and barley seedlings.

A number of enzymes catalysing the phosphorolysis of NDP-sugars have been detected in plant tissues. For instance, an enzyme from wheat germ breaks down ADP-D-glucose into ADP and G-1-P in an essentially irreversible reaction in which orthophosphate is incorporated into the terminal position in ADP [eq. (7.12)].

$$\text{A-R-P-P-G} + \overset{*}{\text{P}} \rightarrow \text{A-R-P-}\overset{*}{\text{P}} + \text{G-1-P} \quad (7.12)$$

This enzyme also catalyses the phosphorolysis of ADP-D-mannose but is less effective with other ADP-sugars and inactive with UDP-D-glucose and GDP-D-mannose. A similar enzyme catalysing the phosphorolysis of GDP-D-mannose and UDP-D-glucose has been found in yeast.

(c) Glycosides

A glycoside is formed when the hydroxyl group on the anomeric carbon atom (i.e. C–1 of aldoses, C–2 of ketoses) of the pyranose or fructose form of a monosaccharide is replaced by a molecule or grouping possessing a nucleophilic atom. The nucleophilic atom is most commonly oxygen ($>$O:) present in an alcohol, phenol or an acid; this gives rise to O-glycosides in which the linkage between the monosaccharide and the aglycone is via an oxygen atom (i.e. C–O–C). When the nucleophilic atom is sulphur ($>$S:, e.g. in compounds with sulphydryl groups) or nitrogen ($>$N:, e.g. in amines), S-glycosides and N-glycosides are formed in which the monosaccharide–aglycone linkages are C–S–C and C–N–C respectively. Less frequently the nucleophilic atom is carbon (i.e. a carbanion); this gives rise to the C-glycosides in which the monosaccharide–aglycone linkage is C–C. The C–C linkage of the C-glycosides differs from the monosaccharide–aglycone linkage of the O-, S- and N-glycosides in not being susceptible to acid-, alkali- or enzyme-catalysed hydrolysis.

Although many different monosaccharides are found as glycosides in plants, D-glucose is the most common. Several glycosides are known in which the monosaccharide moiety is a monosaccharide de-

rivative (e.g. D-glucuronic acid) rather than a true monosaccharide. Some glycosides have more than one monosaccharide moiety. These can be divided into two classes: those in which the monosaccharide moieties are linked to each other forming a single oligosaccharide chain and those in which the aglycone has more than one nucleophilic atom (e.g. hydroxyl groups) and can therefore form glycosidic linkages with more than one mono- or oligosaccharide.

It should be noted that aldose 1-phosphates are a metabolically important type of O-glycoside. Moreover all oligo- and polysaccharides contain the C–O–C type of glycosidic linkage; they can be regarded as a special type of O-glycoside in which the aglycone moiety is a saccharide (if one excuses the contradiction implicit in the terms 'aglycone' and 'saccharide'). Nucleosides such as adenosine are metabolically important N-glycosides. These glycosides apart, the function of glycosides in plants is obscure. However, some of them have a marked physiological effect in animals, for example the cardiac glycosides (e.g. digitonin, XIV) which stimulate the heart muscle to contract more powerfully.

Two interesting groups of plant glycosides whose metabolism has been investigated are the cyanogenic glycosides and the glucosinolates (mustard oil glucosides). The cyanogenic glycosides are capable of liberating hydrogen cyanide when acted upon by certain enzymes. They are widely distributed in the

D-Xyl$_p$ (β1 → 3) D-Glc (β1 → 4) D-Gal (β1 →) O

$$\begin{array}{c} 2 \\ \uparrow \\ \beta 1 \end{array}$$

D-Glc (β1 → 3) D-Gal

Digitonin (D-Xyl$_p$ = D-xylopyranose; D-Gal = D-galactose; D-Glc = D-glucose) (XIV)

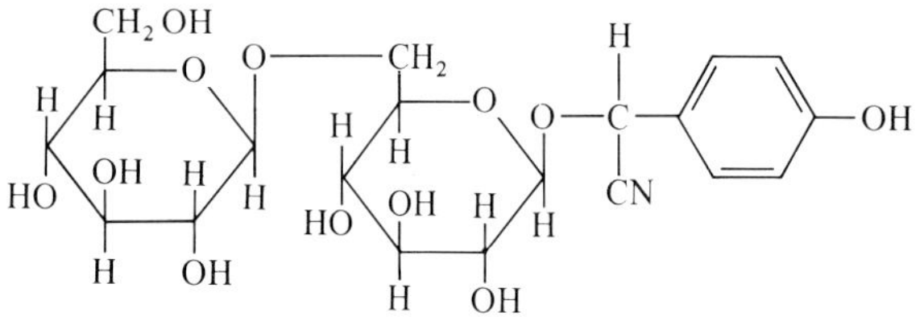

Amygdalin (XV)

plant kingdom and often occur in seeds (e.g. amygdalin, XV in *Prunus amygdalus* seeds). They have the general structure

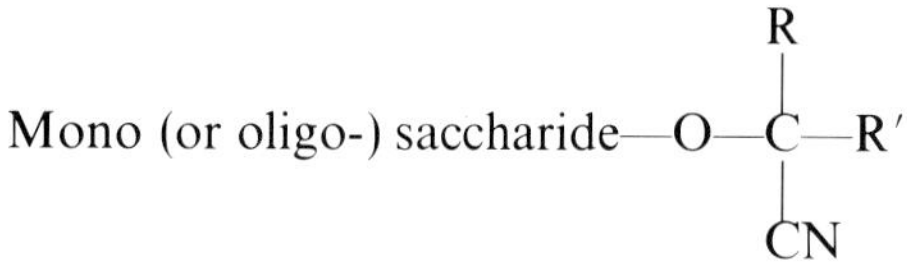

$$\text{Mono (or oligo-) saccharide} - O - \underset{\underset{CN}{|}}{\overset{\overset{R}{|}}{C}} - R'$$

The aglycone moiety is derived from an amino acid, the carboxyl group of which is removed by decarboxylation. The cyano group ($-C\equiv N$) is formed from the α-carbon atom and the amino nitrogen atom of the amino acid. Although the details of the biosynthesis of these compounds is by no means clear, it seems likely that an aldoxime is formed by decarboxylation of the N-hydroxy amino acid which is then dehydrated to the corresponding nitrile. The nitrile then appears to be converted into its cyanohydrin which is then glycosylated. The glycosylating agent has been shown to be a NDP-sugar in some cases. The outline of the biosynthetic pathway of dhurrin, a cyanogenic glycoside present in *Sorghum vulgare*, is shown in Fig. 7.3.

The catabolism of cyanogenic glycosides in plants is shown in Fig. 7.4 and involves initial hydrolysis of the glycosidic linkage by a glycosidase followed by oxidation of the resulting cyanohydrin to the corresponding aldehyde or ketone and HCN. Whilst HCN is very toxic to most organisms (see Chapter 6), plants, especially those containing cyanogenic glycosides, are not only tolerant of it but also utilize it. For instance, higher plants may catalyse the formation of β-cyanoalanine from HCN and L-serine or D-cysteine; the

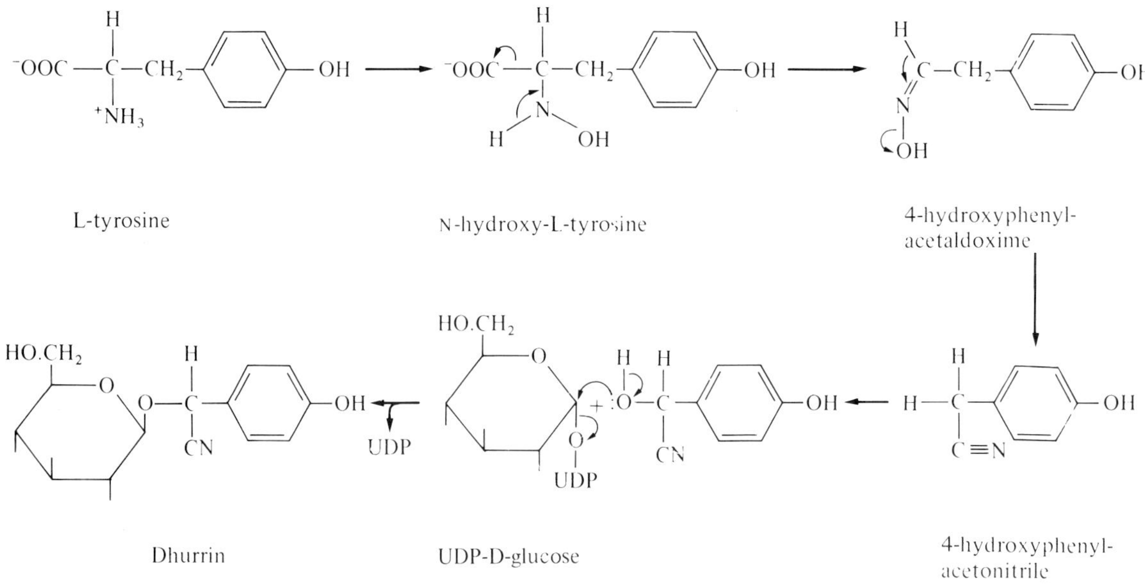

L-tyrosine   N-hydroxy-L-tyrosine   4-hydroxyphenyl-acetaldoxime

Dhurrin   UDP-D-glucose   4-hydroxyphenyl-acetonitrile

Fig. 7.3. Biosynthesis of the cyanogenic glycoside, dhurrin.

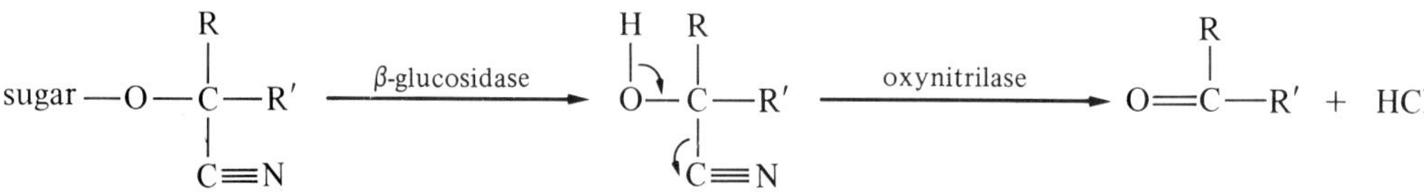

Fig. 7.4. Catabolism of cyanogenic glycosides.

β-cyanoalanine may then be converted into L-asparagine or other compounds. In *Vicia* spp., however, β-cyanoalanine accumulates.

The glucosinolates are found in some plants of the Cruciferae, Capparidaceae, Moringaceae, Resedaceae, Euphorbaceae, Phytolaccaceae and Tropaeolaceae. When the leaves of these plants are rubbed the characteristic smell of mustard oils which is obtained results from the enzymic degradation of glucosinolates. The general structure of glucosinolates is

where X^+ is usually K^+ and R is a moiety derived from an amino acid. The details of the biosynthesis of these compounds has not been completely elucidated. However, the biosynthesis of benzylglucosinolate in the leaves of *Tropaeolium magus* appears to follow the route shown in Fig. 7.5; once again the glycosylating agent is an NDP-sugar, UDP-D-glucose.

The conversion of glucosinolates into mustard oils in plants is catalysed by the enzyme thioglucosidase[20]. This enzyme is located in special cells from which it is released when the leaf tissue is damaged (e.g. when it is rubbed). Some plants also contain an enzyme which catalyses the isomerization of the mustard oils, which are isothiocyanates, into rhodanids, which are thiocyanates (see Fig. 7.6).

(ii) MONOSACCHARIDE INTERCONVERSIONS

The interconversions which monosaccharide derivatives (and in some cases the free monosaccharides) undergo can be divided into three broad classes: (a) internal rearrangements; (b) transfer of C_2 and C_3 units; (c) oxido-reduction reactions.

(a) *Internal rearrangements*

Three classes of internal rearrangements can be grouped under this heading.

Mutation. This first class, involving the repositioning of the phosphate group in the monosaccharide phosphate molecule, has already been referred to in Section B.3(i)(b). This type of reaction is catalysed by mutases, of which phosphoglucomutase[21] is the best known example. Phosphoglucomutase, which has been purified from *Phaseolus radiatus* and potatoes, catalyses the interconversion of α-D-glucose 1-phosphate and α-D-glucose 6-phosphate [eq. (7.13)]. At equilibrium at 30°C the reaction mixture contains 95%

α-D-glucose 1-phosphate α-D-glucose 6-phosphate
$$(\Delta G^{\circ\prime} = -7\cdot4 \text{ kJ mol}^{-1})$$

G-6-P. The mechanism of the reaction has been investigated independently by Leloir and Najjar. Leloir showed that α-D-glucose 1,6-diphosphate was required as a cofactor and Najjar showed that phosphoglucomutase is a phosphoenzyme, the phosphate of which is attached to a serine residue and can be reversibly transferred to either the hydroxyl group at C–1 or C–6 of D-glucose. The phosphoenzyme is relatively unstable and subject to hydrolysis to the dephosphoenzyme and orthophosphate ($\Delta G^{\circ\prime} = -16.4$ kJ mol^{-1}). Because of this a phosphoglucokinase[21] catalyses the formation of G-1,6-diP which is then available for the rephosphorylation of the enzyme.

Other mutases have been found which catalyse similar reactions with the phosphates of D-ribose, D-mannose and D-galactose [eqs. (7.14), (7.15), (7.16)]; catalytic amounts of the diphosphates are required in each case, indicating that the reaction mechanism is the same as that of the phosphoglucomutase-catalysed reaction.

D-ribose 1-phosphate$\rightleftharpoons$D-ribose 5-phosphate
$$(7.14)$$

D-mannose 1-phosphate$\rightleftharpoons$
         D-mannose 6-phosphate   (7.15)

D-galactose 1-phosphate$\rightleftharpoons$
         D-galactose 6-phosphate   (7.16)

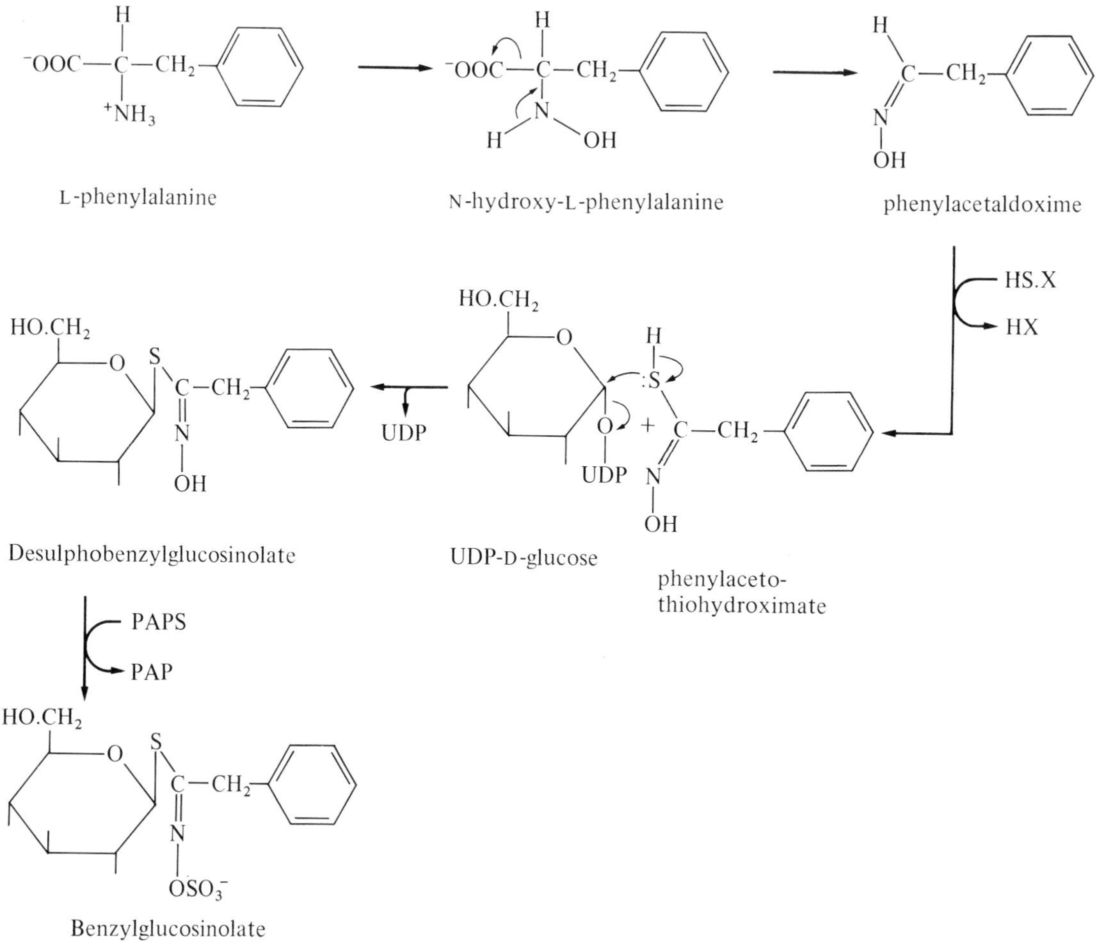
L-phenylalanine
N-hydroxy-L-phenylalanine
phenylacetaldoxime
HS.X
HX
Desulphobenzylglucosinolate
UDP
UDP-D-glucose
phenylaceto-
thiohydroximate
PAPS
PAP
Benzylglucosinolate

Fig. 7.5. Biosynthesis of the mustard oil glycoside, benzylglucosinolate (PAPS = 3′-phosphoadenosine-5′-phosphosulphate).

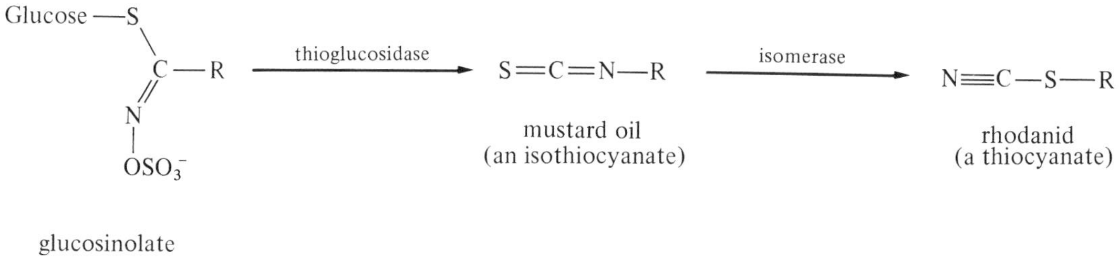
Glucose
thioglucosidase
isomerase
mustard oil
(an isothiocyanate)
rhodanid
(a thiocyanate)
glucosinolate

Fig. 7.6. Catabolism of glucosinolates (mustard oil glycosides).

Epimerization. The second class of internal rearrangements are epimerizations which involve the inversion of the configuration at one of the asymmetric carbon atoms of the monosaccharide. These reactions are catalysed by enzymes known as epimerases, some of which have monosaccharide phosphates as their substrate whilst others have UDP derivatives. An epimerase which has a monosaccharide phosphate as its substrate is ribulose phosphate 3-epimerase[22] which occurs widely in Nature and which participates in the operation of the Calvin cycle (see Chapter 5, Section D) and the

$$
\begin{array}{c}
CH_2OH \\
| \\
C=O \\
| \\
H-C-OH \\
| \\
H-C-OH \\
| \\
CH_2O\,\text{\textcircled{P}}
\end{array}
\quad\rightleftharpoons\quad
\begin{array}{c}
CH_2OH \\
| \\
C=O \\
| \\
HO-C-H \\
| \\
H-C-OH \\
| \\
CH_2O\,\text{\textcircled{P}}
\end{array}
\qquad (7.17)
$$

D-ribulose 5-phosphate D-xylulose 5-phosphate

oxidative pentose phosphate pathway (see Chapter 6, Section F.3).

Epimerases catalysing the epimerization of UDP-sugars are typified by UDP-glucose 4-epimerase[23], UDP-arabinose 4-epimerase[24] and UDP-glucuronate 4-epimerase[25], which catalyse eqs. (7.18), (7.19) and (7.20) respectively. Notice that UDP-arabinose 4-epimerase catalyses the interconversion of a D- and an L-sugar.

UDP-glucose 4-epimerase contains one molecule of tightly bound NAD per molecule of enzyme. This

NAD undergoes oxidation and reduction during reaction (7.18). A study of this reaction using the yeast and E. coli enzymes indicates that the mechanism is that shown in Fig. 7.7; it is likely that this is also the mechanism in plants.

Aldose–ketose interconversions. Thirdly, there are isomerases which catalyse aldose–ketose interconversions. The substrates for isomerases in plants are usually monosaccharides which have their primary hydroxyl group phosphorylated. Three isomerases which are widespread in plants are glucose 6-phosphate isomerase[26], ribose 5-phosphate isomerase[27] and triose phosphate isomerase[28] which catalyse eqs. (7.21), (7.22) and (7.23) respectively. The mechanism of these reactions involves the intermediate formation of an enediol; when G-6-P is isomerized by glucose 6-phosphate isomerase in the presence of deuterated water a deuterium atom enters the resulting F-6-P at C–1, a result which is consistent with an enediol mechanism.

UDP-D-glucose ⇌ UDP-D-galactose   (7.18)

UDP-L-arabinose ⇌ UDP-D-xylose   (7.19)

UDP-D-glucuronic acid ⇌ UDP-D-galacturonic acid   (7.20)

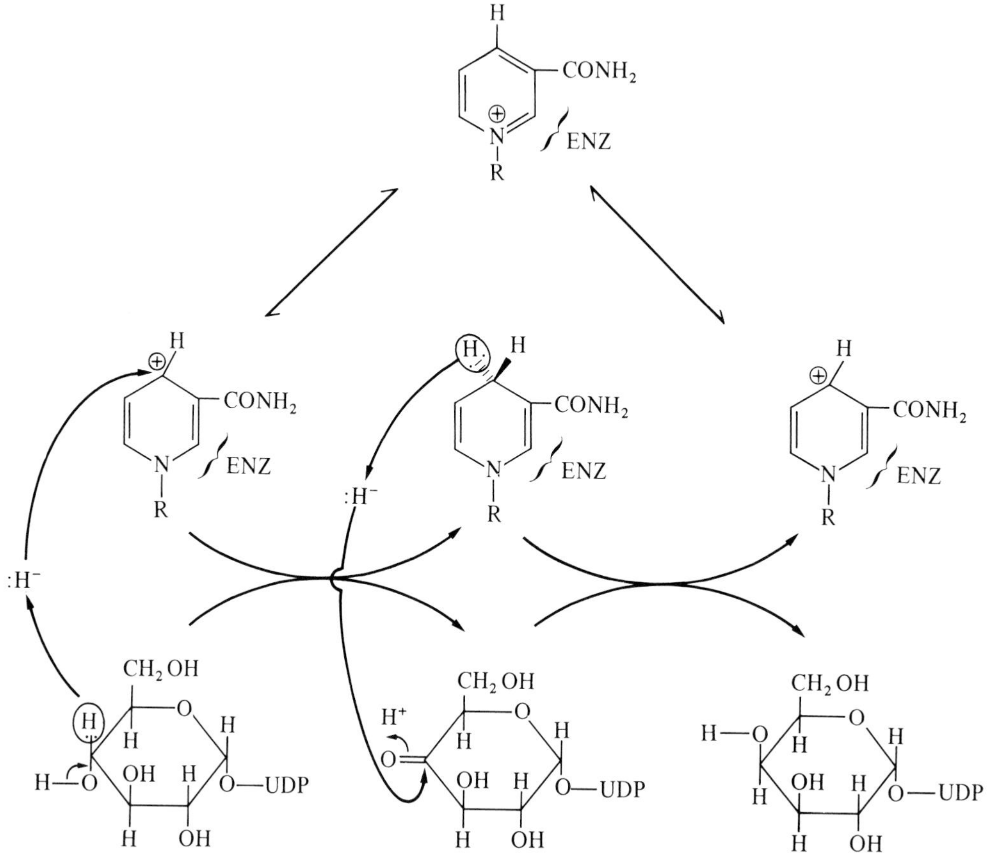

Fig. 7.7. Mechanism of reaction catalysed by UDP glucose 4-epimerase (the reaction is shown proceeding from left to right; it is, however, reversible; $\leftharpoonup$ = resonance).

(b) Transfer of C_2 and C_3 units

Two important enzymes which are of widespread occurrence, transketolase and transaldolase, catalyse the transfer of C_2 and C_3 units respectively from one monosaccharide to another.

Transketolase. Transketolase[29] catalyses the transfer of a ketol moiety $(CH_2OH \cdot CO \cdot)$ from a ketose to an aldose [eq. (7.24)]. The products of the reaction are an aldose derived from, but two carbon atoms shorter than, the donor ketose and a ketose derived from, but two carbon atoms longer than, the acceptor aldose. The donor ketose is a D-sugar with a $3S,4R$ ($\equiv threo$) configuration, as is the ketose resulting from the reaction. Transketolase requires

thiamine pyrophosphate (TPP) and Mg^{2+} for activity. The TPP acts as a go-between in the reaction, receiving the ketol group from the ketose donor and handing it on to the recipient aldose; the ketol group is transiently covalently bound via its carbonyl carbon atom to C–2 of the thiazole ring of TPP during the reaction (see Fig. 7.8). The reaction is freely reversible since the product ketose has all the structural requirements of a donor ketose. D-Xylulose 5-phosphate, D-fructose 6-phosphate and D-sedoheptulose 7-phosphate can act as ketol donors whilst D-ribose 5-phosphate, D-glyceraldehyde 3-phosphate and D-erythrose 4-phosphate can act as acceptor aldoses. Transketolase plays an important role in the Calvin cycle (see Chapter 5, Section D.2) and the oxidative pentose phosphate pathway (see Chapter 6, Section F.3).

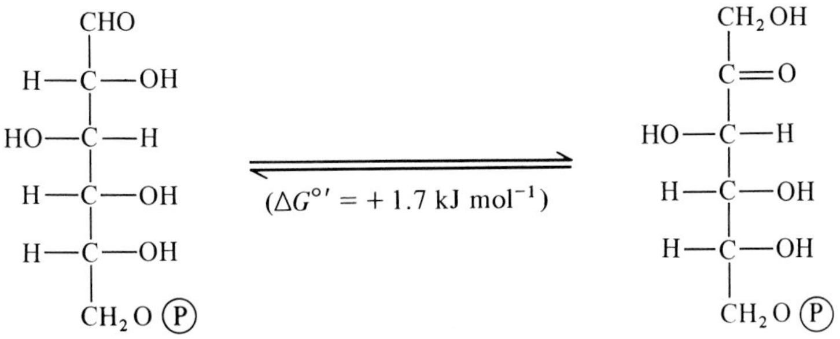

$$
\begin{array}{ccc}
\text{CHO} & & \text{CH}_2\text{OH} \\
\text{H—C—OH} & & \text{C}{=}\text{O} \\
\text{HO—C—H} & & \text{HO—C—H} \\
\text{H—C—OH} & \underset{(\Delta G^{\circ\prime} = +1.7 \text{ kJ mol}^{-1})}{\rightleftharpoons} & \text{H—C—OH} \\
\text{H—C—OH} & & \text{H—C—OH} \\
\text{CH}_2\text{O} \; \text{P} & & \text{CH}_2\text{O} \; \text{P}
\end{array}
\qquad (7.21)
$$

D-glucose 6-phosphate 	 D-fructose 6-phosphate

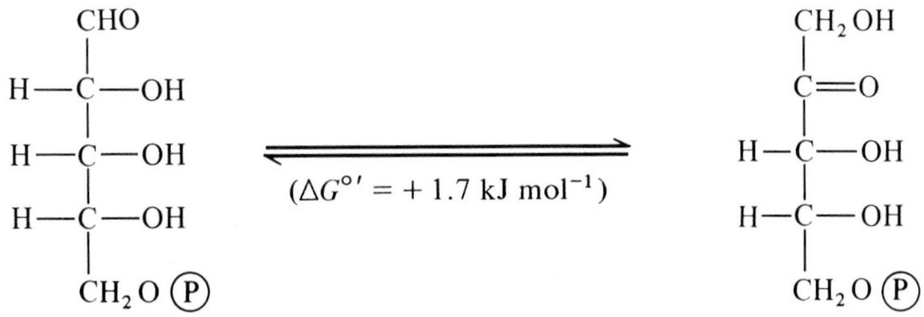

$$
\begin{array}{ccc}
\text{CHO} & & \text{CH}_2\text{OH} \\
\text{H—C—OH} & & \text{C}{=}\text{O} \\
\text{H—C—OH} & \underset{(\Delta G^{\circ\prime} = +1.7 \text{ kJ mol}^{-1})}{\rightleftharpoons} & \text{H—C—OH} \\
\text{H—C—OH} & & \text{H—C—OH} \\
\text{CH}_2\text{O} \; \text{P} & & \text{CH}_2\text{O} \; \text{P}
\end{array}
\qquad (7.22)
$$

D-ribose 5-phosphate 	 D-ribulose 5-phosphate

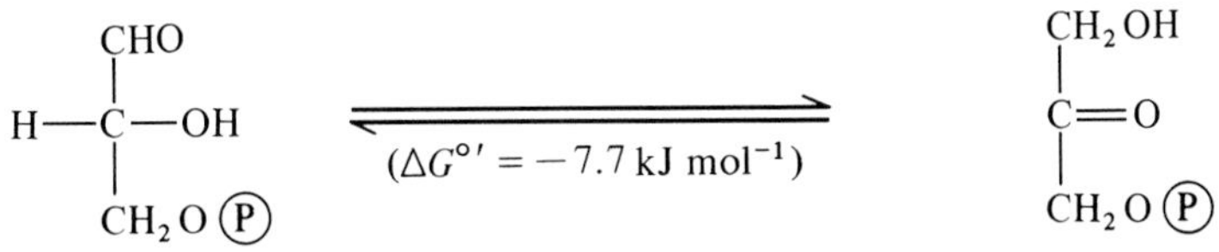

$$
\begin{array}{ccc}
\text{CHO} & & \text{CH}_2\text{OH} \\
\text{H—C—OH} & \underset{(\Delta G^{\circ\prime} = -7.7 \text{ kJ mol}^{-1})}{\rightleftharpoons} & \text{C}{=}\text{O} \\
\text{CH}_2\text{O} \; \text{P} & & \text{CH}_2\text{O} \; \text{P}
\end{array}
\qquad (7.23)
$$

D-glyceraldehyde 	 Dihydroxyacetone
3-phosphate 	 phosphate

ketol $\left\{ \begin{array}{l} \end{array} \right.$

3S.4R $\left\{ \begin{array}{l} \end{array} \right.$

D-sugar $\left\{ \begin{array}{l} \end{array} \right.$

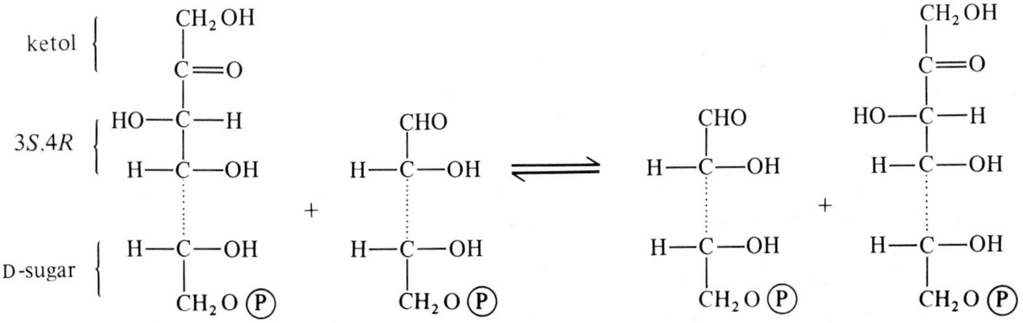

$$
\begin{array}{ccccc}
\text{CH}_2\text{OH} & & & & \text{CH}_2\text{OH} \\
\text{C}{=}\text{O} & & & & \text{C}{=}\text{O} \\
\text{HO—C—H} & \text{CHO} & & \text{CHO} & \text{HO—C—H} \\
\text{H—C—OH} & \text{H—C—OH} & \rightleftharpoons & \text{H—C—OH} & \text{H—C—OH} \\
& + & & & + \\
\text{H—C—OH} & \text{H—C—OH} & & \text{H—C—OH} & \text{H—C—OH} \\
\text{CH}_2\text{O} \; \text{P} & \text{CH}_2\text{O} \; \text{P} & & \text{CH}_2\text{O} \; \text{P} & \text{CH}_2\text{O} \; \text{P}
\end{array}
\qquad (7.24)
$$

ketol 	 ketol
donor 	 acceptor
(a D-ketose 	 (a D-aldose
phosphate) 	 phosphate)

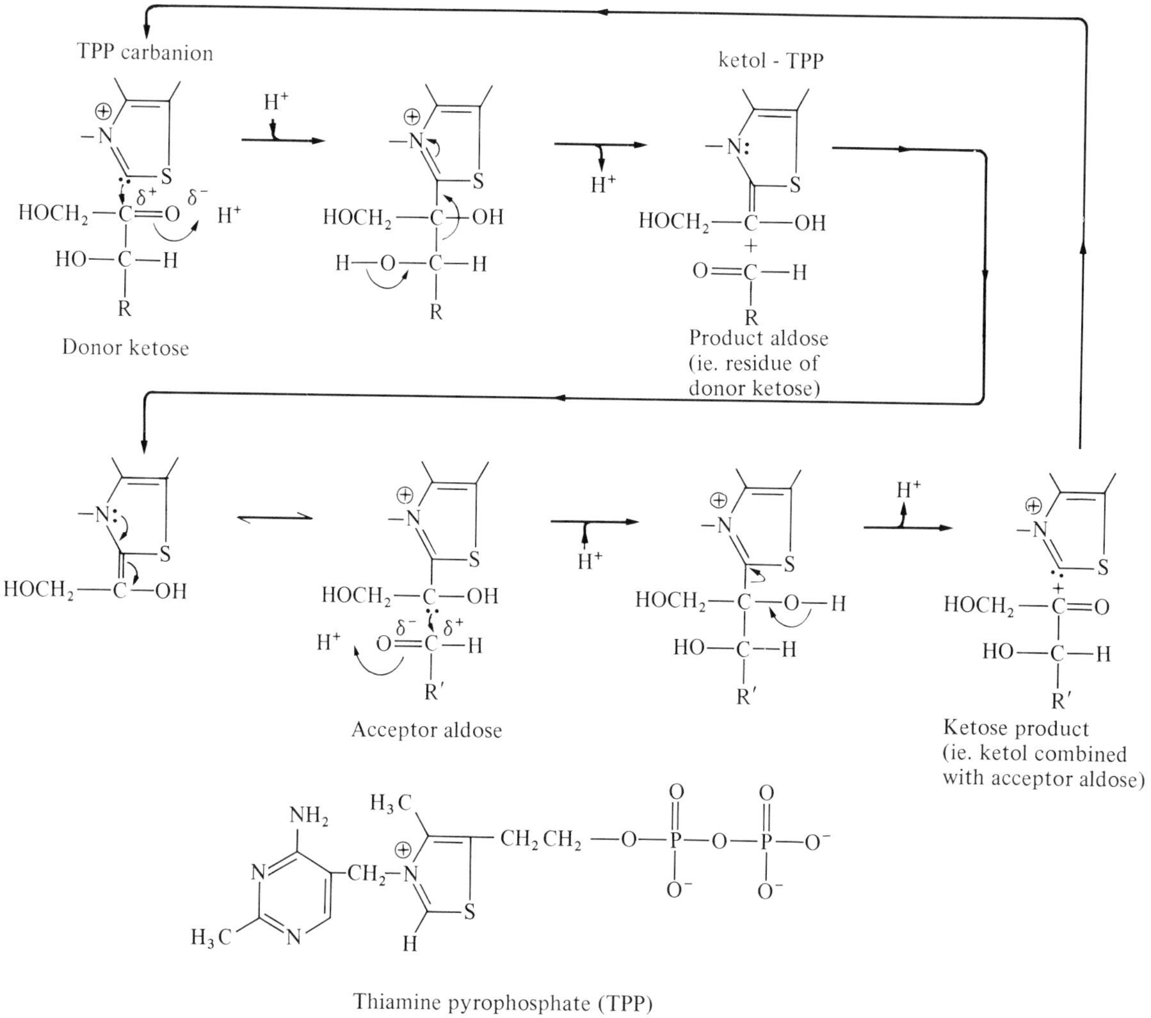

TPP carbanion   ketol - TPP

$HOCH_2$—C=O $\quad$ H$^+$ $\qquad$ $HOCH_2$—C—OH $\qquad$ $HOCH_2$—C—OH

HO—C—H $\qquad\qquad$ H—O—C—H $\qquad\qquad\qquad$ O=C—H

R $\qquad\qquad\qquad\qquad$ R $\qquad\qquad\qquad\qquad$ R

Donor ketose

Product aldose
(ie. residue of
donor ketose)

$HOCH_2$—C—OH $\qquad$ $HOCH_2$—C—OH $\qquad$ $HOCH_2$—C—O—H $\qquad$ $HOCH_2$—C=O

$\qquad\qquad$ H$^+$ $\quad$ O=C—H $\qquad\qquad$ HO—C—H $\qquad\qquad$ HO—C—H

$\qquad\qquad\qquad$ R' $\qquad\qquad\qquad$ R' $\qquad\qquad\qquad$ R'

Acceptor aldose

Ketose product
(ie. ketol combined
with acceptor aldose)

$$NH_2 \quad H_3C \qquad\qquad\qquad O \qquad O$$

$$\cdots CH_2 CH_2 —O—P—O—P—O^-$$

$$H_3C \qquad\qquad H \qquad\qquad O^- \qquad O^-$$

Thiamine pyrophosphate (TPP)

Fig. 7.8. Participation of thiamine pyrophosphate in the reaction catalysed by transketolase.

Transaldolase. Transaldolase[30] catalyses the transfer of a dihydroxyacetone residue ($CH_2OH \cdot CO \cdot CHOH \cdot$) from a ketose to an aldose [eq. (7.25)]. The products of the reaction are an aldose derived from, but three carbon atoms shorter than, the donor ketose, and a ketose derived from, but three carbon atoms longer than, the acceptor aldose. The donor ketose is a D-sugar with a 3S,4R ($\equiv$*threo*) configuration, as is the ketose resulting from the reaction. Unlike transketolase, transaldolase does not utilize a coenzyme. The enzymic mechanism involves the formation of an intermediate Schiff base between the carbonyl of the dihydroxyacetone moiety and the ε-amino group of

a lysine residue at the active site of the enzyme. Transaldolase cannot use free dihydroxyacetone or its phosphate as substrates. The reaction is freely reversible since the product ketose has all the structural requirements of a donor ketose.

D-Fructose 6-phosphate and D-sedoheptulose 7-phosphate can act as dihydroxyacetone donors and D-erythrose 4-phosphate, D-glyceraldehyde 3-phosphate and D-ribose 5-phosphate may act as acceptor aldoses. Transaldolase functions in the oxidative pentose phosphate pathway (see Chapter 6, Section F.3).

An enzyme of equal importance to transketolase and transaldolase is aldolase[31] which can be

Dihydroxy-
acetone

3S,4R

D-sugar

$$
\begin{array}{ccccc}
\text{CH}_2\text{OH} & & & & \text{CH}_2\text{OH} \\
| & & & & | \\
\text{C}=\text{O} & & & & \text{C}=\text{O} \\
| & & & & | \\
\text{HO}-\text{C}-\text{H} & \text{CHO} & & & \text{HO}-\text{C}-\text{H} \\
| & | & & & | \\
\text{H}-\text{C}-\text{OH} & \text{H}-\text{C}-\text{OH} \rightleftharpoons \text{CHO} & & \text{H}-\text{C}-\text{OH} \\
+ & + & & + & \\
\text{H}-\text{C}-\text{OH} & \text{H}-\text{C}-\text{OH} & \text{H}-\text{C}-\text{OH} & \text{H}-\text{C}-\text{OH} \\
| & | & | & | \\
\text{CH}_2\text{O}\,\text{P} & \text{CH}_2\text{O}\,\text{P} & \text{CH}_2\text{O}\,\text{P} & \text{CH}_2\text{O}\,\text{P}
\end{array} \qquad (7.25)
$$

DiHO-acetone
donor
(a D-ketose
phosphate)

DiHO-acetone
acceptor
(a D-aldose
phosphate)

regarded as a rather special transaldolase. It catalyses the reversible combination of D-glyceraldehyde 3-phosphate and dihydroxyacetone
phosphate to form D-fructose 1,6-diphosphate
[eq. (7.26)]; this is a key reaction in glycolysis [see

Dihydroxyacetone
phosphate

$$
\begin{array}{cc}
\text{CH}_2\text{O}\,\text{P} & \text{CH}_2\text{O}\,\text{P} \\
| & | \\
\text{C}=\text{O} & \text{C}=\text{O} \\
| & | \\
\text{CH}_2\text{OH} & \text{HO}-\text{C}-\text{H} \\
+ & | \\
\text{CHO} & \text{H}-\text{C}-\text{OH} \\
| & | \\
\text{H}-\text{C}-\text{OH} & \text{H}-\text{C}-\text{OH} \\
| & | \\
\text{CH}_2\text{O}\,\text{P} & \text{CH}_2\text{O}\,\text{P}
\end{array} \qquad (7.26)
$$

D-glyceraldehyde
3-phosphate

D-fructose
1,6-diphosphate

Chapter 6, Section E.1(i)]. Like the transaldolase-
catalysed reaction, the mechanism of reaction (7.26)
involves the formation of an intermediate Schiff
base between the carbonyl of the dihydroxyacetone
moiety and the ε-amino group of a lysine residue at
the active site of the enzyme. Aldolase will also
catalyse the combination of free D-glyceraldehyde
with dihydroxyacetone phosphate to form
D-fructose 1-phosphate.

(c) *Oxido-reduction reactions*

Aldose sugars may be oxidized at C–1 to yield
aldonic acids or at C–6 to yield alduronic acids.
Monosaccharides may also be reduced to yield
polyhydroxy alcohols.

Formation of aldonic acids. The formation of an
aldonic acid is typified by the conversion of
D-glucose 6-phosphate into D-glucono-δ-lactone-
6-phosphate, the δ-lactone of D-gluconic acid
6-phosphate [eq. (7.27)]. This reaction is catalysed
by glucose 6-phosphate dehydrogenase[32] and is the
initial step in the oxidative pentose phosphate
pathway (see Chapter 6, Section F.3). The hydrogen, removed, as a hydride ion (H⁻), from C–1 of
G-6-P, is transferred by the enzyme to become the 4
pro-S hydrogen on the nicotinamide ring of the
NADP coenzyme. The incipient carbonium ion
resulting from the removal of this hydrogen is
stabilized by loss of H⁺ from the hydroxyl group on
C–1 so forming a carbonyl group. The hydrolysis of
D-glucono-δ-lactone 6-phosphate to yield
D-gluconic acid 6-phosphate [eq. (7.28)] can proceed spontaneously but is normally catalysed by
gluconolactonase[33].

Formation of alduronic acids. Alduronic acid
formation is exemplified by the oxidation of C–6 of
UDP-D-glucuronic acid. This reaction (see Fig. 7.9)
is catalysed by UDP-glucose dehydrogenase[34] and
proceeds in four steps. The first involves the

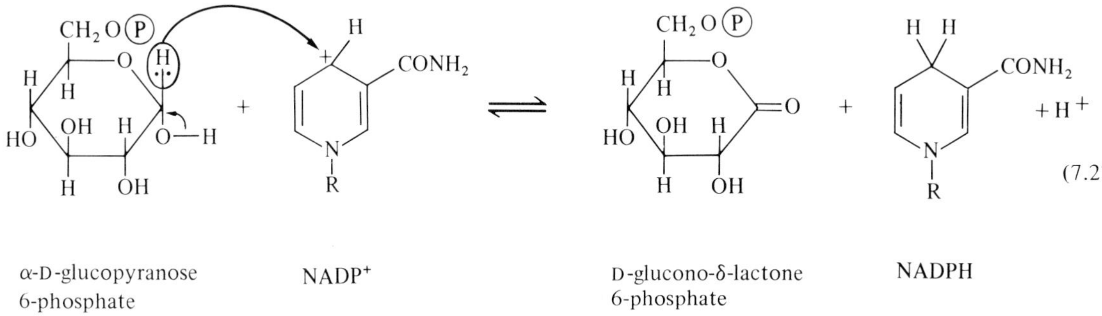

α-D-glucopyranose 6-phosphate   NADP⁺   D-glucono-δ-lactone 6-phosphate   NADPH

$$(7.27)$$

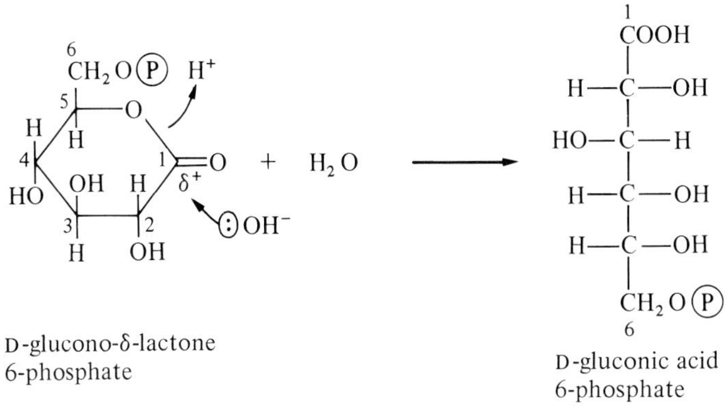

D-glucono-δ-lactone 6-phosphate   D-gluconic acid 6-phosphate

$$(7.28)$$

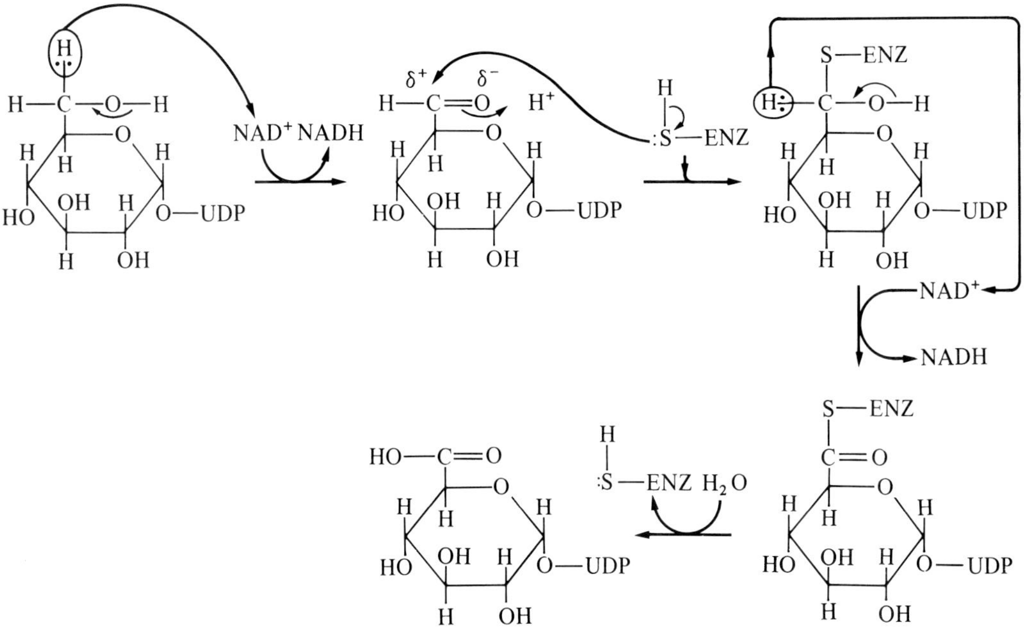

Fig. 7.9. Conversion of UDP-D-glucose into UDP-D-glucuronic acid (HS-ENZ = UDP-glucose dehydrogenase).

oxidation of the C–6 primary alcohol group of UDP-G into an aldehyde group; the 6 pro-*R* hydrogen of the —CH_2OH group is transferred as a hydride ion (H^-) to NAD^+ and the incipient carbonium ion which results is stabilized by loss of H^+ from the hydroxyl group. This is then followed by a reaction between the C–6 aldehyde group and a sulphydryl group at the active site of the enzyme with the formation of a hemithioacetal. The hemithioacetal then undergoes a second dehydrogenation step analogous to the first with NAD^+ again the oxidant and a thioacyl derivative as the product. Hydrolysis of the thioacyl derivative releases the enzyme and UDP-D-glucuronic acid; this final step confers irreversibility on the four-step process.

An alternative route to D-glucuronic acid exists in plants via *myo*-inositol (see Fig. 7.10). An NAD-dependent D-glucose 6-phosphate cycloaldolase catalyses the conversion of G-6-P into *myo*-inositol. The suggested pathway shown in Fig. 7.10 is initiated by the abstraction of a hydride ion (H^-) from C-5 of G-6-P by the NAD^+ of the enzyme. This causes the electron shifts shown which result in the formation of 5-ketoglucose 6-phosphate. This then undergoes an internal aldol condensation to form

D-*myo*-inose-2 1-phosphate which is then reduced by enzyme-bound NADH to *myo*-inositol 1-phosphate. The NADH remains tightly bound to the enzyme throughout and does not exchange with free NADH; moreover, the suggested intermediates (in parenthesis in Fig. 7.10) also remain bound to the enzyme. A specific phosphatase then hydrolyses the phosphate ester bond and so produces *myo*-inositol. The mechanism of the conversion of *myo*-inositol into D-glucuronic acid has not been elucidated; the reaction is catalysed by *myo*-inositol oxygenase[35] which is present in higher plants and yeast. The physiological role of this route for the formation of D-glucuronic acid has yet to be fully evaluated; however, it is clear that in some plant species, at least, it is of significance in the formation of uronic acids since it does not involve the net reduction of NAD^+ (as does the UDP-glucose dehydrogenase route) which ties in nicely with the fact that uronic acids are known to be synthesized in these plants without the net reduction of NAD^+ or $NADP^+$.

Formation of polyhydroxy alcohols (polyols). Polyhydroxy alcohols are formed when the aldehyde group of an aldose or the keto group of a

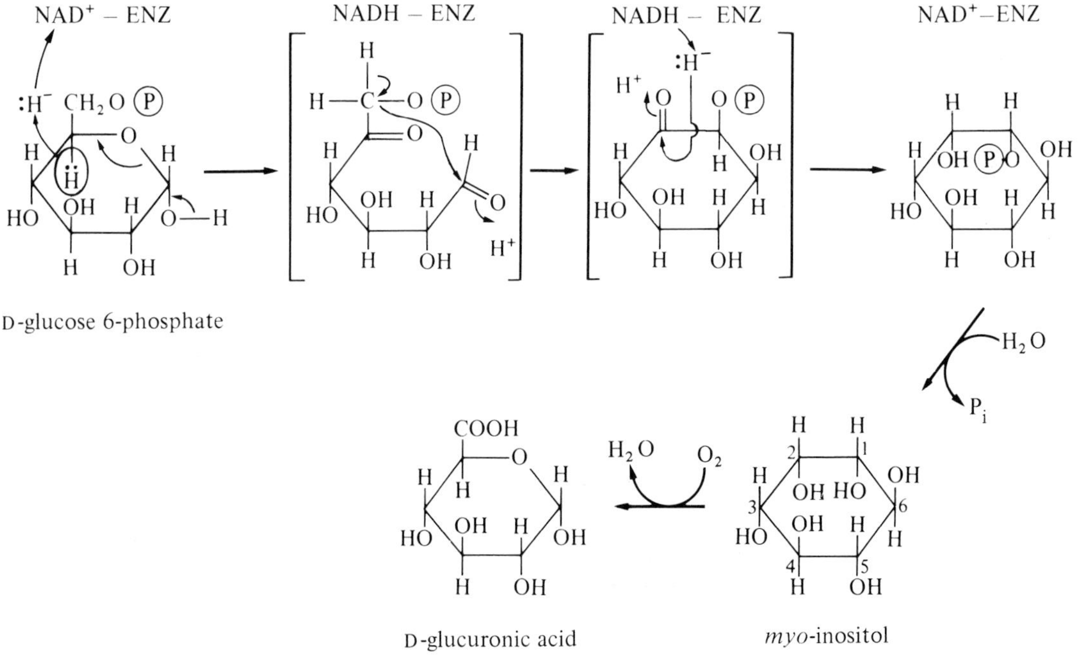

Fig. 7.10. Formation of D-glucuronic acid from D-glucose 6-phosphate via *myo*-inositol.

ketose is reduced. They are found in both the free and combined forms throughout the plant kingdom. The most common are D-mannitol (formed from D-mannose) and D-glucitol (also called sorbitol, formed from either D-glucose or D-fructose). D-glucitol occurs in a large number of plants ranging from algae to spermatophytes; it was discovered in the juice from the berries of the mountain ash (*Sorbus aucuparia*). D-mannitol, unlike D-glucitol, often occurs in plant exudates; it is also present in all brown seaweeds. Lichens (the symbiotic union of an alga and a fungus) accumulate relatively large quantities of polyhydroxy alcohols including D-ribitol, D-arabinitol, D-mannitol and glycerol.

The reduction of the aldehyde and keto groups of aldoses and ketoses to yield polyhydroxy alcohols is presumably brought about by NAD(P)H-requiring reductases; an example of such a reduction is seen in the formation of L-ascorbic acid [see Section B.3(iii)] when D-glucuronic acid is converted into L-gulonic acid and D-galacturonic acid into L-galactonic acid.

(iii) BIOSYNTHESIS OF L-ASCORBIC ACID

Animals, with the exception of man, other primates, the guinea pig, the red vented bubul (*Pycnonotus cafer*) and the Indian fruit bat (*Pteropus medius*), are able to synthesize L-ascorbic acid. They do so by the route from D-glucuronic acid shown in Fig. 7.11. The enzyme glucuronate reductase[36] catalyses the reduction of the aldehyde group at C–1 of D-glucuronic acid forming L-gulonic acid which is then converted by aldolactonase[37] into L-gulono-γ-lactone. The latter is then oxidized by molecular oxygen under the catalytic influence of L-gulonolactone oxidase[38] to H_2O_2 and a 3-keto derivative which tautomerizes to yield L-ascorbic acid. In this reaction sequence there is reversal of the carbon skeleton such that C–1 and C–6 of D-glucuronic acid become C–6 and C–1 respectively of L-ascorbic acid. Man and the other non-ascorbic acid synthesizing animals do not appear to possess L-gulonolactone oxidase and thus the pathway cannot proceed beyond L-gulono-γ-lactone. For these animals L-ascorbic acid is a vitamin (vitamin C) and must be present in their

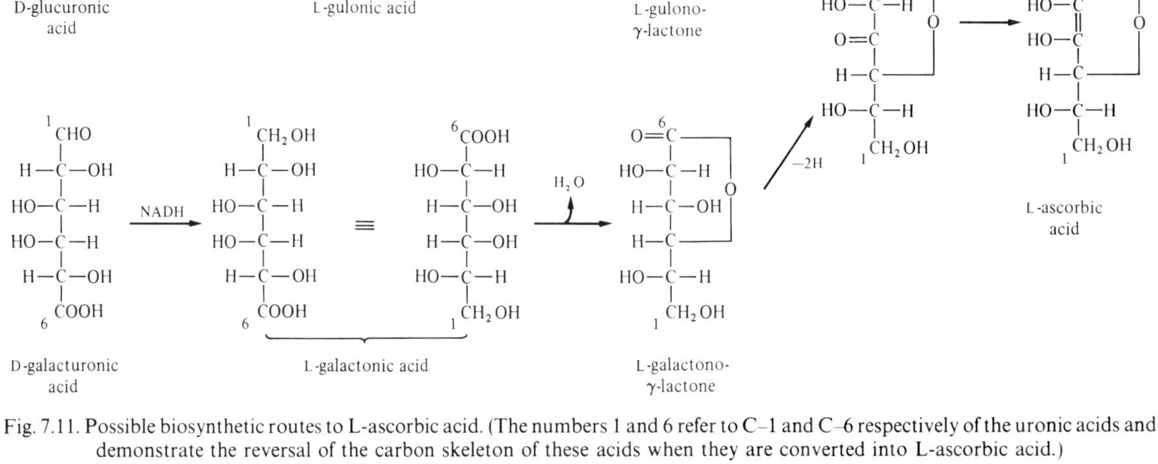

Fig. 7.11. Possible biosynthetic routes to L-ascorbic acid. (The numbers 1 and 6 refer to C–1 and C–6 respectively of the uronic acids and demonstrate the reversal of the carbon skeleton of these acids when they are converted into L-ascorbic acid.)

diet; the source of their dietary L-ascorbic acid is plant tissue.

Plants are able to synthesize L-ascorbic acid very well but the route by which they do it remains somewhat uncertain. Although D-glucuronic acid can be converted into L-ascorbic acid when administered to plants, D-galacturonic acid is a much better substrate. Similarly L-galactono-γ-lactone is a better precursor than L-gulono-γ-lactone. Moreover, experiments in which substrates specifically labelled with ^{14}C were administered to plants showed that inversion of the carbon skeleton had taken place during their conversion into L-ascorbic acid. These results strongly suggest that L-ascorbic acid is biosynthesized in plants preferentially from D-galacturonic acid by a route analogous to that operating in animals (see Fig. 7.11).

However, contradictory results have come from the laboratory of Loewus who found that [6-^{3}H]-D-glucose was converted into [6-^{3}H]-L-ascorbic acid when administered to ripening strawberries. This result clearly indicates that the hydroxymethyl group of D-glucose is retained as such and does not undergo oxidation to a carboxyl group during the conversion of D-glucose into L-ascorbic acid. Moreover, it shows that neither D-glucuronic acid nor their respective γ-lactones can be on the route from D-glucose to L-ascorbic acid and that the route does not involve the reversal of the carbon skeleton.

C. BIOSYNTHESIS OF OLIGOSACCHARIDES

1. Disaccharides (see Fig. 7.12)

(i) Sucrose (α-D-glucopyranosyl-β-D-fructofuranoside, XVI)

Sucrose is the most widely found disaccharide in Nature. It is the principal form in which fixed carbon and energy are translocated in plants. It is composed of an α-D-glucose residue in its pyranose ring form linked by an α1→β2 glycosidic linkage to β-D-fructose in its furanose form; since the anomeric carbon atom of both monosaccharides is involved in the glycosidic linkage their hemiacetal groups are blocked and neither ring may open. Thus sucrose is not a reducing sugar (i.e. it will not reduce Fehling's or Benedict's reagents) and, apart from its extreme sensitivity to acid-catalysed hydrolysis, is chemically inert. It is very soluble in water and has a sweet taste. Note that when D-fructose forms a glycoside (i.e. a fructoside) it always assumes the furanose form; however, free D-fructose in aqueous solution exists as a 4:1 mixture of the pyranose and furanose forms which are in equilibrium with traces of the open-chain form.

There are two enzymes in plants which are capable of catalysing the synthesis of sucrose, namely sucrose synthase[39] and sucrose phosphate synthase[40] which catalyse eqs. (7.29) and (7.30) respectively. The latter enzyme can only utilize UDP-G whilst the former can utilize UDP-G, ADP-G and GDP-G. Reaction (7.30), which already has a more favourable equilibrium constant than reaction (7.29) (3250 versus 5), is pulled even further in the left to right direction by the action of sucrose phosphatase[41] which catalyses the hydrolysis of sucrose 6^F-phosphate to free sucrose [eq. (7.31)].

$$\begin{array}{ccc} \text{UDP-D-glucose} & & \text{UDP} \\ \text{or} & + \text{D-fructose} \rightleftharpoons \text{sucrose} + & \text{or} \\ \text{ADP-D-glucose} & & \text{ADP} \\ & \Delta G^{\circ\prime} = -4 \text{ kJ mol}^{-1} & \end{array} \quad (7.29)$$

$$\text{UDP-D-glucose} + \text{D-fructose 6-phosphate} \rightleftharpoons \\ \text{sucrose } 6^F\text{-phosphate} \\ \Delta G^{\circ\prime} = -20 \text{ kJ mol}^{-1} \quad (7.30)$$

$$\text{Sucrose } 6^F\text{-phosphate} + \text{H}_2\text{O} \rightleftharpoons \\ \text{sucrose} + \text{orthophosphate} \quad (7.31)$$

Present evidence indicates that sucrose biosynthesis in plants occurs by a combination of reactions (7.30) and (7.31) rather than by reaction (7.29). Despite the fact that sucrose biosynthesis has been reported to occur in chloroplasts it now appears clear that it takes place in the cytoplasm of photosynthetic cells from UDP-G and F-6-P derived from dihydroxyacetone phosphate photosynthetically generated in and exported from the chloroplast (see Chapter 5, Section D.6). In nonphotosynthetic tissues (e.g. endosperm of germinating castor bean) sucrose formation from UDP-G and F-6-P probably also takes place in the cytoplasm of the cell.

The enzyme sucrose synthase is thought to be

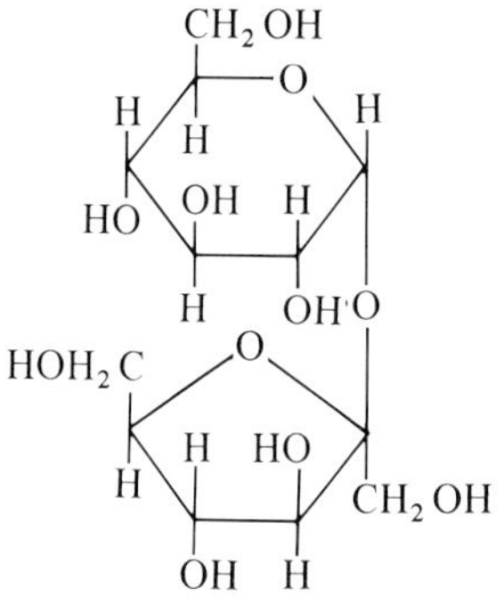

Sucrose (XVI)

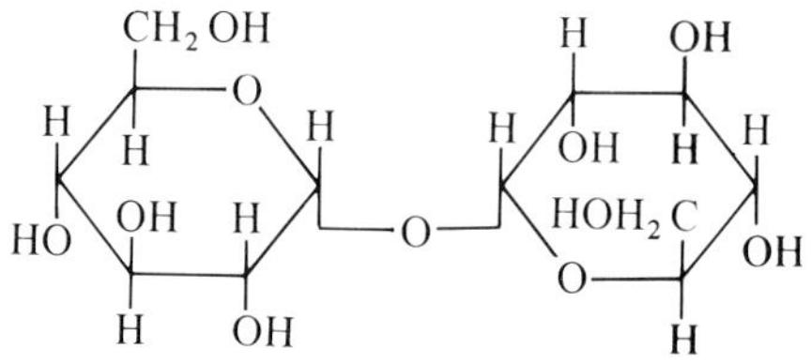

α,α-Trehalose (XVII)

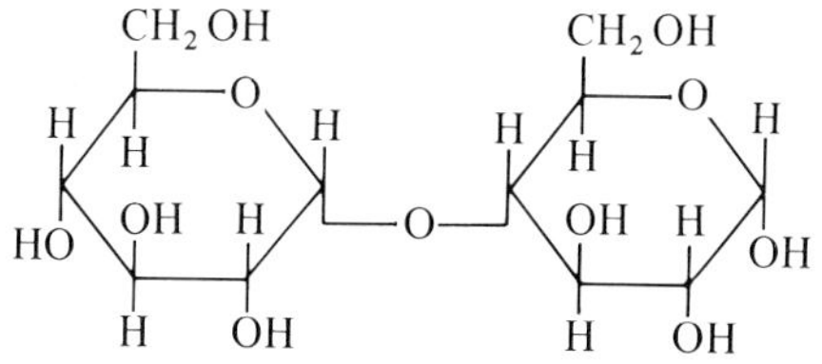

Maltose (XVIII)

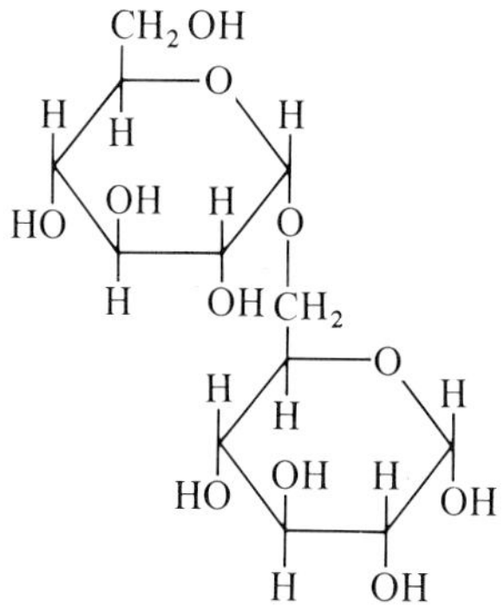

Isomaltose (XIX)

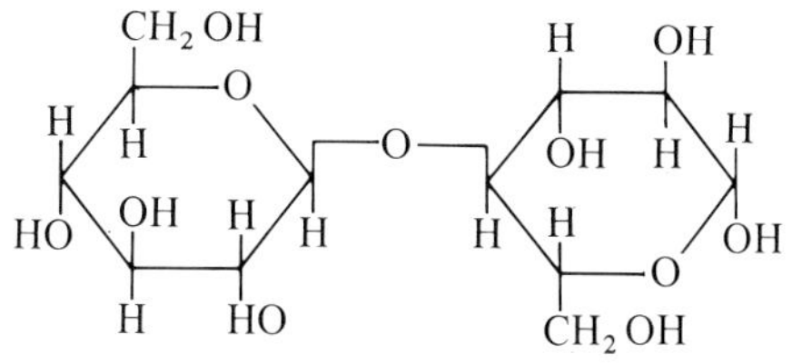

Cellobiose (XX)

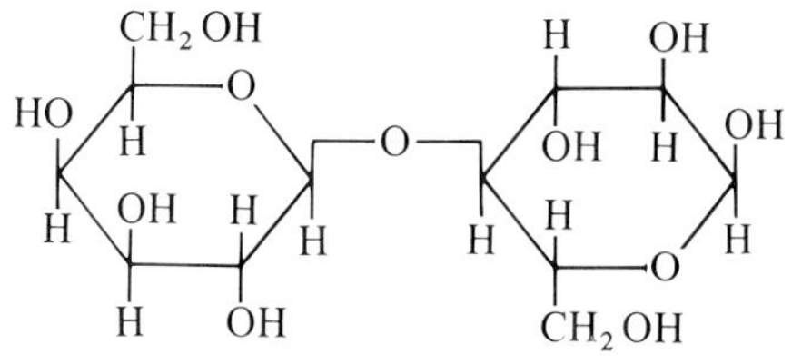

Lactose (XXI)

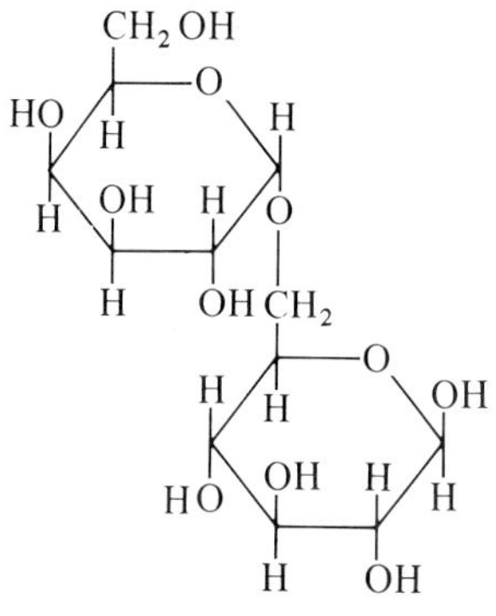

Melibiose (XXII)

Fig. 7.12. Structure of some important disaccharides.

concerned not in the biosynthesis of sucrose but rather with its conversion into starch when it catalyses reaction (7.29) in the right to left direction thereby generating UDP-G (or ADP-G) from sucrose [see Section D.1(ii)].

Sucrose may be hydrolysed in plant tissues under the catalytic influence of invertase[42] [eq. (7.32)].

$$\text{Sucrose} + H_2O \rightleftharpoons \text{D-glucose} + \text{D-fructose} \quad (7.32)$$

(ii) α,α-TREHALOSE (α-D-glucopyranosyl-α-D-glucopyranoside, XVII)

α,α-Trehalose is composed of two α-D-glucose

residues, both in the pyranose form, linked by an α1→α1 glycosidic linkage. It is one of three possible trehalose isomers, the other two being α,β-trehalose (α1→β1 linkage) and β,β-trehalose (β1→β1 linkage). α,α-Trehalose occurs in fungi, blue-green and red algae, the pteridophytes *Selaginella* spp. and *Botrychium lunaria* and the spermatophytes *Echinops persicus, Carex brunescens* and *Fagus silvatica*. α,β- and β,β-trehalose, except for rare cases, do not appear in Nature.

α,α-Trehalose is biosynthesized by a combination of reactions (7.33) and (7.34) which are catalysed by α,α-trehalose phosphate synthase[43] and trehalose phosphatase[44] respectively.

$$\text{UDP-D-glucose} + \text{D-glucose 6-phosphate} \rightleftharpoons$$
$$\text{α,α-trehalose 6-phosphate} + \text{UDP} \quad (7.33)$$

$$\text{α,α-Trehalose 6-phosphate} + H_2O \rightleftharpoons$$
$$\text{α,α-trehalose} + \text{orthophosphate} \quad (7.34)$$

α,α-Trehalose is hydrolysed under the catalytic influence of trehalase[45] [eq. (7.35)] which has been found in the pollen of several species of higher plants.

$$\text{α,α-Trehalose} + H_2O \rightleftharpoons 2 \text{ D-glucose} \quad (7.35)$$

In *Selaginella martensii* labelled α,α-trehalose was present in considerable quantity after photosynthesis in the presence of $^{14}CO_2$; moreover, the labelling increased continuously over many hours. It has been suggested, therefore, that α,α-trehalose replaces sucrose in *Selaginella* spp. as the translocator of fixed carbon and energy; this is supported by the fact that sucrose is found in only minor amounts in *Selaginella* spp.

(iii) MALTOSE   (4-*O*-α-D-glucopyranosyl-D-glucoside, XVIII)

Maltose is composed of two D-glucopyranose units linked together by an α1→4 glycoside linkage. It is found in many plants but usually in small amounts. It probably arises principally as a result of the hydrolysis of starch by the amylase enzymes. However, a little may be formed *de novo* by the transglucosidation activity of α-glucosidase[59]. It may be used as a primer for the enzyme starch synthase[46] or be further hydrolysed by α-glucosidase (maltase) to D-glucose.

(iv) ISOMALTOSE (6-*O*-α-D-glucopyranosyl-D-glucoside, XIX)

Isomaltose is composed of two D-glucopyranose units linked together by an α1→6 glycosidic linkage. It is a structural component of amylopectin and arises during the hydrolysis of starch by α-amylase[47]. It can be hydrolysed to D-glucose by the enzyme oligo-1,6-glucosidase[48] (isomaltase).

(v) CELLOBIOSE   (4-*O*-β-D-glucopyranosyl-D-glucoside, XX)

Cellobiose is composed of two D-glucopyranose units linked together by a β1→4 glycosidic linkage. It is the repeating unit of cellulose from which it is released by the enzyme cellulase[49]. This enzyme, along with the cellobiose-degrading β-glucosidase[50] (cellobiase), is secreted by many microorganisms which live saprophytically on dead vegetation. It is also present in germinating seeds where it assists in the degradation of the cell walls of the parenchymatous cells of the endosperm. Cellobiose is not synthesized from monosaccharide units in plants and only exists in the free state in plant tissues after cellulolysis, a process which only appears to occur in germinating seeds.

(vi) LACTOSE   (4-*O*-β-D-galactopyranosyl-D-glucoside, XXI)

Lactose is composed of a D-galactopyranose unit linked to C-4 of a D-glucopyranose unit by a β1→4 glycosidic linkage. It is a relatively rare and unexpected disaccharide in plants, being present in the chicle plant (*Achras sapota*) and in the anthers of *Forsythia* flowers. Little is known of the mechanism of its formation in plants. It can only be hydrolysed to its component monosaccharides by α-galactosidase[51] which is of widespread occurrence in plants and normally has galactomannans and galactolipids as its substrate.

(vii) MELIBIOSE (6-*O*-α-D-galactosyl-β-D-glucopyranose, XXII)

Melibiose is composed of a D-galactopyranose unit linked to C-6 of a D-glucopyranose unit by an α1→6 glycosidic linkage. It is found in many plant exudates. It is present in the nectaries of a number of plants. It may arise by the action of invertase on the trisaccharide raffinose.

2. Tri-, tetra- and pentasaccharides

A large number of tri-, tetra- and pentasaccharides have been isolated from a considerable range of plants and from various plant tissues. Generally they are present in small quantities. However, the trisaccharide raffinose is an exception being present in cotton-seed meal and sugar beet in considerable quantities. Raffinose can be considered as the trisaccharide member of a family of oligosaccharides whose structure is based upon sucrose plus one or more D-galactose residues (see Table 7.1). It has been suggested that they constitute a storage form of D-galactosyl, D-glucosyl and D-frucosyl residues in the plant for they occur in the seeds of many plants. Raffinose occurs in cereal grains, cotton seeds, soybean and many legumes. Stachyose occurs in many genera of the Leguminosae and the Labiateae; the seeds of *Glycine max* (*Soja hispida*) and the rhizomes of *Stachys sieboldii* are particularly rich in stachyose. Verbascose is present in soybeans, seeds of lucerne, vetch and meadow sage and the rhizomes of *Teucrium canadense*.

The biosynthesis of the raffinose family of oligosaccharides involves the successive transfer of a D-galactosyl residue from galactinol (*O*-α-D-galactopyranosyl [1→1] *myo*-inositol) to the C–6 hydroxyl group of the terminal D-glucosyl or D-galactosyl residue of the preceding oligosaccharide of the sequence. Galactinol is formed from UDP-D-galactose and *myo*-inositol as shown in eq. (7.36); the enzyme catalysing the reaction is UDP-D-galactose:*myo*-inositol galactosyltransferase, which requires Mn^{2+} for activity. The formation of raffinose, stachyose and verbascose from sucrose and galactinol by successive galactosyl transfer then follows as shown in eqs. (7.37), (7.38) and (7.39):

$$sucrose + galactinol \rightleftharpoons raffinose + \textit{myo}\text{-inositol} \tag{7.37}$$

$$raffinose + galactinol \rightleftharpoons stachyose + \textit{myo}\text{-inositol} \tag{7.38}$$

$$stachyose + galactinol \rightleftharpoons verbascose + \textit{myo}\text{-inositol} \tag{7.39}$$

Other oligosaccharides are also based upon sucrose. Gentianose (*O*-β-D-glucopyranosyl-(1→6)-*O*-α-D-glucopyranosyl-(1→2)-β-D-fructofuranoside) is found in the rhizomes of many species of *Gentiana*. Melezitose (*O*-α-D-glucopyranosyl-(1→3)-*O*-β-D-fructofuranosyl-(2→1)-α-D-glucopyranoside) occurs in the exudates which form on the leaves of several species of tree after insect attack. It is probable that the exudate does not come from the leaves but is the honeydew secretion of the

Table 7.1. The raffinose family of oligosaccharides

Oligosaccharide	Constituent monosaccharides and glycosidic linkages
Sucrose	$glu_p(\alpha1 \rightarrow \beta2)fru_f$
Raffinose	$gal_p(\alpha1 \rightarrow 6)glu_p(\alpha1 \rightarrow \beta2)fru_f$
Stachyose	$gal_p(\alpha1 \rightarrow 6)gal_p(\alpha1 \rightarrow 6)glu_p(\alpha1 \rightarrow \beta2)fru_f$
Verbascose	$gal_p(\alpha \rightarrow 6)gal_p(\alpha1 \rightarrow 6)gal_p(\alpha1 \rightarrow 6)glu_p(\alpha1 \rightarrow \beta2)fru_f$

(glu_p = D-glucopyranosyl; gal_p = D-galactopyranosyl; fru_f = D-fructofuranosyl)

UDP-D-galactose   *myo*-inositol   galactinol

$$\tag{7.36}$$

insects; the melezitose then arises as a result of the transglucosylation action of the insect invertase.

Two oligosaccharides composed entirely of D-glucose residues are maltotriose (O-α-D-glucopyranosyl-(1→4)-O-α-D-glucopyranosyl-(1→4)-D-glucopyranoside) and panose (O-α-D-glucopyranosyl-(1→6)-O-α-D-glucopyranosyl-(1→4)-D-glucopyranoside). Maltotriose is probably formed in the hydrolysis of starch by α-amylase whilst panose has been shown to arise in the fungus *Aspergillus niger* by transglycosylation from D-maltose [eq. (7.40)].

$$\text{Maltose} + \text{maltose} \rightleftharpoons \text{panose} + \text{D-glucose} \qquad (7.40)$$

D. BIOSYNTHESIS OF POLYSACCHARIDES

Polysaccharides are synthesized in plants by successive transglycosylation reactions involving a large number of glycosyl donor molecules and an acceptor molecule, frequently referred to as a primer molecule. The primer molecule is elongated by the successive transfer of glycosyl moieties from the donor molecules to one end of the primer molecule (see Fig. 7.13). The overall process is summarized in eq. (7.41). The beauty of this mechanism is that, although the formation of a particular polysaccharide may require a thousand or more glycosyl transfers, the same enzyme catalyses them all. This enzyme specifies the identity of the glycosyl moiety

$$n\text{Glycose-X} + \text{Acceptor} \rightleftharpoons$$

<u>DONOR</u>

$$(\text{Glycose})_n - \text{Acceptor} + n\text{X} \qquad (7.41)$$

transferred and the nature of the glycoside bond formed. In forming the new glycosidic bond the anomeric carbon atom of the glycosyl unit is bonded to a nucleophilic atom in the acceptor molecule; thus the last glycosyl unit to be added to the polysaccharide constitutes the non-reducing end of the chain. The original acceptor molecule (i.e. the acceptor molecule as it exists before the first transfer has taken place) for the biosynthesis of linear homopolysaccharides is frequently a small oligosaccharide whose structure is identical or very close to that of the polysaccharide to be formed. For example, maltose can act as the acceptor for amylose formation and sucrose for fructosan formation.

The most important glycosyl donors are nucleoside diphosphate sugars [see Section B.3(i)(b)] although disaccharides such as sucrose act as donors of fructofuranosyl moieties in the synthesis of fructosans. The importance of the NDP-sugars in polysaccharide biosynthesis is largely due to their high free-energy content. The $\Delta G^{\circ\prime}$ of hydrolysis of NDP-glycose ($\rightarrow$NDP+glycose) is $-33.5\ \text{kJ mol}^{-1}$ in contrast to the $\Delta G^{\circ\prime}$ of hydrolysis of one glycose unit from a polyglycose chain which is in the order of $-21\ \text{kJ mol}^{-1}$. The reverse of the latter reaction, the addition of one glycose unit to a polyglycose chain, therefore has an approximate $\Delta G^{\circ\prime}$ of $+21\ \text{kJ mol}^{-1}$. Thus if the

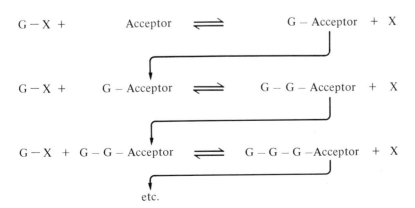

Fig. 7.13. General mechanism for polysaccharide biosynthesis (G = glycosyl moiety; X = other component of glycosyl donor molecule).

glycose is transferred from NDP-glycose to the polyglycose chain the $\Delta G^{\circ\prime}$ is approximately equal to $-33.5 + (+21) = -12.5$ kJ mol^{-1}. With NDP-glycose as the donor, reaction (7.41) proceeds with a considerable loss in free energy so that the equilibrium position is well to the right, favouring polysaccharide formation.

1. Storage polysaccharides

(i) STARCH

Starch acts as both a long-term and a short-term storage polysaccharide in plants. It is stored on a long-term basis in many seeds, tubers and rhizomes, only being utilized when these structures germinate or sprout. However, it forms only a short-term store in chloroplasts during periods of rapid photosynthesis, being mobilized and exported from the leaves as sucrose during an ensuing period of darkness.

Starch is always formed and stored as grains in

plastids; in non-photosynthetic tissues the plastid is an amyloplast whereas in photosynthetic tissues it is a chloroplast. Starch grains are highly organized structures whose size and shape vary but are frequently characteristic of the plant species. The shape may be spherical, ovoid, lens-shaped or irregular and the size may be 1–100 μm in diameter. The grains contain up to 20% water, of which 10% is chemically bound to the starch, and are composed of a series of concentric layers of starch arranged round a centrally or excentrically located spot called the hilum. Present evidence indicates that starch grains are built up by apposition of layers; starch grains from developing potato tubers and waxy maize seeds, examined after the parent plants had been exposed to $^{14}CO_2$, contained ^{14}C in the peripheral layer only, indicating that the newly formed [^{14}C] starch had been laid down on top of existing layers. However, the molecular architecture of these layers of starch remains unknown.

Starch is a mixture of two polysaccharides, amylose (XXIII) and amylopectin (XXIV and Fig. 7.14). Amylose molecules are long, unbranched

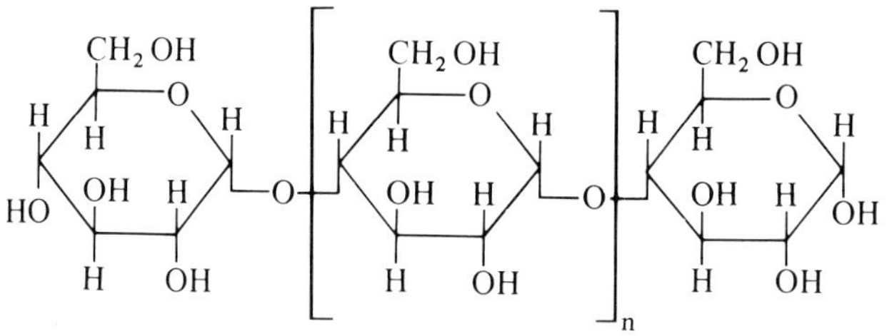

Amylose (XXIII)

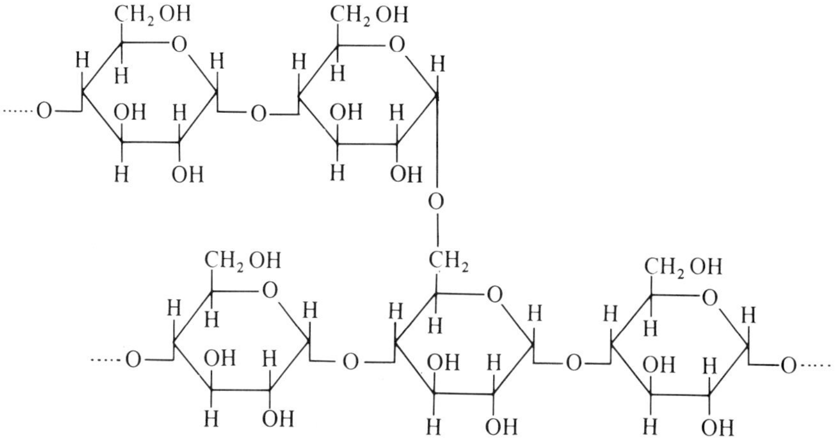

Section of amylopectin molecule (XXIV) (see also Fig. 7.14).

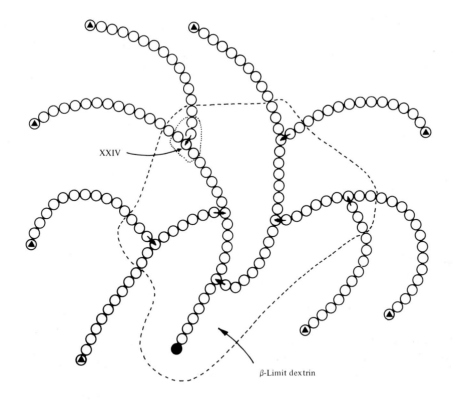

XXIV

β-Limit dextrin

● = reducing end i.e.

⬤ = non-reducing end i.e.

◯◯ = α-1,4-glucosidic linkage

◯⬤ = α-1,6-glucosidic linkage

β-Limit dextrin = residue remaining after β-amylase-catalysed hydrolysis of amylopectin

XXIV = segment of amylopectin shown in molecular structure XXIV

Fig. 7.14. Simplified, two-dimensional structural representation of amylopectin.

chains of about a hundred to several thousand α-D-glucopyranose units linked together by $\alpha1\rightarrow4$ glycosidic bonds. X-ray diffraction studies have shown that amylose molecules have a helical structure with a diameter of 1.3 nm and six successive glucose residues per turn. Amylose molecules are soluble in hot water but the resulting solution is unstable and eventually spontaneous precipitation (known as retrogradation) occurs. Retrogradation is irreversible and arises because of the tendency of the long thin amylose molecules to align themselves side by side and then to form insoluble aggregates by hydrogen bonding.

Amylopectin molecules have highly branched tree-like structures (Fig. 7.14) containing up to 50,000 α-D-glucopyranose residues which are predominantly linked together by $\alpha1\rightarrow4$ glycosidic linkages. However, at each branch point an α-D-glucopyranose residue is involved in an $\alpha1\rightarrow6$ glycosidic linkage (via its C–6 hydroxyl oxygen atom to C–1 of the first glucose residue of the branch) in addition to two $\alpha1\rightarrow4$ linkages (XXIV). The $\alpha1\rightarrow6$ glycosidic linkages constitute about 5% of the total glycosidic linkages in amylopectin molecules. Each amylopectin molecule has one reducing end (i.e. a glucose residue with an anomeric carbon atom not involved in a glycosidic linkage) and many non-reducing ends. The molecular structure is far more complex than that shown in Fig. 7.14 in two ways: firstly, there are many more branch points and secondly, the structure is three-dimensional with branches emerging in all directions giving a spherical shape. The average length of the exterior chains (i.e. from non-reducing end to the first branch point) is 16 glucose units whilst the average length of the interior chains (i.e. from branch point to branch point) is 7 glucose residues. About 50–60% of the glucose residues of the molecule are present in the exterior chains. The structure of amylopectin is very similar to that of the animal storage polysaccharide glycogen, the main difference being that there are more branch points in glycogen thus rendering the exterior and interior chains shorter.

The ratio of amylose to amylopectin in starch depends upon the plant species and is under genetic control. The starch grains synthesized by most plants contain 15–25% amylose and 85–75% amylopectin. However, certain varieties of maize and rice, known as waxy or glutinous varieties, produce starch grains which consist almost entirely of amylopectin; these varieties are homozygous with respect to the recessive waxy gene. At the other extreme a variety of maize known as amylomaize and a variety of wrinkled pea (var. Steadfast) have been bred which produce starch grains with a 50% and 80% amylose content respectively.

The synthesis of starch is a two-step process. Amylose, being the simpler component of starch, is synthesized first and then a proportion of it undergoes rearrangement to form the basis of the more complex amylopectin; amylose is thus the precursor of amylopectin. This is indicated by the fact that the amylose of wheat plants exposed to $^{14}CO_2$ or [^{14}C]sucrose becomes labelled well before amylopectin.

The enzyme catalysing the formation of amylose is called starch synthase[52]. At least two starch synthase isoenzymes are present in plant tissues. The first to be discovered is tightly bound to developing starch grains; it was originally found in starch grains in immature dwarf bean seeds but has since been found in a wide variety of plants. The second starch synthase is a soluble enzyme which is present in amyloplasts in developing seeds, tubers, etc., and chloroplasts in photosynthetic tissues. Both enzymes catalyse the transfer of an α-D-glucopyranosyl residue from an NDP-D-glucose to the non-reducing end of an α-1,4-D-glucan, the acceptor or primer molecule, and attach it thereto by means of an $\alpha1\rightarrow4$ glycosidic linkage [eq. (7.42)]. The term α-1,4-glucan refers to an oligo- or polysaccharide composed of α-D-glucopyranose residues linked together by $\alpha1\rightarrow4$ glycosidic bonds.

The starch synthase which is tightly bound to starch grains can utilize both UDP-D-glucose and ADP-D-glucose in eq. (7.42) although the latter is the more efficient glucosyl donor. The soluble starch synthase, however, can only utilize ADP-D-glucose. Both enzymes can use a wide variety of α-1,4-glucan chains as the primer, ranging from maltose to amylose molecules composed of thousands of glucose units and including the relatively short stretches of α-1,4-glucan chains found in 'branched α-1,4-:α-1,6-dextrins' and the exterior chains of amylopectin. The term 'branched α-1,4-:α-1,6-dextrin' refers to an oligosaccharide composed of α-D-glucopyranose residues linked together mainly

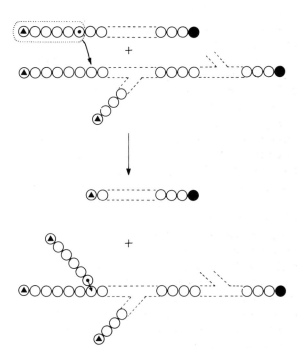

NDP-glucose — Non-reducing end of primer

(7.42)

by $\alpha 1 \rightarrow 4$ glycosidic linkages but which has one or more branch points due to the presence of $\alpha 1 \rightarrow 6$ glycosidic bonds. When the primer is an α-1,4-glucan, starch synthase catalyses the formation of amylose. However, since branched α-1,4-:α-1,6-dextrins can act as primers, starch synthase is also concerned in amylopectin biosynthesis.

Amylopectin is formed by the combined activity of starch synthase, which catalyses the linking together of α-D-glucopyranose residues by $\alpha 1 \rightarrow 4$ glycosidic linkages, and the 1,4-α-glucan branching enzyme[53] (originally called the Q-enzyme) which catalyses the transfer of an oligosaccharide segment from the non-reducing end of an α-1,4-glucan chain (e.g. an amylose molecule) to a non-terminal glucosyl residue in an adjacent α-1,4-glucan chain. The inserted segment is attached to the non-terminal glucosyl residue by means of an $\alpha 1 \rightarrow 6$ linkage (see Fig. 7.15). The acceptor α-1,4-glucan chain may be another amylose molecule or an exterior chain of a growing amylopectin molecule; whichever is the case a new non-reducing end group has been added to the acceptor molecule, which can then be further elongated by starch synthase. The elongated branch may then be further branched by receipt of a further segment from amylose under the catalytic influence of the 1,4-α-glucan branching enzyme. It is thought likely that an external chain of a growing amylopectin molecule can also act as the donor of an oligosaccharide segment as well as being a recipient. Repetition of the branching–elongation sequence ultimately builds up an amylo-pectin molecule.

Starch can be degraded in plant tissues to its component monosaccharide in a process which involves several enzymes, all of which are hydrolases except for starch phosphorylase[54] (sometimes called the P-enzyme) which is a glucosyltransferase.

Fig. 7.15. Reaction catalysed by the 1,4-α-glucan branching enzyme in amylopectin biosynthesis.

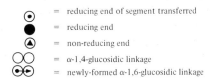

⊙ = reducing end of segment transferred
● = reducing end
▲ = non-reducing end
○○ = α-1,4-glucosidic linkage
⊙○ = newly-formed α-1,6-glucosidic linkage

The hydrolases include α-amylase[55], β-amylase[56], amylopectin 6-glucanohydrolase[57], oligo-1,6-glucosidase[58] and α-glucosidase[59]. Although starch phosphorylase, the α- and β-amylases and amylo-pectin 6-glucanohydrolase will all begin the de-gradation of starch *after it has been extracted from plant tissues*, only α-amylase is capable of this process in starch as it occurs in starch grains. It is therefore assumed that starch degradation *in vivo* is initiated by α-amylase, after which the other enzymes may also participate if they are present.

α-Amylase, as befits its role as the sole initiator of starch breakdown *in vivo*, is widely distributed in plants (moreover, salivary and pancreatic amylases, which are responsible for the digestion of dietary starch in animals, are also α-amylases). It catalyses the essentially random hydrolysis of non-terminal α1→4 glycosidic linkages in the linear amylose and the branched amylopectin molecules. It is therefore able to by-pass the α1→6 glycosidic linkages of amylopectin. Amylose is initially degraded to a mixture of α-1,4-dextrins whilst amylopectin is initially degraded to a mixture of α-1,4-dextrins and branched α-1,4-:α-1,6-dextrins; there is therefore rapid depolymerization of both components of starch. The α-1,4-dextrins and branched α-1,4-:α-1,6-dextrins are further degraded; however, as the hydrolysis proceeds, the degradation becomes less random since there is a significant increase in the relative proportion of α1→4 linkages which are adjacent to branch points or to a terminal glucose residue, for which α-amylase has a much lower affinity. The normal end-products from amylose are maltose and glucose and from amylopectin, maltose, glucose and small oligosaccharides containing an α1→6 glycosidic linkage; the major end product from both amylose and amylopectin, however, is maltose.

Several α-amylase isoenzymes have been found in plant tissues but their physiological significance is not yet clear. Ungerminated cereal seeds contain little or no α-amylase (although they do contain β-amylase which, however, is unable to degrade the starch in the starch grains in the absence of an initial attack by α-amylase). However, when germination commences there is a dramatic increase in the synthesis of α-amylase in the aleurone layer of cells which surrounds the endosperm. This increase in α-amylase synthesis appears to be stimulated by gibberellin which is synthesized in the scutellum after the seed has imbibed water and which is translocated to the aleurone layer. The newly synthesized α-amylase is then secreted into the endosperm, along with a complex cocktail of other enzymes including some which degrade the cell walls of the parenchymatous endosperm cells, where starch degradation takes place.

β-Amylase is much less widespread in plants. It appears to occur only in cereal seeds and certain other higher plants such as sweet potatoes and soybeans. It catalyses the hydrolysis of the penultimate α1→4 glycosidic linkage from the non-reducing end of an α-1,4-D-glucan, thus clipping off maltose which appears as β-maltose. The removal of a maltose unit produces another non-reducing end, so that the process can be repeated many times. β-Amylase will degrade amylose completely to maltose but with amylopectin degradation is incomplete. β-Amylase can attack the non-reducing ends of the many exterior chains of amylopectin and clip off successive maltose units until it gets to within two to three glucose residues of a branch point; here it stops and thus leaves behind a large, branched-chain polysaccharide known as β-limit dextrin (see Fig. 7.14) which is effectively amylopectin minus its exterior chains. In this process β-amylase catalyses the hydrolytic removal of 50–60% of the glucose residues originally present in amylopectin. Several β-amylase isoenzymes have been detected in wheat and barley seeds but their physiological significance is unknown.

Starch phosphorylase is widespread in the leaves and most types of storage organs of higher plants; again several isoenzymes have been detected but their significance remains to be elucidated. This enzyme catalyses the transfer of the D-glucopyranosyl residue at the non-reducing end of an α-1,4-D-glucan to orthophosphate, producing D-glucose 1-phosphate. This transfer produces another non-reducing end so that the process can be repeated many times. Amylose is degraded completely to G-1-P but amylopectin is incompletely degraded because the enzyme cannot by-pass the branch points. It attacks the non-reducing ends of the many exterior chains of amylopectin and degrades each chain to within five to six glucose residues of the first branch point; here it stops and leaves behind a large, branched-chain polysaccharide known as phosphorylase-limit dextrin. In this process it catalyses the removal of about 40% of the glucose residues originally present in amylopectin.

Amylopectin 6-glucanohydrolase occurs widely in plants. It catalyses the hydrolysis of α1→6 glycosidic linkages in amylopectin and β-limit- and phosphorylase-limit dextrins producing a complex mixture of the much smaller α-1,4-dextrins and branched α-1,4-:α-1,6-dextrins. It does not readily catalyse the hydrolysis of the α1→6 glycosidic linkages in the branched α-1,4-:α-1,6-dextrins.

Oligo-1,6-glucosidase also occurs widely in plants. It catalyses the hydrolysis of α1→6 glycosidic linkages in isomaltose and the branched α-1,4-:α-1,6-dextrins produced by amylopectin 6-glucanohydrolase and α-amylase. The products of this hydrolysis are glucose (from isomaltose) and short-chain α-1,4-dextrins (from the branched α-1,4-:α-1,6-dextrins) which can be further degraded to maltose by the amylases or to G-1-P by starch phosphorylase.

α-Glucosidase is widely distributed in plants and several isoenzymes have been found. It catalyses the hydrolysis of α1→4 glycosidic linkages in maltose and α-1,4-linked glucose oligosaccharides, producing D-glucose. It therefore represents the final step in the hydrolytic degradation of starch.

Thus the end products of starch degradation in plants are D-glucose and D-glucose 1-phosphate. The former arises by the collective action of the hydrolases whilst the latter arises by the collective action of the hydrolases, which cleave α1→6 glycosidic linkages, and phosphorylase.

(ii) SUCROSE–STARCH INTERCONVERSION

The interconversion of sucrose and starch occurs widely in plants. It is hardly surprising that this is so since sucrose is the main transporter of fixed carbon and energy in the plant and starch is the most usual storage form of fixed carbon and energy.

In developing seeds the direction of conversion is of sucrose into starch; the sucrose has been imported from the photosynthetic tissues and is converted into starch which is stored as grains in the seed. When the seed germinates the process is reversed and the resulting sucrose is translocated to the rapidly growing meristems of the seedling.

In the photosynthetic cells of leaves the direction of conversion is of starch into sucrose; the starch has been synthesized during active photosynthesis and its conversion into sucrose, for export purposes, takes place in an ensuing dark period.

In ripening fruit (e.g. apple, banana) the direction of conversion is also of starch into sucrose (and other sugars); hence the sweetness of fully ripe fruit.

The pathway by which starch is converted into sucrose in photosynthetic cells is shown in Fig. 5.37 (Chapter 5) whilst that in germinating seeds is probably that outlined in Fig. 7.16.

The pathway by which sucrose is converted into starch in developing seeds has been the subject of considerable study in recent years. Present evidence indicates that the enzyme sucrose synthase plays an important role in it. As indicated in Section C.1(i), the reversible reaction [eq. (7.29)] catalysed by this enzyme operates in vivo in the direction of conversion of sucrose and NDP into NDP-D-glucose and D-fructose [eq. (7.43)] for which the $\Delta G^{\circ\prime}$ is $+4$ kJ mol^{-1} and NDP may be UDP or ADP.

$$\text{Sucrose} + \text{NDP} \rightarrow \text{NDP-D-glucose} + \text{D-fructose} \tag{7.43}$$

It was then thought that the NDP-D-glucose so formed could act as the glucosyl donor for starch synthase so forming starch [see Section D.1(i)] in a reaction [eq. (7.42)] whose $\Delta G^{\circ\prime}$ is -12.5 kJ per mole of glucosyl units transferred. The overall $\Delta G^{\circ\prime}$ for the conversion of sucrose into starch by the combination of reactions (7.43) and (7.42) is there-

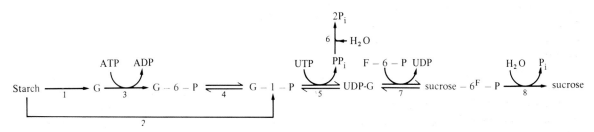

Fig. 7.16. Conversion of starch into sucrose in germinating seeds. (1 = cocktail of hydrolases; 2 = starch phosphorylase plus hydrolases cleaving α1→6 linkages; 3 = hexokinase; 4 = phosphoglucomutase; 5 = UDP-G pyrophosphorylase; 6 = pyrophosphatase; 7 = sucrose phosphate synthase; 8 = sucrose phosphatase.)

fore $(+4)+(-12.5)=-8.5\,\text{kJ mol}^{-1}$. Thus the slightly unfavourable equilibrium of the sucrose synthase-catalysed reaction is overcome by the exergonic reaction (7.42).

This rather elegant two-step incorporation of the glucosyl moiety of sucrose into starch has an advantage to the plant over any mechanism that involves the hydrolysis of sucrose into D-glucose and D-fructose because both ATP and UTP are conserved. If sucrose were hydrolysed, the resulting D-glucose would have to be converted into UDP-G or ADP-G so that it could be incorporated into starch by reaction (7.42); this conversion requires that D-glucose be first converted to G-6-P by the ATP-utilizing hexokinase, then isomerized to G-1-P by phosphoglucomutase before being converted into UDP-G or ADP-G by the appropriate UTP- or ATP-utilizing pyrophosphorylase. By avoiding this conversion two molecules of NTP are conserved (i.e. not used) per glucose residue incorporated from sucrose into starch.

However, it now appears unlikely that the conversion of sucrose into starch in developing seeds is quite so simple as the two-step process just described. The reason for this stems from two facts. Firstly, sucrose synthase utilizes UDP almost ex-

clusively in reaction (7.43) in spite of the fact that both ADP-G and UDP-G can act as its substrates for the reverse reaction: this follows from the observations that (i) the sucrose synthase from maize and mung beans has a high affinity for UDP but a very low affinity for ADP and (ii) its ability to catalyse reaction (7.43) with ADP as a substrate is strongly inhibited when UDP is also present. Secondly, the starch synthases present in developing seeds will either only utilize, or markedly prefer, ADP-G [see Section D.1(i)]. Thus if sucrose synthase only produces UDP-G and starch synthase predominantly utilizes ADP-G, it is clear that the sucrose–starch conversion requires a mechanism for converting UDP-G into ADP-G; moreover, this mechanism should not involve a net conversion of ATP or UTP to their diphosphates if the NTP-conservation advantage of the process is not to be lost. The way this is now thought to be accomplished is shown in Fig. 7.17. The UDP-G produced by sucrose synthase is converted into G-1-P and UTP by reversing the freely reversible reaction (7.7), catalysed by UDP-G pyrophosphorylase. The resulting G-1-P is then converted into ADP-G by ADP-G pyrophosphorylase [reaction (7.7) operating in the normal left to right direction]. The

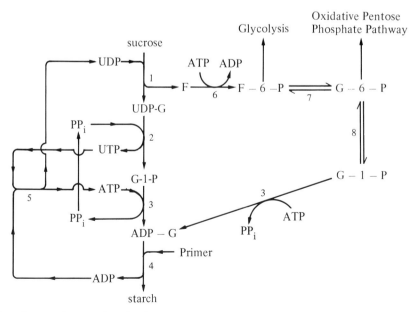

Fig. 7.17. Conversion of sucrose into starch in developing seeds. (1 = sucrose synthase; 2 = UDP-G pyrophosphorylase; 3 = ADP-G pyrophosphorylase; 4 = starch synthase; 5 = nucleoside diphosphate kinase; 6 = hexokinase; 7 = glucose 6-phosphate isomerase; 8 = phosphoglucomutase.)

pyrophosphate formed in the latter reaction is used to drive the former reaction. The glucose residue of ADP-G is then incorporated into starch by starch synthase. The incorporation of the glucosyl moiety of sucrose into starch by this four-step process does not cause the net conversion of NTP into NDP because of the activity of the enzyme nucleoside diphosphate kinase[60] which uses the UTP generated in the UDP-G phosphorylase-catalysed step to convert the ADP generated in the starch synthase-catalysed step into the ATP and UDP which drive the ADP-G pyrophosphorylase-catalysed and sucrose synthase-catalysed steps respectively.

This mechanism is in accord with the increases in the levels of sucrose synthase, the pyrophosphorylases, ADP-G specific starch synthase and hexokinase which are seen during periods of intense starch formation in the endosperm of developing maize seeds.

The D-fructose which is produced by sucrose synthase is probably phosphorylated by hexokinase to yield F-6-P which may be catabolized by glycolysis (see Chapter 6, Section E.1) to generate ATP or isomerized by glucose 6-phosphate isomerase to G-6-P. The G-6-P may be catabolized by the oxidative pentose phosphate pathway (see Chapter 6, Section F.3) to generate NADPH or pentose phosphates for other biosynthetic purposes within the developing seed, or isomerized by phosphoglucomutase to G-1-P which can be converted into ADP-G and used for starch synthesis. If the D-fructose is used for starch synthesis two ATP molecules are converted into ADP for each D-fructose incorporated. Since the incorporation of only the D-fructose component of sucrose into starch utilizes ATP, the complete conversion of sucrose into starch by the pathway shown in Fig. 7.17 requires two ATPs per sucrose molecule.

(iii) PHYTOGLYCOGEN

Maize (*Zea mays*) differs from most plants by synthesizing a glycogen-like polysaccharide, called phytoglycogen, in addition to the usual two-component starch (amylose plus amylopectin). Phytoglycogen has a higher degree of branching than amylopectin; about 10% of its glycosidic linkages are $\alpha1\rightarrow6$ in contrast to amylopectin where

about 5% are $\alpha1\rightarrow6$. In order to synthesize phytoglycogen, maize has an enzyme which has been called amylopectin-branching glycosyltransferase. This enzyme is readily separable from the normal 1,4-α-glucan branching enzyme (Q-enzyme) which catalyses the formation of amylopectin (but not phytoglycogen). It can catalyse the introduction of additional branches into amylopectin thereby converting it into phytoglycogen. Maize also has an additional hydrolase called isoamylase[61] to assist the usual hydrolases [see Section D.1(i)] to degrade phytoglycogen. Isoamylase, in contrast to amylopectin 6-glucanohydrolase, is able to catalyse the hydrolysis of the $\alpha1\rightarrow6$ linkages in phytoglycogen.

(iv) FRUCTOSANS

Fructosans are polymers of β-D-fructofuranose. There are two main types, the inulin-type (XXV), in which the fructofuranose residues are linked together by $\beta2\rightarrow1$ glycosidic linkages, and the levan-type (XXVI), in which the fructofuranose units are linked by $\beta2\rightarrow6$ glycosidic linkages. Both types have an α-D-glucopyranose residue linked to a terminal fructofuranose residue by an $\alpha1\rightarrow\beta2$ glycosidic linkage, which is the same type of linkage as is found in sucrose; thus inulins and levans have a terminal sucrose unit. Notice that neither end of the polysaccharide chain of these fructosans can be described as a reducing end because both terminal monosaccharide residues have their anomeric carbon atoms involved in a glycosidic linkage.

The inulin-type fructosans are present as storage polysaccharides in the underground storage organs of plants of the Compositae and Campanulaceae families (e.g. Dahlia tubers, Jerusalem artichoke rhizomes); in these plants inulin is stored rather than starch. The most important member of this group of fructosans is inulin itself which, like other polysaccharides, is not a single molecular entity but rather a group of molecular species having roughly the same molecular size; inulin is usually regarded as being composed of 30–35 monosaccharide residues (MW 4900–5700). Inulin is frequently accompanied by a range of shorter chain members of this group; it is believed that these are biosynthetic intermediates of inulin.

The levan-type fructosans are found in the leaves, stems and roots of a number of monocotyledonous

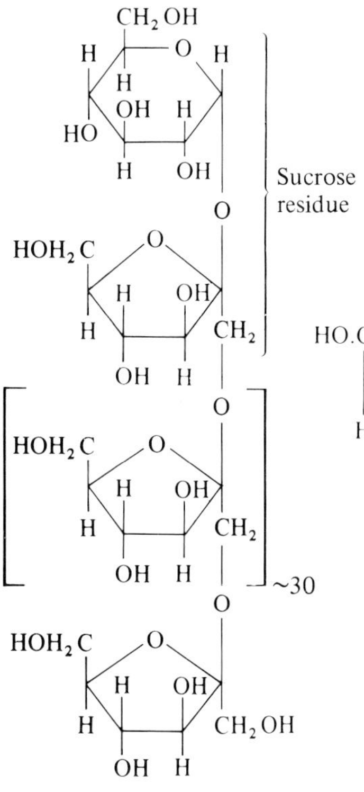

Fig. 7.18. Structure of fructosans.

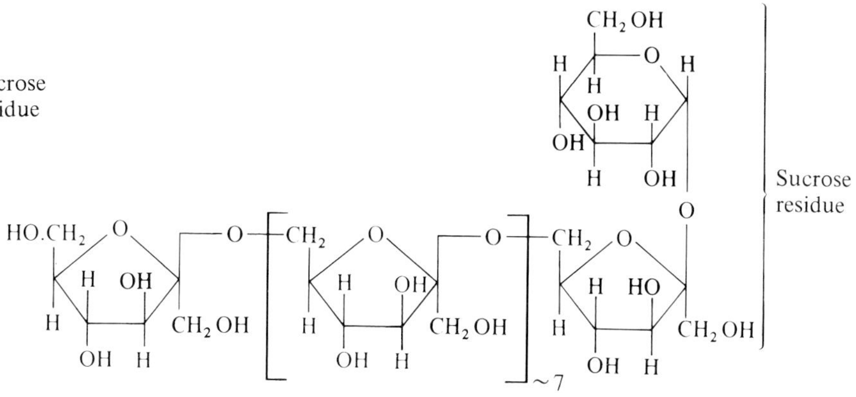

plants, of which the Gramineae constitute the most important family. In these plants they function principally as temporary storage polysaccharides. In general levans are linear molecules with a shorter chain than inulin, for instance the levan of perennial rye grass (*Lolium perenne*) has a linear chain of 7–8 fructofuranose units. However, some levans have a low degree of branching, for instance the levan of Italian rye grass (*Lolium multiflorum*) and cocksfoot (*Dactylis glomerata*) consists of 2–3 chains, each of which has 13–24 fructofuranose residues.

The inulins and levans appear to be biosynthesized from sucrose. The fact that all fructosans have a terminal glucopyranose residue linked to the first fructofuranose residue by a $\beta 2 \rightarrow 1$ linkage indicates that sucrose is the acceptor molecule in eq. (7.41). By analogy with the biosynthesis of amylose one would expect that the fructofuranose donor molecule would be an NDP-D-fructose but no convincing evidence that this is so has yet been

found. All the evidence points to sucrose being the fructofuranose donor. This is not at all unreasonable from a thermodynamic point of view, for the $\Delta G^{\circ\prime}$ of hydrolysis of sucrose is -29.3 kJ mol^{-1} and, as seen earlier, about 20 kJ mol^{-1} are required to add one glycosyl unit to a polyglycose chain; thus the $\Delta G^{\circ\prime}$ for the transfer of a fructofuranosyl residue from sucrose to a sucrose acceptor molecule or to a growing fructosan chain would be in the order of $(-29.3) + (+20) = -9.3$ kJ mol^{-1}. The equilibrium of the reaction would therefore be favourable to fructosan synthesis.

Two enzymes appear to be intimately concerned in fructosan synthesis, inulosucrase[62] and levansucrase[63]. Inulosucrase catalyses the removal of the fructofuranose residue of sucrose and its attachment to the terminal fructofuranose residue of the acceptor molecule by a $\beta 2 \rightarrow 1$ glycosidic linkage; the acceptor molecule may be sucrose or any inulin-type fructosan from a trisaccharide upwards (see

Fig. 7.19). Levansucrase catalyses an identical reaction save that $\beta2{\rightarrow}6$ glycosidic linkages are formed (see Fig. 7.19).

The biosynthesis of inulin in artichoke rhizomes has been investigated more thoroughly than that of most fructosans. The pathway (see Fig. 7.20) involves at least two enzymes. The first, inulo-sucrase, catalyses the formation of the trisac-charide 1^F-fructosylsucrose by the mechanism shown in Fig. 7.19. The second enzyme, a trans-fructosylase, catalyses the transfer of fructofuranose

residues from 1^F-fructosylsucrose to oligosaccha-rides having the inulin structure, the first of which is probably 1^F-fructosylsucrose itself.

The degradation of fructosans involves various β-fructofuranosidases which are presumably spec-ific to either $\beta2{\rightarrow}1$ or $\beta2{\rightarrow}6$ fructoside linkages.

(v) OTHER STORAGE POLYSACCHARIDES

Galactomannans are stored in the seeds of plants of the Leguminosae whilst glucomannans are

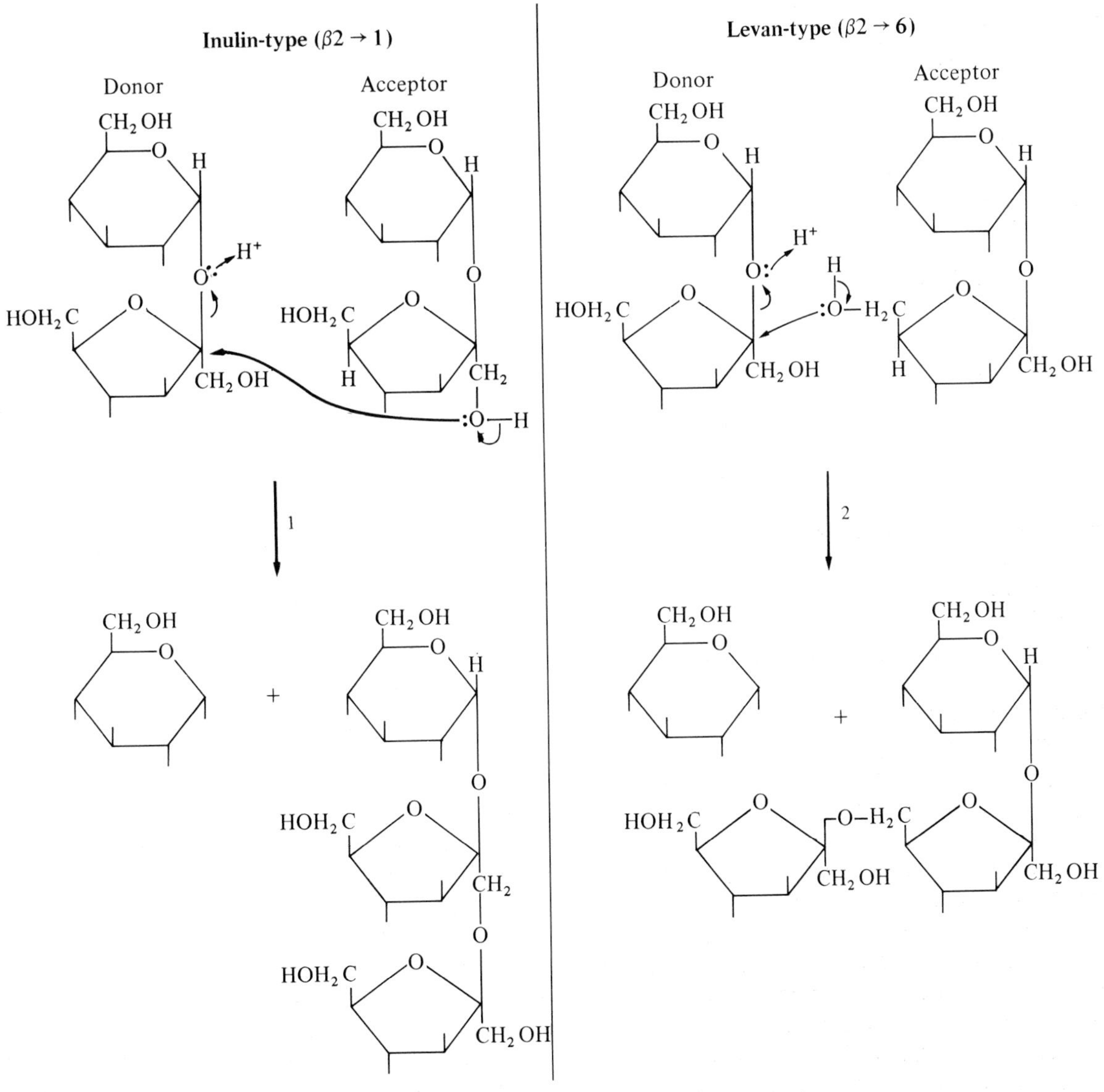

Fig. 7.19. Transfructosylation reactions involved in fructosan biosynthesis. (1 = Inulosucrase; 2 = levansucrase.)

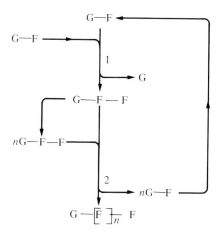

Fig. 7.20. Biosynthesis of inulin in artichoke rhizomes. (G-S = sucrose; G-F-F = 1^F-fructofuranosylsucrose; G[F-]$_n$F = inulin; 1 = Inulosucrase; 2 = transfructosylase.)

stored in the tubers of a small number of plants (e.g. ivory nut, *Amorphophallus tuberosus* and some members of the Orchidaceae). The structure of these compounds has not been fully elucidated and little is known of their biosynthesis or degradation. It is, however, clear that they act as a reserve polysaccharide for they are used up during germination or sprouting.

The storage polysaccharides of algae are mainly glucans (i.e. polymers of D-glucose) which fall into two classes: (a) those which have a backbone of $\alpha 1 \rightarrow 4$ linked D-glucopyranose residues with side chains joined by $\alpha 1 \rightarrow 6$ linkages (cf. amylopectin) and (b) those which have a backbone of $\beta 1 \rightarrow 3$ linked D-glucopyranose residues with side chains joined by $\alpha 1 \rightarrow 6$ linkages. Paramylon, which accumulates in *Euglena gracilis*, falls into the latter category and has been the subject of considerable study. The structurally similar β-1,3-glucans of algae of the families Phaeophyceae (brown), Chrysophyceae (yellow-brown) and Bacillariophyceae (diatoms) have been given different names, laminaran, chrysolaminaran and leucosin respectively. These β-1,3-glucans are similar to the structural polysaccharide, callose, which is produced by higher plants [see Section D.2(v)].

2. Structural polysaccharides

Structural polysaccharides are mainly those which are to be found in the walls of plant cells. Their structure and distribution have been discussed in Chapter 4. Remarkably little is known about the biosynthesis of these polysaccharides. This is in some ways surprising; after all, since cellulose is the most abundant organic compound in Nature it would seem reasonable to assume that its biosynthesis had been fully elucidated. In other ways it is not surprising, since the detailed structure of many of the cell-wall polysaccharides is not known.

It does appear fairly certain that the precursors of these structural polysaccharides are NDP-sugars and that transferase enzymes catalyse the transfer of their monosaccharide moiety to an acceptor molecule; the overall process is therefore described by eq. (7.41). The way in which the various NDP-sugars required for the formation of the known range of cell-wall polysaccharides are formed is largely known. However, precious little is known about the transferases or the nature of the acceptor molecules.

The experimental procedure used to investigate the transferase-catalysed reactions which are supposed to lead to the synthesis of cell-wall polysaccharides *in vivo* has usually involved the incubation of a fairly crude, cell-free enzyme system from young, actively growing (and therefore cell-wall synthesizing) plant tissue with an NDP-[^{14}C]sugar. It is usually assumed that the enzyme system contains, not only the appropriate transferase, but also the appropriate primer. During the incubation it is hoped that the transferase will catalyse the successive transfer of the [^{14}C]sugar to the primer and so produce a [^{14}C]polysaccharide. The [^{14}C]compound produced during the incubation is identified as a polysaccharide and separated from residual NDP-[^{14}C]sugar (and any other [^{14}C]metabolites which may have been formed) by its relative insolubility or by its immobility when subjected to electrophoresis in sodium tetraborate buffer. It is then degraded to a mixture of ^{14}C-labelled di- and trisaccharides by partial hydrolysis or acetolysis. These are then used to identify the nature of the glycosidic bond which links their component monosaccharides together.

This procedure, though very useful, has several limitations. Firstly, it only gives information about the [^{14}C]sugar incorporated into the polysaccharide and not about any unlabelled sugars or non-sugars which may be present. Secondly, it does not distinguish between the *de novo* synthesis of a polysaccharide and the mere addition of a few monosaccharide residues on to the end of a pre-existing polysaccharide. Thirdly, it does not prove that the experimentally produced polysaccharide is identical with the naturally occurring polysaccharide. Moreover, users of this procedure have usually supplied the enzyme system with one or at most two species of NDP-sugar at a time; whilst this is appropriate for homopolysaccharides like cellulose it is inappropriate for the polysaccharides of the plant cell-wall matrix which are heteropolysaccharides frequently containing several different monosaccharides (e.g. the xyloglucan of dicotyledonous primary cell walls which contains L-fucopyranose, D-galactopyranose and L-arabopyranose in addition to D-glucopyranose and D-xylopyranose).

(i) CELLULOSE

In 1964 Hassid and his colleagues were the first to succeed in demonstrating the synthesis of a cellulose-like polysaccharide; they accomplished this by incubating GDP-D-glucose with a particulate enzyme system prepared from mung beans (*Phaseolus aureus*). The enzyme catalysing this synthesis was tightly bound within this preparation. A similar enzyme has since been detected in the roots and hypocotyls of pea, maize and string bean and in the immature seed hairs of cotton. The mung bean enzyme was solubilized and partially purified in 1970. It appears to catalyse reaction (7.44).

nGDP-D-glucose + Primer $\rightleftharpoons$

$\quad [(\beta1 \rightarrow 4)$-D-glucose$]_n$ − Primer + nGDP   (7.44)

It was later found that this enzyme, when supplied with GDP-D-mannose, would catalyse the synthesis of a glucomannan.

The nature of the primer in eq. (7.44) is not known. Moreover, it is likely that the product was a relatively short β-1,4-glucan rather than cellulose; indeed its relationship to the cellulose produced *in vivo*, which is synthesized as microfibrils (see Chapter 4), is somewhat remote. Several authors have, within the last few years, concluded that the claims of *in vitro* cellulose synthesis have been premature. Albersheim in 1978 maintained that no product synthesized in a cell-free enzyme system had, up to that time, been chemically characterized

as cellulose. He further suggested that the main reason why there has been such difficulty in achieving the cell-free synthesis of cellulose is that the synthesizing enzymes are part of a closely integrated complex which is disrupted in making the cell-free enzyme system. As discussed in Chapter 4 the cellulose-synthesizing enzymes are built into the outer surface of the plasmalemma and produce bundles of cellulose molecules (microfibrils) which immediately become incorporated into the adjacent, inner surface of the cell wall. Thus, when, during enzyme preparation, the plasmalemma is pulled away from the cell wall the cellulose-synthesizing enzymes are pulled out of the plasmalemma and their structure is grossly disrupted.

(ii) PECTINS

The major pectin is a rhamnogalacturonan [see Chapter 4, Section B.1(iii)(b)]. Its biosynthesis has not been demonstrated. However, a particulate enzyme system prepared from mung beans catalyses the incorporation of the D-galacturonic moiety of UDP-D-galacturonic acid into an α-1,4-D-galacturonan. The transferase in this enzyme system presumably catalyses reaction (7.45).

nUDP-D-galacturonic acid + Primer $\rightleftharpoons$

$\quad [(\alpha1 \rightarrow 4)$-D-galacturonic acid$]_n$ − Primer + nUDP

(7.45)

The nature of the primer is unknown and the yield of galacturonan is very low. A homogalacturonan, having at least 25 α1→4 linked D-galacturonic acid residues and therefore conforming to the basic structure of the *in vitro* product, has been found in the pectin of primary cell walls. However, since the enzyme system has not been incubated with NDP-L-rhamnose along with UDP-D-galacturonic acid it is not known whether it is capable of synthesizing a rhamnogalacturonan.

Some of the carboxyl groups of the rhamnogalacturonan occur as methyl esters. The methyl groups are derived from the *S*-methyl group of methionine and are introduced by a transmethylation reaction in which 'active' methionine, *S*-adenosylmethionine is the methyl donor. The recipient of these methyl groups appears to be the rhamnogalacturonan itself rather than the UDP-D-galacturonic acid because it was found that the mung bean enzyme system, mentioned above, was incapable of incorporating the carboxymethyl-galacturonic acid moiety of UDP-D-carboxymethyl-D-galacturonic acid into the galacturonan.

Little is known of the biosynthesis of the other pectin component(s) which, in dicotyledonous primary cell walls, is either an arabinogalactan or an arabinan plus a galactan [see Chapter 4, Section B.1(iii)(b)]. The particulate mung bean enzyme system mentioned above was also found to be capable of catalysing the incorporation of the D-galactopyranose moiety of UDP-D-galactose into a galactan, although the yield was very low. A similar result was obtained a few years later with a partially purified enzyme system from mung bean hypocotyls; this system could utilize only the UDP derivative of D-galactose and catalysed the addition of about two galactose residues to a galactan which was shown to have $\beta1\rightarrow4$ linkages. Presumably the enzyme catalysed reaction (7.46).

nUDP-D-galactose + Primer $\rightleftharpoons$

$[(\beta1\rightarrow4)$-D-galactose$]_n$ – Primer + nUDP   (7.46)

A crude particulate preparation from mung bean shoots has been described which catalyses the incorporation of the L-arabinose moiety of UDP-L-arabinose into an arabinan. However, insufficient arabinan was synthesized for it to be fully characterized.

Alginic acids are the functional equivalents of the galacturonans in the matrix of the cell walls of brown algae. They are heteropolysaccharides, being made up of stretches of $\beta1\rightarrow4$-linked D-mannuronic acid residues, stretches of $\alpha1\rightarrow4$-linked L-guluronic acid residues and stretches where the two types of uronic acid residues occur in an almost regularly alternating sequence. It has been assumed that alginic acids are biosynthesized from uronic acid residues derived from NDP-D-mannuronic acid and NDP-L-guluronic acid. Indeed, a cell-free preparation from *Fucus gardneri* was shown to catalyse the incorporation of ^{14}C into a polyuronide fraction when incubated with GDP-D-[^{14}C]mannuronic acid. Moreover, small amounts of GDP-L-guluronic acid have been tentatively identified in *Fucus*. However, it is possible that alginic acids arise from a homo-D-mannuronan by epimerization of some of the

D-mannuronic acid residues in it at C–5 so converting them into L-guluronic acid residues. Evidence for this mechanism has come, not from the brown algae, but from the bacterium *Azotobacter vinelandii* which produces an extracellular polysaccharide with the same chain structure as the algal alginates. A protein fraction, presumably enzymic acid secreted by the bacterium, has been isolated from the *Azotobacter* growth medium. It has been shown to catalyse the conversion of a homo-D-mannuronan into an alginic acid which is indistinguishable from a typical algal alginate. When the incubation was carried out in tritiated water, tritium was located almost entirely in the newly formed L-guluronic acid residues. The obvious conclusion to be drawn from these experiments is that the C–5 epimerization took place at the level of the polymer. It remains to be seen whether this is the mechanism by which the alginic acids of the brown algae are formed.

(iii) HEMICELLULOSES

Even less is known about the biosynthesis of the various polysaccharide species which are classified as hemicelluloses. A detergent-solubilized enzyme system has been obtained from mung bean hypocotyls and a particulate system from the third internode of young pea plants which catalyse the formation of a glucomannan. Both enzyme preparations catalyse (i) the formation of a homo-β-1,4-mannan when incubated with GDP-D-mannose, (ii) the formation of a homo-β-1,4-glucan when incubated with GDP-D-glucose and (iii) the formation of a β-1,4-glucomannan when incubated with UDP-D-mannose and UDP-D-glucose. The transferase enzymes catalysing the formation of this glucomannan appear to be located in the Golgi bodies.

A particulate enzyme system prepared from immature corncobs (developing maize seeds) has been described which, when incubated with UDP-D-xylose, catalyses the formation of an arabinoxylan composed of L-arabofuranose residues linked to a β-1,4-linked chain of D-xylopyranose residues. The enzyme system was shown to contain UDP-arabinose 4-epimerase [see Section B.3(ii)(a)] which catalysed the conversion of some of the added

UDP-D-xylose into UDP-L-arabinose. It was assumed that the enzyme system also contained two distinct transferases which then catalysed the formation of the arabinoxylan from the two UDP-sugars. However, this assumption fails to explain how the arabinose moiety is changed from the pyranose form, in which it exists in UDP-L-arabinose, to the furanose form, in which it exists in the arabinoxylan.

(iv) INTERCONVERSIONS OF NDP-SUGARS AND THEIR PROBABLE BIOSYNTHETIC RELATIONSHIP WITH CELL-WALL POLYSACCHARIDES

Figure 7.21 shows those interconversions of NDP-sugars which have been demonstrated in plants and which appear from Section D.2 to provide the immediate precursors of cell-wall polysaccharides. Four of the enzymes catalysing these interconversions are epimerases [see Section B.3(ii) (a)], two are dehydrogenases which catalyse the conversion of a UDP-sugar into its corresponding UDP-sugar uronic acid using two molecules of NAD^+ in the process (see Fig. 7.9) and one is an NAD^+-requiring-UDP glucuronate decarboxylase[64] which catalyses the decarboxylation of UDP-D-glucuronic acid with the formation of UDP-D-xylopyranose.

In the mid-1960s an alternative route from G-6-P to UDP-D-glucuronic acid was suggested which bypassed G-1-P and UDP-glucose. This route involved the conversion of G-6-P into free D-glucuronic acid via myo-inositol (see Fig. 7.10)

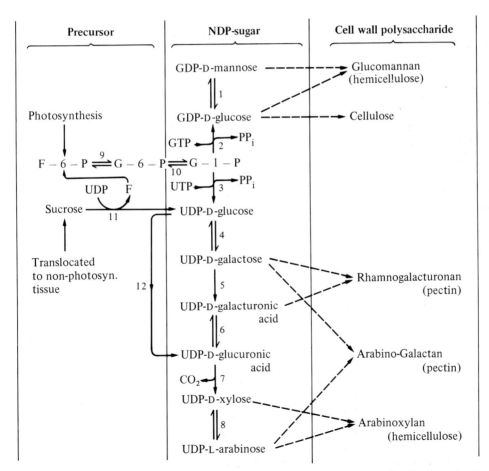

Fig. 7.21. Interconversions of NDP-sugars and their probable biosynthetic relationship with cell-wall polysaccharides. (1 = GDP mannose 2-epimerase; 2 = GDP-G pyrophosphorylase; 3 = UDP-G pyrophosphorylase; 4 = UDP glucose 4-epimerase; 5 = UDP galactose dehydrogenase; 6 = UDP galacturonate 4-epimerase; 7 = UDP glucuronate decarboxylase; 8 = UDP arabinose 4-epimerase; 9 = glucose 6-phosphate isomerase; 10 = phosphoglucomutase; 11 = sucrose synthase; 12 = UDP glucose dehydrogenase.)

and the subsequent formation of UDP-glucuronic acid by a pyrophosphorylase-catalysed reaction. The attraction of this alternative route was that it could be envisaged as a means of regulating the relative amounts of glucose being converted into glucans like cellulose and into pectins and hemicelluloses. By switching on the *myo*-inositol route glucose would be diverted away from glucan synthesis and towards pectin and hemicellulose synthesis and by switching it off the pattern of synthesis would be reversed. However, it now appears that this rather elegant regulatory hypothesis is not tenable because the *myo*-inositol route has been shown to be of no more than minor importance in the biosynthesis of cell-wall polysaccharides. This is evident from a study using a *myo*-inositol-dependent culture of *Fraxinus pennsylvanica* which shows that only 0.5% of the total galacturonic acid in the cell wall is derived from *myo*-inositol.

(v) CALLOSE

The callose molecule is composed of β-D-glucopyranose residues linked together by β1→3 glycosidic linkages (XXVII). This polysaccharide is widely distributed in higher plants and is characteristically found in phloem sieve tubes where it appears to be important in the formation of the sieve plate. It is also produced during the healing of damaged tissue.

Callose closely resembles the storage β-1,3-glucans of algae [see Section D.1(v)].

A soluble enzyme preparation from mung bean catalyses the incorporation of the glucose residue of UDP-D-glucose into a water-insoluble polysaccharide with a structure similar to that of callose.

(vi) THE INVOLVEMENT OF LIPID-LINKED SUGARS IN POLYSACCHARIDE BIOSYNTHESIS

During the past few years convincing evidence has accumulated which indicates that lipid-linked sugars are involved in the biosynthesis of polysaccharides in bacteria, yeast and animals. The lipid component of these compounds is a polyprenol (XXVIII) whose structure shows a degree of variation in terms of the total number of isoprene units present and how many of them are saturated or have a *cis*-configuration. The lipid-linked sugars are of two types and are formed by reactions (7.47) and (7.48). Both types exhibit sugar transfer potentials of the same order as NDP-sugars and can, therefore, function as sugar donors. The major difference between their capability as a sugar donor and that of the NDP-sugar is that they can function in the hydrophobic environment of a membrane because of the long, hydrocarbon polyprenol chain whereas

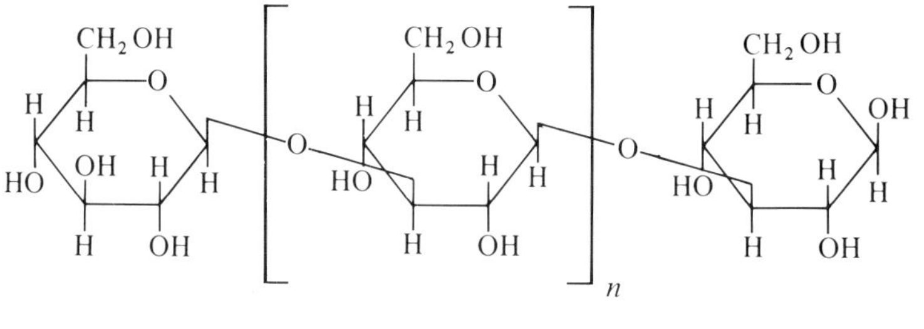

Callose (XXVII)

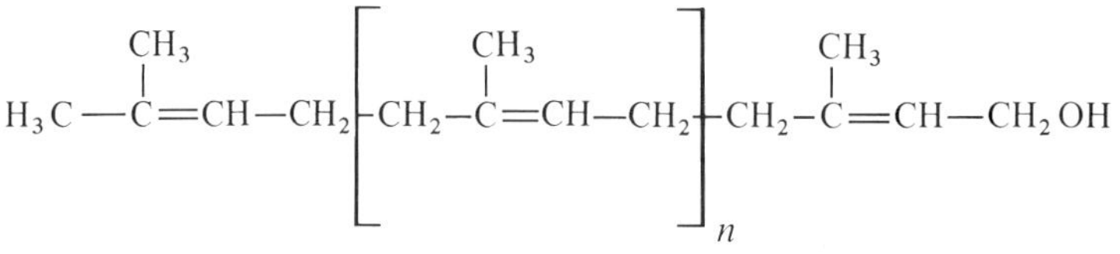

Polyprenol (XXVIII)

(bacteria, $n = 9$; mammalian liver, $n =$ mainly 18; higher plants, $n = 4 - 11$)

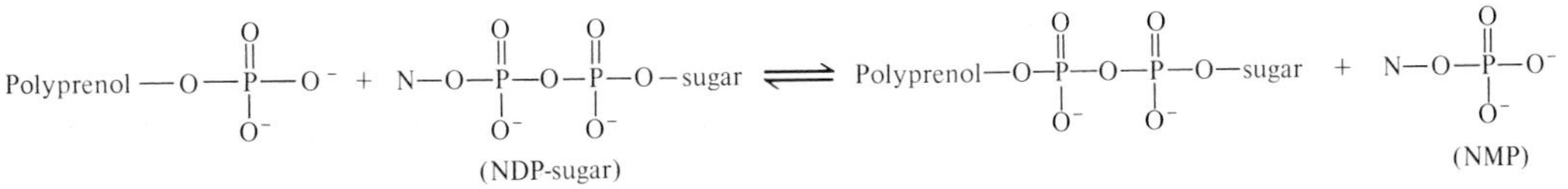

$$\text{Polyprenol}-O-\overset{\overset{\displaystyle O}{\|}}{\underset{\underset{\displaystyle O^-}{|}}{P}}-O^- \;+\; N-O-\overset{\overset{\displaystyle O}{\|}}{\underset{\underset{\displaystyle O^-}{|}}{P}}-O-\overset{\overset{\displaystyle O}{\|}}{\underset{\underset{\displaystyle O^-}{|}}{P}}-O-\text{sugar} \;\rightleftharpoons\; \text{Polyprenol}-O-\overset{\overset{\displaystyle O}{\|}}{\underset{\underset{\displaystyle O^-}{|}}{P}}-O-\overset{\overset{\displaystyle O}{\|}}{\underset{\underset{\displaystyle O^-}{|}}{P}}-O-\text{sugar} \;+\; N-O-\overset{\overset{\displaystyle O}{\|}}{\underset{\underset{\displaystyle O^-}{|}}{P}}-O^- \qquad (7.47)$$

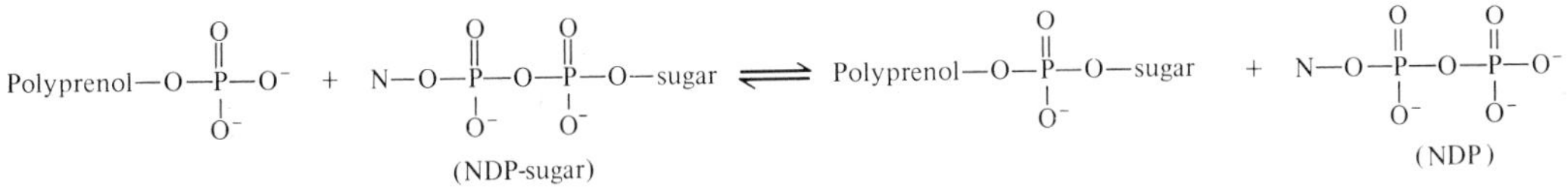

(NDP-sugar)   (NMP)

$$\text{Polyprenol}-O-\overset{\overset{\displaystyle O}{\|}}{\underset{\underset{\displaystyle O^-}{|}}{P}}-O^- \;+\; N-O-\overset{\overset{\displaystyle O}{\|}}{\underset{\underset{\displaystyle O^-}{|}}{P}}-O-\overset{\overset{\displaystyle O}{\|}}{\underset{\underset{\displaystyle O^-}{|}}{P}}-O-\text{sugar} \;\rightleftharpoons\; \text{Polyprenol}-O-\overset{\overset{\displaystyle O}{\|}}{\underset{\underset{\displaystyle O^-}{|}}{P}}-O-\text{sugar} \;+\; N-O-\overset{\overset{\displaystyle O}{\|}}{\underset{\underset{\displaystyle O^-}{|}}{P}}-O-\overset{\overset{\displaystyle O}{\|}}{\underset{\underset{\displaystyle O^-}{|}}{P}}-O^- \qquad (7.48)$$

(NDP-sugar)   (NDP)

the NDP-sugars cannot. They are therefore admirably suited to act as sugar donor molecules in polysaccharide biosynthesis occurring in or on membranes. This is how they function in animal and bacterial cells, since they are concerned in the biosynthesis of antigenic glycoproteins located on the outer surfaces of the plasma membranes.

Plant biochemists have been interested to see whether similar lipid-linked sugars are involved in the biosynthesis of cell-wall polysaccharides. The notion that they might be was an attractive one. It could have gone some way to explaining the difficulties encountered in demonstrating the biosynthesis of cell-wall polysaccharides *in vitro* [see Section D.2(i)–(iii)], in that the wrong sugar donors (i.e. NDP-sugars) were being used. Additionally it is known that the biosynthesis of cell-wall polysaccharides is intimately associated with membranes; the cellulose-synthesizing complex is built into the outer surface of the plasmalemma and the matrix polysaccharides are synthesized within the Golgi cisternae. However, despite numerous investigations there is precious little evidence that lipid-linked sugars are concerned in any way with the biosynthesis of cell-wall polysaccharides.

Nevertheless it is clear that lipid-linked sugars are concerned with the biosynthesis of other polysaccharides in plants; these are the often quite large oligosaccharide moieties of certain glycoproteins. Several classes of glycoproteins can be recognized in higher plants, namely the lectins (or phytoagglutinins), storage glycoproteins, some enzymes and some toxins and allergens.

The lectins are a fascinating group of compounds about which little is known at the present time but which are stimulating a great deal of investigation. They have the ability to bind tightly to and cause the precipitation of, specific polysaccharides and glycoproteins. This enables them to bind to the antigenic glycoproteins on the surface of cells thus causing the cells to agglutinate. They are able to agglutinate red blood cells, in certain cases with such a high degree of specificity that some are used in the typing of human blood. Particularly interesting, and potentially of enormous importance, is their ability to agglutinate malignant cells preferentially. The function of lectins in the plant is still a matter for speculation. It has been suggested that they may be antibodies to counteract soil bacteria or to protect against fungal attack. Alternatively they may be involved in the specific recognition of beneficial organisms such as symbiotic rhizobia by legumes.

In the majority of those plant species which have been shown to have them, the lectins are located in the highest concentration in the seed but they may

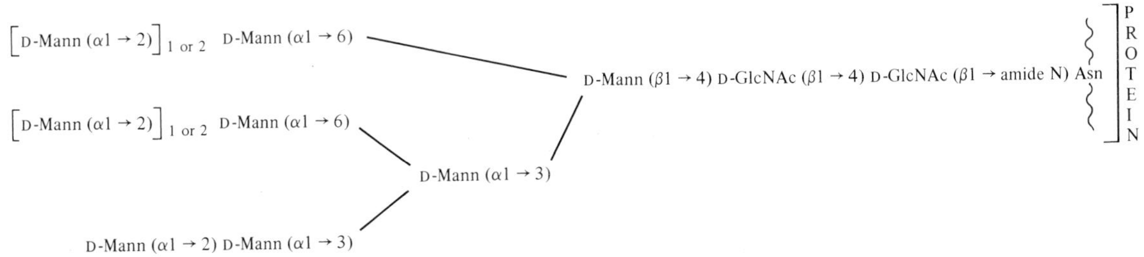

Fig. 7.22. Structure of the oligosaccharide moiety of soyabean agglutinin. (D-Mann = D-mannopyranose; D-GlcNAc = D-N-actyl glucosamine; Asn = asparagine.)

$$\left[\text{D-Mann}\right]_n\text{---D-Mann}\ (\beta 1 \rightarrow 4)\ \text{D-GlcNAc}\ (\beta 1 \rightarrow 4)\ \text{D-GlcNAc}\ (\beta 1 \rightarrow \text{amide N})\ \text{Asn} \quad \left.\begin{array}{c} \\ \\ \end{array}\right] \begin{array}{c} P \\ R \\ O \\ T \\ E \\ I \\ N \end{array}$$

Fig. 7.23. Common structural feature of plant glycoproteins. (D-Mann = D-mannopyranose; D-GlcNAc = D-N-acetylglucosamine; Asn = asparagine.)

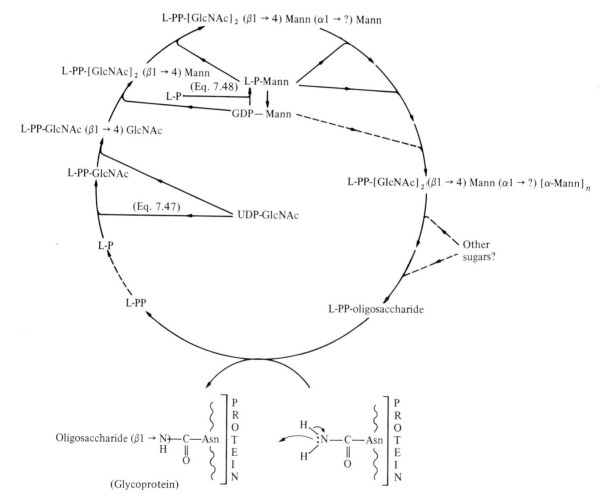

Fig. 7.24. Outline of the probable pathway for the biosynthesis of glycoproteins. (L = polyprenol; Mann = D-mannopyranose; GlcNAc = N-acetylglucosamine; Asn = asparagine; ----▶ = reactions not yet demonstrated; ? in ($\alpha 1 \rightarrow$?) indicates that there is variation in the nature of this glycosidic bond.)

lso be present to a lesser extent in the leaves, roots nd stems. Within the seed the lectins appear to be ocalized in the cytoplasm of the cotyledon(s) and mbryonic cells. It is thought that lectins are ynthesized in the leaves and immediately transfer-ed to the developing seed; the evidence for this is hat the rise in lectin concentration in the de-veloping seed is paralleled by a rapid decrease in its oncentration in mature leaves.

Not all lectins are glycoproteins (e.g. concana-ralin A is 100% protein) but the majority appear to be. Only two of them have been extensively nvestigated to date, soyabean agglutinin and that rom *Phaseolus vulgaris*. The structure of the oligosaccharide component of soyabean agglutinin s shown in Fig. 7.22.

Storage glycoproteins are typified by vicillin, one of the two main storage proteins in the seeds of egumes. Glycoprotein enzymes are typified by pineapple stem bromelain, a proteolytic enzyme esembling papain and ficin (which are, however, not glycoproteins) in activity, and horseradish peroxidase. The glycoprotein toxins and allergens re exemplified by ricin, the toxin of castor beans, nd the pollen allergen of rye grass.

All of the plant glycoproteins studied so far have a ommon structural feature; the oligosaccharide omponent is composed of a series of D-mannopyranose residues, often linked together by $\alpha 1 \rightarrow 2$ glycosidic linkages, linked by a $\beta 1 \rightarrow 4$ inkage to two β-D-N-acetylglucosamine residues which are in turn glycosidically linked to the amide nitrogen atom of an asparagine residue in the protein moiety (see Fig. 7.23). This structural unit is lso found in many glycoproteins of animals and pacteria. The outline of the way in which it is ynthesized in animals and bacteria is now clear; moreover present indications are that this same pathway (see Fig. 7.24) is also operative in plants. The pattern of its biosynthesis is the same as that of ll polysaccharides in that glycosyl residues are ransferred successively from glycosyl donor mole-ules to an acceptor (or primer) molecule. However, n this case the glycosyl donors are polyprenyl phosphate sugars rather than NDP-sugars and the primer is a polyprenyl phosphate rather than an oligosaccharide. Thus the oligosaccharide compo-ent of these glycoproteins is built up on a long, hydrophobic polyprenyl chain. Notice that when

the oligosaccharide is complete the polyprenyl pyrophosphate changes its role from glycosyl ac-ceptor to glycosyl donor because the oligosaccha-ride is transferred to the amide nitrogen of an asparagine residue in the protein moiety of the glycoprotein under construction. The resulting linkage between the terminal N-acetylglucosamine of the oligosaccharide and the asparagine amide nitrogen atom is a $\beta 1 \rightarrow$ amide N-glycosidic linkage; the glycoprotein can thus be regarded as an N-glycoside [see Section B.3(i)(c)].

SUGGESTIONS FOR FURTHER READING

Miller, L. P. (ed.) (1973) *Phytochemistry*, Vol. 1, *The Process and Products of Photosynthesis*, Chapters 6–11, pp. 145–375. Van Nostrand Reinhold Co., New York, London, Toronto, Melbourne.

Briggs, D. E. (1973) 'Hormones and carbohydrates metabolism in germinating cereal grains', in *Biosynthesis and its Control in Plants*, Proc. of Phytochemical Society, Symposium No. 9 (Milborrow, B. V., ed.), pp. 219–277. Academic Press, London & New York.

Pridham, J. B. (ed.) (1974) *Plant Carbohydrate Biochemistry*, Proc. of Phytochemical Society, Symposium No. 10. Chapters 3–6, 8, 9, 12–16. Academic Press, London and New York.

Turner, J. F. and Turner, D. H. (1975) 'The regulation of carbohydrate metabolism', *Ann. Rev. Plant Physiol.* **26**, 159–186.

Banks, W. and Greenwood, C. T. (1975) *Starch and its Components*. Edinburgh University Press, Edinburgh.

Liener, I. E. (1976) 'Phytohaemagglutinins (phytolectins)', *Ann. Rev. Plant Physiol.* **27**, 291–319.

Goldstein, I. J. and Hayes, C. E. (1978) 'The Lectins: carbohydrate-binding proteins of plants and animals', in *Advances in Carbohydrate Chemistry and Biochemistry* (Tipson, R. S. and Horton, D., eds.), Vol. 35, pp. 128–340. Academic Press, London and New York.

Dey, P. M. (1978) 'Biochemistry of plant galactomannans', in *Advances in Carbohydrate Chemistry and Biochemistry* (Tipson, R. S. and Horton, D., eds.), Vol. 35, pp. 341–376. Academic Press, London and New York.

Gabriel, O. and Van Lenten, L. (1978) 'The interconversion of monosaccharides', in *Biochemistry of Carbohydrates*, II (Manners, D. J., ed.) (*International Review of Biochemistry*, Vol. 16, Kornberg, H. L. and Phillips, D. C., Series eds.), pp. 1–36. University Park Press, Baltimore.

Elbein, A. D. (1979) 'The role of lipid-linked saccharides in the biosynthesis of complex carbohydrates', *Ann. Rev. Plant Physiol.* **30**, 239–272.

ENZYMES

1. 3-Phospho-D-glycerate carboxy-lyase (dimerizing), EC 4.1.1.39.
2. Orthophosphate: oxaloacetate carboxy-lyase (phosphorylating), EC 4.1.1.31.
3. ATP: glycerol 3-phosphotransferase, EC 2.7.1.30.
4. sn-Glycerol 3-phosphate: NAD$^+$ 2-oxidoreductase, EC 1.1.1.8.
5. ATP: oxaloacetate carboxy-lyase (transphosphorylating), EC 4.1.1.49.
6. ATP: D-fructose 6-phosphate 1-phosphotransferase, EC 2.7.1.11.
7. Succinate: (acceptor) oxidoreductase, EC 1.3.99.1.
8. L-Malate: NAD$^+$ oxidoreductase, EC 1.1.1.37.
9. D-Fructose 1,6-bisphosphate 1-phosphohydrolase, EC 3.1.3.11.
10. ATP: pyruvate 2-O-phosphotransferase, EC 2.7.1.40.
11. L-Malate: NAD$^+$ oxidoreductase (decarboxylating), EC 1.1.1.39.
12. ATP: D-hexose 6-phosphotransferase, EC 2.7.1.1.
13. ATP: D-glucuronate 1-phosphotransferase, EC 2.7.1.43.
14. ATP: D-galacturonate 1-phosphotransferase, EC 2.7.1.44.
15. ATP: D-xylulose 5-phosphotransferase, EC 2.7.1.17.
16. ATP: L-arabinose 1-phosphotransferase, EC 2.7.1.46.
17. ATP: D-fructose 6-phosphotransferase, EC 2.7.1.4.
18. ATP: D-ribulose 5-phosphate 1-phosphotransferase, EC 2.7.1.19.
19. Pyrophosphate phosphohydrolase, EC 3.6.1.1.
20. Thioglucoside: glucohydrolase, EC 3.2.3.1.
21. ATP: D-glucose 1-phosphate 6-phosphotransferase, EC 2.7.1.10.
22. D-Ribulose 5-phosphate 3-epimerase, EC 5.1.3.1.
23. UDP-D-glucose 4-epimerase (EC 5.1.3.2).
24. UDP-L-arabinose 4-epimerase (EC 5.1.3.5).
25. UDP-D-glucuronate 4-epimerase (EC 5.1.3.6).
26. D-Glucose 6-phosphate ketol-isomerase (EC 5.3.1.9).
27. D-Ribose 5-phosphate ketol-isomerase (EC 5.3.1.6).
28. D-Glyceraldehyde 3-phosphate ketol-isomerase (EC 5.3.1.1).
29. D-Sedoheptulose 7-phosphate: D-glyceraldehyde 3-phosphate glycolaldehyde-transferase, EC 2.2.1.1.
30. D-Sedoheptulose, 7-phosphate: D-glyceraldehyde 3-phosphate dihydroxyacetone-transferase, EC 2.2.1.2.
31. D-fructose 1,6-bisphosphate: D-glyceraldehyde 3-phosphate lyase, EC 4.1.2.13.
32. D-Glucose 6-phosphate: NADP$^+$ 1-oxidoreductase, EC 1.1.1.49.
33. D-Glucono-δ-lactone hydrolase, EC 3.1.1.17.
34. UDP-D-glucose: NAD$^+$ oxidoreductase, EC 1.1.1.22.
35. myo-Inositol: oxygen oxidoreductase, EC 1.13.99.1.
36. L-Gulonate: NAD$^+$ 1-oxidoreductase, EC 1.1.1.19.
37. L-Gulono-γ-lactone hydrolase, EC 3.1.1.18.
38. L-Gulono-γ-lactone: oxygen 2-oxidoreductase, EC 1.1.3.8.
39. UDP-D-glucose: D-fructose 2-α-glucosyltransferase, EC 2.4.1.13.
40. UDP-D-glucose: D-fructose 6-phosphate 2-α-glucosyl transferase EC 2.4.1.14.
41. Sucrose 6^F-phosphate phosphohydrolase, EC 3.1.3.24.
42. β-D-Fructofuranoside fructohydrolase, EC 3.2.1.26.
43. UDP-D-glucose: D-glucose 6-phosphate 1-α-glucosyltransferase, EC 2.4.1.15.
44. Trehalose 6-phosphate phosphohydrolase, EC 3.1.3.12.
45. α,α-Trehalose glucohydrolase, EC 3.2.1.28.
46. ADP-D-glucose: 1,4-α-D-glucan 4-α-glucosyltransferase EC 2.4.1.21.
47. 1,4-α-D-Glucan glucanohydrolase, EC 3.2.1.1.
48. Dextrin 6-α-glucan glucanohydrolase, EC 3.2.1.10.
49. 1,4-(1,3;1,4)-β-D-Glucan 4-glucanohydrolase, EC 3.2.1.4.
50. β-D-Glucoside glucohydrolase, 3.2.1.21.
51. α-D-Galactoside galactohydrolase, EC 3.2.1.22.
52. ADP-D-glucose: 1,4-α-D-glucan 4-α-glucosyltransferase EC 2.4.1.21.
53. 1,4-α-D-Glucan: 1,4-α-D-glucan 6-α-(1,4-α-glucano)-transferase, EC 2.4.1.18.
54. 1,4-α-D-Glucan: orthophosphate α-glucosyltransferase, EC 2.4.1.1.
55. 1,4-α-D-Glucan glucanohydrolase, EC 3.2.1.1.
56. 1,4-α-D-Glucan maltohydrolase, EC 3.2.1.2.
57. Amylopectin 6-glucanohydrolase, EC 3.2.1.41.
58. Dextrin 6-α-glucanohydrolase, EC 3.2.1.10.
59. α-D-Glucoside glucohydrolase, EC 3.2.1.20.
60. ATP: nucleoside monophosphate phosphotransferase, EC 2.7.4.6.
61. (Phyto)glycogen 6-glucanohydrolase, EC 3.2.1.68.
62. Sucrose: 2,1-β-D-fructan 1-β-fructosyltransferase, EC 2.4.1.9 2.4.1.9.
63. Sucrose: 2,6-β-D-fructan 6-β-fructosyltransferase, EC 2.4.1.10 2.4.1.10.
64. UDP-glucuronate carboxy-lyase, EC 4.1.1.35.

CHAPTER 8

Lipid Metabolism

CONTENTS

A. INTRODUCTION

Lipids constitute one of the four major classes of compounds which are found in living tissues, the others being carbohydrates, proteins and nucleic acids.

The term lipid is rather difficult to define. It does not describe a group of compounds with a common, readily identifiable structural feature. It is a term which includes a structurally diverse range of compounds which have two features in common: (i) their presence in living tissues and (ii) their general solubility in organic solvents and insolubility in water. They are extracted as a complex mixture from fresh tissue by organic solvents or solvent mixtures which are wholly or significantly miscible with water, such as ethanol, acetone or chloroform:methanol (2:1, v/v); miscibility with water is required because upwards of 70% of fresh tissue is water. This property of the extracting solvent is not required, however, when lyophilized tissue is extracted.

This definition of the term lipid, based as it is, largely on solubility properties, includes many isoprenoid compounds such as sterols, carotenoids and even chlorophylls. However, since these compounds are discussed in other chapters they are omitted from this one. This leaves as the subject matter of this chapter a broad category of lipids whose structure is characterized by the presence of fatty acid moieties and which are best described as 'acyl lipids'.

It is possible to classify acyl lipids in a number of different ways. The method used in this chapter initially subdivides them into 'neutral' and 'polar'. This subdivision is convenient because it relates directly to their behaviour in the laboratory. Neutral acyl lipids are more readily soluble in non-polar hydrocarbon solvents such as light petroleum and benzene whilst polar acyl lipids are much more soluble in polar solvents like ethanol. Moreover, neutral acyl lipids are held much less strongly on chromatography columns of aluminium oxide and silicic acid and are therefore eluted by non-polar solvents or relatively non-polar solvent mixtures; polar acyl lipids are held much more tightly on such columns and require quite polar solvents or solvent mixtures to elute them.

Neutral lipids are subdivided into glycerides (also called acylglycerols) and waxes; glycerides are fatty acid esters of the trihydroxy alcohol glycerol whilst waxes are predominantly fatty acid esters of long-chain monohydroxy alcohols.

Polar lipids may be subdivided into phospholipids which are diesters of orthophosphoric acid, and glycolipids which contain one or more monosaccharide residues but no orthophosphoric acid.

Phospholipids may be further subdivided into glycerophospholipids and sphingophospholipids, compounds in which one of the two residues bonded to the orthophosphoric acid is a glycerol derivative and a sphingosine derivative respectively.

Glycolipids may be further subdivided into galactosyldiglycerides, cerebrosides and sulpholipids.

Of these acyl lipids, glycerides accumulate as food reserves in seeds and/or the fleshy part of fruits and, apart from their obvious value to the plant, are of enormous commercial importance. Waxes, phospholipids and glycolipids have structural functions. Waxes occur in the cuticular layer that covers the epidermal cells and helps prevent the excessive passage of water either into or out of the plant. Phospholipids and glycolipids are structural components of membranes and as such are distributed throughout the plant.

The nomenclature of acyl lipids was last revised by the IUPAC-IUB Commission on Biochemical Nomenclature in 1976 (see, for example, *Eur. J Biochem.* **79**, 11–21, 1977). Many of the older and more familiar lipid names are still used in this chapter but, where appropriate, a compound is also given its new name; this appears in square brackets after the old name on the first, subsequent occasion that it occurs in the chapter.

B. ACYL LIPID STRUCTURE AND DISTRIBUTION

1. Fatty acids

Over 200 different fatty acids have been identified in the plant kingdom. The majority of these may be classed as 'unusual' since they occur in only a few plant species. Of the remaining widely distributed fatty acids, a few are abundant and occur in most acyl lipids; these may be termed 'major' fatty acids. The rest of the widely distributed fatty acids occur in much smaller quantities and may be termed 'minor' fatty acids.

(i) MAJOR FATTY ACIDS

The major fatty acids are all saturated or unsaturated monocarboxylic acids with an unbranched, even-numbered carbon chain. The unsaturated fatty acids have *cis* double bonds and in the case of the polyunsaturated acids these are arranged in a methylene-interrupted system [see Fig. 8.19(a)].

Table 8.1 lists the major plant fatty acids; these acids are usually found in the lipids from all parts of all plants. Palmitic, oleic and linoleic acids often predominate but in general the saturated acids are less abundant than the unsaturated acids. Chloroplast lipids are particularly rich in α-linolenic acid.

(ii) MINOR FATTY ACIDS

The minor fatty acids fall into two main categories: (i) the saturated- and the *cis*-Δ^9-monounsaturated acids and (ii) the polyunsaturated acids.

The saturated acids generally have an even

Table 8.1. The major plant fatty acids

Common name	Symbol[a]	Structure
Lauric acid	12:0	$CH_3(CH_2)_{10}COOH$
Myristic acid	14:0	$CH_3(CH_2)_{12}COOH$
Palmitic acid	16:0	$CH_3(CH_2)_{14}COOH$
Stearic acid	18:0	$CH_3(CH_2)_{16}COOH$
Oleic acid	18:1 (9c)	$CH_3(CH_2)_7CH{=}CH(CH_2)_7COOH$
Linoleic acid	18:2 (9c, 12c)	$CH_3(CH_2)_3(CH_2CH{=}CH)_2(CH_2)_7COOH$
α-Linolenic acid	18:3 (9c, 12c, 15c)	$CH_3(CH_2CH{=}CH)_3(CH_2)_7COOH$

[a] The number before the colon is the number of carbon atoms in the fatty acid whilst that after the colon is the number of double bonds present; each of the numbers in parenthesis is the lowest numbered carbon atom of the two connected by a given double bond whilst the *c* or *t* which immediately follows it indicates that the double bond has the *cis* or *trans* configuration respectively.

number of carbon atoms and are shorter (e.g. C_6, C_8 and C_{10}) or longer (e.g. C_{20}–C_{30}) than the major saturated acids. Saturated acids with an odd number of carbon atoms are rare but C_{15} and C_{17} acids have been found.

Of the *cis*-Δ^9-monounsaturated acids, palmitoleic (16:0, 9c) and myristoleic (14:0, 9c) are the most common.

Most of the minor polyunsaturated acids belong to one of three biosynthetic families of fatty acid, designated ω9, ω6 and ω3. These designations utilize a strictly biochemical (and non-chemical) method of numbering the fatty acid molecule. The carbon chain is numbered from the methyl, or ω end, rather than the carboxyl end; thus the methyl or ω carbon is C–1 rather than the carboxyl carbon. An ω9 fatty acid, therefore, has a double bond between carbons 9 and 10 when the molecule is numbered from the ω end. This system is of value to the biochemist because fatty acids are elongated *in vivo* by the successive addition of 2C units to the carboxyl end [see Section C.1(i)(c)]; thus the position of a double bond moves with respect to the carboxyl end during elongation but not with respect to the methyl end. Hence the system allows the biosynthetic relationships of unsaturated fatty acids to be seen at a glance. It should be noted that the use of the ω-system is discouraged by the IUPAC-IUB Commission on Biochemical Nomenclature who wish to replace it with what may be called the '*n* minus *x*' system in which *n* is the number of carbons in the fatty acid molecule and *x* is the lower of the two carbon atoms linked together by the double bond, when numbered from the methyl end.

Of the ω9-fatty acids, 11-eicosenoic acid (20:1,

11c) is present in some seed oils, erucic acid (22:1, 13c) is present in the seed oils of the Cruciferae and Tropaeolaceae (particularly in that of rape, *Brassica napus*) and higher homologues are present in other seed oils (e.g. *Ximenia caffra*).

The ω6-fatty acids are frequently found in phytoplankton but less so in higher plants; 11,14-eicosadienoic acid (20:2, 11c, 14c) and γ-linolenic acid (18:3, 6c, 9c, 12c) have been found in members of the Boraginaceae. Arachidonic acid (20:4, 5c, 8c, 11c, 14c) has been found in algae, mosses and ferns but not in higher plants.

The ω3-fatty acids are particularly common in lower members of the plant kingdom; they are generally of chain length C_{16}–C_{22} with methylene-interrupted systems of 3–6 double bonds. However, 7,10,13-hexadecatrienoic acid (16:3, 7c, 10c, 13c) has been found in rape leaves.

(iii) UNUSUAL FATTY ACIDS

This group includes fatty acids which have (i) non-conjugated double bonds which are *trans* or in an unusual position, (ii) conjugated double bond systems, (iii) allenic double bonds, (iv) triple bonds, (v) oxygen functions, e.g. hydroxy, keto, epoxy groups and (vi) branched chains. A selection of these acids is shown in Table 8.2.

2. Neutral lipids

(i) GLYCERIDES

Glycerides [acylglycerols] are fatty acid esters of the trihydroxy alcohol, glycerol. They are the main

Table 8.2. Some unusual plant fatty acids

Unusual feature	Structure[a] (and trivial name)	Source[b]
Trans and/or non-conjugated double bonds	16:1 (3*t*)	Chloroplasts
	18:2 (5*c*, 9*c*)	Conifer leaves
	18:3 (5*t*, 9*c*, 12*c*)	*Thalictrum* spp. (40–50%)
Conjugated double bonds	18:3 (9*c*, 11*t*, 13*c*)	*Punica granatum* (71%)
	18:4 (9*c*, 11*t*, 13*t*, 15*c*)	*Impatiens edgeworthii* (48%)
Allenic double bonds	18:2 (5, 6*al*) (Labellenic acid)	*Leonotis hepetaefolia* (16%)
	18:3 (5, 6*al*, 16*t*)	*Lamium purpureum*
Triple bonds	18:2 (9*c*, 12*a*) (Crepenynic acid)	*Crepis foetida* (60%)
	18:1 (9*a*) (Stearolic acid)	Santalaceae
Hydroxy group	D(+)-12*h*-18:1 (9*c*) (Ricinoleic acid)	*Ricinus communis* (90%)
	D-2*h*-16:0	*Pisum sativum* leaves
Keto group	15*k*-24:1 (18*c*)	*Cuspidaria pterocarpa* (5%)
	17*k*-26:1 (20*c*)	*Cuspidaria pterocarpa* (13%)
Epoxy group	12,13*ep*-18:1 (9*c*) (Vernolic acid)	*Vernolia anthelmintica* (72%)
	15,16*ep*-18:2 (9*c*, 12*c*)	*Camelia sativa*
Branched chain	9,10-methylene-18:1 (9*c*) (Sterculic acid)	*Stercula foetida* (53%)
	13-(2-cyclopentenyl)-13:0 (Chaulmoogric acid)	*Hydnocarpus wightiana* (27%)

[a] The designation of the structure of the fatty acids follows that of Table 8.1; additional abbreviations are: *al* = allenic (-CH=C=CH-); *a* = acetylene (-C≡C-); *h* = hydroxy; *k* = keto; *ep* = epoxy.
[b] A figure in parenthesis gives the percentage by weight of the fatty acid in the seed oil of that species relative to the total weight of fatty acids present.

constituents of natural fats and oils (note that in this context the only difference between a fat and an oil is that the former is a solid at room temperature whilst the latter is a liquid; this is due to the predominance of saturated fatty acid residues in the fat and of unsaturated fatty acid residues in the oil). Natural fats and oils contain minor quantities of other lipids in addition to glycerides.

The vast majority of glycerides in Nature have all three of the glycerol hydroxyl groups esterified with fatty acids and are called triglycerides [triacylglycerols], a term that is often used synonymously with the term 'fat'. Monoglycerides [monoacylglycerols] and diglycerides [diacylglycerols], having one and two of the glycerol hydroxyl groups esterified respectively, do not occur in appreciable amounts in living organisms although they are important metabolic intermediates.

A triglyceride is said to be 'simple' when all the fatty acid residues in the molecule are the same, and 'mixed' when two or more of them are different. Natural fats and oils are largely composed of mixed triglycerides. In theory each of the three glycerol hydroxyl groups may be esterified with any fatty acid. There is, therefore, the possibility of an enormous number of triglyceride species. In practice, however, this is limited by the relatively small number of abundant fatty acids in a given tissue. But even then a surprisingly large number of species are possible; for example, even if only three different fatty acids are available for triglyceride synthesis, say palmitic, oleic and linoleic, eighteen positional isomers can be formed. Moreover, with three different fatty acid residues in the molecule optical isomerism exists because the central carbon of the glycerol residue is asymmetric; there are eighteen enantiomorphic pairs (i.e. 36 optical isomers) of this particular triglyceride. Once again, however, Nature has simplified the situation by (i) synthesizing only one of each pair of enantiomers and (ii) exercising some degree of specificity in respect of which fatty acids may esterify with a given glycerol hydroxyl group; unsaturated fatty acids tend to predominate on the C–2 hydroxyl for instance.

In an attempt to resolve long-standing confusion over the nomenclature of asymmetric glycerol derivatives (this includes triglycerides, glycerophospholipids, and some glycolipids) the IUPAC-IUB Commission on Biochemical Nomenclature has introduced the 'stereospecific numbering' system. The carbon atom that appears at the top of the carbon chain when the glycerol molecule is written as *a Fischer projection with the secondary hydroxyl group (i.e. that on the central carbon) to the left* is designated C–1. To distinguish this numbering system from all others the prefix 'sn' (for stereospecific *numbering*) is used. The application of this system to a triglyceride, a glycerophospholipid and glycerol phosphate (a biosynthetic precursor of both) is shown in Fig. 8.1 (I–IV).

Triglycerides are widely distributed in the plant kingdom. In higher plants they are found in both vegetative and reproductive tissues; however, the

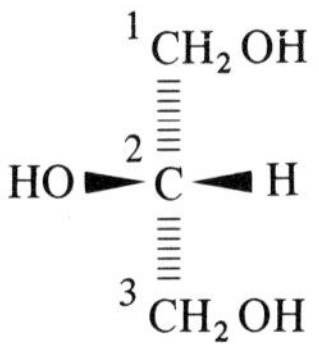

$$^1CH_2OH$$

$$HO \blacktriangleright \overset{2}{C} \blacktriangleleft H$$

$$^3CH_2OH$$

(I) Glycerol
(with its carbon atoms
stereospecifically numbered)

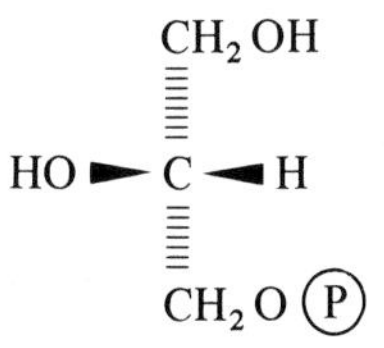

$$CH_2OH$$

$$HO \blacktriangleright C \blacktriangleleft H$$

$$CH_2O\,\text{\textcircled{P}}$$

(II) *sn*-Glycerol 3-phosphate
($\equiv$ L-α-glycerophosphate,
L-glycerol-3-phosphate,
D-glycerol 1-phosphate)*

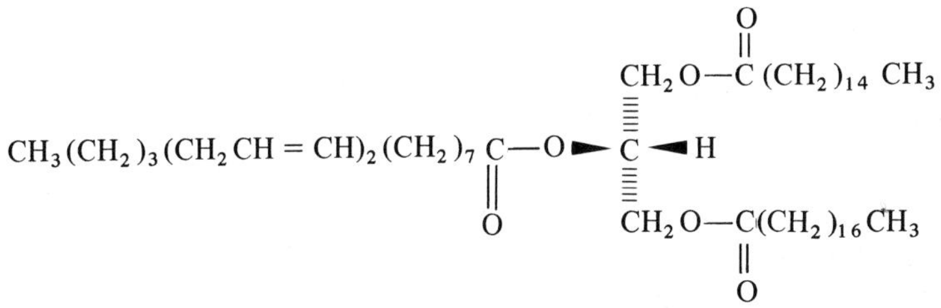

$$CH_2O-\overset{\overset{\textstyle O}{\|}}{C}(CH_2)_{14}\,CH_3$$

$$CH_3(CH_2)_3(CH_2\,CH=CH)_2(CH_2)_7\,\underset{\|}{\overset{}{C}}-O\blacktriangleright C\blacktriangleleft H$$
$$O$$

$$CH_2O-\underset{\|}{\overset{}{C}}(CH_2)_{16}\,CH_3$$
$$O$$

(III) 1-Palmityl-2-linoleyl-3-stearyl-*sn*-glycerol
(a triglyceride or triacylglycerol)

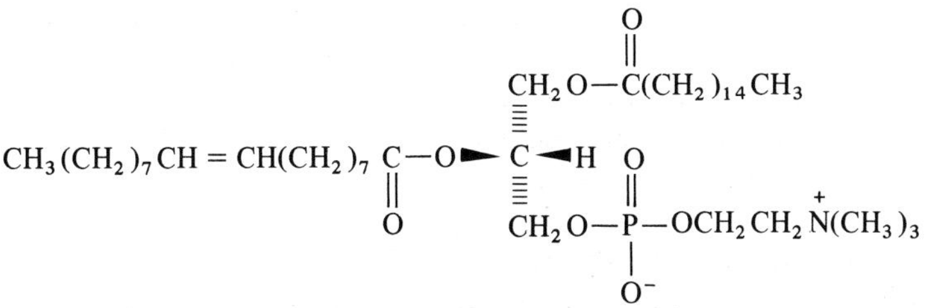

$$CH_2O-\overset{\overset{\textstyle O}{\|}}{C}(CH_2)_{14}CH_3$$

$$CH_3(CH_2)_7\,CH=CH(CH_2)_7\,\underset{\|}{\overset{}{C}}-O\blacktriangleright C\blacktriangleleft H \quad O$$
$$O \qquad CH_2O-\underset{|}{\overset{\|}{P}}-OCH_2\,CH_2\,\overset{+}{N}(CH_3)_3$$
$$O^-$$

(IV) 1-Palmityl-2-oleyl-*sn*-glycero-3-phosphocholine
(or alternatively 3-*sn*-phosphatidylcholine)

Fig. 8.1. Application of the stereospecific numbering system to glycerol and some of its derivatives. (* Older names whose use is now discouraged.)

former contain only small amounts whilst the latter (e.g. fruits and seeds) contain relatively large amounts, particularly in some plant species (e.g. avocado pear mesocarp, castor bean) and frequently represent important food-storage materials for the germinating seedling.

A variation in triglyceride structure occurs in those molecular species which contain a hydroxy fatty acid. Here the hydroxyl group of the fatty acid may also form an ester with another fatty acid. This type of glyceride, however, is rare in plants. The seed oil of *Sapium sebiferum* contains the tetra-acid glyceride shown in Fig. 8.2 (V).

A further variation in triglyceride structure which

has become apparent in recent years involves glycerides which contain an acyl group linked to glycerol by an ether rather than an ester linkage. These compounds are called 'ether triglycerides'. The ether-linked alkyl group is normally on the *sn*-1 carbon of glycerol and either has or has not a Δ^1-double bond (see Fig. 8.2, VI and VII). Only very small quantities of these compounds are found in plants.

Some seed oils contain small amounts of a class of neutral lipids which are not glycerides but which have similar properties; these are usually called 'diol lipids'. They are fatty acid esters of 2C, 3C and 4C dihydroxy alcohols. A typical example is *erythro-*

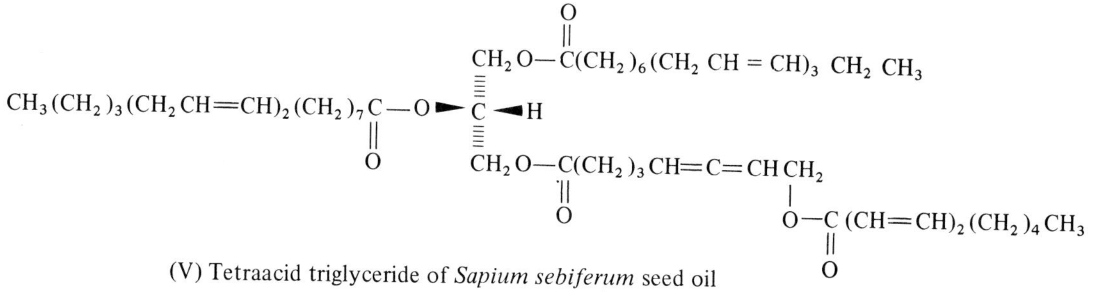

(V) Tetraacid triglyceride of *Sapium sebiferum* seed oil

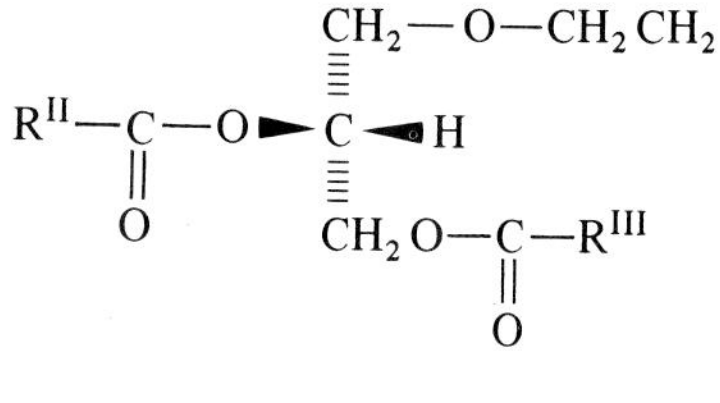

(VI) 1-Alkyl-2,3-diacyl-*sn*-glycerol
(an ether triglyceride)

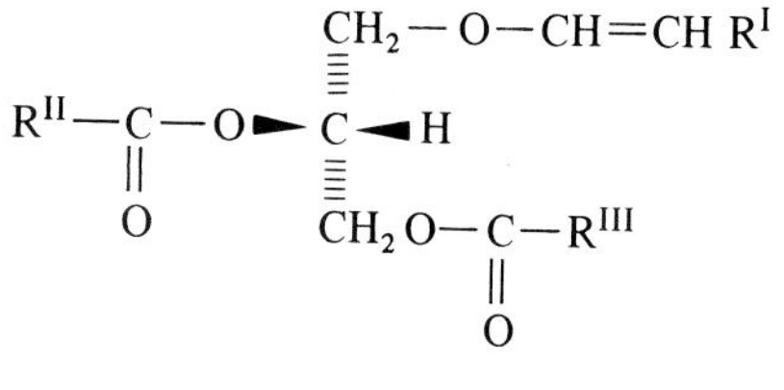

(VII) 1-Δ^1-Alkenyl-2,3-diacyl-*sn*-glycerol
(an ether triglyceride)

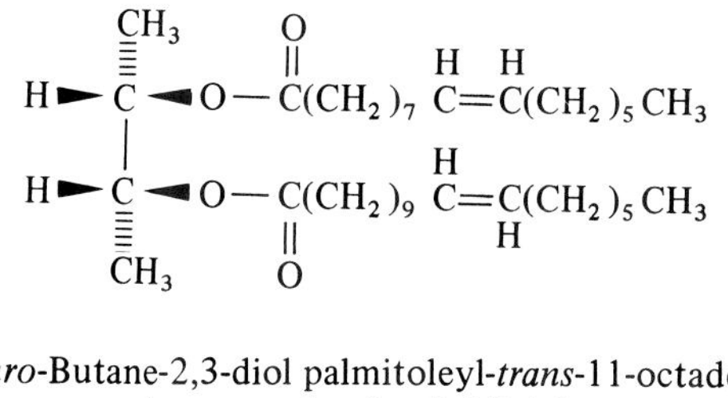

(VIII) *erythro*-Butane-2,3-diol palmitoleyl-*trans*-11-octadecenoate
(an example of a diol lipid)

Fig. 8.2. Structural variants of triglycerides which occur only in minor amounts in plants, particularly in seed oils. (R^I–R^{III} are hydrocarbon chains derived from fatty acids or fatty alcohols.)

butane-2,3-diol palmitoleyl, *trans*-11-octadeceno-ate from *Coix lachrima* seed oil (see Fig. 8.2, VIII).

(ii) Waxes

Waxes occur in the cuticles of plants [see Chapter 4, Section B.5(i)]. They are complex mixtures of highly non-polar lipids. The composition of this mixture is different in different plants and frequently quite characteristic.

The most abundant components of the mixture are wax esters. These are esters of *n*-alkan-1-oic acids and *n*-alkan-1-ols. In the acyl portions, even-numbered, C_{20}–C_{24} carbon chains tend to predominate whilst the alcoholic portion is principally composed of C_{24}–C_{28} carbon chains.

In most plants the wax esters are accompanied by significant amounts of unesterified fatty alcohols and fatty acids. The chain-length distribution of the alcohols is similar to that of the alcohols in the wax esters. The chain-length distribution of the fatty acids, however, is substantially longer than that of the wax ester acids. Fatty aldehydes, with chain lengths similar to those of the alcohols, are present in the wax of some plants.

Hydrocarbons are very common components of waxes. They are predominantly *n*-alkanes with an odd number of carbon atoms in the range C_{25}–C_{35}, of which C_{29} and C_{31} are the most abundant in higher plants. However, there are many exceptions to this generalization.

Oxygenated derivatives of alkanes, such as ketones and secondary alcohols, occur as minor components in the wax of many plants. The oxygen function is usually to be found near the middle of the carbon chain and the chain length distribution of these compounds usually matches that of the alkanes; nonacosane-15-one, for instance, occurs in the wax of plants in which the C_{29} *n*-alkane predominates.

3. Polar lipids

(i) Phospholipids

Phospholipids are diesters of orthophosphoric acid and have the general structure IX (Fig. 8.3). They may be subdivided into glycerophospholipids and sphingophospholipids which have general structures X and XI (Fig. 8.3) respectively.

(a) *Glycerophospholipids*

Glycerophospholipids are, as can be seen from their general structure (X) (Fig. 8.3), esters of 3-*sn*-phosphatidic acid (XII, Fig. 8.4). They can be subdivided into five groups: (i) phosphatidyl-cholines [1,2-diacyl-*sn*-glycero-3-phosphocholine], (ii) phosphatidylethanolamines [1,2-diacyl-*sn*-glycero-3-phosphoethanolamine], (iii) phospha-tidylserines [1,2-diacyl-*sn*-glycero-3-phospho-serine], (iv) phosphatidylinositols [1-(3-*sn*-phos-phatidyl)inositol] and (v) phosphatidylglycerols which include the monophosphatidylglycerols [1,2-diacyl-*sn*-glycero-3-(phospho-1-*sn*-glycerol)] and diphosphatidylglycerols or cardiolipins [1,3-*bis*(3-*sn*-phosphatidyl) glycerol], on the basis of the alcohol esterified to the 3-*sn*-phosphatidic acid. The structures of these alcohols (XIII–XVII) is shown in Fig. 8.4. Within each of the glycerophospholipid subgroups there is a range of molecular species which differ from one another in the fatty acids esterified at the *sn*-1 and *sn*-2 positions of the glycerol residues.

Free phosphatidic acid is seldom a major lipid component of plant tissues but it is an important metabolite. All the subclasses of glycerophospho-lipids are found in plant photosynthetic tissue; the approximate proportion of each, expressed as a percentage of the total phospholipid present, is phosphatidylcholines, 45%; phosphatidylethanol-amines, 10%; phosphatidylserines, trace; phospha-tidylinositols, 8%; monophosphatidylglycerols, 35% and diphosphatidylglycerols, 2%.

Chloroplasts contain monophosphatidylgly-cerols, phosphatidylcholines, phosphatidylinositols and phosphatidylethanolamines in the approximate proportions (expressed as a percentage of the total phospholipid) 60, 25, 10 and 5% respectively. Phosphatidylcholines are found in the membranes of the chloroplast envelope and thylakoid system. Chloroplastidic phosphatidylethanolamines, how-ever, appear to be located solely in the envelope membranes. The fatty acid *trans*-Δ^3-hexadecenoic acid is found only in chloroplasts; it is located

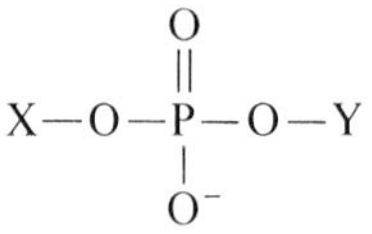

$$\text{X}-\text{O}-\overset{\overset{\displaystyle O}{\|}}{\underset{\underset{\displaystyle O^-}{|}}{\text{P}}}-\text{O}-\text{Y}$$

Where X and Y are alcoholic residues esterified to orthophosphoric acid

(IX) General structure of a phospholipid

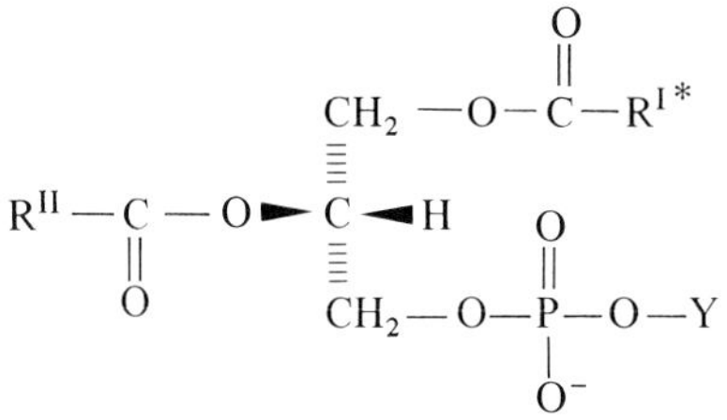

$$R^{II}-\overset{\overset{\displaystyle}{}}{\underset{\underset{\displaystyle O}{\|}}{C}}-O\blacktriangleright\ C\blacktriangleleft H \qquad \begin{array}{l}CH_2 \ —O—\overset{\overset{\displaystyle O}{\|}}{C}—R^{I*}\\[4pt] \\ CH_2—O—\overset{\overset{\displaystyle O}{\|}}{\underset{\underset{\displaystyle O^-}{|}}{P}}—O—Y\end{array}$$

Where the X in (IX) above is an sn-1,2-diglyceride (sn-1,2-diacylglycerol) and Y is some other alcoholic residue, eg. choline; R^I and R^{II} are fatty acid residues

(X) General structure of a glycerophospholipid (*Plasmalogens, rare in plants, have a Δ^1-alkenyl residue in the sn-1 position, as in structure VII)

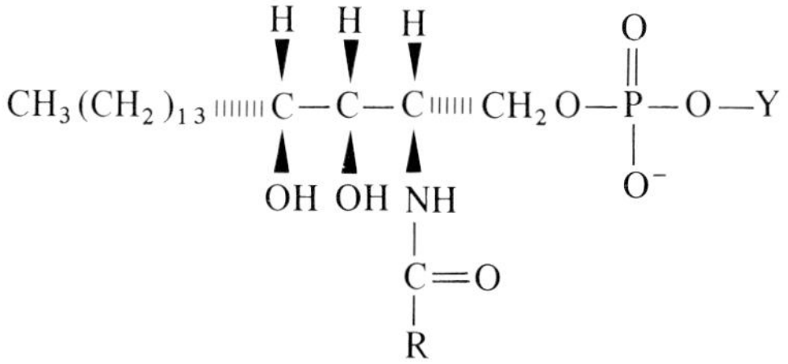

$$CH_3(CH_2)_{13}\text{||||||}\overset{\overset{\displaystyle H\ \ H\ \ H}{\blacktriangledown\ \blacktriangledown\ \blacktriangledown}}{\underset{\underset{\displaystyle OH\ OH\ NH}{\blacktriangle\ \ \blacktriangle\ \ \blacktriangle}}{C-C-C}}\text{||||||}CH_2O—\overset{\overset{\displaystyle O}{\|}}{\underset{\underset{\displaystyle O^-}{|}}{P}}—O—Y$$

$$\underset{\underset{\displaystyle R}{|}}{\overset{\overset{\displaystyle}{}}{C=O}}$$

Where the X of (IX) above is an N-acyl derivative of phytosphingosine (4D-hydroxysphinganine) and Y is an oligosaccharide; R is a fatty acid residue

(XI) General structure of a plant sphingophospholipid (NB. Sphingomyelins, in which X and Y of (IX) above are N-acylsphingosine (N-acyl-trans-4-sphingenine) and choline (or ethanolamine) respectively, are not found in plants)

Fig. 8.3. Structures of phospholipids.

specifically in the sn-2 position of the monophosphatidylglycerols.

The major phospholipids of mitochondria are phosphatidylcholines and phosphatidylethanolamines but significant amounts of phosphatidylinositols are also present. It has recently been found that diphosphatidylglycerol is present only in mitochondria; moreover, within the mitochondria it is localized in the inner membrane.

The glycerophospholipids are only minor components of the lipids of seeds, where triglycerides predominate.

(b) *Sphingophospholipids*

The key structural features of sphingophospholipids are the residues of (i) orthophosphoric acid and (ii) a sphinganine derivative. The latter, however, is not confined to sphingophospholipids; it is also a structural component of cerebrosides [see Section B.3(ii)(c)] which in this chapter are classified as glycolipids since they are not orthophosphoric acid esters. Alternative systems of classification combine the sphingophospholipids and cerebrosides as a distinct class known as sphingolipids

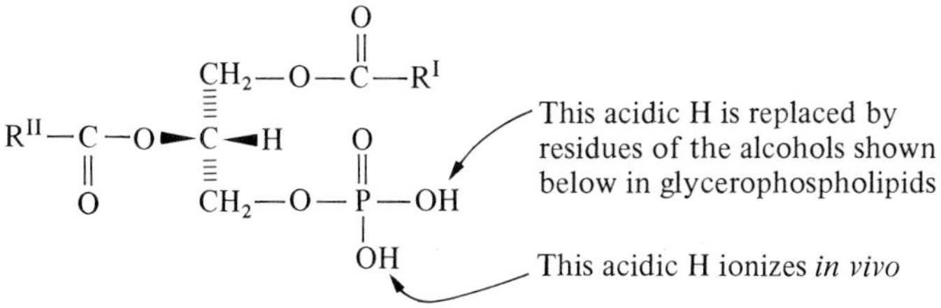

$$CH_2-O-\overset{\overset{\displaystyle O}{\|}}{C}-R^I$$

$$R^{II}-\overset{\|}{\underset{O}{C}}-O\blacktriangleright\overset{\equiv}{\underset{\equiv}{C}}\blacktriangleleft H \qquad \overset{\displaystyle O}{\|}$$

$$CH_2-O-\overset{\displaystyle}{\underset{\displaystyle OH}{P}}-OH$$

This acidic H is replaced by residues of the alcohols shown below in glycerophospholipids

This acidic H ionizes *in vivo*

(XII) 3-*sn*-Phosphatidic acid (formerly L-α-Phosphatidic acid)

Glycerophospholipid	Alcohol	Residue (attached to XII; ionized *in vivo*)
Phosphatidylcholine	$HOCH_2CH_2\overset{+}{N}(CH_3)_3$ (XIII) Choline	$-CH_2CH_2\overset{+}{N}(CH_3)_3$
Phosphatidylethanolamine	$HOCH_2CH_2NH_2$ (XIV) Ethanolamine	$-CH_2CH_2\overset{+}{N}H_3$
Phosphatidylserine	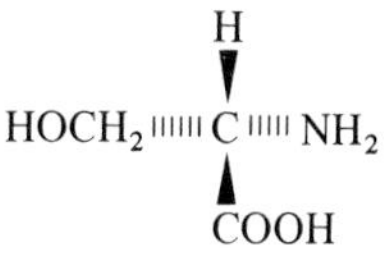 $HOCH_2\overset{\overset{\displaystyle H}{\uparrow}}{\underset{\underset{\displaystyle COOH}{\uparrow}}{C}}NH_2$ (XV) L-Serine	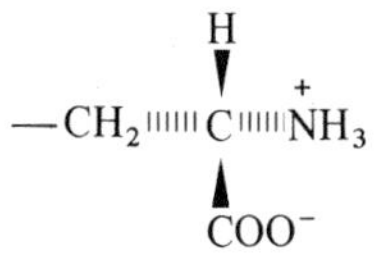 $-CH_2\overset{\overset{\displaystyle H}{\uparrow}}{\underset{\underset{\displaystyle COO^-}{\uparrow}}{C}}\overset{+}{N}H_3$
Phosphatidylinositol	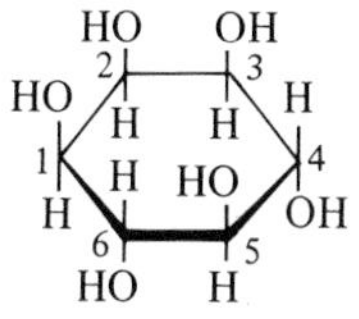 (XVI) *myo*-Inositol	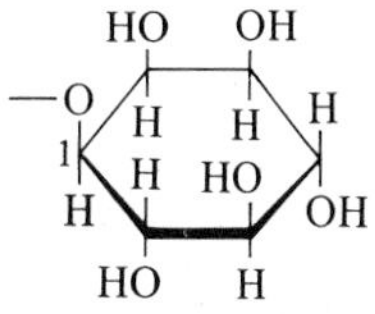
Monophosphatidyl-glycerol	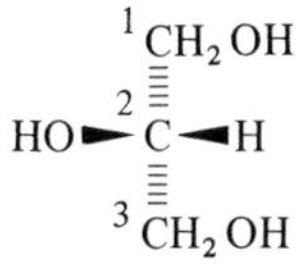 $^1CH_2OH$ $HO\blacktriangleright\overset{2}{C}\blacktriangleleft H$ $^3CH_2OH$ (XVII) Glycerol	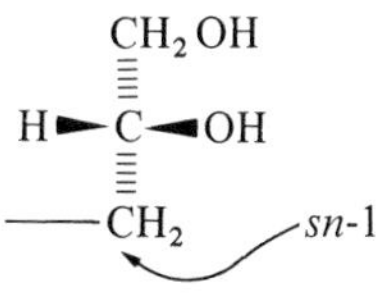 CH_2OH $H\blacktriangleright C\blacktriangleleft OH$ $-CH_2$ *sn*-1
Diphosphatidylglycerol (Cardiolipin)		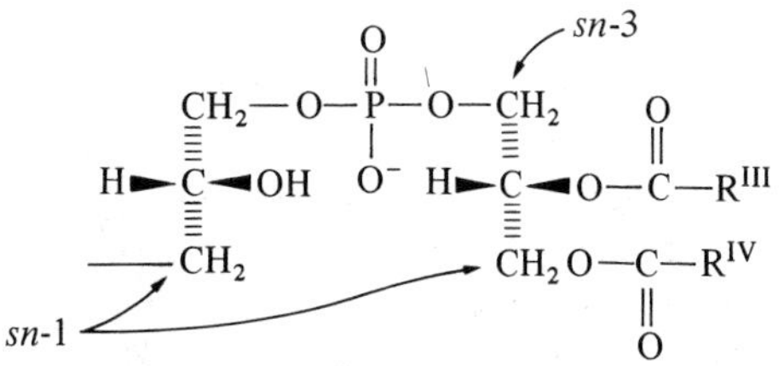

Fig. 8.4. Structures of glycerophospholipids. (R^I–R^{IV} are fatty acid residues.)

which are characterized by the presence of a sphinganine derivative.

The structure of the parent compound, sphinganine (2*S*, 3*R*-2-amino-1,3-octadecanediol), and its main derivatives, sphingosine [*trans*-4-sphingenine] and phytosphingosine (4D-hydroxysphinganine], are shown in Fig. 8.5 (XVIII–XX respectively).

A group of sphingophospholipids having the general structure XI (Fig. 8.3) has been found in the seeds of several plant species. They differ from one another in the nature of the oligosaccharide linked to the orthophosphoric acid residue; however, the way in which some of the monosaccharide components are linked together in the various oligosaccharide residues remains to be elucidated. The partial structures of the sphingophospholipids from maize seeds (*Zea mays*) and peanut (*Arachis hypogea*) are shown in Fig. 8.6 (XXI and XXII respectively). Similar compounds have been reported in the yeasts, *Candida utilis* and *Saccharomyces cerevisiae*, and the green alga, *Scenedesmus obliquus*.

It should be noted that the main sphingophospholipids of animals, the sphingomyelins, are not present in plants; the general structure of the sphingomyelins is given in Fig. 8.6 (XXIII) for comparison purposes.

(ii) GLYCOLIPIDS

Glycolipids are structurally characterized by the presence of one or more monosaccharide residues and the absence of a phosphoric acid residue. They are *O*-glycosides [see Chapter 7, Section B.3(i)(c)] of either glycerol or a sphinganine derivative.

(a) *Galactosyldiglycerides*

Monogalactosyldiglycerides (MGDG) and digalactosyldiglycerides (DGDG) (XXIV and XXV, Fig. 8.7) were originally isolated from wheat flour; their structures were elucidated by Carter in 1961. A trigalactosyldiglyceride (TGDG) (XXVI, Fig. 8.7), isolated from potato tubers, was characterized by Galliard in 1969 and it is possible that even higher homologues exist in plants.

MGDGs and DGDGs have now been shown to be present in a wide variety of higher plant tissues. Early analyses of subcellular fractions led to the belief that they were located solely in the chloroplasts. Subsequent work has shown that whilst most of the cellular MGDG and DGDG is located in the chloroplasts, some is also present in mitochondria. Nevertheless the galactosyldiglycerides are the predominant lipid of chloroplasts. Within the chloroplasts they are located in the membranes; the ratio of MGDG:DGDG in the membranes of the envelope is about 0.9:1 whilst that in the membranes of the thylakoid system is about 2:1.

MGDGs and DGDGs are also present in algae, including the blue-green algae (cyanobacteria).

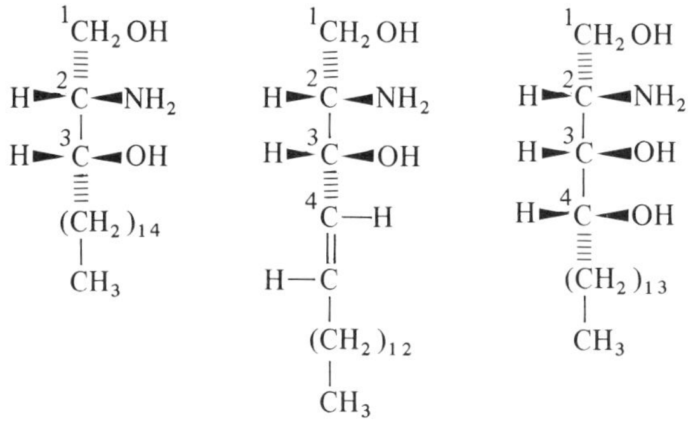

(XVIII) Sphinganine, (2*S*,3*R*)-2-amino-1,3-octadecanediol.

(XIX) Sphingosine, [4-sphingenine], (2*S*,3*R*,4*E*)-2-amino-4-octadecene-1,3-diol.

(XX) Phytosphingosine, [4D-hydroxysphinganine], (2*S*,3*S*,4*R*)-2-amino-1,3,4-octadecanetriol.

Fig. 8.5. Structure of sphinganine and its main derivatives.

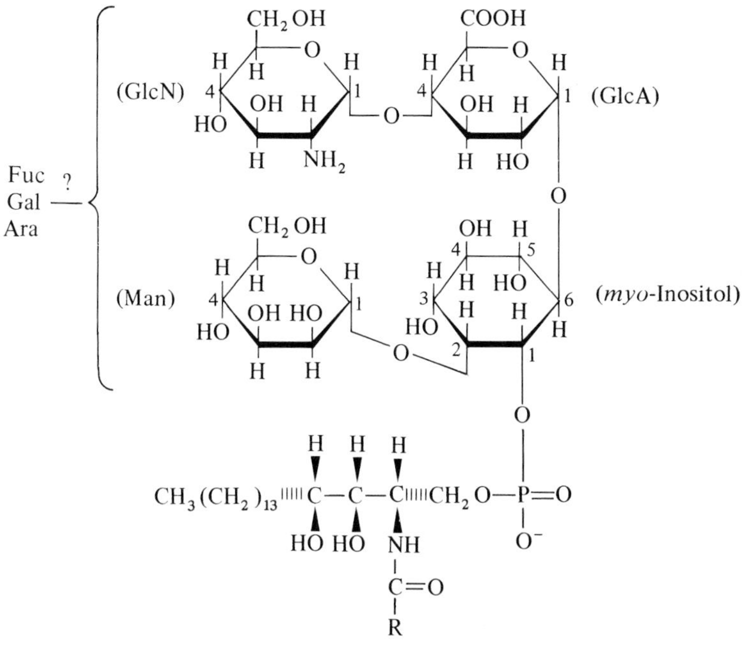

(XXI) Partial structure of the sphingophospholipid from maize seeds

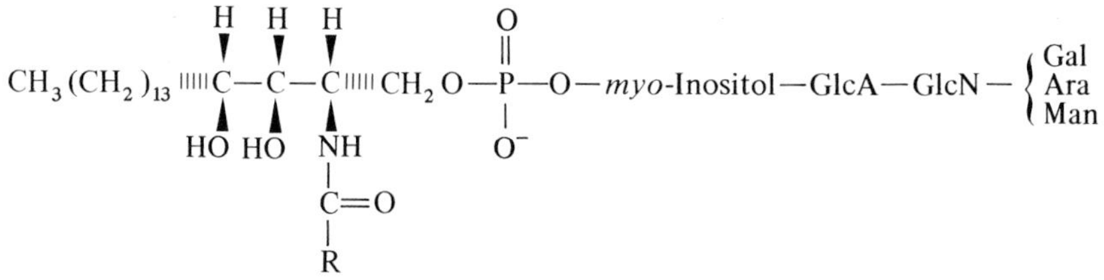

$$CH_3(CH_2)_{13}\text{''''}C-C-C\text{''''}CH_2O-P-O-myo\text{-Inositol}-GlcA-GlcN-\begin{cases}Gal\\Ara\\Man\end{cases}$$

(XXII) Partial structure of the sphingophospholipid from peanuts

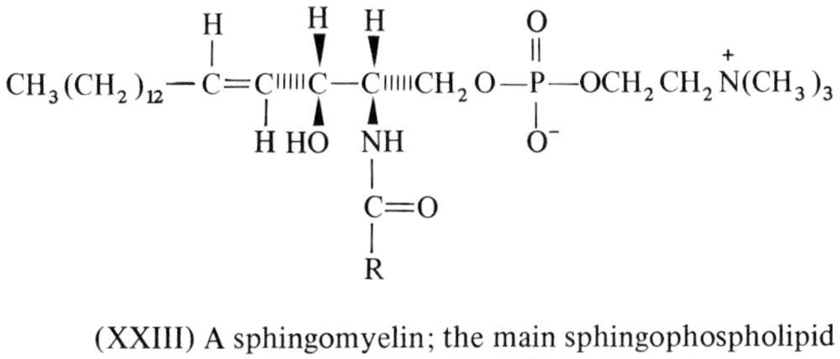

$$CH_3(CH_2)_{12}-C=C\text{''''}C-C\text{''''}CH_2O-P-OCH_2CH_2N(CH_3)_3^+$$

(XXIII) A sphingomyelin; the main sphingophospholipid of animals but which is not present in plants

Fig. 8.6. Structures of sphingophospholipids. (GlcN = D-glucosamine; GlcA = D-glucuronic acid; Man = D-mannose; Gal = D-galactose; Ara = L-arabinose; Fuc = L-fucose; R = a fatty acid residue.)

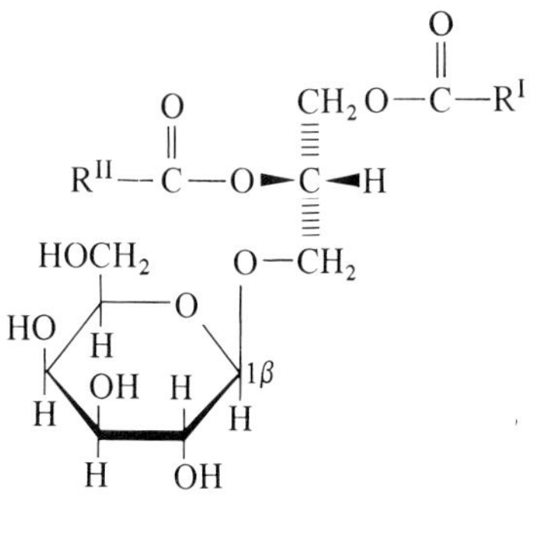

(XXIV) A monogalactosyldiglyceride

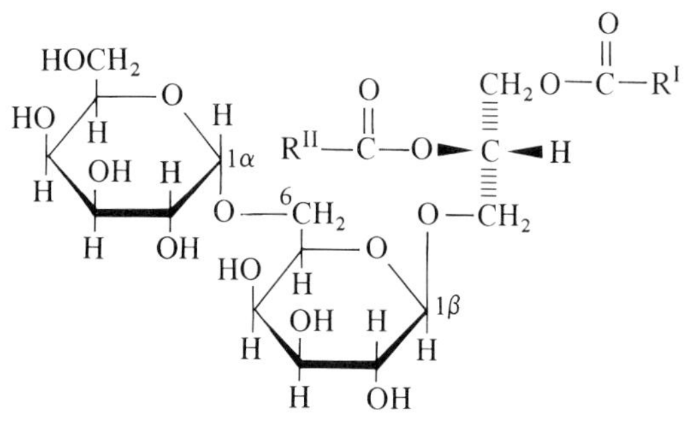

(XXV) A digalactosyldiglyceride

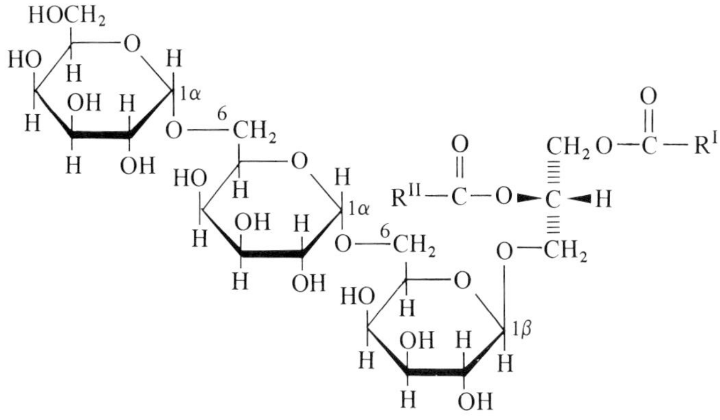

(XXVI) A trigalactosyldiglyceride

Fig. 8.7. Structures of galactosyldiglycerides. (R^I and R^{II} are fatty acid residues.)

Analysis of the fatty acid moieties of higher plant galactosyldiglycerides allows the following generalizations to be made: (i) although both MGDGs and DGDGs are highly unsaturated lipids, the former are the most unsaturated, (ii) hexadecatrienoic acid, when present, occurs mainly in MGDGs whilst palmitic acid occurs mainly in DGDGs, (iii) unsaturated fatty acids predominate at the sn-2 position in both MGDGs and DGDGs.

(b) *Sulpholipids*

The only sulpholipid present in higher plant tissues is the sulphonolipid, often called the plant sulpholipid (XXVII, Fig. 8.8), discovered by Benson in 1959. The glycosyl moiety, 6-sulpho-6-deoxy-D-glucose (6-sulpho-D-quinovose), is linked by an α-glycosidic bond to the sn-3 position of 1,2-diacyl-sn-glycerol. It should be noted that the glycosyl moiety is a sulphonate (C—S bond) and not a sulphate ester (C–O–S linkage). The predominant fatty acid is α-linolenic acid but more saturated fatty acids, such as palmitic acid, may also be present.

In photosynthetic tissues the sulphonolipid is localized in the chloroplasts, predominantly in the membranes of the thylakoid system. It is also present in etiolated tissues, presumably in the etioplasts, and in tissues containing only proplastids, though in lower amounts. Tiny amounts (~1% of total lipid) have been found in tissues such as potato tuber and apple fruit.

Although the sulphonolipid is clearly a chloroplastidic lipid it is much less abundant than the galactosyldiglycerides; in a survey of twenty plant species the sulphonolipid:galactosyldiglyceride ratio ranged from 1:8 to 1:46.

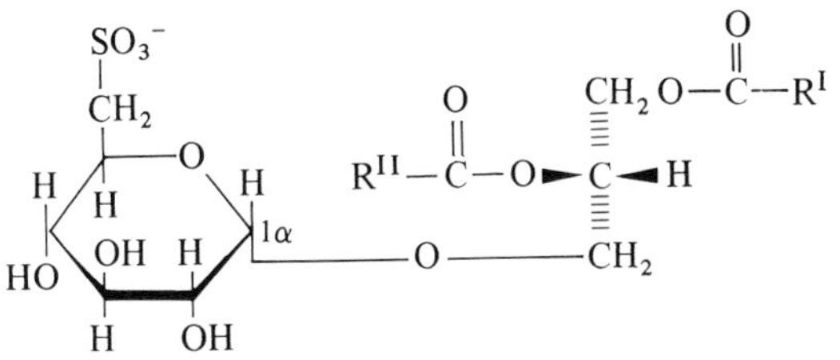

(XXVII) The plant sulpholipid (sulphonolipid)

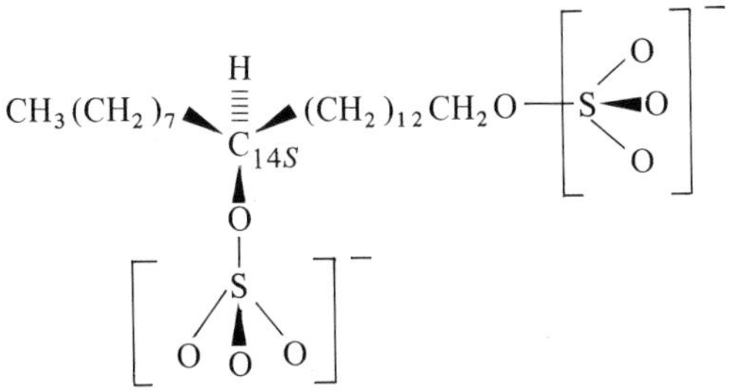

(XXVIII) *n*-Docosane-1,14(S)-diol disulphate

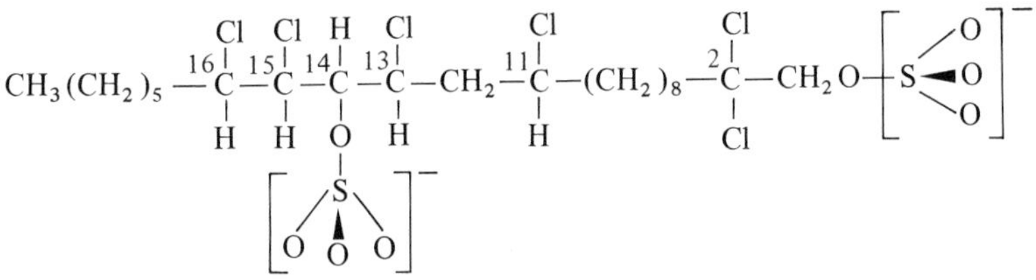

(XXIX) 2,2,11,13,15,16-Hexachloro-*n*-docosane-1,14-diol disulphate

Fig. 8.8. Structures of sulpholipids. (R^I and R^{II} are fatty acid residues.)

An interesting class of sulpholipids, called chloro-sulpholipids, was discovered in 1969 in the Chrysophyte alga, *Ochromonas danica*, in which it constitutes about 15% of the total lipid. These lipids have since been shown to occur in a wide range of fresh-water algae, though in much smaller amounts, but not in higher plants. Though they are not glycolipids, it is convenient to discuss them in this section.

There are two series of chlorosulpholipids, one based on *n*-docosane-1,14-diol disulphate (XXVIII, Fig. 8.8) and the other, much less abundant, based on *n*-tetracosane-1,15-diol disulphate. The 13-chloro-, 11,15-dichloro-, 2,2,11,13,15-pentachloro- and 2,2,11,13,15,16-hexachloro- (XXIX, Fig. 8.8)

derivatives of *n*-docosane-1,14-diol disulphate have been identified, along with several partly character-ized tri- and tetrachloro derivatives. The 14-chloro-, 2,12,14,16,17-pentachloro- and 2,2,12,14,16,17-hexachloro derivatives of *n*-tetracosane-1,15-diol disulphate have also been detected along with partly characterized di- and trichloro derivatives.

(c) *Cerebrosides*

Cerebrosides (XXX, Fig. 8.9) are composed of a monosaccharide residue glycosidically linked to C–1 of an *N*-acylated sphinganine derivative, known as a ceramide. In plants the monosaccharide

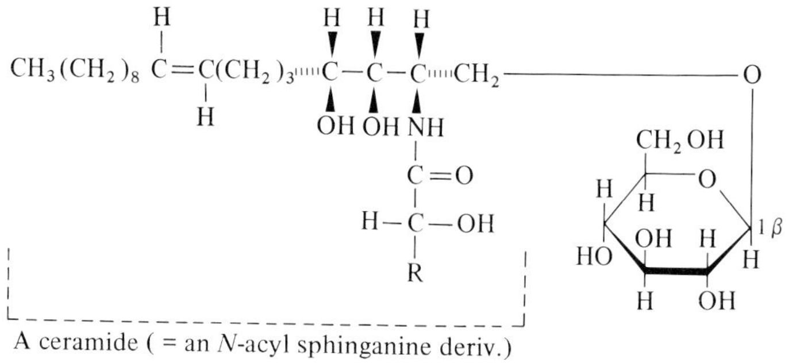

$$CH_3(CH_2)_8 \; \overset{\displaystyle H}{\underset{\displaystyle H}{C}}{=}C(CH_2)_3\text{''''}\overset{\displaystyle H}{\underset{\displaystyle OH}{C}}{-}\overset{\displaystyle H}{\underset{\displaystyle OH}{C}}{-}\overset{\displaystyle H}{\underset{\displaystyle NH}{C}}\text{''''}CH_2 \longrightarrow O$$

$$\overset{|}{C}{=}O$$

$$H{-}\overset{|}{\underset{|}{C}}{-}OH$$

$$R$$

A ceramide (= an *N*-acyl sphinganine deriv.)

Fig. 8.9. Structure of a plant cerebroside (XXX). (R is a 2-hydroxy fatty acid residue.)

residue is always D-glucose; this contrasts with animal cerebrosides where it is usually D-galactose. The sphinganine derivative is mainly *trans*-8-dehydrophytosphingosine along with phytosphingosine (XX, Fig. 8.5); cerebrosides with C_{16}, C_{19} and C_{20} analogues of phytosphingosine have also been reported. The fatty acid component is usually a saturated C_{16}, C_{22} or C_{24} 2-hydroxy acid.

Cerebrosides were first demonstrated in plants by Carter who isolated them from wheat flour in 1961. Sastry and Kates discovered them in runner bean leaves in 1964 and since then they have been found in many photosynthetic tissues, in potato tubers, apple fruit and fungi.

C. METABOLISM OF LIPIDS

1. Metabolism of fatty acids

(i) FATTY ACID BIOSYNTHESIS

An outline of the biosynthetic route to long chain, saturated and unsaturated fatty acids is shown in Fig. 8.10. The route may be divided into three distinct phases. Firstly, acetyl-CoA is carboxylated to yield malonyl-CoA; this reaction is catalysed by acetyl-CoA carboxylase[1]. Secondly, C_2 units derived from malonyl-CoA condense together to form a fatty acid of intermediate chain length which is usually palmitic acid (C_{16}). This is a multistep process and is catalysed by a multienzyme complex called the fatty acid synthase. For the condensation process to begin acetyl-CoA is required as a primer.

The methyl and carbonyl carbons of the acetyl residue become the ω- and next-to-ω carbons respectively of the fatty acid product of the synthase, all the other carbons being derived from malonyl residues (see Fig. 8.10). Thirdly, the range of very long chain, saturated and unsaturated fatty acids are derived from the palmitic acid by the concerted action of fatty acid elongation and desaturation systems.

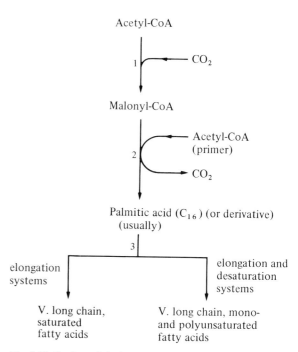

Fig. 8.10. Outline of the biosynthetic route to very long-chain, saturated and unsaturated fatty acids.

(a) *Formation of malonyl-CoA*

Malonyl-CoA is the key intermediate in fatty acid biosynthesis. It may be formed in two different ways in plants. The most important of these is the carboxylation of acetyl-CoA [eq. (8.1)]; the other is via the reaction catalysed by malonate thiokinase [eq. (8.2)].

Acetyl-CoA carboxylase, the enzyme catalysing eq. (8.1), is a multienzyme complex. In prokaryotic organisms (e.g. the much-investigated bacterium, *Escherichia coli*), this complex readily dissociates into three different proteins, two enzymes and a biotin-containing coenzyme (MW 22,500) called biotin carboxyl carrier protein (BCCP). The two enzymes are biotin carboxylase[2] (MW 98,000) and a carboxyltransferase (MW 130,000) which catalyse reactions (8.3) and (8.4) respectively.

In animal cells acetyl-CoA carboxylase is more complex and its structure appears to be different in different tissues. The chicken liver enzyme may exist in two forms. In the absence of citrate it occurs as an inactive monomer (MW 475,000–500,000) which is composed of four different proteins tightly bound together; one of these proteins (MW 117,000) contains covalently bound biotin and is therefore BCCP. Two of the others are biotin carboxylase and carboxyltransferase whilst the fourth has been referred to as the citrate-binding protein. In the presence of citrate about twenty inactive monomers bind together to form the catalytically active form of the enzyme.

In higher plants acetyl-CoA carboxylase resembles that of *E. coli* more closely than that of animals. However, there appear to be at least two different acetyl-CoA carboxylases in higher plants. The enzyme from wheat germ has been purified. Like that of *E. coli*, it can be separated into BCCP and the two enzymes; however, unlike that of *E. coli*, the three components tend to reaggregate readily and the two enzymes have much higher molecular

weights. The enzyme from spinach chloroplasts is peculiar in that the BCCP component is tightly bound to the thylakoid membranes whilst the two enzymes are present in the stroma. The biotin carboxylase component is activated by Mn^{2+} and is quite stable whilst the carboxyltransferase is activated by Mg^{2+} and is labile.

The mechanism of the reaction catalysed by biotin carboxylase [eq. (8.3)] has not been fully elucidated yet. Two postulated mechanisms are shown in Fig. 8.11. The concerted reaction (Fig. 8.11; mechanism 1) has been generally accepted for some time but the more recently suggested route via the transient intermediates *O*-phosphobiotin-BCCP and 1'-*N*-carbonylphosphate biotin-BCCP is also attractive (Fig. 8.11; mechanism 2). Present experimental evidence is consistent with both mechanisms.

The mechanism of the reaction catalysed by the transcarboxylase [eq. (8.4)] has also not been fully elucidated yet. Two postulated mechanisms are shown in Fig. 8.12. The usually accepted view is that the carboxyl group of 1'-*N*-carboxybiotin-BCCP is transferred to the methyl carbon of acetyl-CoA following the nucleophilic attack of an enolate anion on the carboxylate carbon (Fig. 8.12; mechanism 2). Recent work with other carboxyltransferases has, however, suggested that the 2'-carbonyl moiety of the 1'-*N*-carboxybiotin-BCCP assists in the removal of a hydrogen from the methyl group of acetyl-CoA thereby facilitating the nucleophilic attack of the methyl carbon on the carboxylate carbon (Fig. 8.12; mechanism 1).

(b) *Utilization of malonyl-CoA to form palmitic acid*

The malonyl-CoA formed by acetyl-CoA carboxylase is utilized to generate a straight-chain, saturated fatty acid with an even number of carbon

$$CO_2 + CH_3 \cdot CO \cdot S \cdot CoA + ATP + H_2O \rightarrow HOOC \cdot CH_2 \cdot CO \cdot S \cdot CoA + ADP + P_i \qquad (8.1)$$

$$HOOC \cdot CH_2 \cdot COOH + HS \cdot CoA + ATP \rightarrow HOOC \cdot CH_2 CO \cdot S \cdot CoA + AMP + PP_i \qquad (8.2)$$

$$CO_2 + BCCP + ATP \rightarrow BCCP \cdot CO_2 + ADP + P_i \qquad (8.3)$$

$$BCCP \cdot CO_2 + CH_3 \cdot CO \cdot S \cdot CoA \rightarrow HOOC \cdot CH_2 \cdot CO \cdot S \cdot CoA + BCCP \qquad (8.4)$$

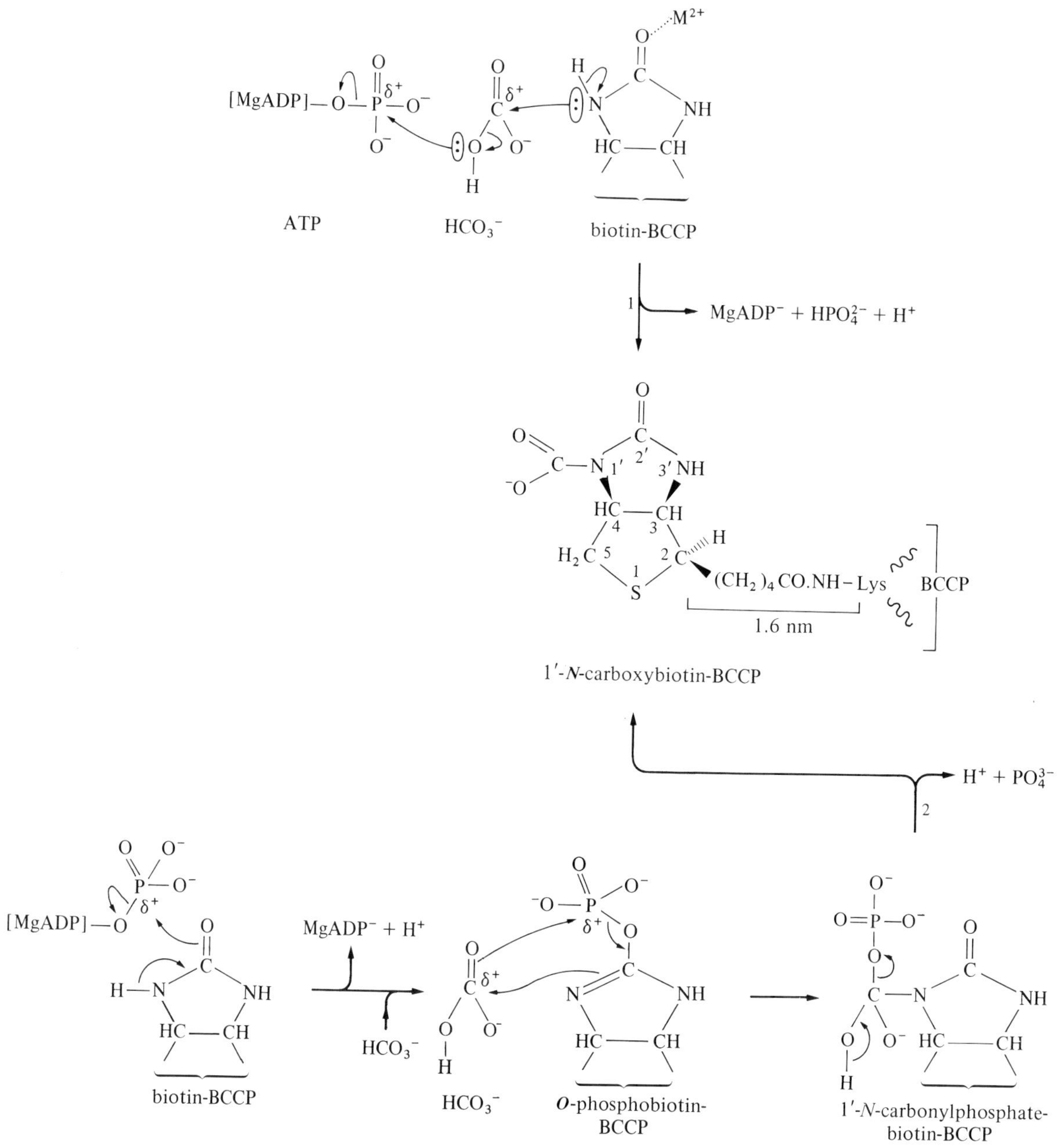

Fig. 8.11. Two postulated mechanisms for the reaction catalysed by the biotin carboxylase component of acetyl-CoA carboxylase (EC 6.4.1.2).

atoms; the usual product is palmitic acid which has sixteen carbon atoms. This is multistep process in which the fatty acid is built up in a stepwise manner by the successive addition of two-carbon units derived from malonyl-CoA to an acyl chain which originates from the acetyl moiety of acetyl-CoA.

Thus acetyl-CoA, as well as malonyl-CoA, is involved in the process; it acts as the primer and contributes the two carbon atoms which reside at the ω end of the palmitic acid molecule.

The process is catalysed by a multienzyme complex called the fatty acid synthase and is

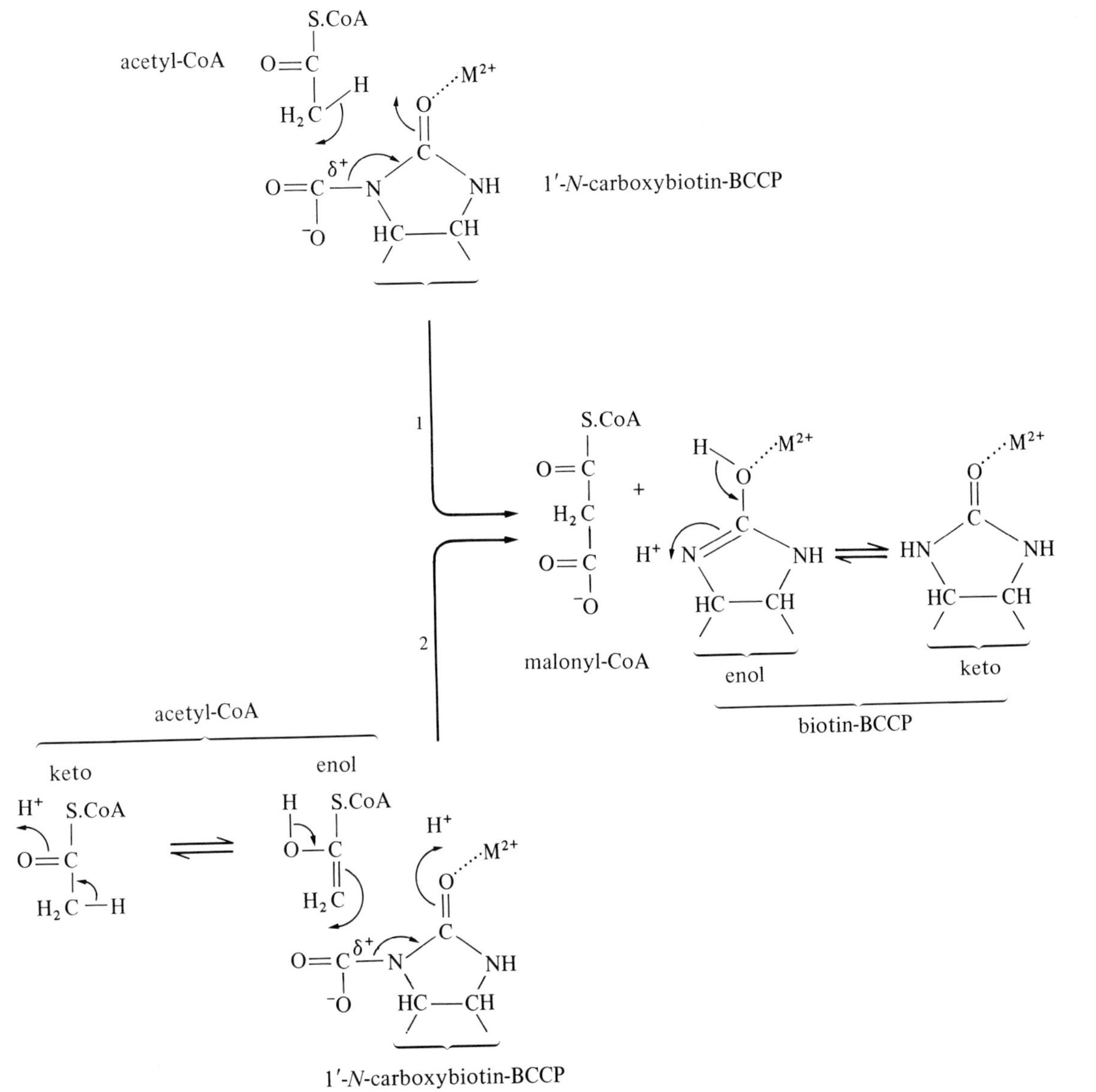

acetyl-CoA

1'-N-carboxybiotin-BCCP

malonyl-CoA

biotin-BCCP

enol   keto

acetyl-CoA

keto   enol

1'-N-carboxybiotin-BCCP

Fig. 8.12. Two postulated mechanisms for the reaction catalysed by the transcarboxylase component of acetyl-CoA carboxylase (EC 6.4.1.2).

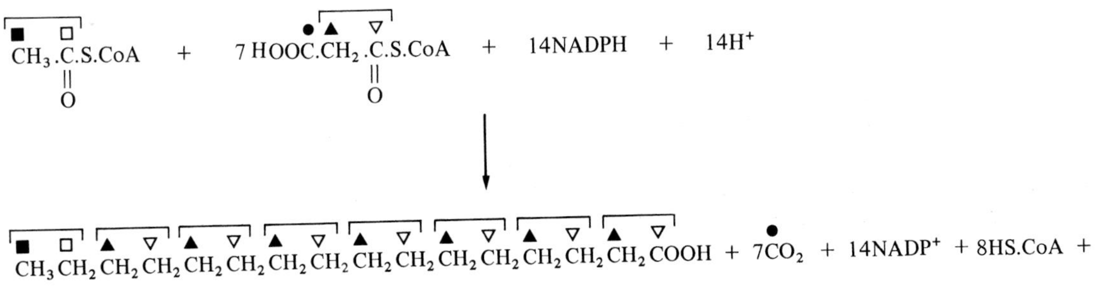

$$CH_3.C.S.CoA \quad + \quad 7\,HOOC.CH_2.C.S.CoA \quad + \quad 14NADPH \quad + \quad 14H^+$$

$$\text{(8.5)}$$

$$CH_3CH_2CH_2CH_2CH_2CH_2CH_2CH_2CH_2CH_2CH_2CH_2CH_2CH_2CH_2COOH \; + \; 7CO_2 \; + \; 14NADP^+ \; + \; 8HS.CoA \; + \; 6H_2O$$

described by the overall reaction (8.5). Labelling experiments have shown that (i) the odd-numbered carbon atoms of palmitic acid in the range 1–13 are derived from the carbonyl carbon of malonyl-CoA, (ii) the even-numbered carbon atoms in the range 2–14 are derived from the methylene carbon of malonyl-CoA, (iii) C–15 and C–16 are derived from the carbonyl and methyl carbons of acetyl-CoA respectively and (iv) the CO_2 eliminated in the process is derived from the carboxyl carbon of malonyl-CoA. The latter finding may, at first sight, seem rather odd, for the carboxyl group of malonyl-CoA which is eliminated as CO_2 is the one which was inserted in eq. (8.1). The cell appears to be indulging in a wasteful exercise, that of using energy (i.e. ATP) to carboxylate acetyl-CoA only to decarboxylate the resulting malonyl-CoA in the next step of fatty acid biosynthesis. That this is not so will be demonstrated later.

There appear to be two types of fatty acid synthase in Nature. The first type is considered to be the most primitive and is typified by that of E. coli. The second type is considered to be more advanced and is typified by that of animal liver. Present evidence indicates that the fatty acid synthase of higher plants resembles that of E. coli more closely than that of animal liver. On the other hand, the fatty acid synthase of yeast resembles that of animal liver more closely than that of E. coli. Euglena gracilis is an alga which displays both plant and animal characteristics. It can grow photoautotrophically in the light with CO_2 as the carbon source like plants and is green because it has abundant chloroplasts. It can also grow heterotrophically in the dark on an organic carbon source and is not green because chloroplasts do not develop. When grown in the dark, E. gracilis contains a fatty acid synthase closely resembling that of animal liver. When it is grown in the light it contains two fatty acid synthases, one resembling that of animal liver and one resembling that of E. coli and higher plants. Not all bacterial fatty acid synthases are like that of E. coli; that of Mycobacterium phlei, an advanced prokaryote, is of the animal type. M. phlei is the least advanced organism so far shown to possess the advanced type of fatty acid synthase.

The fatty acid synthase of E. coli, typical of the primitive type, is presumed to exist within the bacterial cell as a loose aggregate of at least seven enzymes and a heat-stable cofactor called the acyl carrier protein (ACP). The seven enzymes catalyse the different reactions involved in the multistep reaction (8.5). The acyl derivatives, formed as intermediates in the multistep process, are covalently bound to ACP which therefore functions by carrying them from enzyme to enzyme within the complex. ACP is analogous in function to coenzyme A which carries the acyl intermediates from enzyme to enzyme in the process of β-oxidation [see Section C.1(ii)(a) and Chapter 6, Section E.2(ii)(a)]. The analogy does not stop there, however, because both ACP and coenzyme A have 4′-phosphopantetheine as their functional component. In ACP the 4′-phosphopantetheine is covalently bound by a phosphate ester bond to the hydroxyl group of a serine residue in a polypeptide chain. The structure of E. coli ACP is shown in Fig. 8.13; it has a molecular weight of 8847 and consists of a polypeptide chain of seventy-seven amino acids of which the serine linked to the 4′-phosphopantetheine occupies position 36. In coenzyme A the 4′-phosphopantetheine is covalently linked by a phosphate diester bond to the 5′-phosphate of adenosine-3′,5′-diphosphate (see Fig. 8.14). In both ACP and coenzyme A the acyl moieties are linked by a thioester bond to the sulphydryl sulphur atom of the 4′-phosphopantetheine residue.

The evidence that the seven enzymes and ACP catalysing reaction (8.5) in E. coli are aggregated together into a multienzyme complex known as a fatty acid synthase is weak. When extracts of E. coli are examined the components of the supposed synthase are not associated with one another as might be expected and can be separated and purified by conventional techniques. Moreover, in vitro they show no tendency to associate into a multienzyme complex. Nevertheless it is believed that in vivo there must be some structural organization of the component enzymes around the coenzyme ACP, to account for the observed efficiency of fatty acid synthesis in E. coli. It is argued that since the process involves such a large number of steps there must be efficient transfer of intermediates from enzyme to enzyme and that this demands structural organization. Present evidence indicates that ACP is located at or near the inner surface of the E. coli plasma membrane and that the synthase enzymes

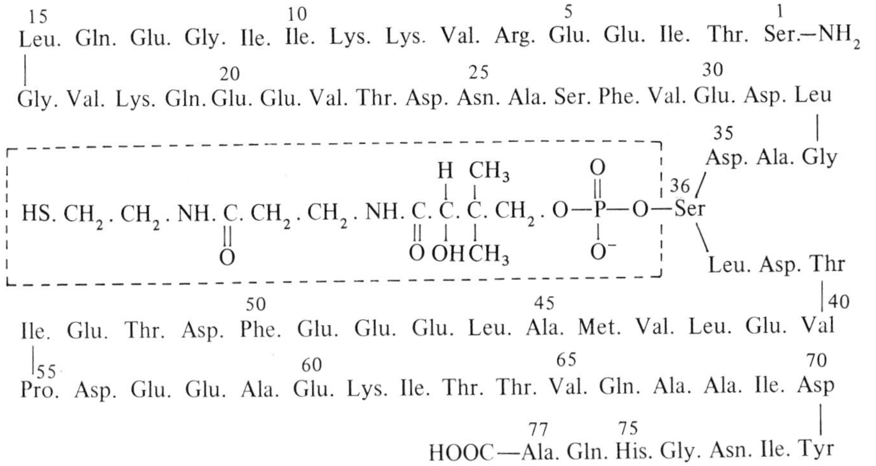

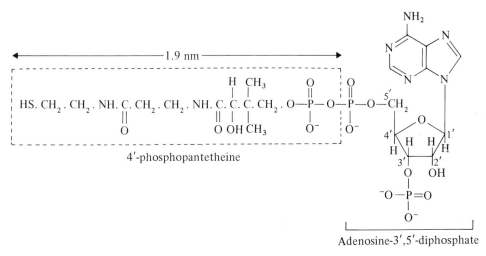

15   10   5   1
Leu. Gln. Glu. Gly. Ile. Ile. Lys. Lys. Val. Arg. Glu. Glu. Ile. Thr. Ser.—NH$_2$

                    20   25   30
Gly. Val. Lys. Gln. Glu. Glu. Val. Thr. Asp. Asn. Ala. Ser. Phe. Val. Glu. Asp. Leu

                                                                      35
                          H CH$_3$ O   Asp. Ala. Gly
                          | | ‖   36 /
HS. CH$_2$. CH$_2$. NH. C. CH$_2$. CH$_2$. NH. C. C. C. CH$_2$. O—P—O—Ser
                ‖   ‖ | | | \
                O   O OH CH$_3$ O⁻   Leu. Asp. Thr
                                                       |40
      50   45
Ile. Glu. Thr. Asp. Phe. Glu. Glu. Glu. Leu. Ala. Met. Val. Leu. Glu. Val

|55   60   65   70
Pro. Asp. Glu. Glu. Ala. Glu. Lys. Ile. Thr. Thr. Val. Gln. Ala. Ala. Ile. Asp

                                77   75 |
                       HOOC—Ala. Gln. His. Gly. Asn. Ile. Tyr

Fig. 8.13. Structure of the acyl carrier protein (ACP) of *Escherichia coli* (fatty acyl intermediates are covalently linked by a thioester bond to the sulphur atom of the 4′-phosphopantetheine residue).

Fig. 8.14. Structure of coenzyme A (HS.CoA).

are arranged in a functionally coherent manner around it.

It is the lack of *in vitro* association of the components of the fatty acid synthase of *E. coli* and higher plants which distinguishes it from that of animals and yeast. There is no tendency for the components of the fatty acid synthase of animals and yeast to dissociate *in vitro* or, indeed, *in vivo*. This structural integrity ensures the efficient transfer of intermediates from enzyme to enzyme during the multistep process and is one of the main reasons why it is thought that animal and yeast fatty acid synthases are more advanced than those of *E. coli* and higher plants.

The multistep process catalysed by the *E. coli* fatty acid synthase is shown in Fig. 8.15. The process is cyclic. Each cycle involves the addition of a 2C unit derived from malonyl-CoA to an acyl unit, so elongating it by two carbon atoms. To build up the 16C palmitic acid seven cycles are required. The first and last cycles are slightly different from the others because they include the initiation and termination steps respectively in addition to the elongation steps which are common to all seven cycles.

Initiation phase (steps 1–3)

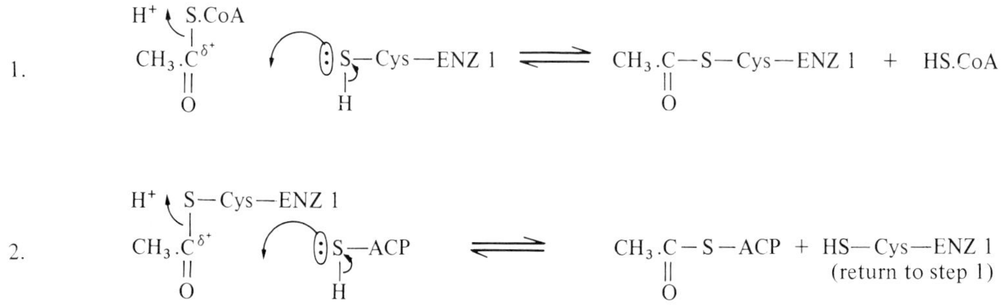

1.

$$CH_3.C\overset{\delta+}{\underset{\parallel O}{|}} \quad \overset{H^+ \ S.CoA}{} \quad :S\overset{H}{\underset{|}{}}-Cys-ENZ\ 1 \rightleftharpoons CH_3.C\underset{\parallel O}{}-S-Cys-ENZ\ 1 \ + \ HS.CoA$$

2.

$$CH_3.C^{\delta+}\underset{\parallel O}{} \quad \overset{H^+ \ S-Cys-ENZ\ 1}{} \quad :S\overset{H}{\underset{|}{}}-ACP \rightleftharpoons CH_3.C-S-ACP \ + \ HS-Cys-ENZ\ 1$$
(return to step 1)

Elongation phase (successive cycles of steps 3–9; step 3 of the first cycle is also part of the initiation phase)

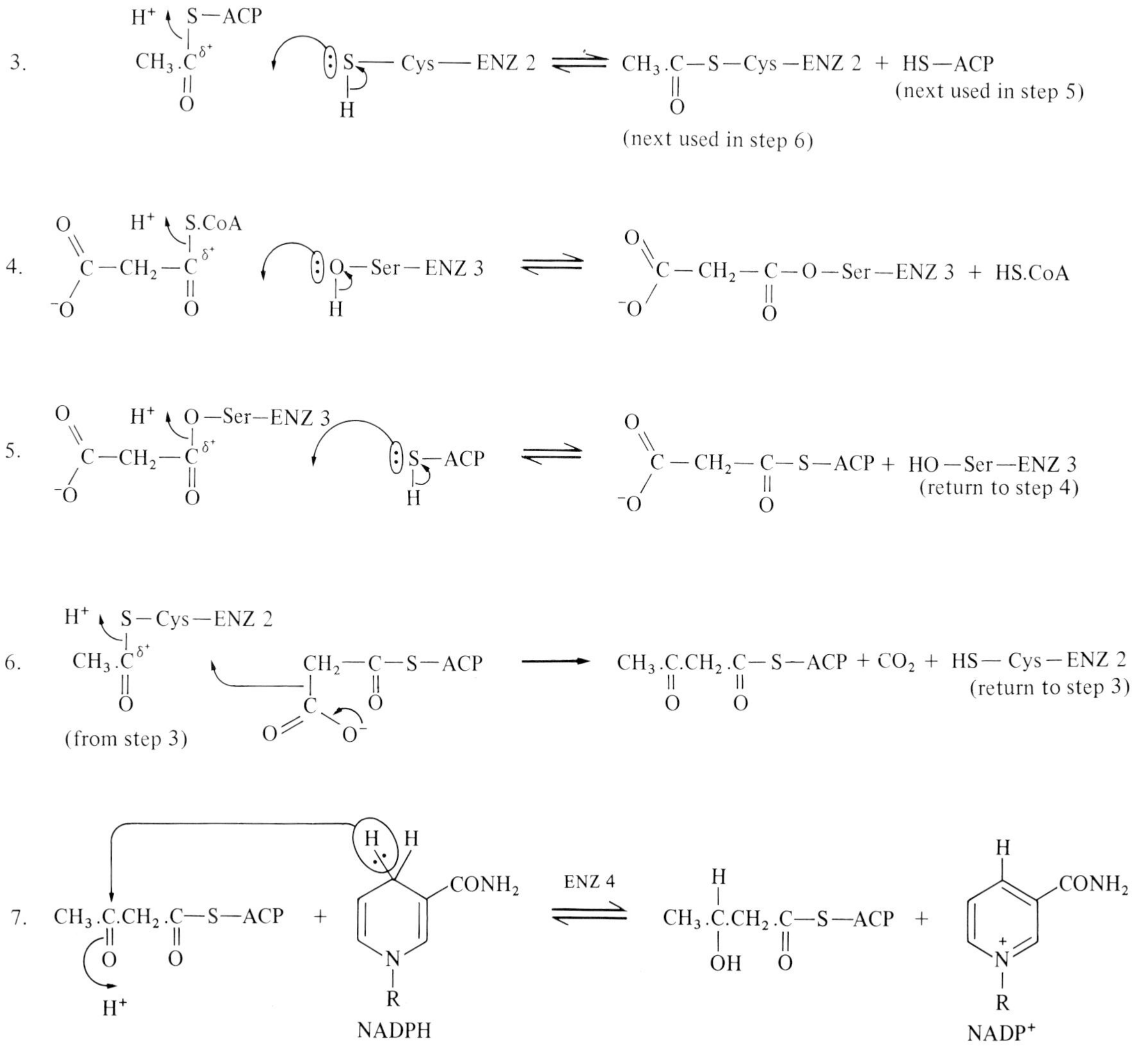

3.

$$CH_3.C^{\delta+}\underset{\parallel O}{} \quad \overset{H^+ \ S-ACP}{} \quad :S\overset{H}{\underset{|}{}}-Cys-ENZ\ 2 \rightleftharpoons CH_3.C-S-Cys-ENZ\ 2 \ + \ HS-ACP$$
(next used in step 5)

(next used in step 6)

4.

$$\overset{O}{\underset{-O}{\parallel}}C-CH_2-C^{\delta+}\underset{\parallel O}{} \quad \overset{H^+ \ S.CoA}{} \quad :O\overset{H}{\underset{|}{}}-Ser-ENZ\ 3 \rightleftharpoons \overset{O}{\underset{-O}{\parallel}}C-CH_2-C-O-Ser-ENZ\ 3 \ + \ HS.CoA$$

5.

$$\overset{O}{\underset{-O}{\parallel}}C-CH_2-C^{\delta+}\underset{\parallel O}{} \quad \overset{H^+ \ O-Ser-ENZ\ 3}{} \quad :S\overset{H}{\underset{|}{}}-ACP \rightleftharpoons \overset{O}{\underset{-O}{\parallel}}C-CH_2-C-S-ACP \ + \ HO-Ser-ENZ\ 3$$
(return to step 4)

6.

$$CH_3.C^{\delta+}\underset{\parallel O}{} \quad \overset{H^+ \ S-Cys-ENZ\ 2}{} \quad CH_2-C-S-ACP \longrightarrow CH_3.C.CH_2.C-S-ACP + CO_2 + HS-Cys-ENZ\ 2$$
(return to step 3)

(from step 3)

7.

$$CH_3.C.CH_2.C-S-ACP \ + \quad \overset{H \ H}{\underset{}{}} \quad \overset{ENZ\ 4}{\rightleftharpoons} \quad CH_3.C.CH_2.C-S-ACP \ + \quad \overset{CONH_2}{}$$

H+

NADPH   NADP+

Fig. 8.15. For legend see facing page.

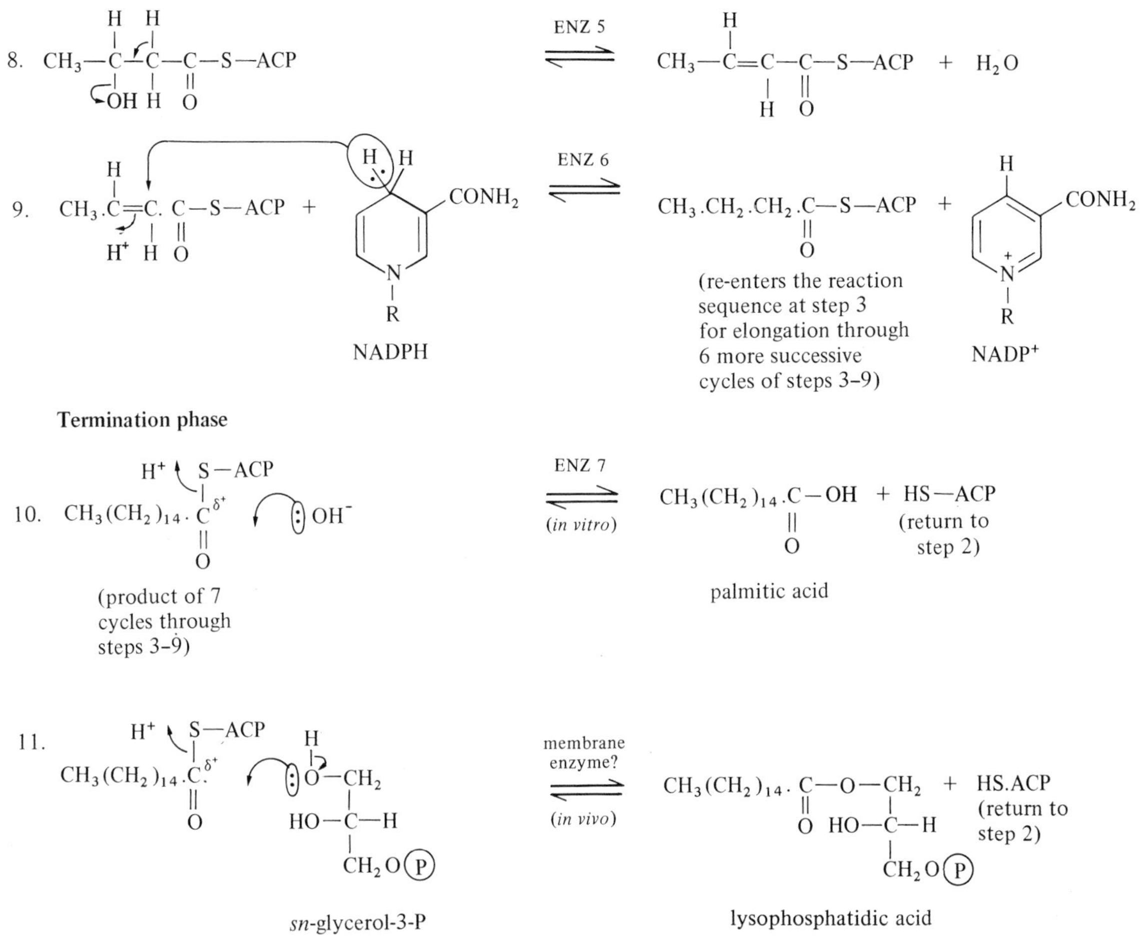

8. $CH_3-\underset{\underset{OH}{|}}{\overset{\overset{H}{|}}{C}}-\underset{\underset{H}{|}}{\overset{\overset{H}{|}}{C}}-\underset{\overset{\|}{O}}{C}-S-ACP$ $\overset{ENZ\ 5}{\rightleftharpoons}$ $CH_3-\underset{\underset{H}{|}}{\overset{\overset{H}{|}}{C}}=C-\underset{\overset{\|}{O}}{C}-S-ACP + H_2O$

9. $CH_3.\underset{\underset{H^+}{|}}{C}=C.\underset{\overset{\|}{O}}{\underset{H}{\overset{|}{C}}}-S-ACP +$ NADPH $\overset{ENZ\ 6}{\rightleftharpoons}$ $CH_3.CH_2.CH_2.\underset{\overset{\|}{O}}{C}-S-ACP +$ NADP⁺

(re-enters the reaction sequence at step 3 for elongation through 6 more successive cycles of steps 3–9)

Termination phase

10. $CH_3(CH_2)_{14}.\overset{\delta^+}{\underset{\overset{\|}{O}}{C}}$ $\overset{H^+ \ S-ACP}{}$ OH^- $\overset{ENZ\ 7}{\underset{(in\ vitro)}{\rightleftharpoons}}$ $CH_3(CH_2)_{14}.\underset{\overset{\|}{O}}{C}-OH + HS-ACP$ (return to step 2)

(product of 7 cycles through steps 3–9)

palmitic acid

11. $CH_3(CH_2)_{14}.\overset{\delta^+}{\underset{\overset{\|}{O}}{C}}$ $\overset{H^+ \ S-ACP}{}$ $\underset{\underset{\underset{CH_2O(P)}{|}}{HO-C-H}}{\overset{H}{\overset{|}{O}-CH_2}}$ $\overset{membrane\ enzyme?}{\underset{(in\ vivo)}{\rightleftharpoons}}$ $CH_3(CH_2)_{14}.\underset{\overset{\|}{O}}{C}-O-CH_2 + HS.ACP$ $\underset{\underset{CH_2O(P)}{|}}{HO-C-H}$ (return to step 2)

sn-glycerol-3-P

lysophosphatidic acid

Fig. 8.15. The multistep process catalysed by the *E. coli* fatty acid synthase.

ENZ 1 = acetyl-CoA: ACP transacylase
ENZ 2 = acyl-ACP: malonyl-ACP condensing enzyme (β-ketoacyl-ACP synthase)
ENZ 3 = malonyl-CoA: ACP transacylase
ENZ 4 = β-ketoacyl-ACP reductase
ENZ 5 = D(−)-β-hydroxyacyl-ACP dehydratase
ENZ 6 = *trans*-$\Delta^{2,3}$-dehydroacyl-ACP reductase
ENZ 7 = palmityl-ACP thioesterase
ACP = acyl carrier protein

The process is initiated by acetyl-CoA; it acts as the primer. Thus in the first cycle the acetyl moiety of acetyl-CoA is the acyl unit which is elongated. However, before this can take place the acetyl unit has to be transferred from coenzyme A to the active site of the acyl-ACP:malonyl-ACP condensing enzyme (ENZ 2; Fig. 8.15). This transfer is called the initiation phase and involves steps 1–3 (Fig. 8.15). The first step involves the transfer of the acetyl unit from the sulphydryl sulphur of coenzyme A to a sulphydryl sulphur of a cysteine residue at the active site of the enzyme, acetyl-CoA:ACP transacylase (ENZ 1; Fig. 8.15). In the second step the acetyl unit is transferred to the sulphydryl sulphur of ACP. In the third step the acetyl unit is transferred to the sulphydryl sulphur of a cysteine residue at the active site of the acyl-ACP:malonyl-ACP condensing enzyme (ENZ 2; Fig. 8.15).

Malonyl-CoA enters the reaction sequence at step 4 in which the malonyl unit is transferred to the hydroxyl group of a serine residue at the active site of malonyl-CoA:ACP transacylase (ENZ 3; Fig. 8.15); the new linkage is a conventional ester bond (i.e. an oxygen ester) rather than the thioester bond of the coenzyme A derivative. In step 5 the malonyl unit is transferred to the sulphydryl sulphur of ACP.

The resulting malonyl-ACP now reacts in step 6 with the acetyl unit linked to the active site of the acyl-ACP:malonyl-ACP condensing enzyme (ENZ 2; Fig. 8.15). In this reaction a new C—C bond is formed between the methylene carbon of malonyl-ACP and the carbonyl carbon of the acetyl unit and the acetyl-enzyme thioester bond is cleaved. The formation of the new C—C bond is facilitated by the presence of the carboxyl group in malonyl-ACP in two ways. Firstly, its presence increases the nucleophilicity of the methylene carbon, thus making a reaction with the electrophilic carbonyl carbon of the acetyl unit more likely. Secondly, by coupling the C—C bond formation with decarboxylation the reaction is driven to completion. This, then, is the reason why the cell goes to the trouble of carboxylating the methyl group of acetyl-CoA only to decarboxylate the resulting malonyl-CoA in the condensation step.

The product of the condensation step is a 4C β-ketoacyl unit (acetoacetyl) linked by a thioester bond to ACP. In step 7 the β-keto moiety is reduced by β-ketoacyl-ACP reductase (ENZ 4; Fig. 8.15) with the production of D($-$)-β-hydroxybutyryl-ACP; the reductant utilized by the enzyme is NADPH. In step 8 the elements of water are removed from the D($-$)-β-hydroxybutyryl-ACP by D($-$)-β-hydroxyacyl-ACP dehydratase (ENZ 5; Fig. 8.15) so producing a 4C $trans$-$\Delta^{2,3}$-dehydro-acyl-ACP (crotonyl-ACP). This $trans$-$\alpha\beta$-double bond is then saturated in step 9 by $trans$-$\Delta^{2,3}$-dehydroacyl-ACP reductase (ENZ 6; Fig. 8.15) which also uses NADPH as the reductant. The product is butyryl-ACP, a 4C saturated acyl-ACP; this is the end product of the first turn of the cycle.

The second cycle begins at step 3. However, the acetyl-ACP used in this step in the first cycle is replaced by the butyryl-ACP just formed. Steps 4–9 follow as before and the end product is hexanoyl-ACP. This then re-enters the reaction sequence at step 3 and the third cycle commences. By the time step 9 of the seventh cycle has been completed palmityl-ACP, a 16C saturated acyl-ACP, has been formed. At this point elongation of the acyl chain by further cycling through steps 3–9 stops. Quite why it stops at the 16C stage is not absolutely clear; it is presumed that one or more of the enzymes cannot handle acyl chains longer than this.

Two different termination reactions have been demonstrated experimentally. Soluble extracts of E. coli contain the enzyme palmityl-ACP thioesterase which catalyses step 10, liberating free palmitic acid and ACP. On the other hand, membraneous preparations from E. coli catalyse the transfer of the palmityl residue from ACP to the sn-l-hydroxyl group of sn-glycerol 3-phosphate (step 11, Fig. 8.15) so forming lysophosphatidic acid which is a precursor of phospholipids known to occur in the E. coli plasma membrane. Clearly both of these termination reactions could occur in vivo but it is thought that step 11 is the predominant one because ACP, and presumably the synthase multienzyme complex, being at or near to the inner surface of the plasma membrane, are ideally located for the biosynthesis of the acyl-lipids required for membrane formation or repair.

Higher plant fatty acid synthases resemble that of E. coli in that the component enzymes and ACP are not tightly associated. They are probably to be found in all cell types within the plant for they are required to produce the fatty acids which are incorporated into membrane lipid components. Within these cells they appear to be located in two sites, the cytosol and the plastid. The fatty acid synthase dissolved in the cytosol is probably concerned in the synthesis of the fatty acids required for the production of the lipid components of non-plastidic membranes. The fatty acid synthase of chloroplasts, which is dissolved in the stroma, is probably concerned in the production of the fatty acids required for the production of the lipid components of the thylakoid membranes. Palmitic acid is the end product of both synthases. The NADPH required in steps 7 and 9 is derived in the chloroplast from the light phase of photosynthesis (see Chapter 5, Section C.6) as is the ATP required for the production of malonyl-CoA by acetyl-CoA carboxylase.

In addition to the synthases of vegetative cells, the cells of certain specialized tissues of seeds (e.g. castor bean endosperm) and fruit (e.g. avocado pear mesocarp) have a fatty acid synthase associated with the organelles known as spherosomes (see Chapter 3, Section B.8) which are distributed in considerable numbers throughout their cytoplasm. These tissues are adapted to store food reserves in the form of triglycerides; to some extent, therefore, they are akin to the adipose tissue of animals. How the spherosomes develop is still in dispute. One attractive hypothesis suggests that they originate from a section of endoplasmic reticular membrane which is pinched off to form a provacuole and that the triglycerides are formed and accumulate between the two halves of the unit membrane. The external half-membrane then surrounds the developing spherosome and thus separates the triglyceride from the cytosol, whilst the internal half membrane eventually disappears. Electron micrographs showing spherosomes surrounded by what appears to be a half-membrane have been published, thus supporting the hypothesis. On this basis the fatty acid synthase would be expected to be built into the external half-membrane along with acetyl-CoA carboxylase and all the other enzymes necessary for the biosynthesis of triglycerides. A membraneous fraction obtained from spherosomes has been shown to catalyse the synthesis of fatty acids from acetyl-CoA and their incorporation into triglycerides.

The fatty acid synthase of yeast is a tightly associated, multienzyme complex with a molecular weight of 2.3×10^6. It exhibits many of the criteria of protein purity; it behaves as a single protein in the analytical ultracentrifuge and when subjected to electrophoresis. However, it is clearly not a single protein because (a) it can be split into enzymically inactive particles with an average molecular weight of 110,000 by treatment with 0.2 M deoxycholate or 6 M urea, (b) under the electron microscope it appears to be made up of oval particles with long and short diameters of 25 nm and 21 nm respectively, (c) it catalyses seven different reactions and (d) work in the mid-1960s showed that it had seven different N-terminal amino acids, each of which was

present to the extent of 3 moles per mole of synthase (this is now known to be incorrect—see later).

A number of differences between the catalytic activity of the yeast synthase and that of E. coli soon became apparent. The end product of the yeast synthase is not free palmitic acid but palmityl-CoA and stearyl-CoA. The pure enzyme is yellow due to the presence of flavin mononucleotide (FMN). The FMN can be dialysed away to yield a colourless but inactive enzyme; activity is restored by adding back the FMN. This indicated that one of the component enzymes is a flavoprotein; subsequent work showed that the only enzymic activity to be lost in the absence of FMN was that of the $trans$-$\Delta^{2,3}$-dehydroacyl-ACP reductase.

The yeast synthase is an SH-enzyme; it is inhibited by reagents which alkylate SH groups. Moreover, it was shown that there were two functionally distinct types of SH groups which were called 'peripheral' and 'central'. The peripheral SH groups are alkylated readily by iodoacetamide indicating that they are readily accessible to the reagent and suggesting that they are located at the outer surface of the synthase. Preincubation of the synthase with acetyl-CoA, but not malonyl-CoA, prevented its inhibition by iodoacetamide, indicating that acetyl residues, but not malonyl residues, become bound to the peripheral SH groups which were presumed to belong to the acyl-ACP:malonyl ACP condensing enzyme. The central SH groups are not alkylated at all by iodoacetamide indicating that they are not accessible to the reagent and suggesting that they are located in the inner regions of the synthase. The central SH groups were considered to bind malonyl residues and, by analogy with the E. coli synthase, to be the yeast equivalent of ACP.

The picture of the yeast synthase that emerged from the evidence just described was of a functional unit, triplicated in the complete particle, which was composed of the seven different enzymes arranged around a centrally placed ACP. The failure to dissociate this complex into catalytically active monofunctional components was attributed to its exceptional tightness and the interdependence of the individual enzymes.

However, this picture of the yeast fatty acid synthase is no longer tenable. This has resulted from a genetic analysis of yeast fatty acid auxotrophs by

Schweizer in the early 1970s which shows that all the enzyme and coenzyme (i.e. ACP) activities of the synthase are coded by only two genetically unlinked gene loci, designated *fas* 1 and *fas* 2. Each of these polycistronic clusters codes a single polypeptide chain which is multifunctional. The two polypeptide chains together account for all the reactions catalysed by the yeast synthase. It is apparent, therefore, that the yeast synthase contains only two different protein components. These have been separated by SDS-polyacrylamide gel electrophoresis; one, designated α, has a molecular weight of 185,000 and is coded by *fas* 2, whilst the other, designated β, has a molecular weight of 180,000 and is coded by *fas* 1. Protein α has the catalytic activities which correspond to the *E. coli* acyl-ACP:malonyl-ACP condensing enzyme and β-ketoacyl-ACP reductase and the coenzyme activity of ACP. Protein β has the catalytic activities corresponding to the *E. coli* enzymes, acetyl-CoA:ACP transacylase, malonyl-CoA:ACP transacylase, D($-$)-β-hydroxyacyl-ACP dehydratase and *trans*-$\Delta^{2,3}$-dehydroacyl-ACP reductase.

How does this new information fit in with the information upon which the old picture of the yeast synthase was based? The established molecular weight of the synthase (2.3×10^6) taken with those of proteins α and β indicate that six copies of each protein is present giving an $\alpha_6\beta_6$ structure. The work which showed that the synthase had seven different *N*-terminal amino acids is clearly incorrect; an $\alpha_6\beta_6$ structure would require a maximum of two different *N*-terminal amino acids. A re-examination revealed no *N*-terminal amino acids at all. This suggested that the *N*-terminal amino acids are blocked. However, the finding of two different *C*-terminal amino acids, lysine for α and valine for β, was consistent with the presence of only two different polypeptide chains. The reason for the earlier finding of seven *N*-terminal amino acids has now become clear. It has been found that great care is required during the extraction and purification of the synthase to prevent proteolysis; this was not realized in the mid-1960s and led to proteolytic artefacts which gave incorrect information about the *N*-terminal amino acids.

Further work has shown that both the peripheral and central SH groups are present in protein α. The peripheral SH group is located on a cysteine residue

in that part of protein α which corresponds to the acyl-ACP:malonyl-ACP condensing enzyme. The central SH group is located on a 4'-phosphopantetheine residue in that part of protein α which corresponds to ACP.

It has also been shown that the initial binding sites of acetyl and malonyl residues transferred from their coenzyme A derivatives are the hydroxyl groups of serine residues located in the acetyl-CoA:ACP transacylase and malonyl-CoA:ACP transacylase regions of protein β respectively. The acetyl and malonyl residues are then transferred in sequence from these serine hydroxyls to the sulphydryl sulphur of the 4'-phosphopantetheine in protein α.

The malonyl-CoA:ACP transacylase region of protein β has been shown to have a second enzymic activity, that of palmityl (or stearyl)-ACP:CoA transacylase; both of these enzymic activities are knocked out by the same genetic lesion. Thus, as well as being involved in the elongation phase (steps 4 and 5; Fig. 8.16), malonyl-CoA:ACP transacylase also functions in the termination phase (steps 10 and 11; Fig. 8.16). It catalyses the transfer of the palmityl (or stearyl) residue formed during the elongation phase from the sulphydryl sulphur of 4'-phosphopantetheine in protein α to that of coenzyme A by way of the serine hydroxyl at its active site. This is, from a chemical point of view, the reverse of steps 4 and 5 (Fig. 8.16).

The multistep process catalysed by the yeast fatty acid synthase is shown in Fig. 8.16; the sequence is cyclic and differs from that of *E. coli* only in steps 1, 2, 9, 10 and 11.

The fatty acid synthases of animal tissues are also composed of two different multifunctional proteins. The liver proteins I and II show a remarkable functional resemblance to the yeast proteins α and β respectively; the major difference is that the *trans*-$\Delta^{2,3}$-dehydroacyl-ACP reductase activity in the liver synthase is in protein I rather than II as would be expected had complete correspondence obtained. Another difference concerns the termination phase. Animal synthases produce free palmitic acid rather than the CoA derivative and thus contain a palmityl-ACP thioesterase (cf. *E. coli*) rather than relying on the double function of malonyl-CoA:ACP transacylase as is the case in the yeast synthase.

Initiation phase (steps 1-3)

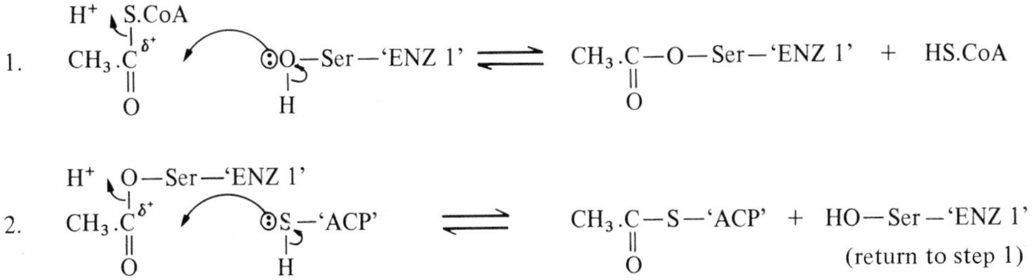

1. $CH_3.C \xrightarrow{H^+ \quad S.CoA} + {}^{\ominus}O-Ser-\text{'ENZ 1'} \rightleftharpoons CH_3.C-O-Ser-\text{'ENZ 1'} + HS.CoA$

2. $CH_3.C \xrightarrow{H^+ \quad O-Ser-\text{'ENZ 1'}} + {}^{\ominus}S-\text{'ACP'} \rightleftharpoons CH_3.C-S-\text{'ACP'} + HO-Ser-\text{'ENZ 1'}$
(return to step 1)

Elongation phase (successive cycles of steps 3–9; step 3 of the first cycle is also part of the initiation phase)

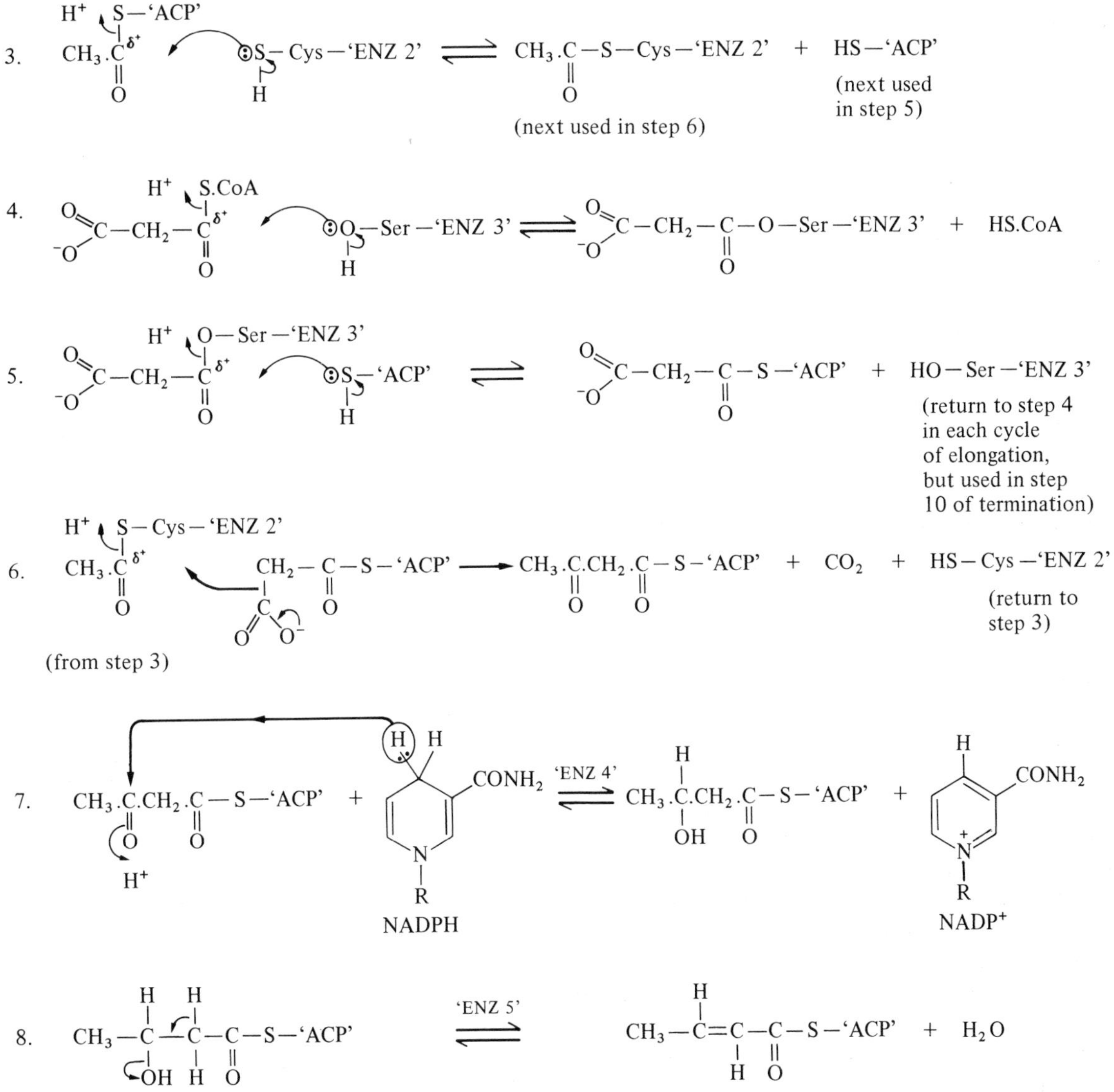

3. $CH_3.C \xrightarrow{H^+ \quad S-\text{'ACP'}} + {}^{\ominus}S-Cys-\text{'ENZ 2'} \rightleftharpoons CH_3.C-S-Cys-\text{'ENZ 2'} + HS-\text{'ACP'}$
(next used in step 6)   (next used in step 5)

4. $\overset{O}{\underset{{}^-O}{>}}C-CH_2-C \xrightarrow{H^+ \quad S.CoA} + {}^{\ominus}O-Ser-\text{'ENZ 3'} \rightleftharpoons \overset{O}{\underset{{}^-O}{>}}C-CH_2-C-O-Ser-\text{'ENZ 3'} + HS.CoA$

5. $\overset{O}{\underset{{}^-O}{>}}C-CH_2-C \xrightarrow{H^+ \quad O-Ser-\text{'ENZ 3'}} + {}^{\ominus}S-\text{'ACP'} \rightleftharpoons \overset{O}{\underset{{}^-O}{>}}C-CH_2-C-S-\text{'ACP'} + HO-Ser-\text{'ENZ 3'}$
(return to step 4 in each cycle of elongation, but used in step 10 of termination)

6. $CH_3.C \xrightarrow{H^+ \quad S-Cys-\text{'ENZ 2'}} + CH_2-C-S-\text{'ACP'} \rightarrow CH_3.C.CH_2.C-S-\text{'ACP'} + CO_2 + HS-Cys-\text{'ENZ 2'}$
(from step 3)
(return to step 3)

7. $CH_3.C.CH_2.C-S-\text{'ACP'} + \text{NADPH} \underset{\text{'ENZ 4'}}{\rightleftharpoons} CH_3.C.CH_2.C-S-\text{'ACP'} + \text{NADP}^+$

8. $CH_3-C-C-C-S-\text{'ACP'} \underset{\text{'ENZ 5'}}{\rightleftharpoons} CH_3-C=C-C-S-\text{'ACP'} + H_2O$

Fig. 8.16. For legend see page 298.

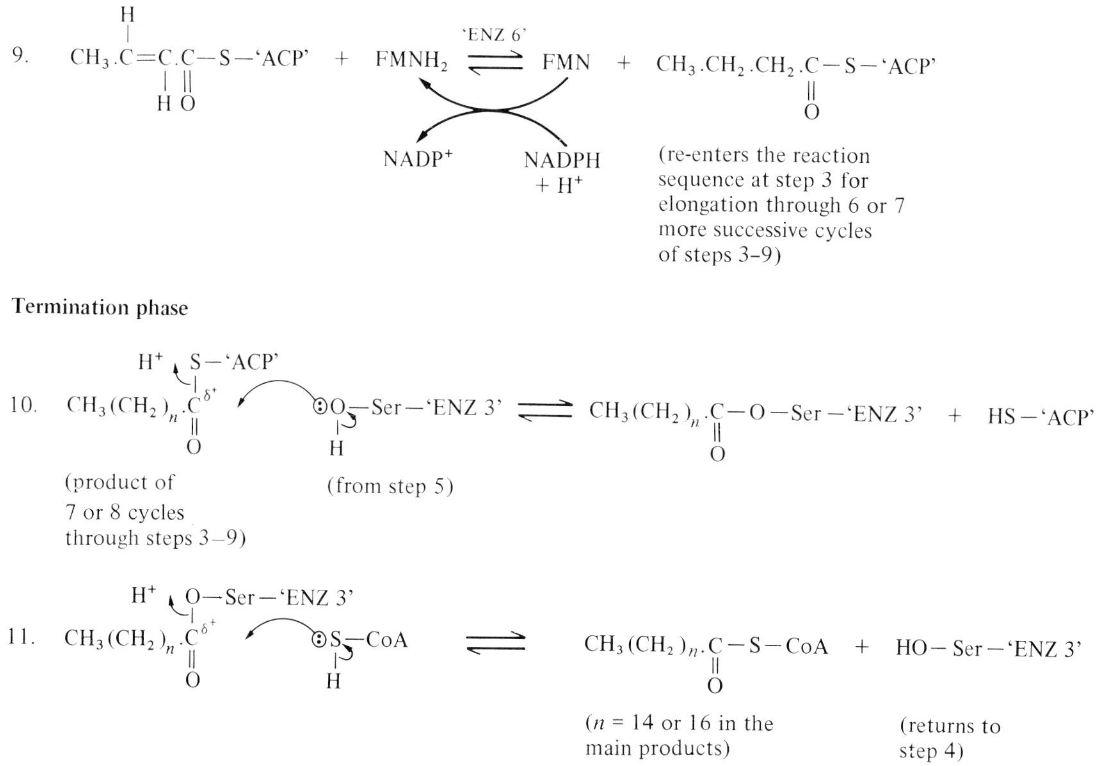

9. $CH_3.\overset{\overset{\displaystyle H}{|}}{C}=\overset{\overset{\displaystyle}{|}}{\underset{\underset{\displaystyle H}{|}}{C}}.\overset{\overset{\displaystyle}{}}{\underset{\underset{\displaystyle O}{\|}}{C}}-S-\text{'ACP'}$ + $FMNH_2$ $\underset{\text{'ENZ 6'}}{\rightleftharpoons}$ FMN + $CH_3.CH_2.CH_2.\underset{\underset{\displaystyle O}{\|}}{C}-S-\text{'ACP'}$

$NADP^+$ $NADPH$
$+ H^+$

(re-enters the reaction
sequence at step 3 for
elongation through 6 or 7
more successive cycles
of steps 3–9)

Termination phase

10. $CH_3(CH_2)_n.\overset{\overset{\displaystyle H^+ \curvearrowright S-\text{'ACP'}}{|}}{\underset{\underset{\displaystyle O}{\|}}{C^{\delta^+}}}$ $\overset{\ominus}{\underset{\underset{\displaystyle H}{|}}{O}}-Ser-\text{'ENZ 3'}$ $\rightleftharpoons$ $CH_3(CH_2)_n.\underset{\underset{\displaystyle O}{\|}}{C}-O-Ser-\text{'ENZ 3'}$ + $HS-\text{'ACP'}$

(product of   (from step 5)
7 or 8 cycles
through steps 3–9)

11. $CH_3(CH_2)_n.\overset{\overset{\displaystyle H^+ \curvearrowright O-Ser-\text{'ENZ 3'}}{|}}{\underset{\underset{\displaystyle O}{\|}}{C^{\delta^+}}}$ $\overset{\ominus}{\underset{\underset{\displaystyle H}{|}}{S}}-CoA$ $\rightleftharpoons$ $CH_3(CH_2)_n.\underset{\underset{\displaystyle O}{\|}}{C}-S-CoA$ + $HO-Ser-\text{'ENZ 3'}$

($n = 14$ or 16 in the   (returns to
main products)   step 4)

Fig. 8.16. The multistep process catalysed by the yeast fatty acid synthase.

'ENZ 1' (Subunit β; fas 1) = catalytic activity of acetyl-CoA:ACP transacylase
'ENZ 2' (Subunit α; fas 2) = catalytic activity of acyl-ACP:malonyl ACP condensing enzyme
'ENZ 3' (Subunit β; fas 1) = catalytic activity of malonyl-CoA:ACP transacylase
'ENZ 4' (Subunit α; fas 2) = catalytic activity of β-ketoacyl-ACP reductase
'ENZ 5' (Subunit β; fas 1) = catalytic activity of D($-$)-β-hydroxyacyl-ACP dehydratase
'ENZ 6' (Subunit β; fas 1) = catalytic activity of trans-$\Delta^{2.3}$-dehydroacyl-ACP reductase
'ACP'   (Subunit α; fas 2) = functional activity of acyl carrier protein

The molecular weight of most animal synthases is in the range 400,000–500,000 which suggests that they have only one copy of each protein giving an $\alpha\beta$ structure.

(c) *Fatty acid elongation systems*

Palmitic acid, produced by the plant fatty acid synthases, can be elongated by the successive addition of 2C units to the carboxyl end to produce a range of straight-chain, saturated fatty acids from C_{18} to $\sim C_{30}$.

Higher plants appear to have two elongation systems, one that elongates palmitic acid to stearic acid (C_{18}) but no further and one that elongates stearic acid and both saturated and unsaturated C_{18+} fatty acids.

Both systems are probably multienzyme complexes resembling the fatty acid synthase in that they catalyse a multistep process.

The $C_{16} \rightarrow C_{18}$ system is soluble; it has been found in the cytosol of cells and in the stroma of chloroplasts. Its substrates are palmityl-ACP and malonyl-ACP, to which it shows absolute specificity. It requires NADPH and its products are stearyl-ACP and CO_2. This process is roughly equivalent to one cycle of the elongation phase (steps 3, 6–9; Fig. 8.15) of the process catalysed by the *E. coli* fatty acid synthase. That this system is quite distinct from the higher plant fatty acid synthase is shown by the following facts: (i) it is more heat-labile and much more easily inhibited by arsenite than the *de novo* synthase, (ii) it is much less sensitive to the antibiotic cerulenin ([2S;3R]-2,3-epoxy-4-oxo-6,10-dodecadienoylamide) than the *de novo* synthase and (iii) the substrate is specifically palmityl-ACP. Since cerulenin has been shown to inhibit the acyl-ACP:malonyl-ACP condensing enzyme, but none of the other enzymes, of the *E. coli* fatty acid synthase it is clear that the equivalent enzyme of the plant $C_{16} \rightarrow C_{18}$ synthase is different from that of the plant *de novo* synthase.

The C_{18} elongation system is bound to endoplasmic reticular membranes. However, beyond the fact that it will handle both saturated and unsaturated C_{18} and C_{18+} fatty acids, virtually nothing is known about it.

(d) Fatty acid desaturation systems

Most eukaryotic organisms possess a Δ^9-desaturase. This is an enzyme system which introduces a *cis*-Δ^9 double bond into a saturated fatty acid regardless of the length of its carbon chain; how-ever, a stearyl (C_{18}) derivative is its most common substrate with the formation of an oleyl (18:1, 9c) derivative. It requires NADPH and molecular oxygen for activity. Two Δ^9-desaturase systems are known; one occurs in vertebrate animals and fungi whilst the other occurs in higher plants.

The animal–fungal system is bound to endoplasmic reticular membranes. It consists of a short electron transport chain which carries electrons from NADPH to the desaturase enzyme. The first component is cytochrome b_5 reductase, a flavoprotein which transfers electrons from NADPH to cytochrome b_5. The reduced cytochrome b_5 then hands its electrons on to the desaturase via the cyanide-sensitive factor (CSF). The CSF is a protein of molecular weight 53,000 with an iron-containing prosthetic group whose activity is inhibited by CN^-. Oxygen enters the reaction sequence at the level of the desaturase (see Fig. 8.17) but the mechanism of the desaturation reaction is not known; intermediates hydroxylated at C–9 or C–10 have not been found. The substrate for the desaturase is a saturated fatty acyl-CoA.

The higher plant Δ^9-desaturase system is soluble; it has been shown to occur in the stroma of the chloroplast and in the cytosol. Its substrate is a fatty acyl-ACP (usually stearyl-ACP) rather than the coenzyme A derivative used by the animal–fungal system. The chloroplast system has been studied by Stumpf. Its structure, as presently conceived, is shown in Fig. 8.18. Electrons may pass from NADPH to ferredoxin via the FAD-containing NADP:ferredoxin oxidoreductase and then on to the desaturase as before. However, an even more efficient way of providing the necessary electrons *in vitro* is to illuminate a broken chloroplast preparation in the presence of ascorbate and dichlorophenolindophenol (DCPIP). Under these circum-

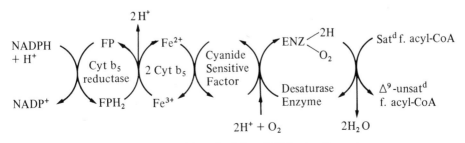

Fig. 8.17. Structure of the animal–fungal Δ^9-desaturation system.

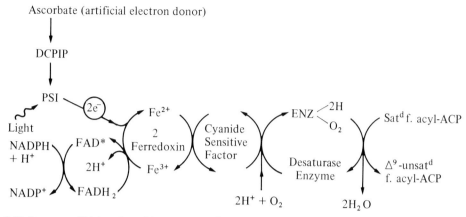

Fig. 8.18. Structure of higher plant chloroplastidic Δ^9-desaturase system. (* = NADP:ferredoxin oxidoreductase.)

stances ascorbate reduces the DCPIP which then hands electrons to a component of the photosynthetic intermediate electron transport chain [see Chapter 5, Section C.8(ii)]; the electrons are then passed directly to ferredoxin by the operation of Pigment System 1 (PS1). The chloroplast system is inhibited by 1 mM KCN, indicating that the CSF is one of its components. The structure of the higher plant cytosol Δ^9-desaturase system has not been elucidated. There is also evidence that there is an active Δ^9-desaturase associated with the spherosomes of developing castor bean seeds.

The fact that Δ^9-desaturases insert the double bond between C–9 and C–10 (counting from the COOH end) regardless of chain length suggests that the fatty acid binds to the enzyme by its carboxyl end (i.e. at a thioester binding site) and that this then positions the C–9 and C–10 methylene carbons adjacent to the active site (i.e. the desaturation site).

The Δ^9-desaturase-catalysed conversion of stearyl-ACP into oleyl-ACP is the main way that oleic acid (18:1, 9c), the main monoenoic fatty acid in Nature, is formed. The chloroplasts of all members of the plant kingdom have another key monoenoic fatty acid, trans-Δ^3-hexadecenoic acid (16:1, 3t), which occurs almost exclusively esterified to the sn-2-hydroxyl group of monophosphatidylglycerol which is to be found in the thylakoid membranes. The biosynthesis of this fatty acid has not been fully elucidated. It is known that palmitic acid is its immediate precursor and that O_2 is required during its formation from palmitate. There

are indications that the desaturation may take place after the palmitate has been incorporated into the monophosphatidylglycerol molecule.

The monoenoic fatty acids produced by the Δ^9-desaturase represent the main starting-point for the information of polyunsaturated fatty acids; oleic acid is particularly important in this respect. Some of the polyunsaturated fatty acids have eighteen carbon atoms and are therefore produced directly from oleic acid by desaturation systems. Others have twenty or more carbon atoms and are produced by the sequential operation of the desaturation and elongation systems. All the desaturation systems concerned in the further desaturation of monoenoic fatty acids that have been studied to date require O_2 and reduced pyridine nucleotide; this suggests that they are structurally similar to the Δ^9-desaturase.

Most of the polyunsaturated fatty acids that are produced in Nature, and certainly the most abundant ones, have a methylene-interrupted double-bond system [see Fig. 8.19(a)]; moreover, the double bonds have the cis configuration.

There is a clear-cut difference between the desaturase systems catalysing polyenoic fatty acid formation in animals and those in members of the plant kingdom. The animal enzymes always introduce the additional double bonds between the pre-existing double bond and the carboxyl group whereas the plant enzymes introduce them between the pre-existing double bond and the terminal methyl group [see Fig. 8.19(b)]. Recently this distinction has been

(a)
$$CH_3.CH_2.\overset{H}{\underset{}{C}}=\overset{H}{\underset{}{C}}.CH_2.\overset{H}{\underset{}{C}}=\overset{H}{\underset{}{C}}.CH_2.\overset{H}{\underset{}{C}}=\overset{H}{\underset{}{C}}.(CH_2)_7.COOH$$

α-Linolenic acid—an example of a
methylene-interrupted double bond system

(b) 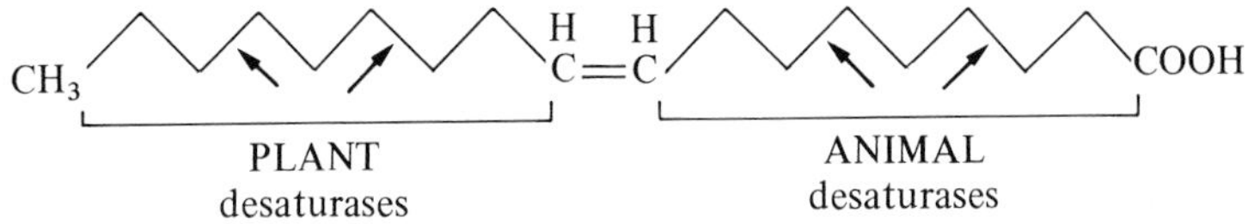 CH₃ ... $\overset{H}{\underset{}{C}}=\overset{H}{\underset{}{C}}$ ... COOH

PLANT
desaturases

ANIMAL
desaturases

Fig. 8.19. Nature of polyunsaturated fatty acids and the desaturases that synthesize them.

blurred at the lower end of the evolutionary scale in both animal and plant kingdoms; in the plant kingdom this concerns the blue-green algae (cyanobacteria), cryptomonad algae, marine dinoflagellates and the much-investigated chrysophyte alga, *Ochromonas danica*. Nevertheless in higher plants oleic acid is successively desaturated to linoleic acid 18:2, 9c, 12c) and α-linolenic acid (18:3, 9c, 12c, 15c) whilst in higher animals it is desaturated to *cis*, *cis*-$\Delta^{6,9}$-octadecadienoic acid. Higher animals cannot, therefore, synthesize linoleic acid although they need it; they obtain it from dietary plant material. They can further desaturate this dietary linoleic acid with a Δ^6-desaturase to form γ-linolenic acid (18:3, 6c, 9c, 12c). This can then be elongated to dihomo-γ-linolenic acid (20:3, 8c, 11c, 14c) which can be further desaturated by a Δ^5-desaturase to yield arachidonic acid (20:4, 5c, 8c, 11c, 14c); both of these C_{20} polyunsaturated fatty acids are prostaglandin precursors in animals.

The mechanism by which higher plants successively desaturate oleic acid to linoleic and γ-linolenic acids is not clear. This is largely because the desaturases are membrane-bound and consequently difficult to purify. There is evidence that the desaturation of oleic acid to linoleic acid takes place after it has been esterified to the *sn*-2 hydroxyl group of phosphatidylcholine; this is catalysed by a desaturation system located in endoplasmic reticular and chloroplastic membranes and the phospholipid so produced is incorporated into these membranes. α-Linolenic acid, the major fatty acid component of the chloroplast galactolipids, has been shown to be formed in two different ways. Mazliak has demonstrated that it can be formed by

direct desaturation of linoleic acid in the chloroplast. Stumpf has shown that it is produced anaerobically in the chloroplast stroma by the successive elongation of a 12:3 (3c, 6c, 9c) acid via 14:3 (5c, 8c, 11c) and 16:3 (7c, 10c, 13c) acids. Stumpf considers that the 12:3 (3c, 6c, 9c) acid is formed in the cytosol of the cell and is then transferred to the chloroplast; precisely how it is formed is not clear.

Desaturases producing linoleic and α-linolenic acids from oleic acid are also associated with the membrane (or half-membrane) surrounding the spherosomes present in developing triglyceride-storing seeds. In the case of safflower seeds, whose triglycerides are particularly rich in linoleic acid, the oleyl-CoA→linoleyl-CoA conversion has been demonstrated.

The desaturases acting on oleic acid in *Chlorella* have been studied by James. Like those of higher plants, the *Chlorella* desaturases insert the additional double bonds between the pre-existing double bond and the methyl terminal end of the fatty acid. However, two types of desaturase have been found. One type inserts the new double bond a fixed distance from the carboxyl end of the fatty acid whilst the other inserts it at a fixed distance from the methyl end.

(e) *Fatty acid hydroxylation*

Two different routes for the formation of hydroxy fatty acids have been observed in the plant kingdom. The first occurs by direct oxygenation of a saturated carbon atom in the fatty acid whilst the second occurs by the hydration of a double bond. In the first route the oxygen of the new hydroxyl group comes from molecular oxygen whilst in the second it comes from water.

Fatty acids with a hydroxyl group at C-2 are widespread but minor components of plant lipids. The 2-D-hydroxy fatty acids arise from the 2-D-hydroperoxy fatty acids which are intermediates in the fatty acid α-oxidation process [see Section C.1(ii)(b)].

Fatty acids with a hydroxyl group at C-3 are also formed in plants. 3-L(+)-Hydroxy fatty acids arise in the normal course of fatty acid β-oxidation due to the hydration of the trans-Δ^2-dehydroacyl-CoA intermediates. 3-D(−)-Hydroxy fatty acids were formed when saturated fatty acyl-CoAs, in the range C_8–C_{12}, were incubated with a soluble preparation from maturing avocado pear mesocarp in the presence of O_2; no requirement for reduced pyridine nucleotide could be demonstrated. The significance of this enzyme system is not known.

The major fatty acid in the triglycerides synthesized and stored in the castor bean is ricinoleic acid (cis-Δ^9-12-D(−)-hydroxy 18:1). This fatty acid is formed from oleic acid. A mixed function oxygenase, utilizing NADH and O_2 and located in the endoplasmic reticular membranes of maturing castor bean seeds, catalyses the hydroxylation of the saturated C-12 of oleyl-CoA. The ricinoleic acid produced by the parasitic ergot fungus (Claviceps purpurea) is formed in a totally different way, namely by the anaerobic hydration of the Δ^{12}-double bond of linoleic acid.

Fatty acids are hydroxylated at the ω-carbon atom and also at non-terminal carbon atoms during the formation of cutin and suberin [see Section C.2(ii) and Chapter 4, Section B.5(i)]. Mixed function oxygenases, requiring NADPH or NADH and O_2, appear to catalyse these hydroxylations; their substrates are usually fatty acyl-CoAs.

(ii) FATTY ACID CATABOLISM

Several different catabolic routes exist for fatty acids in plants. Of these the β-oxidation spiral is probably the most important. However, the processes of α-oxidation and ω-oxidation also play a significant role in the breakdown and alternative utilization of fatty acids in plant tissues.

(a) β-Oxidation

The mechanism of the β-oxidation spiral has been discussed in Chapter 6, Section E.2(ii)(a).

Two functionally and physically separate roles exist for the β-oxidation spiral in higher plants. The β-oxidation spiral operates in the mitochondria of all plant cells. Here fatty acids are broken down into acetyl-CoA which may then be oxidized in the TCA cycle. The NADH and reduced flavin produced in these processes is reoxidized by the mitochondrial electron transport chain and ATP is concomitantly generated [see Chapter 6, Sections B and E.2(ii)]. The process is therefore very like that of animal mitochondria. However, the major differences between animals and plants is that in animals fatty acids are a major source of energy (i.e. ATP) whereas in plants they are not. It is likely that the fatty acids catabolized in the mitochondria of plant cells arise from the turnover of membrane lipids and that β-oxidation functions mainly to break them down into small units which can be re-used; the concomitant ATP production is, therefore, a bonus rather than a primary aim. After all plants generate their ATP by photosynthesis whilst animals, which do not, have to generate ATP by the oxidation of sugars or fatty acids.

In animal cells fatty acids enter the mitochondria in the form of fatty acyl O-carnitine derivatives. This is necessary because the inner membrane of the mitochondrion is impermeable to free fatty acids and to their coenzyme A derivatives. There exists, in the inner membrane of animal mitochondria, a shuttle mechanism (see Fig. 8.20) involving carnitine to overcome this difficulty. The carnitine shuttle has not yet been demonstrated in plant mitochondria but the occurrence of relatively large amounts of carnitine in mitochondria-rich plant tissues suggests that it does operate.

The second location of the β-oxidation spiral is in the glyoxysomes present, during germination, in the fat-storing cells of the seeds of those plants that produce fat-storing seeds (e.g. the endosperm cells of castor bean seeds). In these cells fatty acids and monoglycerides derived from the triglyceride stored in the spherosomes (oil bodies) pass into the glyoxysomes and are catabolized by the β-oxidation spiral (see Chapter 7, Section B.2). The resulting acetyl-CoA is then converted into succinate by the

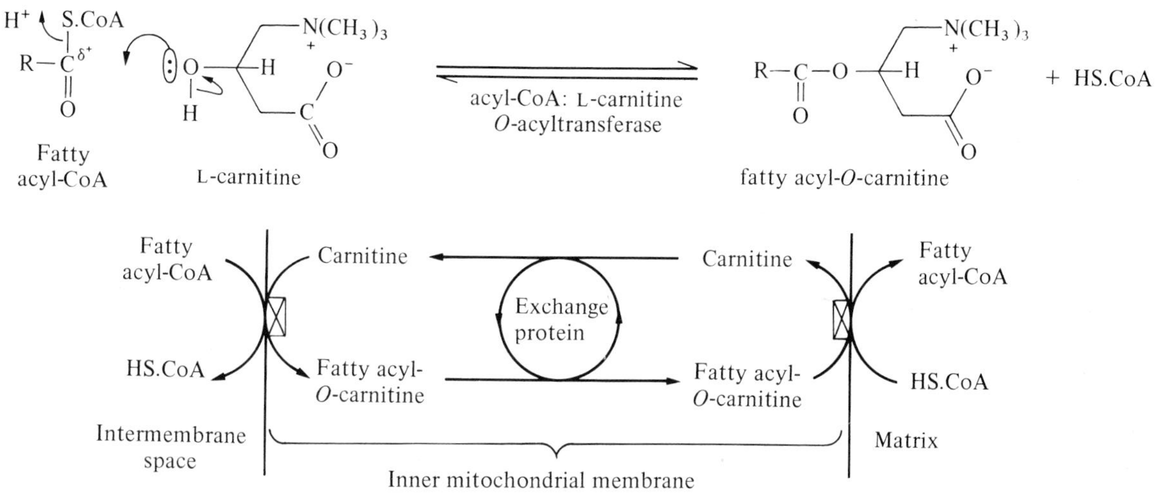

Fig. 8.20. Transfer of fatty acyl residues across the inner membrane of animal mitochondria by the carnitine shuttle. (Exchange protein = carnitine:acylcarnitine exchange protein which brings about a 1:1 exchange of carnitine and acylcarnitine across the inner membrane; ⊠ = acyl-CoA:L-carnitine O-acyltransferase.)

glyoxylate cycle [see Section C.1(iii)]. The succinate then passes out of the glyoxysome and is converted into carbohydrates (see Chapter 7, Section B.2). In this case the β-oxidation spiral is participating in a process, known as gluconeogenesis, in which the stored fat is converted into the carbohydrate which is required for the growth of the seedling until it becomes fully photosynthetic.

It is doubtful whether the carnitine shuttle is involved in the transportation of fatty acids across the glyoxysome membrane.

(b) α-Oxidation

The mechanism of α-oxidation has been discussed in Chapter 6, Section E.2(ii)(b).

The physiological role of α-oxidation in plants is not yet fully established. It has been suggested that it may be concerned in the degradation of branched chain fatty acids as has been demonstrated in animal tissues in the case of the phytanic acid formed from the phytol moiety of dietary chlorophyll. Phytanic acid has a methyl group on its β-carbon (C–3) atom which blocks the β-oxidation sequence of reactions. However, since the α-carbon (C–2) is unsubstituted, phytanic acid is susceptible to α-oxidation and is converted thereby to pristanic acid which is one carbon shorter. The offending

methyl group now resides on the α-carbon (C–2) atom of pristanic acid and the β-carbon (C–3) is unsubstituted. Pristanic acid is therefore susceptible to β-oxidation. The α-methyl group is by-passed by the β-oxidation enzymes and propionyl-CoA is eliminated by the first turn of the spiral instead of acetyl-CoA. Moreover, the two remaining methyl branches of the molecule also reside on α-carbon atoms as the β-oxidation spiral continues and they, too, are by-passed with the production of propionyl-CoA.

The catabolism of the propionyl-CoA formed by the β-oxidation of branched chain and odd-numbered fatty acids is described in Chapter 6, Section E.2(ii)(a). It has also been suggested that propionyl-CoA is of quantitative importance as a precursor of β-alanine which, in turn, is required for coenzyme A and ACP biosynthesis.

α-Oxidation is clearly the main source of the odd-numbered fatty acids and their derivatives which occur in some plant lipids.

It is thought likely that the 2-D-hydroxy fatty acids which are the major acyl components of cerebrosides in higher plants are formed by the offshoot of the α-oxidation spiral (see Fig. 6.44).

(c) ω-Oxidation

Fatty acids with oxygen functions (alcoholic,

carbonyl or carboxyl) at the methyl terminal end (ω-end) are formed by ω-oxidation and frequently occur as components of cutin and suberin [see Section C.2(ii) and Chapter 4, Section B.5(i)].

The initial oxidation step which brings about the conversion of an ω-methyl fatty acyl-CoA into an ω-hydroxymethyl fatty acyl-CoA is catalysed by a mixed function oxygenase. All such enzymes require molecular oxygen, reduced pyridine nucleotide and a non-haem iron protein. These requirements have been demonstrated for the higher plant system; however, in yeast cytochrome P-450 is also involved.

The conversion of the ω-hydroxymethyl fatty acid, formed by hydrolysis of its coenzyme A derivative, into the corresponding α,ω-dicarboxylic acid has been shown to be catalysed by an NADP-specific dehydrogenase in a supernatant fraction from *Vicia faba* epidermal tissue.

(d) Lipoxygenase[3]

The term lipoxygenase is given to a group of enzymes which catalyse the conversion of polyunsaturated fatty acids with a *cis,cis*-1,4-pentadiene system (e.g. linoleic and α-linolenic acids) into the corresponding acid with a *cis, trans*-1,3-butadienehydroperoxide system.

Lipoxygenases were thought to be confined to the seeds of legumes and some cereals but are now known to be widely distributed amongst higher plants. Particularly rich sources are potato tubers and the fruit of the egg plant. They have been demonstrated in leaves and in green algae. Lipoxygenases from different sources differ from one another in several ways, including substrate specificity, pH optima, effects of inhibitors and the isomeric form of their products. In some plants, e.g. soyabean, several lipoxygenase isoenzymes exist.

The physiological function of the lipoxygenases is not known. However, they have recently become the subject of active research because it has been realized that they cause the production of both desirable and undesirable flavour components in edible plant products.

Like most enzymes catalysing a reaction between molecular oxygen and an organic molecule, lipoxygenase contains a transition metal, namely iron; this

occurs in lipoxygenase to the extent of one atom per molecule of enzyme. The reason for this probably follows from the fact that oxygen has a triplet ground state (it is a diradical, having two unpaired electrons) whereas the stable organic reactants and products of the lipoxygenase-catalysed reaction are singlets, having all their electrons paired. Now the direct reaction of a triplet molecule with a singlet to give singlet products will not occur readily because it is a spin-forbidden process. In theory this difficulty could be overcome by converting oxygen from its triplet ground state into a singlet state. However, this would require the input of at least 93 kJ mol^{-1} of energy and enzymes are not capable of supplying electronic energy of this magnitude. One of the ways which oxygenases have evolved to surmount this problem is to complex the triplet oxygen with an enzyme-bound transition metal, which itself has unpaired electrons. The resulting complex can react with a singlet organic compound to give singlet oxidized products, for it is a spin-allowed process. The presence of iron (in ionic form) suggests that lipoxygenase utilizes this solution to the problem. The nature of the bonding of the iron to the enzyme is not known, beyond the fact that it is not present as haem. The possible participation of tryptophan residues within the enzyme in facilitating the binding of oxygen has been indicated.

The details of the mechanism of the lipoxygenase-catalysed reaction are unknown at the present time. Figure 8.21 gives an idea of the process as applied to linoleic acid. It appears to be initiated by the stereospecific removal of hydrogen from the C–11 methylene group. Removal of the 11-pro-S-hydrogen results in the subsequent stereospecific formation of 13R-hydroperoxy-$\Delta^{9(cis),11(trans)}$-octadecadienoic acid; this reaction is catalysed by soyabean lipoxygenase operating at pH 9.0. Removal of the 11-pro-R-hydrogen, however, results in the stereospecific formation of 9S-hydroperoxy-$\Delta^{10(trans)12(cis)}$octadecadienoic acid; this reaction is catalysed by corn germ lipoxygenase operating at pH 6.6. The free radical which is thought to result from hydrogen abstraction at C–11 rearranges as shown in Fig. 8.21; when the 11-pro-S-hydrogen is removed the unpaired electron comes to reside on C–13 and oxygen attacks from behind the plane of the page whereas when the 11-pro-R-hydrogen is removed the the unpaired electron comes to reside on C–9 and

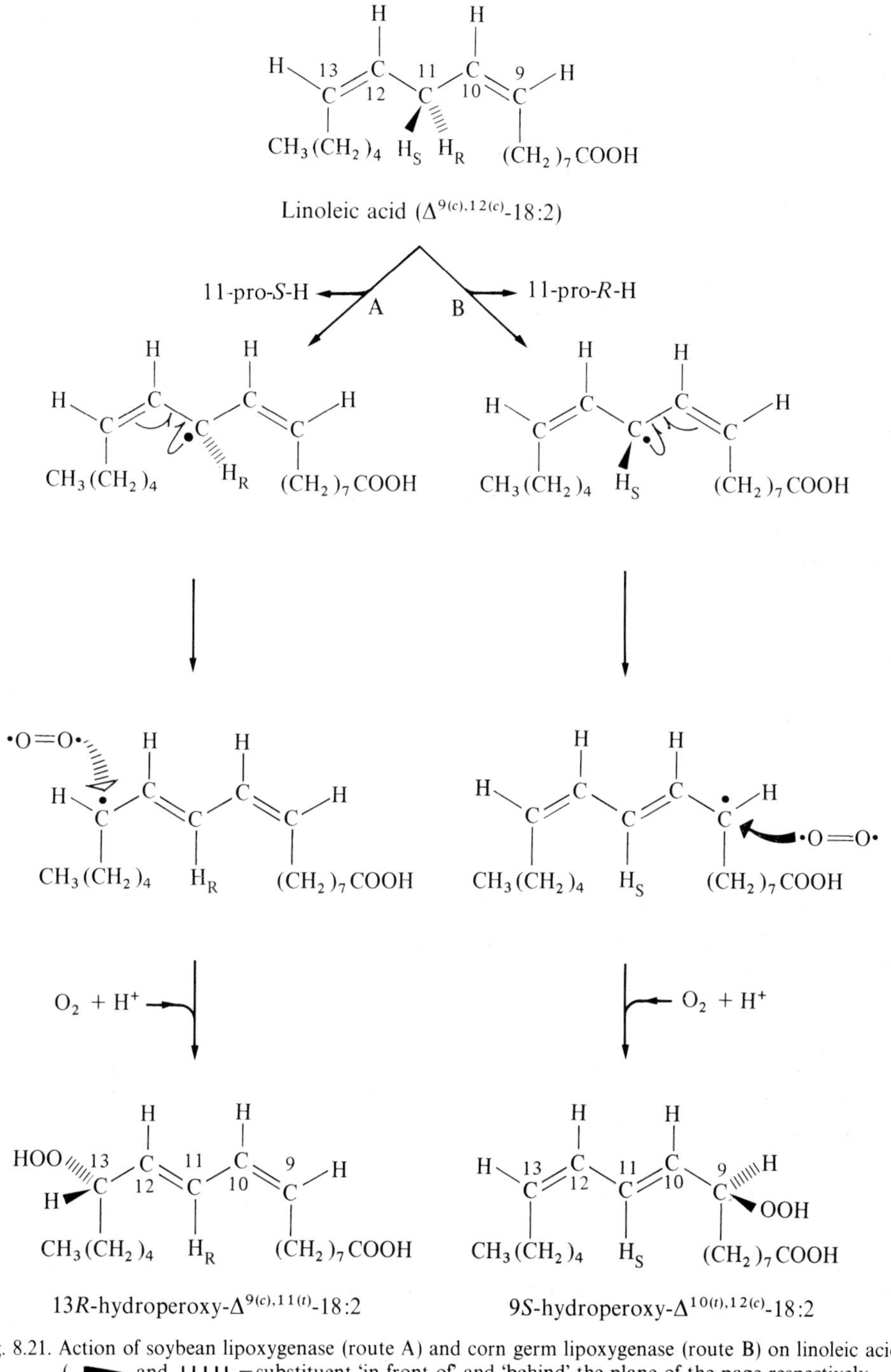

Fig. 8.21. Action of soybean lipoxygenase (route A) and corn germ lipoxygenase (route B) on linoleic acid. (▶ and ııııı = substituent 'in front of' and 'behind' the plane of the page respectively, ◀ and ₋ııɪɪ◁ = attack from 'in front of' and 'behind' the plane of the page respectively.)

oxygen attacks from in front of the plane of the page.

(iii) THE GLYOXYLATE CYCLE

The glyoxylate cycle enables living tissues to bring about the *net* conversion of fat (triglyceride) into carbohydrate; this is an example of gluconeogenesis (see Chapter 7, Section B.2). In the absence of the glyoxylate cycle the *net* conversion of fat into carbohydrate is not possible for reasons which will become apparent later. Not all living organisms have a glyoxylate cycle. It occurs in many bacteria, algae and specific tissues of the seeds, *during germination*, of certain higher plants. It does not occur in animals or in the vegetative tissue (i.e. leaf, stem, root) or the *developing* seeds of higher plants.

The glyoxylate cycle is of particular importance to those plants that store fat, as their main food reserve, in their seeds. During germination of these seeds the fat is utilized to support growth until the seedling is fully photosynthetic. This, of course, involves the conversion of the fat into carbohydrate, for example, in the synthesis of cell-wall polysaccharides.

Most of the fat-storing seeds are dicotyledonous. In many of these the fat is stored in the cells of the cotyledons, e.g. sunflower (*Helianthus annuus*), rape (*Brassica napus*), abyssinian kale (*Crambe abyssinica*) and peanuts (*Arachis hypogaea*) whilst in others it is present in the cells of the endosperm, e.g. castor beans (*Ricinus communis*). Within the cells of these tissues the fat is stored in organelles called spherosomes (see Chapter 3, Section B.8) which are distributed extensively throughout the cytoplasm. During the germination process another type of organelle, called the glyoxysome (see Chapter 3, Section B.7), develops in considerable numbers in these same cells. The glyoxysomes contain the enzymes of the glyoxylate cycle. When the fat reserve of the seed becomes depleted towards the end of germination, the number of glyoxysomes drops rapidly to zero.

Glyoxysomes, possessing an active glyoxylate cycle, are also present, during germination, in some seeds which do not store fat as their main food reserve. They are, for instance, present in the cells of the cotyledons of jojoba (*Simmondsia chinensis*) seeds which store wax esters; these wax esters

constitute about 60% of the dry weight of the cotyledons and are found in organelles, called wax bodies, which are similar to spherosomes. During germination the wax esters are hydrolysed by a lipase with an alkaline pH optimum which is bound to the membrane of the wax bodies. The resulting fatty acids and fatty alcohols then pass to the glyoxysomes where the key steps in their conversion into carbohydrates takes place.

Certain cereal seeds which store starch in their endosperm as their main food reserve also have glyoxysomes. In barley (*Hordeum distichon*) and wheat (*Triticum vulgare*) the glyoxysomes are present, during germination, in the aleurone cells which form a thin layer surrounding that part of the endosperm not adjacent to the scutellum (the single cotyledon). It is significant that the aleurone layer is the only tissue in these seeds that is rich in fat; in barley 3–4.6% of the dry weight of the seed is lipid of which 90%, mainly fat, is in the aleurone layer. In both these seeds the development of the glyoxysomes during germination appears to be initiated by gibberellin produced by the embryo. As germination progresses the fat content of the aleurone layer decreases and all the available evidence indicates that it is being converted into carbohydrate in a process involving the glyoxysomes. Although the quantitative importance of this process in these cereal seeds cannot be as great as that of the fat-storing (or wax-storing) seeds, it clearly has a significance which is not yet apparent.

The conversion of fat into carbohydrate and the role of glyoxysomes in it has been most extensively studied in germinating castor beans, particularly by Beevers and his colleagues. The details of the overall process, with the exception of the section that occurs within the glyoxysome, have been described in Chapter 7, Section B.2 and will not be repeated here. Free fatty acids and monoglycerides, derived from the fat stored in the spherosomes of the castor bean endosperm, enter the glyoxysomes. A membrane-bound, monoglyceride-specific lipase with an alkaline pH optimum then hydrolyses the monoglycerides producing free fatty acids and glycerol. The glycerol passes into the cytosol and is utilized for glucose formation.

The free fatty acids within the glyoxysome are converted into their coenzyme A derivatives by acyl-CoA synthetases. The fatty acyl-CoAs are then

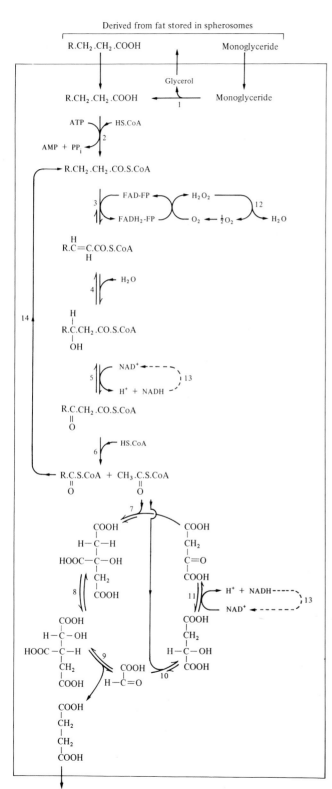

Fig. 8.22. Metabolism of fatty acids in the glyoxysome. (1 = lipase (alkaline pH optimum), EC 3.1.1.3; 2 = acyl-CoA synthetase, EC 6.2.1.3; 3 = acyl-CoA dehydrogenase, EC 1.3.99.3; 4 = enoyl-CoA hydratase, EC 4.2.1.17; 5 = 3-hydroxyacyl-CoA dehydrogenase; 6 = acetyl-CoA acyl-transferase, EC 2.3.1.16; 7 = citrate synthase, EC 4.1.3.7; 8 = aconitase, EC 4.2.1.3; 9 = isocitrate lyase, EC 4.1.3.1; 10 = malate synthase, EC 4.1.3.2; 11 = malate dehydrogenase, EC 1.1.1.37; 12 = catalase, EC 1.11.1.6; 13 = re-oxidation in mitochondrion (see text); 14 = note that additional enzymes are present to cope with the β-oxidation of unsaturated and hydroxy fatty acids (see text).)

Derived from fat stored in spherosomes

Further metabolism in mitochondria

degraded by the enzymes of the β-oxidation spiral into acetyl-CoA [see Section C.1(ii)(a) and Chapter 6, Section E.2(ii)(a)]. All the β-oxidation spiral enzymes have been shown to be present in the glyoxysome. This includes the $\Delta^{3-cis} \rightarrow \Delta^{2-trans}$-enoyl-CoA isomerase and the D($-$)-3-hydroxyacyl-CoA epimerase that are required to handle unsaturated fatty acids. The castor bean glyoxysome must also include additional enzymes to cope with the hydroxylated fatty acyl-CoA intermediates derived from ricinoleic acid (cis-Δ^9-12-D($-$)-hydroxy-18:1) since it has been shown that this acid is degraded completely to acetyl-CoA within the glyoxysome; ricinoleic acid is the most abundant fatty acid in the triglycerides stored in the spherosomes of the castor bean endosperm. The acyl-CoA dehydrogenase and enoyl-CoA hydratase appear to be loosely bound to the inner surface of the glyoxysome membrane.

The acetyl-CoA now enters the glyoxylate cycle, which resembles a TCA cycle minus the two decarboxylation steps. The first two reactions of the glyoxylate cycle are identical to those of the TCA cycle [see Chapter 6, Section E.1(ii)]. In the first acetyl-CoA is condensed with oxaloacetate to form citrate under the catalytic influence of citrate synthase[4]. In the second citrate is isomerized to 2S, 3R-isocitrate by the enzyme, aconitase[5]. The next reaction is peculiar to the glyoxylate cycle. In it the enzyme isocitrate lyase[6] catalyses the cleavage of the C–2 $\rightarrow$ C–3 bond in 2S,3R-isocitrate with the formation of glyoxylate and succinate [eq. (8.6)].

The glyoxylate then reacts with a molecule of acetyl-CoA (this is the second acetyl-CoA to enter the cycle) to form L-malate [eq. (8.7)]. The enzyme catalysing this reaction, malate synthase[7], is also peculiar to the glyoxylate cycle. The L-malate is then oxidized to oxaloacetate by NAD$^+$ under the catalytic influence of malate dehydrogenase[8]. This reaction is identical with the final reaction of the TCA cycle. Since it regenerates oxaloacetate, one of the two starting compounds, it is also the final reaction of the glyoxylate cycle.

The product of the glyoxylate cycle is the succinate formed in eq. (8.6). Since two molecules of acetyl-CoA are introduced into each turn of the glyoxylate cycle and one molecule of succinate is produced, the cycle can be thought of as a means of combining two acetyl (2C) units to form succinate (4C). The glyoxysome cannot metabolize the succinate further because it does not have the relevant enzymes (e.g. succinate dehydrogenase[9], fumarate hydratase[10]). Succinate, therefore, leaves the glyoxysome and enters the mitochondrion where it is converted into oxaloacetate via fumarate and L-malate by the appropriate TCA cycle enzymes. The oxaloacetate is then converted into phosphoenolpyruvate by PEP carboxykinase[11] [eq. (7.3)]. The phosphoenolpyruvate is then converted into D-glucose via a slightly modified glycolysis sequence operating in reverse (see Fig. 7.1).

The five reactions of the glyoxylate cycle, catalysed in sequence by the enzymes, citrate synthase,

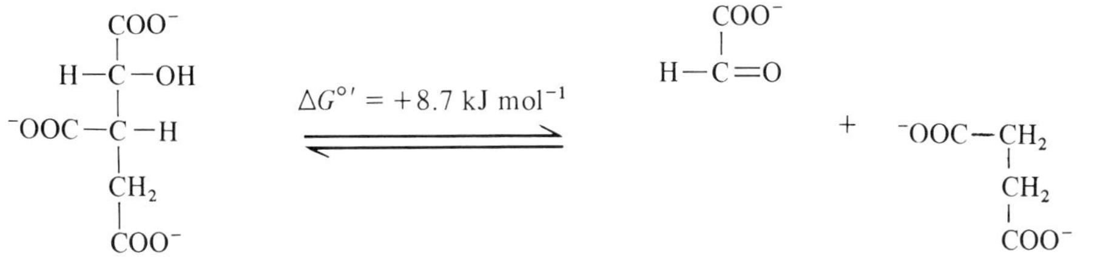

$$
\begin{array}{l}
\text{COO}^- \\
| \\
\text{H–C–OH} \\
| \\
{}^-\text{OOC–C–H} \\
| \\
\text{CH}_2 \\
| \\
\text{COO}^-
\end{array}
\qquad
\xrightleftharpoons{\quad \Delta G^{\circ\prime} = +8.7 \text{ kJ mol}^{-1} \quad}
\qquad
\begin{array}{l}
\text{COO}^- \\
| \\
\text{H–C=O}
\end{array}
\; + \;
\begin{array}{l}
{}^-\text{OOC–CH}_2 \\
| \\
\text{CH}_2 \\
| \\
\text{COO}^-
\end{array}
\qquad (8.6)
$$

$$\text{2S,3R-isocitrate} \qquad\qquad\qquad\qquad \text{glyoxylate} \qquad\qquad \text{succinate}$$

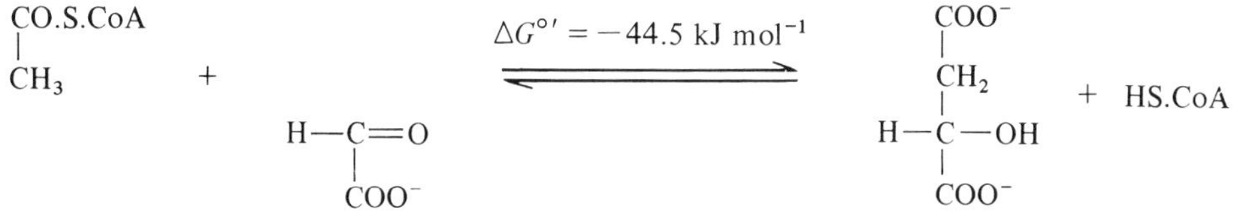

$$
\begin{array}{l}
\text{CO.S.CoA} \\
| \\
\text{CH}_3
\end{array}
\; + \;
\begin{array}{l}
\text{H–C=O} \\
| \\
\text{COO}^-
\end{array}
\qquad
\xrightleftharpoons{\quad \Delta G^{\circ\prime} = -44.5 \text{ kJ mol}^{-1} \quad}
\qquad
\begin{array}{l}
\text{COO}^- \\
| \\
\text{CH}_2 \\
| \\
\text{H–C–OH} \\
| \\
\text{COO}^-
\end{array}
\; + \; \text{HS.CoA} \qquad (8.7)
$$

aconitase, isocitrate lyase, malate synthase and malate dehydrogenase have $\Delta G^{\circ\prime}$ values of -31.4, $+6.7$, $+8.7$, -44.5 and $+27.9$ kJ mol^{-1} respectively. This gives an overall $\Delta G^{\circ\prime}$ value of -32.6 kJ mol^{-1} which corresponds to an equilibrium constant of 5.14×10^5. Thus the combining of two acetyl units to form succinate by the glyoxylate cycle is thermodynamically favourable.

The operation of the β-oxidation spiral and the glyoxylate cycle in the glyoxysome generates the reduced coenzymes, FADH$_2$ and NADH. The former arises from the operation of the acyl-CoA dehydrogenase of the β-oxidation spiral. The latter arises from two reactions, that catalysed by 3-hydroxyacyl-CoA dehydrogenase in the β-oxidation spiral and that catalysed by malate dehydrogenase in the glyoxylate cycle. The FADH$_2$ and NADH have to be reoxidized to keep the glyoxysome processes going. However, this presents a problem because, unlike the mitochondrion, the glyoxysome does not possess an electron transport chain. The way this problem is overcome has not been fully elucidated. It is fairly certain that the FADH$_2$ of the acyl-CoA dehydrogenase is re-oxidized directly by oxygen. The resulting H$_2$O$_2$ is then converted into H$_2$O and oxygen by the enzyme, catalase[12], which is very abundant in the glyoxysome. The NADH is not reoxidized in the glyoxysome. It appears to be reoxidized in the mitochondrion. It is not known whether this is accomplished directly by the mitochondrial NADH dehydrogenase which is specific for exogenous NADH (see Chapter 6, Section C) or by a shuttle mechanism, in which reducing equivalents are transferred from one organelle to the other instead of NADH itself. It has been suggested that the reducing equivalents participating in such a shuttle might be the products of the transaminases (serine-glyoxylate transaminase[13], glutamate-glyoxylate transaminase) known to be present in both glyoxysomes and mitochondria.

It was stated earlier in this section that but for the glyoxylate cycle the *net* conversion of fat into carbohydrate would be impossible. The reason for this is that, in the absence of the glyoxylate cycle, the acetyl units of acetyl-CoA would be oxidized to two molecules of CO$_2$ by the TCA cycle operating in the mitochondrion. Thus the acetyl units derived from the stored fat would be lost as CO$_2$ and could not be

used to synthesize carbohydrate. The glyoxylate cycle is rather like a TCA cycle minus its two decarboxylation reactions, catalysed by isocitrate dehydrogenase[14] and α-oxoglutarate dehydrogenase[15]; these are the reactions which in the TCA cycle effectively convert the two carbon atoms of the acetyl unit into CO$_2$. In bacteria, which do not have any intracellular organelles, the glyoxylate cycle enzymes are not compartmentalized from the TCA cycle enzymes. Thus the glyoxylate cycle is more like a by-pass of the TCA cycle decarboxylation reactions than a distinct cycle and the rate at which it operates has to be regulated allosterically, isocitrate lyase being strongly inhibited by phosphoenolpyruvate. In the seed tissues of higher plants the glyoxylate and TCA cycles are physically separated from one another by being in different organelles. This means that the enzymes that they have in common must be duplicated within the cell and glyoxysome and mitochondrial isoenzymes of citrate synthase, malate dehydrogenase and certain transaminases have been demonstrated. The physical separation of the glyoxylate and TCA cycles in seed tissues and the fact that the β-oxidation enzymes are located in the glyoxysomes ensure that virtually all the fatty acid derived from the stored fat is utilized for carbohydrate formation. The main carbohydrate to be formed initially is sucrose which is translocated from the endosperm cells to the developing seedling where it is converted into the complex range of proteins, lipids and carbohydrates required for growth.

2. Metabolism of neutral lipids

(i) METABOLISM OF GLYCERIDES

(a) *Glyceride biosynthesis*

The biosynthesis of glycerides may be subdivided into three phases: (i) the formation of the fatty acyl residues, (ii) the formation of the glycerol moiety and (iii) the linking of the former to the latter. Since fatty acid biosynthesis has been described earlier [Section C.1(i)], this section will deal primarily with phases (ii) and (iii).

The study of glyceride biosynthesis in plants has

been confined mainly to those tissues which synthesize and store large quantities of triglyceride, namely the endosperm of castor bean seeds and the cotyledons of the seeds of such plants as sunflower, peanut and abyssinian kale. During the development of these seeds triglycerides accumulate in organelles called spherosomes (or oil bodies; see Chapter 3, Section B.8) which eventually occupy most of the volume of the cells of the endosperm or cotyledons.

The spherosomes not only store triglycerides but also play a major role in their synthesis. They appear to possess all the enzymes, presumably built into their bounding membrane, necessary for the biosynthesis of fatty acids from acetyl-CoA and their incorporation into triglycerides. However, the origin of the acetyl-CoA and of the glycerol moieties of the triglycerides is sucrose imported into the developing seed from the photosynthetic tissues of the plant. The overall process and its relationship to the biosynthesis of glycerophospholipids and galactosyldiglycerides is shown in Fig. 8.23.

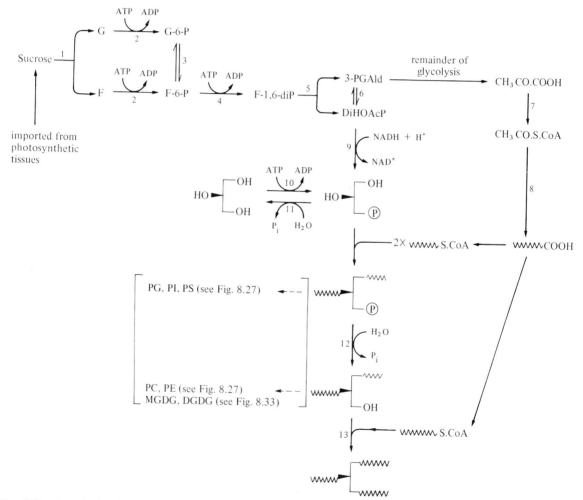

Fig. 8.23. Biosynthesis of triglycerides in fat-storing seeds. ($\blacktriangleright[$ = glycerol residue; $\wedge\wedge\wedge\wedge$ = fatty acid residue; $\textcircled{P}$ and P = orthophosphoric acid residues; P_i = orthophosphate; G = D-glucose; F = D-fructose; 3-PGAld = 3-phosphoglyceraldehyde; DiHOAcP = dihydroxyacetone phosphate; PG, PI, PS, PC and PE = phosphatidyglycerol, -inositol, -serine, -choline and -ethanolamine; MGDG and DGDG = mono- and digalactosyldiglyceride; 1 = invertase, EC 3.2.1.26; 2 = hexokinase, EC 2.7.1.1; 3 = glucosephosphate isomerase, EC 5.3.1.9; 4 = 6-phosphofructokinase, EC 2.7.1.11; 5 = fructose-bisphosphate aldolase, EC 4.1.2.13; 6 = triose phosphate isomerase, EC 5.3.1.1; 7 = pyruvate dehydrogenase complex; 8 = acetyl-CoA carboxylase, EC 6.4.1.2, fatty acid synthase plus elongation and desaturation systems; 9 = glycerol-3-phosphate dehydrogenase (NAD$^+$), EC 1.1.1.8; 10 = glycerol kinase, EC 2.7.1.30; 11 = glycerophosphatase; 12 = phosphatidate phosphatase, EC 3.1.3.4; 13 = diacylglycerol acyltransferase, EC 2.3.1.20.)

The imported sucrose is presumably hydrolysed to D-glucose and D-fructose under the catalytic influence of invertase[16]. The two monosaccharides are then phosphorylated by hexokinase[17] and are catabolized by the glycolysis enzymes in the cytoplasm of the endosperm or cotyledon cells. The resulting pyruvate is then oxidatively decarboxylated to yield acetyl-CoA which is converted into a range of fatty acyl-CoAs by the spherosome enzymes.

Some dihydroxyacetone phosphate (DiHOAcP) is tapped off the glycolysis pathway and reduced by an NAD-specific glycerol 3-phosphate dehydrogenase[18] to sn-glycerol 3-phosphate. The intracellular location of this enzyme is not known with certainty; work with animal tissues suggests that it is likely to be cytoplasmic. Within the spherosome the sn-glycerol 3-phosphate is acylated to form 3-sn-phosphatidic acid. The acyl residues are derived from two fatty acyl-CoAs. Their transfer to the hydroxyl groups of sn-glycerol 3-phosphate is catalysed by acyltransferases. It is not known which of the two hydroxyl groups (sn-1 or sn-2) is acylated first but whichever it is a lyso-3-sn-phosphatidic acid (a monoacyl-sn-glycerol 3-phosphate) is formed as an intermediate. Saturated fatty acyl groups tend to be transferred to the sn-1 position and unsaturated ones to the sn-2 position.

The phosphate ester bond in 3-sn-phosphatidic acid is then hydrolysed under the catalytic influence of phosphatidate phosphatase[19]. The resulting sn-1,2-diglyceride (sn-1,2-diacylglycerol) is then converted into a triglyceride by the transfer of an acyl residue from a third fatty acyl-CoA catalysed by diacylglycerol acyltransferase[20].

Free glycerol may be formed in the cytoplasm of the cell from sn-glycerol 3-phosphate by glycerophosphatase and rephosphorylated by glycerokinase[21].

Little is known of the biosynthesis of the rare 'ether triglycerides' and 'diol lipids' in plants. Ether triglycerides in animal tissues are formed by a variation of the dihydroxyacetone phosphate pathway. DiHOAcP is acylated from fatty acyl-CoA to yield an acyl-DiHOAcP; this is then immediately followed by the exchange of the acyl residue for a fatty alcohol with the formation of an alkyl-diHOAcP. The keto group is then reduced to yield 1-alkyl-sn-glycerol 3-phosphate. The sn-2 hydroxyl group is then acylated from fatty acyl-CoA. The resulting 1-alkyl-2-acyl-sn-glycerol 3-phosphate is then successively dephosphorylated and acylated in a manner analogous to reactions 12 and 13 of Fig. 8.23.

(b) Glyceride catabolism

The catabolism of triglycerides may be subdivided into three phases: (i) the hydrolytic cleavage of the three ester bonds to yield glycerol and fatty acids, (ii) the catabolism of the glycerol and (iii) the catabolism of the fatty acids.

Phase (i) is catalysed by a group of enzymes known as lipases[22]. Phase (ii) is initiated by glycerokinase (reaction 10, Fig. 8.23); the resulting sn-glycerol 3-phosphate is then oxidized to DiHOAcP which may then be catabolized to CO_2 via glycolysis and the tricarboxylic acid cycle (see Chapter 6, Section E.1) or utilized for carbohydrate synthesis via a reversed, modified glycolysis pathway (see Chapter 7, Section B.2). The processes involved in fatty acid catabolism [phase (iii)] have been described earlier [Section C.1(ii) and Chapter 6, Section E.2(ii)].

The term lipase has been used rather loosely in the past to describe any enzyme that liberated fatty acids from acyl lipids. However, the term should now only be used to describe an enzyme that catalyses the hydrolysis of one or more of the ester bonds in a triglyceride whilst acting at a triglyceride–water interface. Thus lipases act at the surface of a fat droplet utilizing the H^+ and HO^- of the surrounding water to effect the cleavage of the ester bonds.

Most of our knowledge of plant lipases comes from studies of germinating, fat-storing seeds. The triglyceride of these seeds, stored in the spherosomes, is utilized to support growth until the seeding is fully photosynthetic; the details of this process have been described in Section C.1(iii) and in Chapter 7, Section B.2. The spherosomes possess a membrane-bound lipase which becomes active during germination. The nature of this enzyme no doubt differs slightly in different plant species. However, all the spherosome lipases studied so far have two properties in common: (i) they have an acid pH optimum, e.g. pH 4.1 for castor bean spherosome lipase, and (ii) they are capable of catalysing the hydrolysis of all the ester bonds in the triglycerides characteristic of their own plant species; however, the castor bean spherosome lipase shows a preference for the primary ester bonds (*sn*-1 and *sn*-3) over the secondary ester bond (*sn*-2). Castor bean spherosome lipase requires for activity a low molecular weight, heat-stable glycoprotein and a cyclic tetramer of ricinoleic acid.

A lipase with an alkaline pH optimum has been detected in the membranes of glyoxysomes isolated from the endosperm cells of germinating castor bean seeds. This lipase catalyses the hydrolysis of the ester bond in monoglycerides imported from the spherosomes. These monoglycerides have been produced by the incomplete action of the spherosome lipase on the stored triglycerides. It is likely that glyoxysomes from other plant species also possess this

enzyme. However, because the substrate of this enzyme is a monoglyceride rather than a triglyceride, Galliard believes that it should not be regarded as a true lipase. The deacylation of monoglycerides by enzymes distinct from true lipases has also been described in wheat seeds, potato tuber and in the leaves of potato and French bean; the full substrate specificities of most of these enzymes have not been determined.

(ii) WAX AND CUTIN METABOLISM

Our knowledge of the biosynthesis of the wax and cutin of plant cuticles owes much to the work of Kolattukudy and his coworkers.

The biosynthesis of the components of wax is outlined in Fig. 8.24. Palmitic acid produced from acetyl-CoA by the catalytic activities of acetyl-CoA

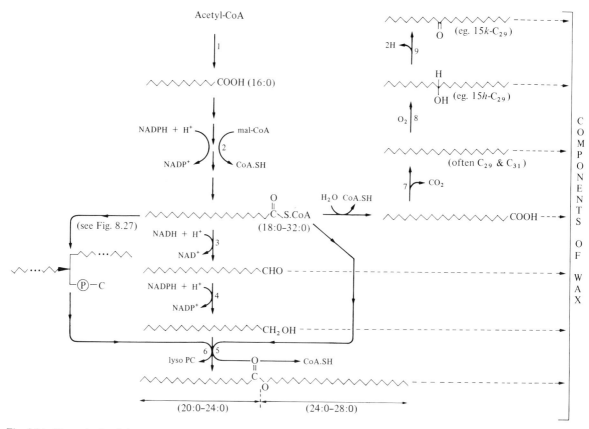

Fig. 8.24. Biosynthesis of the components of wax. (mal-CoA = malonyl-CoA; lyso PC = lysophosphatidylcholine; 1 = acetyl-CoA carboxylase and fatty acid synthase; 2 = fatty acid elongation system(s); 3 = fatty acyl-CoA reductase; 4 = fatty aldehyde reductase; 5 = fatty acyl CoA : fatty alcohol acyltransferase; 6 = phosphatidylcholine : fatty alcohol acyltransferase; 7 = fatty acid decarboxylase; 8 = a hydroxylase; 9 = an oxidoreductase; $\wedge\wedge\wedge\wedge$ = a hydrocarbon chain.)

carboxylase [Section C.1(i)(a)] and the fatty acid synthase [Section C.1(i)(b)] is elongated to give a range of saturated, even-numbered fatty acids with 18–32 carbon atoms. Little is known of the elongation systems involved save that (i) a particulate preparation from pea leaf epidermis utilized malonyl-CoA as the source of 2C units and NADPH as the reductant and (ii) there is evidence that several different systems are operative.

The fatty acyl-CoAs produced by the elongation systems are modified in a number of different ways to form all the components of wax. Free fatty acids are produced by the hydrolytic activity of fatty acyl-CoA thioesterases. Fatty aldehydes are formed by the action of an NAD-specific fatty acyl-CoA reductase. Fatty alcohols are formed from the fatty aldehydes by the action of an NADP-specific fatty aldehyde reductase. The latter two enzymes have been partly purified from an acetone powder of young leaves of broccoli (*Brassica oleracea*); the

fatty aldehyde reductase has an SH group at its catalytic site. Similar enzymes have been demonstrated in dark-grown *Euglena gracilis* which synthesizes large quantities of wax esters.

Wax esters, the principal components of cuticular waxes, may be formed in two different ways: (i) by transfer of an acyl moiety from a phospholipid such as phosphatidylcholine to a fatty alcohol and (ii) by transfer of the acyl moiety of a fatty acyl-CoA to a fatty alcohol. Acetone powders of young leaves have been shown to catalyse both of these reactions. An acyl-CoA:fatty alcohol acyltransferase with a broad pH optimum (6.0–9.0) has been partly purified by ammonium sulphate fractionation of an acetone powder of young broccoli leaves followed by gel filtration on Sephadex G100.

The hydrocarbon components of wax, odd-numbered *n*-alkanes, are formed by decarboxylation of the next higher, even-numbered free fatty acid; thus a C_{32}-fatty acid yields a C_{31}-*n*-alkane.

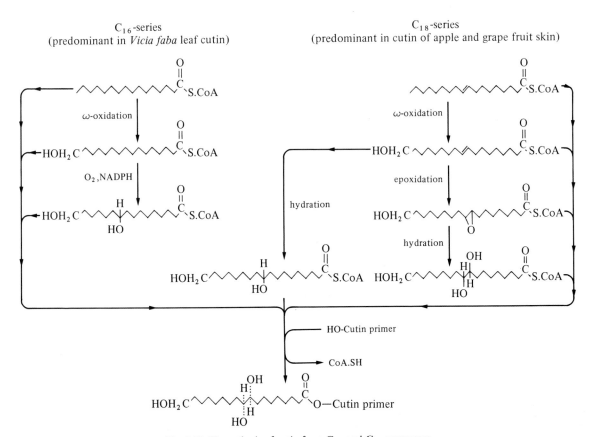

Fig. 8.25. Biosynthesis of cutin from C_{16} and C_{18} precursors.

Oxidation of the hydrocarbons yields secondary alcohols with the hydroxyl group on a carbon atom near the centre of the molecule; thus exogenous nonacosane (n-C_{29}) was shown to be converted by broccoli leaves into a mixture of nonacosan-15-ol and nonacosan-14-ol in the ratio 3:2, which is identical to the ratio of these compounds in the wax of these leaves.

Subsequent oxidation of the secondary alcoholic function yields a ketone. This oxidation step appears to be highly specific with respect to the position of the secondary hydroxyl group; for instance, broccoli leaves oxidize only the nonacosan-15-ol because virtually all the C_{29}-ketone in the wax is the 15-keto isomer.

The biosynthesis of cutin can be divided into two phases: (i) the formation of cutin monomers which are a range of C_{16}- and C_{18}-fatty acids commonly hydroxylated at the ω- or a central carbon and (ii) incorporation of the monomers into cutin which is a polymer of interesterified hydroxy fatty acids [see Chapter 4, Section B.5(i)]. The overall process is shown in Fig. 8.25.

The C_{16}-family of hydroxy fatty acids which are predominant in the cutin of broad bean (*Vicia faba*) leaves are derived from palmityl-CoA. The initial step is an ω-oxidation [see Section C.1(ii)(c)] which yields 16-hydroxypalmityl-CoA. A mixed function oxygenase then introduces a hydroxyl group at C–10 to form 10,16-dihydroxypalmityl-CoA.

The C_{18}-family of hydroxy fatty acids which are predominant in the cutin of the skin of the fruits of the apple and grape are derived from oleyl-CoA. The initial step is again an ω-oxidation. It yields 18-hydroxyoleyl-CoA which undergoes either hydration to form 10,18-dihydroxystearyl-CoA or epoxidation to form 18-hydroxy-9,10-epoxy-stearyl-CoA. The latter may also hydrate to form 9,10,18-trihydroxy-stearyl-CoA.

During phase (ii), the polymerization phase, the acyl residues of the C_{16}- and C_{18}-families of fatty acids are transferred to hydroxyl groups in a growing cutin molecule. The cutin hydroxyl groups reside either on the ω-carbon or one of the central carbon atoms of a hydroxylated fatty acid already in the cutin molecule. The nature of this process closely

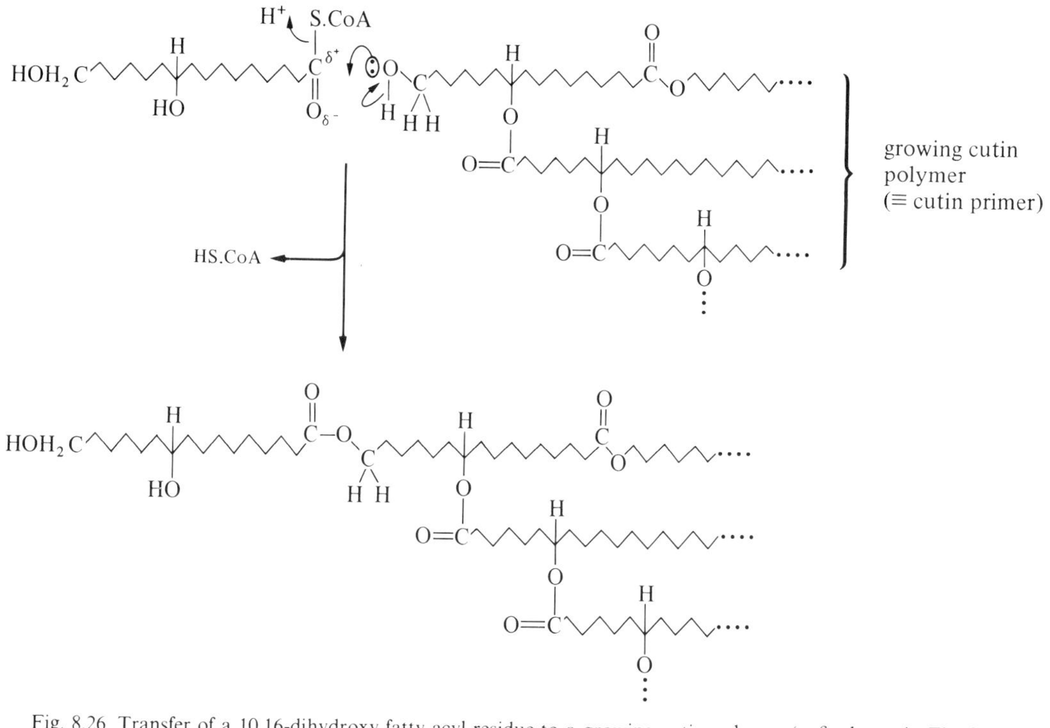

Fig. 8.26. Transfer of a 10,16-dihydroxy fatty acyl residue to a growing cutin polymer (= final step in Fig. 8.25).

resembles that by which polysaccharides are formed [see Chapter 7, Section D, eq. (7.41)]; the donor molecule is an acyl-CoA, the acceptor molecule, or primer, is a growing cutin molecule. Although the transfer of hydroxy-fatty acyl residues from their CoA derivatives into endogenous cutin has been demonstrated with a particulate preparation from *V. faba* leaf epidermis, it is not yet clear what the initial primer molecule (i.e. the equivalent of maltose in amylose formation) is.

The mechanism of the acyl transfer reaction is shown in Fig. 8.26; it is presumably catalysed by an acyl-CoA:cutin acyltransferase.

The biosynthesis of the wax and cutin components of the cuticles of higher plants occurs only in the epidermal layer of cells. It is, in fact, highly likely that the main metabolic activity of the epidermal cells is the biosynthesis of the cuticle.

3. Metabolism of polar lipids

(i) BIOSYNTHESIS OF PHOSPHOLIPIDS

(a) *Glycerophospholipid biosynthesis*

The biosynthesis of glycerophospholipids (see Fig. 8.27) has in common with the biosynthesis of triglycerides (see Fig. 8.23) the fact that *sn*-glycerol 3-phosphate is a key intermediate. In both biosynthetic pathways the *sn*-glycerol 3-phosphate is formed by the reduction of DiHOAcP under the catalytic influence of glycerol-3-phosphate dehydrogenase. However, the immediate origin of the DiHOAcP can be different in the two pathways. In the triglyceride pathway which only takes place to any great extent in the cells of developing, fat-

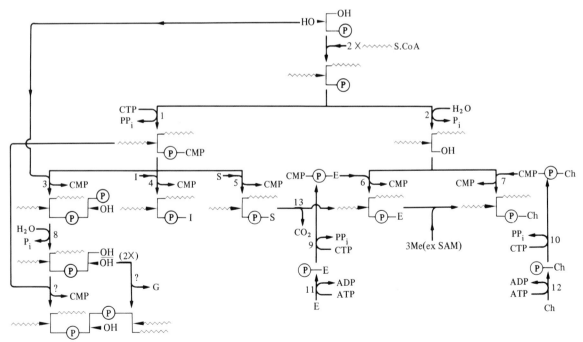

Fig. 8.27. Biosynthesis of glycerophospholipids in plants. (►─[= glycerol residue; ∿∿∿ = fatty acid residue; ℗ = orthophosphoric acid residue; P_i = orthophosphate; PP_i = pyrophosphate; CTP = cytidine 5′-triphosphate; CMP = cytidine 5′-monophosphate; Ch = choline residue; E = ethanolamine residue; S = L-serine residue; I = *myo*-inositol residue; G = glycerol; Me = methyl group; SAM = S-adenosylmethionine; 1 = phosphatidate cytidylyltransferase, EC 2.7.7.41; 2 = phosphatidate phosphatase, EC 3.1.3.4; 3 = phosphatidylglycerol phosphate synthase; 4 = phosphatidylinositol synthase, EC 2.7.8.11; 5 = phosphatidylserine synthase, EC 2.7.8.8; 6 = ethanolaminephosphotransferase, EC 2.7.8.1; 7 = cholinephosphotransferase, EC 2.7.8.2; 8 = phosphatidylglycerophosphatase, EC 3.1.3.27; 9 = ethanolaminephosphate cytidylyltransferase, EC 2.7.7.14; 10 = cholinephosphate cytidylyltransferase, EC 2.7.7.15; 11 = ethanolamine kinase, EC 2.7.1.82; 12 = choline kinase, EC 2.7.1.32; 13 = phosphatidylserine decarboxylase, EC 4.1.1.65.)

storing seeds, it originates from sucrose imported from photosynthetic tissue. In the glycerophospholipid pathway which takes place in all cells since its products are required for membrane formation, the DiHOAcP may originate from imported sucrose in non-photosynthetic cells or directly from the chloroplast in photosynthetic cells. In the latter case, when photosynthesis is taking place, the DiHOAcP is tapped off from the Calvin cycle and diffuses across the chloroplast envelope into the cytoplasm (see Chapter 5, Section D.6; Fig. 5.36) whilst, when photosynthesis is not taking place during the hours of darkness, it is formed by the intra-chloroplastidic breakdown of starch (see Fig. 5.37).

The sn-glycerol 3-phosphate is acylated to form 3-sn-phosphatidic acid as in the triglyceride biosynthetic pathway. Although the transfer of acyl residues from fatty acyl-CoAs to sn-glycerol 3-phosphate and to lyso-3-sn-phosphatidic acid has been demonstrated in vitro using microsomal enzyme preparations from spinach leaves, the possibility that fatty acyl-ACPs could also act as acyl donors cannot be ruled out.

The glycerophospholipid biosynthetic pathway now bifurcates. If phosphatidylcholine or -ethanolamine are to be formed, the 3-sn-phosphatidic acid is hydrolysed to sn-1,2-diglyceride by phosphatidate phosphatase, as in the triglyceride pathway. If, on the other hand, phosphatidylglycerol, -inositol or

-serine are to be formed, the 3-sn-phosphatidic acid reacts with cytidine 5′-triphosphate (CTP) to form cytidine 5′-diphosphate diglyceride (CDP-diglyceride) and inorganic pyrophosphate (see Fig. 8.28); the reaction is catalysed by phosphatidate cytidylyltransferase,[23] an enzyme whose activity has been demonstrated in cell-free preparations from the leaves of spinach and maize and from cauliflower inflorescence. It should be noted that of the two orthophosphoric acid residues in CDP-diglyceride, one originates from the 3-sn-phosphatidic acid, the other from CTP. Moreover, CDP-diglyceride functions as a donor of 3-sn-phosphatidic acid in the biosynthesis of mono- and diphosphatidylglycerols and of phosphatidylinositol and -serine as shown in Fig. 8.29. Therefore from a biosynthetic point of view CDP-diglyceride is better described as CMP-phosphatidic acid; this point is emphasized in the way CDP-diglyceride is represented in Fig. 8.27.

Kates has demonstrated the presence of a phosphatidylserine synthase[24] in a particulate fraction from spinach leaves; this enzyme catalyses the transfer of the phosphatidic acid moiety of CDP-diglyceride to L-serine, so forming phosphatidylserine and CMP.

A phosphatidylinositol synthase[25] has been demonstrated in a mitochondrial preparation from cauliflower inflorescence by Mudd and in a spinach

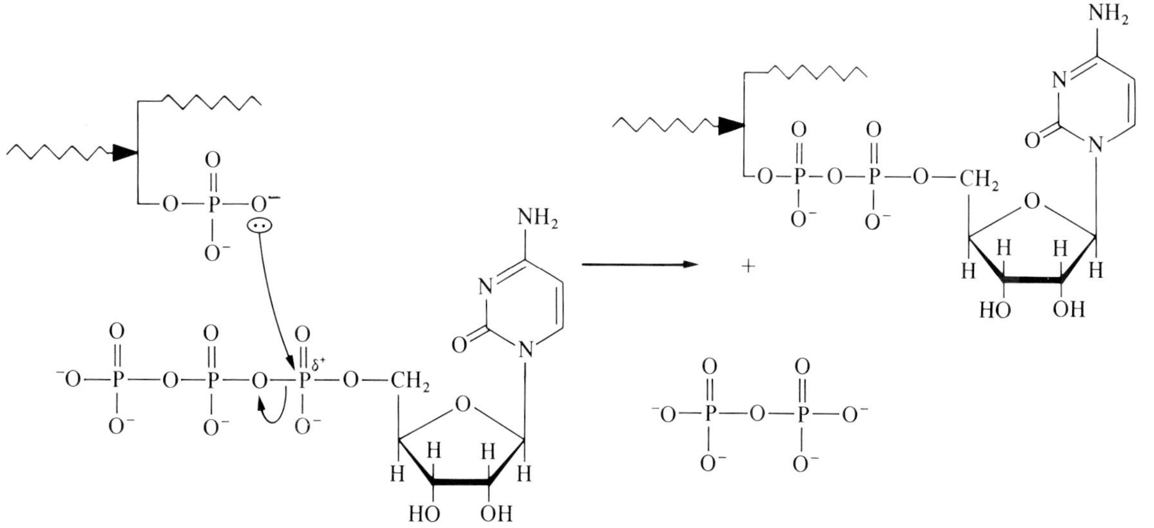

Fig. 8.28. Formation of CDP-diglyceride, catalysed by phosphatidate cytidylyltransferase (reaction 1, Fig. 8.27).

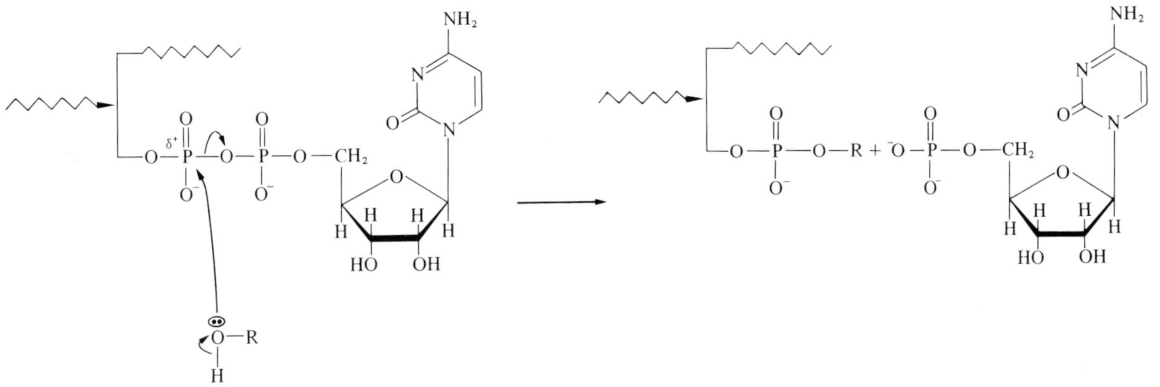

Fig. 8.29. General reaction in which CDP-diglyceride acts as a donor of 3-*sn*-phosphatidic acid. (R.OH = *sn*-glycerol 3-phosphate, monophosphatidylglycerol, *myo*-inositol or L-serine.)

leaf particulate preparation by Kates. This enzyme catalyses the transfer of the phosphatidic acid residue of CDP-diglyceride to *myo*-inositol in the presence of Mn^{2+} ions, so forming phosphatidylinositol and CMP.

Monophosphatidylglycerol is formed in a two-reaction sequence. In the first, the phosphatidic acid residue of CDP-diglyceride is transferred to the *sn*-1 hydroxyl group of *sn*-glycerol 3-phosphate, so forming monophosphatidylglycerol phosphate. In the second, the orthophosphoric acid residue esterified to the *sn*-3 hydroxyl group of the newly introduced *sn*-glycerol 3-phosphate moiety is hydrolysed. Evidence that this reaction sequence operates in cauliflower inflorescence and in the endoplasmic reticular membranes of spinach leaves has been provided by Douce and Kates respectively.

Diphosphatidylglycerol (cardiolipin) is known to be formed in animal tissues by the transfer of a phosphatidic acid residue from CDP-diglyceride to the free *sn*-3 hydroxyl group of monophosphatidylglycerol. Bacteria (e.g. *E. coli, Staphylococcus aureus*), however, utilize a different reaction in which a phosphatidic acid residue is transferred from one molecule of monophosphatidylglycerol to the free *sn*-3 hydroxyl group of a second; free glycerol is formed as a by-product. It is not yet clear which of these routes is responsible for the biosynthesis of diphosphatidylglycerol in plants.

Phosphatidylcholine and -ethanolamine are formed by the transfer of phosphocholine and phosphoethanolamine respectively from CDP-cho-

line and CDP-ethanolamine to the *sn*-3 hydroxyl group of *sn*-1,2-diglyceride; these two transfer reactions are catalysed by distinct enzymes but are mechanistically identical (see Fig. 8.30).

CDP-choline and CDP-ethanolamine are formed by analogous routes. The initial step is the phosphorylation by ATP of the amino alcohol; choline kinase[26] and ethanolamine kinase[27] are the enzymes involved. The resulting phosphocholine and phosphoethanolamine then react with CTP under the catalytic influence of distinct enzymes to form CDP-choline and CDP-ethanolamine.

As with CDP-diglyceride, CDP-choline and CDP-ethanolamine are better described as CMP-phosphocholine and CMP-phosphoethanolamine since this reflects their biosynthesis and their function as phosphoaminoalcohol donors in glycerophospholipid formation; this point is emphasized in the way they are represented in Fig. 8.27.

Work in the laboratories of Kates and Mudd has demonstrated that the enzymes catalysing the incorporation of choline and ethanolamine into phosphatidylcholine and -ethanolamine respectively occur in the microsomal fraction of spinach leaves.

Phosphatidylcholine can also be formed from phosphatidylethanolamine in the fungus, *Neurospora crassa*, the yeast, *Saccharomyces cerevisiae*, the chrysophyte alga, *Ochromonas malhamensis*, and probably in both photosynthetic and non-photosynthetic higher plant tissues. This pro-

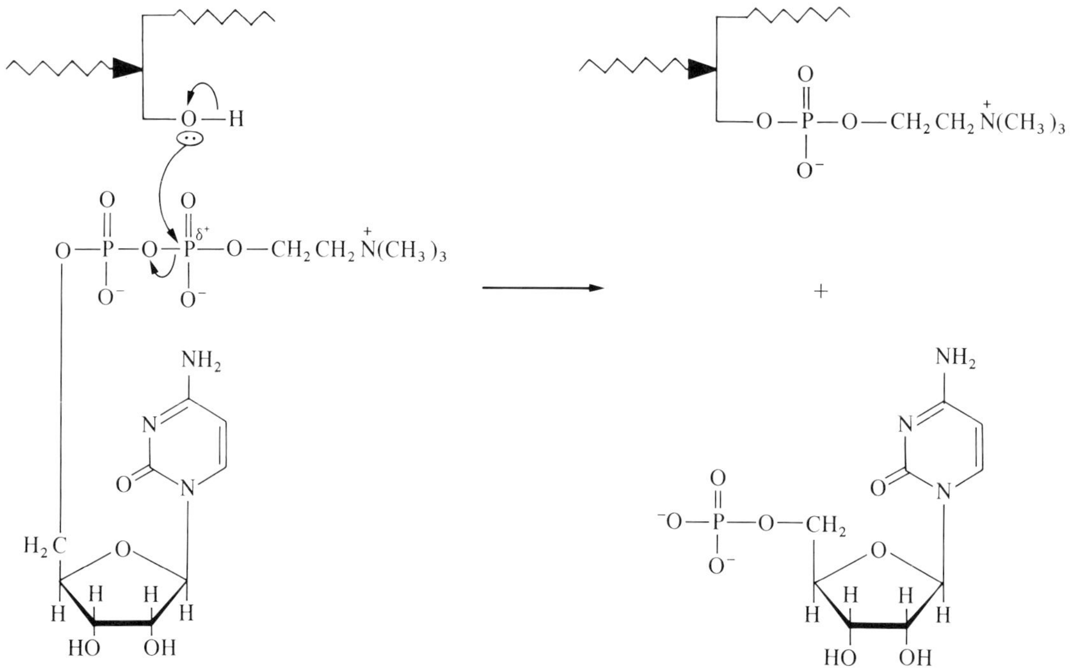

Fig. 8.30. Transfer, catalysed by cholinephosphotransferase, of the phosphocholine residue from CDP-choline to sn-1,2-diglyceride with the formation of phosphatidylcholine (reaction 7, Fig. 8.27).

cess involves the stepwise introduction of three methyl groups on to the amino nitrogen atom of phosphatidylethanolamine; the methylating agent is S-adenosylmethionine. Evidence from N. crassa suggests that separate enzymes catalyse the first (see Fig. 8.31) and subsequent methylations.

Phosphatidylethanolamine can be formed by de-

carboxylation of the serine moiety of phosphatidylserine. Phosphatidylserine decarboxylase[28] activity has been clearly demonstrated in spinach leaf homogenate fractions by Kates and there is indirect evidence of its occurrence in Chlorella vulgaris, tomato root and pea seedlings.

The subcellular site of glycerophospholipid bio-

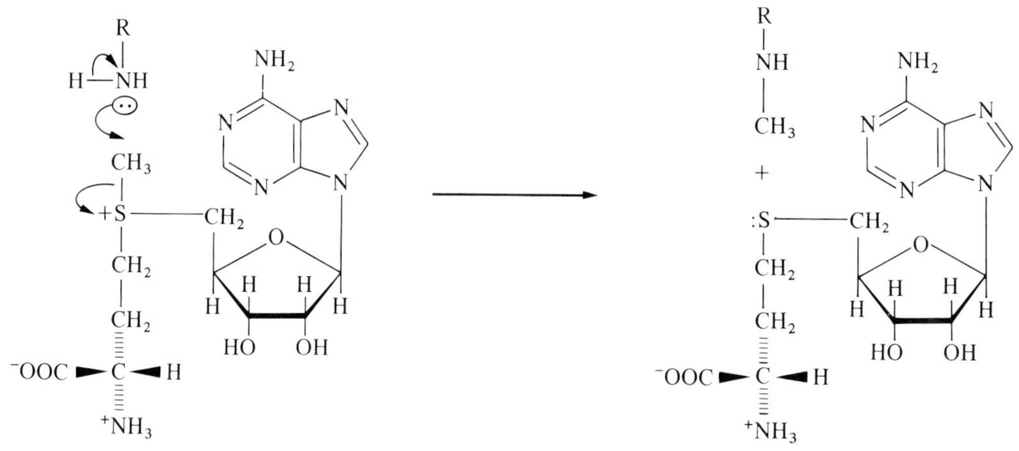

Fig. 8.31. Mechanism of the first methylation by S-adenosylmethionine of the amino nitrogen atom of phosphatidylethanolamine, catalysed by phosphatidylethanolamine methyltransferase, EC 2.1.1.17. (R = rest of phosphatidylethanolamine molecule.)

synthesis appears at the present time to be the endoplasmic reticular membranes. However, in view of the endomembrane concept [see Chapter 3, Section B.2(vi)], future work may show that the glycerophospholipids of the inner membrane of the mitochondrion and of the thylakoid system of the chloroplast are synthesized within these organelles.

(b) *Sphingophospholipid biosynthesis*

Virtually nothing is known about the biosynthesis of the plant sphingophospholipids. However, an examination of their partial structures (see Fig. 8.6, XXI and XXII) shows that they are composed of an inositol-containing oligosaccharide linked via the C–1 hydroxyl group of inositol to the C–1 hydroxyl group of a phytosphingosine-containing ceramide. Therefore the most likely alternative biosynthetic routes are: (i) the transfer of the preformed oligosaccharide to the ceramide, (ii) the transfer of the preformed oligosaccharide to phytosphingosine followed by N-acylation, (iii) the stepwise build-up of the oligosaccharide on the ceramide and (iv) the stepwide build-up of the oligosaccharide on phytosphingosine followed by N-acylation.

The only aspect of these alternatives upon which there is any information is the formation of ceramides and even here much of it comes from investigations on animal tissues. Figure 8.32 gives the outline of the biosynthesis of sphingosine and ceramide in animal tissues. Serine palmityltransferase[29], catalysing the first reaction, is a pyridoxal-phosphate-protein. The resulting (2S)-1-hydroxy-2-amino-octadecan-3-one is then reduced by an NADP-specific oxidoreductase to sphinganine into which a trans-Δ^4 double bond is then introduced by an FAD-specific flavoprotein. The amino nitrogen atom of the resulting sphingosine is then acylated to form a ceramide; a fatty acyl-CoA is the acyl donor and sphingosine acyltransferase is the enzyme.

Studies with cell-free, particulate enzyme systems from the yeast, *Hansenula ciferri*, show that in this organism sphinganine is formed by the pathway shown in Fig. 8.32 but that sphingosine is formed from trans-Δ^4-hexadecenoyl-CoA by a distinct but

parallel pathway. It is not known whether these two routes are also operative in higher plants.

Phytosphingosine could arise in two ways: (i) hydroxylation at C–4 of sphinganine by a mixed function oxygenase or (ii) by hydration of the trans-Δ^4 double bond of sphingosine; it is not known which is operative.

(ii) Biosynthesis of glycolipids

(a) *Galactosyldiglyceride biosynthesis*

Mono- and digalactosyldiglycerides (MGDG and DGDG) are the predominant lipid of chloroplasts and have been shown to be synthesized by the membranes of the chloroplast envelope. The more recent discovery that they are also present in mitochondria suggests that they may also be synthesized there. Their chloroplastidic biosynthetic route is outlined in Fig. 8.33.

The diglyceride moiety of the galactosyldiglycerides is derived from sn-glycerol 3-phosphate by way of 3-sn-phosphatidic acid and sn-1,2-diglyceride (cf. phosphatidylcholine and -ethanolamine biosynthesis, Fig. 8.27). The fatty acyl residues of the sn-1,2-diglyceride may be derived from fatty acyl-CoA or fatty acyl-ACPs which are presumably synthesized within the chloroplast. The sn-glycerol 3-phosphate is formed within the chloroplast by reduction of DiHOAcP which is tapped off from the Calvin cycle during daylight or formed by intra-chloroplastidic starch breakdown during the hours of darkness. (See Chapter 5, Section D.6; Figs. 5.36 and 5.37).

The sn-1,2-diglyceride then reacts with UDP-D-galactose under the catalytic influence of UDP galactose-1,2-diacylglycerol galactosyltransferase[30]. This enzyme catalyses the transfer of the D-galactose moiety of UDP-D-galactose to the sn-3-hydroxyl oxygen of the sn-1,2-diglyceride with the formation of a $\beta1 \rightarrow sn$-3-O-glycosidic linkage. The product of the reaction is MGDG.

MGDG may itself react with UDP-D-galactose to form DGDG. This reaction is catalysed by a distinct enzyme which may be called UDP-galactose-MGDG galactosyltransferase. It catalyses the transfer of the D-galactose residue of UDP-D-galactose to the oxygen of the C–6 hy-

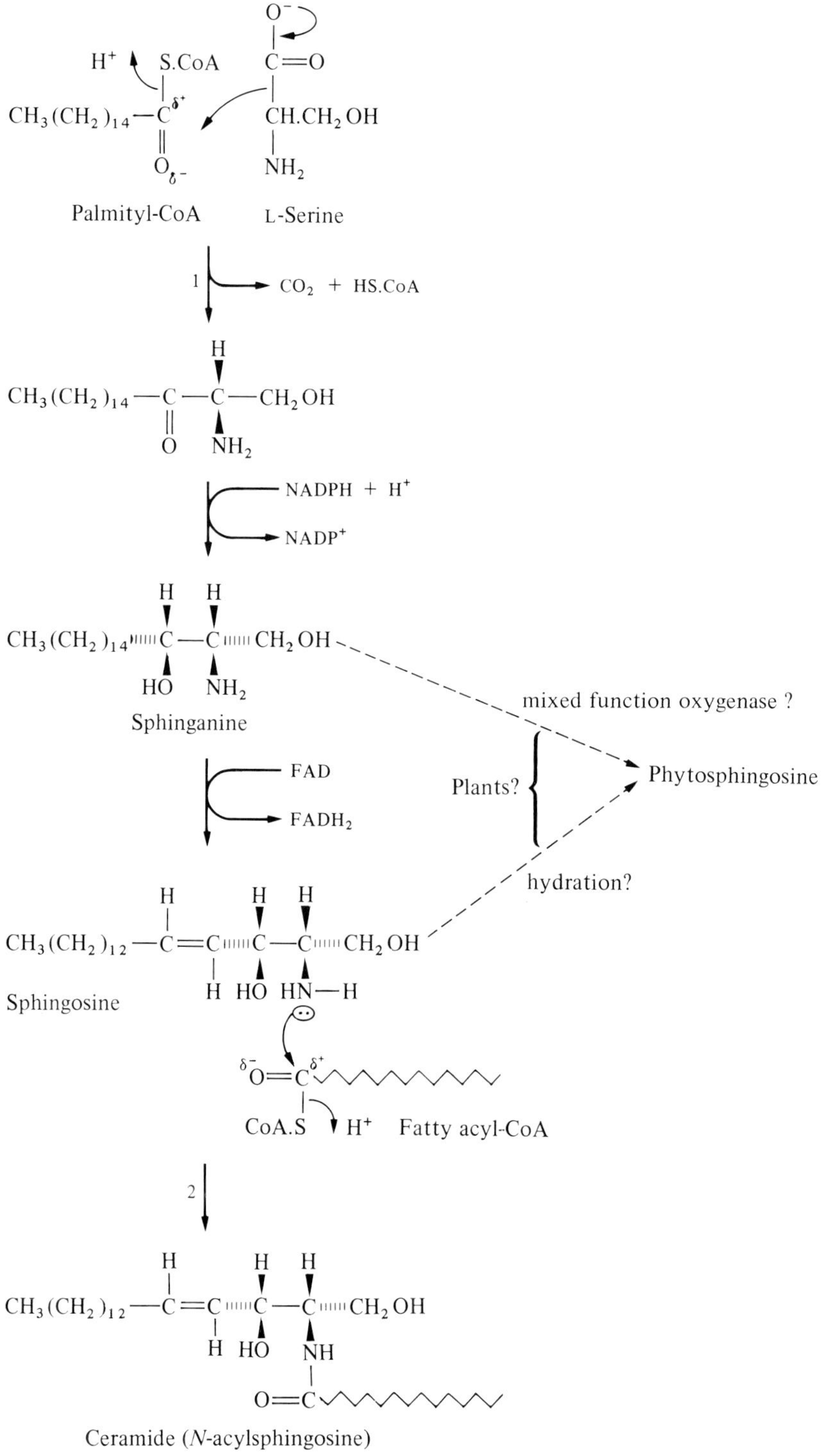

Palmityl-CoA   L-Serine

$CH_3(CH_2)_{14}$—C—C—CH_2OH
with O and NH_2

Sphinganine

mixed function oxygenase ?

Phytosphingosine

Plants?

FAD

$FADH_2$

hydration?

Sphingosine

CoA.S H^+ Fatty acyl-CoA

Ceramide (*N*-acylsphingosine)

Fig. 8.32. Biosynthesis of sphingosine and ceramides in animals. (1 = serine palmityltransferase, EC 2.3.1.50; 2 = sphingosine acyltransferase, EC 2.3.1.24.)

droxyl group of the MDGD D-galactose residue with the formation of an α1→6 O-glycosidic linkage.

A limited degree of acyl exchange takes place at the level of MGDG and DGDG. A fatty acyl residue is removed by enzyme-catalysed hydrolysis so forming a lyso-MGDG or a lyso-DGDG. It is then replaced by a different acyl residue derived from fatty acyl-CoAs presumably formed within the chloroplast. It is thought possible that this acyl exchange may account for the characteristic fatty acid composition of the galactosyldiglycerides.

(b) Sulpholipid biosynthesis

Our knowledge of the biosynthetic pathway of the plant sulpholipid is characterized more by speculation than by hard experimental evidence.

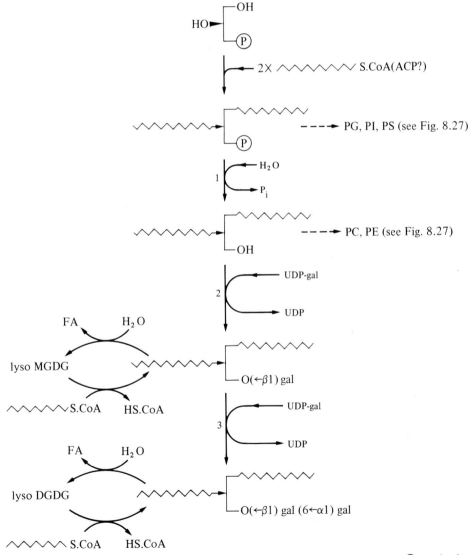

Fig. 8.33. Biosynthesis of galactosyldiglycerides. (►⎣ = glycerol residue; ⋁⋁⋁ = fatty acid residue; Ⓟ = orthophosphate residue; P_i = orthophosphate; UDP = uridine 5′-diphosphate; gal = D-galactopyranose; Lyso-MGDG = lysomonogalactosyldiglyceride; lyso DGDG = lysodigalactosyldiglyceride; FA = fatty acid; PG, PI, PS, PC and PE = phosphatidylglycerol, -inositol, -serine, -choline and -ethanolamine; 1 = phosphatidate phosphatase, EC 3.1.3.4; 2 = UDP galactose-1,2-diacylglycerol galactosyltransferase, EC 2.4.1.46; 3 = UDP-galactose-MGDG galactosyltransferase.)

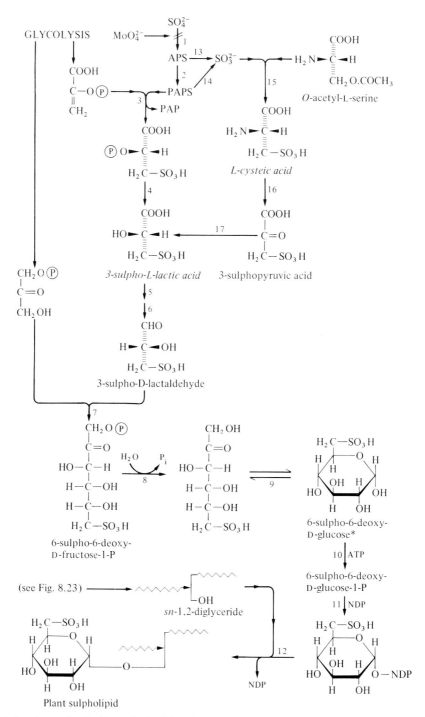

Fig. 8.34. Postulated pathways for the biosynthesis of the plant sulpholipid (redrawn from Davies *et al.*, 1966; Haines, 1973; and Harwood, 1980). (1 = sulphate adenylyltransferase; 2 = adenylylsulphate kinase; 3 = a sulphotransferase; 4 and 8 = phosphatases; 5 and 6 = a reductase and an epimerase; 7 = an aldolase; 9 = a ketol isomerase; 10 = a kinase; 11 = a pyrophosphorylase; 12 = an NDP-6-sulpho-6-deoxyglucose-1,2-diacylglycerol glucosyltransferase; 13 = APS reductase (photosynthetic organisms); 14 = PAPS reductase (non-photosynthetic organisms); 16 = a transaminase (e.g. asparate aminotransferase); 17 = a dehydrogenase (e.g. malate dehydrogenase); compounds whose names are italicized are readily incorporated into the plant sulpholipid; * this sugar is also known as 6-sulphoquinovose.)

The main hypotheses have been combined in Fig. 8.34. These hypotheses assume that the formation of the C—S bond occurs at the level of a 3C compound. This is based upon: (i) the finding, by Benson in 1961, of sulpholactate, sulpholactaldehyde and sulphopropanediol in *Chlorella pyrenoidosa*, (ii) the demonstration by Davies in 1966 that L-cysteic acid, but not L-cysteine, was incorporated into the plant sulpholipid in *Euglena gracilis* even more readily than sulphate ions and (iii) the confirmation of the latter in germinating alfalfa and brussels sprout seeds by Harwood in 1975 coupled with the finding that sulpholactate diluted out the incorporation of label from $^{35}SO_4^{2-}$ into the plant sulpholipid in these tissues.

Davies in 1966 also showed that molybdate ions (MoO_4^{2-}) inhibited the incorporation of $^{34}SO_4^{2-}$ into the plant sulpholipid. Since molybdate was known to inhibit sulphate adenylyltransferase[31], a key enzyme in the formation of 'active sulphate'

(PAPS; see Fig. 8.35), Davies took this as evidence that PAPS was the direct donor of the sulphur of the plant sulpholipid and postulated that 6-sulpho-6-deoxy-D-glucose (6-sulphoquinovose) was formed by steps 1–9 (Fig. 8.34). Cysteic acid, he suggested, was not on the direct pathway but could be converted into 3-sulpholactate by steps 16 and 17.

The main weakness of this pathway is the fact that PAPS is known as a sulphate donor forming a C—O–S linkage rather than the producer of a sulphonic acid which has a C—S bond.

Haines in 1973 got over this difficulty by proposing that the immediate source of the sulphonic acid residue is sulphite (SO_3^{2-}) derived from APS in photosynthetic tissue (step 13, Fig. 8.34) or from PAPS in non-photosynthetic tissue (step 14, Fig. 8.34). The sulphite then effects a nucleophilic displacement of *O*-acetyl from *O*-acetyl-L-serine with the formation of L-cysteic acid which is then

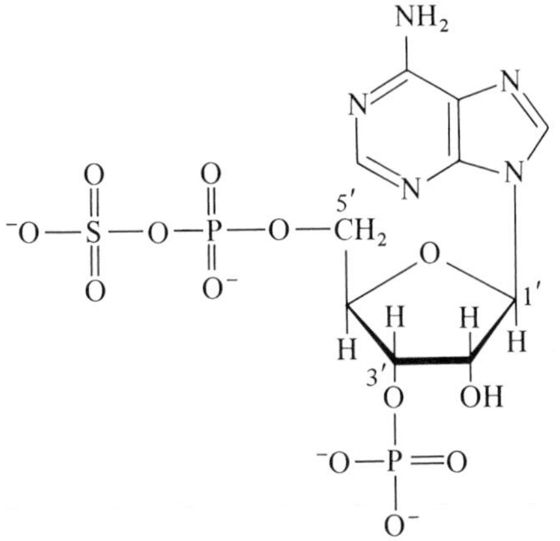

3'-phosphoadenosine-5'-phosphosulphate (PAPS)

$$(i) \quad ATP + SO_4^{2-} \; \underset{\Delta G^{\circ\prime} = +46 \text{ kJ mol}^{-1}}{\overset{1}{\rightleftharpoons}} \; APS + PP_i$$

$$(ii) \quad PP_i + H_2O \; \underset{\Delta G^{\circ\prime} = -33.5 \text{ kJ mol}^{-1}}{\overset{2}{\rightleftharpoons}} \; P_i + P_i$$

$$(iii) \quad APS + ATP \; \underset{\Delta G^{\circ\prime} \text{ very negative}}{\overset{3}{\rightleftharpoons}} \; PAPS + ADP$$

Fig. 8.35. Structure and biosynthesis of PAPS ('active sulphate'). (APS = adenosine-5'-phosphosulphate; 1 = sulphate adenylyltransferase, EC 2.7.7.4; 2 = inorganic pyrophosphatase, EC 3.6.1.1; 3 = adenylylsulphate kinase, EC 2.7.1.25.)

converted into 6-sulphoquinovose via steps 17, 5–9 (Fig. 8.34). This route not only explains the inhibition by MoO_4^{2-} but also has L-cysteic acid on an intermediate rather than on a side pathway.

Whichever of these pathways operates an inversion of configuration must occur on the carbon that becomes C–5 in 6-sulphoquinovose; this must take place somewhere between 3-sulpho-L-lactate and 6-sulphoquinovose and in Fig. 8.34 is shown as taking place at steps 5 or 6, catalysed by an epimerase.

Little is known of the final stages of the pathway, the conversion of 6-sulphoquinovose into the plant sulpholipid. By analogy with the galactosyldiglycerides (see Fig. 8.33), it is generally assumed that it will involve the transfer of the 6-sulphoquinovose residue from a nucleoside diphosphate-6-sulphoquinovose to the sn-3-hydroxyl group of sn-1,2-diglyceride (step 12, Fig. 8.34). Mudd in 1975 has pointed out that this would require that the C–1 hydroxyl group of 6-sulphoquinovose be phosphorylated (step 10, Fig. 8.34) so that the NDP-6-sulphoquinovose may be formed (step 11, Fig. 8.34) by an appropriate pyrophosphorylase.

The biosynthesis of chlorosulpholipids in the chrysophyte alga, Ochromonas danica, has been studied mainly by Haines, Vagelos and Elovson. The route by which the docosane series is formed is outlined in Fig. 8.36.

The elucidation of this pathway has been greatly hindered by the difficulty of obtaining catalytically active cell-free enzyme systems from O. danica. It is thought that the chlorosulpholipids themselves are responsible for this. They are powerful detergents and have been shown to inactivate a number of enzymes (e.g. fatty acid synthase). It is presumed that during the homogenization of O. danica cells, chlorosulpholipids are released from their normal intracellular location, probably the membranes, and inactivate a range of enzymes including those catalysing their own synthesis. Progress has been possible, without the availability of cell-free enzyme systems, however, because the cell wall free cells of O. danica have the ability to take in a wide range of inorganic and organic putative chlorosulpholipid precursors from their growth medium.

The C_{22} hydrocarbon skeleton of the docosane series of chlorosulpholipids is formed by way of long-chain fatty acids. Acetyl-CoA is carboxylated

to form malonyl-CoA; both of these compounds are then utilized by the fatty acid synthase to form palmitic acid which is then elongated by the fatty acid elongation systems to docosanoic acid (22:0). Docosanoic acid is then hydroxylated by a mixed

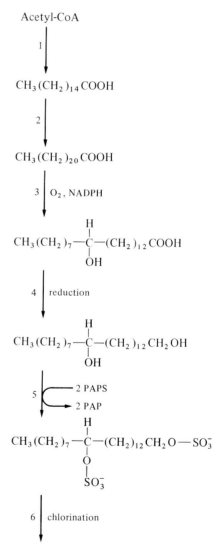

Fig. 8.36. Biosynthesis of the docosane series of chlorosulpholipids (1 = acetyl-CoA carboxylase and fatty acid synthase; 2 = fatty acid elongation systems; 3 = a mixed function oxygenase; 4 = an oxidoreductase; 5 = a sulphotransferase; 6 = a chlorination system, possibly chloride peroxidase; PAPS = 3′-phospho-adenosine-5′-phosphosulphate; PAP = 3′-phospho-adenosine-5′-phosphate; N.B. the fatty acid intermediates shown above probably participate as thioesters of CoA or ACP.)

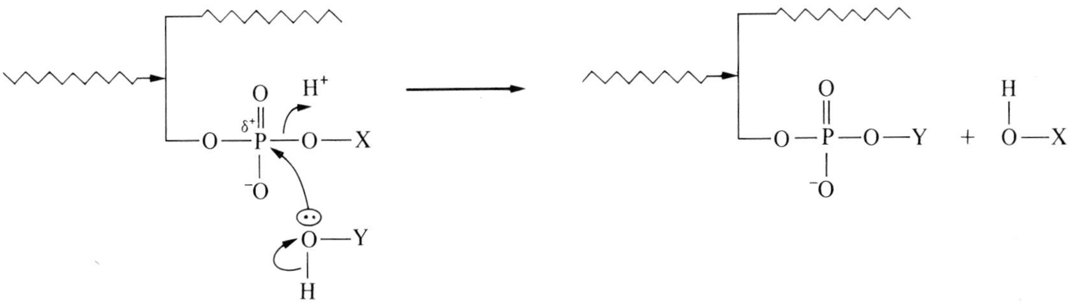

Fig. 8.37. Reaction catalysed by phospholipase D (EC 3.1.4.4). (∿∿→[∿∿∿=an *sn*-1,2-diglyceride residue; X = choline, ethanolamine, serine, glycerol but not inositol; Y—OH = water or an alcohol.)

function oxygenase at C–14 to form 14-hydroxy docosanoic acid; the oxygen of the 14-hydroxy group is labelled when 14-hydroxy docosanoic acid is formed in the presence of $^{18}O_2$ but not in the presence of $H_2^{18}O$. The carboxyl group of 14-hydroxy docosanoic acid is then reduced to a primary alcohol group, so forming *N*-docosane-1,14-diol.

The two hydroxyl groups of *N*-docosane-1,14-diol are then sulphated by PAPS (see Fig. 8.35) to form *N*-docosane-1,14-diol disulphate. The latter is then chlorinated; the mechanism by which the chlorine atoms are introduced has not been elucidated but it has been suggested that a chloride peroxidase[32], catalysing eq. (8.8), may be involved.

$$2RH + 2Cl^- + H_2O_2 \rightarrow 2RCl + 2H_2O \qquad (8.8)$$

It has proved possible in the laboratory to induce *O. danica* to synthesize brominated chlorosulpholipids. This involves depleting the cells of chloride by repeated subculturing on a chloride-free medium and then inoculating them into a medium containing bromide ions. Iodide and fluoride, however, will not substitute for chloride.

(c) *Cerebroside biosynthesis*

Little is known of the biosynthesis of cerebrosides in plants. An examination of their structure suggests two possible alternative routes: (i) glucosylation of a phytosphingosine-containing ceramide or (ii) glucosylation of the phytosphingosine followed by *N*-acylation.

The biosynthesis of phytosphingosine and ceramides have been discussed in Section C.3(i)(b). For cerebrosides the acyl donors would presumably be

2-hydroxy fatty acyl-CoAs formed from the 2-D-hydroperoxy fatty acyl CoAs that are intermediates in the α-oxidation process [see section C.1(ii)(b)].

By analogy with the galactosyldiglycerides [see Section C.3(ii)(a)] the glucose donor would be expected to be UDP-D-glucose.

(iii) Catabolism of polar lipids

(a) *Enzymes attacking the acyl groups of polar lipids*

There is considerable confusion in the general area of acylhydrolases acting on polar lipids. Depending upon the substrate chosen for assay, various acylhydrolase activities have been described as phospholipases, galactolipases, sulpholipases, monoglyceride lipases, lipases and esterases. However, the names cannot usually be justified because the substrate specificities of these enzymic activities have not been determined; thus an enzyme described as a galactolipase because it has been shown to catalyse the hydrolysis of ester bonds in a galactosyldiglyceride may subsequently be shown to be even more active with lysophosphatidylcholine and monoglycerides.

Many of these different enzyme activities may be due to one and the same enzyme, the non-specific acylhydrolase. This enzyme was discovered by Galliard in 1970 in potato tubers but has since been found in a wide range of plant tissues. It catalyses the hydrolysis of ester bonds in monoglycerides and lysoglycerophospholipids very well, those in monogalactosyldiglycerides quite well and those in diglycerides, digalactosyldiglycerides and glycero-phospholipids fairly well; it will not catalyse the hydrolysis of ester bonds in triglycerides, wax esters or sterol esters.

The non-specific acylhydrolase not only has a broad specificity, but it is also an extremely active enzyme. In potato tubers, for instance, it is so active that most of the endogenous membrane-bound polar lipids are destroyed immediately the tissue is disrupted, even at $0°C$. It is quite likely that this enzyme is responsible for the rapid release of free fatty acids that is observed in homogenates of many plant tissues. Its activity is inhibited at high pH values and by the inclusion of bovine serum albumin in the homogenizing medium.

(b) Enzymes attacking the polar groups of polar lipids

Phospholipase D[33] is the only fully characterized enzyme from plant tissues that attacks the polar groups of polar lipids. It catalyses the general reaction shown in Fig. 8.37. It will act on most of the common glycerophospholipids with the exception of phosphatidylinositols, utilizing water or an alcohol as the attacking species. Phospholipase D therefore catalyses both hydrolysis and trans-phosphatidylation reactions. With the former the products are 3-sn-phosphatidic acid and the ex-alcoholic moiety (e.g. choline) whilst with the latter the phosphatidyl residue of a glycerophospholipid molecule is transferred to another species of alcohol, as shown in eq. (8.9). The physiological significance of transphosphatidylation reactions in glycerophos-pholipid biosynthesis is questionable because the product of eq. (8.9), for example, is phosphatidyl-rac-1'-glycerol, whereas the naturally occurring form is phosphatidyl-1'-sn-glycerol. The transphosphatidylation reaction is, however, re-sponsible for the formation of artefacts such as phosphatidylmethanol during the extraction of lipids from tissues 'killed' by homogenization in methanolic solvents.

Phosphatidylcholine + glycerol →

$$\text{phosphatidylglycerol} + \text{choline} \quad (8.9)$$

The physiological role of phospholipase D is not clear. Its activity is highest in storage tissues such as the stalks of cabbage and celery, the roots of carrot, the seeds of pea and marrow and the rhizomes of Jerusalem artichoke; however, this generalization is not absolute for it is absent or low in potato tubers and many fruits. Its intracellular location is not fully established but there is evidence to suggest that it is present in lysosomal vacuoles [see Chapter 3, Section B.2(iv)].

Evidence for enzymes in plants, other than phospholipase D, attacking the polar groups of polar lipids is virtually non-existent. Phospholipase C[34], an animal and bacterial enzyme, catalysing the hydrolytic cleavage of glycerophospholipids into sn-1,2-diglyceride and phosphoalcohols has not been characterized in plants.

SUGGESTIONS FOR FURTHER READING

General texts

Hitchcock, C. and Nichols, B. W. (1971) *Plant Lipid Biochemistry*. Academic Press.
Erwin, J. A. (ed.) (1973) *Lipids and Biomembranes of Eukaryotic Microorganisms*. Academic Press.
Galliard, T. and Mercer, E. I. (eds.) (1975) *Recent Advances in the Chemistry and Biochemistry of Plant Lipids*. Academic Press.
Tevini, M. and Lichtenthaler, H. K. (1977) *Lipids and Lipid Polymers in Higher Plants*. Springer-Verlag.

Specialist reviews

McMurray, W. C. and Magee, W. L. (1972) 'Phospholipid metabolism', *Ann. Rev. Biochem.* **41**, 129–160.
Mazliak, P. (1973) 'Lipid metabolism in plants', *Ann. Rev. Plant Physiol.* **24**, 287–310.
Fulco, A. J. (1974) 'Metabolic alterations of fatty acids', *Ann. Rev. Biochem.* **43**, 215–241.
van den Bosch, H. (1974) 'Phosphoglyceride metabolism', *Ann. Rev. Biochem.* **43**, 243–277.
Bloch, K. and Vance, D. (1977) 'Control mechanisms in the synthesis of saturated fatty acids', *Ann. Rev. Biochem.* **46**, 263–298.

ENZYMES

1. Acetyl-CoA: carbon dioxide ligase (ADP-forming), EC 6.4.1.2.
2. Biotin-carboxyl-carrier-protein: carbon dioxide ligase (ADP-forming), EC 6.3.4.15.
3. Linoleate: oxygen oxidoreductase, EC 1.13.11.12.
4. Citrate oxaloacetate-lyase (pro-3S-CH$_2$COO$^-$ →acetyl-CoA), EC 4.1.3.7.
5. Citrate (isocitrate) hydro-lase, EC 4.2.1.3.

6. *threo*-D$_s$-Isocitrate glyoxylate-lyase, EC 4.1.3.1.
7. L-Malate glyoxylate-lyase, EC 4.1.3.2.
8. L-Malate: NAD$^+$ oxidoreductase, EC 1.1.1.37.
9. Succinate: (acceptor) oxidoreductase, EC 1.3.99.1.
10. L-Malate hydro-lyase, EC 4.2.1.2.
11. ATP: oxaloacetate carboxy-lyase (transphosphorylating), EC 4.1.1.49.
12. Hydrogen peroxide: hydrogen peroxide oxidoreductase, EC 1.11.1.6.
13. L-Serine: glyoxylate aminotransferase, EC 2.6.1.45.
14. *threo*-D$_s$-Isocitrate: NAD$^+$ oxidoreductase (decarboxylating), 1.1.1.41.
15. 2-Oxoglutarate: lipoamide oxidoreductase (decarboxylating and acceptor-succinylating), EC 1.2.4.2.
16. *β*-D-Fructofuranoside fructohydrolase, EC 3.2.1.26.
17. ATP: D-hexose 6-phosphotransferase, EC 2.7.1.1.
18. *sn*-Glycerol-3-phosphate: NAD$^+$ 2-oxidoreductase, EC 1.1.1.8.
19. L-*α*-Phosphatidate phosphohydrolase, EC 3.1.3.4.
20. Acyl-CoA: 1,2-diacylglycerol *O*-acyltransferase, EC 2.3.1.20.
21. ATP: glycerol *sn*-3-phosphotransferase, EC 2.7.1.30.
22. Triacylglycerol acylhydrolase, EC 3.1.1.3.
23. CTP: phosphatidate cytidylyltransferase, EC 2.7.8.8.
24. CDP diacylglycerol: L-serine *O*-phosphatidyltransferase, EC 2.7.8.8.
25. CDP diacylglycerol: *myo*-inositol 3-phosphatidyltransferase, EC 2.7.8.11.
26. ATP: choline phosphotransferase, EC 2.7.1.32.
27. ATP: ethanolamine *O*-phosphotransferase, EC 2.7.1.82.
28. Phosphatidylserine carboxy-lyase, EC 4.1.1.65.
29. Palmitoyl-CoA: L-serine *C*-palmitoyltransferase (carboxylating), EC 2.3.1.50.
30. UDP galactose: 1,2-diacyglycerol 3-*O*-galactosyltransferase, EC 2.4.1.46.
31. ATP: sulphate adenylyltransferase, EC 2.7.7.4.
32. Chloride: hydrogen peroxide oxidoreductase, EC 1.11.1.10.
33. Phosphatidylcholine phosphatidohydrolase, EC 3.1.4.4.
34. Phosphatidylcholine cholinephosphohydrolase, EC 3.1.4.3.

CHAPTER 9

Nitrogen Fixation, Amino Acid Biosynthesis and Proteins

CONTENTS

A. INTRODUCTION

A fundamental requirement for life to persist on this planet is that the nitrogen cycle should continue to function. For this to happen a complex interplay between all forms of life is necessary and this is outlined in Fig. 9.1. The present interest is the fixation of atmospheric nitrogen (N_2, dinitrogen) into organic compounds, particularly amino acids and eventually proteins, in higher plants. This sector of the cycle is achieved by close collaboration between micro-organisms and higher plants. The first stages of nitrogen fixation, conversion of nitrogen into ammonia, are achieved symbiotically in the root nodules of leguminous plants which contain the nitrogen-fixing bacteria of the genus *Rhizobium*. Some non-leguminous plants such as alder and sea-buckthorn and their associated symbiotic bacteria are also nitrogen-fixers. Free-living bacteria such as *Klebsiella pneumoniae* and the photosynthetic blue-green bacteria also fix nitrogen but this aspect is outside the scope of this chapter.

The process of nitrogen-fixation is controlled by a set of genes named (Nif) genes which have been studied in detail in *Klebsiella pneumoniae*. In this bacterium and in certain free-living *Rhizobium* spp. these genes are repressed but in the root nodule *Rhizobia* they are normally derepressed even in the presence of NH_4^+. The significance of this for man's survival is enormous and much work is in progress in the attempt to introduce (Nif) genes into higher plants which do not normally fix nitrogen. For plants which do not fix atmospheric nitrogen the preferred external source of inorganic nitrogen is NO_3^- so a mechanism must be available to reduce

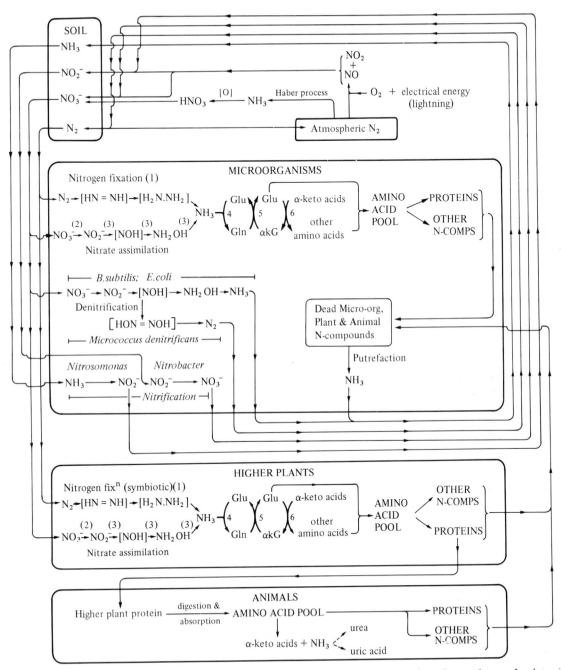

Fig. 9.1. The nitrogen cycle. (1 = nitrogenase; 2 = nitrate reductase; 3 = nitrite reductase; 4 = glutamine synthetase; 5 = glutamine (amide):2-oxoglutarate aminotransferase (oxidoreductase, NADP or Fd) = GOGAT; 6 = aminotransferases; Glu = L-glutamic acid; Gln = L-glutamine; αkG = α-ketoglutarate.)

this to NH_4^+ and this is discussed in Section B.2. The two final sections of the chapter deal with the assimilation of NH_4^+ into an organic form and the formation of the various amino acids found in plants, those required for protein synthesis and those of rather specific structure which are found widely distributed in the free form.

B. NITROGEN FIXATION

1. Formation of NH_3

Electron microscope studies (Fig. 9.2) of soya bean nodules show a system of double membranes, apparently originating from the host cell wall, which surrounds a group of bacteroids (nodule bacteria). The bacteroids lack a rigid cell wall and are osmotically labile. Nitrogen fixation occurs in these nodules and cell-free extracts prepared under argon (oxygen rapidly inactivates the enzyme) and in the presence of polyvinyl pyrrolidone (to remove phenolic compounds which can act as enzyme inhibitors) will fix nitrogen in the presence of ATP, a divalent metal cation and reduced ferredoxin. This complex, known as *nitrogenase*[1], consists of two components which are now known as Protein 1 (formerly called molybdoferredoxin) and Protein 2 (formerly called azoferredoxin). Protein 2 is an extremely oxygen-sensitive iron–sulphur protein which is a dimer of two identical peptides, MW 30,000; the dimer contains four iron atoms, four

sulphur atoms (labile S) and twelve titratable thiol groups. Protein 1 is made up of two different peptide chains, MW 51,000 and 60,000, associated as a mixed $(\alpha_2\beta_2)$ tetramer. The tetramer contains two molybdenum atoms, about twenty-four iron atoms and twenty-four sulphur atoms (labile S), and about thirty titratable thiol groups, possibly in the form of three Fe_4S_4 clusters [see Chapter 6, Section B.2(ii)(c)]. The active nitrogenase complex consists of two Protein 2 dimers associated with one Protein 1 tetramer. The current working model for the complex is outlined in Fig. 9.3.

Six electrons are required for the reduction of N_2 to NH_3 and these are utilized in three steps [eq. (9.1)] with diimide and hydrazine as intermediates.

$$N{\equiv}N \xrightarrow[2H^+]{2e} HN{=}NH \xrightarrow[2H^+]{2e} \cdot$$

$$H_2N{-}NH_2 \xrightarrow[2H^+]{2e} 2NH_3 \qquad (9.1)$$

The electron transfer from NADPH involves firstly Protein 2 and then Protein 1 which is directly coupled with nitrogen [eq. (9.2)]. The point of involvement of ATP was demonstrated by the observation that the EPR signal, characteristic of iron–sulphur centres, of Protein 2 changed on addition of ATP and Mg^{2+} whereas that of Protein 1 was unaltered; ATP also binds tightly to Protein 2, lowering its redox potential $[E_0'$ (pH 7.5)] from -0.29 V to -0.40 V. The role of ATP in the electron transfer is probably to provide energy to drive electrons up a potential gradient (i.e. $+ve \rightarrow -ve$) which would be the equivalent of

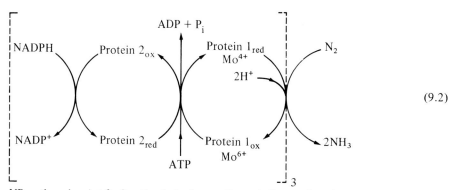

$$(9.2)$$

NB — the subscript 3 after the dashed parentheses indicates that three successive $2e$ steps of the type shown are required to reduce N_2 to $2NH_3$.

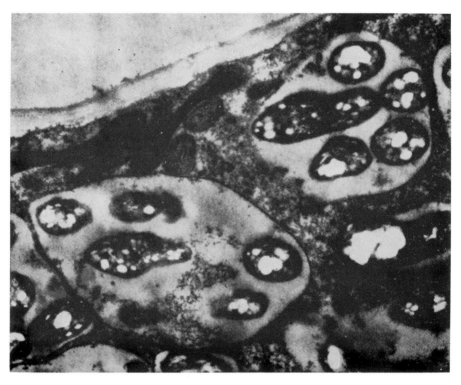

Fig. 9.2. Electron micrograph of soya-bean nodules showing the final stage of the symbiotic structure between host and bacteroid in which a system of double membranes surrounds a group of bacteroids (× 46,000). (Photograph kindly supplied by Drs. D. J. Goodchild and F. J. Bergersen.)

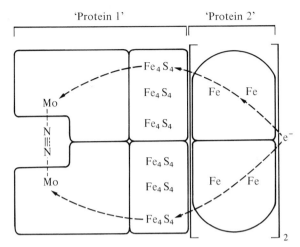

'Protein 1'   'Protein 2'

Fig. 9.3. Hypothetical structure of the N_2-fixing enzyme, nitrogenase. (Protein 1 is an $\alpha_2\beta_2$ tetramer; Protein 2 is an α_2 dimer; Fe_4S_4 = iron–sulphur cluster; Protein 1 and Protein 2 are present in the ratio 1:2.)

reversed electron flow in respiration; however, the need for two to five molecules of ATP per electron pair transferred to nitrogen is not immediately clear. On thermodynamic grounds N_2 should be reduced spontaneously by reduced Protein 2, the $\Delta G^{\circ\prime}$ (pH 7.0) of the reaction being $-89.3\,\mathrm{kJ\,mol^{-1}}$. However, the need for the large amount of ATP is probably connected with the fact that nitrogenase can also reduce protons and that this reaction is closely coupled with nitrogen fixation. The probable stoichiometry of the two reactions is given in Table 9.1. An explanation for these simultaneous reactions is still awaited.

The exact mechanism involved in the reduction is still not clear but in Fig. 9.3 the assumption is that nitrogen receives electrons from both molybdenum atoms. However, recent work suggests that the N_2 binding site contains a cofactor (FeMo-co) having Fe, Mo and S in the ratio 8:1:6.

A nodular component which has attracted much attention is *leghaemoglobin*, which is synthesized on the plant genome. It plays no direct part in nitrogen fixation but it has a very high affinity for oxygen. It occurs in the bacteroid membrane and thus protects the nitrogenase inside the bacteroid from the detrimental effects of oxygen whilst at the same time providing adequate oxygen in the bacteroid membrane to allow respiration to proceed to generate the ATP required for nitrogen fixation.

2. Conversion of nitrate into ammonia

Nitrate, which is the preferred source of exogenous inorganic nitrogen for most plants, arises in the soil mainly from NH_4^+ liberated from decaying organic matter. The conversion is an important aspect of bacterial activity in soil and is called *nitrification*. Two bacteria are involved: *Nitrosomonas* converts NH_4^+ into NO_2^- [eq. (9.3)] and *Nitrobacter* oxidizes NO_2^- to NO_3^- [eq. (9.4)].

It should be emphasized that NH_4^+ can be absorbed by most plants but this process is not as quantitatively significant as NO_3^- absorption. Nitrate is taken up into the plant cell by the action of a permease which is induced by external nitrate. The internal nitrate is sequestered into two pools, one for storage and one for enzyme induction. The process of nitrate reduction (*nitrate reductase*[2]) takes place in two steps: (i) reduction of nitrate to nitrite and (ii) reduction of nitrite to ammonia (*nitrite reductase*[3]).

$$NH_4^+ + 1.5O_2 \rightarrow NO_2^- + 2H^+ + H_2O \quad (9.3)$$
$$\Delta G^{\circ\prime}\,(\mathrm{pH}\ 7) = -272\ \mathrm{kJ\ mol^{-1}}$$

$$NO_2^- + 0.5O_2 \rightarrow NO_3^- \quad (9.4)$$
$$\Delta G^{\circ\prime}\,(\mathrm{pH}\ 7) = -76\ \mathrm{kJ\ mol^{-1}}$$

Table 9.1. *Suggested stoichiometry for the two reactions catalysed by nitrogenase*

1* N_2	+	12ATP	+	$6e^-$	+	$8H^+$	$\xrightarrow{\mathrm{Mg^{2+}}}$			$2NH_4^+$	+	12ADP	+	$12P_i$
2*		4ATP	+	$2e^-$	+	$2H^+$	$\xrightarrow{\mathrm{Mg^{2+}}}$			H_2	+	4ADP	+	$4P_i$
Total: N_2	+	16ATP	+	$8e^-$	+	$10H^+$	$\xrightarrow{\mathrm{Mg^{2+}}}$			$2NH_4$	+	H_2 + 16ADP	+	$16P_i$

* Calculated on the assumption that 75% of the electron flow is through reaction 1. (This percentage can vary under different conditions.)

(i) NITRATE REDUCTASE

Thorough study of this enzyme complex in both higher plants and micro-organisms indicates that two steps are involved which carry out the reactions outlined in eq. (9.5).

The molecular weight for the higher plant enzyme, which is a soluble molybdoflavoprotein, is about 200,000 although reported values of 500,000 suggest that dimerization may occur. The intact complex is an 8S unit in which the NADPH-cytochrome b_{557} reductase [catalysing reactions A and B in eq. (9.5)] exists as a separate (3.7–4.5S) sub-unit which is bound to the nitrate reductase [catalysing reactions C and D in eq. (9.5)] by a molybdenum binding protein of MW 10,000–20,000. There is yet no evidence as to how the molybdate oxyanion MoO_4^{2-} (MoVI) is inserted into the enzyme. It should be emphasized that nitrate reductase is the only higher plant Mo-enzyme known at the time of writing.

Induction of the synthesis of nitrate reductase is stimulated by NO_3^- in many plant tissues and phytochrome and plant hormones also mediate complex stimulatory effects. The enzyme is located in the cytoplasm and may be associated with the chloroplast outer envelope; it has a rapid turn-over, half-life values from 1 to 24 hr have been reported.

Nitrate reductase levels vary with species and correlate with the organic-N: NO_3^-–N ratio in xylem exudates. Species with low nitrate reductase activity, e.g. cotton (*Gossypium*), produce xylem exudates with more than 95% of their nitrogen as free nitrate; on the other hand, species such as radish have high nitrate reductase activity and little or no free nitrate appears in the xylem exudate.

The nitrate reductase from blue-green algae (cyanobacteria) cannot accept electrons directly from NAD(P)H; reduced ferredoxin is the *in vivo* electron donor.

(ii) NITRITE REDUCTASE

This enzyme carries out the six-electron reduction involved in converting nitrite into NH_3 with the probable intermediate formation of nitroxyl (NOH) and hydroxylamine (NH_2OH) [eq. (9.6)]. The enzyme from spinach and marrow has a molecular weight of 61,000–70,000 and probably exists as two sub-units. An iron-porphyrin *sirohaem* (Fig. 9.4) is also present and appears to mediate the six-electron transfer. The immediate source of electrons is ferredoxin and two-electron carriers cannot couple to the system. The site of location of nitrite reductase in green tissues is the chloroplast where it is probably present in the exterior surface of the thylakoid membrane.

3. Assimilation of ammonia

The major route for the assimilation of ammonia

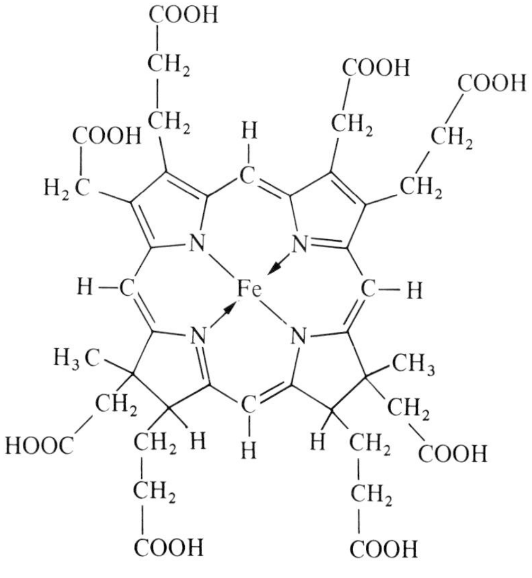

Fig. 9.4. Structure of sirohaem.

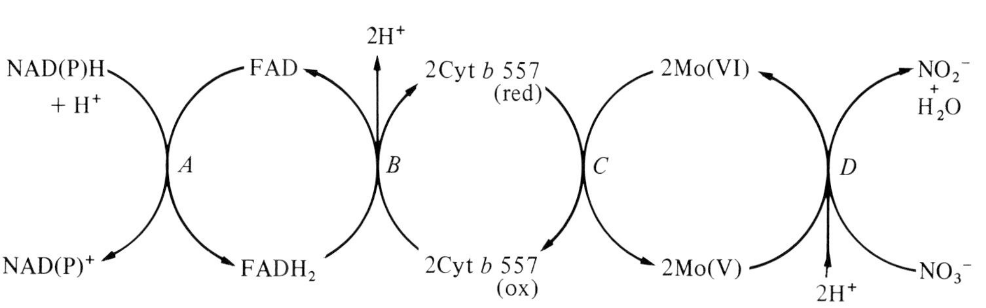

$$ (9.5) $$

$$NO_2^- \xrightarrow[\;OH^-\;]{2e + 2H^+} [NOH] \xrightarrow{2e + 2H^+} NH_2OH \xrightarrow[\;H_2O\;]{2e + 2H^+} NH_4^+ \qquad (9.6)$$

$$\Delta G^{\circ\prime} (pH\ 7) = -433\ kJ\ mol^{-1}$$

into organic nitrogen is the result of the collaborative activity of two enzymes—glutamine synthetase (GS)[4] and glutamine(amide):2-oxoglutarate aminotransferase (oxidoreductase NADP) (GOGAT).† GS catalyses an ATP-dependent incorporation of NH_4^+ into the amide position of glutamate [eq. (9.7)]. It has been purified from soya bean root nodules, has a molecular weight of 330,000 and is made up of eight monomers arranged in two parallel planar tetramers. The activity of the enzyme is the same in the presence of either Mg^{2+} or Mn^{2+} but the pH optima are substantially different being 8.0 in the presence of Mg^{2+} and about 5.0 in the presence of Mn^{2+}. GOGAT[5] from leaves, legume nodules, roots and legume seeds is ferredoxin-dependent and catalyses eq. (9.8). It has also been found in green and blue-green algae. The NADPH-linked GOGAT[6] is found in bacteria, roots and developing seeds but not apparently in leaves [eq. (9.9)].

If eqs. (9.7) and (9.8) are summated the net result

† GOGAT has now been renamed and renumbered by the Enzyme Commission of the IUB. The systematic name of the ferredoxin-dependent enzyme is L-glutamate:ferredoxin oxidoreductase (transaminating), EC 1.4.7.1. The systematic name of the pyridine-nucleotide-dependent enzyme is L-glutamate:NADP$^+$ oxidoreductase (transaminating), EC 1.4.1.13. The abbreviation GOGAT is, however, used in the text.

of the two reactions is the production of one molecule of glutamate from one molecule of α-ketoglutarate and one of NH_4^+.

An additional enzyme glutamate dehydrogenase (GDH)[7] [eq. (9.10)] is widely distributed but is not significantly involved in NH_4^+ assimilation. This conclusion is based on the following observations: (i) the K_m (NH_4^+) for GS is much lower than for GDH so that as plants generally have low cellular levels of NH_4^+, GDH is unlikely to operate significantly in competition with GS, (ii) the nitrogen in NO_3^-, in the amide group of glutamine and in the amino group of glutamate should be in equilibrium if the GS/GOGAT pathway predominates and this has been demonstrated using appropriate substrates labelled with ^{15}N, (iii) methionine sulphoxime, a specific inhibitor of GS, inhibits nitrogen fixation at the level of NH_4^+, which would not be the case if GDH were implicated and (iv) azaserine, a specific inhibitor of GOGAT, causes the accumulation of both α-ketoglutarate and NH_4^+ and the rate of NH_4^+ production is equivalent to the rate of NO_3^- reduction.

About 50% of the GS in leaf tissue occurs in the chloroplast whereas essentially all the GOGAT is chloroplastic, which accounts for the ferredoxin requirement of the reaction. The regulation of these enzymes is not fully understood but some are

$$\begin{array}{c} COOH \\ | \\ (CH_2)_2 \\ | \\ H-C-NH_2 \\ | \\ COOH \end{array} + NH_4^+ + ATP \underset{}{\overset{Mg^{2+}/Mn^{2+}}{\rightleftharpoons}} \begin{array}{c} CO.NH_2 \\ | \\ (CH_2)_2 \\ | \\ H-C-NH_2 \\ | \\ COOH \end{array} + ADP + P_i + H^+ \qquad (9.7)$$

$$\begin{array}{c} CO.NH_2 \\ | \\ (CH_2)_2 \\ | \\ H-C-NH_2 \\ | \\ COOH \end{array} + 2Fd_{(red)} + \begin{array}{c} COOH \\ | \\ (CH_2)_2 \\ | \\ C=O \\ | \\ COOH \end{array} \rightleftharpoons 2\ \begin{array}{c} COOH \\ | \\ (CH_2)_2 \\ | \\ H-C-NH_2 \\ | \\ COOH \end{array} + 2Fd_{(ox)} \qquad (9.8)$$

$$
\begin{array}{c} CO.NH_2 \\ | \\ (CH_2)_2 \\ | \\ H\!-\!C\!-\!NH_2 \\ | \\ COOH \end{array}
\;+\; NAD(P)H \;+\;
\begin{array}{c} COOH \\ | \\ (CH_2)_2 \\ | \\ C\!=\!O \\ | \\ COOH \end{array}
\;\rightleftharpoons\; 2\;
\begin{array}{c} COOH \\ | \\ (CH_2)_2 \\ | \\ H\!-\!C\!-\!NH_2 \\ | \\ COOH \end{array}
\;+\; NAD(P)^+ \qquad (9.9)
$$

$$
\begin{array}{c} COOH \\ | \\ (CH_2)_2 \\ | \\ C\!=\!O \\ | \\ COOH \end{array}
\;+\; NH_4^+ \;+\; NAD(P)H \;\rightleftharpoons\;
\begin{array}{c} COOH \\ | \\ (CH_2)_2 \\ | \\ H\!-\!C\!-\!NH_2 \\ | \\ COOH \end{array}
\;+\; NAD(P)^+ \;+\; H_2O \qquad (9.10)
$$

discussed under storage and assimilation (Section C).

4. Secondary assimilation of ammonia

Recycling of NH_4^+ released from glycine during photorespiration (Chapter 6, Section F.1) is essential to conserve nitrogen within the plant. The key reaction in this process is the acceptance of NH_4^+, produced by glycine metabolism in the mitochondria, by glutamate in the presence of GS [eq. (9.7)]. The glutamine so formed moves into the chloroplast where two molecules of glutamate are formed by the action of GOGAT [eq. (9.8)]. One molecule is used to continue the GS reaction and the second moves to the peroxisome to donate its amino group to glyoxylate to form glycine. This pathway minimizes the energy requirement by normal metabolic pathways because the major endergonic reaction is coupled to the light-dependent reduction of the ferredoxin used in eq. (9.8).

As indicated in Fig. 6.45, two molecules of glycine give rise to one molecule of serine (see also Section E.4), and as the serine/glycine route may be a direct route to sucrose synthesis in many plants it is an important pathway. A scheme involving collaboration of a number of organelles has recently been proposed to integrate this reaction within the general metabolic economy of the cell (Fig. 9.5).

C. TRANSPORT AND STORAGE OF NITROGEN

1. General aspects

In general the chief nitrogen transport compounds in plants are asparagine, aspartate, glutamate, glutamine and arginine; but variation between species is considerable. In temperate legumes, for example, asparagine and glutamine make considerable contributions whereas in some tropical legumes ureides such as allantoin (I, Fig. 9.6) and allantoic acid (II, Fig. 9.6) are the compounds involved. Arginine is involved in long-distance transport in trees. Occasionally a non-protein amino acid is involved in transport; such a case is γ-methylene glutamine (Table 10.1) in peanuts (*Arachis*). The transport compound can also vary according to the nitrogen available to the root. In soya bean (*Glycine max*), for instance, nodulated plants export ureides whereas non-nodulated plants absorbing exogenous NO_3^- export a much smaller quantity of ureides but much more asparagine.

The storage pool of nitrogen in the stems of plants is often similar in composition to that used in transport, but this is not so in leaves as exemplified by *Lycopersicon* where large amounts of alanine and γ-aminobutyric acid are found in the leaves. In leaves of C_3 plants (see Chapter 5, Section D.2) most

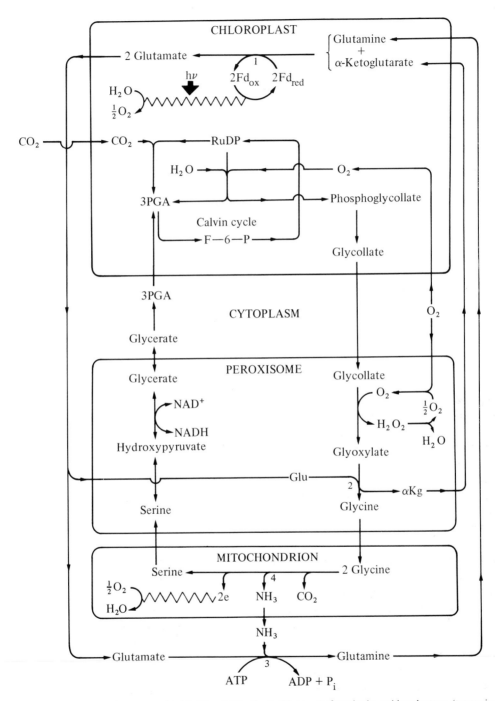

Fig. 9.5. The photorespiratory nitrogen cycle (cf. Fig. 6.45). (1 = L-Glutamate:ferredoxin oxidoreductase (transminating), EC 1.4.7.1 = GOGAT; 2 = glutamate:glyoxylate aminotransferase; 3 = glutamine synthetase, EC 6.3.1.2; 4 = serine hydroxymethyltransferase, EC 2.1.2.1; remaining enzymes see Fig. 6.45.)

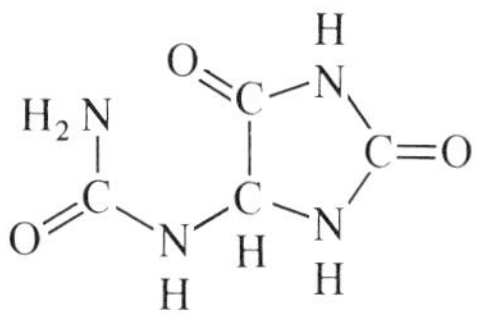

(I) Allantoin 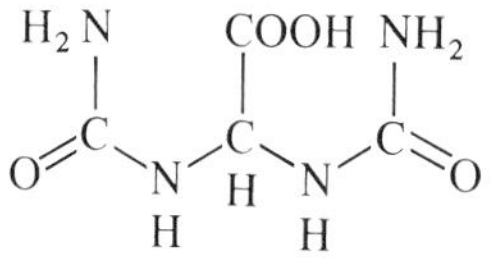   (II) Allantoic acid

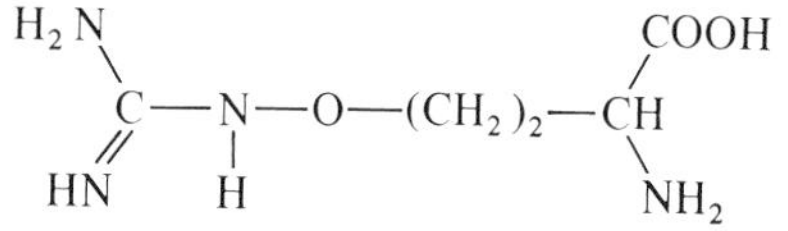

(III) Canavanine

Fig. 9.6. Structures of some compounds involved in the transport or storage of nitrogen.

of the nitrogen is present as Fraction 1 protein. In seeds non-protein amino acids are frequently the main storage product, for example canavanine (III, Fig. 9.6) represents 8% of the dry weight of jack bean seeds. It should also be pointed out that NO_3^- is the major storage pool in some species such as some Cruciferae; in these cases the amino acid pool is much reduced.

In both storage and transport of nitrogen an important criterion is the N:C atomic ratio in the compound involved, the higher the better; in asparagine it is 1:2 and in canavanine 4:5. All storage and transport amino acids are closely related metabolically to either the α-ketoglutarate or oxaloacetate (see Section E.2) carbon skeleton which are easily diverted away from or into the tricarboxylic acid cycle.

Another important requirement of transport and storage amino acids is that their formation and metabolism should be quickly and effectively controlled and react to excess or lack of available NH_4^+. Asparagine synthetase[8] (Section E.2) is inhibited by glucose, light and ATP. With plenty of carbohydrate available to a plant both ATP and α-ketoglutarate would be in good supply. Thus NH_4^+ would be converted by the GS/GOGAT pathway into glutamate which could be used in the synthesis of other amino acids whilst at the same time

asparagine synthesis would be inhibited. When carbohydrate is in short supply ATP and α-ketoglutarate levels would also be low, thus the substrate for synthesis via GS/GOGAT would not be available and at the same time the inhibition of asparate synthetase by ATP would also be removed.

2. Utilization of stored amino acids in developing seeds

The amide nitrogen of asparagine is the chief available source of nitrogen in developing legume seeds. The nitrogen from the amide group is transferred to the carbon residues of other amino acids just before attachment to tRNA (see Chapter 10) because experiments with [^{15}N-amide]-asparagine show very little distribution of ^{15}N in the amino acid pool of the phloem but considerable distribution in the newly synthesized seed protein. Asparaginase[9] is probably the major enzyme concerned with removal of the amide group. Although its K_m(asparagine) is high (~ 10 mM) this is offset by the high concentration of asparagine (~ 30 mM) in the phloem. On the other hand the carbon residue of asparagine is utilized only sluggishly under these circumstances.

In cereals the amino acids are transported as such to the developing seed, there to be incorporated into protein.

D. NATURE OF AMINO ACIDS IN PLANTS

Amino acids synthesized by plants fall into two main categories: (i) those twenty or so which are usually constituents of protein and (ii) those which occur free and are known as non-protein amino acids. The latter constitute a very large group of over 200, and represent a unique feature of amino acid metabolism in plants. It is not possible to discuss all of them in detail but examples of various types are given in Table 9.2. They have been chosen to exemplify the structural features most commonly encountered. One single free amino acid can frequently account for a significant part of the nitrogen pool or reserve as indicated in Section C.

Table 9.2. Examples of non-protein amino acids in higher plants

Name	Structure	Major source
Dicarboxylic acid γ-Methylene-L-glutamic acid	$\begin{array}{c} COOH \\ | \\ C{=}CH_2 \\ | \\ CH_2 \\ | \\ H{-}C{-}NH_2 \\ | \\ COOH \end{array}$	Peanuts, tulips
Imino acids Azetidine 2-carboxylic acid	(azetidine ring with COOH)	Liliaceae
Basic amino acids L-α,γ-diaminobutyric acid	$\begin{array}{c} CH_2NH_2 \\ | \\ CH_2 \\ | \\ H{-}C{-}NH_2 \\ | \\ COOH \end{array}$	*Lathyrus* spp.
Albizziine (L-α-amino-β-ureidopropionic acid)	$\begin{array}{c} H_2N \\ | \\ C{=}O \\ | \\ HN \\ | \\ CH_2 \\ | \\ H{-}C{-}NH_2 \\ | \\ COOH \end{array}$	Mimosaceae
γ-Hydroxyhomoarginine	$\begin{array}{c} H_2N \\ | \\ C{=}NH \\ | \\ HN \\ | \\ (CH_2)_2 \\ | \\ CHOH \\ | \\ CH_2 \\ | \\ H{-}C{-}NH_2 \\ | \\ COOH \end{array}$	*Lathyrus tinigitanus*

(continued overleaf)

Table 9.2 (cont.)

Name	Structure	Major source
Neutral aliphatic amino acids 5-Hydroxynorleucine	CH_3 $\vert$ $CHOH$ $\vert$ $(CH_2)_2$ $\vert$ $H\!-\!C\!-\!NH_2$ $\vert$ $COOH$	*Crotalaria juncea*
2-Amino-4-methylhex-4-enoic acid	CH_3 $\vert$ CH $\Vert$ $C\!-\!CH_3$ $\vert$ CH_2 $\vert$ $H\!-\!C\!-\!NH_2$ $\vert$ $COOH$	*Aesculus californica*
Hypoglycin A	CH_2 $\Vert$ C CH_2 CH $\vert$ CH_2 $\vert$ $H\!-\!C\!-\!NH_2$ $\vert$ $COOH$	*Blighia sapida*
Sulphur-containing amino acids S-Methyl-L-cysteine	$S\!-\!CH_3$ $\vert$ CH_2 $\vert$ $H\!-\!C\!-\!NH_2$ $\vert$ $COOH$	*Phaseolus vulgaris*
Selenium-containing amino acids Se-Methyl-L-cysteine	$Se\!-\!CH_3$ $\vert$ CH_2 $\vert$ $H\!-\!C\!-\!NH_2$ $\vert$ $COOH$	*Astragalus* spp.

Table 9.2 (cont.)

Name	Structure	Major source
Aromatic amino acids 2,4-Dihydroxy-6-methylphenylalanine		Corn cockle
Heterocyclic amino acids Mimosine		*Mimosa pudica*
Acetylated amino acids N-Acetyl-L-ornithine		Widely distributed

E. BIOSYNTHESIS OF AMINO ACIDS PRESENT IN PROTEINS

The biosynthetic activities of plants in synthesizing protein amino acids are best studied in families which arise from either pyruvate, oxaloacetate, α-ketoglutarate, shikimate or directly from an intermediate in the Calvin cycle. It will become apparent as the chapter proceeds that there are many metabolic cross-links between these families.

1. Amino acids derived from pyruvate

(i) LEUCINE, ISOLEUCINE AND VALINE†

As indicated in Fig. 9.7 leucine and valine share

† All reaction letters in this section refer to Fig. 9.7.

the same biosynthetic pathway as far as α-ketoisovaleric acid ($A \rightarrow B \rightarrow C$). The first step ($A$) involves acetolactate synthase[10] which is a thiamine pyrophosphate (TPPP)-requiring enzyme. An active hydroxyethyl-TPP complex is first produced (see Fig. 6.39), then a second pyruvate molecule accepts the hydroxyethyl residue to yield α-acetolactate [eq. (9.11)]. The next enzyme ketol-acid reductoisomerase[11] carries out a two-step reaction (B); the first step involves a methyl shift in the formation of enzyme-bound α-keto-β-hydroxy-isovaleric acid and the second involves a NADPH-linked reduction of this substrate to α, β-dihydroxy-isovaleric acid. The isomerization is similar to the non-enzymic pinacol-pinacolone rearrangement [eq. (9.12)]. The methyl group is transferred to the *re* face of the trigonal centre at C–3 of α-acetolactate. The final common step is dehydration (C), the

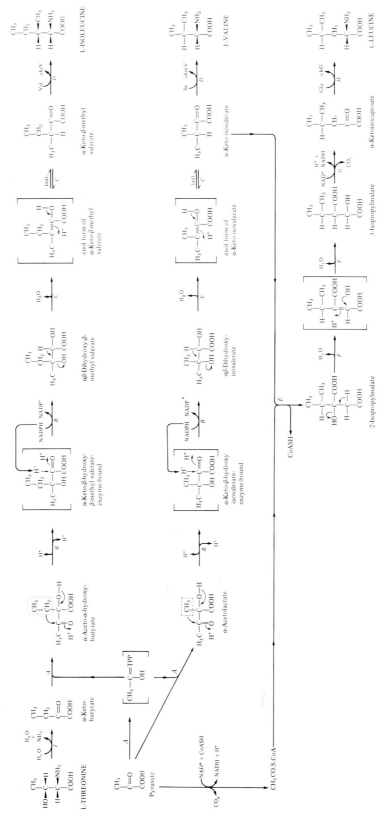

Fig. 9.7. Biosynthesis of leucine, isoleucine and valine in plants. (A = Acetolactate synthase, EC 4.1.3.18; B = ketol-acid reductoisomerase, EC 1.1.1.86; C = dihydroxyacid dehydratase, EC 4.2.1.9; D = valine aminotransferase, EC 2.6.1.32; E = 2-isopropylmalate synthase, EC 2.6.1.32; F = 3-isopropylmalate dehydratase, EC 4.2.1.33; G = 3-isopropylmalate dehydrogenase, EC 1.1.1.85; H = leucine aminotransferase, EC 2.6.1.6; I = threonine deaminase, EC 4.2.1.16; taut = tautomerism; Val = L-valine; αkiV = α-ketoisovalerate; Ile = L-isoleucine; Glu = L-glutamate; αkG = α-ketoglutarate; TPP = thiamine pyrophosphate.) αkmV = α-keto-β-methyl valerate; Glu = L-glutamate; αkG = α-ketoglutarate; TPP = thiamine pyrophosphate.)

$$\text{(structure: thiazolium with } H_3C-C-OH) + CH_3COCOOH \longrightarrow CH_3C(OH)COOH + \text{(thiazolium ylide)} \tag{9.11}$$

with the product group:
$$\begin{array}{c} CH_3\,C(OH)COOH \\ | \\ C{=}O \\ | \\ CH_3 \end{array}$$

$$\begin{array}{c} H_3C \quad CH_3 \\ | \quad\;\; | \\ H_3C-C-C-CH_3 \\ | \quad\;\; | \\ OH \;\; OH \end{array} + H^+ \rightleftharpoons \begin{array}{c} H_3C \quad CH_3 \\ | \quad\;\; | \\ H_3C-C-C-CH_3 \\ | \quad\;\; | \\ OH \;\; {}^+OH_2 \end{array} \xrightarrow{H_2O} \begin{array}{c} H_3C \quad CH_3 \\ | \quad\;\; | \\ H_3C-C-C-CH_3 \\ | \quad\;\; + \\ C \\ | \\ H \end{array} \xrightarrow[H^+]{} \begin{array}{c} CH_3 \\ | \\ H_3C-C-C-CH_3 \\ || \quad\;\; | \\ O \;\; CH_3 \end{array} \tag{9.12}$$

enzyme involved being dihydroxy acid dehydratase[12]. The enzyme from higher plants differs from enzymes from other sources in its pH optimum (8.0–8.2) and its cation requirement (Mg^{2+}). Transamination (valine aminotransferase[13]) leads to valine (D) whilst the conversion of α-ketoisovaleric acid to leucine proceeds through three reactions which have similarities to some of those occurring in the Krebs cycle. Reaction E, the condensation of α-ketoisovalerate with acetyl CoA to form 2-isopropylmalate and its isomerization to 3-isopropylmalate (F), are analogous to the condensation of acetyl-CoA with oxaloacetate to form citrate and its isomerization to isocitrate respectively. Reaction G is analogous to the action of isocitric dehydrogenase to form α-ketoglutarate from isocitrate. The final step H leading to leucine is transamination involving leucine aminotransferase[14].

Enzymes A, B, C and D are quite remarkably involved in the synthesis of another branched chain amino acid, isoleucine, by acting on substrates which are homologues of those involved in valine formation. The homologues arise by the condensation of hydroxyethyl-TPP, not with pyruvate but with α-ketobutyrate, which arises by the deamination of threonine (H). Here we see the interesting situation that a key intermediate for the synthesis of one amino acid is derived from another amino acid not produced from pyruvate (threonine is derived from oxaloacetate—see Section E.2).

Feed-back control of these pathways is exerted by leucine and valine acting co-operatively on acetolactate synthase, by leucine acting on 2-isopropylmalate synthase[15] and by isoleucine acting on threonine deaminase[16]. On the other hand, valine and leucine activate threonine deaminase.

(ii) LYSINE AND ALANINE†

Although lysine is not a branched chain amino acid it is formed by a pathway reminiscent of the valine pathway (Fig. 9.7) because a key intermediate arises by the condensation of pyruvate with a derivate of an amino acid itself not derived from pyruvate. In this case the intermediate is β-L-aspartate semialdehyde (Fig. 9.8) and the product 2,3-dihydrodipicolinate.

Aspartate semialdehyde is formed from aspartate via aspartyl phosphate; the first enzyme involved, aspartate kinase[17] (A), is an obvious control point for lysine biosynthesis but the nature of the control varies with plant species; inhibition can be observed with lysine alone, lysine plus threonine or threonine alone. These differences are probably due to the existence of two isoenzymes, one sensitive to lysine only and the other to threonine only. The second enzyme involved is aspartate semialdehyde dehydrogenase[18] (B). The β-aspartyl semialdehyde so formed then condenses with pyruvate to form 2,3-dihydrodipicolinate in the presence of dihydrodipicolinate synthase[19] (C). The next enzyme D hydrogenates 2,3-dihydrodipicolinate to yield Δ^1-piperideine 2,6-dicarboxylate which then undergoes a novel reaction (E) in which the ring opens as the amino group is succinylated by succinyl-CoA. Succinyldiaminopimelate aminotransferase[20] (F) then forms N-succinyl-L,L-2,6-diaminopimelate.

† All reaction letters in this section refer to Fig. 9.8.

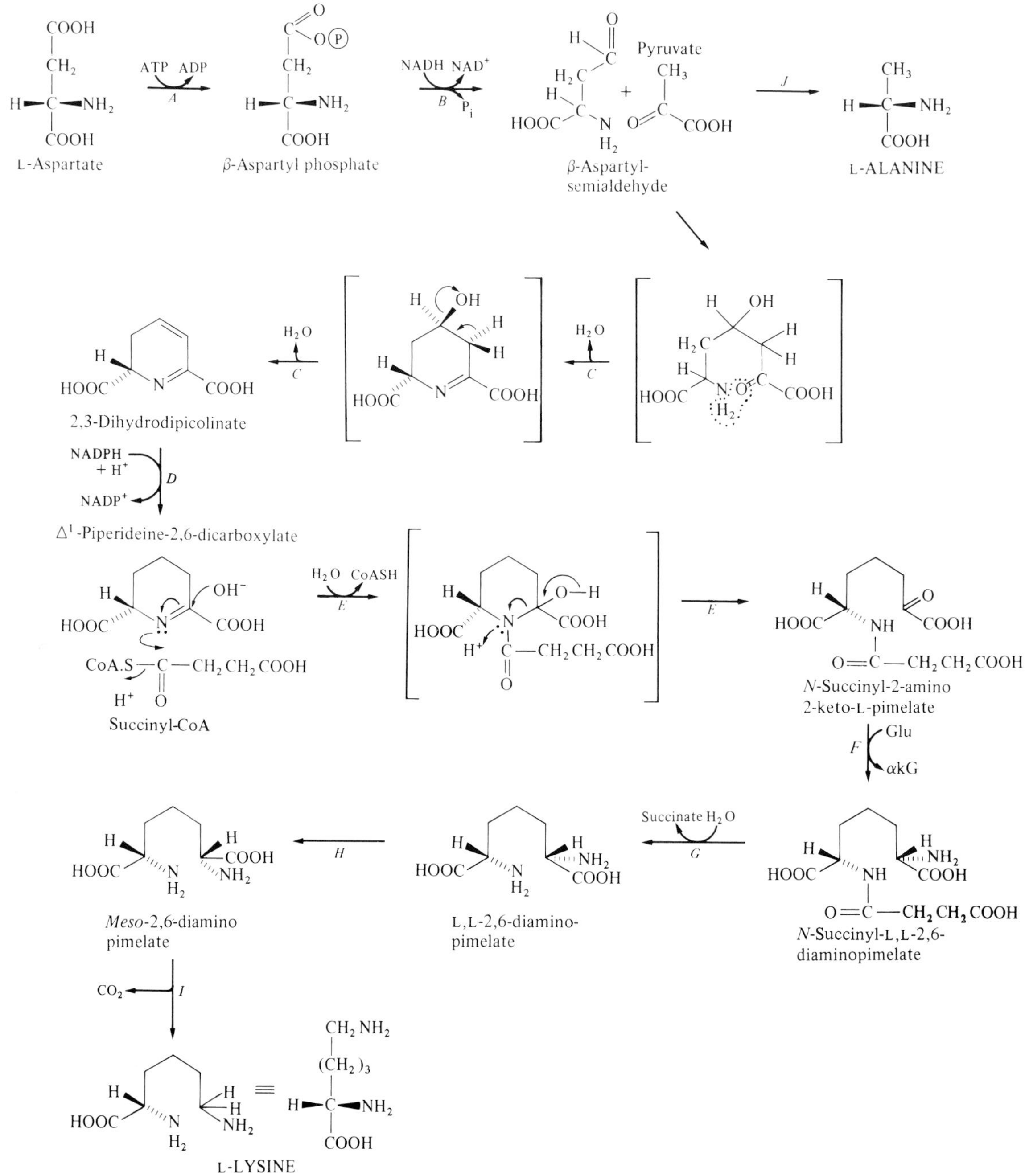

Fig. 9.8. Biosynthesis of lysine and alanine in plants. (A = Aspartate kinase, EC 2.7.2.4; B = aspartate semialdehyde dehydrogenase, EC 1.2.1.11; C = dihydrodipicolinate synthase, EC 4.2.1.52; D = dihydrodipicolinate reductase, EC 1.3.1.26; E = Δ^1-piperideine-2,6-dicarboxylate:succinyl-CoA N-succinyltransferase (ring opening); F = succinyl-diaminopimelate aminotransferase, EC 2.6.1.17; G = succinyl-diaminopimelate desuccinylase, EC 3.5.1.18; H = diaminopimelate epimerase, EC 5.1.1.7; I = diaminopimelate decarboxylase, EC 4.1.1.20; J = alanine aminotransferase, EC 2.6.1.2.)

This is hydrolysed to L,L-2,6-diaminopimelate by succinyl-diaminopimelate desuccinylase[21] (*G*). Diaminopimelate epimerase[22] (*H*) forms *meso*-2,6-diaminopimelate which is the substrate for diaminopimelate decarboxylase[23] (*I*) which produces L-lysine. This pathway is confined to plants and bacteria and a different route is followed in fungi.

Figure 9.8 also indicates how alanine can be formed directly from pyruvate by transamination. The amino donor for the reaction, which is catalysed by alanine aminotransferase[24] (*J*), is glutamate.

2. Amino acids derived from oxaloacetate†

Aspartic acid is the amino acid derived directly from axaloacetate by transamination (*A*), but other amino acids are formed from aspartic acid itself. These are lysine, discussed in the previous section (see Fig. 9.8), threonine and methionine (Fig. 9.9).

In the biosynthesis of threonine, β-aspartylsemialdehyde is reduced by homoserine dehydrogenase[25] (*E*) to L-homoserine which is phosphorylated by homoserine kinase[26] (*F*) in the presence of ATP to L-homoserine-*O*-phosphate. This is then converted by threonine synthase (*G*) into threonine. The reaction involves intramolecular rearrangement of the hydroxyl group and a loss of phosphate. Pyridoxal phosphate is a necessary cofactor and the reaction, which involves the formation of a Schiff's base, is probably that indicated in Fig. 9.10.

L-Homoserine *O*-phosphate is also the branch point for methionine synthesis. The enzyme homoserine *O*-phosphate:cysteine *O*-cysteinyltransferase (*H*) condenses this intermediate with cysteine to form cystathionine with the elimination of inorganic phosphate. The reaction appears to be specific for plants, the substrate in bacterial systems being *O*-acylhomoserine. Cystathionine β-lyase[27] (*J*) then splits cystathionine at its β-linkage into homocysteine, pyruvate and NH_4^+. Finally methionine is formed by methyl transfer to homocysteine from N^5-methyltetrahydrofolate triglutamate (IV, Fig. 9.11) by the appropriate transferase[28] (*K*). Cysteine

itself can also arise from cystathionine by the action of cystathionine γ-lyase[29] (*I*) which produces in addition α-ketobutyric acid and ammonia.

The control of this pathway at the aspartate kinase level has already been discussed in the previous section. However, control is also exerted at homoserine dehydrogenase which exists as a series of isoenzymes, some of which are inhibited by threonine. Threonine synthase is strongly stimulated by active methionine (*S*-adenosylmethionine, SAM, V, Fig. 9.11) and inhibited by cysteine which is a precursor of methionine. Thus only when sufficient methionine is present to give the required amount of SAM is the pathway to threonine activated.

The main route for synthesis of asparagine is by transfer of the amide nitrogen from glutamine in the presence of ATP and asparagine synthetase[30] (*B*). Direct amidation from NH_4^+ is not significant *in vivo*.

3. Amino acids derived from α-ketoglutarate‡

The formation of glutamate from α-ketoglutarate has already been discussed in Section B.3 on the assimilation of ammonia but further metabolism of glutamate leads to the formation of proline and arginine and related compounds.

(i) THE ARGININE GROUP

The first step is the *N*-acetylation of glutamate with either acetyl-CoA or *N*-acetylornithine as donor, the enzymes involved being acetyl-CoA: glutamate *N*-acetyl transferase[31] (*A*) and glutamate acetyltransferase[32] (*E*), respectively (Fig. 9.12). This protects the amino group in the classical manner of organic chemistry and allows the required reactions to proceed without cyclization. Next *N*-acetylglutamate phosphokinase[33] (*B*) produces *N*-acetyl-γ-glutamyl phosphate which is reduced with release of phosphate by the appropriate dehydrogenase (*C*). Transamination (*D*) yields α-*N*-acetylornithine, which, as just stated, can transfer its acetyl group to glutamate (*E*). The latter is an

---

† All reaction letters in this section refer to Fig. 9.9.   ‡ All reaction letters in this section refer to Fig. 9.12.

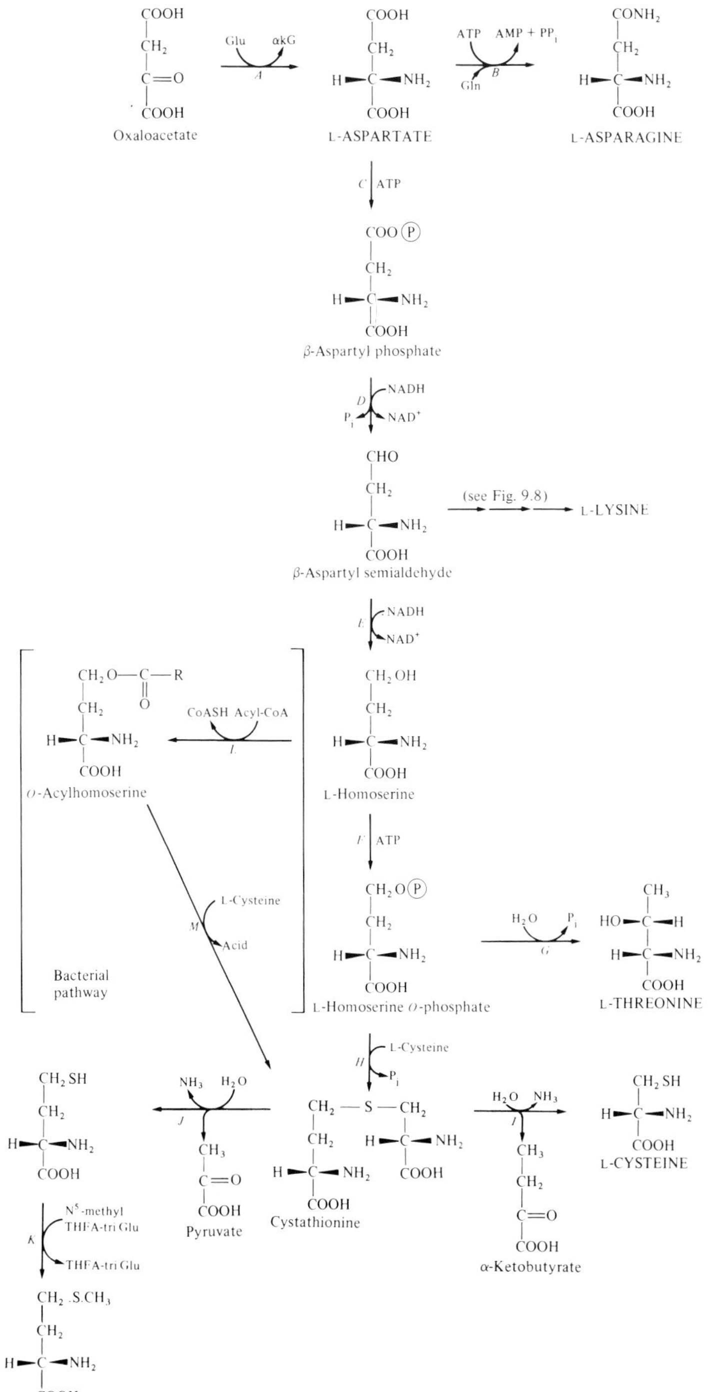

Introduction to Plant Biochemistry

COOH | CH₂ | C=O | COOH
Oxaloacetate

Glu   αkG
A

COOH | CH₂ | H—C—NH₂ | COOH
L-ASPARTATE

ATP   AMP + PP₁
B
Gln

CONH₂ | CH₂ | H—C—NH₂ | COOH
L-ASPARAGINE

C | ATP

COO(P) | CH₂ | H—C—NH₂ | COOH
β-Aspartyl phosphate

D   NADH
P₁   NAD⁺

CHO | CH₂ | H—C—NH₂ | COOH
β-Aspartyl semialdehyde

(see Fig. 9.8) → → L-LYSINE

E   NADH
NAD⁺

CH₂OH | CH₂ | H—C—NH₂ | COOH
L-Homoserine

CoASH   Acyl-CoA
L

CH₂O—C—R | O | CH₂ | H—C—NH₂ | COOH
O-Acylhomoserine

Bacterial pathway

L-Cysteine
M
Acid

F | ATP

CH₂O(P) | CH₂ | H—C—NH₂ | COOH
L-Homoserine O-phosphate

H₂O   P₁
G

CH₃ | HO—C—H | H—C—NH₂ | COOH
L-THREONINE

H   L-Cysteine
P₁

CH₂—S—CH₂ | CH₂ | H—C—NH₂ | COOH   H—C—NH₂ | COOH
Cystathionine

H₂O   NH₃
I

CH₂SH | H—C—NH₂ | COOH
L-CYSTEINE

CH₃ | CH₂ | C=O | COOH
α-Ketobutyrate

NH₃   H₂O
J

CH₃ | C=O | COOH
Pyruvate

CH₂SH | CH₂ | H—C—NH₂ | COOH

N⁵-methyl THFA-tri Glu
K
THFA-tri Glu

CH₂.S.CH₃ | CH₂ | H—C—NH₂ | COOH
L-METHIONINE

Fig. 9.9. Biosynthesis of aspartate, asparagine, threonine and methionine in plants. (A = Aspartate aminotransferase, EC 2.6.1.1; B = asparagine synthetase (glutamine hydrolysing), EC 6.3.5.4; C = aspartate kinase, EC 2.7.2.4; D = aspartate semialdehyde dehydrogenase, EC 1.2.1.11; E = homoserine dehydrogenase, EC 1.1.1.3; F = homoserine kinase, EC 2.7.1.39; G = threonine synthase, EC 4.2.99.2; H = homoserine O-phosphate:L-cysteine O-cysteinyltransferase (phosphate hydrolysing); I = cystathionine γ-lyase, EC 4.4.1.1; J = cystathionine β-lyase, EC 4.4.1.8; K = tetrahydropteroyltriglutamate methyltransferase, EC 2.1.1.14; L = homoserine acyltransferase; M = O-acylhomoserine (thiol)-lyase.)

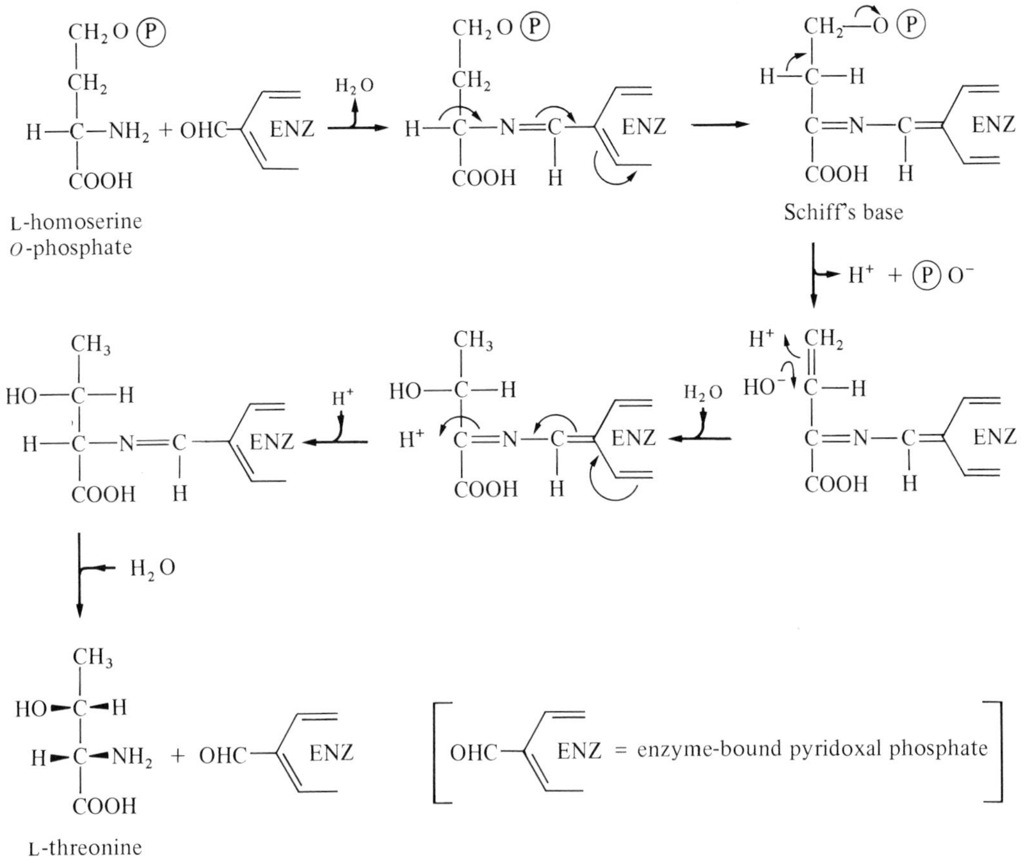

CH₂O (P)   CH₂O (P)   CH₂—O (P)

L-homoserine
O-phosphate

Schiff's base

H⁺ + (P) O⁻

L-threonine

[OHC—ENZ = enzyme-bound pyridoxal phosphate]

Fig. 9.10. Probable mechanism of the threonine synthase-catalysed reaction.

important energy-conserving step because this continuous recycling of acetyl groups allows the continued synthesis of N-acetylglutamate with only the expenditure of the amount of acetyl-CoA necessary to prime the reaction sequence.

A modified urea cycle functions in plants in that ornithine condenses with carbamoyl phosphate to form citrulline (H) which can be converted into arginine via argininosuccinate (I,J); arginine can be converted back into ornithine with the liberation of

urea (K). The carbamoyl phosphate arises from glutamine, itself formed from glutamate by glutamine synthetase (F) (Section B.3). Carbamoyl phosphate synthetase (G) involves the fixation of CO₂ and the utilization of ATP; glutamate is regenerated in the process [eq. (9.13)]. The plant enzyme has not been examined in detail but in E. coli it is biotin-dependent [see Chapter 8, Section C.1(i)(a) for a discussion of biotin-function in CO₂-fixing reactions]. It contains two sub-units,

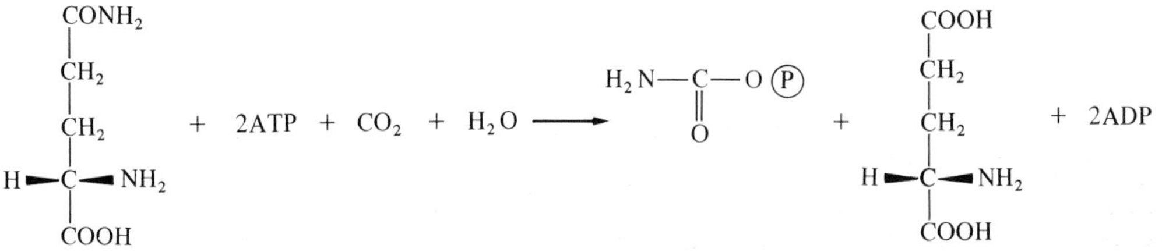

CONH₂
|
CH₂
|
CH₂   + 2ATP + CO₂ + H₂O ⟶
|
H—C—NH₂
|
COOH

     H₂N—C—O (P)
         ‖
         O

 +

COOH
|
CH₂
|
CH₂   + 2ADP + Pᵢ   (9.13)
|
H—C—NH₂
|
COOH

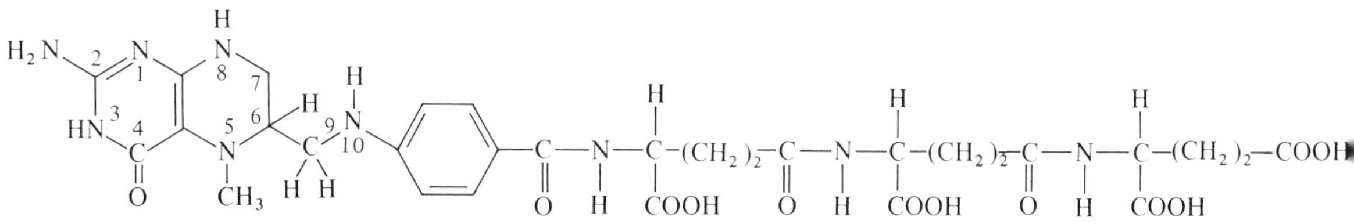

(IV) N^5-methyl-tetrahydrofolate triglutamate

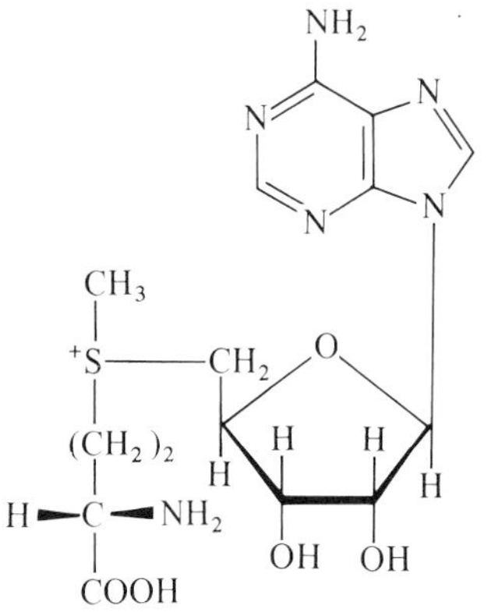

(V) S-adenosyl methionine (SAM)

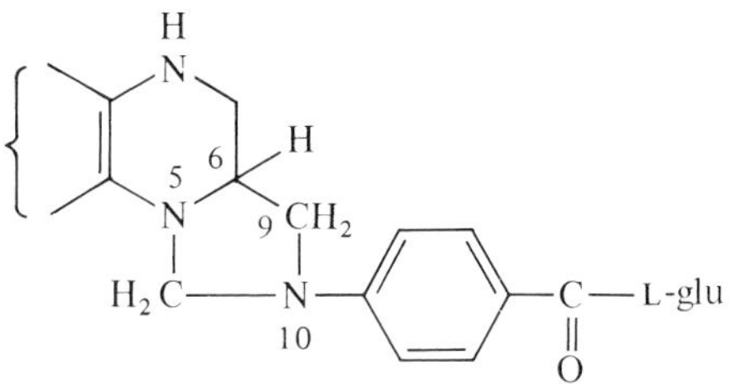

(VI) N^5, N^{10}-methylene THFA
(see section E4)

Fig. 9.11. Structures of some compounds participating in the biosynthesis of oxaloacetate-derived amino acids.

MW 130,000 and 42,000; the smaller has glutaminase activity and presumably liberates ammonia at the active site of the complex. One ATP is used in forming a carboxy-biotin intermediate which carboxylates ammonia to an enzyme-bound carbamate which is phosphorylated by the second molecule of ATP.

Carbamoyl phosphate synthetase in plants produces this substrate for pyrimidine synthesis as well as ornithine synthesis (see Chapter 10) so that some delicate control is necessary to balance the two pathways. In the first place the K_m for carbamoyl phosphate for ornithine carbamoyltransferase[34] (H) is ten times greater than that for asparagine carbamoyltransferase[35], the enzyme concerned with pyrimidine biosynthesis (see Chapter 10). This means that high levels of carbamoyl phosphate are required for significant synthesis of citrulline to occur. UMP (uridine monophosphate, see Chapter 10) inhibits both carbamoyl phosphate synthetase and aspartate carbamoyltransferase, whereas ornithine activates carbamoyl phosphate synthetase. This means that (a) synthesis of carbamoyl phosphate is blocked by UMP until sufficient ornithine accumulates to overcome this inhibition and (b) the pyrimidine pathway is shut off until the UMP levels drop to a low value. Thus a situation exists which permits rapid transfer of nitrogen flow into pyrimidines. Mammals and fungi have solved this particular problem in a different way: two distinct carbamoyl phosphate synthetases exist, one specific for each pathway; that for citrulline synthesis is located in the mitochondria whereas that for pyrimidine biosynthesis is cytoplasmic.

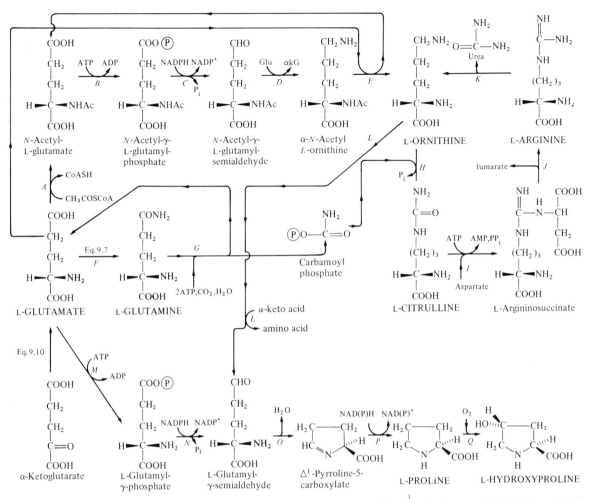

Fig. 9.12. Biosynthesis of arginine, proline and related compounds. ($Ac=CH_3—\overset{|}{C}=O$; A = acetyl-CoA:L-glutamate N-acetyltransferase, EC 2.3.1.1; B = acetylglutamate kinase, EC 2.7.2.8; C = N-acetyl-γ-glutamyl phosphate reductase, EC 1.2.1.38; D = acetylornithine aminotransferase, EC 2.6.1.11; E = glutamate acetyltransferase, EC 2.3.1.35; F = glutamine synthetase, EC 6.3.1.2; G = carbamoyl-phosphate synthetase (glutamine-hydrolysing), EC 6.3.5.5; H = ornithine carbamoyltransferase, EC 2.1.3.3; I = argininosuccinate synthetase, EC 6.3.4.5; J = argininosuccinate lyase, EC 4.3.2.1; K = arginase, EC 3.5.3.1; L = ornithine-oxo-acid aminotransferase, EC 2.6.1.13; M = glutamate kinase, EC 2.7.2.11; O = spontaneous reaction; P = pyrroline 5-carboxylate reductase, EC 1.5.1.2; Q = proline hydroxylase, EC 1.14.11.2.)

The citrulline so formed by ornithine carbamoyl-transferase is converted into argininosuccinate by condensing with aspartate (I). Hydrolysis (J) of this intermediate yields arginine and fumarate. The latter can be recycled by feeding into the TCA cycle whereas arginase[36] (K) converts arginine into urea and ornithine which continue the urea cycle, or are converted into proline [see Section E.3(ii)]. The nitrogen from urea is recycled as ammonia following the action of urease[37].

(ii) PROLINE AND HYDROXYPROLINE

The biosynthesis of proline can take place in the early stages in two ways: (a) via acylated intermediates just discussed for the formation of ornithine which can be deaminated to glutamyl-γ-semialdehyde (L) and (b) via the non-acylated intermediate glutamyl-γ-phosphate (M) which is reduced and dephosphorylated to glutamyl-γ-semialdehyde (N). Which of these two pathways predominates in vivo is not clear. The glutamyl-γ-

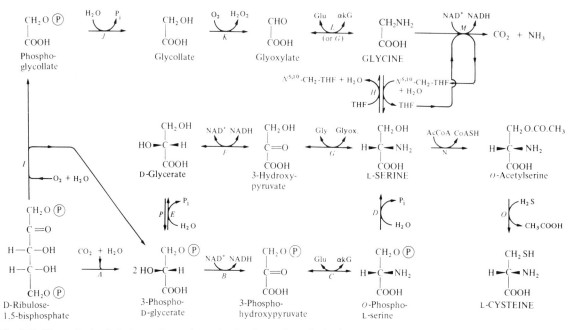

Fig. 9.13. Biosynthesis of glycine, serine and cysteine in plants. (A = Ribulosebisphosphate carboxylase (in carboxylase mode), EC 4.1.1.39; B = phosphoglycerate dehydrogenase, EC 1.1.1.95; C = phosphoserine aminotransferase, EC 2.6.1.52; D = phosphoserine phosphatase, EC 3.1.3.3; E = 3-phosphoglycerate phosphatase, EC 3.1.3.38; F = glycerate dehydrogenase, EC 1.1.1.29; G = serine-glyoxylate aminotransferase, EC 2.6.1.45; H = serine hydroxymethyltransferase, EC 2.1.2.1; I = ribulosebisphosphate carboxylase (in oxygenase mode), EC 4.1.1.39; J = phosphoglycollate phosphatase, EC 3.13.18; K = glycollate oxidase, EC 1.1.3.1; L = glutamate-glyoxylate aminotransferase; M = glycine synthase, EC 2.1.2.10; N = serine acetyltransferase, EC 2.3.1.30; O = cysteine synthase, EC 4.2.99.8; P = glycerate kinase, EC 2.7.1.31; THF = tetrahydrofolic acid; $N^{5,10}$-CH$_2$-THF = $N^{5,10}$-methylene THF; AcCoA = acetyl-CoA; Glyox = glyoxylate; Gly = glycine.)

semialdehyde, formed by either route, spontaneously cyclizes (O) to Δ^1-pyrroline-5-carboxylate which is reduced (P) to yield proline.

Proline is converted into 4-hydroxyproline (Q) only when it is present in a polypeptide chain elongating on a ribosome (see Chapter 10).

4. Amino acids formed from products of the Calvin cycle†

The amino acids so formed are glycine, serine and cysteine with serine occupying a central position. Three possible pathways are available for serine formation (Fig. 9.13). Two start from 3-phosphoglycerate, derived from D-ribulose 1,5-bisphosphate (A), and involve either a phosphorylated or non-

phosphorylated pathway. In the phosphorylated pathway, which occurs in germinating seedlings, the intermediates are 3-phosphohydroxypyruvate, produced by reaction B, and phosphoserine, produced by reaction C; the latter is then dephosphorylated (D) to serine. The non-phosphorylated pathway is found in leaves and the high level of the first enzyme in the pathway, 3-phosphoglycerate phosphatase[38], (E) in C_4 plants [Chapter 5, Section D.3(i)] should be noted. The glycerate formed in this step is oxidized to hydroxypyruvate (F) which is converted into serine by a specific aminotransferase[39] (G) which uses glycine as the specific amino donor.

Glycine is formed from serine by a reversible reaction catalysed by serine hydroxymethyltransferase[40] (H) in which formaldehyde is transferred to the acceptor tetrahydrofolate (THF) with the production of $N^{5,10}$-methylene tetrahydrofolate (VI, Fig. 9.11); pyridoxal phosphate is also involved.

Under conditions conducive to photorespiration

† Reaction letters in this section refer to Fig. 9.13.

the enzyme ribulose 1,5-bisphosphate carboxy-lase/oxygenase[41] acts in its oxygenase mode (Chapter 6, Section F.1) and produces phosphoglycollate (*I*) from ribulose 1,5-bisphosphate; this is then dephosphorylated to glycollate (*J*). These two reactions occur in the chloroplast whence the free glycollate moves into the peroxisomes where it is oxidized to glyoxylate (*K*) which is transaminated to glycine (*L*). Glycine moves out of the peroxisomes and that not required for protein synthesis is transferred to the mitochondria where two molecules are used to form one molecule of serine; one glycine is used to generate $N^{5,10}$-methylene-THF (*M*) which is then used to convert the other into serine (*H*). The reversal of the non-phosphorylated serine pathway ($G \rightarrow F \rightarrow P$) is also important in the overall nitrogen economy of C–3 plants, where it is claimed to be the main pathway of sucrose formation. The amount of ammonia thus liberated in these reactions must be considerable. The importance of its reassimilation has already been discussed (Section B.4).

Serine is converted into cysteine by a two-stage process (Fig. 9.13). The first reaction produces *O*-acetylserine (*N*) and the second involves the uptake of a sulphide group from a carrier molecule and the liberation of acetate (*O*). The enzyme *O*-acetylserine (thiol)-lyase or cysteine synthase[41] (*O*) is a pyridoxal-dependent enzyme. The sulphide arises from the reduction of SO_4^{2-} according to Fig. 9.14. Activated sulphate, adenylyl sulphate, transfers its

sulphuryl group to the SH group of a carrier. This group is then reduced (6-electron reduction) by a ferredoxin-dependent reductase; finally a sulphide group is transferred with the generation of the reduced carrier to continue the cycle.

5. Amino acids derived from shikimate†

The amino acids formed from shikimate are the aromatic amino acids phenylalanine, tyrosine and tryptophan. The pathway to shikimate, the first specific precursor on the route to these aromatic amino acids, has been elucidated by elegant studies on bacteria but little is known about its formation in higher plants. However, there is no reason to believe that it is significantly different from that in bacteria which has been described in Chapter 4 and outlined in Fig. 4.22.

The conversion of shikimate into chorismate (Fig. 9.15), the branch point of tyrosine and tryptophan synthesis, involves the intermediate formation of 5-phosphoshikimate (*A*) 5-phospho-3-enoylpyruvyl shikimate (*B*). The final step is the removal of orthophosphate to form chorismate (*C*). The most interesting reaction in this series is the addition-elimination which results in the formation of 5-phospho-3-enoylpyruvylshikimate (*B*). The key evidence for the mechanism

† Reference letters in this section refer to Figs. 9.15 and 9.16.

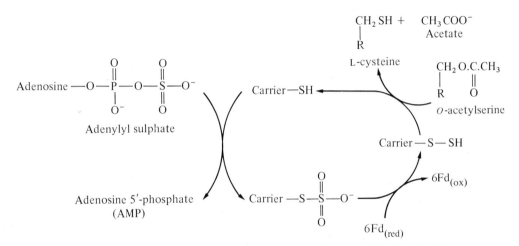

Fig. 9.14. Mechanism of incorporation of sulphur into L-cysteine (Fd = ferredoxin).

outlined in Fig. 9.15 was provided by experiments carried out in ^3H_2O which demonstrated the appearance of tritium in the methylene group of the enoylpyruvate residue. The chorismate so formed is channelled either into tyrosine/phenylalanine pathway (Fig. 9.15) or into the tryptophan pathway (Fig. 9.16).

In the first route isomerization of chorismate to prephenate occurs in the presence of chorismate mutase[42] (D). At least two isoenzymes of this enzyme exist and their regulatory role is considered later in this section. Prephenate can then be converted into tyrosine by two pathways both of which occur in higher plants. They differ in the points at which aromatization occurs. In one pathway prephenate is aromatized to 4-hydroxy-phenylpyruvate (E) which is then converted into tyrosine by the appropriate transaminase (F). In the other pathway, which also occurs in blue-green algae, transamination occurs first and results in the formation of pretyrosine (G); this is then aromatized (H) to tyrosine by a mechanism like that by which prephenic acid is converted into 4-hydroxy-phenylpyruvate (E). Prephenate has yet a third metabolic future: the enzyme prephenate dehydratase[43] (I) produces phenylpyruvate which can be transaminated (J) to phenylalanine.

The second amino acid biosynthetic pathway in which chorismate participates (see Fig. 9.16) leads to the formation of tryptophan and involves firstly the formation of anthranilic acid with the liberation of pyruvate and the utilization of the amide nitrogen of glutamine (A). Further steps (B, C and D) lead to N-(-5'-phosphoribosyl)-anthranilate, 1-(o-carboxy-phenylamino)-1'-deoxyribulose-5'-phosphate and indole 3-glycerolphosphate respectively. The second of these (C) involves a so-called Amadori rearrangement during which C–1 of the ribose residue is reduced and C–2 is oxidized to a carbonyl group; the resulting open chain compound then decarboxylates and ring closes (D) to yield indole-3-glycerol phosphate. The mechanism of these reactions (C and D) is shown in Fig. 9.16 with the hypothetical intermediates placed in parenthesis.

Indole 3-glycerol phosphate is then converted into L-tryptophan by a β-replacement reaction involving L-serine and catalysed by tryptophan synthase[44]. This enzyme is composed of two subunits, α and β, which can be dissociated from one

another. The α-subunit can catalyse the removal of the C_3 side chain from indole 3-glycerol phosphate with the formation of indole and 3-phosphoglyceraldehyde. The β-subunit contains bound pyridoxal phosphate and can catalyse the formation of tryptophan from indole and serine; presumably it first generates from serine the Schiff's base of aminoacrylic acid and then transfers it to C–3 (βC) of indole. When the two subunits act together as tryptophan synthase it is not clear whether free indole is an intermediate or whether there is condensation of indole-3-glycerol phosphate with the aminoacrylate-Schiff's base prior to aldol cleavage as is shown in Fig. 9.17.

The control point located at the chorismate mutase reaction is a particularly good example of isoenzyme function in metabolic control. Three isoenzymes CM_1, CM_2 and CM_3 exist in plants. CM_1 and CM_3 but not CM_2 are inhibited by phenylalanine and tyrosine and activated by tryptophan; tryptophan acts by increasing the substrate affinity for the enzyme. Thus channelling into the two pathways for the requirements for protein synthesis is effectively controlled. The isoenzyme CM_2, on the other hand, is probably used for the production of the phenylalanine and tyrosine required for the synthesis of compounds which are the precursors of lignin and coumarins.

6. Histidine

The biosynthesis of this heterocyclic acid, which plays a significant role at the active site of many enzymes, has not been studied in higher plants. However, there is no reason to believe that the pathway is greatly different from that in bacteria, where it has been studied in great detail. An outline of the pathway is given in Figs. 9.18 a and b. It should be noted that ATP and α-5-phosphoribosyl-l-pyrophosphate (PRPP) constitute the starting-point and that only one nitrogen of the ATP finds its way into the histidine molecule. The numbering of the adenine ring of ATP (1–9) and PRPP (1'–5') is carried through the pathway so as to clarify the biosynthetic origin of the various atoms of histidine. Reaction D involves an Amadori rearrangement identical to that in tryptophan biosynthesis (see step

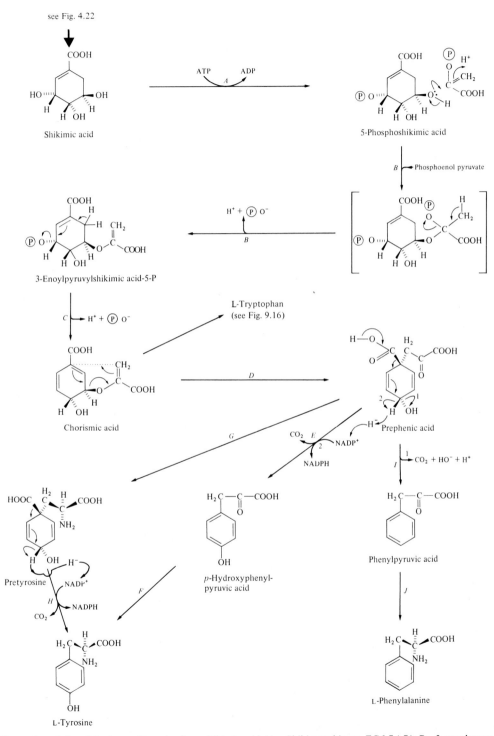

Fig. 9.15. Formation of phenylalanine and tyrosine from shikimic acid. (A = Shikimate kinase, EC 2.7.1.71; B = 3-enoylpyruvylshikimic acid 5-phosphate synthase; C = chorismate synthase, EC 4.6.1.4; D = chorismate mutase, EC 5.4.99.5; E = prephenate dehydrogenase, EC 1.3.1.13; F = tyrosine aminotransferase, EC 2.6.1.5; G = prephenate aminotransferase; H = pretyrosine dehydrogenase; I = prephenate dehydratase, EC 4.2.1.51; J = phenylpyruvate aminotransferase.)

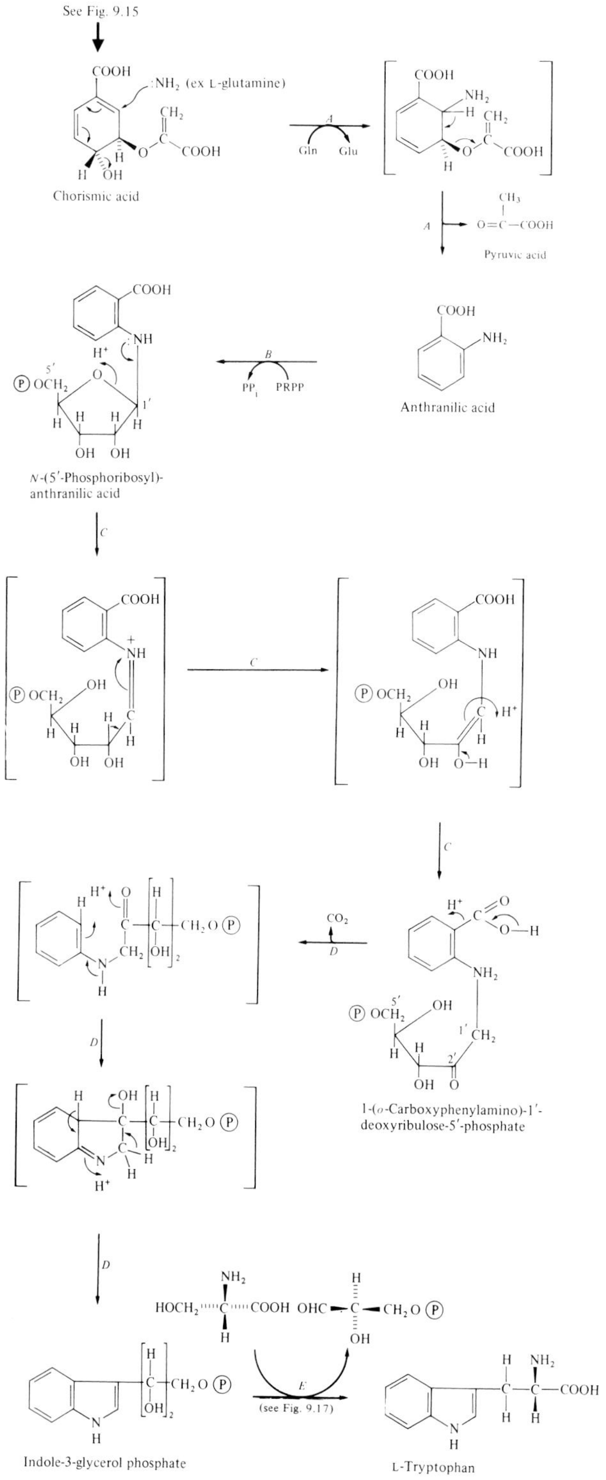

Fig. 9.16. Formation of tryptophan from chorismic acid. (A = Anthranilate synthase, EC 4.1.3.27; B = anthranilate phosphoribosyltransferase, EC 2.4.2.18; C = N-5′-phosphoribosyl-anthranilate ketol-isomerase (an Amadori rearrangement); D = indole-3-glycerol-phosphate synthase, EC 4.1.1.48; E = tryptophan synthase, EC 4.1.2.20.)

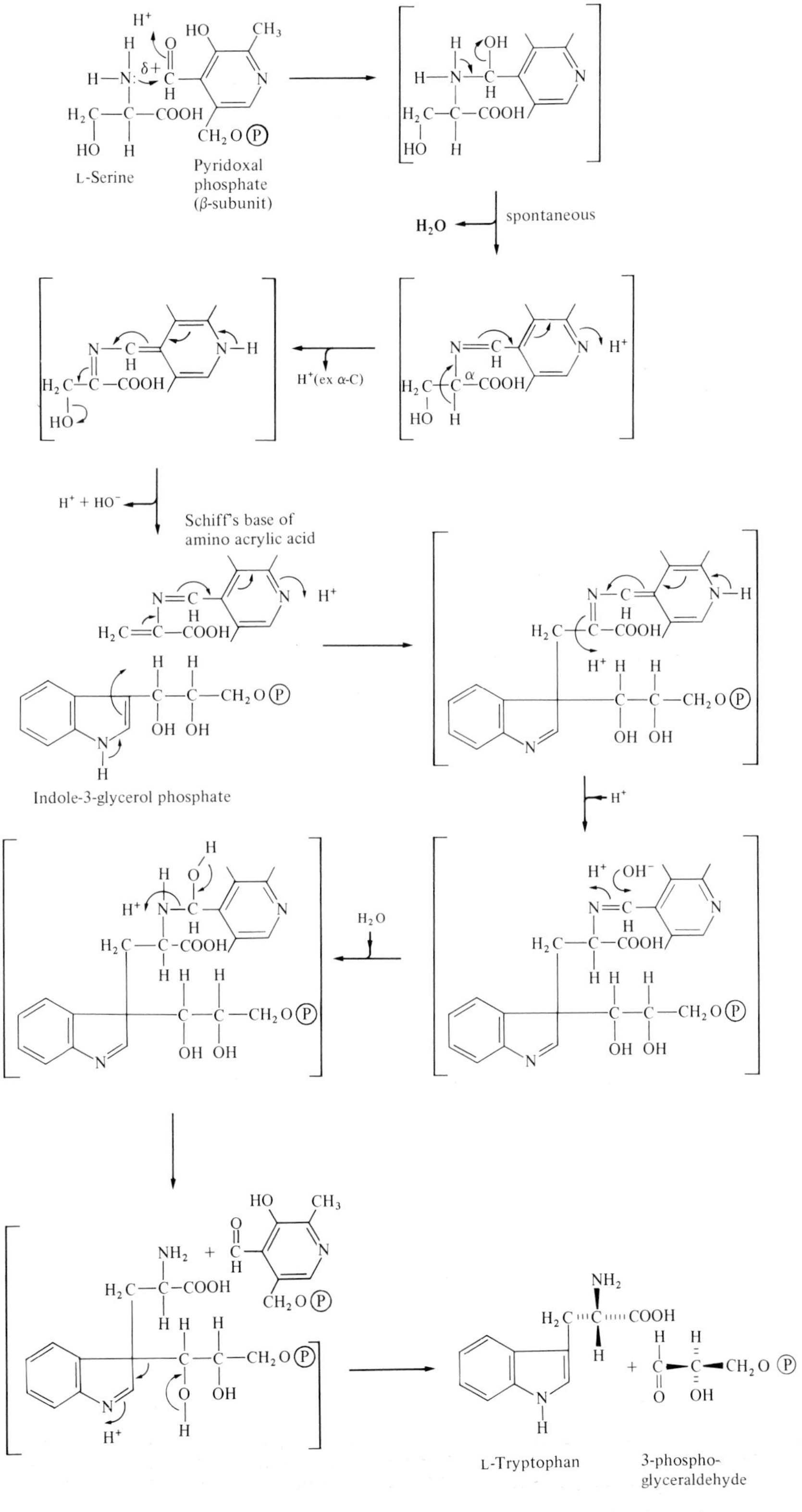

Fig. 9.17. Possible mechanism of the tryptophan synthase-catalysed reaction.

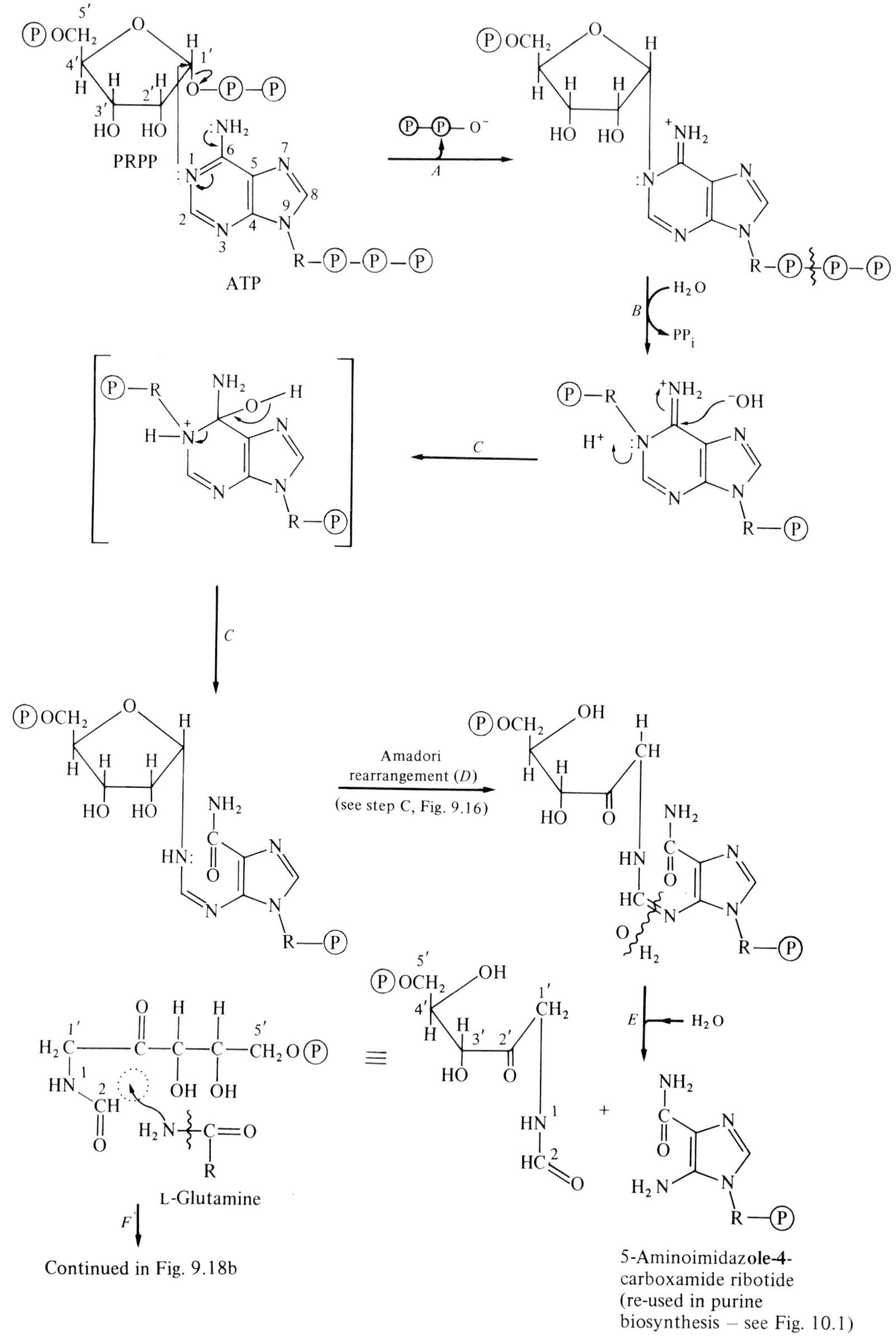

Fig. 9.18a. Biosynthesis of L-histidine (continued in Fig. 9.18b). (*A–F*: see legend of Fig. 9.18b.)

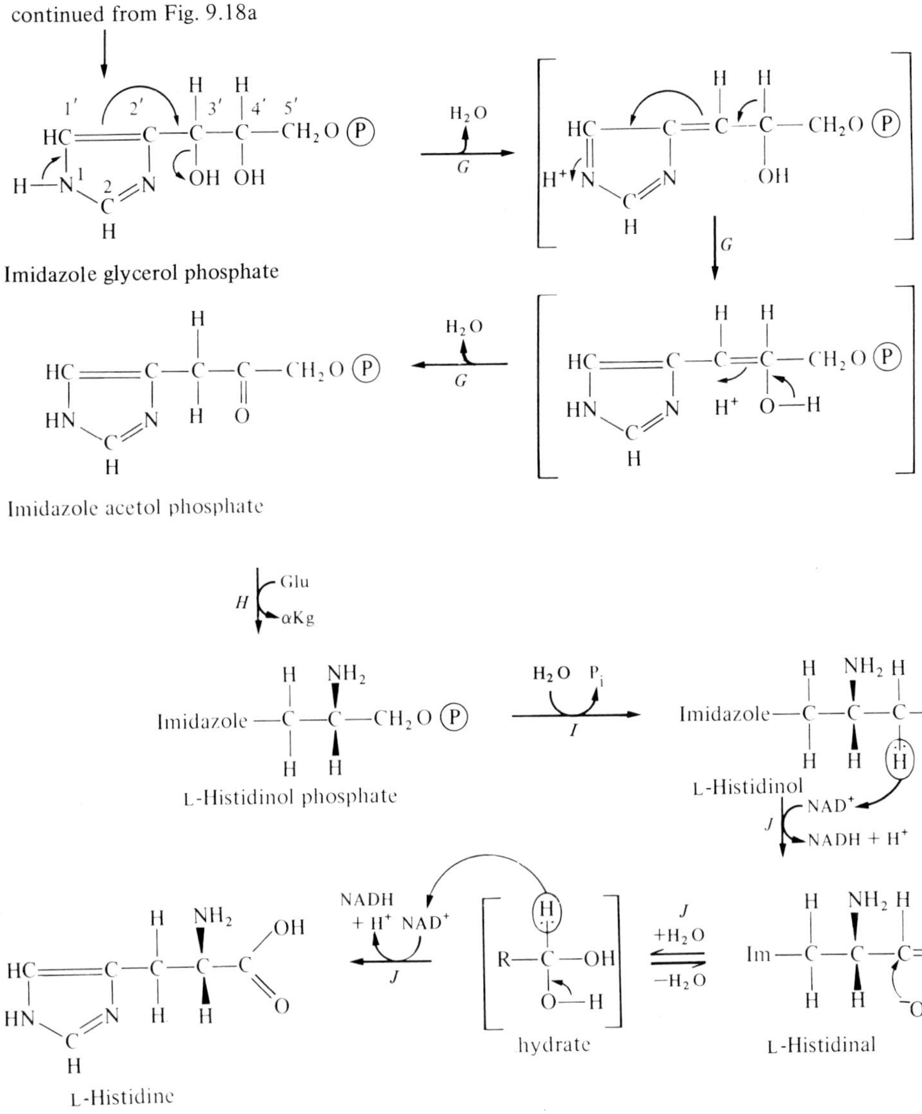

continued from Fig. 9.18a

Imidazole glycerol phosphate

Imidazole acetol phosphate

Fig. 9.18b. Biosynthesis of L-histidine (continued from Fig. 9.18a). (A = ATP phosphoribosyltransferase, EC 2.4.2.17; B = N^1-(5'-phosphoribosyl)-ATP pyrophosphatase; C = phosphoribosyl-AMP cyclohydrolase, EC 3.5.4.19; D = N-(5'-phospho-D-ribosylformimino)-5-amino-1-(5''-phosphoribosyl)-4-imidazole carboxamide isomerase, EC 5.3.1.16; E & F = probably two distinct enzymes, as yet not named; G = imidazoleglycerol-phosphate dehydratase, EC 4.2.1.19; H = histidinol-phosphate aminotransferase, EC 2.6.1.9; I = histidinol phosphatase, EC 3.1.3.15; J = histidinol dehydrogenase, EC 1.1.1.23.)

C, Fig. 9.16). The reaction sequence E and F is complex and imperfectly understood at the present time; in bacteria two genes are involved suggesting that it is catalysed by two distinct enzymes. Control of the pathway is achieved by the end-product histidine which allosterically inhibits enzyme A catalysing the first committed step.

F. BIOSYNTHESIS OF NON-PROTEIN AMINO ACIDS

Not a great deal is known about the formation of members of this large group of amino acids in plants but known pathways can involve either

(i) modification of amino acids which are components of proteins, (ii) modification of the pathways on the route to protein-amino acids or (iii) unique biosynthetic pathways. One example of each will be described. (i) 5-Hydroxytryptophan is formed in legumes by hydroxylation of tryptophan. (ii) β-pyrazol-l-yl-alanine (Formula 9.1) is formed in

$$N-CH_2 CHNH_2 COOH$$

(9.1)

cucurbits by a reaction similar to that catalysed by tryptophan synthase, in which a β-pyrazole derivative rather than indole 3-glycerol phosphate is the base which condenses with serine (see Fig. 9.16). (iii) Lathyrine is formed in *Lathyrus* spp. by the further metabolism of lysine via homoarginine and γ-hydroxyhomoarginine [eq. (9.14)] but alternative routes probably exist because label from [^{14}C]orotate and [^{14}C]uracil is incorporated into it.

(9.14)

γ-Hydroxyhomoarginine   Lathyrine

G. CYANOGENIC GLYCOSIDES

1. Nature and distribution

An important group of nitrogen-containing compounds produced by plants are carbohydrate derivatives of cyanohydrins (2-hydroxynitriles). These compounds will release hydrogen cyanide (HCN) when plants containing them (cyanophoric plants) are crushed and they come into contact with enzymes released from other regions of the plant.

The twenty or so cyanogenic glycosides so far reported in higher plants are all based on the general formula (Formula 9.2). The sugar residue is almost always D-glucose joined by an O-β-D-glucosyl linkage; R_1 is generally an aliphatic or aromatic

(9.2)

(9.3)

group and R_2 in the majority of cases is H. As R_1 and R_2 are usually different, the carbinol carbon is chiral and epimers do occur in Nature but usually not in the same plant species.

A good classification of cyanogenic glycosides is based on the amino acid which is the source of R_1 and examples are given in Table 9.3. Linamarin is one of the few non-chiral cyanogenic glycosides whilst taxiphyllin, from *Taxus* spp. and dhurrin, constitute the (R)- and (S)- isomers respectively of an epimeric pair. The final structural characteristic to be noted is the occurrence of cyclopentene rings in compounds such as gynocardin (Formula 9.3) found in *Gynocardia* spp.

The cyanogenetic glycosides are widespread and are particularly found in the families Rosaceae, Leguminosae, Gramineae, Araceae, Compositae, Euphorbiaceae and Passifloraceae.

2. Biosynthesis

Ideas on the general pattern of the biosynthesis of cyanogenic glycosides developed rapidly after the discovery that the amino acids indicated in Table 9.3 were the precursors of the glycosides and that ^{14}C–^{15}N studies indicated that the carbon–nitrogen bond of the amino acid was being incorporated intact. The general biosynthetic pathway is indicated in Fig. 9.19. The existence of an aldoxime [B, Fig. 9.19] as an intermediate was suggested by the observation that it accumulated in the presence of DL-O-methylthreonine which inhibited the formation of linamarin (Table 9.3) in flax seedlings. The formation of a N-hydroxyamino acid and its oxidative decarboxylation to the aldoxime (A and B, Fig. 9.19) have been demonstrated and the conversion to the cyanohydrin has been achieved in preparations of microsomes of *Sorghum* with L-tyrosine as substrate. Steps B, C and D are indicated by the incorporation of [^{14}C]-isobutyraldoxime,

Table 9.3. Examples of cyanogenic glucosides which arise from different amino acids

Amino acid	Cyanogenic glycoside		Typical source
Valine	Linamarin	H_3C and H_3C on carbon bearing $O-\beta Glc^*$ and $C\equiv N$	*Phaseolus lunatus*
Isoleucine	Lotaustralin	H_3C and $H_3C.H_2C$ on carbon bearing $O-\beta Glc$ and $C\equiv N$	*Lotus tenuis*
Leucine	Heterodendria	$(CH_3)_2CH.CH_2$ group, carbon bearing $O-\beta Glc$ and $C\equiv N$	*Acacia* (African)
Phenylalanine	Amygdalin	phenyl, H, carbon bearing $O-\beta Gen^\dagger$ and $C\equiv N$	*Cotoneaster* spp.
Tyrosine	Dhurrin	*p*-hydroxyphenyl (HO–), H, carbon bearing $O-\beta Glc$ and $C\equiv N$	*Sorghum*

* Glc = D-glucose.
† Gen = gentiobiose (Glc[$\beta 1 \rightarrow 6$]Glc).

isobutyronitrile and α-hydroxyisobutyronitrile into linamarin in flax seedlings and the isolation of these and analogous intermediates after feeding [14C] valine to flax seedlings and [14C]-tyrosine to cherry laurel leaves. Glycosylation of the α-hydroxynitrile is the final step in the pathway (*E*, Fig. 9.19).

3. Catabolism

The most important reactions which cyanogenic glycosides undergo are those leading to the liberation of HCN. These are most obvious when plant tissues are crushed; the glycosides released from the vacuoles come into contact with a β-glucosidase which removes the sugar yielding the α-hydroxynitrile (*A*, Fig. 9.20). This can spontaneously hydrolyse to release HCN but the reaction is greatly accelerated by the enzyme hydroxynitrile lyase (*B*).

In intact cyanophores the enzymic liberation of HCN also occurs, but much more slowly. However, in this case the cyanide is very efficiently fixed into asparagine (Fig. 9.21). The fixing enzyme β-cyanoalanine synthase uses cysteine as the second substrate and H_2S is liberated (*A*, Fig. 9.21). The β-cyanoalanine is then hydrated by the enzyme β-cyanoalanine hydroxylase (*B*, Fig. 9.21) to L-alanine.

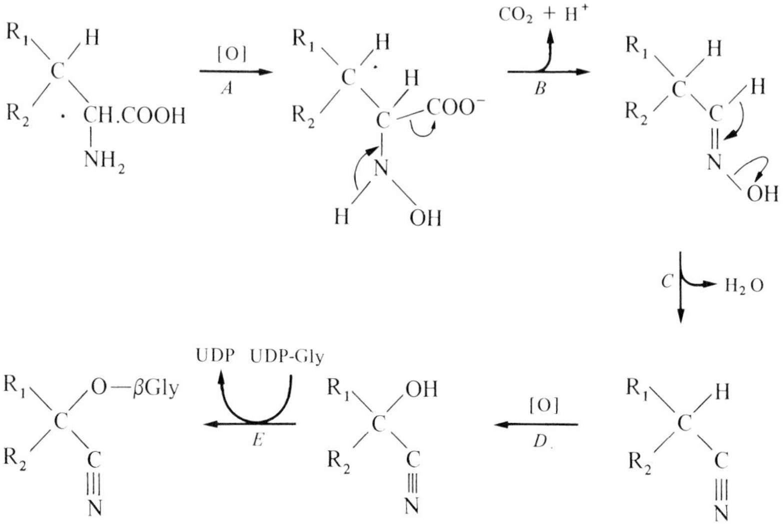

Fig. 9.19. Generalized pathway for the biosynthesis of a cyanogenic glycoside from an amino acid (see also Fig. 7.3).

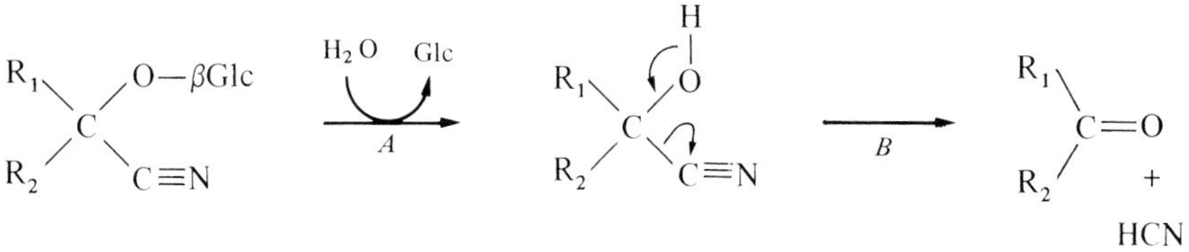

Fig. 9.20. Enzyme-catalysed breakdown of cyanogenic glycosides (Enzyme $A = \beta$-D-glucosidase; Enzyme $B =$ oxynitrilase or hydroxynitrile lyase).

H. PROTEINS†

The major source of amino acids in the cell is, of course, in the form of polymers in which the amino acids are joined covalently by peptide linkage to produce proteins. It is interesting that the classical sub-division of proteins into various groups, which is based mainly on variations in solubility, was developed in the 1930s from studies on plant proteins, particularly seed proteins. The properties of these groups are summarized in Table 9.4. This is still a good usable classification although newer analytical techniques make it difficult to allocate some proteins unequivocally to any one group. Another basic classification is into simple and conjugated proteins which contain in addition a non-protein component such as carbohydrate, lipid or metalloporphyrin. Other classifications include those based on function and on localization. None is fully satisfactory and as we are concentrating on seed and leaf proteins we shall use the classification given in Table 9.4.

1. Seed proteins

Enzymes, structural proteins and storage proteins are all found in seeds. Storage proteins often exist as discrete protein bodies which are engulfed in some plants by a semi-permeable membrane; in others they may be attached to the endoplasmic reticulum. All seed proteins are characteristic of and confined to the seeds; very sensitive immunoelectrophoretic tests have failed to demonstrate their occurrence in other parts of the plant, even the seed coat.

† The biosynthesis of proteins in plants is considered in Chapter 10.

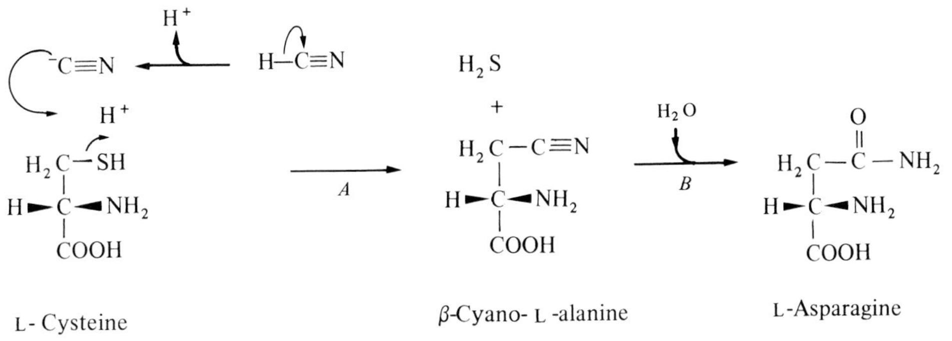

Fig. 9.21. Fixation of HCN in higher plants (Enzyme A = β-cyanoalanine synthase, EC 4.4.1.9; Enzyme B = β-cyanoalanine hydrolase).

Table 9.4. *Classification of plant proteins according to solubility*

Group name	Solubility
Albumins*	Sol. in H_2O and dilute salt
Globulins	Insol. in H_2O; sol. in dilute salt
Glutelins	Sol. in dilute acid and bases; insol. in neutral solvents
Prolamins	Sol. in 70–80% aqueous ethanol; insol. in H_2O

* Also heat-coagulable.

Table 9.5. *Quantitative distribution of amino acids in barley seed proteins*

Fraction	Higher than average	Lower than average
Albumin	glutamic acid aspartic acid leucine	amide-N
Globulin	glutamic acid glycine arginine	amide-N
Prolamin	amide-N proline	lysine
Glutelin	amide-N proline	—

If we confine ourselves to a consideration of seeds from cereals and legumes we find that the distribution pattern is different. In many cereals about 70% of the protein is gluten which consists of about equal amounts of prolamin and glutelin. The remaining protein is made up of about equal amounts of albumin and globulin. Oats is a significant exception in that globulin is the major group present.

Quantitatively the various fractions differ in the amounts of the component amino acids. The main differences in barley proteins from an average composition have been summarized in Table 9.5 and similar conclusions can be drawn for other cereals.

In all legume seeds so far examined the albumin-globulin fraction predominates and can represent more than 80% of the total protein present with globulins preponderating. Frequently prolamins and glutelins cannot be detected. In many non-leguminous dicotyledons, particularly those producing seeds rich in oil, globulins are the major proteins present.

Except in the case of glutelins the other types of seed proteins can be separated into numerous components. The albumin fractions are particularly heterogeneous, which is understandable when it is realized that almost all the enzymes of a seed are concentrated within the albumin fraction. Fewer globulins are present; four have been isolated from wheat seeds and in legumes only two major components, vicilin and legumin, are present. Vicilin probably consists of three peptide chains each with an average MW of 58,000 whereas legumin has twelve components each of average MW 33,000. Legumin has more acidic and basic amino acid residues and three times as many tryptophan residues as vicilin. Wheat glutelin is a very large molecule $(2-3 \times 10^6$ MW) but is a polymer of a small basic unit of about 20,000.

2. Leaf proteins

By far the major part of the protein of leaves is located in the chloroplast in the form of Fraction I protein which is intimately concerned with the fixation of CO_2 [see Chapter 5, Section D.2 (i)]. Chloroplast pigments are present as chromoprotein and many other proteins unique to chloroplasts are enzymes which catalyse reactions which only occur in the chloroplast. Proteins, mainly enzymes, are also present in all other organelles and in the cytoplasm of leaf cells.

3. Isoenzymes

Isoenzymes are enzymes with the same catalytic activity but with different structure. They are generally made up of a number of subunits and it is the varying combination of the subunits which gives rise to enzymes with the same catalytic activity but which can be separated by modern methods of protein chemistry. A major function of isoenzymes is in the control of metabolic activities in the cell and some of these have already been indicated in earlier sections of this chapter on amino acid biosynthesis. Frequently different isoenzymes exist in different organelles of the plant cell, but the full significance of this distribution has yet to be established.

4. Problems associated with investigations on plant proteins

A number of factors peculiar to plant tissues often make plant proteins much more difficult to deal with experimentally than animal or microbial proteins. Such factors include: (a) liberation of a highly acidic vacuolar sap on macerating the tissues: this can be counteracted to some extent by macerating in an appropriate buffer; (b) liberation of highly active oxidative enzymes on maceration, such as polyphenol oxidases which convert phenols into

quinones which then react with proteins to produce brown pigments: this can be overcome by extracting under nitrogen, by adding reducing agents such as ascorbic acid or by adding adsorbents such as insoluble polyvinylpyrrolidone; (c) liberation of proteolytic enzymes on maceration which can reduce the trichloracetic acid–precipitable protein in extracts of some young leaves by as much as 40% within 2 hrs: this can be minimized by working quickly at low temperatures; (d) precipitation of proteins by formation of complexes with tannins and phytates: this is dealt with by adding a suitable adsorbent.

SUGGESTIONS FOR FURTHER READING

Amino acids—nature and distribution

Fowden, L. (1973) in *Phytochemistry*, Vol. II (L. P. Miller, ed.), p. 1. Van Nostrand, Reinhold.
Bell, E. A. (1980) in *Encyclopaedia of Plant Physiology*, Vol. 8. p. 403. Springer.

Nitrogen fixation and amino acid biosynthesis

Miflin, B. J. and Lea, P. J. (1977) *Ann. Rev. Plant Physiology*, **28**, 299.
Shanmugam, K. T., O'Gar, F., Andersen, K. and Valentine, R. C. (1978) *Ann. Rev. Plant Physiology*, **29**, 263.

Proteins

Boulter, D. (1973) in *Phytochemistry*, Vol. II (L. P. Miller, ed.), p. 30. Van Nostrand, Reinhold.

ENZYMES

1. Reduced ferredoxin:dinitrogen oxidoreductase (ATP hydrolysing), EC 1.18.2.1.
2. NAD(P)H:nitrate oxidoreductase, EC 1.6.6.2.
3. NAD(P)H:nitrite oxidoreductase, EC 1.6.6.4.
4. L-Glutamate:ammonia ligase (ADP-forming), EC 6.3.1.2.
5. L-Glutamate:ferredoxin oxidoreductase (transaminating), EC 1.4.7.1.
6. L-Glutamate:NADP$^+$ oxidoreductase (transaminating), EC 1.4.1.13.
7. L-Glutamate:NAD(P)$^+$ oxidoreductase (deaminating), EC 1.4.1.3.
8. L-Aspartate:ammonia ligase (ADP-forming), EC 6.3.1.4.
9. L-Asparagine amidohydrolase, EC 3.5.1.1.
10. Acetolactate pyruvate-lyase (carboxylating), EC 4.1.3.18.
11. 2,3-Dihydroxy-isovalerate NADP$^+$ oxidoreductase (isomerizing), EC 1.1.1.86.

12. 2,3-Dihydroxyacid hydro-lyase, EC 4.2.1.9.
13. L-Valine:3-methyl-2-oxovalerate aminotransferase, EC 2.6.1.32.
14. L-Leucine:2-oxoglutarate aminotransferase, EC 2.6.1.6.
15. 3-Hydroxy-4-methyl-3-carboxyvalerate 2-oxo-3-methyl: butyrate-lyase (CoA-acetylating), EC 4.1.3.12.
16. L-Threonine hydro-lyase (deaminating), EC 4.2.1.16.
17. ATP:L-aspartate 4-phosphotransferase, EC 2.7.2.4.
18. L-Aspartate-β-semialdehyde:NADP$^+$ oxidoreductase (phosphorylating), EC 1.2.1.11.
19. L-Aspartate-β-semialdehyde hydro-lyase (adding pyruvate and cyclizing), EC 4.2.1.52.
20. N-Succinyl-L-2,6-diaminopimelate:2-oxoglutarate aminotransferase, EC 2.6.1.17.
21. N-Succinyl-L,L-2,6-diaminopimelate amidohydrolase, EC 3.5.1.18.
22. 2,6-L,L-Diaminopimelate 2-epimerase, EC 5.1.1.7.
23. meso-2,6-Diaminopimelate carboxy-lyase, EC 4.1.1.20.
24. L-Alanine:2-oxoglutarate aminotransferase, EC 2.6.1.2.
25. L-Homoserine:NAD(P)$^+$ oxidoreductase, EC 1.1.1.3.
26. ATP:L-homoserine O-phosphotransferase, EC 2.7.1.39.
27. Cystathionine L-homocysteine-lyase (deaminating), EC 4.4.1.8.
28. 5-Methyltetrahydropteroyl-tri-L-glutamate:L-homocysteine S-methyltransferase, EC 2.1.1.14.
29. L-Cystathionine cysteine-lyase (deaminating), EC 4.4.1.1.
30. L-Asparate:L-glutamine amido-ligase (AMP forming), EC 6.3.5.4.
31. Acetyl-CoA:L-glutamate N-acetyltransferase, EC 2.3.1.1.
32. N^2-Acetyl-L-ornithine:L-glutamate N-acetyltransferase, EC 2.3.1.35.
33. ATP:N-acetyl-L-glutamate 5-phosphotransferase EC 2.7.2.8.
34. Carbamoylphosphate:L-ornithine carbamoyltransferase, EC 2.1.3.3.
35. Carbamoylphosphate:L-aspartate carbamoyltransferase, EC 2.1.3.2.
36. L-Arginine amidinohydrolase, EC 3.5.3.1
37. Urea amidohydrolase, EC 3.5.1.5.
38. D-Glycerate-3-phosphate phosphohydrolase, EC 3.1.3.38.
39. L-Serine:glyoxylate aminotransferase, EC 2.6.1.45.
40. 5,10-Methylene-tetrahydrofolate: glycine hydroxymethyltransferase, EC 2.1.2.1.
41. O-Acetyl-L-serine acetate-lyase (adding hydrogen sulphide), EC 4.2.99.8.
42. Chorismate pyruvatemutase, EC 5.4.99.5.
43. Prephenate hydro-lyase (decarboxylating), EC 4.2.1.51.
44. L-Serine hydro-lyase (adding indoleglycerol phosphate), EC 4.2.1.20.

CHAPTER 10

Purines, Pyrimidines, Nucleic Acids, Protein Synthesis

CONTENTS

A. INTRODUCTION

Purines and pyrimidines exist in plants not only as the free bases, but also as nucleosides (*N*-glycosides and occasionally *C*-glycosides of the bases) and nucleotides (phosphate esters of nucleosides). The fundamental importance of the polynucleotides DNA and RNA have perhaps led to lack of appreciation of the biological significance of simple nucleotides other than ATP but they play an essential part in metabolism as components of many 'activated' molecules, for example 'active sulphur' (8.35) and coenzyme-A (8.14). This chapter is divided into three main sections, one dealing with the bases, their nucleosides and nucleotides, the second with the polynucleotides, DNA and RNA, and the third with protein biosynthesis.

B. PURINES

1. Nature and distribution

(i) BASES

Purines which are components of nucleic acids

are adenine (10.1) and guanine (10.2).† However, it is not surprising to find other purines in plants occurring in considerable amounts; these include the *N*-methylated bases with importance for man because of their pharmacological properties; examples are theobromine (10.3), which represents up to 3% of cocoa (*Theobroma cacao*) beans, and caffeine (10.4), which can make up 1.5% of coffee beans. All these compounds are derivatives of xanthine (10.5), which also exists in coffee beans.

† Oxygen-containing purine (e.g. guanine) and pyrimidine bases exist as an equilibrium mixture of keto (lactam) and enol (lactim) forms in an aqueous environment. The position of the equilibrium depends upon the pH; the lactam form predominates at the neutral or mildly acidic pHs of plant tissues.

Perhaps the most interesting free purines are the isopentenyl derivatives of adenine some of which are also integral components of *t*RNA (Section D.8). Two typical examples are zeatin (10.6) and *N*[6]-(*trans*-4-hydroxy-3-methylbut-2-enyl)-2-methylthioadenine (10.7) which interestingly contains a sulphydryl residue. These and related compounds promote cell division but also affect other physiological functions (see Chapter 15, Section D).

Uric acid (10.8), allantoin (9.1) and allantoic acid (9.2), which have long been established as end products of nucleic acid metabolism in mammals, are widely distributed in higher plants. However, as indicated in the previous chapter allantoin and allantoic acid are considered important storage

(Keto or lactam form)   (Enol or lactim form)

(10.1)   (10.2)

(10.3)   (10.4) ✓   (10.5)

(10.6)   (10.7)   (10.8)

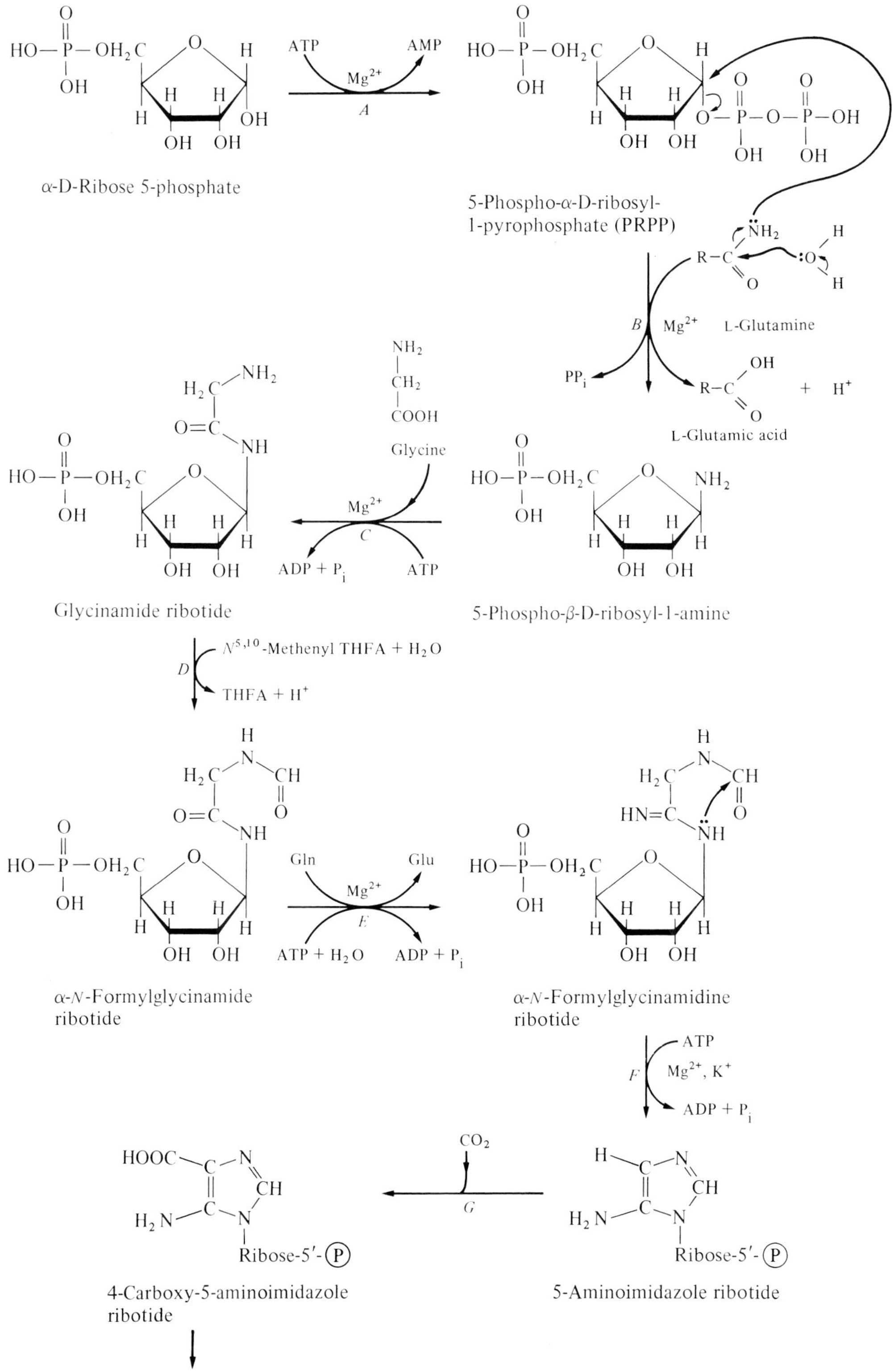
α-D-Ribose 5-phosphate
ATP
AMP
Mg2+
A
5-Phospho-α-D-ribosyl-1-pyrophosphate (PRPP)
B
Mg2+
L-Glutamine
PPi
L-Glutamic acid
+ H+
5-Phospho-β-D-ribosyl-1-amine
Glycine
C
ADP + Pi
ATP
Glycinamide ribotide
D
N5,10-Methenyl THFA + H2O
THFA + H+
α-N-Formylglycinamide ribotide
Gln
Glu
E
ATP + H2O
ADP + Pi
α-N-Formylglycinamidine ribotide
F
ATP
Mg2+, K+
ADP + Pi
5-Aminoimidazole ribotide
CO2
G
4-Carboxy-5-aminoimidazole ribotide

continued on facing page

Fig. 10.1. For legend see facing page.

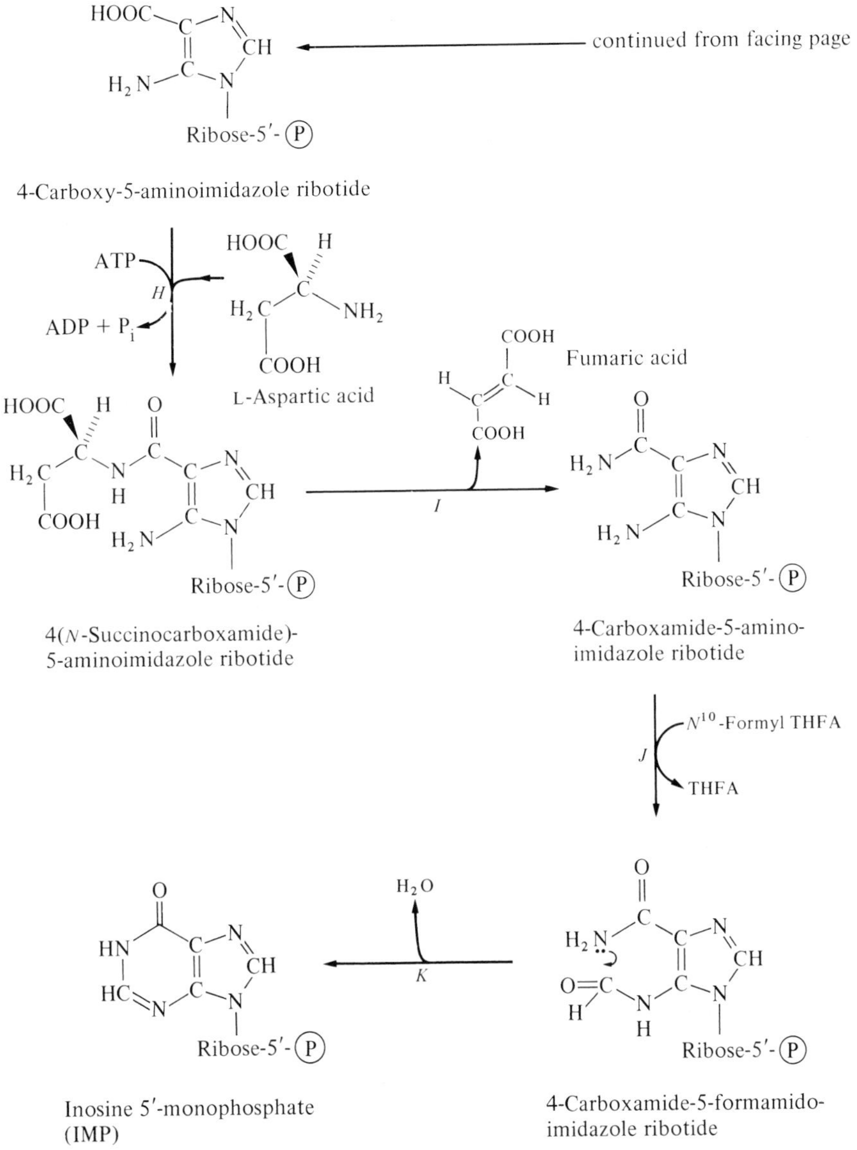

4-Carboxy-5-aminoimidazole ribotide

4(N-Succinocarboxamide)-
5-aminoimidazole ribotide

4-Carboxamide-5-amino-
imidazole ribotide

Inosine 5'-monophosphate
(IMP)

4-Carboxamide-5-formamido-
imidazole ribotide

Fig. 10.1. Biosynthesis of inosine 5'-monophosphate (IMP).

Biosynthesis of inosine 5'-monophosphate (IMP). (The enzymes catalysing reactions *A–K* are:

A = Ribosephosphate pyrophosphokinase, EC 2.7.6.1.
B = Amidophosphoribosyltransferase, EC 2.4.2.14.
C = Phosphoribosylglycinamide synthetase, EC 6.3.4.13.
D = Phosphoribosylglycinamide formyltransferase, EC 2.1.2.2.
E = Phosphoribosylformylglycinamidine synthetase, EC 6.3.5.3.
F = Phosphoribosylaminoimidazole synthetase, EC 6.3.3.1.
G = Phosphoribosylaminoimidazole carboxylase, EC 4.1.1.21.
H = Phosphoribosylaminoimidazole succinocarboxamide synthetase, EC 6.3.2.6.
I = Adenylosuccinate lyase, EC 4.3.2.2.
J = Phosphoribosylaminoimidazolecarboxamide formyltransferase, EC 2.1.2.3.
K = IMP cyclohydrolase, EC 3.5.4.10.
      Abbreviations: Glu = L-glutamic acid; Gln = L-glutamine; THFA = tetrahydrofolic acid—see Fig. 9.11.)

forms of nitrogen in many plant species, as for example in maple.

(ii) NUCLEOSIDES

A typical purine nucleoside which occurs in higher plants is crotonoside (the N^9-riboside of isoguanine) (10.9). The function of such compounds is unknown but a number of antibiotics produced by fungi are nucleosides; an important one is puromycin (10.10).

(iii) NUCLEOTIDES

Many purine nucleotides have a key role to play in the metabolic activities of higher plants. Those already discussed in earlier chapters include ATP, NAD, FAD, GTP and sugar nucleotides such as ADP-glucose. An important regulatory nucleotide, over the existence of which in higher plants there has been some controversy, is cyclic AMP (10.11) (adenosine-3',5'-cyclic phosphate). However, it now seems clear that it is present in plants, but its metabolic role remains undecided.

2. Biosynthesis

(i) RIBOSIDE MONOPHOSPHATES†

Purines are synthesized *de novo* at the nucleoside

† All reaction letters in this section apply to Fig. 10.1.

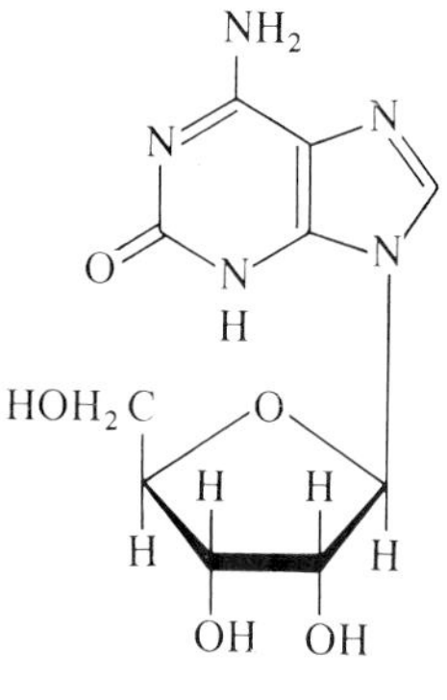

monophosphate level and the first compound formed with the purine ring structure is inosine 5'-monophosphate (IMP). The pathway leading to IMP is outlined in Fig. 10.1. After the conversion of ribose-5-phosphate into phosphoribosyl pyrophosphate (PRPP) (A), the first committed step is a glutamine-dependent step in which pyrophosphate is replaced by the amide-NH_2 of glutamine (B). In this reaction C–1 of PRPP undergoes inversion from the α- to β-configuration and this configuration is then retained throughout the biosynthetic sequence. A reaction similar to (B) is also involved in reaction (E) in the formation α-N-formylglycinamide ribotide. Reaction (C) involves the ATP-dependent attachment of glycine to 5-phosphoribosylamine to produce glycinamide ribotide; a C_1 transfer involving 5,10-methenyl tetrahydrofolic acid then yields α-N-formylglycinamide ribotide (D); after the addition of NH_2 already mentioned (E) an ATP-dependent reaction brings about ring closure to form 5-aminoimidazole ribotide (F). CO_2 fixation then occurs to yield 4-carboxy-5-aminoimidazole ribotide. The newly introduced carboxyl group is then aminated to form 4-carboxamide-5-aminoimidazole ribotide in a two-step process (H and I); L-aspartic acid contributes the amino group (H) and is converted into fumaric acid (I). The amination is mechanistically similar to the L-citrulline→L-arginine conversion (Fig. 9.12, reactions I and J) in the modified urea cycle described in Chapter 9. This is followed

(10.9)

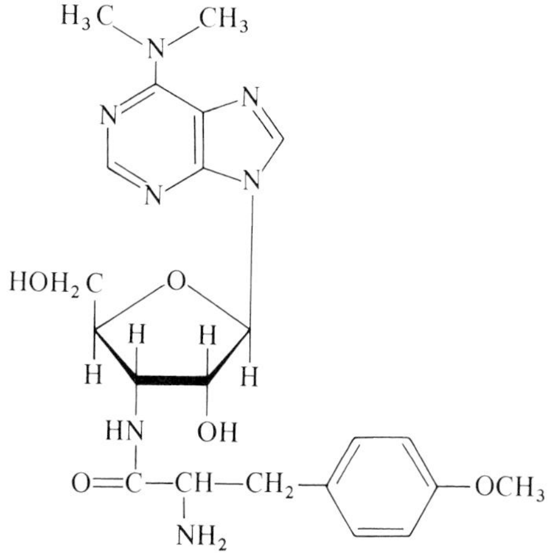

(10.10)

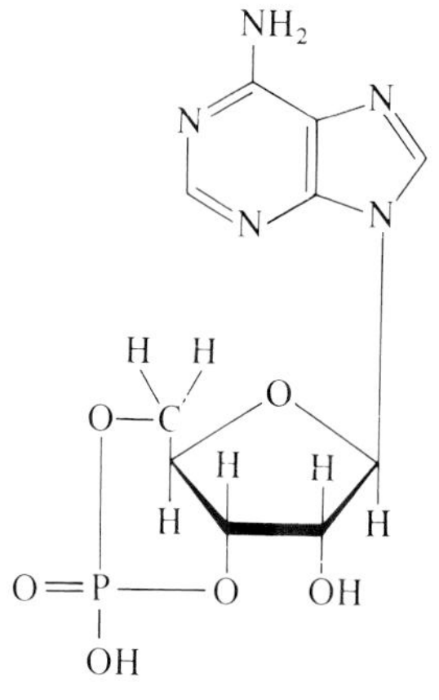

(10.11)

by the addition of another C_1 unit, this time from N^{10}-formyl tetrahydrofolic acid, to yield 4-carboxamide-5-formamidoimidazole ribotide (J). The latter spontaneously ring closes with the elimination of water to form inosine 5'-monophosphate (IMP, inosinic acid) (K); the reaction is accelerated by the enzyme IMP cyclohydrolase[1].

The pathway accounts completely for the labelling pattern found in purines in classical experiments in which the incorporation of different substrates was investigated (Fig. 10.2). Control of the pathway is exerted at two points; the enzyme catalysing reaction (A) is allosterically inhibited by ADP and GDP whilst that catalysing reaction (B), the first committed step, is allosterically inhibited by purine nucleotides, as would be expected.

The bases adenine and guanine are formed, at the ribotide level, from IMP (Fig. 10.3). Adenosine 5'-monophosphate (AMP) is formed by the amination of C–6 of IMP in a two-step process, involving L-aspartic acid, which is analogous to steps H and I of Fig. 10.1; however, in this instance GTP rather than ATP is required as the energy source. The formation of guanosine 5'-monophosphate (GMP) involves firstly the NAD^+-dependent oxidation of C–2 of IMP to yield xanthosine 5'-monophosphate (XMP). This is followed by the ATP-dependent amination of C–2 of XMP, the amino group being derived from the amide group of L-glutamine.

(ii) RIBOSIDE TRIPHOSPHATES

In order to make purines available for synthesis of RNA they need to be in the form of the triphosphates ATP and GTP. The reactions leading to the formation of these compounds involve the enzymes nucleoside monophosphate kinase[2] [eq. (10.1)] and nucleoside diphosphate kinase[3] [eq. (10.2)], where B represents adenosine or guanosine.

$$BMP + ATP \rightleftharpoons BDP + ADP \qquad (10.1)$$
$$BDP + ATP \rightleftharpoons BTP + ADP \qquad (10.2)$$

$$\text{Sum } BMP + 2ATP \rightleftharpoons BTP + 2ADP$$

(iii) DEOXYRIBOTIDES

Purine bases must also be available in the form of deoxynucleotides for the synthesis of DNA. Very little is known about this conversion in plants but it is clear that in animals and bacteria the reduction occurs at the nucleotide level because when a nucleotide is labelled with ^{14}C both in the base (in the classical case, a pyrimidine) and in the ribose and fed to an organism, then in the deoxynucleotide from the isolated DNA the relative amount of ^{14}C in the base and deoxyribose was the same as in the starting material.

In a well-investigated pathway in E. coli it is clear that the reduction takes place at the diphosphate

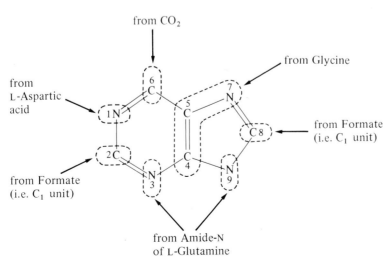

Fig. 10.2. Biosynthetic source of the atoms of the purine ring system.

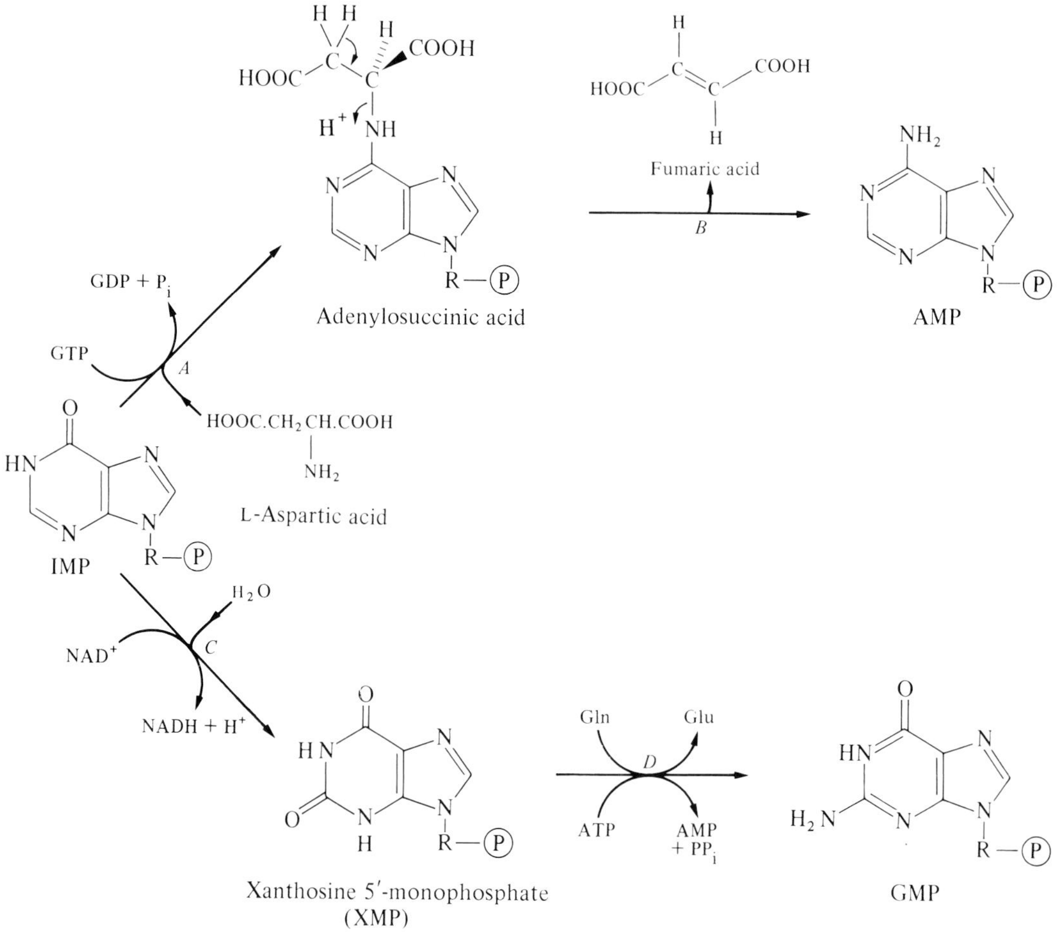

Fig. 10.3. The formation of adenosine 5'-monophosphate (AMP) and guanosine 5'-monophosphate (GMP) from IMP. (The enzymes catalysing reactions A–D are:
A = Adenylosuccinate synthetase, EC 6.3.4.4.
B = Adenylosuccinate lyase, EC 4.3.2.2.
C = IMP dehydrogenase, EC 1.2.1.14.
D = GMP synthetase.
Abbreviations: Glu = L-glutamic acid; Gln = L-glutamine.)

level [eq. (10.3)] and is catalysed by a multienzyme complex. One component is a small acidic, heat-stable, dithiol protein, thioredoxin. It has a MW of 12,000, contains 108 amino acid residues and has an active centre consisting of

$$\overline{S\text{---}S}$$
-Trp-Cys$_{32}$-Gly-Pro-Cys$_{35}$-Lys-Met-

which is invariant in thioredoxin from E. coli, yeast and ox liver. It is the immediate source of reducing power for deoxynucleotide formation. The second

component is ribonucleoside diphosphate reductase[4], which consists of two subunits and requires Mg^{2+} to carry out the reaction indicated in eq. (10.3).

$$BDP + Thioredoxin(-SH\ HS-) \xrightarrow{Mg^{2+}}$$
$$2'\text{-deoxy-BDP} + Thioredoxin(-S\text{---}S-) + H_2O \tag{10.3}$$

Oxidized thioredoxin is reduced to the —SH form by NADPH in the presence of another component of the multienzyme system, a flavoprotein called thioredoxin reductase[5] [eq. (10.4)].

$$Thioredoxin(-S-S-)+NADPH+H^+ \rightleftharpoons$$
$$Thioredoxin(-SH\ HS-)+NADP^+$$
$$(10.4)$$

Other systems found in bacteria appear to require the nucleotide triphosphates (BTP) rather than the diphosphates (BDP) as their substrates.

C. PYRIMIDINES

1. Nature and distribution

(i) BASES

The pyrimidines that are components of nucleic acids are uracil (10.12), cytosine (10.13) and thymine (10.14). Methylcytosine (10.15) is a fairly common minor component in plant nucleic acids.

(ii) NUCLEOSIDES

Nucleosides of pyrimidines are not very widespread in plants and indeed have only been clearly identified in certain *Streptomyces* spp. and sponges.

An example is spongothymidine (10.16). 5-*O*-glycosides are rather more frequently encountered in higher plants as, for example, vicine, which is 2,4-diamino-6-oxopyrimidine-5-*O*-(β-D-glucopyranoside) (10.17), and perhaps of more interest, pseudo-uridine (ψ-uridine, 10.18) which is a normal constituent of transfer RNAs from many plant sources.

2. Biosynthesis†

As with purine nucleotides the first pyrimidine nucleotide formed is a mononucleotide (uridine monophosphate, UMP), but the sugar phosphate is added later in the pathway and not at the beginning (Fig. 10.4).

In the first step (*A*) one molecule each of ammonia and carbon dioxide combine to form carbamoyl phosphate; two ATPs are required. The next step is the first committed step in which carbamoyl phosphate is transferred to aspartate to produce carbamoyl asparate (*B*). Dihydro-orotase (*C*) catalyses ring closure with the elimination of water to form dihydro-orate. Oxidation to orotate

† All letters in this section apply to Fig. 10.4.

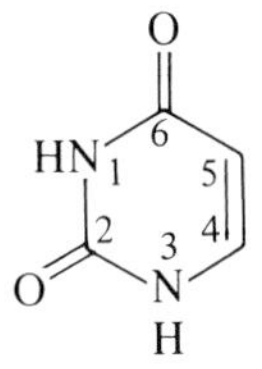

(10.12) 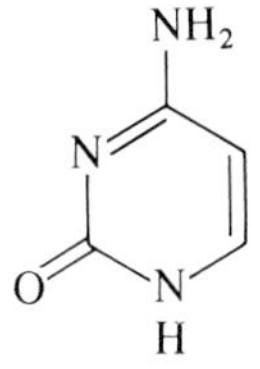   (10.13) 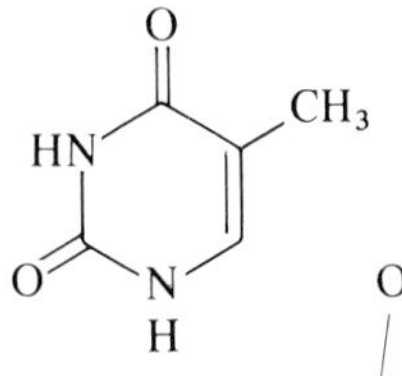   (10.14) 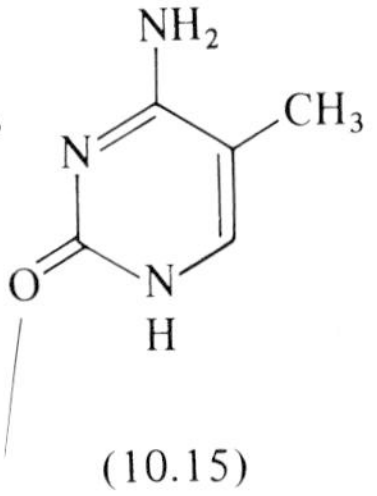   (10.15)

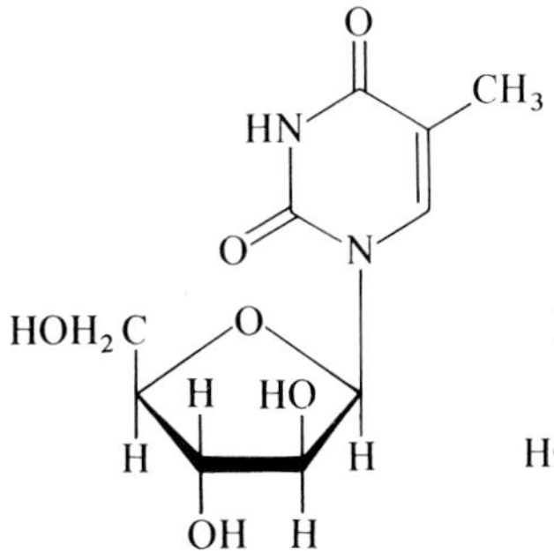

(10.16) 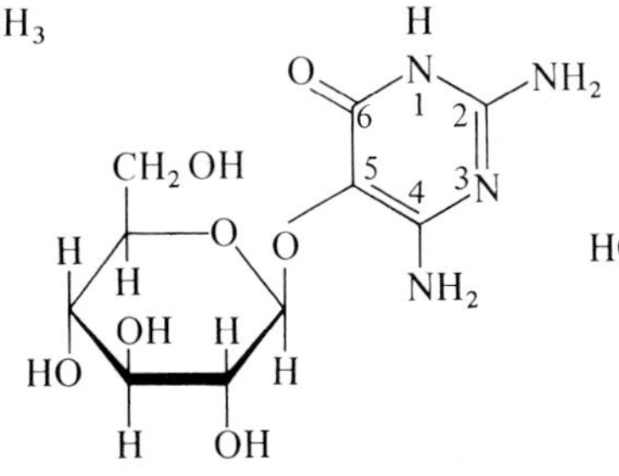   (10.17) 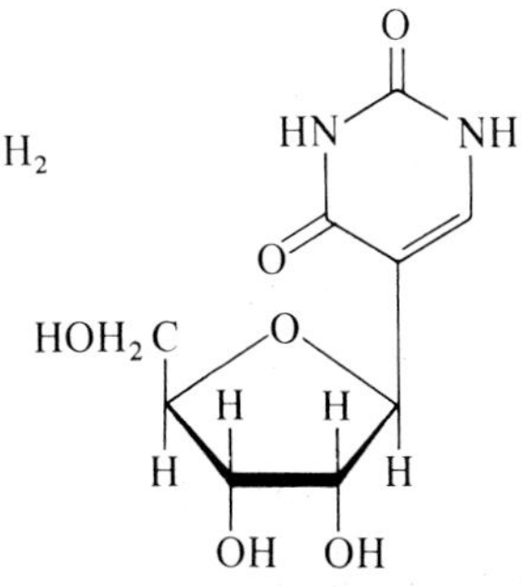   (10.18)

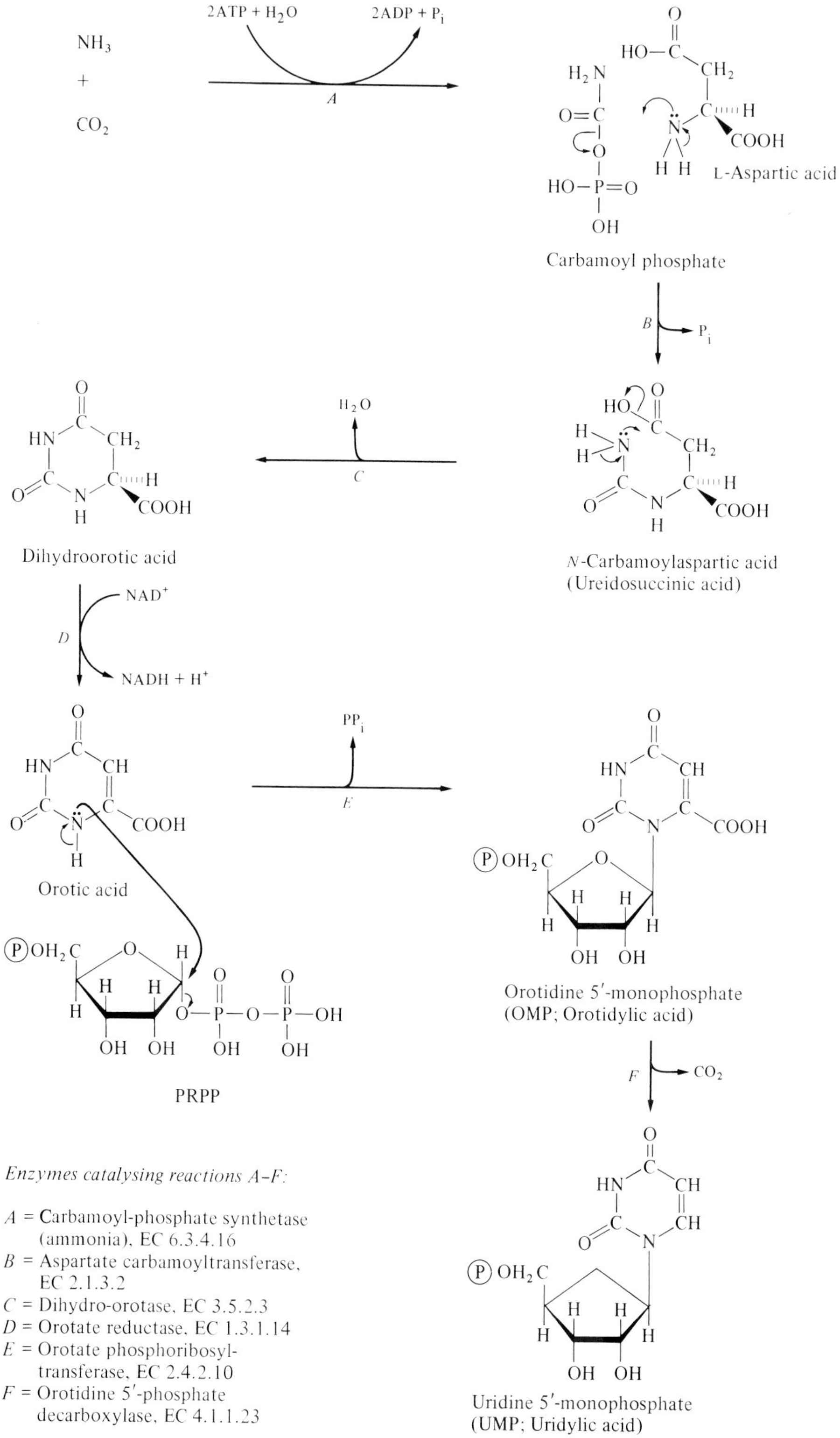

Enzymes catalysing reactions A–F:

A = Carbamoyl-phosphate synthetase
    (ammonia), EC 6.3.4.16
B = Aspartate carbamoyltransferase,
    EC 2.1.3.2
C = Dihydro-orotase, EC 3.5.2.3
D = Orotate reductase, EC 1.3.1.14
E = Orotate phosphoribosyl-
    transferase, EC 2.4.2.10
F = Orotidine 5'-phosphate
    decarboxylase, EC 4.1.1.23

Fig 10.4. Biosynthesis of uridine 5'-monophosphate (UMP).

(D) is followed by transfer of the phosphoribosyl group from PRPP to form orotidine 5′-phosphate (OMP, orotidylic acid) (E). As in the case of purine nucleotide synthesis the configuration at C–1 of PRPP is inverted to form a β-linkage with the base. Removal of CO_2 (F) results in the formation of UMP. For its further conversion into UDP and UTP reactions indicated in eqs. (10.1) and (10.2) are utilized.

Control of the synthesis is at step B, but the modulators vary according to the organism studied; in some higher plants it is UMP whereas in *Pseudomonas fluorescens* it is UTP. In some bacteria no control can be observed at this stage.

The formation of cytidine triphosphate involves not the kinases used for UTP formation but a specific reaction catalysed by CTP synthetase[6] [eq. (10.5)] for which UTP is the substrate.

$$UTP + NH_3 + ATP \xrightarrow{Mg^{2+}} CTP + ADP + P_i$$
$$(10.5)$$

Deoxy-CTP is formed as indicated for the purine deoxyribotides; as is well known, uracil does not occur in DNA but is replaced by thymine (5-methyl-uracil). The substrate is deoxy-UMP, formed by dephosphorylation of deoxy-UDP, and the methyl donor is $N^{5,10}$-methylene-THFA; the enzyme is thymidylate synthase[7] [eq. (10.6)]. Thymidine 5′-monophosphate (TMP) is converted into thymidine 5′-triphosphate (TTP) by the usual reactions [i.e. eqs. (10.1) and (10.2)]. The formation of ψ-uridine takes place *in situ* in the *t*RNA molecule, but the mechanism involved is not known.

D. DNA AND RNA

1. Introduction

The discovery that the polynucleotide DNA is the universal genetic material and the unravelling of the major details of its structure and of how it is replicated represent an enormous achievement of the last 20 years or so. However, when one considers that in addition we now understand how the information is transcribed by specifying the base sequence in an RNA molecule which is then used to translate the unique sequence of amino acids into each individual protein, we can safely say that we are contemplating the greatest scientific achievements of the second half of the twentieth century. These achievements have revolutionized biological thinking to at least as great an extent as Darwin did in the nineteenth century with his *Origin of Species*. In a textbook of this kind it is a problem to strike the right balance between the enormous amount of general information on the structure and function of DNA and RNA and that which is specific to plants. To contain this chapter within reasonable size we will assume somewhat more knowledge of the basic facts of this area of biochemistry than we have done in other areas and give references to more general texts at the end of the chapter. In the first sections of this chapter we have already discussed the basic building units of the polynucleotides so we can start immediately with a consideration of DNA.

$N^{5,10}$-methylene THFA

Dihydrofolic acid

$$(10.6)$$

dUMP

TMP

2. Structure of DNA

DNA was first discovered in 1869 by F. Miescher but indications that it carried genetic information only began to appear in the 1940s, firstly by Avery in studying smooth and rough colonies of a pneumococcus bacterium and later in the 1950s by the famous experiments of Hershey and Chase. It rapidly became a reality following the elucidation of the double helix structure of DNA by Watson and Crick. The primary structure of the component polynucleotides is a series of deoxynucleotides joined through $3' \rightarrow 5'$ phosphodiester linkages. Figure 10.5 gives the structure of a polynucleotide and indicates the shorthand representation used in most discussions of nucleic acid structures. The sequencing of the nucleotides in the DNA is now well advanced in many species (such that the entire sequence of some small genomes has now been deduced) but it will not be dealt with in detail here. The double helix, which is the secondary structure of DNA, is represented by two polynucleotide chains coiled helically in a right-handed configuration

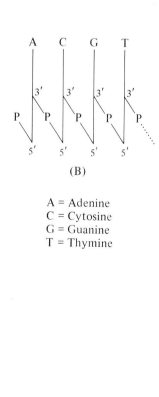

A = Adenine
C = Cytosine
G = Guanine
T = Thymine

(A)

Fig. 10.5. The primary structure of a polydeoxyribonucleotide chain (A) and its shorthand representation (B).

around the same axis. The two chains are *antiparallel*, that is their $3' \rightarrow 5'$ phosphodiester linkages run in opposite directions. The two chains are *plectonemic*, that is they can only be separated by uncoiling the helix. The bases are stacked on the inside of the helix with their planes parallel to each other and perpendicular to the long axis. Only certain base pairs can fit inside this structure so that they can bind to each other by hydrogen bonds. The permitted pairs are *A*denine and *T*hymine, and *G*uanine and *C*ytosine: the exact way thymine and adenine fit is illustrated in Fig. 10.6. The bases are stacked at a centre-to-centre distance of 0.34 nm from each other and there are ten residues per complete turn of the double helix. The stability of the helix is achieved by base stacking forces which are the result of the hydrophobic nature of the bases themselves. This makes contact with the aqueous environment outside the molecule energetically unfavourable and contact is reduced still further by base stacking itself. The hydrogen bonding between bases does confer some additional stability but its significance is mainly in the structural specificity it achieves between complementary bases.

The nature of the double helix demands that in any DNA molecule the sum of A + G equals that of T + C and that A equals T and G equals C. Indeed it was this type of pattern revealed by diligent analytical chemistry which provided the experimental basis of the double-helix concept. In wheat germ DNA, for example, A + G = 50.0, T + C[+ 5-methylcytosine (*10.15*)] = 49.0; A = 27.3, T = 27.1; G = 22.7, C = 16.8, 5-methyl-C = 6.0, moles per 100 g atoms P. Wheat germ DNA falls into the 'A–T rich' type of molecule which is a much commoner type than the 'G–C rich' type. The wide recorded variation in the proportions of bases in DNAs from different sources is to be expected, but the total DNA extracted from the same species always has the same base ratio.

It is very difficult to extract native DNA from chromosomes but intact molecules have been isolated from bacteria, viruses, mitochondria and chloroplasts. These molecules show that the double helix can be further constrained to form tertiary structures such as the double-stranded, circular DNA which has been isolated from chloroplasts. Figure 10.7 indicates three main types of cyclic DNA, (i) *A* is a covalently closed cyclic DNA which contains twelve superhelical turns, (ii) *B* is an open cyclic DNA with only one strand breakage at X, and (iii) *C* is a combination of *A* and *B* in the form of a cantenane. The supercoiled DNA (*A*) can be uncoiled or 'relaxed' by breaking one covalent link in the strand; this is called 'nicking'.

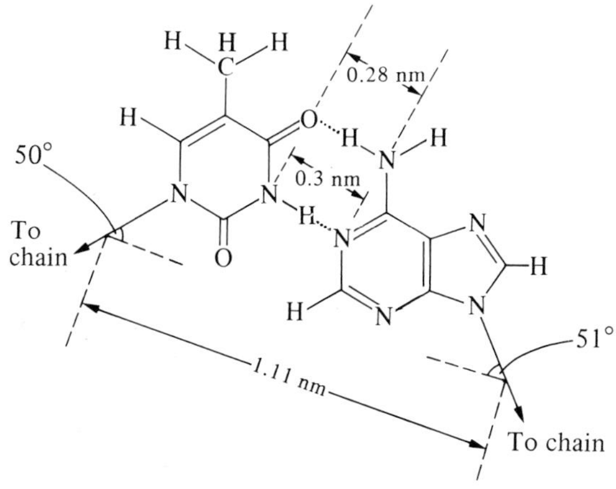

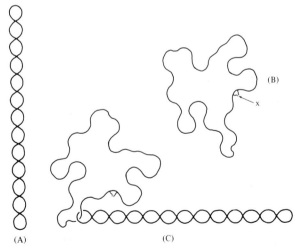

Fig. 10.7. Some typical cyclic DNA molecules.
(*A* = Covalently closed duplex in the form of a superhelix containing twelve turns.
B = Open cyclic molecule with one strand broken at X.
C = A cantenane containing type *A* linked with type *B*.)
Reproduced, with permission, from Davidson, J. N., *Biochemistry of the Nucleic Acids*, 8th edition, revised by Adams, R. L. P., Burdon, R. H., Campbell, A. M. and Smellie, R. M. S., Chapman & Hall.

Fig. 10.6. Scale drawing of the adenine–thymine hydrogen-bonded base pair.

(i) NUCLEAR DNA

The number of chromosomes in a nucleus varies according to species, for example there are fourteen in peas and twenty in maize. Each of these carries a large DNA molecule (~ 2 nm diam.) combined with protein to give a 3 nm diam. fibre which can be aggregated and condensed to give thicker and tighter complexes. In interphase nuclei these DNA–protein complexes are known as chromatin (see Chapter 3, Section B.4). The proteins involved are to a great extent the basic histones. The DNA itself can be considered as composed of (a) unique sequences of bases, (b) 'moderately' repeated sequences and (c) highly repeated sequences. Type (a) sequences appear to code for proteins with specialized functions. A small proportion of type (b) sequences codes for histones and ribosomal RNA whilst type (c) sequences have no clearly delineated biological function at present. In some mammalian cells these highly repeated sequences are found in blocks of several units about 6–8 base pairs long. As these sequences often have a different $G + C$ content from that of the other components, they separate from them on a caesium chloride gradient to give rise to satellite DNA. A final important feature of nuclear DNA is that it contains a considerable number of *palindromic regions*, that is regions with inverted repetitions thus:

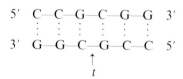

Point *t*, known as the turnabout point, represents the symmetry axis of a structure which possesses two-fold rotational symmetry.

(ii) MITOCHONDRIAL DNA

Mitochondrial DNA is different from that found in the nucleus in that it is usually a small double-stranded circular molecule (MW $\sim 10^7$) and it has a different base composition and different buoyant density ($\delta = 1.706$, compared with 1.698 in pea nuclei). It represents about 10% of the total cellular DNA and codes for some mitochondrial proteins and for mitochondrial ribosomal RNA and transfer RNA.

(iii) CHLOROPLAST DNA

It was in the 1960s that it became clear that chloroplasts contained a unique species of DNA (*Ct*DNA) with a buoyant density (δ) different from that of nuclei. The *Ct*DNA δ value is generally constant in higher plants at about 1.694–1.698. This can be compared with an average value 1.695 for the corresponding nuclear DNA; however, nuclear DNA δ values cover the range 1.691–1.702. The buoyant density suggests a constant GC content for *Ct*DNAs from different species of plant but this does not necessarily mean a common coding structure. Fragmentation of various *Ct*DNAs with EcoRI endonuclease produced no common fragments. Two other important properties of *Ct*DNA is the absence of methylated bases in the molecule and its relative ease of renaturation. *Ct*DNA is a large circular molecule with molecular weight varying according to origin from 85×10^6 (*Sphaerocarpus*; liverwort) to 143×10^6 (*Chlamydomonas*; alga) with higher plant values falling within the range 90–100×10^6. The circular molecule can exist as a covalently closed super-helix (Fig. 10.7) and contains in the case of the material from peas approximately eighteen *ribo*nucleotides. This discovery was based on the fact that there are alkali-labile sites, characteristic of ribonucleotide linkages (see Section D.3), in the closed circular DNA; furthermore, the action of pancreatic RNase and RNase T_1, which nick the molecule and convert covalently closed circular DNA into open circular DNA, indicates that the ribonucleotides are covalently inserted. The significance of this important finding remains to be elucidated. Further discussion of the chloroplast genome is given in Section E.9.

3. Structure of RNA

There are various types of polyribonucleotides in cells: ribosomal (*r*RNA), transfer (*t*RNA), and messenger (*m*RNA). These have very characteristic structures which are best discussed later in the chapter when protein synthesis is being described. However, they all have a common basic structure; they consist of a series of ribonucleotides joined through a $3' \rightarrow 5'$-phosphodiester linkage. That is they have the same basic primary structure as DNA.

This was rather more difficult to establish than in DNA because the deoxynucleotides lack a functional hydroxyl group at C–2'. A key property which led to the elucidation of the structure was the discovery of the nature of the reaction which rendered the RNA molecule alkali-labile. The first step in the reaction is the production of a cyclic nucleoside 2',3'-phosphate which then hydrolyses to yield a mixture of nucleoside 2'-phosphates and nucleoside 3'-phosphates (Fig. 10.8). A nucleotide which cannot form a cyclic 2',3'-phosphate is therefore stable to alkali. This is why the DNA molecule is stable.

The nature of the secondary and tertiary structures of RNA will be discussed later.

4. DNA replication

Cell division produces new cells containing an exact copy of the original genetic material, that is the new DNA is an exact replica of the original DNA. This complex phenomenon has been studied with great intensity during the past 20 years and although the overall picture is clear enough many subtle details of the process, particularly in plants, are still obscure. There is no doubt from the classical experiments of Meselson and Stahl that replication is *semiconservative*. These investigators showed that a DNA double helix can be untwisted to form two separate chains each of which can act as a template for the synthesis of a new complimentary strand to produce two daughter DNA double helices, with one strand of each daughter being derived from the original DNA helix and the other being a product of new synthesis. The Meselson–Stahl experiment consisted of growing E. coli in $^{15}NH_4Cl$ (96.5% ^{15}N) for fourteen generations which resulted in the DNA being almost completely labelled with ^{15}N. The cells were then transferred to a medium containing normal NH_4Cl and the constituent DNA was examined for several generations on a caesium chloride gradient. The observed pattern (Fig. 10.9) showed that after one generation hybrid $^{15}N,^{14}N$-DNA was present which separated into ^{14}N and ^{15}N strands on heating; in later generations the ^{14}N strands

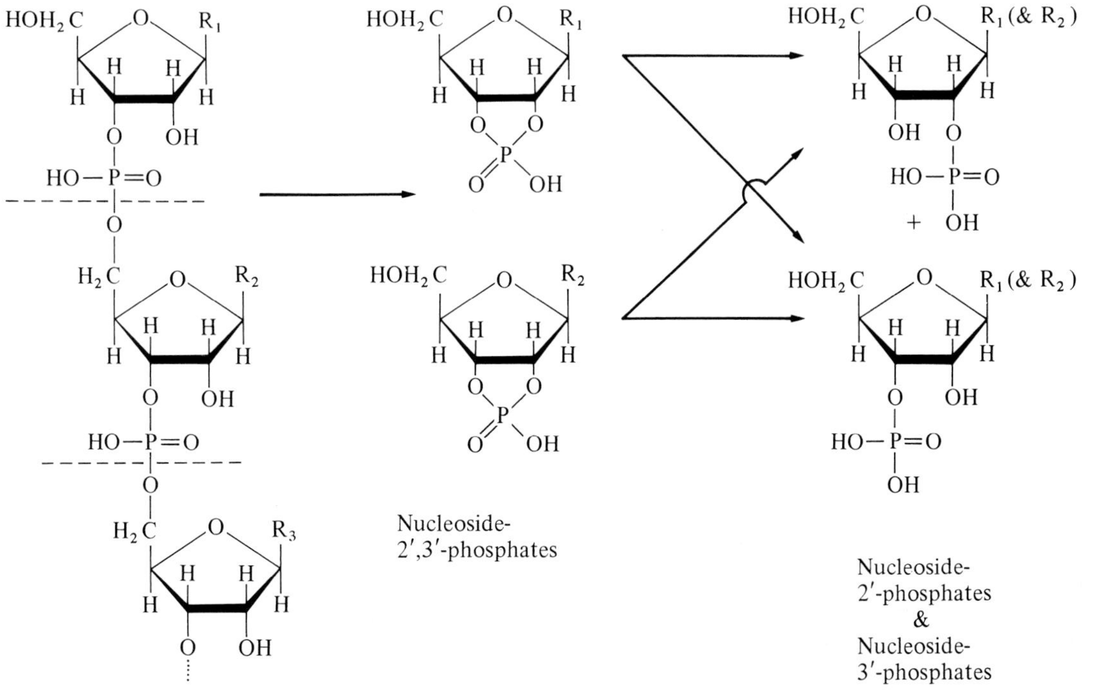

Fig. 10.8. The action of dilute alkali on a polyribonucleotide.

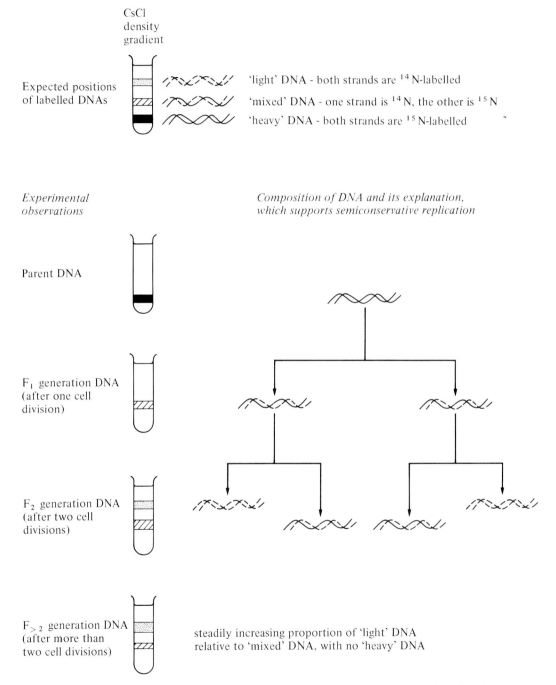

CsCl density gradient

Expected positions of labelled DNAs

'light' DNA - both strands are ^{14}N-labelled

'mixed' DNA - one strand is ^{14}N, the other is ^{15}N

'heavy' DNA - both strands are ^{15}N-labelled

Experimental observations

Composition of DNA and its explanation, which supports semiconservative replication

Parent DNA

F_1 generation DNA (after one cell division)

F_2 generation DNA (after two cell divisions)

$F_{>2}$ generation DNA (after more than two cell divisions)

steadily increasing proportion of 'light' DNA relative to 'mixed' DNA, with no 'heavy' DNA

Fig. 10.9. The Meselson–Stahl experiment proving the semiconservative replication of DNA.

preponderated. This was confirmed in cells of bean seedling roots by feeding tritiated thymidine during periods of DNA synthesis; both daughter chromatids (see Chapter 3, Section B.4) were labelled at cell division. The tritiated thymidine was then withdrawn and after the next cycle each of these chromosomes produced one labelled and one unlabelled chromatid.

(i) INITIATION AND DIRECTION OF REPLICATION

The onset and exact initiation of DNA replication must be controlled by some signal but little is known about its nature. Electron microscopy reveals the initiation sites as double-stranded bubbles, and the sites may be regions of specific base sequences or exist in a certain localized conformation perhaps dictated by the base sequence. Initiation can take place at different times; for example, in *Arabidopsis* two families of initiation sites exist, in one of which DNA synthesis begins 36 min (at 22°) before it starts in the other. Replication is also generally bidirectional.

(ii) MECHANISM OF REPLICATION†

Although many unsolved problems still exist the general picture of the mechanism of replication in prokaryotes is as outlined in Fig. 10.10. The situation in eukaryotes, particularly higher plants, is less clear but may be similar.

The first step (*A*) is the unwinding of the double helix which is driven by energy derived from ATP hydrolysis and catalysed by an enzyme known as helicase. The separated strands are then stabilized by interacting with several molecules of the *single-stranded DNA binding protein* (SS-binding protein). Such proteins have greater binding affinity for single-stranded DNA than for the double-stranded material, thus causing unwinding to form the replication fork (Fig. 10.10) revealed as a bubble in the DNA molecule in the electron microscope. Thus the DNA does not unwind completely before synthesis of new DNA begins. As the parent DNA strands are antiparallel it follows that the structure of one strand runs from 5′ to 3′ and the other from 3′ to 5′. The strand in the replication fork which

runs from 5′→3′ is known as the leading strand and that from 3′→5′ as the lagging strand.

As it is known that all DNA polymerases synthesize DNA in the 5′→3′ direction and that DNA synthesis takes place on both strands of the replication fork, the question arises as to how synthesis takes place on the lagging strand. A short RNA primer (about ten nucleotides) is synthesized at a special primer region by pairing with the bases of the single-stranded DNA (*B*). The enzyme involved, primase, is a DNA-directed RNA polymerase with the direction of synthesis from 5′- to the 3′-terminus in the growing strand of the DNA. DNA polymerase III then elongates the primer in the 5′→3′ direction with the parent single-stranded DNA used as a template to produce *Okazaki fragments* which contain both ribo- and deoxyribonucleotides (*C*). DNA-polymerase I then comes into action; it also has a 5′→3′ exonuclease activity which permits it to catalyse the removal of the primer RNA (*D*); it then exerts its polymerase activity and proceeds to extend the DNA fragments from the adjacent 3′-terminus (*D*). DNA polymerase I also exhibits 3′→5′ exonuclease activity which is considered to represent a proof-reading role because it can remove nucleotides in the direction opposite to that in which replication is taking place; thus it could remove bases which have been inserted by mistake. A DNA polymerase II exists which is similar to DNA polymerase I but lacks its 5′→3′ exonuclease activity.

The DNA fragments eventually formed by the polymerase on the lagging strand are then joined by linking the 5′-phosphate from one fragment to the 3′-OH group of an adjacent segment by the enzyme DNA ligase[8,9] (*E*). The reaction involves the formation of an ADP derivative at the 5′-terminal of one fragment and it is this which condenses with a 3′-OH terminal of the second fragment (Fig. 10.11). The source of the AMP residue is NAD^+ in *E. coli* and ATP in mammals; the source in plants is not known. This enzyme will also join 'nicks' and sew in extraneous pieces of DNA. It is in fact the key enzyme in recombinant DNA (genetic engineering) work. DNA synthesis along the leading strand is essentially continuous.

Three eukaryotic DNA polymerases, α, β and γ, have been described; their detailed action is still under study; but it is considered that α is concerned

† Reaction letters in this section refer to Fig. 10.10.

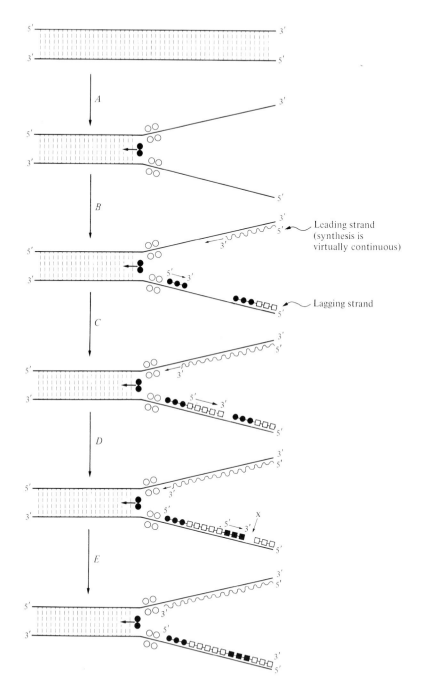

Fig. 10.10. Probable mechanism of DNA replication.
(A = Unwinding of the double helix by helicase, ●●, and stabilization of the single strands by several molecules of the SS-binding
   protein, ○.
B = RNA primers, ●●●, synthesized by primase.
C = Okazaki fragments ●●●□-□-□-□-□, containing both DNA □ and RNA ●, synthesized by DNA polymerase III.
D = DNA strands, □-□-□-■-■-■, produced by removal of the RNA primers by $5' \rightarrow 3'$ exonuclease action of DNA polymerase I, followed
   by the addition of deoxyribonucleotides successively to the adjacent 3'-terminus by the polymerase action of the same enzyme.
E = Linking together of the DNA strands, ■-■-■-□-□-□, at the position marked X by DNA ligase.
→ = Direction of synthesis or of enzyme action.)

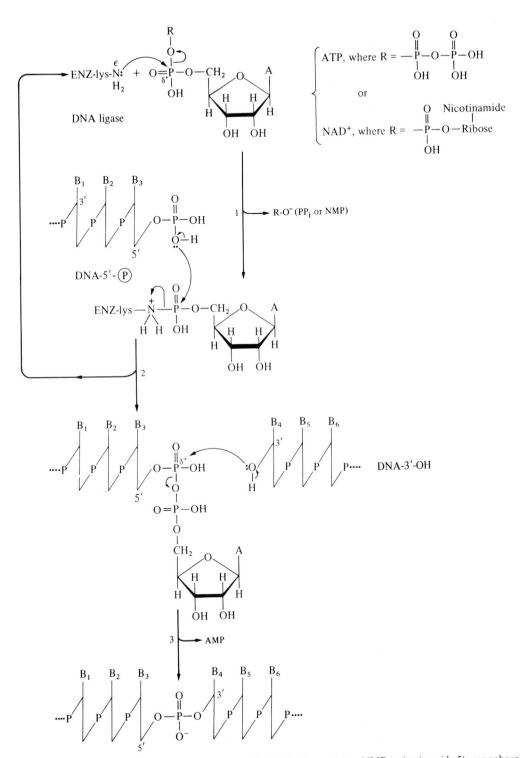

Fig. 10.11. Mechanism of action of DNA ligases (EC 6.5.1.1 and EC 6.5.1.2). (A = adenine; NMP = nicotinamide 5'-monophosphate; B_1, B_2, etc. = purine or pyrimidine bases of DNA; EC 6.5.1.1 utilizes ATP in step 1 and is found in animals and bacteriophage whilst EC 6.5.1.2 utilizes NAD$^+$ and is found in bacteria.)

with replication, β with repair and γ with synthesis of mitochondrial DNA. They have the same basic properties as the prokaryotic enzymes; e.g. they require a template, a 3'-hydroxy primer and the four constituent triphosphates for activity; but they have been reported to have the other catalytic properties.

5. Transcription

The information in a genome (DNA molecule) is not transmitted directly to a protein synthesizing system. It is first transferred to RNA by a process of *transcription*. Basically the process, which, unlike DNA synthesis, does not require a primer, is brought about by a DNA-dependent RNA-polymerase[10], which catalyses the formation of 3',5'-phosphodiester linkages between ribonucleoside triphosphates with the simultaneous release of inorganic pyrophosphate. The sequence of bases in the new RNA is determined by the base sequence in the template DNA and the RNA molecule grows from its 5' end (i.e. in the 5'→3' direction) and proceeds from the 3' end of the template (i.e. in the 3'→5' direction), that is there is antiparallel synthesis.

Detailed information is available on prokaryotic RNA polymerase but less is known about the eukaryotic enzymes. The *E. coli* enzyme is a zinc-containing protein complex MW 500,000, consisting of α_2, β, β' and σ subunits, of MW 39,000, 155,000, 165,000 and 95,000 respectively.

σ (sigma factor) may be considered a recognition factor. It is needed for the correct selection of the promotor site (i.e. a specific starting point for transcription) on the DNA template but is not needed for chain elongation. About seven to eight base pairs are required to provide a unique initiation site. The initial DNA–RNA polymerase complex is considered to be a closed complex because it is formed when the DNA is still a double helix. A conformational change which causes local unwinding of the double helix then takes place at a site closer to the RNA initiation site and leaves the bases of the single-stranded DNA ready to pair with the incoming ribonucleotide triphosphates. The RNA chain is initiated probably by either GTP or ATP reacting with the next nucleotide specified by the template. The G(A)TP retains its triphosphate residue; thus all newly transcribed RNA chains start with a pppG or a pppA. After initiation the σ-factor is released (Fig. 10.12). The RNA chain is elongated by the polymerase core enzyme $(\alpha_2\beta\beta')$ according to the information provided by the template. It must be emphasized that only one DNA strand is used in

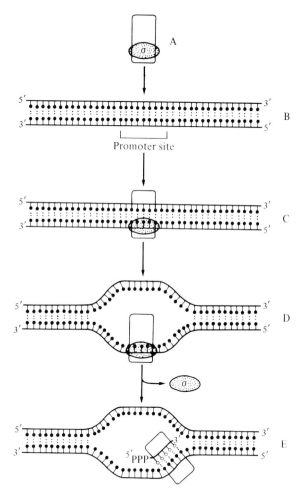

Fig. 10.12. The probable mechanism of initiation of RNA synthesis.
(A = RNA polymerase composed of the σ-factor plus the α_2, β and β' proteins.
B = Double helix of DNA.
C = Recognition of the promoter site by the σ-factor and the binding of the polymerase to it.
D = Opening of the promoter site to give two separate DNA strands.
E = Release of σ-factor and synthesis of an RNA polymer coded by one of the single strands of DNA. The code is read in the 3'→5' direction whilst synthesis occurs in the 5'→3' direction catalysed by the residual components of the polymerase. The second DNA strand is not transcribed.)

transcription. Termination of RNA synthesis is achieved either by specific base sequences in the DNA or by a ρ (rho) factor which is a protein. The final nucleotide in the RNA chain has a free 3'-hydroxyl group. The overall mechanism of elongation of the RNA molecule is outlined in Fig. 10.13.

Three types of RNA polymerase are known in eukaryotes. Those of class I are found in the nucleolus and presumably transcribe pre-rRNA genes (see Section D.9); those of class II transcribe mRNA precursors while class III polymerases transcribe small RNAs, e.g. transfer RNA and 5S RNA.

6. Post-transcriptional processing

Although the base sequences of the various RNA species present in a cell are established by the base sequence of the DNA template, some can undergo further modifications known as *post-transcriptional processing*. This can involve removal of nucleotides

by hydrolysis (*splicing or trimming*), addition of nucleotides (e.g. poly A) to terminal groups, and chemical modification such as methylation. The last reaction is particularly important in plants.

7. Messenger RNA(*mRNA*)

Messenger RNA (*mRNA*) is one of the various types of RNA synthesized by a cell as indicated earlier. Each *mRNA* is a specific RNA which carries information to direct the synthesis of specific proteins. It is short-lived, between several seconds and a few minutes in bacteria, but longer-lived in eukaryotes possibly because it has to move out of the nucleus to the site of protein synthesis in the cytoplasm.

If it is assumed that the average plant *mRNA* molecule contains about 1200 bases, that there are 15,000 genes per haploid genome and that each unique RNA arises from one gene, then 1.8×10^7 base pairs of DNA would be involved in specifying

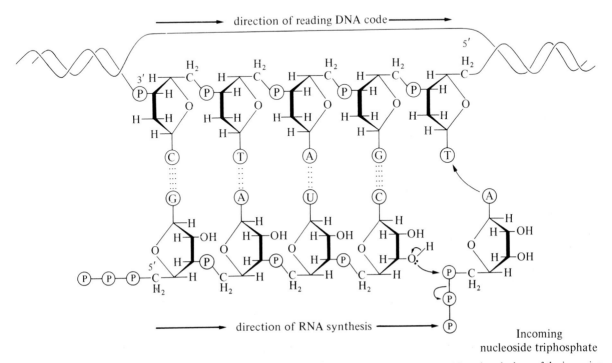

Fig. 10.13. The formation of RNA using one of the strands of a DNA double helix as a template. (Note that the base of the incoming nucleoside 5'-triphosphate must hydrogen bond to the template base before the electron shifts leading to the formation of the new 3',5'-phosphodiester bond and the concomitant elimination of pyrophosphate can take place.)

the *m*RNAs present in one cell. As nuclear genomes from plants can contain between 2×10^8 and 8×10^{10} base pairs it follows that only a small amount of the nuclear DNA is used for coding *m*RNAs.

Two characteristic features of eukaryotic *m*RNAs are (i) the 5'- end is frequently *capped* by 7-methylguanosine attached by a triphosphate group to C–5' of the terminal base: (ii) the 3'-end has a tail of poly A units varying in length from 60–200 residues. Such a tail is, however, missing from the histone *m*RNAs

and although a tail has recently been detected in prokaryotic *m*RNAs in general it is absent. The generalized structure of a plant nuclear *m*RNA is thus that given in Fig. 10.14. The two additional features, capping and adding the poly-A chain, are examples of post-transcriptional processing. Capping renders *m*RNAs more stable by protecting their 5'-ends from phosphatases and nucleases and also appears to enhance their translation. The poly-A tail also increases the stability of *m*RNAs although it does not affect their translation.

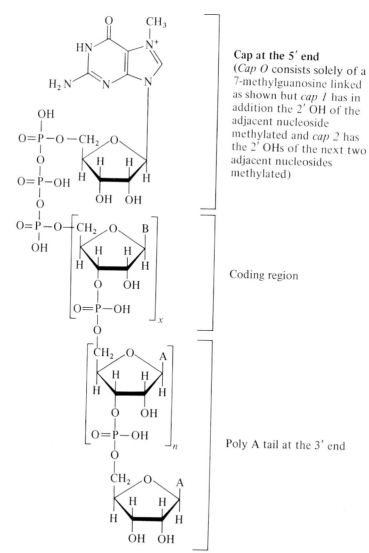

Cap at the 5' end
(*Cap O* consists solely of a 7-methylguanosine linked as shown but *cap 1* has in addition the 2' OH of the adjacent nucleoside methylated and *cap 2* has the 2' OHs of the next two adjacent nucleosides methylated)

Coding region

Poly A tail at the 3' end

Fig. 10.14. Generalized overall structure of eukaryotic *m*RNA. (B=adenine, guanine, cytosine or uracil; A=adenine; $x = \sim 1000$, $n = 60$–200.)

In eukaryotes, including plants, a major fraction of the RNA present in the nucleus is known as *heterogeneous nuclear RNA* (*hn*RNA). This is a mixture of RNAs that are primary transcripts of nuclear genes and as such are *m*RNA precursors. They arise because all eukaryotic nuclear genes so far mapped, with the exception of those for the histones, are discontinuous in that the coding sequence of nucleotides is interrupted by one or more non-coding sequences (i.e. sequences that do not subsequently result in an amino acid sequence in the encoded protein). The coding sequences of the gene are known as *exons* (i.e. expressed regions) and the non-coding sequences are called *introns* (i.e. intervening regions). The initial transcription of the gene includes both the exons and the introns. The resulting RNA thus requires further processing before it becomes the *m*RNA corresponding to the gene; it is a species of *hn*RNA. This processing begins with the addition of a cap and a poly-A tail and is completed by an enzyme-catalysed splicing process in which the non-coding sequences are excised and the cut ends of the coding sequences accurately joined together. The end product is a *m*RNA with a structure corresponding to that shown in Fig. 10.14. In prokaryotes genes are continuous and the post-transcriptional processing just described is not necessary.

8. Transfer RNA (*t*RNA)

Transfer RNA (*t*RNA), carrying an activated amino acid, interacts with *m*RNA and ribosomes during protein synthesis (see Section E). The *t*RNAs have similar but not identical sizes and structure and contain about 74–93 bases. The secondary structure of a plant (algal) cytoplasmic *t*RNA (Fig. 10.15) illustrates the characteristic features of such a molecule. It is a cloverleaf type structure with helical regions with bases paired by hydrogen bonds and loops of unpaired bases. One such loop carries the *anticodon* by which the *t*RNA interacts with *m*RNA (Section E.2). A small extra arm or variable loop is also present. The 3'-end of the *t*RNA is on the amino acid acceptor stem, which in all species of *t*RNAs ends with the base sequence CCA. The tertiary structures of some *t*RNAs are now known.

Many mitochondrial *t*RNAs display structural features which do not correlate with the generalized structure; for example, human mitochondrial *t*RNA^{Ser} lacks completely one of the arms.

The molecules transcribed from the genes for *t*RNA are much larger than the functional molecule. These precursors may contain one or more different *t*RNA molecules. The extra nucleotides are removed by specific nucleases in an orderly manner to yield the mature *t*RNAs. A number of other post-transcriptional modifications occur, for example methylation can occur in the base or in the ribose at C–2', reduction can occur at C–4,5, and in uridine sulphur can replace oxygen at C–2 and C–6 (Fig. 10.16). In the case of methylation of uridine at C–5 the product is ribothymidine. Furthermore, an internal rearrangement to produce ψ-uridine (10.18) can also occur. Purines undergo similar methylations and thiolations and adenine can add an isopentenyl residue to the N at C–6. Such molecules, when free, act as plant hormones (see Section A and Chapter 15, Section D).

9. Ribosomal RNA (*r*RNA)

The ribosomal RNA component of the cytoplasmic ribosome (Chapter 3, Section B.3) (*r*RNA) is transcribed within the nucleolus in eukaryotes. The first molecule transcribed sediments at 45S and has a molecular weight of about 4.1×10^6. This undergoes post-transcriptional changes, the first being methylation at C–2' of some nucleosides and probably conversion of some uridine residues into ψ-uridine residues. Then the molecule is hydrolysed to give units some of which (28S, 18S and 5.8S) are incorporated into ribosomes while others (spacers) are discarded. The splitting, which has been followed in the electron microscope, is outlined in Fig. 10.17. The RNAs are then ready to be incorporated into functional ribosomes. Ribosomal proteins synthesized in the cytoplasm move to the nucleolus where the characteristic 80S eukaryote ribosome is assembled by attaching the 18S, 28S and 5.8S *r*RNA together with a 5S RNA. This last species originates independently in the non-nucleolar region of the nucleus. The intact ribosome can itself

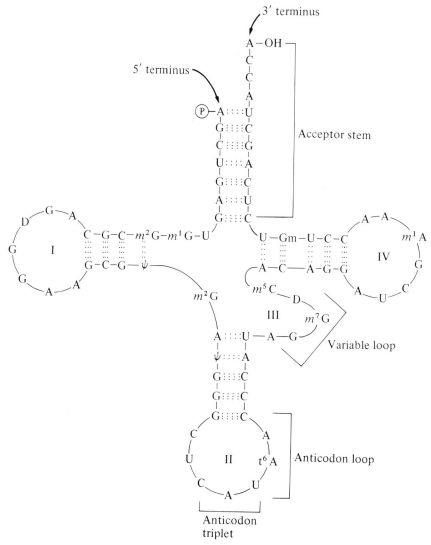

Fig. 10.15. Secondary structure of *Scenedesmus obliquus* cytoplasmic initiator *t*RNA (i.e. *t*RNA$_i^{Met}$) (kindly provided by Dr. D. S. Jones). (Gm = 2′-methylribosylguanine; D = 5,6-dihydrouridine; ψ = pseudouridine; m^{5}C = 5-methylcytidine; m^{1}A = 1-methyladenosine; m^{1}G = 1-methylguanosine; m^{2}G = N^2-methylguanosine; m^{7}G = 7-methylguanosine; t^{6}A = N^6-(*N*-threonylcarbonyl)adenosine.)

be dissociated into two subunits 40S and 60S; the former contains the 18S *r*RNA plus protein whilst the latter contains the 25S, 5.8S and the 5S *r*RNAs plus protein. Associated with the 40S subunit are about thirty proteins whereas about forty are associated with the 60S subunit. The size of the eukaryotic ribosome is greater than that of the prokaryotic ribosome which is a 70S unit, dissociating into 30S and 50S subunits. It is important to emphasize that the chloroplast ribosome is a 70S,

prokaryote type. Mitochondrial ribosomes also resemble prokaryote ribosomes but probably contain more protein.

During protein synthesis ribosomes aggregate to form polysomes (Section E.4).

We have now discussed the various types of RNAs present in a cell and the structure of the ribosomes on which protein synthesis occurs. We can now proceed to discuss the details of protein synthesis.

O replaced by S
Methylation
Hydrogenation
Methylation

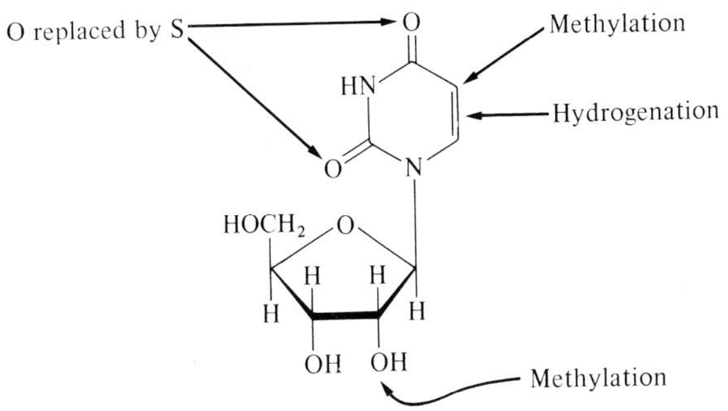

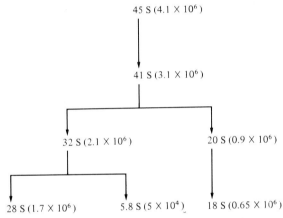

Fig. 10.16. Some types of post-transcriptional modifications which uridine can undergo in *t*RNA.

45 S (4.1 X 10^6)
41 S (3.1 X 10^6)
32 S (2.1 X 10^6)
20 S (0.9 X 10^6)
28 S (1.7 X 10^6)
5.8 S (5 X 10^4)
18 S (0.65 X 10^6)

Fig. 10.17. Origin of eukaryotic *r*RNA species from a 45S precursor. (The numbers in parenthesis are approximate molecular weights.)

E. PROTEIN SYNTHESIS

1. Translation

Protein synthesis involves transfer of information stored in the sequence of bases in a specific *m*RNA to a system which will synthesize a peptide with an amino acid sequence dictated by the *m*RNA. This means that the four bases present in the *m*RNA must be so organized to direct synthesis in a system which has twenty amino acids available for incorporation. This process is termed *translation* and the *controlling signal* is a specific sequence of three bases read in the $5' \rightarrow 3'$ direction. Each sequence of base triplets is known as a *codon*. They are read consecutively and are non-overlapping. Sixty-four triplet codons exist and there is no need here to go into details of how the genetic code was cracked; suffice it to say that the assignment of the various codons to different amino acids is indicated in Table 10.1. The code is degenerate in that more than one codon can specify one amino acid. Codons in Table 10.1 labelled *Term*, e.g. UAA, UGA and UAG, end the synthesis of a peptide chain and are known as *termination codons* (see Section E.7). Earlier they were known as nonsense codons. Other triplets, e.g. AUG and, in prokaryotes, GUG, which code for methionine, are known as initiator codons (see Section E.5). There are no punctuation marks in the code other than full stops (termination codons).

2. Pairing of codons and anticodons

The codon of an *m*RNA must pair with the anticidon of the corresponding *t*RNA. In detail, however, this view has to be modified. The first two bases of a codon, starting at the 5′-hydroxyl end, are mandatory determinants, whereas the final base (at the 3′-hydroxyl end) can vary. This degree of latitude accounts to a great extent for the degeneracy of the code, for example the four codons specific for alanine all begin with GC but can vary in the final base. This aspect of degeneracy has been described as *wobble*. The anticodons of some *t*RNAs are known and frequently contain modified nucleosides, such as inosine and 5-*N*-methylaminomethyl 2-thiouridine (10.19). It is interesting that these modified nucleosides are at the 5′ end of the anti-codon: that is they are the bases corresponding to the wobble position in the codon. This is illustrated for alanine in Fig. 10.18.

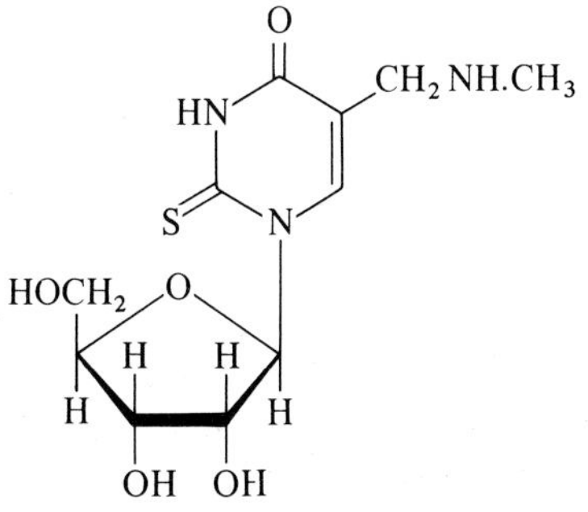

(10.19)

Table 10.1. *The genetic code (mRNA codon base sequences for amino acids)*

5'-OH-terminal base	Middle base				3'-OH-terminal base
	U	C	A	G	
U	Phe	Ser	Tyr	Cys	U
	Phe	Ser	Tyr	Cys	C
	Leu	Ser	Term	Term	A
	Leu	Ser	Term	Trp	G
C	Leu	Pro	His	Arg	U
	Leu	Pro	His	Arg	C
	Leu	Pro	Gln	Arg	A
	Leu	Pro	Gln	Arg	G
A	Ile	Thr	Asn	Ser	U
	Ile	Thr	Asn	Ser	C
	Ile	Thr	Lys	Arg	A
	{ Met / fMet	Thr	Lys	Arg	G
G	Val	Ala	Asp	Gly	U
	Val	Ala	Asp	Gly	C
	Val	Ala	Glu	Gly	A
	{ Val / fMet	Ala	Glu	Gly	G

Phe	= Phenylalanine		Ile	= Isoleucine
Leu	= Leucine		Met	= Methionine
Ser	= Serine		fMet	= N-Formylmethionine
Tyr	= Tyrosine		Thr	= Threonine
Cys	= Cysteine		Asn	= Asparagine
Trp	= Tryptophan		Lys	= Lysine
Pro	= Proline		Val	= Valine
His	= Histidine		Ala	= Alanine
Gln	= Glutamine		Glu	= Glutamic acid
Arg	= Arginine		Gly	= Glycine
Asp	= Aspartic acid		Term	= Termination codons

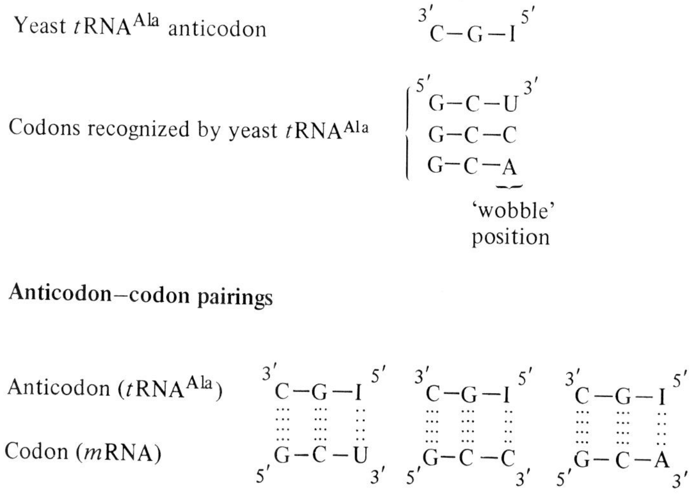

Yeast tRNA^{Ala} anticodon

$$^{3'}\text{C}-\text{G}-\text{I}^{5'}$$

Codons recognized by yeast tRNA^{Ala}

$$\begin{cases} ^{5'}\text{G}-\text{C}-\text{U}^{3'} \\ \text{G}-\text{C}-\text{C} \\ \text{G}-\text{C}-\underset{\sim}{\text{A}} \end{cases}$$

'wobble' position

Anticodon–codon pairings

Anticodon (tRNA^{Ala})

Codon (mRNA)

$$^{3'}\text{C}-\text{G}-\text{I}^{5'} \quad ^{3'}\text{C}-\text{G}-\text{I}^{5'} \quad ^{3'}\text{C}-\text{G}-\text{I}^{5'}$$
$$_{5'}\text{G}-\text{C}-\text{U}_{3'} \quad _{5'}\text{G}-\text{C}-\text{C}_{3'} \quad _{5'}\text{G}-\text{C}-\text{A}_{3'}$$

Fig. 10.18. Anticodon–codon pairings for alanine illustrating 'wobble'. (I = inosine, whose structure is shown in Fig. 10.1.)

3. Activation of amino acids

Once the overall apparatus is available for protein biosynthesis the first step is the binding of the amino acid to the appropriate enzyme as an aminoacyl-adenylate complex [eq. (10.7)]. The complex then must recognize the appropriate tRNA and bind to it. How this is achieved is still essentially a mystery. There follows a transfer of the amino acid to either the 2'- or 3'-hydroxy position at the CCA end of the tRNA [eq. (10.8)].

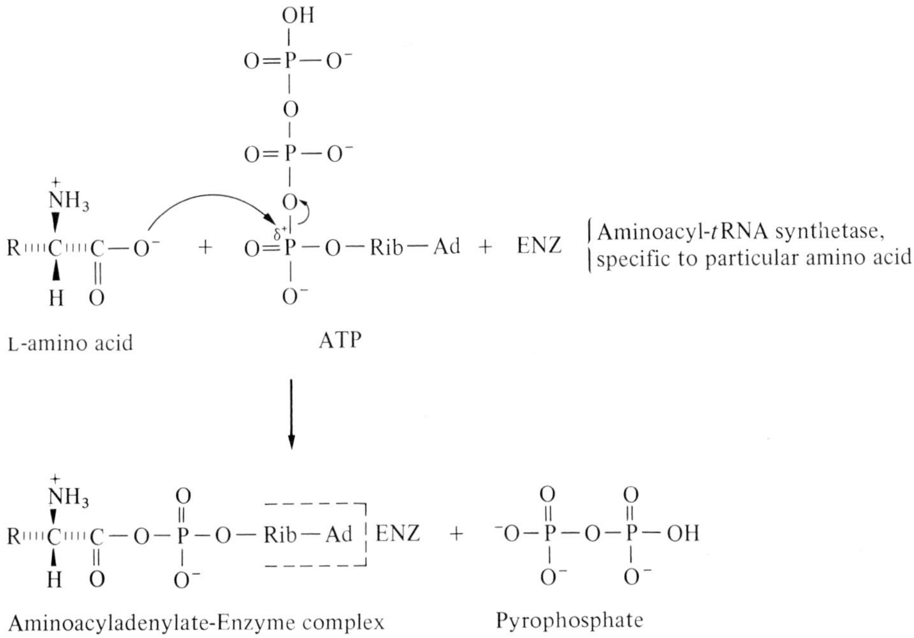

(10.7)

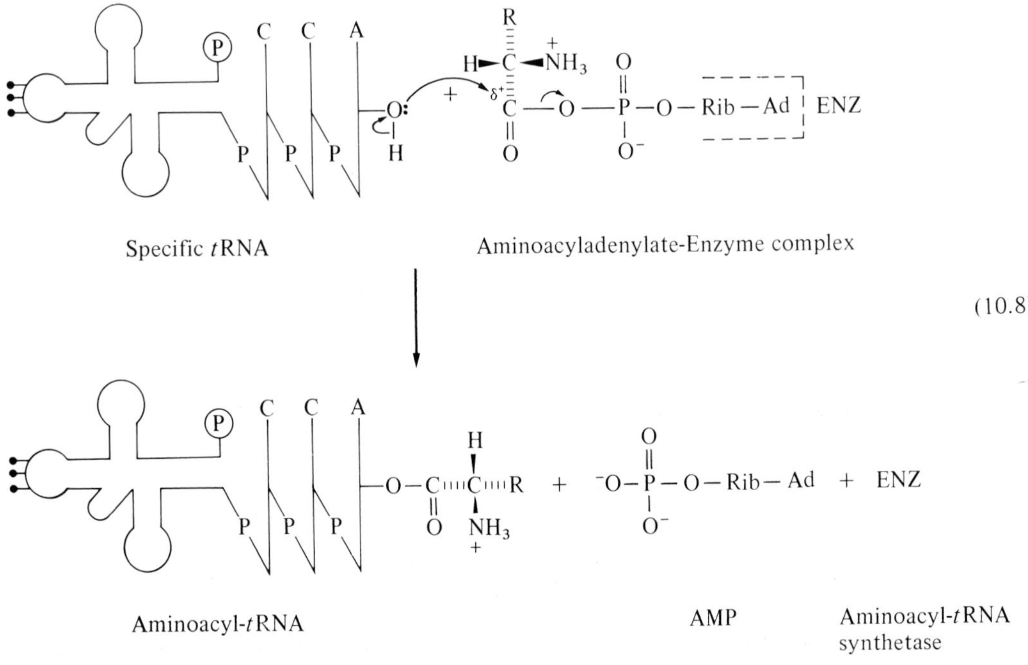

(10.8)

4. Basic reactions of protein synthesis

mRNA, after its formation, becomes bound to ribosomes which move along its length translating it into a polypeptide chain. During this process tRNAs bring amino acids to the mRNA-ribosome complex and position them correctly for linking together in the correct sequence. Protein synthesis therefore takes place on the ribosome, which must be capable of binding both mRNA and tRNA. There are two sites for tRNA binding, known as the 'P' site and the 'A' site. Bound to the P site (peptidyl or donor site) is the tRNA carrying the growing polypeptide chain whilst to the A site is bound the tRNA carrying the next amino acid to be added to the chain. The mRNA is translated in the $5' \rightarrow 3'$ direction and the first amino acid introduced is at the amino terminus of the completed polypeptide. A static picture of a ribosome synthesizing a protein is given in Fig. 10.19. However, in vivo a number of

ribosomes are attached to a single mRNA to form a polysome. These represent ribosomes at various distances from the starting point and they will have attached to them increasingly long threads of polypeptide as they approach the termination point where the protein and ribosome are released, the latter to continue the cycle of protein synthesis (Fig. 10.20). The maximum density of ribosomes on mRNA is about one per eighty nucleotides.

5. Initiation of protein synthesis†

The process of initiation of protein synthesis discussed in this section and the processes of elongation and termination discussed in Sections 6 and 7 respectively have been studied more intensively in bacteria (i.e. prokaryotes) such as E. coli

† All letters in this section refer to Fig. 10.21.

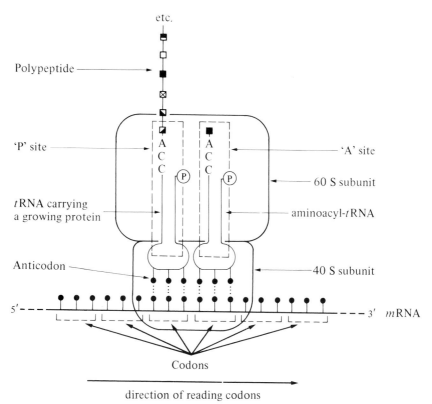

Fig. 10.19. Diagram indicating how mRNA is attached to the ribosome and to tRNAs at the P (peptidyl) and A (aminoacyl) sites in the cytoplasm of eukaryotic cells. (■ □ ▨ ◣ ⊠ and ▬ represent different amino acids.)

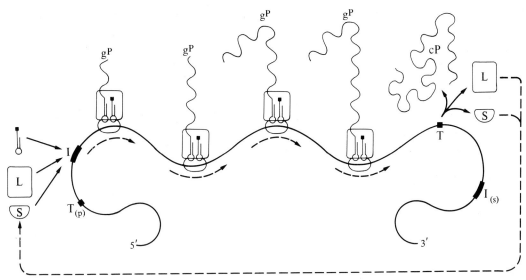

Fig. 10.20. Diagrammatic representation of a polyribosome consisting of four ribosomes strung along a cistron (i.e. a section coding for one protein) in a polycistronic *m*RNA molecule. (S = small ribosomal subunit, 30S in prokaryotes, mitochondria and chloroplasts but 40S in the cytoplasm of eukaryotes; L = large ribosomal subunit, 50S in prokaryotes, mitochondria and chloroplasts but 60S in the cytoplasm of eukaryotes; gP = growing polypeptide chain; cP = completed protein; I = initiator sequence consisting of the initiator codon preceded by a purine rich sequence recognized by the *r*RNA component of S; I(s) = initiator sequence of succeeding cistron; T = terminator codon; T(p) = terminator codon of preceding cistron; = aminoacyl-*t*RNA. Note that eukaryotic *m*RNAs are monocistronic whilst many prokaryotic *m*RNAs are polycistronic.)

than in any other protein-synthesizing entity. As a result the prokaryotic processes of initiation, elongation and termination are best understood. It must be emphasized, however, that they are not fully understood; this will not be achieved until the function of the numerous protein components of the ribosomal subunits has been elucidated. Enough is known of the mechanism of these processes in the cytoplasm, the mitochondria and the chloroplasts of eukaryotic cells to say that they are in essence the same as in prokaryotic cells, although there are minor differences. The most obvious of these is that the ribosomes in the cytoplasm of eukaryotic cells are 80S structures (40S subunit + 60S subunit) whereas those in the mitochondria and chloroplasts are, like those of bacteria, 70S structures (30S subunit + 50S subunit). Indeed there is mounting evidence that mitochondrial and chloroplastidic protein synthesis resembles that of prokaryotes more closely than that of the cytoplasm in which the organelles reside; this ties in nicely with the endosymbiont hypothesis of the evolutionary origin of mitochondria and chloroplasts [see Chapter 3, Sections B.5 and B.6(viii)].

Because most is known about the bacterial

processes of initiation, elongation and termination, they will be used as the basis of the descriptions in this and Sections 6 and 7; where the eukaryotic cytoplasmic processes differ, this will be pointed out.

In prokaryotes, mitochondria and chloroplasts the first step in the initiation process is the formation of an *N*-formyl-L-methionine-transfer RNA complex (fMet-*t*RNA$_f^{Met}$). This complex is able to recognize the AUG (or GUG) initiator codon which signals the start of a protein coding sequence (a cistron) within a polycistronic *m*RNA and can differentiate it from an internal AUG, which also codes for methionine (or an internal GUG, which codes for valine). How is this accomplished? The answer lies in the bringing together of two facts: (i) fMet-*t*-RNA$_f^{Met}$ can *only* bind to the P site of the ribosome (see Fig. 10.19) whereas all other aminoacyl-*t*RNAs can *only* bind at the A site and (ii) the initiator AUG (or GUG) is specifically positioned at the P site when it binds to the ribosome as a result of the hydrogen bonding of a short sequence of purine nucleotides near to, and on the 5′ side of, the initiator codon to a complementary sequence of pyrimidine nucleotides near to the 3′ end of the 16S *r*RNA of the 30S ribosomal subunit. Thus only fMet-tRNA$_f^{Met}$ can hydrogen bond to the initiator codon.

The tRNA that takes N-formylmethionine to the ribosome is different from the tRNA that inserts methionine into an internal position; the former is referred to as tRNA$_f^{Met}$, the latter as tRNA$_m^{Met}$. Methionine is linked to both species of tRNA by the same aminoacyl-tRNA synthetase [eqs. (10.7) and (10.8)] so producing Met-tRNA$_f^{Met}$ and Met-tRNA$_m^{Met}$. A specific transformylase[11] then catalyses the transfer of the formyl group from N^{10}-formyl THFA to the amino-N of Met-tRNA$_f^{Met}$ producing fMet-tRNA$_f^{Met}$; this transformylase will not utilize free methionine or Met-tRNA$_m^{Met}$ as substrates.

In the eukaryotic cytoplasmic initiation process the initiator codon is recognized by a methionine-tRNA complex rather than a formylated methionine-tRNA. Again there are two tRNAs specific to methionine, one taking methionine to the initiator codon, the other taking it to internal positions. The former is referred to as tRNA$_i^{Met}$, the latter as tRNA$_m^{Met}$. Only met-tRNA$_i^{Met}$ can bind to

the initiator codon which, as in the prokaryotic system, is specifically located at the P site. Mammalian cytoplasmic met-tRNA$_i^{Met}$ can be readily formylated by the bacterial transformylase but that from plants either cannot be formylated or can be formylated only with difficulty, depending upon species. The reason for this is unclear since eukaryotic cytoplasmic tRNA$_i^{Met}$ from different sources, plant or animal, show considerable nucleotide sequence homology.

With the formation of fMet-tRNA$_f^{Met}$ the prokaryotic initiation process continues as shown in Fig. 10.21. In the first step (A) the 30S initiation complex is formed by the precise association of the 30S ribosomal subunit, mRNA, fMet-tRNA$_f^{Met}$ and GTP in the presence of three proteinaceous initiation factors termed IF-1, IF-2 and IF-3. IF-3 participates in the binding of mRNA to the 30S ribosomal subunit and also prevents the binding together of the 50S and 30S ribosomal subunits in the absence of a mRNA molecule with the consequent formation of a functionless 70S ribosome.

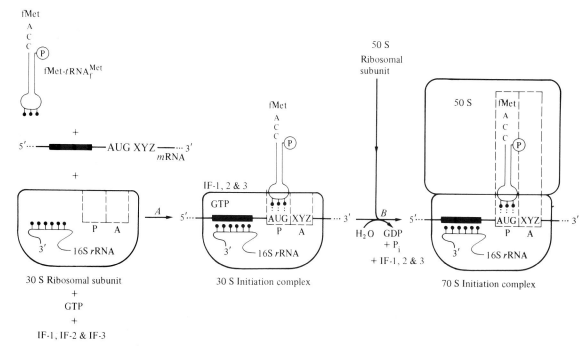

Fig. 10.21. Diagrammatic representation of the probable mechanism of the initiation of protein synthesis in prokaryotes. (fMet = N-formyl-L-methionine; AUG = initiator codon; XYZ = codon on the 3′ side of the initiator codon; ▬■▬ = short sequence of purine nucleotides near to and on the 5′ side of the initiator codon which hydrogen bonds to a complementary sequence of pyrimidine nucleotides near the 3′ end of the 16S rRNA of the 30S ribosomal subunit and positions the initiator codon at the P site; P = P site; A = A site; IF-1, IF-2 and IF-3 = proteinaceous initiation factors.)

IF-2 complexes with the fMet-tRNA$_f^{Met}$ and GTP and then binds to the mRNA-30S ribosomal subunit complex. IF-1 also appears to enhance the binding of fMet-tRNA$_f^{Met}$ in the 30S initiation complex but its precise role is not yet clear.

In the eukaryotic cytoplasmic initiation process step A results in the formation of a 40S initiation complex. In this step it is known that Met-tRNA$_i^{Met}$ binds to the 40S ribosomal subunit before the mRNA and that three initiation factors, designated eIF-1, eIF-2 and eIF-3, are involved.

Prokaryotic initiation continues (B) with the binding of the 50S ribosomal subunit to form the 70S initiation complex. In this step the bound GTP is hydrolysed to GDP and P$_i$ which are then released, along with the initiation factors, to the surrounding medium.

Step B of the eukaryotic cytoplasmic initiation process involves the binding of the 60S ribosomal subunit to form an 80S initiation complex.

6. Elongation of the peptide chain†

With the formation of the 70S initiation complex, the peptide elongation process, depicted diagrammatically in Fig. 10.22, can begin. This consists of three stages: (i) the hydrogen bonding of the appropriate aminoacyl-tRNA to the free codon at the A site of the ribosome (A), (ii) peptide bond formation (B) and (iii) translocation (C and D).

In the first stage (A) the aminoacyl-tRNA whose anticodon is complementary to the codon at the empty A site is delivered and hydrogen bonded to that codon whilst complexed to a cytosolic, protein elongation factor EF–Tu which itself contains bound GTP. The positioning of the aminoacyl-tRNA leads to the liberation of the EF–Tu and the hydrolysis of the bound GTP to GDP. The released GDP–EF–Tu complex then binds to another cytosolic protein elongation factor known as EF–Ts with release of GDP. GTP then binds to the EF–Tu component of the binary complex, causing the release of EF–Ts. This regenerates GTP–EF–Tu

which can then deliver another aminoacyl-tRNA to the mRNA-ribosome complex.

The GTP–EF–Tu can bind to all aminoacyl-tRNAs except fMet-tRNA$_f^{Met}$. Thus the latter cannot be delivered to the A site and accounts for the fact that internal AUG (or GUG) codons are not read by fMet-tRNA$_f^{Met}$.

Now that both P and A sites on the ribosome are occupied stage (ii) of the elongation process, peptide bond formation, can occur (B). This reaction is catalysed by peptidyltransferase[12], an enzyme that is one of the protein components of the 50S ribosomal subunit. The N-formylmethionine residue is transferred from its tRNA$_f^{Met}$ at the P site to the amino-N of the aminoacyl-tRNA at the A site [see eq. (10.9)].

Phase (iii) of the elongation process, translocation, then takes place (C and D) in the presence of a third elongation factor, known as EF–G or translocase. This factor, containing bound GTP, binds to the ribosome and causes the release of the tRNA$_f^{Met}$ from the P site (C). It then appears to facilitate the transfer of the peptidyl-tRNA from the A site to the P site and the simultaneous movement of the ribosome along the mRNA in the $5' \rightarrow 3'$ direction a distance of one codon. These movements are accompanied by the release of EF–G to the surrounding medium and the hydrolysis of the GTP it carried to GDP and P$_i$ (D).

Translocation frees the A site of the ribosome and completes the first cycle of the elongation process in which N-formylmethionine has been elongated to a dipeptide. The second cycle then follows producing a tripeptide and further elongation cycles occur until the protein under construction is complete. The termination process then ensues.

7. Termination of protein synthesis‡

Peptide chain elongation continues until the ribosome, moving codon by codon along the mRNA molecule, reaches the point where a terminator codon occupies the A site. There are three terminator codons UAG, UAA and UGA. Since normal cells do not have any tRNAs with anti-

---

† All letters in this section refer to Fig. 10.22.   ‡ All letters in this section refer to Fig. 10.22.

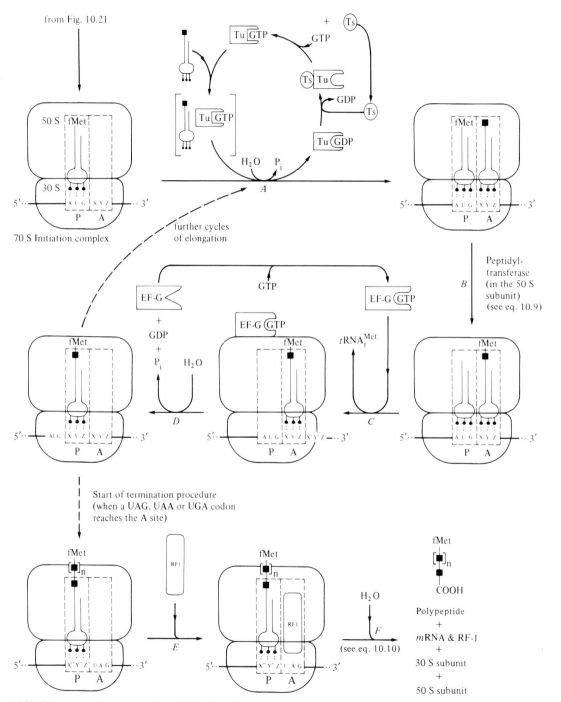

Fig. 10.22. Diagrammatic representation of the elongation and termination phases of protein synthesis in prokaryotes. (Tu and Ts = elongation factors referred to as EF–Tu and EF–Ts; EF–G = another elongation factor, often called 'translocase'; RF-1 = release factor 1, which recognizes terminator codons UAG and UAA: a second release factor, RF-2, recognizes UGA and UAA; AUG = initiator codon; XYZ, X'Y'Z' and X"Y"Z" = various 'in chain' mRNA codons; ■ = amino acid; fMet = N-formyl-L-methionine; P = P site; A = A site; ⬭ = tRNA.)

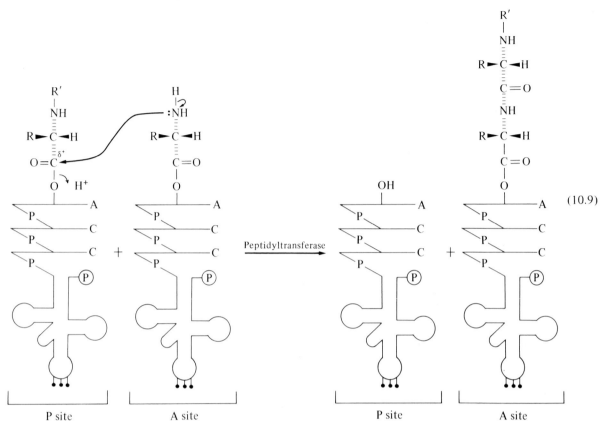

(R = amino acid side chain; R' = peptidyl residue or H or CHO if R is the side chain of the initiator amino acid, methionine)

codons complementary to the terminator codons, there is no aminoacyl-*t*RNA to occupy the A site which was left free after the translocation step of the previous elongation cycle.

The terminator codon at the A site is, however, recognized by one or other of two protein release factors, designated RF–1 and RF–2. The former recognizes UAG and UAA, the latter UAA and UGA. The release factor binds to the terminator codon (*E*); here is an example of a protein recognizing a nucleotide sequence with great precision. This binding apparently alters the specificity of the peptidyltransferase in the 50S ribosomal subunit such that it catalyses the transfer of the peptidyl residue from the peptidyl-*t*RNA at the P site to water instead of the amino-*N* of the aminoacyl-*t*RNA which usually occupies the A site (*F*). Thus it effectively catalyses the hydrolysis of the ester

bond linking the peptidyl residue to *t*RNA [see eq. (10.10)].

The liberation of the protein is closely followed by the dissociation of the release factor, *m*RNA, and the ribosomal subunits (*F*); the latter may be either recycled or degraded.

Termination in the cytoplasm of eukaryotic cells appears to be identical to that in prokaryotes save that only one release factor has been identified.

8. Post-translational modification of proteins

Many of the polypeptides produced by the translation of *m*RNA undergo post-translational modification. In prokaryotes the formyl group is removed from the *N*-terminal *N*-formylmethionine

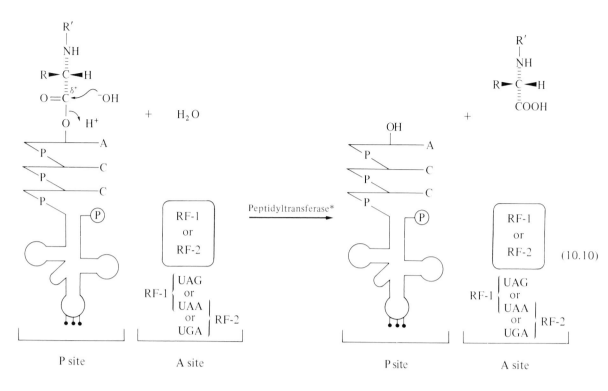

(10.10)

(R = side chain of C-terminal amino acid; R' = peptidyl residue; * = enzyme specificity modified by the proteinaceous release factor, RF-1 or RF-2, bound at the A site)

by a deformylase and in both prokaryotes and eukaryotes the N-terminal methionine is sometimes hydrolysed off while the rest of the polypeptide chain is being synthesized. Aminopeptidases may even remove several amino acids from the N-terminal end. Some proteins, particularly those designated for export from the cell or for transfer from one intracellular compartment to another, are synthesized with a special 'signal' segment at the N-terminal end which is removed during or after the translocation process. The side chains of some amino acids within the polypeptide chain may be modified. Examples of such modifications are (i) the oxidation of two cysteine residues to give a disulphide bridge, (ii) the hydroxylation of proline residues in the cell wall protein, extensin [see Chapter 4, Section B.3(ii)], (iii) the attachment of sugar residues to the side chains of asparagine in the formation of glycoproteins [see Chapter 7, Section D.2(vi)], (iv) the phosphorylation of the hydroxyl group of serine in some enzymes, e.g. phosphogluco-

mutase, and (v) the covalent attachment of prosthetic groups such as biotin in BCCP and coenzyme A in ACP [see Chapter 8, Section C.1(i)].

9. The chloroplast genome

(i) CODING CAPACITY

The chloroplast genome occurs in multiple copies whereas the nuclear genome is haploid; this means that although there is much less genetic material in a chloroplast than in a nucleus (only 10^{-4} the amount) there is even less genetic information available (0.3–1.0×10^{-5} the amount). In spite of this, appropriate calculation reveals that there is sufficient information to code for rRNA, tRNA and some 100–300 proteins with average molecular weights of 4.0×10^4. In fact in higher plant chloroplasts two genes exist for 23S rRNA and for 16S rRNA and they are non-overlapping. Only one gene

Table 10.2. *A comparison of some properties of rRNAs from the chloroplast and cytoplasm of higher plants*

	Chloroplastidic *r*RNA	Cytoplasmic *r*RNA
Size	23S, MW 1.1×10^6	25S, MW 1.3×10^6
	16S, MW 0.6×10^6	18S, MW 0.7×10^6
	5S[a], MW 0.4×10^5	5.8S, MW 0.5×10^5
	4.5S, MW 0.3×10^5	5S[a], MW 0.38×10^5
GC content	55.9%	54.6%
Size of precursor	23S, 1.16×10^6	—
	16S, 0.65×10^6	—

[a] These two species are different.

exists for each of most chloroplast *t*RNAs. Two proteins are clearly established as coded by *Ct*DNA, the large subunit of RuDP carboxylase[13] and photosystem 1 chlorophyll protein, so it may be that most of the *Ct*DNA is regulatory in function.

(ii) BIOSYNTHESIS OF *Ct*DNA

The mode of replication of *Ct*DNA in higher plants is not yet settled but in *Chlamydomonas* and *Euglena* it is semi-conservative as demonstrated by Meselson–Stahl experiments.

Results of experiments with pea and corn chloroplast DNA suggest that the mechanism of replication involves displacement loops (D-loops) and rolling circle intermediates.

(iii) CHLOROPLAST RIBOSOMES

The chloroplast ribosome is, as indicated earlier, a 70S ribosome, whereas the cytoplasmic ribosome in plants is 80S. Other differences between 70S and 80S ribosomes in plants exist; e.g. differential sensitivity to antibiotics: protein synthesis on 70S ribosomes is inhibited by chloramphenicol but not by cycloheximide whereas the reverse is true with 80S ribosomes. Chloroplast ribosomes dissociate into 50S and 30S units at low Mg^{2+} concentration and reassociate on raising the Mg^{2+} concentration; prolonged dialysis in the absence of Mg^{2+} is needed for dissociation of 80S ribosomes and no reassociation occurs on the addition of Mg^{2+}. Chain initiation of protein synthesis on 70S ribosomes requires *N*-formylmethionine and not methionine itself.

About fifty proteins are associated with the 70S ribosomes, about 20–30 on the 30S and 30–40 on the 50S subunits. These proteins are different from those found in the 80S ribosomes from the same plant. Although the *r*RNA of the 70S ribosome is transcribed from the *Ct*DNA, many of its constituent proteins are products of the nuclear genome and have thus been synthesized by the 80S cytoplasmic ribosomes. The mechanisms involved in the integration of the cellular activities which allows the fully functional 70S ribosome to be constructed remain to be investigated.

(iv) CHLOROPLAST *r*RNA

The main differences between chloroplast *r*RNA and plant cytoplasmic *r*RNA are summarized in Table 10.2.

There is now evidence from DNA/RNA hybridization experiments and from biosynthetic studies that isolated chloroplasts can synthesize their own *r*RNA and it now seems that larger precursor molecules are first formed as in the synthesis of cytoplasmic *r*RNA (Section D.9). The pattern of processing is indicated in Fig. 10.23.

The chloroplast RNA polymerase has now been isolated but its detailed properties remain to be securely based.

(v) CHLOROPLAST *m*RNA

The existence of chloroplast *m*RNA, at least that which directs the synthesis of the large subunits of RuDP carboxylase, is well established (see Chapter 5, Section D.2). It is only translated on 70S

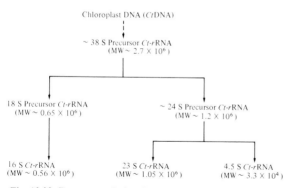

Fig. 10.23. Post-transcriptional processing of ribosomal RNA in spinach chloroplasts.

ribosomes which suggests that the genetic information in the chloroplast is processed, i.e. translated and transcribed, wholly within the chloroplast. There is still some doubt whether or not the chloroplast *m*RNA is similar to eukaryote *m*RNA in containing poly-A sequences although some of the recent evidence is tending to indicate that such sequences are present in some chloroplast *m*RNAs.

What is the mechanism which allows chloroplast *m*RNA to recognize and utilize 70S ribosomes and reject 80S ribosomes? The same question but in reverse must be asked when cytoplasmic *m*RNA reacts with 80S but not 70S ribosomes. A possible

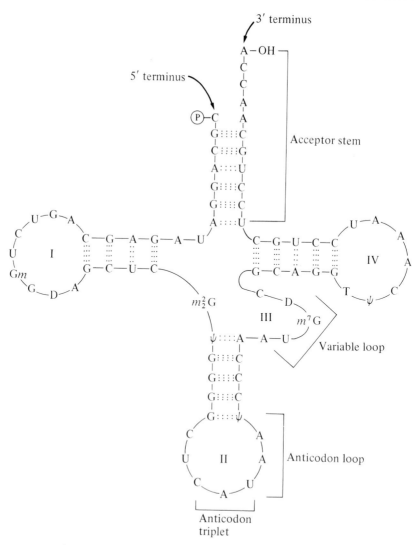

Fig. 10.24. Secondary structure of *Scenedesmus obliquus* chloroplast initiator *t*RNA (i.e. *t*RNA$_f^{Met}$) (kindly provided by Dr. D. S. Jones). (m_2^2G = N^2-dimethylguanosine; T = ribothymidine; all other nucleoside abbreviations are as for Fig. 10.15.)

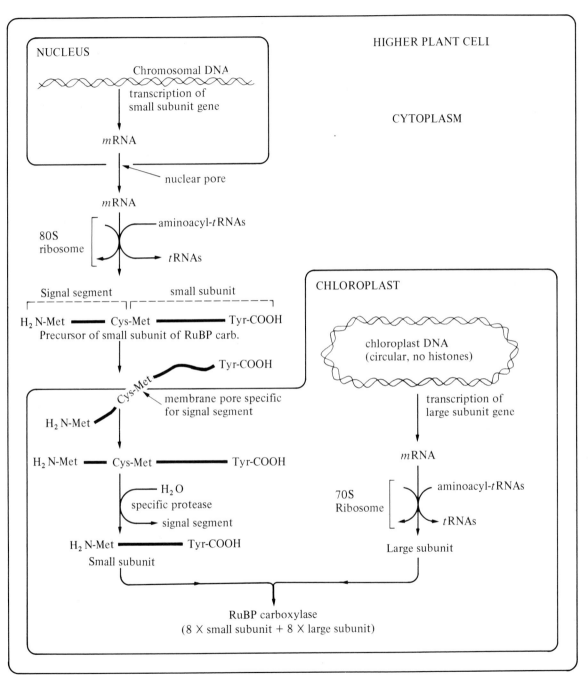

Fig. 10.25. Formation of RuBP carboxylase (EC 4.1.1.39; Fraction 1 protein) in pea leaves (redrawn, with permission, from R. J. Ellis (1979) *Trends in Biochemical Sciences*, **4**, 241–244).

explanation arises from work with *E. coli*, in which it seems that *m*RNA attaches to the 16S *r*RNA at the 3′ end during formation of the initiation complex for protein synthesis. As the sequence in this region of the 16S *r*RNA is CCUCCUUA whereas in 18S *r*RNA from yeast (a eukaryote) it is GAUCAUUA, this may represent two different recognition sites. It will be interesting to see what the sequence in this area is for 16S *r*RNA from chloroplasts and 18S *r*RNA from the cytoplasm of the same plant species.

The *t*RNAs of the chloroplast are somewhat different in structure from cytoplasmic *t*RNAs and are apparently coded for by *Ct*DNA; for example 16 *t*RNA cistrons have been demonstrated in *Ct*DNA from maize. The structure of the $tRNA_f^{Met}$ from the chloroplasts of *Scenedesmus obliquus* (Fig. 10.24) resembles that of $tRNA_f^{Met}$ from prokaryotic organisms more than the initiator *t*RNA from the cytoplasm of *S. obliquus* (Fig. 10.15). It contains all the usual invariant nucleotides of prokaryotic *t*RNA including the important GTψC in loop IV. This is absent from all cytoplasmic initiator *t*RNAs (see Fig. 10.15). Another structural characteristic of chloroplast and prokaryotic initiator *t*RNAs is that they lack a base pair between the first nucleotide at the 5′ end and the nucleotide fifth from the 3′ end. It is interesting that in some *Ct*DNA genes there are intervening sequences (introns). Intervening sequences have only previously been detected in eukaryotic nuclear genes.

The aminoacyl-*t*RNA synthetases of the chloroplast are different from those in cytoplasm. In the case of methionyl-, valyl- and leucyl chloroplastidic *t*RNAs it has been shown that they are only aminoacylated by the chloroplast enzyme; the cytoplasmic enzyme will not recognize them.

(vii) INITIATION, ELONGATION, TERMINATION

As is the case with prokaryotes *N*-formylmethionyl-*t*RNA functions as the initiator of protein synthesis in chloroplasts, but little is known about the initiation factors involved. Elongation of the peptide chain requires at least three factors: one, involved in binding of the aminoacyl-*t*RNA to the ribosome, is probably activated by a second factor

and the third is concerned with the translocation of peptidyl-*t*RNA from the P site to the A site. Nothing detailed is known about the termination reaction.

(viii) PRODUCTS OF THE CHLOROPLAST GENOME

One fully authenticated product of the chloroplast genome is the large sub-unit of RuDP carboxylase; its *m*RNA is present in the chloroplast and it is synthesized on chloroplast ribosomes (inhibited by chloramphenicol but not by cycloheximide). The small subunit, on the other hand, is synthesized as a precursor protein on cytoplasmic ribosomes. The precursor has a 'signal' sequence of amino acids (44 in that of *Chlamydomonas reinhardtii*) at its *N*-terminal end which is apparently recognized by a specific 'pore' in the chloroplast membrane. Once the precursor has entered the chloroplast through the pore a specific chloroplastidic protease catalyses the removal of the signal segment. Eight of the resulting small subunits then combine with eight large subunits to form the functional enzyme (see Fig. 10.25).

Other proteins that are coded by *Ct*DNA are PSI chlorophyll protein, cytochrome *f*, some thylakoid membrane proteins and some subunits of the coupling factor, CF-1.

SUGGESTIONS FOR FURTHER READING

Kung, S-d. (1977) 'Expression of chloroplast genomes in higher plants', *Ann. Rev. Plant Physiol.* **28**, 401.
Hall, T. C. and Davies, J. W. (1979) *Nucleic Acids in Plants*, Vols. 1 and 2. C.R.C. Press, Florida, U.S.A.
Flavell, R. (1980) 'The molecular characterization and organization of plant chromosomal DNA sequences', *Ann. Rev. Plant Physiol.* **31**, 569.
Davidson, J. N. (1981) *Biochemistry of the Nucleic Acids*, 9th edn. (revised by Adams, R. L. P., Burdon, R. H., Campbell, A. M. and Smellie, R. M. S.). London, Chapman & Hall.
Ellis, R. J. (1981) 'Chlorophyll proteins; synthesis, transport and assembly', *Ann. Rev. Plant Physiol.* **32**, 111.

ENZYMES

1. IMP 1,2-hydrolase (decyclizing), EC 3.5.4.10.
2. ATP: nucleosidemonophosphate phosphotransferase, EC 2.7.4.4.

3. ATP: nucleosidediphosphate phosphotransferase, EC 2.7.4.6.
4. 2'-Deoxyribonucleosidediphosphate: oxidized thioredoxin 2'-oxidoreductase, EC 1.17.4.1.
5. NADPH: oxidized-thioredoxin oxidoreductase, EC 1.6.4.5.
6. UTP: ammonia ligase (ADP-forming), EC 6.3.4.2.
7. 5,10-Methylene-tetrahydrofolate: dUMP C-methyltransferase, EC 2.1.1.45.
8. Poly (deoxyribonucleotide): poly (deoxyribonucleotide) ligase (AMP forming), EC 6.5.1.1.
9. Poly (deoxyribonucleotide): poly (deoxyribonucleotide) ligase (AMP forming, NMN-forming), EC 6.5.1.2.
10. Nucleosidetriphosphate: RNA nucleotidyltransferase, EC 2.7.7.6.
11. 10-Formyltetrahydrofolate: L-methionyl-tRNA N-formyltransferase, EC 2.1.2.9.
12. Peptidyl-tRNA: aminoacyl-tRNA N-peptidyltransferase, EC 2.3.2.12.
13. 3-Phospho-D-glycerate carboxy-lyase (dimerizing), EC 4.1.1.39.

CHAPTER 11

Terpenes and Terpenoids

A. INTRODUCTION

The tremendous biosynthetic potentialities of plants are not better illustrated than by the group of natural products known as terpenoids. In general the term *terpene* is used to denote compounds containing an integral number of 5C units, whether or not they contain other elements such as oxygen. *Terpenoids* are compounds with varying numbers of carbon atoms which are clearly derived from 5C units. Probably more individual terpenes and terpenoids exist than any other group of plant products. At one time along with alkaloids and phenolics (see Chapters 13 and 14 respectively) terpenes were classed as secondary plant products; this was presumably on the assumption that they were manifestations of the unplanned biosynthetic prodigality of plants. Although there are still many terpenes which appear to be produced by the channelling of key precursors into metabolic culs-de-sac, it is now clear that the essential of plant life, photosynthesis, depends on the existence of certain terpenes and terpenoid derivatives (carotenoids and chlorophylls) (see Chapter 5). Many plant hormones are also terpenoids (see Section D and Chapter 15).

B. NATURE AND DISTRIBUTION

1. Introduction

Chemically all terpenes and terpenoids can be considered to be derived from a basic branched 5C unit:

$$\begin{array}{c} C \\ \diagdown \\ \diagup \\ C \end{array} C-C-C$$

and they are classified according to the number of such units present in the molecule. Table 11.1 indicates this classification and gives a basic example in each class. Except for the hemiterpenes, sesterterpenes and rubber, the compounds illustrated in Table 11.1 are those which, generally in the form of their pyrophosphates, are important biosynthetic intermediates; this will become obvious in Section C. The usual shorthand formula used is also given in Table 11.1. Sometimes the isoprenoid residue is indicated by *ip*. Thus farnesol is *ip ip ip*, indicating that it is formally made up of three *ip* residues joined *head to tail*. Note that the branched end of the isoprene unit is regarded as the 'head' and the unbranched end the 'tail'. Although the head to tail junction is the most common method of linking isoprene units together, *tail to tail* junctions also occur and are indicated by *ip pi*. This type of junction is to be found at the centre of tri- and tetraterpene molecules; squalene, for example (see Table 11.1), is formally composed of two farnesol residues joined tail to tail and is represented by *ip ip ip pi pi pi*.

2. Hemiterpenes

The only free hemiterpene which is widespread in plants is isoprene (see Table 11.1) and generally it only exists in extremely small amounts. It needs highly sophisticated mass spectrometry to detect isoprene in leaves, so its wide distribution has only recently come to light. The phosphorylated hemiterpenes dimethyallyl pyrophosphate (Δ^2-isopentenyl pyrophosphate) and isopentenyl pyrophosphate (Δ^3-isopentenyl pyrophosphate) (see Fig. 11.2) are key biosynthetic intermediates and thus exist only in trace amounts. However, hemiterpenes are frequently found combined with non-terpenoid moieties as *mixed terpenoids*. Some have already been discussed in Chapter 10; these are purine derivatives which occur in both the free state, when they exert cytokinin activity (see Chapter 15, Section D), and also in *t*RNA. Another typical example is *osthol* (11.1).

It must be emphasized that some 5C branched chain compounds are not isoprenoid in origin; an obvious example is the amino acid valine, which from the evidence in Chapter 9 is clearly not formed via the isoprenoid biosynthetic pathway.

3. Monoterpenes

Monoterpenes can be classified into three basic types: (i) acyclic, (ii) cyclohexanoid which can be mono-, bi- or tricyclic and (iii) cyclopentanoid. Examples are given in Table 11.2. Some *irregular* monoterpenes, such as artemesia ketone (11.2), are composed of two isoprene units which are not joined together in either the normal (*ip ip*) or even the rarer (*ip pi*) ways; these will be discussed later.

Monoterpenes are very widespread in higher plants and their strong smells have made them of great importance to the perfumery industry. They often occur in special secretory glands (e.g. mint) as important components of *essential oils*, which are produced by subjecting plants to steam distillation. The essential oils occur mainly in the Labiatae, Pinaceae and Umbelliferae. It would seem that the only plants which do not produce significant amounts of monoterpenes are in the orders of the class Violales. Monoterpene glycosides, which are

(11.1)   (11.2)

Table 11.1. *General classification of terpenes*

Type	No. of C atoms	Example Structure	Name
Hemiterpene	5		Isoprene
Monoterpene	10		Geraniol
		OR	
Sesquiterpene	15		Farnesol
Diterpene	20		Geranylgeraniol
Sesterterpene	25		Ophiobolin A
Triterpene	30		Squalene
Tetraterpene	40		Phytoene
Polyterpene	$\sim 7.5 \times 10^3$ to $\sim 3 \times 10^5$		Rubber

* Note the conventional shorthand representation of terpene structures.

Table 11.2. *Classification of monoterpenes*

Type		Name	Structure	Typical source
Acyclic		Myrcene		*Rhus* spp.
Cyclohexanoid	Monocyclic	Limonene		*Citrus* spp.
	Bicyclic	α-Pinene		*Pinus* spp.
		Thujyl alcohol	HO	*Artemesia absinthium*
		Borneol	OH	Widespread
	Tricyclic	Teresantalol	—CH₂OH	*Santalum album*
Cyclopentanoid	Iridane-derived	Loganin	H₃C H O—β—D—Glc HO O H O=C—OCH₃	*Menyanthes trifoliata*

Glc = d-glucopyranosyl.

Table 11.3. *Classification of sesquiterpenes*

Type	Name	Structure	Source
Acyclic	Nerolidol		Widespread
Monocyclic	γ-Bisabolene		Widespread
	Elemol		*Canaricum luzonicum*
	Humulene		*Humulus lupulus*
	Germacrone		*Geranium* spp.
Bicyclic	α-Cadinene		*Cedrus* spp.
	α-Eudesmol		*Eucalyptus* spp.
	Guaiol		*Eucalyptus maculata*
	β-Vetivone		*Veteveria zizanioides*

Table 11.3 (*cont.*)

Type	Name	Structure	Source
Bicyclic (*cont.*)	Maaliol		*Canarium samoense*
	Caryophyllene		*Eugenia caryophyllata*
	Eremophilone		*Eremophila mitchelli*
	Iresin		*Iresine celosioides*
	Drimenol		*Drymis winteri*
	β-Santalene		*Santalum album*
	Cuparene		*Cupressus* spp.
Tricyclic	Ledol		*Ledum palustre*
	Patchouli alcohol		*Pogostemon patchouli*
	Tricyclovetivene (= zizaene)		*Vetivera zizanioides*

non-volatile and do not appear in steam distillates, are also widespread.

4. Sesquiterpenes

The sesquiterpenes are the largest group of terpenes known, not only from the point of view of the number of compounds found in Nature (several thousand) but also from the view point of structural variation and innovation (almost 200 different skeletons). Table 11.3 gives examples of some of the main types of structures encountered. They frequently co-exist with monoterpenes in essential oils. In these cases they are found in specific cellular structures. The orders Magnoliales, Rutales, Cornales and Asterales are the most prodigal in the number and type of sesquiterpenes produced, whereas the Ranunculales seem to be the only class which accumulate no sesquiterpenes. They must produce the sesquiterpene farnesyl pyrophosphate as an intermediate in sterol biosynthesis but presumably cannot channel it into the usual sesquiterpenes. One particularly interesting compound is an aromatic sesquiterpene dimer, gossypol (11.3), found in cotton. Non-volatile sesquiterpene glycerides, esters and alkaloids are also known.

(11.3)

5. Diterpenes

The characteristic structures, with examples, of diterpenes are given in Table 11.4. The existence should be noted of two tetracyclic groups with the same basic structure but with different configurations, namely the kaurane and *ent*-kaurane group; derivatives of the latter are steviol (Table 11.4) and

the gibberellins (see Chapter 15, Section C). Over 500 different diterpenes have been reported.

6. Sesterterpenes

Until fairly recently it had been assumed that, with the exception of rubber, terpenes higher than diterpenes (20C) were structurally head to tail dimers such as the sterols (*ip ip ip pi pi pi*) and carotenoids (*ip ip ip ip pi pi pi pi*). However, recently this has been shown not to be necessarily so with the discovery of polyprenols (see Section B.10) followed by the 25C sesterterpenes. The best known sesterterpenes are all members of the ophiobolane family and are produced only by fungi. The numbering of the parent molecule is given in (11.4). A typical isolate is ophiobolin A (Table 11.1). However, five other groups are known.

(11.4)

7. Triterpenes and triterpenoids

The triterpenes and their derivatives [for example the sterols (triterpenoids)] represent another vast group of isoprenoid compounds. The main groups consist of tetracyclic derivatives based on the parent hydrocarbons lanostane (11.5), cycloartane (11.6), dammarane (11.7), and euphane (11.8) and pentacyclic compounds based on ursane (11.9), oleanane (11.10), lupane (11.11) and hopane (11.12). The numbering of the lanostane C-skeleton is given in 11.5; carbons 28 and 29 are reserved for the supernumerary carbons which appear on C–24 in plant sterols. Examples of naturally occurring compounds, which are derivatives of these basic hydrocarbons, are given in Table 11.5. It will

Table 11.4. *Some characteristic diterpenoids*

Type	Name	Structure	Source
Acyclic	Phytol		Universally distributed as a component of chlorophylls
Monocyclic	α-Camphorene		*Cinnamomum camphora*
Bicyclic	Agathic acid		*Agathis alba*
Tricyclic	Abietic acid		*Pinus palustris*
	Cassaic acid		*Erythrophleum guinaense*
Tetracyclic	A = Kaurene B = *ent*-Kaurene		*Agathis australis*
	Steviol		*Stevia rebaudiana*

Table 11.5. *Some characteristic cyclic triterpenoids*

Ring system	Name	Structure	Source
Lanostane (11.5)	Lanosterol		Fungi
Cycloartane (11.6)	Cycloartenol		*Euphorbia* spp.
Dammarane (11.7)	Dammarenediol I (20R)		Dammar resin[a]
5α-Euphane (11.8)	Euphol		*Euphorbia* spp.
Ursane (11.9)	Ursolic acid		Widely distributed in leaves and skins of fruit, e.g. apples and pears

Table 11.5 (cont.)

Ring system	Name	Structure	Source
Oleanane (11.10)	β-Amyrin		Elemi resin[b]
Lupane (11.11)	Lupeol		*Lupinus luteus*
Hopane (11.12)	Hydroxyhopanone		Dammar resin[a]

[a] Dammar is the name given to a group of natural resins originating from trees of the Dipterocarpaceae family which occur in Malaysia and Indonesia.

[b] Elemi is the name given to an oleoresin obtained from the species *Canarium* of which *C. luzonicum* is the principal: this is known as Manila elemi and comes mainly from the Philippines.

become clearer later in this chapter, when biosynthesis is described, that lanostane derivatives are only rarely found in photosynthetic organisms; they are replaced by cycloartane derivatives. Structural changes which are observed in triterpenoids include (i) ring contraction, as in ceanothic acid (11.13), which is derived from lupane (11.11), ring expansion as in serratenediol (11.14), and ring scission as in nyctanthic acid (11.15), which has a seco-oleanane ring system. More complex triterpenoids include the bitter principles from citrus fruit, the limonoids and, from the Cucurbitaceae, the cucurbitacins. Typical examples are limonin (11.16) and cucurbitacin A (11.17). Finally there exist tetracyclic products, such as α-onocerin (11.18), which has been known for over 100 years, in which cyclization appears to have taken place from each end of the molecule simultaneously. Triterpenoid alkaloids, such as cycloprotobuxine A (11.19) have recently been isolated from the Buxaceae; they are derivatives of cycloartenol.

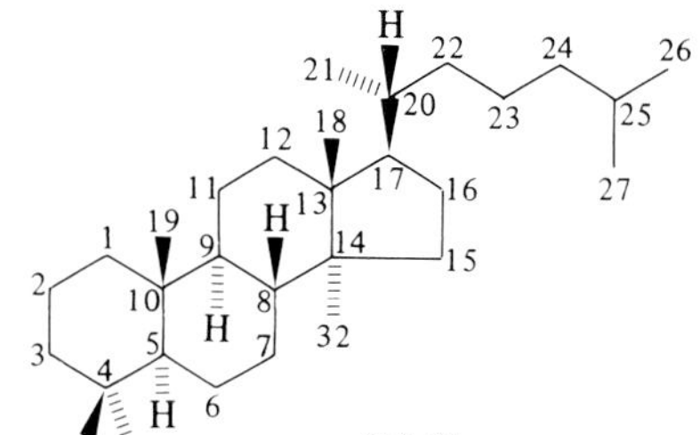

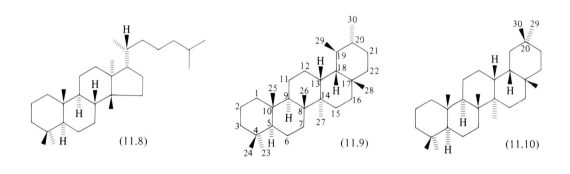

(11.5) 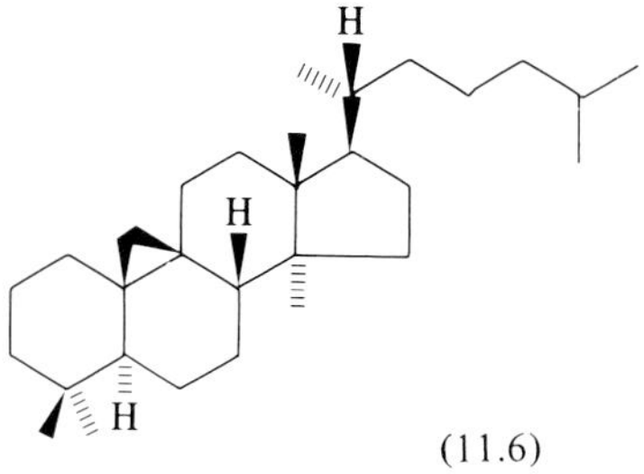   (11.6) 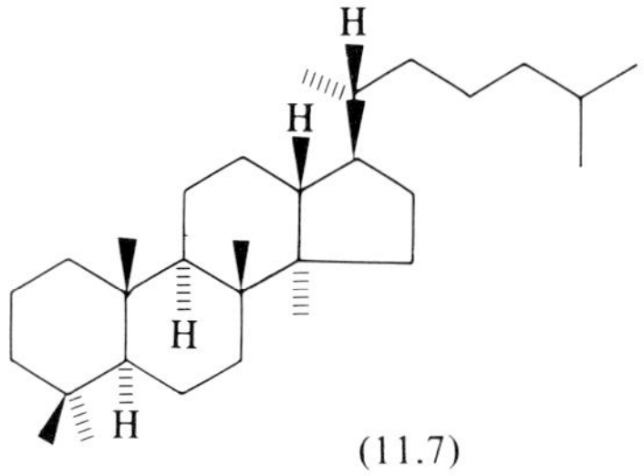   (11.7)

(11.8)   (11.9)   (11.10)

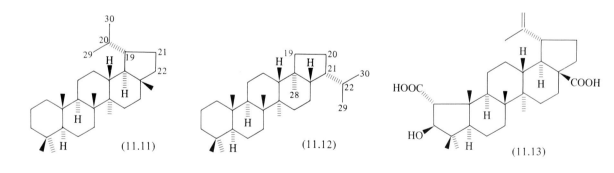

(11.11)   (11.12)   (11.13)

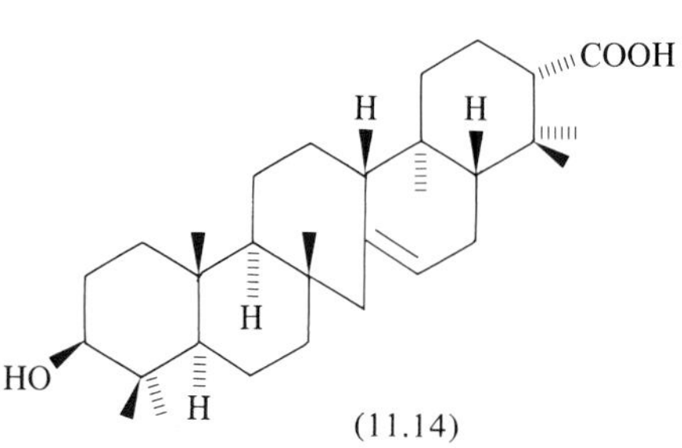

(11.14) 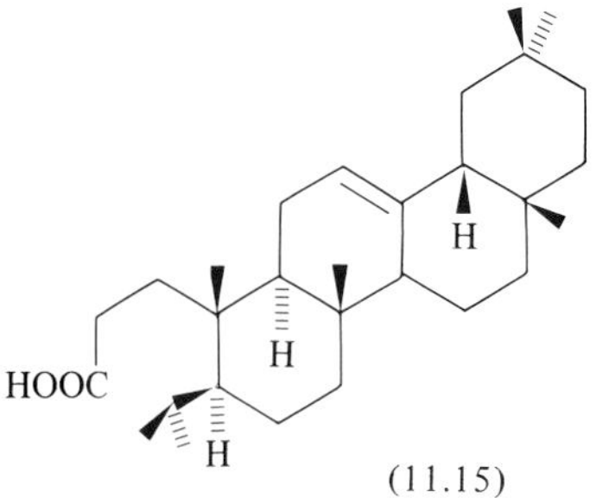   (11.15)

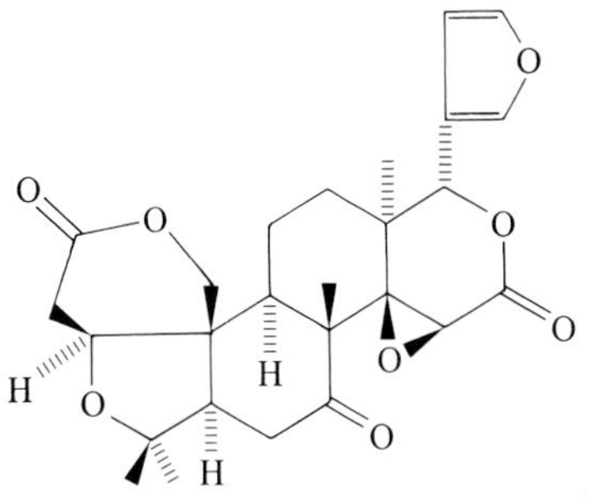

(11.16)

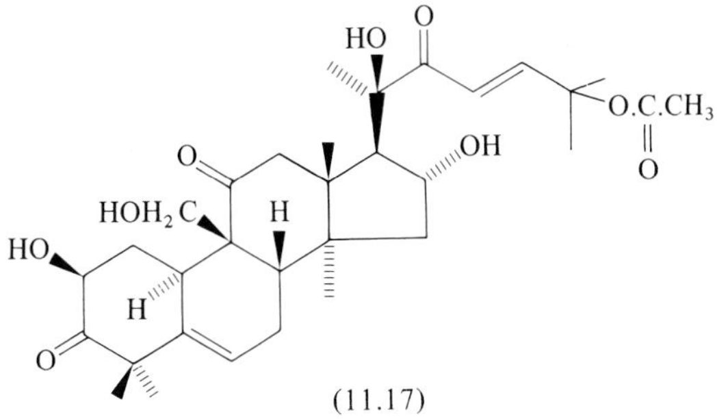

(11.17)

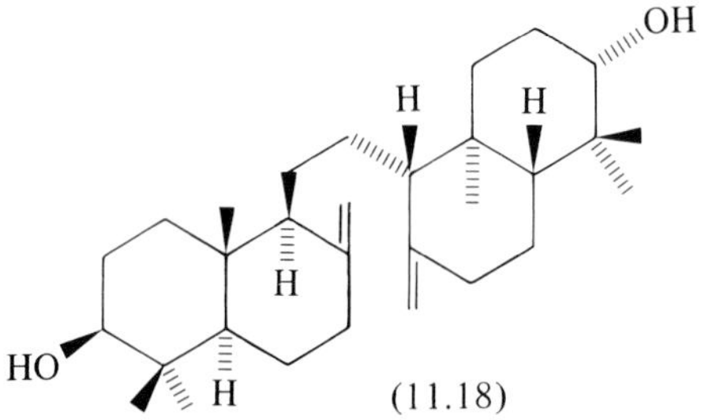

(11.18)

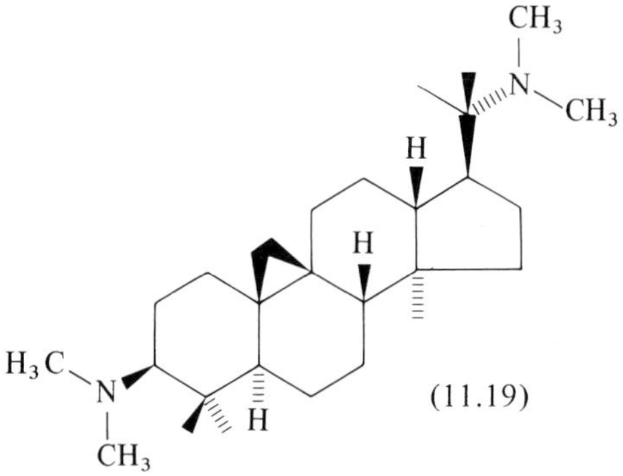

(11.19)

8. Steroids

(i) STEROLS

An examination of their molecular structure reveals that sterols are formally derivatives of tetracyclic triterpenes.

Mammalian and fungal sterols are formally derived from lanostane (11.5). This is consistent with their biosynthesis, for they arise from lanosterol (see Table 11.5) by loss of three methyl groups, two on C-4 and one on C-14, along with other structural alterations.

The sterols of higher plants and algae are also formally derived from lanostane but are biosynthetically formed from a cycloartane (11.6) derivative, cycloartenol (see Table 11.5). This involves loss of the equivalent three methyl groups and the opening of the $9\beta,19$-cyclopropane ring.

Plant sterols in general, whether from higher plants, fungi or algae, are characterized by the presence of a one-carbon or a two-carbon substituent in the side chain at C-24. These substituents are introduced by transmethylation reactions [see

Section C.6(b)]. Table 11.6 shows the various types of C-24 substituents in plant sterols. It should be noted (i) that ethylidene substituents may have either the Z or the E configuration and (ii) that methyl and ethyl substituents may have α- or β-chirality. In general 24α-methyl and 24α-ethyl sterols preponderate in higher plants; however, the great improvements in recent years in methods for separating and identifying sterols with minimal structural differences have demonstrated that among the 24-methyl sterols small amounts of 24β-epimers frequently accompany the 24α-epimers. In contrast to the sterols of higher plants, the sterols of fungi and algae usually have 24β substituents. In algae these are composed of one or two carbon atoms whilst in fungi they generally have only one carbon atom.

Sterols having no extra C-24 substituents are also found in plants. For example, cholesterol (11.20), the main sterol of mammals, is present in trace amounts in higher plants and is the predominant sterol in some red algae.

Another characteristic of plant sterols is the frequent presence of a trans-Δ^{22} double bond in the

Table 11.6. Types of C–24 substitution found in the side chain of plant sterols

No substitution at C - 24

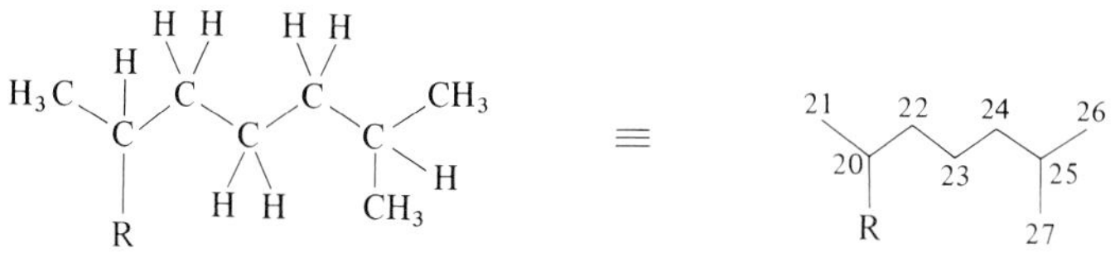

Substitution at C - 24 with a one - carbon unit

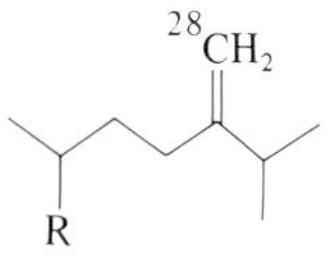

24 - Methylene 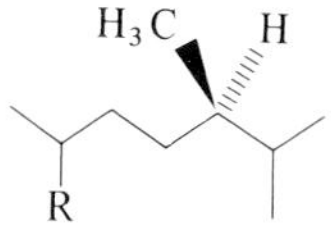   24 α - Methyl 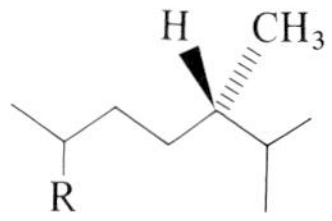   24 β - Methyl

Substitution at C - 24 with a two - carbon unit

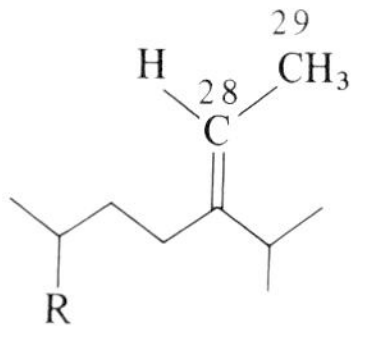

Z - 24 - Ethylidene* 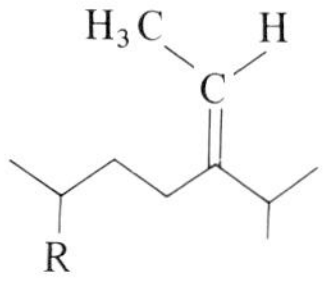   E - 24 - Ethylidene* 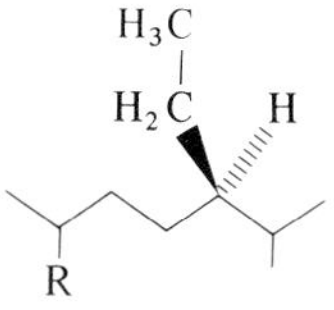   24 α - Ethyl 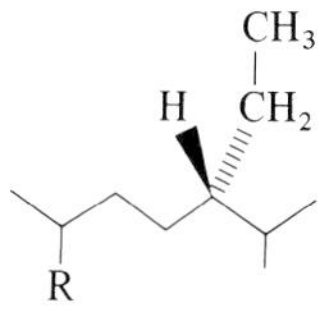   24 β - Ethyl

R = sterol nucleus

* The designations Z (from the German 'zusammen' meaning 'together') and E (from the German 'entgegen' meaning 'opposite') define the geometrical configuration around the C-24, C-28 double bond and roughly correspond to the older terms *cis* and *trans* respectively. In order to use this nomenclature the sequence-rule-preferred atom or group attached to one of the pair of atoms joined by the double bond is compared with that attached to the other; if the two sequence-rule-preferred atoms or groups are on the same side of the double bond the configuration is described as Z, whereas if they are on opposite sides it is described as E. Note that the sequence rule referred to is that of the Cahn-Ingold-Prelog (R/S) convention – see Appendix 1.

NB: The designation of alkyl substituents at C-24 as α or β is preferable to the use of the Cahn-Ingold-Prelog (R/S) convention because the latter gives opposite designations for identically oriented 24-alkyl substituents in saturated side chains and those having a Δ^{22} double bond: this may be summarized in the following way, 24α ≡ 24R (saturated side chain) ≡ 24S (Δ^{22} side chain) and 24β ≡ 24S (saturated side chain) ≡ 24R (Δ^{22} side chain). This results from a change in the priorities of the atoms and groups attached to C-24 when a Δ^{22} double bond is inserted.

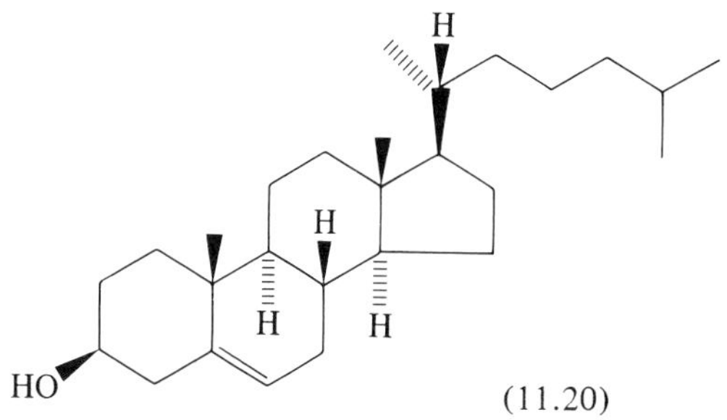

(11.20)

side chain; this feature is never seen in mammalian sterols.

The structural features of the cyclopentanoperhydrophenanthrene nucleus of plant sterols are very like those of mammalian sterols. There is always a 3β-hydroxyl group and Δ^5, Δ^7 or $\Delta^{5,7}$ double bond systems are common. In general Δ^5-sterols predominate. Table 11.7 shows the structures of some typical plant sterols.

Table 11.7. *Some typical plant sterols*

Name	Structure	Source
24-Methylene-cholesterol ($\Delta^{5,24(28)}$-sterol)		Pollen
Brassicasterol ($\Delta^{5,22}$-24β-methyl sterol)		*Brassica rapa*
Campesterol (Δ^{5}-24α-methyl sterol)		Widespread
7-Dehydroporiferasterol ($\Delta^{5,7,22}$-24β-ethyl sterol)		*Ochromonas malhamensis*
Spinasterol ($\Delta^{7,22}$-24α-ethyl sterol)		*Spinaceae oleraceae*
Fucosterol (*E*-24-ethylidene-Δ^{5}-sterol)		Brown algae (esp. *Fucus* spp.)
Δ^{7}-Avenasterol (*Z*-24-ethylidene-Δ^{7}-sterol)		*Avena sativa*

Note that the configuration at C–20 is R.

Sterols which contain one or two methyl groups at C–4, and which are normally biosynthetic intermediates between cycloartenol and 24-alkyl sterols (see Section C.6), sometimes accumulate in plant tissues in larger amounts than would be expected. They are frequently called 4α-methyl sterols and 4,4-dimethyl sterols and can be readily separated from each other and the 4-demethyl sterols, the usual end products of sterol biosynthesis, by thin layer chromatography. Examples are lophenol (11.21), first isolated from the cactus *Lophocereus schotti*, and cyclolaudenol (11.22), isolated from the opium poppy *Papaver somniferum*.

Derivatives of sterols that occur naturally are esters with palmitic, oleic, linoleic and α-linolenic acids as the usual acyl component. The 4α-methyl- and 4,4-dimethyl sterols are also encountered as fatty acid esters. However, only the 4-demethyl sterols occur as β-O-glycosides or β-O-acyl-glycosides (in the latter a hydroxyl group of the sugar residue, commonly that on C–6, is esterified to a fatty acid).

The distribution of sterols, steryl esters, steryl glycosides and steryl acylglycosides in a number of representative plant tissues is given in Table 11.8. It

Table 11.8. *Distribution of sterol and sterol derivatives in some plant tissues*

	S	SE	SG	ASG
Potato tuber	6.5	5.0	12.2	72.2
Apple fruit	68.8	7.2	18.5	5.5
Soyabean seed	59	15	11	15
Wheat endosperm				
(a) Aragon 03	13.8	82.1	1.6	2.5
(b) Mara	66.7	27.6	2.9	2.8
Tobacco leaf	64.3	13.5	18.1	4.1

S = Sterol.
SE = Steryl esters.
SG = Steryl glycosides.
ASG = Acylated steryl glycosides.

is interesting that the sterol distribution in the endosperm of two strains of wheat is markedly different.

The intracellular distribution of sterol in *Calendula* leaves and maize shoots is recorded in Table 11.9. It is clear that the bulk of the sterol is in the microsomes (endoplasmic reticular membranes) and mitochondria. The distribution of steryl esters, normally constituting about 1% of the total sterol, is

Table 11.9. *Distribution of sterol amongst intracellular organelles*

	% of total present			
Tissue	Cell walls	Chloroplasts	Mitochondria	Microsomes
Calendula leaf	4	5	24	66
Maize shoots				
(a) Free sterol	8.4	7.4	27.3	56.9
(b) Esterified sterol	73.0	8.7	9.0	9.3

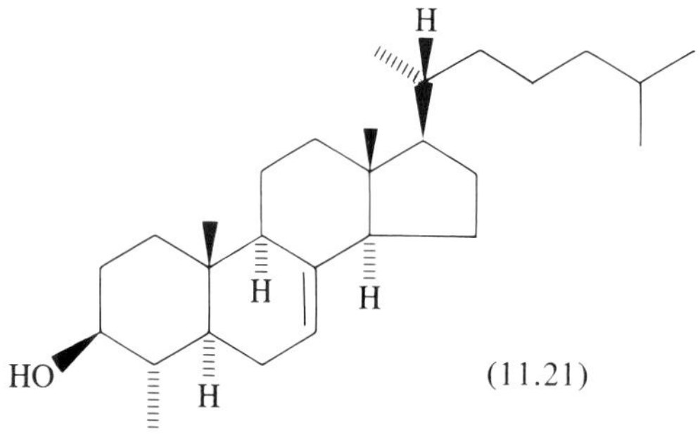

(11.21)

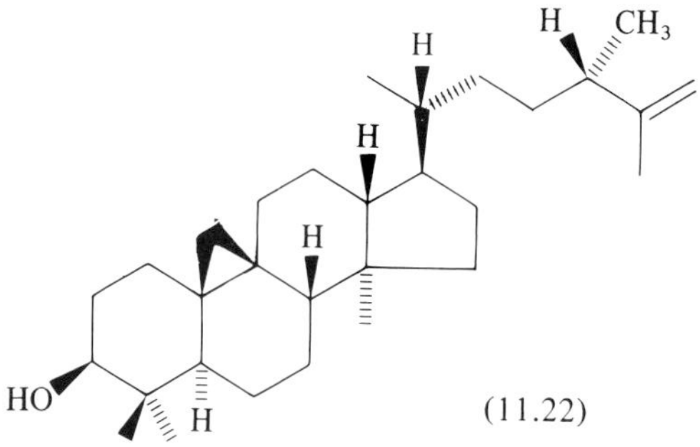

(11.22)

markedly different in that most is associated with the cell-wall fraction.

(ii) SAPOGENINS

Sapogenins are C–27 sterols in which the side chain has undergone metabolic changes to produce a spiroketal. Such a compound is tigogenin (11.23). Sapogenins occur naturally as the saponins, which are 3-O-glycosides of the parent steroid. A well-known example is digitonin (11.24) from *Digitalis*. Saponins are detergents and disintegrate membranes, hence their ability to cause haemolysis of red blood cells. They are widely distributed.

(iii) CARDIAC GLYCOSIDES

Steroid glycosides with important pharmacological effects on heart muscle are found in about eleven plant families. They are either 23-C compounds, called cardenolides, or 24-C compounds which are called bufadienolides because they were originally found in toad venom (Latin: *bufo*, toad). They arise by loss of carbon atoms from the sterol side chain. Both types do not occur together in one plant. The glycosidic parts of these molecules are often complex and contain unique sugars; however, typical

aglycones are digitoxigenin (11.25) (cardenolide) from *Digitalis* and hellebrigenin (11.26) (bufadienolide) from the rhizomes of the Christmas rose.

(iv) STEROID HORMONES

A number of steroids, which are known to be hormones in animals, are somewhat widespread in plants but their biological activity, if any, in plants has yet to be clearly demonstrated. The animal hormones are derived from cholesterol after partial or complete removal of the side chain and this is probably how they arise in plants (see Section C.6). The more common of these compounds that are produced by plants are outlined in Table 11.10.

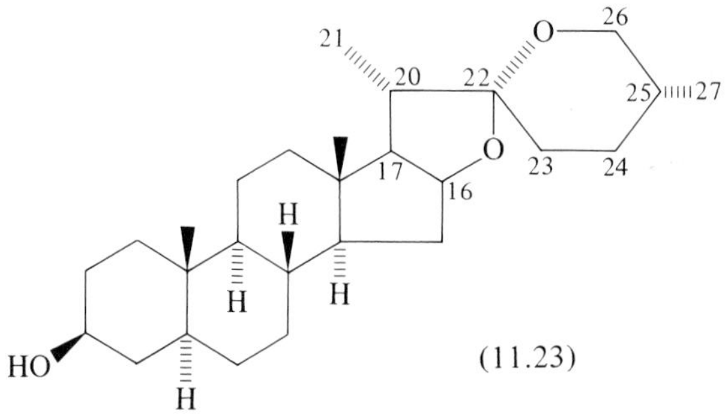

(11.23)

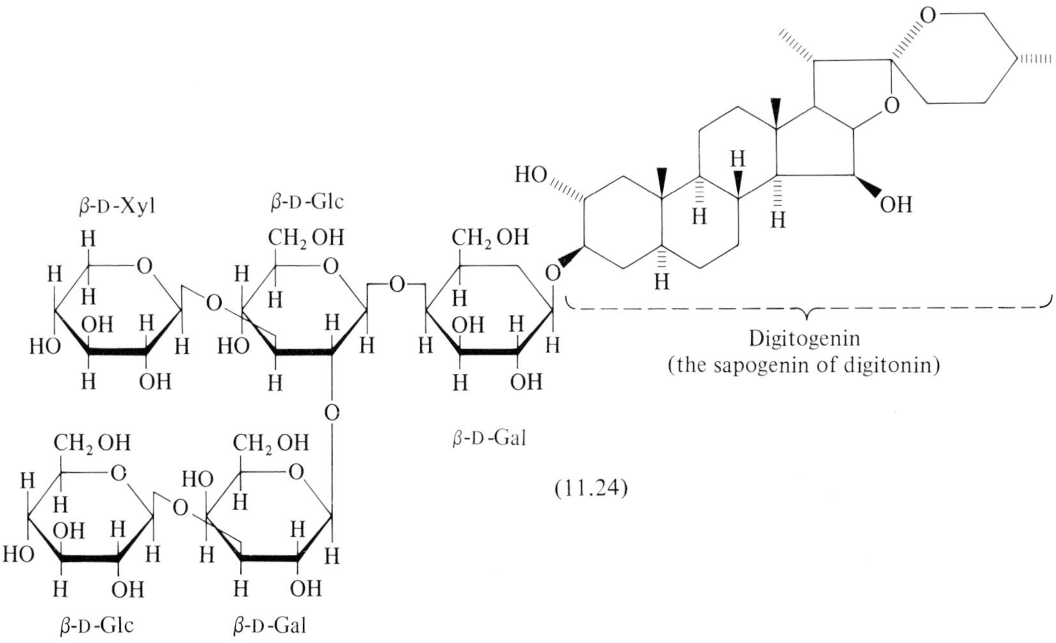

β-D-Xyl   β-D-Glc

Digitogenin
(the sapogenin of digitonin)

(11.24)

β-D-Glc   β-D-Gal

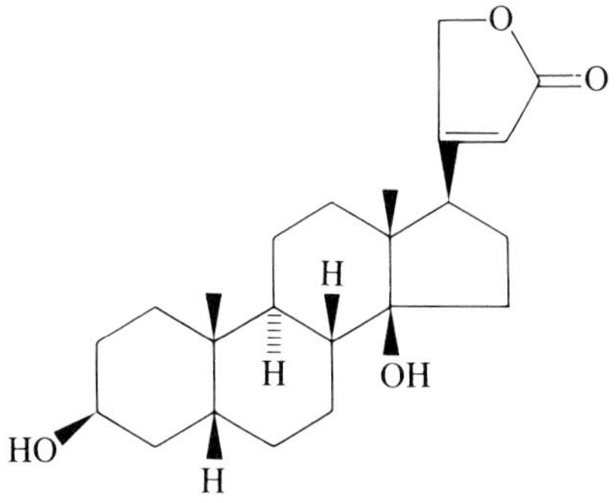

(11.25)

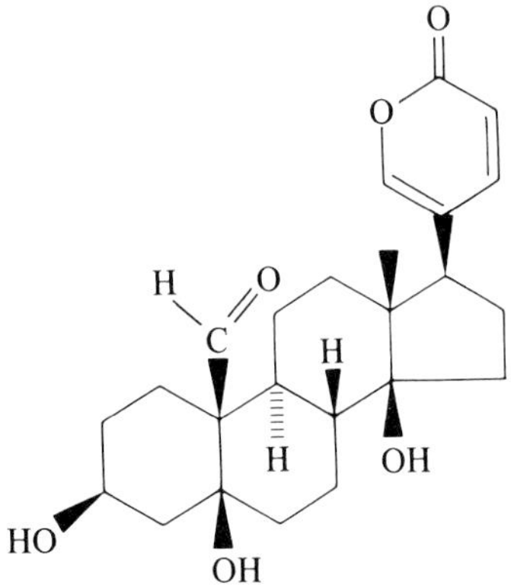

(11.26)

Table 11.10. *Occurrence of some animal steroid hormones in plants*

Name (animal source)	Structure	Plant source
Progesterone (corpus luteum of the ovary; placenta)	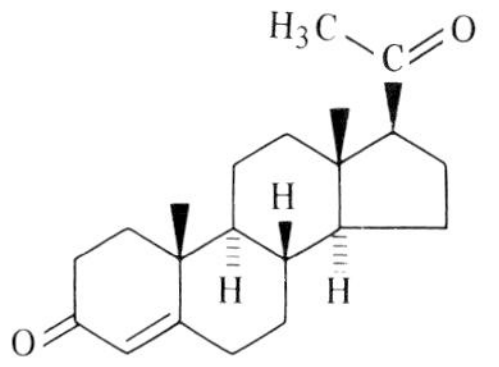	*Hollarrhena floribunda*
Deoxycorticosterone (adrenal cortex)	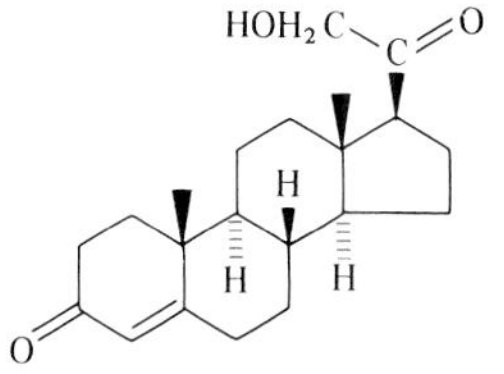	*Digitalis lanata*
Androstanetriol[a] (testes)	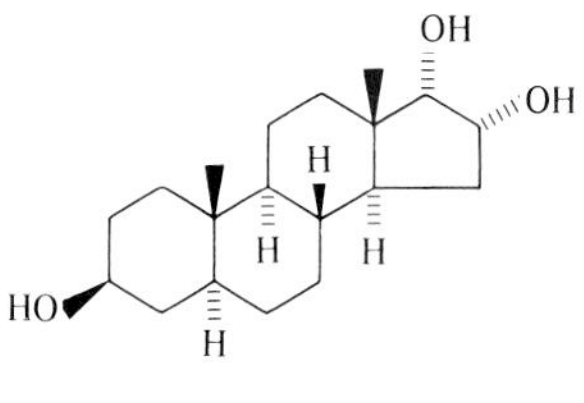	*Haplopappus heterophyllus*

[a] Androstanetriol is not technically a testicular hormone but rather a metabolite of testosterone.

These compounds are extremely polar substances which have only recently been discovered in plants. They have the same basic structure as the insect moulting hormones and are not widely distributed (in about 80 families so far). Typical examples are ecdysone (11.27) isolated from bracken (*Pteridium*) rhizomes and ponasterone A, which is 20R-hydroxyecdysone, from the conifer *Podocarpus*.

Two main groups of steroid alkaloids exist, the 27-C alkaloids, which appear to be formed directly from cholesterol, and the 21-C alkaloids presumably produced from pregnenolone (11.28).

Many of the 27-C alkaloids are nitrogen analogues of the 27-C sapogenins; solasodine (11.29) is a typical example. Other groups, as exemplified by rubijervine (12α-hydroxysolanidine) (11.30), contain nitrogen as part of the six-membered condensed ring system of the molecule. Rubijervine is obtained from *Veratrum* roots, which also produce poly-hydroxy triterpenoid alkaloids such as germine (11.31) which occurs naturally as various esters with

acids such as α-methylbutyric acid and acetic acid (see also Section B.7).

A number of 21-C alkaloids have been recently isolated from the Apocynaceae and Buxaceae. For example, holaphyllamine (11.32) has been isolated from *Holarrhena* (Apocynaceae).

9. Tetraterpenes—Carotenoids

Tetraterpenes consist of only one group, the carotenoid pigments. About 500 structures are known but only a limited number are found in green leaves (see Chapter 5). They are localized in the chloroplasts; their protective presence is essential for the survival of chloroplasts under light/aerobic conditions and they also contribute to light harvesting in photosynthesis [see Chapter 5, Sections C.2 (ii), C.3 and C.4].

Some information about carotenoid structures has already been given in Chapter 5 (Fig. 5.5); here the subject is taken further but reference to Fig. 5.5 will be necessary. The colour of carotenoids is due to the presence of a long series of conjugated double

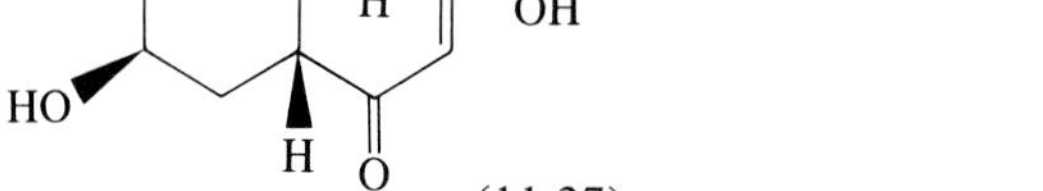

(11.27)

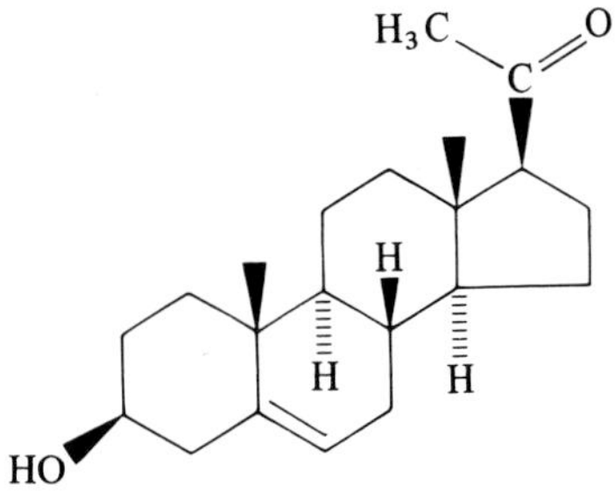

(11.28)

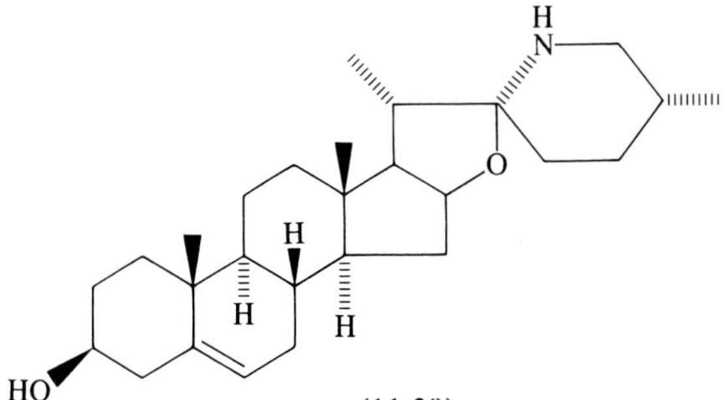

(11.29)

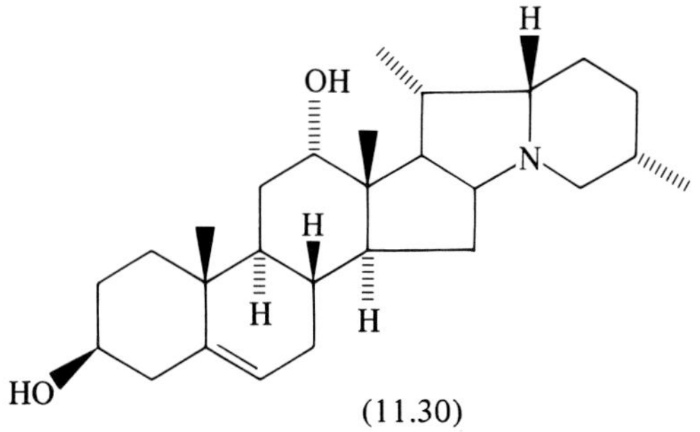

(11.30)

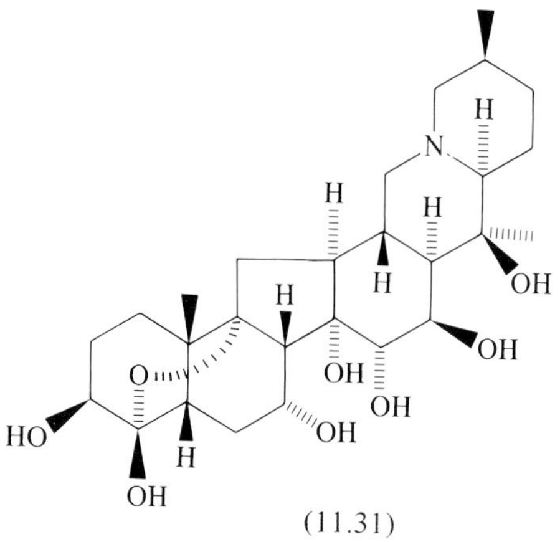

(11.31)

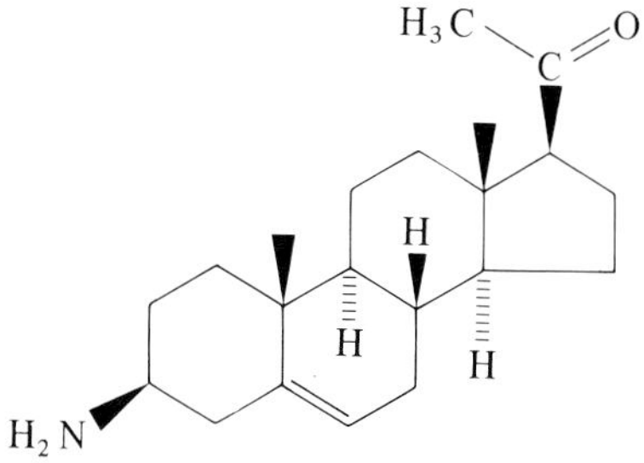

(11.32)

bonds (7–13) present in the molecule and the additional individual pigments of higher plants are the result of variations on the basic structure of the acyclic lycopene (11.33), the characteristic carotenoid of tomato fruit. Cyclization at one end of the lycopene molecule can result in the formation of a β-ring (end group) as in γ-carotene (11.34), an ε-ring (end group) as in δ-carotene (11.35) or a γ-ring (end group) as in (11.36). Cyclization at both ends can yield pigments such as β-carotene (11.37), α-carotene (11.38), or ε-carotene (11.39). The acyclic end group, as in lycopene (11.33), is termed a ψ-group. The numbering of the carotenoid molecule is illustrated for α-carotene (11.38). The rule is that the plain numerals are assigned to that half of the molecule which carries the end group characterized by the Greek letter nearest the beginning of the alphabet (i.e. $\beta > \varepsilon > \gamma > \psi$). Hydrocarbon tetraterpenes are termed *carotenes*; those containing oxygen functions are termed *xanthophylls*; oxygen functions involved include hydroxy groups as in lutein (3,3′-dihydroxy-α-carotene) (IX, Fig. 5.5), epoxide groups as in violaxanthin (5,6,5′6′-diepoxyzeaxanthin, X, Fig. 5.5) [zeaxanthin is 3,3′-dihydroxy-β-carotene (VIII, Fig. 5.5)], and furanoid oxides as in 5,8-epoxy-β-carotene (11.40).

Other important structural variations include the appearance of allenic groups, as in neoxanthin (XII,

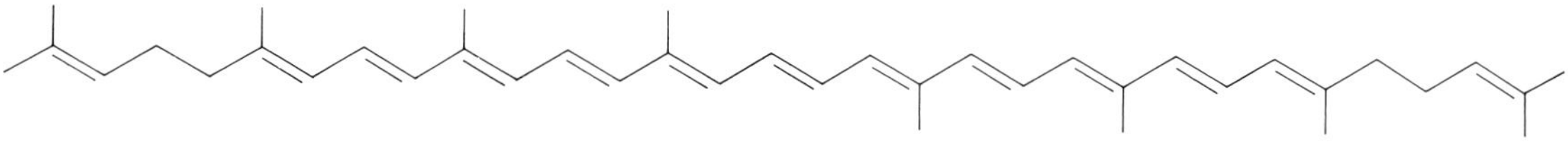

(11.33)

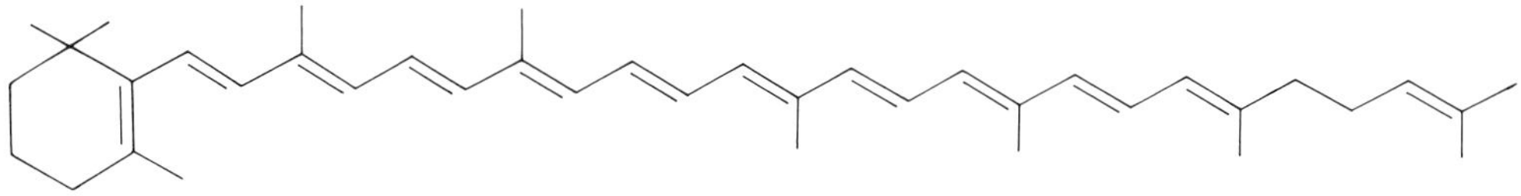

(11.34)

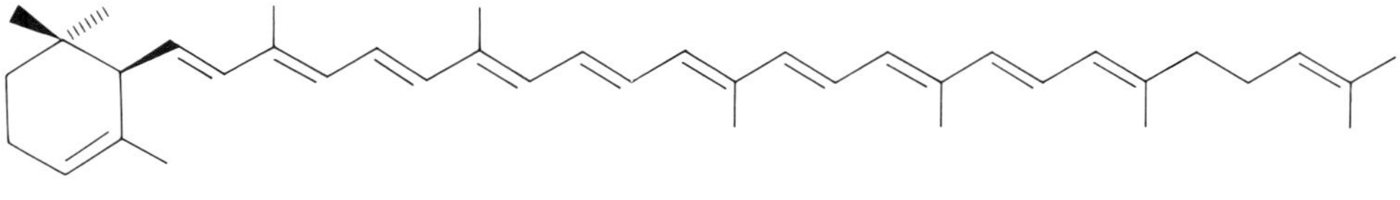

(11.35)

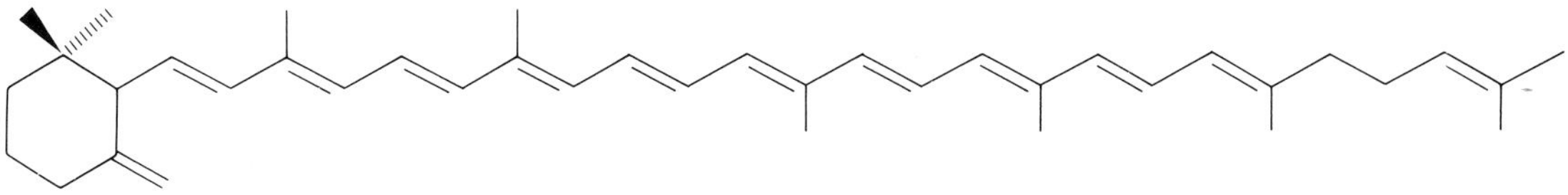

(11.36)

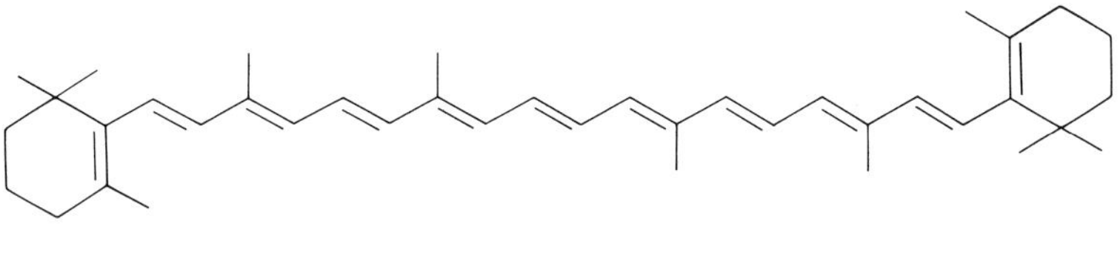

(11.37)

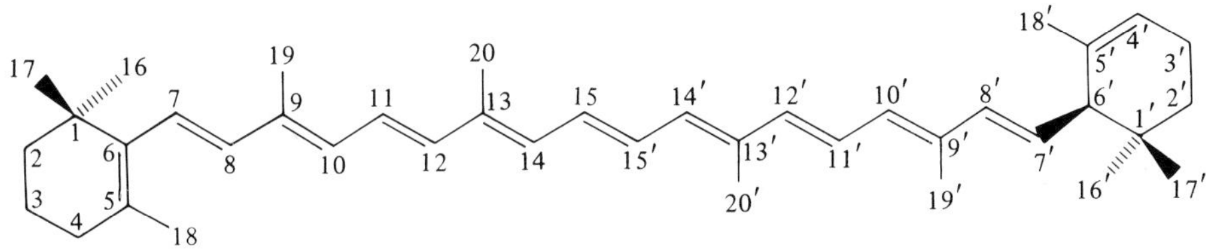

(11.38)

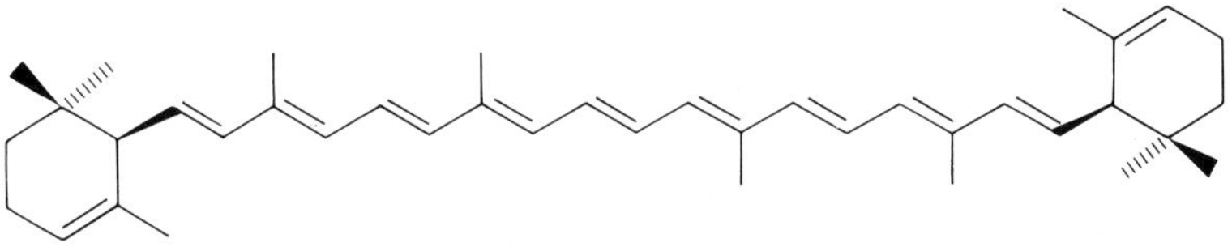

(11.39)

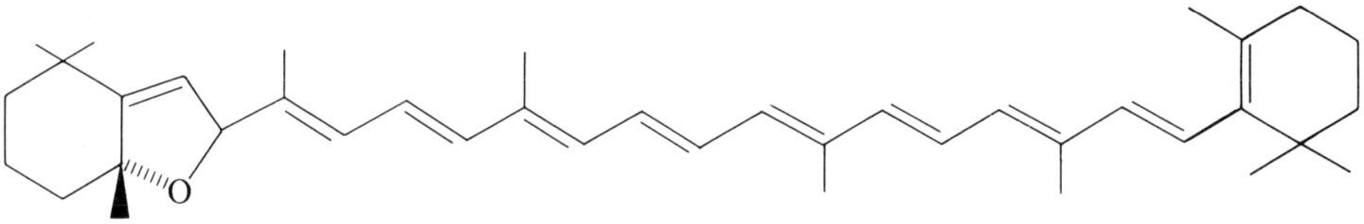

(11.40)

Fig. 5.5), and acetylenic bonds as in alloxanthin (XV, Fig. 5.5). It will be noted that all the double bonds in the structures so far considered have the *trans* configuration. However, pigments with a *cis* configuration are occasionally encountered, as, for example, prolycopene (11.41) (7,9,7′,9′-tetra-*cis*-lycopene) which is present in the fruit of certain tomato mutants.

Apart from being present in all photosynthetic tissues, including algae and photosynthetic bacteria, carotenoids are widespread, but not universally present, in non-photosynthetic organs of higher plants, and in fungi and non-photosynthetic bacteria.

The group of carotenoids present in green leaves is qualitatively the same throughout the higher plant kingdom (mainly β-carotene, lutein, violaxanthin and neoxanthin). However, in organs which are carotenogenic such as petals and fruit, wide structural variations are encountered. Carotenogenic flowers fall into three main groups according to whether they accumulate (i) highly oxidized pigments such as auroxanthin (5,8,5′,8′-diepoxyzeaxanthin), (ii) hydrocarbons such as lycopene or β-carotene, or (iii) highly specific carotenoids such as eschscholtzxanthin (11.42) from the californian poppy. The pigments accumulate in chromoplasts which develop from chloroplasts. Carotenoids also

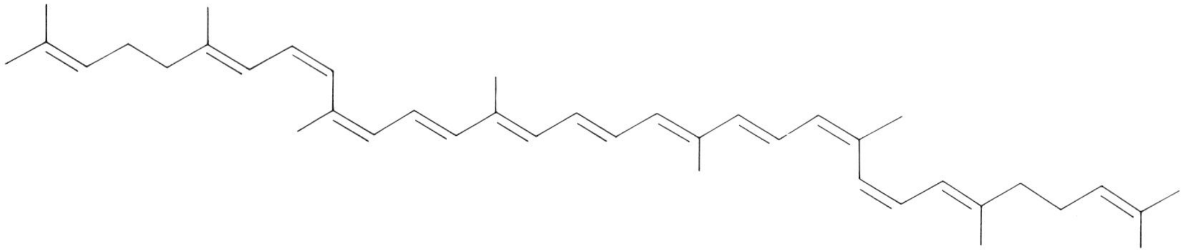

(11.41)

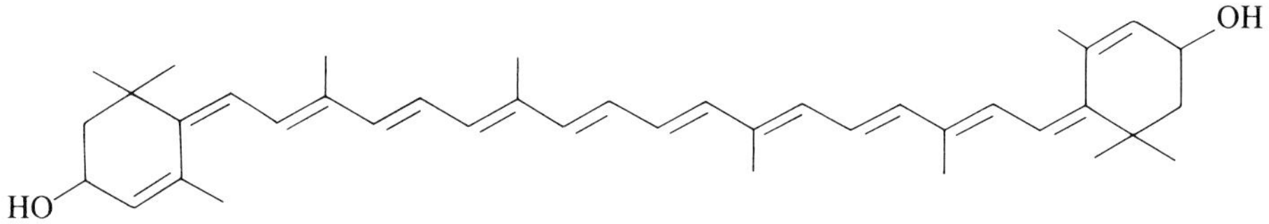

(11.42)

accumulate in the chromoplasts in fruit. Fruit can be divided roughly into nine groups according to the carotenoids they accumulate; typical examples are collected in Table 11.11. The brilliant red pigment of peppers is capsanthin (11.43) in which one of the usual cyclohexene end groups has contracted to a cyclopentane ring. The secocarotenoids, such as semi-β-carotenone (11.44), have had one cyclohexane ring split between C–5 and C–6 and the apocarotenoids which are characteristic of citrus fruit (Group IX; Table 11.11) have a portion of one end of the molecule chopped off. β-Citraurin is 3-hydroxy-β-8′-apocarotenal (11.45).

The domestic carrot is, of course, the classical source of 'carotene' (a mixture of all the carotenes but with β-carotene generally comprising more than 90% of the total).

10. Polyprenols

An acyclic diterpenoid alcohol, phytol (Table 11.4), has been known for a long time in esterified linkage with chlorophyll (Chapter 12). During the past 25 years, however, much larger prenols have been found free in plant tissues; the first to be fully described was the 45C, nonaprenol, solanesol (11.46) isolated from tobacco leaves. This has all its double bonds in the *trans* configuration. Families of polyprenols with mixed (*cis* and *trans*) double

bonds exist; typical examples are the castaprenols from horse chestnut (*Aesculus hippocastanum*) and the betulaprenols from the silver birch (*Betula verrucosa*). The convention for describing these mixed polyprenols can be considered in relation to the castaprenols which are represented as ω–T–T–T–[C]$_n$–C–OH. This means that the three isoprene units following the ω-isoprene residue are *trans* whilst the remainder are *cis*. In the castaprenols n can vary between 4 and 8.

Table 11.11. *Summary of carotenoid distribution in fruit*

Group	Example	Carotenoids
I	*Pyracantha rogersiana*	Traces only
II	*Sambucus nigra*	Chloroplast carotenoids
III	*Lycopersicon* (tomato)	Lycopene series
IV	*Mangifera indica* (mango)	β-Carotene series
V	*Crataegus pratensis*	Epoxides
VI	*Capsicum annuum* (pepper)	Species-specific; capsanthin (11.43)
VII	*Pyracantha angustifolia*	Procarotenes (poly *cis*); prolycopene (11.41)
VIII	*Murraya exotica*	Secocarotenoids; semi-β-carotenone (11.44)
IX	*Citrus* spp.	Apocarotenoids β-citraurin (11.45)

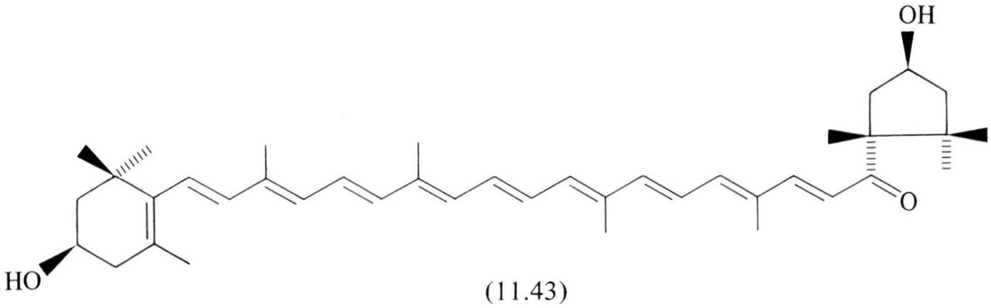

(11.43)

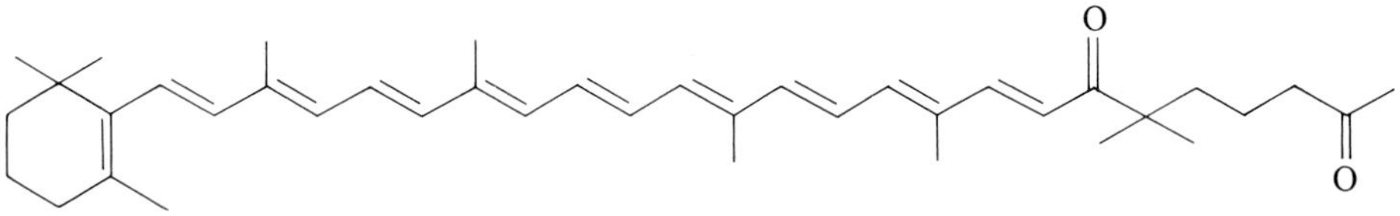

(11.44)

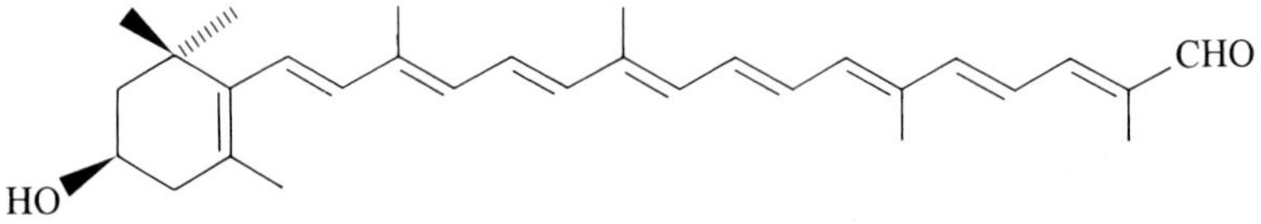

(11.45)

All-*trans* polyprenols are found attached to a number of metabolically important quinones, such as the plastoquinones (11.47), the ubiquinones (11.48) and phylloquinone (vitamin K_1) (11.49). Chromanols derived from plastoquinone by cyclization of the first isoprene residue of the side chain, e.g.

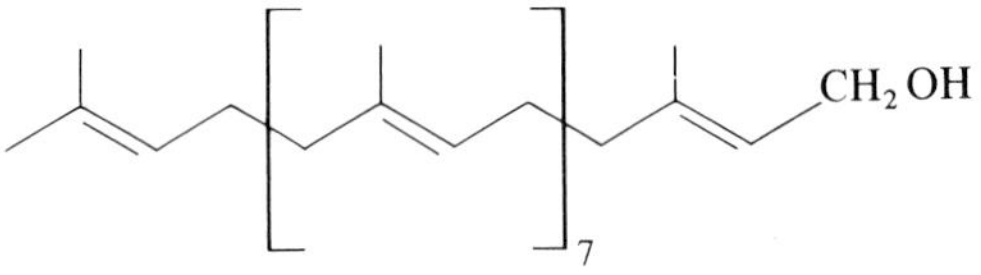

(11.46)

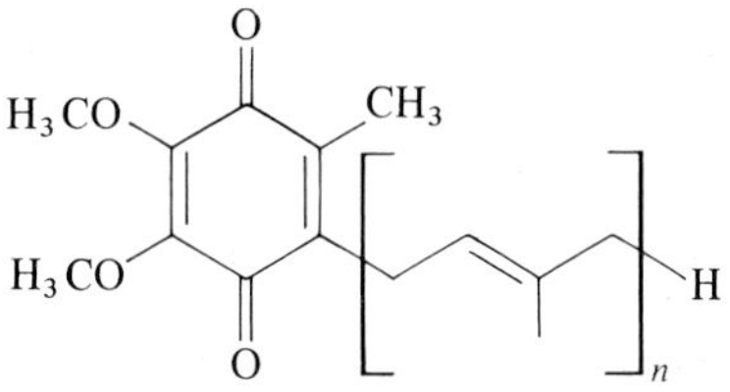

n is most commonly 9 or 10 in higher plants

(11.48)

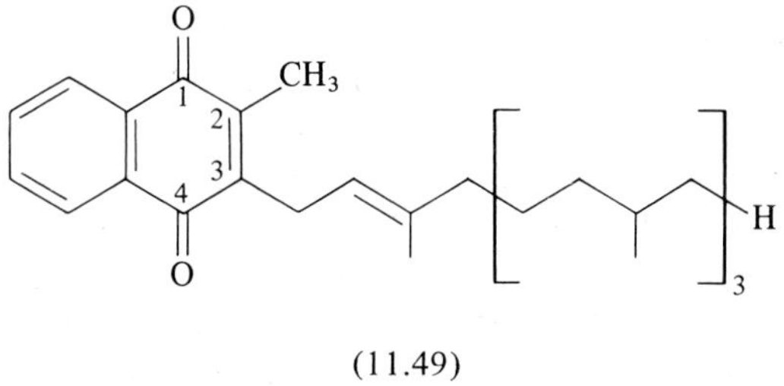

plastochromanol-8 (11.50), also exist in plants. The ubiquitous tocopherols (vitamins E) are also chromanols (11.51); however, in these compounds the isoprene units of the side chain are saturated.

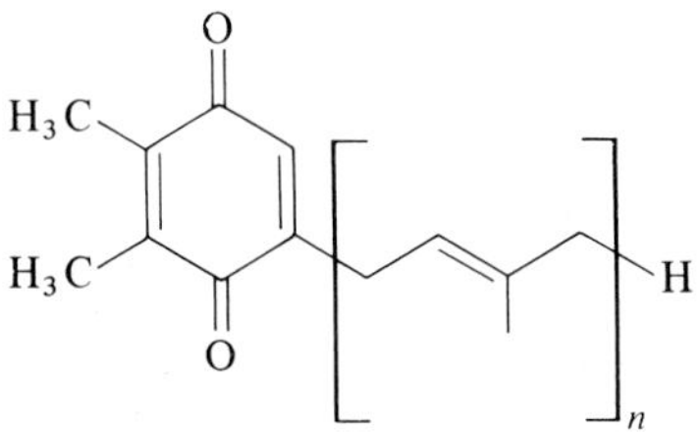

n is most commonly 9 in higher plants

(11.47)

(11.49)

(11.50)

(11.51)

	R_1	R_2
α-Tocopherol	CH_3	CH_3
β-Tocopherol	CH_3	H
γ-Tocopherol	H	CH_3
δ-Tocopherol	H	H

11. Rubber, gutta and chicle

(i) RUBBER

Rubber is a high molecular weight polyisoprene produced in the latex of about 300 genera of Angiosperms. Only *Hevea brasiliensis* is used commercially and high-yielding clones have been developed in Malaysia which produce at least 1500 pounds of rubber per acre per year. The plant breeders have bred out to a great extent the regulatory mechanism which controls the channelling of photosynthate into rubber, so that far more rubber is synthesized than would be produced in normal strains of *Hevea*. The effect of this high rubber production on growth is considerable in young trees from high-yielding clones; the reduction

in girth increment during the first 8 months of tapping can be as great as 70%.

The latex which contains the rubber particles accumulates in specialized cells or vessels known as lactifers. In *Hevea* rubber is formed and stored in the rings of lactifers in the bark. Anastomoses between adjacent vessels in a ring allow the latex from a large area of the cortex to drain on tapping.

Latex contains many organelles in addition to rubber particles; these include nuclei, mitochondria, fragments of endoplasmic reticulum and ribosomes as well as two specialized particles, the lutoids and Frey-Wyssling particles. Lutoids, a major component of latex, are osmotically active spheres which are 1–3 μm in diam., grey in colour and bounded by a unit membrane. The lipids of the latter are rich in phosphatidic acid and saturated fatty acyl residues. The name lutoid (=yellow) turned out to be a misnomer because as originally isolated they were contaminated with the yellow Frey-Wyssling particles, which contain β-carotene.

Frey-Wyssling particles represent a minor component of latex; they are 4–6 μm in diam., are bounded by a double membrane and contain many membranes and tubular structures as well as β-carotene.

An electron micrograph of an ultra-thin section of young latex vessels is given in Fig. 11.1. Rubber is a *cis*-1,4-polyisoprene (11.52) with a wide spectrum of molecular weights ranging from $\sim 1 \times 10^5$– $\sim 4 \times 10^6$ which are equivalent to polymers of ~ 1500–$\sim 60,000$ isoprene residues.

(ii) GUTTA

Gutta is a *trans*-1,4-polyisoprene (11.53) with a lower molecular weight range than rubber. It is produced by various trees of the genus *Palaquium* (Sapotaceae) growing in Malaysia. The latex does not flow as readily as rubber latex and so harvesting involves the destruction of the trees. This has led to the virtual extinction of the former main source of gutta, *P. gutta*. It is now produced commercially on a small scale from guayule (*Parthenium argentatum*), a desert shrub of the family Compositae, native in Mexico and Texas. Gutta is thermoplastic: it is a tough, hard, inelastic solid below 65°C but at this temperature it changes to a soft, pliable, but still inelastic, form.

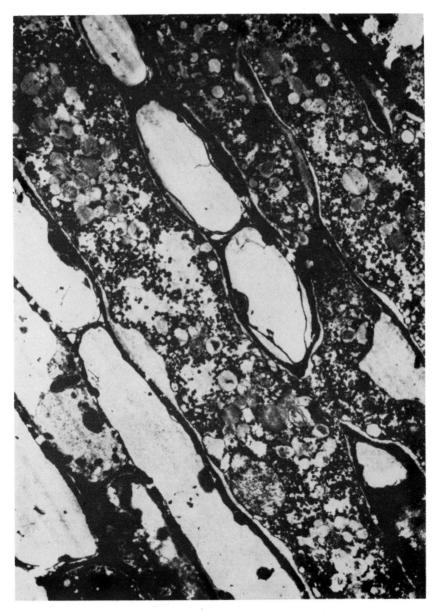

Fig. 11.1. Electron micrograph of an ultra-thin longitudinal section of young latex vessels from a green branch region of a 5-year-old *Hevea brasiliensis* tree (kindly provided by Dr. P. B. Dickenson.)

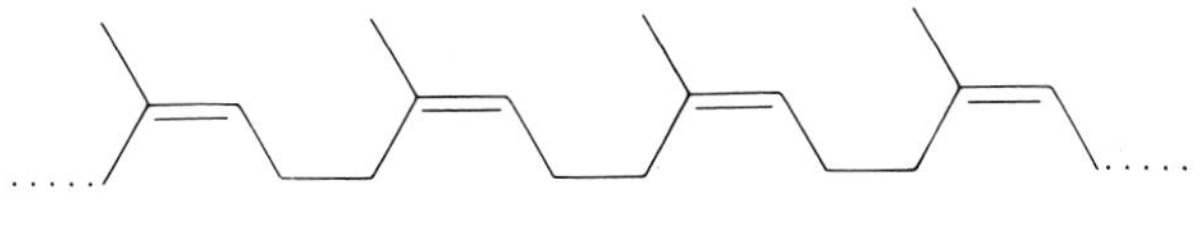

(11.52)

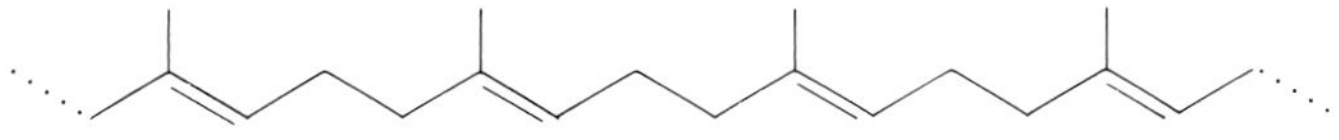

(11.53)

(iii) CHICLE

Chicle, the original chewing-gum base, is obtained from *Achras sapota*. It is a mixture of relatively low molecular weight *cis*- and *trans*-1,4-polyisoprenoids together with acetone-soluble resins.

C. BIOSYNTHESIS

1. Basic reactions

As long ago as 1887 Wallach proposed the 'isoprene rule'. This, in essence, stated that many naturally occurring substances were built up of isoprene C-skeletons and therefore probably arose by the polymerization of isoprene. The ideas behind this rule were extended by Ruzicka who developed the *biogenetic isoprene rule* and proposed that all terpenoids were synthesized from a precursor called 'active isoprene'. This became a biochemical reality when Lynen identified this substance as Δ^3-isopentenyl pyrophosphate (IPP). The key precursor of IPP was shown by Folkers in 1956 to be the 6-C hydroxy acid, (3*R*)-mevalonic acid (11.54).

The basic precursor of terpenoid biosynthesis is acetyl-CoA and the pathway from this to IPP is outlined in Fig. 11.2.

IPP is then converted into all the different terpenes and terpenoids found in Nature according to the scheme shown in Fig. 11.3. IPP is firstly isomerized by the sulphydryl-enzyme, IPP isomerase[1], to dimethylallyl pyrophosphate (DMAPP; also known as Δ^2-isopentenyl pyrophosphate) [eq. (11.1)]. Either of these compounds

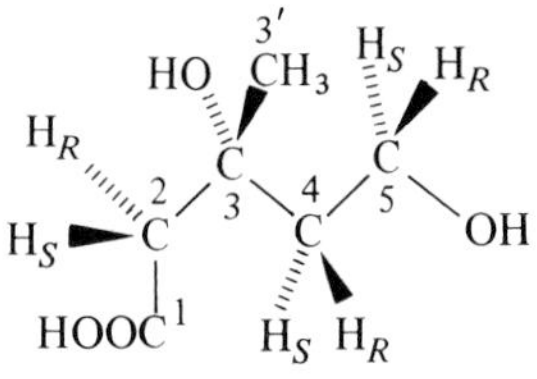
HO
3′
CH3
HS
HR
HR
2
3
4
5
OH
HS
HOOC 1
HS HR

(11.54) (+)-*R*-Mevalonic acid (MVA)

(The designation of the hydrogen atoms on Cs 2, 4 and 5 as *R* or *S* indicates that they are pro-*R* or pro-*S* hydrogens – see Appendix 1)

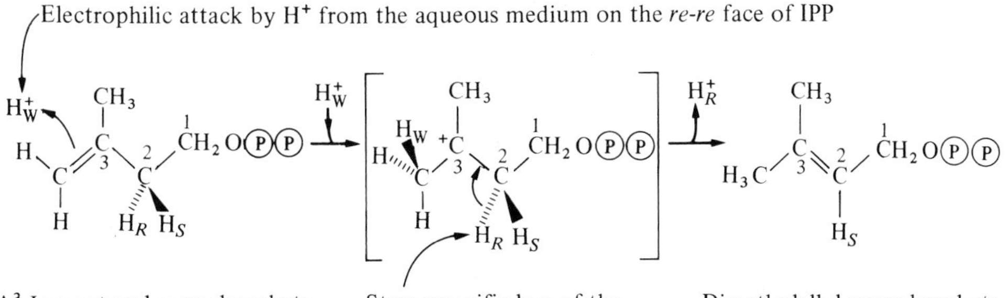
Electrophilic attack by H+ from the aqueous medium on the re-re face of IPP
CH2O(P)(P)
HR HS

(11.1)

Δ^3-Isopentenyl pyrophosphate (IPP)

Stereospecific loss of the 2 pro-*R* hydrogen of IPP which is the former 4 pro-*S* hydrogen of MVA

Dimethylallyl pyrophosphate (DMAPP) (≡ Δ^2-Isopentenyl pyrophosphate)

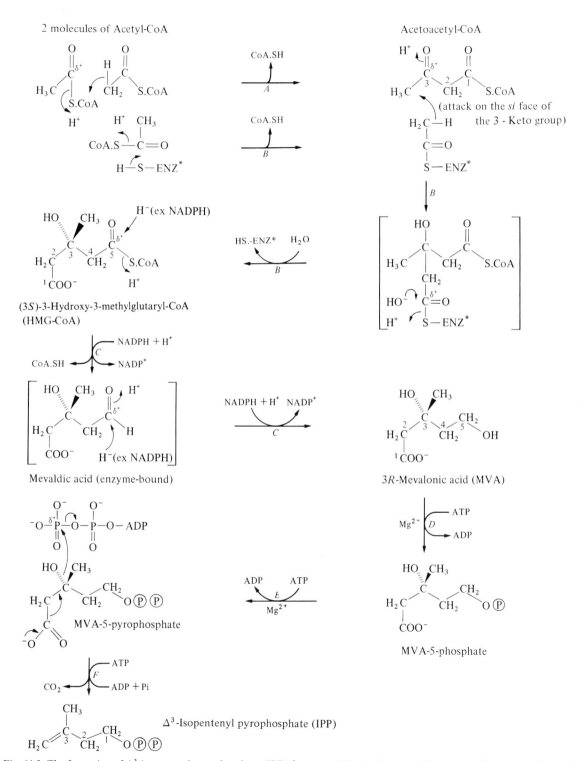

Fig. 11.2. The formation of Δ^3-isopentenyl pyrophosphate (IPP) from acetyl-CoA. (A = Acetyl-CoA .acetyl-CoA C-transferase, EC 2.3.1.9; B and ENZ* = hydroxymethylglutaryl-CoA synthase, EC 4.1.3.5; C = hydroxymethylglutaryl-CoA reductase (NADPH), EC 1.1.1.34; D = mevalonate kinase, EC 2.7.1.36; E = phosphomevalonate kinase, EC 2.7.4.2; F = pyrophosphomevalonate decarboxylase, EC 4.1.1.33.)

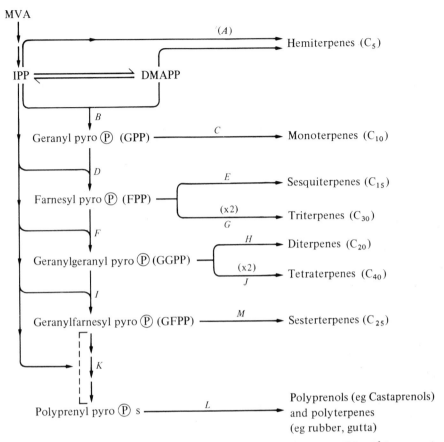

Fig. 11.3. General scheme of terpenoid biosynthesis. (MVA = (3*R*)-mevalonic acid; IPP = Δ^3-isopentenyl pyrophosphate; DMAPP = dimethylallyl pyrophosphate; X2 = tail-to-tail dimerization.)

can then be converted into hemiterpenes (A, Fig. 11.3). DMAPP also acts as a starter for chain elongation and under the influence of the enzyme prenyltransferase[2] condenses with a molecule of IPP to form the C_{10}-compound, geranyl pyrophosphate (GPP) (B, Fig. 11.3 and Fig. 11.4). Mechanistically DMAPP plays the same role in this reaction as does H^+ in eq. (11.1). The GPP can then either be channelled into monoterpene biosynthesis (C, Fig. 11.3) or condense with another IPP molecule to form the C_{15} compound, farnesyl pyrophosphate (FPP) (D, Fig. 11.3). FPP has three possible metabolic fates, to be channelled into sesquiterpenes (E, Fig. 11.3), to undergo chain extension to form the C_{20} compound, geranylgeranyl pyrophosphate (GGPP) (F, Fig. 11.3), or to undergo 'tail-to-tail' dimerization with the formation of the C_{30} triter-

penes (G, Fig. 11.3). GGPP has the same type of metabolic options as FPP, diversion to diterpenes (H, Fig. 11.3), chain elongation to the C_{25} compound, geranylfarnesyl pyrophosphate (GFPP) (I, Fig. 11.3) and dimerization to the C_{40} tetraterpenes (J, Fig. 11.3). GFPP generally undergoes further multiple chain extension to yield polyprenyl pyrophosphates (K, Fig. 11.3) and ultimately polyprenols (e.g. castaprenols) and polyterpenes (e.g. rubber) (L, Fig. 11.3): occasionally it is converted into sesterterpenes (M, Fig. 11.3).

When the reactions from mevalonic acid to geranyl pyrophosphate and farnesyl pyrophosphate are considered in detail, certain important stereochemical questions emerge, which have been answered by Cornforth and Popják and their colleagues by sophisticated experiments based on

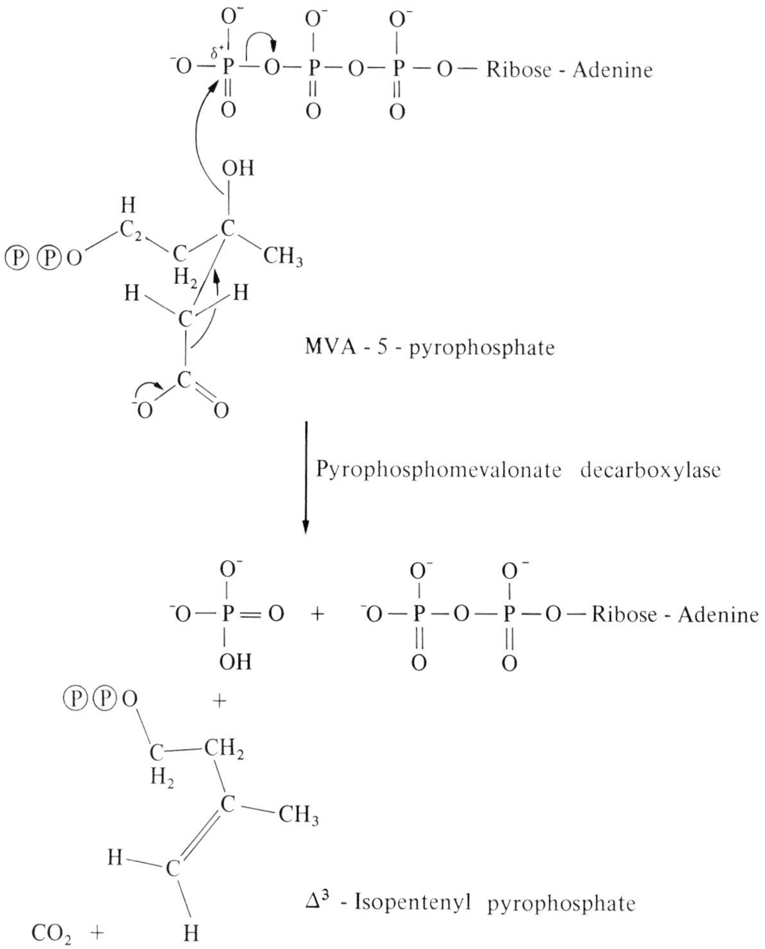

MVA - 5 - pyrophosphate

(11.2)

Pyrophosphomevalonate decarboxylase

Δ^3 - Isopentenyl pyrophosphate

the synthesis of six stereospecifically labelled species of mevalonic acid. From (11.54) it will be seen that mevalonic acid has three prochiral centres, that is carbon atoms with two identical ligands, in this case hydrogen, at C–2, C–4 and C–5. These hydrogens are labelled R or S in the formula and are called pro-R or pro-S hydrogens according to whether, when replaced by another ligand, a chiral centre with the configuration R or S is produced (see Appendix 1 for a discussion of the R, S convention for describing absolute configuration). Although not distinguishable chemically these hydrogens are distinguished by enzymes and the hydrogen at each chiral centre that is involved in the various steps of terpene biosynthesis is now known.

The conversion of MVA-5-pyrophosphate into IPP (see Fig. 11.2) involves the elimination of the

C–1 carboxyl group and the C–3 hydroxyl group. The enzyme catalysing this reaction, pyrophosphomevalonate decarboxylase[3], requires ATP for activity and specifies a *trans* elimination reaction [eq. (11.2)], presumably by holding the MVA-5-pyrophosphate at its active site such that the COOH and OH groups are on opposite sides of the molecule. The elimination is believed to be the concerted process depicted in eq. (11.2).

The isomerization of IPP to DMAPP [eq. (11.1)] involves an electrophilic attack by H^+ from the aqueous medium on the *re–re*† face of the IPP double bond. This produces a transient carbonium ion which is stabilized by the stereospecific elimination of the 2 pro-R hydrogen. (Note that this

† See Appendix 1 on configuration for explanation of this term.

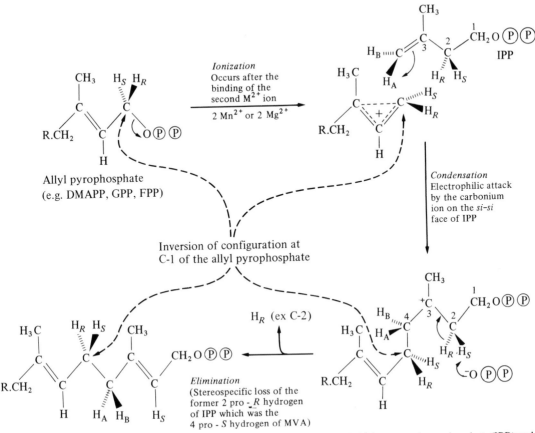

Fig. 11.4. The proposed mechanism of the condensation of an allyl pyrophosphate with Δ^3-isopentenyl pyrophosphate (IPP) under the catalytic influence of prenyltransferase (EC 2.5.1.1). (DMAPP = dimethylallyl pyrophosphate; GPP = geranyl pyrophosphate; FPP = farnesyl pyrophosphate; H_A and H_B of IPP are the former 2 pro-R and 2 pro-S hydrogens of MVA respectively; the carbonium ion is envisaged as a 'bridged' or 'non-classical' carbonium ion to account for the observed inversion of configuration at C–1 of the allyl pyrophosphate.)

hydrogen, i.e. the 2 pro-R hydrogen of IPP, is, by the R, S-convention, identical to the 4 pro-S hydrogen of MVA.) As the entering and leaving H$^+$ ions are on opposite sides of the molecule the isomerization is considered to be a concerted process.

The reactions of the chain elongation process, which begin after sufficient DMAPP has been formed to act as the starter, involve a condensation between an allyl residue (i.e. dimethylallyl, geranyl, farnesyl) and IPP. Present evidence indicates that the reaction is initiated by the ionization of the allyl pyrophosphate to yield a bridged or non-classical carbonium ion (see Fig. 11.4) and pyrophosphate. This ionization requires the binding of two divalent metal ions (Mn^{2+} or Mg^{2+}) per molecule of allyl pyrophosphate at the active site of prenyltransfer-

ase[4], ionization occurs only after the second ion is bound. The allyl carbonium ion, being an electrophilic species like H$^+$ in eq. (11.1), attacks the double-bond system of IPP. However, unlike H$^+$ in eq. (11.1), it attacks the si–si† face. The carbonium ion which results from this condensation is then stabilized by the stereospecific elimination of the 2 pro-R hydrogen. This hydrogen is identical to that which was eliminated from IPP in eq. (11.1) (i.e. the former 4 pro-S hydrogen of MVA). Because the entering and leaving ionic species are on the same side of the IPP molecule the overall reaction cannot be a concerted process. It has been shown that there is inversion of configuration at C–1 of the allyl

† See Appendix 1 on configuration for explanation of this term.

pyrophosphate during the reaction. Since ionization occurs prior to condensation, this inversion of configuration has been taken to indicate that the allyl carbonium ion is non-classical (i.e. the positive charge is spread over adjacent carbons, thereby producing a rigid planar structure with distinct faces) rather than classical (i.e. the positive charge is located at one carbon, thereby allowing free rotation about the $C—\overset{+}{C}$ bond). The overall reaction catalysed by prenyltransferase (see Fig. 11.4) is thus conceived as an ionization–condensation–elimination reaction.

The ionization–condensation–elimination reaction just described, however, produces only the *trans*-polyisoprenoid chain that is found in the majority of naturally occurring terpenoids. The equivalent reaction leading to the production of the *cis*-polyisoprenoid chain that occurs in rubber is catalysed by a different prenyltransferase which has a different stereospecificity (see Section C.10).

The mechanism by which chain elongation is controlled is as yet unknown.

2. Hemiterpenes

Isoprene appears to be the only true naturally occurring hemiterpene. It is formed in the chloroplast, but the pathway involved is not clear. Compounds like isoamylalcohol (11.55) or isovaleraldehyde (11.56) are formed directly from the metabolism of valine and not from HMG-CoA or mevalonate.

3. Monoterpenes†

The basic precursor of the monoterpenes is geranyl pyrophosphate but this has been difficult to prove in many cases because the incorporation rate is low, particularly in whole plants, where the products appear to be compartmentalized, and in crude enzyme systems, which are rich in phosphohydrolases. The mechanism of formation of nerol, the *cis*-analogue of geraniol, is not yet clear although nerol would, on the face of it, appear to be

(11.55)

(11.56)

a necessary intermediate in the formation of cyclic monoterpenes. However, studies with purified γ-terpinene synthetase in which nerol pyrophosphate and geranyl pyrophosphate are equally active has led to the view of the formation of a common enzyme-bound cyclic intermediate (step A). This, when followed by proton loss from C–5 and a 1,2-hydride shift with expulsion of the enzyme, would yield γ-terpinene (step B). A similar mechanism (step C) with the loss of a proton from C–6 and the formation of a cyclopropane ring could yield α-thujene, which is a minor product of the action of γ-terpinene synthetase on geranyl pyrophosphate. The cyclization reaction, the displacement of an allylic pyrophosphate by an electron pair from a double bond to form a —C—C— bond, is very similar to the reaction outlined in Fig. 11.4, and the type of ionized intermediate proposed is similar. This proposed mechanism is significant in that it explains why geranyl pyrophosphate can be a precursor of a cyclic monoterpene without having first to be converted into neryl pyrophosphate.

There is good evidence that γ-terpinene is a precursor of the aromatic monoterpenes *p*-cymene and thymol [eq. (11.3)].

The biosynthetic pathway for camphor shown in eq. (11.4) has recently been amended to accommodate the observation that bornyl pyrophosphate is the first product of cyclization.

γ-Terpinene   *p*-Cymene   Thymol

(11.3)

† Letters in this section refer to Fig. 11.5.

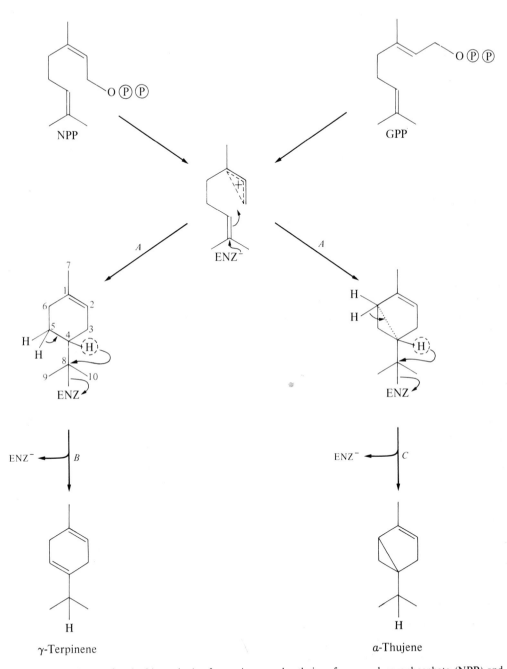

Fig. 11.5. Proposed mechanism for the biosynthesis of γ-terpinene and α-thujene from neryl pyrophosphate (NPP) and geranyl pyrophosphate (GPP).

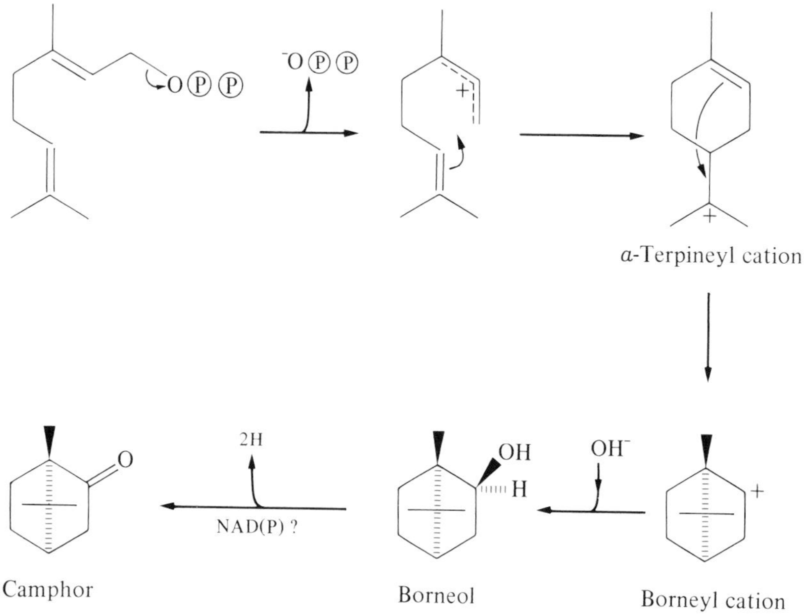

(11.4)

a-Terpineyl cation

Camphor   Borneol   Borneyl cation

There is not space to discuss all the cyclization reactions involved in other cyclic monoterpene syntheses, but it is clear that a small number of basic cyclic structures undergo extensive secondary changes in order to produce the vast array of known monoterpenes. A good example of this is the various reactions which occur in peppermint following the formation of piperitenone (Fig. 11.6).

4. Sesquiterpenes†

In spite of the impressive number of sequiterpenes known in higher plants very little is known of their detailed biosynthesis. Figure 11.7 gives an important example in which it was shown that *trans, trans*-farnesyl pyrophosphate forms a cyclic intermediate cation with the loss of inorganic pyrophosphate (step *A*); this then rearranges by a 1,3-hydride shift specifically involving the pro-*S* hydrogen at C–1 of farnesyl pyrophosphate (step *B*); a final rearrangement based on proton attack on the exocyclic double bond yields γ-cadinene (step *C*).

† Letters in this section refer to Fig. 11.7.

5. Diterpenes

A most important group of diterpenes is, as already indicated, the gibberellins. Their biosynthesis from geranylgeranyl pyrophosphate has been studied in detail and it has been decided that this would best be included in Chapter 15, where plant hormones are discussed.

6. Triterpenes

(i) BASIC REACTIONS—SQUALENE FORMATION

In studying triterpene biosynthesis a new concept arises. This is that the 30C triterpene precursor is synthesized by the tail-to-tail junction of two 15C farnesyl residues derived from farnesyl pyrophosphate (FPP). The first stable compound formed is presqualene pyrophosphate. Its formation is catalysed by a membrane-bound enzyme called presqualene synthase[5] which, when in its polymeric form, also catalyses the second reaction, the reduction of presqualene pyrophosphate by NADPH to squalene. These reactions, first described and studied in liver and yeast, have now been demonstrated in higher plants.

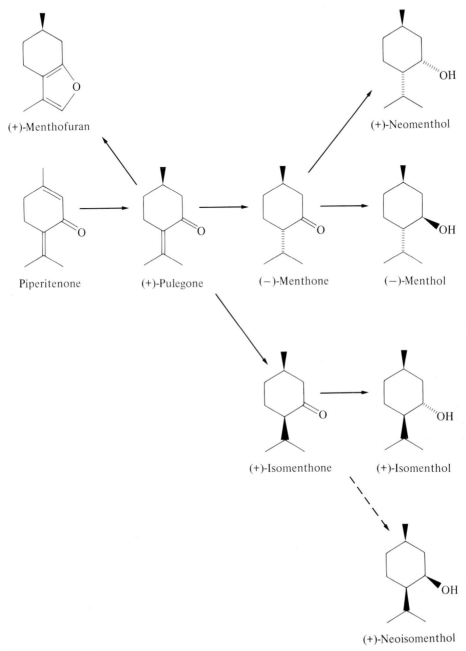

(+)-Menthofuran

(+)-Neomenthol

Piperitenone   (+)-Pulegone   (−)-Menthone   (−)-Menthol

(+)-Isomenthone   (+)-Isomenthol

(+)-Neoisomenthol

Fig. 11.6. Metabolic interconversions of monoterpenes in peppermint (solid arrows indicate reactions which have been demonstrated in cell-free preparations.)

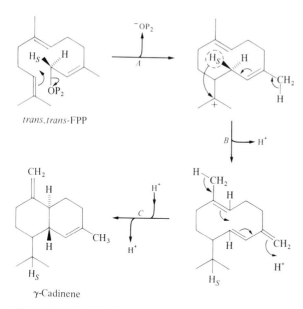

Fig. 11.7. Mechanism for the conversion of *trans*, *trans*-farnesyl pyrophosphate into γ-cadinene.

The stereochemistry of the two-step conversion of two molecules of FPP into squalene has been elucidated by Popják and Cornforth. It may be summarized as follows: (1) one of the participating FPP molecules loses a hydrogen from its C–1, the other FPP does not lose hydrogen from its C–1, (ii) the hydrogen lost from C–1 of FPP is the 1 pro-*S* hydrogen, (iii) the eliminated 1 pro-*S* hydrogen is replaced by the 4 pro-*S* hydrogen of the nicotin-amide ring NADPH (i.e. the hydrogen from the *B* face of the nicotinamide ring), (iv) the 4 pro-*S* hydrogen from NADPH becomes the 12 pro-*R* hydrogen of squalene and (v) there is inversion of configuration at C–1 of the FPP that does not lose hydrogen from its C–1 during squalene formation.

A mechanism for the conversion of 2FPP into squalene which is consistent with the stereochem-istry and the intermediacy of presqualene pyrophos-phate is shown in Fig. 11.8. The reaction sequence commences with the activation of one of the participating FPP molecules (designated FPP_2 in Fig. 11.8); this is accomplished by the attack of a nucleophilic group, designated X^- in Fig. 11.8 and quite probably a component of presqualene syn-thase, on C–3′. The activated species then carries out an S_N2 attack on C–1 of the other FPP (designated FPP_1 in Fig. 11.8). This results in the

loss of pyrophosphate from and the inversion of configuration of C–1 and the formation of a bond between C–1 and C–2′ (*A*, Fig. 11.8). This is followed by a stereospecific 1,3-elimination reaction in which the 1 pro-*S* hydrogen of FPP_1 is lost as H^+ along with X^- from FPP_2. Coincident with these losses a new bond is formed between C–1 and C–3′ so creating a cyclopropane ring between Cs 1, 2′ and 3′ and forming presqualene pyrophosphate (*B*, Fig. 11.8).

Presqualene pyrophosphate then loses pyrophos-phate and rearranges so as to form a cyclopropyl carbinyl cation (*C*, Fig. 11.8); the reaction is formulated as an S_N2 displacement in which the attacking nucleophile is the electron pair constitut-ing the C–1—C–3′ bond and the leaving group is pyrophosphate. This would account for the ob-served inversion of configuration at C–1′ of FPP_2 (i.e. the FPP that does not lose H from its C–1). The carbonium ion is stabilized by the nucleophilic attack of a hydride ion derived from the B face of the nicotinamide ring of NADPH (designated H_B^- in reaction D, Fig. 11.8) on C–1 and the simultaneous shifting of the electron pair from the C–1—C–2′ bond to form a double bond between C–2′ and C–3′. The hydride attack is formulated as an S_N2 displace-ment which therefore causes inversion of configur-ation at C–1 (C–12 of the newly formed squalene) and thereby nullifies the inversion caused at this carbon by the S_N2 displacement in step A. This results in H_B from NADPH becoming the 12 pro-*R* hydrogen of squalene; equally the 1 pro-*R* hydro-gen of FPP_1 becomes the 12 pro-*S* hydrogen.

A further stereochemical point to notice is that at each end of the squalene molecule there is a prochiral centre with, in this case, two methyl groups attached to one carbon atom. As with the prochiral centres involving hydrogen, these geminal methyls retain their individuality: the methyl that is *trans* with respect to the rest of the squalene molecule arises exclusively from C–2 of mevalonic acid.

(ii) STEROL FORMATION

Squalene is firstly converted by squalene epoxi-dase[5], a mixed function oxygenase requiring O_2 and NADPH, into squalene-2,3-oxide in which the configuration at C–3 is *S*.

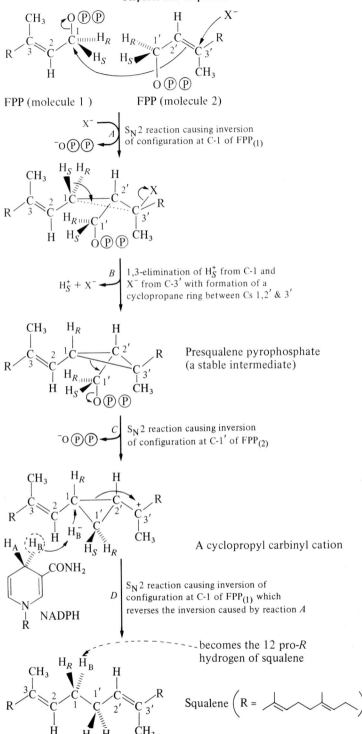

Fig. 11.8. Mechanism for the conversion of two molecules of farnesyl pyrophosphate (FPP) into squalene which is consistent with the known stereochemistry of the process and the intermediacy of presqualene pyrophosphate. (The numbering of all the participating molecular species is that of FPP; in squalene the carbons numbered 1 and 1′ above become 12 and 12′, respectively, in the squalene numbering system; it is important to note that unless stated otherwise, the designation of hydrogens as pro-R (i.e. H_R) or pro-S (i.e. H_S) refer *only* to their spacial orientation in MVA and FPP which in this instance are identical.)

The next step is the cyclization of (3S)-squalene-2,3-oxide. This is a reaction which, apart from photosynthesis itself, distinguishes photosynthetic organisms from non-photosynthetic organisms. In photosynthetic organisms (e.g. higher plants, algae) the cyclization product is cycloartenol (11.57) whilst in non-photosynthetic organisms (e.g. fungi, animals) it is lanosterol (11.58).

The cyclization of squalene-2,3-oxide to cycloartenol is catalysed by squalene-2,3-oxide:cycloartenol cyclase. The mechanism of cyclization is shown in Fig. 11.9. It is initiated by an electrophilic attack by H$^+$ on the epoxy oxygen whilst the squalene-2,3-oxide is held in the chair–boat–chair–boat-unfolded conformation on the active site of the enzyme. This leads to a wave of cyclization reactions (A, Fig. 11.9) running forward through the molecule which results in the formation of four interlinked rings, three cyclohexane and one cyclopentane. The forward cyclization stops when the electron deficiency reaches C–20 because the rest of the molecule is not folded into a potential chair or boat which would bring the remaining double bond sufficiently close to allow further electron shifts.

The transient carbonium ion produced by the forward cyclization then undergoes a backward rearrangement. This takes the form of a series of Wagner–Meerwein shifts. A Wagner–Meerwein shift may be regarded as the movement of a negatively charged atom or group (e.g. H$^-$ or H$_3$C$^-$) from one carbon to an immediately adjacent carbon (i.e. a 1,2 shift) when the latter carries a positive charge. Such a shift actually occurs by way of a bridged carbonium ion (see Fig. 11.10) and is only allowed when the appropriate groups (i.e. the moving group and the group that leaves to create the carbonium ion) are *trans* and coplanar. Thus in a concerted series of Wagner–Meerwein rearrangements, the migrating groups move in a sequentially *trans* manner with inversion of configuration at each of the carbons involved. The sequence of Wagner–Meerwein shifts occurring in cycloartenol formation is: (i) movement of the 17β hydrogen as a hydride ion, H$^-$, to C–20; the spatial orientation taken up by the hydrogen is such that the configuration at C–20 is *R*, (ii) movement of the 13αH so as to become the 17αH, (iii) movement of the 14β methyl as H$_3$C$^-$ so as to become the 13β methyl, (iv) movement of the 8α methyl so as to become the 14α

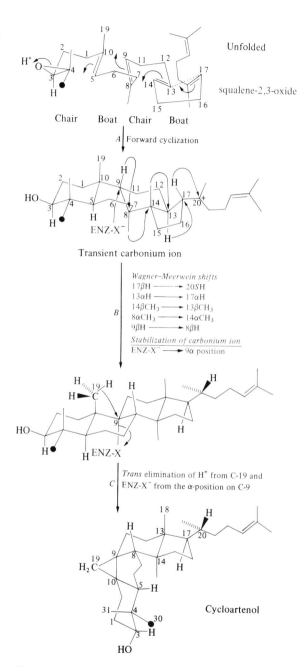

Fig. 11.9. Mechanism of the cyclization of squalene-2,3-oxide to cycloartenol. (The numbering throughout the figure is the sterol numbering system; substituents that project above the general plane of the ring system are said to be β whilst those that project below are said to be α; α- and β-substituents are represented in two-dimensional formulae by dashed (...) and full or wedge-shaped (— or ▶) bond lines respectively; the methyl group on C–4 indicated by a full circle (●) is derived from C–2 of mevalonic acid.)

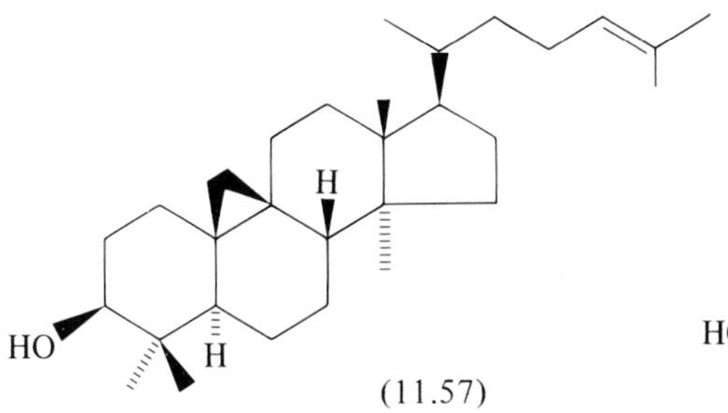

(11.57) 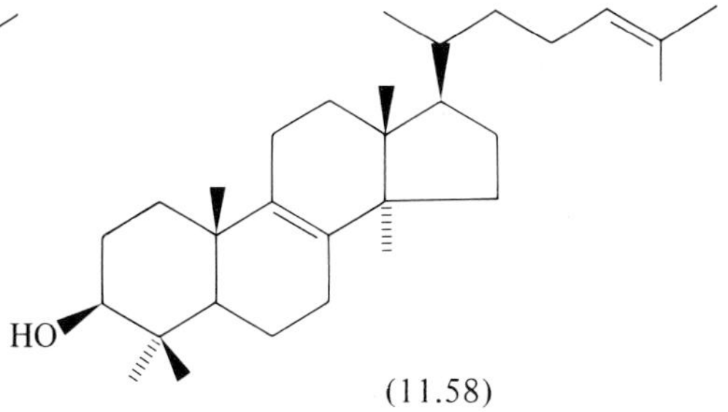   (11.58)

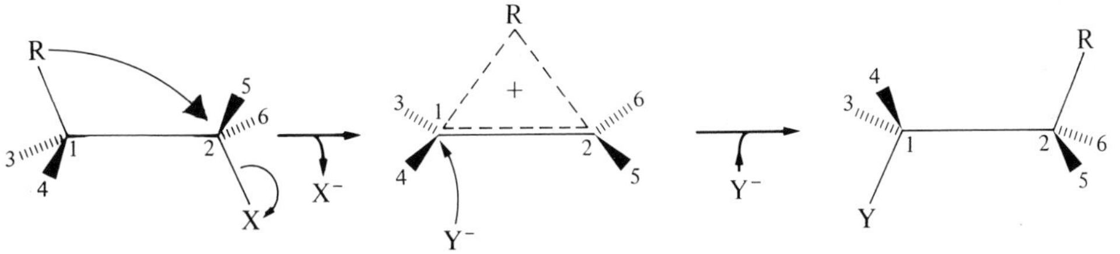

Bridged carbonium ion

Fig. 11.10. Mechanism of a Wagner–Meerwein shift.

methyl, and (v) movement of the 9βH so as to become the 8βH. At this point the carbonium ion is stabilized by the addition of a nucleophilic group (⁻X-Enzyme) at C–9 in the α-orientation. This is then followed by the *trans* elimination of a hydrogen, as H⁺, from the methyl group (C–19) attached to C–10 and of ⁻X-Enzyme from the 9α position; this results in the formation of a cyclopropane ring between Cs 9, 10 and 19. The participation of ⁻X-Enzyme allows the final step to be a *trans* elimination as is required by the biogenetic isoprene rule. Without this the final elimination would have been *cis* which is electronically unacceptable.

In non-photosynthetic organisms such as fungi and animals squalene-2,3-oxide is converted into lanosterol by a different cyclizing enzyme, squalene-2,3-oxide:lanosterol cyclase. The mechanism of cyclization follows the same route as that for cycloartenol until the electron deficiency resides at C–8 following the Wagner–Meerwein shift of the 8α methyl group. This electron deficiency is made good (i.e. the carbonium ion is stabilized) by loss of the 9β hydrogen as H⁺ so forming a double bond between C–8 and C–9.

In both cycloartenol and lanosterol the methyl group on C–4 that has the α-orientation (i.e. is below the general plane of the ring system and indicated in

Fig. 11.9 by a closed circle) is derived from C–2 of mevalonic acid.

As indicated earlier in the chapter, alkylation at C–24 is characteristic of plant sterols and involves either a single or double methylation with S-adenosylmethionine as the methyl donor. Since single methylation of cycloartenol is the next step in the formation of the sterols of photosynthetic organisms it is appropriate to deal with alkylation at this point.

The basic mechanism is similar in all cases but the details vary according to (i) the final chirality observed at C–24, i.e. whether a 24α-methyl or ethyl or a 24β-methyl or ethyl group results and (ii) the organism, in some cases. The variations in the mechanism of single alkylation are shown in Fig. 11.11 and in that of the second alkylation in Fig. 11.12.

The evidence upon which the alkylation routes are based comes from the integration of the results of four lines of investigation: (i) biosynthesis of the sterol in the presence of L-[methyl-²H₃]methionine followed by mass spectral determination of the number of deuterium atoms in the C–24 substituent; for example, the 24-methyl groups of the sterols produced by routes 1, 2 and 3 of Fig. 11.11 would have 3, 2 and 2 deuterium atoms respectively,

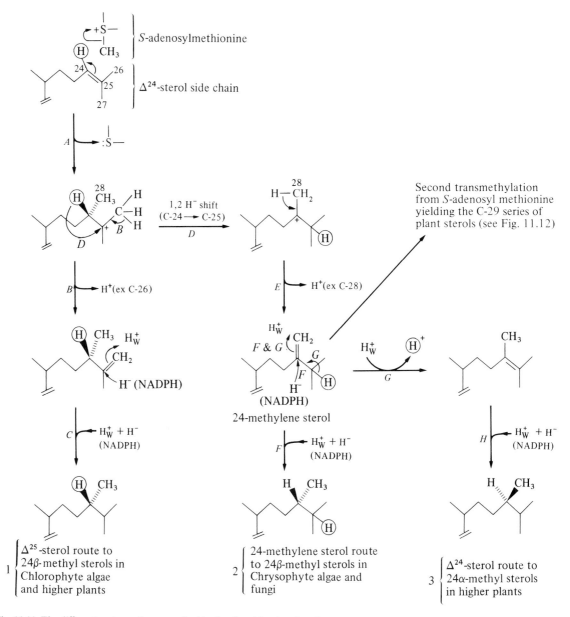

Fig. 11.11. The different routes so far recognized in the plant kingdom for alkylation at C–24 leading to the formation of 24α- and 24β-methyl sterols. (Ⓗ is the hydrogen originally at C–24 of the Δ^{24}-sterol side chain which would be a tritium if the sterol had been synthesized from $[(4R)\text{-}4\text{-}^3H_1]$ mevalonic acid; H_W^+ is a hydrogen ion from the aqueous environment.)

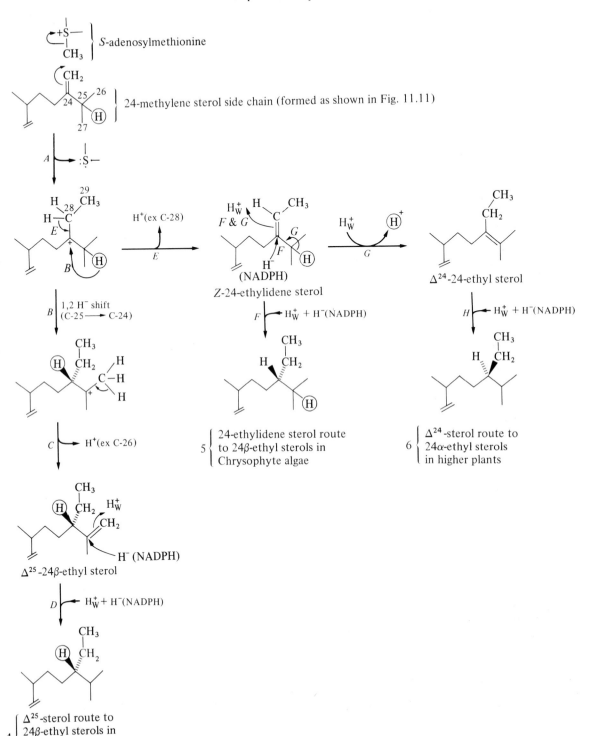

Fig. 11.12. The different route so far recognized in the plant kingdom for further alkylation of the side chain leading to the formation of 24α- and 24β-ethyl sterols. (Ⓗ is the hydrogen originally at C–24 of the Δ²⁴-sterol side chain which would be a tritium if the sterol had been biosynthesized from [(4R)-4-³H₁] mevalonic acid; H_W^+ is a hydrogen ion from the aqueous environment.)

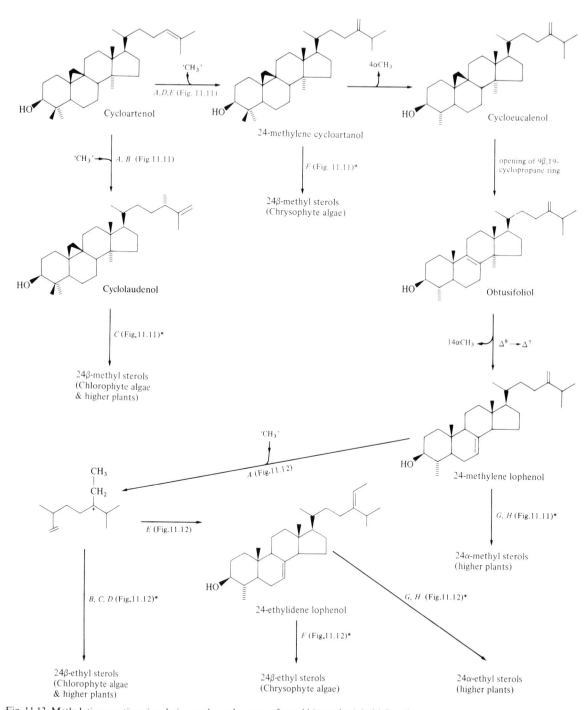

Fig. 11.13. Methylation reactions in relation to the early stages of sterol biosynthesis in higher plants and algae. (The letters *A–H* refer to the reaction steps shown in either Fig. 11.11 or Fig. 11.12; the asterisks indicate that changes in the sterol nucleus and sometimes the side chain occur in addition to the reactions indicated by the letters.)

(ii) biosynthesis of the sterol in the presence of [(4R)-4-³H₁]mevalonic acid—this places a tritium atom on C–24 of the Δ²⁴-sterol undergoing the initial methylation—followed by an examination of the side chain to see whether there was a tritium atom at C–24 as in route 1 (Fig. 11.11), a tritium atom at C–25 as in route 2 (Fig. 11.11) or no tritium atoms at either C–24 or C–25 as in route 3 (Fig. 11.11), (iii) identification of the minor sterols in a given organism and comparison of their side chains with the structures proposed in the numerous theoretically possible alkylation routes and (iv) administration of labelled putative sterol precursors with critical side-chain structures to a given organ-

ism to see whether they could be converted into the major sterol(s) of that organism.

The key features of the initial step of both the first and second alkylations (*A* in Figs. 11.11 and 11.12) are (i) the positively charged sulphur atom (sulphonium ion) of the methyl donor, S-adenosyl-methionine and (ii) the presence of an appropriately placed double bond in the methyl recipient, the Δ²⁴-double bond of cycloartenol in the first alkylation and the Δ²⁴⁽²⁸⁾-double bond in the 24-methylene sterol in the second alkylation. These lead to the transfer of the methyl group and the formation of a carbonium ion, the rearrangement and stabilization of which give the various alkylation routes.

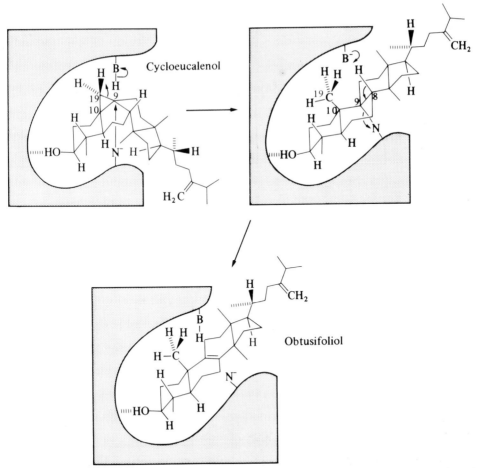

Fig. 11.14. Mechanism of opening the 9β,19-cyclopropane ring in cycloeucalenol. (B and N are active groups at the active site of the enzyme catalysing the reaction; the H which is transferred from B to C–19 of the sterol can exchange with H⁺ in the aqueous environment—thus when the reaction is carried out in D₂O one deuterium atom is incorporated into the 10β-methyl group of obtusifoliol.)

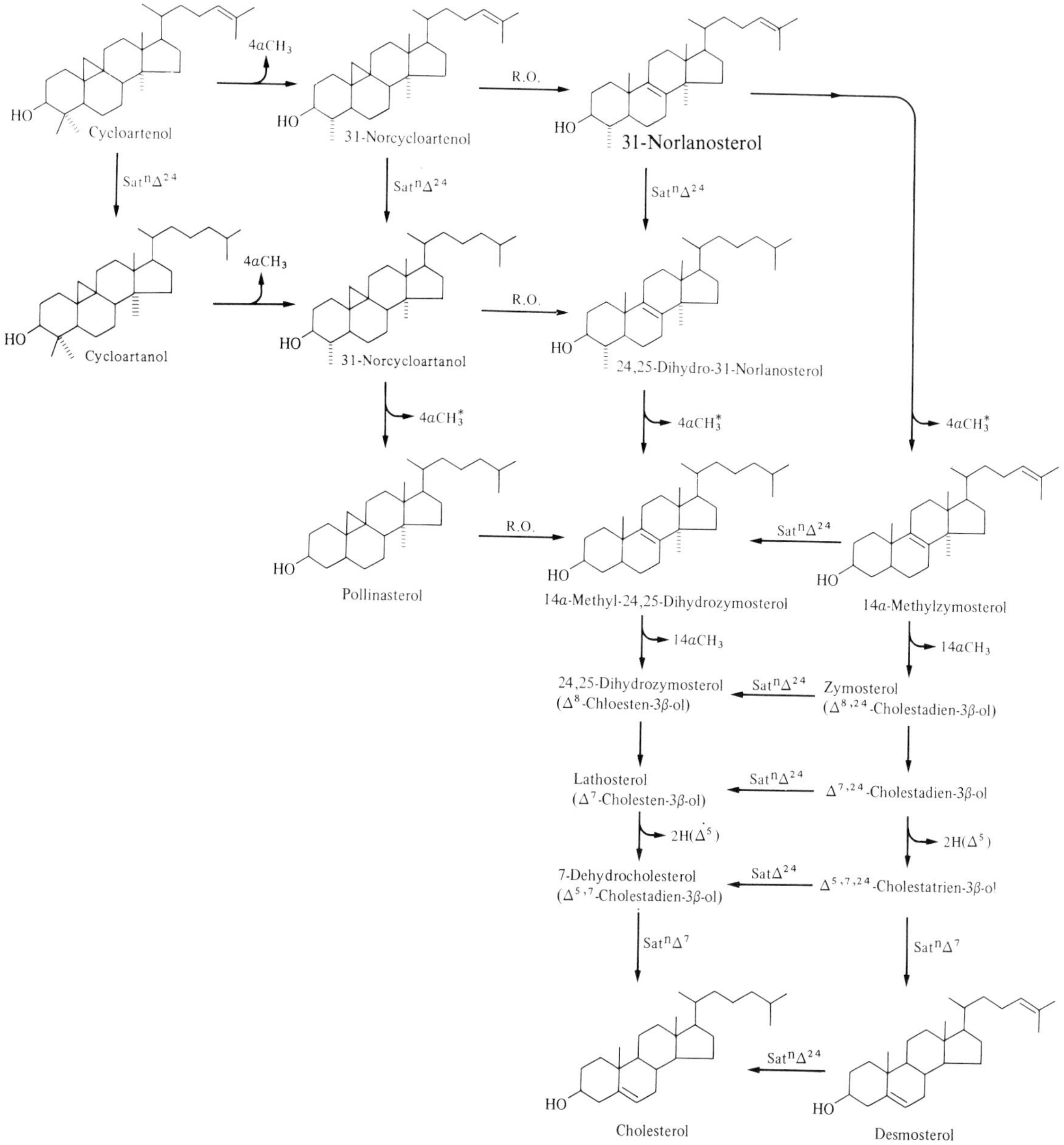

Fig. 11.15. Possible biosynthetic pathways from cycloartenol to cholesterol in higher plants and algae. (R.O. = opening of the 9β,19-cyclopropane ring; the asterisk indicates that although the methyl group has the α-orientation, it was originally the 4β-methyl group of cycloartenol.)

The carbonium ion formed as a result of step *A* in the first methylation (Fig. 11.11) has its positive charge formally located at C–25. In route 1, leading to 24β-methyl sterols in green algae and higher plants, this carbonium ion is stabilized (*B*, Fig. 11.11) by loss of a proton from the C–26 methyl group. The Δ^{25}-double bond that results is then reduced by NADPH to yield the fully saturated 24β-methyl type of side chain (*C*, Fig. 11.11). In this reduction a hydrogen ion (H_W^+) from the aqueous environment is added to C–26 and the hydride ion (H^-) from NADPH to C–25; in such reductions it is usual for H_W^+ to add to the least substituted carbon and H^- to the most highly substituted carbon. In route 1 it should be noted that the chirality at C–24 is determined by the fact that the methyl group is introduced from behind the plane of the Δ^{24}-double bond as depicted in Fig. 11.11.

In routes 2 and 3 the carbonium ion produced in step *A* (Fig. 11.11) undergoes rearrangement. The C–24 hydrogen undergoes a 1,2 hydride shift to C–25 so producing a new carbonium ion with the positive charge formally located at C–24 (*D*, Fig. 11.11). This is then stabilized by loss of a proton

from the newly introduced methyl group and the consequent production of a 24-methylene type of side chain with a $\Delta^{24(28)}$-double bond (*E*, Fig. 11.11).

In route 2, leading to 24β-methyl sterols in Chrysophyte algae (e.g. *Ochromonas*) and fungi (e.g. *Saccharomyces cerevisiae*, *Phycomyces blakesleeanus*, *Mucor pusillus*), the $\Delta^{24(28)}$-double bond is reduced by NADPH, H_W^+ adding to C–28 and H^- to C–24 (*F*, Fig. 11.11). In this route it is the direction of this reduction reaction that determines the chirality at C–24; H^- is added from in front of the plane of the $\Delta^{24(28)}$-double bond as depicted in Fig. 11.11. It is important to remember that in the case of the fungi the Δ^{24}-sterol side chain undergoing the initial methylation (i.e. step *A*, Fig. 11.11) in route 2 is lanosterol and not cycloartenol.

In route 3, leading to the more abundant 24α-methyl sterols in higher plants, it is believed (although it has still to be confirmed) that the 24-methylene side chain is isomerized to a Δ^{24}-side chain (*G*, Fig. 11.11). This involves the electrophilic attack of H_W^+ on the methylene carbon and concomitant loss of H^+ from C–25 to stabilize the

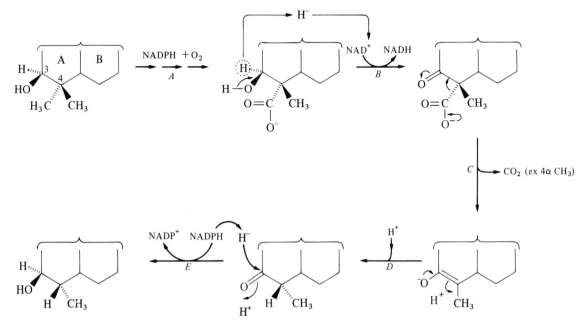

Fig. 11.16. Mechanism for the removal of the 4α-methyl group from a 4,4-dimethyl sterol. (Reaction *A* is a 3-step process involving the successive intermediacy of 4α-hydroxymethyl- and 4α-formyl sterols—each step requires NADPH and O_2; note that the 4β-methyl group of the 4,4-dimethyl sterol is inverted during the demethylation and becomes the 4α-methyl group of the product; the removal of the 4α-methyl group from the 4α-methyl sterol to yield a 4-demethyl sterol occurs in a manner analogous to that shown above.)

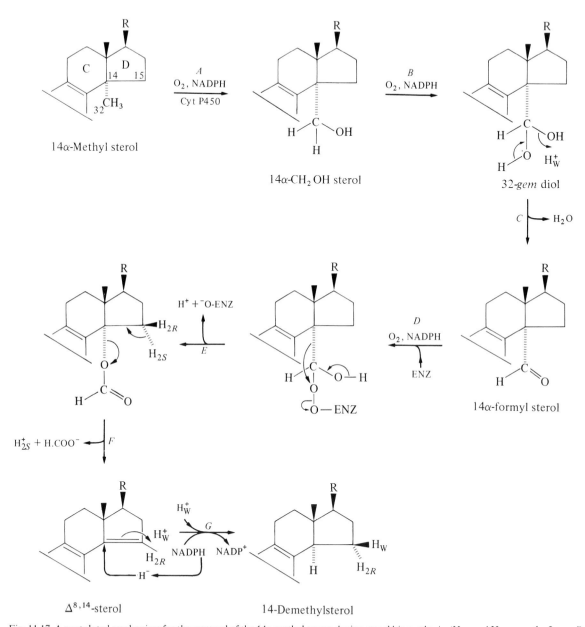

Fig. 11.17. A postulated mechanism for the removal of the 14α-methyl group during sterol biosynthesis. (H_{2R} and H_{2S} were the 2-pro-*R* and 2-pro-*S* hydrogens of mevalonic acid; H_W^+ is a hydrogen ion in the aqueous environment; Cyt P450 = a cytochrome P450 mixed function oxygenase; ENZ is postulated to be a flavoprotein mixed function oxygenase catalysing a Baeyer–Villiger type oxidation, by analogy with that participating in the removal of the 10β-methyl group in oestrogen ring A aromatization in animals.)

incipient C–24 carbonium ion. The Δ^{24}-double bond is then reduced by NADPH (H, Fig. 11.11); in this instance it is not known which of the two double-bonded carbons is attacked by H_W^+ and H^-, nor is it possible to predict this on the basis of their relative degrees of substitution. However, it is the direction of this reduction that must determine the chirality at C–24.

The 24-methylene side chain produced by steps A, D and E (Fig. 11.11) is the starting-point for the second methylation which leads by routes 4, 5 and 6 (Fig. 11.12) to 24α- and 24β-ethyl side chains. It should be noted that a second methylation is very rare in fungi.

The initial step in the second methylation (A, Fig. 11.12) produces a carbonium ion with its positive charge formally located at C–24; the latter also now carries an ethyl group.

In route 4, leading to the 24β-ethyl sterols in green algae and higher plants, the carbonium ion is rearranged by the 1, 2 shift of a hydride ion from C–25 to C–24 (B, Fig. 11.12). The resulting carbonium ion is the 24-ethyl analogue of the carbonium ion produced by step A in Fig. 11.11 and the way in which it is converted into a 24β-ethyl side chain (C and D, Fig. 11.12) is analogous to that by which 24β-methyl side chains are formed in green algae and higher plants (B and C, Fig. 11.11). There is therefore a great deal of similarity between routes 1 (Fig. 11.11) and 4 (Fig. 11.12). However, they differ in the way they establish chirality at C–24; in route 1 it is determined by the initial methylation whilst in route 2 it is determined by the stereospecificity of the 1,2-hydride shift.

In routes 5 and 6 the carbonium ion produced in step A (Fig. 11.12) is stabilized by loss of H^+ from C–28 and the consequent production of a 24-ethylidene type of side chain with a $\Delta^{24(28)}$-double bond (E, Fig. 11.12).

In route 5, leading to 24β-ethyl sterols in Chrysophyte algae (e.g. *Ochromonas malhamensis*, *Monodus subterraneus*), the $\Delta^{24(28)}$-double bond is reduced by NADPH, H_W^+ adding to C–28 and H^- to C–24 (F, Fig. 11.12). As with route 2 (Fig. 11.11), to which route 5 is analogous, the directionality of this reduction determines the chirality at C–24; H^- is added from front of the plane of the $\Delta^{24(28)}$-double bond as depicted in Fig. 11.12.

In route 6, leading to the more abundant 24α-

ethyl sterols in higher plants, the Z-24-ethylidene side chain is isomerized to a Δ^{24}-side chain (G, Fig. 11.12) in a manner analogous to that believed to occur in route 3 (G, Fig. 11.11). Moreover, as with route 3, the Δ^{24}-double bond is reduced by NADPH (H, Fig. 11.12). Thus the direction of the reduction determines the chirality at C–24.

In brown algae (e.g. *Fucus* spp.) the most abundant sterol is the 24-ethylidene sterol, fucosterol, which has the opposite configuration about the $\Delta^{24(28)}$-double bond to that shown in Fig. 11.12 (i.e. it is an E-24-ethylidene sterol). Evidence indicates that it is formed by route 5 (Fig. 11.12) and it is presumed that the opposite configuration is the result of the different stereospecificity of the enzyme catalysing step E.

In the foregoing discussion the actual sterols acting as substrates for the methylation enzymes

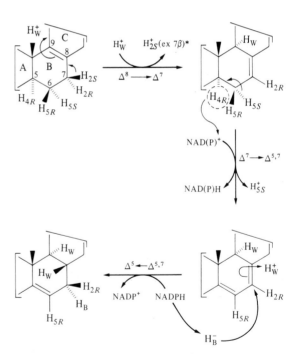

Fig. 11.18. Stereospecificity of the $\Delta^8 \to \Delta^7 \to \Delta^{5,7} \to \Delta^5$ sterol transformations. (The designation of a hydrogen as 2R, 2S, 4R, 4S, 5R or 5S indicates that it was the pro-R or pro-S hydrogen of the carbon of that number in mevalonic acid; H_W^+ is a hydrogen ion in the aqueous environment; H_W is a hydrogen derived from water; H_B is a hydrogen derived from the B face of the nicotinamide ring of NADPH, i.e. the 4-pro-S-hydrogen; the asterisk draws attention to the fact that although the 7β-hydrogen is lost in higher plants and *Ochromonas malhamensis* and during cholesterol biosynthesis in animals, it is the 7α-hydrogen that is lost in yeast.)

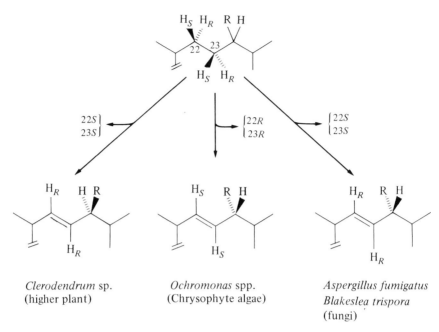

Fig. 11.19. Stereospecificity of hydrogen elimination in the formation of the Δ^{22} double bond of some higher plant, algal and fungal sterols. (The designation of a hydrogen as R or S indicates that it is the pro-R or pro-S hydrogen on the carbon to which it is bound; note that C-22 is derived from C-2 of mevalonic acid and C-23 from C-5.)

were not considered. It is now clear from numerous studies, including work with isolated enzymes, that in photosynthetic organisms cycloartenol is the substrate for the first methylation and that 24-methylene lophenol is the substrate for the second methylation. On the other hand, in fungi (in which only one methylation commonly occurs) the substrate is usually lanosterol. However, in yeast (*S. cerevisiae*) methylation of lanosterol does not occur and is delayed until some or all of the 4,4,14-methyl groups have been removed; zymosterol ($\Delta^{8,24}$-

cholestadien-3β-ol) appears to be the most efficient substrate for methylation.

In Fig. 11.13 the methylation reactions are related to the early stages of sterol biosynthesis as it occurs in higher plants and algae. The enzymes catalysing the initial methylation produce cyclolaudenol (by steps A and B, Fig. 11.11) or 24-methylene cyclo-artanol (by steps A, D and E, Fig. 11.11); both these compounds then undergo further transformations (in the sterol nucleus and, in some instances, the side chain) to yield 24β-methyl sterols. The pathway to

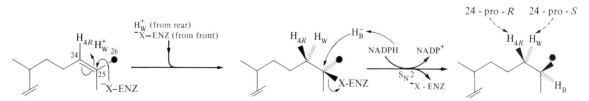

Fig. 11.20. Saturation of the Δ^{24} double bond during cholesterol biosynthesis in animal tissues. (H_{4R} is the former 4-pro-R hydrogen of mevalonic acid which becomes the 24-pro-R hydrogen of cholesterol; H_W^+ is a hydrogen ion in the aqueous medium; H_B^- is a hydride ion derived from the B-face of the nicotinamide ring of NADPH, i.e. the 4-pro-S hydrogen, which carries out an S_N2 displacement of $^-$X-ENZ by attacking C-25 from the rear; $^-$X is a nucleophilic moiety on the enzyme catalysing the overall reaction; ● = the former C-2 of mevalonic acid.)

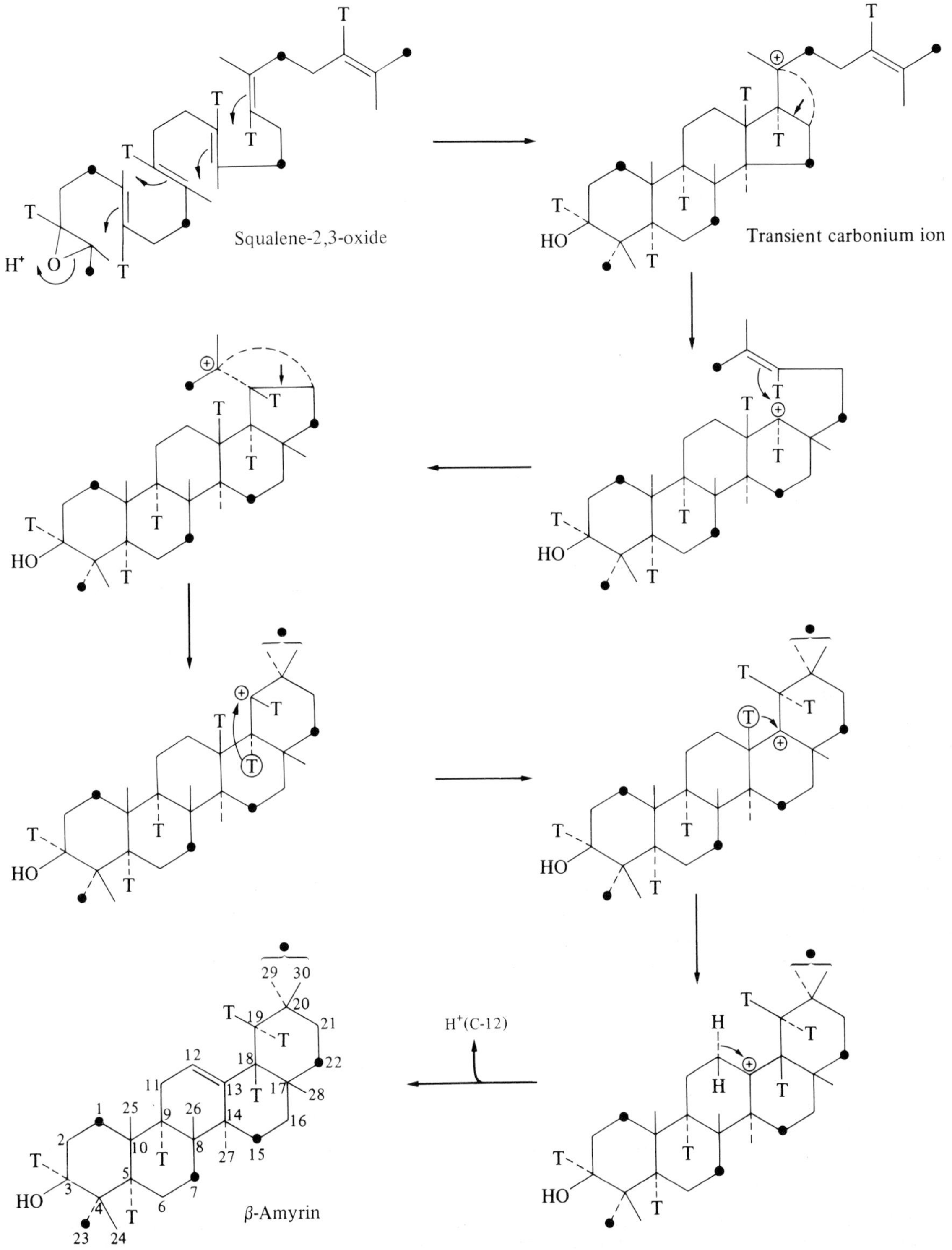

Fig. 11.21. Conversion of squalene-2,3-oxide into β-amyrin. (● = ^{14}C from C-2 of MVA; T = ^{3}H from 4-pro-R position of MVA; a curved, dashed line between two carbons indicates that they become joined by a new bond after the cleavage of the adjacent bond indicated by an arrow.)

24α-methyl sterols involves loss of the 4α-methyl group from 24-methylene cycloartanol giving cyclo-eucalenol which then undergoes opening of the 9β,19-cyclopropane ring to produce obtusifoliol (see Fig. 11.14). The latter is converted into 24-methylene lophenol; this involves loss of the 14α-methyl group and the switching of the Δ^8-double bond to the Δ^7 position. The cyclopropane ring-opening enzyme has considerable specificity in that it acts only on 4α-methyl sterols such as cyclo-eucalenol. This specificity explains the absence of lanosterol from plant tissues; if cycloartenol had been a substrate for this enzyme the product would have been lanosterol. 24-Methylene lophenol is converted into 24α-methyl sterols (by steps G and H, Fig. 11.11) and is the substrate for the second methylation. The enzymes catalysing the second methylation produce 24β-ethyl sterols (by steps A–D, Fig. 11.12) or 24-ethylidene lophenol (by steps

A and E, Fig. 11.12). The latter is then transformed into 24β-ethyl sterols in Chrysophyte algae (by step F, Fig. 11.12 and changes in the nucleus and, in some instances, the side chain) and into 24α-ethyl sterols in higher plants (by steps G and H, Fig. 11.12, and changes in the nucleus and, in some instances, the side chain).

The changes in the sterol nucleus and side chain referred to in the previous paragraph involve (i) loss of the second 4α-methyl group, (ii) conversion of Δ^7-sterols into Δ^5-sterols and (iii) insertion of a Δ^{22}-double bond; all these transformations are discussed in the next section. It is difficult to combine these changes into one clear-cut pathway (indeed, it is quite possible that such a pathway does not exist) for until their enzymology has been worked out, side pathways which vary both qualitatively and quantitatively from plant to plant obtrude. There is therefore little point at this stage in

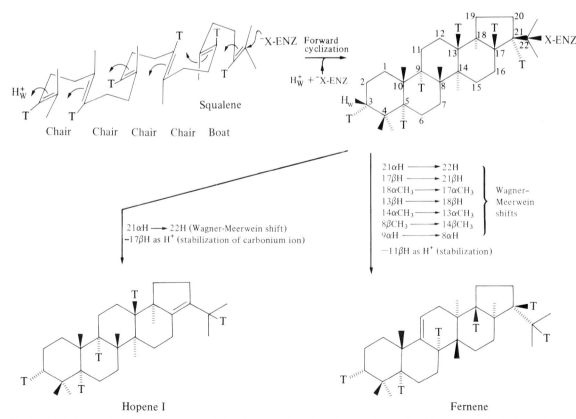

Fig. 11.22. The intramolecular rearrangements during the conversion of squalene into hopene I and fernene. (H_W^+ is a hydrogen ion in the aqueous environment; ^-X is a nucleophile group on the enzymes catalysing the two different cyclizations; T is a tritium atom from [(4R)-4-$^3H_1$] mevalonic acid.)

suggesting a generalized pathway for the later stages of plant sterol synthesis.

Cholesterol, as indicated earlier, is always present to a greater or lesser extent in members of the plant kingdom (particularly the red algae), but its biosynthesis has not been studied in detail. However, Fig. 11.15 shows possible pathways from cycloartenol which are based upon (i) the known occurrence in plants of most of the compounds shown, (ii) analogy with the pathway from lanosterol to cholesterol in mammals and (iii) some observed reactions in plants.

(iii) SOME STEREOCHEMICAL AND MECHANISTIC CONSIDERATIONS

As in squalene, the geminal methyl groups on C–4 of cycloartenol retain their identity, the 4α-methyl

arising specifically from C–2 of mevalonic acid. The removal of these methyl groups occurs in two distinct, but mechanistically similar, reaction sequences with the 4α being the first to be lost. However, in plants these reaction sequences are separated in time by the removal of the 14α-methyl group. Thus the methyl-removal sequence in plants is 4α, 14α and 4β; however, as will become apparent shortly, the 4β-methyl group has assumed the 4α-orientation in the sterol from which it is removed.

The mechanism of removal of the 4α-methyl group is shown in Fig. 11.16. It is firstly oxidized to a carboxyl group in three steps each catalysed by a mixed function oxygenase (A, Fig. 11.16); thus sterols with 4α-CH_2OH and 4α-CHO groups are intermediates. The decarboxylation which follows is facilitated by the prior oxidation of the secondary alcoholic moiety at C–3 to a ketone (B, Fig. 11.16).

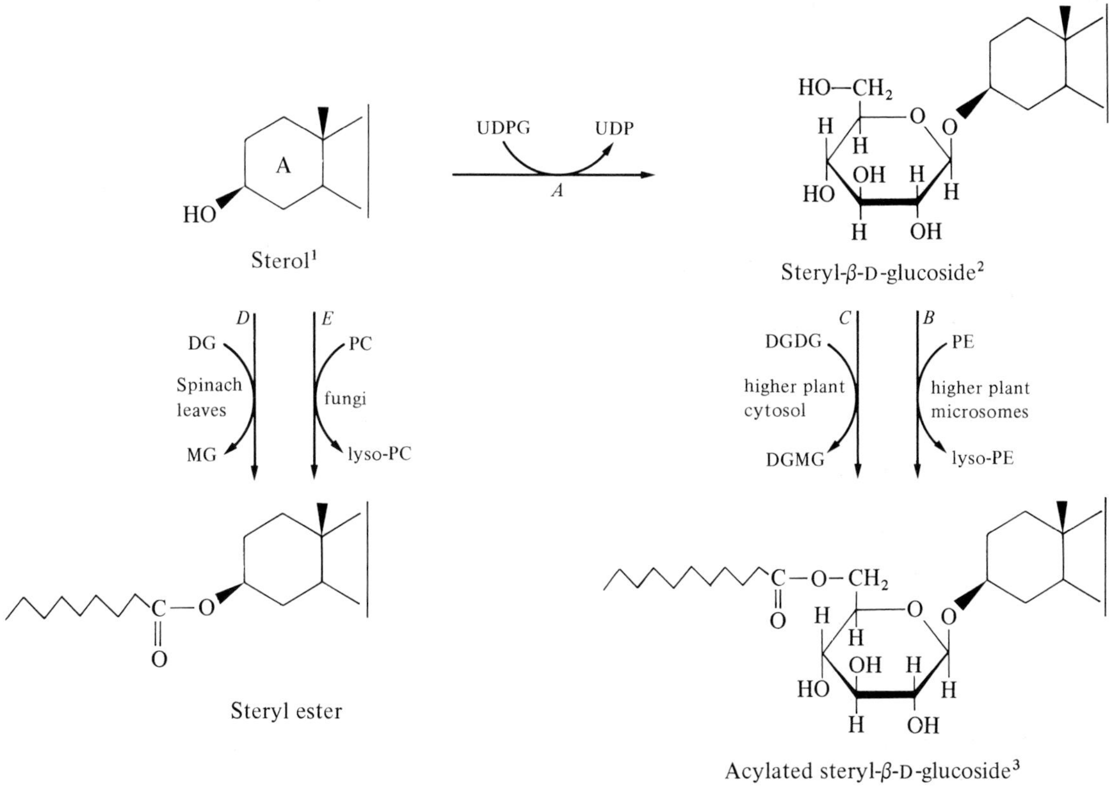

Fig. 11.23. Pathways of biosynthesis of steryl esters, glycosides and acylated glycosides in members of the plant kingdom. (DG = diacylglycerol; MG = monoacylglycerol; PC = phosphatidylcholine; PE = phosphatidylethanolamine; DGDG = digalactosyldiglyceride; DGMG = digalactosylmonoglyceride; 1 = all the sterols in a given tissue probably form esters whereas only the 4,14-demethyl sterols appear to form glycosides; 2 = the most common steryl glycosides are glucosides but mannosides, galactosides and gentiobiosides are known; 3 = acylation usually occurs at the 6′ hydroxyl group of the sugar residue; ∿∿∿∿C— = fatty acid residue.)

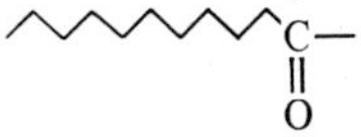

The removal of the carboxyl group forms a 4-methyl-3-enol (C, Fig. 11.16) which on ketonization (D, Fig. 11.16) results in the methyl group taking up the more stable equatorial, 4α orientation. The 3-keto group is then reduced by NADPH so as to re-establish the 3β-hydroxy function.

The mechanism of removal of the second methyl group from C–4 (i.e. the 4β-methyl, which has now become the 4α-methyl) occurs typically at the 4α-methyl-Δ^7-sterol level. The mechanism is identical to that shown in Fig. 11.16 save that the 4β-methyl has been replaced by a 4β-hydrogen which, during the course of the reaction sequence, is inverted to the 4α orientation; the incoming hydrogen ion from the aqueous environment (D, Fig. 11.16) takes up the 4β position.

The removal of the 14α-methyl group has been studied most intensively in cholesterol biosynthesis in animal tissues but its mechanism has still to be fully elucidated. Figure 11.17 shows a postulated mechanism which is consistent with the established stereochemistry and other experimental observations. The 14α-methyl group is firstly oxidized by a cytochrome P450-containing mixed function oxygenase to a 14α-hydroxymethyl group; this step (A, Fig. 11.17) is inhibited by carbon monoxide and a number of fungicides (e.g. triadimefon) that possess

nitrogen-heterocycles. The 14α-hydroxymethyl intermediate is then oxidized to a 14α-formyl sterol by a second mixed function oxidase which, however, is not associated with cytochrome P450 (it is not inhibited by CO). This oxidation is thought to proceed in two stages (B and C, Fig. 11.17); firstly C–32 is further hydroxylated to produce a 32-*gem* diol which then dehydrates to yield the formyl group. The 14α-formyl intermediate is then converted into a $\Delta^{8,14}$-sterol, the formyl group being lost as formate which is subsequently oxidized to CO_2 by formate dehydrogenase[6]. The removal of the formyl group is catalysed by a mixed function oxygenase; it is not inhibited by CO indicating that cytochrome P450 is not involved. Because of the analogy with 10β-demethylation during the aromatization of ring A in oestrogen biosynthesis, it has been postulated that the enzyme is a non-haem flavoprotein catalysing a Baeyer–Villiger type of oxidation (D, E and F, Fig. 11.17). During this reaction, work with [(2R)-2-$^3H_1$]- and [(2S)-2-$^3H_1$]mevalonic acid has shown that the 15α-hydrogen of the sterol intermediate (i.e. the former 2-pro-S hydrogen of MVA) is stereospecifically removed. The Δ^{14}-double bond is then reduced by an NADPH-requiring Δ^{14}-sterol reductase (G, Fig. 11.17); in this reduction H^- from NADPH becomes the 14α-hydrogen whilst H^+

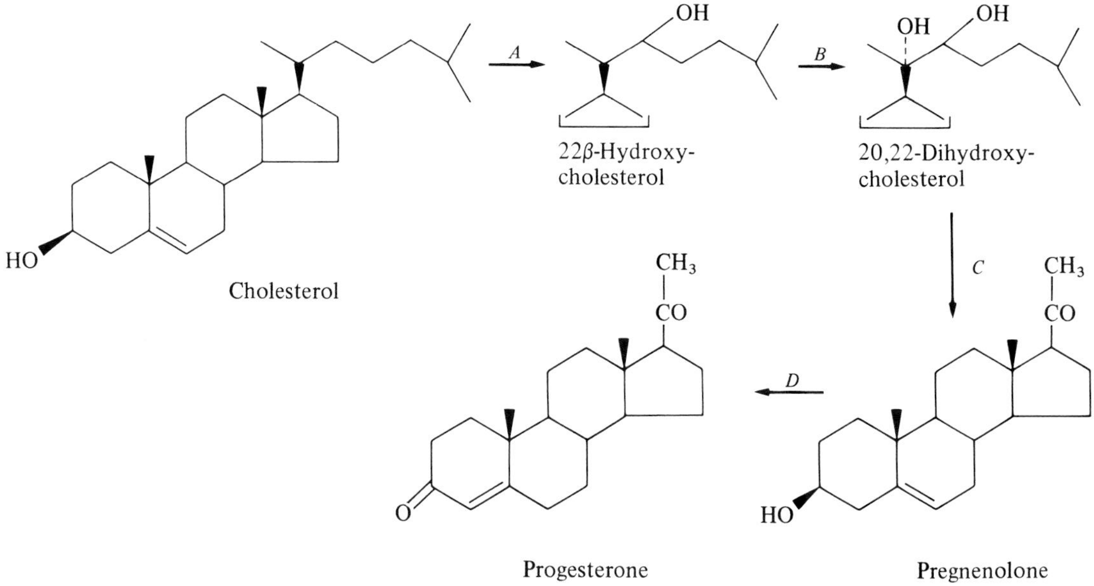

Fig. 11.24. Biosynthesis of progesterone from cholesterol in plants.

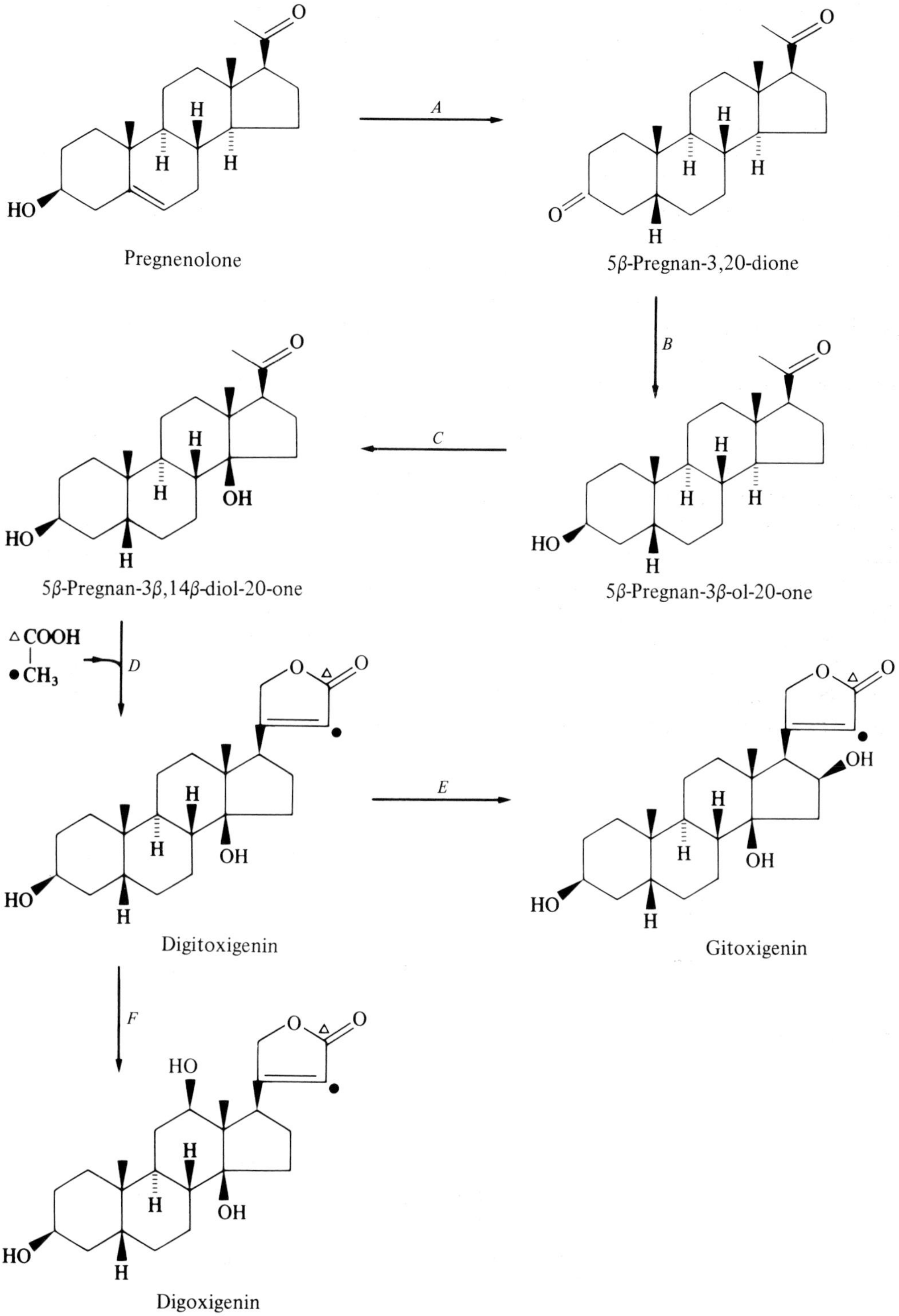

Fig. 11.25. Known steps in the conversion of pregnenolone into cardenolides.

from the aqueous environment becomes the 15β hydrogen of the 14-demethyl sterol.

The conversion of the Δ^8-sterol intermediate into a Δ^5-sterol generally proceeds through the sequence $\Delta^8 \rightarrow \Delta^7 \rightarrow \Delta^{5,7} \rightarrow \Delta^5$. Although the enzymes catalysing these reactions have not been isolated and studied as individual entities, a great deal is known of their stereospecificity through the use of mevalonic acid species stereospecifically labelled with tritium at Cs 2, 4 and 5. This is shown in Fig. 11.18.

Many higher plant, algal and fungal (but not animal), sterols possess a $trans$-Δ^{22}-double bond. This is introduced into the sterol side chain after alkylation at C–24 is complete. Since C–22 and C–23 are the former C–2 and C–5 of mevalonic acid respectively, the use of mevalonic acid species stereospecifically labelled with tritium at C–2 and C–5 has allowed the stereospecificity of hydrogen elimination during Δ^{22}-double bond insertion to be studied. The results of these studies are shown in Fig. 11.19.

The formation of cholesterol requires that the Δ^{24}-double bond be saturated. Nothing is yet known of this process in the plant kingdom but Fig.

11.20 shows its stereochemistry and likely mechanism in animal tissues.

(iv) PENTACYCLIC TRITERPENES

Rather few pentacyclic triterpenes have been studied biochemically but the pathway from squalene-2,3-oxide to β-amyrin, outlined in Fig. 11.21, is supported by considerable experimental data. The Ts in squalene-2,3-oxide and β-amyrin represent the locations in these molecules of the former 4-pro-R hydrogen of mevalonic acid. The intermediate stages in the pathway represent mechanistically acceptable 1,2-hydride shifts and final proton loss which lead to the experimentally observed tritium labelling in β-amyrin.

The pentacyclic triterpenes diploptene (11.59), serratene (11.60), hopene and fernene (Fig. 11.22), produced by the fern *Polypodium vulgare*, are interesting because they are hydrocarbons and, in consequence, cyclization is initiated by proton (H$^+$) attack on squalene itself and not squalene-2,3-oxide. Tracer experiments with mevalonic acid labelled in the 4R position with tritium (T) support the

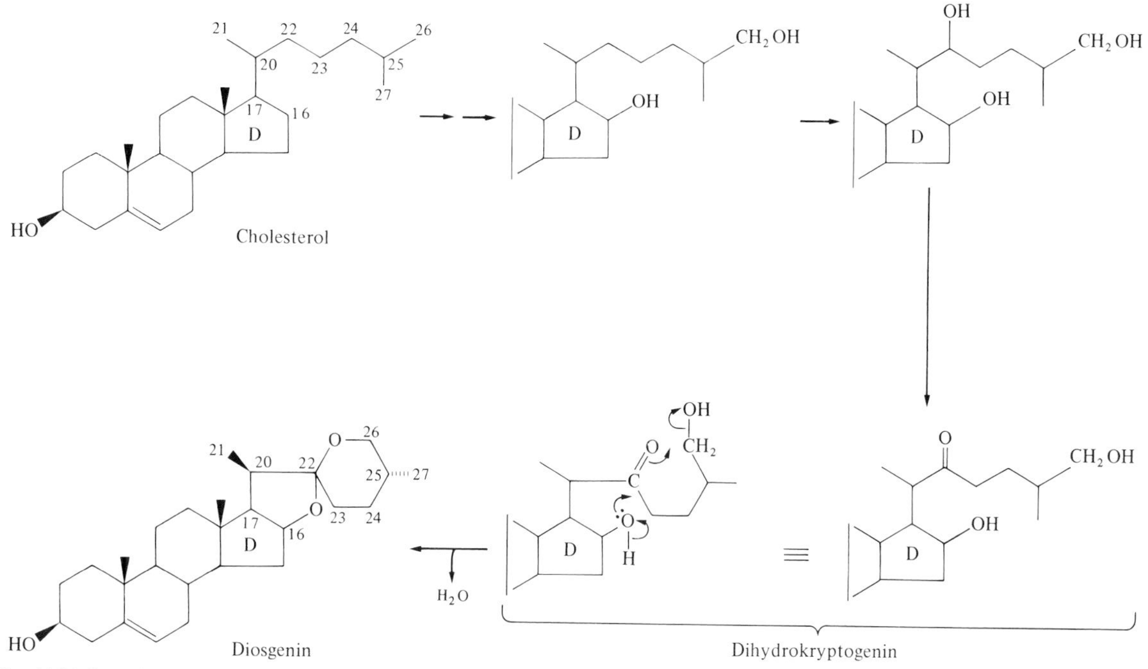

Fig. 11.26. Postulated pathway for the conversion of cholesterol into the sapogenin, diosgenin. (Note that the sterol side chain is not cleaved as is the case with the cardenolides.)

biosynthetic pathway outlined in Fig. 11.22 for hopene I and fernene.

(v) STERYL GLYCOSIDES AND ESTERS†

Figure 11.23 summarizes what is known about the biosynthesis of these sterol derivatives. The steryl glucosides are synthesized in the Golgi apparatus by glucose transfer from UGP-glucose (step A, Fig. 11.23). These glucosides can be further metabolized by being acylated by either particulate or soluble preparations in the presence of a suitable acyl donor such as phosphatidyl ethanolamine (step B) in particulate preparations and digalactosyldiglyceride (step C) in soluble preparations. The acyl donor for sterol esterification in spinach leaves is diacylglycerol (step D) but is phosphatidylcholine in the fungus *Phycomyces blakesleeanus* (step E).

(vi) STEROID HORMONES‡

Progesterone is synthesized in higher plants from cholesterol by a pathway similar to that in animals (Fig. 11.24). The removal of the side chain is preceded by hydroxylation at C–22 and C–20 (steps A and B). The resulting product, pregnenolone (step C), is then oxidized to progesterone (step D). The more abundant sterols with an alkyl group at C–24 can replace cholesterol as the starting material in many plants.

(vii) CARDENOLIDES§

The established steps in the formation of the cardenolides in *Digitalis* is indicated in Fig. 11.25.

† All letters in this section refer to Fig. 11.23.
‡ All letters in this section refer to Fig. 11.24.
§ All letters in this section refer to Fig. 11.25.

Pregnenolone is reduced to the 5β isomer (step A) (*cis* fusion of rings A and B of steroid nucleus) and then, after reduction of the 3-keto group (step B), the 14β hydroxyl group is inserted with inversion of the steroid ring C/D junction (step C) before acetate is added to form the cardenolide (αβ-unsaturated-γ-lactone) ring and thus digitoxigenin (step D). Further hydroxylation can then yield gitoxigenin (step E) and digoxigenin (step F). Reaction A can occur with production of 5α-pregnane derivatives, but these are probably not intermediates. The introduction of the 14β-hydroxyl group does not involve double-bond production at either C–8 or C–15, but the detailed mechanism of the reaction is unknown. The formation of the glycoside linkage which is always β in the cardenolides has not been studied in detail.

(viii) SAPOGENINS

Sapogenins are formed from cholesterol without side-chain cleavage via a pathway (Fig. 11.26) which does not involve pregnenolone as an intermediate.

(ix) ECDYSONES

Ecdysones (11.27) in plants are synthesized from cholesterol but problems which still require elucidation are the mechanism of the formation of the A/B *cis* ring junction and the point at which the 6-keto and the additional hydroxyls are inserted. There is evidence that the Δ^7-double bond is formed early on in the pathway and that it is the 7β and 8β hydrogens which are removed. The A/B *cis* ring junction is not formed via a Δ^4-3-keto derivative as in the cardenolides [Section C.6(vii)] and the proposed mechanism involves a 5α,6α-epoxide (Fig. 11.27). After the formation of the epoxide

(11.59)

(11.60)

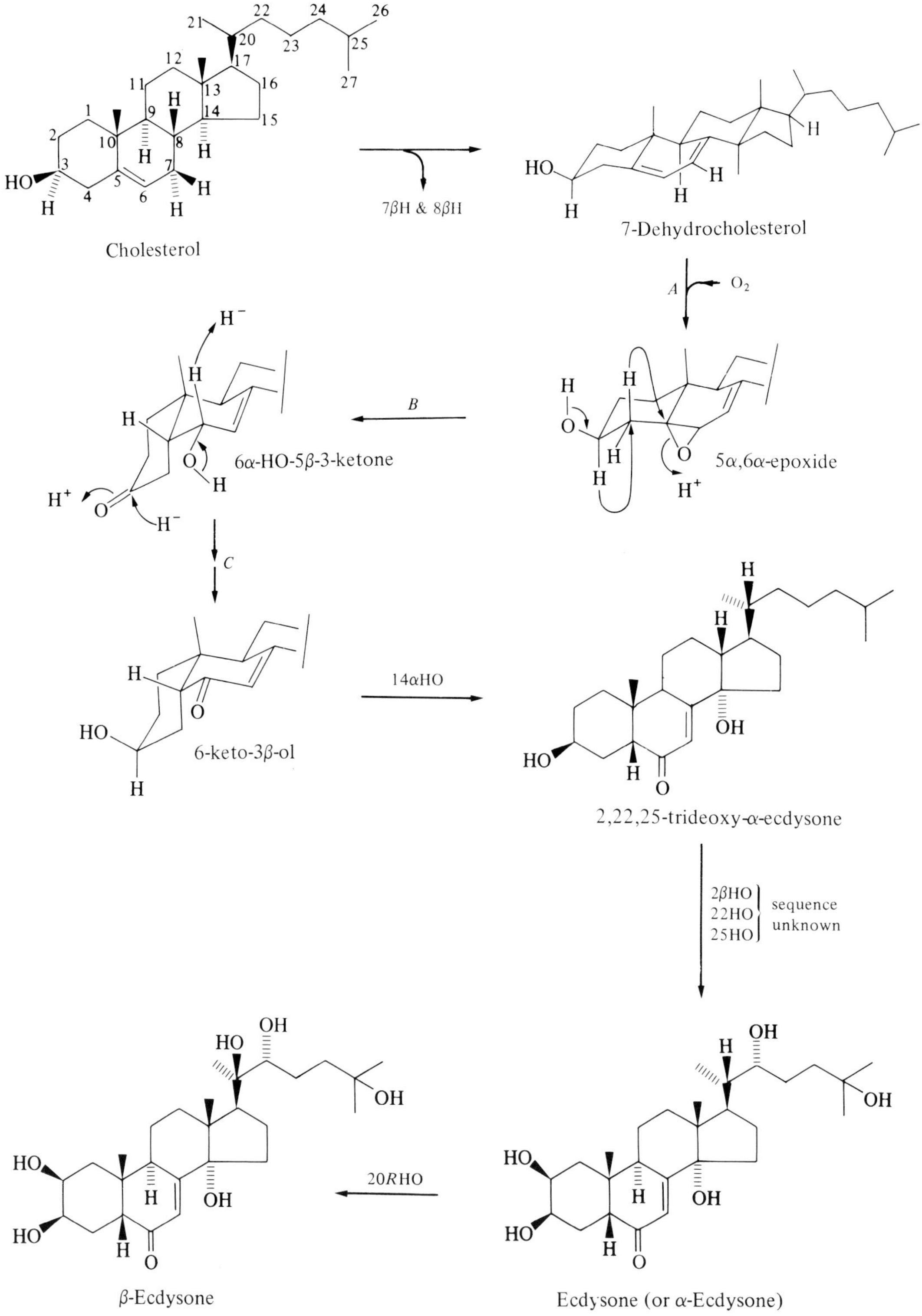

Fig. 11.27. A plausible biosynthetic pathway for ecdysones in *Polypodium vulgare*.

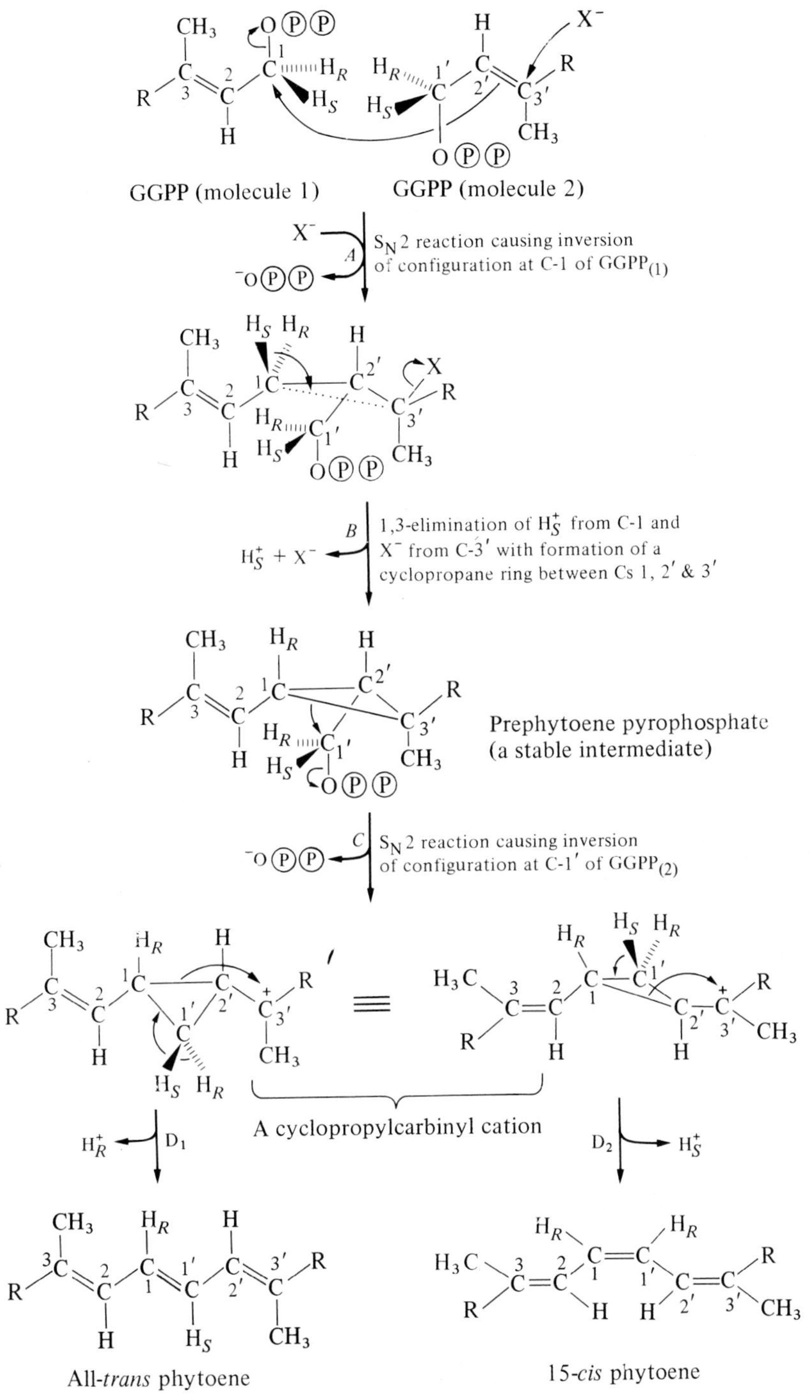

Fig. 11.28. Mechanism for the conversion of two molecules of geranylgeranyl pyrophosphate (GGPP) into all-*trans* phytoene in a *Mycobacterium* sp. and into 15-*cis* phytoene in higher plants and the fungus *Phycomyces blakesleeanus*. (The numbering of all the participating molecular species is that of GGPP; in the phytoene species the carbons numbered 1 and 1' above become 15 and 15', respectively; it is important to note that unless stated otherwise the designation of hydrogens as pro-*R* (i.e. H_R) or pro-*S* (i.e. H_S) refers only to their spacial orientation in MVA and GGPP which in this instance are identical; the two cyclopropylcarbinyl cations are the same species but have been drawn in different conformations so as to demonstrate the production of the *cis* and *trans* isomers of phytoene R = ⌁⌁⌁.)

group (step *A*) proton attack allows the intra-molecular movement of hydride from C–4 to C–5 and from C–3 to C–4 with expulsion of a proton from the hydroxyl group at C–3 (step *B*). The resulting 6α-hydroxy-5β-ketone is then converted into a 3β-hydroxy-6-keto compound (step *C*) which by addition of hydroxyl groups at C–14(α), C–2(β), C–25 and C–22(α) yields ecdysone. Further hydroxylation at C–20 produces β-ecdysone.

(x) STEROID ALKALOIDS

This subject will be considered in Chapter 13.

7. Tetraterpenes—Carotenoids

(i) PHYTOENE FORMATION†

The first 40-C hydrocarbon precursor of carotenoids is phytoene, which is analogous to the 30-C sterol precursor squalene except that the central bond is unsaturated in phytoene. It is formed (see Fig. 11.28) from prephytoene pyrophosphate which is produced by the tail-to-tail junction of two geranylgeranyl residues derived from geranylgeranyl pyrophosphate in exactly the same manner that presqualene pyrophosphate is formed (see Fig. 11.8) from two farnesyl pyrophosphate molecules. The reaction sequence (steps *C* and *D*, Fig. 11.28) to produce phytoene can then proceed in one of two directions depending upon whether all-*trans* or 15-*cis*-phytoene is produced. This follows from the fact that the stereoelimination of the hydrogens at C–15 and C–15′ is different according to the isomer produced. The prephytoene pyrophosphate loses its pyrophosphate moiety as an anion (step *C*) and the resulting cyclopropylcarbinyl cation is stabilized by loss of a proton to produce either all-*trans* phytoene (step *D*1) or 15-*cis* phytoene (step *D*2).

The hydrogen at C–15 in both species of phytoene was originally the 1-pro-*R* hydrogen of GGPP, whereas the hydrogen at C–15′ originated from the 1-pro-*S* hydrogen of GGPP if all-*trans* phytoene is formed and from the 1-pro-*R*-hydrogen of GGPP if 15-*cis*-phytoene is formed.

† The letters in this section refer to Fig. 11.28.

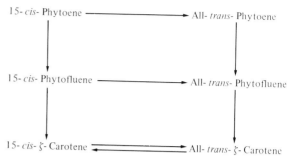

Fig. 11.29. Interconversions of *cis*- and *trans*-carotenoid precursors.

(ii) DESATURATION OF PHYTOENE

The step-wise desaturation of phytoene leads in most higher plants to all-*trans*-lycopene (11.33) so the question arises, at what stage in the desaturation process is the central *cis* double bond in 15-*cis*-phytoene, the normal product of phytoene synthetase in higher plants, isomerized to a *trans* double bond? The situation is not at all clear but the possibilities, all of which have been observed in some individual organisms, are indicated in Fig. 11.29. There is no doubt that from ζ-carotene (see Fig. 11.29) onwards, *trans* intermediates are involved. If the problem of the central *cis* double bond

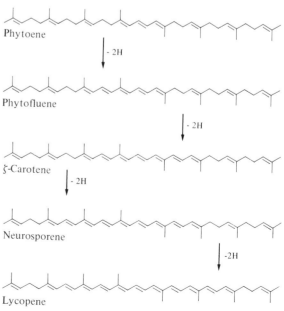

Fig. 11.30. Pathway of conversion of phytoene into lycopene in higher plants.

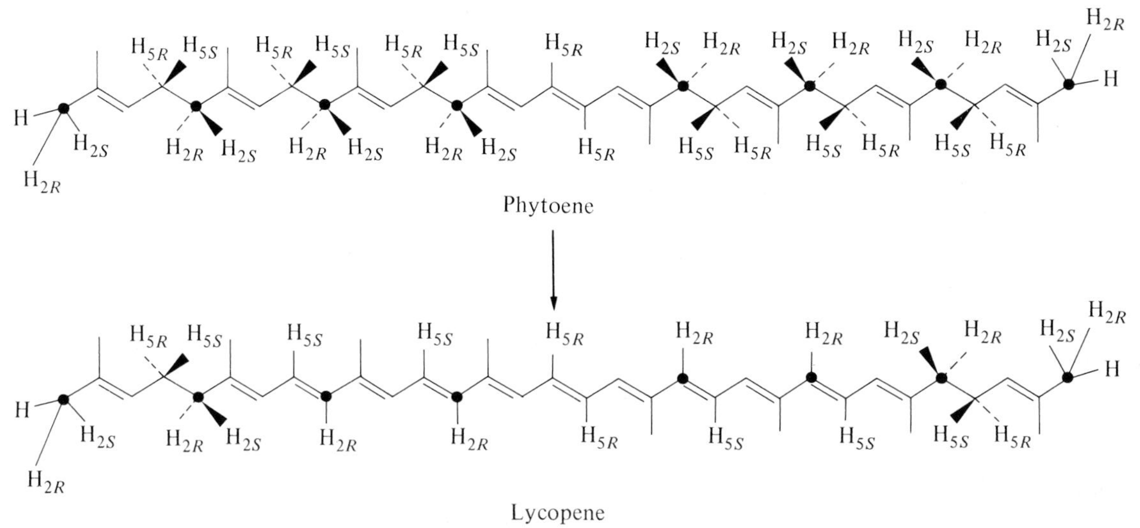

Fig. 11.31. Stereochemistry of the desaturation of phytoene to lycopene. ● indicates C–2 of mevalonate; H_{2R}, H_{2S}, H_{5R} and H_{5S} indicate the pro-2R-, pro-2S-, pro-5R- and pro-5S-hydrogen atoms of mevalonate respectively.

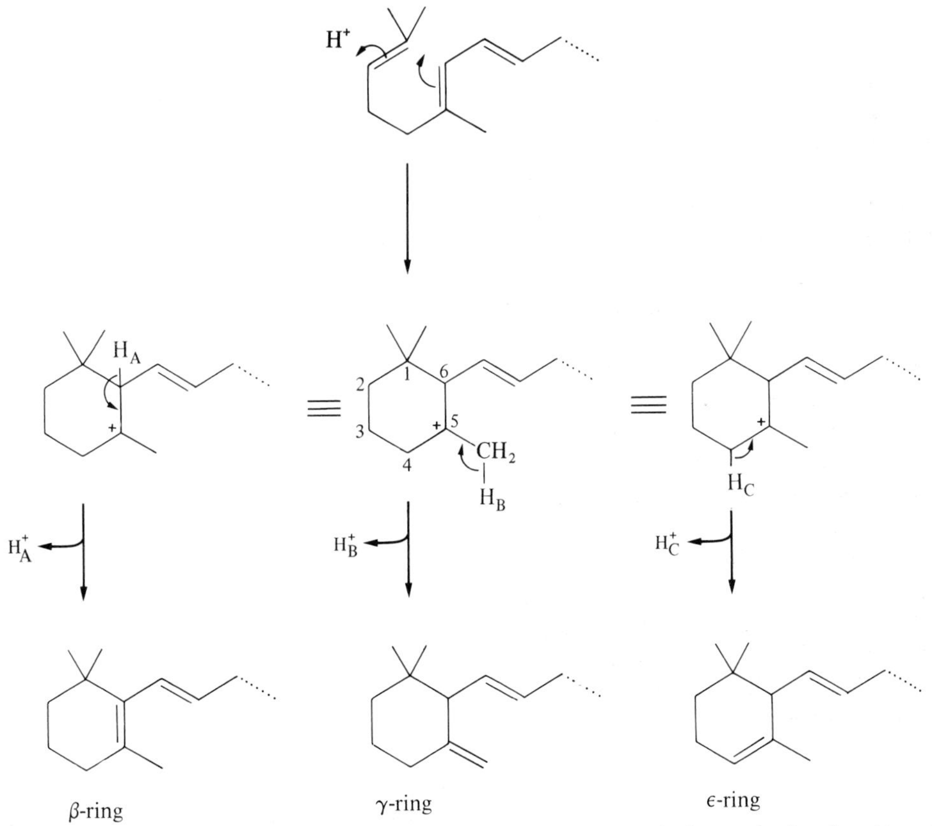

Fig. 11.32. The mechanism of cyclization of an acyclic carotene such as lycopene to form the β-, γ- and ϵ-rings found in cyclic carotenes.

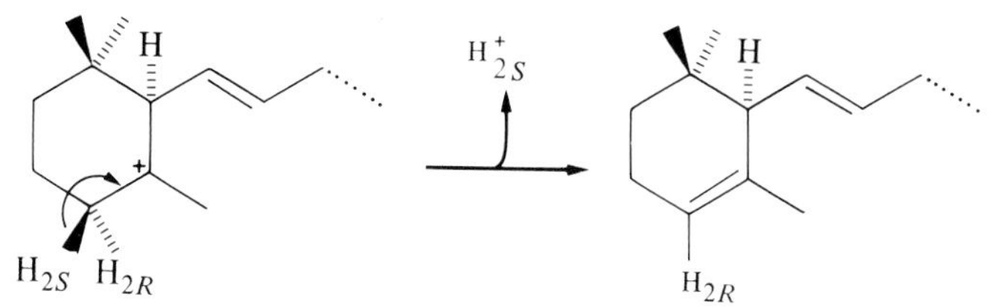

Fig. 11.33. Stereospecificity of hydrogen elimination during the formation of ε-rings (H_{2S} and H_{2R} are the former 2-pro-S and 2-pro-R hydrogens of MVA.)

is ignored then the stepwise desaturation in higher plants takes the pathway indicated in Fig. 11.30. Each desaturation involves the removal of two hydrogens, one originating from C–2 of one mevalonic acid molecule and the other from C–5 of an adjacently incorporated mevalonic acid. As indicated in Fig. 11.31 these two prochiral centres are treated in a stereospecific manner and at each desaturation it is the 2-pro-S and the 5-pro-R hydrogens which are lost.

(iii) Cyclization

The cyclic carotenes are formed from lycopene by proton attack at the C–1 double bond; the resulting intermediate, indicated in Fig. 11.32 as a C–5 carbonium ion, is then stabilized in one of three ways, according to the enzyme involved, by loss of (i) H_A to produce a β-ring, (ii) H_B to produce the exocyclically double bonded γ-ring or (iii) H_C to produce the ε-ring. Once they are formed there is little if any interconversion of rings. A close look at

C–4 reveals that it is a prochiral centre and, as expected, only one specific hydrogen, the former 2-pro-S hydrogen of MVA, is removed in formation of the ε-ring (Fig. 11.33). The stereochemical situation in the formation of the β-ring involves attack by the incoming proton at the re, re face of the double bond and its localization at the β- or pro-S position at C–2; at the same time the geminal methyl groups retain their individuality so that the methyl arising from C–2 of mevalonic acid takes up the α-position at C–1 and therefore becomes C–16 (Fig. 11.34). In this context α and β substituents are defined as projecting down or up respectively from the general plane of the cyclohexenoid ring when the normal sequence of numbering of the carbon atoms of the latter is anticlockwise.

(iv) Xanthophyll formation

In the production of xanthophylls in higher plants, the main reactions are the insertion of a hydroxyl group at C–3 and an epoxy group across

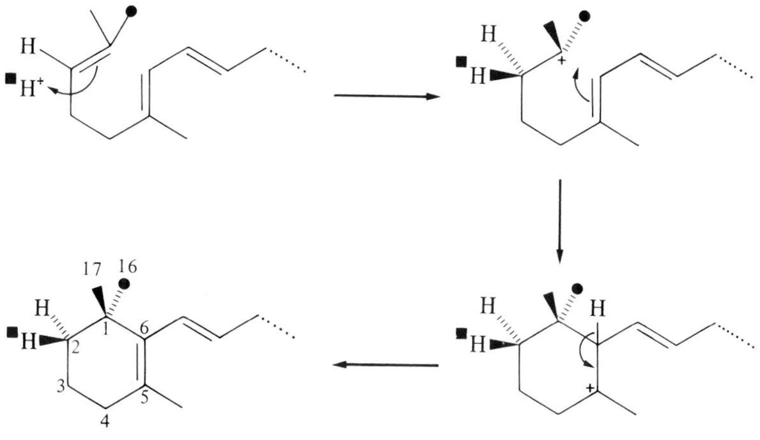

Fig. 11.34. Stereochemistry of protonation at C–2 during the formation of β-rings. ($\bullet$ = carbon derived from C–2 of MVA.)

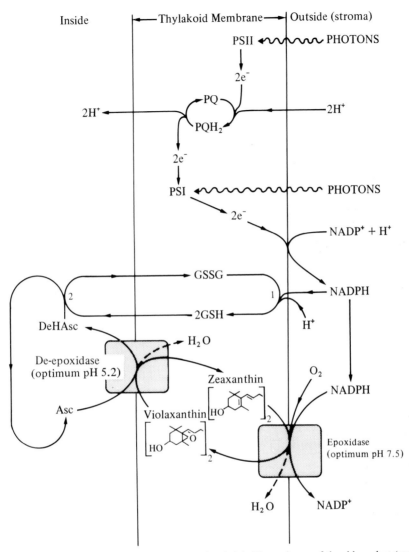

Fig. 11.35. Mechanism and location of the Xanthophyll Cycle in the thylakoid membrane of the chloroplast (modified from Hager, 1975). (PS I = Pigment System I; PS II = Pigment System II; PQ = plastoquinone; GSH = reduced glutathione; GSSG = oxidized glutathione; Asc = ascorbic acid; DeHAsc = dehydroascorbic acid; 1 = glutathione reductase, EC 1.6.4.2; 2 = glutathione dehydrogenase (ascorbate), EC 1.8.5.1.)

the C–5, C–6 double bond of the β-rings. The first reaction involves the direct insertion of a hydroxyl group with retention of configuration [eq. (11.5)]. The epoxidation involves a series of reactions known as the Xanthophyll Cycle (Fig. 11.35). In leaves epoxide (e.g. violaxanthin, a diepoxide) levels decrease on illumination under anaerobic conditions and are re-established in the dark in the presence of oxygen. The enzymes involved are violaxanthin de-epoxidase (MW 54,000, opt. pH 5.2) and zeaxanthin epoxidase, a mono-oxygenase with an optimum pH of 7.5. The enzymes are present in the chloroplast thylakoids and their optimum pH values suggest that the de-epoxidase is located on the inner surface of the thylakoid membrane and the epoxidase on the outer surface.

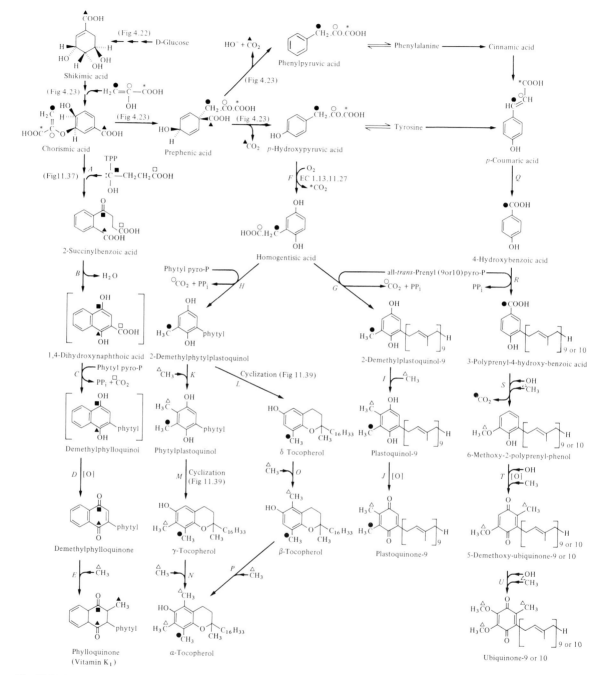

Fig. 11.36. Probable biosynthetic routes to the terpenoid quinones from shikimic acid and prenyl pyrophosphates. (Methyl groups labelled with Δ are derived from S-adenosylmethionine by the mechanisms shown in Fig. 11.38;

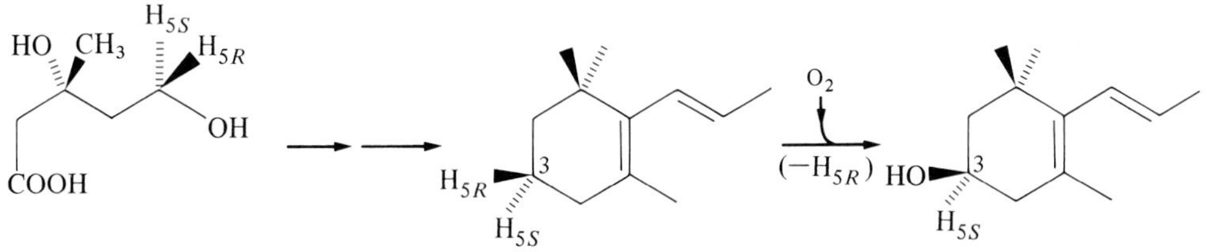

$$(11.5)$$

MVA

3R-hydroxyxanthophyll

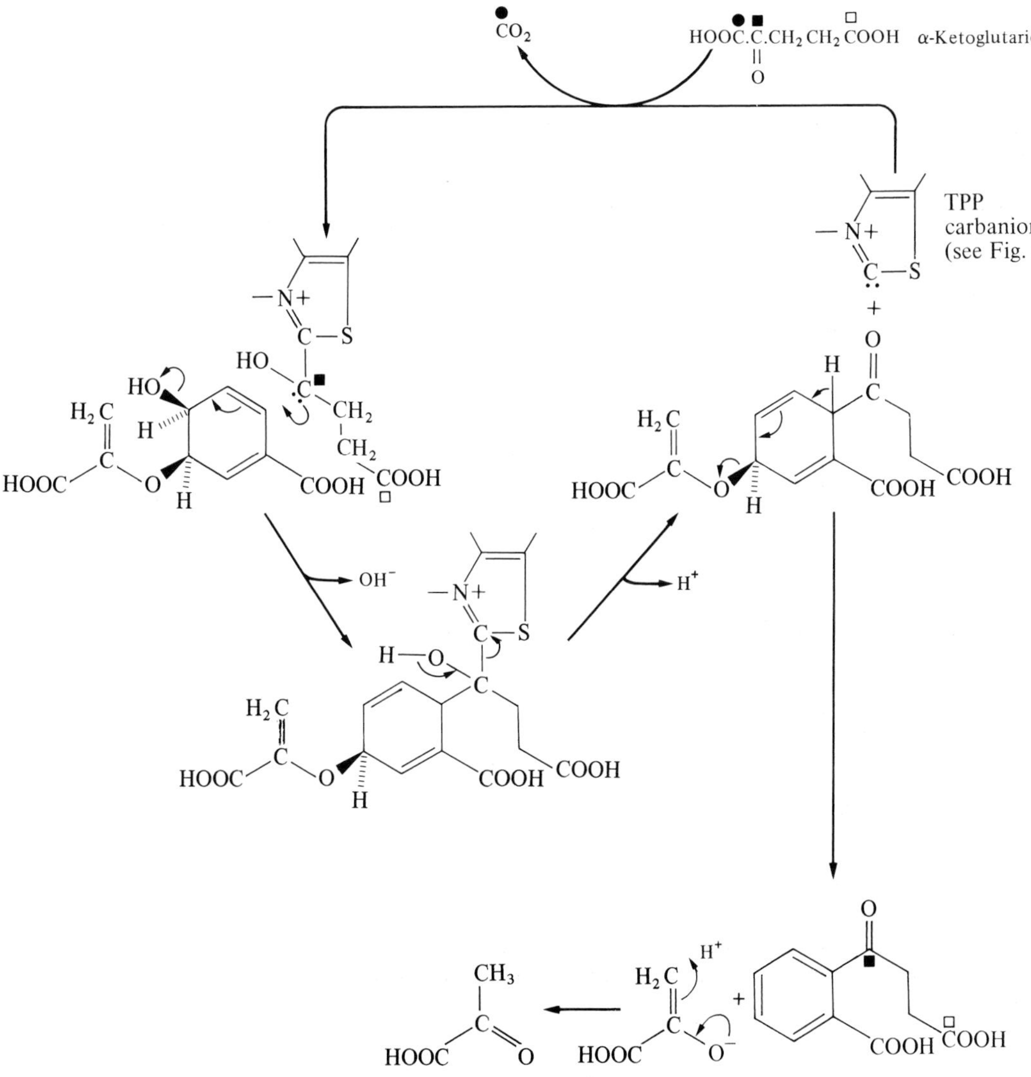

Pyruvic acid   2-Succinylbenzoic acid

Fig. 11.37. Postulated mechanism for the formation of 2-succinylbenzoic acid during phylloquinone biosynthesis.

The probable spatial separation of the enzymes has led to the concept of the Xanthophyll Cycle. Because of the markedly different pH optima of the enzymes the functioning of the cycle depends on the light-induced H^+ ion gradient across the membrane (see Chapter 5, Section C.6) and the translocation of carotenoids across the membrane. The latter has still to be demonstrated experimentally. The reductant, NADPH, generated by light is used in the mono-oxygenase reaction and also to maintain the levels of reduced ascorbic acid which is, in the *in vitro* system at least, the immediate reductant of the de-epoxidase. The intermediate between NADPH and ascorbic acid is the NADPH-dependent glutathione reductase[7]. A clear-cut function for this cycle is still not obvious.

8. Polyprenols

The biosynthesis of polyprenols such as the castaprenols (see Section B.10) begins with the usual synthesis of farnesyl pyrophosphate which are all-*trans* molecules; these compounds are then elongated by *cis*-prenyltransferases which catalyse the stepwise addition of isoprene residues from IPP in such a manner that the double bond in each of the newly introduced isoprene units is *cis*. In these reactions the 2-pro-S-hydrogen of the incoming IPP is lost; this contrasts with the elongation reaction catalysed by the *trans*-prenyltransferase described in Section C.1 where the double bond in the newly introduced isoprene unit is *trans* and the 2-pro-R hydrogen is eliminated.

9. Terpenoid quinones and chromanols†

Terpenoid quinones and chromanols such as phylloquinone (vitamin K_1), the plastoquinones, the ubiquinones and the tocopherols are formed in the plant tissues by the condensation of an appro-

† All letters in this section refer to Fig. 11.36.

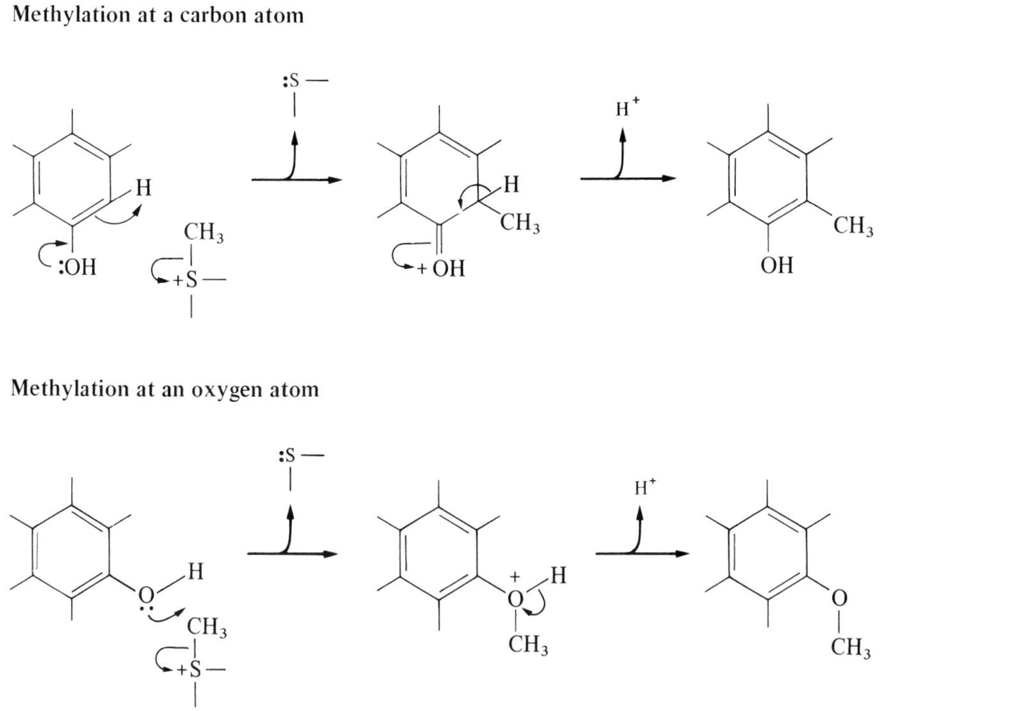

Methylation at a carbon atom

Methylation at an oxygen atom

Fig. 11.38. Mechanism of the methylation reactions occurring in the biosynthesis of terpenoid quinones. (The methylating agent is S-adenosylmethionine.)

priate aromatic residue with an appropriate poly-prenyl residue. The former is derived from chorismic acid which in turn is formed via the shikimic acid pathway (see Figs. 4.22 and 4.23); the latter is derived from a polyprenyl pyrophosphate. Figure 11.36 outlines the probable biosynthetic routes from shikimic acid and polyprenyl pyrophosphates to this group of compounds.

In the case of phylloquinone, chorismic acid is first converted into 2-succinylbenzoate (A); this involves the transfer of the succinol (C$_4$) unit from succinol-thiamine pyrophosphate to the aromatic ring of chorismic acid and the subsequent loss of the hydroxyl and enolpyruvate residues as is shown in Fig. 11.37. The succinol residue is derived from Cs 2–4 of α-ketoglutaric acid; the α-carboxyl group of the latter is eliminated as CO$_2$ during the formation of the succinol–TPP complex. The 2-succinylbenzo-ate is then believed to ring close with the elimination of water to form 1,4-dihydroxynaphthoic acid (B) which then condenses with the phytyl moiety of phytyl pyrophosphate, CO$_2$ and inorganic pyro-phosphate being liberated in the process. The involvement of 1,4-dihydroxynaphthoic acid in reactions B and C has still to be proved experiment-ally, hence it is included in Fig. 11.36 within square brackets indicating its hypothetical status. The same is also true of demethylphylloquinol which is

shown as being the product of reaction C. Indeed it is not yet clear at what stage oxidation of the quinol to the quinone takes place; however, in Fig. 11.36 it is postulated that demethylphylloquinol is formed first and then oxidized to demethylphylloquinone (D). The latter then undergoes C-methylation by the mechanism shown in Fig. 11.38 to yield phyllo-quinone.

In the case of the plastoquinones and toco-pherols, chorismic acid is converted into p-hydroxy-pyruvic acid (see Fig. 4.23) which is then oxidized and decarboxylated by 4-hydroxyphenylpyruvate dioxygenase[8] to yield homogentisic acid (F). The latter then undergoes polyprenylation and simul-taneous decarboxylation; when the polyprenylating agent is all-trans-nonaprenyl pyrophosphate, 2-demethylplastoquinol-9, a plastoquinone precursor results (G) whereas when it is phytyl pyrophos-phate, 2-demethylphytylplastoquinol, a tocopherol precursor, is formed (H).

2-Demethylplastoquinol-9 subsequently under-goes C-methylation (see Fig. 11.38) to yield plastoquinol-9 (I) which is then oxidized to plasto-quinone-9 (J).

2-Demethylphytylplastoquinol undergoes either C-methylation (see Fig. 11.38) to yield phytylplasto-quinol (K) or cyclization (see Fig. 11.39) to yield δ-tocopherol (L). Phytylplastoquinol then undergoes

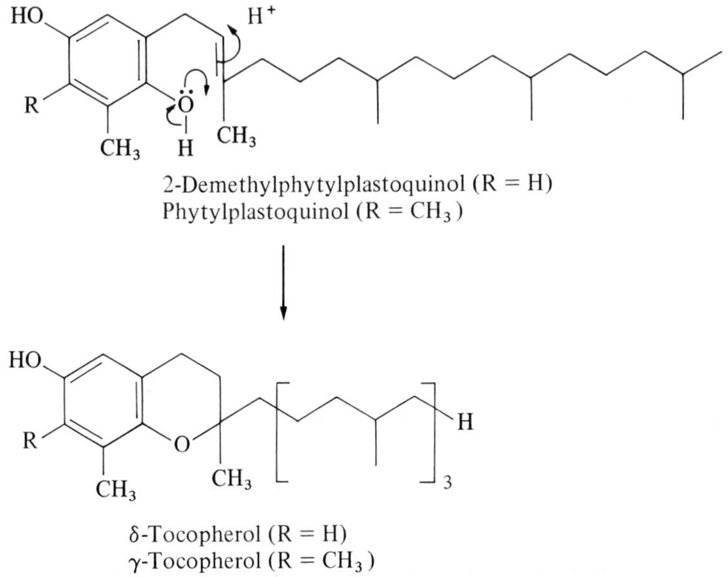

2-Demethylphytylplastoquinol (R = H)
Phytylplastoquinol (R = CH$_3$)

δ-Tocopherol (R = H)
γ-Tocopherol (R = CH$_3$)

Fig. 11.39. Mechanism of the cyclization reaction during tocopherol biosynthesis.

cyclization (see Fig. 11.39) to form γ-tocopherol (M) which is converted into α-tocopherol by further C-methylation (N). The δ-tocopherol is converted into β-tocopherol (O) and possibly α-tocopherol (P) by C-methylations.

In the case of the ubiquinones, chorismic acid is converted into p-coumaric acid via prephenic acid and L-phenylalanine or L-tyrosine. The p-coumaric acid is then oxidized to 4-hydroxybenzoic acid (Q) which reacts with all-*trans*-nona (or deca) prenyl pyrophosphate yielding 3-nona (or deca) prenyl-4-hydroxybenzoic acid (R). This then undergoes hydroxylation, O-methylation (see Fig. 11.38) and decarboxylation (probably in that order) to yield 6-methoxy-2-nona (or deca) prenylphenol (S). The latter is then converted into 5-demethoxyubiquinone-9 (or 10) by further hydroxylation, C-methylation and oxidation of the quinol ring to the quinone ring (T). Further hydroxylation and O-methylation (U) yields ubiquinone-9 (or 10).

Looking at the overall picture of terpenoid quinone and chromanol biosynthesis (Fig. 11.36) it is evident that one methyl group on the ring systems of the plastoquinones and tocopherols arises from a carbon atom in the aromatic precursor (i.e. the β-carbon of p-hydroxyphenylpyruvic acid) whilst all the methyl groups in the ring systems of phylloquinone and the ubiquinones arise by C or O-methylations from S-adenosylmethionine.

The intracellular location of the biosynthetic sequences outlined in Fig. 11.36 are probably as follows: (i) phylloquinone and the plastoquinones are formed in the chloroplast, (ii) the ubiquinones are formed in the mitochondrion and (iii) α-tocopherol is formed both inside and outside the chloroplast.

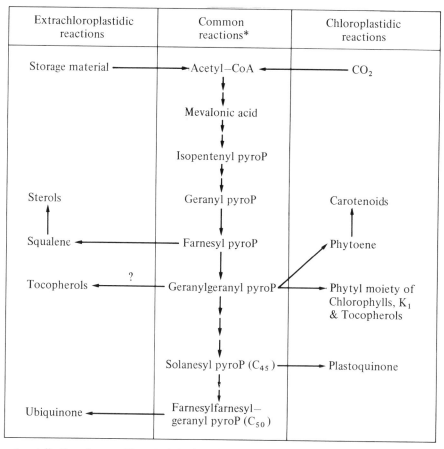

Fig. 11.40. Compartmentalization of terpenoid synthesis in the cells of the photosynthetic tissues of higher plants. (* = reactions that occur both inside and outside the chloroplast.)

10. Rubber

Isolated latex from *Hevea* will incorporate HMG-CoA and IPP very effectively into rubber, apparently by extending the molecules on the surface of the rubber particles. The formation of the *cis* double bonds in rubber is brought about by a specific *cis*-prenyltransferase. It will be recalled that in the formation of *trans*-prenyl pyrophosphates (see Section C.1), it is the 2-pro-*R* hydrogen of IPP which is stereospecifically removed by the *trans*-prenyltransferase; in rubber (11.52) the formation of the *cis*-double bonds results in the stereospecific removal of the 2-pro-*S* hydrogen of IPP.

11. Localization of synthesis

There is no doubt that the chain elongation reactions for terpenoid synthesis can take place both inside and outside the chloroplasts (Fig. 11.40). However, the final specific reactions are localized in organelles. Carotenoids and the mixed terpenoids such as the plastoquinones, phylloquinone and chlorophyll are synthesized exclusively in the chloroplasts although in the case of the pigments, the enzymes concerned are under nuclear control (see Chapter 10). Ubiquinone, on the other hand, is formed in the mitochondria and sterols in the endoplasmic reticulum. There is evidence that a small amount of tocopherol is extraplastidic in origin.

D. FUNCTION

Many terpenoids are considered 'secondary plant products' which are produced in profusion (over 10,000) by plants but which have no known function. This may well be true for many terpernoids but some members of all groups of terpenoids do have important functional activities. Why so many related 'useless' compounds are also synthesized at the same time is not known.

Many terpenoids have hormonal functions and these are discussed in detail in Chapter 15. It is sufficient here to remember that they include the cytokinins, gibberellins, abscisic acid and, possibly, the xanthoxins. Allelopathic roles are often ascribed to mono- and sesquiterpenes [allelopathy is the detrimental effect of one plant (donor) on another (recipient)].

Sterols are localized in plant membranes and presumably function there in the same way as cholesterol does in animal membranes. The reason for the requirement of the side-chain alkyl groups in sterols of plant membranes is not clear. It is assumed that the sterols stabilize membranes and control their permeability. Experiments which support the latter function include those in which the plants are exposed to ozone; this treatment increases membrane permeability and reduces the sterol levels of the membranes; cytokinins protect against the increased permeability and also prevent loss of sterols. In spite of much effort there is yet no proof that endogenous sterols have any hormone function in higher plants.

Carotenoids function by protecting against photodynamic sensitization and in light-harvesting in photosynthesis (see Chapter 5, Section C).

Mixed terpenoids also play key roles in plant metabolism. Chlorophyll without its phytyl side chain is not effective and plastoquinone plays an essential role in photosynthetic electron transport [Chapter 5, Section C.6(ii)] as does ubiquinone in mitochondrial electron transport [Chapter 6, Section B(ii)].

Polyprenyl pyrophosphates function in glycosylation in cell-wall formation [Chapter 7, Section D.2(vi)].

SUGGESTIONS FOR FURTHER READING

Lower terpenes

Banthorpe, D. V. and Charlwood, B. V. (1980) in *Encyclopaedia of Plant Physiology*, Vol. 8, p. 185. Springer, Heidelberg.
Loomis, W. D. and Croteau, R. (1980) in *Biochemistry of Plants* (Stumpf, P. K. and Conn, E. E., eds.), Vol. 4, p. 364. Academic Press, New York.

Steroids

Nes, W. R. and McKean, M. L. (1977) *Biochemistry of Steroids and Other Isopentenoids*. University Park Press, Baltimore, London.

Goodwin, T. W. (1980) in *Biochemistry of Plants* (Stumpf, P. K. and Conn, E. E., eds.), Vol. 8, p. 485. Academic Press, New York.

Grunwald, C. (1980) in *Encyclopaedia of Plant Physiology*, Vol. 8, p. 221. Springer, Heidelberg.

Carotenoids

Goodwin, T. W. (1979) *Ann. Rev. Plant Physiol.* **30,** 369–404.

Goodwin, T. W. (1980) *Biochemistry of Carotenoids*, 2nd edn., Vol. 1. Chapman & Hall, London.

Spurgeon, S. L. and Porter, J. W. (1980) in *Biochemistry of Plants* (Stumpf, P. K. and Conn, E. E., eds.), Vol. 4, p. 419. Academic Press, New York.

Polyprenols and terpenylquinones

Threlfall, D. R. (1980) in *Encyclopaedia of Plant Physiology*, Vol. 8, p. 289. Springer, Heidelberg.

Archer, B. L. (1980) in *Encyclopaedia of Plant Physiology*, Vol. 8, p. 309. Springer, Heidelberg.

Function of terpenoids

Goodwin, T. W., ed. (1978) *The Biochemical Functions of Terpenoids in Plants.* The Royal Society, London.

ENZYMES

1. Isopentenyldiphosphate Δ^3-Δ^2-isomerase, EC 5.3.3.2.
2. Dimethylallyldiphosphate: isopentenyldiphosphate dimethylallyltransferase, EC 2.5.1.1.
3. ATP: 5-diphosphomevalonate carboxy-lyase (dehydrating), EC 4.1.1.33.
4. Farnesyl-diphosphate: farnesyl-diphosphate farnesyltransferase, EC 2.5.1.21.
5. Squalene, hydrogen-donor: oxygen oxidoreductase (2,3-epoxidizing), EC 1.14.99.7.
6. Formate: NAD^+ oxidoreductase, EC 1.2.1.2.
7. NADPH: oxidized glutathione oxidoreductase, EC 1.6.4.2.
8. 4-Hydroxyphenylpyruvate: oxygen oxidoreductase (hydroxylating, decarboxylating), EC 1.13.11.27.

CHAPTER 12

Chlorophylls and Haems

The general structures of the plant chlorophylls are given in Fig. 5.4, but that of chlorophyll a is repeated here (12.1) for ease of reference. The biogenetic implications of the structure are evident if we consider that the result of four pyrrole residues (12.2) joined by methylene bridges at their α-carbon atoms is porphyrinogen (12.3). If the methylene bridges are oxidized to methine bridges porphin (12.4) is formed. Appropriate substitution at carbons 1–8 yields protoporphyrin IX (12.5), a key intermediate in chlorophyll and haem synthesis.

If iron is inserted into protoporphyrin IX (12.5) protohaem (12.6) is formed; insertion of Mg^{2+}, however, yields magnesium protoporphyrin IX which has to undergo a number of changes before chlorophyll a (12.1) is obtained. These are (i) reduction of vinyl at C–4 to ethyl, (ii) formation of a fifth, carbocyclic ring, resulting from oxidation of the side chain at C–6, (iii) methylation of the carboxyl group of the C–6 side chain, (iv) phytylation of the carboxyl group of the side chain at C–7 and (v) reduction of ring D. The remainder of this chapter will be mainly concerned with the formation of chlorophyll.

A. INTRODUCTION

The biologically important porphyrin derivatives in plants are the metalloporphyrins formed by chelation with iron or magnesium. Iron porphyrins are prosthetic groups of the cytochromes (Chapter 6), haemoglobins in root nodules (Chapter 9) and enzymes such as catalase. The magnesium porphyrins are the chlorophylls and these exist in the photosynthetic tissues of plants at about 50 times the concentration of that of the haems (Table 12.1).

Table 12.1. *Levels of chlorophylls and haems in leaf tissue of higher plants*

Source	Concn (mμ moles/mg fresh wt.)	
	Chlorophylls	Haems
Laminium album	2.8	0.046
Stellaria sp.	2.7	0.041
Triticum vulgare	3.7	0.052

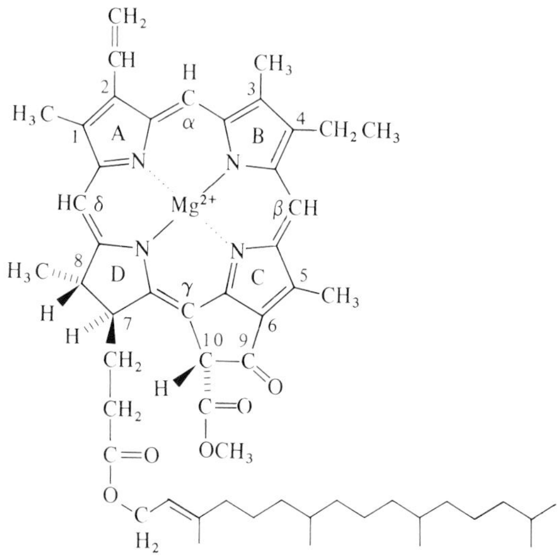

(12.1)

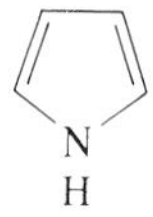

(12.2)

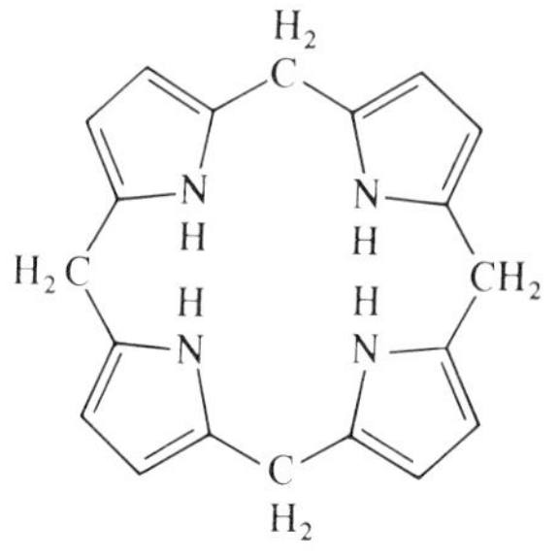

(12.3)

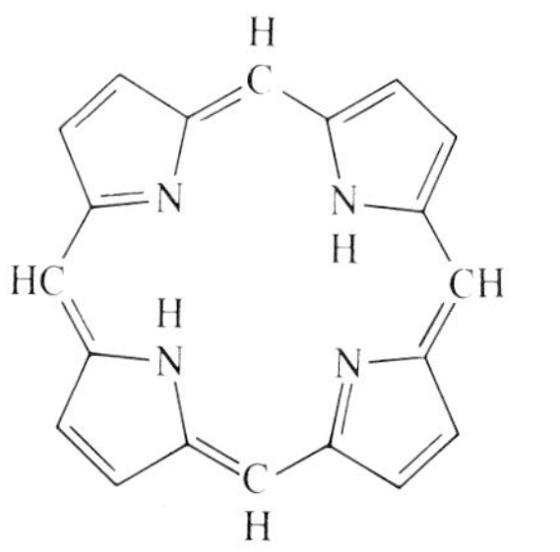

(12.4)

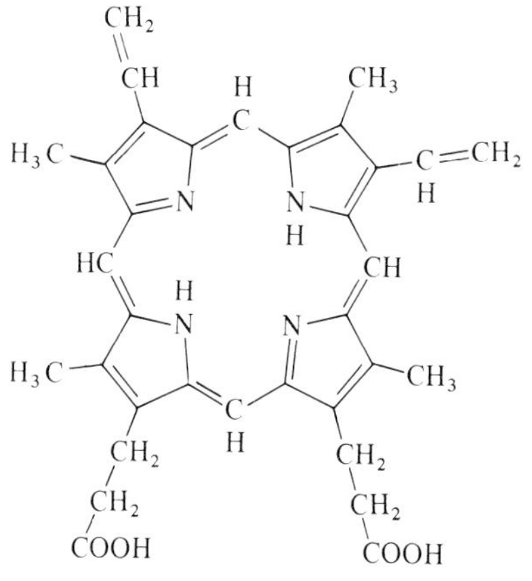

(12.5)

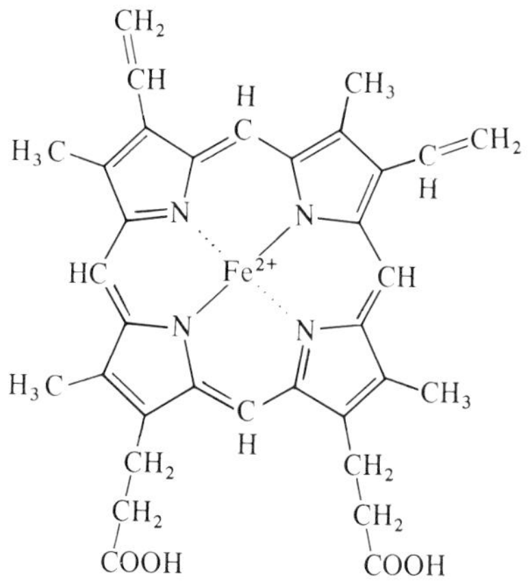

(12.6)

B. BIOSYNTHESIS OF PORPHYRINS

1. Formation of δ-aminolaevulinic acid†

It was the study of haem biosynthesis which eventually revealed that the first specific precursor of the porphyrin molecule was δ-aminolaevulinic acid (ALA) synthesized by ALA-synthase[1] from glycine and succinyl-CoA, in the presence of pyridoxal phosphate. The overall reaction is outlined in

† Letters used in this section refer to Fig. 12.1.

eq. (12.1). This reaction has always been difficult to demonstrate in higher plants and it now seems that ALA synthesized in the chloroplast for chlorophyll production is formed by a different route. It is formed from the intact glutamate molecule according to the pathway outlined in Fig. 12.1. The first two steps (A, B) are reactions analogous to the metabolism of glutamic acid in its conversion into other amino acids (see Chapter 9) except that it is the 1-semialdehyde and not the 5-semialdehyde which is involved. The enzyme for reaction C has only relatively recently been purified from plastids and the reaction represents a novel intramolecular

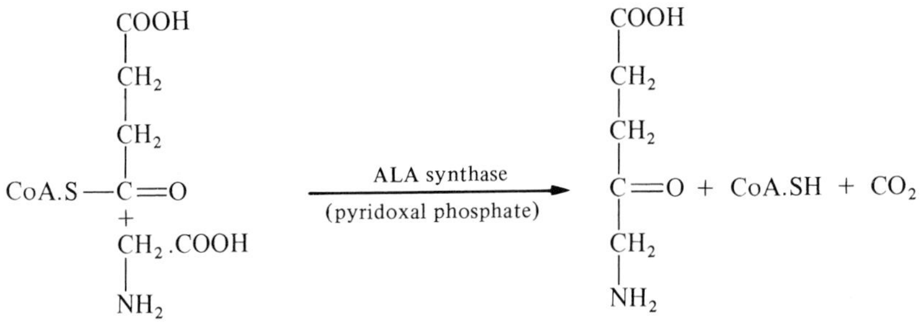

$$
\begin{array}{c}
\text{COOH} \\
| \\
\text{CH}_2 \\
| \\
\text{CH}_2 \\
| \\
\text{CoA.S}-\text{C}{=}\text{O} \\
+ \\
\text{CH}_2.\text{COOH} \\
| \\
\text{NH}_2
\end{array}
\quad
\xrightarrow[\text{(pyridoxal phosphate)}]{\text{ALA synthase}}
\quad
\begin{array}{c}
\text{COOH} \\
| \\
\text{CH}_2 \\
| \\
\text{CH}_2 \\
| \\
\text{C}{=}\text{O} + \text{CoA.SH} + \text{CO}_2 \quad (12.1) \\
| \\
\text{CH}_2 \\
| \\
\text{NH}_2
\end{array}
$$

amino transfer. It is fascinating that such a marked bifurcation of synthesis between higher plants and animals exists at this crucial biosynthetic step. One wonders whether the small amounts of haem derivatives present in plants growing in the dark arise from ALA synthesized in this way or whether the animal route is used but is difficult to detect because of the low levels of substrate involved.

2. Formation of porphobilinogen†

The condensation of two molecules of δ-amino-laevulinic acid (ALA) in the presence of the enzyme porphobilinogen synthase[2] (ALA dehydratase) produces the first pyrrole precursor of the metallo-porphyrins, porphobilinogen. The postulated mechanism of this reaction is shown in Fig. 12.2. The first molecule of ALA (ALA_1) reacts with the ε-amino group of a lysine residue at the active site of the

enzyme (steps A–C) to form a Schiff's base. This then undergoes an aldol-type condensation with a second molecule of ALA (ALA_2, step D). The carbanion of the Schiff's base-ALA which attacks the carbonyl carbon of ALA_2 is generated by deprotonation by B^-, whilst the protonation of the carbonyl oxygen involves BH. B^- and BH are probably the thiolate anion (S^-) and thiol (SH) group respectively of two cysteine residues at the active site; it is known that thiol groups are essential for activity because cysteine alkylating agents such as N-ethylmaleimide and p-chloromercuribenzoate inhibit the enzyme. The initial status of these groups is then regenerated (step E) by the mediation of an imidazole bridge which involves a histidine residue at the active site; it has recently been shown that a histidine residue in the enzyme is essential for activity. There then follows a second deprotonation/protonation sequence (step F) which leads to the elimination of water. B^-, BH and the eliminated water molecule then participate in the ring-closure reaction (step G). In step H the imidazole bridge facilitates another $\text{B}^- {\rightleftharpoons} \text{BH}$ interchange which

† Letters in this section refer to Fig. 12.2.

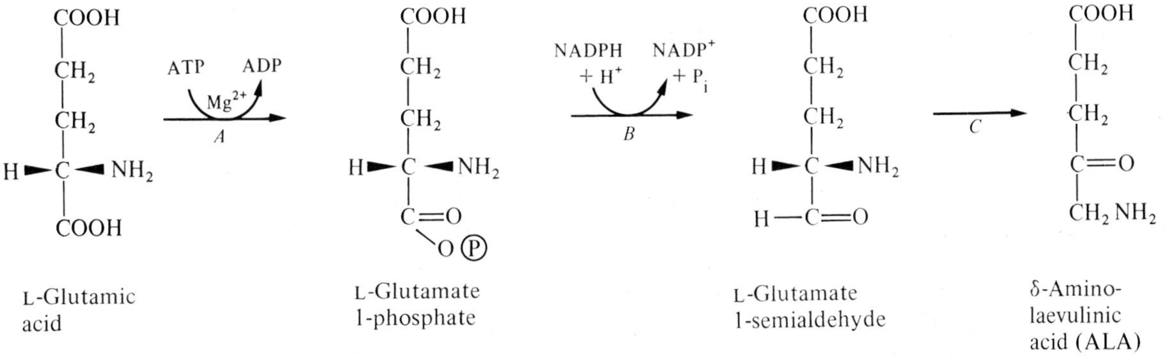

Fig. 12.1. The conversion of L-glutamic acid into δ-aminolaevulinic acid (ALA) in higher plants. (Enzymes involved: A = ATP: L-glutamate 1-phosphotransferase; B = L-glutamate 1-semialdehyde: NADP⁺ oxidoreductase (phosphorylating); C = L-glutamate 1-semialdehyde aminotransferase.)

allows for the dehydrogenation of the pyrrole ring and the cleavage of the resulting porphobilinogen from the lysine nitrogen (step I). The dehydrogenation of the immediate precursor of porphobilinogen is known to be stereospecific.

3. Uroporphyrinogen III formation

The mechanism of the conversion of four molecules of porphobilinogen into uroporphyrinogen III (urogen III) (12.7), the porphyrinogen precursor of chlorophyll and haem, is complex and has only recently come near to elucidation. The complexity arises because a simple head-to-tail condensation of four molecules of porphobilinogen produces uroporphyrinogen I (urogen I) (12.8) in which the acetic and propionic acid residues on the pyrrole rings alternate; this contrasts with uroporphyrinogen III in which the acetic and propionic acid residues on ring D are reversed.

Two enzymes, first isolated from spinach leaves and *Chlorella* and later from animal tissues, are required to form uroporphyrinogen III from porphobilinogen. They are uroporphyrinogen I synthase[3] (sometimes called porphobilinogen deaminase), a relatively heat-stable enzyme with a molecular weight in the range 40,000–60,000, and uroporphyrinogen III cosynthase, a heat-labile enzyme with a molecular weight of 210,000 (in cultured soya bean cells). Uroporphyrinogen I synthase, operating alone, catalyses the conversion of porphobilinogen into uroporphyrinogen I. However, when it operates in the presence of uroporphyrinogen III cosynthase, uroporphyrinogen III is produced. Uroporphyrinogen III cosynthase has no action on porphobilinogen in the absence of uroporphyrinogen I synthase.

Research in 1979 by Battersby and his coworkers and by Burton and Jordan and their coworkers has shown that when porphobilinogen is incubated with uroporphyrinogen I synthase a precursor of both uroporphyrinogens I and III, termed pre-porphyrinogen, is formed. When pre-porphyrinogen is incubated with uroporphyrinogen III cosynthase it is converted into uroporphyrinogen III. However, in the absence of the cosynthase it spontaneously forms uroporphyrinogen I. The structure of pre-porphyrinogen is not yet clear; ^{13}C NMR data are consistent with three alternative tetrapyrrolic structures, one being the spiro compound shown in Fig. 12.3 and another having an open chain form. It has been shown that the sequence of assembly of the pyrrole rings constituting the uroporphyrinogen III macro cycle is $A \rightarrow A-B \rightarrow A-B-C \rightarrow A-B-C-D$ and that it is ring D which then undergoes rearrangement. The scheme shown in Fig. 12.3 is consistent with present knowledge.

4. Formation of protoporphyrin IX

Urogen III is converted into coproporphyrinogen III (coprogen III) by the stepwise decarboxylation of the acetic acid residues at C–8, C–1, C–3 and C–5 (steps A, B, C, and D, Fig. 12.4). The same enzyme, uroporphyrinogen decarboxylase[4], is involved in all cases. The pro-R and pro-S hydrogens of the methylene group retain their individuality in the resulting methyl group which means that the

$A = -CH_2\,COOH$
$P = -CH_2\,CH_2\,COOH$

(12.7)   (12.8)

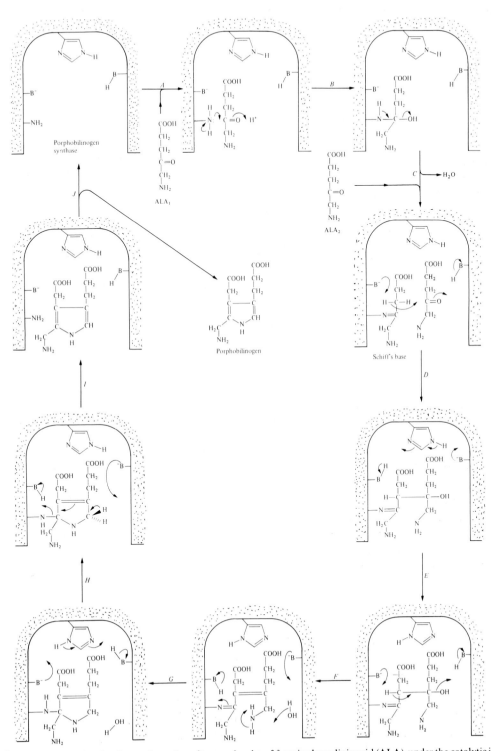

Fig. 12.2. Postulated mechanism for the condensation of two molecules of δ-aminolaevulinic acid (ALA), under the catalytic influence of porphobilinogen synthase (ALA dehydratase), to form porphobilinogen. (NH_2 is the ε-amino group of a lysine residue at the enzyme active site, whilst B^- and BH are probably S^- and SH groups of cysteine residues; the imidazole ring belongs to a histidine residue.)

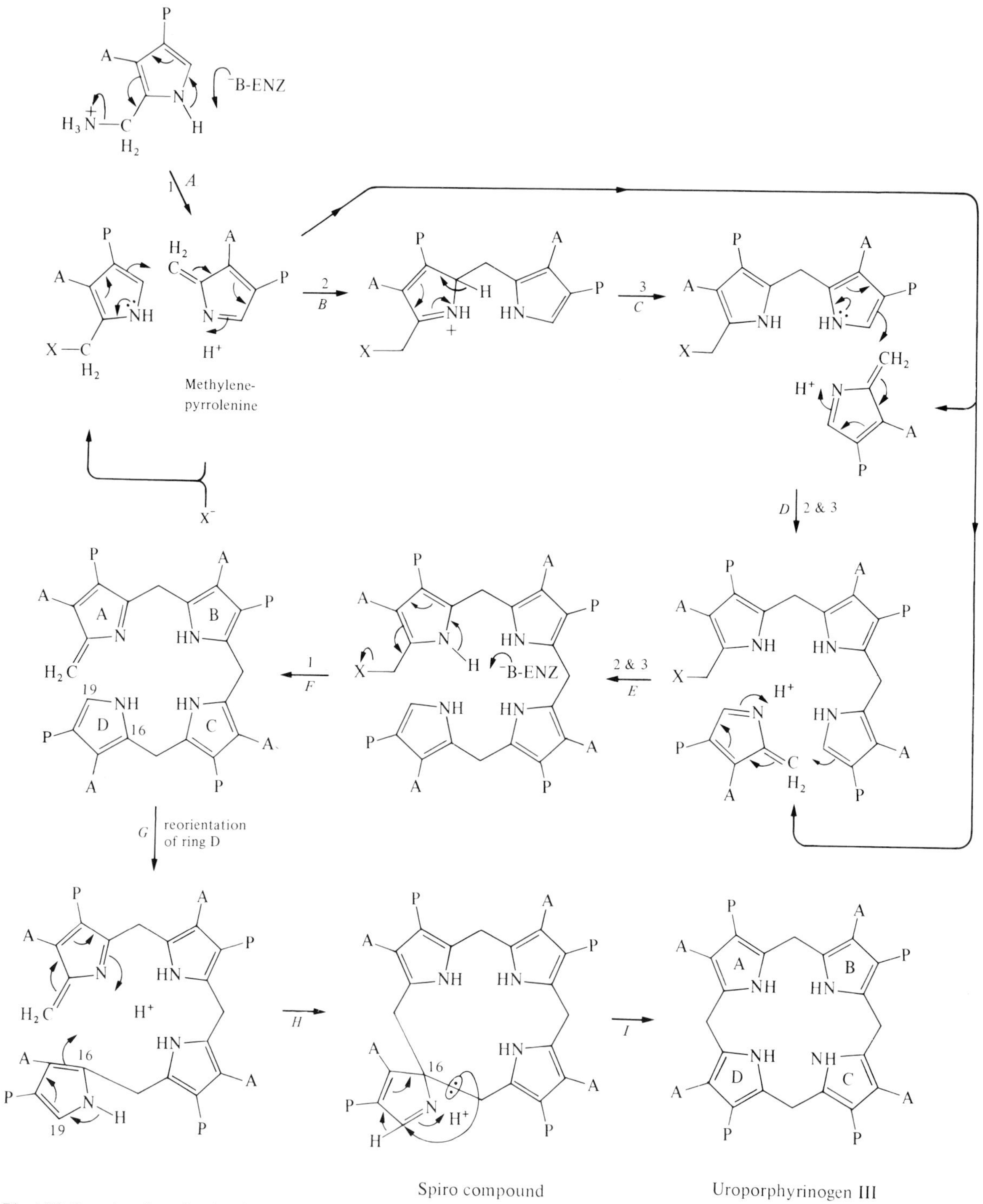

Fig. 12.3. Postulated mechanism for the formation of uroporphyrinogen III from porphobilinogen via pre-uroporphyrinogen III (after Battersby *et al.*, 1979). (A=CH$_2$COOH; P= −CH$_2$CH$_2$COOH; X=enzyme; reaction 1=a pyrrolenine-forming elimination; reactions 2 and 3 are repeated as each methylenepyrrolenine moiety is added; steps *A–F* are catalysed by uroporphyrinogen I synthase; step *G* is simply the redrawing of a structure; steps *I* and possibly *H* are catalysed by uroporphyrinogen III cosynthase.)

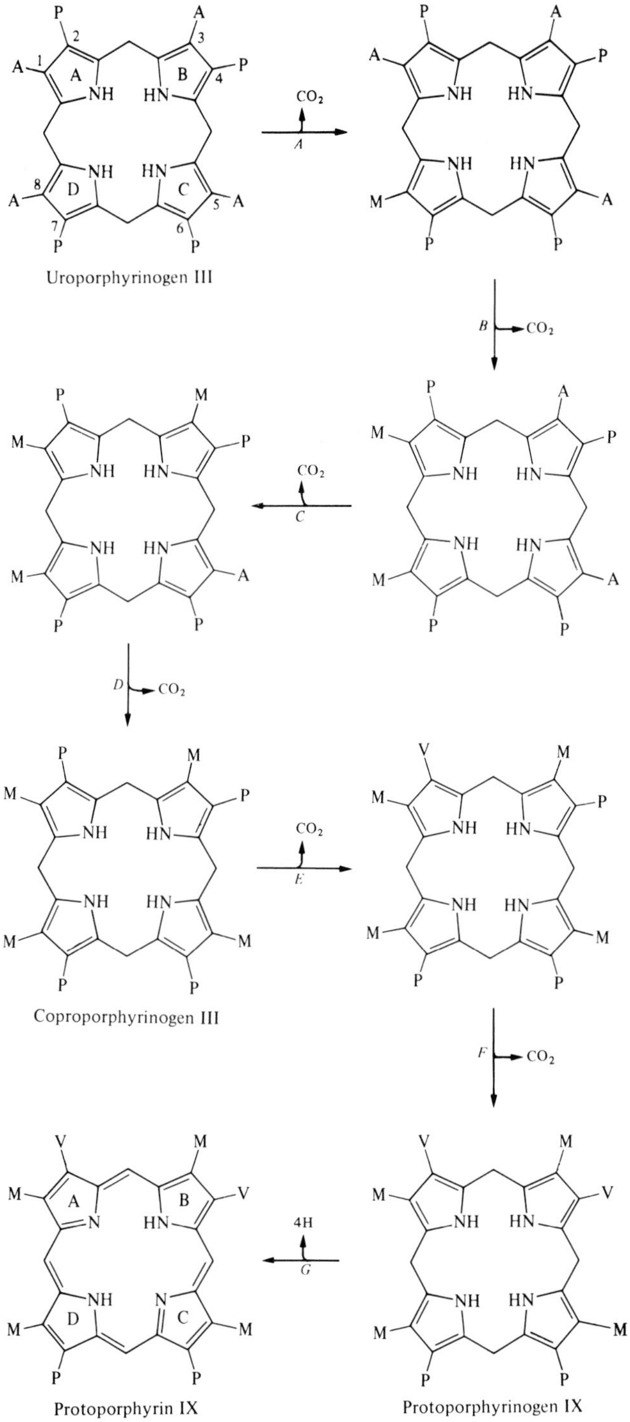

Fig. 12.4. The stepwise conversion of uroporphyrinogen III into protoporphyrin IX via coproporphyrinogen III and proto-porphyrinogen IX. (Steps A–D are catalysed by uroporphyrinogen decarboxylase, EC 4.1.1.37, steps E and F by coproporphyrinogen decarboxylase, EC 1.3.3.3, and step G by protoporphyrinogen oxidase, EC 1.3.3.4; $A = -CH_2COOH$; $P = -CH_2CH_2COOH$; $M = -CH_3$; $V = -CH=CH_2$.)

COOH is replaced by hydrogen without inversion of configuration.

The final steps leading to protoporphyrin IX are the oxidative decarboxylation of the propionic residues at C–2 (E) (Fig. 12.4) and C–4 (F) (Fig. 12.4) of urogen III to yield protoporphyrinogen IX (protogen IX). The same enzyme, coproporphyrinogen decarboxylase[5], is involved in both cases. The reaction involves retention of both hydrogens on the α-carbon atom of the propionic side chain but the pro-S hydrogen is removed stereospecifically from the β-carbon atom. The mechanism proposed (Fig. 12.5) involves the removal of H_S with the formation of a Schiff base intermediate (A, Fig. 12.5), whose electron-withdrawing property facilitates decarboxylation to yield the vinyl derivative (B, Fig. 12.5). In aerobic organisms the presence of oxygen is mandatory. The enzyme has been purified from tobacco leaves. Protogen IX will not chelate with metals and has first to be oxidized to protoporphyrin IX (G, Fig. 12.4). This reaction can proceed spontaneously in the presence of oxygen but it goes much faster in the presence of protoporphyrinogen oxidase[6], which has been obtained from many sources including yeast but has not yet been demonstrated in higher plants.

C. FORMATION OF METALLOPORPHYRINS

1. Iron porphyrins

It is at the protoporphyrin IX stage that the biosynthetic pathway branches—one leads to iron porphyrins (the haems) and the other leads to magnesium porphyrins (the chlorophylls). Both pathways exist in plants but only the former in animals.

To start with the iron porphyrin pathway, an enzyme ferrochelatase[7] has been obtained from both plastids and plant mitochondria which effectively converts protoporphyrin IX into protohaem (haem b) which is the prosthetic group of b type cytochromes (see Fig. 6.19) and, of course, haemoglobin in animals. The pathways leading to haem a (cytochromes a) and haem d (chlorin) are still not known. Sirohaem (Fig. 9.4), present in nitrite reductase[8], has eight carboxyl groups and is clearly a derivative of uroporphyrin IX. This means that in this particular case an enzyme exists which can oxidize uroporphyrinogen III to uroporphyrin IX which is a substrate for a unique ferrochelatase.

2. Magnesium porphyrins

No enzyme has yet been clearly demonstrated which will incorporate Mg^{2+} into protoporphyrin IX in vitro. The evidence that it does occur at this stage is indirect; chlorophyll-deficient mutants accumulate this compound and methylation at the propionic residue at C–6 occurs enzymically much more quickly with Mg protoporphyrin IX than with protoporphyrin IX itself. It should be emphasized that chemically the insertion of Mg^{2+} into protoporphyrin is much more difficult than is the insertion of Fe^{2+}; whereas the latter can proceed spontaneously the former requires magnesium perchlorate and anhydrous, air-free, boiling pyridine.

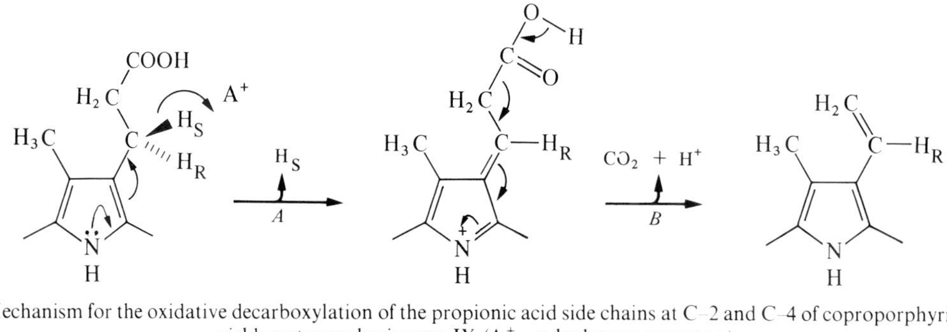

Fig. 12.5. Mechanism for the oxidative decarboxylation of the propionic acid side chains at C–2 and C–4 of coproporphyrinogen III to yield protoporphyrinogen IX (A$^+$ = a hydrogen acceptor.)

D. BIOSYNTHESIS OF CHLOROPHYLLS

1. Formation of protochlorophyllide†

The conversion of Mg protoporphyrin IX into protochlorophyllide follows the pathway outlined in Fig. 12.6. The first step (A), esterification of the propionic residue at C–6, is achieved in the presence of a methyl transferase with S-adenosylmethionine as the methyl donor. The steps involving ring closure (B, C, D, E) can take place with Mg 2,4-divinyl protoporphyrin IX methyl ester as substrate with the resulting Mg 2,4-divinylphaeoporphyrin a_5 methyl ester being reduced to protochlorophyllide (F). The reactions can also take place after Mg 2,4-divinylprotoporphyrin IX methyl ester has been reduced to Mg vinyl phaeoporphyrin IX methyl ester (G); in this case the final product is protochlorophyllide itself. The steps in the production of the carbocyclic ring proceed through the acrylate (B), β-hydroxypropionate (C) to β-ketopropionate (D) (methylation, A, protects this compound from spontaneous decarboxylation) which ring closes on to the γ-methine carbon (E). The evidence for these steps is genetic: compounds with these substituents at C–6 have been isolated from *Chlorella* mutants. However, in all these compounds the vinyl group C–4 was reduced, whereas the first carbocyclic compound formed appears in some cases to be Mg-2,4-divinyl phaeoporphyrin a_5 monomethyl ester, which can be reduced to Mg-2-vinyl phaeoporphyrin a_5 (F), which is protochlorophyllide, by an NADPH-dependent enzyme. The doubt about the exact pathway is indicated in Fig. 12.6 and full enzymic study which will clarify the situation is awaited.

2. Conversion of protochlorophyllide into chlorophyll

With protochlorophyllide a new situation is encountered in that reduction to chlorophyllide‡ is

† Letters in this section refer to Fig. 12.6.
‡ Chlorophyllides are chlorophylls lacking the terpenoid (usually phytol) side chain.

light-catalysed and is carried out on a pigment–protein complex known as protochlorophyllide holochrome. The reduction can be carried out *in vitro* and no external hydrogen donor is required, so it must be provided by the holochrome itself and is probably bound NADPH. An enzyme NADPH-protochlorophyllide oxidoreductase has been purified; it is a single polypeptide, MW 36,000, which binds two or three protochlorophyllide molecules. It seems that this is not the intact holochrome, because the spectral changes noted *in vivo* do not occur with this preparation. Thus the holochrome itself will probably eventually be revealed as a much more complex entity. The reduction in ring D is stereospecific and results in the 7S,8S configuration [eq. (12.2)].

$$(M = -CH_3 ; \quad P = -CH_2 CH_2 COOH)$$

The final step is the attachment of phytol to chlorophyllide.

In some cases esterification may occur before photoreduction, that is protochlorophyll is first formed and then reduced to chlorophyll. An enzyme chlorophyllase[9] which removes phytol from chlorophyll has been known for many years and many experiments have been undertaken in attempts to show that it works reversibly and is the enzyme concerned with converting chlorophyllide into chlorophyll. However, this seems unlikely, and the enzyme and substrate are still not clearly defined. As photoconvertible protochlorophyll esterified with geranylgeraniol can be found in etiolated barley seedlings and chlorophyll esterified with geranylgeraniol is present in newly greened horse chestnut leaves, esterification may first take place at this oxidation level with geranylgeranyl pyrophosphate; reduction to phytol then occurs presumably stepwise after attachment to the chlorophyll molecule [eq. (12.3)].

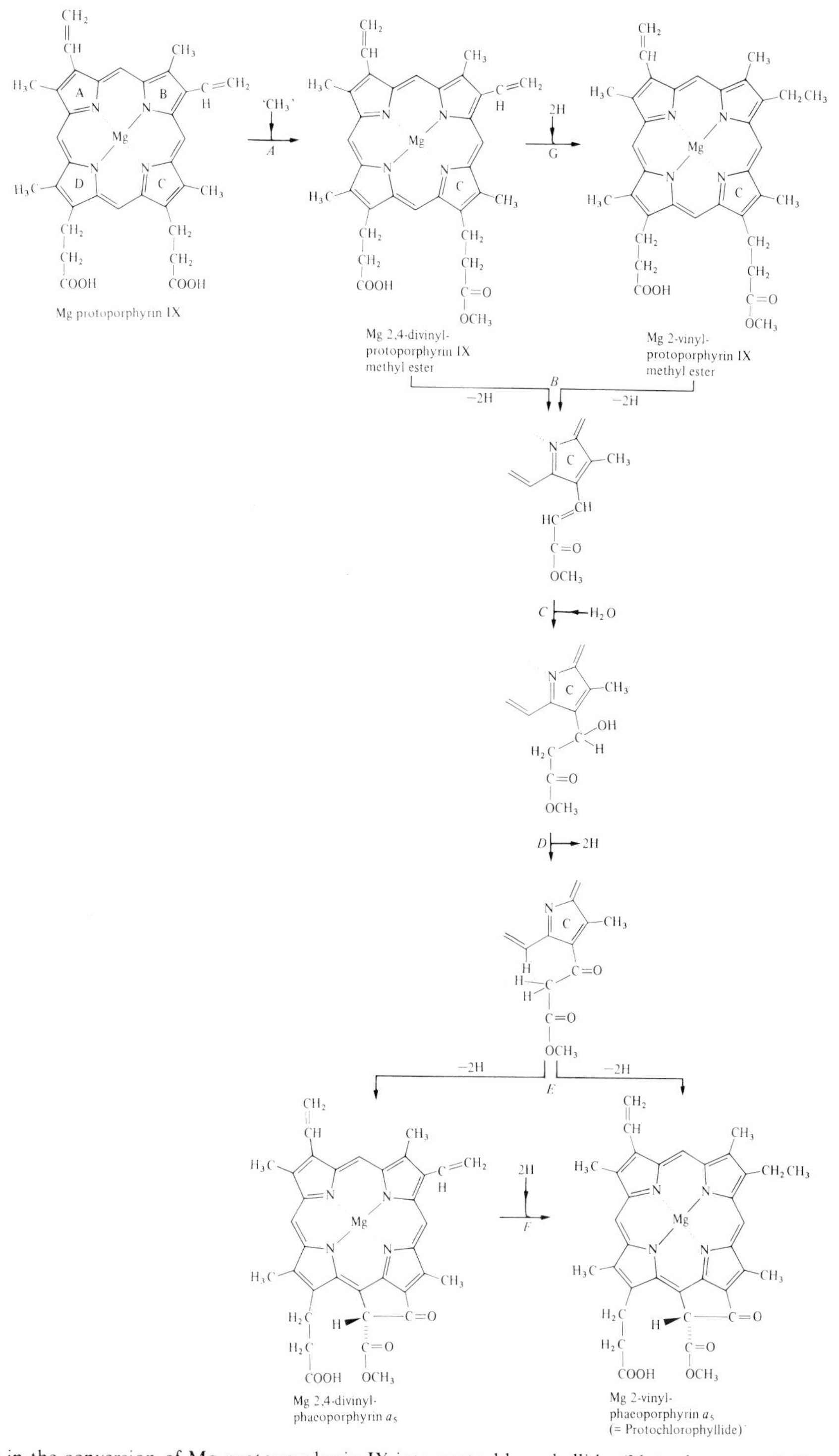

Fig. 12.6. Steps in the conversion of Mg protoporphyrin IX into protochlorophyllide. (Note that steps *B–E* can proceed from two different substrates: 'CH$_3$' indicates transmethylation from *S*-adenosylmethionine.)

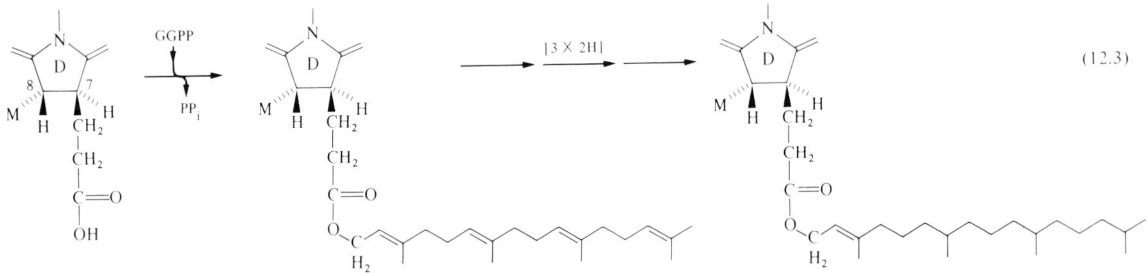

(12.3)

E. GREENING OF ETIOLATED SEEDLINGS†

Etiolated seedlings contain small amounts of protochlorophyllide-holochrome and on brief illumination there is a rapid stoichiometric reduction of protochlorophyllide to chlorophyllide which is then more slowly esterified and converted into chlorophyll *a*. If the seedlings are then returned to darkness about the same amount of protochlorophyllide is formed as originally present and this is also converted into chlorophyllide on illumination. Thus by exposing etiolated seedlings to short (as short as 10^{-4} s) flashes interspersed with 10–15 min dark periods considerable amounts of chlorophyll can be accumulated.

Certain spectral changes can be observed on illuminating etiolated seedlings (Fig. 12.7). Protochlorophyllide absorbs at 650 nm when attached to its holochrome (Pchlide$_{650}$); this maximum corresponds to the maximum for the action spectrum of photoreduction. Photoreduction results in the appearance after 1 s of a band at 678 nm which is bound chlorophyllide *a* (Chlide$_{678}$, *A*, *B*); an intermediate stage which appears after less than 1 s, Chlide$_{668}$ (*A*) may be a *mixed conversion unit* containing both protochlorophyllide and chloro-

† Letters in this section refer to Fig. 12.7.

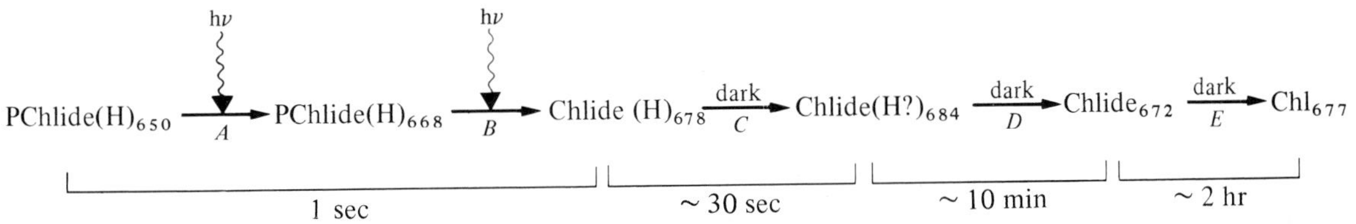

phyllide because evidence is accumulating that protochlorophyllide exists *in vivo* in aggregates. Other spectral changes after 30 s (*C*) and 10 min (*D*) sequentially yielding Chlide$_{684}$ and Chlide$_{672}$ have not been fully explained but lead to the release of chlorophyllide from the holochrome. The final spectral change to 677 nm (*E*), which takes up to 2 hours to be complete, is generally assumed to be due to the phytylation of chlorophyllide *a* to produce chlorophyll *a*. Another peak is observed *in vivo* at 632 nm and this is due to free protochlorophyllide (Pchlide$_{632}$) which is not photoreducible because only limited amounts are needed to charge the holochrome. This is the species of protochlorophyllide which accumulates in the etioplasts of etiolated leaves when the latter are fed ALA. The amounts are often sufficient to give the etiolated leaves a pronounced green colour. The scheme shown in Fig. 12.8 attempts to correlate the spectral changes shown in Fig. 12.7 with the mechanism of the light-dependent conversion of protochlorophyllide into chlorophyllide.

During a lag period, which can last up to 2 hr, illuminated etiolated leaves synthesize chlorophyll at about the same rate as protochlorophyllide is synthesized if the leaves are returned to darkness. Then the rate of protochlorophyllide formation and its conversion into chlorophyllide are increased by at least 10 times. This is associated with the rapid

Fig. 12.7. Time scale of the changes in the red region of the visible spectrum undergone by protochlorophyllide holochrome when an etiolated seedling is illuminated. (PChlide(H)=protochlorophyllide holochrome; Chlide(H)=chlorophyllide holochrome; Chlide=chlorophyllide; Chl=chlorophyll *a*.)

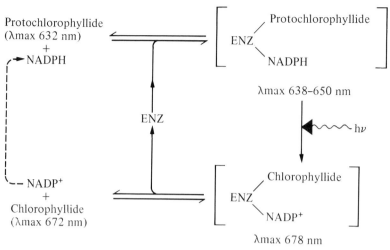

Fig. 12.8. Suggested mechanism of the light-dependent conversion of protochlorophyllide into chlorophyllide in higher plants. (ENZ = protochlorophyllide oxidoreductase.)

formation of the lamellar structure of the chloroplast. If seedlings in the stage of rapid chlorophyll synthesis are returned to darkness, synthesis stops completely and there is no build up of precursors. This brings us to the problem of regulation of chlorophyll synthesis which is considered in detail in Section G.

F. CHLOROPHYLL SYNTHESIS IN THE DARK

Although most higher plants require light for chlorophyll synthesis, some, e.g. conifer seedlings, can synthesize it in the dark, as can many algae, e.g. brown and blue-green algae. The reason why the final reductive step is not light-dependent is not known but in the case of conifers, an as yet uncharacterized factor is involved.

The developing cotyledons synthesize significant amounts of chlorophyll in the dark only so long as they are in contact with the megagametophyte, which normally is until the megagametophyte disappears. Separation from the megagametophyte results in very much lowered synthesis of chlorophyll even if the cotyledons are placed in a nutrient solution. Thus some material in the megagametophyte is intimately involved in chlorophyll synthesis in the dark.

G. REGULATION OF CHLOROPHYLL SYNTHESIS†

The obvious point of regulation would be at the point of ALA synthesis because ALA is the first committed substrate for tetrapyrrole synthesis and it is certainly the rate-limiting step in the process because the addition of ALA to etiolated seedlings in the dark results in an increase in the amount of protochlorophyllide present by at least 10 times. This observation also makes it clear that the enzymes from ALA to protochlorophyllide are constitutive and not induced. The observation that chloramphenicol (inhibitor of proteins synthesized on 70S, plastid ribosomes; see Chapter 10) mimics darkness indicates that the enzyme or enzymes concerned with ALA synthesis are inducible; in fact are photoinducible. Furthermore, ALA formation must be inhibited by protochlorophyllide holochrome because at a critical level of the charged holochrome synthesis of ALA, and therefore later intermediates, is inhibited. The rapid inhibition, as exemplified by the complete cessation of chlorophyll synthesis when greening tissues are placed in darkness, also means that the stability of the ALA synthesizing system must be low, and this is so since its half-life $(T_{\frac{1}{2}})$ is of the order of 30 min. The protochlorophyllide, or a mediator controlled by it,

† Letters in this section refer to Fig. 12.9.

is thought to function by repressing enzyme synthesis by its effect on a stable *mRNA* (*A*, Fig. 12.9). Preliminary work suggests that the required stable *mRNA* is present because actinomycin D, an inhibitor of RNA synthesis, does not inhibit greening of an illuminated etiolated seedling.

It would be reasonable to assume that, as there is apparently only one pool of intermediates for haem and chlorophyll synthesis, there should be control at the bifurcation of the two pathways, that is at the point of insertion of the metal (*B*, *C*). The control of insertion of magnesium by protochlorophyllide 650 (*D*) is indicated by the fact that in isolated etioplasts ALA is incorporated well into intermediates up to magnesium protoporphyrin and badly into those coming after. Control at the protohaem step (*E*) is suggested by experiments with the iron chelator α,α-dipyridyl. It inhibits the formation of haem and therefore the control on the synthesis of the ALA enzyme is removed and the ALA, as observed experimentally, is channelled into porphyrins and magnesium porphyrins.

H. CHLOROPHYLL *b*

Chlorophyll *b* in which C–3 has a CHO group and not a CH_3 group as in chlorophyll *a*, is present along with chlorophyll *a* in all higher plants and the *a/b* ratio in one species is rather constant. In spite of a large amount of work on what would, on the face of it, seem a simple problem, the biosynthetic route to chlorophyll *b* is still not clear. The obvious route, the

oxidation of chlorophyll *a*, is difficult to establish but no other route, if one exists, has yet clearly emerged.

I. OPEN (LINEAR) TETRAPYRROLES

The metabolism of haem in animals results in the removal of iron followed by the splitting of the porphyrin macrocycle by oxidizing away the α-methine carbon. The resulting compounds are open tetra-pyrroles, such as bilirubin (12.9), which are secreted in the bile and are known as bile pigments. The accumulation of such pigments in large amounts occurs only in some algae and blue-green bacteria and these are discussed in detail in Chapter 5, Section C.2(iii). They do not accumulate in higher plants.

Phytochrome

Trace amounts of a bile pigment are, however, probably present in all higher plants as the prosthetic group of the chromoprotein, *phytochrome*.

M   V   M   P   P   M   M   V

O⟨⟩N—C=C—N—C—N—C=C—N⟨⟩O
   H   H H_2 H   H   H

(12.9) M = $-CH_3$; V = $-CH=CH_2$; P = $-CH_2CH_2COOH$

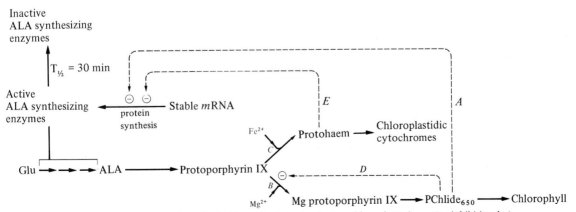

Fig. 12.9. Regulation of porphyrin biosynthesis in developing chloroplasts (---→ ⊖ = inhibition by).

This molecule mediates many physiological responses which are initiated by the absorption of a low dose of red light (660 nm) in an overall dark environment, and reversed by a similar low dose of far-red light (730 nm) under the same conditions.

Its importance is that it is the only known pigment in plants which *senses* light signals for the control of plant development. Many physiological responses have been reported which are phytochrome-mediated and most of them are concerned with morphogenesis.

The inactive form of phytochrome (P_R) has an absorption maximum at 660 nm; on illumination with red light (660 nm) it is converted into its active form (P_{FR}) which absorbs light at 730 nm. P_{FR} can, in turn, be converted back to P_R by illumination with 730 nm light [eq. (12.4)]. This reversible reaction can be repeated *in vivo ad infinitum*, but it is always the last light treatment which controls the eventual physiological response.

Examples of enzymic systems which *in vivo* appear to be under the control of phytochrome are:

(i) The formation of the ALA-synthesizing system capable of producing ALA in response to light (see Sections B.1 and G); here two light-sensitive processes are involved, only one of which is controlled by phytochrome (i.e. red light stimulated/far red light reversed).

(ii) The synthesis of PAL, the enzyme catalysing the conversion of phenylalanine into cinnamic acid, a key precursor in the biosynthesis of flavonoids (see Chapter 14, Section I.3), lignin (see Chapter 4, Section D.2) and some alkaloids [see Chapter 13, Section C.6(ii)(b)].

(iii) The conversion of L-tyrosine into DOPA and its subsequent incorporation into betacyanins and/or betaxanthins [see Chapter 13, Section C.6(iii)].

Elucidation of the processes involved in these systems is hampered by the effect of feedback controls. However, a pointer to what may prove to be the general biochemical action of phytochrome is the recent finding that the synthesis of protein in the light-harvesting chlorophyll protein (LHCP; see Chapter 5, Section C.4) appears to be controlled by

phytochrome. A short flash of low-intensity red light causes the formation of a new *mRNA* which is translated by cytoplasmic ribosomes into a protein which, after structural modification, is rapidly incorporated into LHCP in the chloroplast provided that chlorophyll is available; in the absence of chlorophyll the protein appears to be rapidly degraded. This would suggest that the following sequence of events takes place: the flash of red light converts P_R into P_{FR} which, in some as yet unknown way, stimulates the transcription of the nuclear gene for the LHCP protein; the resulting *mRNA* is then translated by cytoplasmic ribosomes into a precursor of the LHCP protein which may have at its *N*-terminal end a 'signal' amino-acid sequence that is recognized by a specific entry site in the chloroplast membranes [cf. the small subunit of RuDP carboxylase—see Chapter 10, Section E.9(viii)]; once inside the chloroplast the signal sequence is removed from the precursor LHCP protein by an appropriate protease; the completed LHCP protein is then rapidly incorporated, along with chlorophyll and xanthophyll molecules, into LHCP in the thylakoid membranes.

Only recently has the prosthetic group of phytochrome been isolated and shown to be an open tetrapyrrole related to phycocyanobilin (Fig. 5.6). The red form (P_R) has the structure (12.10) and it is converted into (12.11) on illumination with 730 nm

(12.10)

(12.11)

$$P_R \underset{730\ nm}{\overset{660\ nm}{\rightleftharpoons}} P_{FR} \qquad (12.4)$$

light. The isomerization which results in an extended system of double bonds can, to some extent, account for the observed wavelength shift 660–730 nm.

Precisely how P_{FR} triggers the biochemical and physiological changes it causes is not known. Indeed it would be surprising considering the variety of effects if there were only one mechanism involved. Possible mechanisms based on the view that phytochrome is present in membranes include (i) control of active transport of ions and molecules across membranes perhaps by regulating ATPase activity, (ii) controlling the activity of membrane-bound hormones such as gibberellin, (iii) modulating the activity of membrane-bound proteins. All these aspects are the subject of active research and considerable clarification of the situation is to be expected in the next few years.

SUGGESTIONS FOR FURTHER READING

Chlorophylls

Granick, S. and Beale, S. I. (1978) *Adv. Enzymol.* **46**, 33.

Bogorad, L. (1976) in *Chemistry and Biochemistry of Plant Pigments* (Goodwin, T. W., ed.), Vol. 1, p. 64. Academic Press, London.
Beale, S. I. (1978) *Ann. Rev. Plant Physiol.* **29**, 95.

Phytochrome

Marmé, D. (1977) *Ann. Rev. Plant Physiol.* **28**, 173.
Schopfer, P. (1977) *Ann. Rev. Plant Physiol.* **28**, 223.

ENZYMES

1. Succinyl-CoA: glycine *C*-succinyltransferase (decarboxylating), EC 2.3.1.37.
2. 5-Aminolaevulinate hydro-lyase (adding 5-aminolaevulinate and cyclizing), EC 4.2.1.24.
3. Porphobilinogen ammonia-lyase (polymerizing), EC 4.3.1.8.
4. Uroporphyrinogen III carboxy-lyase, EC 4.1.1.37.
5. Coproporphyrinogen: oxygen oxidoreductase (decarboxylating), EC 1.3.3.3.
6. Protoporphyrinogen IX: oxygen oxidoreductase, EC 1.3.3.4.
7. Protohaem ferro-lyase, EC 4.99.1.1.
8. Nitrite: (acceptor) oxidoreductase, EC 1.7.99.4.
9. Chlorophyll chlorophyllidohydrolase, EC 3.1.1.14.

CHAPTER 13

The Alkaloids

CONTENTS

A. INTRODUCTION

The alkaloids represent yet another large group of so-called secondary plant products. Many, but not all, exhibit important pharmacological properties and some of these have been known for hundreds of years. For example, cinchona alkaloids, which are present in the bark of *Cinchona* spp. and *Remijia* spp., plants indigenous to the high eastern slopes of the Andes, have as their main constituent quinine, which has been known since 1639 as an effective antimalarial agent. However, it is within the last 100 years or so that well over 2000 alkaloids have been isolated from plants and their structures determined. When it is realized that less than 5% of the world's plant species have been examined up to now, there is clearly still a vast field in existence for organic chemists to screen for this class of compound.

All alkaloids contain nitrogen, frequently as part of a heterocyclic ring, and most are, as the general name (alkaloid = vegetable alkali) implies, basic. The nomenclature of alkaloids is not very exact but those with heterocyclic rings are frequently called *True Alkaloids* and those without such rings *Protoalkaloids*. Their distal biosynthetic precursors are almost always amino acids; other multicarbon units, e.g. acetate, are also incorporated into the final structure of some alkaloids. Alkaloids, both with and without heterocyclic rings, that are not derived from amino acids are called *Pseudoalka-loids*. These are almost entirely represented by alkaloids in which the carbon skeleton is isoprenoid-derived.

True alkaloids and protoalkaloids are classified either according to the heterocyclic ring system that they contain, as exemplified by Table 13.1, or according to the amino acids from which they are derived. Although a formal table utilizing the latter system of classification is not present in this chapter the relevant information for its construction is included in Table 13.1.

Examples of pseudoalkaloids are given in Table 13.2. No alkaloids have been reported so far which are derived entirely from hemi-, sester-, tetra- or polyterpenes.

B. DISTRIBUTION AND LOCALIZATION OF TRUE AND PROTOALKALOIDS

A generalized scheme for alkaloid distribution in plants is given in Table 13.3. Its limitations are clear when, as indicated in Section A, 95% of plant species still remain to be examined for alkaloids.

Alkaloids accumulate mainly in four types of tissue: (i) actively growing tissue, (ii) epidermal and hypodermal cells, (iii) vascular sheaths and (iv) latex vessels. They are present in vacuoles and so do not appear in young cells until vacuolation occurs. They rarely occur in dead tissue; even in *Cinchona* bark, which can contain up to 12% (w/w) of alkaloids, they are present only in the parenchyma.

Alkaloids are frequently stored in tissues other than their site of synthesis. The best known example of this is nicotine which is synthesized in the roots of tobacco plants but is translocated to the leaves where it is stored; other examples will arise later in the chapter. Secondary structural modifications often occur at sites other than that where primary synthesis took place, for example, the basic ring system of the tropane alkaloids (see Table 13.1) is assembled in the roots of *Datura* spp. and transported to the leaves where it undergoes considerable modification [see Section C.2(ii)].

C. BIOSYNTHESIS OF TRUE AND PROTOALKALOIDS

1. General aspects

Much information has been gained by the judicious use of labelled (often specifically labelled) putative precursors in experiments designed to test the plausible hypothetical biogenetic schemes put forward by organic chemists which were based largely on (i) visual dissection of the complex molecular skeletons into simple units and (ii) known organic reactions. Many of these ideas were remarkably accurate and have been confirmed by the labelling experiments. Consequently we now know in outline the biosynthetic pathway of many alkaloids. However, in biochemical terms, it is true to say that no pathway has been fully delineated because the enzyme(s) concerned in each step has not been isolated and studied. Nevertheless progress in this area is being made and during the following sections reference will be made to the enzymes which have already been reported.

The general approach to elucidating the biosynthetic pathway of a given alkaloid has been to administer a specifically labelled putative precursor to a plant known to be actively synthesizing that alkaloid at the time of the experiment; after allowing sufficient time for incorporation, the alkaloid is extracted from the plant tissue, purified and then chemically degraded in a manner which will allow the position of the labelled atoms within the molecule to be elucidated. The administration of the labelled precursor has been accomplished in a number of different ways: (i) the labelled compound has been added to the hydroponic medium in which the plant was growing, (ii) it has been injected, in aqueous solution, into plant cavities, (iii) it has been placed in a drop of solution on the leaf surface which is then punctured gently several times at that point, (iv) it has been painted on to the leaf surface in a suitable solvent and often oversprayed with silicone oil after the solvent has evaporated, (v) it has been administered by 'wick feeding', a technique in which strands of mercerized cotton are threaded through the plant such that both ends emerge close together and dip into a small volume of solution containing

Table 13.1. *Examples of alkaloid types*

Alkaloid type (based on structure of *N*-heterocycle)	Example		Plant source	Biosynthetic precursor of *N*-heterocycle
	Name	Structure		
1. Pyrrolidine (i) Simple	Stachydrine		*Medicago sativa*	L-ornithine
(ii) Tropane	*l*-Hyoscyamine (*dl*-Hyoscyamine = Atropine[a])		*Atropa belladonna, Datura stramonium* and other Solanaceae	L-ornithine
	Hyoscine (= Scopolamine)		*Datura metel, Scopolia carniolica* and other Solanaceae	L-ornithine
	Cocaine		*Erythroxolon coca*	L-ornithine
2. Piperidine	Anabasine		*Anabasis aphylla* *Nicotiana glauca*	L-lysine
	Sedamine		*Sedum* spp., e.g. *S. sarmentosum*	L-lysine

[a] See page 486.

Table 13.1 (*cont.*)

Alkaloid type (based on structure of *N*-heterocycle)	Example Name Structure	Plant source	Biosynthetic precursor of *N*-heterocycle
2. Piperidine (*cont.*)	Lobeline	*Lobelia* spp., e.g. *L. inflata*	L-lysine
	Coniine	*Conium maculatum*	$4 \times CH_3COOH$ + a N-source
3. Pyridine	Nicotine	*Nicotiana* spp., e.g. *N. tabaccum*, *N. rustica*, *N. glutinosa*	L-aspartic acid + a 3C unit
	Tenellin	*Beauvaria tenella*	L-phenylalanine
4. Pyrrolizidine	Retronecine	*Senecio* spp., e.g. *S. isatidens*, *S. douglasii* *Crotalaria* spp., e.g. *C. spectabilis*	L-ornithine
5. Phenanthroindolizidine	Tylophorine	*Tylophora asthmatica*	L-ornithine (pyrrolidine ring)
6. Quinolizidine	Lupinine	*Lupinus luteus* *Sarothamnus scoparius*	L-lysine

Table 13.1 (*cont.*)

Alkaloid type (based on structure of *N*-heterocycle)	Example Name Structure	Plant source	Biosynthetic precursor of *N*-heterocycle
6. Quinolizidine (*cont.*)	(−)-Sparteine	*Lupinus luteus* *Sarothamnus scoparius*	L-lysine
7. Quinoline	Quinine	*Cinchona* spp., e.g. *C. officinalis*	L-tryptophan
	Dictamnine	*Dictamnus albus*	Anthranilic acid
8. Isoquinoline			
(i) Morphine (or Papaverine alkaloids)	Morphine	*Papaver somniferum*	L-tyrosine
(ii) Amaryllidaceae alkaloids	Lycorine	*Lycoris radiata* *Narcissus tazetta* *Amaryllis belladonna*	L-tyrosine + L-phenylalanine
9. Quinazoline	Vasicine (= Peganine)	*Adhatoda vasica* *Peganum harmala*	Anthranilic acid

Table 13.1 (*cont.*)

| Alkaloid type (based on structure of *N*-heterocycle) | Example | | Plant source | Biosynthetic precursor of *N*-heterocycle |
	Name	Structure		

10. Indole

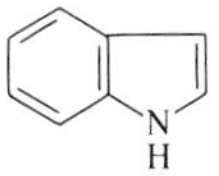

(i) β-Carboline alkaloids | Ajmalicine | | *Corynanthe johimbe* *Rauwolfia serpentina* | L-tryptophan

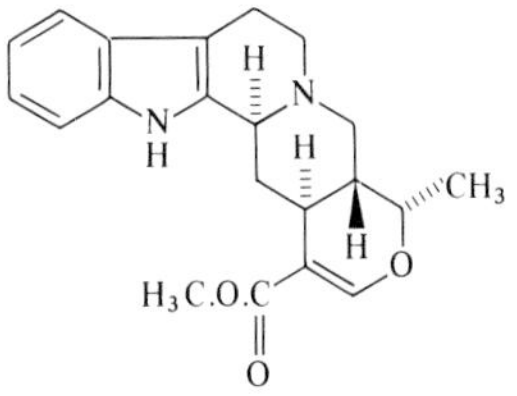

(ii) Ergoline alkaloids[b] | Agroclavine | | *Claviceps purpurea* (ergot fungus[c]) | L-tryptophan

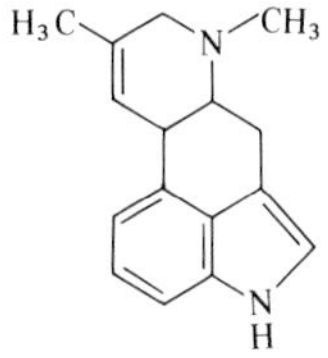

11. Dihydroindole (Betalains) | Betanidin | | Red beet root and other Centrospermae | L-tyrosine

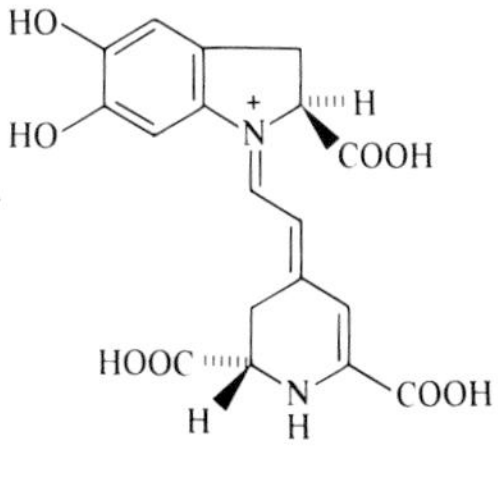

12. Imidazole | Ergothioneine | | *Neurospora crassa* | L-histidine

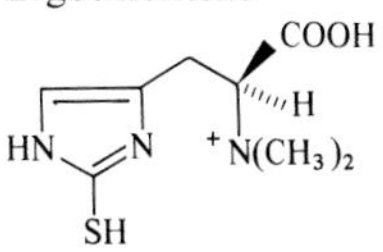

b, c See page 486.

Table 13.1 *(cont.)*

Alkaloid type (based on structure of *N*-heterocycle)	Example Name Structure	Plant source	Biosynthetic precursor of *N*-heterocycle
13. Acridine	Rutacridone 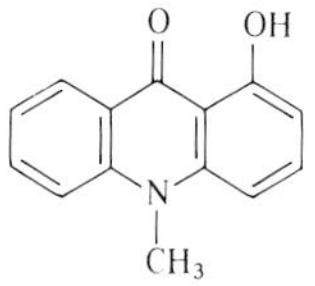	*Ruta graveolens*	?

[a] Partial racemization of *l*-hyoscyamine takes place during extraction from the plant tissue to give the *dl*-mixture which is known as atropine.

[b] Ergoline alkaloids also occur in higher plants, e.g. *Ipomoea rubro-caerulea*, but most biochemical studies have used the ergot fungus.

[c] *C. purpurea* grows parasitically on rye (*Secale cereale*) and other grasses but can be grown saprophytically on defined nutrient media when the alkaloids it produces can be extracted from the mycelium and the culture medium.

the labelled compound and (vi) it has been administered by 'vacuum infiltration' in which the plant is placed in contact with a solution of the labelled compound and then subjected to a moderate decrease in pressure for a few minutes by means of a vacuum pump; this removes from the plant tissues interstitial gases which are replaced by the solution as the partial vacuum is subsequently slowly released. An important prerequisite in all these techniques is the prevention of microbial infection.

Three important general reactions occur many times in alkaloid biosynthesis. These are (i) the oxidative coupling of phenols, (ii) Mannich-type reactions and (iii) Schiff base formation. Because these reactions occur in several of the pathways discussed in later sections they will be considered at this point.

(i) OXIDATIVE COUPLING OF PHENOLS

The hydrogen of the hydroxyl group of a phenol is readily removed by *one electron transfer* oxidizing agents, such as ferric chloride and alkaline potassium ferricyanide, to yield highly reactive mesomeric phenolate free radicals [see eq. (13.1)] which quickly disappear mainly by coupling (dimerization); the coupling of these phenolate radicals may be *ortho–ortho* (*A*), *para–para* (*B*) or *ortho–para* (*C*) (see Fig. 13.1). In living tissues phenolate ions are produced by the catalytic activity of enzymes such as the laccase involved in lignin formation (see Chapter 4, Section D.2) and the phenolases responsible for flavonoid polymerization [see Chapter 14, Section I.3(viii)(f)].

(ii) THE MANNICH REACTION

The generalized Mannich reaction (*A*, Fig. 13.2) involves a carbonyl compound, usually though not necessarily an aldehyde, an amine and a compound that can provide a carbanion, often by loss of an acidic hydrogen. The mechanism involved in the condensation of these three reactants is given in

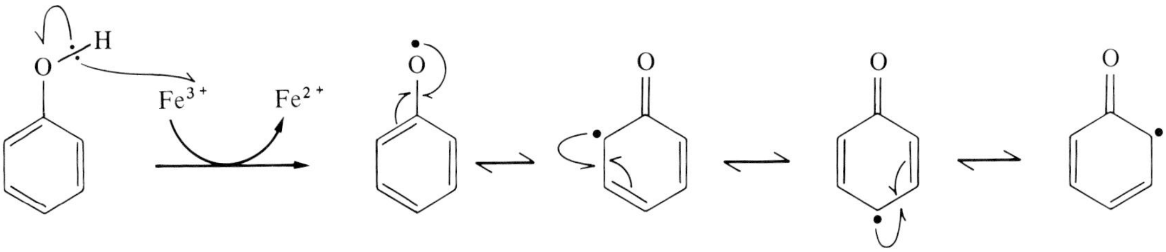

(13.1)

Table 13.2. *Examples of terpenoid alkaloids (pseudoalkaloids)*

Type	Example	Plant source	Precursors
1. Monoterpenoid	Actinidine	*Actinidia polygama*	Mevalonic acid plus a N-source
2. Sesquiterpenoid	Dendrobine	*Dendrobium nobile*	Mevalonic acid plus a N-source
3. Diterpenoid	Atisine	*Aconitum* spp., e.g. *A. heterophyllum*	Mevalonic acid plus a N-source
4. Triterpenoid	Cycloprotobuxine A	*Buxus* spp., e.g. *B. sempervirens*	Mevalonic acid plus a N-source
5. Steroid	Holaphyllamine	*Holarrhena* spp., e.g. *H. congolensis*	Mevalonic acid plus a N-source
	Solasodine	*Solanum* spp., e.g. *S. aviculare*	Mevalonic acid plus a N-source

Table 13.3. General distribution of alkaloids in the plant kingdom

THALLOPHYTA	Absent except in some fungi
BRYOPHYTA	Absent
PTERIDOPHYTA	Limited distribution, e.g. *Equisetum, Lycopodium*
SPERMATOPHYTA	
(i) *Gymnosperms*	Limited distribution
(ii) *Angiosperms*	
(a) Monocotyledons:	mainly in Gramineae and Liliaceae
(b) Dicotyledons:	mainly in Rubiaceae (e.g. coffee)
	Papaveraceae (e.g. poppy)
	Solanaceae (e.g. tobacco)
	Leguminosae (e.g. pea, bean)
	Apocynaceae (e.g. dogbane)
	Fumariaceae
	to a lesser extent in:
	Rosaceae (e.g. rose)
	Compositae (e.g. aster)
	Labiatae (e.g. mint)

B (Fig. 13.2); in many biosynthetic Mannich reactions the substituent R is a —SCoA residue thus making the carbonyl reactant part of a thioester linkage.

(iii) THE FORMATION OF SCHIFF BASES

Compounds with primary amino groups can react, sometimes spontaneously, with carbonyl compounds to form Schiff bases (otherwise called azomethines). When the carbonyl compound is an aldehyde the Schiff base product is an aldimine, when it is a ketone a ketimine results (*A*, Fig. 13.3). The mechanism of the reaction is shown in *B* (Fig. 13.3).

2. Alkaloids derived from L-ornithine

(i) SIMPLE PYRROLIDINES†

The biosynthesis of the pyrrolidine ring of nicotine has been studied in detail and the basic pathway is outlined in Fig. 13.4. Nicotine also contains a second *N*-heterocycle, the pyridine ring, and its pathway of formation and linking up with the pyrrolidine ring precursor is also shown in Fig. 13.4. However, the details of the synthesis of the pyridine ring will be considered in Section C.3(i).

[2-^{14}C]ornithine (the ^{14}C is labelled ▲ in

† All letters in this section refer to Fig. 13.4.

Fig. 13.4) is converted into the pyrrolidine ring of nicotine such that the latter is labelled at C-2′ and C-5′. This means that a symmetrical intermediate must be involved; this is putrescine which is formed by the action of the enzyme ornithine decarboxylase[1] which is found in tobacco root tissues where the alkaloid is synthesized. Putrescine is presumably released from the enzyme before putrescine *N*-methyltransferase[2], in the presence of *S*-adenosylmethionine, can convert it into *N*-methylputrescine (*B*). If the pyrrolidine ring were synthesized on a multienzyme complex putrescine might well have been held on the complex by, for example, the formation of a Schiff base with an enzyme carbonyl group, which would have made it essentially an unsymmetrical molecule for the enzyme-catalysed *N*-methylation. Another way that the symmetry of the reaction may be removed would be to methylate L-ornithine in the *δ-N*-position before decarboxylation. This actually occurs in the biosynthesis of other alkaloids [e.g. hyoscyamine—see Section C.2(ii)], but not in this case.

The next step (*C*) is the oxidative deamination of *N*-methylputrescine by *N*-methylputrescine oxidase to 4-methylaminobutanal which exists in aqueous solution in equilibrium with *N*-methyl-Δ^1-pyrroline (*D*). This has been shown to be an extremely good precursor of nicotine and the proposed coupling mechanism involves 3,6-dihydronicotinic acid and the *N*-methyl-Δ^1-pyrrolinium cation in a concerted

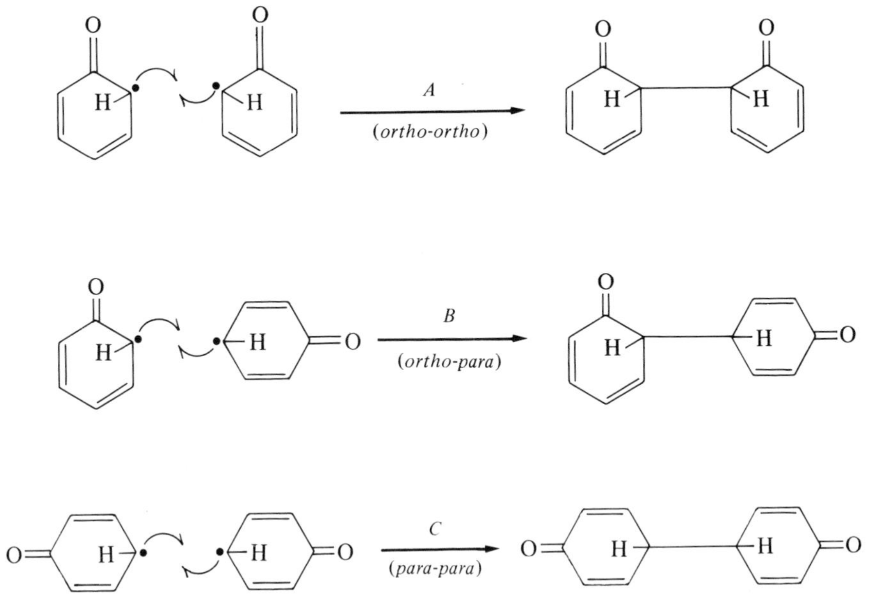

Fig. 13.1. Coupling of phenolate free radicals.

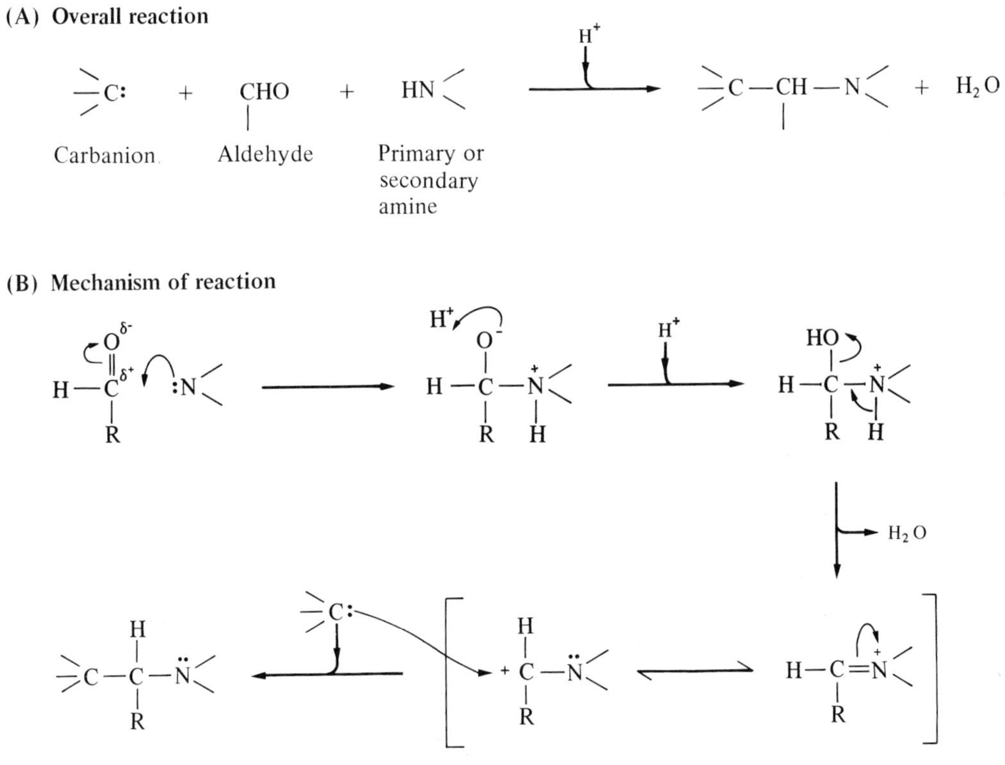

(A) Overall reaction

(B) Mechanism of reaction

Fig. 13.2. The Mannich reaction.

(A) Overall reaction

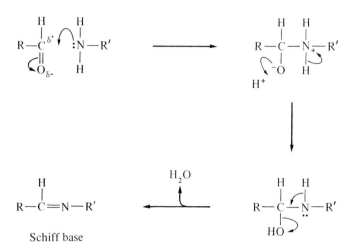

Fig. 13.3. Formation of a Schiff base.

decarboxylation (E) yielding 2′-S-dihydronicotine. This is followed by the stereospecific removal of the pro-S hydrogen from C–6 of the dihydropyridine ring (indicated as T in Fig. 13.4), probably to an electrophilic acceptor site on the appropriate enzyme, and the loss of the hydrogen on C–3 as H^+ to the aqueous environment.

Confusion arose in studying the biosynthesis of nicotine when it was found that $[\alpha\text{-}^{15}N,2\text{-}^{14}C]$ornithine yielded the pyrrolidine ring labelled only with ^{14}C. This was not observed when $[\delta\text{-}^{15}N,2\text{-}^{14}C]$ornithine was the substrate. The pathway involving putrescine should have equilibrated the α- and δ-nitrogens of L-ornithine as it did C–2 and C–5. The explanation of this lack of equilibration is that L-ornithine rapidly equilibrates with α-keto-δ-aminovaleric acid (L), which means that an unlabelled pool of this compound would quickly dilute out the ^{15}N in the α-amino group to an undetectable level.

(ii) TROPANE ALKALOIDS†

The pathway of formation of the pyrrolidine residue of the tropane ring contrasts sharply with that of the pyrrolidine residue of nicotine discussed in the previous section. The molecule of hyoscyamine (Table 13.1) is labelled at C–1 but not at C–5 when it is synthesized from $[2\text{-}^{14}C]$ornithine in *Datura stramonium*. This is a result of a specific methylation of the δ-amino group of L-ornithine as the first step in the biosynthetic sequence (A, Fig. 13.5). The following steps (B, C and D) are analogous to those involved in the synthesis of the pyrrolidine ring of nicotine (Fig. 13.4). Two acetate residues, presumably as acetyl-CoA, provide the carbon atoms for the formation of hygrine from the N-methyl-Δ^1-pyrrolinium salt (E and F). Hygrine itself has been shown to be a precursor of the tropane residue of hyoscyamine, but the reasonable reaction

† All letters in this section refer to Fig. 13.5.

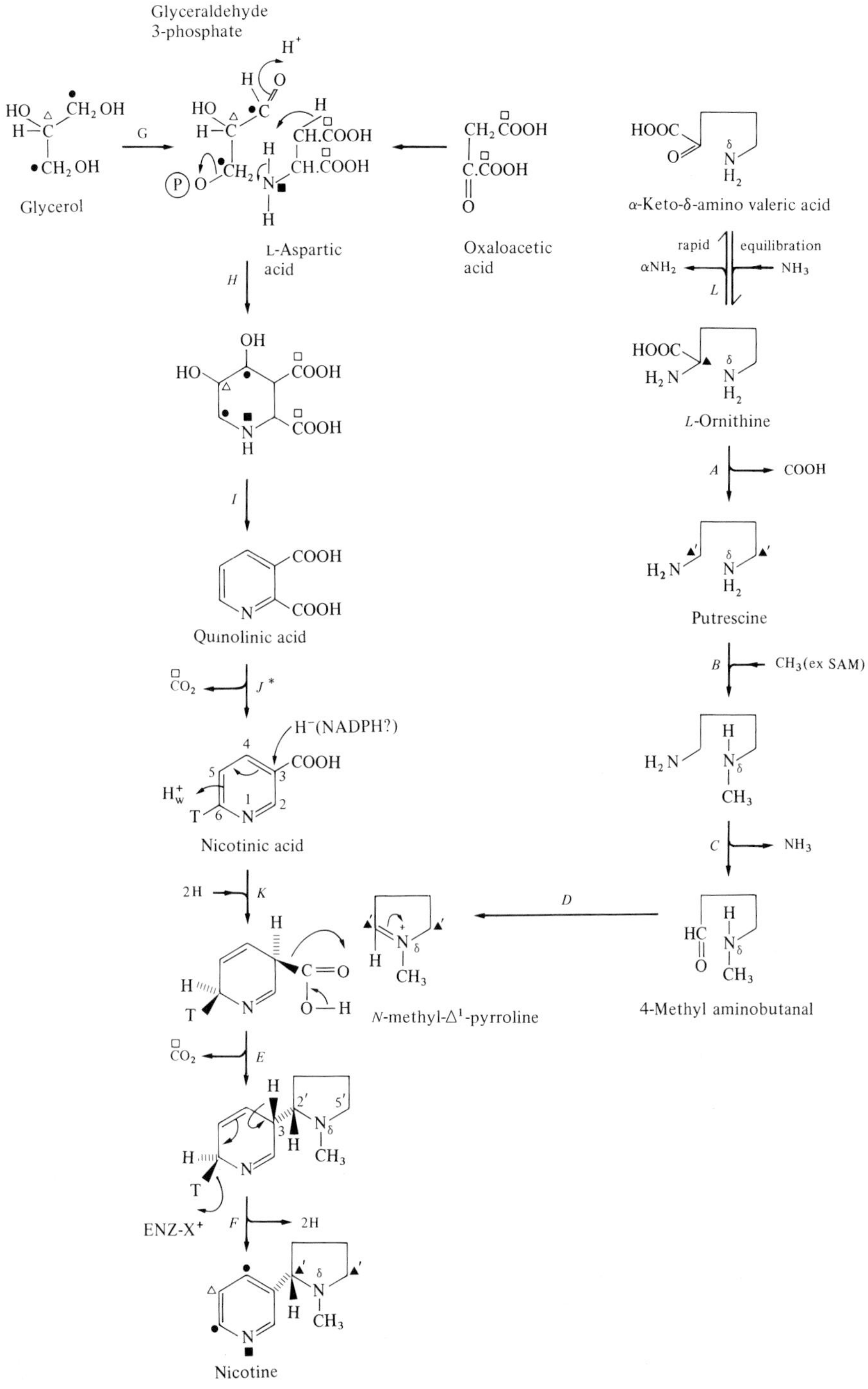

Fig. 13.4. Postulated biosynthesis of nicotine. (●, ○, ■, □, ▲ and △ indicate atoms labelled in precursor molecules; ▲′ indicates that half the label originally present as ▲ in the precursor is now present at that atom in the product—when two atoms in a product molecule are labelled ▲′ randomization of the ▲ label has occurred via a symmetrical intermediate, e.g. putrescine; note that the αNH₂ of L-ornithine is not incorporated into nicotine—this is believed to be due to the rapid equilibration of L-ornithine and α-keto-δ-aminovaleric acid in the plant tissues; note that [6-³H₁] nicotinic acid loses tritium during its conversion into nicotine. Enzymes involved: A = ornithine decarboxylase, EC 4.1.1.17; B = putrescine N-methyltransferase, EC 2.1.1.53; C = N-methylputrescine oxidase.
* Nicotinic acid mononucleotide is probably an intermediate in this reaction, see Section C.3(i).)

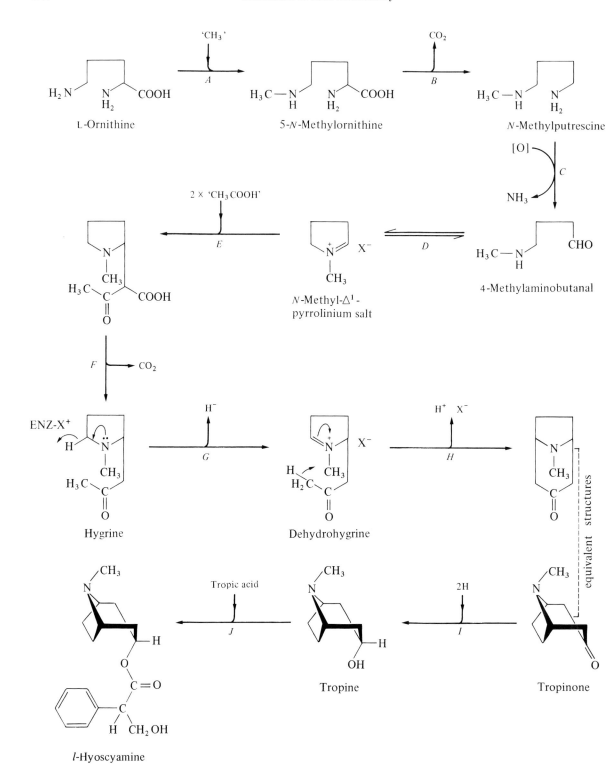

Fig. 13.5. Probable biosynthetic pathway from L-ornithine to the tropine alkaloid, *l*-hyoscyamine. (Enzyme for step J = Tropine esterase, EC 3.1.1.10.)

sequence (G, H and I) leading to tropine has not yet been demonstrated experimentally. Tropine is then esterified with L-tropic acid (13.1) to yield *l*-hyoscyamine (J). The enzyme tropine esterase has been detected in the roots of many Solanaceae. Tropic acid racemizes easily, thus *l*-hyoscyamine is easily converted into *dl*-hyoscyamine, which is atropine. Tropic acid arises from L-phenylalanine. Labelling experiments indicate that the carboxyl group of L-phenylalanine remains as the carboxyl group of tropic acid. This means that an intramolecular rearrangement must take place with movement of the carboxyl group from C–2 to C–3 of the L-phenylalanine skeleton.

(iii) PYRROLIZIDINES†

Most pyrrolizidine alkaloids exist as esters in which the base is known as a necic base and the esterifying acids are either one or two monocarboxylic acids or one dicarboxylic acid. These acids, which generally have branched carbon chains, are known as necic acids. Illustrative examples of such naturally occurring pyrrolizidine alkaloids are lindelofine (13.2), heliosupine (13.3) and retrorsine (13.4).

† All letters in this section refer to Fig. 13.6.

The necic base retronecine (Fig. 13.6) is also derived from L-ornithine and the labelling pattern after feeding [2-^{14}C]- and [5-^{14}C]ornithine is consistent with the first step (A, Fig. 13.6) being the production of putrescine in which C–1 and C–4 become equivalent. The proposed pathway involves the oxidative deamination of putrescine to form 4-aminobutanal (B) by the action of the copper-containing enzyme amine oxidase[3]. Two molecules of 4-aminobutanal combine to form an aldimine (Schiff base—see Fig. 13.3) (C) which undergoes reduction of the C=N double bond and oxidative deamination (D and E). This is followed by an intramolecular Mannich-type reaction (F, G and H; cf. Fig. 13.2) which results in the closure of both five-membered heterocyclic rings. The final reactions (I) evidently include dehydrogenation at Cs 1 and 2, reduction of the formyl group to a primary alcohol group and a stereospecific hydroxylation at C–7; all these reactions are chemically feasible but none has been examined enzymically.

At first sight the necic acids released from pyrrolizidine alkaloids by hydrolysis of the ester linkage, e.g. dicrotalic acid from dicrotaline [eq. (13.2)], would appear to be isoprenoid-derived. This is not the case for they are formed from branched chain amino acids. This has been particularly clearly demonstrated in the formation of

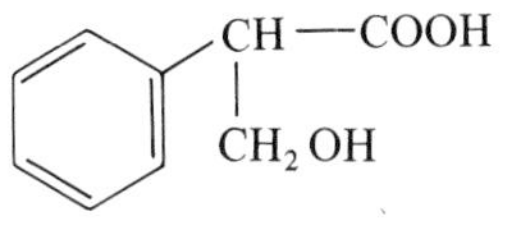

(13.1)

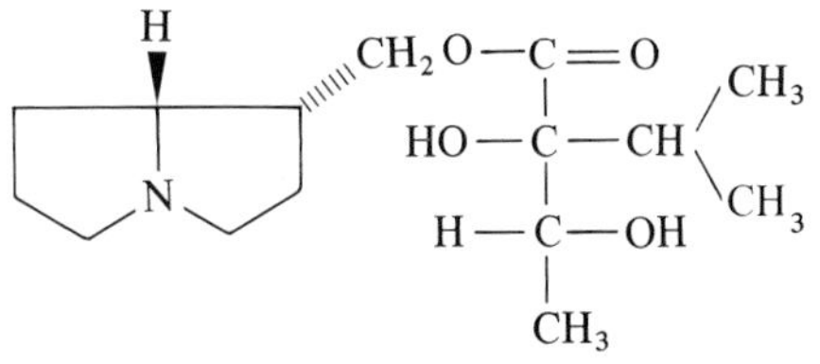

(13.2)

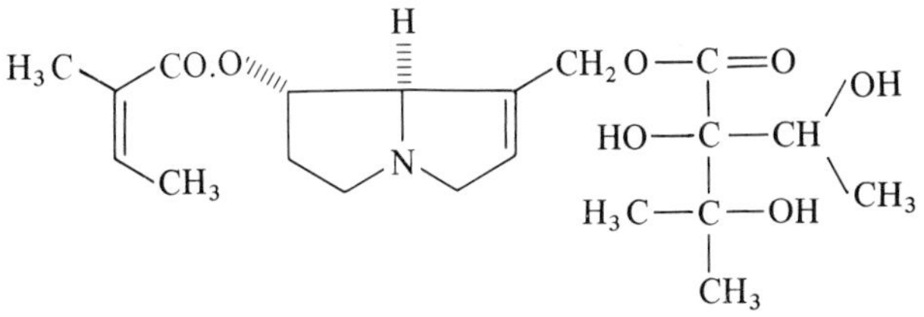

(13.3)

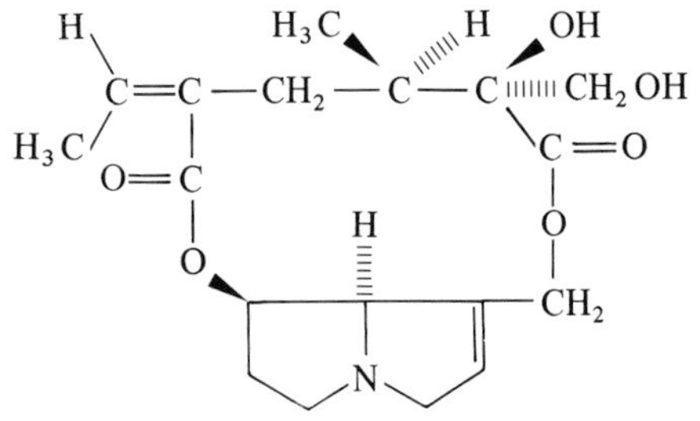

(13.4)

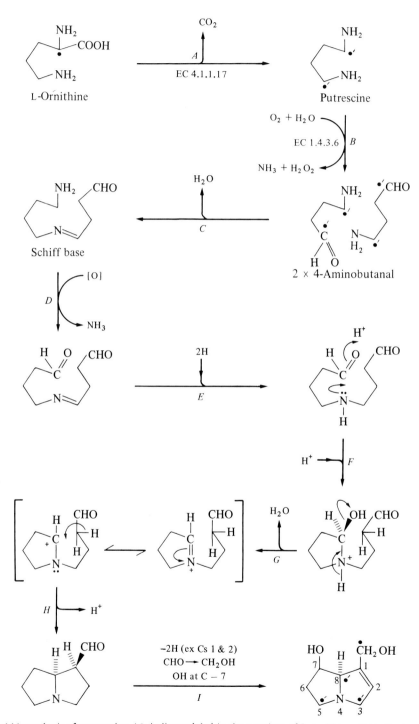

Fig. 13.6. Postulated biosynthesis of retronecine. (● indicates label in the α-carbon of L-ornithine; ●′ indicates that half the label originally presented in the α-carbon of L-ornithine is now located at that atom in retronecine or its precursor—this is due to randomization of the ● label brought about by the participation of symmetrical intermediates; EC 4.1.1.17 = ornithine decarboxylase; EC 1.4.3.6 = amine oxidase (copper containing).)

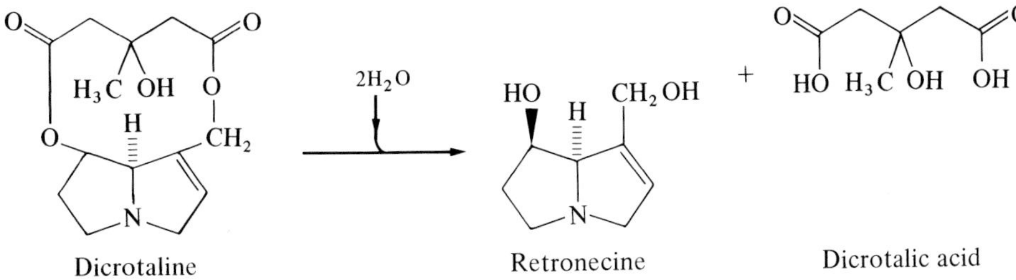

(13.2)

Dicrotaline   Retronecine   Dicrotalic acid

the echimidinic acid residue of heliosupine (Fig. 13.7). The pathway involves the formation of ketovaline (3-methyl-2-ketobutyric acid) which condenses with a C_2 compound (probably active acetaldehyde) to yield a C_7 compound that is eventually converted into echimidinic acid.

3. Alkaloids derived from L-aspartate via nicotinic acid

(i) PYRIDINES

As already indicated in Fig. 13.4 nicotinic acid is a precursor of the pyridine ring of nicotine. The distal precursor of nicotinic acid is L-aspartic acid and tracer experiments indicate that the non-aspartate carbons arise from a metabolite of glycerol, probably D-glyceraldehyde 3-phosphate (G, Fig. 13.4). Condensation produces a heterocyclic dicarboxylic acid (H, Fig. 13.4) which is aromatized to quinolinic

L-Valine   Ketovaline

'CH₃CHO' → B

2H [O]

C

Echimidinic acid

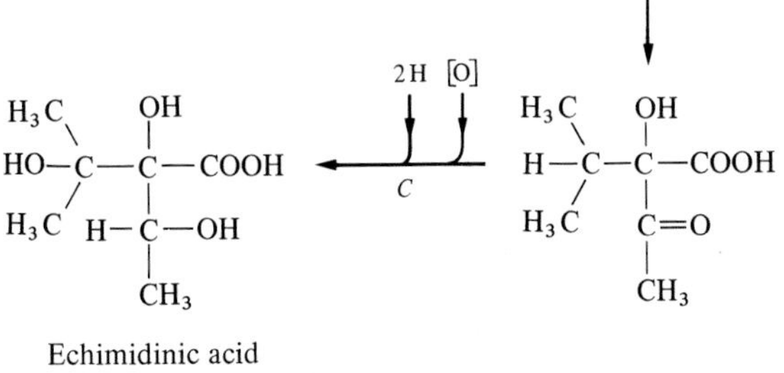

Fig. 13.7. The formation of echimidinic acid from L-valine and a 2C unit, probably 'active acetaldehyde'.

acid (I, Fig. 13.4) which then undergoes decarboxylation to yield nicotinic acid (J, Fig. 13.4). Nicotinic acid mononucleotide is probably an intermediate in this reaction, formed by the simultaneous addition of ribose 5'-phosphate from PRPP and loss of CO_2; the mononucleotide is then hydrolysed to nicotinic acid. Nicotinic acid is then reduced (K, Fig. 13.4) to 3,6-dihydronicotinic acid which condenses with N-methyl-Δ^1-pyrroline to yield the immediate precursor of nicotine.

(ii) ISOQUINUCLIDINES†

The pyridine ring in the isoquinuclidine alkaloid dioscorine (Fig. 13.8) arises from nicotinic acid; the feeding of [2-¹⁴C]- and [5,6-¹³C]nicotinic acid to dioscorine-synthesizing plants yielded dioscorine labelled with ¹⁴C at C–3 and ¹³C at C–1 and C–7. These observations are compatible with the pathway shown in Fig. 13.8. It is assumed that, as in the case of nicotine synthesis, the condensing species is 3,6-dihydronicotinic acid which undergoes a concerted decarboxylation and condensation (A) with the C_8 acetate-derived acyclic precursor. The resulting compound, after reduction and double bond rearrangement in the heterocyclic ring (B), undergoes an aldol condensation with decarboxylation (C) to yield a compound which is converted into dioscorine by lactonization (D).

4. Alkaloids derived from L-lysine

(i) PIPERIDINES‡

The piperidine rings of anabasine (13.5), N-methylpelletierine (13.6), anaferine (13.7) and adenocarpine (13.8) arise from L-lysine, but in all cases

† All letters in this section refer to Fig. 13.8.
‡ All letters in this section refer to Fig. 13.9.

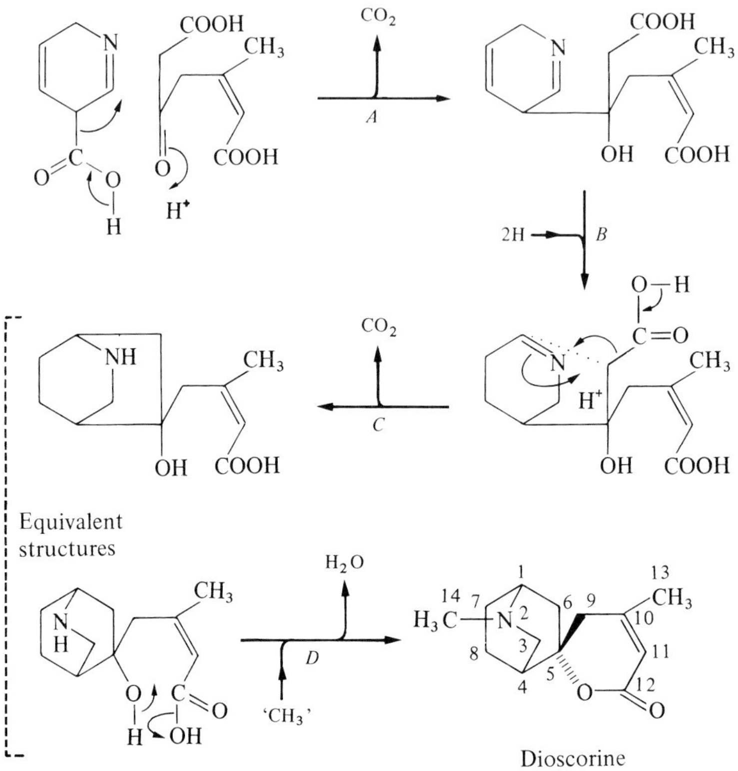

Fig. 13.8. Biosynthesis of dioscorine from 3,6-dihydronicotinic acid and an acetate-derived C_8 acyclic precursor.

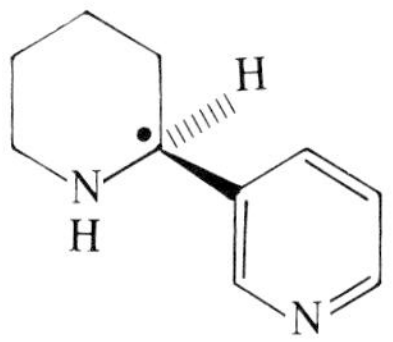

(13.5)

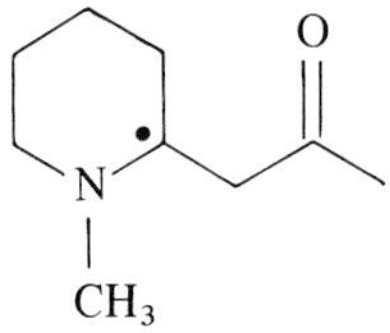

(13.6)

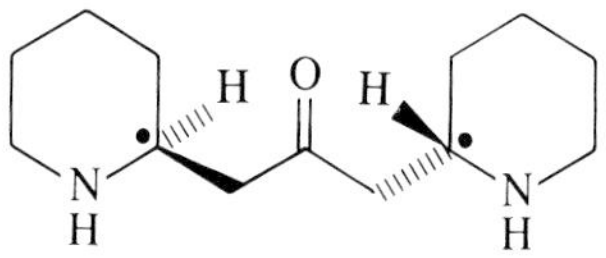

(13.7)

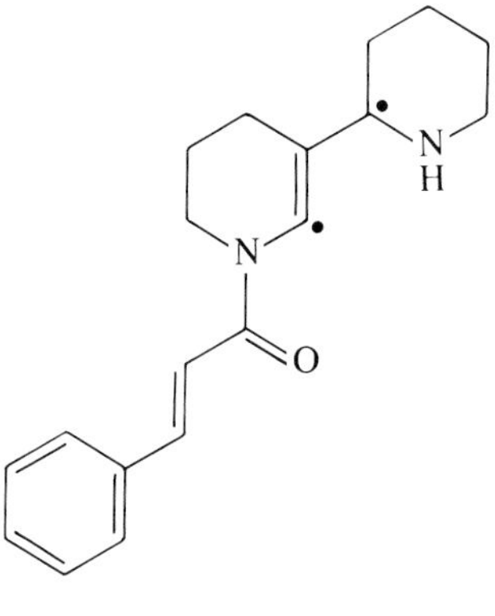

(13.8)

C–2 of L-lysine retains its individuality and thus its symmetrical decarboxylation product, cadaverine, cannot be an intermediate. The location of C–2 of L-lysine in these alkaloids is indicated thus ● in formulae (13.5)–(13.8) and it will be noted that anaferine represents a piperidine dimer. The intermediate which prevents randomization of the C–2 label of L-lysine is a Schiff base formed from L-lysine and pyridoxal phosphate (A, Fig. 13.9) which undergoes decarboxylation to the Schiff base of cadaverine (B) so that C–1 and C–4 of cadaverine never become equivalent. Hydrolysis of the Schiff base yields pyridoxamine phosphate and 5-amino-pentanal (C); the latter undergoes ring closure to

Δ^1-piperideine (D) which appears to be the key precursor of almost all piperidine alkaloids. The pyridoxamine phosphate liberated in step C is converted into pyridoxal phosphate and re-used in step A.

Anabasine, for example (see Table 13.1), is formed by the condensation of Δ^1-piperideine with 3,6-dihydronicotinic acid in an analogous fashion to the condensation of N-methyl-Δ^1-pyrroline with 3,6-dihydronicotinic acid in nicotine biosynthesis (E, Fig. 13.4).

N-methylpelletierine (13.6) results from the condensation of Δ^1-piperideine with acetoacetate, probably as its CoA derivative, followed by methyl-

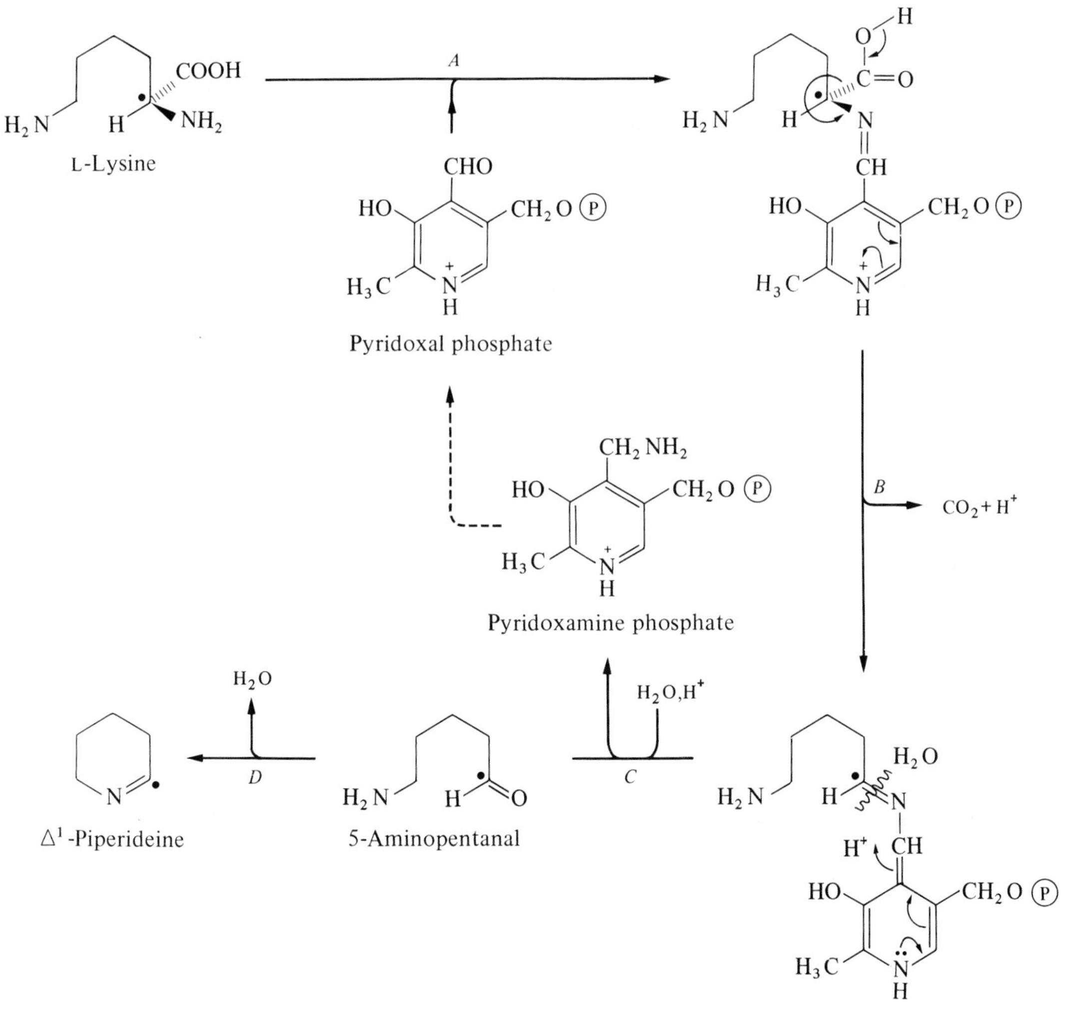

Fig. 13.9. Biosynthesis of Δ^1-piperideine from L-lysine.

ation. Anaferine (13.7) probably results from the condensation of pelletierine with a second molecule of Δ^1-piperideine. Adenocarpine (13.8) appears to result from the dimerization of two Δ^1-piperideine molecules, a reaction which occurs easily *in vitro*.

It is important to note, however, that the hemlock alkaloids, e.g. coniine, which have a piperidine ring, are biosynthesized not from L-lysine but from four molecules of acetate. The observed labelling pattern in coniine derived from [1-^{14}C]acetate is shown in Fig. 13.10. The reduction of γ-coniceine to coniine has been demonstrated enzymically, the 4-pro-S hydrogen (see Appendix 1) of the nicotinamide ring of NADPH being utilized in the process.

(ii) Quinolizidines†

The quinolizidine ring in alkaloids such as lupinine arises from L-lysine (see Fig. 13.11) but, in contrast to the formation of the piperidine ring, C–2 of L-lysine is randomized and thus free cadaverine can be considered a true intermediate and label from

† All letters in this section refer to Fig. 13.11.

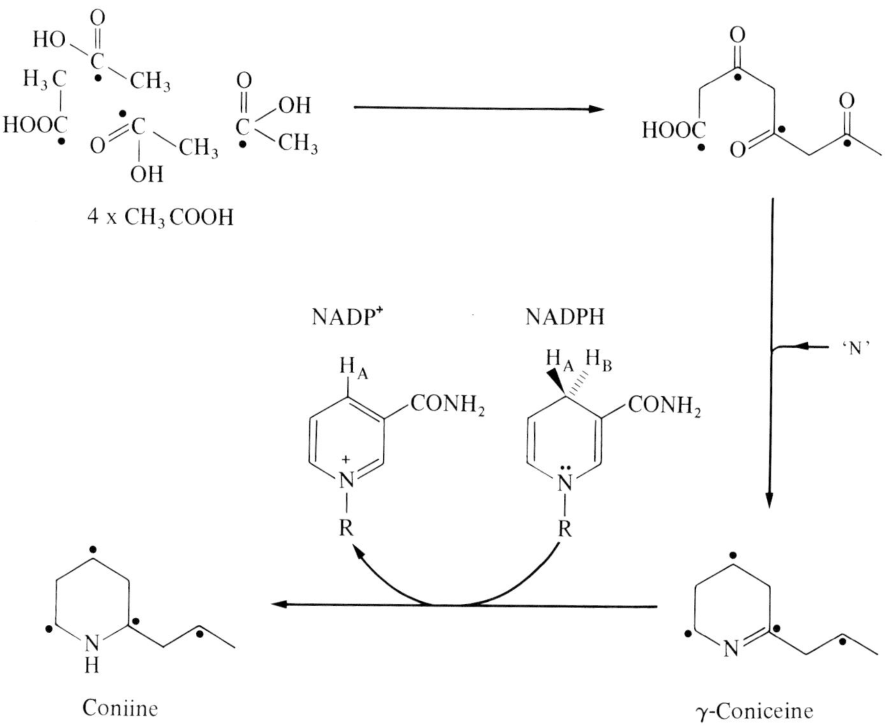

[2-^{14}C]lysine is found at C–1 and C–5 after the action of lysine decarboxylase[4] (A). Cadaverine undergoes oxidation to 5-aminopentanal (B), two molecules of which condense, probably after one has been oxidatively deaminated to 4-formylbutanal, to yield a Schiff base (C) which, after reduction of the C=N double bond (D), undergoes an intramolecular Mannich-type reaction (E, F, G and H; cf. Fig. 13.2 and Fig. 13.6) resulting in the closure of both six-membered heterocyclic rings. Reduction of the aldehyde group (I) to a primary alcohol group then yields lupinine.

The label that would be expected in lupinine on the assumption that there is randomization of C–2 of L-lysine via cadaverine is indicated thus ● in Fig. 13.11 and corresponds exactly to that found experimentally.

Lycopodine (13.9) from *Lycopodium* spp. also arises from L-lysine via a symmetrical intermediate and the observed labelling from [2-^{14}C]lysine is indicated thus ● in formula (13.9). Mechanisms for the biosynthesis of this compound have been proposed but are not discussed here.

Fig. 13.10. The biosynthesis of coniine from four molecules of acetate. (● = labelling from [1-^{14}C]CH$_3$COOH; H$_A$ = 4-pro-R hydrogen of NADPH; H$_B$ = 4-pro-S hydrogen of NADPH.)

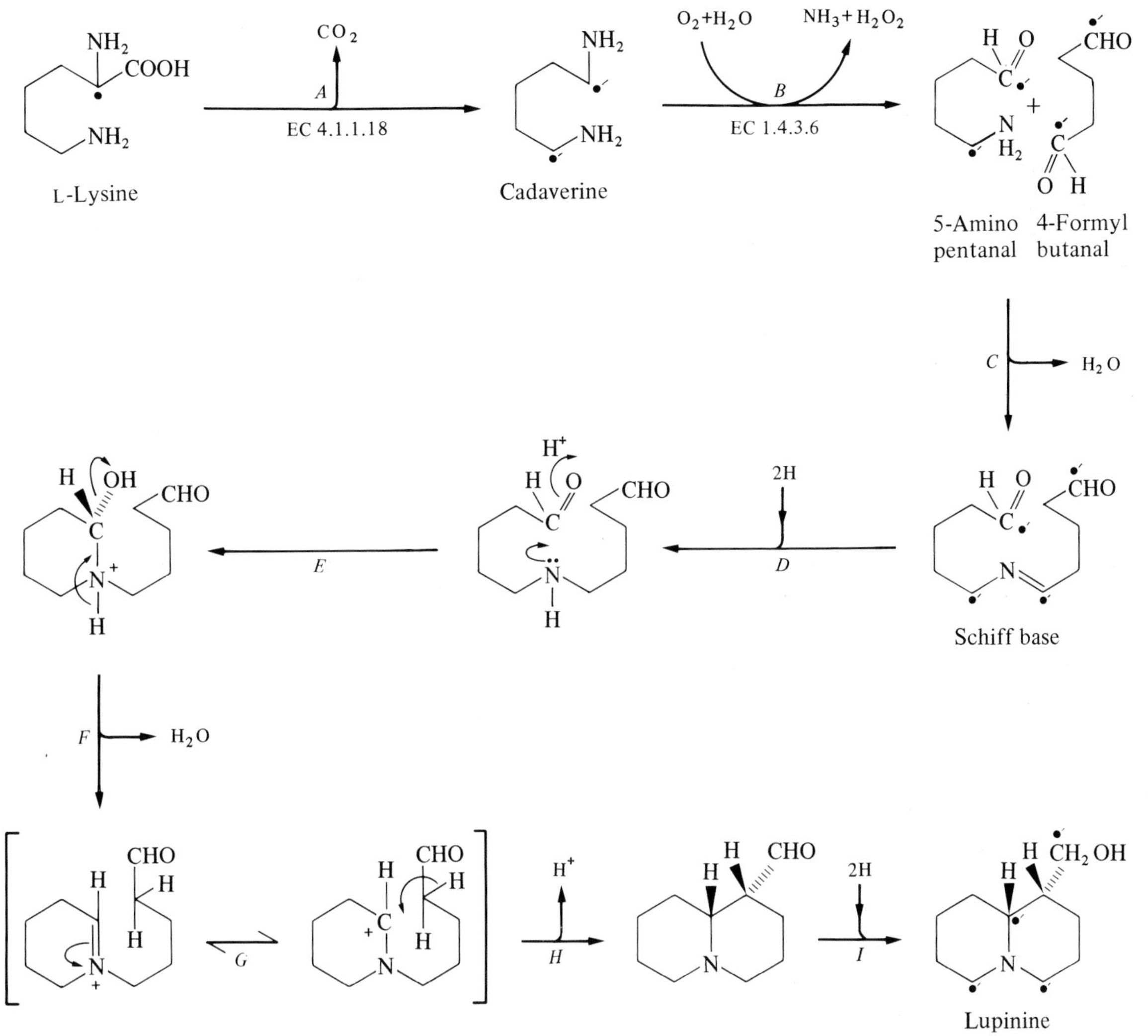

Fig. 13.11. Postulated biosynthesis of the quinolizidine alkaloid, lupinine. (EC 4.1.1.18=lysine decarboxylase; EC 1.4.3.6=amine oxidase (copper containing); ● indicates label in the α-carbon of L-lysine; ●′ indicates that half the label originally present in the α-carbon of L-lysine is now located at that atom in lupinine or its precursors—this is due to randomization of the ● label by the participation of symmetrical intermediates.)

5. Alkaloids derived from L-phenylalanine

L-phenylalanine *per se* does not give rise to many alkaloids by contributing *both* the aromatic ring system and the α-amino group to the alkaloid structure. However, it is often the source of aromatic rings in alkaloids and this aspect of its metabolism will be discussed when it arises [see, for example,

Section C.6(i)(b)]. Experiments with [β-^{14}C]phenylalanine indicate that it is the precursor of ephedrine (13.10) in *Ephedra* spp. [● represents the position of the ^{14}C from [β-^{14}C]phenylalanine in (13.10)]. The mechanism for the formation of this alkaloid is still in doubt because of unexpected results showing that although ^{15}N from [^{15}N]phenylalanine is incorporated, ^{14}C from [α-^{14}C]phenylalanine is not.

(13.9)

(13.10)

6. Alkaloids derived from L-tyrosine

Unlike L-phenylalanine, L-tyrosine provides both an aromatic ring and a nitrogen atom for many complex alkaloids. Only a selected few of these can, however, be considered here.

(i) ISOQUINOLINES†

The formation of isoquinoline alkaloids involves the hydroxylation of the aromatic ring of L-tyrosine followed by decarboxylation and then ring closure with another compound, which provides an additional one or two carbon units, occurs. A typical example of this procedure is seen in the formation of anhalonidine by the peyote cactus (Fig. 13.12). It is assumed that 3,4-dihydroxyphenylalanine (DOPA), formed by 3-hydroxylation of L-tyrosine, is converted into 3-hydroxy-4,5-dimethoxyphenyl-ethylamine, known to be present in the cactus *Lophophora williamsii*, by a series of steps (*B, C, D* and *E*), all of which are chemically acceptable, but which are, apart from *B*, still unproven. However, all the intermediates have been found in plants. The 3-hydroxy-4,5-dimethoxyphenylethylamine then undergoes ring closure (*F*) with pyruvate to form peyoruvic acid which is then decarboxylated (*G*) to anhalonidine. Evidence for the final steps (*F* and *G*) comes from (i) the incorporation of [2-¹⁴C]tyrosine into anhalonidine with the label in the position indicated thus ● in Fig. 13.12 and (ii) the very efficient conversion of 3-hydroxy-4,5-dimethoxy-phenylethylamine into anhalonidine. The proposed intermediate, peyoruvic acid, has been isolated from *L. williamsii*. The analogue of peyoruvic acid, peyoxylic acid, is formed by the condensation of the

precursor amine with glyoxylic acid (*H*); peyoxylic acid can also be incorporated into peyote alkaloids (*I*). The *N*-methylated isoquinoline alkaloids are formed by *N*-methylation of a precursor amine. Just where along the biosynthetic line this methylation occurs is not clear but *N*-methyl-3-hydroxy-4,5-dimethyoxyphenylethylamine, doubly labelled with ¹⁴C and ³H, has been shown to be incorporated intact and efficiently into pellotine (*J, K*). In Fig. 13.11, therefore, *N*-methylation is assumed to occur at *J* but it may happen earlier in the pathway.

(ii) BENZYLISOQUINOLINES

(a) *Benzylisoquinolines derived from two molecules of L-tyrosine*‡

Our example in this group is the formation of the morphine alkaloids in the opium poppy (Fig. 13.13). The first step (*A*) involves the conversion of two molecules of L-tyrosine into two molecules of DOPA. One molecule of DOPA is converted into dopamine (*B*) whilst the other is converted into 3,4-dihydroxyphenylpyruvate (*C*). These two compounds then combine with the elimination of water and carbon dioxide in a Mannich-type reaction (see Fig. 13.2) to yield a molecule of norlaudanosoline (*D*). *O*- and *N*-methylation (*E*) then lead to (−)-reticuline (sometimes called protothebaine). The next step involves oxidative phenol coupling [see Section C.1(i)]. In the presence of phenol oxidase⁵ (−)-reticuline yields a diradical (*F*), of which one resonance form undergoes *o,p'* coupling (*G*) to produce (+)-

---

† All letters in this section refer to Fig. 13.12.

‡ All letters in this section refer to Fig. 13.13.

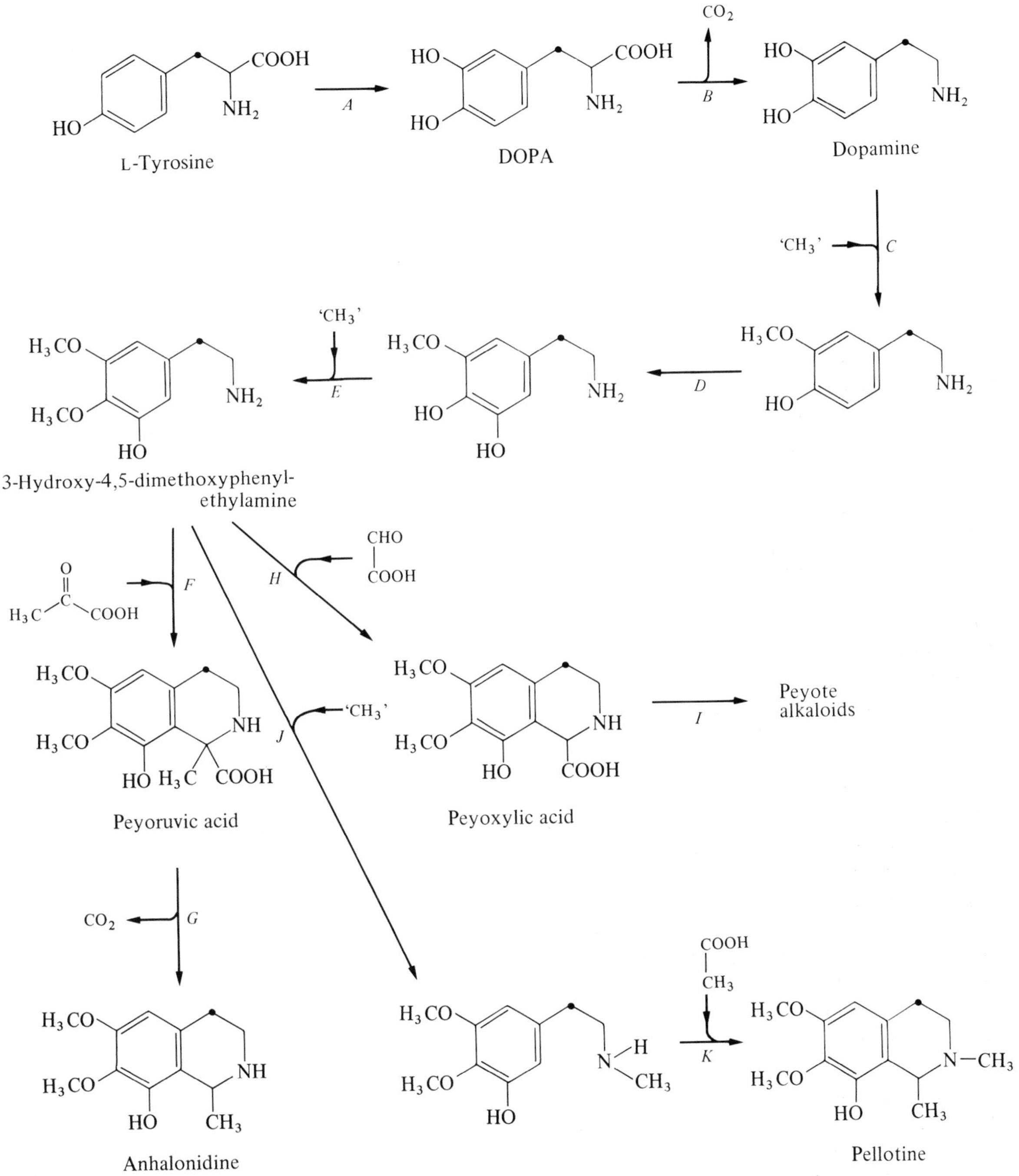

Fig. 13.12. The formation of peyote alkaloids from L-tyrosine.

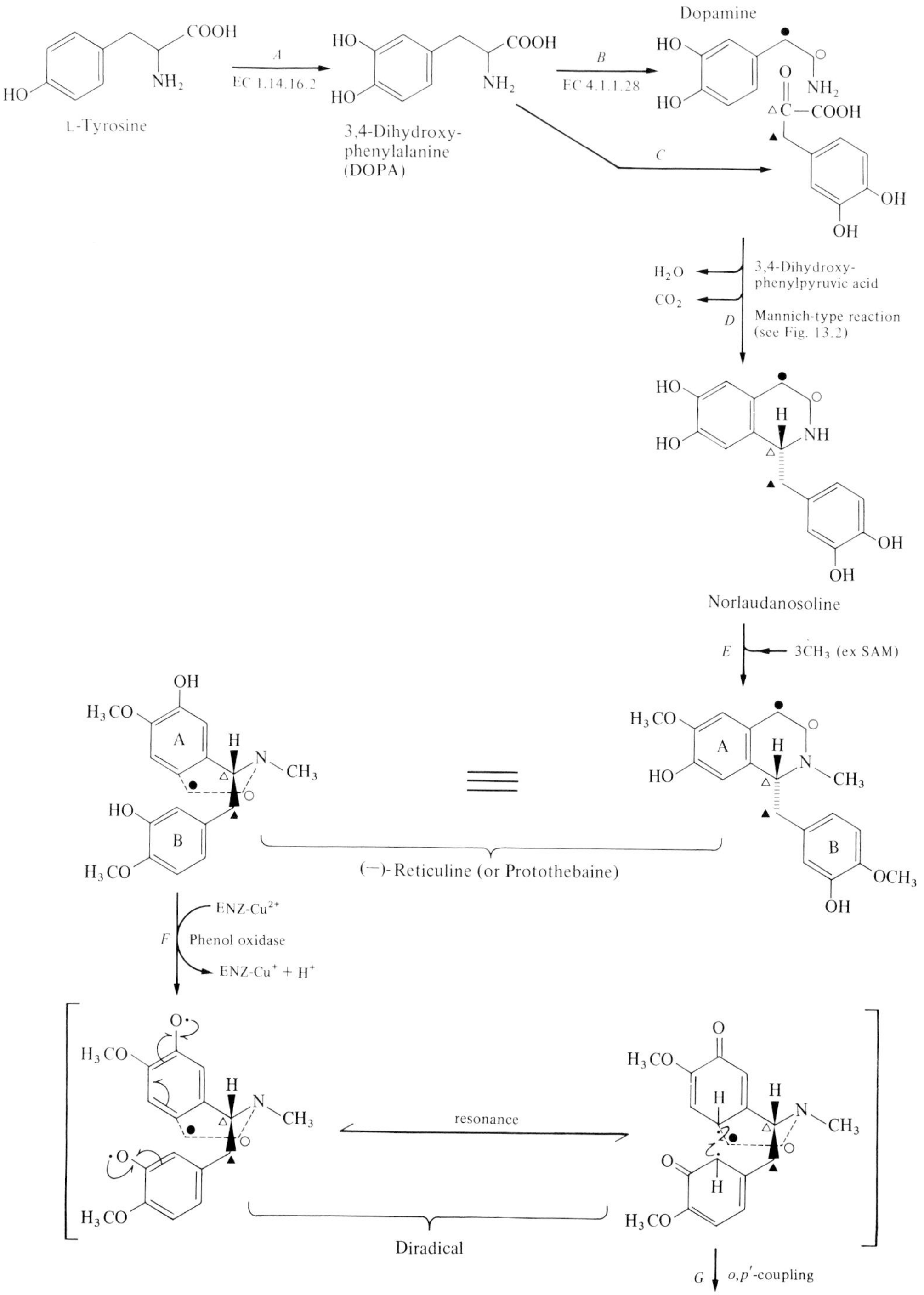

Fig. 13.13. Postulated biosynthesis of examples of the hydrophenanthrene series of isoquinoline alkaloids found in opium poppy (*Papaver somniferum*). (EC 1.14.16.2 = tyrosine 3-monooxygenase; EC 4.1.1.28 = aromatic L-amino acid decarboxylase; SAM = *S*-adenosylmethionine; >S: = *S*-adenosylhomocystein?; ⇀ = movement of an electron pair in the direction indicated; ⇁ = movement of a single electron in the direction indicated; *o,p*-coupling = the coupling of an *ortho*-phenolate radical with a *para*-phenolate radical; ○ and ● = C–1 and C–2 of dopamine; ○ and △ both originate from C–2 of tyrosine; △ and ▲ = C–2 and C–3 of 3,4-dihydroxyphenylpyruvic acid; ● and ▲ both originate from C–3 of tyrosine.)

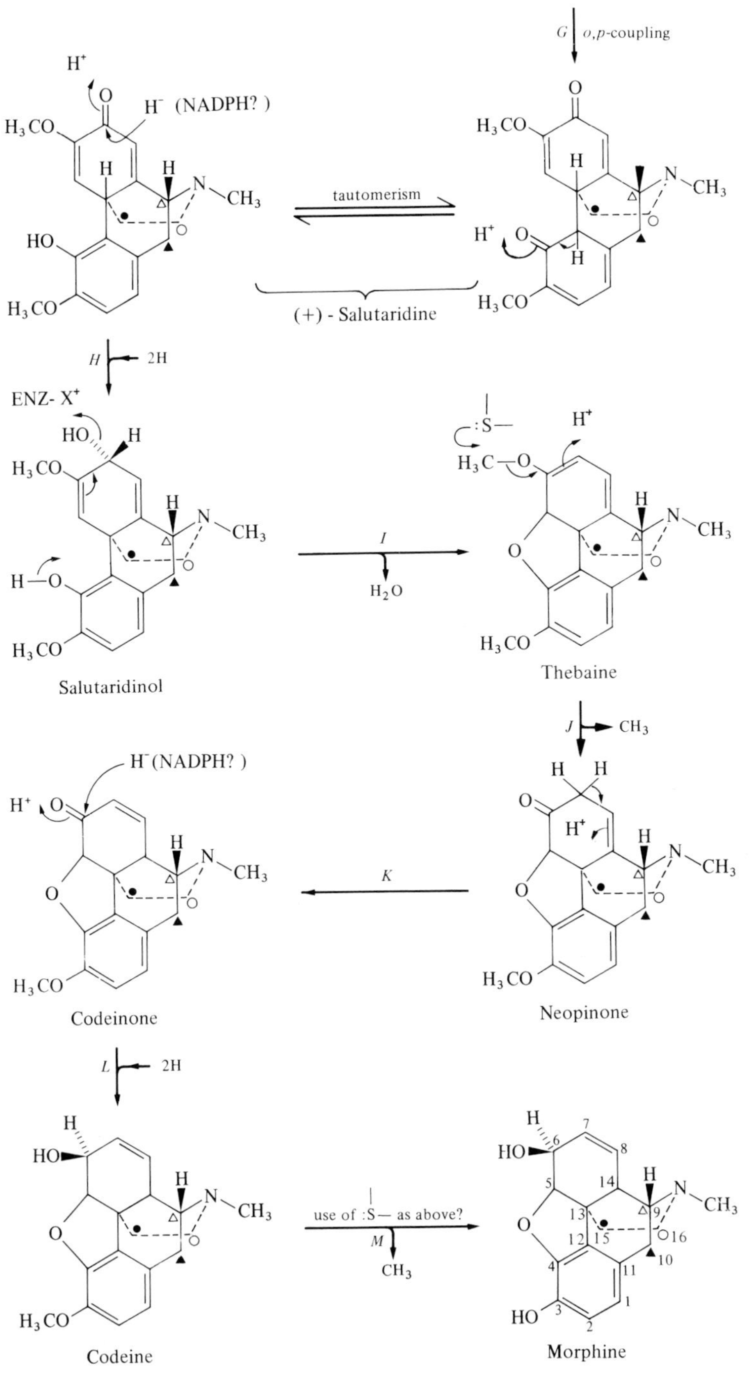

Fig. 13.13 (*cont.*).

salutaridine, a known alkaloid. Reduction to saluta-ridinol (*H*) is followed by a dehydration to yield thebaine (*I*). Demethylation of thebaine yields neopinone (*J*) which isomerizes to codeinone (*K*). Finally reduction of the codeinone keto group yields codeine (*L*) which on demethylation produces morphine (*M*).

Evidence for the reality of this pathway includes: (i) the fact that ^{14}C from $[\beta\text{-}^{14}C]$tyrosine is incorporated into C–10 and C–15 of morphine, (ii) the fact that ^{14}C from $[\alpha\text{-}^{14}C]$dopamine is incorporated only into C–16 of morphine, (iii) the fact that labelled laudanosoline, (−)-reticuline and (+)-salutaridine all yield morphine labelled in the predicted positions and (iv) the fact that (−)-reticuline labelled in four positions is converted into thebaine labelled in the equivalent four positions [eq. (13.3)].

(b) Benzylisoquinolines derived from one molecule of L-tyrosine and one molecule of L-phenylalanine†

The most interesting group of alkaloids in this category are those formed by plants of the Amaryllidaceae. This group divides naturally into three series of alkaloids: (i) the galanthamine (or dibenzofuran) series, (ii) the lycorine (or pyrrolo-phenanthrene) series and (iii) the haemanthamine (or ethanophenanthridine) series (see Fig. 13.14). Alkaloids of the three series are all synthesized from the same alkaloid precursor, *p′-O*-methylnorbel-ladine, the final product depending upon the type of oxidative phenolate coupling [see Section C.1(i)] involved.

† All letters in this section refer to Fig. 13.14.

Experiments with $[2\text{-}^{14}C]$- and $[3\text{-}^{14}C]$tyrosine demonstrated that *p′-O*-methylnorbelladine and its derivatives, which are composed of a $C_6\text{--}C_2\text{--}N\text{--}C_1\text{--}C_6$ skeleton, obtain the $C_6\text{--}C_2\text{--}N$ residue from L-tyrosine via tyramine (*A*). Labelled phenylalanine is not incorporated into this residue but it does provide the $C_1\text{--}C_6$ residue; it has been shown that the compound which condenses with tyramine in step *G* is 3-hydroxy-4-methoxybenzaldehyde formed via *trans*-cinnamic acid (*B*), *trans*-coumaric acid (*C*), caffeic acid (*D*) and 3,4-dihydroxybenzaldehyde (*E*). The condensation of 3-hydroxy-4-methoxybenzaldehyde with tyramine yields a Schiff base (*G*) which, on reduction, forms *p′-O*-methylnorbelladine.

The copper-containing enzyme phenol oxidase, acting on *p′-O*-methylnorbelladine, produces a phenolate free radical (I) which, according to the environment in which it finds itself, can undergo *o′,p*- or *o,p′*- or *p,p′*-coupling yielding a precursor of galanthamine, lycorine or haemanthamine respectively.

To deal with galanthamine formation first, coupling occurs between positions *ortho* to the hydroxyl of the ring contributed by L-phenylalanine and *para* to the hydroxyl of the L-tyrosine-derived ring (J_1, K_1). Methylation (L_1) and tautomerization (M_1) lead to a dienone which rearranges to yield narwedine (N_1); a final reduction step then produces galanthamine (O_1).

When coupling occurs between the positions *para* to the hydroxyl of the ring contributed by L-phenylalanine and *ortho* to the hydroxyl of the L-tyrosine-derived ring a dione results (J_2, K_2) which undergoes ring closure to produce a heterocyclic ring (L_2), tautomerism (M_2) and reduction (N_2) to

(−)-Reticuline → Thebaine   (13.3)

yield norpluviine. Removal of water accompanied by ring closure yields caranine (O_2) which is stereospecifically hydroxylated at C–2 with concomitant loss of the 2β-hydrogen to form lycorine (P). The final reaction (P) was demonstrated by the loss of tritium from $[2\beta\text{-}^3H_1]$caranine when it was converted into lycorine.

When coupling occurs between the positions *para* to the hydroxyl of the rings contributed by both L-phenylalanine and L-tyrosine two isomers are possible, one with the bridge being *endo*, as in haemanthamine (see Fig. 13.14) and its derivatives, and one with the bridge being *exo* as in buphanisine (13.11).

In Fig. 13.14 we illustrate only the coupling leading to the *endo* bridge and thus to haemanthamine. The steps in this route (J_3, K_3, L_3, M_3 and N_3) are analogous to those described in the formation of galanthamine and lycorine. In the final step (O_3), hydroxylation involves the stereospecific loss of the 3-pro-R hydrogen of tyrosine.

Colchicine (13.12), from autumn crocus and well known for its effect on mitosis (see Chapter 3,

Section B.9), is characterized by containing the seven-membered aromatic tropolone ring. It is in fact a modified phenylethylisoquinoline alkaloid formed by p,p' coupling of derivatives of L-phenylalanine and L-tyrosine. The tropolone ring arises from the L-tyrosine residue by ring expansion; this was demonstrated by the incorporation of ^{14}C from $[\beta\text{-}^{14}C]$tyrosine and $[4\text{-}^{14}C]$tyrosine into C–12 [labelled ● in (13.12)] and C–9 [labelled ○ in (13.12)] of colchicine respectively.

(iii) BETALAINS

Betalains are alkaloids that are unique in being highly coloured. They occur widely in the Centrospermae, including the Cactaceae, and are divided into the *betacyanins* and the *betaxanthins*, which are acylated and nonacylated glycosides of aglycones such as betanidin [(13.13), a betacyanin] and indicaxanthin [(13.14), a betaxanthin]. Isobetanidin, with the opposite chirality (R) at C–15 from that in betanidin (15S), also exists in Nature. A typical glycoside is betanin (13.15). In acyl

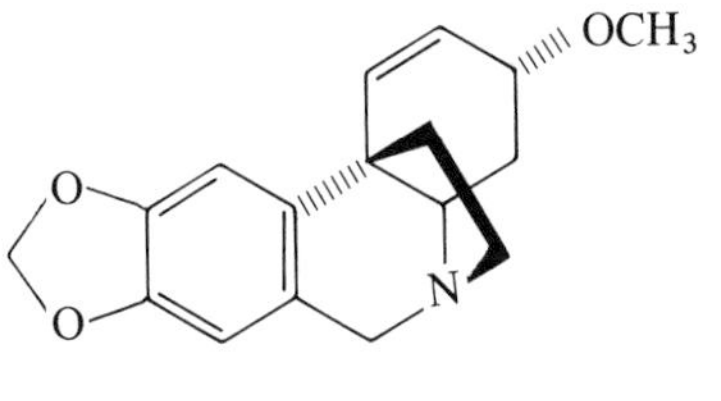

(13.11)

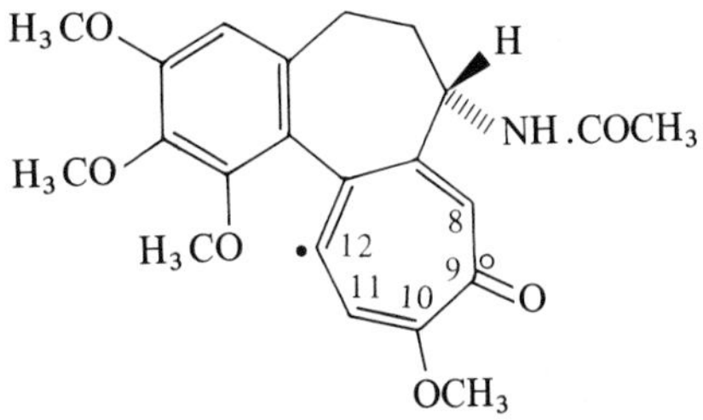

(13.12)

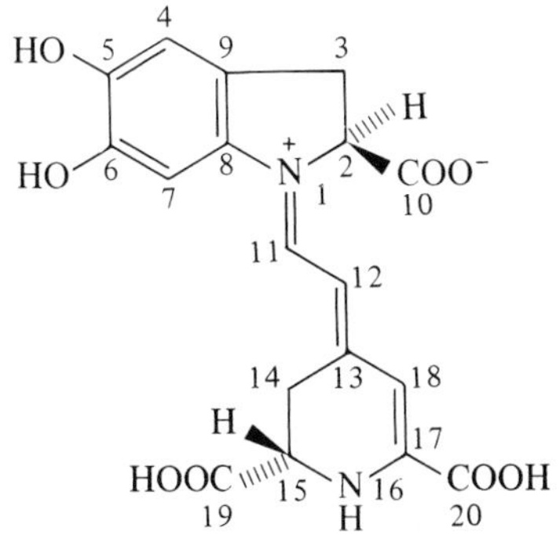

(13.13)

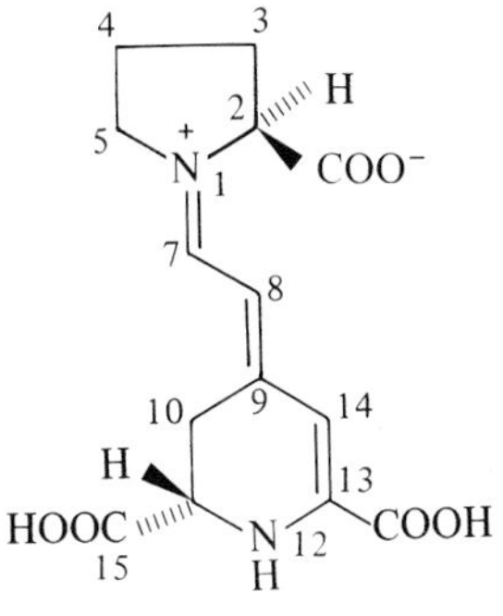

(13.14)

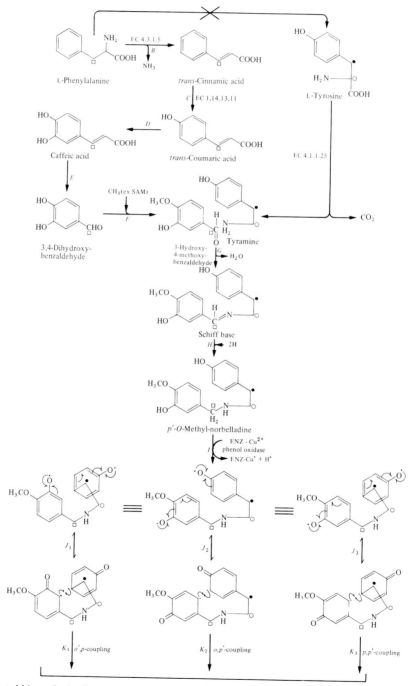

Fig. 13.14. Postulated biosynthesis of examples of the three major series of isoquinoline alkaloids found in the Amaryllidaceae. (EC 4.3.1.5 = L-phenylalanine: ammonia lyase; EC 1.14.13.11 = *trans*-cinnamate, NADPH: oxygen oxidoreductase (4-hydroxylating); EC 4.1.1.25 = L-tyrosine carboxy-lyase; SAM = *S*-adenosylmethionine; ⌒ = movement of an electron pair in the direction indicated; ⌒ = movement of a single electron in the direction indicated; ↔ = resonance. Note that L-phenylalanine and L-tyrosine are not interconvertible in the Amaryllidaceae.)

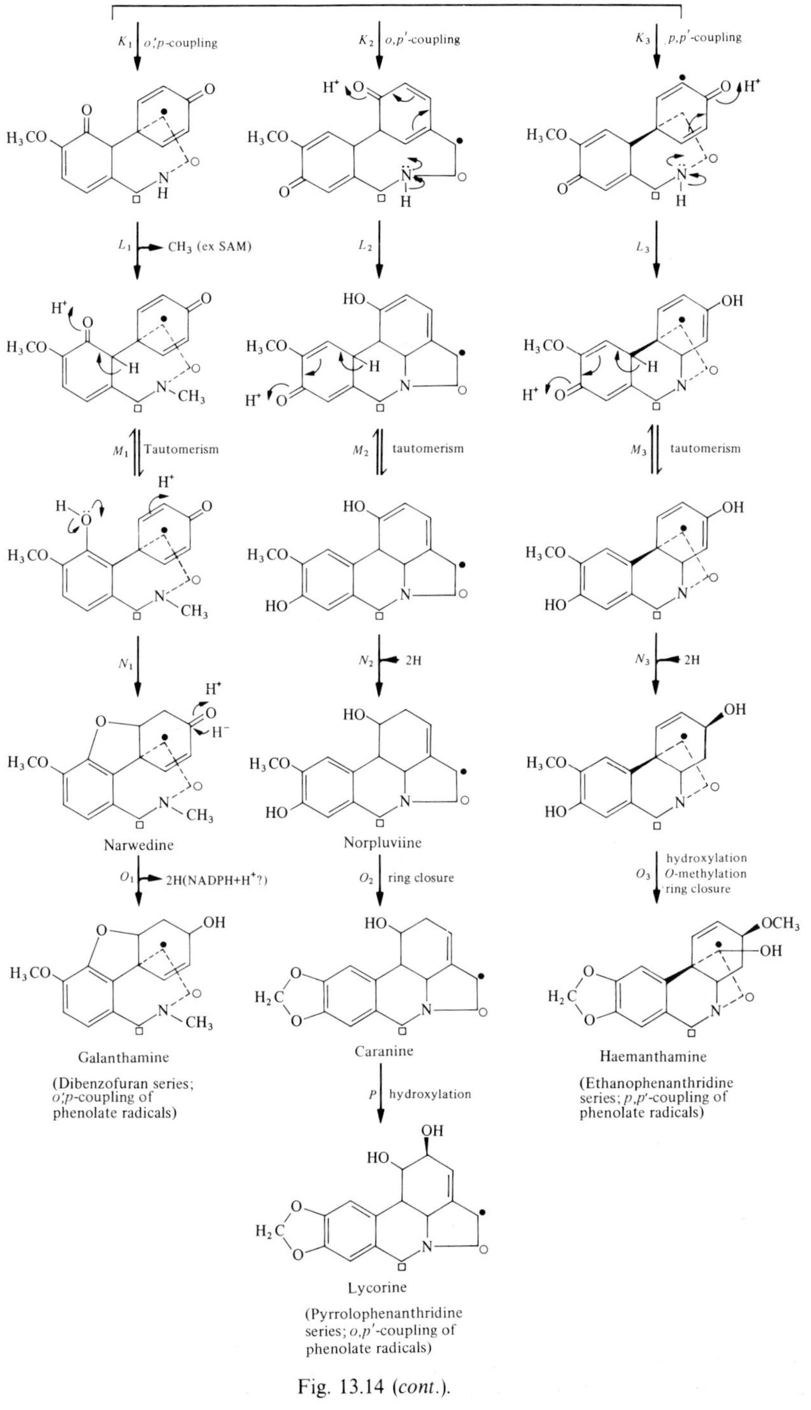

K_1 | o,p-coupling K_2 | o,p'-coupling K_3 | p,p'-coupling

L_1 — CH$_3$ (ex SAM) L_2 L_3

M_1 Tautomerism M_2 tautomerism M_3 tautomerism

N_1 N_2 — 2H N_3 — 2H

Narwedine   Norpluviine

O_1 — 2H(NADPH+H$^+$?) O_2 | ring closure O_3 | hydroxylation *O*-methylation ring closure

Galanthamine

(Dibenzofuran series;
o,p-coupling of
phenolate radicals)

Caranine

Haemanthamine

(Ethanophenanthridine
series; p,p'-coupling of
phenolate radicals)

P | hydroxylation

Lycorine

(Pyrrolophenanthridine
series; o,p'-coupling of
phenolate radicals)

Fig. 13.14 (*cont.*).

derivatives the acyl group is always linked to the sugar residue.

The betalain pigments are localized in the cell vacuole and impart red-violet (betacyanins) or yellow (betaxanthins) colours to the plants which contain them. The structural difference between the two is that the dihydroindole ring system in the betacyanins is replaced by that of L-proline in the betaxanthins.

Also present in betalain-containing plants is betalamic acid (see Fig. 13.16) which is, as will become apparent later in this section, a key biosynthetic intermediate in betalain formation. Betalains are also found in the mushroom *Amanita muscara* as, for example, musca-aurin I (13.16) and muscapurpurin (13.17). Note should be taken of the occurrence in *Amanita* and *Hygrocybe* of a betalain-like pigment called muscaflavin (13.18) which contains the seven-membered dihydroazepine ring

system. The close relationship to the betalains is emphasized when it is realized that muscaflavin (13.18) and betalamic acid (see Fig. 13.16) are isomers and are both biosynthesized from DOPA (see later).

Labelling experiments have shown that DOPA is effectively incorporated into the betalamic acid residue of betaxanthins; the labelling pattern [see eq. (13.4)] is consistent with a 4,5-extradiol cleavage of DOPA. [5-T]tyrosine yields indicaxanthin labelled in C–7 which arises from the aldehyde group of betalamic acid. Similarly [^{14}COOH]tyrosine labels C–15 of indicaxanthin which indicates that the carboxyl carbon of L-tyrosine appears in the carboxyl group indicated thus ● in betalamic acid in eq. (13.4). This labelling pattern can only arise by the 4,5-extradiol cleavage of DOPA [A, eq. (13.4)] followed by recyclization [B, eq. (13.4)].

The close relationship between betalamic acid

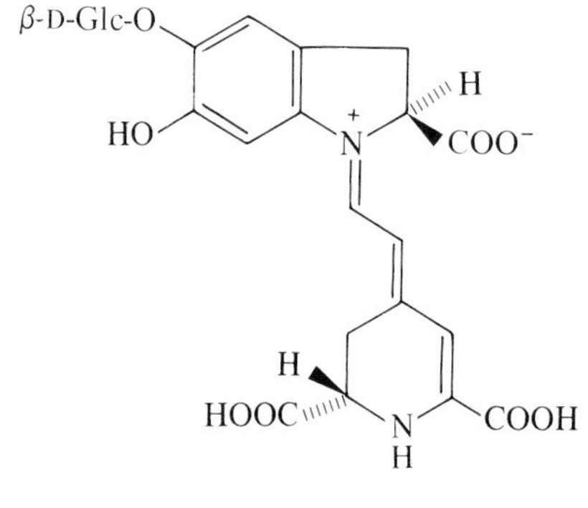

(13.15)

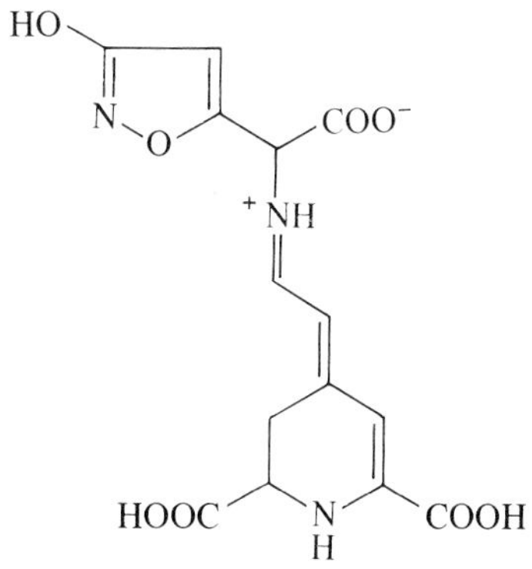

(13.16)

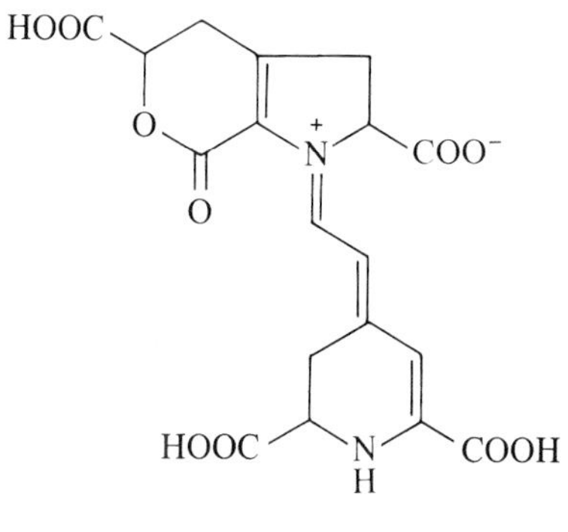

(13.17)

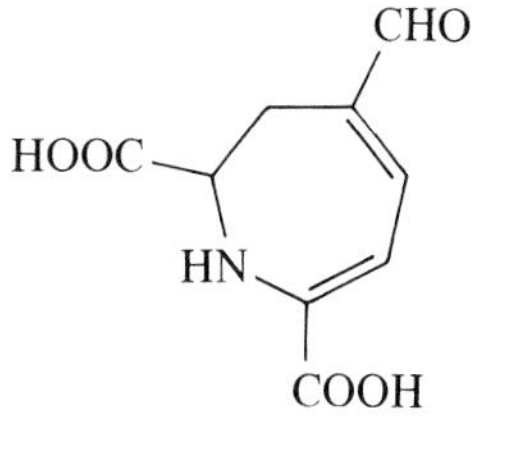

(13.18)

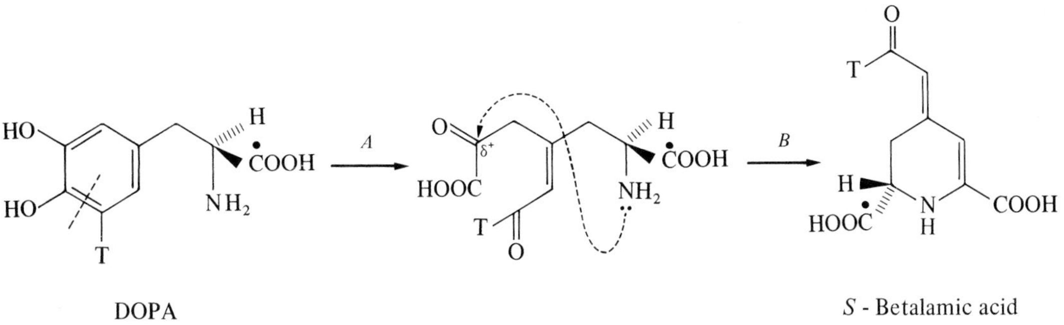

$$(13.4)$$

DOPA S - Betalamic acid

and muscaflavin is clear when it is realized that a 2,3-extradiol cleavage of DOPA, followed by re-cyclization, leads to muscaflavin (Fig. 13.15). This 2,3-extradiol ring cleavage is catalyzed *in vitro* by some plant oxygenases but they are not specific; for example they will also catalyze a 3,4-intradiol cleavage.

Experiments with [^{15}N]tyrosine showed that this amino acid provides the nitrogen for both the heterocyclic rings of betanin and further experiments with [^{14}C]tyrosine showed that it was the sole source of the dihydroindole ring. The latter is considered to arise from DOPA via *S*-cycloDOPA (*A*, Fig. 13.16). Thus betanidin arises from two molecules of DOPA, one yielding *S*-cycloDOPA

and the other *S*-betalamic acid [*B*, Fig. 13.16 and *B*, eq. (13.4)]. These two metabolites then condense to form betanidin (*C*, Fig. 13.16). The efficient conversion of betanidin into betanin in the fruit of *Opuntia dilleni* suggests that glycosylation is a late step in the biosynthetic pathway (*D*, Fig. 13.16).

However, in a variety of *Celosia plumosa S*-cycloDOPA-5-*O*-β-D-glucoside proved to be a far better precursor of amaranthin (see Fig. 13.16) than was betanidin suggesting that, although glycosylation of the betacyanin aglycone can occur, it normally takes place at the *S*-cycloDOPA level (*E*, *F* and *G*, Fig. 13.16).

Apart from the fact that betalamic acid is the precursor of the dihydropyridine ring of betaxan-

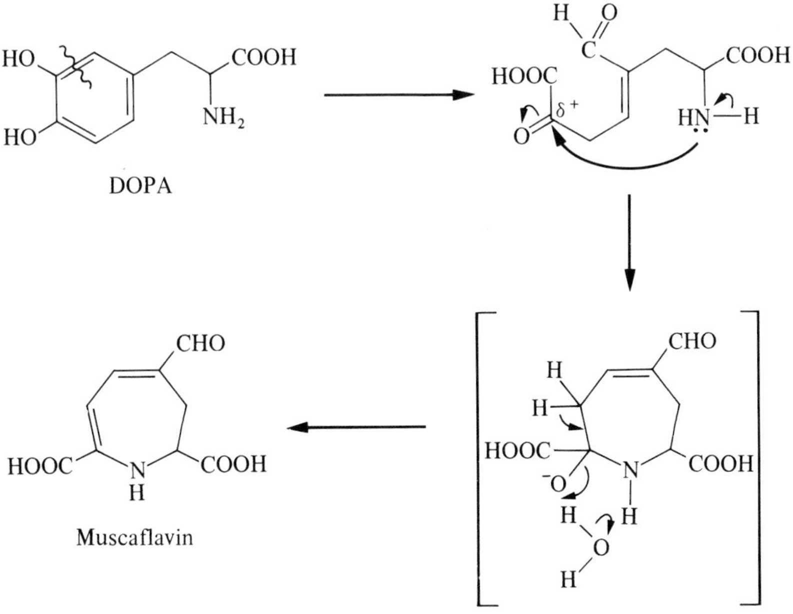

Fig. 13.15. Conversion of DOPA into muscaflavin by 2,3-extradiol cleavage followed by recyclization.

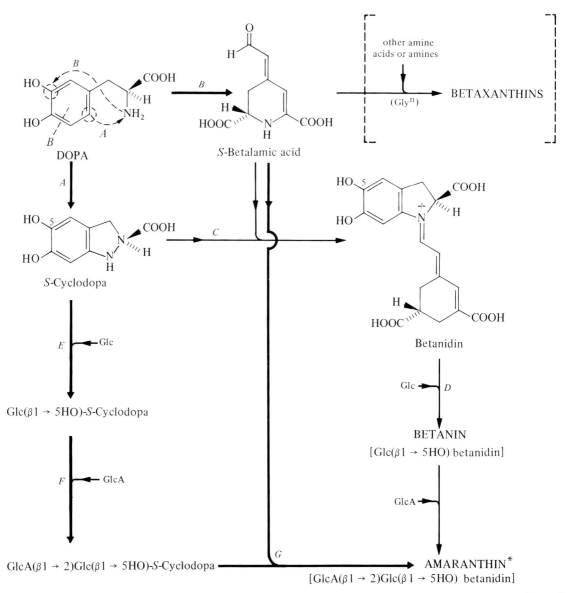

Fig. 13.16. Formation of betacyanins from two molecules of DOPA. (*Also called Amarantin; Glc = D-glucose; GlcA = D-glucuronic acid; Gly^n = glycosylation; ➡ = probable major route; → = probable minor route; note that the glycosylation steps probably involve NDP-sugars.)

thins, nothing is known of their biosynthesis. Presumably they result from the condensation of amino acids or amines with betalamic acid, a reaction which is known to occur.

Light is an absolute requirement for betalain synthesis in many species and has a quantitative effect on synthesis in all the plants so far examined. The synthesis may also be controlled by phytochrome and by plant hormones (e.g. cytokinins).

7. Alkaloids derived from L-tryptophan

(i) SIMPLE BASES

Tryptamine (see Fig. 13.17) and its N-methyl and N-dimethyl derivatives are, according to the generally accepted definition, protoalkaloids. They are widely distributed in plants, as are their hydroxylated derivatives, e.g. serotonin and psilocine (see Fig. 13.17).

The biosynthetic pathway to psilocine involves the initial conversion of L-tryptophan into tryptamine (A, Fig. 13.17) and the reaction is catalyzed by the pyridoxal-dependent enzyme, tryptophan decarboxylase[5]. Two N-methylations of tryptamine then follow (B and C, Fig. 13.17) and the resulting N,N-dimethyltryptamine is then hydroxylated at C-4 of its aromatic ring (D, Fig. 13.17).

In the formation of serotonin hydroxylation (E, Fig. 13.17) occurs before decarboxylation (F, Fig. 13.17), with 5-hydroxy-L-tryptophan being the intermediate compound.

A biosynthetically more interesting protoalkaloid derived from L-tryptophan is gramine (Fig. 13.18) which is found in relatively large amounts in barley. Detailed isotope labelling experiments with $[\beta\text{-}^3H_2,\beta\text{-}^{14}C]$tryptophan have shown that the methylene group of gramine arises intact from the methylene group of L-tryptophan and that the amino group of L-tryptophan is the source of the gramine nitrogen atom. The intramolecular movement of NH_2 from the αC to the βC

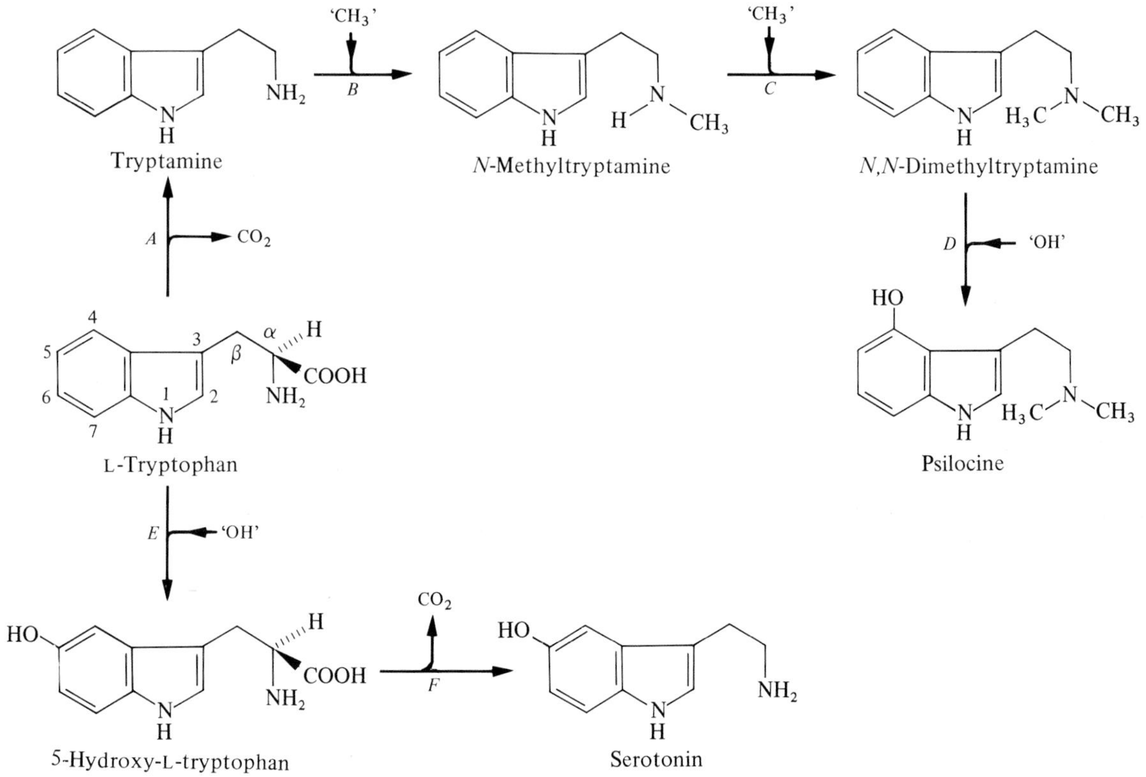

Fig. 13.17. Biosynthesis of serotonin and psilocine from L-tryptophan.

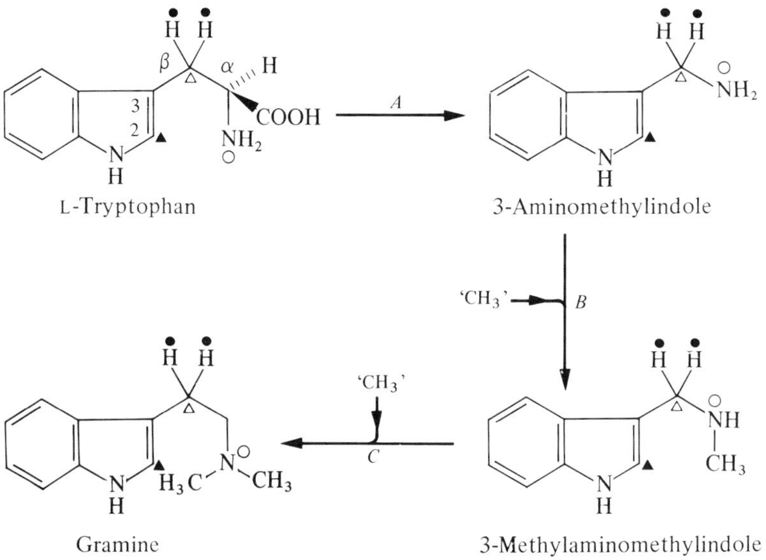

Fig. 13.18. The biosynthesis of gramine from L-tryptophan. ($\triangle = \beta C$ of tryptophan; $\blacktriangle = C$–2 of tryptophan; $\bullet$ = Hs attached to βC of tryptophan; $\bigcirc$ = N at αC of tryptophan.)

of the L-tryptophan side chain occurs before N-methylation because 3-aminomethylindole (A, Fig. 13.18) and 3-methylaminomethylindole (B, Fig. 13.18) occur naturally in trace amounts and are rapidly incorporated into gramine (C, Fig. 13.18). The mechanism of the intramolecular movement of the amino group (A, Fig. 13.18) is unknown.

(ii) SIMPLE β-CARBOLINES

About fifteen plant families contain β-carboline alkaloids which can be divided into three classes according to whether they contain (i) an aromatic pyridine ring, (ii) a dihydroaromatic ring or (iii) a tetrahydroaromatic ring. The best known example is harmine (see Fig. 13.19). Experiments with L-tryptophan and tryptamine doubly labelled with ^{14}C and ^{15}N have shown that the carboxyl group of the former is lost in the biosynthesis of harmine although both compounds are good precursors. The first step in harmine biosynthesis thus appears to be the conversion of L-tryptophan into tryptamine (A, Fig. 13.19). The next step is the N-acetylation of tryptamine (B, Fig. 13.19). Ring closure of the resulting N-acetyltryptamine then yields harmalan (C, Fig. 13.19) which is finally oxidized to harmine (D, Fig. 13.19).

(iii) ERGOLINE ALKALOIDS†

Ergoline (13.19) alkaloids are therapeutically important compounds produced mainly by the *Claviceps* genus of fungi which are parasitic on rye and some wild grasses. They can be divided into two groups: (i) the clavine alkaloids which are derivatives of 6,8-dimethylergoline and (ii) the peptide alkaloids which are derivatives of lysergic acid (see Fig. 13.20) and exemplified by ergotamine (see Fig. 13.20). Ergolines are also found in higher plants particularly in the seeds of *Rivea* and *Ipomoea* species; they are the hallucinogenic compounds found in Mexican Ololinqui, a mixture of seeds of *R. corymbosa* and *I. violacea*.

These alkaloids are formed from a precursor

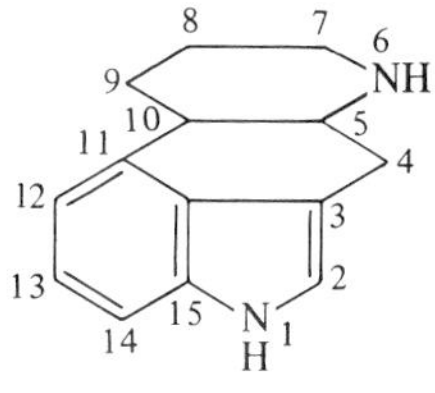

(13.19)

† All letters in this section refer to Fig. 13.20.

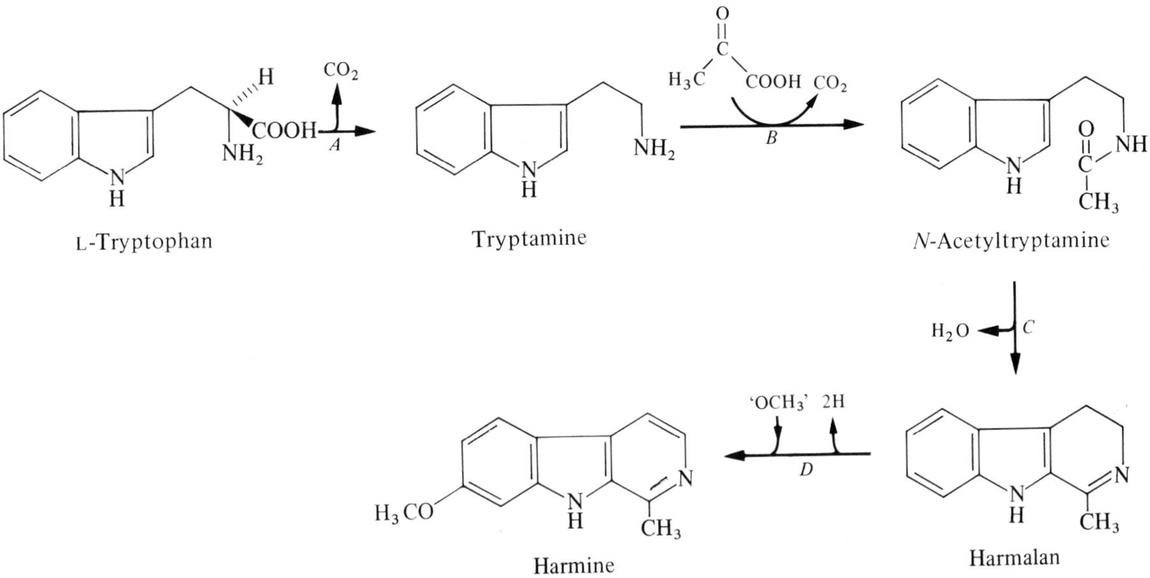

Fig. 13.19. Biosynthesis of harmine from L-tryptophan. (Some evidence suggests that methoxylation may occur at the tryptamine stage, i.e. prior to reaction B.)

which is derived from the condensation of L-tryptophan with the hemiterpene dimethylallyl pyrophosphate synthesized from mevalonic acid (A) (see Chapter 11, Section C.1) in the presence of DMAPP:L-tryptophan dimethylallyltransferase (B). The product, 4-dimethylallyltryptophan, undergoes oxidation of the methyl group arising from C–3′ of mevalonic acid to a hydroxymethyl group (C), decarboxylation, ring closure and N-methylation (D, E) to yield chanoclavine I; although Fig. 13.20 gives a sequence for these reactions the exact order of them is not known, apart from the fact that methyl oxidation takes place before cyclization. An interesting *trans→cis* isomerization occurs during reactions A–E. This follows from the experimental observation that the 4-pro-R hydrogen of mevalonic acid is retained in dimethylallyltryptophan, indicating that the isoprenoid double bond in the latter has the *trans* configuration (see Chapter 11, Sections C.1 and C.10), and also in chanoclavine I in which the isoprenoid double bond undoubtedly has the *cis* configuration. In Fig. 13.20 the *trans→cis* isomerization is shown as step D. The oxidation step (C) is catalysed by a mixed function oxygenase. Chanoclavine I is converted into agroclavine by ring closure but here there is a *cis→trans* isomerization (F), reversing the effects of the earlier one, before cyclization (G). Thus a superficial investi-

gation would have led to the conclusion that no isomerization steps occur between dimethylallyl-tryptophan and agroclavine.

A three-step process converts agroclavine into lysergic acid (H) after which combination with various peptides yields ergot alkaloids. The example given (I) is ergotamine. Virtually nothing is known of the biochemistry of these final steps.

(iv) COMPLEX INDOLE ALKALOIDS

Alkaloids of this class are synthesized from L-tryptophan and a monoterpene (C_{10}) unit as building blocks. They represent the majority of tryptophan-derived alkaloids and are found mainly in eight plant families of which the Rubiaceae, Loganiaceae and Apocynaceae are the best sources. Three main structural types can be discerned: (i) the *Corynanthe* type (e.g. corynantheine, 13.20), (ii) the *Aspidosperma* type (e.g. vindoline, (13.21)] and (iii) the *Iboga* type [e.g. catharanthine, (13.22)].

Once it was proved that the C_9 or C_{10} non-tryptophan residue of these alkaloids arose from the monoterpene geraniol it was clear that chemically acceptable rearrangements of this residue could lead to the three different structural skeletons. This is illustrated in Fig. 13.21 where geraniol, arising from

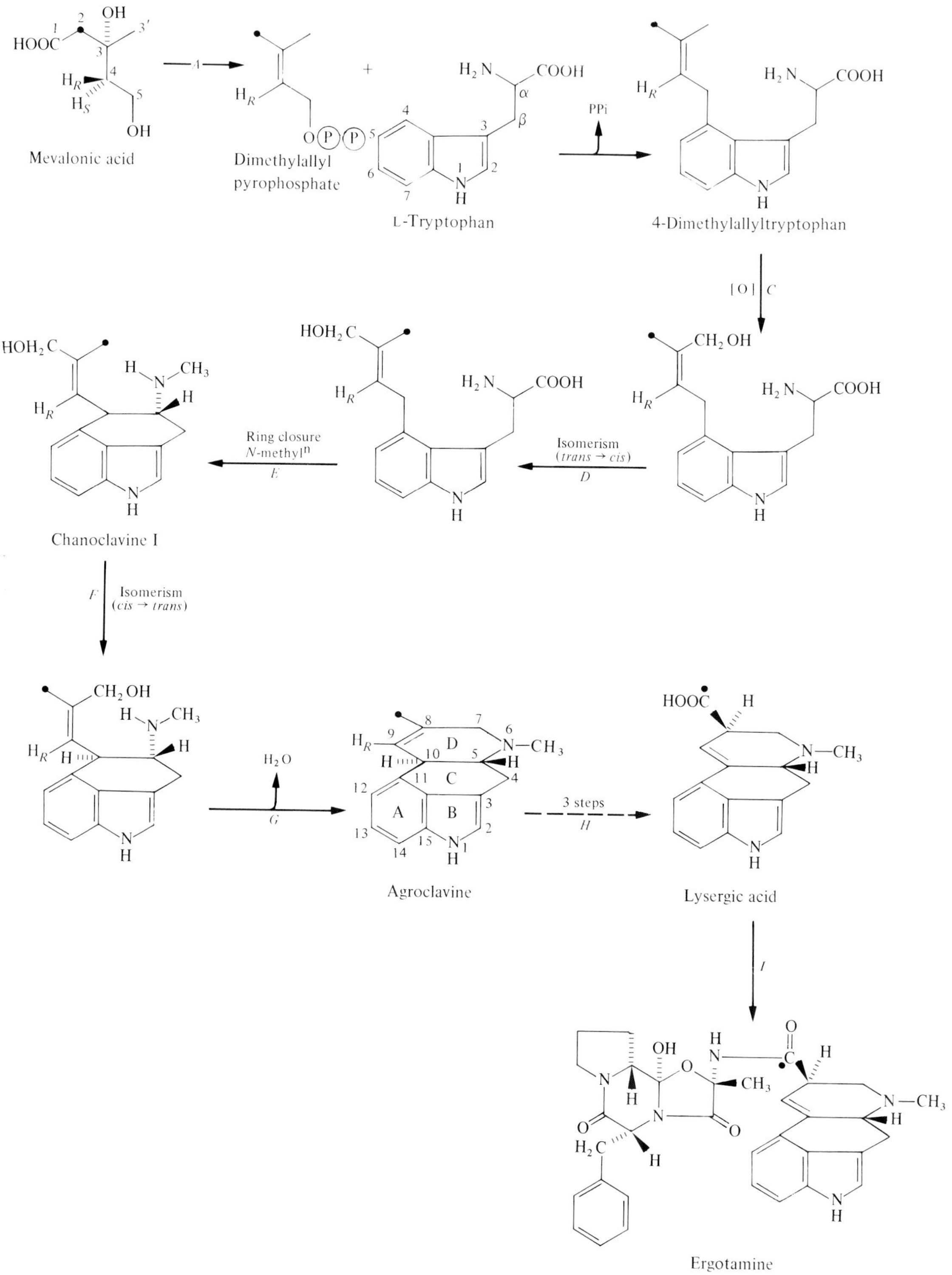

Fig. 13.20. The biosynthesis of ergot alkaloids from L-tryptophan and mevalonic acid. (H_R and H_S are the 4-pro-R and 4-pro-S hydrogen atoms of mevalonic acid or are derived from them; ● = the carbon arising from C–2 of mevalonic acid.)

(13.20)   (13.21)   (13.22)

two molecules of mevalonic acid (*A*, Fig. 13.21), is considered to produce a cyclopentane intermediate by forming a C—C bond between the former C–4s of the precursor mevalonic acid molecules. A cleavage between C–2′ and C–3′, followed by rotation around the C–4,4′ bond, will then give rise to the *Corynanthe* type of structure (*B*, *C*, Fig. 13.21). An intramolecular movement of the branched 3C unit (Cs 2, 3 and 6) from C–4 to C–5 (*D*(a), Fig. 13.21) yields the *Iboga* type of structure whilst movement of this unit from C–4 to C–4′ (*D*(b),

Fig. 13.21) yields the *Aspidosperma* type of structure.

Experiments involving the administration of different species of ^{14}C-labelled mevalonic acids to appropriate plants followed by chemical degradation of the various alkaloids to determine their labelling pattern, all confirmed the postulates of Fig. 13.21. The next important step was the identification of the proposed cyclopentane intermediate as the monoterpene glycoside, loganin (see Fig. 13.22), which is formed from geraniol. The latter is first

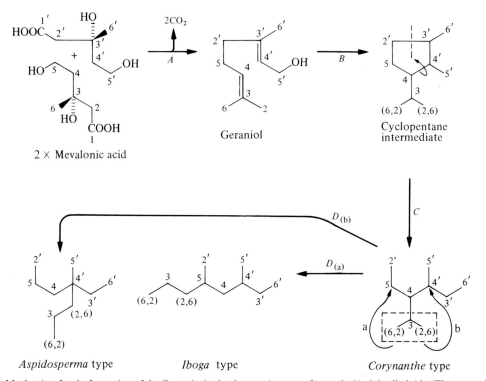

Fig. 13.21. Mechanism for the formation of the C_{10} units in the three main types of 'complex' indole alkaloids. (The source is mevalonic acid, two molecules of which give rise to a monoterpene, geraniol, whose carbon skeleton is rearranged as indicated.)

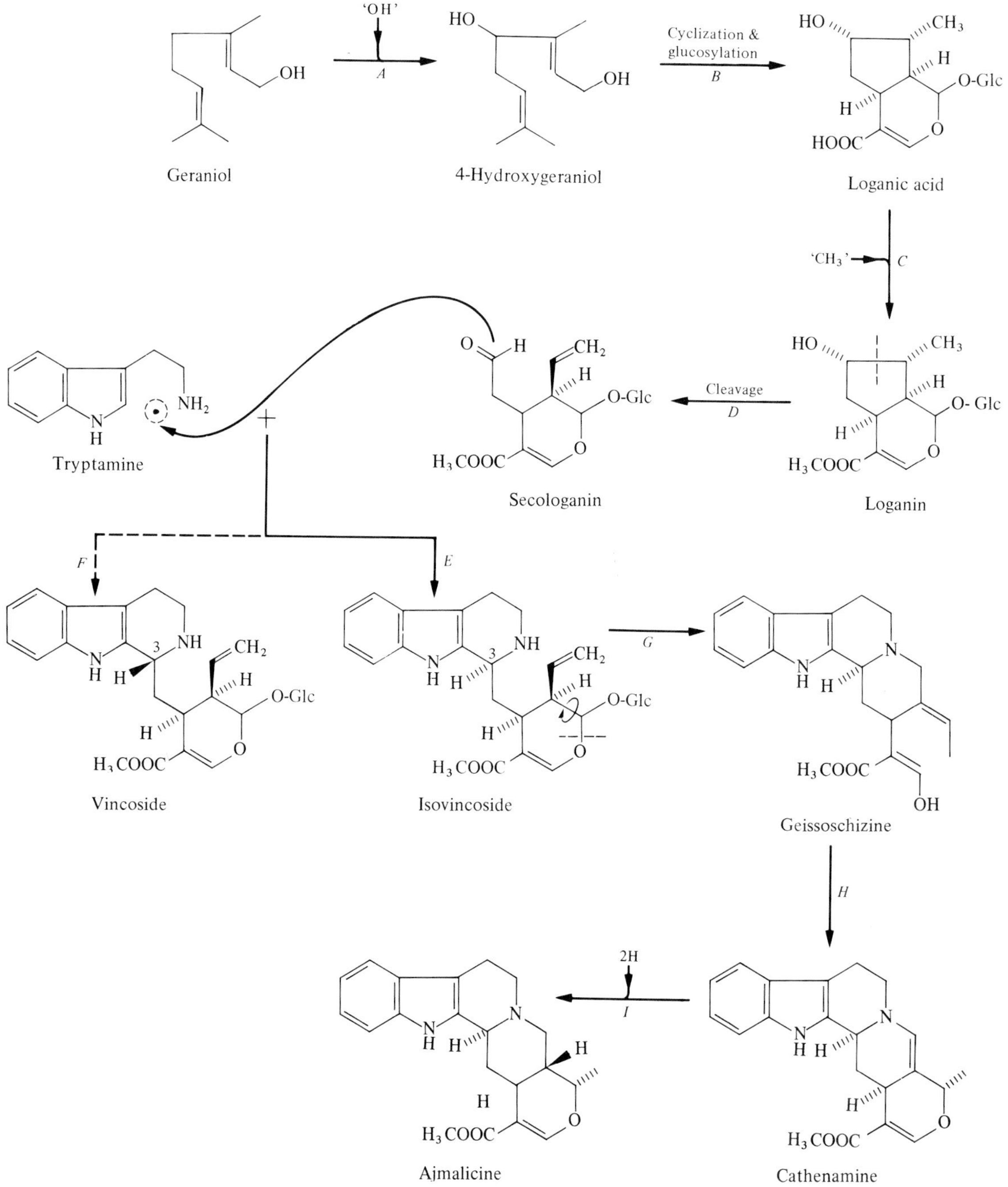

Fig. 13.22. Biosynthesis of ajmalicine (*Corynanthe* type) from one molecule of tryptamine and one of a monoterpene (secologanin).

converted into 4-hydroxygeraniol (A, Fig. 13.22) by a mixed function oxygenase present in microsomes prepared from *Catharanthus roseus* seedlings. Cyclization of 4-hydroxygeraniol yields loganic acid (B, Fig. 13.22) which is methylated by a methyltransferase found in leaves and roots of *C. roseus*, giving loganin (C, Fig. 13.22).

Loganin is then converted into secologanin by C—C bond cleavage (D, Fig. 13.22; cf. C, Fig. 13.21). The secologanin condenses with tryptamine to form isovincoside (strictosidine) (E, Fig. 13.22) although earlier experiments indicated that vincoside, which is the opposite chirality at C–3, was the product (F, Fig. 13.22). Tryptamine arises conventionally from L-tryptophan. Cell-free systems from *C. roseus* convert tryptamine and secologanin into isovincoside and some preparations will also convert isovincoside into geissoschizine (G, Fig. 13.22) and, in the absence of NADPH, to cathenamine (20,21-dehydroajmalicine) (H, Fig. 13.22); addition of NADPH to these preparations allows the reaction to continue to ajmalicine (I, Fig. 13.22). In order to carry out the reactions involving secologanin D-δ-gluconolactone must be added to the crude enzyme systems to inhibit the glucosidases present which would otherwise degrade the alkaloid and its precursors.

It seems clear from labelling experiments and from a study of the sequential appearance of alkaloids in germinating seedlings of *C. roseus* that alkaloids of the *Corynanthe* type are converted into those of the *Aspidosperma* type, which are themselves converted into alkaloids of the *Iboga* type. Figure 13.23 outlines the probable route from geissoschizine (*Corynanthe* type) via preakuammicine (*Corynanthe* type; A, Fig. 13.23) and stemmadenine (*Corynanthe* type; B, Fig. 13.23) to tabersonine (*Aspidosperma* type; C, D, Fig. 13.23) and to catharanthine (*Iboga* type; C, E, Fig. 13.23).

8. Alkaloids derived from anthranilic acid

Anthranilic acid is a key intermediate in the biosynthesis of L-tryptophan (see Chapter 9, Section E.5) but in plants it is also the precursor of a number of alkaloids *but not because of its being a precursor of L-tryptophan*. During its conversion into L-tryptophan anthranilic acid is decarboxylated; however, in the formation of alkaloids the carboxyl group is either directly involved in the formation of a *N*-heterocycle or is retained as such. In mammals anthranilic acid is also a key catabolite of L-tryptophan but this route, even if it exists in higher plants, is of no quantitative significance.

(i) PROTOALKALOIDS

Damascenine, found in *Nigella* spp. stored in the papillate cells of the seed epidermis, is formed from anthranilic acid (see Fig. 13.24). Anthranilate 3-monooxygenase[6] inserts a hydroxyl group at C–3 (A, Fig. 13.24) and then methylation of the hydroxyl, carboxyl and amino groups occurs giving rise to damascenine (B, Fig. 13.24); the sequence of methylation is not known. A key isotope experiment in the elucidation of this pathway showed that the label from [carboxy-^{14}C; ^{15}N]anthranilic acid was incorporated into damascenine.

(ii) QUINOLINES

Quinoline alkaloids such as the *Cinchona* alkaloids (e.g. quinine, Fig. 13.25) generally occur along with the *Corynanthe* alkaloids [see Section C.7(iv)] and are derived therefrom (see Fig. 13.25). These alkaloids therefore have L-tryptophan as their distal precursor. However, a large number of quinoline alkaloids present in Rutaceae are derived directly from anthranilic acid. Experiments with labelled precursors showed that anthranilic acid provided the aromatic ring and C–4 of dictamnine (see Fig. 13.26), that the carboxyl group of acetate provided C–10 and that C–4 and C–5 of mevalonic acid provided C–2 and C–3. Thus condensation of anthranilic acid with 'acetate' would yield 4-hydroxy-2-quinolone (A, Fig. 13.26) which on isoprenylation by isopentenyl pyrophosphate derived from mevalonate yields 4-hydroxy-3-prenyl-2-quinolone (B, Fig. 13.26). After methylation (C, Fig. 13.26) a cyclization reaction, involving loss of water, yields platydesmine (D, Fig. 13.26) from which the isopropyl residue is cleaved to form dictamnine (E, Fig. 13.26). Further hydroxylation and methylation in an, as yet, undetermined order yields skimmianine (F, Fig. 13.26).

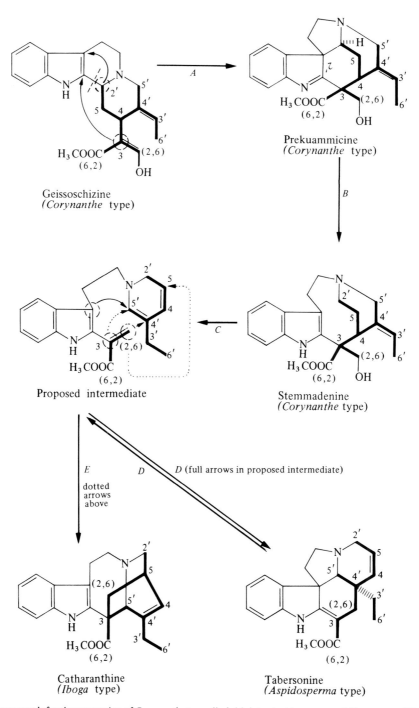

Fig. 13.23. Outline proposals for the conversion of *Corynanthe* type alkaloids into *Aspidosperma* and *Iboga* types. (The disposition of the monoterpene carbon atoms is indicated by the heavy lines; the numbering of these carbon atoms should be compared with those forecast and illustrated in Fig. 13.21.)

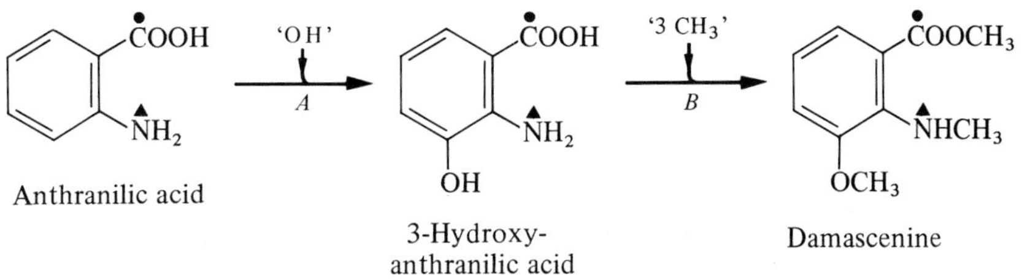

Fig. 13.24. Biosynthesis of damascenine from anthranilic acid.

(iii) QUINAZOLINES

Three main types of quinazoline alkaloids can be distinguished: (i) simple derivatives of quinazol-4-one, such as glycorine (13.23) from *Glycosmis arborea*, (ii) the pyrrolidinoquinazolines, such as peganine (13.24) (also called vasicine) from *Adhatoda vasica* and (iii) β-carbolinoquinazolones, such as evodiamine (13.25) from *Evodia rutaecarpa* (these alkaloids are confined to Rutaceae).

Biosynthetic studies have been mainly focused on peganine. Anthranilic acid is incorporated *in toto* into peganine in *A. vasica* whilst C–1, C–2 and N-11

[see (13.24)] probably arise from L-aspartic acid and C–3 and C–10 from an as yet unspecified C_2 unit.

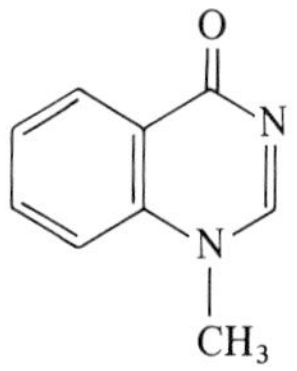

(13.23)

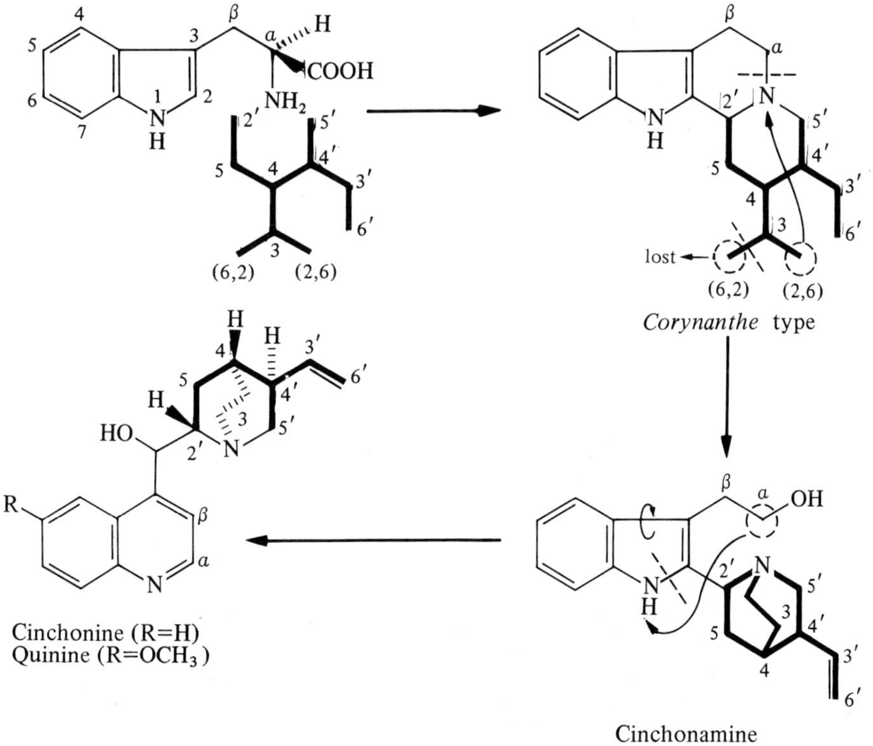

Cinchonine (R=H)
Quinine (R=OCH$_3$)

Corynanthe type

Cinchonamine

Fig. 13.25. The formation of *Cinchona* alkaloids by rearrangement of a *Corynanthe* type alkaloid. (The disposition of the monoterpene carbon atoms is indicated by the heavy lines and their numbering relates back to that given in Fig. 13.21.)

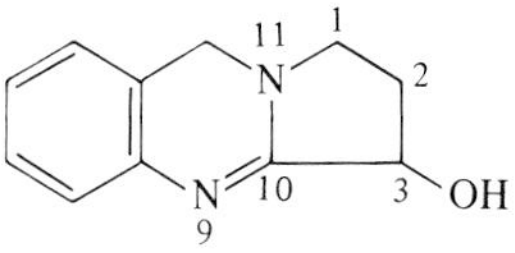

(13.24)

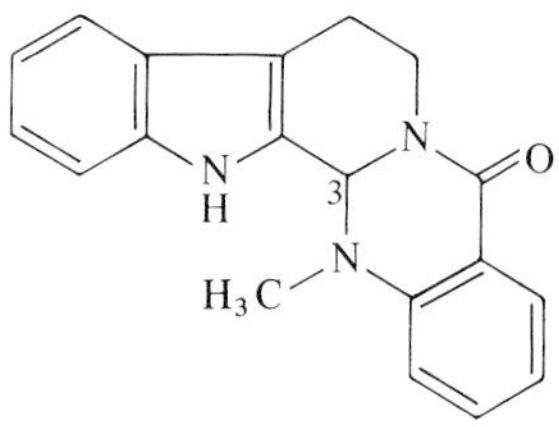

(13.25)

Although we have placed evodiamine (13.25) in this section it could equally well have been placed in Section 7(iii) because the tryptamine residue arises from L-tryptophan. Rather unexpectedly C–3 arises from the methyl of methionine whilst the N-methyl group arises, as expected, from the same source. The remainder of the molecule arises from anthranilic acid.

9. Alkaloids derived from L-histidine

Histidine (see Fig. 13.27) is the precursor of ergothioneine in a number of microorganisms including *Neurospora crassa*. The evidence for the pathway outlined in Fig. 13.27 centres around the observation that $[^{14}C$-N-methyl,2-$^{14}C]$hercymine

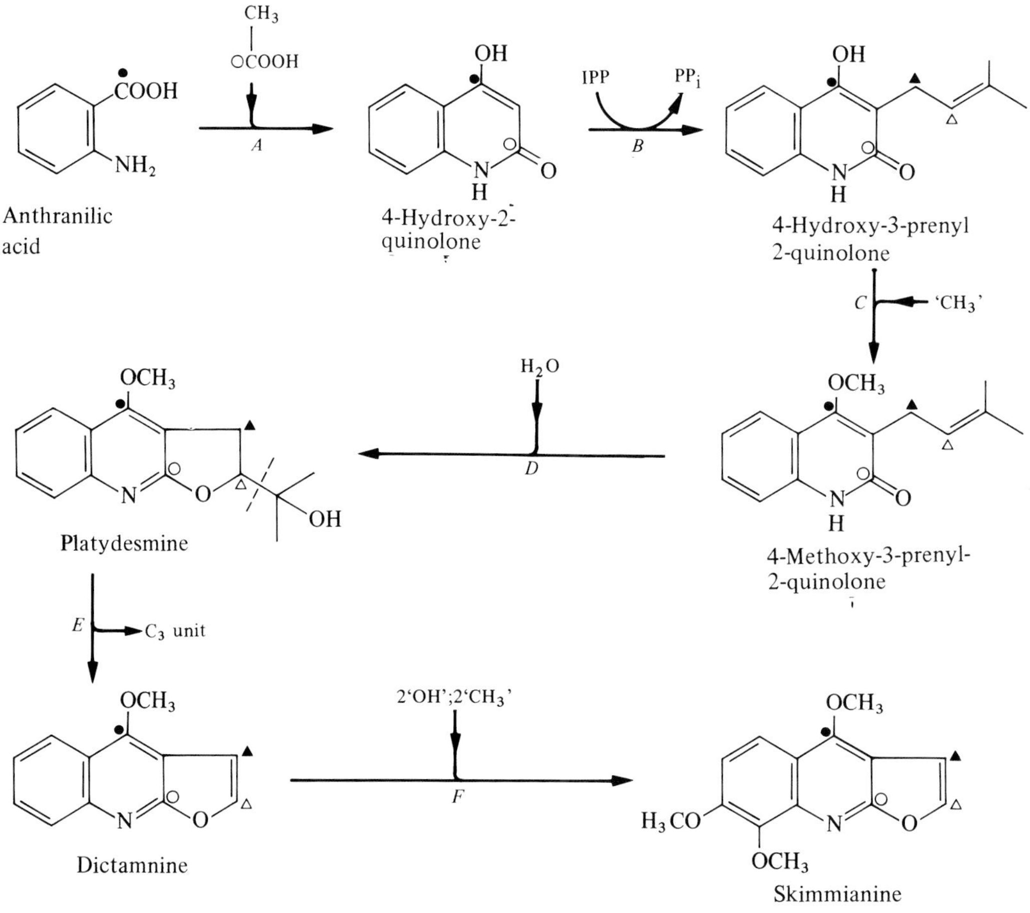

Fig. 13.26. Formation of quinoline alkaloids from anthranilic acid, acetate and isopentenyl pyrophosphate (IPP). ($\bullet$ = label from carboxyl carbon of anthranilic acid; $\bigcirc$ = label from carboxyl carbon of acetate; $\blacktriangle$ = label from C–1 of IPP $\equiv$ C–5 of mevalonic acid; $\triangle$ = label from C–2 of IPP $\equiv$ C–4 of mevalonic acid.)

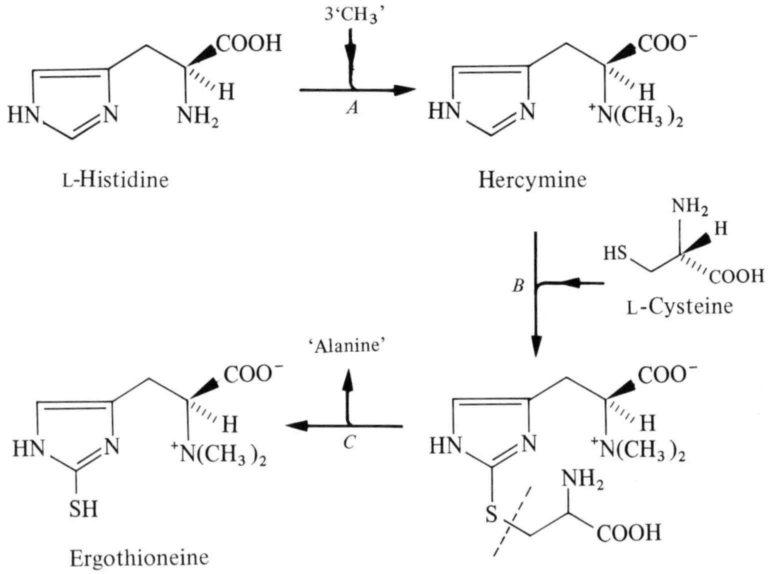

Fig. 13.27. Biosynthesis of ergothioneine from L-histidine in *Neurospora crassa*.

is incorporated into ergothioneine such that the labelled atoms occupy the corresponding positions. This demonstrates that methylation (*A*, Fig. 13.27) precedes thiolation (*B*, *C*, Fig. 13.27) which is presumed to involve the addition of cysteine followed by the loss of its alanine residue.

D. ISOPRENOID ALKALOIDS (PSEUDOALKALOIDS)

1. General aspects

In the previous sections we have dealt with alkaloids in which virtually the entire skeleton of an amino acid is incorporated into the alkaloid molecule; now we shall consider the pseudoalkaloids in which nitrogen is inserted into a non-amino acid residue, which is, in general, isoprenoid in character. The importance of hemi- and monoterpene residues in the biosynthesis of true alkaloids has already been stressed [Sections C.7(iv) and C.8(ii)]. The isoprenoid residues involved in the formation of pseudoalkaloids are mono-, sesqui-, di- or triterpenoid in character. So far no pseudoalkaloids have been reported which, apart from the nitrogen, arise from hemi-, sester-, tetra- or polyterpenoids.

2. Nature and distribution

(i) MONOTERPENOID ALKALOIDS

The structure of all monoterpenoid alkaloids is based on that of actinidine (Table 13.2) although the pyridine ring can be replaced by the piperidine ring as in skytanthine (13.26) from *Skytanthus acutus*.

(ii) SESQUITERPENOID ALKALOIDS

Four main groups of sesquiterpenoid alkaloids have been described. These are based, not upon chemical structure, but upon the genera of the plants from which they were first isolated (Table 13.4). In addition to these, waterlilies of the *Nuphar* genus synthesize unique alkaloids which contain sulphur as well as nitrogen. It will be realized that neothiobinupharidine (13.27), one of

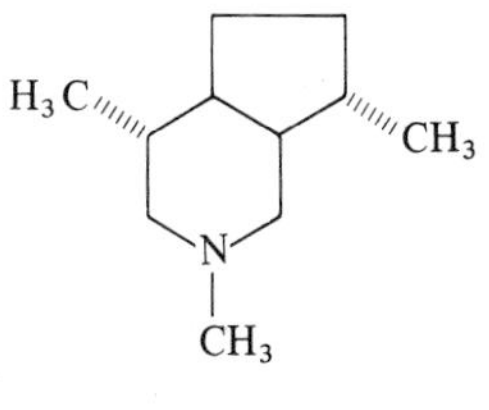

(13.26)

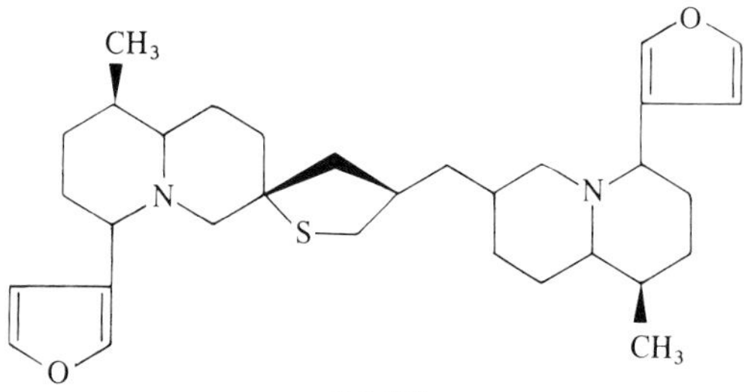

(13.27)

these alkaloids, is a dimer of two deoxynupharidine molecules (see Table 13.4) joined by a sulphur bridge.

(iii) DITERPENOID ALKALOIDS

Diterpenoid alkaloids can be divided into two groups: (i) the aconitines from *Delphinium* and *Aconitum* spp., e.g. aconitine (13.28), and (ii) the atisines, e.g. veatchine (13.29) which is found in *Garrya* spp. The derivation of the latter from kaurene

[see Chapter 15, Section C.2(i)] is clear. However, hydrolysis of aconitine yields a C–19 alkamine indicating that one carbon atom of the diterpene precursor has been lost in forming the aconitine skeleton. It is interesting to note that although aconitine is extremely toxic to man, the hydrolysis product is relatively non-toxic.

(iv) TRITERPENOID ALKALOIDS

These alkaloids are not very widespread but a number have been obtained from *Daphniphyllum macropodum*. The two major types are the C_{32} compounds based on daphniphylline (13.30) and the C_{27} compounds based on yuzurimine (13.31). A structurally distinct group has been isolated from *Buxus* spp. Alkaloids of this group are structurally closely related to cycloartenol, the first cyclic triterpenoid precursor of the plant sterols [see Chapter 11, Section C.6(ii)]. A typical example is cycloprotobuxine A (see Table 13.2). Another characteristic structure on which triterpenoid al-

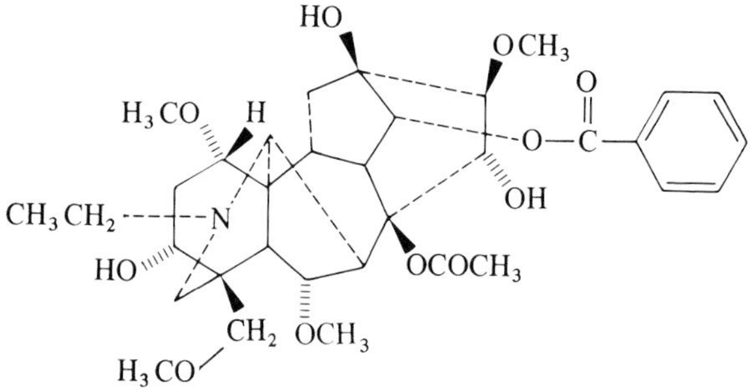

(13.28)

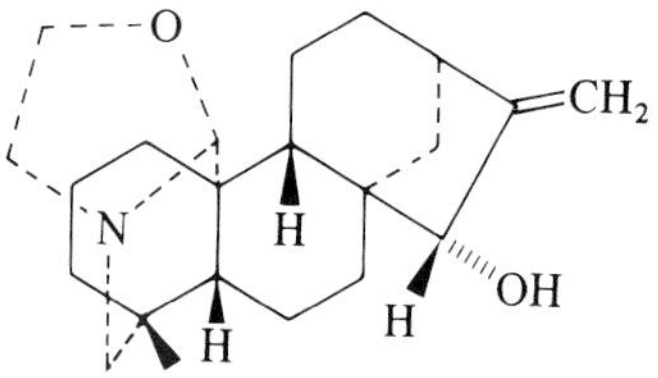

(13.29)

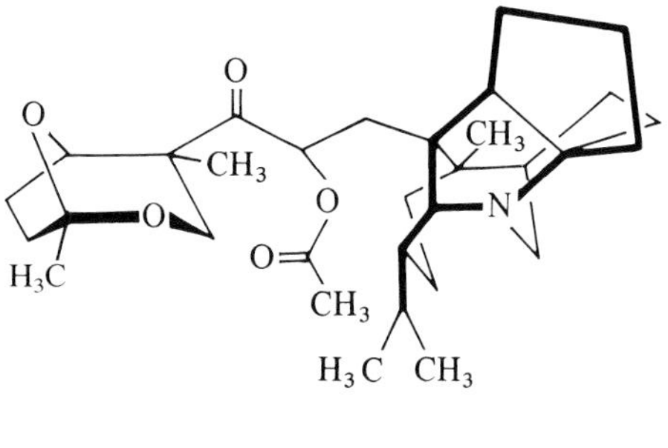

(13.30)

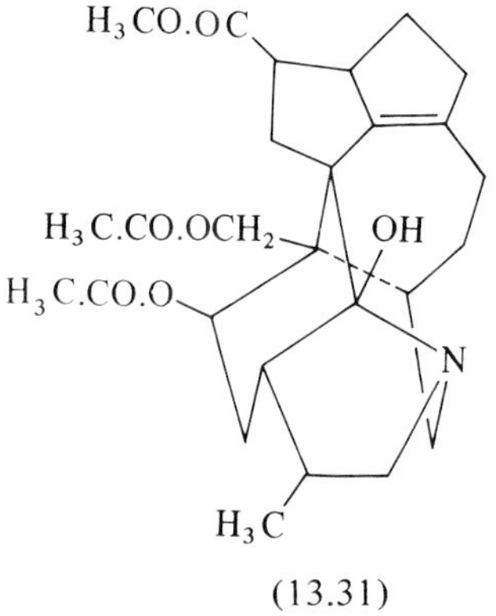

(13.31)

Table 13.4 *Classification of sesquiterpene alkaloids according to their distribution in the plant kingdom*

Plant source	Name of example	Structure
Nuphar luteum (water lily)	Deoxynupharidine	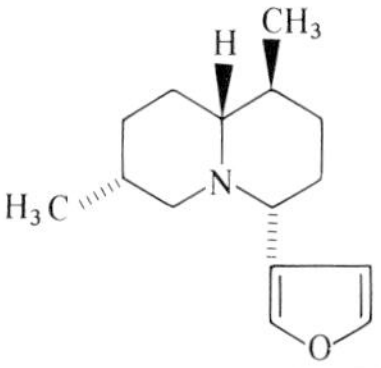
Dendrobium nobile (orchid)	Dendrobine	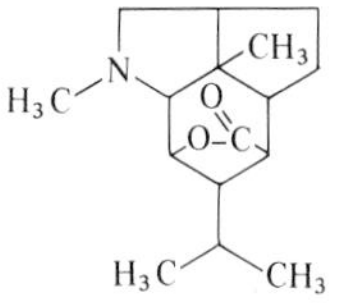
Pogostemon patchouli	Patchoulipyridine	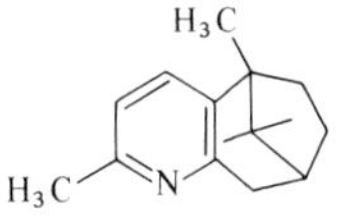
Fabiana imbricata	Fabianine	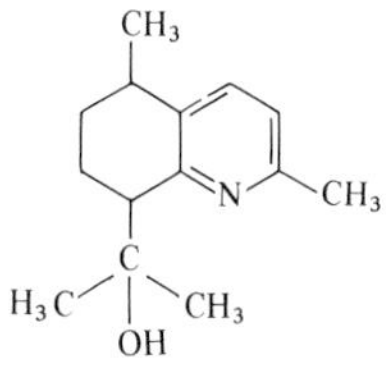

kaloids are based is one in which ring B has been expanded to include the cyclopropane ring of the basic cycloartenol skeleton. This is exemplified in buxamine-E (13.32). It will be noted that in this compound and in cycloprotobuxine A the side chain has been reduced from the eight carbon atoms of the parent compound to two.

(v) STEROIDAL ALKALOIDS

(a) *Cholestane* (C_{27}) *derivatives*

Alkaloids derived from the cholestane skeleton are found mainly in the Solanaceae and for this reason are often called Solanum alkaloids. They occur naturally as glycosides and can be divided into two major groups in which the basic structures are spirosolane and solanidine. An important aglycone in the spirosolane group is solasodine (13.33) and a naturally occurring glycoside of this compound is solasonine (13.34); another example is tomatidine (13.35) and its glycoside α-tomatine (13.36). The important stereochemical difference at C–22 between solasodine and tomatidine must be emphasized. The aglycone solanidine (13.37) is found glycosylated with the same sugar residues as solasodine: an example is α-solanine (13.38). A structural variant is demissine (13.39) which has a saturated ring system.

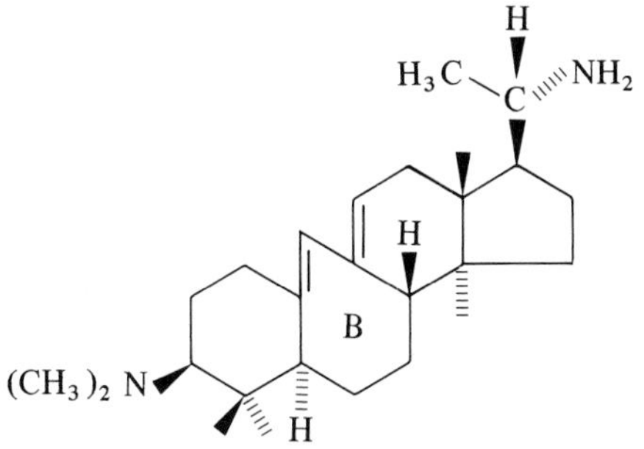

(13.32)

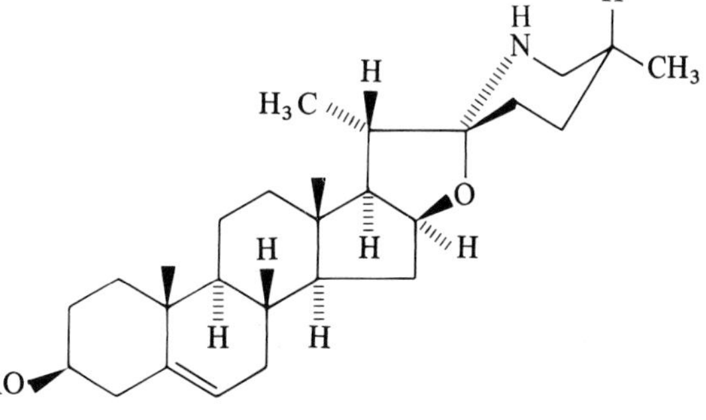

(13.33) (R = H)
(13.34) (R = Glc(β1 → 3)Gal(β1 →)
2
↑
β1
L-Rha

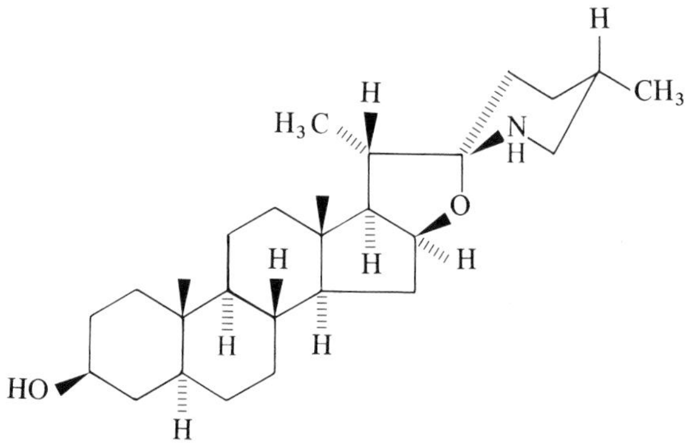

(13.35)

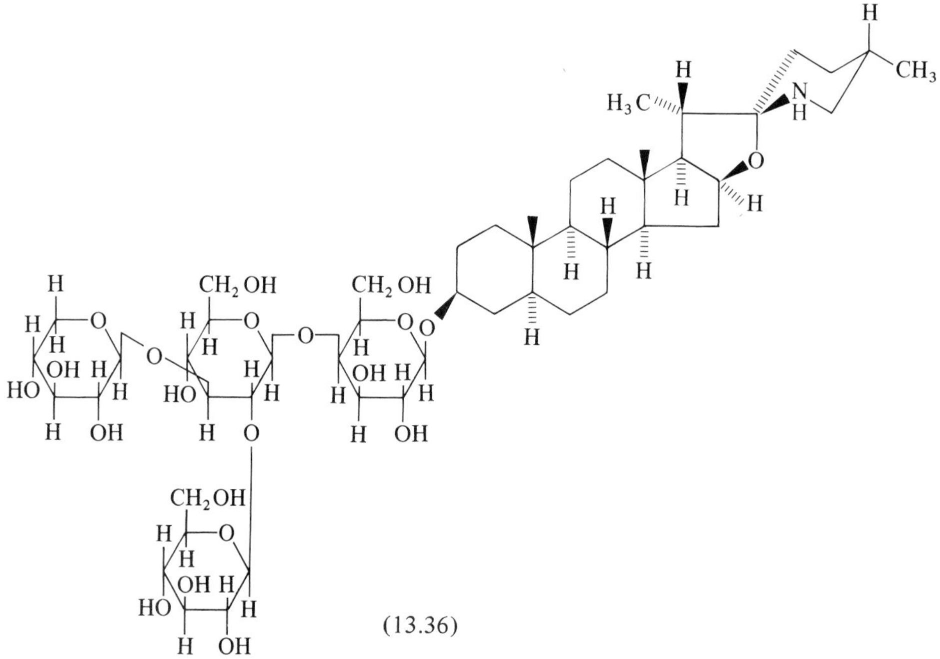

(13.36)

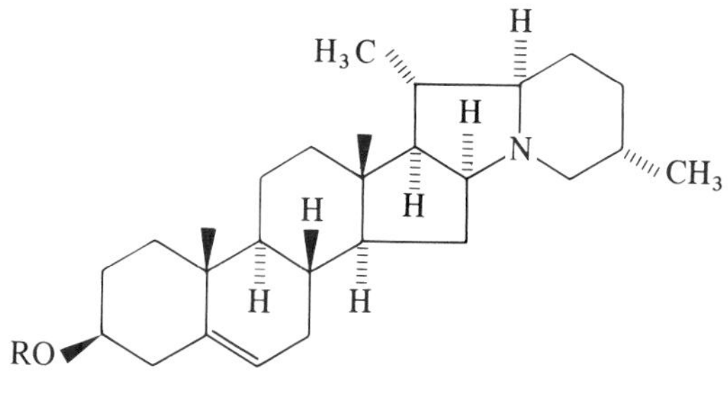

(13.37) R = H
(13.38) R = Rha-Glc-Gal

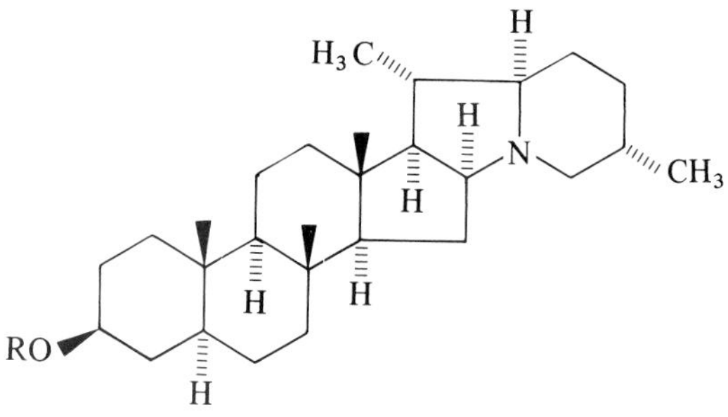

(13.39) R = Xyl-2xGlc-Gal

Compounds with no ring E, such as tomatillidine (13.40), have also been found in the Solanaceae; they probably represent biosynthetic intermediates [see Section D.3(iii)].

These alkaloids probably accumulate in the vacuole of the cell and, in the case of α-tomatine at least, are synthesized in both root and shoot, with greatest synthesis taking place in the meristematic tissues of these organs.

(b) *Pregnane (C$_{21}$) derivatives*

Three main types of C$_{21}$ pseudoalkaloids can be discerned which are derived structurally from the pregnane skeleton. One group [e.g. holaphyllamine (13.41)] carries an amino group at C–3. A second group has the amino group in the two carbon side chain as in funtuphyllamine A (13.42). In the third group the nitrogen is incorporated into a pyrolidine ring as in conessine (13.43). These alkaloids tend to be confined to the Apocynaceae.

(c) *C-nor-D-homosteroidal alkaloids*

These pseudoalkaloids, in which the angular methyl group on the C-ring (i.e. C–18) has disappeared and the D-ring has expanded by one carbon atom, are confined to the Liliaceae, especially the *Veratrum* spp., hence their general name *Veratrum alkaloids*. Two main groups can be distinguished on the basis of their characteristic oxygenation patterns. Examples are jervine (13.44) and germine (13.45).

3. Biosynthesis

(i) GENERAL ASPECTS
Not a great deal of work has been reported on the biosynthesis of the terpenoid alkaloids, but two general points can be made here to prevent constant repetition in the later stages of Section D.3: (i) these compounds are mevalonic acid-derived and (ii) when *O*- and *N*-methyl groups occur their origin is the conventional one of *S*-adenosylmethionine. If very little more than this has been reported for certain alkaloid groups, they are not specifically discussed in this section. A remarkable ignorance of the source of the nitrogen in terpenoid alkaloids is to be observed.

(ii) TRITERPENOID ALKALOIDS
[^{14}C]Squalene is incorporated into daphniphylline (13.30) in *Daphniphyllum macropodum*. Experiments with [2-^{14}C,4R-^{3}H$_1$]mevalonic acid in *Buxus sempervirens* yielded a ^{3}H:^{14}C atomic ratio of 3:4 in cyclovirobuxine D (13.46) and a ratio of 3:3 in cyclobuxine D (13.47). This indicates that the location of the labelled atoms is as indicated, on the assumption that cycloartenol is the precursor of these compounds and that it is formed in the conventional way (see Chapter 11, Section C.6(ii)). These results indicate that (i) amination at C–3 and C–20 involves a keto intermediate because tritium would normally be found at these carbons in cycloartenol and (ii) the unique methylene group at C–4 of cyclobuxine D is derived from the 4β-methyl of cycloartenol rather than the 4α-methyl since the former, like the C–4 methylene group, is not labelled with ^{14}C whilst the latter would have been.

(iii) STEROIDAL ALKALOIDS
Cycloartenol is converted into tomatidine (13.35) probably via cholesterol. There is a suggestion that the nitrogen of solanidine in *Veratrum grandiflorum* is derived from L-arginine and that glycosylation of the aglycones occurs by steps involving UDP-glycosyltransferase enzymes.

Pregnenolone (13.48) and cholesterol, but not, as might be expected, progesterone (13.49), are precursors of holaphyllamine (13.41).

E. BIOLOGICAL FUNCTION

The widespread distribution of alkaloids in all parts of plants has stimulated searches for a function of these compounds in the general metabolism of plants. Many suggestions have been made, but apart from the ecological observation that they offer protection against predators, little compelling evidence has been collected for any unequivocal biochemical or physiological function. Amongst the less bizarre proposals the following can be noted:

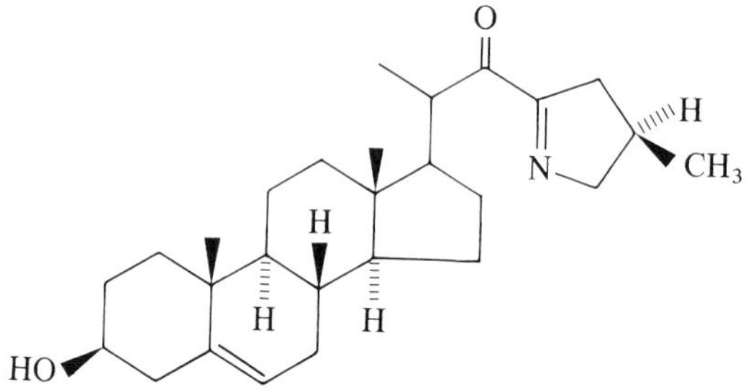

(13.40)

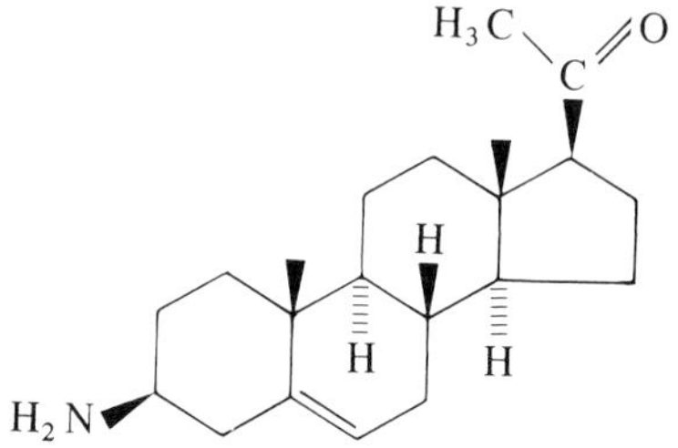

(13.41)

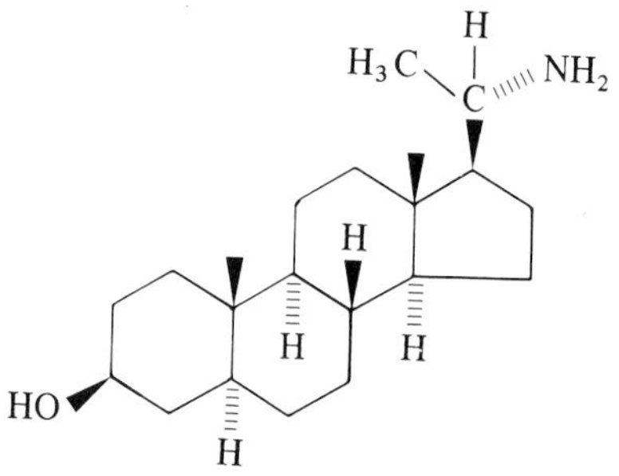

(13.42)

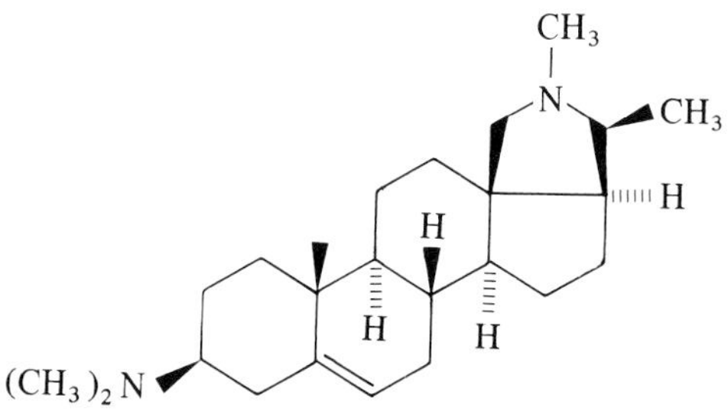

(13.43)

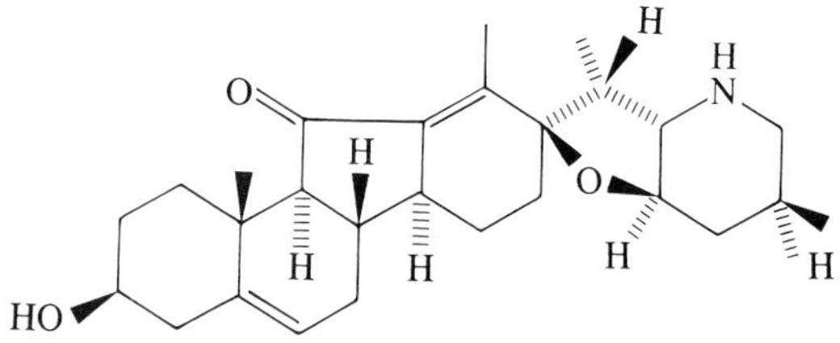

(13.44)

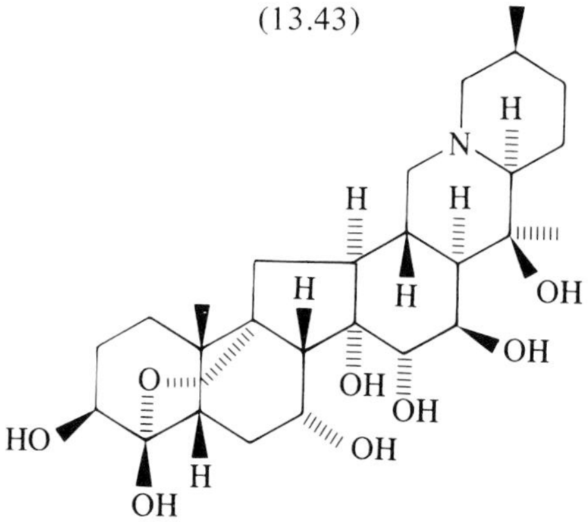

(13.45)

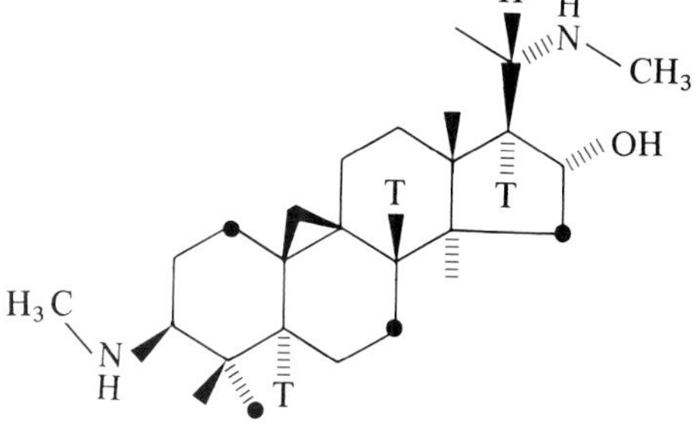

(13.46) (● = ^{14}C, T = tritium
from [2-^{14}C,4R-^{3}H$_1$] MVA)

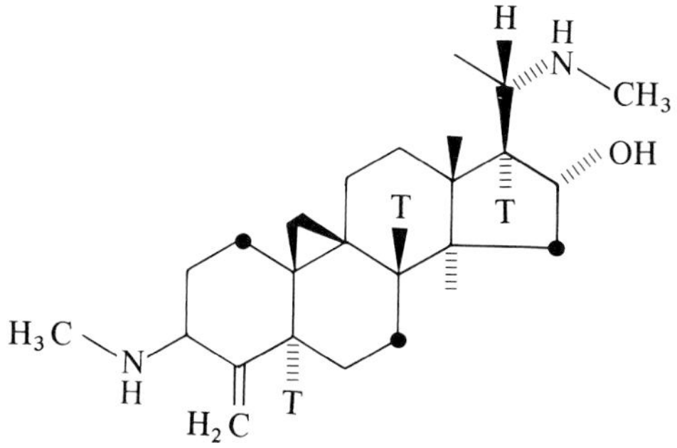

(13.47) (● = ^{14}C, T = tritium
from [2-^{14}C,4R-^{3}H$_1$] MVA)

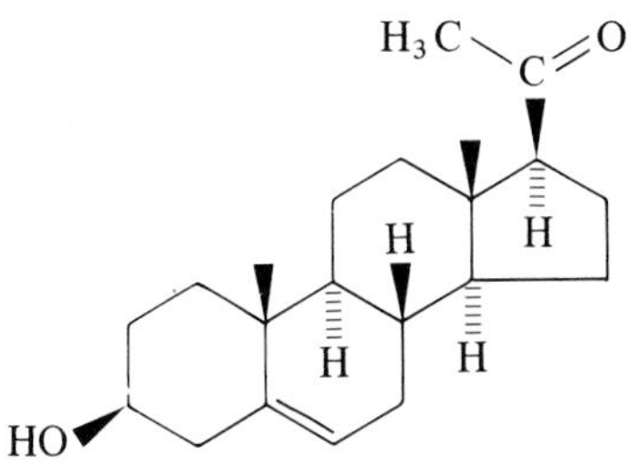

(13.48)

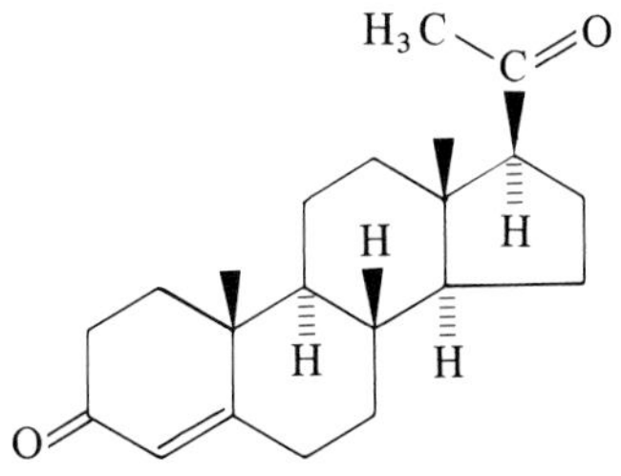

(13.49)

(i) they serve as *N*-excretory products in the same way as urea and uric acid do in animals;

(ii) they act as a nitrogen reserve; but there is little evidence that they are utilized under conditions of nitrogen deficiency;

(iii) they act as growth regulators, in particular as germination inhibitors;

(iv) they help to maintain ionic balance by virtue of their chelating power.

Alkaloids have many profound effects on animals and animal tissues and are used and abused as neurotrophic agents. This is not the place to discuss this aspect of alkaloids and the reader is referred to pharmacological texts for further information.

SUGGESTIONS FOR FURTHER READING

General texts

Manske, R. H. F. and Holmes, H. L. (eds.) (1950–54) *The Alkaloids*, Vols. I–IV. Academic Press, London.
Manske, R. H. F. (ed.) (1955–77) *The Alkaloids*, Vols. V–XVII. Academic Press, London.

Specific reviews

Leete, E. (1980) 'Alkaloids derived from ornithine, lysine and nicotinic acid' in *Encyclopaedia of Plant Physiology* (Bell, E. A. and Charlwood, B. V., eds.), Vol. 8, pp. 65–91. Springer-Verlag, Heidelberg.
Fodor, G. B. (1980) 'Alkaloids derived from phenylalanine and tyrosine', *ibid.*, pp. 92–127.
Gröger, D. (1980) 'Alkaloids derived from tryptophan and anthranilic acid', *ibid.*, pp. 128–159.
Fodor, G. B. (1980) 'Alkaloids derived from histidine and other precursors', *ibid.*, pp. 160–166.
Roddick, J. G. (1980) 'Isoprenoid alkaloids', *ibid.*, pp. 167–184.
Mabry, T. J. (1980) 'Betalains', *ibid.*, pp. 513–534.
Robinson, T. (1981) *The Biochemistry of Alkaloids*, 2nd edition. Springer-Verlag, Heidelberg.

ENZYMES

1. L-Ornithine carboxy-lyase, EC 4.1.1.17.
2. *S*-Adenosyl-L-methionine: putrescine *N*-methyltransferase, EC 2.1.1.53.
3. Amine: oxygen oxidoreductase (deaminating) (copper-containing), EC 1.4.3.6.
4. L-Lysine carboxy-lyase, EC 4.1.1.18.
5. Aromatic-L-amino acid carboxy-lyase, EC 4.1.1.28.
6. Anthranilate, tetrahydropteridine: oxygen oxidoreductase (3-hydroxylating), EC 1.14.16.3.

Plant Phenolics

A. INTRODUCTION

Plants produce many thousands of compounds which contain one or more phenolic residues. These compounds can be divided into major groups according to the number of carbon atoms in their skeleton (Table 14.1). The first impression of an overwhelming array of structures becomes less bewildering when it is realized that all except flavonoids arise from a common biosynthetic intermediate, phenylalanine or its close precursor shikimic acid. In the case of the flavonoids one aromatic ring and its C_3 side chain arises from phenylalanine whilst the other arises from acetyl-CoA via the polyketide pathway [see Section I.3(i)]. An outline of the synthesis of the plant phenolics is given in Fig. 14.1, and the details will be painted in later.

Table 14.1. *The major classes of plant phenolics*

No. of C atoms	Basic skeleton	Class	Example	Source
6	C_6	Phenols	Catechol	*Gaultheria* leaves
7	C_6-C_1	Phenolic acids	*p*-Hydroxybenzoic acid	Widespread
8	C_6-C_2	Phenylacetic acids	2-Hydroxyphenylacetic acid	*Astilbe* leaves
		Hydroxycinnamic acids	Caffeic acid	Ubiquitous
		Phenylpropenes	Myristicin	*Myristica fragrans* (nutmeg)
9	C_6-C_3	Coumarins	6,7-Dimethoxycoumarin	*Dendrobium densiflorum* (orchid)
		Isocoumarins	Hydrangenol	*Hydrangea macrophylla*

Table 14.1 (*cont.*)

No. of C atoms	Basic skeleton	Class	Example	Source
9 (*cont.*)	C_6-C_3	Chromones	Eugenin	*Eugenia aromatica*
10	C_6-C_4	Naphthoquinones	Juglone	*Juglans nigra* (walnuts)
13	$C_6-C_1-C_6$	Xanthones	Mangiferin (a *C*-glucoside)	Widespread
14	$C_6-C_2-C_6$	Stilbenes	Lunularic acid	Liverworts (e.g. *Lunularia cruciata*)
		Anthraquinones	Emodin	Rhubarb
15	$C_6-C_3-C_6$	Flavonoids	(see Section I and Table 14.3 for details)	
		Lignans	Pinoresinol	Conifers
18	$[C_6-C_3]_2$			

Table 14.1 (*cont.*)

No. of C atoms	Basic skeleton	Class	Example	Source
18 (*cont.*)	$[C_6-C_3]_2$	Neolignans	Eusiderin	Heartwood of Magnoliaceae, *Eusideroxylon zwageri*
30	$[C_6-C_3-C_6]_2$	Biflavonoids	Amentoflavone (= 3′,8″-biapegenin)	Gymnosperms
	$[C_6-C_3]_n$	Lignins	(see Chapter 4, Sections B.2 and D.2 and Fig. 4.14)	Cell walls
n	$[C_6]_n$	Melanins	(see Section M for details)	
	$[C_6-C_3-C_6]_n$	Condensed tannins (flavolans)	(see Section L.1(ii) for details)	

B. PHENOLS

1. Nature and distribution

Simple phenols such as catechol (Table 14.1) are not widely distributed, the most common being hydroquinone (14.1). Derivatives are occasionally encountered and interesting examples are urushiol (14.2) (catechol with a C_{15} side chain), which is the toxic principle of poison ivy (*Toxicodendron radicans*) and Δ^1-tetrahydrocannabinol (also called Δ^9-tetrahydrocannabinol) (14.3), a resorcinol derivative which, as its name implies, is the hallucino-

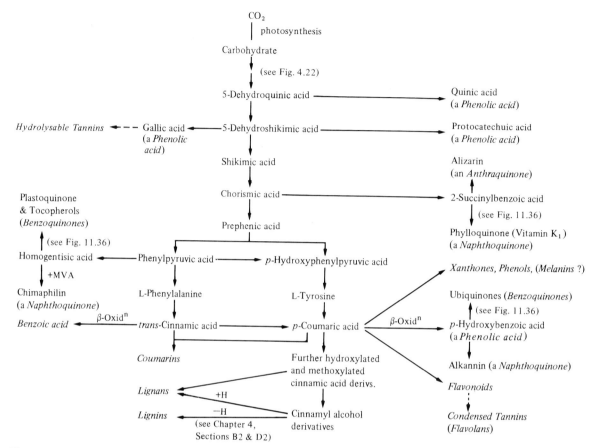

Fig. 14.1. An outline of the biosynthesis of plant phenolics showing the involvement of the shikimic acid pathway. (N.B. Plastoquinone, phylloquinone, ubiquinones and tocopherols are usually classed as terpenoid quinones or chromanols (see Chapter 11, Section C.9) rather than plant phenolics but are included in this figure for comparison purposes because other quinones are regarded as plant phenolics.)

genic principle of *Cannabis sativa*. Plant melanins (see Section M) are probably catechol polymers.

2. Biosynthesis

Many labelling experiments have confirmed the view that phenols arise from phenylalanine via *p*-coumaric acid [the formation of *trans-p*-coumaric acid has already been discussed in Chapter 4 (see Fig. 4.23)] which is converted into a benzoic acid (see Section C.2) which is oxidatively decarboxylated to yield a phenol.

C. PHENOLIC ACIDS

1. Nature and distribution

Phenolic acids are widespread throughout the plant kingdom and *p*-hydroxybenzoic acid (Table 14.1), protocatechuic acid (14.4), vanillic acid (14.5) and syringic acid (14.6) are universally present in the angiosperms so far examined. Gallic acid (14.7) is usually present in polymerized forms as soluble tannins (see Section L). Aldehydes and alcohols corresponding to the parent carboxylic acids are also known; important examples are vanillin (14.8)

Structures (14.4), (14.5), (14.6), (14.7), (14.8), (14.9)

from the *Vanilla* pod and salicyl alcohol (14.9) from the willow. The carboxyl and hydroxyl groups may be methylated as in the acetophenone 3-acetyl-6-methoxybenzaldehyde (14.10) from the desert shrub *Encelia farinosa*.

2. Biosynthesis

Benzoic acid is formed from *trans*-cinnamoyl-CoA by β-oxidation followed by oxidative decarboxylation, a reaction analogous to that described for *p*-hydroxybenzoic acid formation from *p*-coumaric acid. Acids with additional hydroxyl groups are formed not by hydroxylation of benzoic or *p*-hydroxybenzoic acid but by a pathway which branches from the main shikimate pathway at 5-dehydroshikimate (Fig. 14.2). The reactions in Fig. 14.2 have been demonstrated in cell-free systems obtained from cell-suspension cultures of mung bean by using [^{14}C]shikimate in the presence of $NADP^+$. The oxidized nucleotide serves two purposes; it is the co-factor of shikimate dehydrogenase[1] which catalyses the reversible reaction (*A*) (Fig. 14.2) and is also required for the conversion of the enol form of 5-dehydroshikimate (*B*) into gallic acid (*C*). The enzyme dehydroshikimate dehydratase converts the enol into protocatechuic acid (*D*).

D. PHENYLACETIC ACIDS

Little is known about the distribution of phenylacetic acids in plants but both 2- and 4-hydroxyphenylacetic acids [Table 14.1 and formula (14.11) respectively] have been found in *Astilbe* leaves and in dandelion (*Taraxacum officinalis*), respectively.

E. HYDROXYCINNAMIC ACIDS, PHENYLPROPENES, COUMARINS AND CHROMONES

1. Nature and distribution

Hydroxycinnamic acids, such as *p*-coumaric acid (14.12) and caffeic acid (Table 14.1) and their methylated derivatives such as ferulic acid (14.13) and sinapic acid (14.14), are apparently universally present in higher plants. They occur free and in a very large range of esterified forms. A list of the main groups esterified with hydroxycinnamic acids is given in Table 14.2. It should be noted that in sugar derivatives the sugar is attached by an ester and not a glycosidic linkage. The range of esters of the hydroxycinnamic acids is far greater than that of any other plant phenols.

Structures (14.10), (14.11), (14.12), (14.13), (14.14)

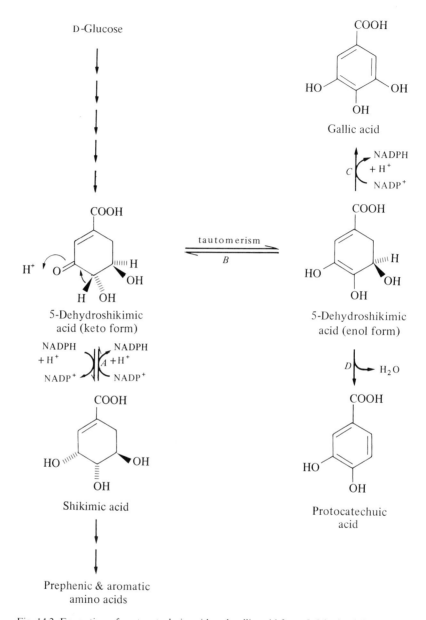

Fig. 14.2. Formation of protocatechuic acid and gallic acid from 5-dehydroshikimic acid.

Phenylpropenes are not widely distributed but occur sporadically in essential oils. For example, myristicin (Table 14.1), which occurs in nutmeg, is also present in essential oils. It is said to have hallucinogenic properties.

Coumarins are lactones which can be considered to be derived formally from *o*-hydroxycinnamic acid by ring closure between the *ortho* hydroxyl group and the carboxylic group of the side chain, after *trans→cis* isomerization of the side chain (the naturally occurring cinnamic acids have the *trans* configuration). Coumarin itself (14.15) probably does not exist free in fresh leaves. In damaged tissues, however, it can be easily formed from *trans-O*-glucosyloxycinnamic acid (14.16), a constant component of fresh leaves. The damage allows

Table 14.2. *Some compounds whose hydroxyl groups may be esterified with the carboxyl group of hydroxycinnamic acids*

Class of compound	Hydroxy compound forming ester	Example	
		Structure	Name
Cyclohexane carboxylic acid	Quinic acid		Chlorogenic acid
Sugar	D-glucose		Caffeyl-β-D-glucose
Amine	Tryptamine		N-Ferulyltryptamine
Phenolic acid	3,4-Dihydroxy-phenyllactic acid		Rosmarinic acid
Terpenoid	Catalpol		10-(4-Hydroxy-cinnamoyl)-catalpol (Scutellarioside II)

* Caff = caffeic acid (see Fig. 14.1); Fer = ferulic acid (14.13); Coum = *p*-coumaric acid (14.12).

access to enzymes which remove glucose and cause *trans*→*cis* isomerization followed by ring closure. These reactions with the resulting formation of the volatile coumarin (14.15) are the cause of the smell of new mown hay. Dicoumarol (14.17), the haemorrhagic factor in spoiled sweet clover, is formed from coumarin by microbial action during spoilage.

Hydroxycoumarins, formally derived from *p*-coumaric acid (14.12), caffeic acid (Table 14.1) and ferulic acid (14.13), are widespread. Examples are umbelliferone (14.18), aesculetin (14.19) and scopoletin (14.20). These compounds are characterized by their intense fluorescence in solution which can be used as the basis of their quantitative assay. Coumarins with a furan ring (the furocoumarins) are occasionally found, as are isoprenylated

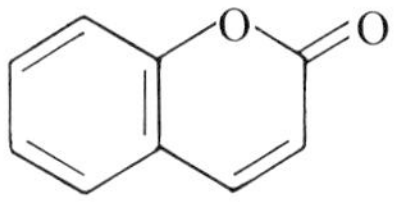

(14.15) 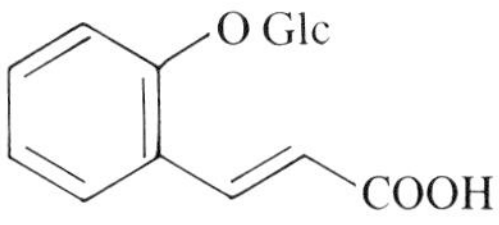   (14.16) 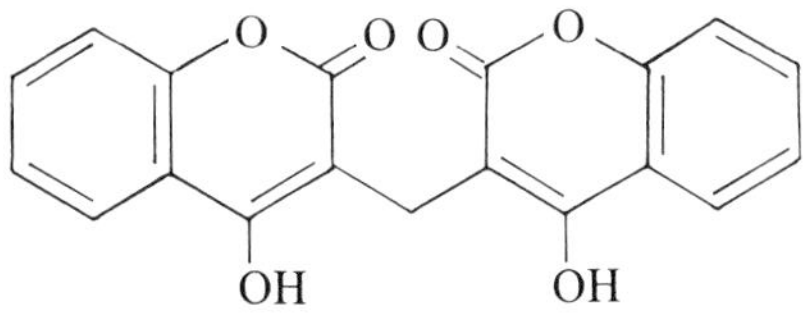   (14.17)

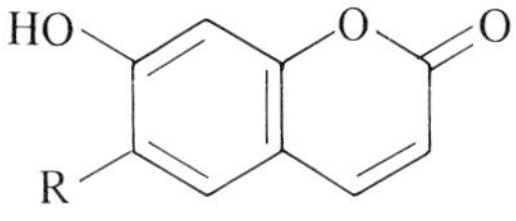

R = H (14.18)   (14.21)   (14.22)
R = OH (14.19)
R = OCH₃ (14.20)

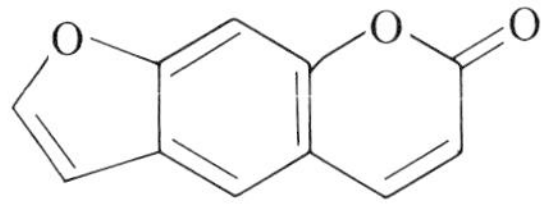

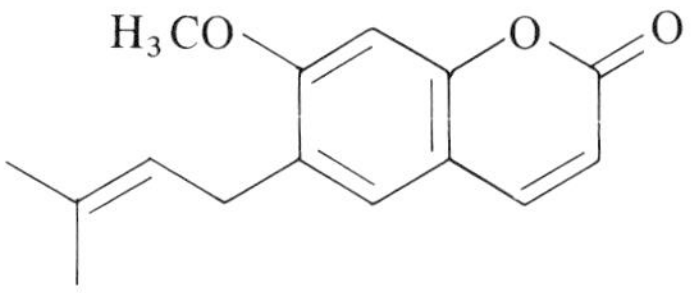

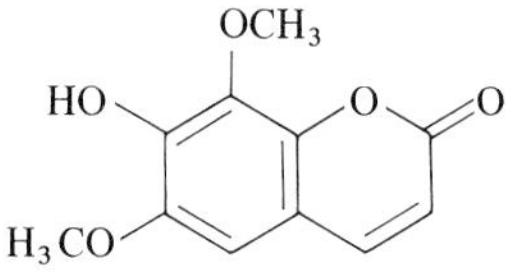

(14.23)

2. Biosynthesis

The key step in the biosynthesis of coumarins is the *ortho* (C–2) hydroxylation of a cinnamic acid which is necessary for later lactonization. The enzyme responsible, which is membrane-bound, has been obtained from chloroplasts. The mechanism involved is not yet known. Tracer experiments have clearly shown that *p*-coumaric acid is a precursor of umbelliferone (*A*, Fig. 14.3) and demonstrate that in this case at least, *p*-hydroxylation occurs before *o*-hydroxylation. However, the conversion of umbelliferone into aesculetin (*B*, Fig. 14.3) shows that hydroxylation can also occur after lactonization. Glycolysation (*C*, Fig. 14.3) is the final step in the pathway resulting in the production of cichoriin (aesculetin 7-*β*-D-glucoside).

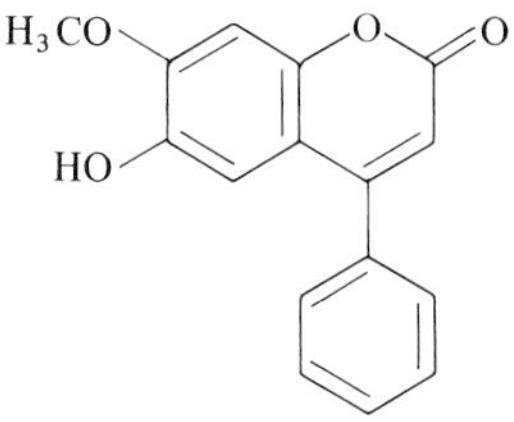

(14.24)

coumarins. Examples are psoralen (14.21) (furo-coumarin) from *Psoralea corylifolia* and suberosin (14.22) (6-dimethylallyl-7-methoxycoumarin) from the bark of *Xanthoxylum suberosum*. In contrast derivates of sinapic acid (14.14) are rare; one example is isofraxidin (14.23) which occurs only in the bark of *Fraxinus*. The so-called neoflavonoids are 4-phenylcoumarins and their existence is probably restricted to the Leguminosae and Guttiferae. Dalbergin (14.24) is a typical example.

The chromones, which are isomeric with coumarins (keto groups at C–4 and C–2, respectively) are not widely distributed but eugenin (Table 14.1) is a well-established natural product.

Another group of compounds isomeric with the coumarins are the isocoumarins. They are not widely distributed; for example, hydrangenol (Table 14.1) has only been found so far in *Hydrangea macrophylla*. The other isocoumarins also tend to contain additional phenyl residues.

F. QUINONES

1. Nature and distribution

Benzoquinone is not known as a plant product but in its reduced form, hydroquinone (14.1), it is widely distributed. Benzoquinone derivates such as plastoquinone (11.47) and ubiquinone (11.48) have important functional roles in photosynthesis and mitochondrial electron transport, respectively. One

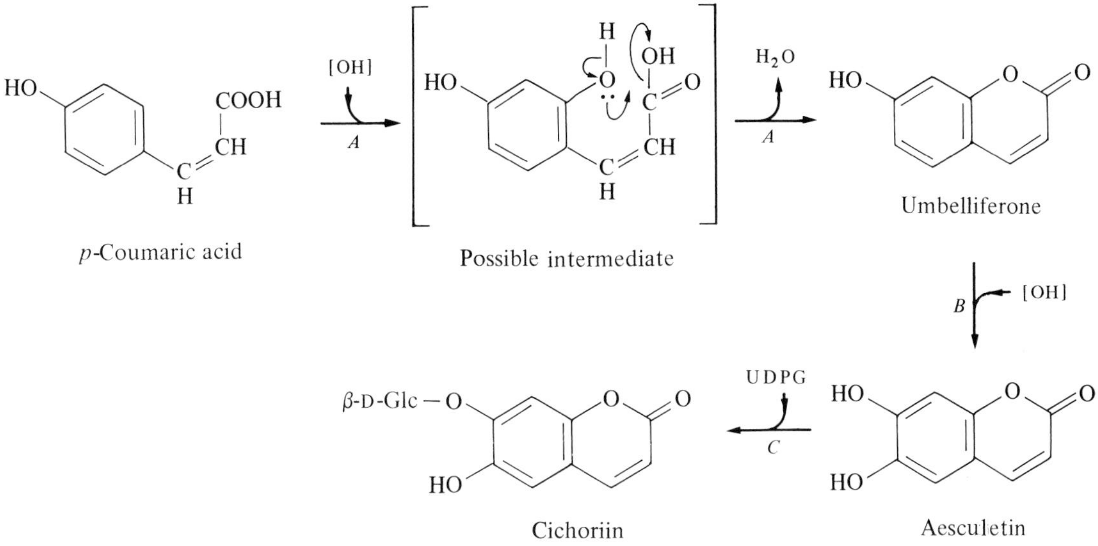

Fig. 14.3. Biosynthesis of the coumarin cichoriin from *p*-coumaric acid.

other interesting derivative is primin (14.25), the dermatitic principle in *Primula obconica*.

Naphthoquinones occur generally in the reduced and glycosylated forms. During extraction they are often oxidized and the isolate is coloured. An example of a plant napthoquinone derivative is the glucoside of reduced juglone (14.26) obtained from walnuts. However, free naphthoquinones occur in the heartwood of trees, often as dimers, trimers and tetramers. Such a compound is the blue diosindigo A (14.27) from *Diospyros*. A naphthoquinone which functions in photosynthesis is phylloquinone (Vitamin K_1) (11.49).

More than 200 anthraquinones have been reported in flowering plants. Apart from simple anthraquinones such as emodin (Table 14.1), mixed dimers, such as palmidin A (14.28), can also exist.

2. Biosynthesis

Benzoquinones are rare in higher plants and when they exist they probably occur in the reduced, quinol, form and the formulation of these compounds has already been considered. In lower fungi, benzoquinones are derived either from phenylalanine, that is via the shikimate pathway, or from acetyl-CoA via the polyketide pathway (see Section I.3). For example, coprinin (14.29) arises from tyrosine and spinulosin (14.30) from a polyketide. The benzoquinone residue in terphenylquinones, however, arises in an entirely different manner from the carbon side chains of either two phenylpyruvate, two phenylalanine, two tyrosine, or, as in the case of volucrisporin (Fig. 14.4), two *m*-tyrosine molecules.

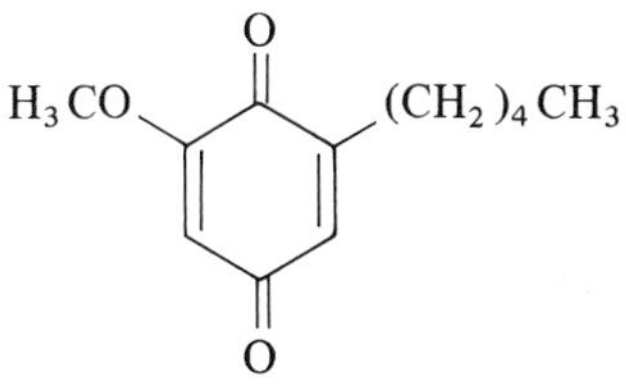

(14.25)

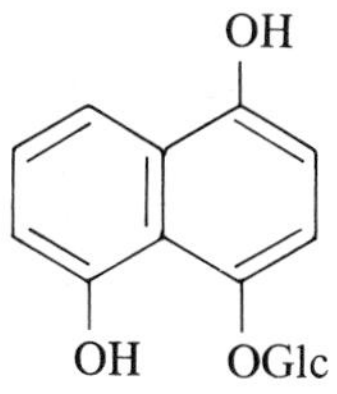

(14.26)

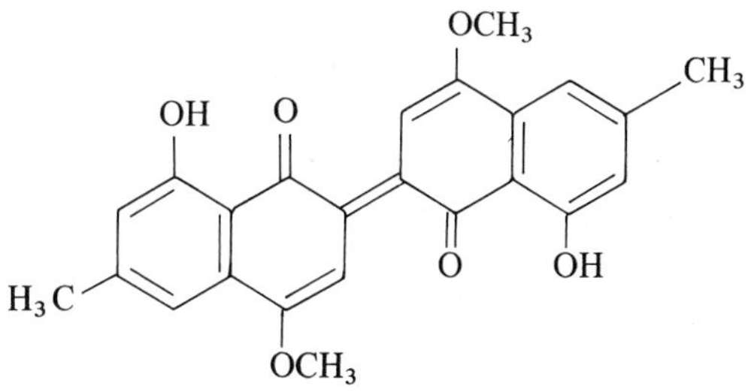

(14.27)

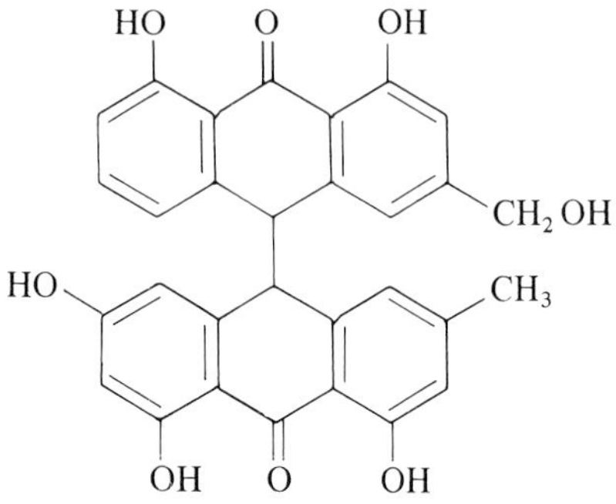

(14.28)

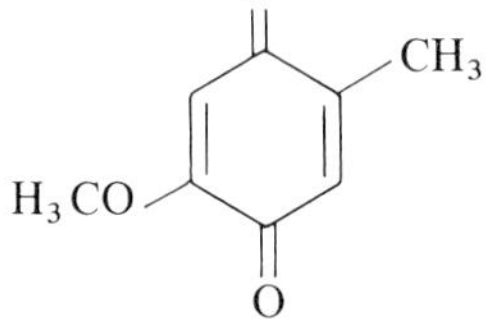

(14.29)

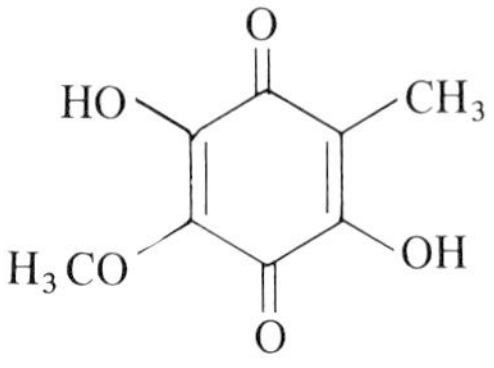

(14.30)

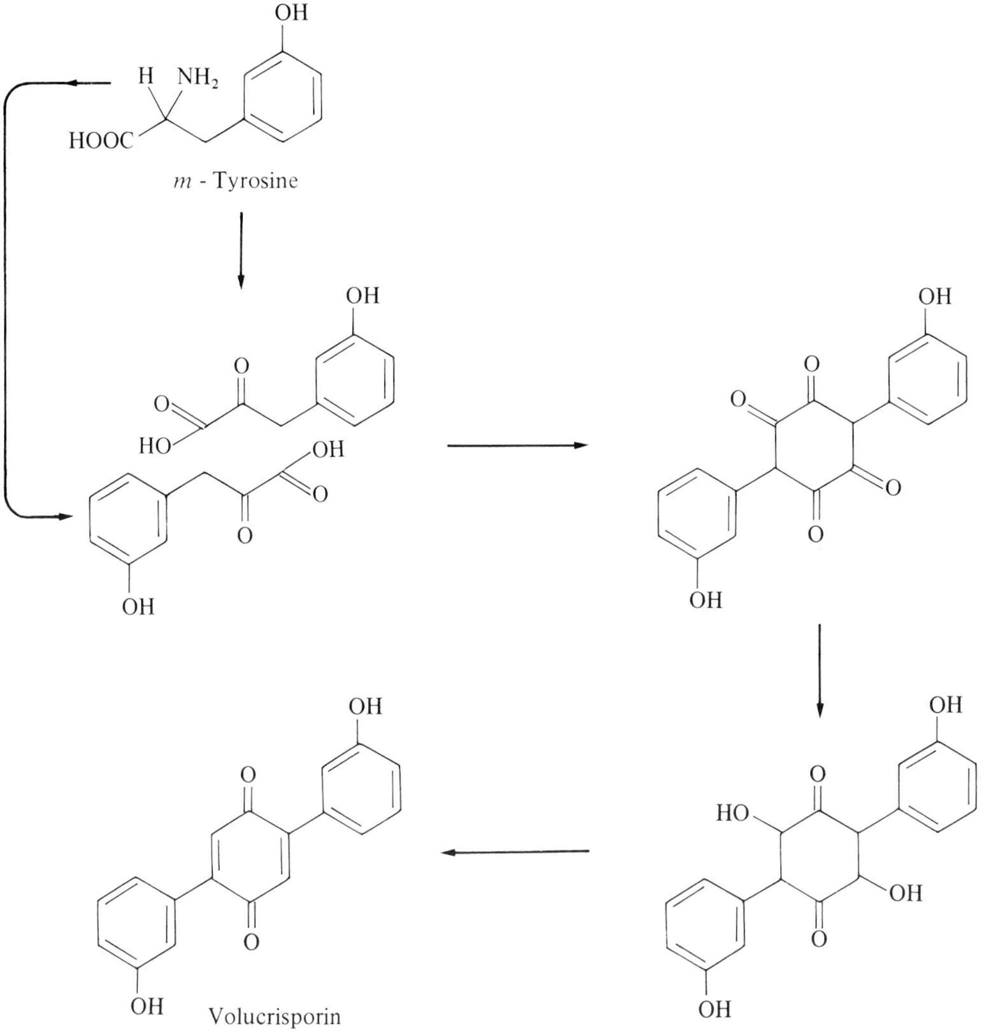

m - Tyrosine

Volucrisporin

Fig. 14.4. Biosynthesis of volucrisporin from *m*-tyrosine.

Naphthoquinones can be biosynthesized in a number of ways. Firstly, via 2-succinoyl benzoate as in the formation of prenyl quinones (Chapter 11, Section C.9) and lawsone; secondly, via the polyketide route as in plumbagin (Fig. 14.5) formed in *Plumbago*; thirdly, from tyrosine via homogentisic acid which couples with mevalonate to yield compounds such as chimaphilin from *Chimaphila umbellata* (Fig. 14.5); and fourthly, from *p*-hydroxybenzoic acid and two molecules of mevalonate, probably as geranyl pyrophosphate, to form alkannin in *Plagiobothrys arizonicus*. The labelling patterns observed in these compounds after feeding various substrates are summarized in Fig. 14.5. It is experiments of this type which have led to the conclusions outlined above; detailed enzymological studies have still to be carried out.

The anthraquinone alizarin is formed from the substrate 2-succinoyl benzoate as are some naphthoquinones, such as lawsone just discussed. However, in this case the naphthalene intermediate, not yet identified but possibly naphthalene 2-carboxylic acid (*A*, Fig. 14.6), is isoprenylated by isopentenyl pyrophosphate arising from mevalonic

Lawsone   Plumbagin

Chimaphilin   Alkannin

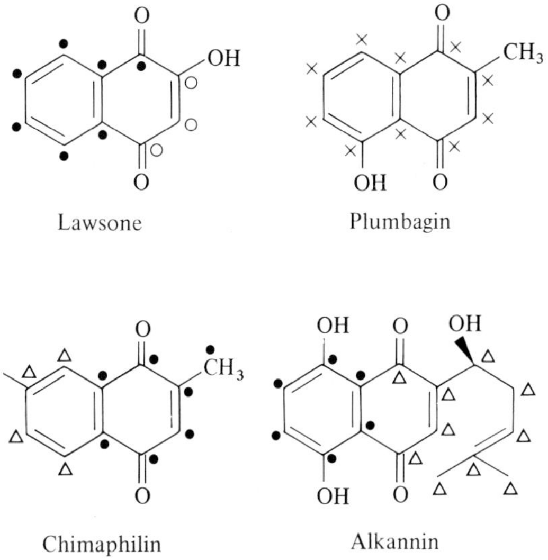

Fig. 14.5. The source of the carbon atoms in four different naphthoquinones. (● = carbons arising from the shikimic acid/phenylalanine/tyrosine pathway; ○ = carbons arising from succinyl-CoA; × = carbons arising from acetic acid; △ = carbons arising from mevalonic acid. Lawsone synthesized by *Impatiens balsamina*. Plumbagin synthesized by *Plumbago* spp. Chimaphilin synthesized by *Chimaphila umbellata*. Alkannin synthesized by *Plagiobothrys arizonicus*.)

Shikimic acid   Succinyl-CoA   1, 4-Dihydroxynaphthalene 2-carboxylic acid (probable intermediate)

Alizarin   Deoxy-lapachol quinol

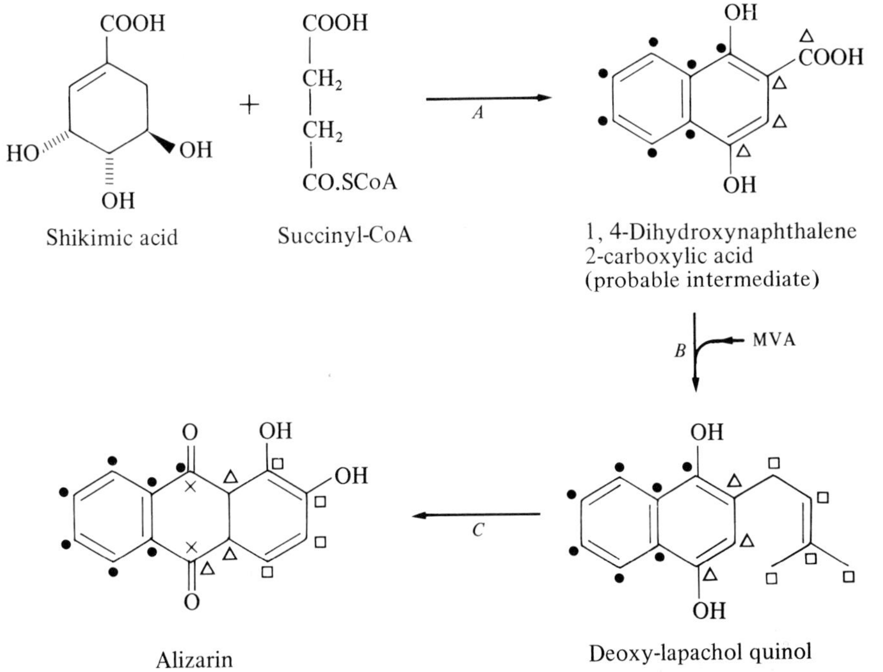

Fig. 14.6. Biosynthesis of alizarin in *Rubia tinctorum*. (● = carbons arising from the shikimic acid pathway; △ = carbons arising from succinyl-CoA; □ = carbons arising from mevalonic acid; × = carbons arising from deoxy-lapachol.)

acid (*B*, Fig. 14.6) and this is followed by ring closure (*C*, Fig. 14.6). On the other hand, emodin (Table 14.1) produced by *Rhamnus frangula* is derived entirely from acetate as indicated in Fig. 14.7. The solitary unlabelled methyl carbon atom in Fig. 14.7 presumably arises by transmethylation.

G. XANTHONES

1. Nature and distribution

The distribution of xanthones is generally re-stricted to the Guttiferae and Gentianaceae. Over seventy compounds which can contain from one [7-hydroxyxanthone (14.31)] to five [4,7-dimethoxy-1,3,8-trihydroxyxanthone, 14.32] hydroxyl groups, have been described. They are found in both the free state and as glycosides in heartwood and roots. Mangiferin (14.33) differs from most other xanthones in that it is widely distributed in the plant world including the ferns.

2. Biosynthesis

A limited number of investigations indicate that one aromatic ring is acetate-derived and the other is shikimate-derived (Fig. 14.8). *p*-Coumaryl-CoA was incorporated intact into mangiferin (14.33) in *Anemarrhena asphodeloides* whilst acetate accounted for the remainder of the carbon atoms in gentisein (Fig. 14.8).

H. STILBENES

1. Nature and distribution

The most common plant stilbene is the dihydro-stilbene lunalaric acid (Table 14.1). It occurs in all liverworts where it acts as a growth inhibitor similar in action to abscisic acid in higher plants (see Chapter 15, Section E). Stilbenes themselves exist as compounds such as pinosylvin (14.34) in *Pinus* and resveratrol (14.35) in *Eucalyptus* spp.

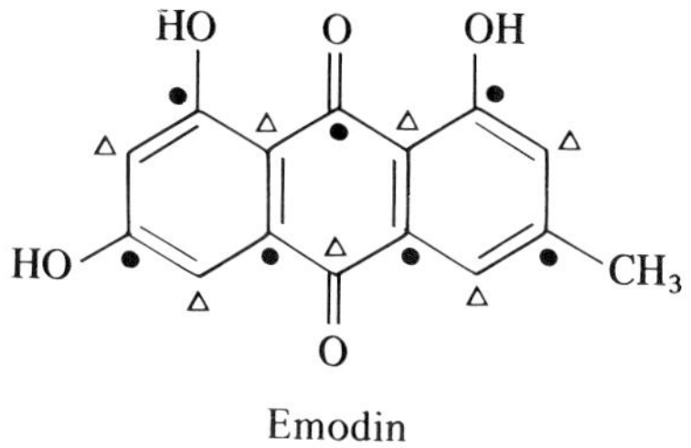

Emodin

Fig. 14.7. The source of the carbon atoms of emodin synthesized by *Rhamnus frangula*. (● = carbons arising from the carboxyl carbon of acetic acid; △ = carbons arising from the methyl carbon of acetic acid. The methyl group of emodin probably arises by transmethylation.)

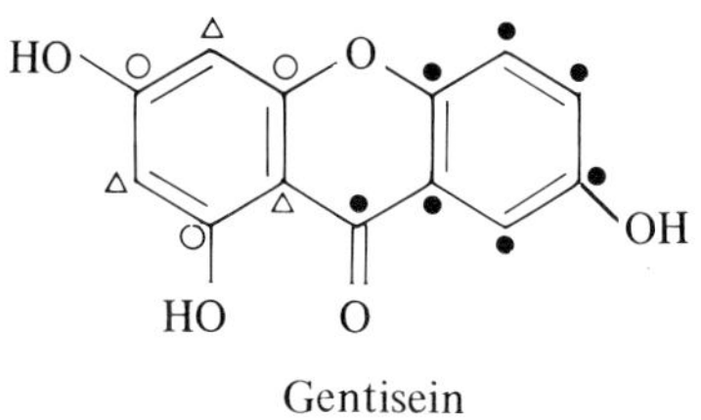

Gentisein

Fig. 14.8. Source of the carbon atoms in the xanthone, gentisein, synthesized by *Gentiana lutea*. (● = carbons arising from the shikimic acid/phenylalanine pathway; △ = carbon atoms arising from the methyl carbon of acetic acid; ○ = carbon atoms arising from the carboxyl carbon of acetic acid.)

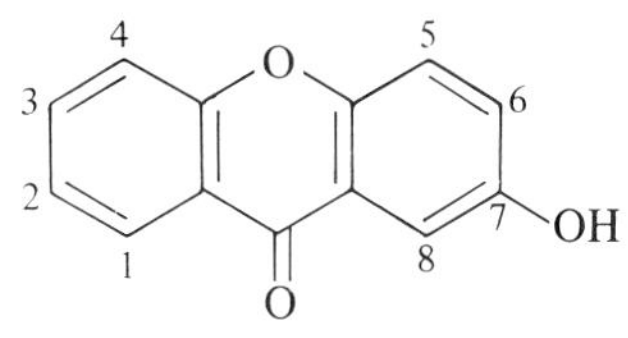

(14.31)

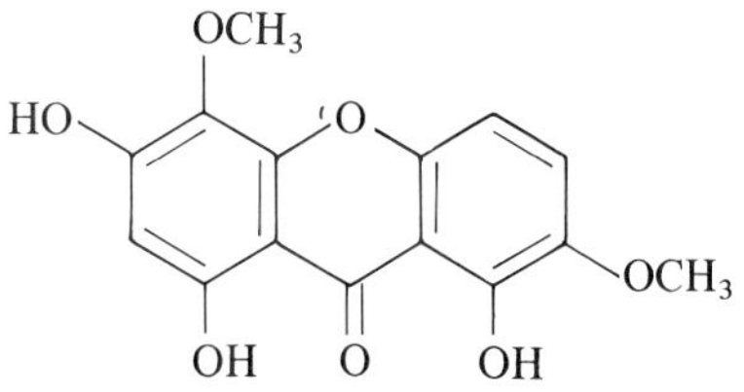

(14.32)

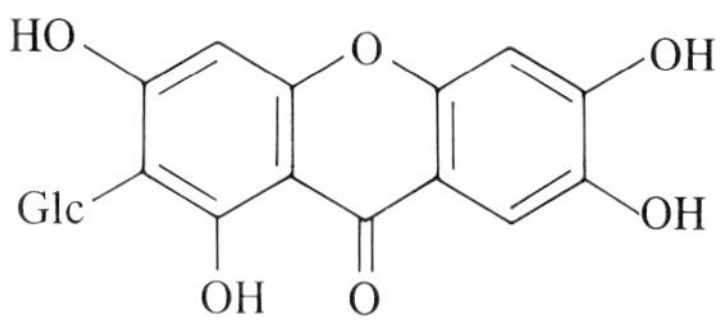

(14.33)

2. Biosynthesis

Little is known about the biosynthesis of stilbene but resveratrol (14.35) is a precursor of dimerized derivatives such as ε-viniferin (14.36) in *Vitis vinifera*.

change is merely a consequence of the contraction of the heterocyclic ring.

The pattern of hydroxylation in the flavonoids can vary considerably but the emergence of eight main types can be discerned. Examples of these are given in Table 14.4. The existence of stereoisomers in the

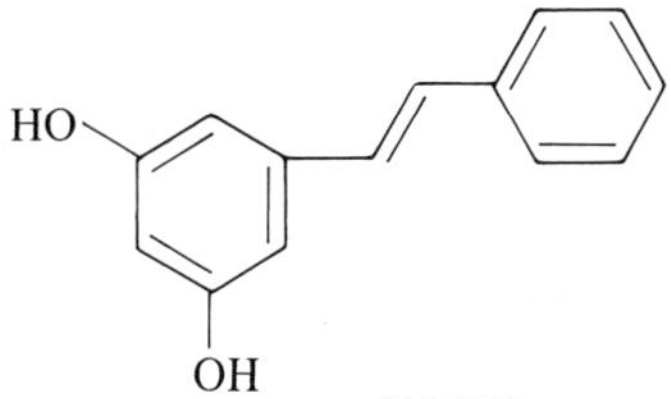

(14.34)

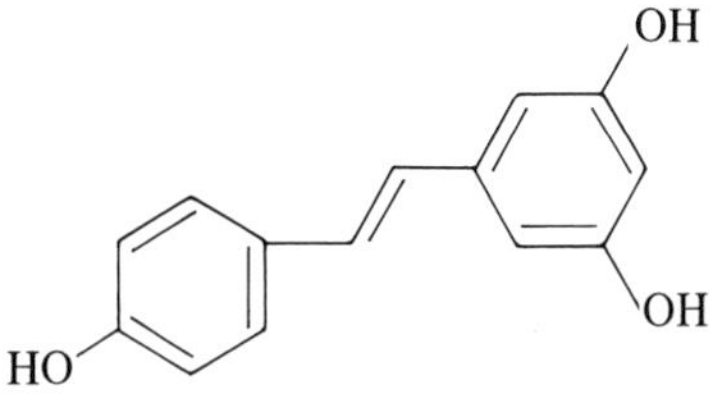

I. FLAVONOIDS

1. Nature

Flavonoids represent a very widespread group of water-soluble phenolic derivatives many of which are brightly coloured being red, crimson, purple or yellow. They are generally found in the vacuoles although some are also found in chromoplasts and chloroplasts. Flavonoids are glycosides and the structures of their aglycones are based on the flavan structure (14.37) which consists of two aromatic rings joined in a chroman structure by a three-carbon unit (C_6–C_3–C_6); thus they are, as are many of the compounds already discussed in this chapter, phenylpropane derivatives. Free aglycones are found in dead woody tissues where they have apparently been formed from flavonoids by enzymic hydrolysis.

The aglycones are classified according to the state of oxidation of the C_3 unit (C–2,3,4) in the molecule (Table 14.3). The last three groups in the table, chalcones, dihydrochalcones and aurones, are not strictly flavonoids but their close relationship, both chemically and biosynthetically, makes it sensible to include them in the flavonoid grouping. However, care must be taken over the numbering of these molecules which is different from that of the flavonoids. The chalcone numbering (14.38) is particularly confusing whereas the aurone (14.39)

(14.35)

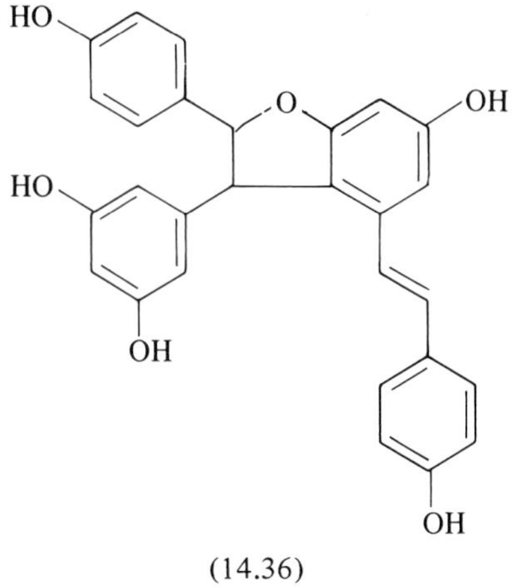

(14.36)

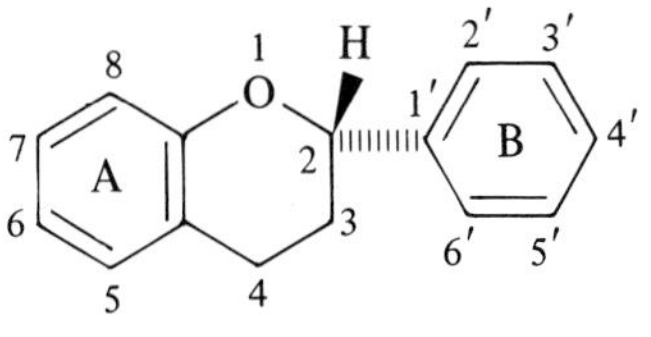

(14.37)

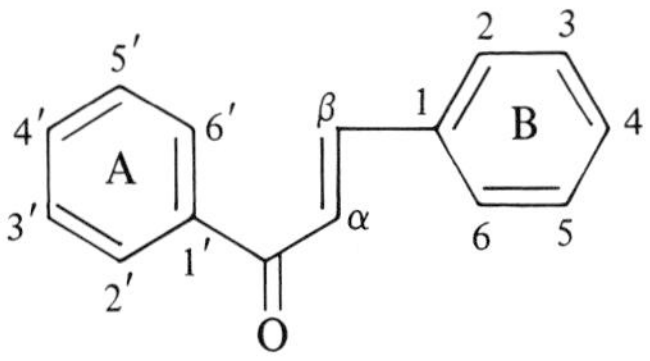

(14.38)

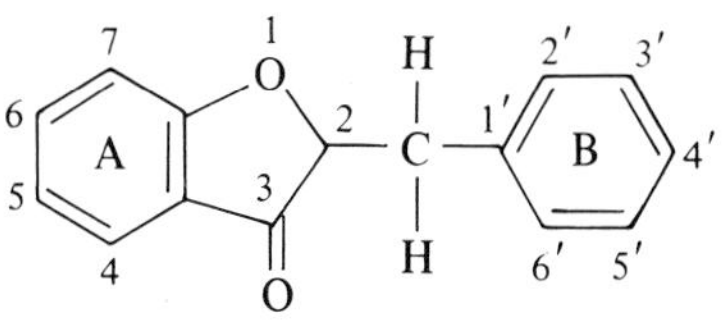

(14.39)

Table 14.3. *Classification of flavanoid aglycones according to the oxidation state of the C$_3$ residue of the phenylpropane unit*

Class	Structure	Example (Nos. indicate positions of hydroxyl groups in the remainder of the molecule)
Flavones		Apigenin (5,7,4′)
Flavonols		Kaempferol (5,7,4′)
Catechins (Flavan-3-ols)	(2R,3S)	Catechin (5,7,3′,4′)
	(2R,3R)	(−)-Epicatechin (5,7,3′4′)
	(2S,3R)	(+)-Epicatechin (5,7,3′,4′)
Flavanones		Naringenin (5,7,4′)

Table 14.3 (*cont.*)

Class	Structure	Example (Nos. indicate positions of hydroxyl groups in the remainder of the molecule)
Dihydroflavonols	(2R,3S)	Taxifolin (5,7,3′,4′)
Flavan-3,4-diols (Proanthocyanidins or Leucoanthocyanidins)	(2R,3R,4R)	Teracacidin (7,8,4′)
	(2R,3R,4S)	Isoteracacidin (7,8,4′)
	(2R,3S,4R)	Leucorobinetinidin (7,3′,4′,5′)
	(2R,3S,4S)	Guibaurtacacidin (7,4′)
Anthocyanidins	[a]	Pelargonidin (5,7,4′)

[a] This is just one of a number of resonating structures of the flavylium cation.

Table 14.3 (cont.)

Class	Structure	Example (Nos. indicate positions of hydroxyl groups in the remainder of the molecule)
Isoflavones		Genistein (5,7,4′)
Neoflavones (4-Phenylcoumarins —see Section E)		Dalbergin (6-hydroxy,7-methoxy)
Chalcones		Butein (3,4,2′,4′)
Dihydrochalcones		Phloretin (4,2′,4′,6′)
Aurones		Sulphuretin (6,3′,4′)

N.B. For details of the numbering systems for the flavonoid, chalcone and aurone carbon skeletons see formulae (14.37), (14.38) and (14.39), respectively.

catechins and flavan-3,4-diols must be emphasized.

The various aglycones have one or more of their hydroxyl groups attached to a sugar by a glycosidic linkage which is usually β except in the case of the L-rhamnosides and L-arabinosides where it is α. The flavonoids containing monosaccharide, disaccharide or trisaccharide residues are termed mono-sides, diosides or triosides, respectively. A compound with two of its hydroxyl groups attached to monosaccharide residues would be a dimonoside. Examples of naturally occurring flavonoids containing different sugars are given in Table 14.5.

Table 14.4. *Main hydroxylation patterns observed in flavonoids*

Flavonoid	R_1	R_2	R_3	R_4	R_5	R_6	R_7	R_8	R_9
5-Hydroxyflavone	OH	H	H	H	H	H	H	H	H
5,8-Dihydroxyflavone	OH	H	H	OH	H	H	H	H	H
Digicitrin	OH	OCH_3	OCH_3	OCH_3	H	OCH_3	OCH_3	OH	OCH_3
Morin	OH	H	OH	H	OH	H	OH	H	OH
Isoetin	OH	H	OH	H	OH	H	OH	OH	H
Gossypetin	OH	H	OH	OH	H	OH	OH	H	OH
Quercetagetin	OH	OH	OH	H	H	OH	OH	H	OH
6-Methoxyluteolin	OH	OCH_3	OH	H	H	OH	OH	H	H
Diosmetin	OH	H	OH	H	H	OH	OCH_3	H	H

2. Flavonoids and plant coloration

The main classes of flavonoids which contribute colour to plants are the anthocyanidins, flavonols, chalcones and aurones, and it is important to discuss this aspect of their biochemistry when studying the general properties of flavonoids.

(i) ANTHOCYANIDINS

Although some twenty-two anthocyanidins are known the three most common ones are pelargonidin (Table 14.3), delphinidin (14.40) and cyanidin (14.42); they differ only in the number of hydroxyl groups in ring B. Because of their ionic character both the intensity and shade of colour of anthocyanidins vary with changes in pH. In acid solution (methanol-HCl) the colours vary from orange-red (pelargonidin, abs. max. 520 nm) through magenta (cyanidin; abs. max. 535 nm) to mauve (delphinidin; abs. max. 545 nm). The differences depend on the number of hydroxyl groups in ring B. If the pH of an acid (orange) solution of an anthocyanidin is raised the solution becomes colourless near pH 7.0 owing to the formation of a colourless pseudobase (*A*, Fig. 14.9); above pH 7.0 the bluer anhydrobases are formed (*B*, Fig. 14.9) and at very high pH values irreversible changes occur which are initiated by ionization of the phenolic hydroxyls (*C*, Fig. 14.9).

Glycosylation of anthocyanidins at C–3 to produce anthocyanins results in a marked hypsochromic effect (~ 15 nm), but this has no marked effect on plant colours *in situ* because virtually all anthocyanidins are glycosylated at C–3. Glycosylation at other positions, which is much more sporadic, has only minor effects on colour. However, the amount of anthocyanin present in a tissue, which can vary from 0.01% to 15.0% of dry matter, has a very marked effect. For example, the anthocyanin concentration in 'normal' blue cornflowers is 0.05% (dry wt.) whereas that in the deep purple varieties is 13–14% (dry wt.).

(a) *Petals*

Variation in the pH of the cell sap of plants is not great and is not the major factor in controlling the colour of the petals of those plants which is based mainly on anthocyanin pigments. More important factors are the phenomenon of co-pigmentation and the ability of the pigments to chelate with metals.

(b) *Co-pigmentation*

Blueness of flowers is due to co-pigmentation between an anthocyanin and a flavone glycoside or hydrolysable tannin [see Section L.1(i)]. For example, the anthocyanin in maroon and mauve

Table 14.5. *Some examples of flavonoids with different sugar residues*

Class	Flavonoid (source)	Aglycone	Sugar	Linkage (sugar→Aglycone HO)
Monoside	Delphinidin-3-rhamnoside (flower petals of *Lathyrus odorata*)	Delphinidin (14.40)	L-Rhamnose (pyr)	$\alpha 1 \to 3HO$
	Myrtillin *a* (*Vaccinium myrtillus*)	Delphinidin	D-Glucose (pyr)	$\beta 1 \to 3HO$
	Quercitrin (bark of *Quercus tinctoria*)	Quercetin (14.41)	L-Rhamnose (pyr)	$\alpha 1 \to 3HO$
	Isoquercitrin (flowers of *Gossypium herbaceum*)	Quercetin	D-Glucose (pyr)	$\beta 1 \to 3HO$
	Hirsutrin (flowers of *Gossypium herbaceum*)	Quercetin	D-Glucose (fur)	$\beta 1 \to 3HO$
	Quercimeritrin (flowers of *Gossypium herbaceum*)	Quercetin	D-Glucose (pyr)	$\beta 1 \to 7HO$
Dioside	Quercetin-3-sophoroside (leaf of *Pisum sativum*)	Quercetin	Sophorose [D-Glc$_p$($\beta 1 \to 2$)D-Glc$_p$]	$\beta 1 \to 3HO$
	Neohesperidoside (leaf of *Typha latifolia*)	Quercetin	L-Rha$_p$($\alpha 1 \to 2$)D-Glc$_p$	$\beta 1(Glc) \to 3HO$
	Rutin (*Ruta graveolens*)	Quercetin	Rutinose [L-Rha$_p$($\alpha 1 \to 6$)D-Glc$_p$]	$\beta 1(Glc) \to 3HO$
Trioside	Quercetin-3-(2^G-Glucosyl-rutinoside (flower petals of *Solanum tuberosum*)	Quercetin	L-Rha$_p$($\alpha 1 \to 6$)⌐ ⌐D-Glc$_p$($\beta 1 \to 2$)⌐ ⌐D-Glc$_p$	$\beta ?1(Glc) \to 3HO$
Dimonoside	Cyanin (petals of *Centaurea cyanus*)	Cyanidin (14.42)	β-D-Glc$_p$ β-D-Glc$_p$	$\beta 1 \to 3HO$ $\beta 1 \to 5HO$
Acylated mono- and diosides	Petunoside (flower petals of *Petunia*)	Kaempferol (Table 14.3)	2'-Ferulyl-sophoroside[a]	$\beta 1 \to 3HO$
	Luteolin-7-(*p*-coumarylglucoside) (flower petals of *Catalpa bignonioides*)	Luteolin (14.43)	*p*-Coumaryl-D-glucoside	$\beta 1(Glc) \to 7HO$

pyr = pyranose; fur = furanose; D-Glc$_p$ = D-Glucopyranose; L-Rha$_p$ = L-Rhamnopyranose.

[a] HO—⟨ ⟩—C=C[C → 2'HO]D—Glc$_p$[$\beta 1$ → 2]—D—Glc$_p$ ∿

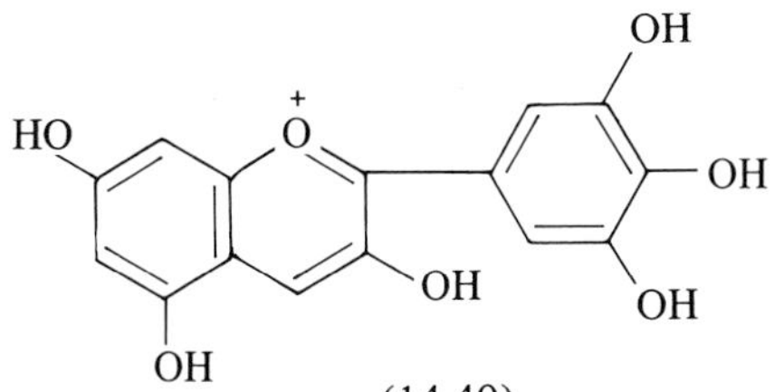

(14.40)

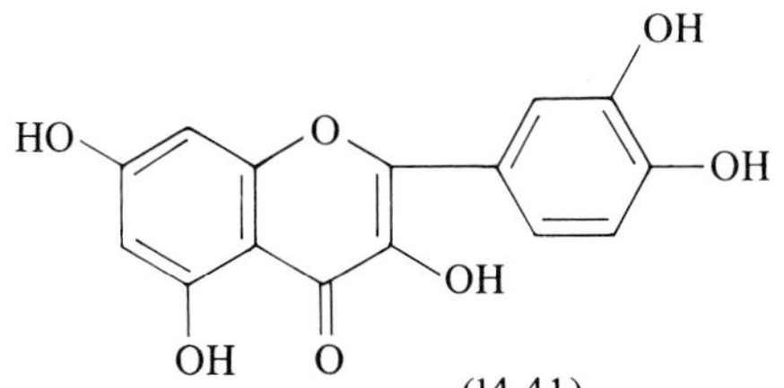

(14.41)

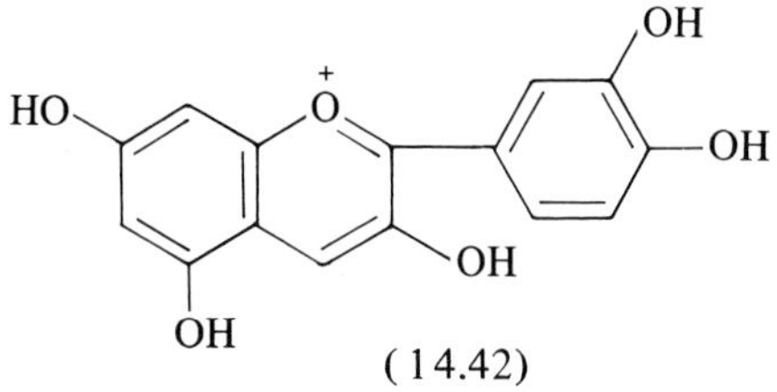

(14.42)

Primula sinensis is malvidin 3-glucoside (14.44) in both cases. However, the difference in colour is the result of a spectral shift of the absorption maximum by 5 nm to higher wavelengths in the mauve variety; this is caused by the co-pigmentation of (14.44) with kaempferol (Table 14.3) glycosides, the concentration of which is 3–5 times greater in the mauve than in the maroon strains. Purple roses contain cyanidin (14.42) 3,5-diglucoside, which is also the main pigment in crimson roses, co-pigmented with large amounts of gallotannin; the spectral shift in this case is from 507 nm to 512 nm. Co-pigmentation is the result of hydrogen bonding between the carbonyl group of the anhydrobase and the aromatic hydroxyl groups in the complexing flavonoid.

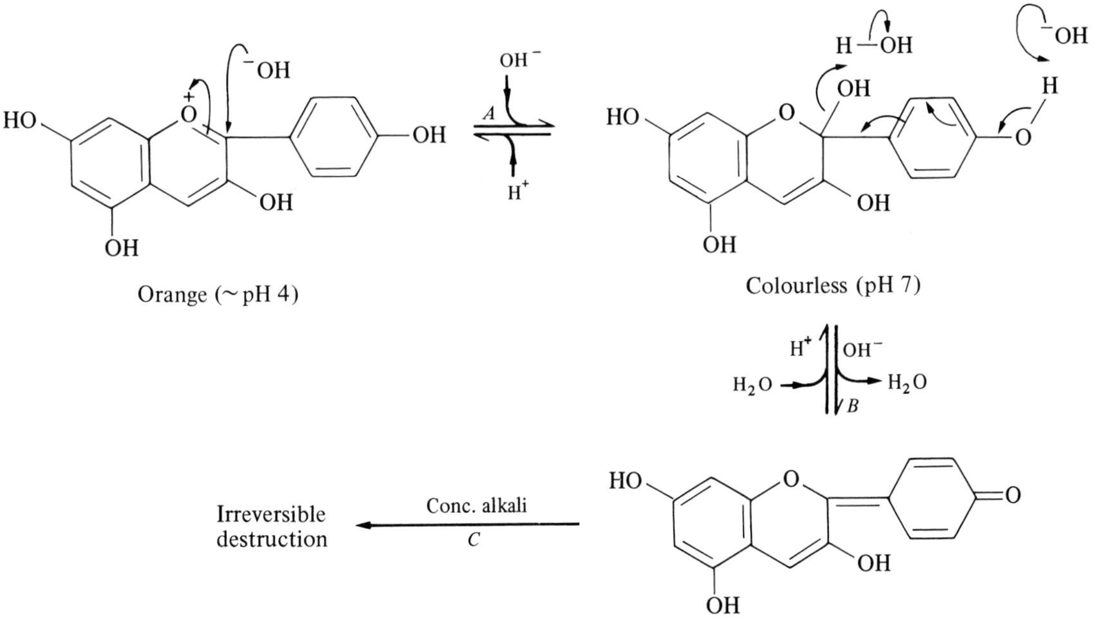

Orange ($\sim$ pH 4)

Colourless (pH 7)

Blue ($\sim$ pH 10)

Fig. 14.9. The effect of varying the pH on anthocyanidins. (The anthocyanidin taken as the example is pelargonidin; the curly arrows refer to electron pair shifts when the pH is raised, but the reader can readily conceive of the electron shifts that would occur when the reversible reactions are driven in the opposite direction with H^+ instead of OH^- as the attacking ionic species.)

OH

HO—⟨⟩—O—⟨⟩—OH

HO   O

(14.43)

OCH₃

HO—⟨⟩—O⁺—⟨⟩—OH

OH   OGlc   OCH₃

(14.44)

(c) *Metal chelation*

The effect of metal chelation on colour of petals is probably best illustrated by comparing the blue colour of a cornflower with that of the red colour of a rose, in which the major anthocyanidin, cyanidin (14.42), is the same in both cases. Procyanin, a blue crystalline iron complex containing four molecules of cyanin (cyanidin 3,5-diglucoside) and three of the flavone apigenin-7-glucuronide-4'-glucoside (14.45), can be isolated from cornflower sap. So in this case the combined effect of metal chelation and co-pigmentation is observed. On the other hand, similar extraction of rose petals yields a metal-free 'genuine red anthocyanin'. If the mineral balance of *Hydrangea macrophylla* is correct the chelating metals aluminium and molybdenum are easily accumulated and the petals turn blue; otherwise they are red.

The brown colour of some petals, in for example the wallflower, is due to a combination of the water-soluble magenta anthocyanin in the vacuole with

GlcAO—⟨⟩—O—⟨⟩—OGlc

OH   O

GlcA = glucuronic acid

(14.45)

the yellow carotenoids in the chromoplasts. Colour pattern is mainly due to local increases in pigment production, as in foxgloves, or to a superimposition of an additional pigment on the background pigment. An example of this is the deep purple coloration at the centre of some poppy flowers which is the result of a high local concentration of cyanidin on a pelagonidin background. A summary

of the major factors controlling anthocyanin flower colours is given in Table 14.6.

(d) *Leaves*

Anthocyanins are the only flavonoids to impart colour to leaves; they contribute to the transient coloration in young leaves, to permanent coloration and to autumn coloration. In nearly all cases the pigment involved is cyanidin 3-glucoside although *Primula* spp. and *Solanum* are unusual in accumulating delphinidin (14.40) in their leaves.

The reason for the transient flush of anthocyanins in the early stages of leaf development is not known but the permanent coloration patterns in the common house plant *Coleus* are under strict genetic control. The copper beech is a sport from the green *Fagus sylvatica*. Sometimes the intense leaf production of anthocyanins spills over into the flowers as in some species of *Antirrhinum*.

The appearance of anthocyanins is particularly striking in the red autumn colours of *Acer* and *Pyrus*. The quantitative production of the pigment (cyanidin 3-glucoside) depends on the climatic conditions in autumn but exact biochemical control points are not known. The yellow and browns of autumn foliage are due mainly to carotenoids (Chapter 11) and tannins (see Section L), respectively.

Table 14.6. *Major factors controlling anthocyanin pigmentation in flower petals*

1. Nature of anthocyanins, particularly the extent of hydroxylation of ring B
2. Concentration of anthocyanin
3. Molar ratio of anthocyanin to co-pigment
4. Metal chelation, particularly with magnesium and iron
5. pH of cell sap

(e) *Fruit*

The anthocyanins represent major contributors to the colours of many edible fruit and a short summary is given in Table 14.7. Variation in colour of different varieties of fruit, e.g. 'black' and 'white' cherries, is usually due to quantitative differences, rather than changes in the nature of the pigments present. The blue colours of many fruit are probably the result of complex formation with a metal and protein and not to co-pigmentation as in petals.

(ii) FLAVONOLS AND FLAVONES

Common flavonols and flavones, although widespread in flower petals, do not contribute to their colour. However, flavonols make some contribution if they are methylated or have unusual glycosidic patterns. For example, syringetin (14.46) (myricetin 3',5'-dimethyl ether) contributes to the yellow colour of the flowers of the meadow pea, *Lathyrus pratensis*, and isorhamnetin (14.47) (quercetin-3'-methyl ether) may also contribute to the petal colour of the common marigold (*Calendula officinalis*) although in this case the major pigmentation is probably due to carotenoids.

8-Hydroxyflavonols such as gossypetin (14.48) account for the distinctive yellow colour of cotton and primula flowers. 6-Hydroxyflavonols appear to be restricted to Compositae where, for example, quercetagetin (14.49) is found in the African marigold (*Tagetes erecta*).

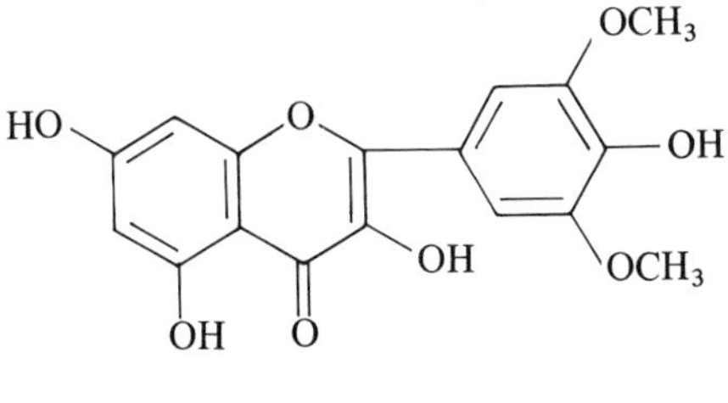

Table 14.7. *Anthocyanidin*[a] *distribution in some edible fruit*

Pigment	Fruit
Cyanidin (14.42)	Apple (skin); cherry; raspberry; red currant
Delphinidin (14.40)	Aubergine (skin); pomegranate (juice)
Pelargonidin (Table 14.3)	Strawberry; passion fruit
Cyanidin plus delphinidin	Blackcurrant; blood orange (juice)

[a] The pigments exist *in vivo* as glycosides (anthocyanins).

It must be emphasized that although flavones do not contribute directly to flower colour, they can act as co-pigments intensifying the colour of yellow flavonols, chalcones and aurones, in particular apigenin 7,4'-diglucuronide (14.50) co-pigments with aureusin (14.51) in some yellow *Antirrhinum* cultivars.

Colourless flavonols and flavones appear to provide 'body' to white, cream and ivory-coloured flowers. This has been demonstrated in an albino mutant of *Antirrhinum majus*, which completely lacks flavones; its flowers can be easily distinguished from those of the native strain. The major contributors to this phenomenon are kaempferol (Table 14.3) and quercetin (14.52).

In general flavonols and flavones make no contribution to fruit colours although they are widely distributed in fruit. One exception is the isoflavone, osajin (14.53), which is responsible to some extent for the colour of the osaje orange (*Maclura pomifera*). Note that this pigment also contains two isoprenoid residues.

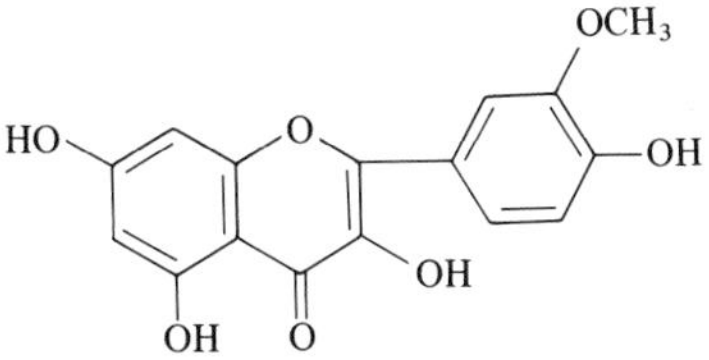

(14.46)

(14.47)

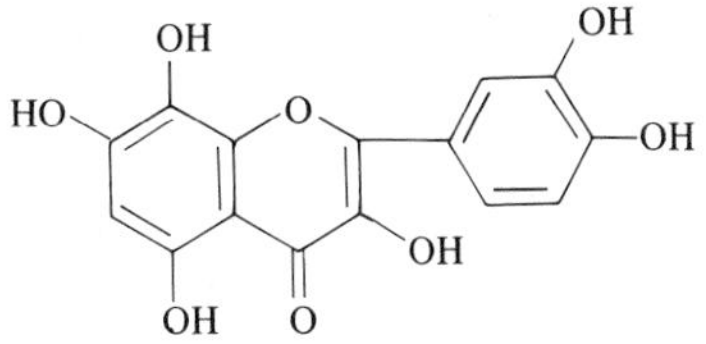

(14.48)

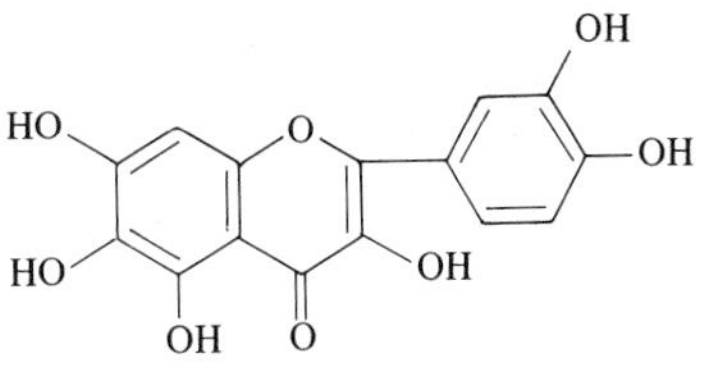

(14.49)

(14.50) (GlcA = β1-linked D-Glucuronic acid)   (14.51)   (14.52)

(iii) CHALCONES AND AURONES

The presence of these yellow pigments can easily be demonstrated in petals by exposing them to ammonia vapour when the colour changes dramatically from yellow to red. For this reason chalcones and aurones are often called *anthochlor* pigments. Their distribution is restricted to about nine plant families. Examples in which these pigments are the major contributors to the yellow colour of petals are the chalcone isosalipurposide (14.54) in yellow carnations and the aurone aureusin (14.51) in yellow *Antirrhinum majalis*. Anthochlor pigments are frequently associated with carotenoids in yellow petals.

Leaves and fruit occasionally produce anthochlor pigments. For example, the methylated chalcone, pedicinin (14.55), in which ring A has been oxidized to the quinonoid form, occurs in the coloured dust which accumulates on the underside of the leaves of *Didymocarpus pedicellata*. The fruit of *Kyllingia brevifolia* contains the bright yellow chalcone okanin (14.56) whereas aureusin (14.51) is present in the fruit of *Gahnia clarkei*.

3. Biosynthesis

(i) FIRST STEPS

The biosynthesis of flavonoids is unique in that the two component aromatic rings arise via different pathways. The phenylpropane residue (ring B and

$C_{2, 3 \text{ and } 4}$) derives from *p*-coumaric acid, itself formed via the shikimate pathway. Ring A, on the other hand, is basically formed from acetate and is a rather special case of polyketide synthesis. The characteristic of polyketides is that they are all assembled from C_2 units derived from acetate. Belonging to this group are the fatty acids already considered in Chapter 8 as well as certain phenols, mainly those produced by fungi.

The first view of phenol synthesis was that four molecules of acetate condensed to form a polyketide (3,5,7-triketo-octanoic acid) (such an intermediate has not yet been isolated) which could cyclize in various ways. One way, appropriate to our discussion, is illustrated in Fig. 14.10. Further investigations revealed that the starter R.COOH was acetyl-CoA and that the other 'acetates' added were in fact malonyl-CoAs. These condensation reactions have been explained in Chapter 8, Section C.1(i). Now, if R.COSCoA is *p*-coumaryl-CoA rather than acetyl-CoA, then the cyclization envisaged in Fig. 14.10 can yield a chalcone or a flavanone (Fig. 14.11). Which is the primary product of the cyclization is still in doubt, mainly owing to the ease with which chalcones and flavanones can interconvert. An enzyme system from parsley cell cultures will convert *p*-coumaryl-CoA and malonyl-CoA into 4,2′,4′,6′-tetrahydroxychalcone which rapidly and spontaneously isomerizes to naringenin (Fig. 14.11). There is also a widespread enzyme chalcone–flavanone isomerase[2] which catalyses the interconversion, and the stereochemistry of the

(14.53)   (14.54)   (14.55)

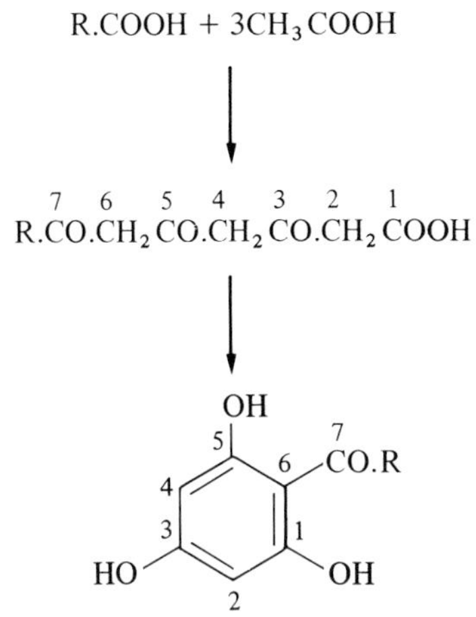

R.COOH + 3CH$_3$COOH

$\downarrow$

$\overset{7}{\text{R.CO.}}\overset{6}{\text{CH}_2}\overset{5}{\text{CO.}}\overset{4}{\text{CH}_2}\overset{3}{\text{CO.}}\overset{2}{\text{CH}_2}\overset{1}{\text{COOH}}$

$\downarrow$

Fig. 14.10. The basic mechanism for forming aromatic rings from polyketides (other possibilities of cyclization are not included; in the simplest example R = CH$_3$.)

reaction has been studied using a chalcone deuterated in the α position as substrate. The resulting flavanone had the S-configuration at C–2 and the deuterium took up the equatorial position at C–3 (A, Fig. 14.12). On the other hand, when the reaction was carried out in D$_2$O with unlabelled substrate, the deuterium occupied the axial position at C–3 (B, Fig. 14.12). This means that cyclization is formally a cis addition to the α,β-double bond.

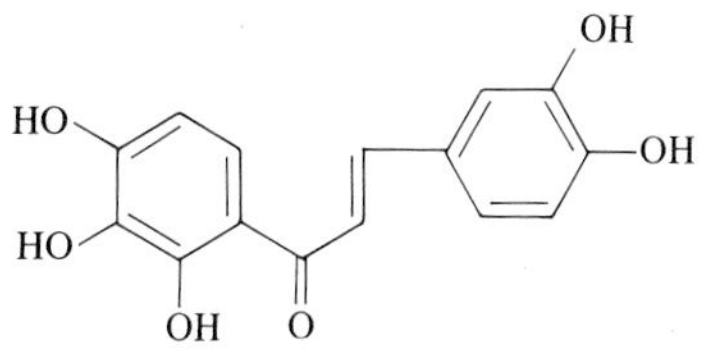

(14.56)

(ii) DIHYDROFLAVONOLS

Feeding experiments indicate that these are formed directly from chalcones, but the mechanism of the hydroxylation involved has not yet been reported (A, Fig. 14.13).

(iii) FLAVONES AND FLAVONOLS

Flavones can be formed by the oxidation of flavanones by flavanone oxidase (B, Fig. 14.13) which, when isolated from young primary leaves of parsley, will oxidize naringenin (Fig. 14.11) to apigenin (Table 14.3). Another possible route to flavones is indicated in Fig. 14.14, in which the oxidation takes place at the chalcone level (C, Fig. 14.14) via a hypothetical chalcone epoxide. Dibenzoylmethanes of the type envisaged in Fig. 14.14 are natural products and can easily be converted chemically into flavones. Chalcone epoxides are easily converted into dibenzoylmethanes photochemically. Dihydroflavonols are probably

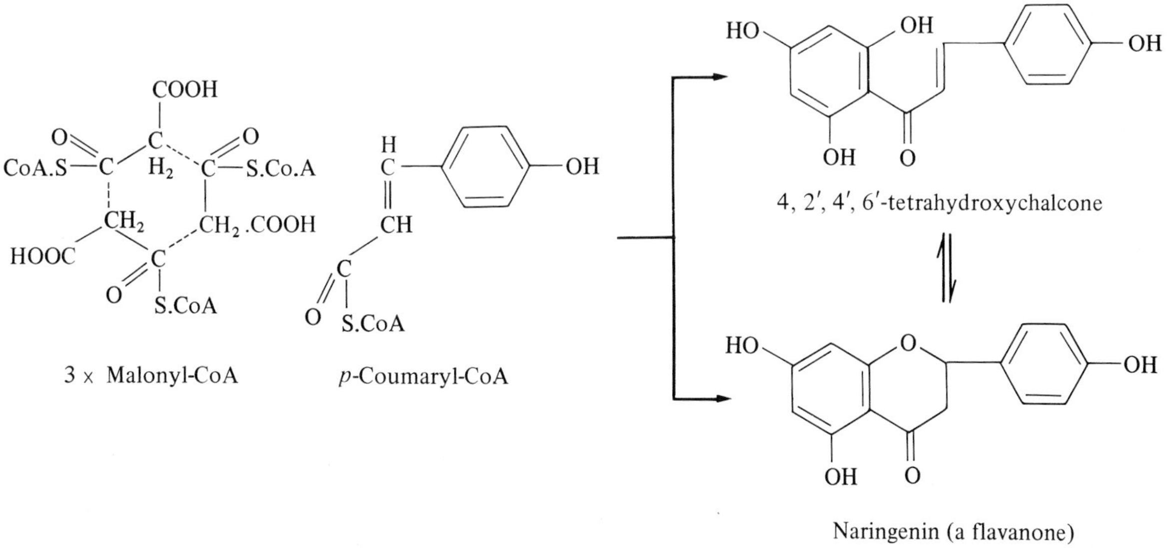

3 × Malonyl-CoA   p-Coumaryl-CoA

4, 2', 4', 6'-tetrahydroxychalcone

Naringenin (a flavanone)

Fig. 14.11. The condensation of three molecules of malonyl-CoA and one molecule of p-coumaryl-CoA to yield a chalcone or a flavanone (it is difficult to decide which is the primary product because of the rapid equilibrium established between the chalcone and flavanone).

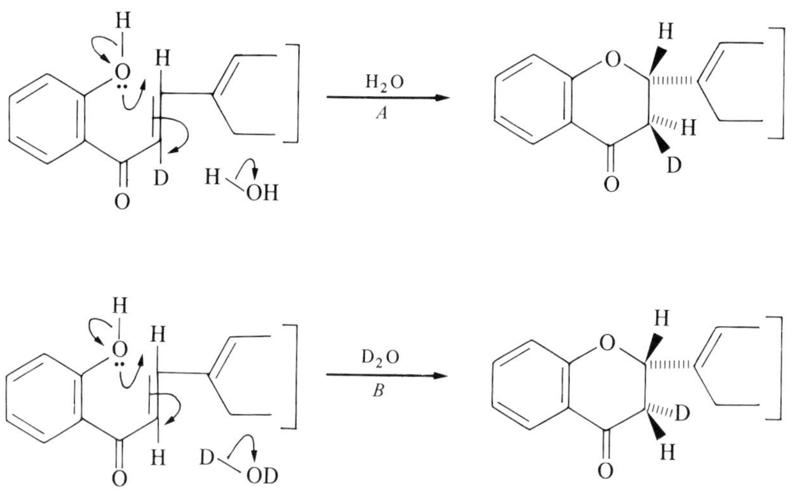

Fig. 14.12. Stereochemistry of the reaction catalysed by chalcone-flavanone isomerase. A = Reaction proceeds in H_2O with the chalcone labelled at the α-carbon atom with deuterium. B = Reaction proceeds in D_2O with unlabelled chalcone.

the precursors of flavonols (D, Fig. 14.13) because isotope experiments, for example, have demonstrated the conversion of dihydroquercetin (dihydroflavonol) into quercetin (flavonol) [eq. (14.1)]. The enzymology of the reaction is not clear but peroxidase-like enzymes may be involved.

(iv) Isoflavones

There is considerable radioisotope evidence that an aryl migration occurs from C–2 to C–3 in isoflavone formation. Experiments with $[^{14}COOH]$-, $[\alpha$-$^{14}C]$- and $[\beta$-$^{14}C]$phenylalanine showed labelling patterns in formononetin produced by red clover indicated in Fig. 14.15. These patterns are consistent with an aryl migration. The in-

volvement of chalcones in this synthesis is demonstrated by the observation that labelled genistein is produced from labelled 4,2′,4′,6′-tetrahydroxy-chalcone [eq. (14.2)] (G, Fig. 14.13). The mechanism by which the aryl migration occurs is not known.

The rotenoids are derivatives of isoflavones and Fig. 14.16 indicates the basic reactions leading to the formation of rotenone from formononetin.

(v) Anthocyanidins

Isotope experiments indicate that dihydro-flavonols are precursors of anthocyanidins (E, Fig. 14.13) in both intact plants and cell-suspension cultures. The mechanism involved is not known but as the two compounds are of the same oxidation

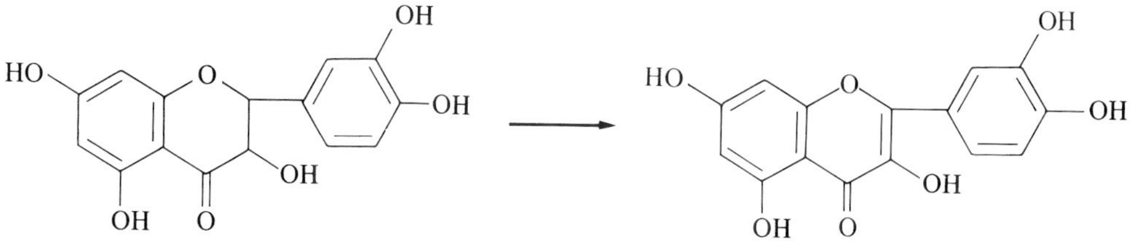

(14.1)

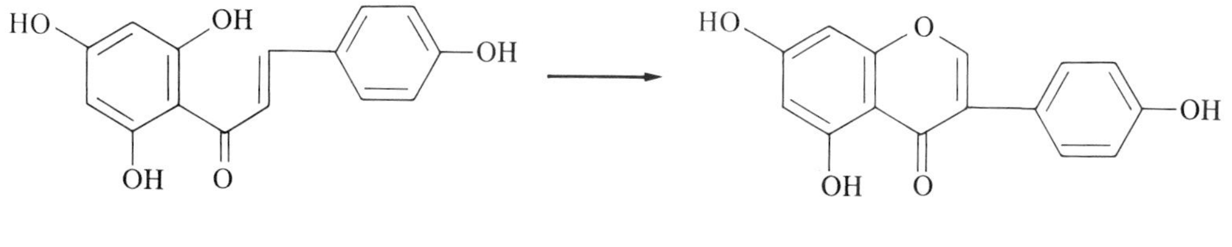

(14.2)

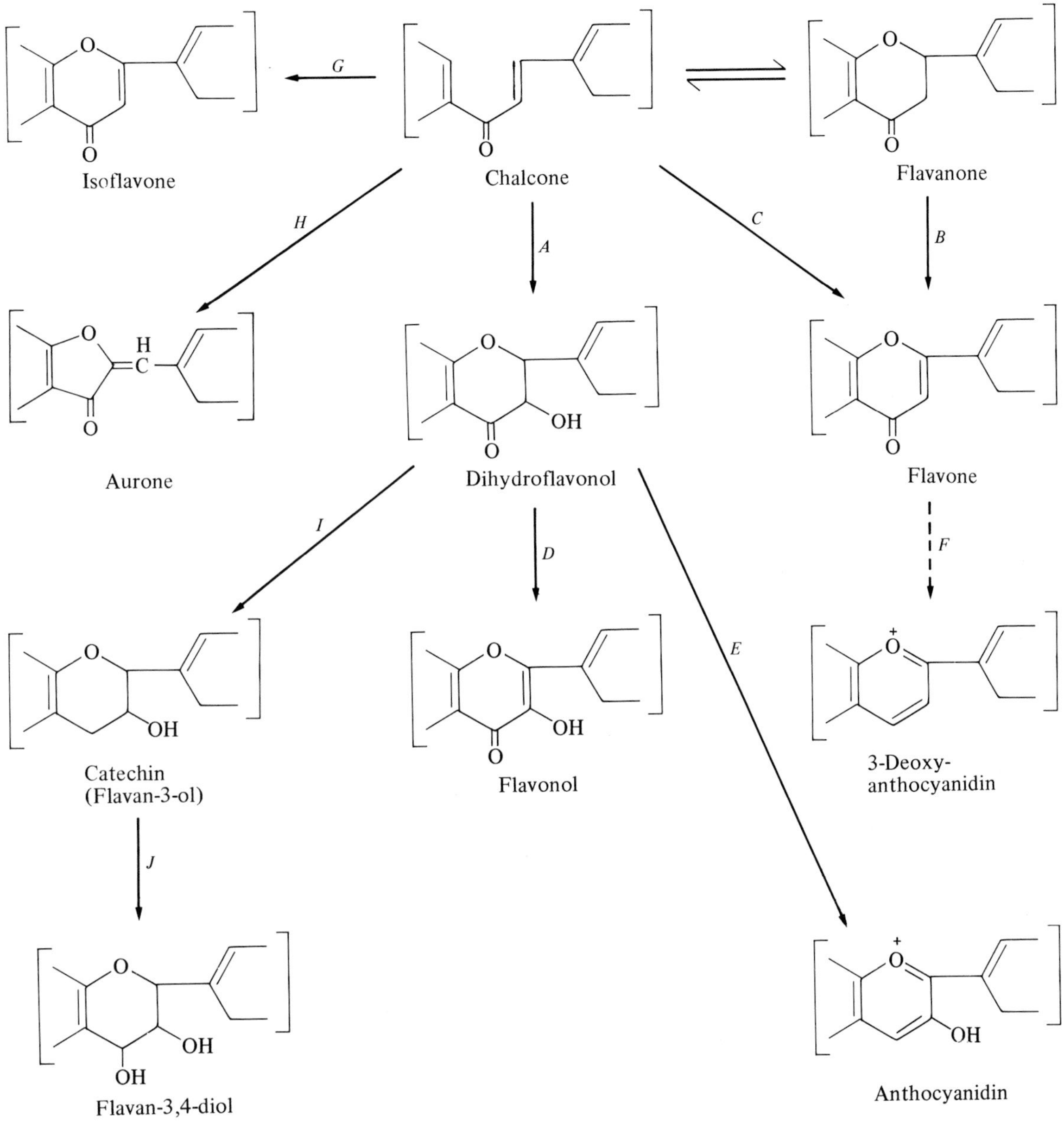

Fig. 14.13. General pattern of flavonoid biosynthesis from chalcones.

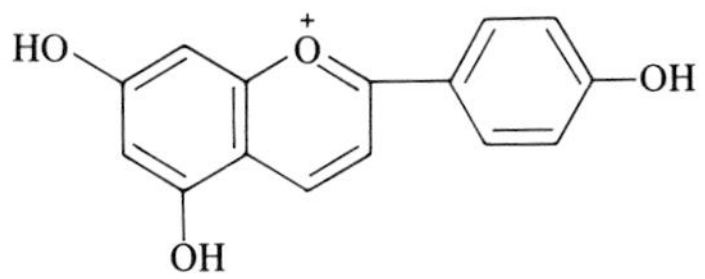

(14.57) 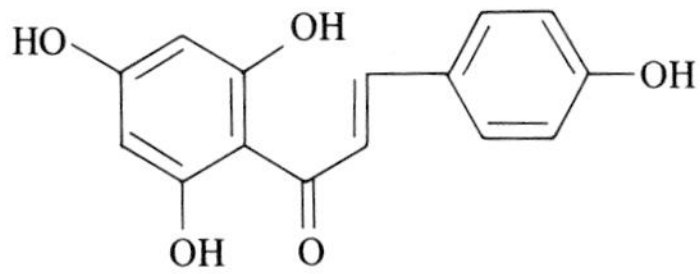   (14.58)

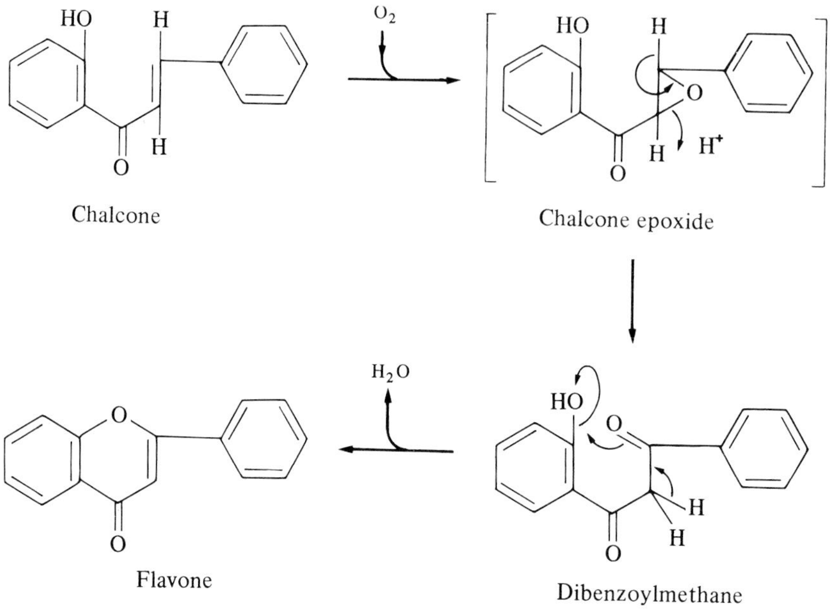

Fig. 14.14. Possible route for the conversion of a chalcone into a flavone.

[¹⁴C] substrate (L-phenylalanine)	Labelling pattern in the isoflavone, formononetin	% of total cpm in formononetin in the position indicated
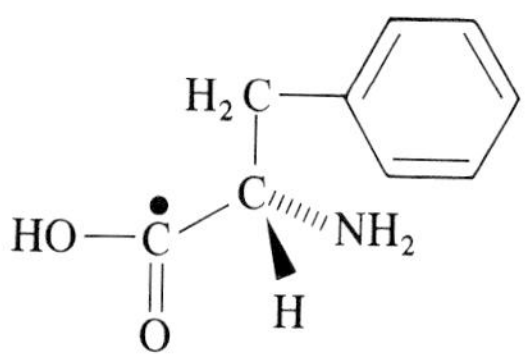	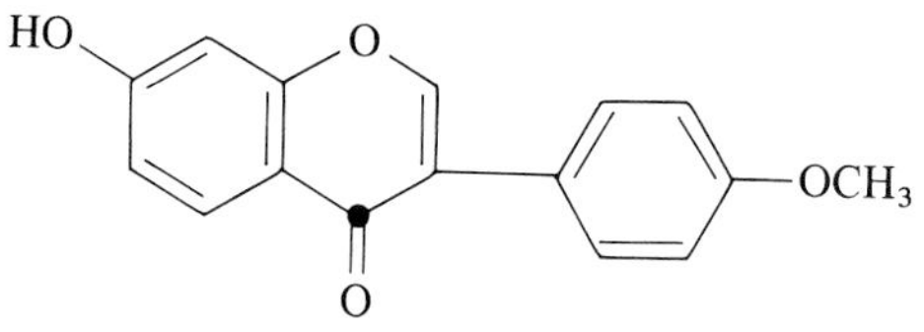	93
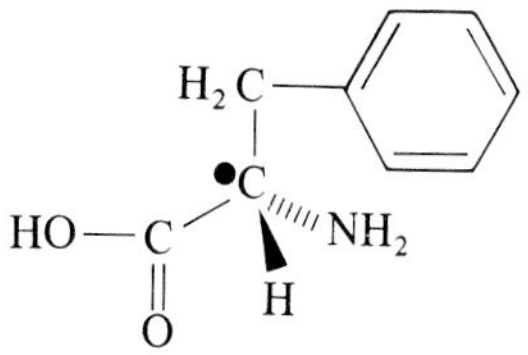	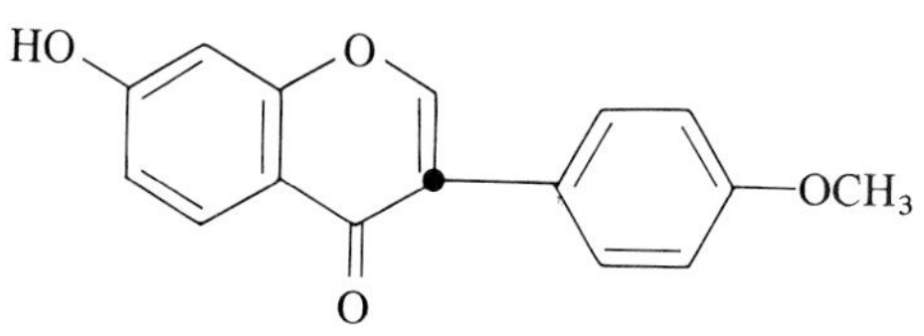	82
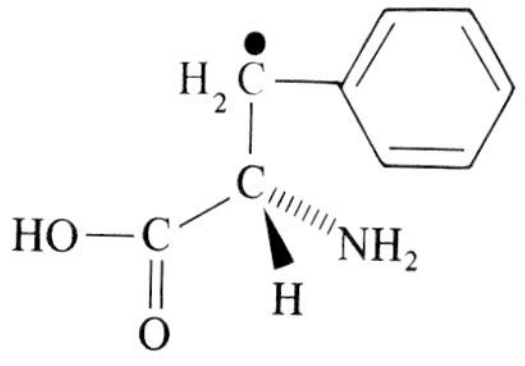	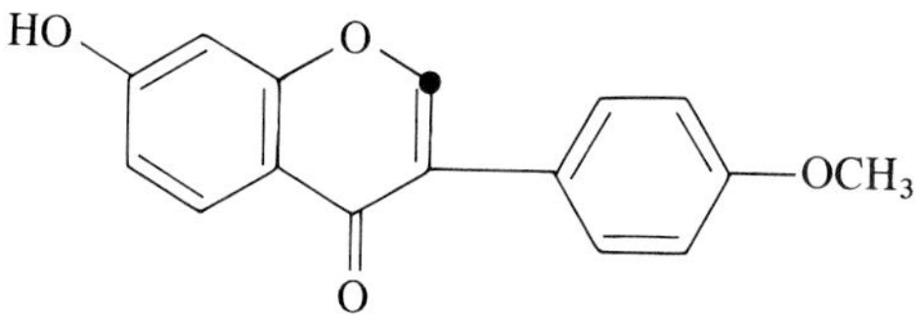	96

Fig. 14.15. Pattern of labelling in the isoflavone, formononetin, produced in red clover from different ¹⁴C-labelled species of L-phenylalanine (● = ¹⁴C).

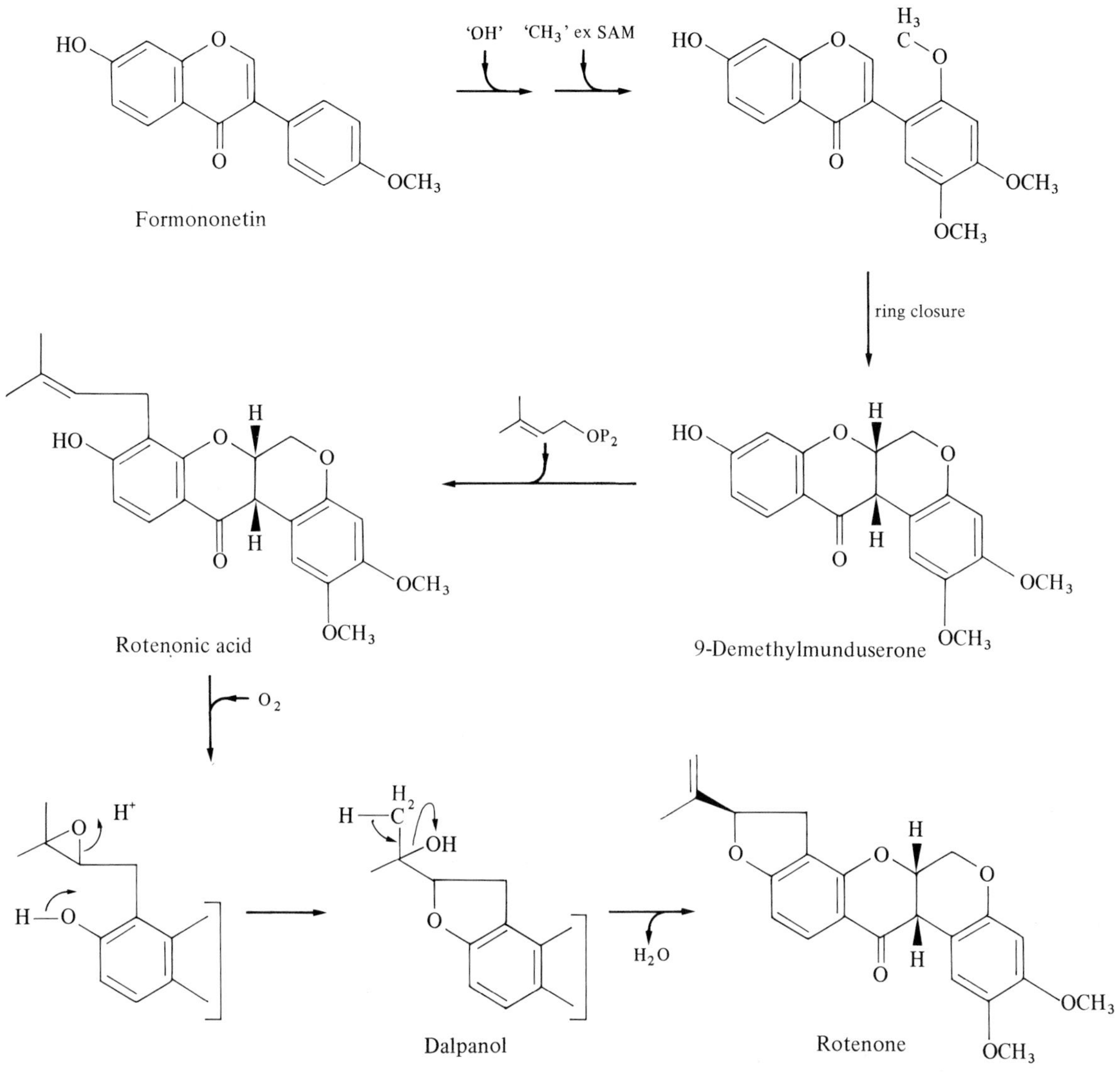

Fig. 14.16. Biosynthesis of rotenone from isoflavones via formononetin (SAM = S-adenosylmethionine).

state a scheme involving non-redox steps has been proposed (Fig. 14.17). The conversion of dihydro-flavonols into anthocyanidins has still to be de-monstrated with enzyme systems. It is reasonable to assume that naturally occurring 3-deoxyantho-cyanidins, such as apigeninidin (14.57), are formed from flavones (*F*, Fig. 14.13) but this has not yet been investigated.

(vi) AURONES

Chalcones are precursors of aurones (*H*, Fig. 14.13) and studies with cell-free systems from

Cicer indicated that a peroxidase-like enzyme is involved, and that the course of the reaction may be that indicated in Fig. 14.18.

(vii) CATECHINS AND FLAVAN-3,4-DIOLS

In contrast to the flavonoids considered pre-viously catechins and flavan-3,4-diols are at a lower oxidation level than chalcones. However, the cate-chins are formed from chalcones via dihydrofla-vonols (*I*, Fig. 14.13). For example, chalcononarin-genin (14.58) is converted into epicatechins, and 1,2-dihydro-kaempferol (Table 14.3) is converted into

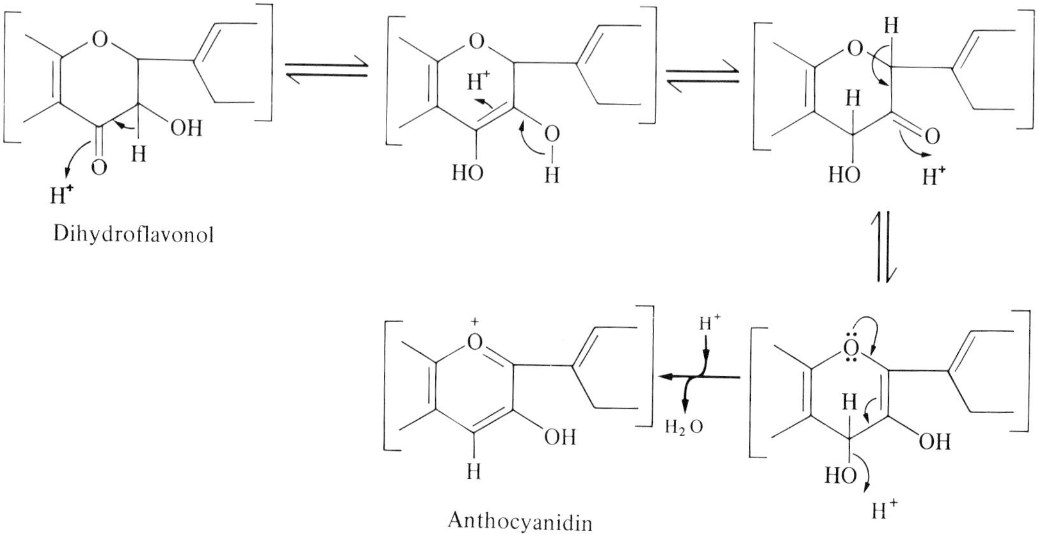

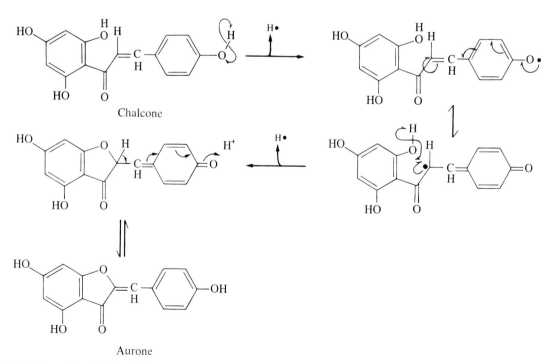

Fig. 14.17. Proposed non-redox mechanism for the conversion of dihydroflavonols into anthocyanidins.

Fig. 14.18. Proposed mechanism for the conversion of chalcones into aurones catalysed by a peroxidase-like enzyme. (⌐ = movement of a single electron in the direction indicated; ⌐ = movement of an electron pair in the direction indicated; H• = hydrogen radical abstracted under the catalytic influence of the enzyme.)

catechins (Table 14.3) in tea leaves. Little is known about the mechanisms involved in producing the different stereoisomers of the catechin family. Similarly little is known about the biosynthesis of flavan-3,4-diols, but because of the various naturally occurring stereoisomers they could be formed by oxidation of catechin epimers (*J*, Fig. 14.13) but probably not directly from flavonols.

The reason why flavan-3,4-diols are sometimes called proanthocyanidins is that they are easily converted into anthocyanidins when heated in the presence of acid; they have no biosynthetic relationship one to the other.

(viii) SECONDARY MODIFICATIONS

(a) *Hydroxylation*

The basic pattern of hydroxylation is illustrated by consideration of the structure of chalcononaringenin (14.58). Hydroxylation at C–4 (chalcone numbering) is almost universal, indicating that *p*-coumaric acid is the usual precursor of the phenyl propane residue of flavonoids. The hydroxylation pattern which is most frequently encountered in ring A (C–2′,4′,6′, chalcone numbering) is that which is expected from its polyketide origin (see Fig. 14.10). Although the basic pattern is widespread compounds with the hydroxyl missing at C–5 (flavonoid numbering) are known, particularly in the Leguminosae; an example is fisetin (14.59) from *Rhus* spp.

Genetic studies suggest that the removal of the C–5 hydroxyl group probably occurs at the polyketide stage and certainly at the chalcone stage at the latest because the presence of the gene controlling the removal results in all flavonoids being equally affected. Furthermore, tracer studies have shown that in normal plants 5-deoxychalcones give rise only to 5-deoxyflavonoids.

Flavonoids with additional hydroxyl groups in ring A are also known; for example, galetin [OH at C–6 (14.60)] and hibiscetin [OH at C–8 (14.61)] from *Galega* and *Hibiscus*, respectively. In contrast to the removal of OH from C–5, the addition at C–6 and C–8 appears to take place after chalcone formation.

In ring B many flavonoids are known with additional hydroxyl groups at C–3′ and C–5′ [e.g. fisetin (14.59) and hibiscetin (14.61)]. Although the matter is by no means settled tracer studies suggest that these additions occur at the chalcone stage although hydroxylated *p*-coumaric acids are also effectively incorporated in some species. The mechanism of hydroxylation *in vivo* is not clear but *in vitro* purified phenolases[3], as for example that from leaves of spinach beet, will hydroxylate 4′-hydroxyflavonoids. However, genetic studies suggest that separate hydroxylation enzymes may exist for different species of flavonoids. For example, gene M in *Antirrhinum* affects the hydroxylation pattern of flavones but not of aurones.

(b) *O-Methylation*

Many *O*-methylated flavonoids are known, for example syringetin (14.46), and there is no doubt that the methylation step occurs late on in the biosynthetic sequence. The reaction is catalysed by conventional methyl transferases with *S*-adenosylmethionine as the methyl donor but the enzymes are specific for hydroxyls in different positions. For example, a mutant of subterranean clover accumulates the isoflavones daidzein (14.62) and genistein (14.63) whereas the native strain contains large amounts of the methylated derivatives formononetin (14.64), biochanin A (14.65) and pratensein (14.66). However, the methylated flavones contain the methyl group at C–3′ and occur to the same extent in the native and mutant strains. These studies demonstrate that methylation occurs

(14.59)

(14.60)

(14.61)

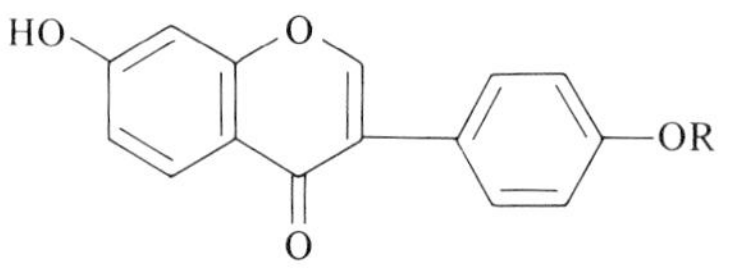

(14.62) R=H
(14.64) R=CH₃

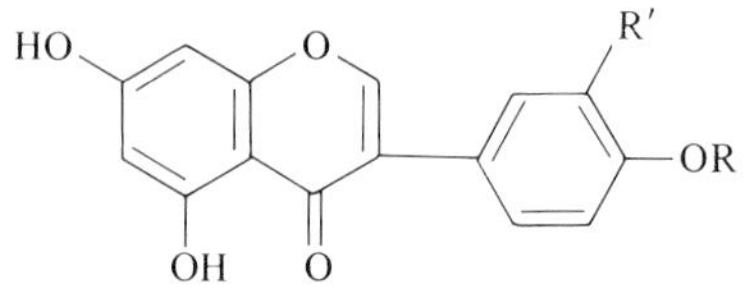

(14.63) R=R′=H
(14.65) R=CH₃; R′=H
(14.66) R=CH₃; R′=OH

late on and that position-specific methylases occur in clover.

(c) O-Glycosylation

Specific glycosyl transferases utilizing UDP-sugars have been obtained from parsley cell suspensions. UDPG:flavone 7-O-glucosyltransferase is specific for the 7-O position and catalyses the first step in the formation of apigenin 7-O-apiosylglu-

coside malonate (Fig. 14.19), which is the production of apigenin 7-O glucoside (A, Fig. 14.19). This in turn is converted into the 7-O-apiosylglucoside in the presence of UDP apiose-flavone apiosyltransferase[4] and UDP-apiose (B, Fig. 14.19). The enzyme is specific for UDP-apiose, which is formed from UDP-glucuronide by decarboxylation and rearrangement [eq. (14.3)]. The final step in Fig. 14.19 is not a glycosylation but the esterification of the glucose residue with malonyl-CoA.

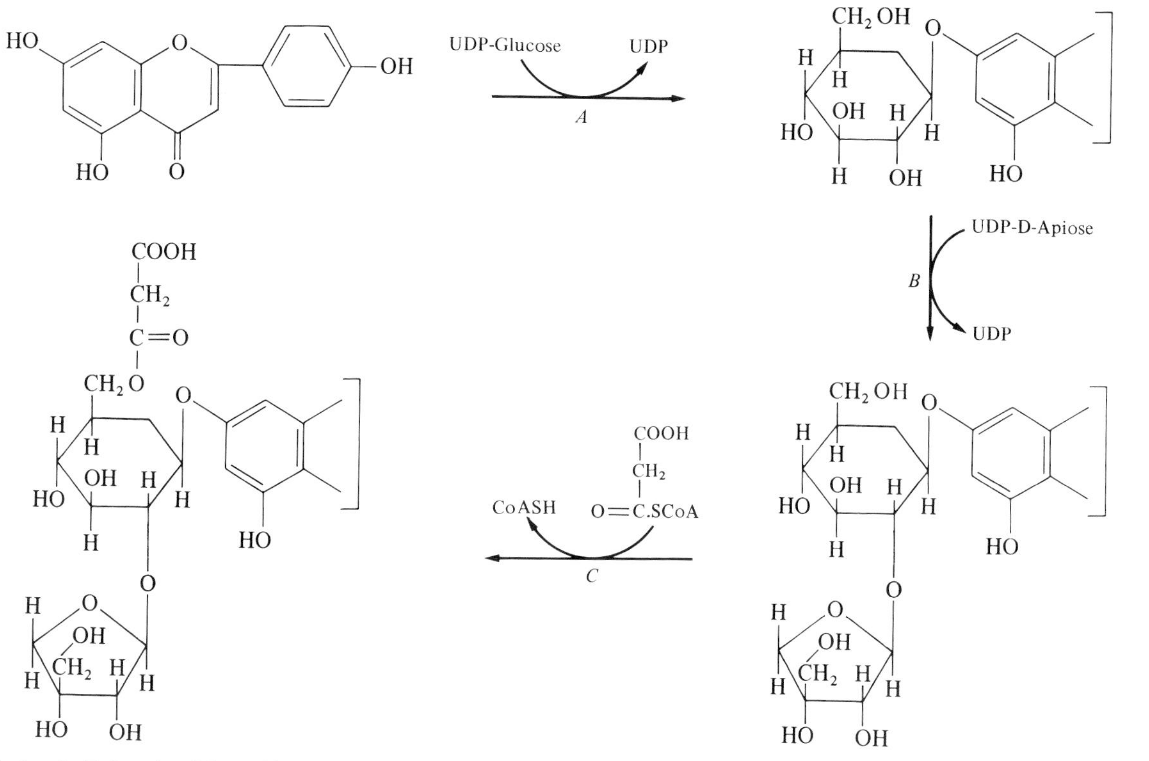

Apigenin 7-O-apiosylglucoside malonate (Apiin)

Fig. 14.19. Steps in the biosynthesis of apigenin 7-O-apiosyl-glucoside malonate, an acylated flavone dioside. (N.B. The exact position of the malonyl residue in the glucose moiety has not been determined but it has been placed in the 6′-O position by analogy with the known occurrence of 6′-O-malonylisoflavone glucosides in *Trifolium* spp.)

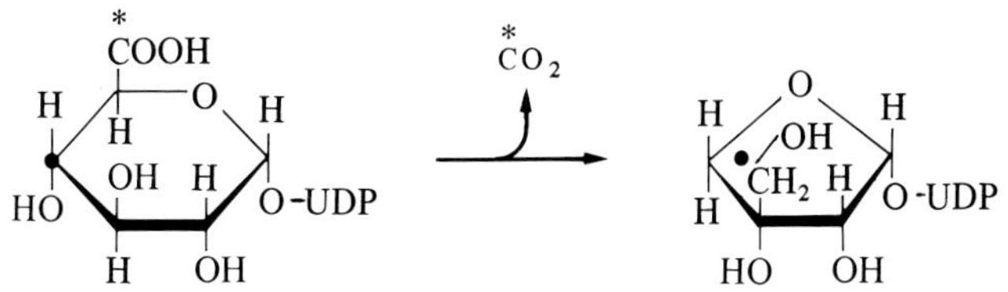

UDP-D-Glucuronic acid   UDP-D-Apiose

$$\text{UDP-D-Glucuronic acid} \quad \xrightarrow{\overset{*}{C}O_2} \quad \text{UDP-D-Apiose} \tag{14.3}$$

(d) *C-Glycosylation*

C-glycosides are found predominantly amongst the flavones and the attachment of the sugar occurs early in the biosynthetic sequence, probably at the chalcone stage. Tracer studies showed that [^{14}C]apigenin was incorporated, in for example *Spirodela*, only into *O*-glycosides but not *C*-glycosides, whereas naringenin (Table 4.3), which is in equilibrium with chalcononaringenin (14.58) (see Fig. 14.13), is incorporated into both *C*- and *O*-glycosides. Furthermore, *C*-glycosides can be *O*-glycosylated. The mechanism of *C*-glycosidation is unknown but probably involves UDP-sugars.

(e) *C-Alkylation*

C-Methylated flavonoids are relatively rare and no biosynthetic studies have been reported. A typical example is sideroxylin (14.67) from *Eucalyptus* spp., which also contains an *O*-methyl group.

Isoprenylation at carbon atoms is also encountered in some sixty flavonoids of which mulberrin (14.68), present in the bark of *Morus alba*, is a typical example. As the formation of terpenes is characterized by C—C bond formation (Chapter 11) it is not unexpected to find *C*-isopentenyl residues in flavonoids. Although no-

thing is known of the mechanism of isoprenylation in flavonoids it can reasonably be assumed that, as in terpene biosynthesis, isopentenyl pyrophosphate is the substrate for a specific prenyl transferase.

(f) *Enzymic polymerization*

Apart from naturally occurring biflavonyls (see Section K), polymerization of flavonoids only occurs after cell damage which allows flavonoids to come into contact with phenolases and similar enzymes. *In vitro* studies with phenolase indicate that catechins are oxidized via a quinone (*A*, Fig. 14.20) and a 6′,8-linked dimer (*B*, Fig. 14.20) to a polymer (*C*, Fig. 14.20). A 6′,8-linked dimer occurs naturally in oak bark. It is such polymerization reactions which cause the rapid browning of damaged plant tissues, by converting colourless precursors into brown polymers.

J. LIGNANS AND NEOLIGNANS

1. Nature and distribution

The name *lignan* was introduced in 1936 to describe a group of naturally occurring phenylpropanoid dimers. The C_6–C_3 units are joined by a

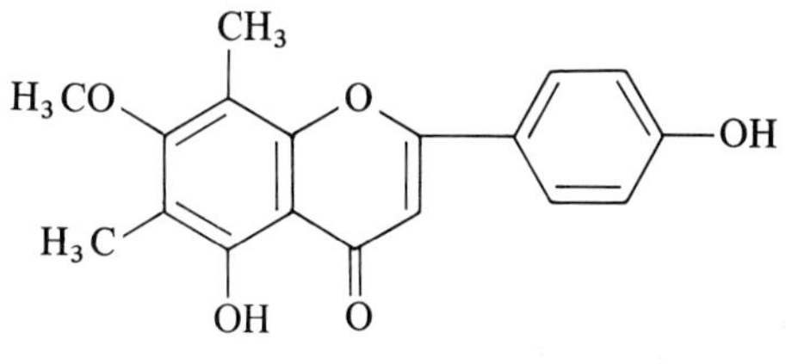

(14.67)

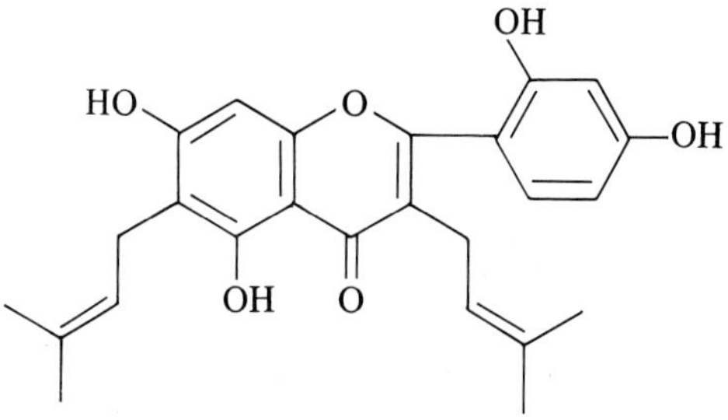

(14.68)

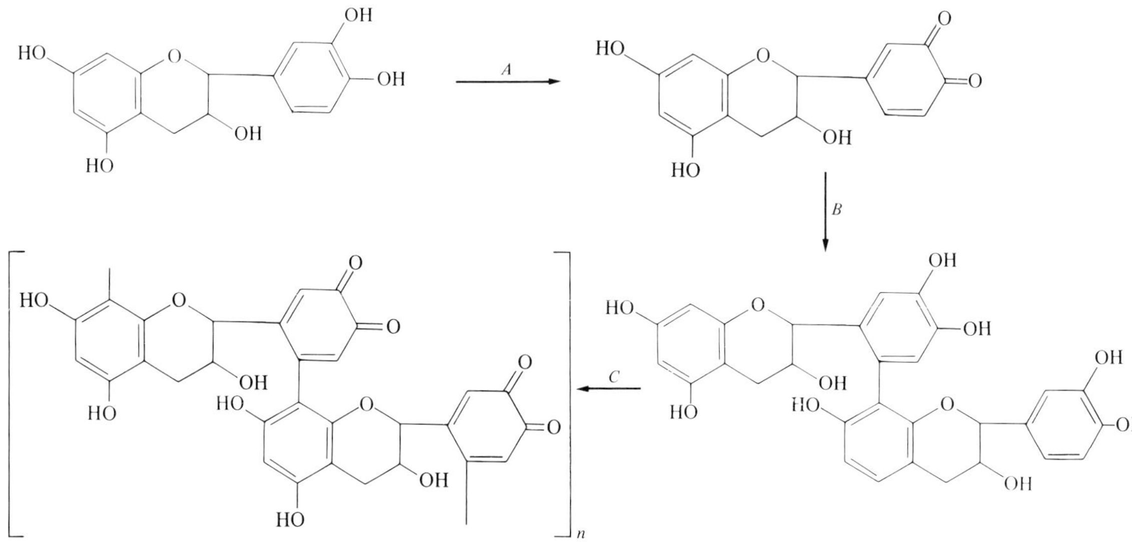

Fig. 14.20. Possible route for the oxidation of catechins to brown polymerized products by phenolases.

carbon–carbon bond between the middle carbons of their side chains (14.69).

Two ways of writing the basic structure are illustrated in (14.70) for dihydroguaiaretic acid; the numbering of the side chain is also indicated. All lignans have chiral centres at C–2 and C–3 and the absolute configuration of many lignans is known. For example, pinoresinol (Table 14.1) is found associated with its epimers diapinoresinol (14.71) and epipinoresinol (14.72). The pinoresinols exemplify a characteristic structural feature of lignans which is the furan ring. Other structural variations occur in olivil (14.73) which makes up 50% of the resinous exudate of the olive tree, and lariciresinol (14.74) from the resin of *Larix decidua* in which one or both methyl groups have been oxidized to hydroxymethylene groups. Some compounds, such as conidendrin (14.75), have an additional carbon–carbon bond between the side chain and one of the aromatic residues.

Neolignans consist of two C_6–C_3 units joined head-to-tail (14.76) instead of tail-to-tail as in lignans (14.69). A typical example is eusiderin (see Table 14.1) which is obtained from the heartwood of the Magnoliaceae, Piperaceae and Lauraceae.

2. Biosynthesis

Little is known of the biosynthesis of lignans. Although they are clearly related to lignins they cannot be synthesized by the free radical mechanism

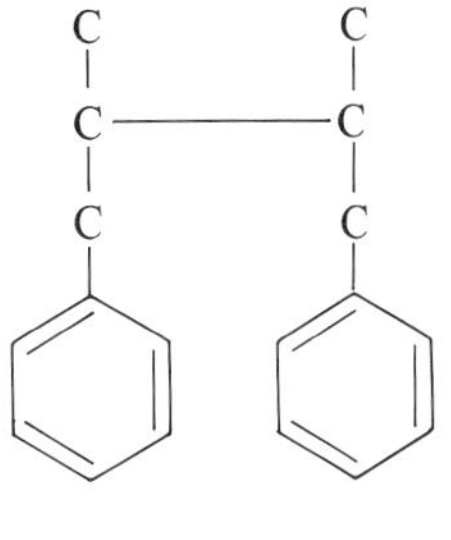

(14.69)

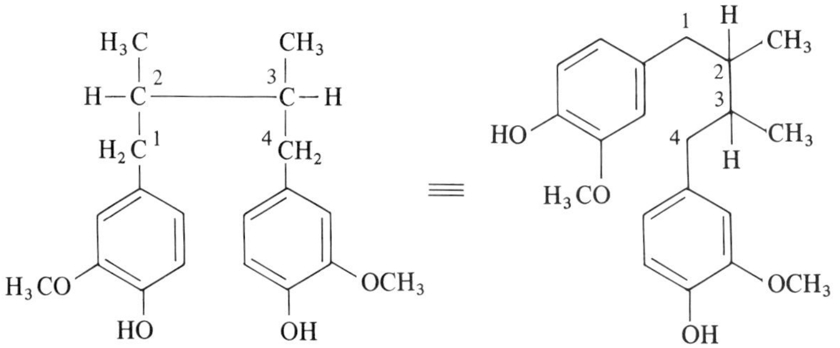

(14.70)

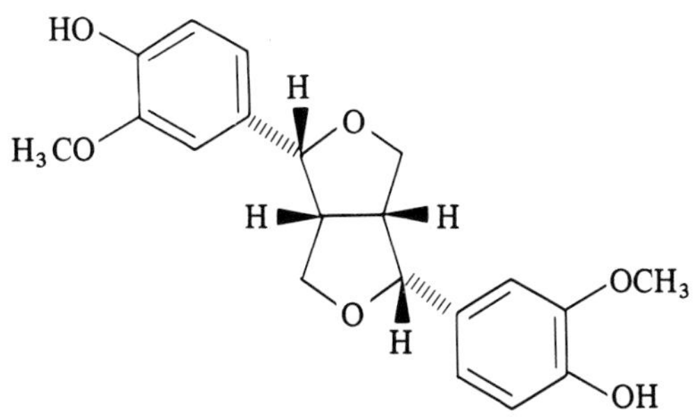

(14.71)

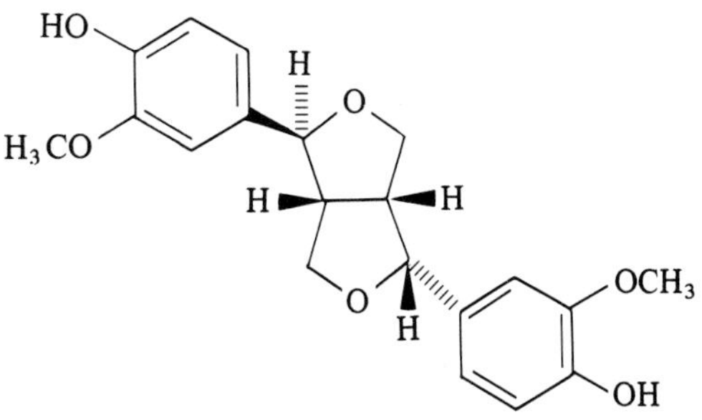

(14.72)

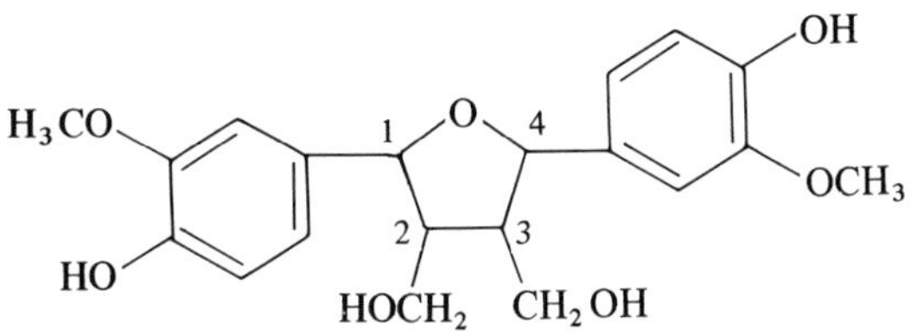

(14.73)

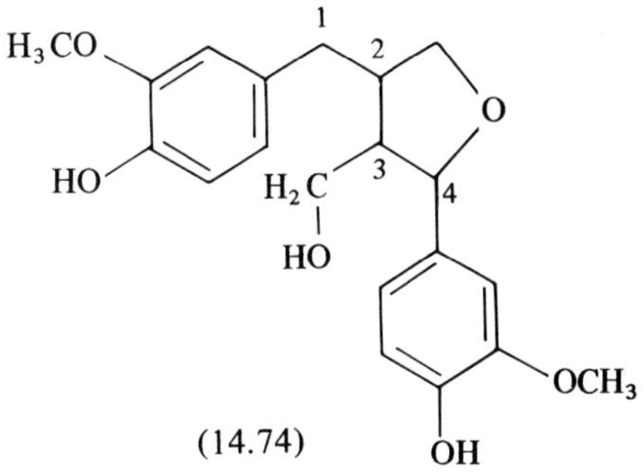

(14.74)

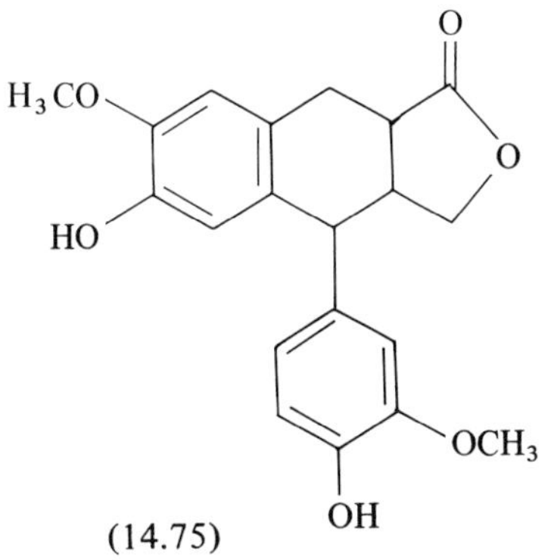

(14.75)

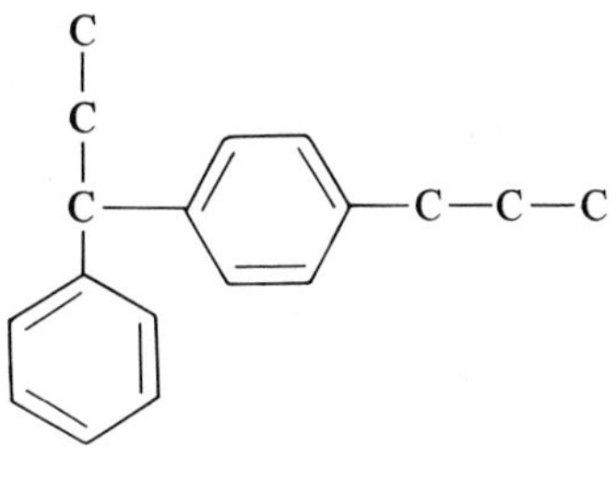

(14.76)

involved in lignin biosynthesis (see Chapter 4, Section D.2 and Figs. 4.28 and 4.29) because they have chiral centres. A possible mechanism (see Fig. 14.21) involves the reductive coupling of two C_6–C_3 units which have unsaturated side chains.

to a great extent confined to the gymnosperms. The most common member of this group is amento-flavone (3',8"-biapegenin) (see Table 14.1). Nothing is known of their formation.

L. TANNINS

1. Nature

K. BIFLAVONOIDS

Many plants have the ability to dimerize flavones by carbon–carbon linkages to form biflavonoids. Over sixty such compounds are known and they are

The term tannin was introduced in 1796 to describe a group of compounds present in some

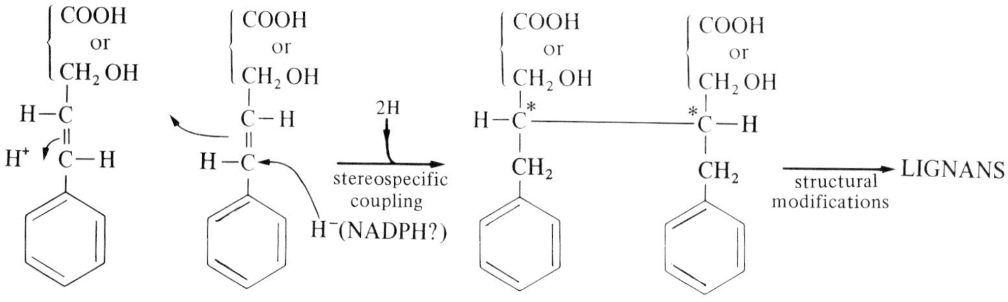

(* = chiral centre)

Fig. 14.21. Possible pathway of lignan biosynthesis involving a stereospecific reductive coupling of C_6-C_3 compounds with unsaturated side chains.

plants, which can *tan* animal skins to produce leather. The word *tan* is derived from the Latin form of a Celtic word for oak, an extract of oak bark being a common tanning agent. To be a good tanning agent a substance should have a molecular weight of between 500 and 3000 and contain sufficient phenolic hydroxyl groups (1–2 per 100 MW) to form effective cross-links with protein. Simple phenolic compounds are too small to form such effective cross-links although they may be adsorbed on to proteins, whereas compounds with very high molecular weights are ineffective because they cannot penetrate between the collagen fibrils of the animal skin.

Tannins and other polyphenols are an unmitigated nuisance to enzymologists working with plant extracts. When they are liberated from one part of the cell during the preparation of plant extracts they *tan* the enzymes liberated at the same time from another part of the cell. This trouble can be avoided to a great extent by adding poly-vinylpyrrolidone in one of its various forms to the extracting media.

Tannins are divided into two main groups,

hydrolysable tannins and *condensed tannins* (non-hydrolysable).

(i) HYDROLYSABLE TANNINS

Hydrolysable tannins contain a core of a polyhydric alcohol, usually, if not always, glucose, which is esterified with either gallic acid (Fig. 14.2) to form the *gallotannins* or with hexahydroxydiphenic acid (14.77) to form the *ellagitannins*. However, it has now been realized that ellagic acid (14.78) is an artefact resulting from the lactonization of hexahydroxydiphenic acid so the name ellagitannin is inappropriate.

The simplest example of a hydrolysable tannin is Chinese Tannin from sumach (*Rhus.* sp.); (14.79) indicates one possible structure and it is easy to appreciate from this how careful hydrolysis with acids and enzymes yields only gallic acid and glucose.

(ii) CONDENSED TANNINS

Condensed tannins are made up only of phenols of the flavone type and are often called flavolans

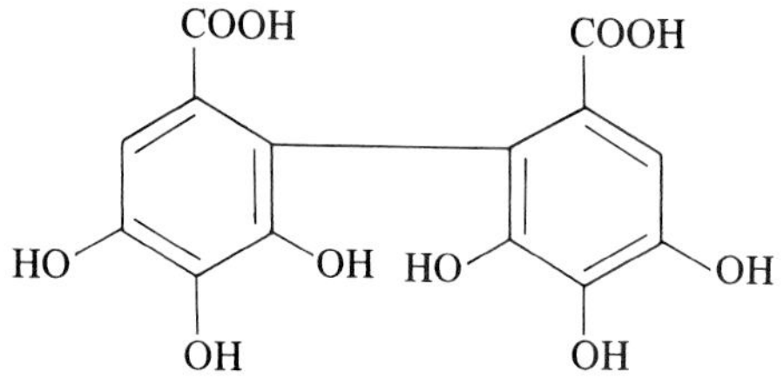

(14.77)

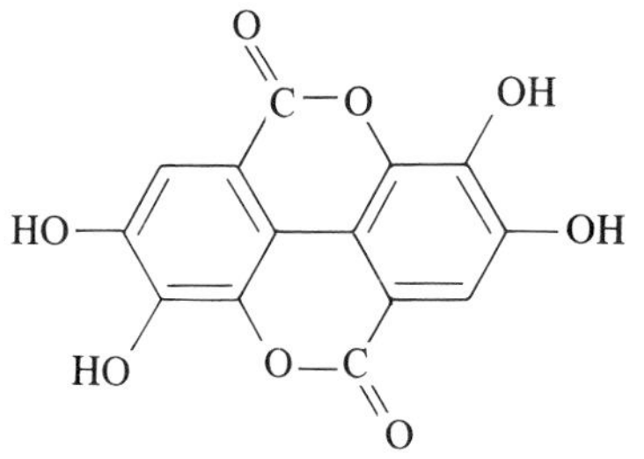

(14.78)

(14.79)

because they are polymers of flavans such as flavan-3-ol (Table 14.3) or flavan-3,4-diols (pro- or leucoanthocyanidins) (Table 14.3). In contrast to the hydrolysable tannins they never contain sugar residues. A typical condensed tannin can be represented by the dimer procyanidin to which further molecules of flavans can be added as indicated (14.80). On treatment with hydrolytic agents condensed tannins yield no significant amounts of compounds of low molecular weight; on the contrary they tend to polymerize, particularly in acid, to yield amorphous compounds, which are often red, called *phlobaphenes*.

2. Distribution

Tannins in small amounts are widespread but large accumulations (45% dry wt.) of hydrolysable tannins are found in the seed pods of Algarobilla (*Caesalpinia brevifolia*) and Divi-divi (*C. coriaria*). Similar concentrations of condensed tannins are found in the bark of eucalyptus and mangrove (*Rhizophora* sp.). Even higher concentrations are found in pathogenic tissues, for example the galls produced on the leaves of *Rhus semialata* are made up of 64% hydrolysable tannins.

3. Biosynthesis

No details are available on the reactions involved in tannin biosynthesis.

M. MELANINS

Melanin is frequently used in a general sense to include all deep brown or black naturally occurring pigments. Within this generalization, however, there exist many different structures. Melanins characteristic of plants are called *catechol melanins* because they yield catechol on alkali fusion. Such compounds have been found in the seeds of *Helianthus annuus* and *Citrullus vulgaris* and in the spores of *Ustilago maydis* and in the ascomycete *Daldinia concentrica*. A possible structure of the melanin from *D. concentrica* is (14.81) which is formed by the oxidation of 1,8-dihydroxynaphthalene because *D. concentrica* produces a black quinone (14.82) as well as (14.81).

Animal melanins, usually termed *eumelanins*, are formed in essence by the polymerization of indole-5,6-quinone (14.83) which arises from the amino acid tyrosine by the action of the tyrosinase enzyme

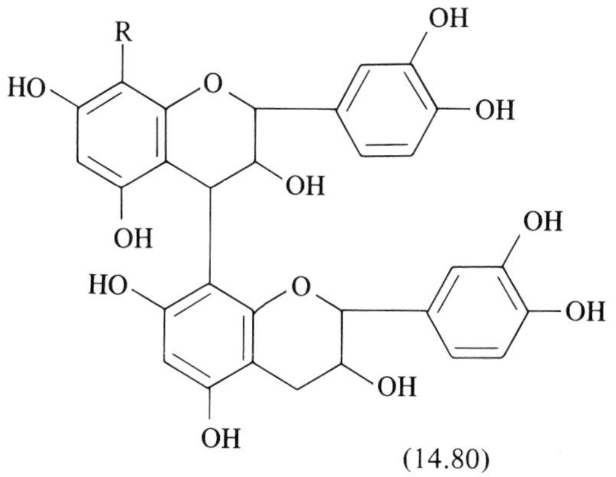

(14.80)

R = [flavan-3-ol]$_n$

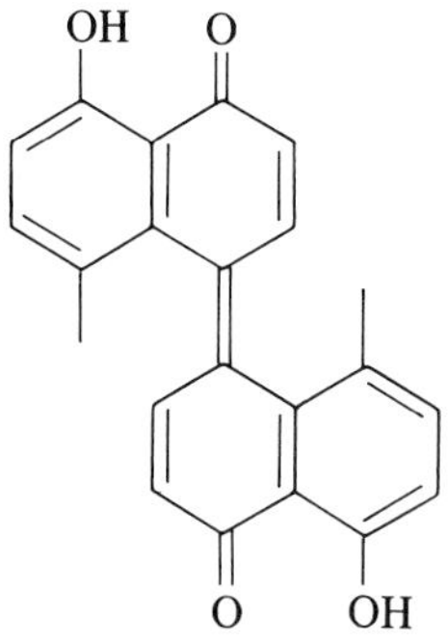

(14.81)

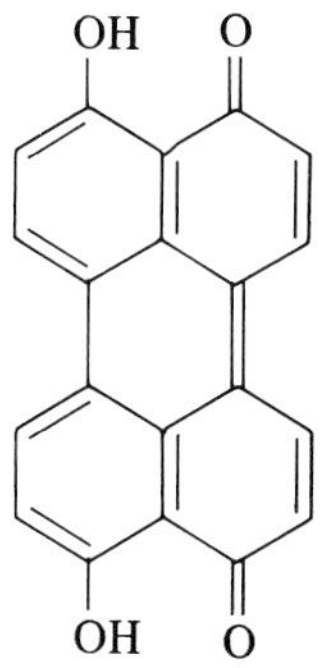

(14.82)

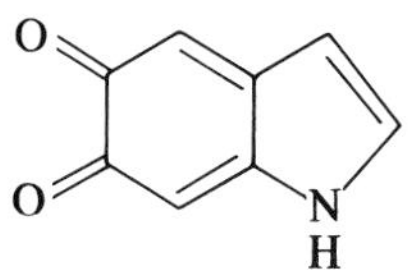

(14.83)

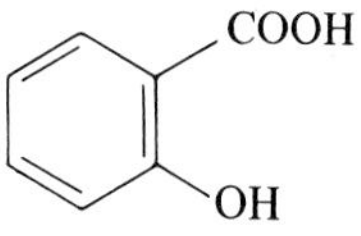

(14.84)

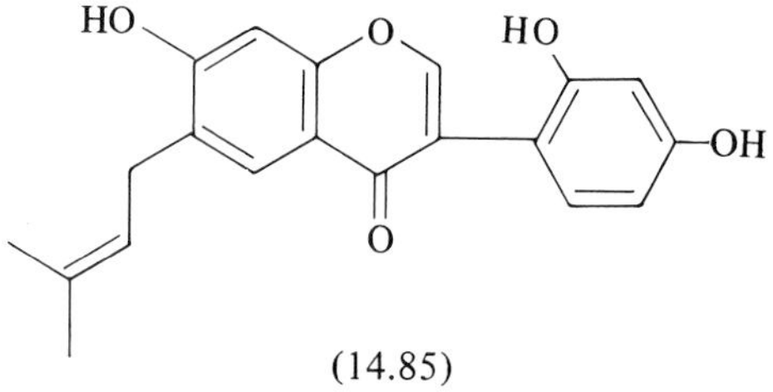

(14.85)

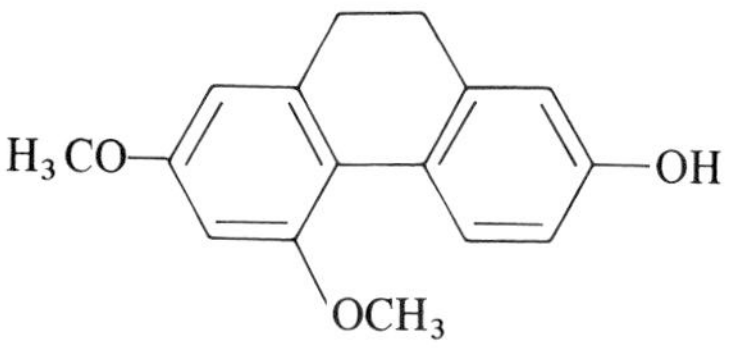

(14.86)

complex. Some plant melanins, particularly those formed after cellular damage, are probably eumelanins but evidence is circumstantial, based on the co-existence of the substrate, enzyme and melanin in the particular part of the plant involved. In contrast to the catechol melanins, eumelanins contain nitrogen.

N. REGULATION OF SYNTHESIS OF PHENOLICS

Biochemical studies on the regulation of any cellular activity can only be fully investigated if the enzymology of the system is clearly defined. The latter has not been accomplished for the synthesis of phenolics in intact higher plants, but the situation is much clearer in plant tissue cultures.

A key enzyme in these studies is phenylalanine ammonia-lyase[5] (PAL), which as indicated earlier yields cinnamic acid, a precursor of lignin (Chapter 4, Section D.2) and some alkaloids [see Chapter 13, Section C.6(ii)(b)] as well as the plant phenolics. The activity of PAL is clearly a phytochrome-controlled reaction (Chapter 12); its activity in buds of etiolated pea seedlings is greatly increased following exposure to red light and this increase does not occur if the red-light treatment is followed by a dose of far-red light. It has been shown that the site of this effect is on the *m*RNA for PAL.

A problem which has to be considered in intact plants is that PAL does not appear to be a rate-limiting enzyme for phenolic biosynthesis because it is present in much greater amounts than required and the observed variations in activity do not parallel changes in phenolic accumulation. The control may be one step back with a key factor being the control of the channelling of phenylalanine into either protein or phenolics according to the physiological requirements of the plant. However, there is no doubt that in some conditions in plant tissue cultures, PAL is certainly rate-limiting. Studies on cultured parsley cells indicate that there are two groups of enzymes which are separately regulated although closely related. The first, Group I, are those concerned with general phenylpropanoid formation, and the second, Group II, are concerned with flavonoid glycoside synthesis. Both are induced by ultraviolet light but only Group I enzymes are induced by diluting the cultures.

Two isoenzymes of 4-coumarate:CoA ligase have been observed in soya bean cultures. The first has a very low K_m value for the substrates of the lignin pathway and is therefore probably concerned with lignin biosynthesis. The second has different substrate specificities and, although its direct participation in flavonoid biosynthesis cannot yet be claimed, it looks highly likely because the single 4-coumarate:CoA ligase present in parsley cell cultures, which specifically synthesizes flavonoids, has properties very similar to the second isoenzyme in soya-bean cultures.

The isomerases are strongly inhibited by AMP and the degree of inhibition depends also on the amount of ATP present; this means that both pathways can be controlled by the energy charge in the cell.

O. FUNCTION OF PHENOLICS

1. Biochemical

There is as yet little evidence that phenolics play a key biochemical role in plant growth and development although there are some pointers in this direction. Apart from lunularic acid (Table 14.1), which appears to replace abscisic acid in liverworts as a phytohormone (Chapter 15, Section E), no phenol plays a clear primary hormonal role. There are many reports, however, of their effect *in vitro* on certain hormones or enzyme systems related to hormone synthesis or action. Whether these effects exist *in vivo* still remains moot. Perhaps the most likely function is in the production of ethylene from methionine (Chapter 15, Section F.4) in which a *p*-coumaric acid ester is a necessary co-factor.

2. Ecological

The importance of flavonoid pigments in flowers and fruit in pollination and seed dispersal by attracting birds and insects by their colour is now well established. However, the colourless flavones and flavonols which are ubiquitous in flowers also play an important part in attracting insects because of their absorption of near ultraviolet radiation which insects can detect. In black-eyed susan (*Rudbeckia hirta*), for example, such flavonoids are present only at the base of the petals but their strong absorption in the spectral region 240–380 nm means that pollinating insects such as bees can accurately locate the centre of the flower where the pollen and nectar are.

Other actions include their activity as allelopathic agents (excretory products which may be autotoxic or affect neighbouring plants), e.g. salicylic acid (14.84) in *Quercus falcata*, as feeding deterrents, e.g. the tannins, as antimicrobial agents, e.g. luteone (14.85) in Lupins, and as phytoalexins, e.g. orchinol (14.86) in *Orchia militaris*.

SUGGESTIONS FOR
FURTHER READING

General

Harborne, J. B. (1980) in *Encyclopaedia of Plant Physiology*, Vol. 8, p. 329. Springer, Heidelberg.
Swain, T. (1976) in *Chemistry and Biochemistry of Plant Pigments*, (Goodwin, T. W., ed.), Vol. 1. Academic Press, p. 425.
Harborne, J. B., Mabry, T. J. and Mabry, H. (1975) (eds.), *The Flavonoids*. Chapman & Hall, London.

Biosynthesis

Wong, E. (1976) in *Chemistry and Biochemistry of Plant Pigments* (Goodwin, T. W., ed.), Vol. 1. Academic Press, p. 464.
Hahlbrock, K. and Grisebach, H. (1979) *Ann. Rev. Plant Physiol.* **30**, 105.

Function

Harborne, J. B. (1976) in *Chemistry and Biochemistry of Plant Pigments* (Goodwin, T. W., ed.), Vol. 1. Academic Press, p. 736.

ENZYMES

1. Shikimate: NADP$^+$ 3-oxidoreductase, EC 1.1.1.25.
2. Flavanone lyase (decyclizing), EC 5.5.1.6.
3. Benzenediol: oxygen oxidoreductase, EC 1.10.3.2.
4. UDP apiose: 7-O-β-D-glucosyl-5,7,4′-trihydroxyflavone apiofuranosyltransferase, EC 2.4.2.25.
5. L-Phenylalanine ammonia-lyase, EC 4.3.1.5.

CHAPTER 15

Phytohormones and Related Compounds

A. INTRODUCTION

The term 'hormone' was coined in 1902 by Bayliss and Starling to describe a substance present in higher animals that is now called secretin. Secretin is a low molecular weight protein synthesized in the cells of the duodenal mucosa. When the acidic contents of the stomach empty into the duodenum these cells are stimulated to secrete secretin into the blood of nearby capillaries. The blood circulatory system then distributes the secretin to all regions of the body. When it reaches the pancreas the acinar cells are stimulated to secrete copious amounts of pancreatic juice, via the pancreatic duct, into the lumen of the small intestine where it participates in the further digestion of the food.

The word 'hormone' is derived from a Greek root meaning 'to excite' or 'to arouse'. In the original case, the hormone 'excited' the pancreas to secrete pancreatic juice. However, it is now known that not all hormones 'excite'; some hormones have the opposite effect, they inhibit rather than stimulate. In fact secretin itself has an inhibitory effect in addition to the stimulatory one just described. It inhibits the secretion of hydrochloric acid by the oxyntic cells of the gastric mucosa. This also shows that a given hormone may stimulate a particular process in one type of cell whilst inhibiting a different process (or, in some cases, the same process) in another type of cell.

The definition of the term 'hormone' in its original (i.e. higher animal) context is that it is a chemical substance that is synthesized in tiny quantities in the cells of one tissue in the body, is then secreted in tiny quantities into the blood stream and is carried by the blood to the cells of distant organs or tissues to modify their structure and/or function. The tissue producing the hormone is called an 'endocrine' tissue and the tissue affected by the hormone is called the 'target' tissue.

The hormones of higher animals fall into three chemical classes: (i) amino acid derivatives, e.g. the thyroid hormones and the catecholamines, (ii) peptides, e.g. the hypothalamic and pituitary hormones and (iii) the steroid hormones, e.g. the adrenocorticoid hormones, the androgens and the oestrogens. However, they fall into two classes on the basis of their action at the molecular level. One class, which includes the catecholamine and peptide hormones, act at the outer surface of the plasma membrane of the cells of their target tissue; they do not pass through the plasma membrane and enter the cell. They reversibly bind to a receptor molecule, usually a protein, located in the outer surface of the plasma membrane. Receptor molecules are present only in the membranes of target cells; each hormone has its own specific receptor. The hormone–receptor complex then activates a nucleotidyl cyclase located in the inner surface of the plasma membrane which catalyses the formation of a $3':5'$-cyclic nucleotide (e.g. $3':5'$-cyclic AMP) in the cytoplasm. This then initiates a cascade of events which amplifies the original hormonal signal and results in the activation of some enzymes and the inactivation of others; the enzymes affected are characteristic of the hormone and of the target cell. The enzymes that are activated (i.e. switched on) and inactivated (i.e. switched off) catalyse rate-limiting steps in important metabolic processes. Thus this class of hormones exerts control over its target cells by switching on some metabolic processes and switch-

ing off others. Since the exercise of this control involves the activation or inactivation of enzyme molecules already present in the cell and does not require that new enzyme molecules be synthesized, response to this type of hormone is rapid.

The second class of higher animal hormones, which include the steroid and thyroid hormones, are able to enter their target cells. In fact they can pass into and out of all cells because their lipophilic nature allows them to cross the lipid bilayer of the plasma membrane. However, only their target cells possess receptor molecules which bind them and allow them to exert their biological effect. These receptor molecules are proteins and, in most cases, are located in the cytosol. Each hormone has its own specific receptor. The hormone–receptor complex is translocated into the nucleus, usually after some slight modification to the protein moiety, where it switches on specific genes. These are then transcribed into mRNAs which pass into the cytosol where they are translated into specific proteins. These proteins may have a structural or an enzymic role. In the latter case they catalyse the rate-limiting steps in important metabolic sequences but in both cases they are frequently concerned with growth (one of the characteristic responses to androgens and oestrogens). Since this class of hormones exerts its control by synthesizing new enzyme molecules, the response that is elicited is relatively slow.

As knowledge of the control exerted by hormones over the bodily functions of higher animals increases it is increasingly clear that a great deal of interplay amongst the hormones exists in which the effect of some hormones is antagonistic to that of others. For instance, the normal level of blood glucose (~ 80 mg/100 ml blood) is the resultant of several hormonal influences, some stimulating a rise in the level (e.g. glucagon, glucocorticoids) and others stimulating a fall in the level (e.g. insulin). Similarly a delicate balance between bone deposition and resorption is maintained by calcitonin and the parathyroid hormone which have antagonistic effects.

Because of its original application the term 'hormone' must be used with care in the plant context. Clearly the definition given earlier cannot be applied to the substances that are known as plant hormones (or phytohormones) because plants do not have a blood circulatory system for their distribution. Nevertheless it is useful in prompting the sort of questions about a given phytohormone that plant physiologists and biochemists ought to answer, namely (i) what internal or external stimulus causes its synthesis and/or secretion?, (ii) in what intracellular organelle, in what cell type, in what part of the plant is it synthesized?, (iii) what is its target tissue?, (iv) how is it transported—from cell to cell or within the vascular tissues?, (v) what is its physiological action? and (vi) how is its physiological action brought about at the molecular level?

Five major hormones (or groups of hormones) are generally recognized as occurring in higher plants: (i) the auxins, (ii) the gibberellins, (iii) the cytokinins, (iv) abscisic acid and (v) ethylene. In general terms, the auxins, gibberellins and cytokinins are plant growth promoters whilst abscisic acid and ethylene are plant growth inhibitors.

The existence of other phytohormones has been postulated and indeed the evidence for some of them is quite persuasive but they have yet to be isolated and identified. Probably the best-known example is florigen, a hormone that is supposedly formed in leaves under favourable daylength conditions and translocated to the shoot meristem where it promotes flowering.

Additionally, there are many substances present in higher and lower plants that have profound stimulatory or inhibitory effects on higher plant growth when supplied exogenously at very low concentrations. Though it is doubtful whether these compounds regulate plant growth *in vivo*, a few are described briefly in Section G. Lower plants also produce hormones, most of which appear to be concerned with the regulation of their sexual reproduction; some of them are also described briefly in Section G.

B. AUXINS

1. Discovery and characterization

The first of the plant hormones to be investigated were the auxins. The history of their discovery goes back to 1880 when Charles Darwin and his son Francis were investigating phototropism, the bending of plants towards a light source. In this work they used seedlings of ornamental canary grass (*Phalaris canariensis*), the coleoptiles of which were strongly phototropic. A coleoptile is a protective sheath covering the first leaves of seedling grasses; it is usually conical with a pore at the tip through which the leaves eventually emerge, but in some primitive grasses it is an open leaf. It grows principally by cell elongation and is characteristically phototropic.

The Darwins showed, by the experiments outlined in Fig. 15.1, A, that light was detected only by the tip of the coleoptile and that the bending response occurred in a lower region of the coleoptile. They concluded that when light is detected 'some influence is transmitted from the upper to the lower part, causing the latter to bend'.

In 1913 Boysen-Jensen demonstrated that the 'influence' postulated by the Darwins was more likely to be chemical rather than physical (e.g. electrical) by the experiment shown in Fig. 15.1, B. He excised the tip from an oat coleoptile, put a thin layer of gelatin over the cut end of the stump and then placed the tip on the gelatin. The coleoptile resumed its growth and when the tip was exposed to light from one side the stump bent towards the light source. This indicated that the 'influence' was able to pass from the tip, through the gelatin layer, to the stump and strongly suggested that it took the form of a diffusible chemical compound.

Paál in 1919 demonstrated that this diffusible chemical had a growth-promoting effect lower down the coleoptile. He excised the tip of the coleoptile (but not that of the rolled-up foliage leaf within), then replaced it asymmetrically on the stump and removed any source of light. He found that if the tip were placed to the left the coleoptile stump bent to the right and vice versa. This indicated that the chemical had diffused from the tip preferentially into the region of the coleoptile immediately beneath it and caused it to grow faster than the region not beneath it. This asymmetric growth caused the coleoptile to bend.

In 1926–8 Went proved that the growth stimulus is a diffusible chemical. The critical experiment is shown in Fig. 15.1, D. Went placed coleoptile tips, immediately after excision, on a layer of agar and left them there for about an hour. He then removed the

A. The Darwins' experiments on phototropism (1880)

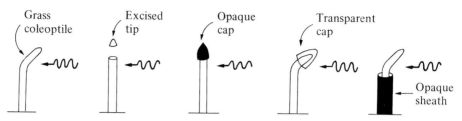

B. Boysen-Jensen's experiment (1913)

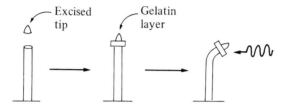

C. Paál's experiment (1919)

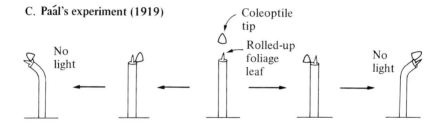

D. Went's experiment (1928)

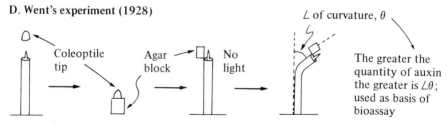

Fig. 15.1. Some of the key experiments in the discovery of auxin. (⬅〰〰 = light.)

tips and cut the agar into small blocks which he placed asymmetrically on freshly cut coleoptile stumps. These were left in the dark. He found that the coleoptile stumps bent to the left if the agar blocks were placed to the right of centre and vice versa. No such bending was caused by blocks of pure agar. This indicated that the growth-promoting influence had diffused from the coleop-

tile tips into the agar and then from the agar blocks into the coleoptile stumps. This clearly showed that it was chemical rather than physical because physical stimuli (e.g. electrical) cannot be stored in agar.

Went called this diffusible chemical 'auxin' (from the Greek word *auxem*, to grow). His experiment was of considerable significance for it was the first proof of the presence of higher plant analogues to the animal hormones and it formed the basis of an auxin bioassay (see Fig. 15.1, D).

The next step was to isolate auxin and to identify it. However, in spite of the availability of a bioassay and a known auxin source, this proved difficult. These difficulties largely stemmed from the lack of any knowledge as to what type of chemical structure auxin had, to its presence in plant tissues in such small amounts and to its destruction by oxidative enzymes in homogenates. The latter was overcome by grinding the tissue in chloroform and then adding 0.2 vol. of 1 M HCl. The resulting chloroform extract on evaporation yielded a lipoidal residue containing auxin activity. This was then extracted into water. The elucidation of the identity of auxin was facilitated by the discovery of three compounds with auxin activity in human urine. One of these, termed heteroauxin at the time, was shown in 1934 to be indole 3-acetic acid (Fig. 15.4, IV). The similarity of its properties to those of the auxin extracted from plant tissues led to the belief that the two were one and the same. However, it was not until 1972 that 'diffusible' auxin (i.e. the auxin diffusing from coleoptile tips into agar) was conclusively identified as indole 3-acetic acid (IAA) by mass spectrometry. Prior to that, however, IAA had been identified by less rigorous methods in a whole range of tissues (e.g. shoots, roots, buds, leaves, cotyledons and fruits). The quantity of IAA, as measured by bioassay of chromatographically purified extracts, in plants is in the range 1–100 μg kg^{-1} wet weight of tissue.

Although IAA is probably the main auxin it certainly is not the only substance present in plants to have auxin activity. Many of the other auxins are indole derivatives closely related to IAA (e.g. indole 3-acetonitrile, IAN; Fig. 15.4, VII) but others are not (e.g. phenylacetic acid). It is because of the existence of more than one compound in plants with auxin activity that plant physiologists often speak of auxins (*plur.*) rather than auxin (*sing.*).

Auxins appear to be synthesized mainly in meristematic tissues such as those of shoot and root apices, developing leaves, flowers and fruits. However, auxins are to be found in most regions of the plant body; this follows from the fact that they are translocated from their site of synthesis to their site of action.

2. Movement of auxin in the plant

(i) LATERAL MOVEMENT

Lateral movement of auxin in shoot and root tissues has been observed.

Lateral movement of auxin in coleoptile tips (a shoot tissue) occurs in response to light. This has been demonstrated by the experiment outlined in Fig. 15.2, A. The tips of two coleoptiles are placed on agar blocks partitioned by a barrier impermeable to auxin. In one tip, (a), the barrier extends only slightly into its base whilst in the other, (b), the barrier divides it completely into two parts. The tips are illuminated from one side only. The quantity of auxin present in the two half-blocks from (a) and (b) are subsequently determined by the Went bioassay procedure. The two half-blocks from (b) have identical quantities of auxin in them. However, the half-block from (a) remote from the light source contains about three times as much auxin as the half-block nearest the light source, which in turn contains about half that of the half-blocks from (b). This indicates that light causes the lateral movement of auxin from the illuminated side of the coleoptile tip to the shaded side. How this is brought about is not clear although it must involve the detection of the light stimulus by a pigment in the tip (carotenes and flavins have been suggested) and the coupling of this to auxin movement.

Lateral movement of auxin in shoots also occurs in response to gravity. When a plant is laid on its side the shoot bends upwards; this is a negative geotropic response for the shoot is turning away from the gravitational pull. It is found that in a horizontally placed plant the concentration of auxin increases in the lower side of the shoot and decreases in the upper side. This causes the lower side to grow faster than the upper side and the shoot to turn upwards. Again the mechanism of this process is not understood. Clearly the meristematic cells in the shoot tip detect the pull of gravity, possibly by its effect on the distribution of intracellular inclusions, and then cause a differential auxin movement.

Lateral movement of auxin in response to gravity also occurs in roots. The roots of a horizontally placed plant bend downwards; this is a positive geotropic response for the roots are bending in the direction of the gravitational pull. It is found that in

A. Lateral movement of auxin in the coleoptile tip

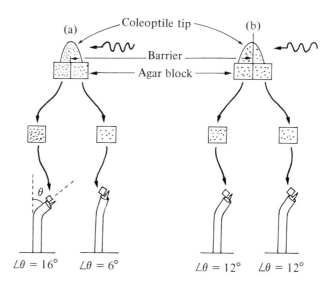

B. Polar movement of auxin in the coleoptile

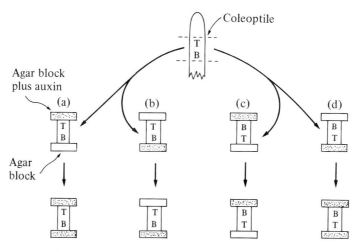

Fig. 15.2. Movement of auxin in the coleoptile. (⟿ = light; T = top; B = bottom; the more dots per unit area, the greater the auxin concentration.)

such a plant the concentration of auxin increases in the lower side of the root and decreases in the upper side, thereby exactly matching its distribution in the shoot. However, the response of the root to this asymmetric auxin distribution is the reverse of that

of the shoot; the side with most auxin grows less than the side with least, causing the root to turn downwards. This different effect of auxin in the root has been explained by its induction of the synthesis of the growth inhibitor ethylene (see Section B.4);

however, recent evidence indicates that abscisic acid may be more important than auxin in root geotropism [see Section E.5(i)]. The detection of gravity by roots appears to involve the starch grains of the amyloplasts in the root cap which redistribute themselves under the gravitational pull.

The movement of auxin in the shoot and root is polarized. The term polarized means that auxins are usually transported along the longitudinal axis of the plant more rapidly in one direction than the other. However, this direction is different in different parts of the plant. In shoot tissues (e.g. coleoptile) auxins move basipetally (i.e. apical→basal). In roots the situation is more complex; over the greater part of the length of the root auxin movement is acropetal (i.e. basal→apical) but near the root apex it is basipetal.

These polar auxin flows are, however, the resultant of two opposite but unequal flows. Thus in the shoot the basipetal flow mentioned above is the resultant of a metabolically driven basipetal flow and an acropetal flow, probably due to passive back diffusion, in the ratio of about 10:1.

The rate of the polar movement of auxin is roughly the same in shoots and roots, namely about 1 cm hr^{-1}. In both shoots and roots auxins appear to be transported mainly in vascular tissues, particularly cambium and newly formed phloem.

The polar movement of auxin in the coleoptile, a shoot tissue, is indicated by the experiment outlined in Fig. 15.2, B. Sections of coleoptile are removed from just beneath the tip. The top (T) and bottom (B) of each section are noted. Two sections [(a) and (b)] are placed vertically the right way up whilst two others [(c) and (d)] are placed vertically upside down. Agar blocks containing auxin (many black dots) are placed on the cut surfaces of (a) and (c) which are now uppermost and on the cut surfaces of (b) and (d) which are now bottommost. Agar blocks containing no auxin (no dots) are then placed against the remaining cut surface of each section. After some hours the quantity of auxin in all the agar blocks is determined by bioassay. Both blocks from (a) and from (d) contain roughly equal amounts of auxin indicating that auxin has moved through the coleoptile section in the direction T→B even though this was against the pull of gravity in the case of (d). Only the agar block that contained auxin at the start possesses any at the end of the experiment in (b) and (c) indicating that auxin has not moved through the coleoptile section in the direction B→T in spite of the fact that the pull of gravity would have assisted such a movement in (c).

The polar transport of auxin appears to involve both active and passive components. The participation of an active transport component is indicated by the inhibition of polar transport by anaerobiosis, cyanide, azide, 2,4-dinitrophenol and other inhibitors of respiration. The participation of a passive diffusion component is indicated by the fact that respiration inhibitors do not completely stop the efflux of auxin in individual cells.

The mechanism of the polar transport of auxin is not completely understood. The theory which held sway until the early 1970s postulated that auxin diffuses passively into the cytoplasm of a given cell (and possibly also into the vacuole) and is actively transported out by means of an energy-requiring carrier located in the plasmalemma at one end of the cell (e.g. at the end remote from the apex in the case of a cell in the shoot).

More recently an alternative mechanism, known as the chemiosmotic theory of polar auxin transport, has been put forward (see Fig. 15.3). This is based upon the following facts: (a) the pH outside the protoplast of a plant cell is about 1 pH unit lower than that of the cytosol, (b) there is an electrical potential gradient of about 0.058 V across the plasmalemma with the outside electropositive relative to the inside and (c) the unionized form of IAA (IAAH), being less polar, can pass through the lipid bilayer of the plasmalemma much more easily than the ionized form (IAA$^-$). To these facts are added the postulates, (a) that respiratory energy is used to generate and maintain the pH and electrical potential gradients by actively pumping H$^+$ ions out of the cell, (b) that IAAH diffuses passively into the cell down an IAAH concentration gradient, (c) that IAA$^-$ is pulled out of the cell passively by the combined pH and electrical potential gradients; the passage through the plasmalemma is probably

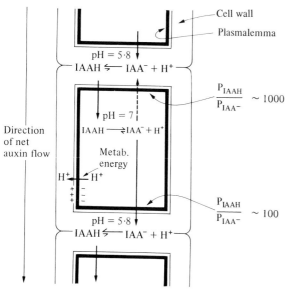

Fig. 15.3. A chemiosmotic theory of polar auxin transport. (P = permeability of plasmalemma to IAA^- and IAAH = ionized and unionized forms, respectively, of indole 3-acetic acid.)

facilitated by a specific carrier system and (d) that the ratio expressing the permeability of the plasmalemma to IAAH relative to that of IAA^- (i.e. P_{IAAH}/P_{IAA^-}) is about 10 times greater at one end of the cell (i.e. the apical end in the shoot) than the other; this effectively means that IAA^- passes out of the cell more readily at one end than the other.

Assuming that the pH in the cell wall and cytosol are 5.8 and 7.0 respectively it can be calculated, knowing that the pK_a of IAA is 4.7, that the ratio IAA^-/IAAH in the cell wall and the cytoplasm are about 12.6 and 200 respectively. Thus a greater proportion of the IAA is as IAAH in the cell wall than in the cytosol. The IAAH in the wall diffuses passively down an IAAH concentration gradient into the cytosol where most of it ionizes because of the higher pH. The resulting IAA^- is then pulled passively out of the cell and into the cell wall by the combined pH and electrical potential gradients. However, more IAA^- will diffuse out at one end of the cell than the other (the basal end in the shoot). In the wall a portion of it will be protonated to IAAH at the lower pH. This can then diffuse down the IAAH concentration gradient into the next cell and the process is repeated.

A number of structurally diverse compounds (e.g. naphthylphthalamic acid, 2,3,5-triiodobenzoic acid, 3,3a-dihydro-2-(p-methoxyphenyl)-8H-pyrazolo-[5,1-a]-isoindol-8-one and morphactins) are specific inhibitors of auxin transport. They appear to block the cellular efflux of auxin but do nothing to stop its influx; this leads to a cellular accumulation of auxin. If the chemiosmotic theory is correct these compounds would appear to block the outward, carrier-mediated diffusion of IAA^- across the plasmalemma in response to the pH and electrical potential gradients; they could do this by binding to the carrier and indeed there is evidence to suggest that they compete with auxin for membrane-bound sites.

3. Metabolism of auxin

(i) BIOSYNTHESIS OF AUXIN

The main endogenous auxin, IAA, is synthesized in the plant from the amino acid L-tryptophan (Fig. 15.4, I) which is present in plants in the free form and combined in proteins. The quantity of free tryptophan in leaves is about a thousandfold greater than that of IAA, being in the order of 20–40 μg g^{-1} wet weight.

Most plants convert L-tryptophan into IAA by the indole 3-pyruvic acid pathway (Fig. 15.4, I→II→III→IV). The conversion of L-tryptophan into indole 3-pyruvic acid (II, a very unstable compound, particularly under alkaline conditions) is catalysed by tryptophan aminotransferase[1]. This enzyme in plants has a fairly broad specificity in that it can catalyse the transamination of all the aromatic amino acids found in proteins and L-aspartic and L-glutamic acids. It is widely distributed in plants and requires pyridoxal phosphate for activity. The conversion of indole 3-pyruvic acid into indole 3-acetaldehyde (III) is catalysed by indolepyruvate decarboxylase which is distinct from classical pyruvate decarboxylase[2]. The conversion of indole 3-acetaldehyde into IAA is catalysed by either an NAD-dependent indoleacetaldehyde dehydrogenase (demonstrated in mung bean seedlings) or an oxygen-requiring indoleacetaldehyde oxidase (demonstrated in oat coleoptiles). Both enzymes are cytosolic.

The shoots of tobacco, tomato and barley plants

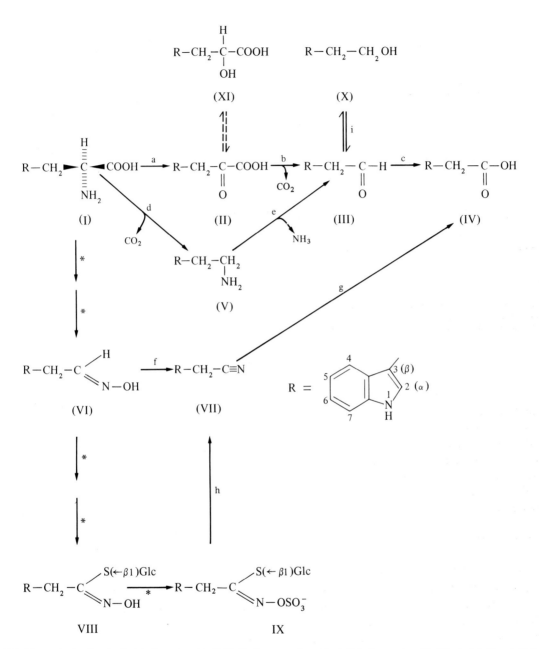

Fig. 15.4. Biosynthesis of auxin (indole 3-acetic acid, IAA). (I = L-tryptophan; II = indole 3-pyruvic acid; III = indole 3-acetaldehyde; IV = indole 3-acetic acid (IAA); V = tryptamine; VI = indole 3-acetaldoxime; VII = indole 3-acetonitrile (IAN); VIII = desulphoglucobrassicin; IX = glucobrassicin; X = tryptophol; XI = indole 3-lactic acid; a = tryptophan aminotransferase, EC 2.6.1.27; b = indolepyruvate decarboxylase; c = indoleacetaldehyde oxidoreductase; d = aromatic-L-amino acid decarboxylase, EC 4.1.1.28; e = tryptamine oxidoreductase (deaminating); f = indoleacetaldoxime dehydratase, EC 4.2.1.29; g = nitrilase, EC 3.5.5.1; h = thioglucosidase, EC 3.2.3.1; i = tryptophol oxidoreductase; * = see Fig. 7.5; Glc = D-glucose.)

and the coleoptiles of oat seedlings are able to convert L-tryptophan into IAA via tryptamine (Fig. 15.4, I→V→III→IV). The first step of this pathway is catalysed by a pyridoxal phosphate-requiring L-aromatic amino acid decarboxylase[3] which is particularly active with L-tryptophan. The resulting tryptamine (V) is then oxidatively de-aminated to indole 3-acetaldehyde under the catalytic influence of a copper-requiring tryptamine oxidoreductase (deaminating).

Members of the Cruciferae can form IAA from indole 3-acetonitrile (VII) which is formed either directly from indole 3-acetaldoxime (Fig. 15.4, I→VI→VII→IV) or indirectly via glucobrassicin (Fig. 15.4, I→VI→VIII→IX→VII→IV). Indole 3-acetaldoxime (VI) has been shown to promote the growth of coleoptile and hypocotyl tissues of all plants capable of converting it into indole 3-acetonitrile. Glucobrassicin is a growth promoter in the coleoptile bioassay. The enzyme, thiogluco-sidase[4] (also known as myrosinase), which catalyses the conversion of glucobrassicin into IAN is found mainly in cruciferous plants. The enzyme, nitrilase[5], which catalyses the conversion of IAN into IAA is present in plants of the families Gramineae and Musaceae as well as the Cruciferae.

Indole 3-lactic acid (XI) and tryptophol (X) are natural constituents of some plants and have weak and strong auxin activity respectively in the coleoptile bioassay. Tryptophol has been shown to be converted into IAA in tomato shoots but the evidence for the conversion of indole 3-lactic acid into IAA is less convincing. The probable metabolic interrelationships of these compounds with IAA is shown in Fig. 15.4.

Plants are colonized by many species of bacteria which are capable of converting L-tryptophan and tryptamine into IAA. Since greater amounts of IAA are found in non-sterile plants than in sterile plants, the suggestion was made in the late 1960s that all the IAA present in plants is synthesized by epiphytic bacteria. It has since been shown that this is definitely not the case; sterile plants do contain IAA and can be shown to synthesize it from L-tryptophan.

(ii) CATABOLISM OF AUXIN

IAA is broken down quite readily in plant tissues.

This is mainly accomplished by oxidase enzymes of which there appear to be two types: (a) peroxidases with IAA oxidase activity and (b) IAA oxidases with no peroxidase activity. The intracellular location of these enzymes is by no means clear. Hinman and Lang in 1965 proposed that IAA is catabolized by the route shown in Fig. 15.5.

IAA in aqueous solution is readily decomposed by light. Since this is accelerated by both synthetic and natural pigments, it has been suggested that photo-oxidation of IAA may occur in plants.

(iii) OTHER METABOLITES OF AUXIN

When labelled IAA is supplied to plants many more labelled compounds can be detected than can be accounted for by the pathways discussed in Figs. 15.4 and 15.5. Some of these are shown in

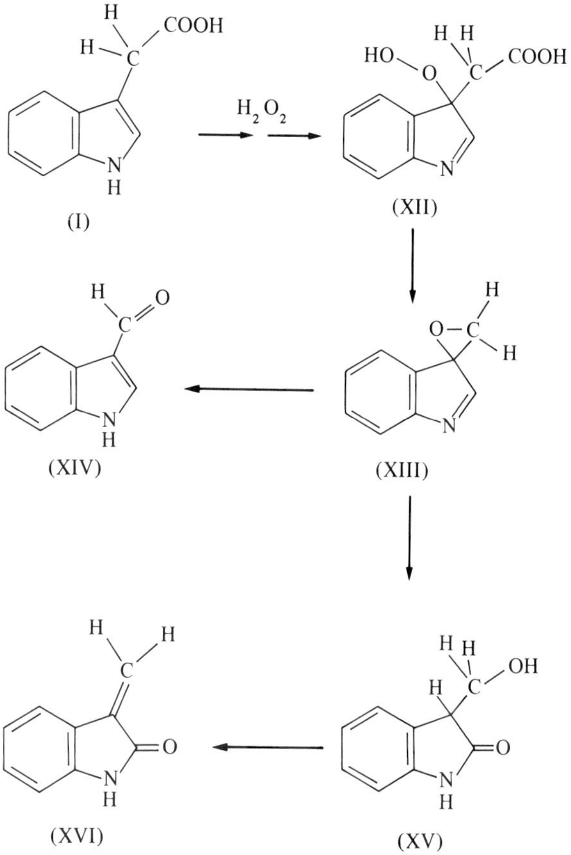

Fig. 15.5. The catabolism of indole 3-acetic acid (Hinman and Lang, 1965). (I=indole 3-acetic acid; XII=indolenine hydroperoxide; XIII=indolenine epoxide; XIV=indolealdehyde; XV=oxindole 3-carbinol; XVI=3-methyleneoxindole.)

Fig. 15.6; 1′-(indole 3-acetyl)-β-D-glucose (XVII) is formed in *Colchicum* and barley leaves and in pea and bean stems and indole acetyl-L-aspartic acid (XVIII) is formed in numerous plant species (see Fig. 15.6).

Other IAA derivatives have been found in seeds. Immature pea seeds contain an interesting set of chlorinated IAA derivatives, 4-chloroindolyl-3-acetic acid (XIX) and its methyl ester (XX) and monomethyl-4-chloroindolyl-3-acetyl-L-aspartic acid (XXI). Maize seeds contain a set of IAA-*myo*-inositol derivatives, IAA-*myo*-inositol arabinosides and IAA-*myo*-inositol galactosides. They also contain a series of compounds in which IAA is ester-linked to a β-1,4-glucan of variable length.

4. Physiological effects of auxins

The physiological effects of auxins are complex. Different tissues respond to it in different ways and the reasons for this are imperfectly understood. Moreover, it is often difficult to resolve primary effects from secondary effects.

The classical effect of auxin is to promote growth by stimulating the elongation of the cells constituting a given tissue. However, the straightforwardness of this effect was soon complicated by the results of dose-response experiments. From the dose-response curves (see Fig. 15.7, full lines) for various tissues it was apparent that (i) as the concentration of IAA increased the stimulatory effect on growth reversed and ultimately became inhibitory and (ii) the concentration of IAA at which there was maximum stimulation of growth was different for different tissues, being $\sim 5 \times 10^{-10}$ M for roots, $\sim 10^{-9}$ M for buds and $\sim 10^{-6}$ M for shoots.

The reason for the reversal in tissue response above a certain IAA concentration puzzled plant physiologists for some time and indeed even now it is not clear in the case of some tissues. However, some light was thrown on the problem in the mid-1960s when Burg and others showed that in a number of tissues, which include the stems of dicotyledonous plants (e.g. pea, sunflower), roots and probably buds, the direct effect of auxin at all concentrations is to stimulate, and never to inhibit, growth. But when the auxin concentration rises to a critical level, which is different for each tissue, ethylene production is induced [see Fig. 15.7,

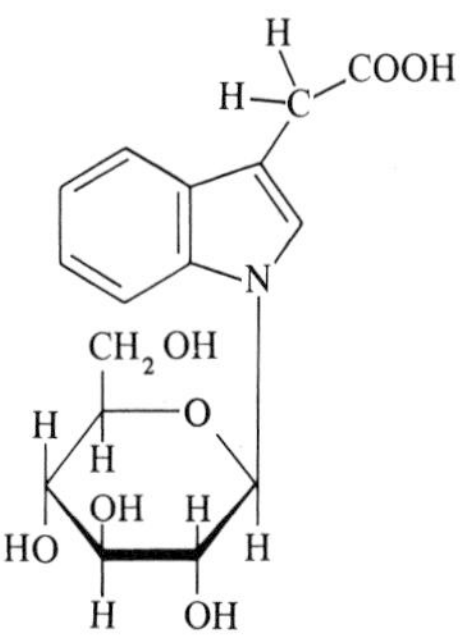

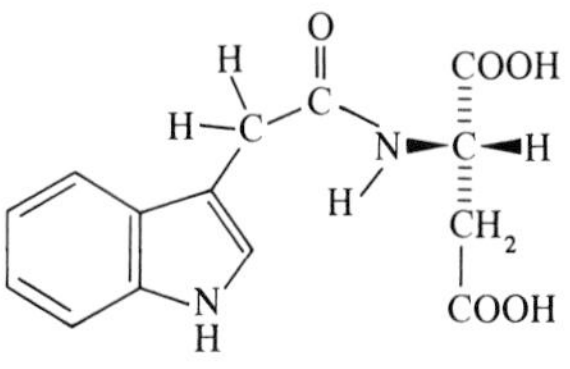

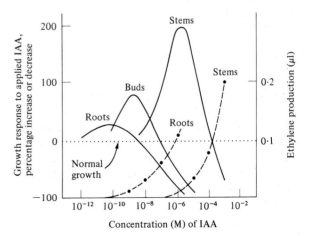

Fig. 15.6. Some metabolites of auxin. (XVII = 1′-(indole 3-acetic acid)-β-D-glucose; XVIII = indoleacetyl-L-aspartic acid; XIX = 4-chloroindolyl-3-acetic acid; XX = methyl-4-chloro-indolyl-3-acetate; XXI = monomethyl-4-chloroindolyl-3-acetyl-L-aspartic acid.)

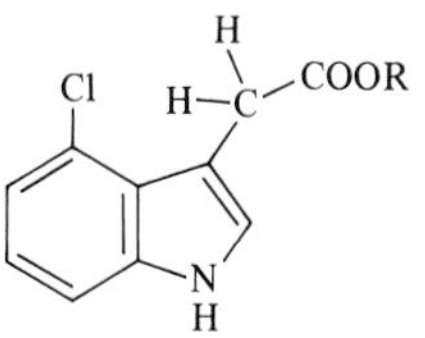

R = H = XIX
R = CH$_3$ = XX

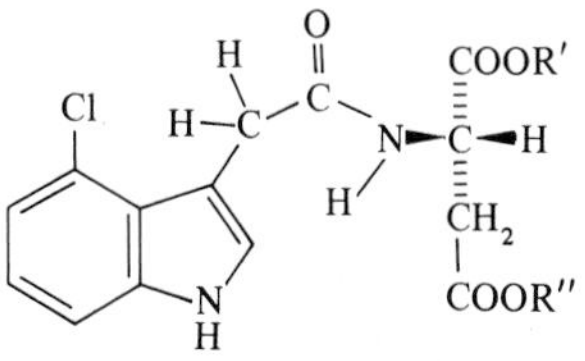

R′ & R″ = H & CH$_3$ or CH$_3$ & H

(XXI)

Fig. 15.7. Effect of exogenous indole-3-acetic acid (IAA) on the growth (full lines) and ethylene production (dashed lines) of roots, buds and stems.

dashed lines, and Section F.4(i)] and it is ethylene that is the direct inhibitor of growth. Thus the inhibition of growth that occurs in these tissues as the auxin concentration exceeds a certain level is due to the inhibitory effect of ethylene overcoming the stimulatory effect of auxin. In other tissues, notably the coleoptiles of monocotyledonous seedlings, the inhibition of growth is not due to ethylene despite the induction of its synthesis by auxin. This is indicated by the observation that growth and ethylene production are maximal in these tissues at the same auxin concentration and decline together as the auxin concentration is raised further.

Until recently it was believed that auxin-induced ethylene was responsible for root geotropism (i.e. the downward growth of the roots of the horizontally oriented plant). It was thought that the greater concentration of auxin in the lower side of the root, produced by the effect of gravity, induced the formation of ethylene which then inhibited growth in this region. The uninhibited growth of the upper side then caused the root to turn downwards. However, it now appears that abscisic acid is the principal mediator of geotropism in roots [see Section E.5(i)].

Auxin is also thought to play a part in the abscission of leaves and flowers. The observed sequence of events leading to leaf abscission commences with a marked decrease in the production of auxin in the leaf in question. This allows progressive senescence of the leaf to take place. The senescing tissues produce ethylene (see Sections F.2 and F.4) which acts upon the abscission zone located at the base of the leaf petiole. The abscission zone is a thin layer of cells which cuts transversely across all but the vascular tissues of the petiole. Ethylene acts particularly on the proximal layer of cells within the abscission zone (i.e. the layer remotest from the leaf blade). It causes them to enlarge and to synthesize and secrete cell-wall degrading enzymes such as cellulase[15] and pectinase[16]. These enzymes attack and weaken the cell walls of the rest of the cells in the abscission zone, commencing with those nearest to their site of origin. This results in a weak region in the petiole which breaks readily under stress (e.g. movement of the leaf by the wind) and causes the leaf to fall from the plant. Thus auxin delays leaf abscission by delaying the onset of senescence. It plays a similar role in flower abscission. When a

flower is fertilized fruit develop in the ovary (or the flower receptacle in some plants). If fertilization does not occur the flower senesces and produces ethylene which causes abscission of the flower stalk. When, however, fertilization does take place senescence and consequent ethylene production are prevented by auxin which is, in some cases, e.g. orchids, initially produced by the pollen grains responsible for fertilization and then by the developing seeds.

In addition to their role in cell elongation and abscission, there is evidence that auxins are involved in cell division. They appear to stimulate cambial activity. It is thought that auxins are involved in the differentiation of vascular tissue in the shoot apical region and lead to the production of the new vascular tissue that occurs in the spring when plant growth resumes in temperate climates. It also appears to be responsible for the initiation of lateral roots which originate from small groups of cambial cells in the pericycle. Thus auxins can be applied to roots to induce lateral branching; however, the concentrations required for this are so high that they inhibit root elongation and a short bushy root system results. Auxins also stimulate the development of adventitious roots from stem and leaf cuttings and are used commercially for this purpose.

In passing it is worth mentioning that two widely used selective herbicides, 2,4-D and 2,4,5-T (see Fig. 15.8) have many, but not all, of the properties of auxins. When used at the correct concentration they kill broad-leaved plants but not grasses or related narrow-leaved monocotyledonous plants. They kill by causing uncoordinated and distorted growth. The basis of their selectivity is not understood.

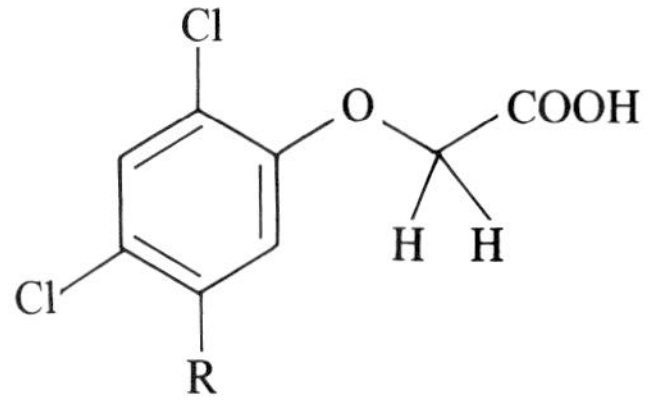

R = H = 2,4-dichlorophenoxyacetic acid (2,4-D)
R = Cl = 2,4,5-trichlorophenoxyacetic acid (2,4,5-T)

Fig. 15.8. Selective herbicides which have many, but not all, of the properties of auxins.

5. Biochemical mode of action of auxin

The main target cells of auxin appear to be those in the process of differentiating close to the various meristems. The general response to auxin is cell elongation. The question therefore arises as to how this response is elicited.

Auxin undoubtedly enters its target cells. Extrapolating from what is known of animal hormones, it would be expected that within the target cell auxin would bind to a specific receptor molecule, probably a protein, and that the resulting complex would switch on specific genes. These would be transcribed into mRNAs which would, in turn, be translated into specific proteins. The latter would be key enzymes whose activity would facilitate the physiological response of cell elongation. Present evidence is only partly consistent with this postulated sequence of events. Two facts that do not fit into it very well are (i) the initial response to IAA occurs within 10–15 min. which is too soon to be explained by new protein synthesis and (ii) no auxin receptors, protein or otherwise, have yet been identified. On the other hand Key and his coworkers have shown that (i) IAA stimulates the synthesis of all types of RNA (mRNA, tRNA and rRNA) in elongating and maturing sections of the hypocotyls of soya bean seedlings and (ii) the ability of IAA to stimulate cell elongation is dependent upon the synthesis of new RNA and protein. Some of these inconsistencies can be reconciled, however, if it is postulated that auxin elicits a rapid response, which is not dependent upon RNA and protein synthesis, and a slower response, which is. An attempt to delineate these two responses is made in the ensuing discussion.

Cell elongation requires that the longitudinal walls stretch. This will occur under the driving force of turgor pressure provided that the longitudinal walls become more plastic. Such an increase in plasticity is thought to involve the breaking of some of the bonds between the polysaccharide cell wall components and the re-making of others. These bonds are of two types: (i) hydrogen bonds, which link the cellulose microfibrils to the hemicellulosic xyloglucans that coat them and (ii) glycosidic linkages which bind the xyloglucans to the arabino-galactans and the latter to the rhamnogalac-

turonans (see Chapter 4, Sections B.1(iii) and C.1). The breaking of hydrogen bonds is potentiated by an increase in the H^+ ion concentration and their re-establishment by a decrease. The breaking and re-making of glycosidic bonds, on the other hand, involves enzymes.

IAA is known to increase cell wall plasticity. Precisely how it accomplishes this is not known with certainty. However, Rayle and Cleland in 1970 postulated that the primary effect of auxin is to cause the cell to lower the pH of the aqueous phase of the cell wall, possibly by stimulating a membrane-bound H^+ ion pump. As mentioned above this increase in H^+ ion concentration would then weaken the hydrogen bonding between the cellulose microfibrils and the xyloglucans and allow them to slide past each other under the pull of turgor pressure. Moreover, it may further aid wall plasticity by activating cell wall bound glycosidases, some of which are known to have low pH optima. The evidence in favour of this hypothesis is: (i) segments of oat coleoptiles, pea and soya bean stems and other tissues elongate more rapidly when exposed to low pHs—this elongation is not suppressed by respiratory inhibitors or O_2 lack; (ii) IAA-induced elongation in oats, peas and maize is preceded by the passage of H^+ ions from the cells into the surrounding medium; (iii) IAA-induced elongation and IAA-induced H^+ ion extrusion are both suppressed by respiratory inhibitors or O_2 lack; (iv) cell wall extension, whether induced directly by H^+ ions or by IAA, only occurs when the turgor pressure of the cells is at or near its maximum. It would appear that H^+ ion-induced stretching of the cell wall constitutes the rapid response to auxin mentioned above.

However, cell elongation involves more than cell wall stretching. This follows from the fact that elongation does not cause the longitudinal cell walls to become thinner as would be the case if only stretching were involved. This implies that the stretching of the cell wall is followed by the introduction of newly synthesized cell wall material into it. This in turn suggests that auxin stimulates the synthesis and deposition of new cell wall polysaccharide as well as increasing wall plasticity, and there is evidence that this is the case. The synthesis of new cell wall polysaccharides could be accomplished by promoting the synthesis of more polysaccharide-synthesizing enzymes or by activating those already present. The former would require the switching on of the appropriate genes whilst the latter would not. It is likely that both processes occur because (i) gene 'activation' would explain the observed increase in RNA synthesis following IAA treatment and the dependence of IAA-induced elongation on the synthesis of new RNA and protein and (ii) enzyme activation has been demonstrated; Ray in 1973 showed that treatment of pea stems with IAA caused a 2- to 4-fold increase in the activity of the particulate cellulose synthase (UDP-forming) in tissue homogenates. It appears that these processes constitute at least part of the slower response to auxin mentioned above.

C. GIBBERELLINS

1. Discovery and characterization

Gibberellins are a large family of closely related tetracyclic diterpenoid compounds. They were discovered as metabolites of an ascomycete *Gibberella fujikuroi* (imperfect state, *Fusarium moniliforme*). This fungus was recognized by Japanese scientists in the 1920s as the cause of the 'bakanae' disease which afflicted rice seedlings. This disease was typified by excessive stem and leaf elongation. The infected seedlings thus became abnormally tall and spindly and usually fell over; hence the name *bakanae* which literally means 'foolish seedling'.

In 1926 Kurosawa showed that when rice seedlings were treated with a sterilized filtrate of the medium in which *G. fujikuroi* had been growing, they developed the typical *bakanae* symptoms. This showed that the fungus secreted some substance into its growth medium which would cause excessive elongation in rice. Moreover, it implied that when the fungus was growing on rice seedlings this same substance was secreted into the tissues and caused the *bakanae* symptoms.

The next task was to isolate this substance from the *G. fujikuroi* growth medium and identify it. By 1939 Yabuta and his coworkers had obtained the active material in crystalline form. They called it 'gibberellin' but did not identify it. No further progress was made until 1954 when 'gibberellin' was

shown to be a mixture and the first pure compound, gibberellic acid (now known as GA_3), was isolated by Cross. The structure of GA_3 was elucidated in 1959 by Cross and coworkers. This facilitated the structural elucidation of several other gibberellins present in much smaller quantities in *G. fujikuroi* growth medium.

Gibberellins were then found to be present in higher plants. Immature seeds were a particularly good source. Indeed the first higher plant gibberellin was isolated from immature bean seeds and shown by MacMillan in 1960 to be identical with GA_1 isolated from *G. fujikuroi*.

Since then about 53 gibberellins have been identified; of these about eight have been found only in *G. fujikuroi*, about 32 only in higher plants and about 13 in both *G. fujikuroi* and higher plants (see Table 15.1). However, no more than 15 GAs have been found in any one higher plant species.

The gibberellins are all carboxylic acids and are therefore called gibberellic acids or GAs. They are distinguished from each other by a subscript numeral (i.e. GA_1, GA_2, ..., GA_n).

The GAs are biosynthetically derived [see Section C.2(i)] from *ent*-kaurene (sometimes called (−)-kaurene) whose parent hydrocarbon is *ent*-kaurane (XXIII, Fig. 15.9). The carbon skeleton of this tetracyclic diterpene is modified during biosynthesis by the contraction of ring B to give a range of C_{20}–GAs, presently about 23 in number, whose parent hydrocarbon is *ent*-gibberellane (XXV, Fig. 15.9). A second range, the C_{19}–GAs is formed by a further skeletal modification, the loss of C–20; the parent hydrocarbon of the C_{19}–GAs is *ent*-20-norgibberellane (XXV, Fig. 15.9 but with H replacing CH_3 on C–10; the prefix 'nor' indicates that a carbon is missing, whilst the number 20 indicates which carbon it is).

The prefix *ent* in *ent*-kaurane and *ent*-gibberellane is an abbreviation of the term 'en-antiomer' and indicates that these compounds are the enantiomers of the 'normal' tetracyclic diterpenes, kaurane (XXII, Fig. 15.9) and gibberellane (XXIV, Fig. 15.9). Enantiomers are optical isomers that are mirror images of each other; this relationship can be seen in the case of the kaurane/*ent*-kaurane and gibberellane/*ent*-gibberellane enantiomeric pairs in Fig. 15.9 if the dotted lines between the perspective formulae are imagined to be mirrors.

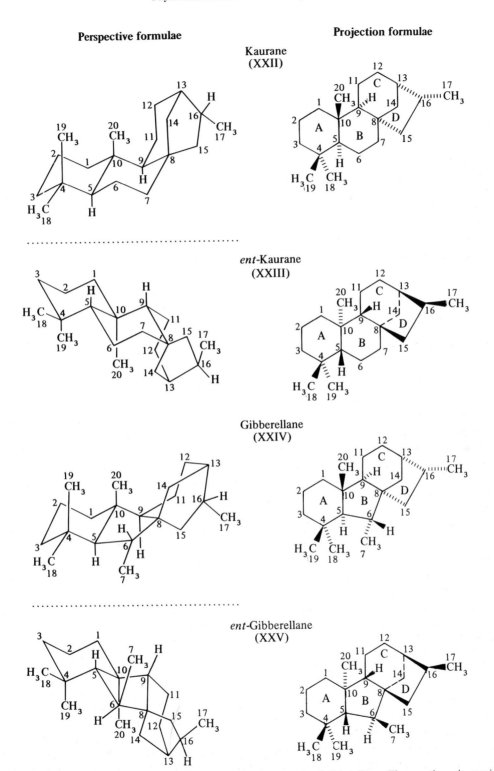

Fig. 15.9. Perspective and projection formulae of the enantiomers of kaurane and gibberellane. (The enantiomeric or mirror-image relationship of these pairs of compounds can be seen if a mirror is imagined in the position of the dotted lines; notice that substituents that project upwards in one enantiomer project downwards in the other.)

Table 15.1. *Structures of the naturally occurring gibberellins*

Gibberellin GA_x[a]	No. of carbon atoms	Type of rings (see below) A	B	C+D	Substituents[b] (see XXV, Fig. 15.9 for the numbering of the ring system)
GA_1(g, p)	19	2	1	1	3βHO, 4βMe, 13HO
GA_2(g)	19	2	1	2	3βHO, 4βMe
GA_3(g, p)	19	2	1	1	Δ^1-en, 3βHO, 4βMe, 13HO
GA_4(g, p)	19	2	1	1	3βHO, 4βMe
GA_5(p)	19	2	1	1	Δ^2-en, 4βMe, 13HO
GA_6(p)	19	2	1	1	$2\beta,3\beta$-epoxy, 4βMe, 13HO
GA_7(g, p)	19	2	1	1	Δ^1-en, 3βHO, 4βMe
GA_8(p)	19	2	1	1	2βHO, 3βHO, 4βMe, 13HO
GA_9(g, p)	19	2	1	1	4βMe
GA_{10}(g)	19	2	1	2	4βMe
GA_{11}(g)	19	4	1	1	4βMe
GA_{12}(g)	20	1	1	1	10αMe
GA_{13}(g)	20	1	1	1	3βHO, 10αCOOH
GA_{14}(g)	20	1	1	1	3βHO, 10αMe
GA_{15}(g)	20	3	1	1	4βMe
GA_{16}(g)	19	2	1	1	1αHO, 3βHO, 4βMe
GA_{17}(p)	20	1	1	1	10αCOOH, 13HO
GA_{18}(p)	20	1	1	1	3βHO, 10αMe, 13HO
GA_{19}(p)	20	1	1	1	10αCHO, 13HO
GA_{20}(p)	19	2	1	1	4βMe, 13HO
GA_{21}(p)	19	2	1	1	4βCOOH, 13HO
GA_{22}(p)	19	2	1	1	Δ^2-en, 4βCH$_2$OH, 13HO
GA_{23}(p)	20	1	1	1	3βHO, 10αCHO, 13HO
GA_{24}(g, p)	20	1	1	1	10αCHO
CA_{25}(g, p)	20	1	1	1	10αCOOH
GA_{26}(p)	19	2	1	1	2βHO, 3βHO, 4βMe, 12 keto
GA_{27}(p)	20	3	1	1	2βHO, 3βHO, 4βMe
GA_{28}(p)	20	1	1	1	3βHO, 10αCOOH, 13HO
GA_{29}(p)	19	2	1	1	2βHO, 4βMe, 13HO
GA_{30}(p)	19	2	1	1	Δ^1-en, 3βHO, 4βMe, 12αHO
GA_{31}(p)	19	2	1	1	Δ^2-en, 4βMe, 12αHO
GA_{32}(p)	19	2	1	1	Δ^1-en, 3βHO, 4βMe, 12αHO, 13HO, 15βHO
GA_{33}(p)	19	2	1	1	1βHO, 3 keto, 4βMe, 12αHO
GA_{34}(p)	19	2	1	1	2βHO, 3βHO, 4βMe
GA_{35}(p)	19	2	1	1	3βHO, 4βMe, 11βHO
GA_{36}(g)	20	1	1	1	3βHO, 10αCHO
GA_{37}(g, p)	20	3	1	1	3βHO, 4βMe
GA_{38}(p)	20	3	1	1	3βHO, 4βMe, 13HO
GA_{39}(p)	20	1	1	1	3βHO, 10αCOOH, 12αHO
GA_{40}(g)	19	2	1	1	2αHO, 4βMe
GA_{41}(g)	20	1	1	2	3βHO, 10αCOOH
GA_{42}(g)	20	1	1	2	3βHO, 10αMe
GA_{43}(p)	20	1	1	1	2βHO, 3βHO, 10αCOOH
GA_{44}(p)	20	3	1	1	4βMe, 13HO
GA_{45}(p)	19	2	1	1	4βMe, 15βHO
GA_{46}(p)	20	1	1	1	2βHO, 10αCOOH
GA_{47}(g)	19	2	1	1	2αHO, 3βHO, 4βMe
GA_{48}(p)	19	2	1	1	2βHO, 3βHO, 4βMe, 12βHO
GA_{49}(p)	19	2	1	1	2βHO, 3βHO, 4βMe, 12αHO
GA_{50}(p)	19	2	1	1	2βHO, 3βHO, 4βMe, 11βHO
Ga_{51}(p)	19	2	1	1	2βHO, 4βMe
GA_{52}(p)	20	3	1	1	2βHO, 3βHO, 4βMe, 12αHO
GA_{53}(p)	20	1	1	1	10αMe, 13HO

For notes [a] and [b] see facing page.

It can be seen from the perspective formulae in Fig. 15.9 that substituents on the carbon atoms constituting the ring system project roughly upwards or downwards from the general plane of the ring system. Upwardly projecting substituents are indicated in the more easily drawn projection formulae (see Fig. 15.9) by solid, wedge-like bond lines and are regarded as being above the plane of the paper whilst downwardly projecting substituents are indicated by broken bond lines and are regarded as being below the plane of the paper.

The convention governing the systematic nomenclature of these compounds decrees that substituents that project upwards from the general plane of the ring system of the **normal**-isomer be termed 'β' substituents whilst those that project downwards be termed 'α'. The key feature of this convention is that the 'β=upward: α=downward'

relationship applies solely to the normal-isomer. The terms 'α' and 'β' are also used in the nomenclature of the *ent*-isomer but must always be related back to the normal-isomer remembering that, because of the mirror image relationship of the two isomers, a substituent that projects upwards in the normal-isomer and is therefore β projects downwards in the *ent*-isomer **but remains β**. The net effect therefore is that in the *ent*-isomer the 'β=upward: α=downward' relationship of the normal isomer is reversed. Thus compound XXVI (Fig. 15.10) has the systematic name *ent*-7α-hydroxykaurenoic acid in spite of the fact that its 7-hydroxy group projects upwards. This follows because (i) the name *ent*-7α-hydroxykaurenoic acid means 'enantiomer of 7α-hydroxykaurenoic acid' and (ii) 7α-hydroxykaurenoic acid (XXVII) has a downwardly projecting 7-hydroxy group which is

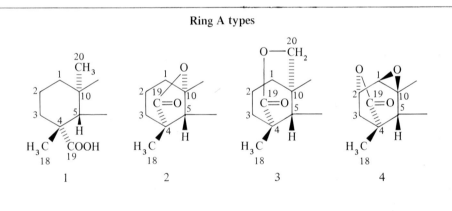

Ring A types

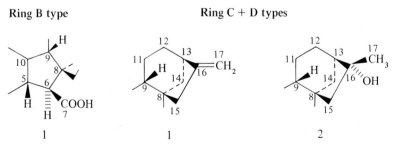

Ring B type   Ring C + D types

[a] Location of GA is indicated by g and p; g=present in *Gibberella fujikuroi*; p=present in some higher plants.
[b] Me=methyl; HO=hydroxyl; COOH=carboxyl; CHO=formyl; CH_2OH=hydroxymethyl; Δ^1-en=a double bond between Cs 1 and 2; Δ^2-en=a double bond between Cs 2 and 3.

The projection formula of each of the GAs in the table can be constructed by combining the appropriate A, B and C+D rings, whose structures are shown below, to form the tetracyclic diterpenoid ring system and then adding the appropriate substituents to it; α-substituents project below the plane of the paper and are shown with broken (ıııııı) bond lines whilst β-substituents project above the plane of the paper and are shown with solid, wedge-like (►) bond lines.

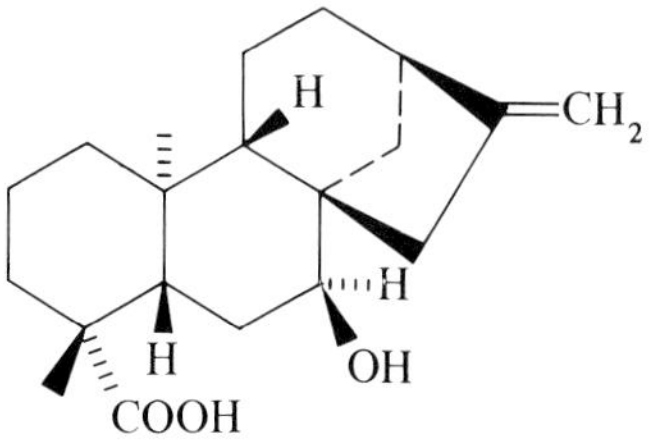

(XXVI)

ent-7α-hydroxykaurenoic acid

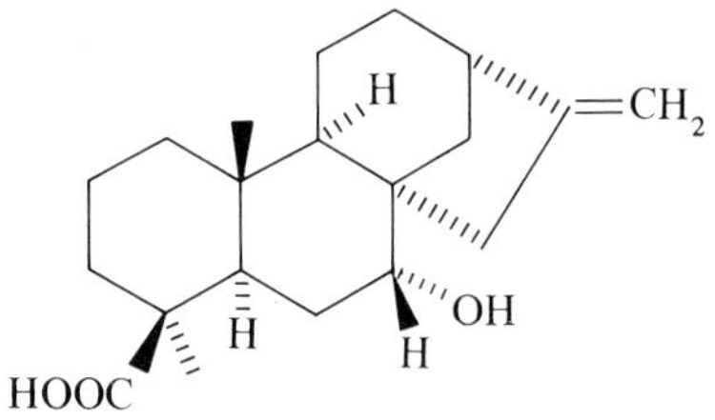

(XXVII)

7α-hydroxykaurenoic acid

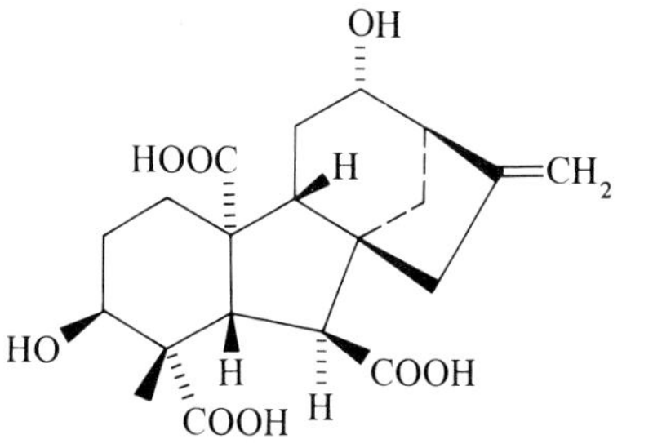

GA$_{39}$ (XXVIII)

ent-3α,12β-dihydroxy-
gibberell-16-en-4β,10β-dioic acid

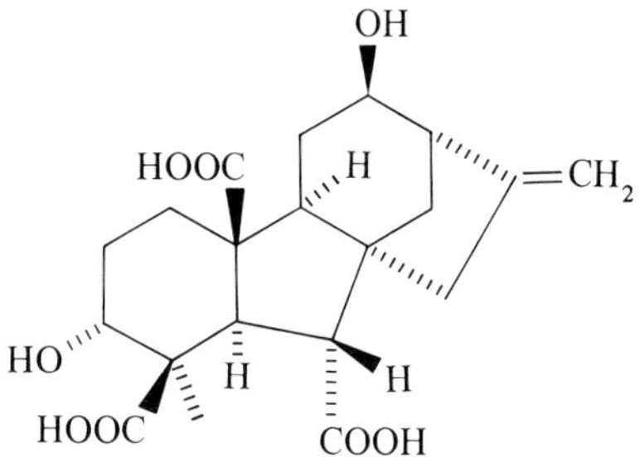

(XXIX)

3α,12β-dihydroxygibberell-
16-en-4β,10β-dioic acid

Fig. 15.10. Examples of nomenclatural problems in the gibberellin field (see text).

termed α because 7α-hydroxykaurenoic acid is for-mally derived from the normal-isomer of kaurane.

Confusion has arisen because this systematic nomenclature has not been applied rigorously in naming the gibberellins. It is in common usage in the case of gibberellin precursors (e.g. ent-7α-hydroxykaurenoic acid—see Fig. 15.12) but not for the GAs themselves. As was mentioned earlier GAs are usually named by a trivial system of nomencla-ture (GA$_1$, GA$_2$, ..., GA$_n$) rather than by the systematic one. In this trivial system the terms 'α' and 'β' are used to describe the orientation of substituents with respect to the general plane of the ring system but the 'β=upward: α=downward' relationship is used in spite of the fact that the parent hydrocarbon of the GAs is ent-gibberellane. Thus GA$_{39}$ (XXVIII, Fig. 15.10) is said to have 3β and 12α hydroxyl groups and 4α and 10α carboxyl groups in spite of the fact that its systematic name is ent-3α,12β-dihydroxgibberell-16-en-4β,10β-dioic acid (i.e. it is the enantiomer of 3α,12β-dihydroxy-gibberell-16-en-4β,10β-dioic acid, XXIX, Fig.15.10).

In the list of GA structures in Table 15.1 the trivial system of nomenclature is used. In the following text systematic nomenclature is used for GA precursors (i.e. ent-kaurene→ent-7α-hydroxykaurenoic acid, see Fig. 15.12) and the trivial system for the GAs themselves.

Common usage has also led to the abandonment of one of the original rules of GA nomenclature, namely that for a compound to be included in the list of GAs and given an identifying number it should not only (i) have the correct carbon skeleton and (ii) be naturally occurring but should also show activity in the specific GA bioassays. Several of the GAs listed in Table 15.1 have little or no biological activity. This is particularly true of those with a 2β-hydroxy group for 2β-hydroxylation appears to be a key method of deactivating GAs in higher plants and is quite probably concerned in the regulation of GA levels. No 2β-hydroxy GAs have been found in G. fujikuroi.

A number of GA derivatives have been identified. The first to be discovered was 3-O-β-acetyl GA$_3$ in

G. fujikuroi; n-propyl esters of GA_1 and GA_3 have since been found in seeds of *Cucumis sativus*. The most abundant GA derivatives in higher plants are, however, the β-D-glucosyl ethers (often called GA-glucosides) in which the D-glucose residue is often glycosidically linked to the 2β-hydroxyl group (e.g. GA_8-$O(2)$-β-D-glucoside, XXX, Fig. 15.11) and occasionally to the 3β-hydroxy group (e.g. GA_3-$O(3)$-β-D-glucoside) and to the 11β-hydroxyl group (e.g. GA_{35}-$O(11)$-β-D-glucoside). *Phaseolus vulgaris* has a number of β-D-glucosyl esters (e.g. GA_1-β-D-glucosyl ester, XXXI, Fig. 15.11). Immature seeds of *Pharbitis nil* contain a more remote GA derivative, gibberethione (formerly called pharbitic acid, XXXII, Fig. 15.11).

GA-activity has been recorded in extracts of all parts of higher plants, including shoots, roots, leaves, flower buds, flower petals, anthers and seeds; it has also been demonstrated in plastids. In general the vegetative parts of the plant contain less GA activity than the reproductive parts. Immature

seeds are particularly rich in GAs (10–100 μg per g wet weight) and have been a favourite tissue for the isolation of GAs and the study of their biosynthesis. As seeds mature their GAs tend to become derivatized.

GA activity is measured by bioassays of which the most used are based upon (i) the increased growth of the seedlings of genetically dwarf cultivars of pea (a multigene mutant), maize (single gene mutants d-1, d-2, d-3, d-5 and an_1) and rice in response to GAs, (ii) the increased growth of the hypocotyls of lettuce and cucumber in response to GAs and (iii) the increased α-amylase activity in the endosperm of cereal seeds (e.g. barley, rice) in response to GAs.

The biological activity of GAs is a function of their structure. The following generalizations can be made: (i) the 7β carboxyl group is required for activity, (ii) C_{19}-GAs are more active than C_{20}-GAs, (iii) 3β-hydroxylation, 3β,13-dihydroxylation or the introduction of a Δ^1-double bond increases the activity of C_{19}-GAs, (iv) the oxidation of the 10α

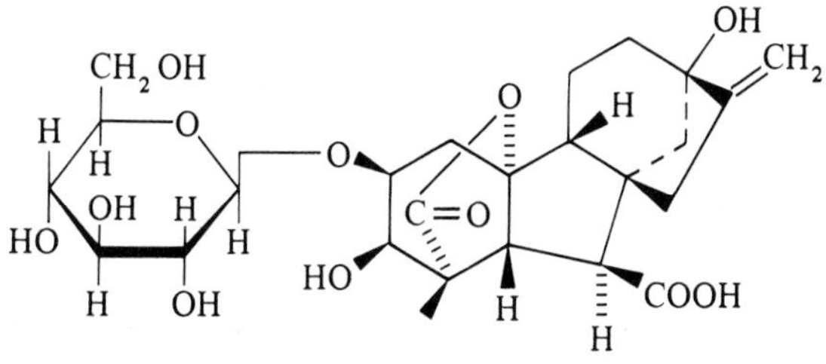

GA$_8$-O(2)-β-D-glucoside (XXX)

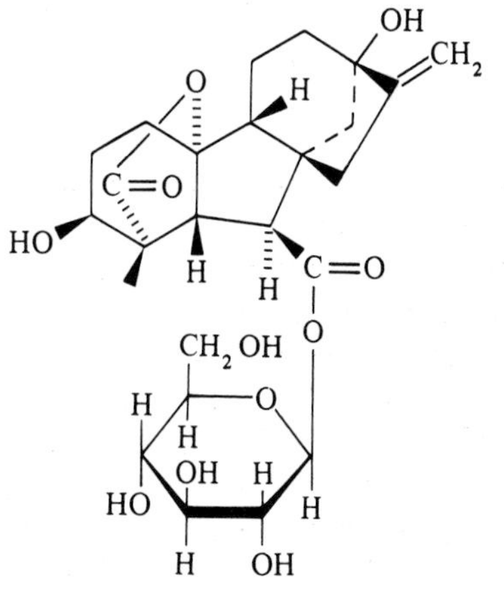

GA$_1$-β-D-glucosyl ester (XXXI)

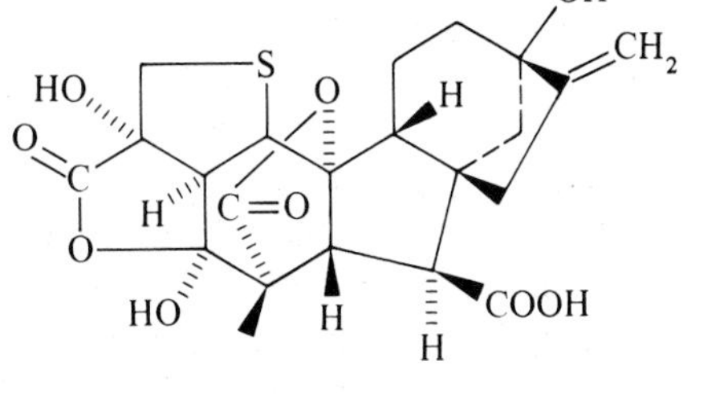

Gibberethione (formerly pharbitic acid, XXXII)

Fig. 15.11. Examples of some naturally occurring GA-derivatives.

methyl group to a 10α aldehyde (formyl) group or the δ-lactonization of ring A (e.g. GA_{15}) increases the activity of C_{20}-GAs and (v) 2β-hydroxylation causes loss of activity in C_{19}- and C_{20}-GAs.

2. Metabolism of gibberellins

(i) BIOSYNTHESIS OF GIBBERELLINS

The first stage in biosynthesis of gibberellins (acetyl-CoA→GA_{12}-aldehyde) is identical in *G. fujikuroi* and higher plants. This stage may be divided into two parts: The first (acetyl-CoA→HMG-CoA→MVA→MVAP→MVAPP→IPP→DMAPP →GPP→FPP→GGPP) is composed of the earliest steps of the polyisoprenoid biosynthetic pathway which have been described in Chapter 11 whilst the second (GGPP→GA_{12}-aldehyde) is peculiar to GA biosynthesis.

The steps in the pathway from GGPP to GA_{12}-aldehyde are shown in Fig. 15.12. The first two steps (GGPP→copalyl pyrophosphate→*ent*-kaurene) are catalysed by a soluble enzyme complex known as *ent*-kaurene synthase which has two catalytic activities, known as A (responsible for GGPP→copalyl pyrophosphate) and B (responsible for copalyl pyrophosphate→*ent*-kaurene). The mechanism of these steps is shown in Fig. 15.13. The cyclization of GGPP to copalyl pyrophosphate is initiated by the electrophilic attack of H^+ on the Δ^{14} double bond and terminated by loss of H^+ from the C–7 methyl group which becomes a methylene group. The further cyclization of copalyl pyrophosphate to *ent*-kaurene is initiated by removal of the pyrophosphate moiety as inorganic pyrophosphate ion followed by cyclization, skeletal rearrangement and finally stabilization (by H^+ loss from C–17) of the resulting carbonium ion. The next three steps involve the sequential oxidation of the C–19 methyl group to give *ent*-kaurenol, *ent*-kaurenal and *ent*-kaurenoic acid. This is followed by hydroxylation at C–7 to give *ent*-7α-hydroxykaurenoic acid. The enzymes catalysing the steps from *ent*-kaurene to *ent*-7α-hydroxykaurenoic acid appear to be membrane-bound cytochrome P450 mixed function oxygenases and therefore require O_2 and NADPH for activity. The conversion of *ent*-7α-hydroxykaurenoic acid into GA_{12}-aldehyde involves the contrac-

tion of ring B from a six- to a five-membered ring by the extrusion of C–7 as an aldehyde group and the formation of a new bond between C–6 and C–8. The use of *ent*-7α-hydroxykaurenoic acid stereospecifically labelled at C–6 with tritium has shown that the 6α-hydrogen (H_b in Fig. 15.14) is lost in this process. A formalized mechanism for the reaction is shown in Fig. 15.14. The enzyme catalysing it is probably membrane bound.

The pathway from GA_{12}-aldehyde to GAs is better understood in *G. fujikuroi* than in higher plants; it is outlined in Fig. 15.15. This fungus is particularly suitable for biosynthetic studies because (i) it grows easily on a synthetic medium, (ii) it absorbs ^{3}H and/or ^{14}C-labelled putative GA precursors from the medium, (iii) it synthesizes large amounts of GAs within 4–6 days and (iv) it secretes these GAs into its growth medium thus greatly facilitating their extraction and purification prior to identification by GC-MS (a combination of gas–liquid chromatography and mass spectrometry) and radioassay. Much of our current knowledge of GA biosynthesis in *G. fujikuroi* has come from the exploitation of its B1-41a mutant which does not synthesize GAs because the pathway is blocked between *ent*-kaurenal and *ent*-kaurenoic acid (see Fig. 15.12). However, the mutant still possesses all the post-blockage enzymes of GA biosynthesis because it will convert post-*ent*-kaurenoic acid GA-precursors into the same range of GAs as the GA-producing *G. fujikuroi* strain from which it originated. The great advantage that the B1-41a mutant has over GA-producing strains in the elucidation of GA biosynthesis is that ^{3}H and/or ^{14}C-labelled GAs and their precursors can be fed to it and the resulting metabolites recovered without the complication of their being diluted by endogenous synthesis (i.e. synthesis from the unlabelled general carbon sources of the medium via acetyl-CoA and MVA).

From such investigations it is clear that the GA biosynthetic pathway bifurcates at GA_{12}-aldehyde, giving rise to the so-called 'early 3β-hydroxylation pathway' and the 'non-3β-hydroxylation pathway'.

The 'early 3β-hydroxylation pathway' leads to probably the best known of all GAs, GA_3, and a range of other 3β-hydroxy GAs. This pathway is initiated by the 3β-hydroxylation of GA_{12}-aldehyde. The resulting GA_{14}-aldehyde is then converted into GA_4. This conversion requires (i) the

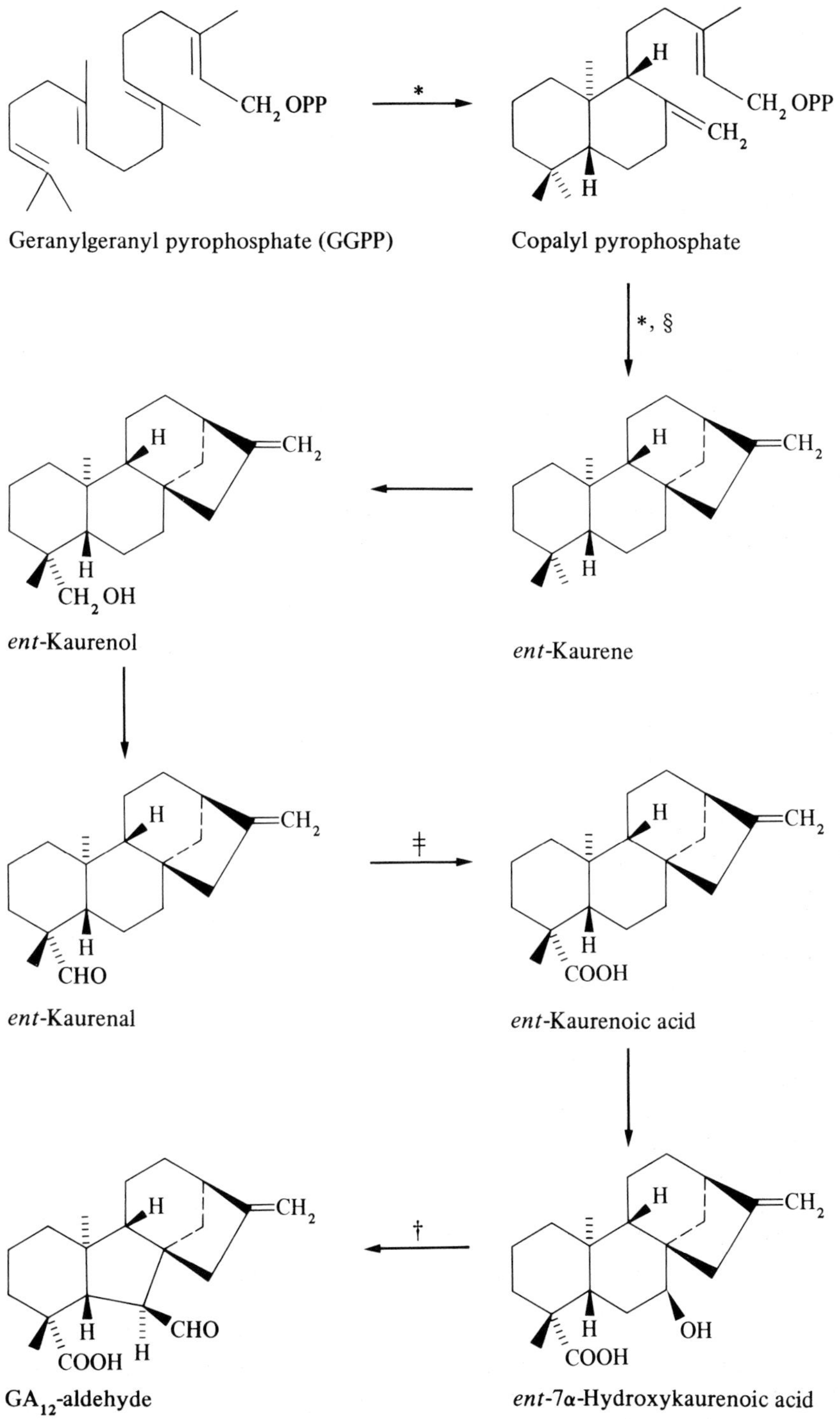

Fig. 15.12. The conversion of GGPP into GA$_{12}$-aldehyde in *Gibberella fujikuroi* and higher plants.
(*See Fig. 15.13 for detailed mechanism.
†See Fig. 15.14 for detailed mechanism.
‡Blocked in the *G. fujikuroi* mutant B1-41a.
§Blocked in the maize mutant *dwarf*-5.)

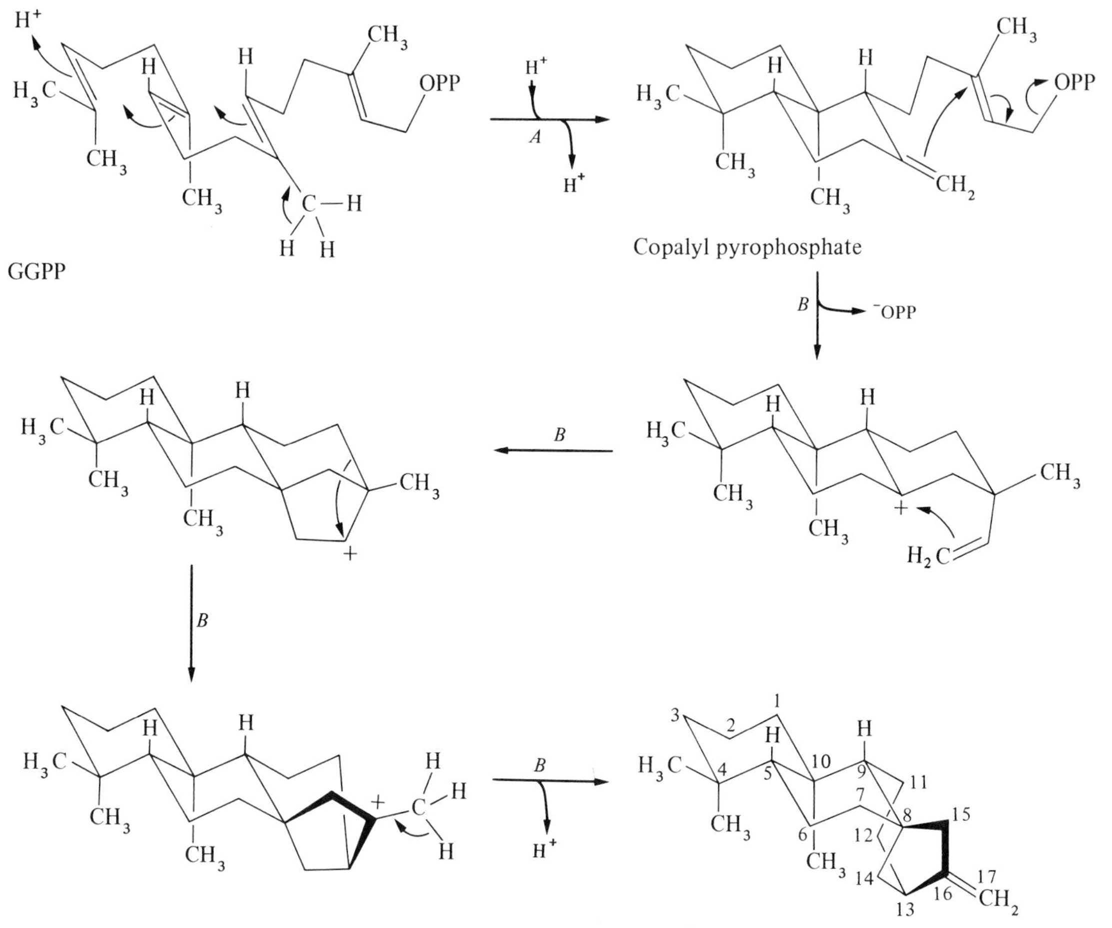

Fig. 15.13. The postulated mechanism of cyclization of geranylgeranyl pyrophosphate (GGPP) to *ent*-kaurene via copalyl pyrophosphate. (A = 'A'-activity' of *ent*-kaurene synthase; B = 'B'-activity' of *ent*-kaurene synthase.)

oxidation of the 7β aldehyde (formyl) group into a 7β carboxyl group and (ii) the loss of the 10α methyl group and the formation of a lactone ring between C–19 and C–10; this is termed the standard lactonization of ring A in Fig. 15.15. It is not yet clear which of these two events occurs first. If the sequence is (i)→(ii) then GA_{14} is the intermediate between GA_{14}-aldehyde and GA_4 but if it is (ii)→(i) the intermediate is GA_4-aldehyde. At first sight the answer seems clear enough because GA_{14} is a major metabolite of GA_{14}-aldehyde in feeding experiments whilst GA_4-aldehyde has not been detected. However, it is necessary to postulate its participation because feeding experiments also show that GA_{14} is converted into GA_4 and other C–19 GAs

more slowly than is GA_{14}-aldehyde. From GA_4 the pathway divides several ways: three of these are hydroxylations, 1α-hydroxylation to give GA_{16}, 2α-hydroxylation to give GA_{47} and 13-hydroxylation to give the major metabolite GA_1, whilst the fourth involves the introduction of a Δ^1-double bond to form GA_7, another major metabolite. The latter is then 13-hydroxylated to yield the main metabolite, GA_3. GA_{14} is also converted into other C–20 GAs. It is hydrated at its Δ^{16}-double bond to give GA_{42} and oxidized at C–20 to different extents to give GA_{36} (10α CHO) and GA_{13} (10α COOH); feeding experiments show that GA_{36} is not oxidized to GA_{13} and that GA_{13} is not decarboxylated to give the C–19, ring A lactone, GA_3.

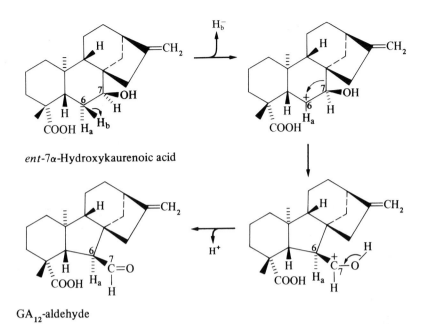

ent-7α-Hydroxykaurenoic acid

GA₁₂-aldehyde

Fig. 15.14. A formalized mechanism for the conversion of *ent*-7α-hydroxykaurenoic acid into GA_{12}-aldehyde.

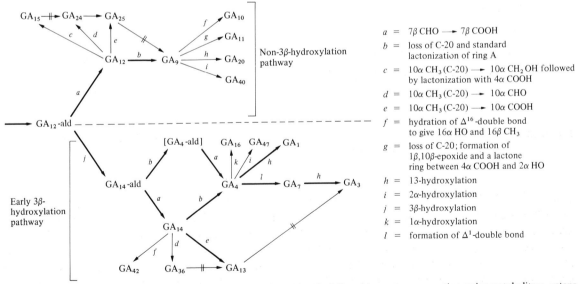

Fig. 15.15. The GA biosynthetic pathway after GA_{12}-aldehyde in *Gibberella fujikuroi*. (➡ = steps connecting major metabolites; → steps connecting minor metabolites; ─╫➤ = steps that cannot be demonstrated; [ ] = hypothetical intermediate.)

The non-3β-hydroxylation pathway yields GA_{12} and GA_9 as the main products. The initial step is the oxidation of the 7β aldehyde (formyl) group of GA_{12}-aldehyde to a 7β carboxyl group so forming GA_{12}. Feeding experiments with GA_{12} show that although its main metabolite is GA_9, GA_{15} (10α CH_2OH followed by lactonization), GA_{24} (10α CHO) and GA_{25} (10α COOH) are also formed as minor metabolites by different degrees of oxidation of the 10α methyl group (C–20). GA_9 is further metabolized to four minor metabolites, GA_{10}, GA_{11}, GA_{20} and GA_{40}.

The conversions of GA_{12} into GA_9 (non-3β-hydroxylation pathway) and of GA_{14} into GA_4 (early 3β-hydroxylation pathway) have the same effect, namely the removal of the 10α methyl group (C–20) and the formation of a lactone ring between C–19 and C–10, and are presumably brought about by the same mechanism. However, the nature of this mechanism is not clear. At first sight the most likely mechanism appeared to involve the oxidation of the 10α methyl group followed by its removal as —CHO or CO_2 with the concomitant production of a $\Delta^{1(10)}$-, $\Delta^{5(10)}$- or Δ^9-double bond which could then be hydrated to give a C–10 hydroxyl group; the latter could then form a lactone ring with the 4α carboxyl group (C–19) by elimination of H_2O. Such a mechanism has been ruled out by labelling experiments which have shown (i) that no hydrogen atoms are lost from Cs 1, 5 and 9 during the conversion of C_{20}-GAs into C_{19}-GAs; this rules out the involvement of $\Delta^{1(10)}$-, $\Delta^{5(10)}$- and Δ^9-double bonds and (ii) that both oxygen atoms of the 4α-carboxyl group (C–19) are present in the lactone; this rules out the formation of a lactone by loss of the elements of water between the 4α carboxyl group and a C–10 hydroxyl group because one of the lactone oxygen atoms would then have been derived from the latter. Point (ii) also rules out a biological type of Baeyer–Villiger reaction and the involvement of a C–19 peracid. It is clear that the C_{20}-GA→ C_{19}-GA conversion involves an intermediate with an electrophilic centre at C–10 which is attacked by the C–19 carboxylate ion. Furthermore, it is quite possible that the latter participates in the loss of C–20; however, it is not clear at what state of oxidation C–20 is lost because the results of feeding experiments with the various possibilities have been inconclusive.

The post-GA_{12}-aldehyde pathway in higher plants is less clear; the present information is fragmentary and scattered amongst several plant species, notably pumpkin (*Cucurbita maxima*), french bean (*Phaseolus vulgaris*) and pea (*Pisum sativum*). Most studies with labelled metabolites have been carried out with developing seeds because they have higher levels of GAs than vegetative tissues.

It appears that both of the *G. fujikuroi* pathways (i.e. 'early 3β-hydroxylation' and 'non-3β-hydroxylation') operate in some higher plants and

that in addition there is a third pathway, the 'early 13-hydroxylation pathway', which is absent from *G. fujikuroi*. Present evidence indicates that the 'early 3β-hydroxylation pathway' operates in bean and pumpkin, the 'non-3β-hydroxylation pathway' in pea and pumpkin and the 'early 13-hydroxylation pathway' in french bean, pea and maize (*Zea mays*). Apart from the 'early 13-hydroxylation pathway' which is absent from *G. fujikuroi*, the details of the higher plant 'early 3β-' and 'non-3β-hydroxylation' pathways are different from those in the fungus. In particular 2β-hydroxylation occurs commonly in higher plants but not in *G. fujikuroi*; this results in loss of biological activity. Moreover, derivatization of GAs giving β-D-glucosyl ethers and esters occurs in higher plants but not in *G. fujikuroi*. These pathways are outlined in Fig. 15.16.

The site of GA synthesis in higher plants is not absolutely clear. It is likely that the main sites are the shoot and root tips, young leaves, flower parts, immature seeds and germinating embryos. However, it is possible that almost all tissues can synthesize GA to some degree. The subcellular sites of GA synthesis are probably the plastids. The evidence for intraplastidic GA synthesis is quite good; it is very strong for the MVA→*ent*-kaurene section of the biosynthetic pathway, fairly strong but limited to barley chloroplasts for the *ent*-kaurenol→*ent*-7α-hydroxykaurenoic acid section and fair for the final stages. However, it is not possible to say at the present time whether it is confined to the plastids.

A number of compounds have been shown to inhibit GA biosynthesis. The best known of these, AMO-1618 (2′-isopropyl-4′-trimethylammonium-chloride)-5′-methylpiperidine carboxylate), CCC (β-chloroethyltrimethyl ammonium chloride) and phosphon D (tributyl-2,4-dichlorobenzylphos-phonium chloride) appear to inhibit the A-activity of *ent*-kaurene synthase (Fig. 15.13). Ancymidol (α-{4 - methoxyphenyl) - α - cyclopropyl - 5 - pyrimidine methanol, EL 531) and the closely related triarimol (α-(2,4-dichlorophenyl)-α-phenyl-5-pyrimidinemeth-anol, EL 273) inhibit all the steps of the following sequence: *ent*-kaurene→*ent*-kauranol→*ent*-kaur-enal→*ent*-kaurenoic acid. These are all oxidations catalysed by cytochrome P450 mixed function oxy-genases and it is thought that these inhibitors

(a) Cell-free enzyme system from pumpkin (*Cucurbita maxima*) endosperm

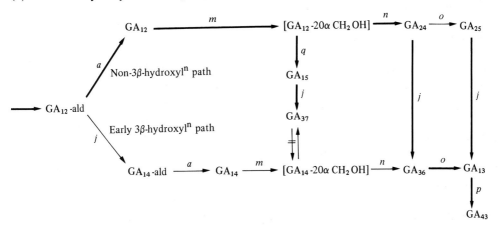

(b) Intact developing bean (*Phaseolus vulgaris*) seeds

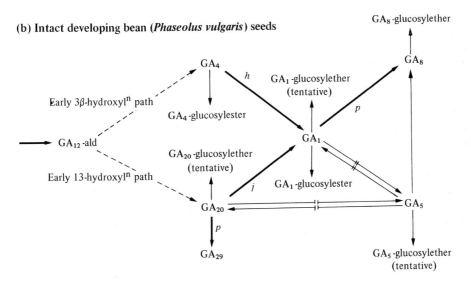

(c) Intact seedlings and developing seeds of pea (*Pisum sativum*)

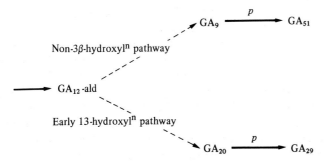

Fig. 15.16. The post-GA_{12}-aldehyde GA biosynthetic pathway as presently seen in some higher plants. ($a–l$, [ ] and the various forms of arrow have the same meaning as in Fig. 15.15; $m = 20\alpha CH_3 \rightarrow 20\alpha CH_2 OH$; $n = 20\alpha CH_2 OH \rightarrow 20\alpha CHO$; $o = 20\alpha CHO \rightarrow 20\alpha COOH$; $p = 2\beta$-hydroxylation; $q =$ lactonization between $20\alpha CH_2 OH$ and $4\alpha COOH$.)

interact with the cytochrome P450 component; this interaction may involve the binding of one of the pyrimidine nitrogen atoms to the protohaem Fe of cytochrome P450 thereby preventing the binding of oxygen to it. These compounds reduce the growth of many higher plants and are used as growth retardants; it is quite likely that this effect is due to the blocking of GA formation.

The introduction of a 2β-hydroxyl group into a GA markedly reduces its biological activity. It is thought that 2β-hydroxylation of GAs, an irreversible process, is widespread in higher plants and that it may serve as a means of regulating the level of biologically active GAs in the tissues although there is no direct evidence for this as yet. There is, however, evidence that the 2β-hydroxylation of GA_{20} to GA_{29} occurs only at a particular stage in the development of pea seeds which is perhaps consistent with a regulatory role.

The β-D-glucosyl ethers and esters of GAs are either low or lacking in biological activity. Their physiological role in higher plants is uncertain but their formation may constitute another deactivation mechanism. They are formed from free GAs as developing seeds mature. On germination the GA-β-D-glucosyl esters appear to be hydrolysed to yield free GAs; this capability is lost as the seedlings age.

Gibberethione (XXXII, Fig. 15.11), formed in immature seeds of *Pharbitis nil* by the oxidation of the 3β-hydroxyl group of GA_3 followed by the addition of mercaptopyruvic acid or L-cysteine to the resulting α,β-unsaturated ketone, has no GA-activity and is also regarded as a GA deactivation product.

3. Movement of gibberellins in the plant

The picture that we have of the movement of endogenous GAs in higher plants at the various stages of their growth is incomplete and far from clear. Numerous experimental approaches have been used to investigate this problem. These include (i) experiments involving the local application of a radiolabelled GA, often GA_3, to the exterior surface of one part of the plant followed by the determi-

nation of its distribution by radioautography of the whole plant some hours later, (ii) experiments involving the standing of excised shoots in a solution of a radiolabelled GA followed by the determination of its distribution by radioautography, (iii) experiments in which an agar block containing a radiolabelled or an unlabelled GA is placed on one of the cut ends of a stem or petiole segment followed by the determination, by radioassay or bioassay as is appropriate, of the amount of GA reaching an agar block placed on the other cut end and (iv) experiments in which the sieve tube (phloem) sap and the bleeding (xylem) sap of several plant species are analysed for GAs. Although these experiments have been criticized in a number of ways they have shown that (i) GAs can move in both the phloem and the xylem, (ii) GAs generally move to young, growing tissues such as the shoot and root tips and immature leaves and (iii) GAs do not exhibit the polarity of transport characteristic of auxin.

The lateral movement or redistribution of GAs in response to phototropic and geotropic stimulation has been studied in a number of different tissues but a simple picture has not emerged.

There is one well-established movement of GAs in the plant. This occurs in the germinating cereal seed where GAs move from the scutellum to the cells of the aleurone layer (see Sections C.4 and C.5).

4. Physiological effects of gibberellins

The most dramatic effect of GAs is their stimulation of stem elongation in (i) dwarf varieties of plants (e.g. the dwarf pea—*Pisum sativum* cv. Meteor), (ii) certain single gene dwarf mutants (e.g. dwarf maize—*Zea mays* mutant d-5) and (iii) rosette plants (e.g. cabbage—*Brassica oleracea*, lettuce—*Latuca sativa*) in which the leaves are tightly clustered around a stem that has undergone little elongation. Exogenously applied GAs do not have a marked stimulatory effect on the elongation of naturally tall plants.

GA-stimulated stem elongation is due to cell elongation rather than increased cell division. This follows from the observation that barley and wheat seedlings in which cell division had been completely inhibited by irradiation with α-rays still elongated when treated with GA. In this respect GAs behave like auxins.

Although there are many examples of a positive correlation between GA and stem elongation all involve exogenously applied GA. The involvement of GA in the process of stem elongation during normal plant growth would be more convincing if there were a positive correlation between the **endogenous level of identified GAs** and stem elongation. To date the limited information that we have is based upon apparent levels of **extractable GA-like materials**, for which there are examples of both positive and negative correlations with growth. The maize mutant d-5 is an example of a positive correlation. This mutant grows to about a fifth of the height of normal maize. No substances with GA-like activity have ever been extracted from it; this contrasts with normal maize, from which GA-like activity has been extracted. Since exogenous application of a number of GA-precursors including *ent*-kaurene and its oxidation products stimulates the mutant to grow to the height of normal maize it appeared that the single gene mutation had adversely affected one of the enzymes catalysing a pre-*ent*-kaurene step in the GA-biosynthetic pathway. Work with cell-free systems has shown this to be the case; the B-activity of *ent*-kaurene synthase (see Fig. 15.12) in the mutant d-5 is only about a fifth that in the normal maize. The corn cockle plant (*Agrostemma githago*) provides an example of a negative correlation between the level of extractable GA-like materials and stem elongation. This plant grows as a rosette in short days and bolts (i.e. the stem elongates) when exposed to long days. The exogenous application of GA to rosetted plants kept under short days also causes bolting. Since GA mimics long days in causing rosetted corn cockle plants to bolt, it might be expected that long days have the effect of raising the endogenous level of GA. However, this has not proved to be the case; the level of extractable GA-like material in rosetted plants is equal to or greater than that in photoperiodically induced bolting plants. Negative correlations between extractable GA-active material and stem elongation such as this do not, however, necessarily rule out the involvement of endogenous GA in the process of stem elongation because it is not difficult to think of possible explanations for

them. A possible explanation of the negative correlation in corn cockle is that under short days GA is compartmentalized in the cells and not available for the stimulation of stem elongation until long days initiate its release; if this were the case then the level of extractable GA-active material would be the same in rosetted and bolting plants.

Most of the other effects of GAs are also brought about by their exogenous application to the plant or part of the plant. GAs stimulate flowering in a wide range of plant species; they are the only group of chemicals known to do this. The effect is almost entirely restricted to long-day plants that grow as rosettes under short days and elongation of the stem (bolting) normally accompanies the flowering response. The corn cockle mentioned above is an example of this; lettuce, spinach (*Spinacia oleracea*) and Darnel grass (*Lolium temulentum*) are others. The dose of GA required to cause flowering ranges from 3 to 100 μg per plant depending upon the species and may have to be repeated at frequent intervals. Most investigations in this area have used GA_3; however, GA_4 and GA_7 have been found to be much more active than GA_3 with *Crepis parviflora*.

Exogenously applied GA overcomes the requirement for vernalization in biennial plants. Normally biennials germinate and grow vegetatively but do not flower during their first year. However, after exposure to the cold temperatures of the ensuing winter they flower in their second year. It is the prolonged exposure to low temperatures (followed by the correct photoperiod) that is necessary for flowering; without it the plants would remain vegetative indefinitely. If a biennial is given an 'artificial' cold treatment during its first year of growth (followed by the correct photoperiod) it will flower in the same year, thus behaving like an annual plant. The acquisition or acceleration of the ability of a plant to flower by chilling the vegetative plant itself is known as vernalization. The locus of perception of vernalization within the plant has been variously ascribed to the growing tip or to any region of dividing cells within the plant. The age at which a plant is sensitive to vernalization is different in different species; for example, cereals may be vernalized during the germination of the seed whilst brussels sprouts (*Brassica oleracea* var. *gemmifera*) require the presence of about 30 leaves. The application of GA to biennials enables them to flower without exposure to cold. However, GA does not replace the cold requirement for flowering in cold-requiring caulescent plants (plants with a visible stalk, cf. rosette plants).

Exogenously applied GA also overcomes the need for stratification in those seeds that normally require it. Stratification is the requirement of the seeds of some plant species (e.g. *Pinus rigida*) to be exposed to low temperatures for a prolonged period before they will germinate.

GA also causes parthenocarpy: this can be accomplished by spraying the flowers of the plant with a GA solution. Parthenocarpy is the process in which fruits form without fertilization having occurred. The resulting fruits (developed from the ovary wall) appear normal but are seedless. Auxin is also capable of inducing parthenocarpic fruit set but GA is more active; GA will, for instance, function with stone fruit which are unresponsive to auxin.

The external application of GA to the mature form of ivy (*Hedera helix*) induces it to revert to the juvenile form which has considerable morphological differences.

There is one well-documented effect of endogenous GA, rather than externally applied GA. This is to be found in the germinating cereal seed (e.g. barley—*Hordeum vulgare*). Shortly after the seed has imbibed water through its micropyle as a first step in germination, the embryo synthesizes GAs (GA$_3$, GA$_7$) in the scutellum. These GAs are then translocated to the cells of the aleurone layer which surrounds the starch-storing cells of the endosperm. The GAs stimulate the synthesis of α-amylase[6] and other enzymes in the aleurone cells (see Section C.5). These enzymes are then secreted into the endosperm where they participate in the breakdown of starch [see Chapter 7, Section D.1(i)] into sugars which are absorbed by the embryo and used to support its growth.

5. Biochemical mode of action of gibberellins

As can be seen from Section C.4, GAs bring about or are involved in several diverse effects in plants. At the present time none of these effects can be completely explained at the molecular level. In fact for only one of them, that involving cereal aleurone

cells, is there any real body of knowledge and even here it is too incomplete to give more than a blurred picture of what is happening.

The overall picture of the effect of GA in the barley aleurone cells has been outlined in Section C.4; the key feature is that it stimulates the secretion of α-amylase and other hydrolases into the endosperm to degrade the stored starch. Although the aleurone cells of the dry, non-germinating seed contain β-amylase[7], probably stored in the aleurone grains, they have no α-amylase at all. The formation of α-amylase is essential during germination because it alone of all the enzymes concerned in starch breakdown can attack starch in the form of starch grains, which is how it is present in the endosperm cells. It has been shown experimentally that GA$_3$ stimulates the *de novo* synthesis of four α-amylase isoenzymes in the aleurone cells. There is, however, a lag of about 8–10 hr between the GA$_3$ application and the first appearance of α-amylase. At the same time that α-amylase appears other newly synthesized hydrolase enzymes appear, including a protease, ribonuclease[8] and to a lesser extent endo-1,3-β-D-glucanase[9]. These enzymes, along with others which, like β-amylase, have previously been stored in the aleurone grains, are then secreted from the aleurone cell into the surrounding endosperm tissue. Some of the hydrolases attack the cell walls of the endosperm cells to facilitate the entry of the starch degrading enzymes.

Although the synthesis and secretion of α-amylase and other hydrolases is the most obvious response by the aleurone cells to GA it is by no means the only one. It is preceded by the proliferation of endoplasmic reticular membranes which begins about 2–4 hr after the application of GA. This is indicated by electron microscopic observation of a progressive increase in rough endoplasmic reticulum, the increased incorporation of precursors of membrane lipids (e.g. $^{32}PO_4^{3-}$, [^{14}C]choline) into the membrane fraction and the increased activity of key enzymes in the biosynthesis of membrane lipids (e.g. cholinephosphate cytidylyltransferase[10] and cholinephosphate glyceridetransferase)[11].

This increased membrane synthesis might be expected in view of the large-scale secretion of enzymes which is to come. Proteins such as these which are destined for export from the cell are segregated from those which are not by the process described in Chapter 4, Section D.3. They are synthesized by ribosomes adhering to the endoplasmic reticular membranes and pass into the lumen of the endoplasmic reticular tubule system where they undergo structural modification under the catalytic influence of 'tailoring' enzymes. They then become incorporated into Golgi cisternae where further structural modification may occur. Finally they are incorporated into Golgi vesicles which pass to the plasmalemma, fuse with it and empty their contents, the enzymes, to the exterior of the cell. The process of enzyme secretion therefore coincidentally leads to the transfer of membrane material from the endoplasmic reticulum to the plasmalemma. This loss of endoplasmic reticulum must be replaced by the synthesis of new endoplasmic reticular membrane. Thus the increased synthesis of endoplasmic reticulum beginning 2–4 hr after GA application could be interpreted as the cell preparing itself for the large-scale transfer of endoplasmic reticulum to plasmalemma that will accompany the secretion of α-amylase and the other hydrolases. The fact that the endoplasmic reticulum is liberally covered with ribosomes (i.e. there is a high proportion of rough endoplasmic reticulum) during its GA-induced proliferation before the onset of α-amylase secretion is probably indicative of the synthesis of protein membrane components rather than of the synthesis of enzymes destined for secretion.

From the foregoing account it is clear that GA has at least two actions in aleurone cells which are separated in time. The first is the switching on of the synthesis of endoplasmic reticular membranes whilst the second is the switching on of the synthesis of α-amylase and the other hydrolases. Precisely how GA accomplishes the first is not known; however, it has been suggested that it directly or indirectly (i.e. via a specific receptor) activates key enzymes in membrane synthesis. There is a little more information about GA's second action. Actinomycin D, an antibiotic derived from some *Streptomyces* spp. which blocks the transcription of DNA into *m*RNA by intercalating its phenoxazole ring system between base pairs of the DNA double helix in regions containing guanine, blocks the GA-induced synthesis of α-amylase; the earlier in the lag phase it is administered the more effective it is. This suggests that GA stimulates the synthesis of the

*m*RNA specific for α-amylase. This *m*RNA appears to be a long-lived species because actinomycin D becomes less and less effective in preventing GA-stimulated α-amylase synthesis the later in the lag phase it is administered and has virtually no effect after the lag phase is over. Exactly how GA stimulates the synthesis of the α-amylase-specific *m*RNA is not known; it is possible that it directly or indirectly (i.e. via a specific receptor which may be different from that mentioned earlier) derepresses the gene for α-amylase. There is a strong possibility that the *de novo* synthesis of the other hydrolases is stimulated in the same way.

GA is not the only plant hormone to be implicated in the secretion of α-amylase from aleurone cells. Abscisic acid (see Section E) inhibits the GA-induced α-amylase secretion regardless of whether it is administered before or after the end of the lag phase; it does this without affecting the overall rate of protein synthesis. This inhibition by abscisic acid can be overcome by cytokinins (see Section D) which, however, have no effect on α-amylase formation when administered by themselves. Ethylene (see Section F) does not have any effect on the synthesis of α-amylase but promotes its release from aleurone cells.

From Section C.4 it is clear that the most dramatic effect of GA is on cell elongation. How GA brings this about is not known. However, the underlying principles of elongation discussed in Section B.5 are equally applicable in this context. It is therefore possible that GA, as well as auxin, is concerned in some way with cell wall plasticity.

D. CYTOKININS

1. Discovery and characterization

Cytokinins were discovered as a result of the extensive work on plant tissue cultures carried out by Skoog in the period 1945–55. Skoog found that when segments of tobacco stem, consisting of cortical, vascular and pith tissues, were cultured on a medium containing auxin, the pith cells proliferated. If, however, pith tissue alone was cultured on the same medium the cells enlarged but did not divide. Cell division, in this medium, was induced if the pith tissue was placed in contact with vascular tissue or if an extract of vascular tissue was added. Clearly the vascular tissue contained a substance which promoted cell division in tobacco pith. Cell-division-promoting substances were also found in coconut milk (liquid endosperm), malt and an autoclaved preparation of herring sperm DNA. The latter proved to be a particularly rich source and in 1956 the active substance, termed kinetin, was isolated from it and identified as 6-(furfurylamino)purine (XXXIII, Table 15.2). Kinetin induced cell division in tobacco pith when present in the culture medium at a concentration of 1 μg ml^{-1}. However, it was realized that kinetin is not a naturally occurring substance; it was formed as an artefact of the autoclaving of DNA by the dehydration of a deoxyribose residue and its migration from the 9-position of adenine to the 6-amino nitrogen (the N^6-position).

Although kinetin does not occur in plant tissues, substances with a similar biological activity were shown to be widely distributed in plants. These substances were initially called kinins but this name has now been replaced by the generally accepted term 'cytokinin'. The definition of this term is physiological not chemical; a cytokinin is said to be a substance that promotes cell division in a cultured plant callus tissue grown on a defined medium containing all necessary organic and inorganic nutrients and growth factors including an optimal exogenous supply of auxin.

Despite the known presence of cytokinins in plant tissues, the first isolation and identification of a naturally occurring cytokinin did not come until 1963, 9 years after the discovery of kinetin. This was largely due to the technical difficulties of purifying minute quantities of biologically active compounds from large amounts of plant material and the determination of their structure with physical methods of limited sensitivity. However, in 1963 Letham identified a cytokinin, now known as zeatin (XXXIV), from immature maize seeds as 6-(4-hydroxy-3-methyl-*trans*-2-butenylamino)purine; in 1966 he also identified zeatin riboside (9-β-D-ribofuranosyl zeatin, XXXV) and in 1973 zeatin riboside 5'-monophosphate (XXXVI) and several other cytokinins from the same source. Table 15.2 gives the structure and occurrence of several of the naturally occurring cytokinins now known.

The identification of the first cytokinin in 1956 as a 6-substituted aminopurine led to the synthesis of a large number of analogues. This work enabled the structure–biological activity relationship of cytokinins to be studied. This may be summarized by saying that: (i) replacement of the N^6 substituent of kinetin with a wide variety of different groups produced a range of compounds with a considerable spectrum of cell division promoting activity; (ii) most of these compounds were less active than kinetin but a few, notably 6-(3-methyl-3-butenylamino)purine (i^{6}Ad, XXXVII), were more active; (iii) in general, the most active of these compounds had an N^6 side chain of 4–6 atoms; (iv) alterations of the purine nucleus led to a large reduction or total loss of activity and occasionally produced an anticytokinin. An example of the latter is 3-methyl-7-*n*-pentylaminopyrazolo[4,3-*d*]pyrimidine (XLV, Fig. 15.17).

In spite of (iv) above, the presence of a purine ring is not essential for cytokinin activity. A considerable number of compounds lacking a purine ring are cytokinins. The most active of these are phenylurea derivatives of which N-3-chlorophenyl-N'-phenylurea and N-3-nitrophenyl-N'-phenylurea (XLVI and XLVII respectively, Fig. 15.17) are good examples. Weak cytokinin activity is shown by some aromatic amide derivatives, azaindene and azanaphthalene derivatives, 4- and 5-substituted benzimidazoles, systemic benzimidazole fungicides such as methyl benzimidazol-3-yl-carbamate and 6-(substituted amino)-8-azapurines. Some phenylurea derivatives, notably N-benzyl-N'-phenylurea and N-benzyl-N'-3,4-dichlorophenylurea (XLVIII and XLIX, Fig. 15.17), are anticytokinins; both these compounds will antagonize the biological action of purine and phenylurea cytokinins. This suggests that purine and phenylurea cytokinins act at the same site in the cell.

Cytokinin research took an interesting turn in 1966 when Zachau and coworkers identified i^{6}Ad as a constituent of two *t*RNA^{Ser} species from yeast. It was subsequently shown that the i^{6}Ad occurred only once in the *t*RNA^{Ser} molecules in a position adjacent to the 3' end of the anticodon. Skoog then identified four cytokinins, namely i^{6}AdR (XXXVIII), ZR (XXXV), MeS^{2}i^{6}AdR (XLIII) and MeS^{2}ZR (XLIV), in a variety of *t*RNA species. It was subsequently found that cytokinins are not

Table 15.2. *Structures of some cytokinins*

	Cytokinin		Substituents of adenine[a]			
Text no.	Chemical name	Abbrev.[b]	$N^{6\ c}$	C–2	N–9	Occurrence
XXXIII	6-(Furfurylamino)-purine	Kinetin	H_2C — (furfuryl group, O)	H	H	Artefact from autoclaved DNA
XXXIV	6-(4-Hydroxy-3-methyl-*trans*-2-butenylamino) purine	Zeatin (Z)	H_2C / CH_2OH / CH_3	H	H	Immature maize seeds, cotton ovules, coconut milk, sycamore sap
XXXV	6-(4-Hydroxy-3-methyl-*trans*-2-butenylamino)-9-β-D-ribofuranosyl-purine	Zeatin riboside (ZR)	H_2C / CH_2OH / CH_3	H	β-D-R_f[d]	Maize seeds, cotton ovules, chicory root, coconut milk, sycamore sap
XXXVI	6-(4-Hydroxy-3-methyl-*trans*-2-butenylamino)-9-(β-D-ribofuranosyl-5′-monophosphate) purine	Zeatin riboside phosphate (ZRP)	H_2C / CH_2OH / CH_3	H	β-D-R_f 5′-Ⓟ	Maize seeds
XXXVII	6-(3-Methyl-2-butenyl-amino) purine	i⁶Ad	H_2C / CH_3 / CH_3	H	H	*Corynebacterium fascians*, cotton ovules
XXXVIII	6-(3-Methyl-2-butenyl-amino)-9-β-D-ribofuranosylpurine	i⁶AdR	H_2C / CH_3 / CH_3	H	β-D-R_f	Tobacco tissue, cotton ovules
XXXIX	6-([3S]-4-Hydroxy-3-methylbutylamino) purine	Dihydro-zeatin (diHZ)	H_2C / CH_2OH / CH_3 / H	H	H	Immature lupin seeds, cotton ovules, sycamore sap
XL	2-Hydroxy-6-(4-hydroxy-3-methyl-*trans*-2-butenyl-amino) purine	HO²Z	H_2C / CH_2OH / CH_3	OH	H	Maize seeds
XLI	6-(3,4-Dihydroxy-3-methylbutylamino) purine	HOZ	H_2C / CH_2OH / OH / CH_3	H	H	Maize seeds
XLII	6-(O-Hydroxybenzylamino)-9-β-D-ribofuranosyl-purine	HOBz⁶AdR	HO— (o-hydroxybenzyl group) H_2C	H	β-D-R_f	*Populus robusta* leaves

(*continued overleaf*)

For notes [a–e] see page 598.

Table 15.2 (cont.)

| Text no. | Cytokinin | | Substituents of adenine[a] | | | |
	Chemical name	Abbrev.[b]	N^{6c}	C–2	N–9	Occurrence
XLIII	6-(3-Methyl-2-butenyl-amino)-2-methylthio-9-β-D-ribofuranosyl-purine	MeS²i⁶AdR		$H_3C \cdot S$	β-D-R$_f$	tRNA of wheat germ, pea shoots, and bacteria
XLIV	6-(4-Hydroxy-3-methyl-cis-2-butenylamino)-2-methylthio-9-β-D-ribofuranosylpurine[e]	MeS²ZR(c)		$H_3C \cdot S$	β-D-R$_f$	tRNA of wheat germ, pea shoots, and Pseudomonas aeruginosa

[a] Adenine numbering:

N^6 position (ie the nitrogen attached to position 6 in the purine ring)

[b] Abbreviation and/or trivial name.
[c] Note that all the N^6 substituents in this table, except those of XXXIII and XLII, have the isoprene carbon skeleton, (see Chapter 11).
[d] β-D-ribofuranosyl.
[e] Mixture of cis and trans isomers in tRNA of pea shoots.

present in all tRNA species; they appear to be confined to tRNA species that respond to mRNA codons with the initial letter U (see Fig. 15.18).

The distribution of cytokinins in tRNA species from different sources is interesting: only i⁶AdR has been found in animal tRNAs; i⁶AdR and MeS²i⁶AdR have been found in bacterial tRNAs; hydroxylated i⁶AdR derivatives appear to be peculiar to plant tRNAs and with one exception have the cis configuration.

Some cytokinins have an effect in animal cells. For example, they have been shown to (i) inhibit blood platelet aggregation, (ii) inhibit the growth of the cells of some tumours, (iii) stimulate the elongation of cultured fibroblasts and (iv) behave as immunosuppressive agents and induce remissions in leukaemia patients.

2. Bioassay and distribution of cytokinins

A considerable number of different cytokinin bioassays have been used. They fall into six main classes: (i) bioassays based on the ability of cytokinins to stimulate the expansion of excised leaf or cotyledon tissue—probably the best of these uses radish cotyledons and is capable of detecting kinetin concentrations as low as 10 ng ml^{-1}; (ii) bioassays based on the stimulation of growth of duckweed species, e.g. Lemna minor, Spirodela—these are capable of detecting kinetin concentrations in the range of 10–60 ng ml^{-1}; (iii) bioassays based on the stimulation of growth of stem or coleoptile sections—these are quick to perform, taking only one day to produce a result, but lack specificity; (iv) bioassays which depend upon the ability of

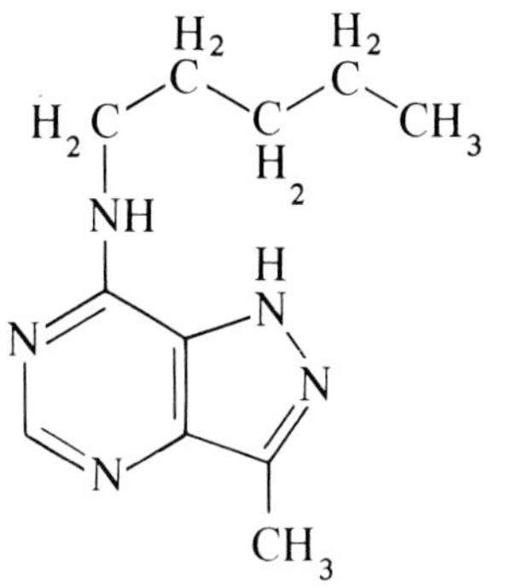

(XLV) 3-methyl-7-*n*-pentylaminopyrazolo [4,3-*d*] pyrimidine (a pseudopurine anticytokinin)

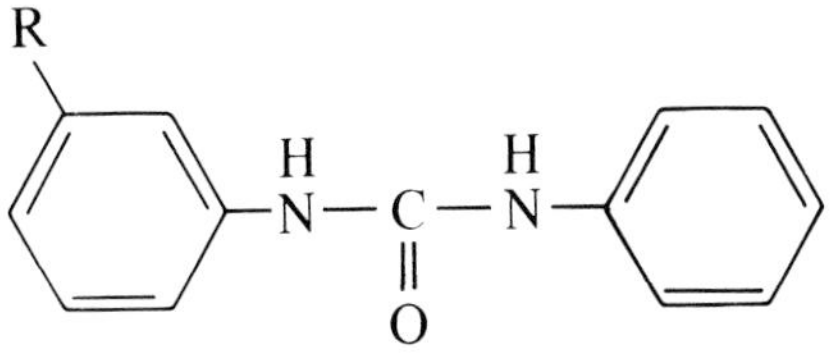

(XLVI) R = Cl; *N*-3-chlorophenyl-*N*′-phenylurea
(XLVII) R = NO₂; *N*-3-nitrophenyl-*N*′-phenyl-urea (two non-purine cytokinins)

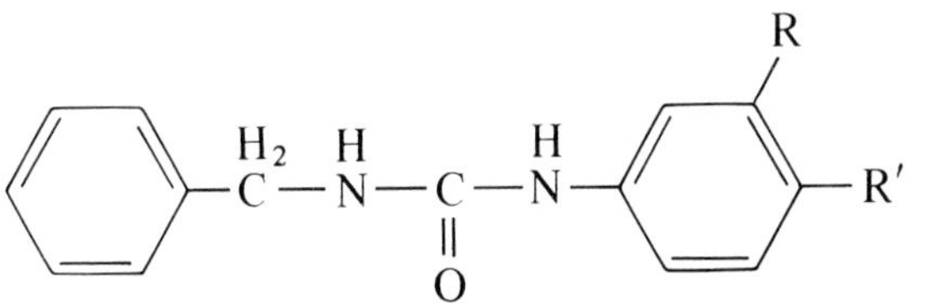

(XLVIII) R = R′ = H; *N*-benzyl-*N*′-phenylurea
(XLIX) R = R′ = Cl; *N*-benzyl-*N*′-3,4-dichlorophenylurea
(two non-pseudopurine anticytokinins)

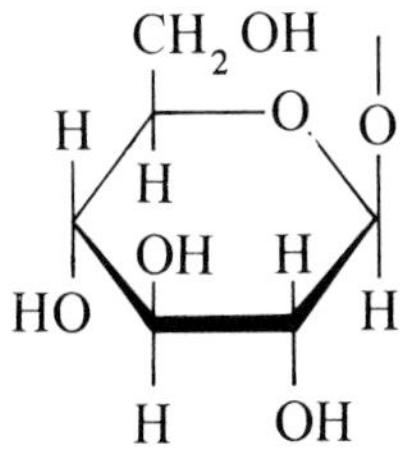

β-D-glucopyranosyl (see structures L–LIV below)

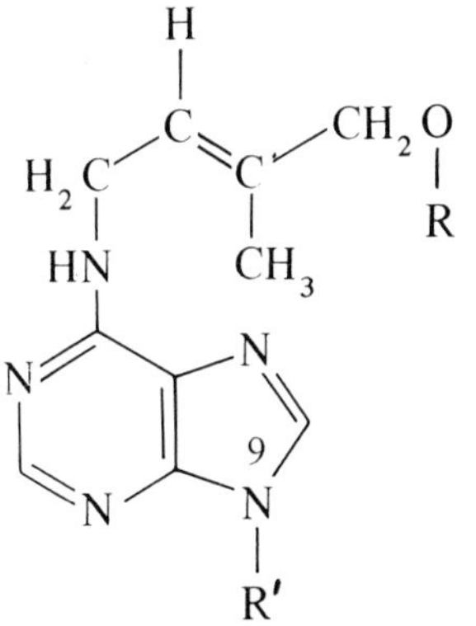

(L) R = *β*-D-glucopyranosyl; R′ = H

(LIII) R = H; R′ = —CH₂ ''''C ''''COOH
with NH₂ above and H below

(LIV) R = H; R′ = *β*-D-glucopyranosyl
(see Fig. 15.21)

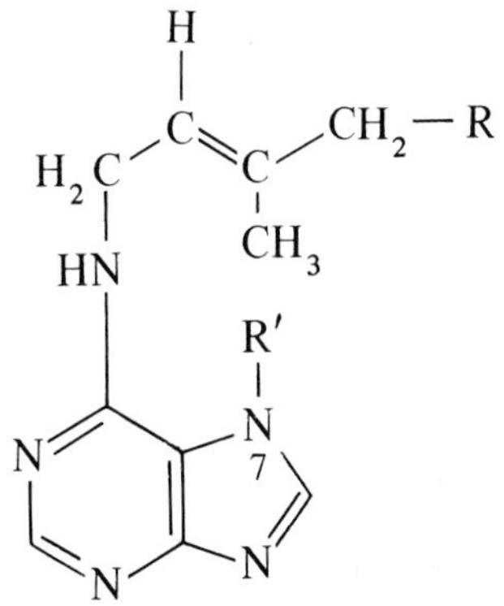

(LI) R = H; R′ = *β*-D-glucopyranosyl
(LII) R = OH; R′ = *β*-D-glucopyranosyl
(see Fig. 15.21)

Fig. 15.17. Some cytokinins, anticytokinins and cytokinin metabolites.

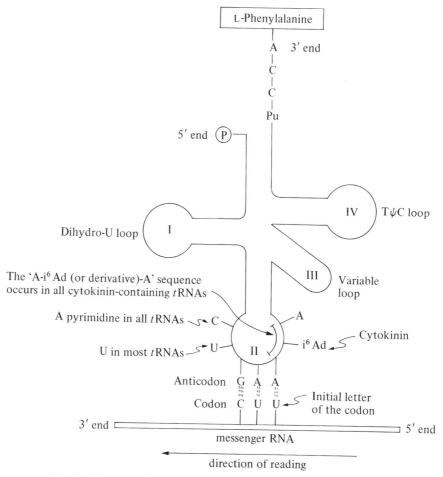

Fig. 15.18. Structure of yeast tRNA^{Phe} showing the position of the cytokinin.

cytokinins to induce cell division in tissue cultures—these have the disadvantage of taking 3 weeks or more to produce a result but are generally very sensitive, often being capable of detecting kinetin concentrations in the range 0.5–3 ng ml^{-1}; (v) bioassays based on the ability of cytokinins to retard leaf senescence—these take about 6 days, are fairly specific and are capable of detecting kinetin concentrations in the range 3–10 ng ml^{-1}; (vi) bioassays based upon the ability of cytokinins to stimulate pigment formation, e.g. chlorophyll in cucumber cotyledons and betacyanin in Amaranthus—these give a result within 1–3 days, have high specificity and are capable of detecting a

kinetin concentration of 1–2 ng ml^{-1}.

Cytokinins have been found in a wide range of plant tissues; they are particularly abundant in root tips, xylem sap, developing fruits, tumour tissue and germinating seeds.

The cytokinins of roots are Z, ZR and possibly ZRP (see Table 15.2) and are located predominantly in the root tip. Root nodules contain unusually high cytokinin levels; the main cytokinin in Vicia faba root nodules is Z whilst Z and ZR are found in those of Pisum sativum. It is not clear how much of the cytokinin in the root nodules is contributed by the symbiotic bacteria. Cytokinins are found in the outer regions and the 'eyes' (i.e. buds) of potato tubers; the level rises markedly towards the end of tuber dormancy.

Xylem sap contains ZR as its principal cytokinin although Z may also be present at certain stages of growth. The phloem sap of a few plants, e.g. *Xanthium, Salix babylonica*, has also been shown to contain cytokinin, probably ZR.

Buds do not contain detectable amounts of cytokinins when dormant but as the time for normal bud break approaches in the spring cytokinins appear and their level rises to a maximum.

Cytokinin levels in leaves vary with the age of the leaf. Total cytokinin activity and diversity of structural species are greatest in young expanding leaves and lowest in mature and senescing leaves. Z and ZR are the principal leaf cytokinins.

Cytokinin-autotrophic tissue (i.e. plant tissues which can be cultured *in vitro* in the absence of exogenously supplied cytokinin) appear to be able to synthesize cytokinins for themselves. A cytokinin-autotrophic strain of soya bean cotyledon callus cells has been shown to produce three cytokinins which appear to be Z, ZR and ZRP.

Cytokinins have also been found in crown-gall tumour tissue. Crown-gall tumours arise when plant cells are infected by *Agrobacterium tumefaciens*. It is possible to obtain crown-gall tumour cells free from bacteria and to culture them on a synthetic medium containing no auxin or cytokinin. Extracts from cells cultured in this way contain cytokinin activity suggesting that they are capable of synthesizing cytokinins for themselves. The cytokinin from crown-gall tissue cultures from *Parthenocissus tricuspidata* and *Vinca rosea* is probably ZR.

Cytokinins have been detected in several marine algae (e.g. *Gymnodinium splendens, Cricosphaera* spp., *Laminaria digitata, Hypnea musciformis*) and certain freshwater algae (e.g. *Volvox carteri*). They are also present in some bacteria (e.g. *Corynebacterium fascians, A. tumefaciens, Rhizobium japonicum*) and in some fungi (e.g. the mycorrhizal fungi, *Rhizopogon roseolus* and *Suillus punctipes*, and the phytopathogen *Monilia* spp.).

Cytokinins have been found in the labial glands of the larvae of *Stigmella* spp. This caterpillar infests the leaves of deciduous trees. In the autumn, as the leaves senesce, turning yellowish-brown, the location of the caterpillar is indicated by a 'green island' which is very rich in cytokinins, mainly Z.

3. Movement of cytokinins in the plant

The picture that we have of cytokinin translocation in plants is unclear. There is evidence to suggest that the cytokinin synthesized in the root tips is transported to the leaves and probably the rest of the plant via the xylem sap. However, the cytokinin that is synthesized in other regions of the plant such as developing seeds does not appear to be translocated. Numerous investigations in which cytokinins have been applied to the external surface of leaves and other organs or to the cut ends of stems or petioles or injected into various parts of the plant have generally produced unsatisfactory and often contradictory results.

Some evidence suggests that it is the conjugated types of cytokinins (e.g. ribosides such as ZR) that are translocated rather than the free purine types (e.g. Z).

4. Metabolism of cytokinins

(i) BIOSYNTHESIS OF CYTOKININS

Cytokinins occur in the free form or in *t*RNA. The question that immediately arises is 'Are these two forms synthesized by independent pathways or is one form derived from the other?' Despite the paucity of information about cytokinin biosynthesis generally, the answer to this question is now clear; in growing plants the two forms arise by independent pathways.

Studies with cell-free plant systems have shown that the Δ^2-isopentenyl side chain is transferred to an adenine residue adjacent to the 3' end of the anticodon in *t*RNA (see Fig. 15.18) at the macromolecular level. This follows from the use of *t*RNA that has been treated with permanganate under mild conditions; such treatment specifically cleaves off the Δ^2-isopentenyl residue from the i[6]Ad leaving adenine. When the *t*RNA, produced by this treatment, was incubated with cell-free preparations from tobacco pith or yeast in the presence of [2-^{14}C]mevalonic acid i[6]Ad was formed in the *t*RNA. Purification of the yeast enzyme showed that the immediate isopentenyl donor was Δ^2-isopentenyl pyrophosphate (i.e. dimethylallyl pyrophosphate) rather than Δ^3-isopentenyl pyrophosphate. Presumably other tRNA cytokinins (e.g. Me^2i^6AdR, $Me^2ZR(c)$, see Table 15.2) are also formed at the macromolecular level by modification of the adenine and/or Δ^2-isopentenyl moieties of i^6Ad.

The evidence that free cytokinins in plants are not formed by hydrolysis of tRNA to the mononucleotide level is very strong. Two powerful examples are: (i) the tRNA of cytokinin-autotrophic tissue (see Section D.2) contains cytokinins but such tissue will not grow on cytokinin-free media—hence physiologically significant amounts of cytokinins are not produced by tRNA breakdown and (ii) tRNA cytokinins such as Me^2i^6AdR, $MeS^2ZR(c)$ and the cis isomer of ZR have never been found in plants in the free form. It is, however, possible that cytokinins may be released from tRNA in dying, autolysing plant cells. Moreover, cytokinins may arise from tRNA in bacteria where the turnover of tRNA is more rapid.

The pathway by which free cytokinins are synthesized in plants is virtually unknown. It seems likely that a Δ^2-isopentenyl residue derived from mevalonic acid is transferred to the N^6 of adenine, adenosine or adenosine 5'-monophosphate and that this is then modified by $trans$-hydroxylation of a C–3 methyl group and/or reduction of the Δ^2 double bond (see Fig. 15.19).

The principal site of biosynthesis of the free cytokinins in the vegetative plant appears to be the root tip. These cytokinins are probably distributed to the rest of the plant via the xylem sap. Free cytokinins are also synthesized in developing buds, developing seeds and in the embryonic axis of germinating seeds.

It is more than likely that the cytokinins that appear in tRNA are formed in every living cell in the plant. Within the living cell they are known to be formed in two distinct compartments, the cytoplasm and the chloroplast, and are probably also synthesized in the mitochondrion although this has yet to be demonstrated. The cytoplasmic tRNAs are formed by transcription of nuclear genes (see Fig. 15.20) whilst the chloroplastidic and mitochondrial tRNAs are formed by transcription of genes in chloroplastidic and mitochondrial DNA respectively. The isopentenylation and other modifications

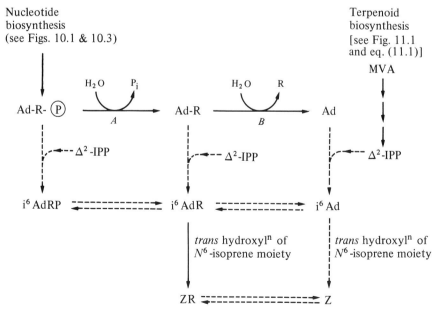

Fig. 15.19. Hypothetical pathway for the biosynthesis of free cytokinins in plants. (A = 5'-Nucleotidase, EC 3.1.3.6; B = adenosine nucleosidase, EC 3.2.2.7; Ad = adenine; R = D-ribose; P_i and $\textcircled{P}$ = orthophosphate; Δ^2-IPP = Δ^2-isopentenyl pyrophosphate = dimethylallyl pyrophosphate; MVA = mevalonic acid; i^6Ad, i^6AdR, i^6AdRP, Z and ZR have the same meanings as in Table 15.2; → = known reaction; ----→ = postulated reaction.)

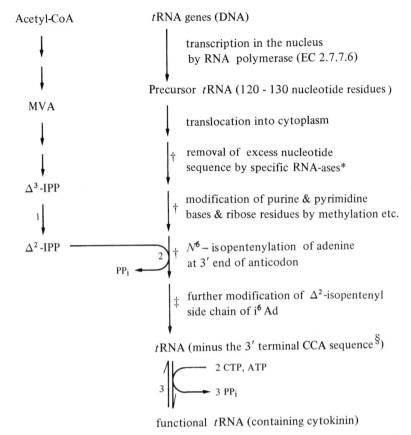

Fig. 15.20. Probable pathway for the biosynthesis of cytoplasmic *t*RNA-cytokinins in plant cells (a similar pathway probably occurs in the stroma of chloroplasts and mitochondria for the biosynthesis of chloroplastidic and mitochondrial *t*RNA-cytokinins.) (1 = isopentenyl diphosphate Δ-isomerase, EC 5.3.3.2; 2 = *t*RNA isopentenyltransferase, EC 2.5.1.8; 3 = *t*RNA cytidylyltransferase, EC 2.7.7.21 and *t*RNA adenylyltransferase, EC 2.7.7.25—it is thought that these two enzymes may be identical.)
* Some trimming of *t*RNA may also occur in the nucleoplasm.
† These steps do not necessarily occur in the order shown.
‡ This step does not occur in the formation of i⁶Ad-*t*RNAs.
§ Whether the 3′ terminal CCA sequence is transcribed from the *t*RNA gene or is added after transcription as shown is controversial; it is quite likely that in some *t*RNAs it is transcribed and that in others it is added later; however, it is known that there is considerable turnover of this sequence.)

to the adenine on the 3′ side of the anticodon that generate the cytokinin in the *t*RNAs occur in the cytoplasm in the case of cytoplasmic *t*RNAs and in the organelle stroma in the case of chloroplastidic and mitochondrial *t*RNAs. The cytokinins of cytoplasmic *t*RNAs tend to be different from those of chloroplastidic *t*RNAs. The former are like the free cytokinins found in plants whilst the latter are like the *t*RNA-cytokinins of bacteria; this fits in rather well with the hypothetical endosymbiont origin of chloroplasts [see Chapter 3, Section B.6(viii)].

(ii) CONJUGATION AND BREAKDOWN OF CYTOKININS

Cytokinins have been shown to be conjugated with D-glucose to form *O*-, 7- and 9-β-D-glucosides in a number of plant tissues (see Fig. 15.21). This glucosylation leads to inactivation of the cytokinins; however, the physiological significance of this in the plant tissues in which it occurs has yet to be explored. Seedlings of *Lupinus luteus* are also capable of conjugating Z with L-alanine to form lupinic acid (see Fig. 5.21).

Plant tissues are also capable of degrading

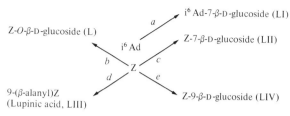

Fig. 15.21. Conjugation of cytokinins with D-glucose and L-alanine. (a = in tobacco cell cultures; b = in soya bean tissue culture, *Populus alba* leaves and *Lupinus luteus* seedlings; c = in *Raphanus sativus* leaves and as the minor metabolite in *Zea mays* plants; d = in *Lupinus luteus* seedlings; e = major metabolite in *Zea mays* plants; the roman numerals refer to structures in Fig. 5.17.)

cytokinins by cleaving off the isopentenyl side chain; this results in total loss of activity. The enzyme catalysing this cleavage has been partly purified from maize seeds. It requires molecular oxygen and has some of the characteristics of a mixed function oxygenase. It is active with i^6Ad and i^6AdR and catalyses their degradation to Δ^2-isopentenyl aldehyde (3-methylbut-2-enal) and adenine or adenosine respectively. It has been given the trivial name cytokinin oxidase although cytokinin oxygenase probably fits it better.

5. Physiological effects of cytokinins

A considerable number of physiological effects of cytokinins have been described. However, they can probably all be placed in one or other of two broad categories: (i) promotion of cell division and differentiation and (ii) retardation of the process of senescence.

The effect of cytokinins in promoting cell division was first seen with plant tissue cultures and indeed is the basis of the definition of a cytokinin (see Section D.1). The specificity in this respect is now unquestioned and it is well established that they act as a specific trigger of mitosis; it is probable that they fulfil the same function in whole plants. Cytokinins also play an important part in cell differentiation and organogenesis in plant tissue cultures. In this their effect is dependent upon their concentration relative to that of IAA and GA. This is typified by the following results with tobacco callus grown with various ratios of cytokinin, IAA

and GA_3 in the agar medium: (i) no hormones—little growth; (ii) IAA only—cell enlargement but no cell division or differentiation; (iii) cytokinin only—growth but no differentiation; (iv) cytokinin and IAA—growth and differentiation; (v) cytokinin/IAA ratio low—growth and differentiation mainly into roots; (vi) cytokinin/IAA ratio high—growth and differentiation mainly into buds and leaves; (vii) cytokinin/GA_3 ratio low—growth and differentiation into slender, etiolated plants with narrow leaves; and (viii) cytokinin/GA_3 ratio high—growth and differentiation into short, green plants with rounded leaves.

Other cytokinin effects that can be put in the 'promotion of cell division and differentiation' category are (i) the promotion of germination of some seeds, (ii) the induction of tuberization in potato stolons and (iii) the promotion of bud development. In the latter case cytokinins appear to play a part in overcoming apical dominance. Apical dominance is the inhibition of the growth of subtending lateral (axillary) buds by the growing shoot apex. Although the mechanism of this phenomenon is not completely understood it appears to involve an antagonism between cytokinins and auxins. Auxins are produced by the shoot tip and move down the stem (see Section B.2) inhibiting the development of the adjacent axillary buds; the nearer an axillary bud is to the shoot tip, the greater is the inhibition and vice versa. This inhibition can be relieved artificially by removing the shoot tip or by applying cytokinin to the axillary buds. The axillary buds synthesize some cytokinin themselves. This is evidently insufficient to overcome the inhibitory effect of auxin in the intact plant but can promote bud development when the auxin supply from the shoot tip has been removed.

The effect of cytokinins in retarding the rate of senescence is best seen in leaves. When a mature leaf is removed from a plant and the petiole is immersed in water, senescence occurs rapidly. The most obvious external manifestation of this is a yellowing due to the destruction of the chlorophyll, whose green colour normally masks the yellow carotenoids. The leaf, however, does not senesce if adventitious roots develop on the petiole. This led Chibnall in the late 1930s to propose that senescence was being prevented by a hormone produced in the roots. The elucidation of the identity of this hormone was brought a little closer when, in 1957, Richmond and Lang found that senescence in detached leaves of cocklebur (*Xanthium pennsylvanicum*) was retarded by kinetin. It was soon shown that this was not an isolated instance but a general phenomenon; kinetin and N^6-benzyladenine (a synthetic cytokinin) delayed the senescence of detached leaves and leaf sections from many species. In 1959 Mothes applied a kinetin solution to some parts of the external surface of a leaf and not to others; he found that the treated areas remained green whilst the untreated areas yellowed. The general picture that emerges is that cytokinins produced in the roots are trans-located via the xylem sap to the leaves where they promote key metabolic processes and thereby maintain the structural and functional viability of the leaves. It follows from this that senescence occurs when the supply of cytokinin to the leaves or the *active* cytokinin content of the leaves falls; however, the experimental data necessary to back up this conclusion are contradictory.

6. Biochemical mode of action of cytokinins

It is not possible at the present time to explain the physiological effects of cytokinins (see Section D.5) at the molecular level. The situation is complicated by the presence of two metabolically distinct types of cytokinin, the free cytokinins and the *t*RNA-cytokinins. This raises a number of questions. Firstly, is only one type or are both types physiologically active? If only one type is physiologically active, which is it? If both groups are physiologically active, do they act in the same or different ways? If they have different physiological actions, what are they? The present indications are that both types are physiologically active and play an important role in metabolic control. It is a fair guess that each of the two types plays a different role. The *t*RNA-cytokinins would appear to be the more primitive type in the evolutionary sense because they have a function (probably an identical function) in bacteria and higher plants whereas the free cytokinins have a function only in higher plants; the free cytokinins of bacteria probably result from *t*RNA turnover and have no separate function. Thus it seems reasonable to suppose that the function of the *t*RNA-cytokinins

is the same in bacteria and higher plants (and indeed in all cellular living organisms) whilst the function of the free cytokinins evolved much later and is peculiar to higher plants. Clearly the *t*RNA-cytokinins are not translocated round the plant and are not phytohormones; the free cytokinins are translocated in the plant and are phytohormones.

The biochemical mode of action of the *t*RNA-cytokinins appears to be concerned with protein biosynthesis at the translational level (see Chapter 10, Section E.1). Evidence for this comes from the observation that chemical modification of the purine component of the i^6Ad of yeast $tRNA^{Ser}$ markedly reduced the binding of serine-$tRNA^{Ser}$ to the *t*RNA–ribosome complex; it did not, however, affect the ability of the $tRNA^{Ser}$ to form serine-$tRNA^{Ser}$. Identical results have been obtained with *E. coli* where it was shown that a $tRNA^{Tyr}$ with an adenine adjacent to the 3' end of the anticodon, bound to the *t*RNA–ribosome complex markedly less well than a $tRNA^{Tyr}$ with i^6Ad adjacent to the 3' end of the anticodon, although both species of $tRNA^{Tyr}$ formed tyrosine-$tRNA^{Tyr}$ equally well.

The biochemical mode of action of free cytokinins in plants is unknown. It is thought that they may bind to some receptor in the target cell. The receptor is usually envisaged as an allosteric protein which is physiologically inactive in the absence of the cytokinin but which is converted to a physiologically active conformation when the cytokinin molecule binds to it. It is presumed that the cytokinin–receptor complex then switches on or off some metabolically important step, the consequence of which is the observed physiological response (e.g. cell division, retardation of senescence). The only evidence along these lines to date is the observation made by Fox and Erion in the period 1975–7 that cytokinins bind with quite high specificity to higher plant ribosomes. They isolated three cytokinin-binding proteins from wheat germ ribosomal pre-parations which they called CBF-1 (MW 93,000), CBF-2 (MW 30,000) and CBF-3 (MW 250,000). CBF-1 and CBF-2 had high affinities for cytokinins whilst CBF-3 had a low affinity. CBF-1 is un-doubtedly a ribosomal protein but it is less certain that the other two are specifically ribosomal. There is one molecule of CBF-1 per ribosome and it binds a wide range of cytokinins and cytokinin analogues with great specificity. The biological function of

CBF-1 is not known; it could conceivably have a role in the regulation of protein biosynthesis.

The mode of action of cytokinins in delaying leaf senescence has received some attention. It has been shown that the protein and the RNA content of leaves declines during senescence. Extrapolating from the results of investigations with senescing tobacco leaf discs, the major RNA species to be depleted is chloroplastidic ribosomal RNA. The decline in protein and RNA content is primarily due to increased breakdown by proteases and RNA-ase respectively rather than to decreased synthesis. There is increased protease synthesis and RNA-ase activity during senescence. Cytokinins have been shown to slow down the rate of protein and RNA loss in detached leaves. They appear to do this by suppressing the synthesis of proteases and preventing any increase in RNA-ase activity.

E. ABSCISIC ACID

1. Discovery and characterization

Abscisic acid was discovered as the result of two independent investigations, one by Addicott and his coworkers in Davis, California, into the naturally occurring compounds that accelerate leaf abscission, the other by Wareing and his coworkers in Aberystwyth into naturally occurring compounds that promote bud dormancy in woody plants.

In 1963 Addicott reported the isolation of a substance from young cotton (*Gossypium hirsutum*) fruit that accelerated the abscission of excised, debladed cotton petioles (cotyledonary petioles were used in the bioassay). This activity was detectable when amounts as low as 0.01 μg were applied externally to each abscission zone. The substance was shown to have a molecular formula of $C_{15}H_{20}O_4$ and was named 'abscisin II' in deference to 'abscisin I', another abscission-accelerating substance that had been isolated from mature cotton burs 2 years earlier by Liu and Carns. (Interest in abscisin I has waned and its structure does not appear to have been elucidated.) Abscisin II was also shown to inhibit the IAA-induced straight growth of *Avena* coleoptiles but to have no GA-activity in the dwarf maize bioassay (see Section C.1).

Also in 1963 Wareing and Eagles isolated a substance from birch (*Betula pubescens*) leaves which caused the buds of growing seedlings to go dormant when applied to them. They named it 'dormin'. To speed up the elucidation of the structure of dormin Wareing enlisted the assistance of Cornforth and his colleagues at Shell Research Ltd., Sittingbourne, and in 1965 they were able to show, from a comparison of physical properties, that dormin and abscisin II were identical. In the same year Addicott proposed a structure for abscisin II and a few months later Cornforth confirmed it by chemical synthesis. It proved to be a sesquiterpenoid (C_{15}, composed of three isoprene units—see Chapter 11) compound with single carboxyl, hydroxyl and keto groups. It had one chiral centre (C-1') and was synthesized as a racemic mixture of the (+)- and (−)-enantiomorphs which have the *S* and *R* configurations at C-1' respectively. The enantiomorphs can be distinguished from each other by their optical rotatory dispersion curves and, with the development of this technique by Milborrow into a sensitive assay system, it had become clear by 1966 that the naturally occurring abscisin II (dormin) was the (+)-(*S*)-enantiomorph (LV, Fig. 15.22).

Possible confusion over the use of two names to describe the same compound was averted when in 1967 Addicot's, Wareing's and Cornforth's groups got together and decided on the name abscisic acid and the abbreviation ABA.

2. Assay and distribution of ABA

ABA has been estimated by bioassays and by physical methods. Bioassays based on the following phenomena have been developed: (i) the acceleration of the abscission of cotyledonary petioles in explants of 14-day-old cotton seedlings, (ii) the inhibition of IAA-induced, straight growth of oat (*Avena*) coleoptiles, (iii) the inhibition of germination of isolated wheat (*Triticum*) embryos, (iv) the inhibition of the growth of a duckweed species (*Lemna minor*), (v) the inhibition of GA-induced synthesis of α-amylase by barley (*Hordeum*) aleurone tissue and (vi) the stimulation of stomatal closure.

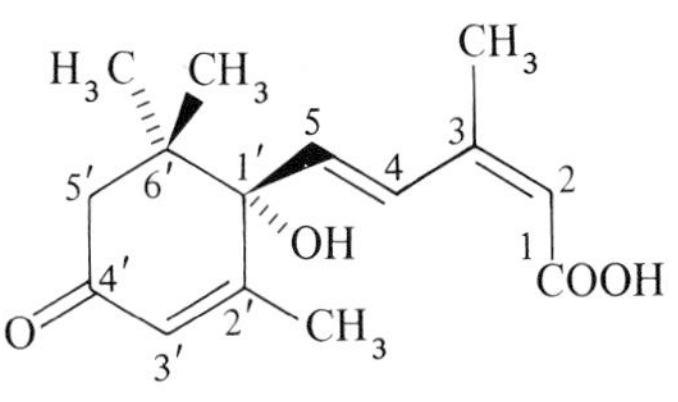

(LV) (+)-(S)-Abscisic acid 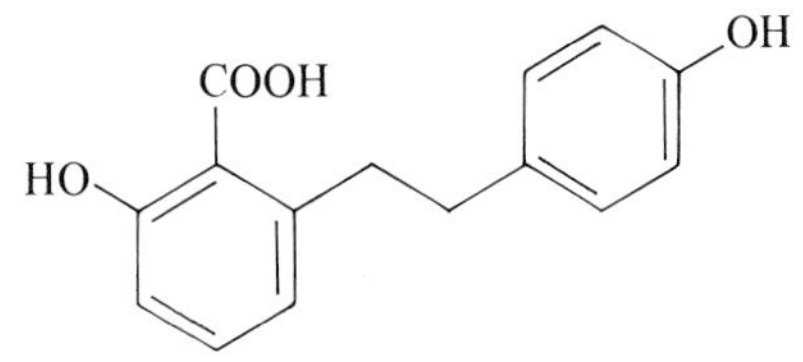   (LVI) Lunularic acid

Fig. 15.22. Structure of abscisic acid and lunularic acid, its possible functional equivalent in liverworts and algae.

Physical methods have included ultraviolet absorption, optical rotatory dispersion and circular dichroism, gas-liquid chromatography of the methyl ester or, less frequently, the trimethylsilyl ester of ABA on various silicone liquid phases and combined gas-liquid chromatography-mass spectrometry. The latter technique gives the most certain identification of ABA; the mass spectral cracking pattern of methyl abscisate is well documented.

ABA has been identified in so many angiosperms, both mono- and dicotyledonous, that it can be safely assumed that it is present in all species. It has also been identified in several gymnosperms (e.g. Yew, Silver fir, Pine, Sitka spruce) and a few Pteridophytes (e.g. *Anemia phyllitidis, Equisetum arvense*). Among the Bryophytes it has been found in a moss but not in any liverworts. It has not been found in any algae. The place of ABA as a growth inhibitor in liverworts and algae appears to have been taken by lunularic acid (LVI, Fig. 15.22).

ABA has been found in many higher plant tissues, namely roots, stems, buds, leaves, fruits and seeds. It has also been found in phloem and xylem sap and in nectar. Its concentration varies from tissue to tissue and also within a given tissue during the course of its development. Most tissues contain 20–100 ng per g wet weight but concentrations of 10 μg and 20 μg per g fresh weight have been found in avocado fruit pulp and dormant cocklebur (*Xanthium*) buds respectively.

3. Movement of ABA in the plant

The picture that we have of the movement of endogenous ABA in higher plants at various stages of growth is not absolutely clear. Numerous approaches have been used to investigate this problem. These include, (i) experiments involving the local application of [¹⁴C]ABA to the exterior surface of one part of the plant followed by determination of the distribution of radioactivity some time later, (ii) experiments involving the application of [¹⁴C]ABA to the cut ends of stem, petiole and root segments followed by the determination of how much radioactivity reached the other end and (iii) the examination of phloem and xylem sap from various vascular organs for the presence of ABA.

From these experiments it appears that (i) externally applied ABA is able to get into tissues rapidly and is distributed freely in all directions within the plant, (ii) cell to cell transport of ABA is slow and non-polar, (iii) ABA is present in the phloem and xylem sap and is probably transported throughout the plant in this way, (iv) ABA, synthesized in the root cap, moves basipetally in the tissues of the stele (central vascular region). There is also some evidence for lateral ABA movement in roots; it has been shown that in the root tips of maize (*Zea mays*) a basipetally moving inhibitor, probably ABA, is redistributed laterally under the influence of gravity and participates in the geotropic response by causing an asymmetric inhibition of root growth.

The form in which ABA is transported is not yet clear; most of the evidence seems to indicate that it is free ABA but the possibility that it is a derivative, such as ABA-β-D-glucopyranoside, cannot be excluded.

4. Metabolism of ABA

(i) BIOSYNTHESIS OF ABA

Most higher plant tissues appear to be capable of synthesizing ABA. Its synthesis from mevalonic acid (MVA) has been demonstrated in fruit tissues, seeds (the embryo and cotyledons of avocado, the endosperm and embryo of wheat), roots (sunflower), stems (avocado) and leaves (avocado).

Within the cells of these tissues it appears likely that most of the ABA is synthesized in the plastids.

The biosynthetic pathway from MVA to ABA is not clear. Two possibilities exist: (i) that ABA is formed at the C_{15} level by cyclization and modification of farnesyl pyrophosphate (FPP) and (ii) that ABA is formed by catabolism of a carotenoid such as violaxanthin.

Although MVA is clearly a precursor of ABA regardless of whether it is formed by pathway (i) or pathway (ii) it was hoped that it would be possible to decide which operated *in vivo* by comparing the incorporation of tritium (^3H) from six stereospecifically tritiated species of MVA, namely $[2\text{-}^{14}C,(2R)\text{-}2\text{-}^3H_1]MVA$, $[2\text{-}^{14}C,(2S)\text{-}2\text{-}^3H_1]MVA$, $[2\text{-}^{14}C,(4R)\text{-}4\text{-}^3H_1]MVA$, $[2\text{-}^{14}C,(4S)\text{-}4\text{-}^3H_1]MVA$, $[2\text{-}^{14}C,(5R)\text{-}5\text{-}^3H_1]MVA$ and $[2\text{-}^{14}C,(5S)\text{-}5\text{-}^3H_1]MVA$, into ABA with that into carotenoids such as violaxanthin. The results of these experiments are shown in Fig. 15.23; although the pathway in this figure is hypothetical the locations of the former pro-*R* and pro-*S* hydrogen atoms on Cs 2, 4 and 5 of MVA in the ABA structure shown in the figure are those that were experimentally found. Unfortunately the results failed to distinguish between the two possible pathways, for the labelling pattern found in ABA was consistent with both. These experiments did, however, show that the Δ^2 double bond of ABA is formed with the *trans* configuration and must therefore be isomerized to the *cis* configuration at some subsequent step in the pathway; this follows from the fact that the C–2 hydrogen of ABA is derived from the 4-pro-*R* hydrogen of MVA rather than from the 4-pro-*S* hydrogen as would have been the case if the Δ^2 double bond had been formed directly with the *cis* configuration (see Chapter 11, Section C.10).

Figure 15.23 shows a hypothetical 'non-carotenoid' pathway for the biosynthesis of ABA. It postulates that farnesol, derived from FPP, is desaturated to give *trans*-Δ^4-dehydrofarnesol (step *a*); experiments with MVA species stereospecifically labelled with ^3H at C–2 and C–5 indicate that this reaction involves the stereospecific loss of hydrogens that were the 2-pro-*S* and 5-pro-*R* hydrogens of MVA respectively.

The pathway then assumes that the *trans*-Δ^2 double bond of *trans*-Δ^4-dehydrofarnesol is isomerized to the *cis*-Δ^2 double bond that occurs in ABA (step *b*); this must take place with retention of the former 4-pro-*R* hydrogen of MVA on the carbon that becomes C–2 of ABA.

This is then followed by a cyclization reaction (step *c*), postulated as being mechanistically similar to that by which ε-rings of carotenoids are formed. This results in the formation of compound LVII. Experiments with MVA stereospecifically labelled with ^3H at C–2 indicate that during the cyclization a former 2-pro-*S* hydrogen of MVA is lost from the carbon that becomes C–3' of ABA.

It is then assumed that there is a stepwise oxidation of the primary alcohol group to a carboxyl group via an aldehyde group (steps *d* and *e*). This would be likely to involve pyridine nucleotide-requiring oxidoreductases and would cause the loss of the former 5-pro-*R* and 5-pro-*S* hydrogens of MVA from the carbon that becomes C–2 of ABA.

The resulting compound LVIII is then assumed to be hydroxylated at what will be C–4' of ABA probably by a mixed function oxygenase (step *f*). The newly inserted hydroxyl group is then oxidized to a keto group, possibly by a pyridine nucleotide-requiring oxidoreductase to produce compound LIX (step *g*) which is then hydroxylated, probably by a mixed function oxygenase, at C–1' to give ABA (step *h*). Step *h* would cause the loss of a former 4-pro-*R* hydrogen of MVA.

It must be emphasized that this pathway is conjectural and there is no evidence that any of the steps from FPP or farnesol occur in higher plants. There are, however, indications that a pathway like it may occur in the ABA-forming fungus *Cercospora rosicola*. Horgan and his coworkers in Aberystwyth have recently shown that compound LIX is formed by this fungus and that it is a precursor of ABA. They have also shown that compounds LVII, LVIII and LIX are converted into ABA in high yield by the

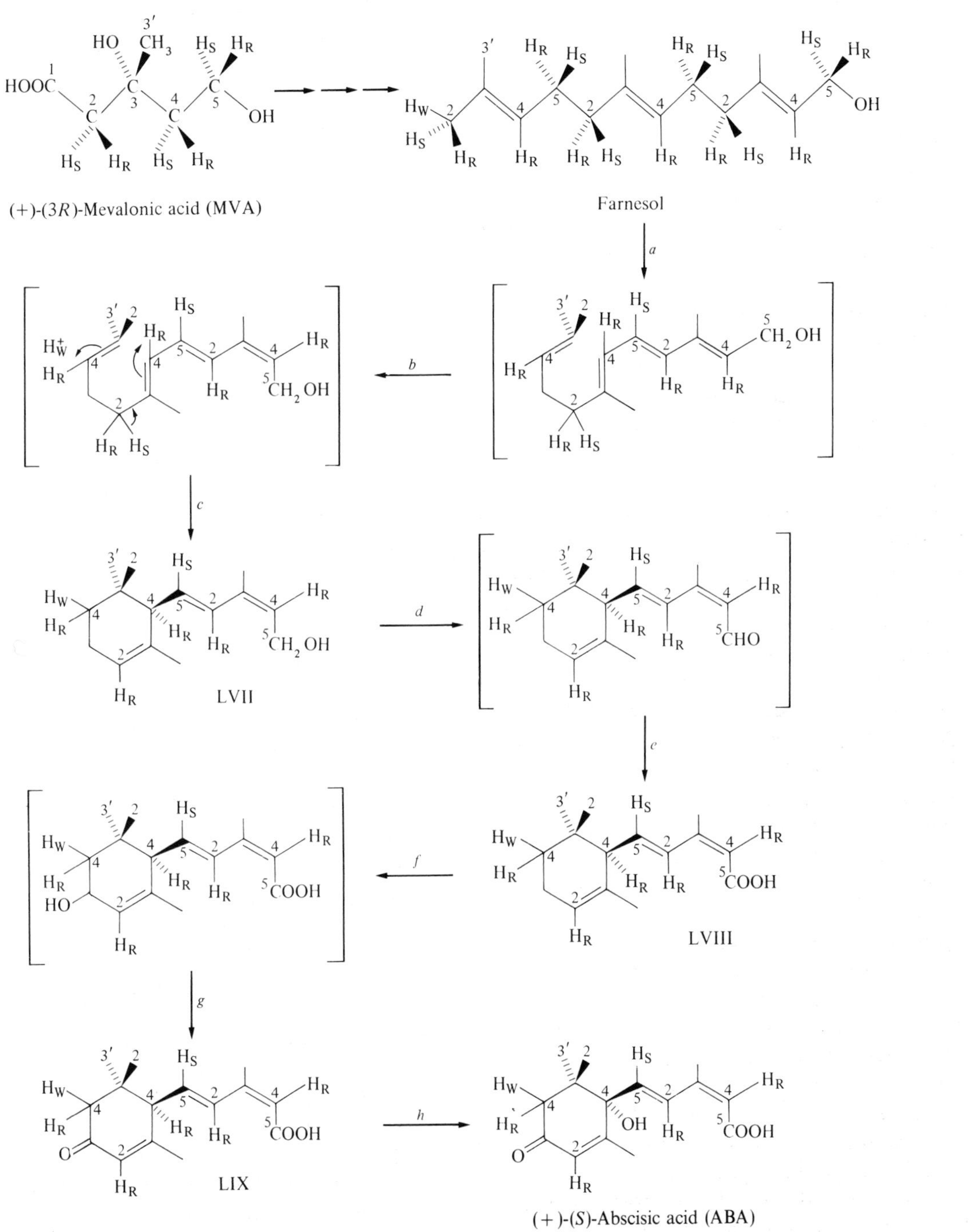

Fig. 15.23. Hypothetical 'non-carotenoid' pathway for the biosynthesis of ABA. (The numbers adjacent to some of the carbon atoms of the intermediates from FPP onwards indicate their derivation from MVA; the designation of a hydrogen as H_R or H_S in the intermediates from FPP onwards indicates that it was the pro-R or pro-S hydrogen of the carbon of that number in MVA; H_w is a hydrogen derived from water; compounds LVII, LVIII and LIX have been shown to be converted into ABA by the fungus *Cercospora rosicola*; compound LIX has also been shown to be present in *C. rosicola*; compounds in parenthesis have yet to be identified as intermediates in the biosynthesis of ABA.) *The authors wish to thank Dr R. Horgan, U.C.W., Aberystwyth, for his help in the preparation of this figure.*

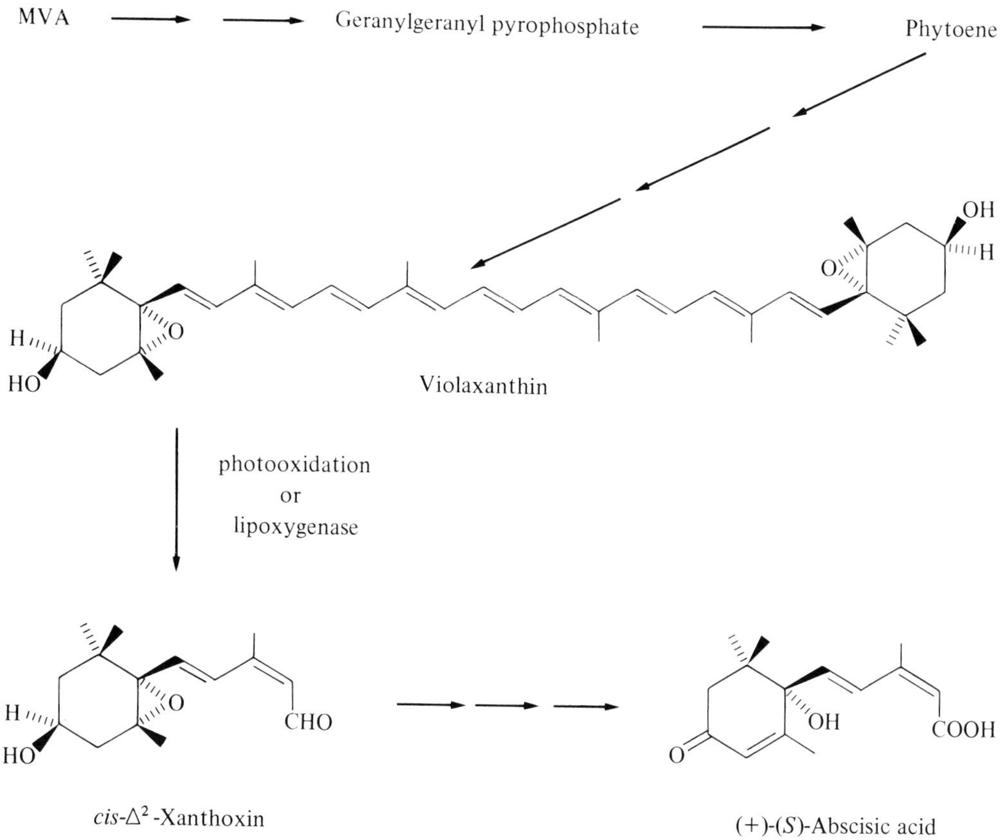

Fig. 15.24. Postulated 'carotenoid pathway' for the biosynthesis of ABA.

fungus although it has yet to be shown that they are endogenous to the fungus.

The 'carotenoid' pathway for the biosynthesis of ABA is shown in Fig. 15.24. It is suggested that a carotenoid such as violaxanthin is converted into cis-Δ²-xanthoxin and that the latter is in turn converted into ABA, a process that involves (i) oxidation of the aldehyde group to a carboxyl group, (ii) oxidation of the hydroxyl group to a keto group and (iii) reduction of the epoxide to a hydroxyl group.

The evidence for this pathway is as follows: (i) violaxanthin is converted into cis-Δ²-xanthoxin when it is subjected to high-intensity light or when it is incubated with lipoxygenase [see Chapter 8, Section C.1(ii)(d)], (ii) cis-Δ²-xanthoxin is rapidly converted into ABA in tomato shoots, (iii) cis-Δ²-xanthoxin has been isolated from the shoots and leaves of a number of higher plants and (iv) cis-Δ²-xanthoxin is a powerful inhibitor of plant growth.

Although at first sight this evidence is persuasive, a closer examination reveals its weaknesses. Firstly, the light intensity required to convert violaxanthin or any other carotenoid into cis-Δ²-xanthoxin is so high that it is doubtful whether it could occur in vivo. Secondly, ABA has been shown to be synthesized in a number of tissues which are devoid of carotenoids. Thirdly, cis-Δ²-xanthoxin has not been found in a number of tissues known to synthesize ABA. Fourthly, cis-Δ²-xanthoxin is more strongly inhibitory to plant growth in some assay systems than can be accounted for by its complete conversion into ABA. Fifthly, Robinson showed that when [14C]phytoene and [2-3H]MVA were simultaneously fed to avocado fruit the ABA that was synthesized was labelled with 3H only in spite of the formation of [14C] carotenoids from the [14C]phytoene.

Thus, although the formation of ABA from carotenoids cannot be said never to occur in plant tissues, it is clear that ABA is normally formed at the C_{15} level. It is worth noting that cis-Δ^2-xanthoxin could equally well be formed at the C_{15} level and may yet prove to be a natural precursor of ABA in some tissues.

(ii) CONJUGATION AND FURTHER METABOLISM OF ABA

The principal metabolic changes undergone by ABA in plants are shown in Fig. 15.25. ABA may be reversibly converted into its D-glucose ester, ABA-β-D-glucopyranoside, or it may be converted into phaseic acid and a mixture of epimeric dihydrophaseic acids, probably via 6'-hydroxymethyl-ABA. This can be regarded as an inactivation process because neither phaseic acid nor the dihydrophaseic acids have much ABA activity. No interconversion of the dihydrophaseic acids appears to take place indicating the absence of an appropriate epimerase.

Phaseic acid and the dihydrophaseic acids may be converted into polar conjugates the identities of which remain to be elucidated; the free acids are readily released from the conjugates by hydrolysis.

It has recently been shown that the methyl group on C–6' of ABA which is incorporated into the furan ring of phaseic acid is the one which is (i) on the same face of the cyclohexenyl ring as the C–1' hydroxyl group and (ii) derived from 3' carbon of MVA.

5. Physiological effects of ABA

Although ABA was discovered as a result of its ability to (i) promote abscission in explants of cotton seedlings and (ii) induce bud dormancy, and has since been implicated in several other processes

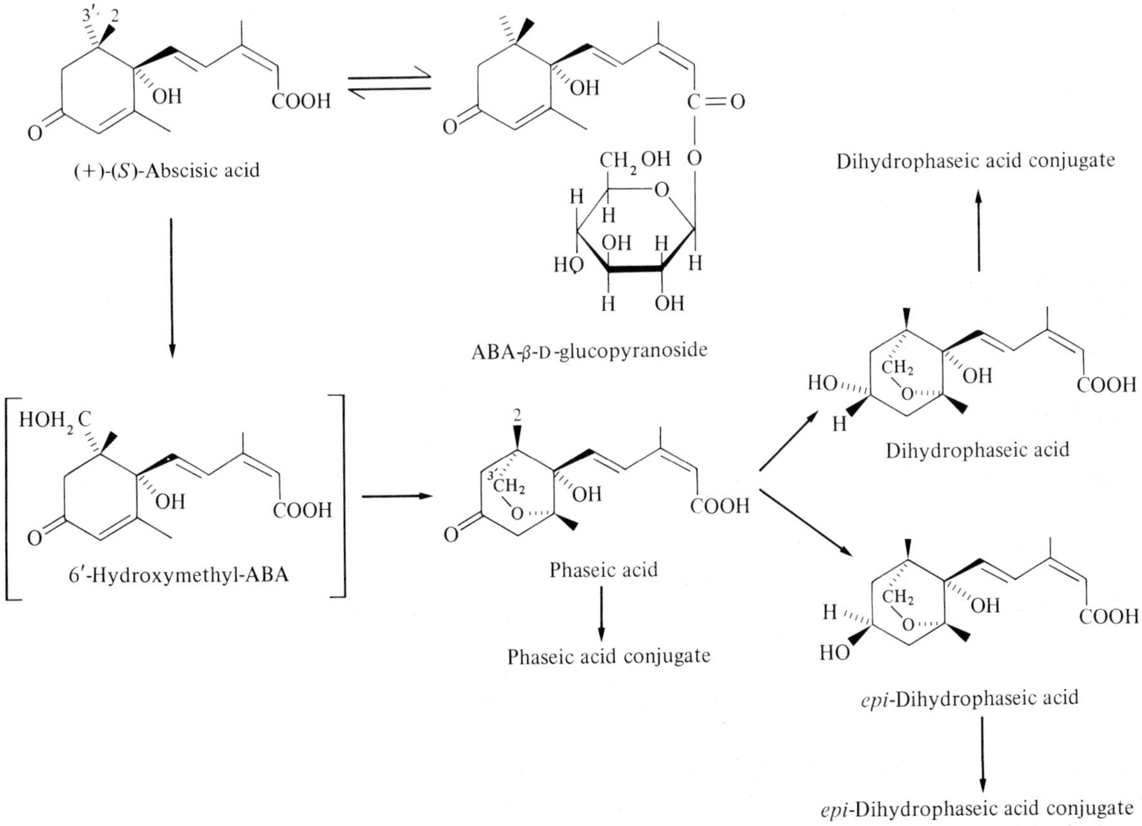

Fig. 15.25. Metabolism of (+)-(S)-abscisic acid. (The numbers 2 and 3' in the formulae of ABA and phaseic acid indicate the derivation of the atoms concerned from Cs 2 and 3' of MVA respectively; the identities of the phaseic acid and dihydrophaseic acid conjugates have yet to be elucidated.)

such as seed dormancy and germination, tuberization, leaf senescence and fruit ripening, the best authenticated examples of its hormonal activity are (i) its role as a growth inhibitor in the downward growth response of roots to gravity and (ii) its role in causing stomatal closure in response to water stress.

(i) ABA AND GEOTROPISM IN ROOTS

The roots of a horizontally held plant grow downwards under the influence of gravity. It has long been held that this is the result of an asymmetric redistribution of IAA in the root (see Section B.4). However, more recent work with maize (*Zea mays*) roots has led to a revision of views on the mechanism of geotropism and implicates ABA in it. It is now thought that ABA, synthesized in the root cap, is normally translocated in the vascular tissue in a basipetal direction but that when the root is held horizontally it is pulled by gravity to the lower side where it inhibits growth. Since growth of the upper side of the root is uninhibited, the result is a downward curvature. The evidence for this is, (i) removal of the root cap destroys the geotropic response, (ii) removal of half the root cap results in curvature towards the side with the remaining half, (iii) ABA has been extracted from root caps and is known to be synthesized in them, (iv) application of ABA to the root surface at concentrations of 10^{-8}–10^{-4} M causes inhibition of growth and (v) asymmetric application of ABA to the root surface causes curvature towards the site of application.

(ii) ABA AND STOMATAL CLOSURE

Land plants protect themselves against excessive water loss by reducing the size of the stomatal openings in what is otherwise a virtually impermeable epidermis. This protection appears to involve two lines of defence. The first line consists of immediate responses to components of the aerial environment such as water vapour and CO_2. When the water vapour pressure of the atmosphere is significantly lower than that of the air in the intercellular spaces of the leaf the stomata close. Similarly an increase in CO_2 concentration around the guard cells causes the stomata to close. The second line of defence comes into action when water

uptake by the roots is still less than water loss by transpiration despite the operation of the first line of defence. The second line of defence is mediated by ABA.

The drop in water potential appears to be detected by the chloroplasts of the mesophyll cells of the leaf, whose photosynthetic activities are probably the most sensitive of all the plant's metabolic processes to water stress. It is suggested that the drop in water potential causes a marked increase in the permeability of the chloroplast membranes to ABA. The ABA synthesized by and stored in the chloroplasts then diffuses into the mesophyll cytoplasm and moves from cell to cell through plasmodesmata to the guard cells where it induces stomatal closure. The drop in the level of stored ABA in the mesophyll chloroplasts induces fresh biosynthesis of more ABA which continues to be translocated to the guard cells as long as the water potential remains low. The release of ABA from the mesophyll chloroplasts stops when the water potential is restored to normal and its rate of synthesis is markedly reduced.

The evidence for this is, (i) the exogenous application of low concentrations (e.g. 1 mM) of ABA to leaves causes stomatal closure within 3–9 min, (ii) the concentration of ABA in water-stressed leaves rises markedly, (iii) mesophyll chloroplasts can synthesize ABA but guard cell chloroplasts cannot and (iv) the ABA of leaves of well-watered spinach has been shown to be contained in the chloroplasts but after 4 hr of water stress the total ABA level was found to have risen 11-fold whilst that of the chloroplasts had risen only 2-fold; this suggests that the newly synthesized ABA is rapidly exported from the chloroplasts.

(iii) ABA AND BUD DORMANCY

'Innate' or 'true' dormancy (i.e. temporarily suspended growth due to internal conditions) is exhibited by the buds of temperate-zone trees during winter, stem tubers (e.g. potato), root tubers (e.g. Dahlia), corms, bulbs and rhizomes. Hemberg in 1949 found that the dormancy of buds of ash (*Fraxinus*) and potato tubers was positively correlated with the concentration in them of unknown growth inhibitors. It was the desire to find out what these growth inhibitors were that ultimately led to the isolation of dormin and its identification as ABA (see Section E.1). The external application of ABA to buds of growing seedlings caused them to become dormant. The extrapolation of these findings was that, in response to some environmental trigger, ABA was synthesized in the plant, travelled to the buds and induced them to go dormant.

This certainly seemed to be true of the buds of temperate-zone trees (e.g. birch—*Betula pubescens*; sycamore—*Acer pseudoplatanus*). Here the environmental trigger was the shorter days of autumn. Short days (photoperiods) were shown to cause increased levels of ABA in leaves and buds. The leaves appeared to play the crucial role in this response and it seemed clear that short days induced the leaves to synthesize more ABA and to export it to the buds. However, in these early experiments the ABA levels were measured in relatively crude extracts by bioassay. When it became possible to determine ABA levels by physical means it was found that they were no higher in leaves that had been exposed to short days than those that had remained under long days. On the other hand, physical assay methods have confirmed the presence of high levels of ABA in the dormant buds of several species and have shown that these levels progressively decline throughout the winter. Thus the role of ABA in bud dormancy is equivocal at the present time.

(iv) ABA AND SEED DORMANCY

ABA only appears to play a role in the dormancy of seeds that require stratification (i.e. seeds that will not germinate unless they are exposed to periods of low temperature) and especially those that are 'innately' dormant (e.g. *Taxus baccata*) (in contrast to those which exhibit 'seed coat-imposed dormancy').

It has been shown that dormant, immature embryos of yew (*Taxus baccata*) can be induced to germinate without stratification if they are placed in a nutrient solution which causes an ABA-like compound to be leached from them. Dormancy can then be reimposed by immersing the leached embryos in an ABA solution. Similar results have been obtained with apple seed embryos.

However, attempts to correlate endogenous ABA levels with the state of dormancy of the seed have given contradictory results. In some species the ABA level is high in the dormant seed and declines during stratification whereas in others no such decline occurs during stratification. Nevertheless it is probable that ABA is important in the maintenance of dormancy in seeds. Recent work with stratification-requiring seeds suggests that dormancy is regulated by an antagonistic interaction between ABA (which inhibits growth and therefore maintains dormancy) and GAs and cytokinins (which promote growth and therefore stimulate germination). The role of GAs and cytokinins in breaking seed dormancy has been indicated by the observations that, (i) externally applied GAs and/or cytokinins stimulate germination in many types of dormant seed and (ii) treatments, such as stratification, that overcome dormancy result in increased levels of GAs and cytokinins in the seed.

(v) ABA AND ABSCISSION

It is now believed that ABA is not concerned in the abscission of leaves but that it may have a regulatory role in the abscission of flowers and fruit. The evidence for this is, (i) ABA, applied externally to the intact plant, accelerates the abscission of the mature fruits of grape, olive, citrus and apple and the flowers of grape but only promotes leaf abscission at very high concentrations (2–4 mM) and (ii) there are two peaks in the endogenous level of ABA in developing cotton fruits, the first coinciding with the dropping of immature fruits, the second with the final dehiscence of mature fruits.

(vi) OTHER EFFECTS OF ABA

ABA appears to be involved in the process of tuberization. Wareing has shown that ABA, formed by the leaf of a one-node cutting taken from a short-day-induced plant of *Solanum andigena* (a wild species of potato which requires about 20 short-day cycles to fully induce tuberization), is exported to the axillary bud and causes it to develop rapidly into a small tuber. Externally applied ABA has also been shown to promote the formation of tubers in Dahlia and Jerusalem artichoke.

ABA may also be involved in senescence and ripening processes. Externally applied ABA accelerates the senescence of detached leaves; however, its effect on the senescence of leaves still attached to a healthy plant is minimal. Externally applied ABA accelerates the ripening of young fruit; moreover, there is a rise in the level of endogenous ABA during the ripening of strawberries and grapes.

ABA may also be involved in increasing the resistance of temperate-zone plants to frost damage; the external application of ABA has been shown to increase the frost hardiness of seedlings of box elder (*Acer negundo*), apple and alfalfa (*Medicago sativa*) and to ameliorate chilling injury in cucumber seedlings.

6. Biochemical mode of action of ABA

Little is known of the way in which ABA acts at the molecular level and none of the physiological effects described in Section E.5 can yet be fully explained.

However, an examination of these physiological effects shows that most of them involve an inhibition of growth which immediately suggests that ABA may block some step in the process in which DNA is transcribed into *m*RNA and the *m*RNA is translated into protein. Accordingly the effects of ABA on nucleic acid and protein synthesis have been studied in a wide variety of systems, involving growing and mature plant tissues. The picture that has emerged is not clear but the following generalizations can be made: (i) ABA does not exert a general inhibition of DNA transcription but may inhibit the synthesis of a limited number of specific *m*RNA species, e.g. the α-amylase *m*RNA of the cells of the aleurone layer in barley seeds, and (ii) ABA exerts its main effect by blocking protein synthesis at a post-transcriptional level such as *m*RNA processing or translation rather than by preventing DNA transcription, e.g. ABA has been shown to inhibit the synthesis of protease and isocitrate lyase[12] in the embryos of cotton seeds even though the appropriate *m*RNAs are present.

Some of the physiological effects of ABA are, however, too rapid to be explained by its inhibition of protein synthesis and there is an increasing body of evidence to suggest that in some tissues ABA acts at membrane sites. It has, for instance, been shown to alter the permeability of some membranes to a variety of ions. The effect of ABA on stomatal closure [see Section E.5(ii)] appears to fall into this category. ABA has been shown to inhibit the efflux of H^+ ions from the guard cells and the influx of K^+ ions into them. Moreover, this ion exchange appears to be directly connected with ABA-stimulated stomatal closure. The currently accepted mechanism of stomatal operation is outlined in Fig. 15.26. It is believed that the plasmalemma of the guard cell contains a proton pump (possibly driven by ATP formed by cyclic photophosphorylation in the guard cell chloroplasts) which actively pumps H^+ ions out of the cell. This is compensated by the passive movement of K^+ ions into the cell. The loss of H^+ ions from the cytosol increases its pH into the range 8–9 which is the optimum for phosphoenolpyruvate carboxylase[13]. This enzyme catalyses the conversion of phosphoenolpyruvate, derived from starch, into oxaloacetate which is then reduced to L-malate by malate dehydrogenase[14]. The L-malate accumulates in the vacuole. The resulting decreased water potential (increased osmotic pressure—in older terminology) of the vacuole draws water into the cell. As the turgidity of the two guard cells increases the stomatal aperture widens. It is believed that ABA interrupts this normal sequence of events by blocking the proton pump. The changes consequent upon this increase the water potential (decrease the osmotic pressure) of the vacuole and water flows out of the guard cells. As the guard cells become more flaccid the stomatal aperture narrows.

F. ETHYLENE

1. Properties and discovery as a phytohormone

Ethylene (ethene according to the IUPAC system of nomenclature) has the formula $H_2C{=}CH_2$ (MW 28.05). It is a gas at STP (m.p. $-169.15°C$; b.p. $-103.71°C$). It is colourless, has an ether-like smell, is lighter than air (density of ethylene at 0°C and 1 atmosphere = 1.260 g l^{-1}; density of dry air at 0°C and 1 atmosphere = 1.293 g l^{-1}) and is highly flammable. It is more soluble in water than O_2 or N_2 but is less soluble than CO_2 (approx. solubilities in water at 0°C and 25°C respectively when the gas is at a pressure of 1 atmosphere, expressed in ppm: ethylene, 315 and 140; oxygen, 70 and 40; nitrogen, 30 and 18; carbon dioxide, 3387 and 1500). The

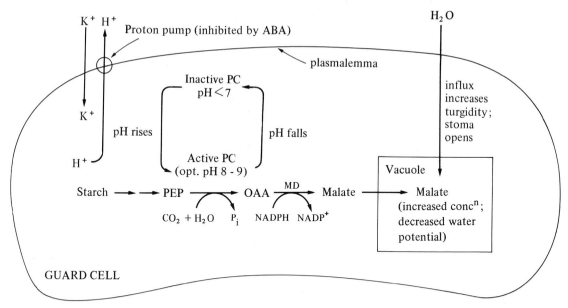

Fig. 15.26. Mechanism of stomatal opening. (PEP = phosphoenolpyruvate; OAA = oxaloacetate; PC = phosphoenolpyruvate carboxylase; MD = malate dehydrogenase; ABA = (+)-(S)-abscisic acid.)

diffusion equilibrium of ethylene at 1 ppm between air and water is about 10,000:1 at 25°C.

Ethylene reacts readily with halogens to form halohydrocarbons, with a variety of metallic ions to form coordination compounds (the coordination complex with Hg^{2+} ions, formed when ethylene reacts with mercuric perchlorate, is a useful method of trapping ethylene in solution; moreover, ethylene can be quantitatively recovered from it by the addition of Cl^- ions) and is easily oxidized (the oxidation of ethylene to ethylene glycol, $HOCH_2 \cdot CH_2OH$, by $KMnO_4$ is used in horticulture to remove it from the atmosphere; fruits and flowers are commonly transported in packages containing alumina or silica gel impregnated with $KMnO_4$ to remove ethylene so as to decrease the rate at which ripening or opening occurs).

The discovery of ethylene as a plant hormone can be traced back to the last century when it was found that coal gas had profound effects on plants—trees shed their leaves prematurely, seedlings grew horizontally, flowers faded quickly. Coal gas was produced by heating bituminous coal to about 700°C. It was initially used to illuminate streets and houses and then as a source of heat. It was a mixture of gases; the principal heat producers were hydrogen

(38–55%, by vol) and methane (22–25%, by vol) whilst the main illuminants (2.5–5%, by vol) were olefins (e.g. ethylene), acetylene and benzene. It was Neljubow, a Russian physiologist, who, in 1901, showed that it was the ethylene component that was responsible for the effects of coal gas on plants. He showed that ethylene, in the concentrations found in coal gas, caused (i) inhibition of stem elongation, (ii) stem swelling and (iii) horizontal stem growth in pea seedlings. He also found that ethylene did not cause a permanent change in growth; when the ethylene was removed from the atmosphere normal growth resumed. In 1912 Sievers and True reported that Californian citrus growers had found that the fumes from kerosene (paraffin) stoves accelerated the ripening of green lemon fruit and Denny in 1924 proved that ethylene was the active component of the fumes.

Up to this point in the story the ethylene affecting the plants has been an abnormal pollutant, introduced into the air by combustion. However, it eventually became apparent that ethylene is also produced by plants themselves. It had long been known by commercial storers and shippers of fruit that overripe and rotting fruit accelerated the ripening of fruit in the immediate vicinity. In 1910

Cousins reported that a volatile compound produced by ripe oranges speeded up the ripening of bananas stored close by. In 1933 Botjes showed that the volatile compound produced by ripe apples that caused the downward curvature of tomato seedlings could be removed from the air by ethylene adsorbants. Other workers obtained similar effects but it was Gane in 1934 who finally proved that ethylene was produced by plants. He passed the air stream containing all the gaseous products of metabolism of 27.2 kg of Worcester Pearmain apples through bromine at $-65°C$ for 4 weeks. This would trap any ethylene as its dibromide. Fractional distillation of the oil derived from the bromine yielded 0.65 g of a product boiling below $140°C$ which would contain ethylene dibromide (b.p. $131°C$) if it were present. That it was present was shown by the formation of crystals identified as N,N'-diphenylethylenediamine when the product was treated with aniline.

In the late 1930s ethylene was shown to be produced not only by fruits but also by several other plant tissues, e.g. flowers, leaves, leafy stems, roots and seeds, principally by Denny and his colleagues. This, taken with its known effects on plant growth, led to the suggestion that ethylene might be an endogenous growth regulator.

Although the list of plant tissues shown to produce ethylene increased over the next two decades interest in it as a growth regulator did not become widespread until about 1960 when Burg, Pratt and others began to use gas chromatography for ethylene analysis. This was not only easier and quicker than the older manometric procedures but also gave a million-fold increase in sensitivity. Gas–solid chromatography on columns of alumina or silica gel and using a flame ionization detector enables as little as 10^{-15} mole of ethylene in a 1–2 ml gas sample to be measured in about 5 min and even now is still the analytical method of choice.

The increase in interest in the physiology and biochemistry of ethylene sparked off by its ease of analysis by gas chromatography produced an increase in knowledge which led to its general acceptance as an important plant hormone.

2. Tissue production of ethylene

Ethylene has been shown to be produced by

angiosperms, gymnosperms, pteridophytes (e.g. *Pteridium*), bryophytes (e.g. *Sphagnum squarrosum, Polytrichum, Funaria*), green algae (e.g. *Chara corallina, Chlorella protothecoides*) and some fungi (e.g. *Mucor heimalis, Penicillium digitatum* Sacc., *Saccharomyces cerevisiae, Candida vartiovaatia*).

It is also produced by the cyanobacterium (blue-green alga) *Anacystis nidulans* and a number of other bacteria (e.g. *Pseudomonas solanacearum, Escherichia coli*). It does not, however, appear to be produced by healthy mammalian tissues.

In angiosperms, all tissues are able to synthesize ethylene. However, the tissues producing the highest levels $(>1 \text{ nl g}^{-1} \text{ hr}^{-1})$ are those which are (i) senescing or ripening or (ii) immature and rapidly dividing or expanding. Mature tissues produce much lower levels $(<0.1 \text{ nl g}^{-1} \text{ hr}^{-1})$. This generalization is complicated by the fact that tissues of all ages are able to produce high levels of ethylene in response to (i) wounding, (ii) stress, e.g. water deficit, low temperatures and (iii) the appropriate hormonal stimulus, e.g. IAA, cytokinin, IAA plus cytokinin.

3. Movement of ethylene in the plant

Ethylene is sufficiently soluble in water to be transported in aqueous solution and sufficiently non-polar to pass through cell membranes readily. However, it is believed that the site of action of ethylene is close to its site of synthesis and that there is no long-range movement of ethylene within the plant.

4. Metabolism of ethylene

(i) BIOSYNTHESIS OF ETHYLENE

The biosynthesis of ethylene has been difficult to elucidate because (i) cell-free systems from higher plant tissues capable of catalysing the synthesis of ethylene have so far proved impossible to prepare and (ii) non-physiological ethylene formation can readily occur. Most investigations have therefore used model systems or tissue segments or slices.

There is general agreement that the key precursor of ethylene in higher plants is L-methionine. Labelling experiments have shown that ethylene is derived from C–3 and C–4 of L-methionine; moreover, during the conversion of L-methionine into ethylene C–1 becomes CO_2, C–2 becomes formic acid (and eventually CO_2) and the CH_3S group is retained intact in the tissue. There is evidence to indicate that the CH_3S group is recycled back to L-methionine. Conservation of the CH_3S group by recycling is necessary because it has been shown that the levels of free and protein methionine in apple fruit cannot account for the observed level of ethylene production. In the biosynthetic pathway proposed by Adams and Yang (see Fig. 15.27) 5′-methylthioadenosine and 5-methylthioribose act as CH_3S carriers. The CH_3S group is finally substituted for the hydroxyl group of L-homoserine, derived from D-glucose via L-aspartate (see Fig. 9.9), to regenerate L-methionine.

In the Adams and Yang pathway L-methionine is converted into S-adenosylmethionine which then breaks down to yield 5′-methylthioadenosine (MTA) and l-aminocyclopropane-l-carboxylic acid (ACC). The 5′-methylthioadenosine is then hydrolysed to 5-methylthioribose (MTR) which reacts with L-homoserine to regenerate L-methionine. The 1-aminocyclopropane-1-carboxylic acid is cleaved under aerobic conditions to ethylene, CO_2, formate and NH_4^+. The evidence for this pathway is (i) when [methyl-^{14}C]- or [^{35}S]methionine were administered to climacteric apple tissue labelled MTR and MTA were produced as major and minor metabolites respectively, (ii) when ethylene production in climacteric apple tissue was inhibited by aminoethoxyvinylglycine (AVG) no labelled MTR or MTA was formed from [methyl-^{14}C]- or [^{35}S]-methionine, (iii) preclimacteric apple tissue, which does not form ethylene, did not produce labelled MTR or MTA from [methyl-^{14}C]- or [^{35}S]methionine, (iv) when [^{35}S]MTA was ad-

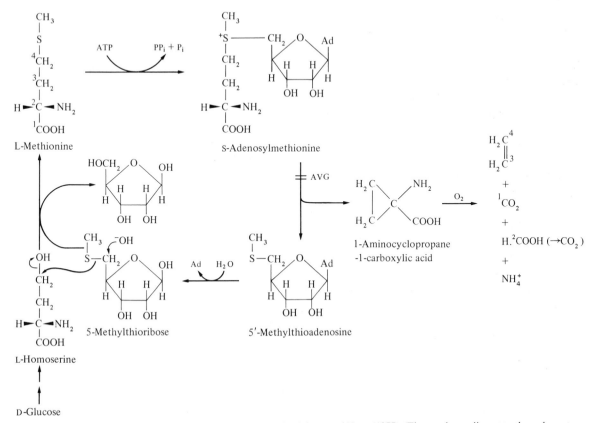

Fig. 15.27. Biosynthesis of ethylene in higher plants as proposed by Adams and Yang (1977). (The numbers adjacent to the carbon atoms of ethylene and the other products of the last reaction indicate their origin from L-methionine; Ad = adenine; AVG = aminoethoxyvinylglycine.)

ministered to climacteric apple tissue [^{35}S]MTR and [^{35}S]methionine were formed, (v) when [^{35}S]MTR was administered to climacteric apple tissue [^{35}S]methionine was formed but not [^{35}S]MTA, (vi) when [U-^{14}C]methionine was administered to climacteric apple tissue under N$_2$ [^{14}C]ACC accumulated, (vii) when [^{14}C]ACC was administered to climacteric apple tissue in air [^{14}C]ethylene was formed although this conversion did not take place in N$_2$ and (viii) ACC has been identified in apple and pear juice.

Ethylene production in higher plant tissues has been shown to be inhibited by, (i) metal chelators, e.g. EDTA, (ii) Co^{2+} ions at 0.1 mM, (iii) enol ether amino acid analogues, e.g. Rhizobitoxine, AVG, MVG (see Fig. 15.28), that are found in the medium after the growth of certain microorganisms, (iv) canaline (α-amino-γ-aminobutanoic acid), (v) free-radical quenching agents, e.g. n-propylgallate, sodium benzoate and (vi) a protein isolated from mung bean (*Phaseolus aureus* Roxb.) hypocotyls. Canaline is a known inhibitor of pyridoxal-dependent enzymes but it is probable that its inhibition of ethylene formation does not involve the inhibition of such an enzyme. This follows because canaline inhibition of ethylene formation is partially reversed by methionine but

not by pyridoxal phosphate. The proteinaceous inhibitor inhibits IAA-induced ethylene formation in subapical hypocotyl segments from mung beans. It has no effect on ethylene production in apple slices. There is a possibility that it does not act directly on the ethylene-forming system but blocks the active site with which IAA must react in order to stimulate the ethylene-forming system.

The intracellular site of ethylene biosynthesis in higher plant tissues is not known. However, recent work suggests that it may be located in the plasmalemma. This is supported by the fact that, although homogenization of cells destroys ethylene-synthesizing ability, gently prepared protoplasts can form ethylene.

The formation of ethylene by higher plant tissues is stimulated by auxin (IAA). This IAA-stimulated increase in ethylene production requires the continuous presence of relatively high levels of IAA (10^{-5}–10^{-3} M), occurs after a lag period of 1–3 hr and is prevented by inhibitors of RNA and protein synthesis. It therefore appears that IAA induces the ethylene-forming enzyme system. Some of the growth inhibitory effects of IAA are believed to be due to its induction of ethylene synthesis. Cytokinins have been shown to enhance IAA-induced ethylene formation.

```
        CH₂OH
         |
        CH.NH₂
         |
        CH₂
         |
         O
         |
    H — C
         ‖
         C — H
         |
  H ►— C ◄ NH₂
         |
        COOH
```

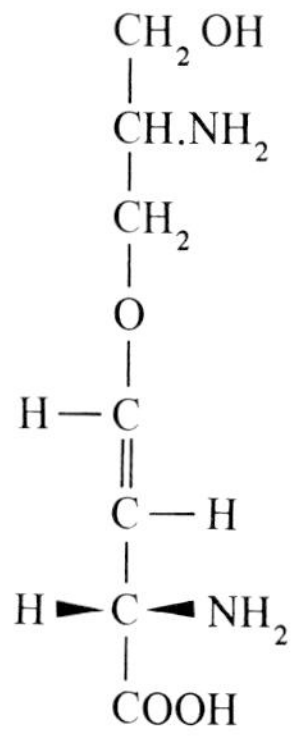

Rhizobitoxine;
L-2-amino-4-
(2-amino-3-
hydroxypropoxy)-
trans-3-butenoic
acid
(ex *Rhizobium
japonicum*)

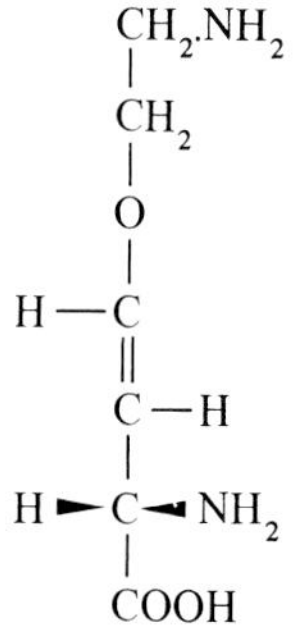

```
        CH₂NH₂
         |
        CH₂
         |
         O
         |
    H — C
         ‖
         C — H
         |
  H ►— C ◄ NH₂
         |
        COOH
```

Aminoethoxyvinyl-
glycine (AVG);
L-2-amino-4-
(2-aminoethoxy)-
trans-3-butenoic
acid
(ex *Streptomyces* sp.)

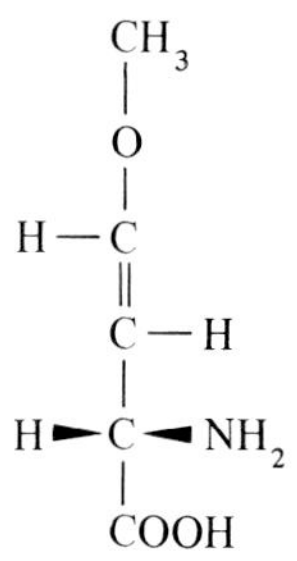

```
        CH₃
         |
         O
         |
    H — C
         ‖
         C — H
         |
  H ►— C ◄ NH₂
         |
        COOH
```

Methoxyvinyl
glycine (MVG);
L-2-amino-4-
methoxy-*trans*-
3-butenoic acid
(ex *Pseudomonas
aeruginosa*)

Fig. 15.28. Structure and origin of some enol ether amino acid analogues that inhibit ethylene biosynthesis.

Ethylene biosynthesis has also been studied in microorganisms. The green mould of citrus fruit, *Penicillium digitatum* Sacc., produces large quantities of ethylene when grown in static culture. Under these growth conditions ethylene is not formed from L-methionine; it appears to be derived from C–3 and C–4 of α-ketoglutarate. If, however, *P. digitatum* is grown in shake culture with L-methionine in the medium a methionine–ethylene-forming system is induced after a lag period. A methionine–ethylene-forming system has also been shown to be induced by L-methionine in shake cultures of *Saccharomyces cerevisiae*.

(ii) CATABOLISM OF ETHYLENE

Little is known of the catabolism of ethylene in higher plants. Early work suggested that very little catabolism of ethylene occurs (i.e. <0.05% of an administered amount); moreover, the catabolic products included compounds such as benzene and toluene. These results were criticized by Beyer in 1975 because (i) the [^{14}C]ethylene used had not been purified and (ii) no precautions had been taken to avoid microbial contamination of the plant tissues used.

Beyer has since shown (1975–8), in experiments carried out under aseptic conditions with pure [^{14}C]ethylene, that ethylene is catabolized in several tissues known to respond physiologically to ethylene, namely etiolated pea seedlings, cut carnation flowers and cut morning glory flowers. In these tissues there appeared to be two distinct routes of ethylene catabolism, a major one leading to CO_2 and a minor one leading to as yet unknown water soluble metabolites. The 'ethylene→CO_2' route was inhibited by high CO_2 (7–10%, by vol) levels but not by Ag^+ ions. In contrast the 'ethylene→water soluble metabolites' route was inhibited by Ag^+ ions but not by high CO_2 levels.

Jerie and Hall in 1978 showed that exogenous ethylene was converted by the cotyledons of developing broad bean (*Vicia faba*) seeds into ethylene oxide in high yield (85–95% conversion in <25 hr). This was a surprising result for two reasons: (i) no previous experiments had demonstrated more than a 1% conversion of the administered ethylene into catabolites and (ii) of all the tissues so far examined only *Vicia faba* has been shown to produce ethylene

oxide from ethylene. However, there is no doubt of its *bona fides* since (i) the ethylene oxide was identified by mass spectrometry, (ii) it has been repeated by other workers and (iii) the enzyme catalysing the conversion has been partly purified. One must conclude, therefore, that this is a catabolic process peculiar to *Vicia faba*, at least amongst the plants examined.

5. Physiological effects of ethylene

A large number of different physiological effects of ethylene have been reported. They include (i) the stimulation of ripening of fleshy fruits, (ii) the stimulation of abscission of leaves, (iii) the so-called 'triple response' of etiolated seedlings such as pea, i.e. the inhibition of stem elongation, the stimulation of radial swelling of stems and the horizontal growth of stems with respect to gravity, (iv) the tightening of the hook of the epicotyl or hypocotyl in etiolated dicotyledonous seedlings, (v) the inhibition of root growth, (vi) the stimulation of adventitious root formation, (vii) the stimulation of flowering in the pineapple, (viii) the stimulation of the fading of some flowers and (ix) the stimulation of leaf epinasty (the downward curvature of the leaf by more rapid growth of the upper side of the petiole).

Of these effects probably the most important are the stimulation of the ripening of fleshy fruits and the stimulation of abscission. However, before dealing with these in more detail it should be pointed out that the effects on root growth [(v) and (vi) above] are probably brought about in Nature by ethylene produced by soil microorganisms, particularly in waterlogged conditions.

The stimulatory effect of ethylene on fruit ripening has been known for many years (see Section F.1). This knowledge has been used since the mid-1930s to prevent the overripening of fruit during storage. The fruit are kept at a low temperature in a gastight room in which the air is freed from ethylene by circulating it through brominated charcoal filters. The respiration of the stored fruit lowers the oxygen content of the air to 5–10% (by vol) and raises the carbon dioxide content to 1–3% (by vol). This assists storage because oxygen is

necessary for the production and the action of
ethylene and carbon dioxide inhibits its action.

The changes associated with fruit ripening are
numerous, complex and imperfectly understood.
They may be regarded as a special form of
senescence the purpose of which is to render the fruit
more palatable to animals and thereby to facilitate
seed dispersal. The most obvious ripening changes
are (i) tissue softening, (ii) hydrolysis of storage
compounds, (iii) changes in pigmentation and
flavour and (iv) changes in respiration rate.

Fleshy fruits can be subdivided into two classes
on the basis of the changes that occur in the
respiration rate following maturity (i.e. attainment
of final size). One class, which includes apples,
pears, bananas, avocados and mangoes, exhibits an
initial decrease in respiration followed by a large
rise, coinciding with ripening, and then a final
decrease as the fruit becomes overripe. The large
increase in respiration is called the 'respiratory
climacteric' and fruit in this class are termed
climacteric fruit. The respiratory climacteric is
thought to be due to cyanide-resistant respiration
(see Chapter 6, Section D). The other class contains
fruit that do not have a respiratory climacteric;
these are known as non-climacteric fruit. This class
is subdivided into (i) fruit such as oranges, lemons
and figs that have a virtually constant respiration
rate during ripening and (ii) fruit, such as peppers,
that show a decrease in respiration during ripening.

There is now convincing evidence that ethylene is
the hormone that initiates fruit ripening. In climacteric fruit the increase in intracellular ethylene to a
level known to cause ripening occurs well in
advance of the respiratory climacteric (which may
be taken as indication of ripening). Moreover,
exogenous ethylene not only stimulates the ripening
of non-climacteric fruit but also induces a respiratory climacteric.

However, there is considerable variation amongst
fruits in their response to ethylene. In most fleshy
fruits a threshold intracellular level of ethylene
(3 ppm in the case of melons) must be present before
ripening will occur; this is achieved by ethylene
synthesis in the fruit itself. Most of these fruit will
ripen on the plant once this level is attained but
some (e.g. avocados, mangoes) will not and require
to be harvested before ripening occurs. In the latter
it is presumed that an inhibitor of ethylene is present

whilst the fruit is attached to the parent plant and
that this gradually disappears after harvesting. The
nature of this inhibitor is unknown and indeed may
be not a substance but rather a particular balance of
other hormones. In other fruit, typified by certain
kinds of bananas, the ripening concentration of
ethylene is present in the immature fruit but the fruit
remains insensitive to it until maturity is reached. In
the fruit of some annual crops (e.g. tomatoes,
melons) ethylene production and ripening will only
occur after a certain physiological age is reached
regardless of whether the fruit remains on the plant
or is harvested.

Ethylene appears to play a key role in leaf and
flower abscission. It is produced as the leaf or flower
tissues senesce and acts on the abscission zone at
the base of the leaf petiole or flower stalk. It causes
the proximal layer of cells within the abscission zone
(i.e. the layer remotest from the leaf blade or flower)
to enlarge and to synthesize and secrete cell wall
degrading enzymes such as cellulase[15] and pectinase[16]. These enzymes attack and weaken the
walls of the rest of the cells in the abscission zone
commencing with those nearest their site of origin.
This results in a weakened region in the petiole or
flower stalk which breaks readily under stress
causing the leaf or flower to fall from the plant.
Auxin inhibits this process (see Section B.4).

Ethylene differs from all the other known phytohormones in its extreme volatility. Osborne has
pointed out that this enables the ethylene produced
by one plant to influence the growth of other plants
close to it. To this extent it is like the insect
pheromones. However, it differs from the pheromones in that it also influences the growth of the
individual that produces it.

6. Biochemical mode of action of ethylene

The mode of action of ethylene at the molecular
level is not clear. However, there are two significant
pointers to the way it may act. Firstly, it is well
established that during the ripening of fruit and the
abscission of leaves ethylene stimulates the *de novo*
synthesis and secretion of cell wall-dissolving enzymes such as cellulase[15] which then cause tissue
softening. Secondly, Hall and his coworkers have

Table 15.3. *Compounds produced by lower plants to regulate their development* (*mainly their sexual reproduction*)

Compound	Source	Physiological effect or function
1. α-Factor (a tridecapeptide composed of protein amino acids)	α-Mating type of *Saccharomyces cerevisiae*	Inhibits mitotic division and causes cell elongation in the *a*-mating type of *S. cerevisiae*
2. Trisporic acid B Trisporic acid: R=COOH Trisporal B: R=CH$_2$OH Methyl dihydrotrisporate B: R=COOCH$_3$; 4-keto group reduced to 4β-hydroxyl	Trisporic acid B is formed in the cells of both (+) and (−) mating types of fungi of the order Mucorales It is formed in the (+) type from trisporol produced by the (−) type It is formed in the (−) type from methyl dihydrotrisporate B produced by the (+) type	Stimulates the switch from asexual to sexual differentiation, characterized by the production of zygospores
3. L-Sirenin	♀ gametes of the water mould *Allomyces* after release from the ♀ gametangia into the surrounding aqueous medium	Attracts ♂ gametes so that they may fuse with the ♀ gametes to produce motile, biflagellate zygotes
4. Antheridiol	Vegetative hyphae of the '♀' mating type of the heterothallic water moulds, *Achlya ambisexualis* and *A. bisexualis*	Induces (i) the formation of antheridial branches in the '♂' mating type, (ii) the growth of these branches towards the source of antheridiol, (iii) the development of antheridia and (iv) the secretion of hormone B (possibly like oogoniol-1 and -2) which induces oogonia formation in the '♀' mating type
5. Ectocarpene	♀ gametes of the marine, brown alga *Ectocarpus siliculosus*	Attracts ♂ gametes
6. Fucoserratene	Eggs of the marine, brown alga *Fucus serratus*	Attracts spermatozoids
7. Multifidene	Eggs of *Cutleria multifida*	Attracts ♂ gametes

(*continued overleaf*)

Table 15.3 (*cont.*)

Compound	Source	Physiological effect or function
8. Antheridogens (e.g. Antheridogen A$_{An}$) 	Prothalli of ferns	Induce the formation of antheridia on fern prothalli

Table 15.4. *Compounds produced by microorganisms that promote the growth of higher plants*

Compound	Source	Physiological effects
1. Helminthosporol 	Mycelium of the pathogenic fungus *Helminthosporium sativum*	Promotes elongation of the leaf sheaths of rice and dwarf maize seedlings; promotes hypocotyl and primary root elongation in intact seedlings; breaks dormancy of artichoke tubers; promotes seed germination and flowering
2. Cotylenins (glycosides of cotylenol—see below) 	Filtrate of culture medium of a fungus believed to be a species of *Cladosporium*	Promotes the growth of Chinese cabbage, radish and cucumber cotyledons
3. Sclerin 	Mycelium of the fungus *Sclerotinia libertiana*	Promotes shoot elongation and root growth in rice and other seedlings; stimulates lipase activity in germinating castor bean seeds and α-amylase activity in germinating rice seeds
4. Pestalotin 	Filtrate of culture medium of the fungus *Pestalotia cryptomeriaecola*	Inactive alone but augments the action of gibberellins

Table 15.4 (*cont.*)

Compound	Source	Physiological effects
5. Fusicoccin 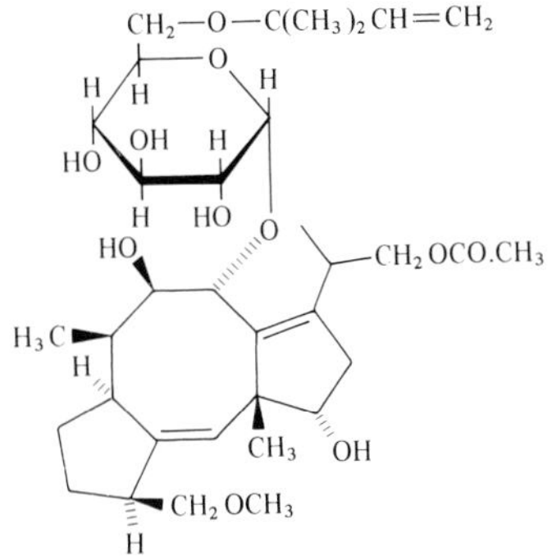	Mycelium of the fungus *Fusicoccum amygdali* which infects almond and peach trees	Causes wilting and dehydration of the shoots on infected almond and peach trees; promotes cell enlargement in a variety of plant tissues—in this it is more potent than auxins; causes the opening of stomata in the presence and absence of light; induces germination of dormant wheat seeds
6. Malformin (cyclo-D-Cys-D-Cys-L-Val-D-Leu-L-Ile)	Mycelium of the fungus *Aspergillus niger*	Promotes cell elongation; causes severe malformations on the stems and petioles of several plant species; causes root curvature

Table 15.5. *Examples of compounds that occur in higher plants and promote or inhibit plant growth when supplied exogenously*

Compound(s)	Source	Physiological effects
1. Phenolic carboxylic acids and *trans*-cinnamic acid 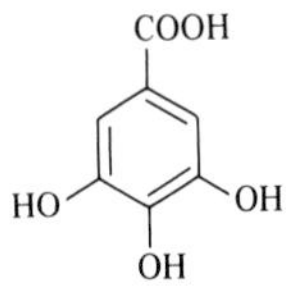 Gallic acid 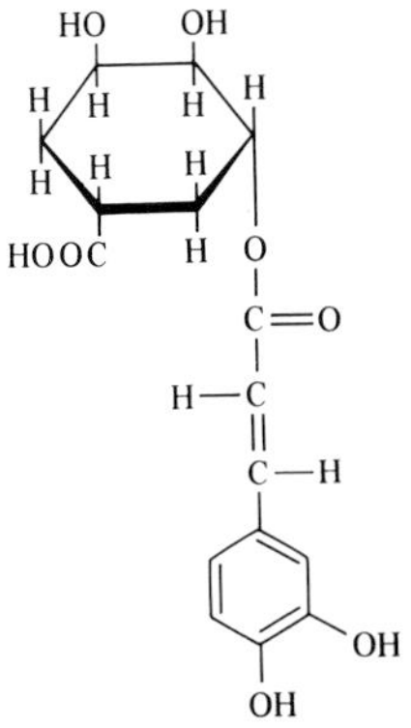 Chlorogenic acid 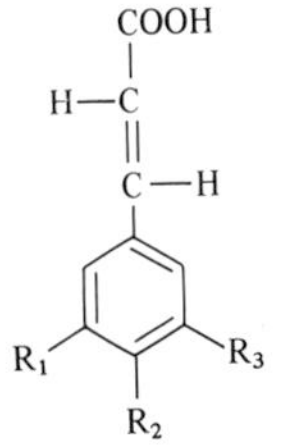 Cinnamic; $R_1 = H$, $R_2 = H$, $R_3 = H$ Coumaric; $R_1 = H$, $R_2 = OH$, $R_3 = H$ Caffeic; $R_1 = H$, $R_2 = OH$, $R_3 = OH$ Ferulic; $R_1 = H$, $R_2 = OH$, $R_3 = OCH_3$ 5-HO-Ferulic; $R_1 = OH$, $R_2 = OH$, $R_3 = OCH_3$ Sinapic; $R_1 = OCH_3$, $R_2 = OH$, $R_3 = OCH_3$	Widely distributed in higher plants in the free form and as glucose esters and amides	Affect IAA-induced growth of sections of coleoptiles, internodes and hypocotyls. In general terms acids with one or no HO group antagonize IAA action probably by stimulating the oxidative decarboxylation of IAA whilst acids with two or three adjacent HO groups enhance the IAA action probably by suppressing the oxidative decarboxylation of IAA. It is uncertain whether these acids regulate auxin levels *in vivo*, but there is some evidence that they do

(*continued overleaf*)

Table 15.5 (*cont.*)

Compound(s)	Source	Physiological effects
2. Flavonoids (see Chapter 14) 	Widely distributed in higher plants mainly as their glycosides; the *O*-glycosides are the most common but *C*-glycosides do occur	Affect IAA-induced growth (e.g. in first internode sections of oats) probably by their effect on the oxidative decarboxylation of IAA. This activity depends on the ring B hydroxylation pattern. In general terms flavonoids with one HO group at the 4′ position inhibit IAA-induced growth by stimulating IAA oxidase whilst flavonoids with HO groups at the 3′ and 4′ positions stimulate it by inhibiting IAA oxidase
3. Coumarins Coumarin; $R_1 = H$, $R_2 = H$, $R_3 = H$ Umbelliferone; $R_1 = H$, $R_2 = OH$, $R_3 = H$ Skimmin; $R_1 = H$, $R_2 = OGlc$, $R_3 = H$	Widely distributed in higher plants, frequently as glucosides	Generally regarded as growth inhibitors but the activity depends on the pattern of substitution. Coumarin inhibits IAA-induced elongation of wheat and oat coleoptile segments, growth of rice, mung bean, lettuce and clover seedlings and the germination of seeds (e.g. lettuce)
4. Other saturated lactones (i) Protoanemonin 	*Ranunculus* spp., also *Clematis flammula*	Inhibitis elongation of oat coleoptile segments; inhibits mitosis in and growth of roots
(ii) Parasorbic acid 	Berries of the mountain ash tree (*Sorbus aucuparia*)	Inhibits seed germination and cell enlargement in oat coleoptiles
(iii) Xanthinin 	Leaves of cocklebur (*Xanthium pennsylvanicum*)	Inhibits growth of oat seedlings and oat coleoptile sections. Inhibits axillary buds when applied to the adjacent leaves

recently demonstrated that ethylene is specifically and reversibly bound to a protein in the endomembrane system (endoplasmic reticular and Golgi membranes) of the cotyledons of developing French bean (*Phaseolus vulgaris*) seeds. This is the first demonstration of an ethylene receptor site in a target cell.

These observations suggest that ethylene enters the target cell, binds to a specific receptor protein located in the endomembrane system and thereby promotes the synthesis of a specific cocktail of enzymes, many of which are secreted from the cell and modify the cell wall structure. At first sight this sequence of events is like that of the steroid hormones in mammals. The steroid hormone enters the cell, binds to a specific protein receptor in the cytosol and then passes, in the bound form, to the nucleus where it switches on specific genes which are transcribed into *m*RNA and then translated into proteins (enzymes). However, there is a fundamental difference: the steroid hormone receptor protein is located in the cytosol and is therefore free to move to the nucleus to switch on the genes whereas the ethylene receptor protein, being membrane-bound, is not. There is therefore an important gap in our knowledge—how does the signal that results from the binding of ethylene to its receptor protein result in the synthesis of specific enzymes?

G. OTHER PLANT-GROWTH REGULATORS

Although auxins, gibberellins, cytokinins, abscisic acid and ethylene are generally regarded as the principal phytohormones they are by no means the only compounds present in plants that have a pronounced effect on plant growth when administered exogenously at low concentrations.

Most of these other physiologically active compounds fall into one or other of the following three categories:

(i) Compounds produced by the lower forms of plant life (i.e. algae, fungi, bryophytes and pteridophytes) to regulate their development.

(ii) Compounds produced by microorganisms which have the ability to affect the growth of higher plants.

(iii) Compounds that are produced by higher plants and that affect plant growth when supplied exogenously to intact plants or plant tissue segments.

Examples of compounds in these three categories, their structures, occurrence and physiological actions are given in Tables 15.3, 15.4 and 15.5. All the compounds in the first category listed in Table 15.3 regulate the sexual development of the organisms producing them. However, to this table could be added lunularic acid which appears to be the physiological equivalent of abscisic acid in liverworts and algae (see Section E.2). All the compounds in the second category listed in Table 15.4 stimulate higher plant growth; however, it should

be noted that there are numerous compounds produced by fungi that are inhibitory or toxic to higher plants. All the compounds in the third category listed in Table 15.5 either stimulate or inhibit plant growth when administered exogenously. However, it is doubtful whether these compounds have a growth regulatory role *in vivo*.

SUGGESTIONS FOR FURTHER READING

General Texts

Letham, D. S., Goodwin, P. B. and Higgins, T. J. V., eds. (1978) *Phytohormones and Related Compounds: a comprehensive treatise*. Vol. I, *The Biochemistry of Phytohormones and Related Compounds*. Vol. II, *Phytohormones and the Development of Higher Plants*. Elsevier/North-Holland Biomedical Press, Amsterdam–Oxford–New York.

Moore, T. C. (1979) *Biochemistry and Physiology of Plant Hormones*. Springer-Verlag, New York–Heidelberg–Berlin.

Wareing, P. F. and Philips, I. D. J. (1981) *The Control of Growth and Differentiation in Plants*, 3rd edn. Pergamon Press, Oxford–New York–Toronto–Sydney–Paris–Frankfurt.

Specialist Reviews

Auxins

Scott, T. K. (1972) 'Auxins and roots', *Ann. Rev. Plant Physiol.* **23**, 235–258.

Goldsmith, M. H. M. (1977) 'The polar transport of auxin', *Ann. Rev. Plant Physiol.* **28**, 439–478.

Gibberellins

Lang, A. (1970) 'Gibberellins: structure and metabolism', *Ann. Rev. Plant Physiol.* **21**, 537–570.

Jones, R. L. (1973) 'Gibberellins: their physiological role', *Ann. Rev. Plant Physiol.* **24**, 571–598.

Rappaport, L. and Adams, D. (1978) 'Gibberellins: synthesis, compartmentation and physiological process', in *The Biochemical Functions of Terpenoids in Plants* (Goodwin, T. W., ed.), pp. 83–101. The Royal Society, London.

Cytokinins

Skoog, F. and Armstrong, D. J. (1970) 'Cytokinins', *Ann. Rev. Plant Physiol.* **21**, 359–384.

Hall, R. H. (1973) 'Cytokinins as a probe of developmental processes', *Ann. Rev. Plant Physiol.* **24**, 415–444.

Horgan, R. (1978) 'Nature and distribution of cytokinins', in *The Biochemical Functions of Terpenoids in Plants* (Goodwin, T. W., ed.), pp. 1–9. The Royal Society, London.

Laloue, M. (1978) 'Functions of cytokinins', *ibid.*, pp. 11–18.

Abscisic acid

Millborrow, B. V. (1974) 'The chemistry and physiology of abscisic acid', *Ann. Rev. Plant. Physiol.* **25**, 259–307.

Mansfield, T. A., Wellburn, A. R. and Moreira, T. J. S. (1978) 'The role of abscisic acid and farnesol in the alleviation of water stress', in *The Biochemical Functions of Terpenoids in Plants* (Goodwin, T. W., ed.), pp. 33–44. The Royal Society, London.

Wareing, P. F. (1978) 'Abscisic acid as a natural growth inhibitor', *ibid.*, pp. 45–60.

Walton, D. C. (1980) 'Biochemistry and physiology of abscisic acid', *Ann. Rev. Plant Physiol.* **31**, 453–489.

Ethylene

Abeles, F. B. (1972) 'Biosynthesis and mechanism of action of ethylene', *Ann. Rev. Plant Physiol.* **23**, 259–292.

Lieberman, M. (1979) 'Biosynthesis and action of ethylene', *Ann. Rev. Plant Physiol.* **30**, 533–591.

Other plant growth regulators

Kefeli, V. I. and Kadyrov, C. S. (1971) 'Natural growth inhibitors, their chemical and physiological properties', *Ann. Rev. Plant Physiol.* **22**, 185–196.

Kochert, G. (1978) 'Sexual pheromones in algae and fungi', *Ann. Rev. Plant Physiol.* **29**, 461–486.

McMorris, T. C (1978) 'Antheridiol and the oogoniols, steroid hormones which control sexual reproduction in *Achlya*', in *The Biochemical Functions of Terpenoids in Plants* (Goodwin, T. W., ed.), pp. 21–32. The Royal Society, London.

Dörffling, K. (1978) 'The possible role of xanthoxin in plant growth and development', *ibid.*, pp. 61–69.

Gooday, G. W. (1978) 'Functions of trisporic acid', *ibid.*, pp. 71–82.

Marrè, E. (1979) 'Fusicoccin: a tool in plant physiology', *Ann. Rev. Plant Physiol.* **30**, 273–288.

Related topics

Wareing, P. F. and Saunders, P. F. (1971) 'Hormones and dormancy', *Ann. Rev. Plant Physiol.* **22**, 261–288.

Evans, L. T. (1971) 'Flower induction and the florigen concept', *Ann. Rev. Plant Physiol.* **22**, 365–394.

Sequeira, L. (1973) 'Hormone metabolism in diseased plants', *Ann. Rev. Plant Physiol.* **24**, 353–380.

Evans, M. L. (1974) 'Rapid responses to plant hormones', *Ann. Rev. Plant Physiol.* **25**, 195–223.

Kende, H. and Gardner, G. (1976) 'Hormone binding in plants', *Ann. Rev. Plant Physiol.* **27**, 267–290.

Wareing, P. F. (1977) 'Growth substances and integration in the whole plant', in *Society for Experimental Biology, Symposium XXXI—Integration and Activity in the Higher Plant* (Jennings, D. H., ed.), pp. 337–365.

ENZYMES

1. L-Tryptophan: 2-oxoglutarate aminotransferase, EC 2.6.1.27.
2. 2-Oxo-acid carboxy-lyase, EC 4.1.1.1.
3. Aromatic-L-amino-acid carboxy-lyase, EC 4.1.1.28.
4. Thioglucoside glucohydrolase, EC 3.2.3.1.
5. Nitrile aminohydrolase, EC 3.5.5.1.
6. 1,4-α-D-Glucan glucanohydrolase, EC 3.2.1.1.
7. 1,4-α-D-Glucan maltohydrolase, EC 3.2.1.2.
8. Plant ribonuclease, EC 3.1.27.1.
9. 1,3-β-D-Glucan glucanohydrolase, EC 3.2.1.39.
10. CTP: cholinephosphate cytidylyltransferase, EC 2.7.7.15.
11. CDPcholine: 1,2-diacylglycerol cholinephosphotransferase, EC 2.7.8.2.
12. *threo*-D_s-Isocitrate glyoxylate-lyase, EC 4.1.3.1.
13. Orthophosphate: oxaloacetate carboxy-lyase (phosphorylating), EC 4.1.1.31.
14. L-Malate: $NADP^+$ oxidoreductase, EC 1.1.1.37.
15. 1,4-(1,3; 1,4)-β-D-Glucan 4-glucanohydrolase, EC 3.2.1.4.
16. Poly(1,4-α-D-galacturonide) glycanohydrolase, EC 3.2.1.15.

Appendix 1

Specification of Absolute Configuration at Chiral Centres Using the Cahn–Ingold–Prelog Convention

The absolute configuration about chiral centres in certain limited groups of compounds has been defined very adequately by a number of internationally agreed conventions. For instance, the terms D and L are used for carbohydrates and amino acids relating them to glyceraldehyde and serine respectively (see Fig. A.1.1) and the terms α and β to substituents of rings A–D of the steroid nucleus relating them to the orientation of the 19-methyl group (see Fig. A.1.2). However, the terms D and L or α and β as defined for these groups of compounds cannot be used as a general system throughout chemistry. The need for such a general system was apparent and this has been met by the Cahn–Ingold–Prelog convention which defines the configuration at a chiral centre as either R or S. This convention was developed during the period 1951–66 by R. S. Cahn, C. K. Ingold and V. Prelog (Cahn, R. S. and Ingold, C. K., *J. Chem. Soc.* 612,

1951; Cahn, R. S., Ingold, C. K. and Prelog, V., *Experientia* **12**, 81, 1956; Cahn, R. S., Ingold, C. K. and Prelog, V., *Angew. Chem., Int. Edit. Engl.* **5**, 385, 1966) and can be applied to all chiral compounds. However, it was not intended to replace 'local' systems, such as those shown in Figs. A.1.1 and A.1.2 but to be used when no 'local' system existed or to assist when (i) local systems conflict or overlap or (ii) new stereochemical factors cannot be readily incorporated into a local system (Cahn, R. S., *J. Chem. Educ.* **41**, 116, 1964).

Before discussing the Cahn–Ingold–Prelog convention the term chiral centre will be discussed. This has been introduced to replace the term asymmetric centre because the latter does not always accurately describe centres which result in optical activity. The essential requirement for optical activity is that a molecule is not identical with its mirror image. It was customary to call this condition asymmetry but

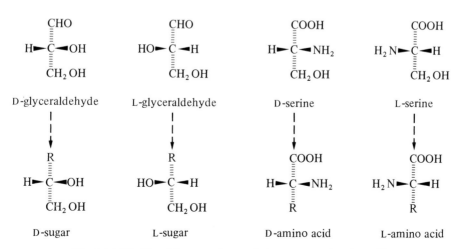

Fig. A.1.1. The D and L convention applied to sugars and amino acids.

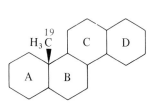

Steroid nucleus showing the 19-methyl projecting up from the plane of the paper and the general plane of the ring system. Such a methyl is termed β (beta) and is shown in structural formulae as being joined to the nucleus by a solid (preferably thickened ▬▬ or wedge-shaped ◀) bond line. Similarly oriented substituents are indicated in the same way. Substituents projecting in the opposite direction are indicated by broken (ⅲⅲⅲⅲ) bond lines. Substituents whose orientation is not known are indicated by wavy (∿∿) bond lines.

Lanostanol showing α- and β-oriented substituents

Fig. A.1.2. The α and β convention for steroid nuclear substituents.

it is now realized that this term is not strictly correct because some molecules exist which are not identical with their mirror images but nevertheless possess symmetry by virtue of axes of rotation. To get over this difficulty all molecules which are not identical to their mirror images, whether they are genuinely asymmetric or have symmetry by virtue of axes of rotation, are now termed chiral or are said to exhibit chirality. Chirality is therefore the necessary and sufficient condition for the existence of optical isomers. The word chiral means 'of or pertaining to the hand', thus chirality means 'handedness'. It should be noted that since the grouping C$xyzw$ has no symmetry whatsoever, the term 'asymmetric carbon atom' is perfectly correct; the asymmetric carbon atom is an example of a chiral centre.

Since the asymmetric carbon atom is the most frequently encountered chiral centre in biochemistry it will be used as the basis for the following description of the Cahn–Ingold–Prelog convention.

The Cahn–Ingold–Prelog convention is composed of two rules, known as the 'Sequence Rule' and the 'Conversion Rule'. The assignment of absolute

configuration to an asymmetric carbon atom is a two-stage process. Firstly, the substituents of the asymmetric carbon atom are given a 'priority' relative to each other; this is accomplished by the application of the Sequence Rule. Secondly, the spatial arrangement of the substituents, in order of priority, is determined; this is determined by application of the Conversion Rule.

Although the Sequence Rule is always applied before the Conversion Rule during the process of assigning configuration the two rules are best explained in reverse order.

1. The Conversion Rule

When the substituents of the asymmetric carbon atom have been placed in order of priority according to the Sequence Rule the molecule is viewed from the direction opposite to the substituent of lowest priority, i.e. the molecule is oriented so that the substituent of lowest priority projects away from the viewer with the other three substituents projecting towards him like the blades of a three-bladed propeller. If the priorities of the three substituents projecting towards the viewer proceed from greatest (i.e. priority [1]) to lowest (i.e. priority [3]) in a clockwise direction the configuration of the asymmetric carbon is said to be R (rectus, right handed); if the priority sequence is anticlockwise the configuration is said to be S (sinister, left handed) (see Fig. A.1.3).

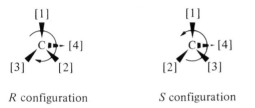

R configuration   *S* configuration

Fig. A.1.3. The Conversion Rule ([1], [2], [3] [4] = substituents of the asymmetric carbon atom with priorities in the sequence [1]>[2]>[3]>[4]: ▶ represents a bond projecting out of the plane of the paper towards the reader; ⟨⟨⟨ represents a bond projecting into the plane of the paper away from the reader).

2. The Sequence Rule

The Sequence Rule assigns the priorities to the four substituents (or ligands) of the asymmetric carbon atom. It is not really a single rule but rather a composite of four sub-rules, of which only two (sub-rules 1 and 2) are required to deal with the vast majority of cases.

Sub-rule 1 states that 'higher atomic number precedes lower'. It is applied initially to the four *atoms* by which the ligands are attached to the asymmetric carbon atom.

The priority of the atoms that are commonly or occasionally found in biochemically important compounds are:

Atom	I	Br	Cl	S	P	F	O	N	C	H
Atomic number	53	35	17	16	15	9	8	7	6	1

→
Decreasing priority

In the following example (Fig. A.1.4) the atoms directly attached to the asymmetric carbon atom are O (in the hydroxyl group), N (in the amino group), C (in the methyl group) and H. These have the priority sequence O>N>C>H (where the

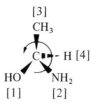

S configuration

Fig. A.1.4. Priority of ligands determined by reference to the atoms directly attached to the asymmetric carbon atom ([1], [2], [3] and [4] = order of priority).

symbol > denotes 'has priority over'). Therefore the priority sequence of the ligands is OH>NH$_2$>CH$_3$>H. To assign the configuration of the asymmetric carbon atom the Conversion Rule must now be applied. The H atom, being the ligand of lowest priority, is placed so that it is furthest away from the viewer (i.e. with the C—H bond projecting into the plane of the paper). The other three ligands now project out of the plane of the paper towards the viewer. Since their priority sequence is anticlockwise the asymmetric carbon has the S configuration.

Frequently the priority of two or more of the ligands attached to the asymmetric carbon atom cannot be determined by considering solely the atoms *directly* attached to it; this is the case in the example given in Fig. A.1.5 where three of the ligands are directly attached to the asymmetric carbon atom by carbon atoms. Sub-rule 1 copes with this situation by saying that where the relative priorities of two (or more) ligands cannot be decided by reference to the atoms directly attached to the chiral centre, the next set of atoms (i.e. those attached to the atoms attached to the chiral centre) must be considered (on the basis of higher atomic number precedes lower), and so on, working outwards from the chiral centre, until a decision can be reached. Where this gives rise to branch points, as it inevitably must, it is necessary to decide which of the branches of a given ligand has the greatest priority using the usual criterion; this branch is known as the 'prior branch'. Then the prior branches must be compared, working outwards until a decision on priority amongst the ligands is obtained. If no decision can be reached by comparing the prior branches of the ligands, then the branches next in order of priority (i.e. priority [2])

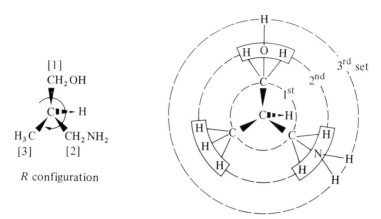

Fig. A.1.5. Priority of three of the ligands determined by reference to the atoms attached to the carbon atoms directly bonded to the asymmetric carbon atom ([1], [2], [3] and [4] = order of priority).

must be compared as before; if these branches do not allow a decision to be reached, then the branches third in order of priority (i.e. priority [3]) must be compared.

In the example shown in Fig. A.1.5 the atoms directly attached to the asymmetric carbon atom are C, C, C and H. The H atom clearly has lowest priority. However, the priority of the other three ligands, CH_3, CH_2NH_2 and CH_2OH, cannot be decided on the basis of the atoms directly attached to the asymmetric carbon atom because they are all the same (i.e. C atoms). Therefore it is necessary to consider the atoms which are attached to these three carbons. They are: [H,H,H] for the methyl C, [H,H,N] for the primary amine C and [H,H,O] for the primary alcohol C. This must be regarded as a case of branching; the methyl C has three Hs as branches, the primary amine C has two Hs and one N as branches whilst the primary alcohol has two Hs and one O as branches. The prior branches of each ligand are: H in the case of the methyl C (because all the branches are H), N in the case of the primary amine C, and O in the case of the primary alcohol C. Since the priority sequence of these atoms is O>N>H, the priority sequence of the ligands is $CH_2OH>CH_2NH_2>CH_3$. Application of the Conversion Rule then shows that the configuration of the asymmetric carbon atom of the molecule in Fig. A.1.5 is R.

The example given in Fig. A.1.6 is rather more complex. In this molecule the H atom and the methyl group are clearly the ligands of priority [4]

and [3] respectively; the problem is to decide which of the other two ligands has priority [1]. The expanded formula shows these two ligands branch at the level of the second set of atoms. It is therefore necessary to decide which of the branches of the ligand on the right is prior and likewise for the ligand on the left. Taking the ligand on the right first, the prior branch must start with one or other of the two carbon atoms in set 2 since they have priority over the hydrogen. A comparison of the atoms in set 3 attached to these carbons shows that the bottom one, with H,C,C, has priority over the top one, with H,H,C. Thus the bottom branch has priority in the case of the ligand on the right. Similar considerations show that it is also the bottom branch that has priority in the ligand on the left. It is now necessary to compare the right-hand bottom branch with the left-hand bottom branch to see whether a decision can be made as to whether the left-hand ligand has priority over the right-hand ligand or vice versa. A comparison of the set 3 atoms of these branches shows that a decision cannot be made at that level because they are both H,C,C. It is therefore necessary to compare the set 4 atoms. These also prove to be identical, being H,H,H and H,H,O in both cases. Thus it is not possible to decide the priorities of the right- and left-hand ligands by comparing their prior branches. It is therefore necessary to compare their branches that are second in order of priority, namely the top, right-hand and top, left-hand branches. Since their set 3 atoms are identical (H,H,C) it is necessary to compare their set

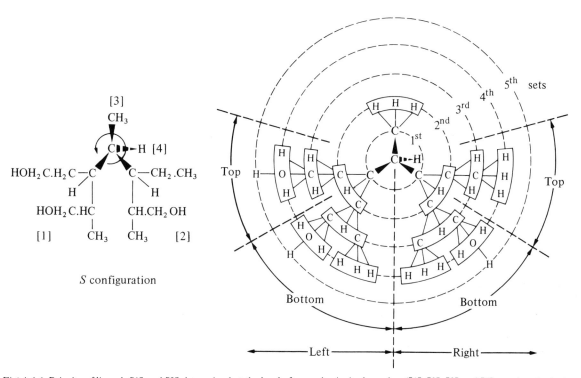

Fig. A.1.6. Priority of ligands [1] and [2] determined at the level of second priority branches ([1], [2], [3] and [4] = order of priority).

4 atoms. Here a decision can be made because the set on the left is H,H,O which has priority over that on the right which is H,H,H. This means that the ligand on the left has priority over the ligand on the right. Application of the Conversion Rule then shows that the configuration of the asymmetric carbon atom is *S*.

Sub-rule 1 also has to cope with ligands containing multiple bonds. Here it is important to note that the treatment of multiple bonds adopted in the 1966 paper supersedes that adopted in the 1951 and 1956 papers. The current treatment is as follows. In the case of a double bond the atom at each end is considered to be duplicated, and then all atoms other than hydrogen are complemented to quadriligancy by notionally adding the necessary number of 'phantom' atoms which are regarded as having an atomic number of zero. The duplicate atoms are shown in parenthesis and the phantom atoms as o.

Thus the $\ce{>C=C<}$ bond becomes

and the $\ce{>C=O}$ bond becomes

Triple bonds are treated similarly; thus the $\ce{-C\equiv C-}$ bond becomes

It should be noted that for the purposes of determining priority duplicate atoms are the same as 'real' atoms of the same element, i.e. (C) = C, and that phantom atoms with a notional atomic number of zero have the lowest priority of all.

Figure A.1.7 gives an example of the use of this treatment of multiple bonds taken in conjunction with ligand branching and complicated by the fact that a ring system is involved.

The example is the 3-hydroxy-β-end group found in such carotenoids as lutein and zeaxanthin. It has an asymmetric carbon at C–3 with the absolute configuration shown in A (Fig. A.1.7). How is this configuration designated according to the Cahn–Ingold–Prelog convention? Of the four ligands attached to C–3, the hydroxyl, by virtue of its oxygen atom, has priority [1] and hydrogen has priority [4]. The problem is therefore to decide which of the two carbon-containing ligands (i.e. C_2–C_1–C_6 ... and C_4–C_5–C_6 ...) has the higher priority. Because of the presence of the 5,6-double

bond it is necessary to carry out the partial expansion shown in B (Fig. A.1.7) and then to consider the dissection (C) derived from it. Comparison of C–2 and C–4 does not allow a priority decision to be made because both their substituents are H,H,C. It is therefore necessary to consider the substituents of C–1 (the prior branch of C–2) and C–5 (the prior branch of C–4). Once again a priority decision cannot be made because both their substituents are C,C,C. It is therefore necessary to compare the prior branch of C–1 with that of C–5. The prior branch of C–1 is C–6 because its substituents are C,C,C whereas the substituents of the other two carbons are H,H,H. The prior branch of C–5 is also C–6 because its substituents are C,C,C whereas the substituents of the other two carbons

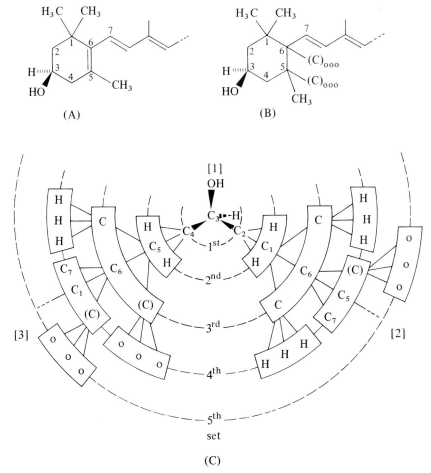

(C)

Fig. A.1.7. Priority of ligands [2] and [3] of C–3 of a 3-hydroxy-β-end group of carotenoids such as lutein and zeaxanthin determined at the level of third-priority branches after the current treatment of the double bond demanded by the Sequence Rule.

are H,H,H and ooo respectively. However, a comparison of these two prior branches does not provide a priority decision because C–6, whether approached via C–4 and C–5 or via C–2 and C–1, has identical substituents. It is therefore necessary to turn to the branches of C–1 and C–5 with priority [2]. Again no priority decision is possible because these branches are methyl groups and therefore have identical substituents, namely H,H,H. It therefore becomes necessary to consider the branches of C–1 and C–5 with priority [3]; these are CH$_3$ and (C)ooo respectively. This allows a priority decision to be made because H,H,H has priority over ooo. Therefore the C–2→C–1→C–6 . . . ligand of C–3 has priority over the C–4→C–5→C–6 . . . ligand. Application of the Conversion Rule shows that the configuration at C–3 of the 3-hydroxy-β-end group of lutein and zeaxanthin is R.

Asymmetry at a carbon atom can also arise as a result of the introduction of isotopically labelled atoms such as deuterium (^{2}H) and tritium (^{3}H). Sub-rule 1 of the Sequence Rule does not cover this eventuality. It is, however, covered by *sub-rule 2* which states that 'higher atomic mass has priority over lower', thus we have the priority sequences ^{3}H$\rangle^2$H$\rangle^1$H$\rangle$ and ^{14}C$\rangle^{13}$C$\rangle^{12}$C.

3. Special points arising from the application of the Sequence Rule

(i) RELATIONSHIP TO REACTION MECHANISMS

The description of an asymmetric carbon atom as R or S has no direct significance in terms of reaction mechanism. Thus in a reaction at an asymmetric carbon atom involving the substitution of one ligand by another ligand, it does not follow that, if configuration is retained in the normal stereo-chemical sense, a compound with an asymmetric carbon atom of the R configuration will give rise to a product whose asymmetric carbon atom is also R. This is demonstrated by the hypothetical reaction shown in Fig. A.1.8. In this reaction configuration is retained (i.e. the four ligands of the asymmetric carbon atom in compound A have the same relative positions in space as the four equivalent ligands of the asymmetric carbon in compound B), one substituent (CH$_3$) being oxidized (to COOH) without rupture of the C—C bond. However, this change alters the priority sequence of the ligands such that, although the asymmetric carbon in A had the R configuration, that in B has the S configuration.

In a similar way it does not follow that in an S$_N$2 substitution reaction involving inversion of configuration at an asymmetric carbon atom, a compound in which the configuration of that atom was R will give rise to a product in which the configuration is S. It may do so, but equally it may not.

Two real examples of this point are:

(a) In the conversion of mevalonic acid (MVA) into Δ^3-isopentenyl pyrophosphate (IPP) (see terpenoid biosynthesis—Chapter 11) it is a consequence of the numbering convention that C–4 of MVA becomes C–2 of IPP. If C–4 of MVA is labelled with tritium (T) such that its configuration is R the IPP biosynthesized from it has the S configuration at C–2 even though inversion of configuration cannot have occurred because none of the bonds linking substituents to the

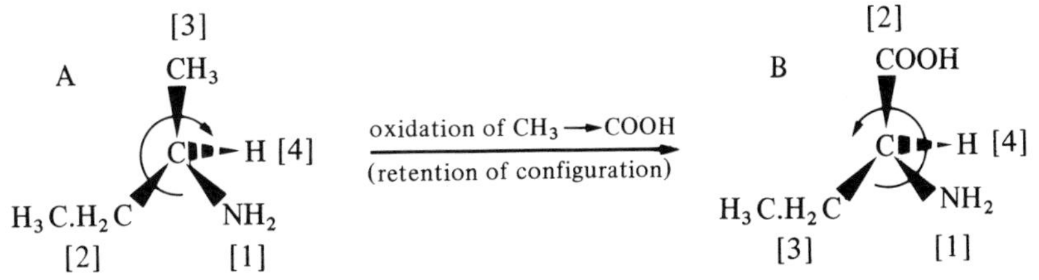

Fig. A.1.8. Hypothetical reaction demonstrating that although the configuration of an asymmetric carbon atom may be retained in the normal stereochemical sense during the reaction, its designation as R or S by the Cahn–Ingold–Prelog convention may be altered owing to a change in ligand priority.

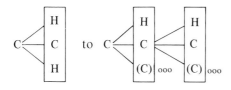

R configuration at C–4 of MVA   S configuration at C–2 of IPP

asymmetric carbon atom (C–4 in MVA, C–2 in IPP) were cleaved during the reaction sequence. The change from R to S at this atom results from a change in the priority sequence of its ligands due to the loss of the hydroxyl group from C–3 of MVA.

(b) The introduction of a Δ^{22} double bond in the sterol side chain alters the configuration at C–24. This is demonstrated below in the case of the conversion of clionasterol (24S) into poriferasterol (24R). The alteration is caused by the change in priorities of C–23 and C–25 from C–25>C–23 in clionasterol to C–23>C–25 in poriferasterol when the Δ^{22} double bond is inserted; the duplication of the carbon atoms required by the Sequence Rule changes C–23 from

which has priority over C–25 which remains

(ii) FAILURE TO RELATE CHEMICALLY OR BIOGENETICALLY SIMILAR SERIES OF COMPOUNDS

Use of the D,L-convention (see Fig. A.1.1) shows that all the amino acids found in proteins in Nature are chemically related by having the L-configuration at their α-carbon atom. Application of the Cahn–Ingold–Prelog convention to this group of compounds fails to show this 'family' relationship because L-cysteine is equivalent to R-cysteine whereas all the other L-amino acids have the S configuration. This is a consequence of the Sequence Rule which gives the CH_2SH ligand, by virtue of its S atom, priority over the COOH ligand in cysteine whereas in all the other amino acids the COOH ligand has priority over the ligand constituting the side chain which consists of the following types:

$$-C(H,H,H), \ -C(C,H,H), \ -C(C,C,H), \\ -C(O,H,H) \ \text{and} \ -C(O,C,H).$$

(iii) IDENTIFICATION OF THE FACES OF A DOUBLE BOND— THE re, si SYSTEM

The two faces of fumaric acid (A and B, Fig. A.1.9), for example, can be distinguished by applying the Sequence Rule to C–2 and C–3 but then ignoring the duplicate atom on the carbon atom under consideration. Expansion of A gives C. When considering the configuration at C–2 the

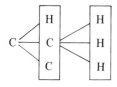

Clionasterol 24S   Poriferasterol 24R

duplicate carbon on C–2 is ignored but the duplicate carbon on C–3 is taken into account. Similarly when considering the configuration at C–3 the duplicate carbon on C–3 is ignored but the duplicate carbon on C–2 is taken into account. The priorities of the ligands of the expanded forms of A and B are in each case COOH > C–2 or C–3 > H. Use of the conversion Rule at C–2 and C–3 of C (i.e. expanded form of A) shows that both configurations are right-handed (clockwise), termed, in this case, *re*. Similarly the configurations at C–2 and C–3 of B are both left-handed (anticlockwise), termed, in this case, *si*. A is thus the *re–re* face of fumaric acid and B is the *si–si* face.

In the case of molecule D the two faces of C–1 cannot be distinguished because of the presence of two hydrogens, but the two faces of C–2 can. Under these circumstances the face of C–1 is specified by reference to that of C–2. The faces of C–1 are named so that the *re* faces of C–1 and C–2 are on the same side of the molecule.

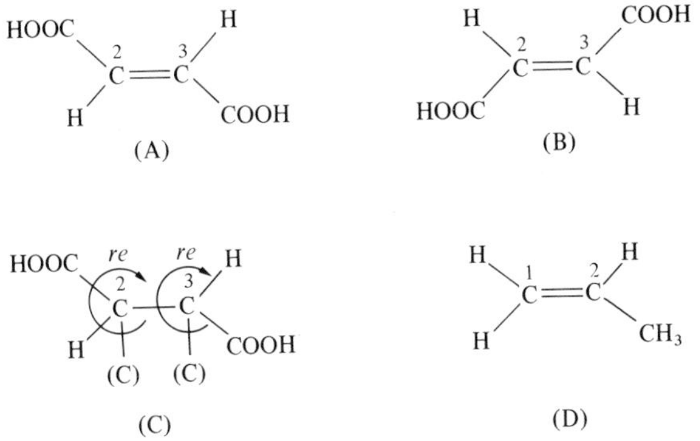

Fig. A.1.9. Identification of the faces of a double bond by the *re, si* system.

(iv) PROCHIRALITY

Prochirality is best explained by taking a simple example of a molecule which is prochiral. Ethanol (Fig. A.1.10) is such a molecule. Its C–1 carries four ligands, H,H,CH$_3$ and OH. Because two of these ligands are identical, C–1 is not an asymmetric carbon atom. If, however, one of the H atoms is replaced by another ligand, a deuterium atom for example, C–1 becomes asymmetric and the molecule exhibits chirality. An achiral molecule, like ethanol, which becomes chiral when one of its ligands is changed is said to be prochiral.

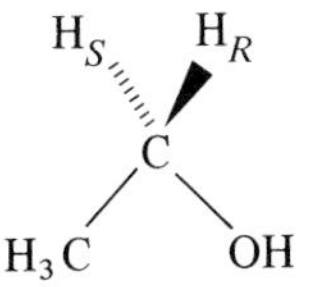

Fig. A.1.10. Pro-*R* and pro-*S* hydrogen atoms of ethanol.

If the hydrogen designated H$_R$ in Fig. A.1.10 is replaced by a higher isotope of hydrogen (i.e. deuterium or tritium), the configuration of C–1 is *R*. If, on the other hand, the hydrogen designated H$_S$ in Fig. A.1.10 is replaced by D or T, the configuration of C–1 is *S*. H$_R$ is said to be the pro-*R* hydrogen of C–1 and H$_S$ the pro-*S* hydrogen. H$_R$ and H$_S$ are therefore said to be pro-ligands of C–1. The general definition of a pro-ligand is that 'it is a ligand which, when replaced, results in chirality'. A pro-ligand which, when replaced, results in *R* chirality is said to be a pro-*R* ligand; similarly a pro-ligand which, when replaced, results in *S* chirality is said to be a pro-*S* ligand. Pro-ligands can have any identity; they do not have to be hydrogens. However, in the biochemical context they frequently are hydrogens. The concept of prochirality and the designation of ligands as pro-*R* or pro-*S* is particularly useful when dealing with the stereospecificity of enzyme action. Enzymes can usually distinguish between two hydrogen atoms at a prochiral carbon atom and remove or replace one of them with absolute specificity. They can also saturate a —C=C— double bond such that the hydrogen 'added' to each carbon atom takes up a specific position in space. It is convenient to be able to say that it is the pro-*R* or the pro-*S* hydrogen that is removed from a given carbon or to say that the hydrogen 'added' to a given carbon becomes the pro-*R* or pro-*S* hydrogen on that carbon. The availability of deuterium and tritium during the past 30 years has allowed stereospecifically deuterated and tritiated compounds to be used to investigate the stereochemistry of reactions and the stereospecificity of enzymes. No area of biochemistry has been investigated more thoroughly in this respect than sterol and carotenoid biosynthesis. This resulted from the pioneering efforts of J. W. Cornforth and G. Popják in the 1960s using various stereospecifically deuterated or tritiated species of mevalonic acid to study the biosynthesis of squalene. Mevalonic acid

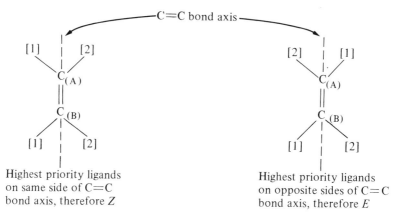

Fig. A.1.11. Specification of the configuration about a double bond using the Z, E system ([1], [2] = order of priority of carbon atom ligands).

has one chiral centre, C–3 (only the 3R isomer is used by the enzymes of terpenoid biosynthesis), and three prochiral centres, namely C–2, C–3 and C–5. Cornforth and Popják synthesized and used species of mevalonic acid stereospecifically labelled with deuterium or tritium at all three prochiral centres in their investigations, the results of which are described in Chapter 11, Section C.1. These labelled MVAs were then used by many other workers to investigate the later stages of the biosynthesis of terpenoids such as sterols, carotenoids, isoprenoid quinones (see Chapter 11) and abscisic acid [see Chapter 15, Section E.4(i)].

(v) CONFIGURATION ABOUT A DOUBLE BOND—THE Z, E SYSTEM

Although it is convenient and, in many cases, satisfactory to describe the arrangement of substituents about a C=C double bond as *cis* or *trans*, there are cases in which this cannot be done

unambiguously. The Sequence Rule has allowed a universal and unambiguous system for the specification of configuration about a double bond to be introduced; this is known as the Z, E system. To use this system it is firstly necessary to determine which of the two ligands attached to each of the two, double bond-linked carbon atoms (i.e. $C_{(A)}$ and $C_{(B)}$ in Fig. A.1.11) has the highest priority using the Sequence Rule criteria. If it is then found that the highest priority ligands are on the same side of the axis of the C=C double bond (i.e. the prior ligand of $C_{(A)}$ is on the same side as the prior ligand of $C_{(B)}$) the configuration is said to be Z. If on the other hand the highest priority ligands are on opposite sides of the C=C bond axis the configuration is said to be E. Thus Z is usually the same as the *cis* arrangement and E is usually the same as the *trans* arrangement (Z and E are derived from the German words *zusammen* meaning 'together' and *entgegen* meaning 'opposite', respectively).

Appendix 2

Units and Dimensions

1. SI units—general picture

Measurements of physical quantity are expressed as a number followed by the unit. The number expresses the ratio of the measured quantity to a fixed standard and the unit is the name of, or symbol for, the standard. The system of units that is now internationally recommended for use is the so-called SI system (Système International d'Unités) and replaces the so-called cgs (centimetre, gram, second system) that was formerly used quite widely.

SI contains three classes of units: (a) base units, (b) derived units and (c) supplementary units.

(a) Base units in SI

There are seven base units:

(i) the metre, the standard of length
(ii) the kilogram, the standard of mass
(iii) the second, the standard of time
(iv) the ampere, the standard of electric current
(v) the kelvin, the standard of temperature
(vi) the candela, the standard of luminous intensity
(vii) the mole, the standard of amount of substance.

(b) Derived units

Derived units can be formed by combining base units. Thus the unit of force can be produced by combining the first three base units. Often derived units are given names, e.g. the unit of force is the *newton*.

(c) Supplementary units

Two supplementary units are at present defined, the radian and the steradian, which are the units for plane and solid angles respectively.

(d) *SI prefixes and multiplication factors*

To obtain multiples and submultiples of units, standard prefixes are used as shown below:

Multiplication factor		Prefix	Symbol
1 000 000 000 000 000 000	$= 10^{18}$	exa	E
1 000 000 000 000 000	$= 10^{15}$	peta	P
1 000 000 000 000	$= 10^{12}$	tera	T
1 000 000 000	$= 10^{9}$	giga	G
1 000 000	$= 10^{6}$	mega	M
1 000	$= 10^{3}$	kilo	k
100	$= 10^{2}$	hecto	h
10	$= 10^{1}$	deca	da
0.1	$= 10^{-1}$	deci	d
0.01	$= 10^{-2}$	centi	c
0.001	$= 10^{-3}$	milli	m
0.000 001	$= 10^{-6}$	micro	μ
0.000 000 001	$= 10^{-9}$	nano	n
0.000 000 000 001	$= 10^{-12}$	pico	p
0.000 000 000 000 001	$= 10^{-15}$	femto	f
0.000 000 000 000 000 001	$= 10^{-18}$	atto	a

2. SI derived units based on length, time and mass

(a) The flow diagram given on page 638 shows the interrelationships of some derived units that are based on length (L), time (T) and mass (M), whilst Table A.2.1 (see page 639) gives a more complete list giving the dimensions of these units in both SI and the older cgs system.

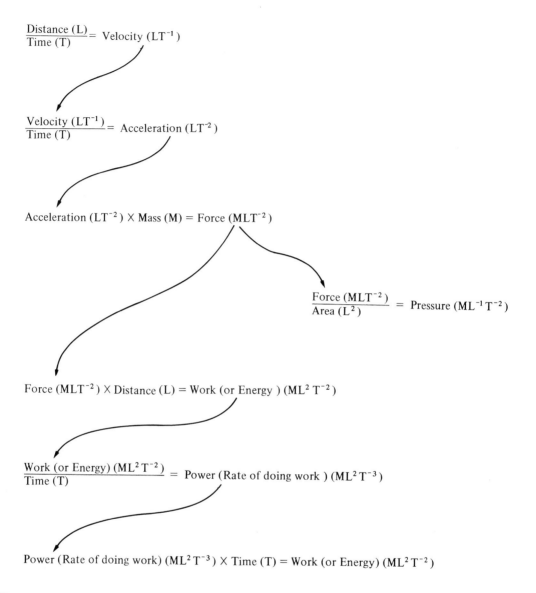

$$\frac{\text{Distance (L)}}{\text{Time (T)}} = \text{Velocity (LT}^{-1})$$

$$\frac{\text{Velocity (LT}^{-1})}{\text{Time (T)}} = \text{Acceleration (LT}^{-2})$$

$$\text{Acceleration (LT}^{-2}) \times \text{Mass (M)} = \text{Force (MLT}^{-2})$$

$$\frac{\text{Force (MLT}^{-2})}{\text{Area (L}^2)} = \text{Pressure (ML}^{-1}\text{T}^{-2})$$

$$\text{Force (MLT}^{-2}) \times \text{Distance (L)} = \text{Work (or Energy) (ML}^2\text{T}^{-2})$$

$$\frac{\text{Work (or Energy) (ML}^2\text{T}^{-2})}{\text{Time (T)}} = \text{Power (Rate of doing work) (ML}^2\text{T}^{-3})$$

$$\text{Power (Rate of doing work) (ML}^2\text{T}^{-3}) \times \text{Time (T)} = \text{Work (or Energy) (ML}^2\text{T}^{-2})$$

(b) UNITS OF LENGTH COMMONLY FOUND IN THE OLDER LITERATURE

The terms micron (μ) and Angstrom unit (Å) have been commonly used as units of length in the measurement of intracellular organelles, chemical bonds, X-ray spacings, etc. They have the following interrelationships:

1 micron = 1 μ = 1 μm = 1×10^{-6} m = 1×10^3 nm = 1×10^4 Å

1 Å = 0.1 nm = 1×10^{-4} μm = 1×10^{-10} m

1 nm = 10 Å = 1×10^{-3} μm = 1×10^{-9} m

(c) THE SI UNIT OF PRESSURE (Pa) RELATED TO OTHER COMMONLY USED PRESSURE UNITS

1 Pa = 9.869 2×10^{-6} Std.Atmos. = 7.500 6×10^{-3} mmHg = 1.450 4×10^{-4} psi.

1 Std.Atmos. = 760 mmHg = 14.696 psi = 1.013 3 $\times 10^5$ Pa.

1 mmHg = 0.491 15 psi = 3.386 4×10^3 Pa = 3.342 1 $\times 10^{-2}$ Std.Atmos.

1 psi = 6.894 8×10^3 Pa = 6.804 6×10^{-2} Std. Atmos. = 51.715 mmHg.

(Pa = Pascal; Std.Atmos. = standard atmosphere; mmHg = millimetre of mercury = Torr; psi = pound force per square inch.)

Table A.2.1. Derived units in SI and the cgs system and their ratios

Quantity	Dimensions	SI unit	cgs unit	Ratio $=\frac{\text{cgs unit}}{\text{SI unit}}$
Length ($=$Distance)	L	metre (m)	centimetre (cm)	10^{-2}
Time	T	second (s)	second (s)	1
Mass	M	kilogram (kg)	gram (g)	10^{-3}
Velocity	LT^{-1}	m s^{-1}	cm s^{-1}	10^{-2}
Acceleration	LT^{-2}	m s^{-2}	cm s^{-2}	10^{-2}
Force	MLT^{-2}	kg m s^{-2}=N (N=Newton)	g cm s^{-2}=dyn (dyn=dyne)	10^{-5}
Work (or Energy)	ML^2T^{-2}	kg m^2 s^{-2}=J (J=Joule)	g cm^2 s^{-2}=erg	10^{-7}
Power	ML^2T^{-3}	kg m^2 s^{-3}=W (W=Watt)	g cm^2 s^{-3}	10^{-7}
Pressure (or Stress)	$ML^{-1}T^{-2}$	kg m^{-1} s^{-2}=Pa (Pa=Pascal)	g cm^{-1} s^{-2}	10^{-1}
Surface Tension	MT^{-2}	kg m^{-2}=Nm^{-1}	g cm^{-2} or dyn cm^{-1}	10^{-3}
Viscosity	$ML^{-1}T^{-1}$	kg m^{-1} s^{-1}	g cm^{-1} s^{-1}=P (P=Poise)	10^{-1}
Frequency	T^{-1}	s^{-1}=Hz (Hz=Hertz)	s^{-1}	1

3. SI supplementary units

(a) *Radian* (rad)=unit for a plane angle (defined as the plane angle between two radii of a circle which cut off on the circumference an arc equal in length to the radius).

Note that:

(i) The degree ($^\circ$) is a unit of angle equal to $\frac{\pi}{180}$ rad.

(ii) The minute of arc (′) is $\frac{1^\circ}{60}$ and is therefore equal to $\frac{\pi}{10,800}$ rad.

(iii) The second of arc (″) is $\frac{1'}{60}$ and is therefore equal to $\frac{\pi}{648,000}$ rad.

(b) *Steradian* (sr)=unit for a solid angle (defined as the solid angle which, having its vertex in the centre of a sphere, cuts off an area of the surface of the sphere equal to that of a square with sides of length equal to the radius of the sphere).

4. Amount of a substance

In SI the amount of a substance is expressed in relation to a fixed mass of the isotope of carbon containing 6 protons and 6 neutrons in its nucleus, namely $^{12}_6C$. The *SI base unit of quantity is the mole* (mol) which is the amount of a substance of a system which contains as many elementary entities as there are atoms in 0.012 kg of $^{12}_6C$. When the term mole is used it is essential to specify the elementary entities under consideration; these may be atoms, molecules, ions, electrons or other particles or groups of particles. It should be noted that in SI Avogadro's constant ($6.022\,174 \times 10^{23}$ mol^{-1}) is defined as the number of atoms in 0.012 kg of $^{12}_6C$ and is therefore the number of entities in a mole of a substance.

5. Concentration of solutions

Molarity (M) = the number of moles of solute per litre of solution.
(The term 'molar' is equivalent to mol l^{-1}.)

Molality (m) = the number of moles of solute per kg of solvent.

%(w/v) = the weight in grams of a solute per 100 ml of solution.

%(w/w) = the weight in grams of a solute per 100 g of solution.

% saturation = the concentration of salt in solution as a percentage of the maximum concentration possible at the given temperature.

6. Units of enzyme activity

The *official unit of enzyme activity is now the Katal* (kat) which is defined as the amount of enzyme that will transform 1 mole of substrate in 1 second. This unit was recommended by the Enzyme Commission of the International Union of Biochemistry in 1972 to bring it in line with SI units and to replace the previous unit of enzyme activity, the International Unit (I.U.) which was defined by the Enzyme Commission in 1961 as the amount of enzyme that will transform 1 μmole of substrate in 1 minute. However, both the katal and I.U. are still in common usage.

It is recognized that the katal is a very large quantity of enzyme activity and is, in most cases, too large for practical use. Therefore it is more usual to find enzyme activity expressed in microkatals (μkat; $= 1 \times 10^{-6}$ kat), nanokatals (nkat; $= 1 \times 10^{-9}$ kat) or picokatals (pkat; $= 1 \times 10^{-12}$ kat) which correspond to the amount of enzyme that transforms 1 micromole (μmol), 1 nanomole (nmol) and 1 picomole (pmol) of substrate respectively per second.

The relationship between the old and new units of enzymic activity is as follows:

$$
\begin{aligned}
1 \text{ kat} \quad &= 1 \text{ mol s}^{-1} \\
&= 60 \text{ mol min}^{-1} \\
&= 60 \times 10^6 \ \mu\text{mol min}^{-1} \\
&= 60 \times 10^6 \text{ I.U.} \\
&= 6 \times 10^7 \text{ I.U.} \\
1 \text{ I.U.} \quad &= 1 \ \mu\text{mol min}^{-1} \\
&= \tfrac{1}{60} \ \mu\text{mol s}^{-1} \\
&= \tfrac{1}{60} \ \mu\text{kat} \\
&= 0.016 \ 67 \ \mu\text{kat} \ (1.667 \times 10^{-2} \ \mu\text{kat}) \\
&= 16.67 \text{ nkat} \\
&= 1.667 \times 10^{-8} \text{ kat}
\end{aligned}
$$

The *specific activity* of an enzyme is defined as the number of katals per kilogram of protein or more usually, from a practical point of view, the number of μkat per mg of protein, under specified conditions of temperature, pH, substrate concentration, etc.

The *molar activity* of an enzyme is defined as the number of katals per mole of enzyme under specified conditions (as above).

The *concentration of enzymic activity* in a solution is defined as activity divided by the volume of solution and is expressed in katals per litre (kat l^{-1}) or suitable multiples thereof.

The term *turnover number* has sometimes been used to express enzymic activity in terms of the number of substrate molecules transformed per minute by one molecule of enzyme. Note that the turnover number is related to molar activity in the following ways:

$$\text{Molar Activity} = \frac{\text{Turnover Number}}{60}$$

and

$$\text{Turnover Number} = 60 \times \text{Molar Activity}$$

7. SI electrical units

(a) *SI base unit*

Ampere (A) = unit of electrical current (defined as that constant current which, if maintained in each of two infinitely long straight parallel wires of negligible cross-section placed 1 metre apart, in a vacuum will produce between the wires a force of 2×10^7 newtons per metre length) which is equivalent to the flow of $6.241 \ 449 \times 10^{18}$ electrons past a given point in 1 second.

(b) *SI derived units*

Coulomb (C) = unit of electrical charge (defined as the quantity of electricity transported per second by a current of 1 ampere).

$$C = A \cdot s \ (= J \cdot V^{-1})$$

Volt (V) = unit of potential difference (defined as that difference in electrical potential between two points of a wire carrying a constant current of 1 ampere when the power dissipation between those points is 1 watt).

$$V = W \cdot A^{-1} \; (= J \cdot s^{-1} \cdot A^{-1} = J \cdot C^{-1})$$

Ohm (Ω) = unit of electrical resistance (defined as the electrical resistance between two points of a conductor when a constant potential difference of 1 volt applied between these points produces in the conductor a current of 1 ampere.

$$\Omega = V \cdot A^{-1}$$

8. SI light units

(a) *SI base unit*

Candela (cd) = unit of luminous intensity (defined as the luminous intensity, in a perpendicular direction, of a surface of 1/600 000 square metre of a full radiator at the temperature of freezing platinum under a pressure of 101 325 newtons per square metre). It replaces the international candle which was defined in terms of the light emitted by a specified electric lamp.

1 candela = 0.982 international candle.

(b) *SI derived units*

Lumen (lm) = unit of luminous flux (defined as the light energy emitted per second within a solid angle of 1 steradian by a uniform point source of luminous intensity equal to 1 candela.

$$lm = cd \cdot sr$$

Lux (lx) = unit of illuminance (defined as the illumination produced by 1 lumen falling on 1 square metre of a perpendicularly placed surface).

$$lx = lm \cdot m^{-2}$$

(Note that 1 lx = 10.764 ft candles.)

Refractive Index (n) = ratio of the velocity of light *in vacuo* to that in the medium in question.

9. Units of temperature

(a) *SI base unit*

Kelvin (K) = unit of thermodynamic temperature (defined as the fraction 1/273.16 of the thermodynamic temperature of the triple point of water).

(b) *Other units*

The Degree Celsius (°C) is based on the Centigrade scale of temperature which uses the ice point as zero and the boiling point of water at 1 standard atmosphere as the upper fixed point set to be 100°C. The Celsius scale of temperature is defined as being the same as the thermodynamic scale with the zero shifted to the ice point which is 273.15 K.

Temp. in °C = Temp. in K − 273.15
Temp. in K = Temp. in °C + 273.15

The Degree Fahrenheit (°F) is based on the Fahrenheit scale of temperature which places the ice point at 32°F and the boiling point of water at 1 standard atmosphere at 212°F.

Temp. in °F = [Temp. in °C × 1.8] + 32
Temp. in °C = [Temp. in °F − 32] × 5/9
Temp. in K = ([Temp. in °F − 32] × 5/9)
$\qquad$ + 273.15
$\qquad$ = [Temp. in °F + 459.67] × 5/9

10. Units of radioactivity

Curie (Ci) = that quantity of a radioactive substance in which 3.7×10^{10} atoms disintegrate per second.
(Thus 1 Ci = 3.7×10^{10} dps =
$\qquad 2.22 \times 10^{12}$ dpm.)

Becquerel (Bq) = that quantity of a radioactive substance in which 1 atom disintegrates per second.
(Thus 1 Ci = 3.7×10^{10} Bq
$\qquad$ 1 Bq = 1 dps = 60 dpm.)

11. Isotopes used in labelling experiments in biochemistry

Isotope	Stable or radioactive	% Natural abundance	Type of decay	$\frac{1}{2}$ life	Maximum energy (MeV)	Average energy (MeV)
^{2}H	Stable	0.015	—	—	—	—
^{3}H	Radioactive	—	β^-	12.43 years	0.018 5	0.005 5
^{13}C	Stable	1.11	—	—	—	—
^{14}C	Radioactive	—	β^-	5730 years	0.156	0.050
^{15}N	Stable	0.36	—	—	—	—
^{18}O	Stable	0.204	—	—	—	—
^{32}P	Radioactive	—	β^-	14.3 days	1.709	0.70
^{35}S	Radioactive	—	β^-	87.4 days	0.167	0.049
^{36}Cl	Radioactive	—	β^-	3.07×10^5 years	0.71	0.30

12. Values of some fundamental physical constants

Physical constant	Symbol	Value (units)
Speed of light *in vacuo*	c	$2.997\,925 \times 10^8$ m s^{-1}
Mass of proton (at rest)	m_p	$1.672\,614 \times 10^{-27}$ kg
Mass of neutron (at rest)	m_n	$1.674\,920 \times 10^{-27}$ kg
Mass of electron (at rest)	m_e	$9.109\,56 \times 10^{-31}$ kg
Energy of proton (at rest)	$m_p c^2$	$1.503\,271 \times 10^{-10}$ J
Energy of neutron (at rest)	$m_n c^2$	$1.505\,343 \times 10^{-10}$ J
Energy of electron (at rest)	$m_e c^2$	$8.187\,26 \times 10^{-14}$ J
Charge of proton	e	$1.602\,192 \times 10^{-19}$ C
Charge of neutron		0
Charge of electron	e	$1.602\,192 \times 10^{-19}$ C
Boltzmann's constant	k	$1.380\,626 \times 10^{-23}$ J K^{-1}
Planck's constant	h	$6.626\,196 \times 10^{-34}$ J s
Gas constant	$R\ (=N \cdot k)$	$8.314\,34$ J K^{-1} mol^{-1}
Faraday's constant	$F\ (=N \cdot e)$	$9.648\,670 \times 10^4$ C mol^{-1}
Avogadro's constant	N	$6.022\,174 \times 10^{23}$ mol^{-1}
Normal volume of ideal gas	V_o	$2.241\,36 \times 10^{-2}$ m^3 mol^{-1}

Index

Reference to chemical formulae is given in italics

664 Index

Plant Index

incorporating references to animals and bacteria

^a animal   ^b bacterium   ^c common name

ᶜWheat (*Triticum vulgare*)   43, 44, 45, 79, 236, 282, 286, 306, 312, 359, 598, 606, 608, 624

Xanthium   601, 607
Xanthium pennsylvanicum   605, 624
Xanthophyceae   98
Xanthoxylum suberosum   536
Ximenia caffra   275

ᶜYeast (*Saccharomyces cerevisiae*)   36, 53, 184, 236, 241, 247, 291, 296, 297, 317, 368, 472, 601, 605
ᶜYew (*Taxus baccata*)   607, 613

Zea mays   fig. 3.10, 143, 261, 282, 590, 592, 604, 607, 612
Zygnema spp.   38
Zygophyllaceae   140, 149

SUGGESTIONS FOR FURTHER REFERENCE

MACURA, P., compiler (1979) *Elsevier's Dictionary of Botany*. Vol. I, *Plant Names*. Elsevier, Amsterdam-Oxford-New York.
CRONQUIST, A. (1981) *An Integrated System for Classification of Flowering Plants*. Columbia University Press, New York.
TAKHTAJAN, A. L. (1980) 'Outline of the classification of flowering plants (Magnoliophyta)', *The Botanical Review* **46,** 225–359.